The Metallurgy of Zinc Coated Steels

Woodhead Publishing Series in Metals and Surface Engineering

The Metallurgy of Zinc Coated Steels

Arnold R. Marder

Frank E. Goodwin

WP
WOODHEAD PUBLISHING
ELSEVIER An imprint of Elsevier

Woodhead Publishing is an imprint of Elsevier
50 Hampshire Street, 5th Floor, Cambridge, MA 02139, United States
The Boulevard, Langford Lane, Kidlington, OX5 1GB, United Kingdom

Notices

Knowledge and best practice in this field are constantly changing. As new research and experience broaden our understanding, changes in research methods, professional practices, or medical treatment may become necessary.

Practitioners and researchers must always rely on their own experience and knowledge in evaluating and using any information, methods, compounds, or experiments described herein. In using such information or methods they should be mindful of their own safety and the safety of others, including parties for whom they have a professional responsibility.

To the fullest extent of the law, neither the Publisher nor the authors, contributors, or editors, assume any liability for any injury and/or damage to persons or property as a matter of products liability, negligence or otherwise, or from any use or operation of any methods, products, instructions, or ideas contained in the material herein.

ISBN: 978-0-323-99984-7 (print)
ISBN: 978-0-323-98489-8 (online)

For information on all Woodhead publications
visit our website at https://www.elsevier.com/books-and-journals

Publisher: Matthew Deans
Acquisitions Editor: Dennis McGonagle
Editorial Project Manager: Howi M. De Ramos
Production Project Manager: Anitha Sivaraj
Cover Designer: Miles Hitchen

Typeset by STRAIVE, India

Dedicated to our wives, Carole Marder and Rosalind Volpe Goodwin, for their infinite patience, understanding, and perseverance.

Contents

Foreword

Galvanizing is one of the world's most important industrial processes. If you give this a moment's thought, you will quickly realize that this is true. The marriage of our most important metallic structural material "steel" with the zinc that sacrifices itself to prevent its destruction by corrosion is one the most beneficial technological relationships that humanity has discovered and applied on a global scale. With the annual cost of corrosion being estimated at USD2.5 trillion, the economic benefits of galvanizing—in all of its forms—cannot be overemphasized. Within the context of the modern concern with sustainability, galvanized advanced steels are significant contributors to the increased fuel efficiency and durability of modern vehicles and are essential enablers for the large-scale deployment of electric vehicles. In addition, the extension of product and infrastructure lifetimes provided by the corrosion prevention of steel through galvanizing has kept millions of tons of carbon dioxide from being emitted in replacement steel production.

This is not to say the road has been an easy one to travel. From the initial discovery of "animal electricity" by Galvani in 1791, to the discovery by Faraday that zinc would sacrificially protect steel, to the development of what we now refer to as the "batch" galvanizing process by Sorel in the 19th century, to the development of electrogalvanizing and the continuous Sendzimir and Cook-Norteman galvanizing processes in the 20th century, the history of galvanizing has been one of continuous technological advancement coupled with significant economic benefits. In recent decades, continuous galvanizing technology, in particular, has undergone transformative technological changes driven by the need to produce increasingly sophisticated, more highly alloyed steels. This path has been documented through the technical literature and is, perhaps, best embodied in the Galvatech series of conferences and the papers of the Galvanizers Association meetings, to which the attentive reader will find many citations within this volume. My students and I have had the privilege of contributing, in our modest way, to this literature and I am acutely aware of the accelerated pace of developments in this corner of the technological society we inhabit since I first entered it at the Galvatech 1995 conference.

You hold in your hands (or are viewing on your screen) a compendium of this journey, written by two of the giants in our field: Professor Arnold Marder and Dr. Frank Goodwin. Although I have never had the opportunity to work with Professor Marder, I have had the privilege to have enjoyed a friendship and worked with Frank Goodwin on various research endeavors for more than two decades now. I cannot think of two more appropriate people to compile our hard-earned knowledge into one place for the benefit of all galvanizing researchers and practitioners.

In 2021, world steel production reached approximately 1.9 billion metric tons, with world zinc production being approximately 12.8 million metric tons. Approximately 50% of this zinc was devoted to galvanizing, which is a lot of steel surface when you appreciate that most zinc coatings are a fraction of the thickness of a human hair. As we move forward, the technological demands surrounding the application of metallic coatings to steel will only increase. Within this volume, the reader will find knowledge for both the researcher and producer, with topics ranging from process fundamentals along with a comprehensive compilation of coating defects to the theory and practice of gas-jet wiping and the galvanizing of the next generation of advanced steels. I strongly encourage the reader to take full advantage of this volume. I know I certainly will.

Joseph R. McDermid
McMaster University, Hamilton, ON, Canada

Preface

It is never sufficient to merely touch the student with the flame of knowledge.
Rather one must "raise up" the flame so that it will grow and nurture the student for a long time.
The task of the teacher is to ensure that the flame will continue to burn on its own, that knowledge will be a lifetime process.

Our effort to create this book is not singular. Like all good scientific progressions, it builds on past reviews, and it is appropriate to acknowledge those accomplishments:

- 1950—H. Bablik, Galvanizing (Hot-Dip), Wiley.
- 1978—D. Horstmann, Reaction Between Iron and Zinc, Zinc Development Association, pp. 20–32.
- 1979—J. Mackowiak, N.R. Short, Metallurgy of Galvanized Coatings, p. 1.
- 1983—G.J. Harvey, P.N. Richards, Zinc Based Coatings for Corrosion Protection of Sheet Steel and Strip, Metals forum, p. 6.
- 2000—A.R. Marder, The metallurgy of zinc-coated steel, Progr. Mater. Sci. 45 (2000) 191–271.
- 2016—V. Kuklik, J. Kudlacek, Hot-Dip Galvanized Steel Structures, Butterworth and Heinemann, Elsevier.

It has been more than 20 years since the publication of my monograph "The Metallurgy of Zinc-Coated Steel" in *Progress in Materials Science*. It was the purpose of that review to supplement and update the excellent reviews of the past. Up to the time of writing the 2000 review, I had the privilege to witness the development of Galvalume 55%Al-Zn (Borzillo and Horton) and the early effort to produce Zn-Al-Mg coatings (Berke and Townsend) as a member of the Research Department at Bethlehem Steel. In a very small way, I also investigated the coating morphology of 55%-Al and ran two collaborative programs with Lehigh University on Si additions to 55%-Al (Selverian and Notis) and electrodeposited Zn coatings (Gu). My interest in coated sheet was piqued even more when T. Sendzimir asked me to review a new coating line concept he was developing. I realized then that coatings had a great future, "the ultimate composite," and that coating metallurgy was a fertile area for research. That exposure led me to continue research on Zn coatings in my new position at Lehigh in 1986. Our group extended the work on electrodeposited Zn coatings and made contributions to the literature on hot-dip Zn coatings (Jordan, Goggins, and Bluni) and the process of galvannealing (Gu and Drewien). In 1997, I took my Sabbatical at the Technion, Israel Institute of Technology, where I taught a course in coatings and had thought-provoking discussions on zinc alloys with Prof. Dan Schectman (later a Nobel laureate in chemistry for his work on quasicrystals). Thus, the idea of writing a review of Zn coatings began.

Since writing the review published in 2000, there has been an exponential rise in the use of zinc coatings particularly in automotive and construction industries. At the same time, there has been an explosion of research from laboratories all over the world on this important product. After retirement from Lehigh in 2007, I spent several years teaching lawyers and testifying to jurors and judges as an expert in Zn coating litigation cases throughout the world. I was well aware of the application of Zn coatings to exposed parts in automobiles, but I did not realize how important these coatings were in construction until I spent time working and touring Australia and New Zealand. I was amazed at the use of Zn-coated steel on the roofs of just about every house and commercial building. In much of my consulting experience, I expended a lot of time reviewing the literature and realized how much the 2000 review was insufficient. I thought a new review of the Zn metallurgy coatings was needed, but I did not think I was familiar enough with all the disparate work that was being done throughout the world. However, I learned that Frank Goodwin had just retired from the International Zinc Association, and as most of us know, Frank has been involved in more Zn research programs than anyone else on the planet. I called Frank to explore the idea of a new review paper and was disappointed when he told me he was already working on a paper. Great idea, ...but it was too late for me. Nevertheless, I was extremely surprised and overwhelmed when Frank called me about six months later and asked if I still wanted to write a review paper.

There is nothing worse than having two retirees think that they are young researchers out to conquer the world again! Well, we decided that a review paper was not enough and that we should write a more comprehensive review: a BOOK.

Wary of the many academics who have started books never to be finished, we decided to solicit publishers with an outline first, before we would begin writing a single word. Fortunately, we received glowing and almost embarrassingly positive reviews and encouragement to start writing. We selected Elsevier publishing for several reasons and especially because of their vast experience in the publication of articles in metallurgy and material science. Thus, our journey began with the similar objective as the prior review, and it is the purpose of this book to supplement and update the vast amount of research in Zn-coated products since my last review in 2000.

It has been established that zinc is one of nature's most vital, essential elements. Humans, animals, plants, and even the smallest microorganisms need zinc to function, and zinc shows the greatest effect on our immune system. Zinc helps fight infections and can even reduce the duration and severity of the common cold. Also zinc may be important for treating COVID-19 infections. The nexus between Zn and COVID-19 gave us an unexpected opportunity. Being trapped at home during the COVID-19 pandemic, we have had plenty of time to review and analyze the literature on Zn coatings and compose this book.

As I started reviewing the history of zinc coatings, it became obvious to me that Bethlehem, Pennsylvania, with Bethlehem Steel and Lehigh University, was one of the major centers of Zn coatings research. Both institutions played essential roles in the research and development of Zn coatings that involved both individual institutional contributions and cooperative efforts. J.W. Richards, Professor of Metallurgy at Lehigh University, from 1890 to 1921, received a patent (US 456,204) in 1891 for using Al in Zn galvanizing baths. A. Borzillo and J. Horton (Beth. Steel) patented Zn-55% Al Galvalume in 1967 and A. Borzillo (Beth. Steel) wrote his MS thesis for W. Hahn (Lehigh), on the inhibition layer in Zn-Al galvanized coatings in 1968. This was followed in 1971 by the work of Ghumen and Goldstein (Lehigh) along with Berke and Townsend (Beth. Steel) in 1983. Many papers were published during this time by Bethlehem Steel, particularly Cleary, Townsend, and Benscoter, as they commercialized Galvalume coatings. Concurrently, in a jointly sponsored research program, J. Selverian (Lehigh), M. Notis (Lehigh), and A. Marder (Beth. Steel) studied the effect of Si in Galvalume coatings. In 1985, M.Gu (Lehigh) and A. Marder (Beth. Steel) began work on electrodeposited Zn-Fe coatings. This was followed in the 1990s by research on galvanizing and galvannealing simulations by C. Jordan (Lehigh) and K. Goggins (Lehigh) along with C. Drewien (Lehigh) on electrodeposited coatings and included Visiting Scholar W. Maschek from Voestealpine. In 1994, R. Fraley (Beth. Steel) studied lead additions in Zn coatings for his Lehigh M.S. thesis. Of utmost importance to the application of zinc coatings in automobiles, the effort was spearheaded by Lehigh University Alumnus, Lee Iacocca at Chrysler Corp. Some 25 miles away in Palmerton, Pennsylvania, the New Jersey Zinc Company operated a zinc mine from the early 1900s through the 1970s. In the 1970s, the company established the New Jersey Zinc Professor Chair in the Department of Materials Science and Engineering. Finally, to emphasize the criticality of the Bethlehem area in Zn coating development, Frank Goodwin, ILZRO and IZA, and my coauthor, was born and raised in Bethlehem!

Zinc-coated steel offers the perfect materials science and engineering paradigm: the interrelationship between structure, processing, and properties. We have attempted to review the structure of Zn coatings in Chapters 3—Fe-Zn phase equilibria; Chapter 5—Interface reactions; Chapter 6—Overlay coating formation; Chapter 8—Galvannealing; and Chapter 9—Electrogalvanizing processes (EG coatings). Processing is reviewed in Chapter 2—Hot-dip coating methods: Continuous processing (CGL); Chapter 4—Zinc bath phenomena; Chapter 7—Hot-dip galvanized coating weight control; Chapter 10—Physical vapor deposited coatings; and Chapter 15—General galvanizing. With some effort, the microstructure of Zn coatings can be investigated in the laboratory, but few realize the difficulty in trying to produce a 10-ton Zn-coated coil. Consider the exacting control of the process that is necessary to run a 6 ft wide (avg. size of a man), 1000 ft sheet (3 football fields long) traveling at 600 ft/min (7 mi/h) and produce a thin uniform 25 μm zinc coating (the size of a human hair) across the entire surface of each side of the coil. The engineering developments to accomplish such a goal are astounding. This process results in very important commercial properties that are reviewed in Chapter 11—Corrosion resistance; Chapter 12—Zn-coated steel product properties; Chapter 13—Zn-coated high-strength steel; and Chapter 14—Defect identification and remediation in zinc-coated steel sheet.

So, armed with our experience, buoyed by the COVID-19 pandemic, and encouraged by our wives to keep busy during our retirement, we were "*galvanized*" *to* undertake this book project on the metallurgy of zinc-coated steel. It is our hope that this effort will serve as a compendium of knowledge on zinc-coated steel, circa 2022. However, as R. Tarfon the great sage once said, "You are not required to complete the task, yet you are not free to withdraw from it." We have tried to document the Zn coating effort to date and have no doubt that it will be further reviewed in the future. Such is scientific progress.

A.R. Marder

Acknowledgments

gal·va·nize (găl′və-nīz′)—to excite about something so that action is taken.

It takes an industry to be **galvanized** to write a galvanizing book. In the combined century of experience of this book's authors, working with the galvanizing industry, we still marvel at the spirit of generous open-handedness with technical information and experience that is thankfully the norm in this industry. It is a very different world from other industries where "secrets" (that one learns in many cases are the same in many companies) are guarded and little is shared. We have so many people to thank for sharing so much with us.

First, we are grateful to the companies that have financially supported the authors in the development of this book. We acknowledge all the contributors in alphabetical order: at ArcelorMittal, Jean-Michel Mataigne, Jonas Staudte, Sylvie Gindre, Narayan Pottore, Weimin Zhong, Stavros Fountoulakis, Henry Hahn, Erika Bellhouse, and Hassan Ghassemi-Armaki; at Cleveland-Cliffs, Paul Janavicius, Steve Boston, and Jerry Arnold; at SSAB, Ari Peltola and Sven-Erik Hörnström; at Steel Dynamics, Joshua Bowne, Margie Farabaugh, and John Nolan; at Teck Metals, Daniel Liu, Graeme Anderson, Paul Kolisnyk, Jim Peterson, and Neil Gao; at Tecnar, Alexandre Nadeau; and at International Zinc Association: Andrew Green and Martin van Leeuwen.

Next, we are grateful to a very special group of people who dedicated many hours to reviewing chapter manuscripts, finding our mistakes or omissions, and patiently providing suggestions for improvement. Each of these people has a passion for our industry, and each has gone beyond our requests for their assistance: Michel Dubois, Gary Dallin, Martin Gagne, Daniel Liu, Edzo Zoestbergen, and Roger Pankert.

Arnold Marder thanks his many graduate students at Lehigh University for their excellent research, much of which has been included in this book. A major thank you goes to John Selverian, Kathy Goggins, Minguan Gu, Celeste Drewien, and Scott Bluni, many of whom were supported by Frank Goodwin and ILZRO. He also thanks Boris Levin for many good discussions on future Zn opportunities and his colleagues at Lehigh—Arlan Benscoter and John DuPont and his students—for the various mini searches of theses, books, papers, and photomicrographs. Arnold Marder's special thanks goes to the late Catherine Jordan, for her excellent pioneering Gleeble work on the fundamentals of galvannealing. Unfortunately, Cathy died this past year of breast cancer, but she will always be remembered not only as a great researcher but also as a wonderful and caring individual. In addition, Arnold Marder thanks his friends for their enthusiastic encouragement as opposed to skepticism that an "old man" could still do something technical. He especially thanks Steve Bauml and Peter Levine. He also thanks Irv Jacobs and Sharon and Larry Chambers for their help in sorting out quotes and historical perspectives for this book.

Frank Goodwin thanks collaborators at universities, industrial, and government laboratories and metals producers and processors globally who are too numerous to be named, in the more than 100 research projects they have done together, always looking over the horizon to find the new knowledge that will take the industry to the next step forward. The constant support of the zinc industry, leveraged by many steel and auto companies, together with their suppliers, provided the platform for the constant learning, and all of the progress made. His special thanks goes to those who cochaired the Galvatech series of conferences in several countries with him over the years: Debanshu Bhattacharya, Josef Faderl, Thomas Koll, Prof. Joseph McDermid, Ruiyu Yin, and Gerhard Hackl. He also acknowledges the support of the staff in industry associations, including International Zinc Association (IZA), American Galvanizers Association (AGA), Association for Iron & Steel Technology (AIST), American Iron and Steel Institute (AISI), Associazione Italiana di Metallurgia (AIM), Austrian Society for Metallurgy and Materials (ASMET), Corrosion Science Society of Korea, European General Galvanizers Association (EGGA), Galvanizers Association (especially remembering Dale Nevison), Iron and Steel Institute of Japan (ISIJ), and Stahlinstitut VDEh. There are many more who responded to his requests, brought people together, and provided invaluable support, and he is very grateful to each one of them.

We thank the Elsevier team who has been with us from the beginning of this project, teaching us the many things there were to learn about publishing a book, which is much more than "just" writing a manuscript. We are especially grateful to Christina Gifford for her efforts to initially set this book project in motion.

Last but not least, we are grateful to our wonderful wives, Carole Marder and Rosalind Volpe Goodwin, for their enduring support, amazing patience, and editing, without whom this book would not have been possible. We greatly appreciate their self-control of our "honey-do" retirement list, allowing us the freedom to write.

Introduction

As you walk around a galvanizing plant, you know there is whole lot going on. Have any one of these questions ever crossed your mind?

- What were the technical innovations that propelled this massive industry to produce these important steel-coated products that impact us daily?
- How is the steel being treated, and how is it reacting to make galvanized steel?
- Steels are becoming more advanced, with higher strengths and more complicated microstructures. How are these steels galvanized to meet both the mechanical property targets and achieve high-quality coatings?
- There are a lot of alloy coatings out there besides zinc, including zinc-aluminum, zinc-aluminum magnesium, and for general galvanizing, zinc-nickel. What is the metallurgy behind these coatings, why do they perform better in many situations?
- I would like to understand not only hot-dip galvanizing but also electrogalvanizing or vapor-deposited coatings—from a metallurgist's point of view. How are these coatings applied and how do they behave? How about zinc alloy electrogalvanized and vapor coatings?
- There are lots of corrosion studies out there, written by corrosion engineers and electrochemists. How do zinc alloy contents and microstructures affect corrosion behavior, from a metallurgist's point of view?
- I need to understand more about forming and welding of zinc-coated sheet steels, especially things that affect galvanneal formability. What is the metallurgical knowledge behind this?
- We know how important good coating weight/thickness control is, both to satisfy customer specifications and for the profitability of the line. What are the main sources of variability in coating weight control? What are the basic coating control relationships? What are triple-spot and single-spot coating weight specifications, and how do these relate to the coating weights we are measuring?
- Our galvanizing bath produces too much dross, or our pot rolls have too short of a life. What is the state of knowledge on these topics? Which operating conditions produce the least amount of dross?
- I need a good guide to coating defect identification and cure. Is there an atlas of typical defects available with this information?
- What is the future of galvanizing, and how is it being integrated into other plant operations as "Industry 4.0" practices are developed? What is the "digital twin" of a galvanized coil, and how does this help me improve quality and reliability?

The answers to these and even more questions about the metallurgy of zinc coated steel can be found in this book. Separate chapters provide an overview of continuous galvanizing line operations and dive deeply into galvanizing bath phenomena and coating weight control. The galvannealing chapter provides a training tool for the operator and is at a high enough scientific level to be a useful reference to the engineering scientist seeking to understand the metallurgy behind many of the gas-metal, zinc-iron, and aluminum-iron reactions that occur during galvanizing. Chapters on phase diagrams and steel-coating interface reactions provide the "road maps" for understanding why zinc, iron, aluminum, and other elements react as they do during the galvanizing process. The knowledge described in these chapters is brought together and invoked in the chapter on identifying and analyzing coating defects.

For the general galvanizer, the latest advances in our knowledge on pretreatment steps including cleaning, pickling, and fluxing are provided, together with a detailed view on exactly what happens when structural steels and other long products are dipped in a zinc bath. For the first time, phase diagrams and bath management data are brought together to assist batch galvanizers refine their procedures for alloy additions and sampling procedures, all explained from a metallurgist's view. Control of steel reactivity in the galvanizing bath is approached using basic metallurgical principles.

For the user or processor of galvanized sheet steel, the metallurgical influence on corrosion is provided both for uncoated and coil coated (continuously painted) sheet steel, in both conventional and alloy coatings, including galvanneal coatings. Details on corrosion of zinc-iron-coated hot press forming steels are included and both automotive and nonautomotive corrosion situations are addressed. Formability of both construction and automotive products is considered for conventional zinc, zinc-aluminum, and zinc-aluminum-magnesium alloy coatings. Recent knowledge on resistance spot welding and gas metal arc brazing includes an introduction to causes of liquid metal embrittlement in advanced high strength steels. A detailed description of coating weight specifications is included at the end of the coating control chapter.

The frontiers of zinc-coated steel are addressed in chapters on galvanizing high strength steels and the metallurgy of hot press-forming steels, and on the constant advances in process control and online characterization that, using increasingly fast computers together with artificial intelligence rather than predictive formulas, are revolutionizing metals processing. The scientific knowledge in this book provides a strong technical platform for the operating engineer, applications engineer, and research scientist to participate in these future advances.

The reader will learn from this book that although there are a great number of steel grades being coated with different zinc alloy coatings and processes, there is a unity of metallurgical knowledge behind it all. Understanding the knowledge presented in this book will enable the reader to see the "big picture," increasing their professional worth together with satisfying their intellectual curiosity about "what is really going on" with zinc-coated steel.

As mentioned in the Preface, this book grew out of the 2000 review paper with the same title as this book, "The Metallurgy of Zinc Coated Steel," written by A.R. Marder, with support from F.E. Goodwin and others. This 80-page paper became one of the most-cited papers in the metallurgical literature in the past 20 years and has been a valuable resource to several generations of students, their professors, and countless numbers of people involved, both producers and users, in the global galvanizing industry. After so many years, it became apparent that an update was required. Also, since 2000, there has been an enormous increase in global galvanizing capacity. As of the time of writing, over five hundred continuous hot-dip galvanizing lines in 73 countries have been identified, nearly doubling the numbers in 2000. There are an additional 90 electrogalvanizing lines [1].

Continuous vapor-coating processing has recently begun, for which capacity growth is also forecast. This has resulted in the involvement of many more people in this industry, all of whom require knowledge at some level of the metallurgy of zinc-coated steel. General galvanizing has kept pace with sheet galvanizing. The update of the 80-page paper quickly grew into a book of about five hundred pages together with more than eight hundred figures and tables, revising and unifying metallurgical knowledge on zinc-coated steels into one volume.

The authors combine over a century of experience working in zinc-coated steel research, production support, and applications support and have been involved in technical leadership in the galvanizing and steel industries for many of those years. Their balance of engineering and scientific experience make them ideal guides for readers who wish to gain a rich understanding of the metallurgy of zinc-coated steel.

Reference

[1] International Lead Zinc Study Group, World Directory: Continuous Galvanizing Lines, 1997 and 2003 editions, International Lead Zinc Study Group, Lisbon, Portugal. www.ilzsg.org.

Chapter 1

History of zinc-coated steel

The history of the development of zinc-coated steel has been well chronicled by Lamesch in The World History of Galvanizing [1]. The purpose of this chapter is to briefly review the highlights of those inventors and inventions that have brought us to this moment in time as summarized in Appendix 1, The Chronological History of Zinc-Coated Sheet Steel. For this review, there is a heavy emphasis on the patent literature that clearly shows the relationship between science and economics. Great scientific ideas are inspired but only enter the world of industry when they are paired with an economic driving force. Throughout the history of zinc coating development, these pairings can be seen. For example, Galvani and Faraday, Sorel and Daniell, Sendzimir and Cook-Norteman, Mayhew and Kohler, and Borzillo and Berke. Practical industrial needs influenced all of these inventions, from iron ships and metal buildings to lightweight energy-efficient automobiles. The history of zinc-coated steel is a testament to human genius and ingenuity.

1.1 Zinc

Zinc, atomic number 30, symbol Zn, has a silvery-grayish appearance and is brittle at room temperature as a result of its hexagonal close-packed crystal structure (Fig. 1.1).

Zinc ranks 24th in abundance in the Earth's crust, and the most common zinc ore is sphalerite (zinc blende), a zinc sulfide mineral. Zinc is economical to produce because zinc deposits are located in concentrated areas in the Earth's crust rather than being widely dispersed, and it can be refined at low cost, compared with rubidium (22nd) or titanium (9th). The production of zinc from recycled sources, mainly from recycling of zinc-coated steel, is expected to increase to 11% of total zinc metal supply by 2025, up from 6% in 2014 [2].

The elemental properties of zinc are given in Table 1.1:

Large-scale production of zinc did not begin until the 12th century in India, although it was known to the ancient Romans and Greeks (300 BCE). Pure zinc production made by the distillation process was found as early as the 9th century in Zawar, in Rajasthan, India, where it continued into the 1830s when the Zawar complex was destroyed during a period of political instability. Zinc was known in China during the 16th century. Paracelsus (1493–1541), a Swiss-born German alchemist, named the metal "zincum" or "zinken" in the 16th century. Before zinc was produced in Europe, it was imported from India beginning about 1600. A Flemish metallurgist and alchemist, P.M. de Respour, extracted metallic zinc from zinc oxide in 1668.

In Bristol, England, William Champion (1709–89) rediscovered the Indian distillation process after a tour of European brass mills that imported Indian zinc and patented a process to extract zinc from calamine (Fe_2O_4Zn) by vapor condensation in a vertical retort furnace in 1738; the process was used for over 100 years through 1851. In 1746, a German Chemist, Andreas Marggraf (1709–82) heated a mixture of charcoal and calamine to obtain zinc, a procedure that became commercially practical by 1752. Marggraf usually gets credit for the discovery of pure metallic zinc, although zinc was distilled from calamine 4 years earlier by a Swedish Chemist, Anton von Swab (1702–68).

Pure metallic zinc oxidizes quickly on exposure to air and corrodes easily; however, in an atmosphere containing carbon dioxide, it can form a thin stable layer of insoluble zinc carbonate, giving the familiar gray color. Zinc has favorable casting properties. It also has good alloying characteristics and is an important alloying element in brass. Zinc oxide is used as an accelerant for vulcanization in the rubber-making industry, and because it is opaque, it is also used to protect polymers and plastics from ultraviolet radiation. Zinc oxide semiconductor properties make it useful in photocopying products and useful extensively as a white pigment in paints and ceramics. As a trace element, zinc is present in all parts of the body, e.g., organs, tissues, and bones, and zinc is essential for growth and reinforcement of the immune system. Zinc's importance is reflected in the Recommended Daily Allowance for zinc by the USA Food and Drug Administration, 11 mg/d for adult men, 8 mg/d for adult women, and 12 mg/d for lactating women [3].

The Metallurgy of Zinc Coated Steels. https://doi.org/10.1016/B978-0-323-99984-7.00011-7

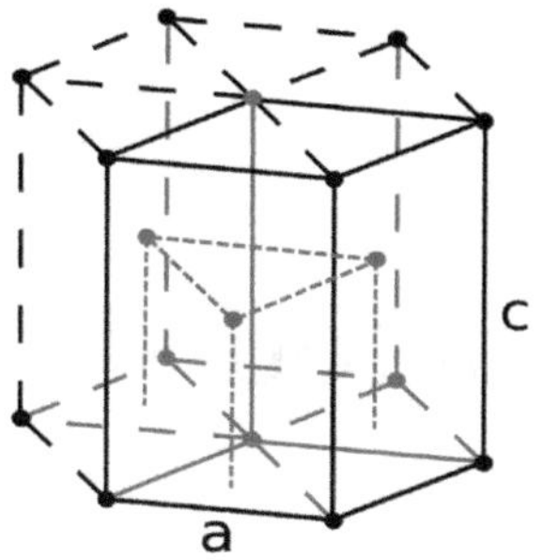

FIG. 1.1 HCP crystal structure.

TABLE 1.1 Elemental properties.

Elemental properties	
Atomic number	30
Atomic weight	65.39
Melting point	420°C (788°F)
Boiling point	907°C (1665°F)
Heat of fusion	7.32 kJ/mol
Heat of vaporization	115 kJ/mol
Density @ R.T.	7.14 g/cm^3
Density @ m.p.	6.57 g/cm^3
Oxidation state	+2
Crystal structure	Hexagonal close-packed (hcp)

Zinc is nonaccumulative in the body, therefore is nontoxic as organisms ingest zinc with food and excrete excess quantities. Zinc supplements are used to fight the common cold, and zinc is thought to improve immune cell function that counters viral infections or by reducing the ability of viruses to multiply, as in the COVID-19 virus [4].

1.2 In the beginning

The history of galvanizing (see Lamesch for a complete review of zinc history, Ref. [1]) involves many scientists that goes back to the first use of zinc as a coating on steel. In 1742, Paul-Jacques Malouin (1701–78) the "father of hot-dip zinc," dipped tin plate into a zinc bath and produced a new coating with rust-proof properties and a replacement for the standard at that time, tin plate [5]. Malouin's discovery was overlooked by many including Antoine-Laurent Lavoisier (1743–94), the founder of modern chemistry, who continued to recommend that laboratory utensils be protected by tin rather than zinc. Little did scientists know that the tin plate in Malouin's experiment acted as a shield between the iron and the oxygen in the air, preventing the iron from oxidizing, forming an oxide film on the surface of the iron, and preventing the zinc from adhering to the iron. However, in 1802, Karl Friedrich Buschendorf (1763–1811) proposed an improved method of zinc plating that instead of using tinplate involves pickling iron sheets in sal ammoniac, better known now as ammonium chloride [6]. Thus, the hot-dip zinc-coated steel process was born. It is interesting that almost no tin-plated items from this era or before can be found. Although tin provides excellent barrier protection, it cannot protect the steel from scratches and cut edges. At these locations, the unprotected iron corrodes, in part to sacrificially protect the tin coating until the article was destroyed. In contrast, one of the great values of zinc coatings are their ability to galvanically protect steel even when the coating is damaged.

True to form, theory followed practice, and it began with the discoveries of Luigi Galvani (1737–98) when in 1791 he published his experiments that caused spasmodic contraction of frog's leg muscles by touching the spinal cord with wires of dissimilar metals [7]. Galvani labeled this "animal electricity." Also, Alessandro Volta (1745–1827) reported his experiments in 1800 proving that electricity could be generated chemically by showing that any pair of dissimilar metals would give rise to an electrical force [8]. A result of his findings enabled Volta to arrange metals in a particular sequence, later to be called the Standard Potential of Elements, which showed the relative reactivity of the elements he studied. For these elements he found the sequence of reactivity to be:

Copper
Tin ↓ **more**
Iron **reactive**
Zinc

The relative position of iron between tin and zinc in this sequence shows the corrosion benefit of coating iron with zinc, the zinc being sacrificial to iron. This eventually led to the development of the battery or voltaic cell [9].

Around the same time in England, Sir Humphrey Davy (1778–1829), working on voltaic cells, was able to show that while producing electricity, one metal corroded protecting the other metal, suggesting that zinc could be used for the protection on iron even outside the voltaic cell [10]. Along with his laboratory assistant at that time, Michael Faraday (1791–1867), in 1829 he found an important property of zinc whereby it protected iron against rust even at a distance. He observed that iron exposed to moisture and air did not rust when placed in physical contact with zinc, leading to the concept of cathodic protection [11]. Faraday continued his studies producing the fundamental laws of electrochemistry, the cathode and anode, which was essential for electro galvanizing.

1.3 Zinc coating processing

Modern galvanizing was born when Stanislas Sorel received a French Patent in May 1837 (Fig. 1.2), followed by the USA Patent (US 510) in December 1837. Sorel built on the works of Davy and Faraday using electrochemical principles that zinc will protect iron from corrosion to produce a zinc-coated iron product. By pickling the surface of the iron prior to dipping into a zinc bath, he was able to show that the zinc was adhesive to the iron and that iron would not rust at the edges or even if filed. He called the process Galvanization of Iron in honor of Luigi Galvani and his frog experiments. The new terms "galvanizing," "to galvanize," and "galvanic" were quickly accepted in the French language and, soon afterward, in other languages.

Actually, the term was first coined from the French "galvanizer" in 1801, "to stimulate by galvanic electricity". The term became part of the local vernacular and was used in the general literature of that time, e.g., by Thomas Hood in his 1845 poem, "Love has no Eyes" [12] and in the novel "Villete" by Charlotte Bronte in 1853 [13]. During the American Civil War, 1861–1865, former Confederate prisoners of war who swore allegiance to the United States and joined the Union Army were called Galvanized Yankees. The term "galvanized" continues to be used metaphorically of any stimulus that results in any activity by a person or group of people, as in "galvanize into action", meaning stimulating a complacent person or group to take action.

Timing is everything. Iron ships were being constructed from 1838, and corrosion in salt water was becoming a significant problem. In the past, copper plates were used to protect wooden hulls but, consistent with its place in the Standard Potential Series, would accelerate corrosion of the iron. Marine paints did not help, so the French naval command encouraged Sorel to galvanize hulls, anchor chains, bolts, and nails with great success. Industrial production of galvanized iron soon followed and so did patents. As an example, C.B. Miller received a US Patent, No. 10,976, in 1854, Fig. 1.3.

Concurrently, based on the experiments of Davy, Volta, and Faraday, electrochemistry became commercially very important. One result was to produce a zinc coating on iron, again for corrosion protection, called electrogalvanizing. Although originally a shortcoming of the process was the limited use of the voltaic cell battery, new battery research by Prof. J.F. Daniell (1790–1845), who invented the Daniell cell in 1840, utilized a semipermeable membrane between the anode and cathode to extend the life of the battery [14]. Sorel obtained an electrogalvanizing patent in 1840. Technological improvements on the electrogalvanizing process soon followed and as an example Charles Bresley in 1862, was awarded a US patent, No. 36,750, for an "Improved Process of Electroplating Iron, steel, …", Fig. 1.4. Sherard O. Cowper-Coles (1866–1936), whose name is also connected with the sherardizing process used for coating of fasteners, obtained a GB Patent 189,419,797 in 1895, which proved impractical. Nevertheless in 1936, T. O. Tapp received a patent (GB445719A) for his continuous electrogalvanizing process. In 1938, Tapp was able to electrogalvanize wide thin sheets with remarkable quality. Development continued in electrogalvanized processing of sheet until it was largely replaced by hot-dip coatings whose quality significantly improved in the 1990s. Electrogalvanizing is still industrially important for the production of small items such as fasteners, where the coating must precisely follow the thread shape.

FIG. 1.2 Original Stanislas Sorel French Patent 93.

In the early stages of electroplating of zinc on steel, both chloride baths (Fr patent 358,622) and cyanide baths were used (US Patents 1,451,543 and 2,080,483). It wasn't until the 1930s when the first bright deposits of zinc were obtained by L.R. Westbrook, with an alkaline cyanide electrolyte (e.g., US patent 2,080,520 and US patent 2,233,500). Smooth electrogalvanized product was obtained in 1940 by J.P.Hibbell using an ammoniacal zinc solution (US Patent 2,200,987). Since that time, much work was carried out on alkaline electrolytes, acidic electrolytes (based on both chlorides and sulfates), and bath compositions in order to produce very bright deposits.

Thus, two zinc coating processes, hot-dip galvanizing and electrogalvanizing, became available to commercially produce zinc coatings. In the 1840s, an English company, Morewood & Rogers, commercially pioneered the use of hot-dip galvanized corrugated sheet for building construction. In particular, their 1843 patent GB 9720, mechanized the production of hot-dip galvanized steel, and is described as an iron frame placed on a zinc pot, which is driven by a steam engine, with supporting rolls designed to propel lengths of sheet through the zinc bath. Morewood & Rogers went on to construct many additional components to their semicontinuous process. In 1846, they proposed a mechanical system for controlling

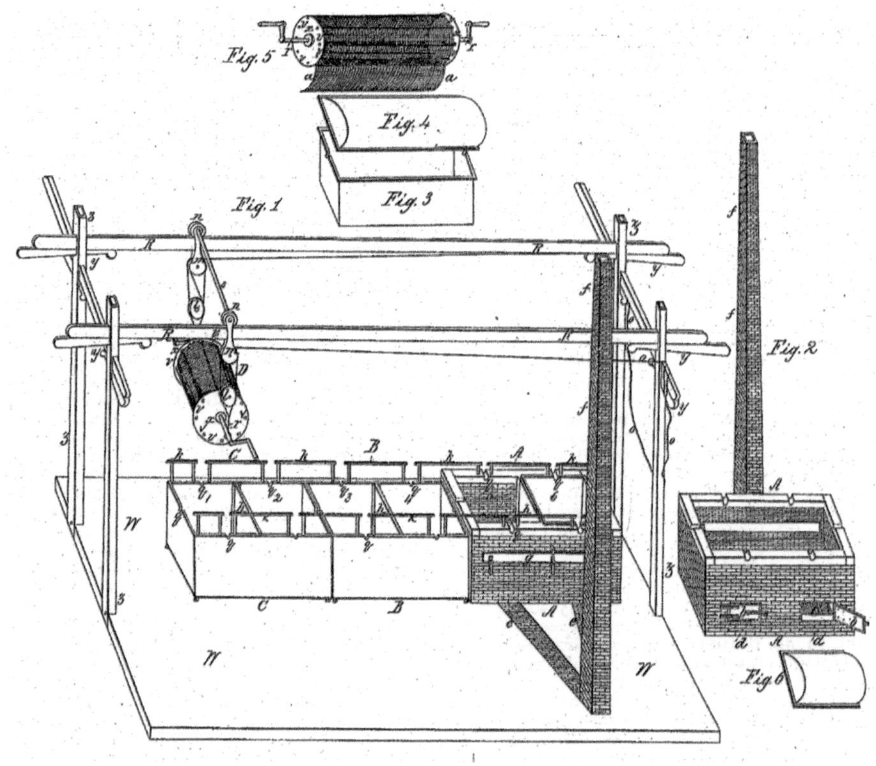

FIG. 1.3 C.B. Miller, US Patent 10,976.

and balancing the thickness of the molten zinc with skimming rolls (GB 11476). In all, they obtained over 20 patents related to the production of zinc-coated steel. Many modifications to the process were made between the years of 1880 and 1920. Of significance was the design of Heathfield in 1879 (US Patent 211,905) and Poppleton at John Lysaght in 1888.

Preparation of the steel surface for galvanizing until this time was based on aqueous treatments. The breakthrough in the development of furnace treatments to prepare the steel surface for hot-dip galvanizing resulted from the inventions of Tadeus Sendzimir (1894–1989). After spending time in Shanghai as a galvanizer and returning to his country of origin, Poland, after WW1, Sendzimir developed the concept of a continuous galvanizing line, to eliminate the sheet-by-sheet processing that had up to this point characterized hot-dip galvanizing. He took his ideas to the United States in 1929, but because of economic circumstances at that time, had to return to Poland. Here, in conjunction with S. Inwald, he was able to build his CGL line and prove his invention.

Sendzimirs' process, using a hydrogen atmosphere to reduce the iron oxides on the surface of the steel sheet, eliminated the need for the pickling-fluxing step and heating of the zinc bath. Eliminating the fluxing salts allowed for the efficient use of aluminum in the zinc bath, ranging between 0.1% and 0.2%. The resulting good-quality product had regular spangles and a very thin intermetallic inhibition layer. The steel strip vertically exited the zinc pot allowing for a uniform zinc layer on both sides of the steel sheet. In the United States, Sendzimir licensed Armco Steel, who acquired the rights to Sendzimir's patents (US Patents 2,110,893, 2,136,957 and 2,197,622), Fig. 1.5. As the Sendzimir concept continued to evolve, it was adapted to additional processes such as aluminizing, continuous annealing, and galvannealing.

Although the Sendzimir process was meant to replace the need for the pickling-fluxing step, it gave rise to the development of a continuous fluxing process that used batch-annealed coils that required no further heat treatment as in Sendzimir, or could be processed "full hard" after cold rolling to give the highest possible strength. The new process, patented in 1958 (US Patents 2,824,020, 2,824,021, and 2,823,641), Fig. 1.6, was developed at Wheeling Steel by N.E. Cook (1900–64) and S.L. Norteman (1913–95). The furnace was therefore limited to drying the fluxing salts and ran at relatively low temperatures, 250°C. As a result, the Cook-Norteman lines led to high throughput, twice the current Sendzimir lines. However, a major problem was the management of flux fumes, despite this, the Cook-Norteman process, or its derivatives, is still used in many countries.

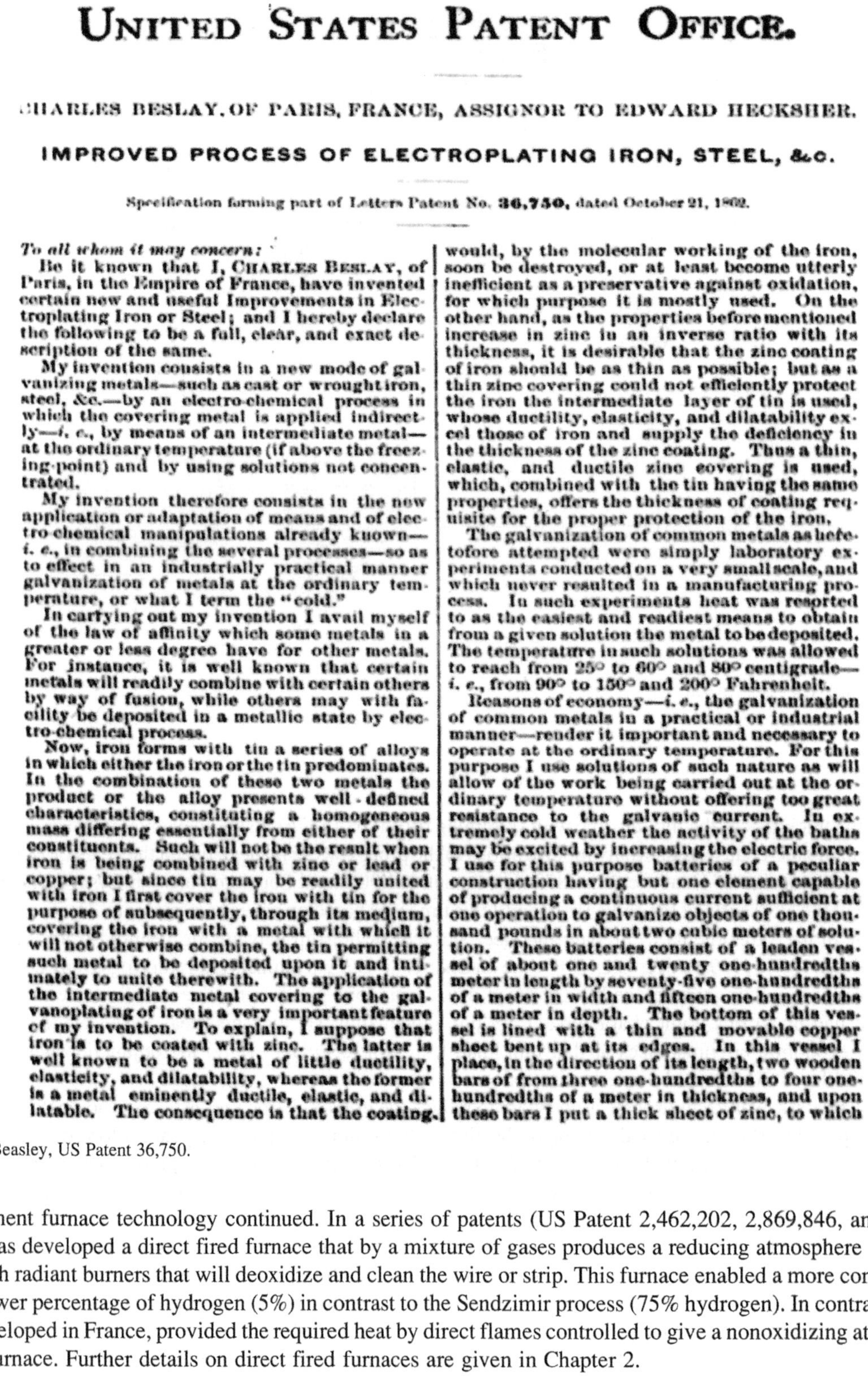

UNITED STATES PATENT OFFICE.

CHARLES BESLAY, OF PARIS, FRANCE, ASSIGNOR TO EDWARD HECKSHER.

IMPROVED PROCESS OF ELECTROPLATING IRON, STEEL, &c.

Specification forming part of Letters Patent No. 36,750, dated October 21, 1862.

To all whom it may concern:

Be it known that I, CHARLES BESLAY, of Paris, in the Empire of France, have invented certain new and useful Improvements in Electroplating Iron or Steel; and I hereby declare the following to be a full, clear, and exact description of the same.

My invention consists in a new mode of galvanizing metals—such as cast or wrought iron, steel, &c.—by an electro-chemical process in which the covering metal is applied indirectly—*i. e.*, by means of an intermediate metal—at the ordinary temperature (if above the freezing-point) and by using solutions not concentrated.

My invention therefore consists in the new application or adaptation of means and of electro-chemical manipulations already known—*i. e.*, in combining the several processes—so as to effect in an industrially practical manner galvanization of metals at the ordinary temperature, or what I term the "cold."

In carrying out my invention I avail myself of the law of affinity which some metals in a greater or less degree have for other metals. For instance, it is well known that certain metals will readily combine with certain others by way of fusion, while others may with facility be deposited in a metallic state by electro-chemical process.

Now, iron forms with tin a series of alloys in which either the iron or the tin predominates. In the combination of these two metals the product or the alloy presents well-defined characteristics, constituting a homogeneous mass differing essentially from either of their constituents. Such will not be the result when iron is being combined with zinc or lead or copper; but since tin may be readily united with iron I first cover the iron with tin for the purpose of subsequently, through its medium, covering the iron with a metal with which it will not otherwise combine, the tin permitting such metal to be deposited upon it and intimately to unite therewith. The application of the intermediate metal covering to the galvanoplating of iron is a very important feature of my invention. To explain, I suppose that iron is to be coated with zinc. The latter is well known to be a metal of little ductility, elasticity, and dilatability, whereas the former is a metal eminently ductile, elastic, and dilatable. The consequence is that the coating would, by the molecular working of the iron, soon be destroyed, or at least become utterly inefficient as a preservative against oxidation, for which purpose it is mostly used. On the other hand, as the properties before mentioned increase in zinc in an inverse ratio with its thickness, it is desirable that the zinc coating of iron should be as thin as possible; but as a thin zinc covering could not efficiently protect the iron the intermediate layer of tin is used, whose ductility, elasticity, and dilatability excel those of iron and supply the deficiency in the thickness of the zinc coating. Thus a thin, elastic, and ductile zinc covering is used, which, combined with the tin having the same properties, offers the thickness of coating requisite for the proper protection of the iron.

The galvanization of common metals as heretofore attempted were simply laboratory experiments conducted on a very small scale, and which never resulted in a manufacturing process. In such experiments heat was resorted to as the easiest and readiest means to obtain from a given solution the metal to be deposited. The temperature in such solutions was allowed to reach from 25° to 60° and 80° centigrade—*i. e.*, from 90° to 150° and 200° Fahrenheit.

Reasons of economy—*i. e.*, the galvanization of common metals in a practical or industrial manner—render it important and necessary to operate at the ordinary temperature. For this purpose I use solutions of such nature as will allow of the work being carried out at the ordinary temperature without offering too great resistance to the galvanic current. In extremely cold weather the activity of the baths may be excited by increasing the electric force. I use for this purpose batteries of a peculiar construction having but one element capable of producing a continuous current sufficient at one operation to galvanize objects of one thousand pounds in about two cubic meters of solution. These batteries consist of a leaden vessel of about one and twenty one-hundredths meter in length by seventy-five one-hundredths of a meter in width and fifteen one-hundredths of a meter in depth. The bottom of this vessel is lined with a thin and movable copper sheet bent up at its edges. In this vessel I place, in the direction of its length, two wooden bars of from three one-hundredths to four one-hundredths of a meter in thickness, and upon these bars I put a thick sheet of zinc, to which

FIG. 1.4 C. Beasley, US Patent 36,750.

Pretreatment furnace technology continued. In a series of patents (US Patent 2,462,202, 2,869,846, and 3,320,085), Fig. 1.7, Selas developed a direct fired furnace that by a mixture of gases produces a reducing atmosphere in the furnace chamber with radiant burners that will deoxidize and clean the wire or strip. This furnace enabled a more compact line and ran much lower percentage of hydrogen (5%) in contrast to the Sendzimir process (75% hydrogen). In contrast, the Heurty process, developed in France, provided the required heat by direct flames controlled to give a nonoxidizing atmosphere in a horizontal furnace. Further details on direct fired furnaces are given in Chapter 2.

Because of the requirements in the automotive industry for smooth, thin zinc coatings, "jet wiping" was invented. Prior to the 1970s, the control of coatings had been managed by coating rolls placed at the exit of the strip from the zinc bath. These had their origins in the mechanically driven wringer rolls used in the early semicontinuous galvanizing machines described earlier. Control of coating weight by coating rolls is still practiced by some lines producing roofing panels and other industrial products.

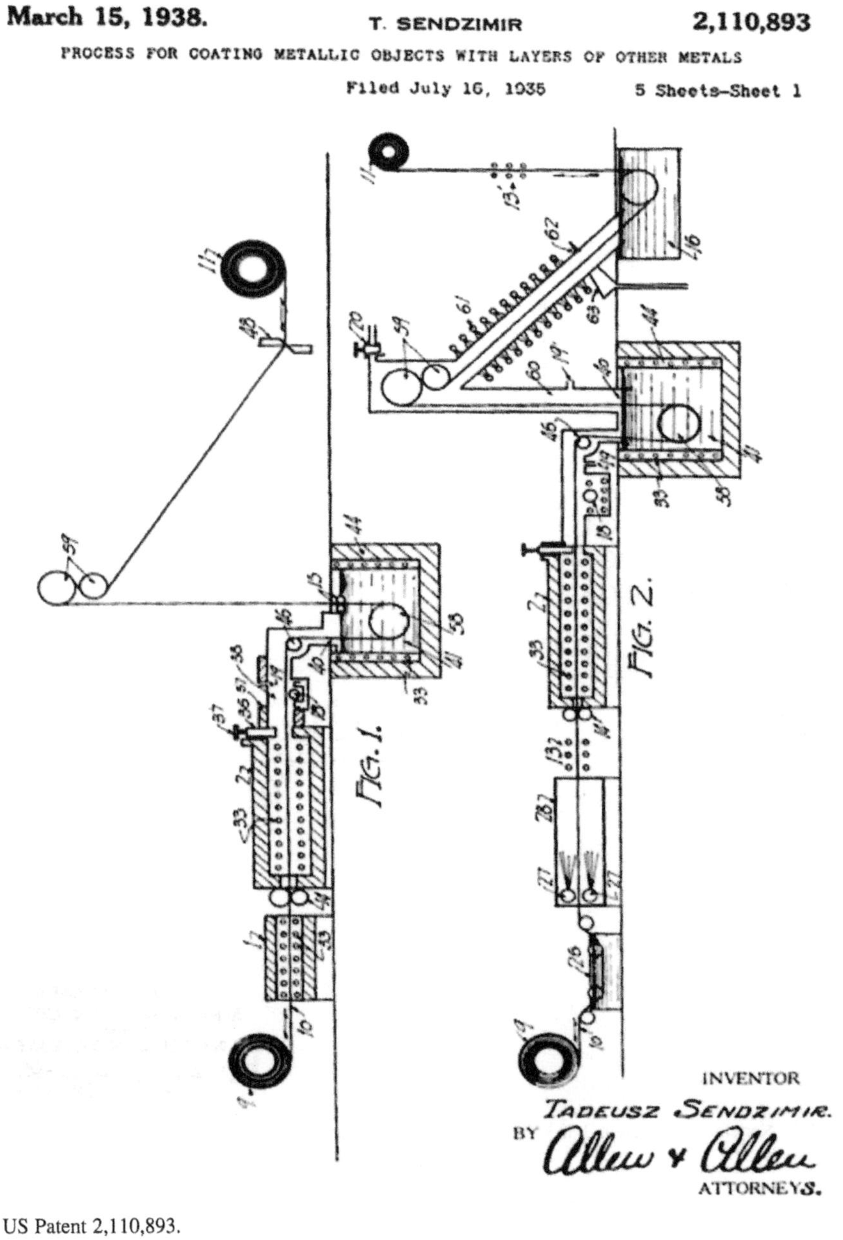

FIG. 1.5 T. Sendzimir, US Patent 2,110,893.

The concept of controlling zinc coating thickness by jet wiping was first developed in 1883 by H.A. Young using compressed air (US Patent 287,076). Because of the process being unsuccessful, it was abandoned, although work continued sporadically until the automotive need appeared. In 1970, after a series of failed patent applications dating back to the 1960s, J.T. Mayhew received a patent (US 3,499,418) for an apparatus that would control coating weight/thickness by using nozzles that would direct gas under pressure onto the surface of the molten zinc after emerging from the zinc bath, Fig. 1.8. This design was similar to those being used in the paper industry at that time. Further details on coating control are given in Chapter 7.

Galvannealing, a post pot heat treatment of the zinc coating in which the zinc coating is transformed to a series of Fe-Zn compounds, has been around since the 1920s. J.I. Herman received a patent (US 1,468,905) in 1923 for producing a heavy zinc coating that was flexible, malleable, and smooth. Commercial galvanneal coatings were also produced by removing as much liquid zinc as possible from a coating produced in a low Al($<0.1235\%$Al at 460°C) bath giving a coating weight of 40 g/m^2. The development of heavy coatings proved difficult because the multiphase microstructure powdered

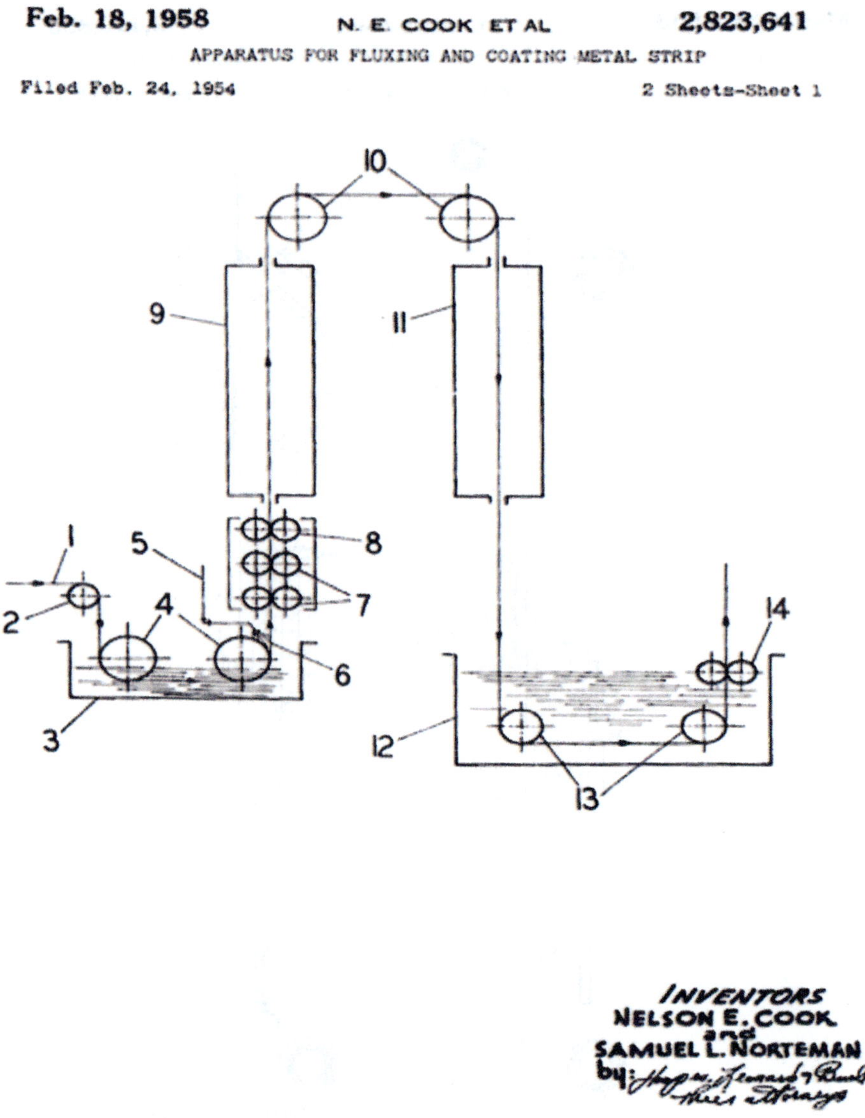

FIG. 1.6 N.E. Cook and S. L. Norteman, US Patent 2,823,631.

easily. Major research in the 1990s on optimum ZnAlFe phase morphology, steel chemistry, and line control resulted in the successful application of Galvaneal coatings with coating thicknesses in excess of 8 μm [15]. Further information on galvanneal coatings is given in Chapter 8.

1.4 Zinc coating alloy development

Additions of alloying elements to the zinc bath for better control of the zinc coated product have long been of interest, particularly for hot-dip zinc coatings. One of the most important elements added to zinc coating is aluminum. J.W. Richards (1891–1918), Professor of Metallurgy at Lehigh University, was a strong proponent of the use of aluminum throughout his career. Yet, in 1891, he received a patent, US No. 456,204, Fig. 1.9, in which he proposed diffusing aluminum throughout a bath of metallic zinc to produce a much more uniform zinc galvanized coating, which is "highly crystalline and permanently brilliant, while at the same time being exceedingly malleable and tenacious, as well as strongly adherent to

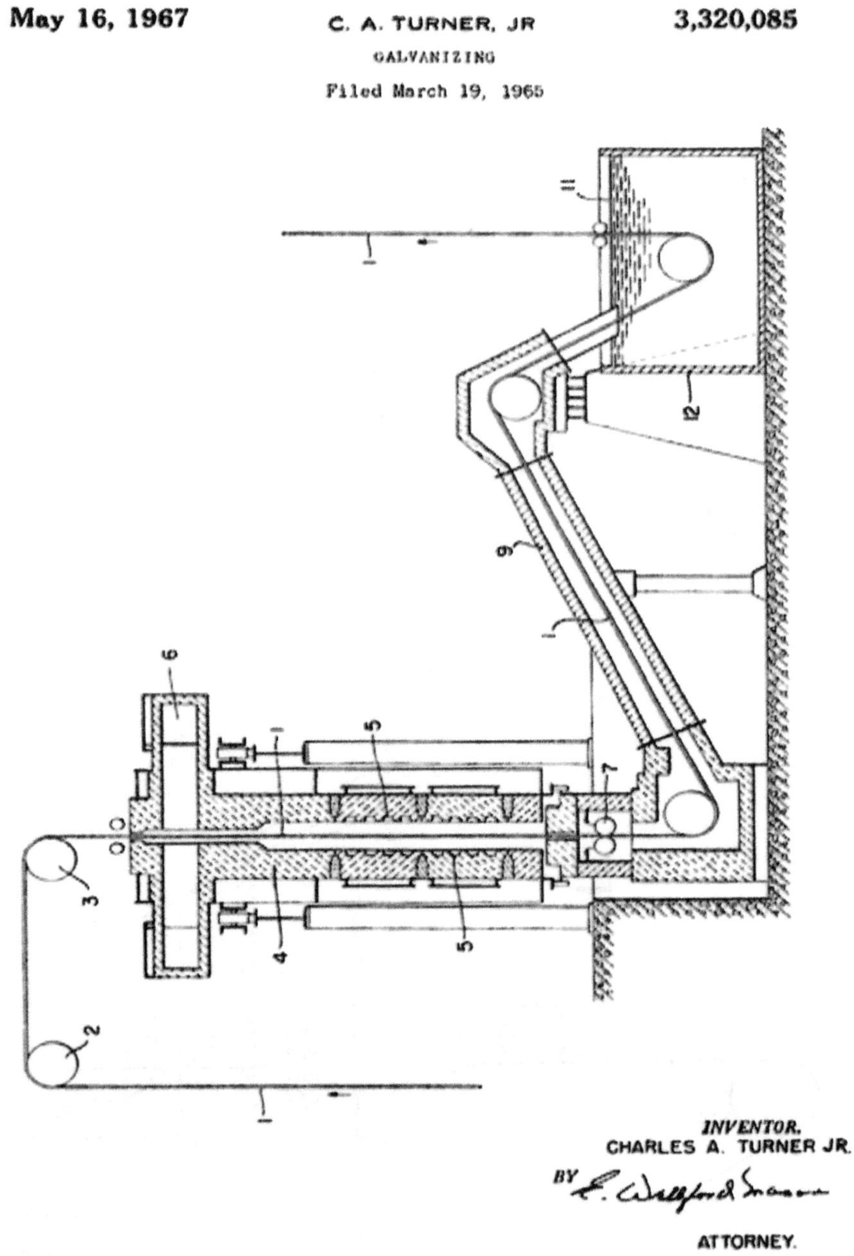

FIG. 1.7 C.A. Turner, US Patent 3,320,085.

the iron." Previously, several investigators experimented with zinc alloy additions such as lead, tin, copper, and antimony (1873, US Patent 144,403), or tin, antimony lead, and bismuth (1885, US Patent 328,239). Data gathered in 1911 and further reported by Cone [16], showed the relationship between the "tightness" (bend cracking resistance) of the coating and total aluminum, this favorable parameter increases rapidly in the Al range between 0.12 and 0.14 wt%, validating the discovery of Prof. Richards.

Zinc-coated iron or steel was coated with small amounts of lead in the zinc (1909, US Patent 933,612), or barium, strontium, and/or beryllium (1933, US Patent 1,939,667). In the 1950s, several investigators were able to show that additions of 0.1%–0.3% Al suppressed the formation of Fe-Zn compounds in the Zn coating [17–19]. This reaction was termed "inhibition." In the 1960s and following years, systematic studies were conducted on Zn-Al alloy-coated steel. In 1967, Borzillo and Horton received patents for a zinc coating containing 25%–70% aluminum and silicon, not less than 0.5% (US Patents 3,343,930 and 3,393,089), Fig. 1.10. These patents were the basis for the Galvalume coatings. In 1970, Ghuman

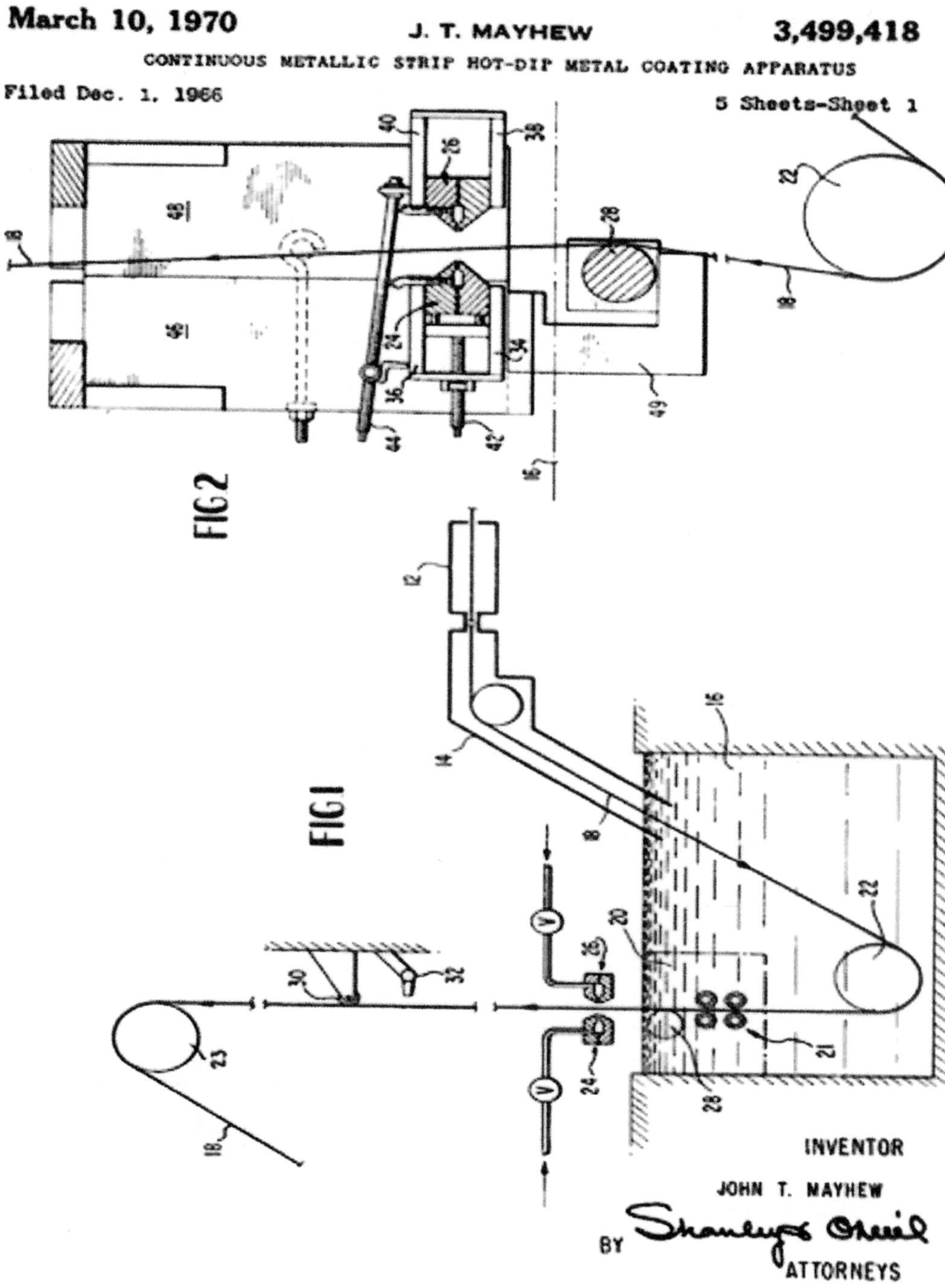

FIG. 1.8 J.T. Mayhew, US Patent 3,499,418.

and Goldstein [20] studied the inhibition effect in 0%–10% Al-Zn baths, followed by research on the effect of silicon content on Zn-55% Al Galvalume coatings by Selverian, et al. [17,21,22]. Following promising work at Inland Steel (now Arcelor Mittal) by H.H. Lee (US Patents 4,029,478 and 4,051,366), Zn-5% Al coatings (Galfan) were developed by CRM in 1978 (US Patent 4,448,748), and in the late 1980s, a great amount of research was conducted on annealing Zn coatings [15]. In 1983, Berke and Townsend received a patent (US 4,401,727) for coatings of Al-Zn-Mg-Si based on the Galvalume composition, although Inland Steel had done research in the 1970s on small additions of 1.5%–5% Mg and at least 0.5% Si (US patents, 3,505,042 and 3,505,043) and further additions of antimony and lead (US Patent 4,056,366) at much lower aluminum contents.

UNITED STATES PATENT OFFICE.

JOSEPH WILLIAM RICHARDS, OF PHILADELPHIA, PENNSYLVANIA.

MANUFACTURE OF GALVANIZED IRON.

SPECIFICATION forming part of Letters Patent No. 456,204, dated July 21, 1891.

Application filed September 27, 1890. Serial No. 366,358. (Specimens.)

To all whom it may concern:

Be it known that I, JOSEPH WILLIAM RICHARDS, of Philadelphia, in the State of Pennsylvania, have invented certain new and useful Improvements in the Manufacture of Galvanized Iron, whereof the following is a specification.

My improvements relate to the process of coating iron with zinc in the manner which is commercially called "galvanizing." As is well known, the ordinary method of galvanizing sheet or other iron consists in dipping it into a bath of melted zinc and allowing it to remain until a layer of zinc of the desired thickness has been deposited thereon and adheres thereto. I have discovered that by the diffusion of a very small percentage of metallic aluminum throughout the bath of melted zinc a much more uniform deposit upon the sheet-iron can be effected, and, moreover, that, as compared with previously-known products, the coating is highly crystalline and permanently brilliant, while at the same time being exceedingly malleable and tenacious, as well as strongly adherent to the iron.

For the conduct of my process in its preferred form I proceed as follows: As the quantity of aluminum required is very small, its diffusion throughout a large bath might be found practically difficult were it used alone or in a pure form. I therefore first form an alumino-zinc alloy by melting in a suitable crucible, say, two per cent. of metallic aluminum and adding thereto ninety-eight per cent. of metallic zinc. When thoroughly combined, the alloy thus formed is cast into small bars for convenience in use. The zinc in this alloy is, however, merely a vehicle for the aluminum contained therein to permit its more ready diffusion throughout the galvanizing-bath, and said alloy is not to be understood as being a product of my process, nor is its use essential to the same. Having prepared the galvanizing-bath in the usual manner by melting a sufficient quantity of zinc, I add the alumino-zinc alloy just described thereto, preferably in the quantity of about four ounces of alloy to the ton of melted zinc in the bath, and after allowing a few minutes for diffusion of aluminum throughout the bath I dip the sheet or other iron which is to be galvanized in the ordinary manner. As the bath is depleted and new additions of zinc are made thereto the alumino-zinc alloy should be added in the same proportions as before. I find that the crystalline formation of the deposited zinc varies with the amount of aluminum in the bath, the size of said crystals increasing (up to a certain limit) with the increased percentage of aluminum; but irrespective of the differences in degree which results from the use of a greater or less percentage of aluminum the advantageous qualities above described as incident to such use will result from the conduct of the galvanizing process in the presence of aluminum in any substantial quantity. Hence I do not limit my claim to the use of the above-mentioned proportions.

It must be understood that my invention does not necessarily involve the deposit of an alumino-zinc alloy recognizable as such in the final product, and I do not wish to be understood as limiting myself to a process which yields such an alloyed coating. Hence in using the words "alumino-galvanizing" as a convenient term to describe the process I do not intend that the coating must essentially be a recognizable alloy, but merely that it possesses the characteristics due to deposit in the presence of aluminum.

Furthermore, though I have described the use of a vehicle or primary alloy of aluminum and zinc as the best method known to me of diffusing the aluminum throughout the galvanizing-bath, I do not mean to restrict my claim to the use of a vehicle for this purpose, since any method of addition which effectually diffuses the metallic aluminum throughout the bath would be the equivalent of the described process.

I claim—

The hereinbefore-described process of alumino-galvanizing iron, which consists in diffusing metallic aluminum throughout a bath of metallic zinc and then dipping the iron in said composite bath and permitting the coating to deposit thereon, substantially as set forth.

JOSEPH WILLIAM RICHARDS.

Witnesses:
JAMES H. BELL,
E. REESE.

FIG. 1.9 J.W. Richards, US Patent 456,204.

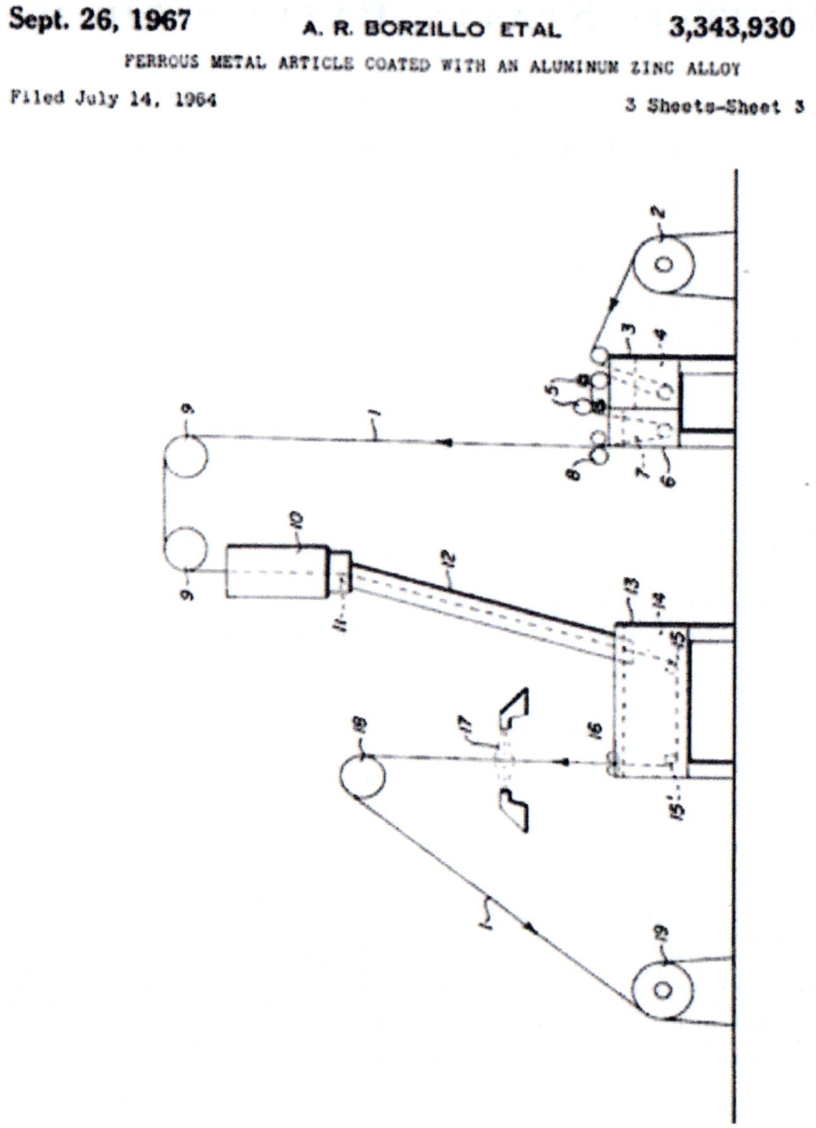

FIG. 1.10 A.R. Borzillo and J.B. Horton, US Patent 1,343,930.

A Appendix

A.1 Chronological history of zinc-coated sheet steel

- **200 BCE**
  Romans used zinc vapor to combine with copper to make brass
- **9th Century**
  Zinc production by the distillation process in India
- **1374**
  Large-scale production of a new metal zinc in India
- **1526**
  Paracelsus names the element "Zink"
- **1668**
  de Respour extracted metallic zinc from zinc oxide
- **1738**
  Champion patented a process for extracting zinc from calamine
- **1742**
  von Swab first distills zinc from calamine in Sweden
- **1742**
  Malouin dipped tinplate into a zinc bath to produce a hot-dip coating
- **1746**
  Marggraf heats calamine and charcoal to produce zinc metal
- **1791**
  Galvani discovers the electrochemical process that occurs between metals
- **1800**
  Volta proved electricity was generated chemically using dissimilar metals.
  Davy discovers corrosion protection of iron by zinc
- **1802**
  Buschendorf produces a zinc coating on a pickled iron substrate
- **1829**
  Faraday proves zinc acts sacrificially in protecting iron from corrosion
- **1837**
  Sorel obtained a French Patent to protect iron with a zinc layer coating
- **1840**
  J.F. Daniell invents the Daniell cell
- **!840**
  Sorel obtains an electro galvanizing patent
- **1843**
  Morewood & Rodgers pioneer the commercial use of zinc hot-dip galvanized corrugated sheet for building construction
- **1854**
  Industrial production of zinc coated French naval components
- **1856**
  John Lysaght buys a small hardware galvanizing business to produce corrugated sheet
- **1883**
  H.A. Young uses compressed air to control coating weight and thickness
- **1891**
  J.W. Richards patents the use of precise aluminum additions to the zinc bath to produce more uniform coatings
- **1936**
  T.O. Tapp receives a patent for a successful continuous electrogalvanizing process
- **1938**
  T. Sendzimir obtains patents for the continuous processing of zinc-coated sheet
- **1940**
  J.P. Hibbell using an ammoniacal zinc solution produces smooth electrogalvanized product
- **1950**
  Bablik published a comprehensive review of Hot-Dip Galvanizing

- **1958**
  Cook and Norteman patented a continuous fluxing process for the production of zinc coatings
- **1967**
  Selas Corp. develops a vertical direct fired furnace that will deoxidize and clean wire and strip. Heurty develops a horizontal direct fired furnace
- **1967**
  Borzillo and Horton patent Galvalume high aluminum zinc coatings
- **1970**
  J.T. Mayhew patents direct gas nozzles to control coating weight/thickness
- **1978**
  Zn-5% Al (Galfan) is developed at CRM
- **1979**
  Mackowiak & Short publish "Metallurgy of Galvanized Coatings"
- **1983**
  Berke and Townsend receive a patent for Al-Zn-Mg-Si hot-dip coatings
- **!983**
  Harvey and Richards publish "Zinc-Based Coatings for Corrosion Protection Of Sheet Steel and Strip"
- **1990s**
  First application of hot-dip coatings and galvanneal coatings to stringent automotive requirements and advanced steel technology
- **2000**
  Marder publishes a review of "The Metallurgy Zinc-Coated Steel"
- **2004**
  Lamesch publishes monumental book on "The World History of Galvannealing"
- **2016**
  Kuklik and Kudlacek publish "Hot-Dip Galvanizing of Steel Structures".

References

[1] J. Lamesch, The World history of Galvanizing, Arcelor Mittal, 2004, ISBN: 2-919969-25-0.

[2] M. Vandermissen, (McKinsey & Co.). "Zinc Market Outlook-What to Expect in Times of Uncertainty", Zinc Metal Roundtable Virtual Event, November 5–6, International Zinc Association, 2020.

[3] Institute of Medicine, Food and Nutrition Board, Dietary Intakes for Vitamin A, Vitamin K, Arsenic, Boron, Chromium, Copper, Iodine, Iron, Manganese, Molybdenum, Nickel, Silicon, Vanadium, and Zinc, Natinal Academy Press, Washington, DC, 2001.

[4] R.H. Schmerling, Do Vitamin D, Zinc and Other Supplements Help Prevent Covid-19 or Hasten Healing?, Harvard Health Publishing, 5 April, 2021.

[5] P.J. Maloun, Memore de L'Academie Royale des Science, 1745, p. 70.

[6] K.F. Buschendorf, Manufaktur, Hanlung, und Mode, J. Fabrik (1802).

[7] L. Galvani, De Viribus Electricitatus in Motu Musculari Commentarius, Annals of the University of Bologna, 1791.

[8] A. Volta, Memorie Sull'electricita Animal, Inserite nel GionarleFisico-Medico del Sig, Brugnatelli, Pavin, 1792.

[9] A. Volta, On the electricity excited by the mere contact of conducting substances of different kinds, Philos. Trans. R. Soc. Lond. (Fr.) 90 (1800) 403–431.

[10] H. Davy, Philos. Trans. R. Soc. (1824). 114, 151, 242, 328.

[11] M. Faraday, Experimental Research in Electricity, vol. 1 and 2, 1855.

[12] T. Hood, Love has not Eyes, 1845.

[13] C. Bronte, Villette, Smith, Elder & Co., London, 1853.

[14] J.F. Daniell, Daniell cell, Philos. Trans. (1836).

[15] C.E. Jordan, A.R. Marder, Morphology development in hot-dip galvanneal coatings, Metal. Mater. Trans. 25A (1994) 937.

[16] C. Cone, Iron and Steel Engineer, Association of Iron and Steel Engineers, March, 1962, p. 80.

[17] H. Bablik, Galvanizing (Hot Dip), third ed., Spon Ltd., London, 1950, pp. 204–223.

[18] M.L. Hughes, Proc. of the International Conference on Hot Dip Galvanizing, vol. 31, Zinc Development Assoc., Copenhagen, 1951.

[19] M.A. Houghton, The effects of aluminum and iron on the structure of galvanized coatings, in: Proc. 2nd Int. Conf. on Hot Dip Galvanizing, 59, Zinc Development Assoc., 1953.

[20] A.R.P. Ghuman, J.I. Goldstein, Reaction mechanisms for the coatings formed during hot dipping of iron in 0-10% Al-Zn Baths at 450° to 700°C, Metal. Trans. 2 (1971) 2903.

[21] J.H. Selverian, A.R. Marder, M.N. Notis, The reaction between solid iron and liquid Al-Zn baths, Metal. Trans. A 19A (1988) 1193–1203.

[22] J.H. Selverian, A.R. Marder, M.R. Notis, The effects of silicon on the reaction between solid iron and liquid 55 Wt pct Al – Zn baths, Metal. Trans. A 20A (1989) 543–555.

Chapter 2

Hot-dip coating methods: Continuous processing (CGL)

2.1 History and overview

As described in Chapter 1, the beginning of the modern era of continuous sheet galvanizing can be dated to Sendzimir's 1938 USA patent 2,110,893 that for the first time describes that a sheet galvanizing process involves producing "a thin and uniform film of oxides of the body (substrate) metal....by treating said...bodies in an oxidizing atmosphere, then under conditions of heat and a reducing atmosphere completely reducing said film so as to form a reduced metal layer in intimate contact and closely adhering to the body metal,......leading them beneath the surface of a bath of ...molten coating metal." This process had many advantages over the Cook-Norteman lines that used wet chemicals to prepare the steel surface for galvanizing, as described in [1]. The genius of Sendzimir's process was that both useful heat treatments of the steel sheet and preparation of a galvanizable surface could be accomplished simultaneously by the continuous galvanizing heat cycles and atmospheres. Most of the sheet steel delivered to modern continuous galvanizing lines (CGLs) is coming from cold rolling mills in full hard condition up to 2 mm in thickness; it is necessary to heat treat it to specified conditions for it to be useful to the customer. For the Cook-Norteman process, such heat treatments, including temper rolling, need to be performed before delivery to the galvanizing line. CGLs also process a wide variety of hot rolled strip up to 4 mm thickness, as described later in this chapter. CGLs have evolved since the time of Sendzimir's patent, handling a thickness range of 0.15–6 mm and strip widths up to 2100 mm with production rates approaching 1 million metric tons of annual production, processing steels that could not be imagined in Sendzimir's time.

Fig. 2.1 shows a modern CGL capable of applying Zn coatings to advanced high-strength steels (AHSS).

As summarized in [1], the basic CGL process steps are precleaning (not shown in Fig. 2.1), heating and soaking in a reducing atmosphere, cooling, and immersion in the Zn coating bath, wiping of the coating after it exits the bath to obtain the desired coating weight, the optional use of a galvannealing furnace when production of galvannealed coatings is required, cooling, normally with the assistance of after-pot coolers before reaching the top roll where the strip temperature must not exceed 360°C, additional cooling, then (not shown in Fig. 2.1) postgalvanizing operations that can include tension leveling, temper rolling, then inspection before coiling or cutting to length of the coated sheet in the delivery section. This chapter explains the advances made in CGL processing technology since 2000 that relate to the metallurgy of Zn-coated steel.

2.2 Cleaning

Rolling mills typical deliver sheet steel to a galvanizing line with total surface residues present at a level of 100–200 mg/m^2, these must be removed so that they do not interfere with the galvanizing reaction, contaminate furnace rolls, or contribute to the dross levels in the galvanizing bath, as described further in Chapter 4. The original Sendzimir process relied on only an open-air direct-fired furnace to clean the steel sheet before entering the high-temperature soaking furnace. This was later changed to heating in a nonoxidizing furnace, because with an oxidizing flame, carbonaceous contaminants can burn on the steel surface, leaving a residue; if they are volatilized in a reducing flame, they can be removed. Iron fines, either in metallic or oxidic form, are commonly found on the surface of rolled sheet. The original Sendzimir process further oxidized these iron fines during the furnace cleaning step, resulting in higher iron levels in the zinc bath compared with a nonoxidizing cleaning step. To reach levels of cleanliness for automotive product, typically 20 g/m^2 per side of iron fines and oil [3], a wet cleaning step prior to furnace entry is almost always used, this involves cleaning with an alkali solution, applied by either spraying or immersion, combined with abrasion by brushes, and in many cases combined with electrolytic action that helps lift oil and iron residues from the surface. A schematic of a typical wet cleaning section is shown in Fig. 2.2.

The Metallurgy of Zinc Coated Steels. https://doi.org/10.1016/B978-0-323-99984-7.00010-5

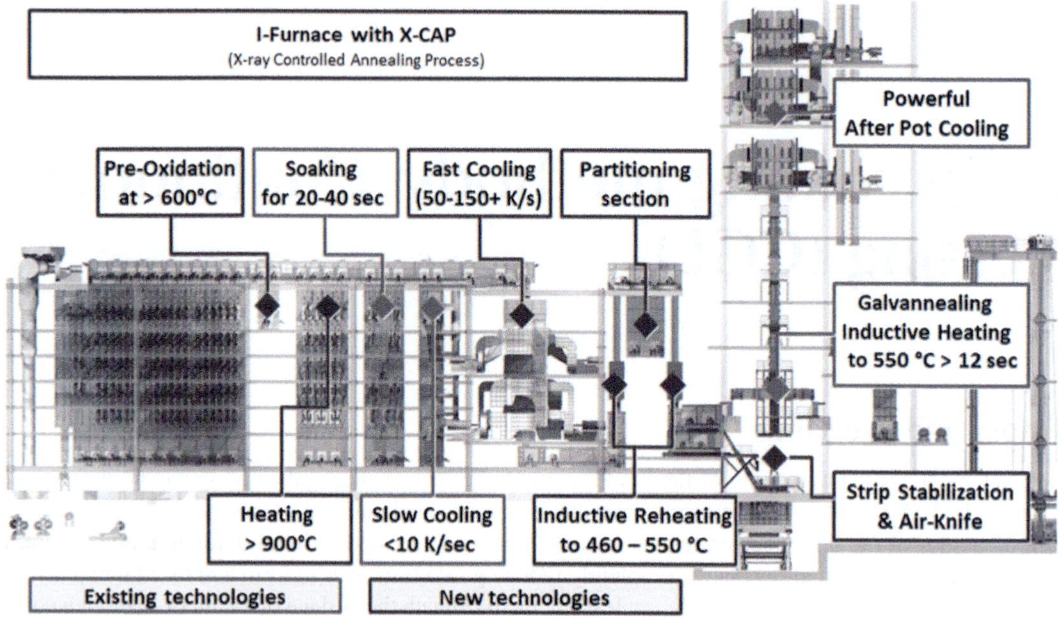

FIG. 2.1 Schematic of a modern continuous galvanizing line [2].

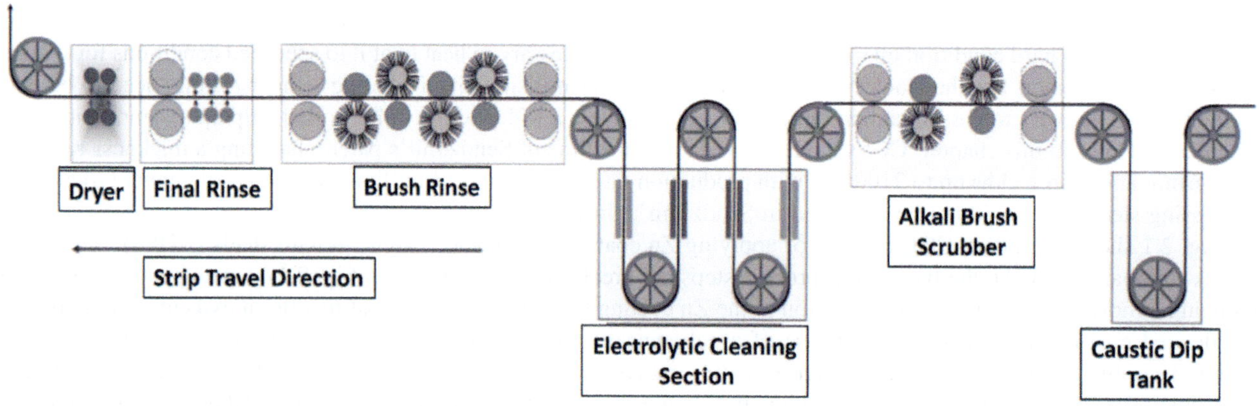

FIG. 2.2 Wet cleaning section of CGL (S. Hesling). *(S. Hesling, "Entry Section Processes" Section 3 "Chemical Cleaning of Strip" Fig. 1, undated.)*

The furnace cleaning step can include up to four furnace zones, with a typical furnace gas temperature of around 1200 °C. This gas temperature maximizes the potential for reducing contaminants and oxides on the strip. As the strip is heated in this section, the carbonaceous contaminants are removed at lower temperatures; surface oxides begin to be reduced in the final sections as the strip temperature reaches the range of 565–675°C. The furnace atmosphere is a mixture of CO from combustion and H_2; the air-fuel ratio in the furnace burners is usually set to around 6% excess combustibles to be sure a reducing atmosphere is present, the sum of the CO and H_2 compositions in the furnace atmosphere is typically 3%. If a neutral flame is used in the burners, soot deposition can result, contaminating the strip surface. The reduction of surface oxides continues in the soaking section that follows. Furnace cleaning that follows the wet cleaning stage has been enhanced in many cases by the use of direct flame impingement (DFI) burners, located a few cm from the strip. This has been found to enhance cleaning, while improving efficiency and productivity, in some cases making prior wet cleaning unnecessary and reducing the required duration of furnace treatment in this process step [4].

2.3 Steel types for continuous galvanizing

The wide variety of steel types processed in modern CGLs include the following major groups, for which Euronorm and ASTM standards are also shown:

- Commercial Steel (CS) ASTM A653, EN 10326
  - C range 0.04%–0.10%, Mn range 0.2%–0.6%, depending on the product being made.
  - steel is cold rolled with 50%–80% reduction.
- Forming Steel (FS) ASTM A653, EN 10142
  - C range 0.04%–0.08%, Mn about 0.25%.
  - to obtain better response to annealing, cold reduction is between 60% and 80%.
- Structural Steel (SS) ASTM A653, EN 10326
  - C range between 0.04% and 0.20%, and Mn range 0.4%–1.6%, depending on the product being made.
  - substrate is cold rolled anywhere from 50% to 70% reduction.
- Deep Drawing Steel (DDS) and Extra Deep Drawing Steel (EDDS) ASTM A653, EN 10292
  - generally made from ultralow carbon stabilized steels: C range 0.0040%–0.0070%, also written as 40–70 ppm, although some DDS is made using extra low carbon (0.015%–0.020%) steel.
  - EDDS and some DDS are produced to be fully stabilized (nonaging) after annealing and coating.
  - To maximize annealing response, cold reduction is generally high at 75% minimum.
- High-Strength-Low Alloy (HSLAS) ASTM A653, EN 10292
  - Typically made from microalloyed low carbon (C range 0.15%–0.25%) steel. The primary alloying element is Nb with a range of 0.005%–0.05%
  - Cold reduction rarely exceeds 60% due to high cold rolling loads.
- Advanced High-Strength Steel (AHSS) ASTM A1088, EN HCT 490X through EN HCT 1180G2
  - Produced using higher levels of alloying elements and carefully controlled annealing and cooling cycles. This category includes solution hardened steel (SHS), bake hardening steels (BHS), dual-phase (DP) steels, transformation-induced plasticity (TRIP) steels, complex-phase (CP) steels, and third-generation (3G) steels
  - Cold reduction rarely exceeds 60% due to high cold rolling loads

Compared with drawing quality conventional C-Mn grades, AHSS alloy compositions include higher levels of elements that have higher oxidation potentials than Fe. These can include Mn, Si, Al, and Cr. For a discussion of the effects of these elements on microstructures and mechanical properties of these grades, the reader is referred to De Cooman and Speer's book [5].

2.4 Galvanizing pretreatment steps for AHSS

The objective of pretreatment is to prepare the steel surface to be reacted with the zinc bath, which depends upon a surface consisting of clean, reduced iron. Simultaneously, the steel must be heat treated to achieve the desired microstructure. Alloying elements and their composition levels in AHSS strongly affect the choice of CGL process steps, if they are to be successfully galvanized. The main issue to be controlled is the surface-selective oxidation of several of these the alloying elements, mainly Mn, Si, Cr, and Al. In a CGL annealing furnace containing an atmosphere that is strongly reducing for Fe, say a partial pressure of oxygen $P_{O_2} = 10^{-26}$ atm, which at a temperature of 675°C has a dew point of −30°C, the concentration of each of these elements in solid solution at and near the steel surface is lowered by oxidation; equilibrium requires that replenishment by diffusion from the bulk (steel subsurface), which is accomplished by solid-state diffusion, particularly along steel grain boundaries. However, if a higher p_{O2} is used in the furnace, say 10^{-24} atm, which at 675°C corresponds to a dew point of +5°C, there is a surplus of oxygen on the surface. Because the diffusion coefficient of oxygen into the steel strip is on the order of 1000 times higher than that of Mn, Si, Al, and Cr, the oxidation of these elements occurs internally rather than on the steel surface when a sufficient oxygen supply is present, improving conditions for galvanizability [6–8]. This internal oxidation process is shown schematically in Fig. 2.3.

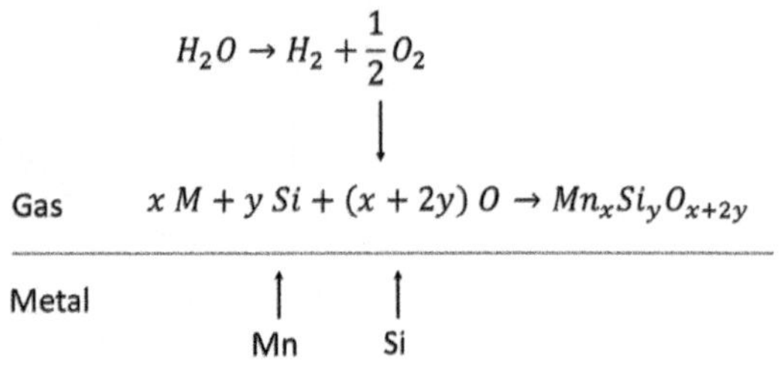

FIG. 2.3 Schematic of diffusion of surface-selective oxidizing elements and oxygen in (Mn,Si) steels.

The transition point between internal and external oxidation for a binary steel composition is given by

$$N^0_{B,\text{Critical}} = \left[\frac{g^*\pi V_{\text{alloy}} N^S_O D_O}{2\nu V_{BO_\nu} D_B}\right]^{\frac{1}{2}}$$

where

$N^0_{B,\text{Critical}}$ = critical mole fraction of alloying element B.
g^* =critical volume fraction of precipitated oxide (usually 0.3 is taken).
ν = stoichiometry of the oxide.
V_{alloy} = molar volume of the alloy.
V_{BO_ν} = molar volume of oxide BO_ν.
D = diffusion coefficient.
N^S_O = mole fraction of oxygen dissolved in A at surface.

Internal oxidation occurs when

$$N^0_B < N^0_{B,\text{Critical}}$$

And external oxidation occurs when

$$N^0_B \geq N^0_{B,\text{Critical}}$$

An example of the use of this equation, for a simple Fe-Mn composition, is shown in Fig. 2.4. The transformation boundary between Mn and MnO is shown by the Mn/MnO line on the left side. The furnace dew points required to reduce MnO to Mn are in the range from −50°C to −7°C even at a soaking temperature of 950°C. The other two lines show the transition boundary between external and internal oxidation for two different models, a simplified version based on the above equations (upper line) and a more generalized version that includes solute interactions and other factors (lower line). "External MnO" indicates the region where external MnO oxidation occurs on the steel surface (to the left side of these two lines), with internal oxidation occurring on the right side. At a constant dew point in the practical operating range for a CGL, i.e., −30°C on the horizontal axis of the graph, solely internal oxidation occurs when Mn < 6 wt% according to the general model. This corresponds to the $N^0_{B,\text{critical}}$ value given above; above this value, external oxides are present, according to the model.

This equation has been broadened to ternary and higher alloyed steels. Interactions between alloying elements can be ignored in the case of low alloy (Mn,Cr) compositions because the solubility products of the oxidation products are very low [10]. A more complicated example is shown in Fig. 2.5, for a C < 0.25%-1.7%Mn-0.1%Si-1.5%Al TRIP composition. Here, as predicted by theory, the oxidation fronts for Al and Si are pushed deeper into the steel as dew point increases. However, Mn is not, because of several competing reactions with FeO (wüstite) and $MnAl_2O_4$. This illustrates some of the complications arising when modeling internal and external oxidation of these steels. However, the formed external

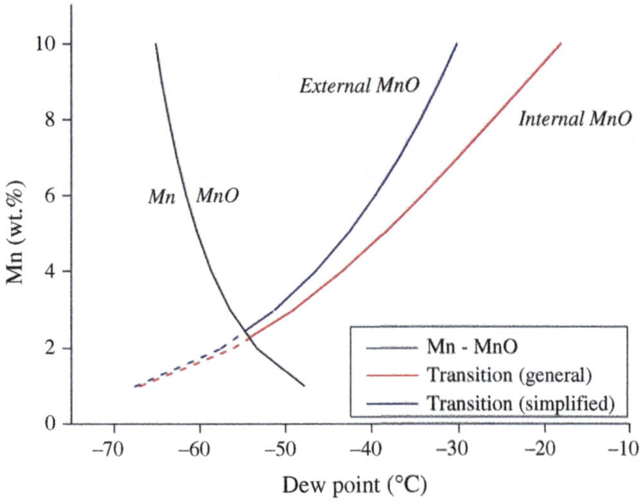

FIG. 2.4 Oxidation domains for binary Fe-Mn alloys oxidizing at 950°C in a N_2-5%H_2 atmosphere. The simplified model is based on the above equations [9]. *(V. Lashgari, Internal and External Oxidation of Manganese in Advanced High Strength Steels (Ph.D. thesis), University of Delft, September 2014.)*

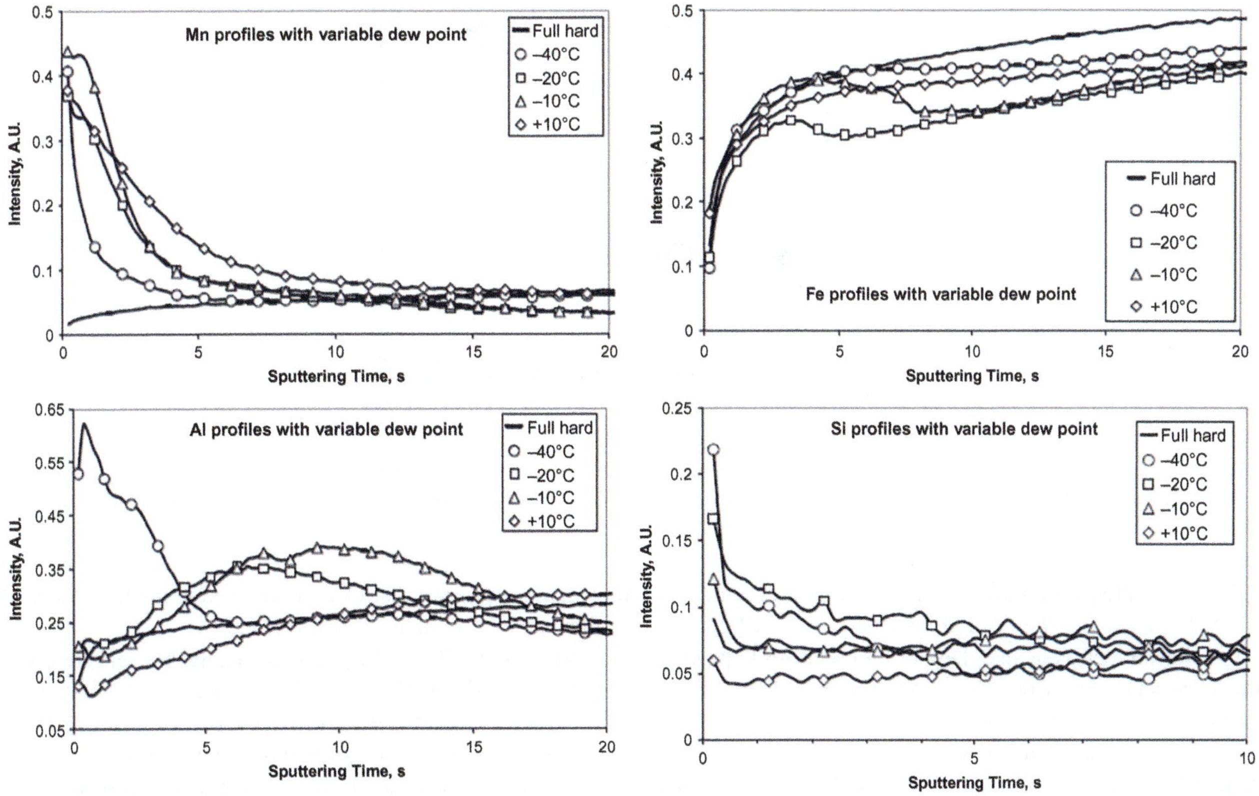

FIG. 2.5 Effect of dew point on near-surface oxidation behavior of an Mn-Al TRIP steel [11]. *(J. Staudte, J.-M. Mataigne, D. Loison, F. Del Frate, Galvanizability of high Mn grade versus mixed Mn-Al and Mn-Si grades, in: Proc. Galvatech 2011, Genoa, Associazione Italiana di Metallurgia.)*

oxides in this case are granular in morphology rather than film-like at high dew points, enabling a satisfactory quality Zn coating to be produced [11]. The internal oxidation layer grows as soaking time increases.

General predictions of the types of oxides to be found on Mn-Si steels are shown in Fig. 2.6A, for external oxides when a low dew point atmosphere is used, and for internal oxides in Fig. 2.6B, when a high dew point furnace is used.

Such steels can also be industrially processed using radiant tube furnaces (RTFs) in which the dew point is high in the first zones, resulting in internal oxidation of the Mn, Si, Al, and Cr with external oxidation of Fe to FeO (wüstite). Subsequent reducing furnace atmosphere zones then transform the surface FeO to metallic Fe, resulting in a surface suitable for

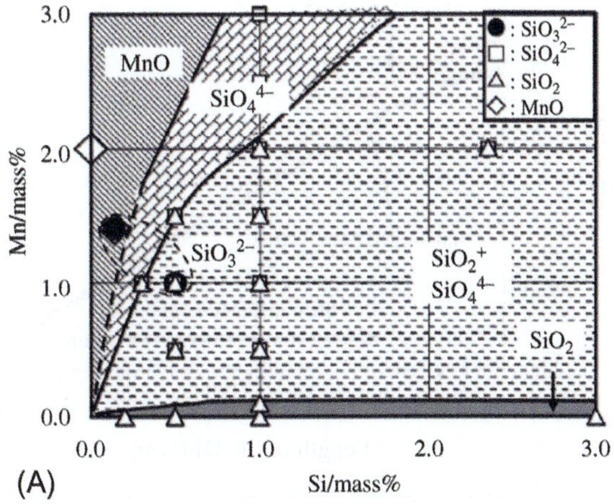

FIG. 2.6A External surface oxides predicted for Mn-Si grades, using a low dew point furnace atmosphere [12]. *(Y. Suzuki, T. Yamashita, Y. Sugimoto, S. Fujita, S. Yamaguchi, Thermodynamic analysis of selective oxidation behavior of Si and Mn-added steel during recrystallization annealing, ISIJ Int. 49 (4) (2009) 564–573.)*

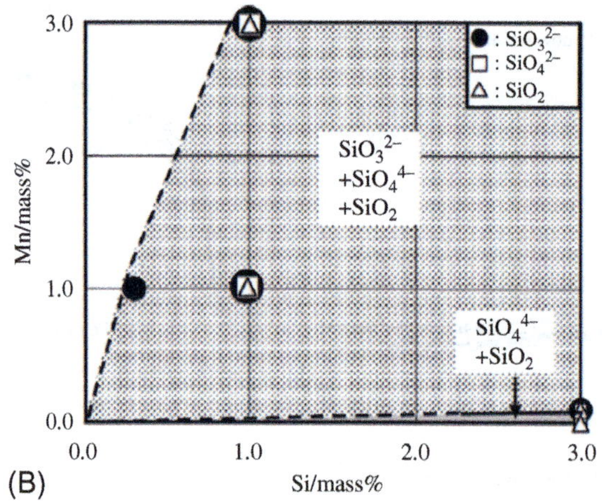

FIG. 2.6B Internal surface oxides predicted for Mn-Si grades, using a high dew point furnace atmosphere [12]. *(Y. Suzuki, T. Yamashita, Y. Sugimoto, S. Fujita, S. Yamaguchi, Thermodynamic analysis of selective oxidation behavior of Si and Mn-added steel during recrystallization annealing, ISIJ Int. 49 (4) (2009) 564–573.)*

galvanizing [13]. Alternatively a preoxidation section, containing an atmosphere of N_2-O_2, is contained in the first zones of an RTF, producing the desired internally oxidized condition, followed by a reducing section [14].

A similar technique uses DFF or DFI sections to produce the desired internally oxidized condition, the burners are set so that excess O_2 is present. This is followed by reducing sections in either a DFF or RTF. Fig. 2.7 shows the effect of λ, the air/fuel ratio, on strip temperature as the strip is heated in the DFF, moving from right to left. The normal burner settings are for a slightly reducing flame, $0.85 < \lambda < 0.96$. A $\lambda = 1.02$ setting increases oxidation of the strip and also its temperature; corresponding furnace temperatures are 500°C higher at any distance along these zones [15]. Surface oxides also increase surface emissivity, further increasing strip temperature at high λ values. External oxide thickness increases as the λ reaches the stoichiometric ratio for complete combustion because excess oxygen is then available to form these oxides, Fig. 2.8. The oxide composition for a Si-TRIP grade containing 1.48%Mn, 1.4wt%Si, and 0.69%Al formed under these oxidizing conditions is shown in Fig. 2.9. The surface has high proportions of both FeO and SiO_2 so that a surface featuring these mixed oxides can be predicted; the lower value of Al suggests that mixed oxides of Al and Si may also be present.

Process control, including control of the surface FeO layer and temperature measurement, has been found to be easier in the all-RTF process [14]. DFI preheating can also be used as the first step of oxidation/reduction treatments used to prepare the more difficult to galvanize AHSS grades for zinc coating. In this case, the surface must be essentially free of volatile

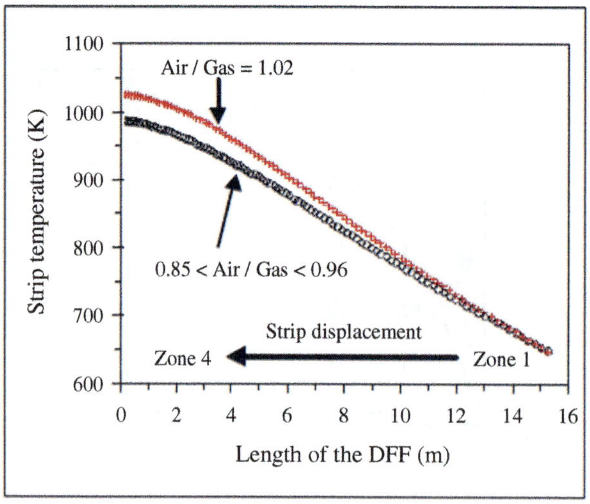

FIG. 2.7 Strip temperature (°K) evolution through the DFF with normal ($0.85 < \lambda < 0.96$) and increased ($\lambda = 1.02$) air/fuel ratio [15]. *(L. Bordignon, G. Angeli, H. Bolt, R. Hekkens, W. Maschek, J. Paavalainen, X. vanden Eynde, Enhanced hot dip galvanising by controlled oxidation in the annealing furnace, in: Proc. 44th Mechanical Working and Steel Processing Conference, vol. XL, AIST, September 2002, pp. 833–844.)*

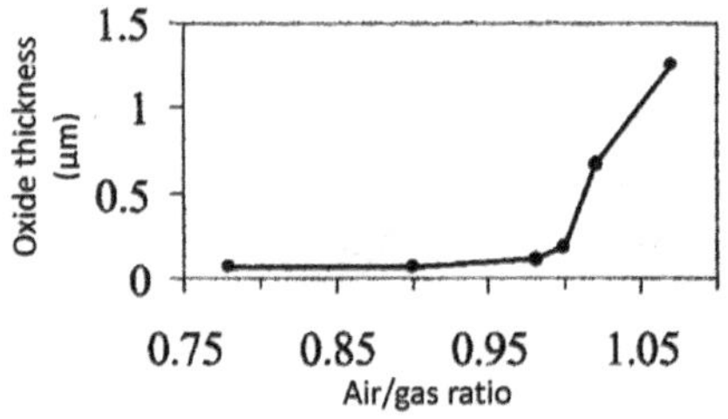

FIG. 2.8 External oxide thickness increases rapidly as λ reaches the stoichiometric ratio for combustion [15]. *(L. Bordignon, G. Angeli, H. Bolt, R. Hekkens, W. Maschek, J. Paavalainen, X. vanden Eynde, Enhanced hot dip galvanising by controlled oxidation in the annealing furnace, in: Proc. 44th Mechanical Working and Steel Processing Conference, vol. XL, AIST, September 2002, pp. 833–844.)*

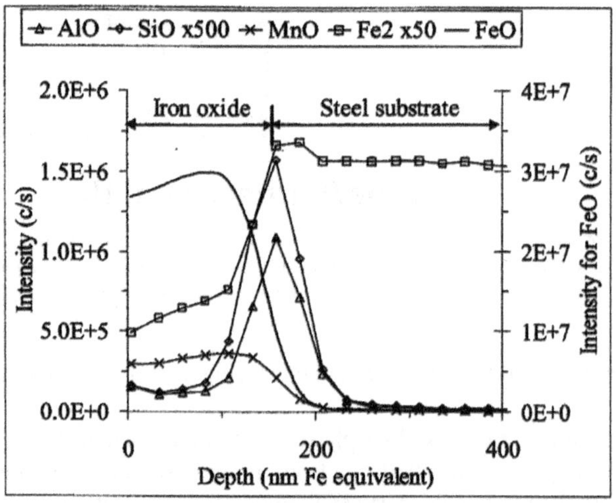

FIG. 2.9 Secondary ion mass spectroscopy surface depth profiles for a 1.48%Mn, 1.4wt%Si, and 0.69%Al TRIP steel preoxidized using the high λ DFF described in the previous figures [15]. *(L. Bordignon, G. Angeli, H. Bolt, R. Hekkens, W. Maschek, J. Paavalainen, X. vanden Eynde, Enhanced hot dip galvanising by controlled oxidation in the annealing furnace, in: Proc. 44th Mechanical Working and Steel Processing Conference, vol. XL, AIST, September 2002, pp. 833–844.)*

contaminants because the DFI burner is run in oxidizing mode; the air/fuel ratio is controlled to give a specified external oxide thickness [16].

The cleanliness of the annealing furnace atmosphere has been enhanced by use of radiant tube heating elements. This also permits more precise control of furnace atmosphere, because burner combustion products are not released into the furnace. RTFs have high thermal inertia and are suitable for long production runs with the same product. Unfortunately, many CGLs are now required to produce short runs of a variety of steel grades (DP, TRIP, IF, etc.) all on the same line, and therefore, rapid changes in furnace temperature profile can be required. In this case, a direct-fired annealing furnace can be more suitable [17].

2.5 Overaging in the CGL

Absent from many CGLs in the years before 2000 was an overaging section after annealing and cooling that is now essential to obtain the desired microstructures and consequent mechanical properties for many steels used today. Aluminum-killed drawing quality steels, introduced in the 1960s, could not be processed on the older lines because of the need to produce a matrix of low-carbon ferrite that contained carbides, mainly cementite, which were initially nucleated during an initial slow cooling stage after high-temperature soaking. After overaging in an offline postannealing furnace, these carbides contained most of the carbon in these grades, enabling their excellent drawing capabilities, for example, transverse r values of 1.8–2 and n values of 0.19–0.2 [18]. Not until the development of ultralow carbon interstitial free (IF) steels in the 1980s was it possible to supply Zn-coated steels meeting automotive forming requirements on existing galvanizing lines, because IF steels need no overaging, in fact keeping some C in solution is required for bake-hardening grades [19]. High-strength low-alloy grades (HSLA) that provided higher strengths by carbide strengthening but less formability also were not

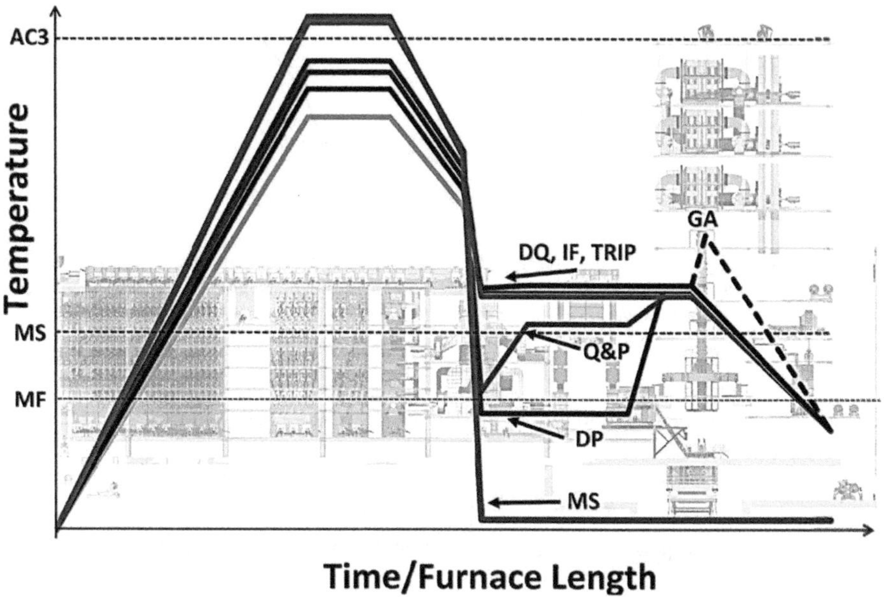

FIG. 2.10 Heat treatment cycles used for various steel grades: DQ = drawing quality, IF = interstitial free, TRIP = transformation induced plasticity, Q&P = quench and partition, DP = dual phase, MS = martensitic [2]. *(F. Brühl, C. Sasse, K. Watson, Annealing and galvanizing of third-generation AHSS grades, in: AISTech 2019 — Proceedings of the Iron & Steel Technology Conference, 6–9 May 2019, Pittsburgh, PA, USA, AIST, Warrendale, PA.)*

necessary to overage. Typical heating, soaking, cooling, and overaging heat treatments used for steels on modern lines are shown in Fig. 2.10.

Overaging is also not required for compatible dual-phase (DP) grades, where the final microstructure of ferrite and martensite can be produced by cooling at a controlled rate from an intercritical annealing temperature (where both ferrite and austenite are present), to below the martensite start (M_s) temperature. If a DP grade has its M_s temperatures below the temperature of strip entry into the galvanizing bath, which is typically 2–3°C above the zinc bath temperature of 460°C, a short heating step between cooling and strip entry into the zinc bath is required; alternatively it can be cooled to the zinc bath temperature, with transformations continuing as the steel is cooled further after coating with zinc. Overaging sections are essential for AHSS in which stable austenite is required in the final product, including TRIP, CP, and 3G steels. These grades have in common a process route that produces, after cooling from an annealing temperature either in or above the intercritical range, a mixture that can include old ferrite, martensite, bainite, and retained austenite. Overaging produces stabilization of the austenite by enabling its enrichment by carbon by reaction with surrounding phases or partitioning. A schematic of this carbon enrichment process for a quench and partition grade containing only austenite and martensite is shown in Fig. 2.11.

It is not unusual to find 15%–50% retained austenite in TRIP and 3G steels, with austenite carbon contents up to 0.5%. Retained austenite is required in these grades to cause the TRIP (transformation-induced plasticity) effect: upon deformation, the retained austenite transforms to martensite, greatly increasing the strain hardening effect while permitting

FIG. 2.11 Microstructure and composition changes occurring in the quench and partition process for a 3G steel, where the partitioning temperature is greater than the quench temperature. Γ = austenite (gamma) phases, m = martensite phase [20]. *(D.K. Matlock, J.G. Speer, Third generation of AHSS: microstructure design concepts, in: A. Haldar, S. Suwas, D. Bhattacharjee (eds.), Microstructure and Texture in Steels, Springer, Basel, 2009, ch. 11.)*

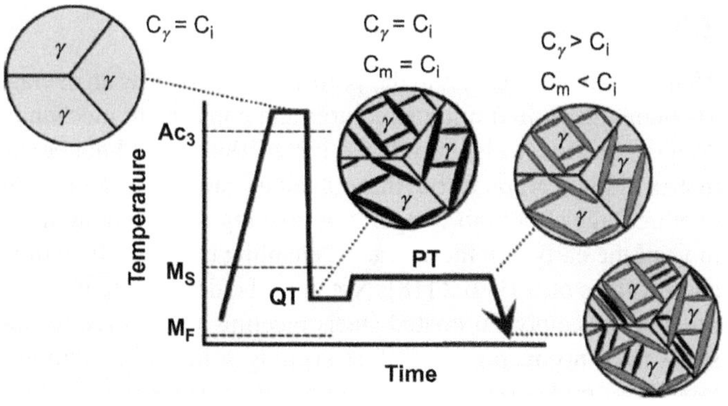

significant ductility. These enhanced stress-strain characteristics greatly increase energy absorption during deformation that is highly sought after by the auto industry to improve crashworthiness.

The process steps inducing internal oxidation in the steel can result in internally oxidized layer that is up to several microns thick. These can result in poor adhesion of the galvanized coating when the sheet is deformed. When the fracture surface is examined, the underside of the separated zinc coating is observed to have a surface composition similar to the internally oxidized layer [21]. Adhesion loss can be minimized if internal oxidation is limited to grain boundaries rather than also reacting in the bulk of the grain, this is controlled by sheet texture, composition, and oxidation treatment process variables [22].

2.6 Metallic coating pretreatments

Another approach to galvanizing steels containing relatively high levels of Mn, Cr, and Al is precoating the steel sheet with a very thin metallic layer. Precoating with Fe, Cu, Ni, and alloys of these metals has been the subject of research investigations [23–26]. Ni was found to affect the nucleation rate of the different intermetallic phases formed at the interface of the coating in steel, including $NiAl_3$ and also NiZn delta and gamma phases, the optimal Ni preplated layer thickness is between 3 and 5 μm and depends upon dipping time in the Zn bath. Use of Ni as an addition to the bath with Al, to form an Ni-Fe intermetallic, has been explored. Fig. 2.12 shows the effect of Fe precoatings, in the range of 3–10 g/m^2 on the oxidation behavior of a 0.004%C-0.8%Si-0.29%Mn steel. An addition of sodium citrate was used for the precoating with 4 mass% oxygen, whereas the other precoating has less than 0.2 mass% oxygen. Fig. 2.11 shows that the surface-selective oxidation behavior of Si was the same in the oxygen-free coating as expected from a sample without precoating. By contrast, the Si combined with the oxygen in the oxygen-containing coating, resulting in internal oxidation that improved hot-dip galvanizability. Because of cost, these pretreatments are rare in industrial practice.

Beyond gas treatment of steel surfaces to influence oxidation behavior, use of a surface treatment technique involving lithium boron oxides has been described [27]. This liquid salt bath, which follows conventional annealing in an RTF, is operated between 650°C and 800°C and was successful in removing surface oxides so that a 1.5%Mn-1.5%Si Si TRIP steel could be successfully galvanized. A 23% Mn TWIP (twinning-induced plasticity) steel was also successfully galvanized. Samples showed good coating adhesion after a 1.5 T-bend test; however, this has not been taken to industrial practice.

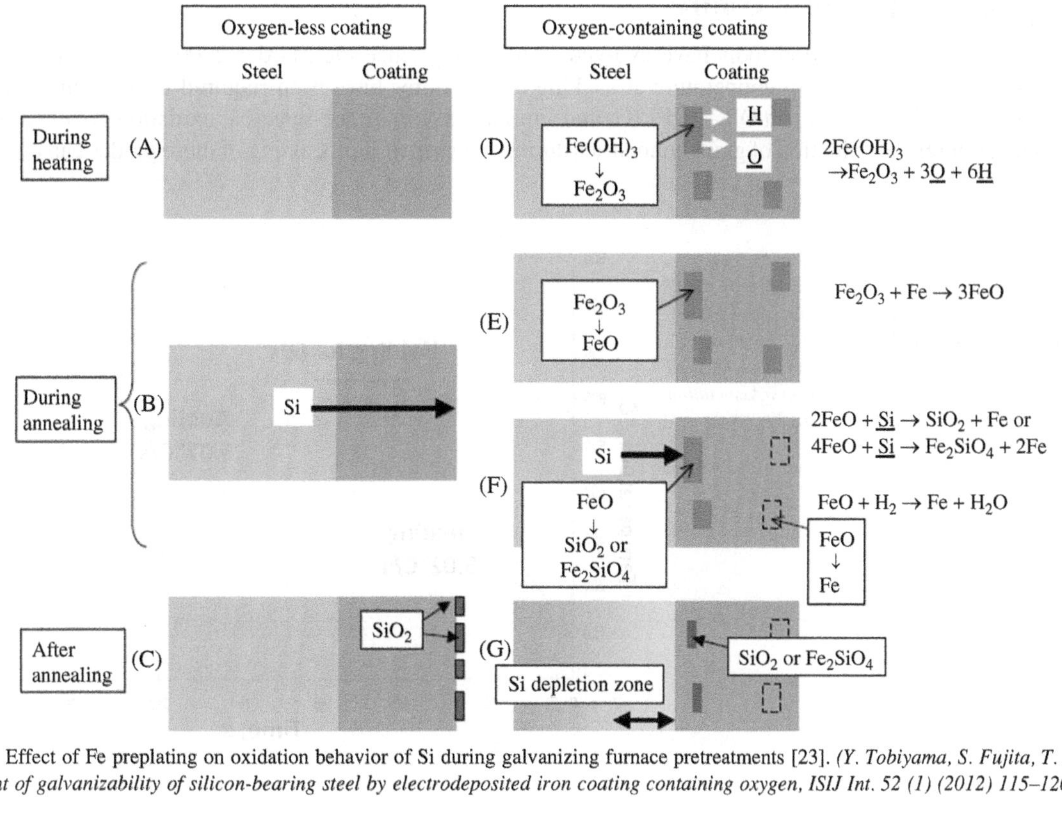

FIG. 2.12 Effect of Fe preplating on oxidation behavior of Si during galvanizing furnace pretreatments [23]. *(Y. Tobiyama, S. Fujita, T. Maruyama, Improvement of galvanizability of silicon-bearing steel by electrodeposited iron coating containing oxygen, ISIJ Int. 52 (1) (2012) 115–120.)*

2.7 Galvanizing of hot rolled strip

Improvements in hot strip mill practice, including continuous casting of steel directly to a thin strip shape, have greatly expanded the quantities of hot-rolled steel being processed on CGLs. Minimum thicknesses of 0.8 mm can be achieved [28]. Most producers carry out hot strip pickling on a separate line; increasingly the trend is to integrate the pickling step integral with the galvanizing process. Before pickling, the hot-rolled grade must be in a metallurgical condition compatible with the treatments it will receive in the CGL and with eventual user requirements. For DP grades, hot strip mill coiling temperatures either above or below the M_s temperature are used, depending upon hardenability. Ferrite-bainite steels are coiled below their B_s temperature to suppress formation of carbides and martensite. TRIP steels are also coiled below B_s to assist in development of desired quantities of austenite, whose C enrichment can be developed during furnace treatments before galvanizing [29]. Because hot-rolled grades have not gone through cold working steps, further heat treatment for many grades to modify microstructures is not required, the furnace heat treatment prior to galvanizing can be limited to a temperature capable of reducing steel surface oxides prior to immersion in the galvanizing bath. CGL strip speeds are usually limited by furnace capacity, so the requirement of a lower-temperature furnace treatment is beneficial for line productivity when processing hot-rolled strip. So-called "heat-to-coat" processing has been applied to precipitation-strengthened and HSLA grades [30,31]. An example of a "heat-to-coat" thermal cycle for a 2.6 mm-thick HSLA grade is shown in Fig. 2.13. This included the following steps:

- Heating in an infrared furnace at 5°C/s to 600°C.
- Holding at 600°C for 60 s or 120 s.
- Cooling at 5°C/s down to the strip entry temperature of 465°C.
- Dipping at 460°C for 4 s. Zinc bath with 0.20% effective Al and Fe saturated.
- Cooling after galvanizing at 3–5°C/s.

The furnace atmosphere used in the first three steps was N_2-20%H_2 with a dew point of −30°C.

There is also increasing use of induction heating to bring hot-rolled strip up to the temperature normally required to reduce surface oxides; DFF is also used and allows increased line speed compared with RTF for these thick sections [28].

Higher-temperature processing and overaging are required to successfully process hot-rolled TRIP grades [32].

2.8 Furnace temperature control

CGL process control and inspection tools have evolved significantly since the Marder 2000 review. Improvements in pyrometry for measurement of strip temperature at soaking temperatures have been required because in steels with the levels of Si, Mn, and other elements in DP, TRIP, 3G, and similar grades, surface-selective oxidation greatly affects emissivity [33]. The preferential oxidation of grain boundaries during hot strip mill processing of these grades can also influence

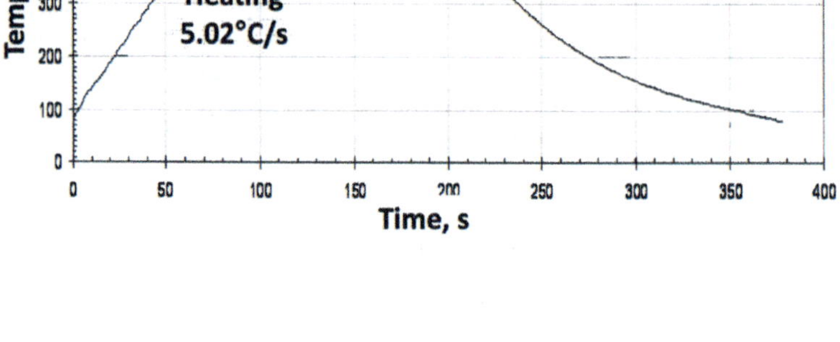

FIG. 2.13 Typical heat-to-coat thermal cycle for hot-rolled steels where annealing is not required [30]. *(V.Y. Gertsman, E. Essadiqi, R. Bouchard, O. Dremailova, J. McDermid, R. Fourmentin, J. Nemes, W. Dabboussi, F. E. Goodwin, Properties of galvanized and galvannealed hot rolled HSLA steel, in: Proc. Galvanizers Association 100th Meeting, Baltimore, MD, October 26–29, 2008.)*

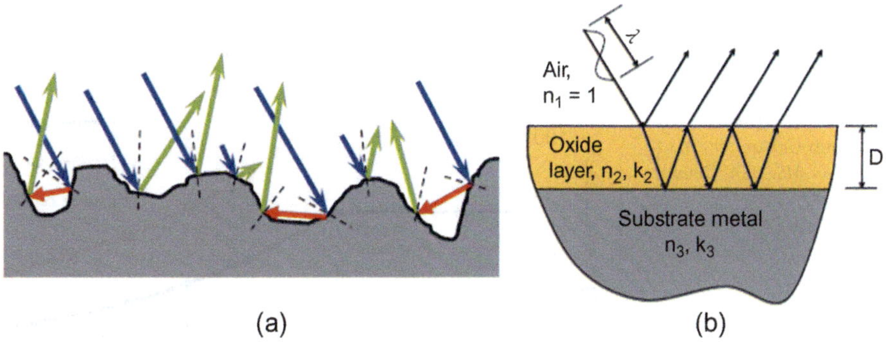

FIG. 2.14 The spectral variation of AHSS radiative properties is caused by two distinct phenomena: (A) scattering by random surface features, including rolling marks, scratches, rolled-in flaps, and oxide nodules; and (B) thin-film interference, from surface oxides, which leads to coherent interference patterns. n = refractive index, k = extinction coefficient (related to reflectance). The surface features in Fig. 2.14A are on the order of microns, while the layer thickness in Fig. 2.14B is on the order of tens of nanometers [35]. *(Final Report for Project ZCO-72, Surface Profile Effects on AHSS Processing, Prof. Kyle J. Daun and Prof. Myriam Brochu, Project sponsored by International Zinc Assn., February 2021.)*

surface roughness [34]. A schematic of the interaction between the steel surface and incident electromagnetic waves is shown in Fig. 2.14. The effect of surface roughness is shown in Fig. 2.14A. Incoming electromagnetic waves, such as those used in pyrometers and including visible wavelengths used to inspect the steel surface, scatter from roughness features, including roll marks, scratches, rolled-in flaps, and oxide features present on the surface at all wavelengths. At shorter wavelengths, significantly less than the dimensions of the roughness features, the electromagnetic waves are subjected to a thin-film interference effect, caused by the oxide layer over the steel surface, Fig. 2.14B, whose thickness is tens or hundreds of nanometers. The thin-film interference effect mainly influences the visible appearance of the steel, including reflectivity and color changes associated with the formation of various oxide layers. These effects occur simultaneously at short wavelengths for cold-rolled strip entering the CGL. Because most steel pyrometry measurements are carried out between 1.5 and 2.5 μm, both effects influence the accuracy of pyrometer measurements.

Emissivity and roughness effects can result in pyrometry errors of more than 20°C. Resulting setpoint errors during, for example, intercritical annealing can result in varying quantities of austenite and ferrite being produced during this step, so that in the case of DP grades, the amount of martensite produced after cooling, and therefore final mechanical properties, can be off-specification. The relationship between surface oxide (mainly Si/Mn) refractive index and extinction coefficient (related to reflectivity) and emissivity is now just beginning to be understood. When the steel surface roughness is close to the wavelength being used for pyrometer measurements, significant interference can also occur. Basic research in these two directions will be used to improve future pyrometer algorithms [36].

To improve the consistency of furnace control regardless of the multiphase steel grades being processed, a novel approach has been taken that measures the austenite content in the ferrite-austenite strip after intercritical annealing before rapid cooling. By obtaining the fraction austenite at this point, the cooling rate can automatically be compensated by use of a closed-loop control. It is claimed that this has saved many coils from downgrades related to mechanical properties issues [2].

2.9 CGL rapid cooling section developments

Requirements for higher cooling capability following soaking have driving development in higher-performance cooling sections. Because H_2-N_2 gas thermal conductivity increases with increasing H_2 content, the H_2 content in the rapid cooling section is increased to 20% or more by injecting pure H_2 into the gas atmosphere. Seals at the entrance and exit of the cooling section prevent mixing with adjacent furnace zones. Cooling rates in excess of 100°C/s, corresponding to heat transfer rates of 100–200 W/m^2°C, can be achieved for strip thicknesses of 1 mm or less, a typical operating curve is shown in Fig. 2.15 for a strip entering the cooling section with a temperature at 800°C and exiting at 450°C. Cooling rates are strongly influenced by strip thickness and line speed, this example shows a three-zone furnace, where additional zones are activated as line speed, related to the shown tons per hour, increases [37]. In terms of tonnage, a coil of 0.8 mm thick steel that is 1200 mm wide and 3000 m long weighs 22.6 metric tons, so the tonnage rates in Fig. 2.15 correspond to processing between three and four such coils per hour or a line speed of 150–200 m/min.

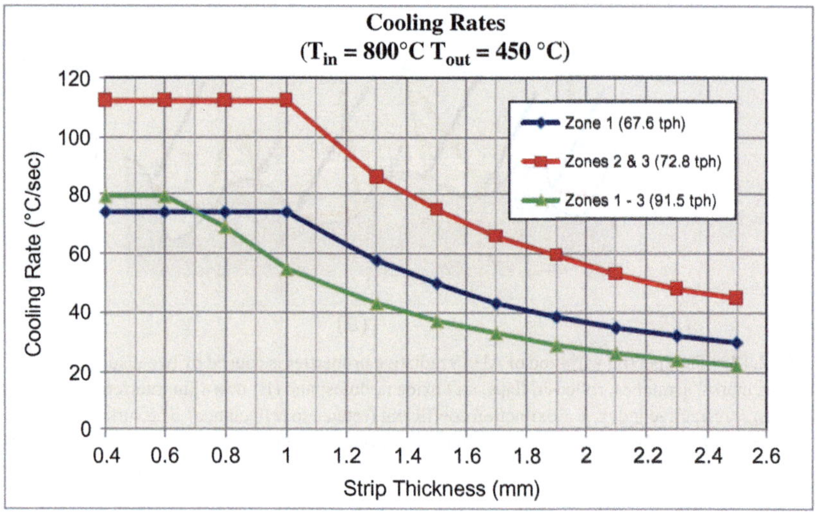

FIG. 2.15 Operating curves for a modern strip rapid cooling section, where up to three furnace zones can be used as cooling demands increase [37]. *(N. Lagios M. Renard, J.-P. Crutzen, J.-M. Raick, New lines for processing advanced high strength steels, in: Proceedings Galvanizers Association 106th Meeting, Jackson, MS, USA, October 5–8, 2014.)*

The benefit of using the zinc bath as a quenching medium for steel strip exiting the furnace section at temperatures up to 650°C has been validated in industrial production at ArcelorMittal's CGL 1 in Asturias, Spain [38]. This process replaces rapid gas cooling; an initial slow gas cooling is used to set the desired phase composition in the strip before quenching in the zinc bath. Improved Zn wettability of the strip is related to a higher soluble Al concentration in the Zn bath adjacent to the immersed strip because of the removal of Fe by precipitation on the bath cooler surface. This encourages the aluminothermic reaction [39] that permits zinc coatings to be produced on steel surface containing MnO:

1. Reduction of MnO by Al in the Zn bath

$$3MnO_{substrate} + 2[Al]_{bath} \rightarrow 2[Mn]_{bath} + Al_2O_3 \tag{2.1}$$

2. Nucleation and growth of $FeAl_3$.

$$Fe_{substrate} + 3[Al]_{bath} \rightarrow Al_2O_3 \tag{2.2}$$

3. $FeAl_3$ transforms to Fe_2Al_5 and growth of Fe_2Al_5

$$FeAl_3 + 2[Al]_{bath} + Fe_{substrate} + x[Zn]_{bath} \rightarrow Fe_2Al_{5-x}Zn_x \tag{2.3}$$

The intermetallic compound $Fe_2Al_{5-x}Zn_x$ is a key component to the galvanizing reaction; further details are given in Chapters 4 and 5.

2.10 Strip entry snout control

The effect of CGL entry snout atmosphere conditions on coating surface quality was explained by Arnold et al. in 1989 [40]. The partial pressure of zinc increases from 10^{-4} to 10^{-3} atm as its temperature is increased from 400°C to 500°C, which is much higher than that of liquid Al. This vaporized zinc can condense in the snout and can be eventually released and fall on the steel strip prior to immersion in the zinc bath. By increasing the partial pressure of oxygen in the snout to a level sufficient to oxidize the liquid zinc layer present in the snout, but below that which would oxidize the metallic Fe surface being introduced into the galvanizing bath, coating defects related to zinc vapor in the snout were controlled, this is commonly termed "wet snout practice." An illustration of this practice is shown in Fig. 2.16. For both Zn and Zn-55Al production, the atmosphere in the snout is isolated from the drier furnace gas in preceding sections; the problem can be more severe in Zn-55Al operations because the higher partial pressure of Zn at higher temperatures as noted before. More details on coating defects related to Zn dust condensation in the snout are described in Chapter 14 and in [41].

2.11 Alloy coating developments

Zinc-coated steel has traditionally consisted of an overlay of zinc-aluminum alloy corresponding to the bath composition that contains 0.13–0.4%Al, over a very thin Al-Fe interface layer, termed "GI." The development of higher Al content Zn-based coatings is described in Chapter 6. The most important new coating alloy system developed since that time has been the Zn-Al-Mg coating family. This family of alloy coatings has improved corrosion resistance compared with Zn and ZnAl coatings with similar Al compositions and was first developed for construction products. Later, ZnAlMg-coated steels were introduced to the auto market, using a lower range of Al (1%–3%) and Mg (1%–2%) compositions [43]. For auto producers, the ability to reduce coating thickness and obtain the same corrosion performance helped improve weldability and formability. Less galling susceptibility is also observed with ZnAlMg automotive coatings, compared with GI coatings [44]. An Mg-containing Zn-55Al coating has also reached the industrial production stage, also because of the improved corrosion performance realized by adding 2% Mg [45]. CGL process developments related to zinc pot operations and coating weight control are discussed in Chapters 4 and 7, respectively, including dross control and the development of practices for coated product transitions in Chapter 4.

2.12 Zinc coating wetting behavior during galvanizing

To achieve a galvanized coating, the liquid zinc or zinc alloy in the coating bath must first wet the prepared steel surface. Wetting angle is the key measure of wetting behavior, and this has been found to strongly depend upon the nature of the steel surface, especially the presence of surface oxides, and the nature of those oxides. As indicated before, the objective of the CGL furnace treatments regarding surface preparation is to produce a thin layer of reduced Fe on the steel surface that can be reacted with the Zn bath. For the simplest carbon steels, this surface is free of surface oxides from other alloy elements. When the sample is withdrawn from the bath and held still, a wetting angle θ can be measured, as shown in Fig. 2.17, this is called the Wilhelmy plate method. The wetting angle is determined by measuring the difference between the free weight of the steel sample and that measured in the static position indicated in Fig. 2.17 according to the formula

$$\cos\theta = \frac{\Delta F_G - \rho zn(T) \cdot V_{zn} \cdot g}{L_l \gamma_1 (Zn)}$$

where T = temperature (°K), V_{Zn} = volume of zinc melt displaced by the dipped sample (m^3), g = gravitational constant, L_1 = immersion depth (mm), and γ(Zn) is the surface tension of liquid zinc. For a GI bath at 460°C: the density of Zn, $\rho_{Zn}(T) = 0.65\,\text{kg}\,\text{m}^{-3}$, the surface tension of liquid zinc $\gamma_l(\text{Zn}) = 0.78\,\text{mN/mm}$, and $g = 9.81\,\text{m/s}^2$. There are other ways of measuring wetting angle, including dispensed droplet [47] and a hot-melt rotator device [48]. The effect of steel composition and resultant surface oxides on wetting angle can be seen in Fig. 2.18. This is for a family of three steels, each with low amounts of alloying elements in which the amount of Si was varied, with levels of 0.1, 0.3, and 0.7 wt% for the steels Si1, Si2, and Si3. These had respective Si/Mn ratios of 0.25, 0.75, and 1.75. Pure zinc was used in these experiments. The wetting behavior of pure iron, on which no oxides were present, is also shown near the vertical axis. The effect of wetting time is also shown: for pure iron, wetting angle decreased from 70 degrees at $t = 0$ to 20 degrees at $t = 5$ s. This is related to

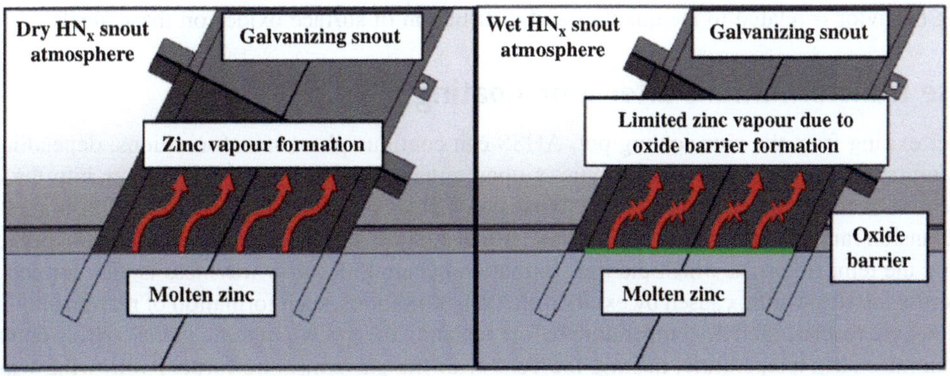

FIG. 2.16 Sketch of a CGL strip entry snout into the zinc bath showing (left) use of conventional dry H_2-N_2 furnace gas and resulting zinc vapor accumulation (right) use of wet furnace gas, producing oxide barrier on the bath surface [42]. *(J. Berry, C.H. Phillips, D.J. Penney, Understanding the role of snout contamination in the formation of an oxide based defect in hot dip galvanised coating, Surf. Coat. Technol. 306 Part B (25) (2016) 397–407.)*

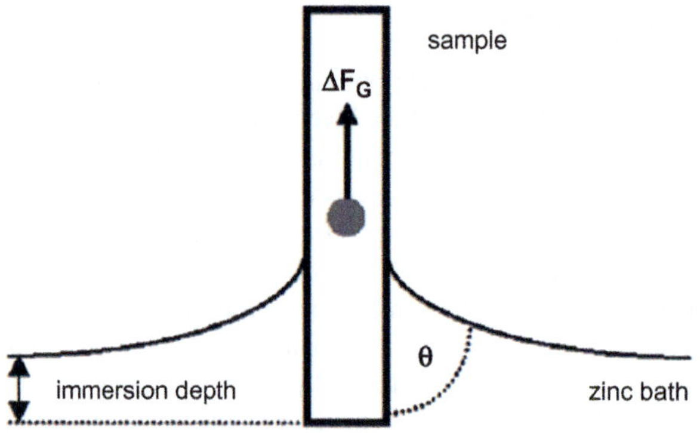

FIG. 2.17 Wilhelmy plate method for measuring wetting angle [46]. *(M. Blumenau, M. Norden, F. Friedel, K. Peters, Wetting force and contact angle measurements to evaluate the influence of zinc metallurgy on galvanizability of high-manganese-alloyed steels, Surf. Coat. Technol. 205 (2010) 828–834.)*

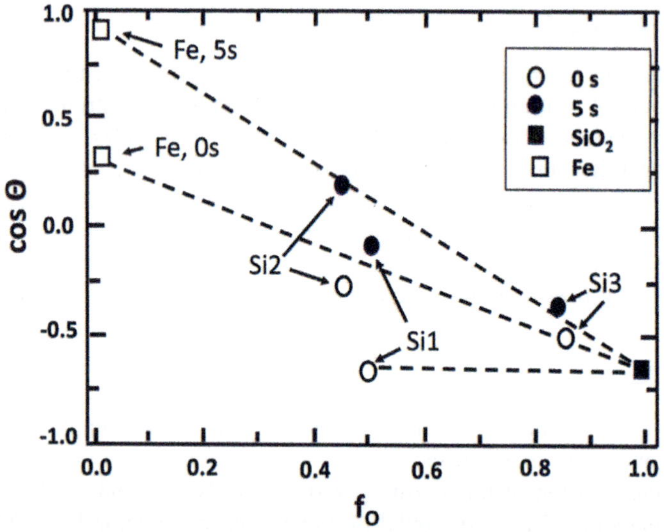

FIG. 2.18 Relationship between cos θ, the cosine of the wetting angle, and the coverage of the surface oxides (f_O) [49]. *(J. Lee, J. Park, S.-H. Jeon, Wetting of low-carbon, interstitial-free steel surfaces with nanostructured oxides by liquid zinc, Met. Mat. Trans. 40B (2009) 1035–1040.)*

the reaction between Fe and Zn that produces intermetallic layers, physically "flattening" the wetting angle. Steels Si1 and Si2 exhibited nearly the same wetting behavior, while the wetting angle of Si3 was higher and decreased much more slowly with time [49].

This wetting behavior is related to the quantity and distribution of surface oxides on these steels.

2.13 Phase transformations after zinc coating

During and after exiting from the zinc coating pot, AHSS can continue phase transformations, depending on their characteristic transformation temperatures. Strip waviness upon entering the zinc pot, and further into the coating control section, has been found to be related to the volume expansion that accompanies the transformation from austenite to martensite. This expansion ranges from 2.0% at 0.19%C to 3.1% at 1.01%C at the M_s temperature largely because of the effect of C on lowering the temperature at which the transformation occurs [50]. Strip waviness is also formed by nonuniform temperatures across the strip width, caused by exothermic transformations, such formation of martensite from austenite, in grades that had not yet reached their M_f temperature before reaching the gas wiping line in the coating control section. The residual heat can raise strip temperature, making the control of the galvannealing furnace following wiping more challenging. This is a particularly difficult problem to detect because when the strip is observed when cool after processing, the transformations have finished and the strip is again flat, the evidence of waviness during wiping is the pattern of coating thickness on the coated sheet [51]. Further details are described in Chapter 7.

CGL line processing advances relating to the galvannealing section for production of the galvanneal "GA" coating are described in Chapter 8.

2.14 Coating solidification developments

For GI-coated sheet, the cooling rate after passing through the coating control (wiping) knives, through the solidification range, greatly influences zinc crystal (spangle) size and the dominant crystallographic orientation (texture) of the zinc coating. The character of the interfacial intermetallic layer is not influenced by the rate of cooling through the zinc solidification temperature range. As shown in Fig. 2.19, conventionally air-cooled coatings in zinc coatings where spangle-formers such as Pb and Sb are not present have a spangle size of 700–800 μm and a roughness R_a of 0.57 μm. Rapid cooling through the solidification range with an electrostatically charged water spray reduced spangle size to 80–200 μm with an R_a of 0.32 μm. The dendrite arm spacing and overall dendrite size also decrease as spangle size decreases, with no dendrites being visible on the surface of the rapidly cooled coating [52].

In this work, the intensity of the basal (0001) plane in the GI coating also increased with increasing cooling rate; with spangle sizes less than 200 μm, Zn crystal growth is much more influenced by existing crystallographic orientation rather than thermal gradients. The (0001) intensity is known to strongly correlate with the strength of adhesion of paint to the GI coating [52,53]. There are conflicting reports about the effect of spangle size on resistance of the GI coating to cracking, whether measured by tension bending or flaking during drawing. The common factor to improved crack resistance is a distribution of crystallographic orientations in the polycrystalline Zn coating, a distribution of Zn crystal orientations is observed to better accommodate deformation and inhibit crack propagation than a single crystal orientation, regardless of spangle size [54].

After completion of solidification by diffusion and intermetallic formation in the case of galvanneal, or conventional solidification in the case of all other metallic coatings, the strip must be cooled to less than 345°C to prevent coating pickup on the top roll that would occur with hotter, softer coatings.

During and just after the solidification of the zinc coating, the surface Al in the liquid, and later, solid solution of the zinc coating oxidizes in preference to the less reactive zinc. Similar to the surface-selective oxidation of AHSS grades described before, the Zn coating surface is enriched by Al, especially at coating grain boundaries. Even in the short time between

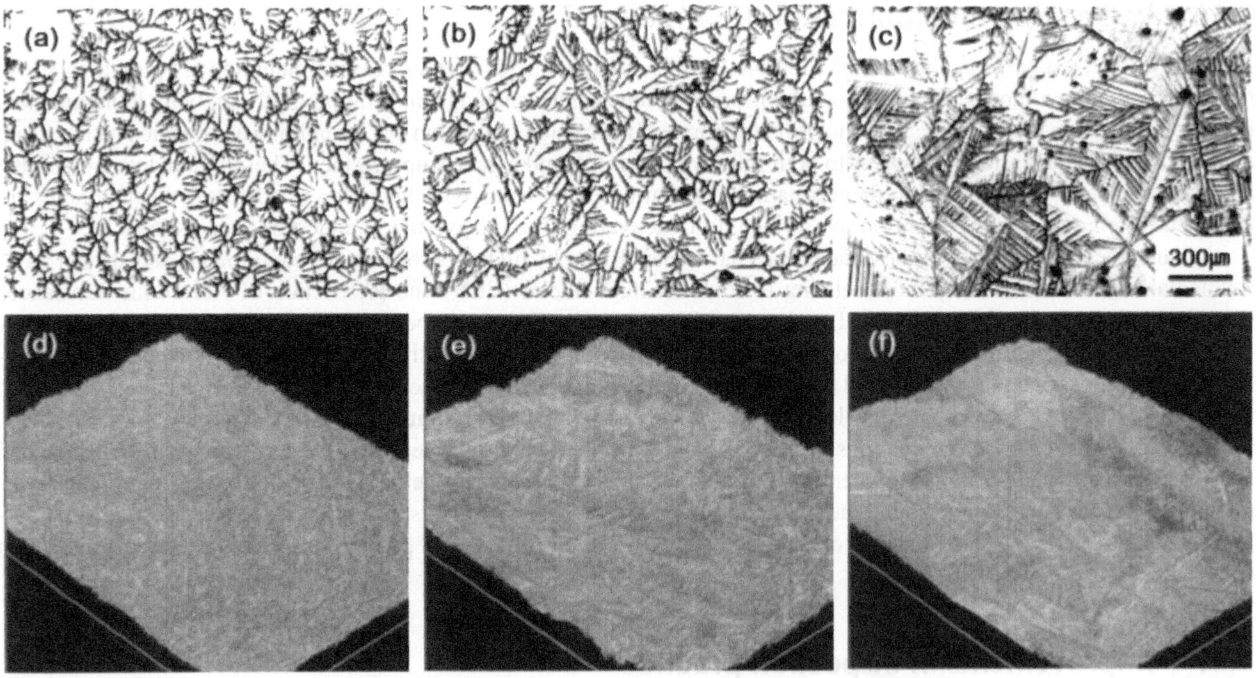

FIG. 2.19 Optical micrographs and roughness profiles of GI steel sheets obtained from (A), (D) small; (B), (E) medium; and (C), (F) large spangles, respectively [52]. *(D.-J. Paik, M.-H. Hong, Sustainable production and technology challenges for galvanized steel sheets, in: Proc. 11th International Conference on Zinc and Zinc Alloy Coated Steel Sheet (Galvatech 2017), November 12–16, 2017, Tokyo, Iron and Steel Institute of Japan, Tokyo, pp. 11–17.)*

solidification and cooling before the top roll, an average surface Al layer weight of 2.5 mg/m^2 can be observed; this influences later temper rolling and chemical pretreatment processes [55].

2.15 Temper rolling and tension leveling

The strip then descends, usually into a water cooling tank where the maximum exit temperature is usually 38°C to prevent diffusion of C and other interstitials in the steel. This is followed by temper rolling, also termed skin passing, which reduces strip thickness by typically 0.5%–1.5%, but up to 2% if the steel is to bake hardened, where a high dislocation density is helpful. Water is used as a lubricant for temper rolling of galvanneal, 1%–2% water-soluble lubricant is added for GI. The temper rolling step has several objectives. The first is to prevent discontinuous yielding, or yield point elongation, associated with the formation of Lüders lines, by introducing dislocations into the steel, to which C and other interstitials are pinned. The second is to condition the strip surface for surface treatments that follow by imprinting a roll surface pattern, or texture onto it. The third is to improve strip shape and flatness. Advances realized since the publication of Marder 2000 have focused on work roll texturing; the operation of the temper rolling mill is still surprisingly manual. Roll textures are either concave, where the depressions are created by grit blasting, electric discharge texturing, electron beam texturing, or laser texturing, or convex, where the surface dimples are created by selective Cr plating (Topochrom) [56–58]. In all cases, the depressions on the surface are intended to trap oil that permits improved formability, or to alter the previously mentioned aluminum oxide surface layer developed on the strip during application of the hot-dip coating, improving reactivity of chemical pretreatments or conversion coatings that follow. The aluminum-rich surface layer is not entirely removed by temper rolling. Hoshino et al. [55] showed that if the temper-rolled surface does not immediately receive a protective treatment such as passivation, additional aluminum oxide is formed on the zinc coating surface that contacted the temper roll. Such an activated and unprotected surface would be vulnerable to moisture-induced staining if improperly stored; this is discussed further in Chapter 11.

As an alternative or supplement to temper rolling, a lower-cost technique called tension leveling, which pulls the strip at a stress above its yield stress, can also eliminate yield point elongation and add surface texture, but cannot change strip shape. If a spangled appearance is to be maintained, tension leveling is used instead of temper rolling, because the temper mill rolls flatten the spangle.

2.16 Passivation and lubrication treatments

The strip is then ready for application of additional surface treatments for various reasons including (1) providing short-term passivation that blocks the formation of corrosion products during storage and shipping; (2) applying lubricants that both facilitate downstream processing and provide short-term corrosion protection or (3) applying an organic coating, from an increasingly wide variety of choices to both improve durability and enhance appearance. These treatments are applied soon after temper rolling to avoid further formation of the surface aluminum oxide layer described above; the treatments are designed to remove oxides present on the surface, fully activating it to receive the intended treatments [59].

Advancements in all three of these categories have been driven by higher performance requirements and stricter environmental regulations, notably those relating to End-of-Life Vehicles (ELV) and Restriction of Hazardous Substances (RoHS) laws. Coatings for the electrical appliance market were the first developed, eliminating the usage of compounds with hexavalent chromium (Cr^{+6}), a known carcinogen. Developments for construction and automotive applications followed. At first, passivation systems, and pretreatment systems for paint, were based on organic resins containing insoluble metal salts. This was followed by development of silanol compounds incorporating inorganic inhibitors [60]. For automotive applications, phosphate coatings utilizing Ti and Zr as a cations have replaced prior formulations that included Ni. For construction applications, trivalent Cr formulations are an alternative, together with a wide variety of organic treatments based on acrylics, silanes, and other types of polymers. No-rinse treatments have also improved environmental performance, these generally use either roll or spray coaters [61]. The performance of painted zinc-coated steels is discussed in Chapter 12.

2.17 Improvements in inspection practice

Before exiting the line, the condition of the processed strip is inspected. The development of machine vision has enabled inspection to be increasingly automated and has provided a wealth of data for use in higher-level control strategies and for the Industry 4.0 data analytics described in Chapter 16. A typical machine vision system applied to galvanized coating

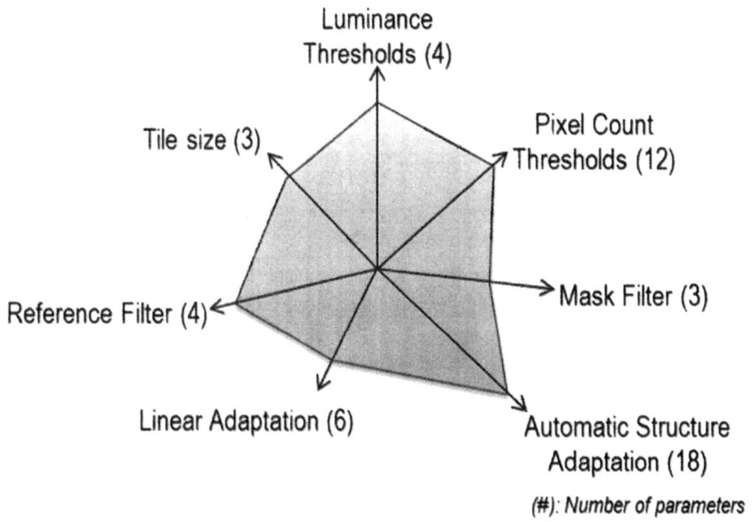

FIG. 2.20 Examples of machine vision inspection system detection parameters to tune for defect detection for each type of zinc-coated surface [62]. *(G.M. Gutmann, S.H. Burkhardt, Quality assurance by reliable automated surface inspection, in: Proceedings Galvanizers Association 109th Annual Meeting, October 15–18, 2017, Troy, MI, Galvanizers Association, Marine City, MI, 2017.)*

inspection tunes the results of selected detection parameters for each coating type. Some of the more important detection parameters are shown in Fig. 2.20.

The zinc coating surface patterns being imaged can have a variety of spangle features, roughnesses, and defect types. To develop independence between reportable types of defects and the underlying variety of coating appearances, 50–100 of the types of parameters in Fig. 2.20 are used. Algorithms are developed and optimized as the system is tuned. Ideally, the Pareto optimum is reached, that is, the vision system cannot improve defect detection performance without identifying pseudo-defects, i.e., false positives. The step beyond this is to correlate the defects with their root cause. In the past, this was done manually, but with the increasing amounts of process data available from each CGL unit, such analyses are entering the realm of Industry 4.0 as described in Chapter 16.

References

[1] A.R. Marder, The metallurgy of zinc-coated steel, Prog. Mater. Sci. 45 (2000) 191–271.

[2] F. Brühl, C. Sasse, K. Watson, Annealing and galvanizing of third-generation AHSS grades, in: AISTech 2019 — Proceedings of the Iron & Steel Technology Conference, 6–9 May 2019, Pittsburgh, PA, USA, AIST, Warrendale, PA, 2019.

[3] F. Chen, R. Patil, An in-depth analysis of various subtle coating defects of the 2000's, in: Proc. 6th International Conference on Zinc and Zinc Alloy Coated Steel Sheet (Galvatech 2004), April 4–7 2004, Chicago, IL, AIST, Warrendale, PA, 2004, pp. 1055–1066.

[4] C. Dulcy, F. Galdo, Recent progress in the process technology of continuous galvanizing lines, in: Proc. 7th International Conference on Zinc and Zinc Alloy Coated Steel Sheet (Galvatech 2007), November 18–22, 2007, Osaka, Japan, Iron and Steel Institute of Japan, Tokyo, 2007, pp. 23–28.

[5] B.C. De Cooman, J.G. Speer, Fundamentals of Steel Product Physical Metallurgy, AIST, Warrendale, PA, 2011, ISBN: 978-1-935117-16-2.

[6] C. Wagner, Reaktionstypen bei der Oxydation von Legierungen, Z. Elektrochem. 63 (1959) 772–782.

[7] R.A. Rapp, The transition from internal to external oxidation and the formation of interruption bands in silver-indium alloys, Acta Metall. 9 (1961) 730–741.

[8] F. Gesmundo, F. Viani, Transition from internal to external oxidation for binary alloys in the presence of an outer scale, Oxid. Met. 25 (5–6) (1986) 269–282.

[9] V.A. Lashgari, Internal and External Oxidation of Manganese in Advanced High Strength Steels (Ph.D. thesis), Technical University of Delft, Netherlands, September 2014.

[10] W. Mao, Y. Ma, W.G. Sloof, Internal oxidation of Fe–Mn–Cr steels, simulations and experiments, Oxid. Met. 90 (2018) 237–253, https://doi.org/10.1007/s11085-018-9836-7.

[11] J. Staudte, J.-M. Mataigne, D. Loison, F. Del Frate, Galvanizability of high Mn grade versus mixed Mn-Al and Mn-Si grades, in: Proc. 8th International Conference on Zinc and Zinc Alloy Coated Steel Sheet (Galvatech 2011), June 21–24, 2011, Genoa, Italy, Associazione Italiana di Metallurgia, Milan, 2011.

[12] Y. Suzuki, T. Yamashita, Y. Sugimoto, S. Fujita, S. Yamaguchi, Thermodynamic analysis of selective oxidation behavior of Si and Mn-added steel during recrystallization annealing, ISIJ Int. 49 (4) (2009) 564–573.

[13] W. Mao, R.W.A. Hendrikx, W.G. Sloof, Prediction of oxide phases formed upon internal oxidation of advanced high-strength steels, Oxid. Met. 89 (2018) 531–549, https://doi.org/10.1007/s11085-017-9815-4].

[14] M. Blumenau, C.O. Gusek, M. Norden, Industrial use of pre-oxidation during continuous hot-dip coating of (high) alloyed steels, in: AISTech 2012 — Proceedings of the Iron & Steel Technology Conference, May 7–12, 2012, Atlanta, GA, AIST, Warrendale, PA, 2012.
[15] L. Bordignon, G. Angeli, H. Bolt, R. Hekkens, W. Maschek, J. Paavalainen, X. van den Eynde, Enhanced hot dip galvanising by controlled oxidation in the annealing furnace, in: Proc. 44th Mechanical Working and Steel Processing Conference, vol. XL, September 9–11, 2002, Orlando, FL, AIST, Warrendale, PA, 2002, pp. 833–844.
[16] K. Kahoul, M. Turchetto, Hot-dip galvanizing line at ARVEDI producing AHSS over 800MPa, in: Proc. 10th International Conference on Zinc and Zinc Alloy Coated Steel Sheet (Galvatech 2015), May 31–June 4, 2015, Toronto, AIST, Warrendale, PA, 2015, pp. 436–443.
[17] G. Angeli, W. Fischer, New galvanizing lines, in: Proc. 8th International Conference on Zinc and Zinc Alloy Coated Steel Sheet (Galvatech 2011), June 21–24, 2011, Genoa, Italy, Associazione Italiana di Metallurgia, Milan, 2011.
[18] E. Buscarlet, J.P. Nauzin, X. Cluzel, S. Mehrain, CAL and CGL metallurgical specifications for AHSS and associated furnaces for automotive deliveries since 1970 and for the future, Proc. 10th International Conference on Zinc and Zinc Alloy Coated Steel Sheet (Galvatech 2015), May 31–June 4, 2015, Toronto, 2015, AIST, Warrendale, PA, 2015, pp. 457–465.
[19] R.P. Foley, D.K. Matlock, G. Krauss, Metallurgical review of processes for obtaining strength and r-value in galvannealed sheet steels, in: Proc. 42nd Mechanical Working and Steel Processing Conf. Vol XXXVIII October 22–25, 2000, Toronto, AIST, Warrendale, PA, 2000, pp. 455–468.
[20] D.K. Matlock, J.G. Speer, Third generation of AHSS: microstructure design concepts, in: A. Haldar, S. Suwas, D. Bhattacharjee (Eds.), Microstructure and Texture in Steels, Springer, Basel, 2009. Ch. 11.
[21] F.E. Goodwin, L. Bordignon, Investigation of hydrogen-free N2-O2 heating and soaking treatments on galvanizability of advanced high strength steels, in: Proceedings Galvanizers Association 104th Meeting, October 21–24, 2012, Detroit, MI, Galvanizers Association, Marine City, MI, 2012.
[22] M. Norden, M. Blumenau, R. Schönenberg, Recent trends in hot dip galvanizing of advanced high strength steel at ThyssenKrupp Steel Europe, in: AISTech 2012 — Proceedings of the Iron & Steel Technology Conference, May 7–12, 2012, Atlanta, GA, 2012.
[23] Y. Tobiyama, S. Fujita, T. Maruyama, Improvement of galvanizability of silicon-bearing steel by electrodeposited Iron coating containing oxygen, ISIJ Int. 52 (1) (2012) 115–120.
[24] T. Matsumoto, Properties of galvannealed steel sheets flash-coated with electroplated layer, CAMP-ISIJ 4 (1991) 1632.
[25] A. Chakraborty, A. Mondal, P. Govardhana, T. Laha, S. Singh, Phase transformation in galvanised coatings on pre-coated steel, in: Proc. 11th International Conference on Zinc and Zinc Alloy Coated Steel Sheet (Galvatech 2017), November 12–16, 2017, Tokyo, Iron and Steel Institute of Japan, Tokyo, 2017, pp. 122–129.
[26] L. Berry, J. Demmel, C. Phillips, D. Penney, A study into the pre-deposition of nickel coatings and their effect on intermetallic formation in hot dip galvanising, in: Proc. 11th International Conference on Zinc and Zinc Alloy Coated Steel Sheet (Galvatech 2017), November 12–16, 2017, Tokyo, Iron and Steel Institute of Japan, Tokyo, 2017, pp. 803–810.
[27] L. Bordignon, M. Larnicol, X. Vanden Eynde, A. Farinha, P. Gerkens, J.F. Noville, J. Smal, M. Bobadilla, Liquid oxide annealing for surface preparation of AHSS, in: Proc. 10th International Conference on Zinc and Zinc Alloy Coated Steel Sheet (Galvatech 2015), May 31–June 4, 2015, Toronto, AIST, Warrendale, PA, 2015, pp. 58–65.
[28] M. Jaenecke, H.-G. Klöckner, M. Cottin, C. Sasse, Economical hot strip galvanizing, in: Proc. 10th International Conference on Zinc and Zinc Alloy Coated Steel Sheet (Galvatech 2015), May 31–June 4, 2015, Toronto, AIST, Warrendale, PA, 2015, pp. 444–451.
[29] Y.S. Jin, Recent progress and development of galvanizing technology in Korea, in: Proc. 9th International Conference on Zinc and Zinc Alloy Coated Steel Sheet (Galvatech 2013), September 23–27, Beijing, Chinese Society for Metals, Metallurgical Industry Press, Beijing, 2013, pp. 25–33.
[30] V.Y. Gertsman, E. Essadiqi, R. Bouchard, O. Dremailova, J. McDermid, R. Fourmentin, J. Nemes, W. Dabboussi, F.E. Goodwin, Properties of galvanized and galvannealed hot rolled HSLA steel, in: Proc. Galvanizers Association 100th Meeting, October 26–29, 2008, Baltimore, MD, Galvanizers Association, Marine City, MI, 2008.
[31] S. Dionne, B. Voyzelle, E. Essadiqi, J. McDermid, E. Baril, F.E. Goodwin, Galvanizing of hot rolled high strength grades: effect of reheat temperature and furnace atmosphere on coating quality, Iron Steel Technol. (2006) 93–101.
[32] J.R. McDermid, S. Dionne, O. Dremailova, B. Voyzelle, E. Essadiqi, E. Baril, F.E. Goodwin, The effect of continuous galvanizing process parameters on the user properties of hot rolled transformation induced plasticity steels, in: 2005 SAE International Congress and Exposition, April 10–15, 2005, Detroit, MI, Paper 05 M-345, SAE International, Warrendale, PA, 2005.
[33] A. Ferraiuolo, A. Fiorini, S. Pantarotto, U. Liuzzo, M. Manaresi, L. Pocaforza, M. Cocco, G. Ruggeri, Optimisation of AHSS annealing cycles by online control of strip surface emissivity and tensile-structure properties, in: Proc. 11th International Conference on Zinc and Zinc Alloy Coated Steel Sheet (Galvatech 2017), November 12–16, 2017, Tokyo, Iron and Steel Institute of Japan, Tokyo, 2017, pp. 904–911.
[34] G. Angeli, J. Faderl, "Galvanized in Europe"—an actual status and challenges for the future, in: Proc. 10th International Conference on Zinc and Zinc Alloy Coated Steel Sheet (Galvatech 2015), May 31–June 4, 2015, Toronto, AIST, Warrendale, PA, 2015, pp. 19–27.
[35] Galvanized Autobody Partnership Project ZCO-72, Surface Profile Effects on AHSS Processing, Progress Report 4, International Zinc Assn., Durham, NC February 2021.
[36] F.K. Suleiman, K. Lin, R. Tsang, T.F. Larrick, K.J. Daun, Impact of evolving spectral emissivity on pyrometry during intercritical annealing, in: Proc. Galvanizers Assn 111th Meeting, October 6–9, 2019, Charleston, SC, Galvanizers Association, Marine City, MI, 2019.
[37] N. Lagios, M. Renard, J.-P. Crutzen, J.-M. Raick, New lines for processing advanced high strength steels, in: Proceedings Galvanizers Association 106th Meeting, October 5–8, 2014 Jackson, MS, Galvanizers Association, Marine City, MI, 2014.
[38] P. Sippola, R. Patil, Unique technology for production of new AHSS and UHSS grades by using the zinc bath as a quenching medium, in: Proc. 10th International Conference on Zinc and Zinc Alloy Coated Steel Sheet (Galvatech 2015), May 31–June 4, 2015, Toronto, AIST, Warrendale, PA, 2015, pp. 172–181.

[39] E.M. Bellhouse, J.R. McDermid, Analysis of the Fe–Zn interface of galvanized high Al–low Si TRIP steels, Mat. Sci. Eng. A 491 (2008) 39–46.
[40] J.L. Arnold, S.L. Boston, F. Caudill, Control of defects caused by zinc vaporization in the snout, in: Proc. 1st International Conference on Zinc and Zinc Alloy Coated Steel Sheet (Galvatech 1989), September 5–7, 1989, Tokyo, Iron and Steel Institute of Japan, Tokyo, 1989, pp. 130–137.
[41] N.-Y. Tang, F.N. Coady, Coating defects: origins and remedies, in: Proc. Galvanizers Association 93rd Meeting, September 17–19, 2001, Portland, OR, Galvanizers Association, Marine City, MI, 2001.
[42] J. Berry, C.H. Phillips, D.J. Penney, Understanding the role of snout contamination in the formation of an oxide based defect in hot dip galvanised coating, Surf. Coat. Technol. 306 (Part B (25)) (2016) 397–407.
[43] G. Angeli, R. Brisberger, M. Bülter, L. Diez, C. Filthaut, W. Fischer, T. Koll, T. Maiwald, C. Pesci, A. Richter, J. Schulz, U. Stellmacher, N. van Landschoot, Zinc-magnesium-aluminium for automotive industry, in: Proc. 9th International Conference on Zinc and Zinc Alloy Coated Steel Sheet (Galvatech 2013), September 23–27, Beijing, Chinese Society for Metals, Metallurgical Industry Press, Beijing, 2013, pp. 632–636.
[44] J. Faderl, Application of continuous galvanized steel in europe: driving forces and game changer, in: Proc. 11th International Conference on Zinc and Zinc Alloy Coated Steel Sheet (Galvatech 2017), November 12–16, 2017, Tokyo, Iron and Steel Institute of Japan, Tokyo, 2017, pp. 36–45.
[45] Q. Liu, J. Williams, D. Nolan, B. Shedden, R. Smith, A. Neufeld, Development of Mg-containing 55%Al-Zn coated steel for building applications, in: Proc. 10th International Conference on Zinc and Zinc Alloy Coated Steel Sheet (Galvatech 2015), May 31–June 4, 2015, Toronto, AIST, Warrendale, PA, 2015, pp. 208–215.
[46] M. Blumenau, M. Norden, F. Friedel, K. Peters, Wetting force and contact angle measurements to evaluate the influence of zinc metallurgy on galvanizability of high-manganese-alloyed steels, Surf. Coat. Technol. 205 (2010) 828–834.
[47] M. Zaïdi, M.-L. Giorgi, H. Duval, J.-B. Guillot, F. Goodwin, Influence of surface oxide particles on wetting in hot dip galvanizing, in: European Congress on Advanced Materials and Processes (Euromat), September 5–8, 2005, Prague, Czech Republic, Federation of European Materials Societies, London, 2005.
[48] S. Frenznick, S. Swaminathan, M. Stratman, M. Rohwerder, A novel approach to determining high temperature wettability and interfacial reactions in liquid metal/solid Interface, J. Mater. Sci. 45 (2010) 2106–2111.
[49] J. Lee, J. Park, S.-H. Jeon, Wetting of low-carbon, interstitial-free steel surfaces with nanostructured oxides by liquid zinc, Met. Mat. Trans. 40B (2009) 1035–1040.
[50] J.M. Moyer, G.S. Ansell, The volume expansion accompanying the martensite transformation in iron-carbon alloys, Met. Mat. Trans. 6A (1975) 1785, https://doi.org/10.1007/BF02642308.
[51] M.M. McCosby, E.A. Silva, Improved properties of galvannealed dual phase 980, in: Proc. Galvanizers Assn 104th Meeting, October 21–24, 2012, Detroit, MI, Galvanizers Association, Marine City, MI, 2012.
[52] D.-J. Paik, M.-H. Hong, Sustainable production and technology challenges for galvanized steel sheets, in: Proc. 11th International Conference on Zinc and Zinc Alloy Coated Steel Sheet (Galvatech 2017), November 12–16, 2017, Tokyo, Iron and Steel Institute of Japan, Tokyo, 2017, pp. 11–17.
[53] H. Leidheiser Jr., D.K. Kim, Crystallographic factors affecting the adherence of paint to deformed galvanized steel, J. Met. 28 (11) (1976) 19–25.
[54] S. Gaignard, C. Esling, J. Wegria, M. Dubois, Optimisation of galvanisation process parameters to improve the cracking resistance in zinc coatings, in: Proc. 12th Int. Conf. on Textures of Materials (ICOTOM-12), August 9–13, 1999, Montreal, Canada, Canadian Institute of Mining, Metallurgy and Petroleum, Montreal, Canada, 1999.
[55] K. Hoshino, K. Oikawa, W. Tanimoto, M. Nagoshi, M. Koba, Segregation mechanism of Al-based oxides on surface of Zn-0.2mass%Al hot-dip galvanized steel sheets, ISIJ Intl. 60 (8) (2020) 1765–1773.
[56] K. Muell, M. Haissig, TOPOCROM®—a new process to treat cold rolling and skin pass rolls, in: Proc. Galvanizers Association 93rd Meeting September 17–19, 2001, Portland, OR, Galvanizers Association, Marine City, MI, 2001.
[57] D.-J. Paik, H.-S. Han, J.-S. Park, S.-H. Jeon, D.-G. Kang, S.-Y. Choun, M.-H. Hong, Effects of POSTEX® texturing parameters on the surface morphologies and roughness changes of galvanized steel sheets, in: Proc. 10th International Conference on Zinc and Zinc Alloy Coated Steel Sheet (Galvatech 2015), May 31–June 4, 2015, Toronto, AIST, Warrendale, PA, 2015, pp. 605–610.
[58] T. Koll, M. Bretschneider, T. Klinkberg, F. Luther, B. Maas, On optimizing the zinc coating surface by an improved temper rolling process, in: Proc. 10th International Conference on Zinc and Zinc Alloy Coated Steel Sheet (Galvatech 2015), May 31–June 4, 2015, Toronto, AIST, Warrendale, PA, 2015, pp. 611–618.
[59] L.T. Jandel, Surface modification by inline thin film coating and organic/inorganic treatment of continuous galvanised steel sheet, in: Proc. 8th International Conference on Zinc and Zinc Alloy Coated Steel Sheet (Galvatech 2011), June 21–24, 2011, Genoa, Italy, Associazione Italiana di Metallurgia, Milan, 2011.
[60] K. Seto, Recent progress and outlook for zinc- and zinc-alloy coated sheets in Japan, in: Proc. 11th International Conference on Zinc and Zinc Alloy Coated Steel Sheet (Galvatech 2017), November 12–16, 2017, Tokyo, Iron and Steel Institute of Japan, Tokyo, 2017, pp. 1–5.
[61] M. Schinzel, M. Dubois, Post treatment products & technology, in: Proc. 9th International Conference on Zinc and Zinc Alloy Coated Steel Sheet (Galvatech 2013), September 23–27, Beijing, Chinese Society for Metals, Metallurgical Industry Press, Beijing, 2013, pp. 592–597.
[62] G.M. Gutmann, S.H. Burkhardt, Quality assurance by reliable automated surface inspection, in: Proceedings Galvanizers Association 109th Annual Meeting, October 15–18, 2017, Troy, MI, Galvanizers Association, Marine City, MI, 2017.

Chapter 3

Zn coating phase equilibria

3.1 The formation of a Zn coating

Before reviewing Fe-Zn phase equilibria, an understanding of the coating formation must be established. As has been discussed in Chapter 2, Hot-Dip Coating Methods, the basic coating process involves (i) immersing a clean surface of iron or steel substrate into a liquid bath of coating alloy (e.g., zinc with specified additions of aluminum and other elements), (ii) withdrawing the steel and liquid coating from the bath, and (iii) rapidly solidifying the coating on the substrate. However, the process involves many fundamental steps to be considered in order to produce a commercially viable coating.

Wettability is the ability of a liquid (Zn) to maintain contact with a solid surface (Fe) and is controlled by the balance between the intermolecular interactions between the adhesive force of the liquid to solid and the cohesive force between the liquid in the coating and the liquid in the bath. Since the substrate steel is being removed from the liquid zinc pot to form the coating at short times, it becomes a dynamic wetting process, which is controlled by diffusion and fluid flow. It is more of a kinetic problem than an equilibrium problem since the dissolution of the substrate steel is also involved. Capillary hydrodynamics along with wettability contribute to the coating formation, but equally important is the nucleation of interfacial intermetallic layer compounds that may be predicted by equilibrium phase diagrams [1]. Finally, solidification of the overlay alloy occurs, which can range from a simple pure zinc coating to a complex coating consisting of intermetallic phases, elemental dendrites, complex inter-dendritic regions, and coating surface oxides. Thus, the basic steps required for coating formation are [2]:

(1) wettability of the substrate by the liquid metal
(2) dissolution of the steel into the liquid bath
(3) formation of intermetallic compounds on the substrate
(4) solidification and growth of the overlay coating

Once the zinc liquid is deposited on the steel substrate, time and temperature-dependent kinetics will control. In the very short term, an intermetallic barrier layer (inhibition layer) will form at the zinc liquid/steel interface as further described in Chapter 5, Interface Reactions. As the inhibition layer breaks down, diffusion of iron and zinc, as well as other added elements in the bath, occurs. Classically, this can be considered a diffusion couple controlled by Fick's Laws, which will eventually, with time, lead to equilibrium and the phases predicted by the equilibrium phase diagrams below. Zn galvanneal coatings are an example of what might be expected in an equilibrium coating, Chapter 8, Heat Treatment.

However, the metallurgical paradigm of "*processing/structure/properties*" will eventually prevail. It remains up to the engineer, given the set of properties required for a particular application, to determine the processing parameters available to produce the optimum microstructure. This depends upon an understanding of phase equilibria along with the kinetics of nucleation of particular phases as well as the kinetics of diffusion of the available elements. This is a very complex problem discussed in Chapter 6, Overlay Coating Formation.

3.2 Zn-based coating systems

Zinc coatings for sheet steel applied by the immersion of steel into a liquid zinc bath can be classified into five categories:

1. Coatings with more than 0.13% Al but less than 1% Al, and usually less than 0.5% Al in a solid solution overlay layer, covering a thin (<150 nm) interface layer of Fe-Al-Zn intermetallic, commonly termed GI coatings, Chapter 5.
2. Coatings with 0%–13% Al and 10%–18% Fe formed by diffusion of liquid Zn into solid Fe in a heat treatment step after the steel exits the hot-dip bath, termed galvanneal or GA coatings, Chapter 8.
3. Coatings with Al greater than 1% and also Zn-Al coatings to which Mg has been added. These include the Zn-Al alloy coatings termed Galfan (Zn-5% Al) and Galvalume (Zn-55% Al). Coatings over the range $1 < Al < 11\%$ and

The Metallurgy of Zinc Coated Steels. https://doi.org/10.1016/B978-0-323-99984-7.00019-1

$1 < Mg < 3\%$ have been separately developed. The Zn-55% Al-Mg coating is usually termed AM while the other Mg-containing coatings are termed ZM, Chapter 6.

4. Coatings produced during the hot press forming process. The precursor coatings are either GI or GA coatings that are subjected to heating of the coated steel to the austenitic range, typically 650–900°C, followed by die quenching. These coatings normally have between 15% and 75% Zn, depending on the steel and coating composition and the process details, Chapter 12.
5. Zinc-coated sheet steel produced by electrogalvanizing (EG) and vapor deposition techniques, Chapter 9.

3.3 Equilibrium phase diagrams

Before discussing the various zinc coating types in the chapters to come, it will be helpful to review the response between the iron substrate and zinc coating components. Reactions that occur between the steel substrate and the Zn alloy coating bath require an understanding of the various equilibrium diagrams defined by the elemental interactions as well as the phase transformation kinetics. Starting with the binary Zn-Fe diagram, the related multicomponent systems required by the alloy content of the bath composition must be reviewed.

3.3.1 Binary systems

3.3.1.1 Fe-Zn

For the formation of a zinc coating in the hot-dip galvanizing process, iron reacts with zinc, with a maximum solubility of 41 at.% Zn at 782°C in the BCC α phase, to form a series of tightly bonded intermetallic compounds: ζ-$FeZn_{13}$, δ-$FeZn_{7-10}$, Γ_1-$Fe_{11}Zn_{40}$, and Γ-Fe_3Zn_{10} [3–5]. Burton and Perrot [6] in 1993 reviewed the many studies on the Fe-Zn equilibria, and since that time several investigators [7–12] have systematically assessed the thermodynamic system by using the CALPHAD (Calculation of PHAse Diagrams) method. Recently, Han, et al. [13] reexamined the Fe-Zn phase equilibria and proposed a new diagram with clearly defined phase boundaries of the intermetallic and liquid phase that allows for a more accurate thermodynamic modeling of the binary system (Fig. 3.1). The phases found in this system are shown in Table 3.1, and a more detailed description has been published previously [14] and has been modified below according to Han, et al. [13]. In addition, Zhu, et al., [15] reexamined the phase equilibria and diffusion coefficients that will provide reliable inputs for future thermodynamic and kinetic modeling in the Fe-Zn binary system.

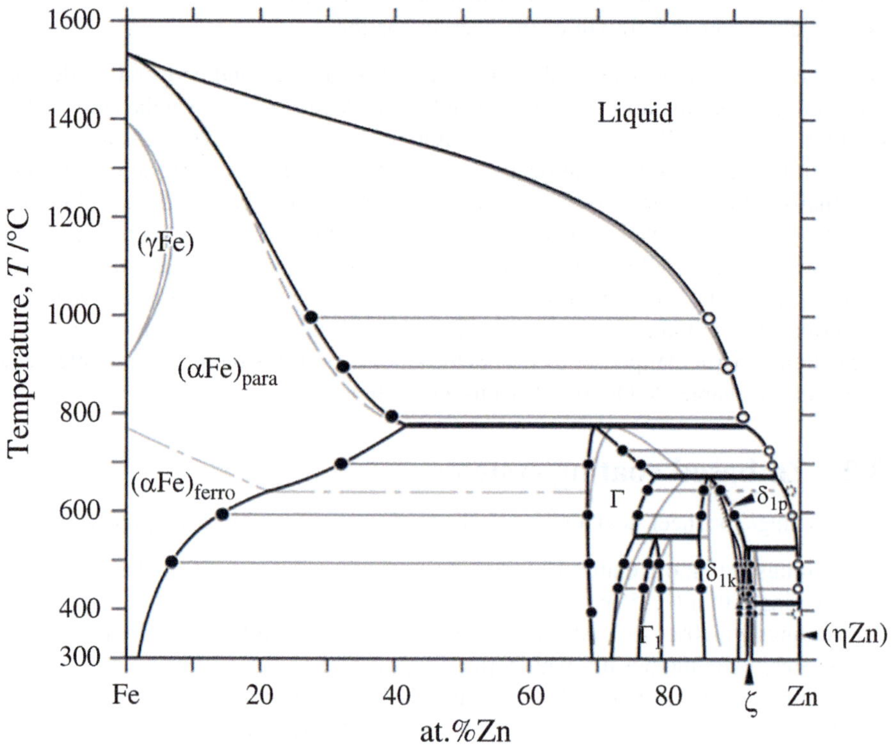

FIG. 3.1 Phase diagram of the Fe-Zn binary system experimentally determined by Han et al. [13]. *(From K. Han, I. Ohnuma, K. Okuda, R. Kainuma, Experimental determination of phase diagram in the Zn-Fe binary system, J. Alloy Compd. 737 (2018) 490–504.)*

TABLE 3.1 Nominal symbols, stoichiometric ratios, and crystal structures of stable phases in the Zn-Fe binary system [13].

Phase	Formula	Pearson symbol	Space group	Proto type
αFe	(Fe, Zn)	*cI*2	*Im*3m	W
Γ	Fe_3Zn_{10}	*cI*52	*I*43m	γ-Cu_5Zn_8
$Γ_1$	$Fe_{11}Zn_{40}$	*cF*408	*F*43m	$Fe_{11}Zn_{40}$
$δ_{1k}$	$FeZn_7$	Three times as large as that of $δ_{1p}$		
$δ_{1p}$	$FeZn_{10}$	*hP*556	*P*6₃/mmc	$FeZn_{10}$
ζ	$FeZn_{13}$	*mC*28	*C*2/m	$FeZn_{13}$
ηZn	(Zn, Fe)	*hP*2	*P*6₃/mmc	Mg

From K. Han, I. Ohnuma, K. Okuda, R. Kainuma, Experimental determination of phase diagram in the Zn-Fe binary system, J. Alloy Compd. 737 (2018) 490–504.

Alpha (α) iron

Alpha or ferritic iron, the common BCC structure found in most iron alloys at moderate temperatures, can have a maximum solubility with zinc of up to 41 at.%.

Gamma (Γ) phase

The gamma (Γ) phase, Fe_3Zn_{10}, is body-centered cubic and has an iron composition range of approximately 69–78 at.% Zn at 672°C. It forms on cooling because of a peritectic reaction at 782°C and 69.3 at.% Zn and has a maximum solubility of Fe in Zn at the delta (δ) phase peritectic temperature of 672°C.

Gamma$_1$ ($Γ_1$) phase

The gamma 1 ($Γ_1$) phase, $Fe_{11}Zn_{40}$, has a face-centered cubic lattice structure with a zinc composition range of approximately 76–79 at.% at 450°C. It forms on cooling as a result of a peritectoid reaction between gamma (Γ) phase and delta (δ) phase at 550°C and approximately 78.5°C.

Delta (δ) phase

The delta (δ) single-phase region in the literature is distinguished into ordered ($δ_{IK}$) and disordered ($δ_{1P}$) phases. The delta ($δ_{1k}$) phase, $FeZn_7$, has an iron composition of 85.1–88.9 at.% zinc at 600°C and forms on cooling from another peritectic reaction, gamma (Γ) and liquid (L) at 672°C and 86.5 at.% zinc. It is HCP and is three times larger than the ($δ_{1p}$) phase. The Delta ($δ_{!P}$) phase field can be found alongside the ($δ_{ik}$) field at higher Zn concentrations. In the past, it was distinguished by its morphology from the ($δ_{1p}$) phase. The ($δ_{1p}$) phase, $FeZn_{10}$, has a palisade morphology, whereas the ($δ_{1k}$) phase is compact. For the short-term nonequilibrium immersion times found in galvanizing, only one delta (δ) phase morphology was reported.

Eta (η) phase

The Eta (η) phase is an HCP terminal solid solution of Zn, with the melting point of pure Zn being 419.3°C. Eta (η) has a maximum solubility of 0.3 at.% Fe in Zn at 417.4°C, the eutectic temperature below which the two-phase zeta (ζ) and eta (η) fields are stable.

3.3.1.2 Zn-Al

Aluminum is added to the zinc bath to either produce a pure zinc overlay for galvanic protection by inhibiting the formation of Fe-Zn phases at the zinc/iron substrate interface or by introducing multiphase microstructures in the overlay coating to improve corrosion resistance. Fig. 3.2 shows the Al-Zn phase diagram [16]. Examples of aluminum added zinc coatings are Zn-5Al (Galfan) and Zn-55Al (Galvalume). An examination of the phase diagram shows that the solid phases present in the Al-Zn system include a solid solution terminal α FCC Al-rich phase with a maximum solubility of approx. 70 wt% Zn and a terminal η HCP Zn-rich phase with a maximum Al content of approx. 1.2 wt% Al. There is a wide solubility β phase, ZnAl,

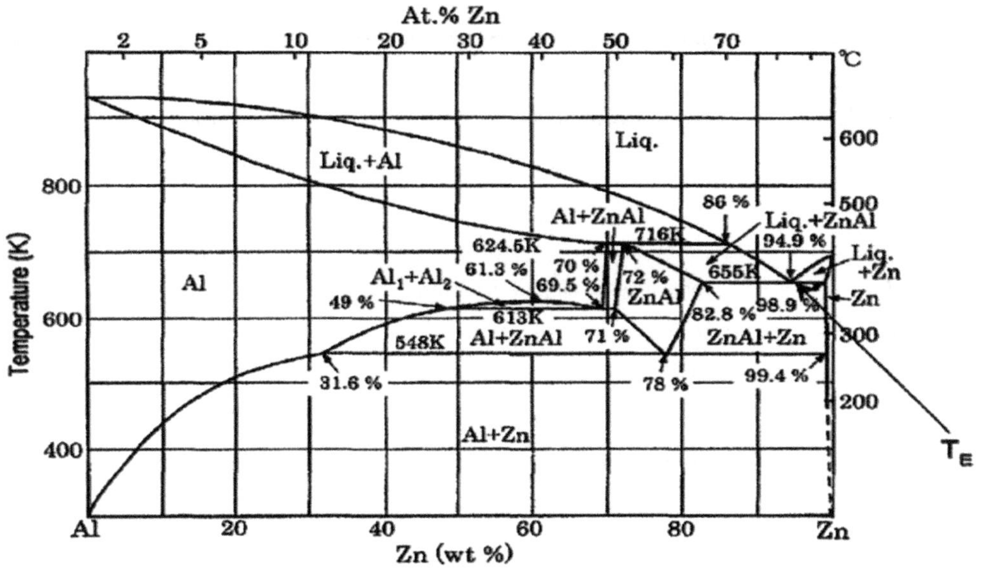

FIG. 3.2 The Al-Zn equilibrium phase diagram [16]. *(From K. Miyamoto, S. Nagagawa, Bonding method, of dissimilar materials from metals and bonding structure thereof, U.S. Patent 8,058,584, November 15, 2011.)*

between 71 wt% and 82.8 wt% that exists from 548 K and 716 K. The diagram also shows a eutectic temperature of 655 K at a composition of 94.9 wt% Zn and a eutectoid transformation located at 78 wt% Zn and 548 K. Diffusion data have been developed as an extension of the CALPHAD approach, known as DICTRA (DIffusion Controlled TRAnsformation) code [17].

3.3.1.3 Al-Fe

Aluminum additions to the zinc bath are deliberately added to form zinc-coated steel with a variety of compositions, e.g., GI coatings (<1% Al), Galfan (Zn-5% Al), Galvalume (Zn-55% Al), as well as a range of Zn-Al-Mg coatings. Besides producing a whole range of overlay microstructures and properties, Al is added to the various Zn baths to "inhibit" brittle Fe-Zn intermetallic compounds from forming. Depending upon the Al content of the bath, inhibition can be temporary, but it can lead to the formation of several Fe-Al intermetallic compounds, Table 3.2 [18].

Although Zn can greatly interact with these Fe-Al intermetallic binary compounds to form ternary Zn-Al-Fe intermetallic compounds, it is instructional to first review the Fe-Al equilibrium phase diagram. Fig. 3.3 shows the three major stoichiometric compounds in this binary system. Takata, et al. [19] studied the microstructure and crystallography of the major intermetallic compound η, Fe_2Al_5, and showed that this intermetallic compound grows toward the solid iron substrate as a columnar structure along the [001] direction, producing a saw tooth morphology at the α-iron/η interface. The stresses produced at the interface are also described and may add to the explanation of inhibition layer breakdown by providing fast diffusion paths.

TABLE 3.2 Stoichiometric ratios and crystal structures of IMC phases in the Al-Fe binary system [18].

Phase	Formula	Lattice	Space group	Pearson symbol	Prototype
ε	Fe_5Al_8	Cubic	$I\bar{4}3m$	cI52	Cu_5Zn_8
ζ	$FeAl_2$	Triclinic	$P\bar{1}(2)$	aP19	Fe_5Al_{12}
η	Fe_2Al_5	Orthorhombic	Cmcm	oS24	$FeAl_{28}$
θ	Fe_4Al_{13}	Monoclinic	C12/m1	mS102	Fe_4Al_{13}

From K. Han, I. Ohnuma, R. Kainuma, Experimental determination of phase equilibria of Al-rich portion of the Al-Fe binary system, J. Alloys Compd. 668 (2016) 97–106.

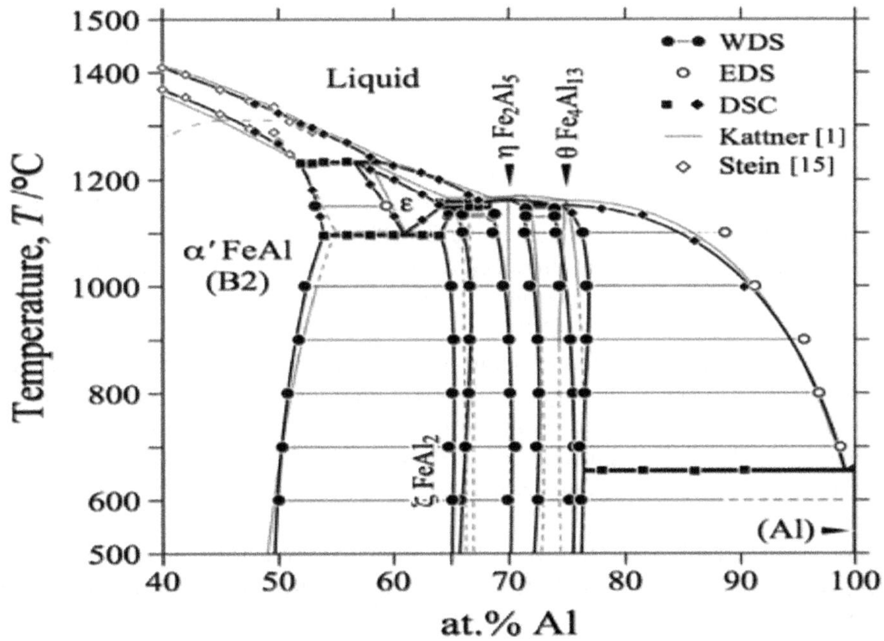

FIG. 3.3 Experimentally determined phase diagram of Al-rich portion of the Al-Fe system [18]. *(From K. Han, I. Ohnuma, R. Kainuma, Experimental determination of phase equilibria of Al-rich portion of the Al-Fe binary system, J. Alloys Compd. 668 (2016) 97–106.)*

3.3.2 Ternary systems

3.3.2.1 Zn-Al-Fe

The Zn-Al-Fe ternary system is important for the optimization of the various coating types described in this chapter. In designing these coatings, the prime objective is to produce a corrosion-resistant coating without deteriorating the formability of the sheet steel being coated. The intermetallic compound layers formed between the steel and the molten bath of the Zn-Al alloys must be optimized. Knowledge of the Zn-Al-Fe system is first based on the binary systems described above and shown in Fig. 3.1: Zn-Al, Fig. 3.2: Al-Zn, and Fig. 3.3 [20] Al-Fe, and summarized in Table 3.3.

TABLE 3.3 Information related to crystal structures of intermetallic compounds in the Fe-Al and Fe-Zn binary systems [20].

Alloy system	Phase	Formula	Lattice	Space group	Pearson symbol
Fe-Al	αFe	(Fe, Al)	Cubic	$Im\bar{3}m$	*cI2*
	ε	Fe_5Al_8	Cubic	$I\bar{4}3m$	*cI52*
	ζ	$FeAl_2$	Triclinic	$P\bar{1}$	*aP19*
	η	Fe_2Al_5	Orthorhombic	*Cmcm*	*oS24*
	θ	Fe_4Al_{13}	Monoclinic	*C12/m1*	*mS102*
	αAl	(Al, Fe)	Cubic	$Fm\bar{3}m$	*cF4*
Fe–Zn	αFe	(Fe, Zn)	Cubic	$Im\bar{3}m$	*cI2*
	Γ	Fe_4Zn_9	Cubic	$I\bar{4}3m$	*cI52*
	$Γ_1$	$Fe_{21.2}Zn_{80.8}$	Cubic	$F\bar{4}3m$	*cF408*
	$δ_{1k}$	$FeZn_7$	Superlattice structure based on $δ_{1p}$		
	$δ_{1p}$	$Fe_{13}Zn_{26}$	Hexagonal	$P6_3/mmc$	*hP556*
	ζ	$FeZn_{13}$	Monoclinic	*C2/m*	*mC28*
	ηZn	(Zn, Fe)	Hexagonal	$P6_3/mmc$	*hP2*

From I. Lee, K. Han, I. Ohnuma, R. Kainuma, Experimental determination of phase diagram at 450°C in the Zn-Fe-Al ternary system, J. Alloys Compds. 854 (2021) 157–163.

Reactions occurring between the steel substrate and the Zn alloy coating bath are described by equilibrium diagrams for the Zn-Fe binary and related multicomponent systems, as shown in Fig. 3.4. A schematic representing the 450°C isotherm of the Zn-Fe-Al system that focuses on the intermetallic compounds, which can be found in galvanizing baths, is seen in Fig. 3.5 [21].

For continuous hot-dip galvanizing, the zinc corner of the ternary phase diagram is extremely important. The most accurate models of Zn-Fe phase equilibria are based on improved descriptions developed for all eight intermediate phases in the Zn-Fe system (intermetallic compounds gamma-one, gamma, delta, and zeta together with FCC and BCC Fe-rich phases and the HCP Zn-rich phase and liquid). These were incorporated in the CALPHAD formalism and correlated with experimental results [8]. The sublattice models derived for the intermetallic compounds of the Zn-Fe diagram were examined to predict which sublattice(s) would be most capable of hosting Al. These were then combined with past assessments of Zn-Al-Fe equilibria and Al activities introduced to the PARROT module of Thermo-Calc. Comparison of calculated quantities experimental results led to an improved thermodynamic model of the Zn-Al-Fe system within the 420–500°C range [22]. The Zn-rich corner of the Zn-Al-Fe phase diagram, of most interest for the galvanizing phenomena, was then characterized in great experimental detail. Results were combined with thermodynamic modeling, permitting mapping of Al and Fe isoactivity contours on isothermal sections [23,24]. A comparison of the Nakano, et al. [22] and the McDermid et al. [23] results is seen in Figs. 3.6–3.9 for various isothermal sections of this corner of the Zn-Al-Fe ternary phase diagram from 450°C to 480°C, respectively [24].

At higher Al contents, 5% Al, Galfan coatings result in the formation of an inhibition layer of $Fe_2Al_5Zn_x$ after short dipping times. Ghuman and Goldstein [25] schematically followed the transformation of this type of alloy through the Fe-Zn-Al ternary phase diagram, e.g., Fig. 3.4, at various dipping temperatures and times. The microstructure of these alloys, especially with additions of Mg, will be discussed in Chapter 6. At even higher Al contents, 55% Al, Galvalume coatings present an even greater challenge in understanding the formation of phases by equilibrium phase diagrams. The Al-rich portion phase diagram in the Zn-Fe-Al system was recently proposed by Lee et al. [20], Fig. 3.10. For these higher aluminum alloys, further elemental additions are made, e.g., Si and Mg, which complicate the phase stability even further, requiring quaternary or higher equilibrium diagrams.

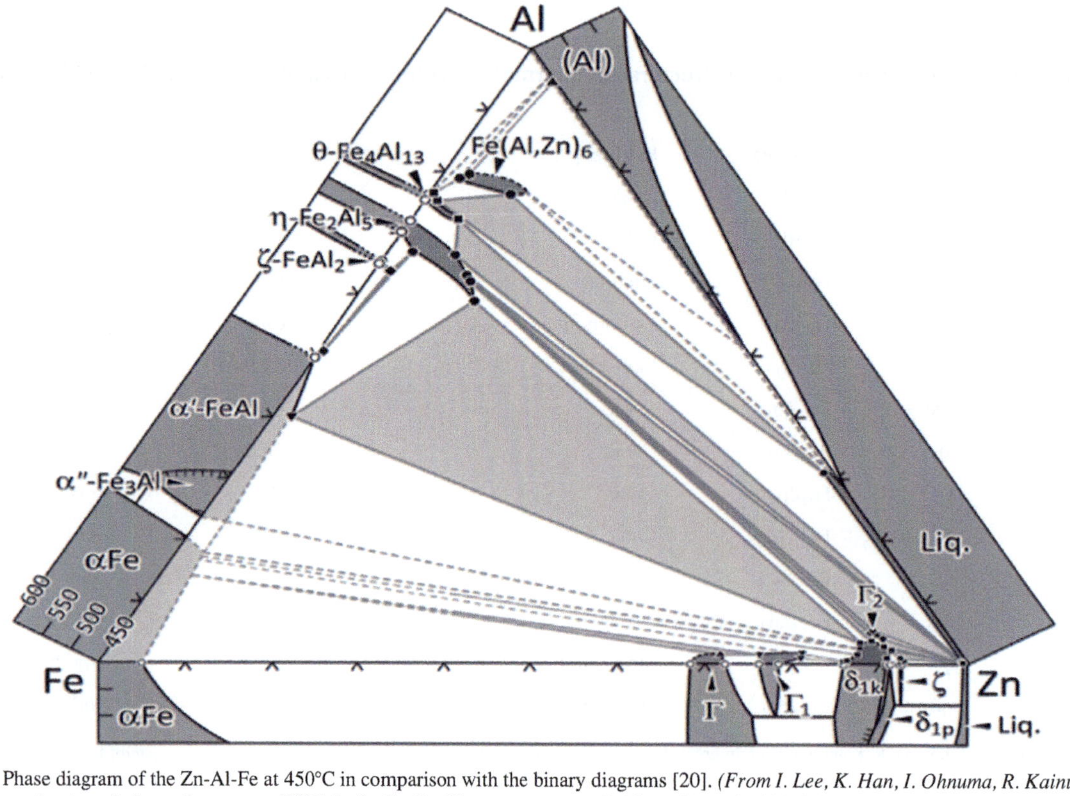

FIG. 3.4 Phase diagram of the Zn-Al-Fe at 450°C in comparison with the binary diagrams [20]. *(From I. Lee, K. Han, I. Ohnuma, R. Kainuma, Experimental determination of phase diagram at 450°C in the Zn-Fe-Al ternary system, J. Alloys Compds. 854 (2021) 157–163.)*

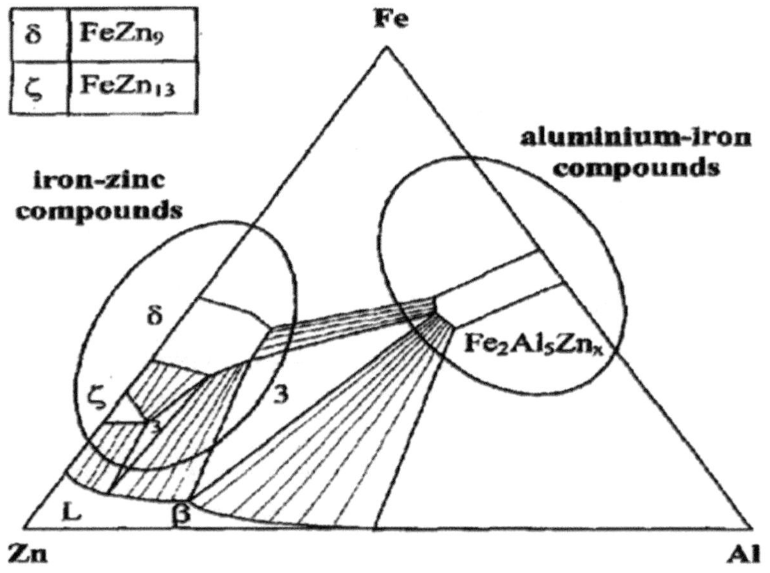

FIG. 3.5 Schematic 450°C isotherm of the zinc-iron-aluminum system [21]. *(From M.L. Giorgi, J. B. Guillot, R. Nicolle, CALPHAD 25 (2001) 461–474.)*

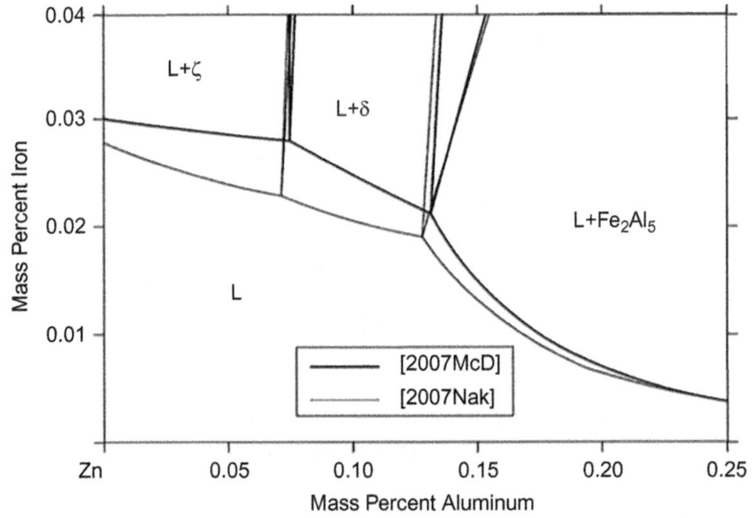

FIG. 3.6 Al-Fe-Zn computed metastable isothermal section at 450°C [24]. *(From V. Raghaven, Al-Fe-Zn (aluminum-iron-zinc), JPEDAV 32 (2011) 143–144.)*

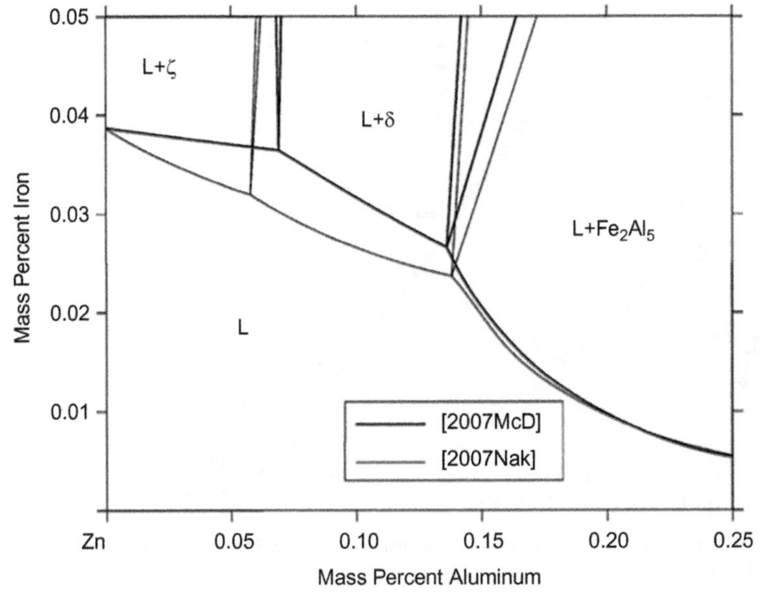

FIG. 3.7 Al-Fe-Zn computed metastable isothermal section at 460°C [24]. *(From V. Raghaven, Al-Fe-Zn (aluminum-iron-zinc), JPEDAV 32 (2011) 143–144.)*

FIG. 3.8 Al-Fe-Zn computed metastable isothermal section at 470°C [24]. *(From V. Raghaven, Al-Fe-Zn (aluminum-iron-zinc), JPEDAV 32 (2011) 143–144.)*

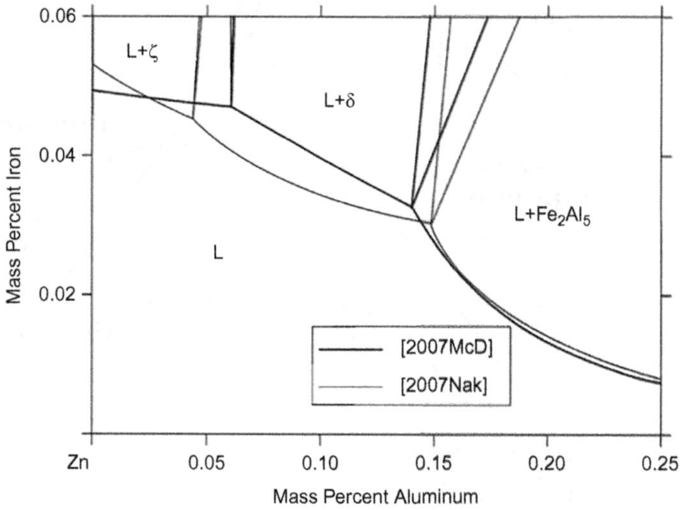

FIG. 3.9 Al-Fe-Zn computed metastable isothermal section at 480°C [24]. *(From V. Raghaven, Al-Fe-Zn (aluminum-iron-zinc), JPEDAV 32 (2011) 143–144.)*

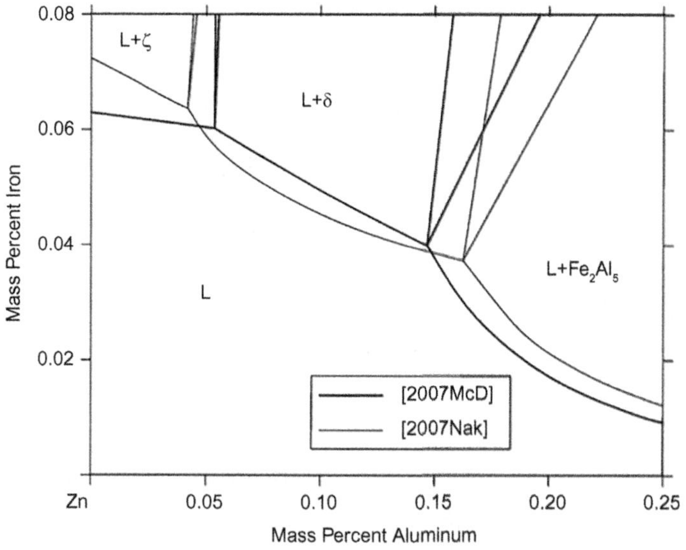

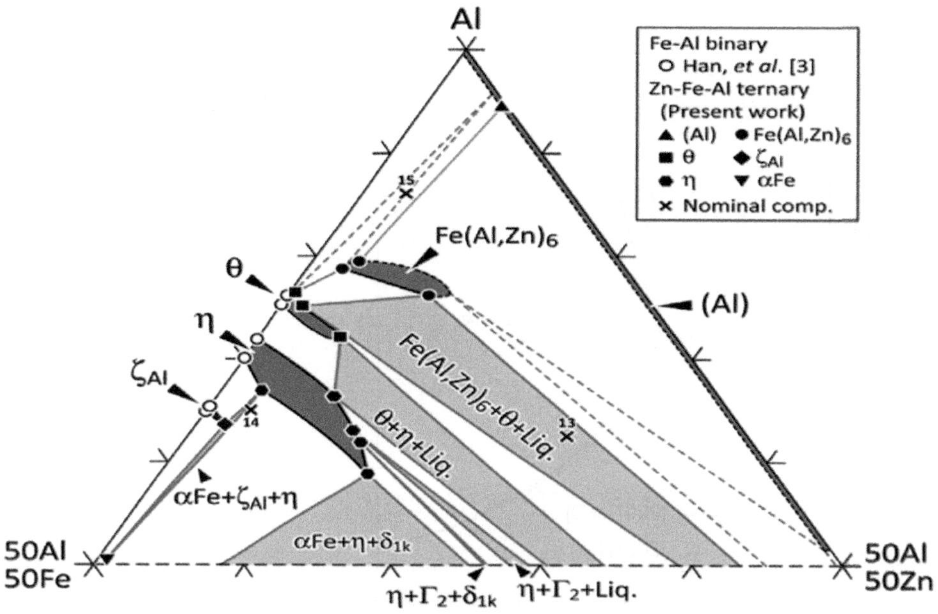

FIG. 3.10 Al-rich portion of the Zn-Fe-Al phase diagram at 450°C [20]. *(From I. Lee, K. Han, I. Ohnuma, R. Kainuma, Experimental determination of phase diagram at 450°C in the Zn-Fe-Al ternary system, J. Alloys Compds. 854 (2021) 157–163.)*

3.3.3 Complex systems

3.3.3.1 Al-Zn-Fe-Si

The Al-Zn-Fe-Si quaternary system comprises four ternary systems, namely Zn-Al-Si, Zn-Al-Fe, Al-Fe-Si, and Zn-Fe-Si, which have been summarized by Liu et al. [26]. The equilibrium phases and phase compositions in the Zn-Al-Fe-Si quaternary system are listed in Table 3.4, and an isothermal section of the Zn-Al-Fe-Si quaternary system at a constant 50 at.% Al is shown in Fig. 3.11 [26]. To sum up, the complexity of the quaternary system is such that it is very difficult to follow the development of the coating structure. Nevertheless, keeping Al constant, studies have been conducted to determine the optimum Si content in 55% Al-Zn Galvalume coatings in order to understand the overlay coating microstructure. Selverian et al. [27] have shown that for high Al-Zn bath compositions, the reaction of an iron sheet placed in the bath is very rapid and strongly exothermic. In bath compositions of 45Al-55Zn, 55Al-45Zn, and 75Al-25Zn, an iron panel was consumed within 2 min by the bath. Borzillo and Horton [28] found that the presence of silicon was required to prevent the exothermic reaction and promote the formation of an adherent coating. Si was found to suppress the formation. Details of these results will be discussed further in Chapter 5, Interface Reactions. Further, it was found that control of $FeAl_3$ by silicon can influence the spangle size, which will be discussed in more detail in Chapter 6 [29].

TABLE 3.4 The equilibrium phases and phase compositions in Zn-Al-Fe-Si quaternary system at 600°C with Al fixed at 50 at.% [26].

		Composition of phases (at.%)			
Specimen	Phase	Al	Si	Fe	Zn
Al	τ_2	55.4	23.0	20.9	0.7
$Al_{50}Zn_{30}Fe_5Si_{15}$	τ_4	49.6	33.9	15.8	0.7
	Si	0.4	98.9	0.2	0.5
	Liq.	42.1	0.5	0.2	57.2
A2	τ_2	61.9	16.2	20.6	1.3
$Al_{50}Zn_{41}Fe_5Si_4$	τ_5	68.4	12.3	18.0	1.3
	$FeAl_3$	66.2	6.0	23.8	4.0
	Liq.	56.8	3.5	0.2	39.5
A3	τ_{10}	56.1	17.5	24.1	2.3
$Al_{50}Zn_8Fe_{27}Si_{15}$	τ_3	53.8	21.3	23.7	1.2
	Fe_2Al_5	61.7	7.9	24.0	6.4
	Liq.	35.7	0.4	0.2	63.7
A4	τ_1	33.4	29.1	37.1	0.4
$Al_{50}Zn_{18.5}Fe_{27.5}Si_4$	τ_{10}	57.1	17.2	24.5	1.2
	Fe_2Al_5	62.3	3.4	25.3	9.0
	Liq.	9.8	1.5	6.0	82.7
A5	τ_2	54.9	23.0	19.1	3.0
$Al_{50}Zn_{15}Fe_{20}Si_{15}$	τ_3	53.5	21.3	23.2	2.0
	Fe_2Al_5	0.1	99.5	0.1	0.3
	Liq.	48.0	7.6	0.3	44.1
A6	τ_4	49.6	33.0	16.9	0.5

Continued

TABLE 3.4 The equilibrium phases and phase compositions in Zn-Al-Fe-Si quaternary system at 600°C with Al fixed at 50 at.% —cont'd

		Composition of phases (at.%)			
Specimen	Phase	Al	Si	Fe	Zn
$Al_{50}Zn_5Fe_5Si_{40}$	αAl	79.5	1.2	0.3	19.0
	Si	0.3	99.3	0.1	0.3
	Liq.	61.5	3.9	0.3	34.3
A7	τ_5	69.5	12.1	17.7	0.7
$Al_{50}Zn_{39}Fe_8Si_3$	$FeAl_3$	67.8	4.7	23.8	3.7
	Liq.	42.2	2.4	0.4	55.0
A8	τ_2	61.6	16.7	20.5	1.2
$Al_{50}Zn_{34}Fe_{10}Si_6$	$FeAl_3$	70.6	5.1	22.6	1.7
	Liq.	46.0	1.5	0.2	52.3
A9	τ_4	48.8	34.0	16.2	1.0
$Al_{50}Zn_{16}Fe_4Si_{30}$	Si	0.1	99.7	0.1	0.1
	Liq.	66.0	10.7	0.2	23.1
A10	τ_2	58.2	19.6	20.3	1.9
$Al_{50}Zn_{43}Fe_2Si_5$	Si	1.3	97.3	0.1	1.3
	Liq.	51.2	3.7	0.3	44.8
All	τ_1	33.2	30.8	35.3	0.7
$Al_{50}Zn_{15}Fe_{30}Si_5$	Fe_2Al_5	64.4	3.4	24.2	8.0
	Liq.	4.1	0.6	4.8	90.5
A12	τ_3	54.2	20.6	24.1	0.8
$Al_{50}Zn_{18}Fe_{22}Si_{10}$	Fe_2Al_5	62.2	8.2	23.9	5.7
	Liq.	11.3	0.4	1.0	87.3
A13	FeAl	46.4	3.7	49.4	0.5
$Al_{50}Zn_{14}Fe_{35}Si_1$	Fe_2Al_5	61.9	0.2	27.6	10.3
	Liq.	2.2	0.1	14.0	83.7
A14	FeAl	48.5	1.5	49.3	0.7
$Al_{50}Zn_3Fe_{45}Si_2$	Fe_2Al_5	62.6	0.2	27.7	9.5
A15	τ_1	35.1	27.9	36.6	0.4
$Al_{50}Zn_3Fe_{37}Si_{10}$	Fe_2Al_5	64.3	1.8	26.3	7.6
A16	τ_{10}	57.5	17.6	24.5	0.4
$Al_{50}Zn_5Fe_{30}Si_{15}$	Fe_2Al_5	61.3	1.8	27.4	9.5

From L. Liu, X. Pan, L. Lu, Y. Liu, X. Su, J. Wang, Phase equilibria of 600°C isothermal section of Zn-Al-Fe-Si quarternary system, JPEDAV 37 (2016) 693–701.

3.3.3.2 Al-Zn-Fe-Mg-Si

Since the review by Marder [14], a new family of Zn-Al-Mg coatings has been developed and industrialized. The first work on Zn-Al-Mg phase equilibria extended the Zn-rich corner of the Zn-Al-Fe phase diagram by adding Mg up to 0.5% to a Zn-0.2% Al bath in a hot-dip galvanizing process simulator and comparing the phases observed with numerical analysis based

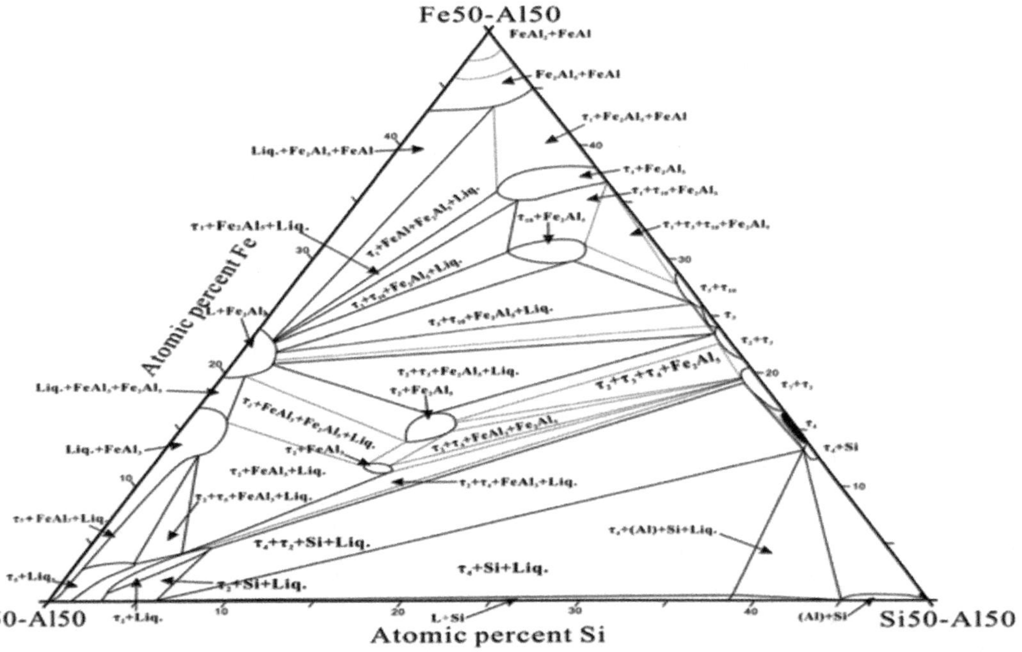

FIG. 3.11 The isothermal section of the Zn-Al-Fe-Si quaternary system at 600°C with Al fixed at 50 at.% [26]. *(From L. Liu, X. Pan, L. Lu, Y. Liu, X. Su, J. Wang, Phase equilibria of 600°C isothermal section of Zn-Al-Fe-Si quarternary system, JPEDAV 37 (2016) 693–701.)*

on data that were, at the time, only available for the Al-rich corner [30]. Differential scanning calorimetry showed the ternary eutectic temperature of this system was 344°C and the peritectic temperature was 359°C. This was extended in the Galvanized Autobody Partnership program ZCO-66 [31], which derived phase diagrams for the same range of Al compositions and Mg additions up to 3%. These diagrams are shown in Fig. 3.12 [31]. It was observed that the Fe-Zn ζ phase is formed at low Al content over the entire temperature range only for 0% and 1% Mg. For 3% Mg, ζ phase forms up to 456°C, and no ζ phase forms for 5% Mg. When increasing Mg content, $FeAl_2$ forms at Al compositions between those corresponding to δ and η (Fe_2Al_5) phases. Fe_2Al_5 forms at temperatures up to 462°C for 1% Mg, up to 472°C for 3% Mg, and up to 479.4°C for 5% Mg. The minimum Al content for which η phase forms is slightly lower for 1% Mg than for 0% Mg and then increases when increasing the Mg content (it remains in the range of 0.11%–0.14% Al).

DeBrucker et al. [30] also examined the addition of Mg in compositions up to Zn-6Al-3Mg. They observed that, contrary to equilibrium predictions, the Mg_2Zn_{11} phase was never observed. The ternary eutectic consisted of Zn-hcp/Al-fcc/$MgZn_2$. The absence of the Mg_2Zn_{11} phase was confirmed in a study of the Zn-11% Al-3% Mg-0.2% Si coating trademarked as SuperDyma [32]. Unlike the lower Al content coatings, in this coating precipitation of the Zn phase from the primary Al phase begins at 308°C. The metastable $MgZn_2$ forms as a Laves phase in the absence of Mg_2Zn_{11} and the associated peritectic reaction.

The solidification path of hypoeutectic Zn-Al-Mg compositions, which are the most widely used Zn-Al-Mg coating compositions, has been described in [33] In the absence of the Mg_2Zn_{11} phase, the composition at the ternary eutectic point is reported as 3.8% Al and 2.2% Mg. In this case, solidification of 1.5% Al and 1.5% Mg alloy (point 1 in Fig. 3.13) begins at the liquidus temperature of 381.5°C. Precipitation of the primary Zn phase continues with the liquid composition following the indicated line from point 1 to point 2, which is at 350°C at a composition of 2.3% Al and 2.8% Mg. Thereafter, the liquid composition follows the univariant line from point 2 to the eutectic at point 5. During this stage, a binary eutectic constituent (Zn + $MgZn_2$) precipitates. At the eutectic point, a ternary eutectic constituent (Zn + $MgZn_2$ + fcc) precipitates isothermally. The hcp Zn phase contains 0.9% Al and 0.1% Mg in solid solution, the $MgZn_2$ phase contains 0.7% Al in solution, and the composition of the fcc phase is 81.5% Zn-18.5% Al-0.06% Mg. For a 3.0% Al-3.0% Mg coating (point 3), solidification begins at the liquidus temperature of 361.5°C. Precipitation of the primary $MgZn_2$ phase continues with the liquid composition following the line from point 3 to point 4, which is 346°C at a composition of 3.1% Al-2.7% Mg.

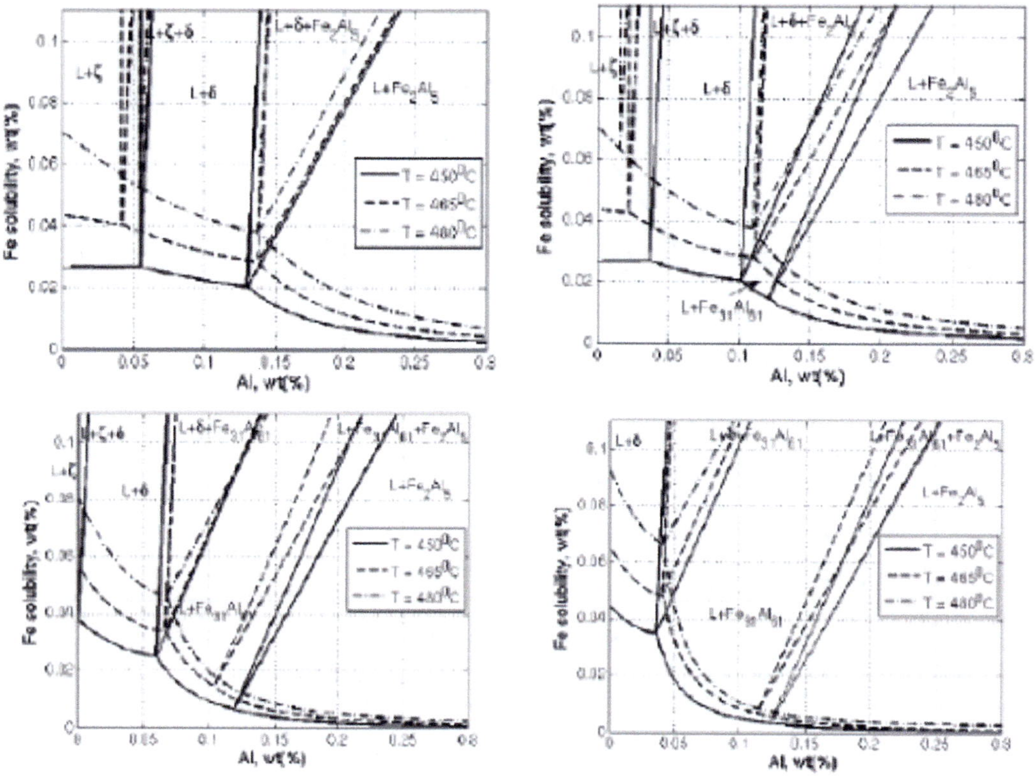

FIG. 3.12 Zn-Al-Fe-Mg phase diagrams for 0%, 1%, 3%, and 5% Mg. Upper left-0% Mg; Upper right-1% Mg; Lower left-3% Mg; Lower right-5% Mg [31]. *(From Galvanized Autobody Partnership Project ZCO-55/66 Final Report, August 2014, 135. Available from www.zinc.org.)*

FIG. 3.13 Liquidus projections of the Zn-rich corner of the Zn-Al-Mg phase diagram, indicating solidification paths of Zn-1.1Al-1.1 Mg alloy (beginning at point 1) and Zn-3% Al-3% Mg alloy (beginning at point 3) [33]. *(From A.D. Pelton, Thermodynamic Calculations of the Steps during Solidification of Certain Zn-Al-mg Alloys, prepared for F. Ajersch, under contract with IZA, 2014.)*

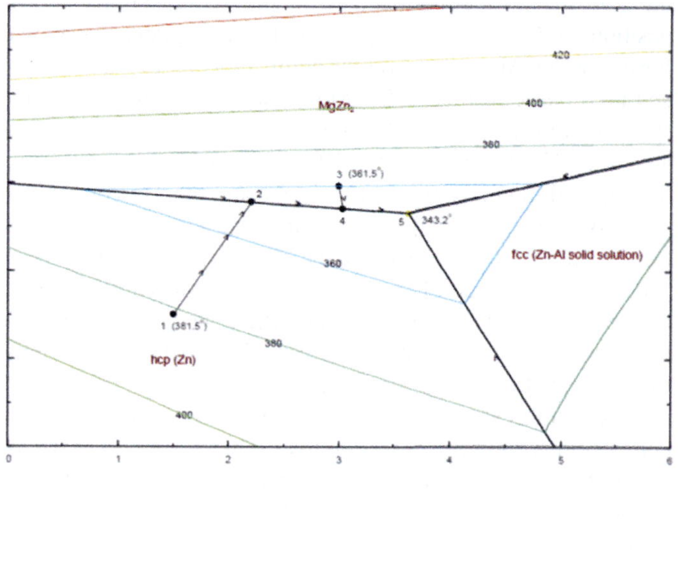

Thereafter, the liquid composition follows the univariant line from point 4 to the eutectic at point 5. During this stage, a binary eutectic constituent (Zn + $MgZn_2$) precipitates. At the eutectic point, a ternary eutectic constituent (Zn + $MgZn_2$ + fcc) precipitates isothermally. The hcp Zn phase contains 0.9% Al and 0.1% Mg in solid solution, the $MgZn_2$ phase contains 0.7% Al in solution, and the composition of the fcc phase is 81.5% Zn-18.5% Al-0.06% Mg. Little difference was seen between equilibrium and Scheil-Gulliver solidification, i.e., no diffusion in the solid.

Most of the industrially used Zn-Al-Mg alloy coatings, particularly those used for automotive applications, have Al and Mg compositions in the range of 1%–3.7% [34]. Because of this, their microstructures are formed in the same way, and they have similar physical and electrochemical properties. The steel-coating interfacial layers formed with coatings made by dipping an IF steel into a Zn-1.0% Al-1.8% Mg bath were compared with coatings made in a Zn-0.28% Al-0.28% Mg bath [35]. The steel entry and bath temperatures were both 460°C and the dipping time was 3s. The lower alloyed bath formed coatings with a $Fe_2Al_{5-x}Zn_x$ interface layer consisting of two morphologies: either a rough, granular layer like that formed in GI baths with crystallographic direction parallel to the normal direction of the steel sheet, or a smooth layer of single crystal $Fe_2Al_{5-x}Zn_x$ grown epitaxially on the substrate. For the higher alloyed bath, the interfacial layer consisted of disk-shape [010]-oriented Fe_4Al_{13} crystals with diameters between 100 and 500nm. No orientation relationship between the Fe_4Al_{13} crystals and α-Fe was found, and no Mg content was observed in any of the intermetallic compounds, a consequence of the mutual insolubility of Fe and Mg. The Zn-Fe-Al ternary phase diagram shows that Fe_4Al_{13} phase can be formed in a bath containing more than 1.3% Al, which is slightly higher than the Al content used in this study [22].

Comparison of a Zn-55% Al-2% Mg-1.6% Si coating interface with that of a Mg-free version of the same coating showed the intermetallic layer of the Mg-added coating to consist of rounded particles with diameters 100–1000nm, but to be largely free of the Si-containing intermetallics normally observed at the interface of the Mg-free Zn-55% Al-1.5% Si coating that has been industrially produced for many years [36]. The microstructure of the Zn-55% Al coating overlay also changed as Mg was added. The Zn-rich interdendritic areas were observed to be more connected in the Al-Zn coating with the Mg addition than in the conventional Zn-55% Al coating. The inter-dendritic areas of the Zn-55% Al coating normally contain isolated islands of Zn-rich particles in an Al-rich matrix. Si-Mg particles were also observed in the Mg-added coating, concentrated mostly in the bottom half of the coating, close to the alloy layer, which will be discussed in more detail in Chapter 6.

References

[1] M. Zaïdi, M.-L. Giorgi, J.-B. Guillot, F. Goodwin, Influence of surface oxide particles on reactive wetting in hot-dip galvanizing, in: Presented at metal European Congress on Advanced Materials and Processes (Euromat), Prague, Czech Republic, September 5–8, Federation of European Materials Societies, London, 2005.

[2] J.B. Newkirk, High-Temperature Materials Coatings and Surface Interactions, Trans Tech Pub, 1980.

[3] J.D. Culcasi, P.R. Sere, C.I. Eisner, A.R. Di Sarli, Control of the growth of zinc-iron phases in hot-dip galvanizing process, Surf. Coat. Technol. 122 (1969) 21–23.

[4] R. Sanguanmoo, E. Nisaratanaporn, Y. Boonyongmaneerat, Hot-dip galvanization with pulse electrodeposited nickel pre-coating, Corros. Sci. 53 (2011) 122–126.

[5] M.H. Razmpoosh, A. Macwan, E. Biro, D.I. Chen, Y. Peng, F. Goodwin, Y. Zhou, Liquid embrittlement in laser beam welding of Zn-coated 22MnB5 steel, Mater. Design 155 (2018) 375–383.

[6] B.P. Burton, P. Perrot, Fe-Zn (iron-zinc), in: H. Okamoto (Ed.), Phase Diagramsof Binary Iron Alloys, ASM International, Materials Park, OH, 1993, pp. 459–466.

[7] C. Reumont, P. Perrot, J.M. Fiorani, J. Hertz, Thermodynamic assessment of the Fe-Zn system, J. Phase Equilib. 21 (2000) 371–378.

[8] X.P. Su, N.Y. Tang, J.M. Toguri, Thermodynamic evaluation of the Fe-Zn system, J. Alloys Compd. 325 (2001) 129–136.

[9] J. Nakano, D.V. Malakhov, G.R. Purdy, A crystallographic consistent optimization of the Fe-Zn system, Calphad 29 (2005) 276–288.

[10] X.P. Su, F.C. Yin, Z. Li, N.Y. Tang, M.X. Zhao, Thermodynamic calculation of the Fe-Zn-Si system, J. Alloys Compd. 396 (2005) 156–163.

[11] W. Xiong, Y. Kong, Y. Du, Z.-K. Liu, M. Selleby, W.-H. Sun, Thermodynamic investigation of the galvanizing systems, 1: refinement of the thermodynamic description for the Fe-Zn system, Calphad 33 (2009) 433–440.

[12] Y. Tang, X. Yuan, Y. Du, W. Xiong, Thermodynamic modeling of the Fe-Zn system using exponential temperature dependence for the excess Gibbs energy, J. Min. Metall. B 47 (2011) 1–10.

[13] K. Han, I. Ohnuma, K. Okuda, R. Kainuma, Experimental determination of phase diagram in the Zn-Fe binary system, J. Alloy Compd. 737 (2018) 490–504.

[14] A.R. Marder, The metallurgy of zinc-coated steel, Prog. Mater. Sci. 45 (3) (2000) 191–271.

[15] L. Zhu, S. Honrao, B. Rijal, R.C. Henning, M.V. Manuel, Phase equilibria and diffusion coefficients in the Fe-Zn binary system, Mater. Design 188 (2020) 1–10.

[16] K. Miyamoto, S. Nagagawa, Bonding Method, of Dissimilar Materials from Metals and Bonding Structure Thereof, U.S. Patent 8,058,584, November 15, 2011.

[17] Y.W. Cui, K. Oikawa, R. Kainuma, K. Ishida, Study of diffusion mobility of Al-Zn solid solution, J. Phase Equilib. Diffus. 27 (2006) 333–342.

[18] K. Han, I. Ohnuma, R. Kainuma, Experimental determination of phase equilibria of Al-rich portion of the Al-Fe binary system, J. Alloys Compd. 668 (2016) 97–106.

[19] N. Takata, M. Nishimoto, S. Kobayashi, M. Takeyama, Crystallography of Fe_2Al_5 at the interface between solid Fe and liquid Al, Intermetallics 67 (2015) 1–11.

[20] I. Lee, K. Han, I. Ohnuma, R. Kainuma, Experimental determination of phase diagram at 450°C in the Zn-Fe-Al ternary system, J. Alloys Compds. 854 (2021) 157–163.
[21] M.L. Giorgi, J.B. Guillot, R. Nicolle, Calphad 25 (2001) 461–474.
[22] J. Nakano, D.V. Malakhov, S. Yamaguchi, G.R. Purdy, A full thermodynamic optimization of the Zn-Fe-Al system within the 420-500°C temperature range, Comput. Coupl. Phase Diagr. Thermochem. 31 (2007) 125–230.
[23] J.R. McDermid, M.H. Kaye, W.T. Thompson, Fe solubility in the Zn rich corner of the Zn-Al-Fe system for use in continuous galvanizing and galvannealing, Met. Mat. Trans. B 38B (2007) 215–230.
[24] V. Raghaven, Al-Fe-Zn (aluminum-iron-zinc), JPEDAV 32 (2011) 143–144.
[25] A.R.P. Ghuman, J.I. Goldstein, Reaction mechanisms for the coatings formed during hot dipping of iron in 0 to 10pct. Al-Zn baths at 450° to 700°C, Met. Trans. 2 (1971) 2903–2914.
[26] L. Liu, X. Pan, L. Lu, Y. Liu, X. Su, J. Wang, Phase equilibria of 600°C isothermal section of Zn-Al-Fe-Si quarternary system, JPEDAV 37 (2016) 693–701.
[27] J.R. Selverian, A.R. Marder, M.R. Notis, The reaction between solid iron and liquid Al-Zn baths, Met. Trans. A 19A (1988) 1193–1203.
[28] A.R. Borzillo, J.B. Horton, U.S. Patent No. 3,343,930, September 26, 1967.
[29] B. Xu, D. Phelan, R. Dippenaar, Role of silicon in solidification microstructure in hot-dipped 55 wt% Al-Zn-Si coatings, Mater. Sci. Eng. A (2008) 76–80.
[30] E. De Bruycker, Z. Zermout, B.C. De Cooman, Zn-Al-mg coatings: thermodynamic analysis and microstructure related properties, Mater. Sci. Forum 539–543 (2007) 1276–1281.
[31] Galvanized Autobody Partnership Project ZCO-55/66 Final Report, August 2014, 135. Available from www.zinc.org.
[32] K. Honda, W. Yamada, K. Ushioda, Solidification structure of coating layer on hot-dip Zn-11%Al-3%mg-0.2% Si coated steel sheet, Mater. Trans. JIM 49 (2008) 1395–1400.
[33] A.D. Pelton, Thermodynamic Calculations of the Steps during Solidification of Certain Zn-Al-mg Alloys, prepared for F. Ajersch, under contract with IZA, 2014.
[34] G. Angeli, R. Brisberger, M. Bulter, L. Diez, C. Filthaut, W. Fischer, T. Koll, T. Maiwald, C. Pesci, A. Richter, J. Schulz, U. Stellmacher, N. van Landschoot, Zinc-magnesium-aluminium for automotive industry, Galvatech (2013) 632–636.
[35] C.W. Hsu, G.L. You, K.K. Wang, L. Chang, D. Gan, L.J. Chiang, Characterization of the Fe-Al intermetallic phases formed in hot-dip Al-Zn-mg coatings, Galvatech (2015) 845–851.
[36] J.X. Li, D.M. Hreso, Morphology of 555Al-Zn coatings and the Al-Zn coating with mg additions, Galvatech (2015) 852–860.

Chapter 4

Zinc bath phenomena

4.1 Introduction

Until the advent of significant automotive applications for zinc-coated steels in the late 1980s, galvanizing baths were run for long campaigns that produced primarily construction and commercial quality products. Line stops typically involved scheduled changes of galvanizing bath hardware or other items whose service lives were measured in weeks. Galvanizing bath target compositions remained in a narrow range, almost always for production of the conventional galvanized (GI) coating with a zinc overlay having the same composition as the bath. Production of the Zn-55%Al coating was then, and continues to be, normally conducted on dedicated lines. There was little motivation to gain knowledge on the nature of transitions from one bath composition to another, or the way that coating formation changed as bath composition changed. Requirements for coating quality were also less strict than automotive standards; defects caused by dross, and variations in coating weights were of concern but not as critical for acceptance. One-sided (unexposed) hot-dip sheet was typically provided for automotive applications until the mid-1990s, two-sided exposed quality could only be obtained with electrogalvanized product. The emphasis was primarily on meeting formability requirements [1]. This all changed with the upgrading of equipment and practices to produce automotive-quality galvanized sheet. Moreover, it was, and is, expected in many cases that the GI and galvannealed (GA) coating would be produced on the same galvanizing line, and that the changeover from high-quality GI to GA, and vice versa, would occur in the least possible time with production of the smallest amounts of galvanizing bath dross and sub-prime coated product. This expectation led to a tremendous increase in knowledge of the nature of the galvanizing bath and how coatings are affected by transitions in bath chemistry, temperature, and line speed, much of it since 2000. This knowledge has led to the development of bath management tools to help optimize practice.

4.2 Typical continuous galvanizing bath characteristics

The arrangement of a typical continuous sheet galvanizing bath and associated equipment is shown in Fig. 4.1.

In this drawing, steel strip from the furnace and cooling sections enters through the snout. A protective gas atmosphere surrounds the strip as it passes from the previous sections until it is immersed in the zinc bath; further details on snout atmosphere are also given in Chapter 2 and later in this chapter. The strip travels out of the snout and around the sink roll. Passage of a strip under tension around a roll at a temperature and tension capable of elongating the outside surface always produces a concave strip shape called crossbow, as illustrated in Fig. 4.2 [3]. To regain strip flatness, a second, smaller roll, called the corrector roll, shown in Fig. 4.1 as the "guide roll," is positioned so that it deflects the strip in the opposite direction, equalizing the deformations imposed on each side.

Because the corrector roll moves the strip position out of the passline, a third roll, the stabilizer roll, is used to bring the strip position back to the passline. It is possible to use only the sink roll and the corrector roll to guide the strip through the bath, but their positions must be set to compensate for the function of the corrector roll; therefore, the three-roll arrangement shown on the left side of Fig. 4.3 is most commonly used. Combinations of roll diameters and positions that minimize crossbow have been calculated using an elastic approximation [4].

Feeding of the zinc to the bath is usually done behind the snout. Large ingots (1 or 2 tons) are gradually fed to the bath, usually by hanging from a chain that lowers the ingot into the bath at a rate that approximates the rate of zinc withdrawal from the bath by the coated strip or by a launder from a premelt pot located close by the snout. Some lines add ingots by lowering them quickly into the bath, which causes temperature, composition, and zinc level fluctuations in the bath that can cause coating defects. In particular, changes of bath composition when transitioning from GA to GI operation, or vice versa, can involve rapid addition of higher-aluminum ingots. The GA to GI transition especially can involve generation of excessive dross because of the rapid rise in aluminum content and consequent lowering of iron solubility as explained later. Smaller (typically 9–25 kg) ingots with various alloy compositions are used to trim the bath composition.

The Metallurgy of Zinc Coated Steels. https://doi.org/10.1016/B978-0-323-99984-7.00013-0

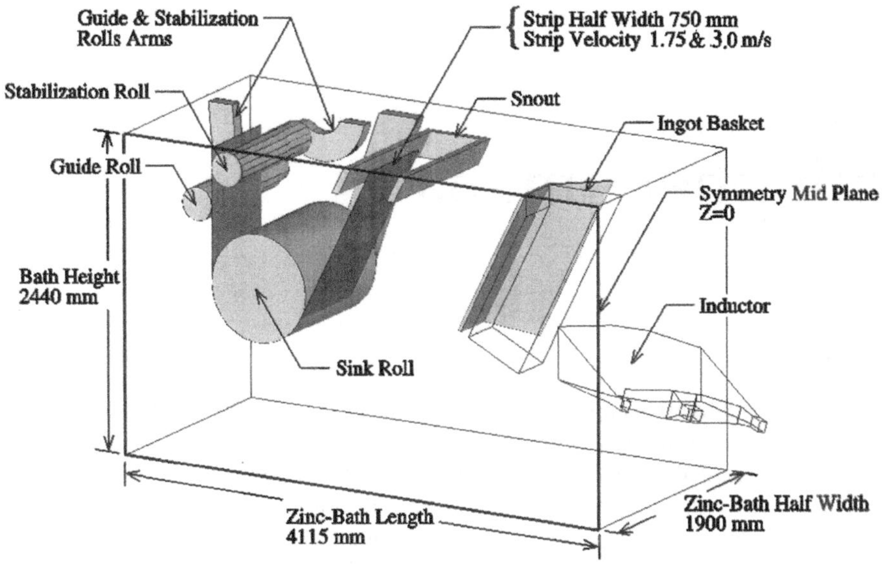

FIG. 4.1 Half-section of a typical continuous sheet galvanizing bath. The symmetry plane is bounded by the lines showing zinc bath length and height [2].

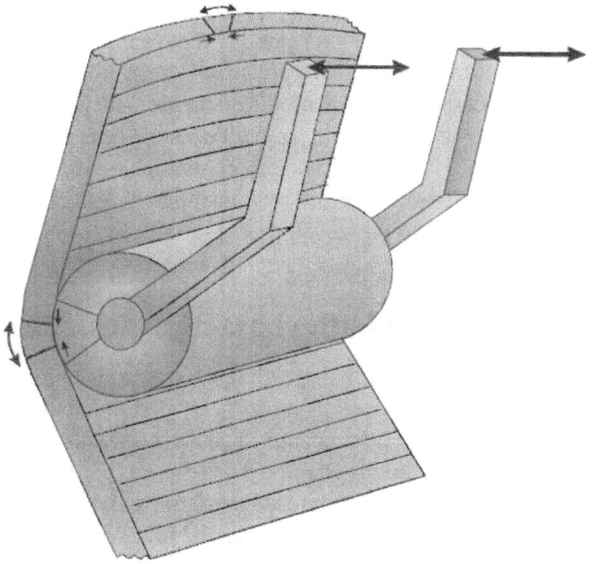

FIG. 4.2 Schematic of formation of crossbow shape as the strip is bent around a roll. Elongation of the outside surface results in bow-shape bending of the strip as it exits contact with the roll [3].

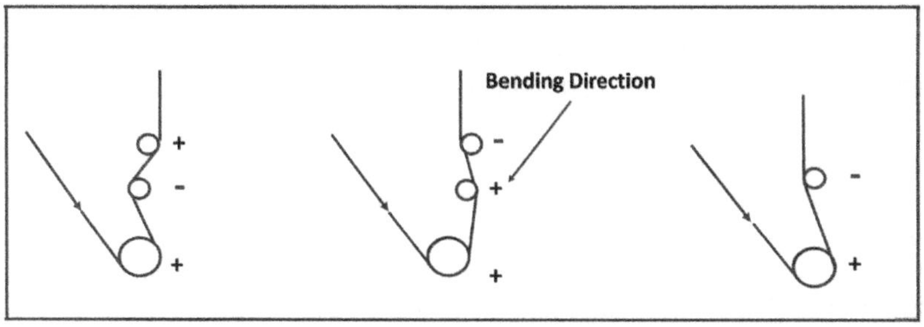

FIG. 4.3 Pot roll arrangements. The sink roll (bottom) induces a tensile force (positive sign) on the outer (front, or bottom) surface of the strip. The strip then travels around the corrector roll above it, which induces a compressive (− sign) or further tensile (+ sign) on this strip surface. The strip then travels around the stabilizer roll (if used) to further change the stress. The objective is to sum the positive and negative stresses to zero after traversing the pot rolls. The left side arrangement is most commonly used [4].

Zinc baths are heated in several ways. Traditionally, side heating by combustion heaters or electric grid panels was practiced. This allowed use of metallic pots similar to those used in general galvanizing as described in Chapter 15. The benefit of much longer lives of ceramic pots led to the adoption of induction heating, which also permits stirring of the bath when inductors are run at sufficient power. Channel inductors are the most widely used, as shown in Fig. 4.1. These inductors are usually located on both sides of the pot behind the snout, below the zinc feeding area. Several lines have used front-and-back inductors and other configurations. The effect of these configurations on bath uniformity and consequent rate of dross generation will be discussed later. Less commonly used are coreless inductor pots. The main benefit of a coreless pot is the ability to pump the pot empty without risk of damage to the channel inductors, which must always be kept filled with zinc, at a bath level at least equal to the height of the top of the channel inductors in the bath. Such complete pumping can speed the transition from one bath chemistry to another.

4.3 Fluid flow and temperature patterns in the galvanizing bath

Flow patterns in the galvanizing bath are mainly influenced by the motion of the strip through the bath, but there can also be significant effects from movement of the pot rolls, metal drawn into and discharged from channel inductors, and thermally driven convection related to temperature differences, mainly near the inductors and ingot liquid metal addition locations. The size and location of bath hardware, inductors, and snout can influence details of fluid flow in the bath. Downward-flowing gas and zinc runback from the wiping rig can strongly influence surface flow. All of these effects will be explained later in this chapter.

An example of fluid flow in a typical galvanizing bath, during a time when no additional zinc is being fed to the bath, is shown in Fig. 4.4. Fig. 4.4A shows the flow pattern 44.5 cm from the bath centerline, Fig. 4.4B shows the flow pattern 103.5 cm from the bath centerline, and Fig. 4.4C is 162.5 cm from the bath centerline. For this example, the strip half-width is 50.8 cm, so Fig. 4.4C is beyond the edge strip, near the edge of the sink and other guide rolls. The half-width of the bath is 190 cm, so Fig. 4.4C is closer to the bath wall than it is to the bath hardware.

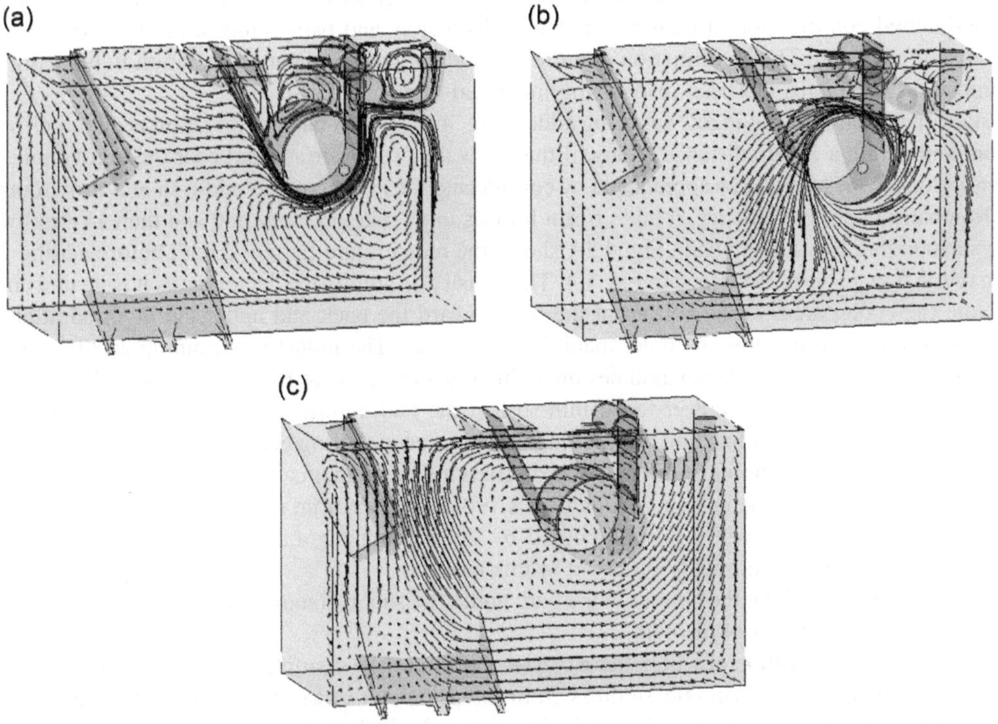

FIG. 4.4 Results of computer modeling of fluid flow in a 250 metric ton galvanizing bath with a snout entry angle of 27 degrees and a line speed of 1.75 m/s. Half-section views are shown with velocity vectors shown for cutting planes moving progressively away from the centerline [2]. (A) Cutting plane 44.5 cm from bath centerline, cutting through the strip and bath hardware. Note circulation in the "V" area above the sink roll. (B) Cutting plane 103.5 cm from bath centerline, just beyond the edge of the bath hardware. (C) Cutting plane 162.5 cm from bath centerline, 27.5 cm away from the bath wall.

During the time modeled in Fig. 4.4 when no zinc is being added, the inductors are running at 20% power, and therefore, most of the heat required to maintain the bath temperature of 460°C is provided by the entering strip, which is at this same temperature. Fig. 4.4A., cutting through the strip, shows the influence of the strip movement on zinc flow adjacent to it; the zinc flow is entrained with the strip surface just below the snout. On the side not contacting the sink roll, customarily called the front or bottom side, this entrained flow follows the direction of strip travel while the zinc flow following the back or top side is mostly ejected at the pinch point where the strip contacts the sink roll. The expelled zinc flow from this pinch point results in a clockwise rotation of zinc flow in the bath volume bounded by the sink roll and the strip before and after it contacts the sink roll, commonly called the "V area" illustrated in Fig. 4.4A. There are two other areas of circular flow: these are driven by the expulsion of entrained metal on the strip front side as it contacts the corrector roll and by the motion of the strip and stabilizer roll. Over most of the width of the strip, these volumes of circular flow can become stagnant with regard to local bath composition and temperature because they only exchange metal with the rest of the bath volume at the strip edges, notably the sideways expulsion of bath metal at points where the strip makes initial contact with the guide rolls. Behind the snout and the volume of metal near the moving strip, flow velocities are much lower. With the inductor on low power, there is little inductor influence on bath flow. The colder bath walls induce a downward natural convection, the zinc near the bottom of the bath behind the snout can be considered a stagnant zone until locations near to the sink roll are approached, where the general direction of flow is to the right side, toward the front of the bath under the moving strip. Moving beyond the strip edge, but still near the edge of the pot roll configuration in Fig. 4.4B., the vectors in the V area point outward, indicating a slow flow of metal toward the side of the bath. The influence of strip and pot roll motion is significant, with the zinc entrained by sink roll motion influencing both the circular flow at the front of the bath and increasing the flow toward the back of the bath.

Finally, near the side wall of the bath, a large circulation loop is observed, mainly moving toward the back of the bath, with forward movement near the bath surface. The movement toward the back along the sides is also supplied with metal from the forward flow of metal under the sink roll, coming from areas nearer to the center of the bath. The small influence of the inductor running at low power is seen in Fig. 4.4C, both the inductor-induced flow entering the inductor at 0.4 m/s and natural convection of the zinc exiting the inductor at a temperature 8°C warmer than the bath cause an upward flow along the bath sidewalls behind the snout. The downward flow at the walls is mainly the result of the cooling effect of the front and back walls and the bath side walls beside the pot rolls. Increasing or decreasing line speed increases the intensity of the observed flow patterns but not the general nature of these patterns. As a general rule, the average zinc flow is 4% of the strip speed [5]. Experimental confirmation of numeric models of bath flow and bath composition has been made with water models utilizing acid injection. The confirmation of numeric modeling of temperature variation in the bath, together with flow and composition variations, has been made using industrial line trials [5,6].

In steady-state operation when no ingots are being added, Fig. 4.5A shows bath temperature is quite uniform, except for the very bottom and within a 10 mm distance of any of the pot walls or top surface and near the ingot chute. The possible temperature variations in the bath can be appreciated by considering the instantaneous addition of a 1 ton room temperature ingot to a 250 ton bath. In normal practice, it takes 6 min for this ingot mass to be heated and then 14 min for it to be fully melted by the surrounding galvanizing bath. Fig. 4.5A shows the temperature field of a bath before adding such an ingot, corresponding to the flow patterns described in Fig. 4.4. The density of liquid zinc decreases as it is heated and increases as it is cooled. Note the cooler areas at the bottom of the bath toward the back and under the ingot addition chute. These correspond to areas of downward flow of cooler metal in these areas. The inductor is running at 20% power. When the ingot is added to the bath, the model shown assumes the inductors will be run at full power, 400 KW, for the 20 min ingot melting period, then returned to 20% power for 40 min so that the bath returns to 460°C after 1 h. Fig. 4.5B shows the temperature field just after the ingot has finished melting. A significant cooling effect can be seen under the ingot chute. Despite the fact that the inductor running at full power is discharging metal that is 20°C hotter with a channel inlet velocity of 0.75 m/s, there is little effect of the inductor action that is 190 cm distant from the bath centerline, except in the shaded zones around the ingot chute. The bath has been cooled by the melting ingot; the cooler zinc flows downward, little influence by the slow flow pattern normally observed in this area.

The forward motion of zinc along the bottom of the bath under the sink roll moves this cooler zinc forward to be mixed with the bath metal in the front part of the bath.

Because the instantaneous addition of a 1 ton ingot can be observed to be disruptive to the bath temperature pattern (also flow and composition patterns) and can release dross from the perimeter of the bath and snout when the liquid level is suddenly raised, it is much more common to lower the ingot into the bath approximately equal to the rate of zinc being consumed. This avoids the "saw tooth" temperature cycle that would be measured by a thermocouple in the vicinity of the ingot addition, replacing it with a more gradual temperature variation. A third approach is to always maintain the mean bath temperature near 460°C, adjusting inductor power so the bath is not overheated, although this means that for much of the period of ingot melting, the bath temperature will be slightly below 460°C. A modeled example is shown in Fig. 4.6.

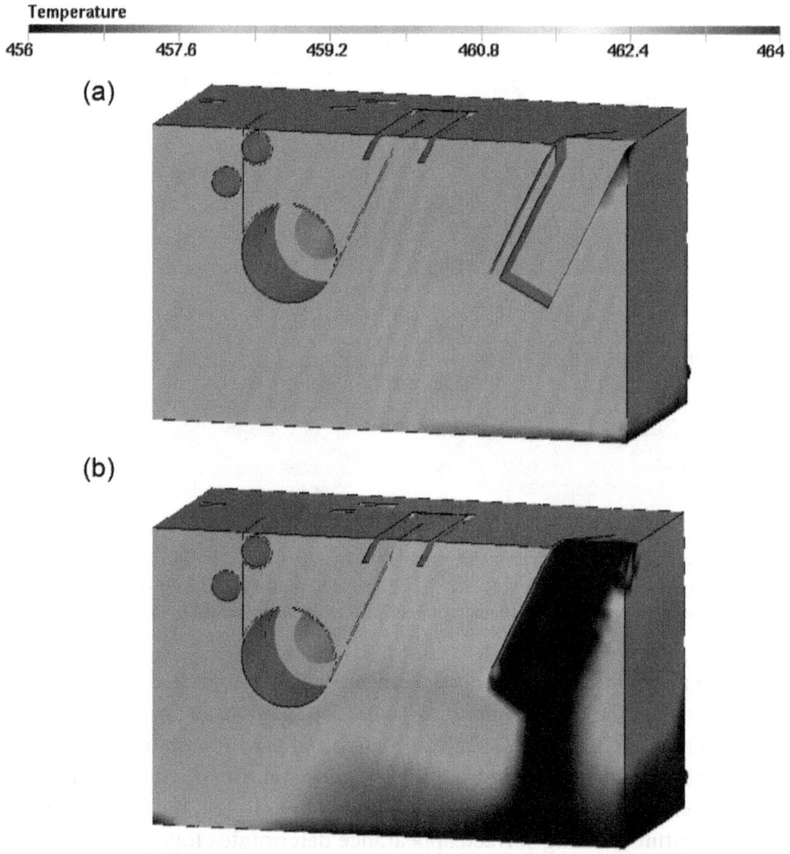

FIG. 4.5 Temperature patterns in a 250 ton galvanizing bath, strip speed 1.75 m/s, strip entry and bath temperature set at 460°C [2]. (A) Galvanizing bath temperature field at a plane through the middle of the bath during a period of time when no ingot is being melted [2]. (B) Temperature pattern at a plane through the middle of the bath 20 min after adding a 1 ton zinc ingot.

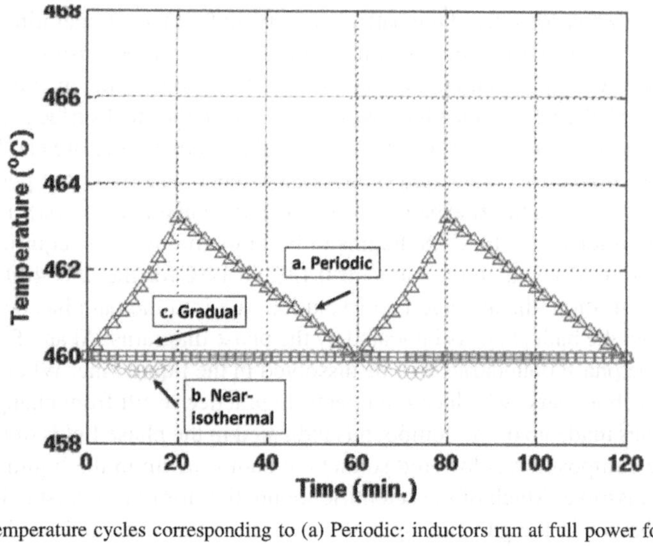

FIG. 4.6 Three galvanizing bath temperature cycles corresponding to (a) Periodic: inductors run at full power for 20 min ingot melting period, 20% power for 40 min with consequent bath overheating. (b) Near isothermal: inductors controlled to achieve mean bath temperature, resulting in slight underheating during 20 min ingot melting period. (c) Gradual: ingot lowered into bath to balance rate of zinc consumption [7].

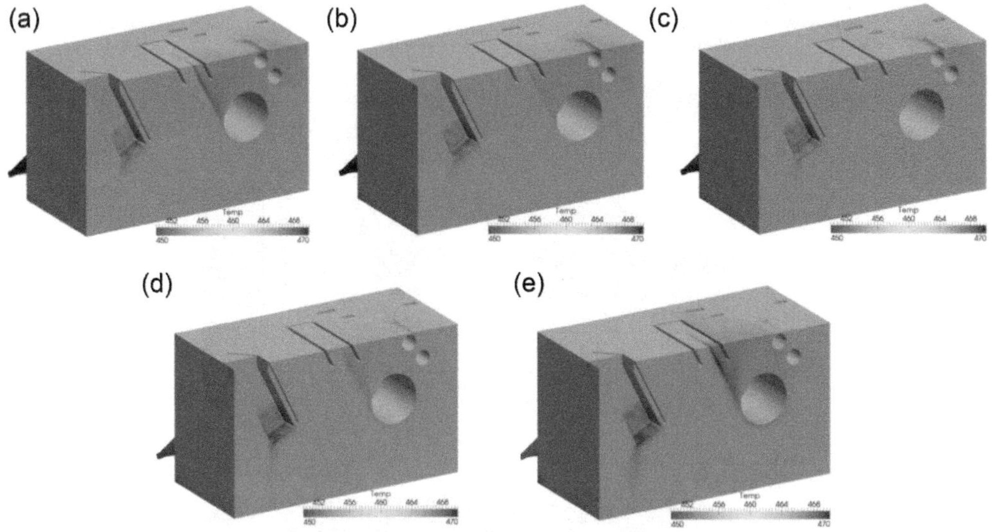

FIG. 4.7 Effect of strip entry temperature to the bath on bath temperature patterns. (A) 420°C, (B) 440°C, (C) 460°C, (D) 480°C, (E) 500°C. Bath temperature is always set at 460°C. Note cool zone at strip exit except in Fig. 6E. because of mixing of bath surface metal caused by strip and stabilizer roll motion, and cool zones in zinc feeding area that decrease in temperature as strip entry temperature is increased [7].

Strip entry temperature also affects bath temperature distribution. Although it is customary practice to set strip entry temperature a few degrees hotter than the bath temperature to reduce inductor energy consumption, it is desirable to minimize the reheating of advanced high-strength steels such as quench and partition grades where the partitioning temperature used to enrich austenite with carbon, as described in Chapter 2, can be significantly below the bath temperature. Reheating of these grades to the bath temperature can reduce formability and strength [8], but if the strip entry temperature is more than 10°C below the bath temperature, coating surface appearance deteriorates together with wettability. The temperature distribution in the bath for five strip entry temperatures is shown in Fig. 4.7 after 2 h of continuous production. The cooling effect of the lower temperature strip is seen in Fig. 4.7A and B. In these two situations, inductor power is needed to maintain the 460°C bath temperature. For the two configurations where the strip entry temperature is higher than the bath temperature, significant heating in the strip entry area can be seen. Because the inductors would be run a little or no power for these two configurations, a cooler ingot melting area can be observed.

4.4 Bath chemistry

None of the description until now has considered the bath chemistry and effects of operating conditions on bath chemistry. After a galvanizing bath is run for even a few minutes, it is saturated in iron, because some iron from the strip is always dissolved as the strip is galvanized. If a galvanizing bath is operated with pure zinc, then according to the phase diagrams shown in Chapter 3, Figs. 4.6–4.9, the Fe-Zn zeta phase will be formed on the steel surface. This will continue to grow with continued immersion in the zinc bath. While coatings with this initial coating structure can be galvannealed, the iron-zinc produced phase interface is unsuitable for production of GI-coated sheet because of the poor formability of the produced coating. To overcome this, Al is added to the zinc bath to form the very planar and compact interface layer consisting of the compound $Fe_2Al_{(5-x)}Zn_x$ that, referring to Fig. 4.8, begins to be in thermodynamic equilibrium at dissolved Al concentrations in a Fe-saturated baths at Al levels in the range of 0.13%–0.14% for the 450–480°C temperature range shown.

With this level of Al concentration, the average Fe dissolution rate into the bath has been measured to be 120 mg per m^2 of steel strip passing through the bath [11]. As indicated by the phase diagrams, Al and Fe combinations in the zinc bath in the liquid (L) field below the phase boundaries will be dissolved in the liquid zinc. When the combination of Al and Fe levels is above the liquid phase boundary, which can in operating practice result from changes in either bath chemistry or temperature, then dross particles made up of the compounds indicated in the phase fields of these diagrams will be formed; if the temperature rises or the composition is lowered so that the bath is again in the liquid phase field indicated by these diagrams, then the dross will dissolve. Much of the scientific foundation for our understanding of Zn-Al-Fe reactions and their relationship to dross behavior in the galvanizing bath is based on the work at Teck Metals of Dr. Nai-Yong Tang.

For dissolved (effective) Al compositions in the range of 0 to about 0.1% at 450°C, the dross formed is zeta phase $FeZn_{13}$ when the Fe present exceeds the solubility limit, which is in the range 0.03%–0.023%, decreasing as Al content increases. The phase diagram shows that below 0.1%Al, iron solubility increases as temperature increases, and that it drops

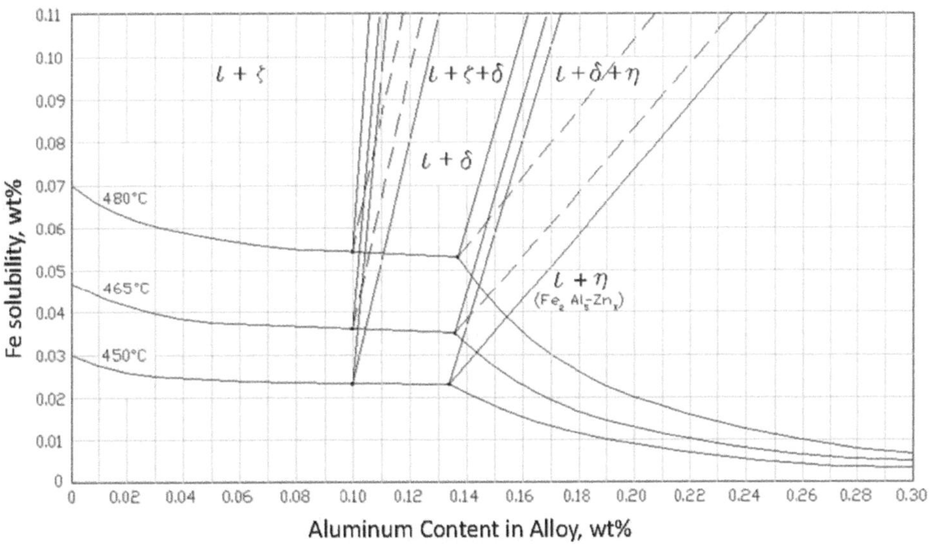

FIG. 4.8 Zn-rich corner of the Zn-Al-Fe phase diagram showing phase boundaries for three temperatures covering the range of galvanizing bath operation [9].

more rapidly as aluminum is added when temperature increases. The zeta phase has a monoclinic (elongated rectangular prism) shape and is likely the only dross constituent of this shape to be encountered in the galvanizing bath and so is easily identified. For dissolved Al compositions from 0.1% to 0.135% at 450°C, the dross formed is the compact delta phase $FeZn_7$. The soluble Fe limit does not change much over this range, remaining at about 0.023%. Delta phase dross particles are more regular (lower aspect ratio) in shape than zeta phase, and portions of the delta particles usually exhibit flat sides, consistent with its cubic crystal structure. Above the dissolved Al level of 0.135% at 450°C, the $Fe_2Al_{(5-x)}Zn_x$ compound is stable. Fe solubility decreases rapidly as Al content increases from this point, from 0.023% to 0.003% at 0.3% dissolved Al. $Fe_2Al_{(5-x)}Zn_x$ particles are similar in appearance to delta phase dross particles. Images of these types of dross particles are shown in Fig. 4.9.

The densities of these dross phases are different from liquid zinc; the minor additions of Al over the range of values in the Fig. 4.8 phase diagram do not significantly change the density of liquid zinc. More information on the densities of liquid Zn-Al alloys is given in Chapter 7.

The rising or settling of dross particles in a zinc bath is described by Stokes's Law, for which the relevant equation in the case of Fe_2Al_5 particles is

$$V_{\text{dross}} = r^2_{\text{Fe}_2\text{Al}_5}\left(\rho_{\text{Zn}} - \rho_{\text{Fe}_2\text{Al}_5}\right)\left(\frac{2g}{9\eta_{\text{Zn}}}\right) \tag{4.1}$$

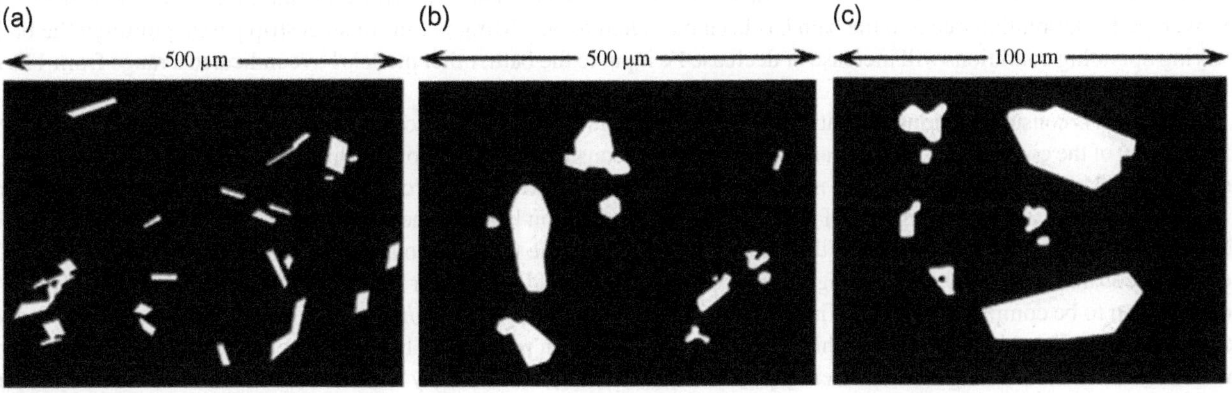

FIG. 4.9 Appearance of dross particles normally encountered in GA and GI production [10]. (A) Zeta phase dross particles. (B) Delta phase dross particles. (C) $Fe_2Al_{(5-x)}Zn_x$ dross particles.

TABLE 4.1 Density of GI and GA galvanizing bath constituents.

		Weight percent			Density	
Name	Formula	Zn	Al	Fe	g/cm^3	Crystal structure
Zeta	$FeZn_{13}$	94	0.5	5.5	7.1	Monoclinic
Delta	$FeZn_7$	92.5	1.5	6	7.2	Hexagonal
Eta	$Fe_2Al_{(5-x)}Zn_x$	35	45	20	4.5	Orthorhombic
Zinc					6.7	Liquid

Derived from T.Q. Coddington, D.C. Cook, Mössbauer spectroscopy of aluminum substituted galvanneal steel, in: Proc. Int'l Conf. on Applications of the Mössbauer Effect, September 10–16, 1995, Rimini, Italy, Paper #14-11 and I. Lee, K. Han, I. Ohnuma, R. Kainuma, Experimental determination of phase diagram at 450°C in the Zn-Fe-Al ternary system, J. Alloys Compd. 854 (2021) 157–163.

where

$r_{Fe_2Al_5}$ = radius of the Fe_2Al_5 particle.
ρ_{Zn} = density of liquid Zn, from Table 4.1.
$\rho_{Fe_2Al_5}$ = density of Fe_2Al_5, from Table 4.1.
g = gravitational force constant $9.8\,m/s^2$.
η = viscosity of liquid Zn, 0.004 Pa or 0.04 dyne-s/cm^2.

A more complete application of Stokes's law to behavior of galvanizing bath dross is given in [12–14]; a more general treatment is given in [15].

Because the density of Fe_2Al_5 is less than zinc, it will rise. This same equation can be used for the zeta and delta phases, but because their density is larger than liquid zinc, the equation will give a negative result, indicating sinking rather than rising behavior. The difference in density between either of the two types of bottom dross, zeta or delta and liquid zinc is much smaller than that between Fe_2Al_5 and liquid zinc; therefore, Fe_2Al_5 top dross particles of the same size will rise much more rapidly than the Fe-Zn bottom drosses sink. The magnitudes of the velocities obtained for representative sizes, between 1 and 100 μm, are much smaller than the steel strip velocity or even the average bath velocity, indicated before as 4% of the strip velocity. For example, using the above equation, a 20 μm diameter Fe_2Al_5 particle has an upward velocity of 0.001 m/s and a 20 μm radius $FeZn_7$ particle has a downward velocity of 0.0003 m/s. Referring to Fig. 4.9, these would be the size of typical particles. Velocity increases with the square of the radius, so that doubling the radius gives a fourfold increase in rising or sinking velocity. In areas of relatively high bath velocity, i.e., anywhere over the average 0.08 m/s bath velocity for a strip speed of 2 m/s, the Stokes velocity is overwhelmed by the velocity of the local bath, it is only in the portions of the bath where low velocity, or near-stagnant, conditions are found where the Stokes equation can be considered. Therefore, drosses in the 20 μm size range tend to be swept through regions of high flow and then rise or sink in areas of slow or near-stagnant flow. At the same time, particle growth or agglomeration can occur when the local bath compositions and temperatures are favorable to dross stability, increasing the Stokes flow velocity. Conversely, they can dissolve when local bath compositions and temperatures are in the liquid domain of the phase diagram, reducing dross size.

The overall composition of the zinc bath is determined by the mass balance for iron and aluminum. As indicated before, the average Fe dissolution rate into the bath has been measured to be 120 mg per m^2 of steel strip passing through the bath. Varying operating conditions will increase or decrease Fe input to the bath; other models have indicated a range from 110 to 250 mg/m^2 [16,17].

Aluminum is consumed during formation of the intermetallic layers, and it also flows out with the liquid zinc to form the zinc overlay of the coating. For both GA and GI coatings, the consumption of Al to form intermetallics on the strip surface is assumed to take place on the first 0.305 m of the strip after entry into the bath (corresponding to 0.2 s. at a strip speed of 1.524 m/s). In particular, the formation of the $Fe_2Al_{(5-x)}Zn_x$ inhibition layer on the strip surface, discussed in more detail in Chapter 5, coats the strip surface within this very short period of time [18]. The overall aluminum concentration of the GI coating is considered to be 0.4% by weight for a coating weight of 200 g/m^2 per side [19]. Other work has considered Al consumption to be complete during the first 0.15 s, giving an Al pickup of 60 mg/m^2 [17], a lower figure than in the model referenced here. The Al pickup in the inhibition layer alone has been reported to be 30 mg/m^2 [16], meaning the rest of the Al taken out with the coating is in the overlay portion of the coating that will vary with coating thickness. This is consistent with a survey of 34 coils taken from four different galvanizing lines made by Dubois [20].

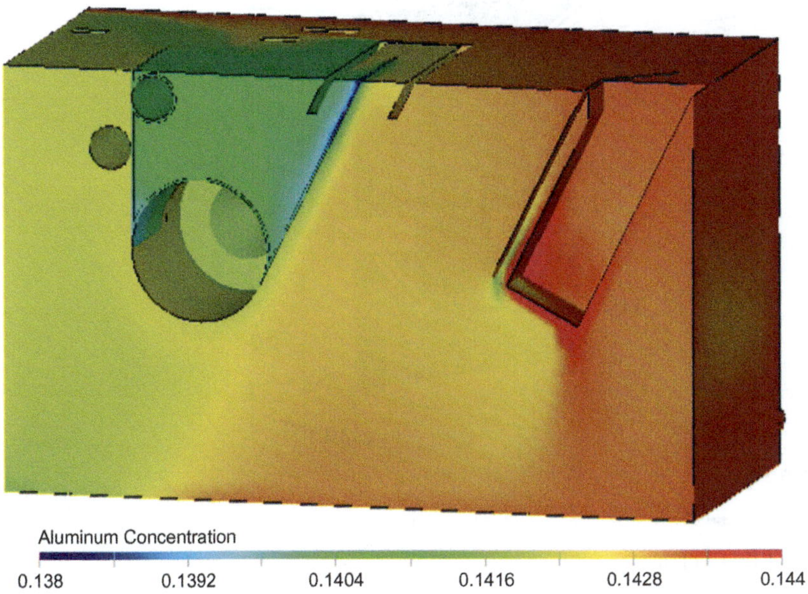

FIG. 4.10 Al composition distribution in the galvanizing bath, 20min after adding a Zn-1%Al 1 ton ingot, strip speed 1.75m/s [2].

With this description of phase equilibria, dross characteristics, and the mass balance for Al and Fe, the patterns of Al and Fe compositions in the galvanizing bath can now be described. Fig. 4.10 shows the same 250ton bath described before, showing the Al composition at the plane of symmetry, 20min after an ingot containing 1%Al has been added. Although we might expect the higher Al concentration melting liquid from the ingot to float upward because of its lower density than pure zinc, the thermal influence on flow described in Fig. 4.5 is far more important. Under the conditions of flow typical in these galvanizing baths, the effect of Al on liquid zinc buoyancy is slight.

Comparing the thermal fields in Fig. 4.5B with the composition field in Fig. 4.10, it is clear that there is a significant portion of the volume where the local bath temperature is lower than average and where the Al composition is higher than average. The phase diagram in Fig. 4.8 tells us that these are locations where dross is expected to precipitate. Figs. 4.4A and B also tell us that the volume under the ingot melting location is also a region of low velocity over much of the width of the bath. It is only in the area next to the inductors, which are running full power during ingot melting, where there is significant bath velocity near the ingot melting location. The dross formed over much of the bath width, from the centerline out to 28cm away from the bath wall, will float upward in the case of the GI bath or downward in the case of a GA bath. Indeed, the floor of this portion of the bath is normally a location where bottom dross accumulation can be found during GA production, the surface of the bath in this area is where top dross accumulation is normally found, where it can be removed by surface skimming. It can also be observed in Fig. 4.10 that the "V" area above the sink roll has a lower Al concentration than the rest of the bath but is near the average bath temperature. The side of the strip facing the sink roll is consuming Al to form intermetallics at the interface, although there is metal circulation in this area, it is not well mixed with the rest of the bath so that Al depletion may occur. Conditions in this portion of the bath are not favorable for dross formation. Indeed, any dross entering the "V" area would be expected to begin dissolution because this region is below Al saturation; moreover, Fe solubility increases with decreasing Al concentration. However, water modeling work has shown that dross particles from outside the "V" area tend to be transported into this area, regardless of their diameter [14].

The Al compositions shown in Fig. 4.10 can be separated into dissolved (or effective) Al and Al tied up in dross. Fig. 4.11 shows the locations in which the history of total Al, temperature, and precipitated Al (Al contained in dross) are described in Figs. 4.12–4.14.

In the first 6min shown in Figs. 4.12–4.14, the ingot is heating up to the melting point. There is no supply of liquid metal to the bath during this time to compensate for the aluminum being consumed by the strip, so Al concentration decreases especially near the strip. After ingot melting begins, a rapid increase in aluminum concentration is observed on the inductor side. On the strip side the increase is delayed by about 3min, corresponding to the time it takes for the melted zinc on the ingot side to travel with bath flow to the front (strip side) of the bath. During ingot melting, the total aluminum concentration continues to increase, with a maximum value near the ingot at about 14min. On the inductor side the differences in Al concentration between different locations are more apparent with larger values on the low-left location and smallest on the up-left location. Flow from the ingot (rich in aluminum) reaches the low-left location first followed by the low-right,

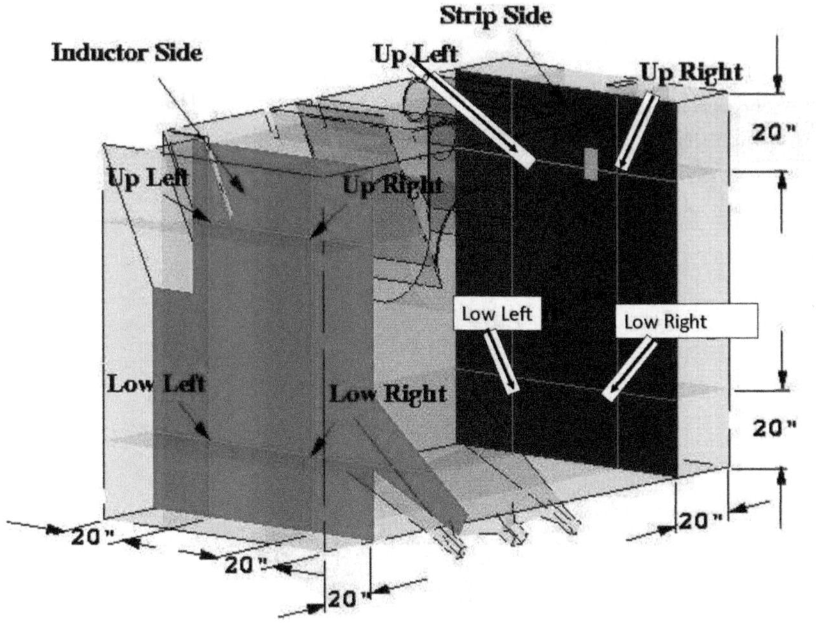

FIG. 4.11 Galvanizing bath locations indicated in Figs. 4.12–4.14 [2].

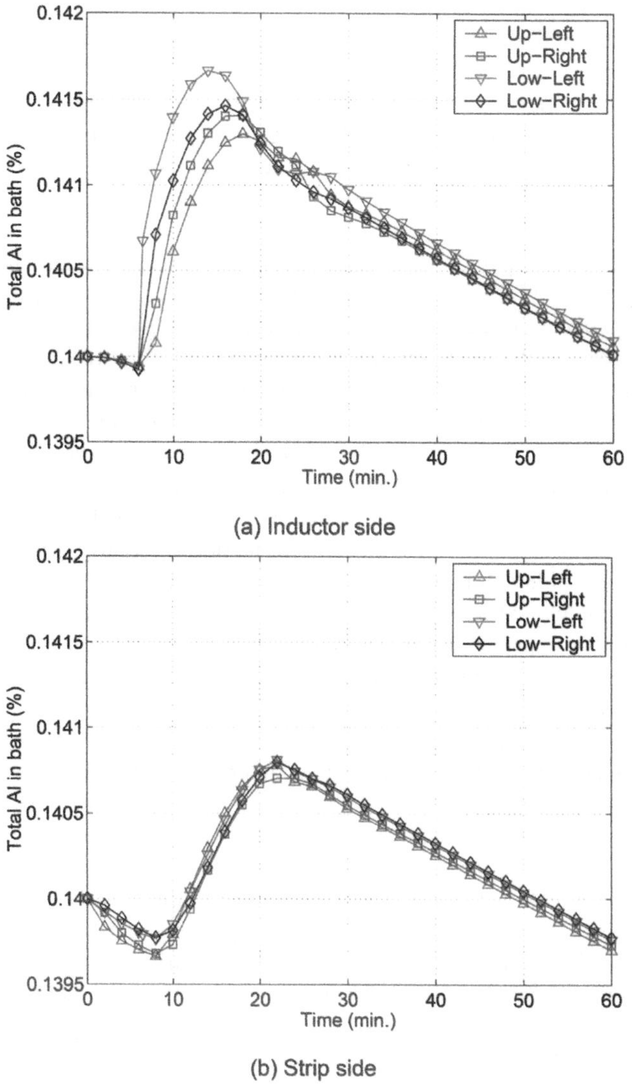

FIG. 4.12 History of total Al (dissolved plus tied up in dross) after adding a 1 ton Zn-1%Al ingot to a 250 ton bath, strip speed 1.75 m/s. (A) Inductor side (back of bath). (B) Strip side (front of bath), referring to Fig. 4.11 locations [2].

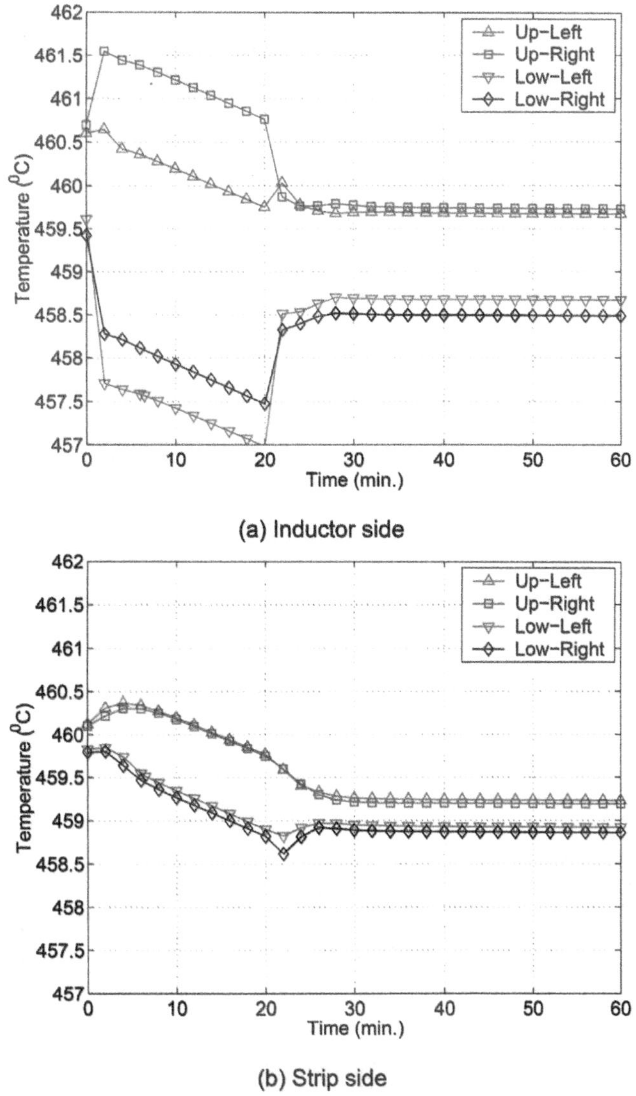

FIG. 4.13 History of bath temperature after adding a 1 ton Zn-1%Al ingot to a 250ton bath, strip speed 1.75m/s. (A) Inductor side (back of bath). (B) Strip side (front of bath), referring to Fig. 4.11 locations [2].

up-right, and finally up-left locations. This is confirmed by the movement of zinc in the bath shown near the inductors in Fig. 4.4C. On the strip side the aluminum concentration is more uniform but remains higher on the ingot side. After 21 min, the aluminum composition on strip side decreases at a constant rate because of the aluminum consumption by the strip.

Fig. 4.13 shows the temperature evolution at the same locations. Temperature differences are observed as soon as the ingot is introduced in the bath at 0min on the timescale. The temperature drops at the inductor/ingot side at the low-left and low-right locations. The temperature rises at the up-right location because inductors are running at full power projecting heated zinc toward the top surface of the bath. Temperature gradients are higher on the inductor side and during the ingot melting period between 0 and 20min. After the ingot is melted and the inductors switched to 20% power, the temperature remains almost constant after an initial stabilization between 20 and 30min. The stratification of temperature in the bath is caused by buoyancy of the higher Al liquid from ingot melting versus lower Al in the overall bath; heat losses to the bath walls and surface also contribute to this temperature pattern.

Fig. 4.14 shows the concentration of aluminum contained in the Fe_2Al_5 dross. This is determined from Fig. 4.8 phase diagram, first calculating the amount of aluminum above the temperature dependent limit of solubility (effective Al) using the local temperature shown in Fig. 4.13 at the given time and then subtracting this from the total Al shown in Fig. 4.12 for the same time. For this analysis, the rate of precipitation and dissolution of dross was taken to be instantaneous. Dross

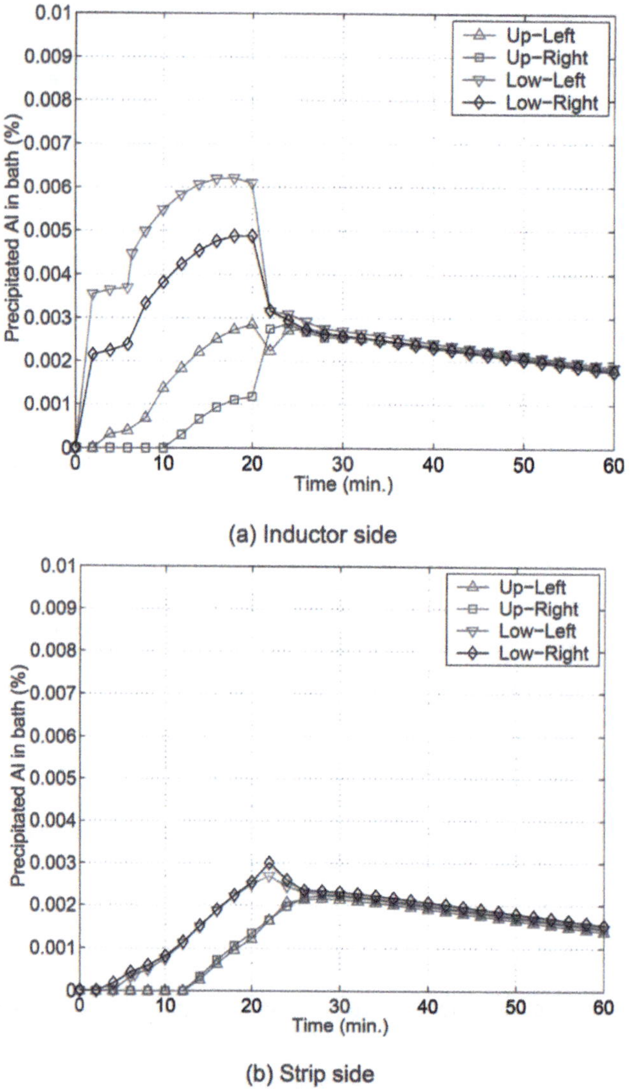

FIG. 4.14 Quantity of Al tied up in Fe_2Al_5 dross after adding a 1 ton Zn-1%Al ingot to a 250 ton bath, strip speed 1.75 m/s. (A) Inductor side (back of bath). (B) Strip side (front of bath), referring to Fig. 4.11 locations [2].

growth and dissolution rates will be described further, in a later section. When the ingot is immersed, we observe a sharp increase in dross, which is highest at the low-left location on the ingot side. This is caused by the sudden drop in temperature when the ingot is immersed. A second wave of dross formation is observed after 6 min, when the ingot begins to melt, introducing aluminum into the bath. Again, larger variations are observed on the inductor side. After 20 min, after the ingot is completely melted, the concentration of dross is rapidly stabilized and becomes more uniform. However, the values on the inductor side are still slightly larger than on the strip side.

4.5 Effect of bath geometry and operating variables on dross production

The terms "dross" and skimmings" are sometimes mistakenly used interchangeably, but they have different meanings. "Dross" refers to the Al-Fe-Zn and Zn-Fe intermetallic compounds formed as a result of Fe reactions in the bath; these are called floating dross and bottom dross, respectively. "Skimmings" means the mixture of surface oxides, floating dross, and bath metal skimmed from the bath surface. Much more details on skimmings are given in Section 4.9.

To give guidance on galvanizing pot equipment design and operating practices that can minimize dross production, 18 different variations from the base configuration were analyzed. The base case was defined as follows:

TABLE 4.2 Base case conditions for dross minimization study [19].

Strip entry angle	27 degrees
Snout depth	20.3 cm (8″)
Strip width	149.86 cm (59″)
Sink roll depth[a]	101.6 cm (40″)
Bath depth	243.84 cm (96″)
Strip entry *T*	460°C
Strip velocity	1.75 m/s
Ingot	0.5%Al
Bath size	250t
Bath composition	0.14% Al, saturated Fe

[a]*Distance from top of sink roll to bath surface.*

The conditions varied were:

TABLE 4.3 Conditions varied in dross minimization study [19].

Case	Condition varied
A.1	Al concentration in ingot 0%
A.2	Al concentration in ingot 1%
A.3	Bath unsaturated with Fe
A.4	Bath with 0.1% Al and saturated with Fe
A.5	Bath with 0.2% Al and saturated with Fe
A.5.1	Bath with 0.2% Al and oversaturated with Fe (0.06% Fe)
A.6	Strip entry temperature 420°C
A.7	Strip deposition rate 100 g/m^2
B.1	100 ton bath
B.2	500 ton bath
C	Side ingot charging
D.1	Gradual ingot charging at bath center
D.2	Gradual ingot charging at bath side
E	Front-back inductors
F.1	30″ sink roll depth
F.2	Asymmetric hardware
G.1	Smaller snout size
G.2	Deeper snout

In the previous section, it was shown that the quantity of Al tied up in dross over time was a good indicator of the amount of dross produced in GI baths. The effects of the variations on this quantity are shown in Table 4.4 for selected configurations. As before, the rate of dross precipitation and dissolution is assumed to instantaneous in this analysis. The base case, Table 4.2, assumes that a one ton ingot is completely submerged in the bath at the beginning of 1 h time period, heating for 6 min, and being completely melted after 20 min. Duration of two 1 h cycles was modeled in this analysis.

The first two cases in Table 4.3 modeled ingot additions into the 0.14% Al bath with Al compositions of 0 and 1%, comparing them with the base case of 0.5%Al. The 1% ingot addition was also described in Section 4.4. For the 0%Al

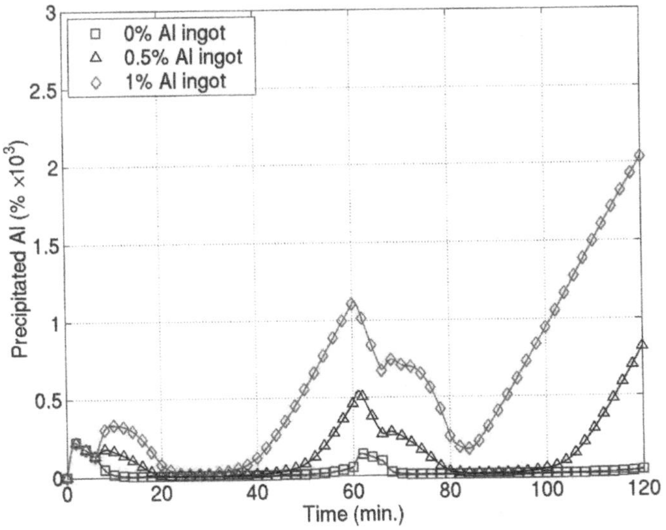

FIG. 4.15 Effect of ingot Al composition on Al tied up in Fe_2Al_5 dross [19].

ingot, some dross forms when the ingot is placed in the bath because of the lowering of temperature; however, as temperature recovers, this dross redissolves. Both Al-containing ingots produce bath Al levels above the solubility limit; the Al tied up in dross increases with increasing Al content in the ingot as shown in Fig. 4.15.

Regarding the effect of bath Al content on dross production, Fig. 4.8 indicates that Al contents below 0.135% at 460°C, down to 0.1%Al, will produce delta phase (for this analysis, using the composition $Fe_3Zn_{24}Al$ to include the Al content indicated in Table 4.1). Also, for this analysis, the iron dissolution rate from the strip is assumed to be the same for each bath composition. The dross concentration is computed by using the dross composition in each case and stipulating that the dross contains no excess Zn, the overall result is shown in Fig. 4.16. The greater amount of dross is found in the 0.1%Al bath and is mainly because this compound contains a high amount of Zn content.

If the strip entry temperature is lowered from 460 to 420°C, there is a strong cooling effect of the strip on the bath temperature. This results in more dross being generated, except for the period after the ingot is fully melted because of the higher inductor power required to melt the ingot, shown in Fig. 4.17. There is also much more stratification of temperature within the bath, so that more dross is formed near the bottom and less near the top than the base case. One exception

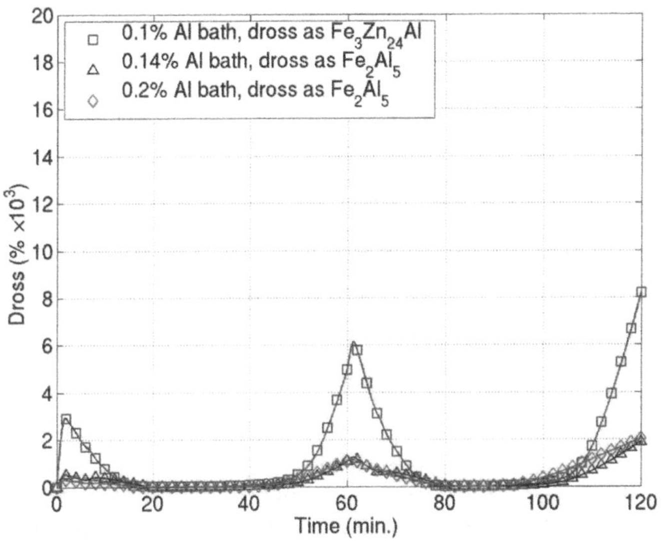

FIG. 4.16 Effect of bath Al composition on dross produced. Note that bath compositions below 0.135%Al at 460°C produce the $Zn_3Fe_{24}Al$ dross compound [19].

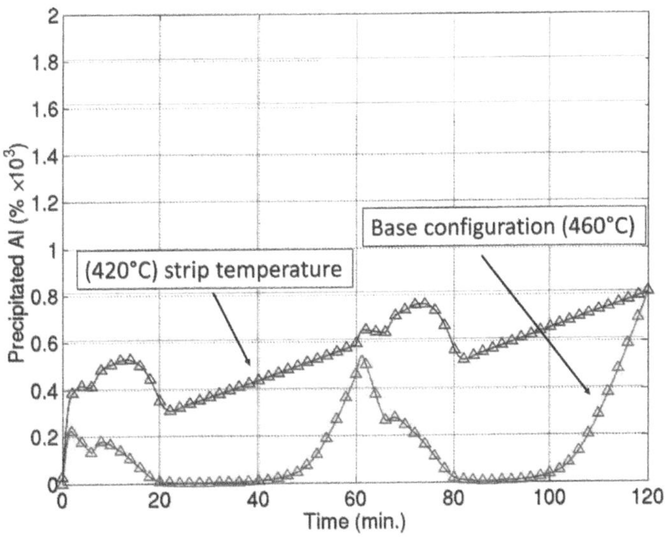

FIG. 4.17 Effect of strip entry temperature on Al tied up in Fe_2Al_5 dross [19].

to this pattern is near the strip exit at the bath centerline, where there is also a cool zone and significant top dross precipitation.

To study the effect of bath size on the amount of dross produced, several operating details needed to be modified. For the 100 ton bath, the depth was reduced by 91.4 cm and the width by 61 cm. The ingot feeding location was moved 89 cm closer to the snout. To maintain the desired thermal cycle, the inductor power was reduced to 60% of maximum power during ingot melting and 15% when an ingot was not being melted. For the 500 ton bath, the dimensions were changed in a similar fashion. The inductor power was held at the maximum during ingot melting and reduced to 50% power when an ingot was not being melted. The dross formation patterns with time were similar to the base case, the only significant effect of bath size was the improved stability of temperature with the larger baths and the lower concentration of dross. The amount of dross generated during ingot melting was the same, but it was distributed in a larger volume.

Increasing coating weight from 60 to 100 g/m^2 per side in the short term, over the 2 h analysis period, decreased the amount of dross produced because more Al is consumed with the higher coating weight; this lowers the bath Al levels at most locations in the bath, decreasing the number of locations where the bath was not in a supersaturated condition with regard to Al and Fe levels together with temperature. For longer periods of time, higher Al input to the bath would be required, for example, by use of a 1%Al ingot rather than the 0.5% composition used in the base case; therefore, in long-term operations, the effect of increase coating weight is likely to have an insignificant effect on dross production during steady-state operation. The converse is true for decreasing coating weights; a lower coating weight reduces Al consumption and therefore increases bath Al, pushing the overall bath state into a more saturated condition. The effect of decreasing coating weight on dross production is shown in Fig. 4.18.

Moving the ingot feeding location closer to one side resulted in insignificant changes in the overall heat and mass transfer inside the bath. The stagnant area near the bottom of the bath under the ingot feeding area is moved to the side away from the basket, but overall dross rates are not changed.

Feeding the ingot at a constant rate to the bath and maintaining inductor power at a constant level rather than running at high power during ingot melting and at low power the rest of the time gives a steadier bath temperature and bath total Al level. Surprisingly, the dross rates are about the same, because with the steady inductor power, the ingot melting area, especially under the melting ingot, is about 2°C cooler than the rest of the bath. With cyclic melting, this region is 3°C cooler, but only for the portion of the operating period when the ingot is being heated and melted, 20 min out of each hour. For the remainder of the period, with no ingot present, this temperature gradient is much less. The amount of dross produced with rapid and gradual ingot immersions is shown in Fig. 4.19.

Adding ingots from the side of the bath behind the snout, rather than at the back of the bath, results in some nonsymmetrical flow, temperature, and composition effects. The flow of the inductor operating beside the feeding position is constrained, giving a higher temperature in this area. This results in a small temperature asymmetry within the snout, also an asymmetry in dross concentration at the strip exit, with less dross present on the ingot feeding side near the strip exit.

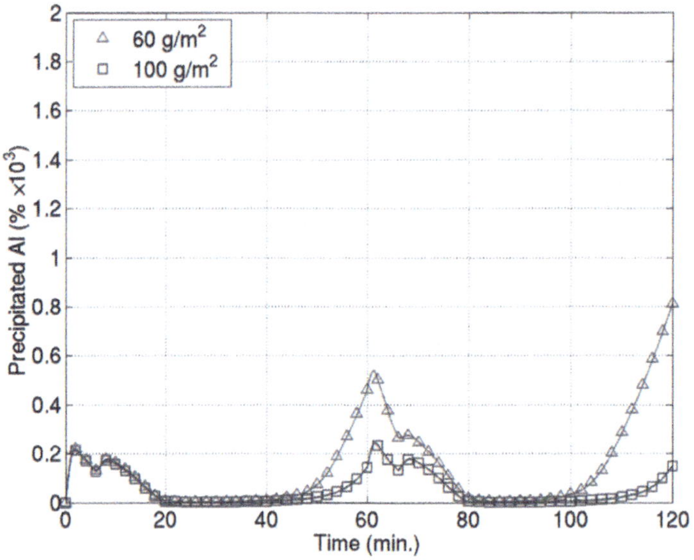

FIG. 4.18 Effect of coating weight on Al tied up in Fe_2Al_5 dross [19].

Positioning the inductors at the front and back of the baths, rather than on the left and right sides toward the back of the bath, gives a slightly higher overall bath temperature at many bath locations. This is because of the more efficient stirring of the bath and consequent lower losses to the bath walls and surface, aided by strip movements when using front and back inductors. However, the temperature in the snout, which is greatly influenced by strip entry temperature, remains the same. The slightly higher temperature reduces the tendency of dross to precipitate.

Positioning the sink roll at a shallower depth in the bath mainly affects conditions at the strip exit side. The shallower sink roll is not effective at stirring the bath as a deeper roll, especially in the "V" volume above the sink roll, where much less metal is then present. The Al level in this "V" is much lower because of the lower amount of mixing that occurs. The volume of metal inside the snout is also affected, the Fe dissolving in the snout on the strip side facing the "V" is not well mixed into the bath volume, this increases local precipitation of dross. However, dross production in the overall bath was not affected by this change of sink roll position.

Moving the sink roll rig 15 cm from the bath centerline resulted in small asymmetric temperature and bath composition patterns. The greatest effect was in the strip exit area, where higher temperatures were seen at the bath centerline, also a higher Fe level was seen inside the snout at the centerline, with lower values on the left and right sides of the snout. Overall dross rates were not affected.

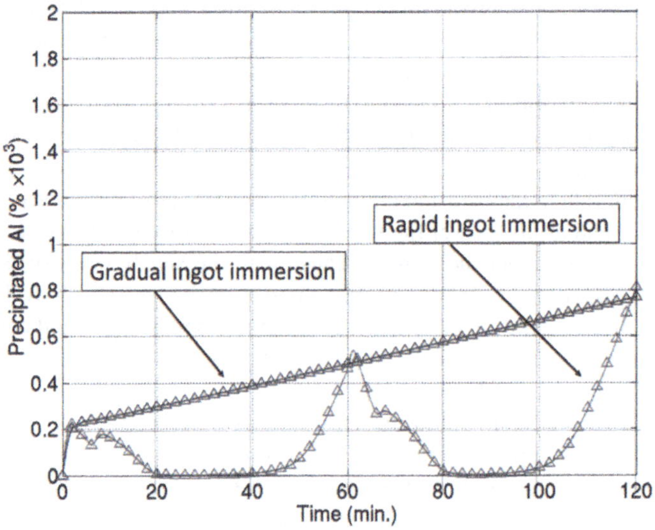

FIG. 4.19 Effect of rate of ingot immersion on Al tied up in Fe_2Al_5 dross [19].

Two variations in snout geometry were analyzed. The base configuration snout is 244 cm long in the transverse strip direction, 42.7 cm long in the strip normal direction and is immersed 20.3 cm into the bath. The effects of a smaller snout 203 × 31.2 cm immersed 20.3 cm into the bath were analyzed, also the base case size immersed deeper, 30 cm, into the bath. Reduced mixing of the flow near the center of the snout was seen with the smaller snout size. This resulted in a lower Al concentration and a corresponding higher Fe concentration at this location for the smaller snout size case, as a result of the Al consumption and Fe dissolution at the strip surface. The net effect on dross production is negligible as the two contributions counterbalance each other. Using the deeper snout gave a a slightly higher temperature for the bath metal, and therefore, the amount of precipitates formed in the bath for this case will be smaller than that for the base configuration. Overall differences in dross production rates were very small.

A summary of these different effects on dross generation rates is shown in Table 4.4.

The dross production rates computed from the ideal situations described in each of these cases are the lowest possible rates that can theoretically be achieved, limited by the iron balance. In an industrial situation, other real-life factors influence temperature, flow, and composition patterns in the bath such as the accuracy of strip and bath temperature control and variations in strip velocity, including line stops. Moreover, these rates are only for the FeZn and FeAl base reaction products resulting from the interaction of the steel strip with the galvanizing bath. Top skimmings, largely composed of

TABLE 4.4 Effect of galvanizing bath details and practices on dross production [19].

Case	Dross (kg)[a]	Comments
Base configuration (BC)	4.55	Dross produced mainly during first minutes of the melting ingot period and at the end of the period without ingot because of the decrease in temperature
Al ingot 0%	0.13	Smaller Al content in the ingot results in a smaller amount of dross formed because of a decrease in the quantity of Al in excess of the solubility limit
Al ingot 1%	11.41	Higher Al content in the ingot results in a higher amount of dross formed because of a larger quantity of Al in excess of the solubility limit
Fe unsaturated	2.55	Unsaturated Fe solution retards formation of dross
Al bath 0.1%	19.90	Larger amount of dross because of $Fe_3Zn_{24}Al$ precipitates having more Zn
Al bath 0.2%	4.93	Similar amount of dross as BC for the case of initial values at solubility limit
Al 0.2%, Fe 0.06%	5.31[b]	The bath is oversaturated with AL and Fe. Initial dross quantity is 258 kg
T_{strip} = 420°C	4.56	Smaller temperature gradients and more gradual dross formation
100 g/m^2 deposition	0.83	Increased Al consumption on the strip resulting in smaller Al content and less dross
100 ton bath	5.22	Higher gradients and smaller bath volume results in higher concentration of precipitates
500 ton bath	3.22	Smaller gradients results in less dross formation
Side ingot	4.35	Non-symmetrical effects more important at the ingot side. Temperature and precipitates concentration higher in the part of the bath containing the ingot
Gradual center	4.31	Smooth variations and smaller gradients. Dross is formed gradually.
Gradual side	3.44	Dross is formed gradually. Non-symmetrical effects more important at the ingot side
Front-back	2.45	Higher temperature gradients, but smaller Al and Fe gradients. Less dross formed because of the higher mean temperature
30″ sink roll	4.50	Less mixing in the snout
Asymm. hardware	3.92	Small asymmetric effects at the ingot side, at the strip exit side for temperature and in the snout for Al and Fe concentrations.
Smaller snout	4.32	Poorer mixing in the snout resulting in larger gradients in this region
Deeper snout	3.80	Poorer mixing in the snout

[a]*Dross generated after a 2 h period.*
[b]*In addition to the initial 258 kg of dross.*

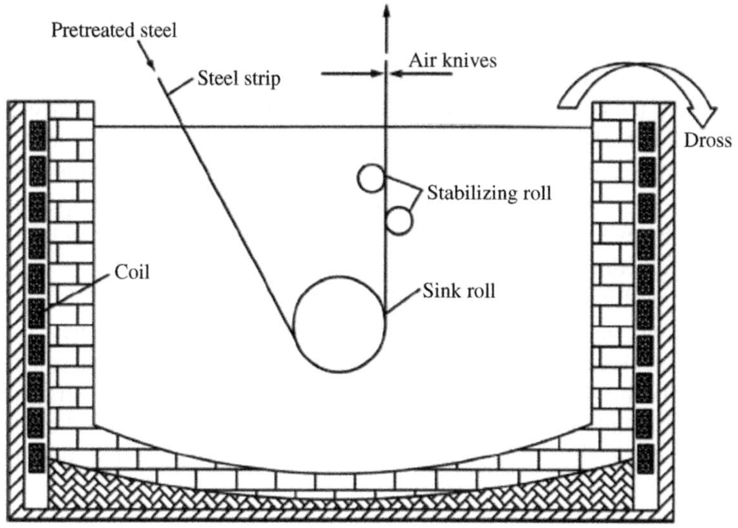

FIG. 4.20 Schematic of a coreless induction heated galvanizing pot [21].

oxides, are also produced and removed with top dross and will be discussed later. Top skimmings can add significantly to the amount of Zn, Al, and Fe removed from the bath by means other than coating of the steel strip.

Galvanizing baths heated by coreless inductors are different from those heated by channel inductors in several ways. As shown in Fig. 4.20, the bath is in the shape of a round bowl, rather than being in the shape of a rectangular prism, so that low-flow areas associated with corners are eliminated. The induction field used to heat the bath also has circular symmetry, also improving the uniformity of fluid flow in the bath. Such flow can also be useful in stirring bottom dross into the bulk of the bath as it is formed, such as in the case of Zn-55%Al coated sheet production. This prevents bottom dross buildup and, if practiced correctly, permits these dross particles to be entrained with strip motion before they grow to a size that affects coating quality. Water modeling of a galvanizing bath in a coreless pot showed that the flow at the strip centerline was dominated by volumes of circular flow behind and in front of the strip, Fig. 4.21, for a strip speed of 2 m/s [21]. Like the flow in channel inductor heated baths, the intensity

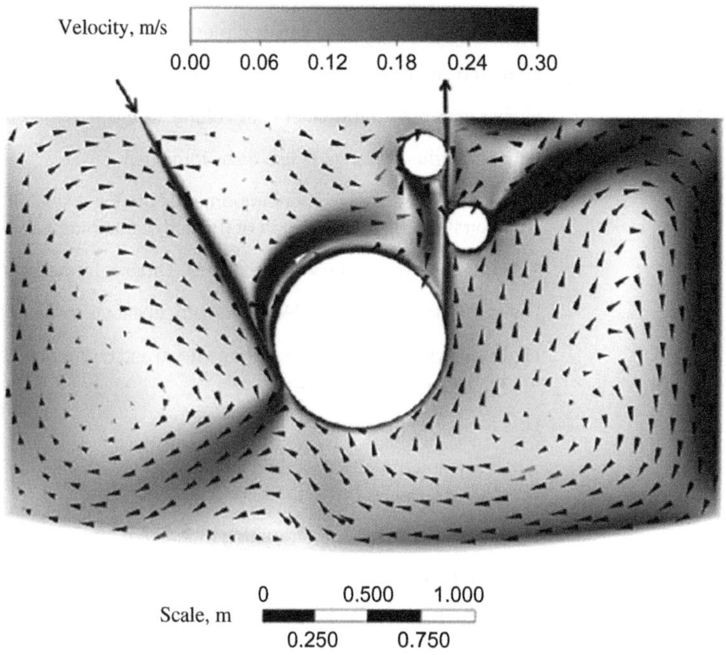

FIG. 4.21 Flow patterns in a modeled coreless inductor pot used for continuous strip galvanizing, at the strip centerline [21].

of the flow patterns changed with strip speed, but the flow patterns were the same. Water modeling was used to understand the nature of dross flow in the bath, using plastic spheres both lighter and heavier than the water bath. The patterns of dross distribution were also affected by strip speed. The floating particles adhered to the strip at low strip speeds but were driven away from the strip at high speed. Increasing the strip speed above 2.3 m/s in this water model began stirring and entraining the denser-than-water dross that had settled to the bottom of the bath, increasing the chance that they would then adhere to the moving strip.

4.6 Dross precipitation and dissolution

Stokes Law describes the velocity of dross particles in a stagnant liquid as a function of relative density and particle size. For the example given in Section 4.4, a 20 μm diameter Fe_2Al_5 top dross particle has an upward velocity of 0.001 m/s and a 20 μm radius $FeZn_7$ bottom dross particle has a downward velocity of 0.0003 m/s. Therefore, in the 2.44 m deep bath shown in Fig. 4.1, a top dross particle of this size at the bottom of the bath would take 41 min to rise to the bath surface in a stagnant area such as near the back of the bath when no ingots were being fed. A bottom dross particle of this size would take 135 min to sink from top to bottom.

Using a bath in the GA range in which a stream of liquid Zn-0.25% Al alloy was introduced, Paik [22] observed that when conditions are favorable for formation of top dross, the first phase to form is a metastable spherical particle of small size, 50-340 nm. This is very different from the polygonal shapes of stable dross particles. Iron content in these particles ranged from 8 to 16%. Al content began near 80% and was lowered to about 15%, decreasing as Zn content increased. These particles were concluded to either transform into stable $Fe_2Al_{(5-x)}Zn_x$ or contribute soluble Al to the bath.

Similarly, Park [23] began with a Zn-0.08%Fe bath and added a liquid Zn-26%Al alloy. When the added amount produced a bath with an overall 0.115%Al composition, local precipitation of the Fe_2Al_5 phase was observed, initially surrounded by an Al-depleted zone (ADZ) in which zeta formed, surrounding the Fe_2Al_5 phase. Because the zeta phase is unstable at Al compositions higher than the zeta stability range, as the surrounding liquid equilibrates in Al, the outer zeta phase begins to decompose, allowing pathways for Al to reach the Fe_2Al_5 core. These phases then dissolved to form the stable FeZn delta phase, but at different rates. The Fe_2Al_5 to delta transformation was complete in about 30 min, whereas the zeta to delta transition took about 2 h. This was interpreted as indicating that there was no supersaturation in the Fe_2Al_5 to delta transformation (and the reverse transformation), and therefore fast kinetics, whereas significant supersaturation existed in the zeta to delta transformation (also the reverse transformation, indicating slow kinetics). However, when growing these crystals directly from saturated baths rather than transforming from one phase to another, zeta has the fastest growth rate, followed by Fe_2Al_5, followed by delta. Moreover, the growth rates of zeta and delta phases are independent of bath Al content [24].

Rather than generating dross by adding Al to the bath as in the above examples, dross generation by Fe additions coming from strip dissolution has been studied. The mass transfer coefficient for iron dissolving from a steel strip was determined by Michal [25] to be in the range of $5–8 \times 10^{-6}$ m/s by applying Fick's second law and using the data of Liu [26] and Toussaint [18], this constant is applied in the formula

$$\frac{dM_{Fe}^{\infty}}{dt} = k_M\left(\frac{\rho}{100}\right)\left(W_{Fe}^{MAX} - W_{Fe}^{\infty}\right) \tag{4.2}$$

where M_{Fe}^{∞} is the mass of dissolved iron per unit area (kg/m^2), ρ is the density of liquid zinc (kg/m^3); W_{Fe}^{MAX} is the maximum solubility of iron in liquid zinc in units of wt. pct. and W_{Fe}^{∞} is the iron concentration in the liquid zinc in wt% far from the steel sheet surface. The mass transfer coefficient was found to vary only slightly with strip speed and bath effective aluminum content but increased as strip temperature increased. The effect of increasing supersaturation was to form larger numbers of fine dross particles, when the bath Fe content in the vicinity of the strip was near or above saturation, the driving force for dissolution was lower, and therefore, fewer dross particles were observed [27].

To better understand dross behavior in industrial conditions, dross samples were taken from the zinc baths of 14 continuous galvanizing baths operated by seven companies, representing a range of GA and GI conditions [28]. Samples were taken from steady-state operations and also from periods of transitions between GA and GI states and the reverse. Strip speeds ranged from 0.21 to 2.91 m/s. The compositions of the 980 dross samples investigated agreed well with the phase diagram shown in Fig. 4.8. Galvanizing baths operating to produce GI coatings, with effective Al contents above 0.16%, were found to have dross made up of only Fe_2Al_5 particles, whereas the GA lines running with bath compositions between 0.13% and 0.14% effective Al had both delta and Fe_2Al_5 dross particles present, and the Fe_2Al_5 particles appeared to be dissolving. GA baths containing only delta particles in the dross were concluded to be operating under stable conditions.

TABLE 4.5 Compositions of dross particles from zinc baths producing GA coatings [28].

			Bath composition	Particle composition		
Code	Type	Temperature of bath (°C)	% Al	% Al	% Fe	% Zn
2-w	Fe_2Al_5	457-460	0,13	45.50	31.22	23.28
2-w	$FeZn_{10}$	457-460	0,13	2.98	4.49	92.54
3	Fe_2Al_5	488-490	–	45.41	30.81	23.79
3	$FeZn_{10}$	488-490	–	2.10	5.28	92.62
4-s	Fe_2Al_5	460-465	0.135 eff	46.59	31.44	21.98
4-s	$FeZn_{10}$	460-465	0.135 eff	2.71	5.12	92.17
5	$FeZn_7$	469	0.135	2.21	5.30	92.5
7-A	Fe_2Al_5	455	0.135	47.28	30.92	21.81
7-A	$FeZn_{10}$	455	0.135	2.70	5.05	92.25
Mean particle composition Fe_2Al_5 (40 spectra)				46.20	31.10	22.71
Standard deviation				0.90	0.29	0.97
Mean particle composition $FeZn_{10}$ (46 spectra)				2.54	5.05	92.42
Standard deviation				0.37	0.33	0.20

Diameter measurements were taken on these dross particles. The diameters of the Fe_2Al_5 particles coming from the baths producing GI coatings ranged from 15 to 65 μm with an overall average diameter of 34.08 μm. For the GA lines in which Fe_2Al_5 particles were found, the Fe_2Al_5 particles were smaller and the ZnFe delta particles were larger.

In this sampling study, the chemical composition of both the Fe_2Al_5 and ZnFe delta particles in GA lines shown in Table 4.5 was much more stable than the Fe_2Al_5 particles from GI baths shown in Table 4.6. This is because zinc baths producing GA coatings operated over a much narrower range of effective Al compositions, 0.13%–0.135% as compared with the GI baths, which varied from 0.17% to 0.28% A1. Table 4.5 shows a standard deviation of 0.90 around a mean on

TABLE 4.6 Compositions of dross particles from zinc baths producing GI coatings [28].

			Bath composition	Particle composition		
Co	Type	Temperature of bath (°C)	% Al	% Al	% Fe	% Zn
1	Fe_2Al_5	456–459	0,21	39.46	36.94	23.60
2-n	Fe_2Al_5	465–468	0.175-0.205	28.20	39.47	32.34
2-p	Fe_2Al_5	453–455	–	40.39	38.13	21.49
4-4	Fe_2Al_5	503	0.17 eff	34.49	36.69	28.82
4-5	Fe_2Al_5	470	0.18 eff	25.91	38.04	36.09
4-7	Fe_2Al_5	455	0.28 eff	37.98	38.34	23.75
6-A	Fe_2Al_5	460–462	0.20-0.23	40.13	37.24	22.33
6-B	Fe_2Al_5	460–462	0.20-0.23	41.14	37.27	21.60
7-I	Fe_2Al_5	458	0.18	46.86	30.96	22.19
Mean particle composition Fe_2Al_5 (92 spectra)				37.21	37.01	25.80
Standard deviation				6.62	2.42	5.34

46.2%Al for Al compositions with lesser variations in Zn and Fe compositions. By contrast, under GI conditions shown in Table 4.6, the mean $Fe_2Al_{(5-x)}Zn_x$ Al composition was 37.21% with a standard deviation of 6.62. Compositions vary slightly with bath temperature and are generally consistent with values predicted by Fig. 4.8 phase diagram and by other published literature. All of the $Fe_2Al_{(5-x)}Zn_x$ compositions in Tables 4.5 and 4.6 exceeded the maximum possible Zn content of 19% determined by Konishi [30]. Because its Zn composition is usually significant, for the remainder of this chapter, the Fe_2Al_5 compound will be referred to as $Fe_2Al_{(5-x)}Zn_x$.

In the Fe_2Al_5 crystal, Zn substitutes into two of the Al sites; when fully substituted the orthorhombic crystal dimensions change only slightly, the *a*-axis is shortened by 0.6% while the *b*- and *c*-axes are lengthened by 1.0% and 0.3%, respectively [31]. The addition of zinc also reduces the hardness and elastic modulus of the $Fe_2Al_{(5-x)}Zn_x$ crystal [32].

An explanation of why the observed dross phases in galvanizing baths do not always correspond to those predicted by Fig. 4.8 phase diagram was given by Nakano [33]. Upon entering the zinc bath, the dissolution of iron is very rapid. The highest concentration of iron dissolved in the bath is very close to the strip, decreasing as the distance from the strip in the bath increases. The iron atom is subject to the body-centered cubic (bcc, ferrite crystal structure)-liquid equilibrium in the ferritic steel considered in this work. The reader may be interested to know that the first realization of the equivalence of behavior of a suspended solid (such as dross) in a liquid and solute atom was indicated in a 1905 paper by Albert Einstein [34]. Nakano also calculated the diffusional velocity of such atoms in many locations of a galvanizing bath that would be overwhelmed by local bath flow effects. With regard to iron dissolution in the bath, the iron atom follows local bath flow movements once it escapes the boundary layer of the moving strip. Unless the volume of the bath it encounters is in a locally unsaturated condition, it will find a more stable equilibrium by combining with zinc or, in many cases, aluminum to form a dross precipitate.

In the region where iron-zinc phase equilibria is expected, the delta phase will preferentially nucleate because it has the same bcc structure as the ferritic iron atom [35]. By considering thermodynamics, the driving force for compositions of interest in GA and GI production, an "isonucleation diagram" can be developed. Fig. 4.22 shows this superposed on the Zn-Al-Fe phase diagram at 450°C. The driving force indicated on the right side vertical axis of this diagram is the Gibbs free energy. The liquidus line has a driving force value of zero, while all points in the liquid phase fields have negative driving forces. What is especially noticeable in this diagram is the wide field where $Fe_2Al_{(5-x)}Zn_x$ is the preferred compound for nucleation, it extends far into the fields of delta and zeta phase stability, agreeing with observations in the bath sampling study described above that $Fe_2Al_{(5-x)}Zn_x$ was found in many GA baths and that the inhibition layer expected to only form under GI coating conditions can be nucleated on steel strip at effective Al levels far below the 0.135%Al phase boundary at 460°C [36].

Several studies have described the growth and dissolution rates of dross particles as temperature and bath composition are varied. The local velocity field has also been found to affect dross growth or dissolution rates; the role of local fluid velocity is particularly important for dross buildup on pot rolls, which will be described later.

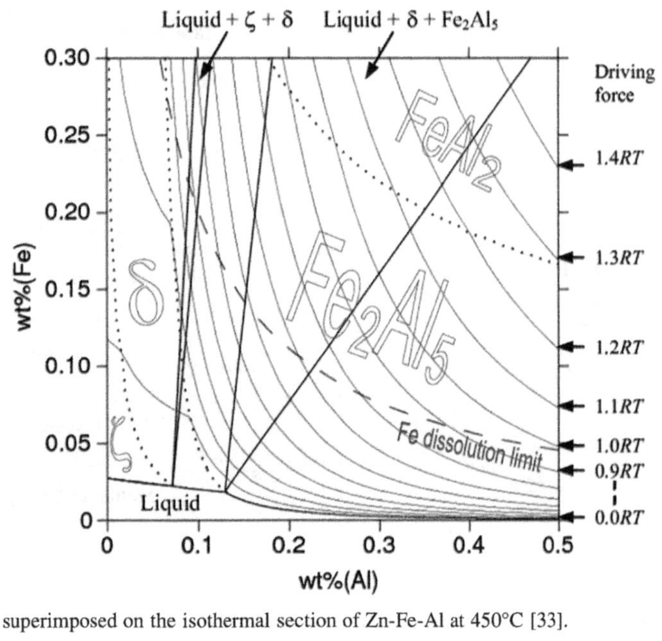

FIG. 4.22 Isonucleation diagram superimposed on the isothermal section of Zn-Fe-Al at 450°C [33].

One investigation of dross growth used a 20 kg zinc bath, stirred with a non-reactive mixer running at either 200 or 500 rpm [10]. To determine the effect of temperature, bath total Al contents of 0.08%, 0.12%, and 0.18% were used, saturated in iron. Bath temperature was lowered from 500°C to 435°C at a linear rate between 0.5°C and 1.0°C/minute and then immediately raised back to 500°C at the same rate. To simulate bottom dross to top dross transition, another set of experiments increased bath total Al content from 0% to 0.18% Al by the addition of Zn-5%Al "brightener bar" alloy at constant 460°C temperature.

The temperature was monitored with a thermocouple. To monitor composition of the dissolved Al and Fe in bath, one of methods of real-time monitoring was used, Laser-Induced Breakdown Spectroscopy (LIBS). Real-time composition monitoring is described further in the section on bath composition transitions later in this chapter. The amount of Al and Fe tied up in dross can be deduced by subtracting the measurements of soluble Al and Fe provided by online monitoring, or by bath sampling followed by chemical analysis, from the known total Al and Fe given by the additions of these elements made to the bath. In the results described below, bath samples were taken and analyzed by Atomic Absorption Spectroscopy (AAS) to provide measures of total Al and total Fe. Bath samples taken during each experiment were also sectioned, polished, and analyzed using a CLEMEX image analysis system. Dross particle sizes are reported as the diameter of a sphere with an equivalent area of the analyzed area exhibited by the dross particle, even though these dross particles are more polygonal than spherical. The aspect ratio is also reported and measured for each particle as the largest projected diameter divided by the smallest projected diameter.

LIBS is an all-optical analytical technique (ref) for the continuous elemental analysis of liquids, solids, and gases. LIBS analysis is carried out by focusing a high power pulsed laser on a material, thereby vaporizing and ionizing a small volume (typically 10^{-8} – 10^{-6} cm^3) of the target to produce high-temperature plasma. Optical emission from this plasma is analyzed with a spectrometer to obtain its atomic composition. Since only optical access is required to generate the plasma and to collect its emitted light, the laser and other potentially vulnerable components can be distanced and completely isolated from the sample. This makes LIBS suitable for analysis in hostile environments where liquids are very hot, corrosive, or in other ways not compatible with alternate measurement technologies. Given its rapidity, simplicity, and the absence of both sample preparation and contact, LIBS is increasingly finding applications in process monitoring.

In a typical LIBS system, brief pulses (about 10 ns duration) of 1064 nm (near-infrared) radiation emitted by a Nd:YAG laser are focused on the sample surface. The high energy flux delivered during each pulse ablates the sample and heats the resulting vapor to create a short-lived (typically a few tens of microseconds) luminous plasma. Depending primarily on the laser power and beam quality, the sample may be located many centimeters, or even meters, from the focusing optics. The experimental setup for these dross growth and precipitation experiments is shown in Fig. 4.23. Light emitted by the plasma was imaged onto an optical fiber bundle, for transmission to a remote Czerny-Turner grating spectrograph. Light dispersed

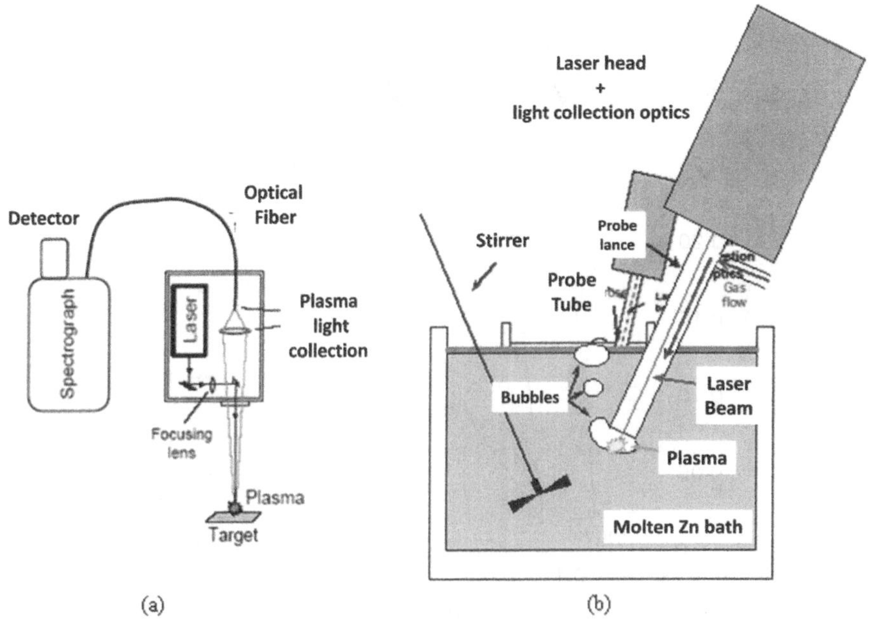

FIG. 4.23 Schematic of the setup for the dross dissolution and growth experiments [10].

by the spectrograph was detected by an intensified charged-coupled device (ICCD) to thereby image a portion of the plasma's emission spectrum, typically 10–40nm wide. The detection of light was timed to select the optimal period in the plasma's lifetime for analytical measurements. Depending on the laser repetition rate and other factors, the above sequence of pulse emission and spectrum creation may be performed several times a second. Thus, subject to the limitations of the required analysis response time, hundreds of spectra may be used to increase the accuracy of elemental analyses derived from peak intensities. The laser head was fitted with a 600mm long mullite probe tube for immersion into the molten zinc. An argon purge flowing at about 1L/min produced bubbles at the tip of the probe tube. Argon was chosen to provide a chemically inert environment for the laser-liquid interaction and plasma formation. The laser was focused at the surface of the bubbles. Since laser pulses typically ablate only tens of nanometers from a targeted material's surface, the dynamic character of bubbles ensured the continuous creation of the fresh and clean molten metal surface required for representative analysis. Radiation from the plasma formed at the bubble surface reached the LIBS head via the probe tube and was spectrally analyzed as described above.

Results for an experiment conducted using a bath total Al of 0.18% are shown in Fig. 4.24A–D. For this composition $Fe_2Al_{(5-x)}Zn_x$ is expected to precipitate, and as described before [22,33], it is still kinetically favorable for $Fe_2Al_{(5-x)}Zn_x$ to precipitate even at the lowest soluble Al content reached in this experiment. After initial rapid nucleation and growth, the particle sizes remain almost constant at about 12μm for both rotational speeds of 200 and 500rpm, with aspect ratios that do not vary greatly between 1.5 and 2.2 at 200rpm and from 2 to 3 at 500rpm. The aspect ratio was typically in the range of 1.6, agreeing with the micrograph shown in Fig. 4.9C. The isolated square data points in Figs. 4.24A and B are the total Al and Fe

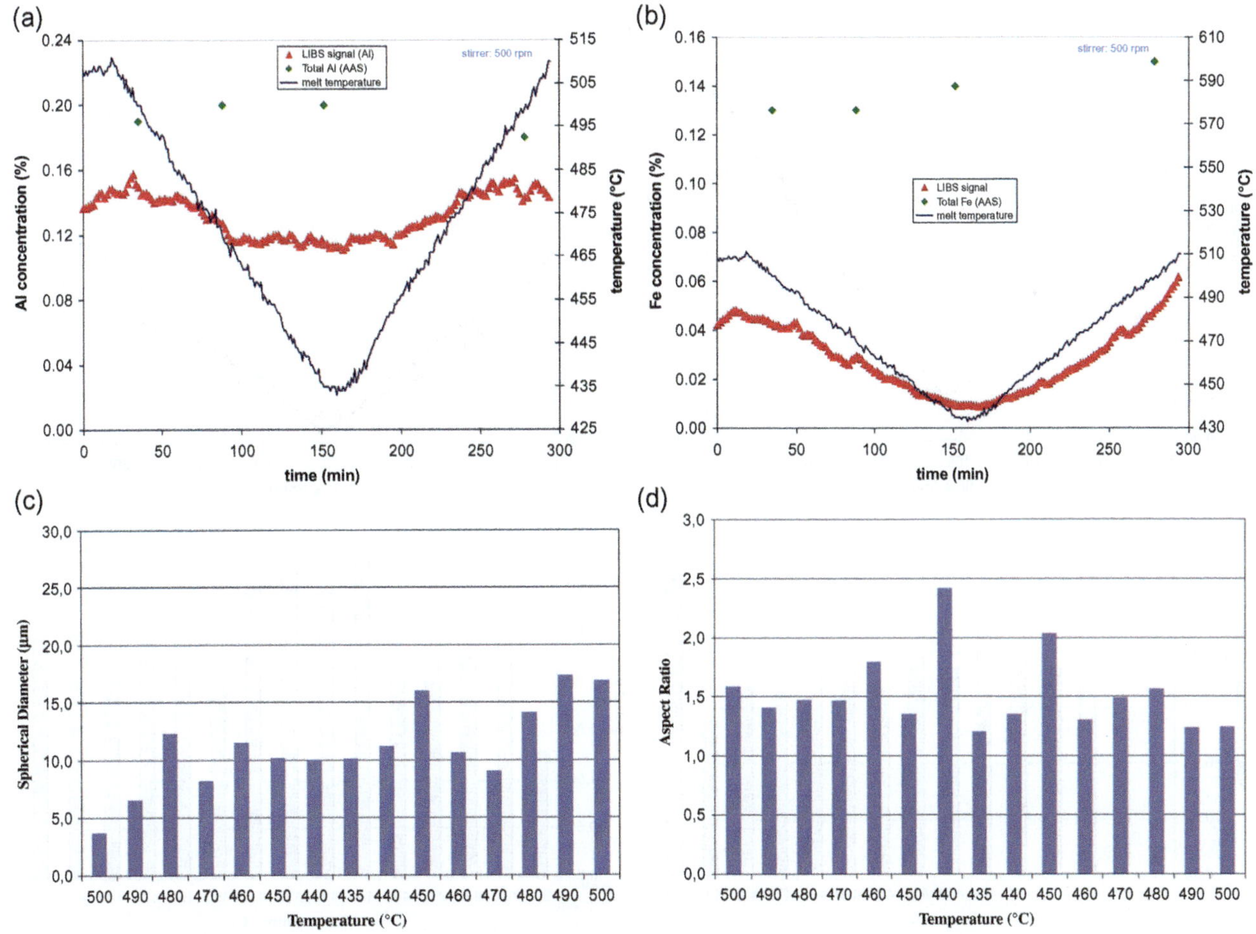

FIG. 4.24 Results of dross precipitation and dissolution experiment for bath with a starting composition of 0.18% total Al, saturated with Fe at 460°C, stirring speed 200rpm [10]. (A) Temperature, effective Al from LIBS, and total Al from AAS for a bath with an initial total Al content of 0.18% [10]. (B) Temperature, effective Fe from LIBS, and total Fe from AAS for a bath with an initial total Al content of 0.18% [10]. (C) Average spherical diameters for dross produced during cooling of a 0.18% total Al bath from 500 to 435°C followed by heating to 500°C at 0.5–1.0°C/s [10]. (D) Aspect ratios for dross produced during cooling of a 0.18% total Al bath from 500 to 435°C followed by heating to 500°C at 0.5–1.0°C/s [10].

contents measured from bath sampling and AAS measurements. The trend of these points is as expected: as more dross is formed, more Al and Fe are tied up in the dross, giving a greater difference between total and effective Al. The dross re-dissolves as the bath is reheated from 435 back to 500°C, as evidenced by the increase in effective Fe content, Fig. 4.24B [10].

Fig. 4.25A–D shows the results of bath cooling experiments for a starting total bath Al of 0.12%, initially saturated in Fe. Here, the dross observed to be formed was a mixture of zeta and delta phase; therefore, the amount of Al lost to dross will be much less than that shown in Fig. 4.24, and indeed the difference between total and effective Al over the course of the temperature cycle shown in Fig. 4.25A is much less than in Fig. 4.24A. It is a different situation for Fe, because of the consumption of soluble Fe from the bath to produce the FeZn drosses, the difference between soluble and total Fe in Fig. 4.25B. grows as temperature decreases and then is reduced as temperature rises.

Also, the effect of fluid flow on the growth rate of the mixed FeZn drosses is much more significant. Fig. 4.25C shows that the particle sizes increase very quickly to about 35 μm for a shown rotational speed of 200 rpm; for a rotation speed of 500 rpm, they grow nearly twice this fast, to about 50 μm. The aspect ratios of the delta phase dross particles are relatively constant between 2 and 2.5, agreeing with the micrograph showing both delta and zeta phase particles shown in Fig. 4.9B. As shown in Fig. 4.25C, the dross particle sizes remain almost constant at about 30 μm for 200 rpm for the remainder of the temperature cycle; larger particles with an average size of about 50 μm when a 500 rpm stirring speed is used.

When a bath starting composition of 0.08%Al, initially saturated in Fe, is used, the precipitated dross is expected to be zeta phase. The counterpart curves to Fig. 4.25A and B are quite similar, because the dross formed is also FeZn, with even less Al tied up in the precipitated dross than in delta phase. Like delta phase, there is a significant effect of flow rate. Fig. 4.26A shows that particles increased in size from about 5 to 30 μm for a rotational speed of 200 rpm, as in the case

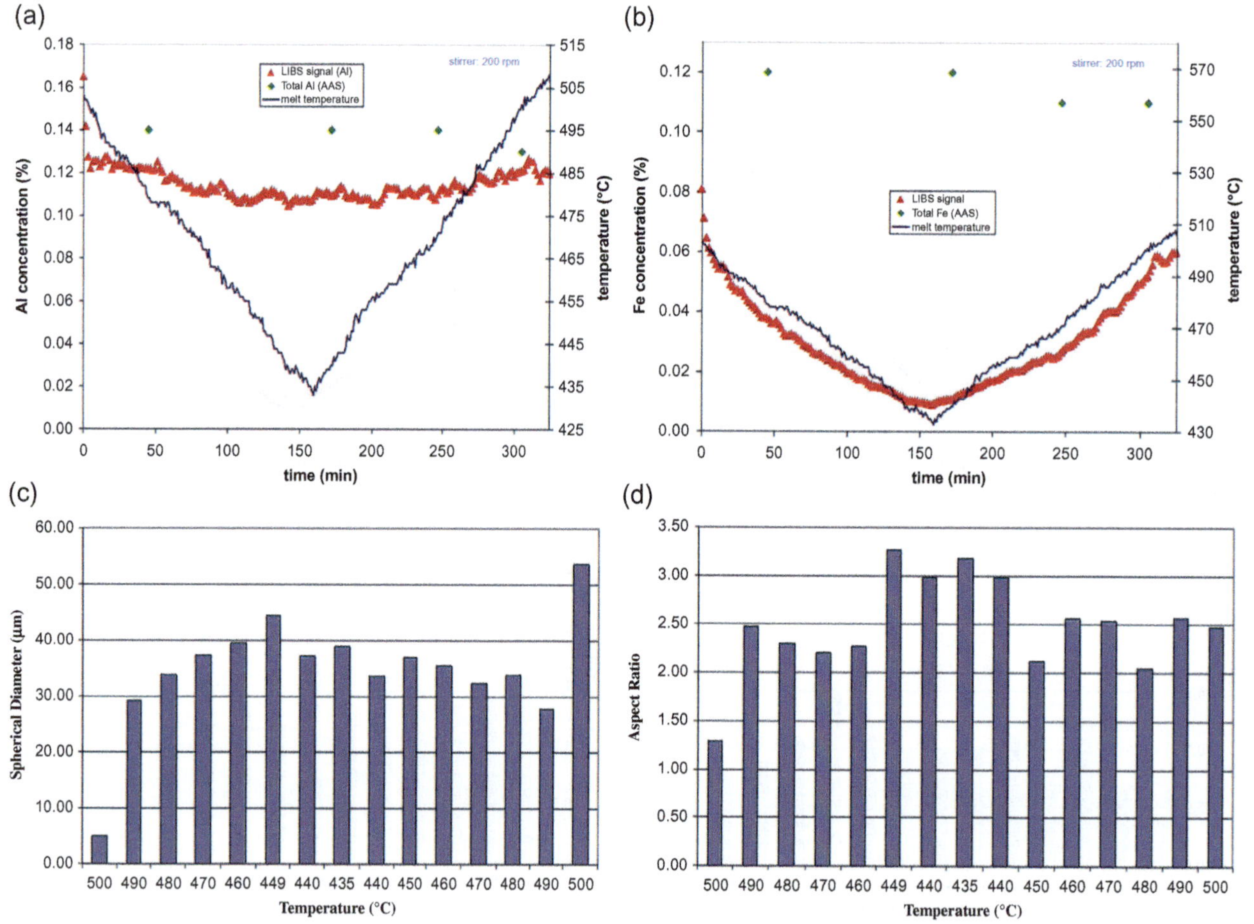

FIG. 4.25 Results of dross precipitation and dissolution experiment for bath with a starting composition of 0.12% total Al, saturated with Fe at 460°C, stirring speed 200 rpm [10]. (A) Temperature, effective Al from LIBS, and total Al from AAS for a bath with an initial total Al content of 0.12% [10]. (B) Temperature, effective Fe from LIBS, and total Fe from AAS for a bath with an initial total Al content of 0.12% [10]. (C) Average spherical diameters for dross produced during cooling of a 0.12% total Al bath from 500 to 435°C followed by heating to 500°C at 0.5–1.0°C/s [10]. (D) Aspect ratios for dross produced during cooling of a 0.12% total Al bath from 500 to 435°C followed by heating to 500°C at 0.5–1.0°C/s [10].

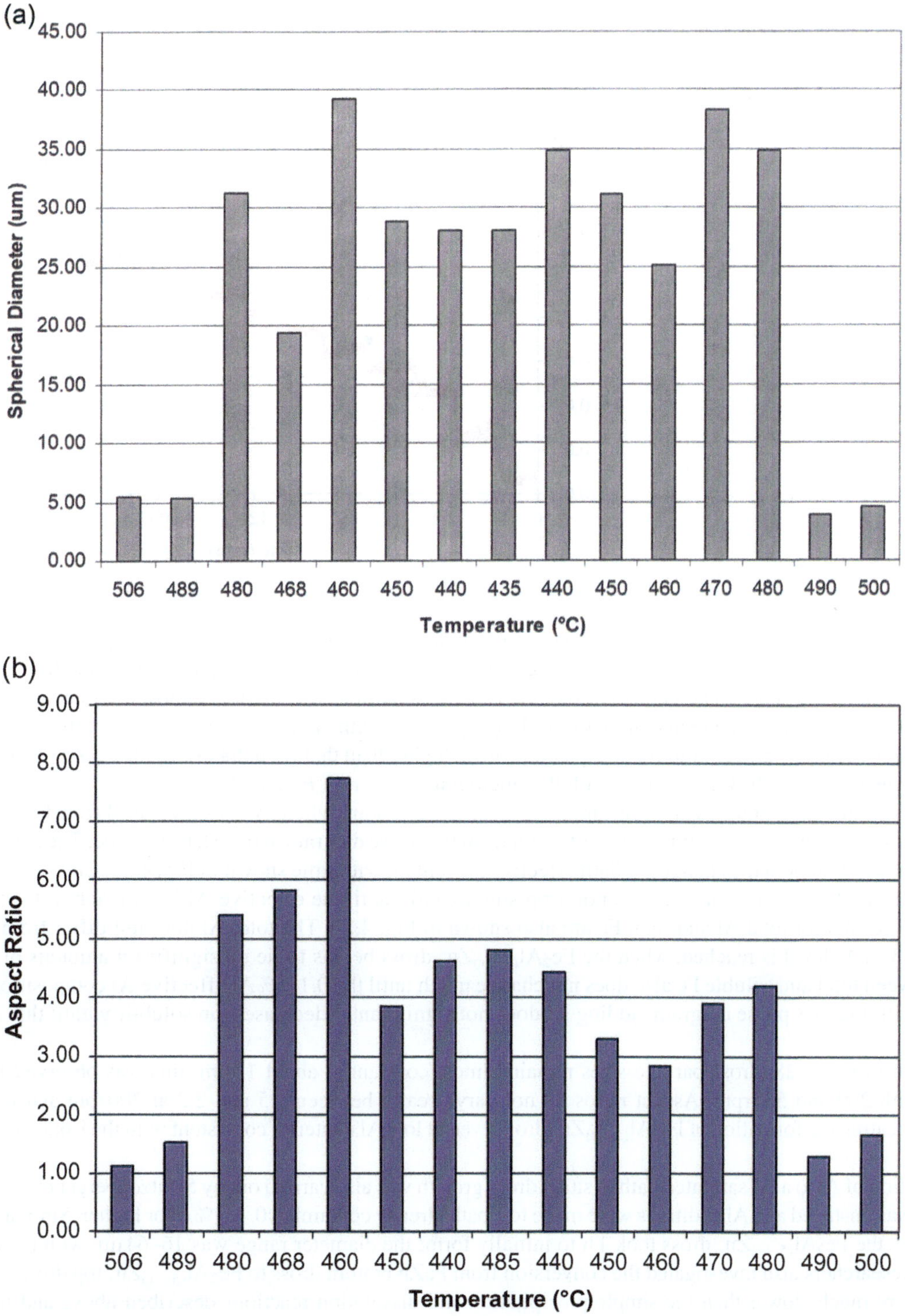

FIG. 4.26 Results of dross precipitation and dissolution experiment for bath with a starting composition of 0.08% total Al, saturated with Fe at 460°C, stirring speed 200 rpm [10]. (A) Average spherical diameters for dross produced during cooling of a 0.08% total Al bath from 500 to 435°C followed by heating to 500°C at 0.5–1.0°C/s [10]. (B) Aspect ratios for dross produced during cooling of a 0.08% total Al bath from 500 to 435°C followed by heating to 500°C at 0.5–1.0°C/s [10].

FIG. 4.27 Effect of Zn-5%Al brightener bar additions into an initially Fe-saturated 0%Al bath. Effective Al and soluble Fe readings from LIBS, total Al, and Fe from AAS readings from bath samples, stirring speed 500 rpm [10].

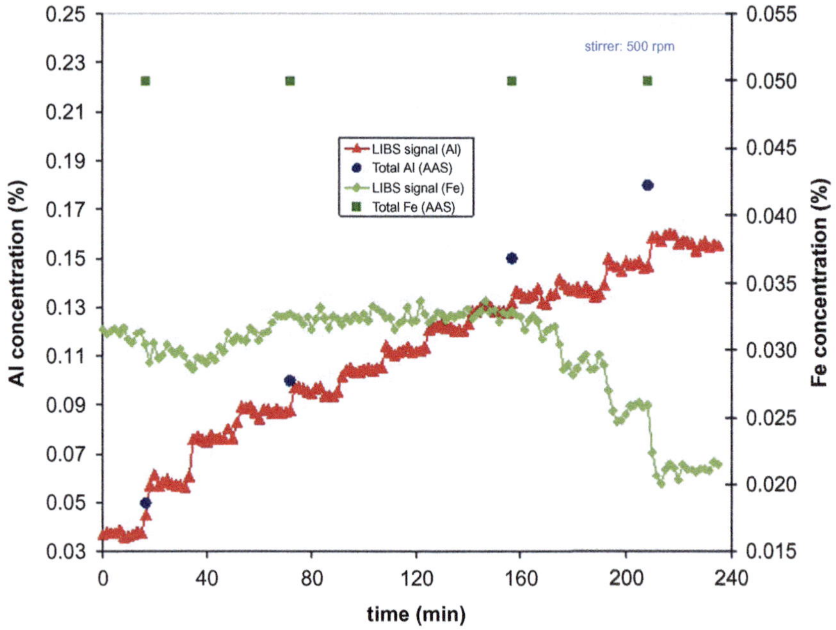

of the mixed zeta and delta phase, the growth rate increases with stirring speed, reaching about 50 μm at 500 rpm during the temperature decrease, increases from about 1 to 5 over this period for a rotational speed of 200 rpm and from about 1 to 6 at 500 rpm. As with mixed zeta and delta phase dross, the zeta dross particle sizes remain almost constant at about 35 μm for 200 rpm for the remainder of the temperature cycle; larger particles with an average size of about 50 μm when a 500 rpm stirring speed is used. The aspect ratio for zeta phase is much higher than that for delta or $Fe_2Al_{(5-x)}Zn_x$, being typical 4 or higher, as shown in Fig. 4.26B, this agrees with the micrograph shown in Fig. 4.9A.

For the experiments conducted at a constant 460°C bath temperature where Al was increased from 0% to 0.18% total Al, the initial Fe content was the saturation level taken from the phase diagram at 0%Al, 0.035%Fe. The Zn-Al brightener bar additions are evident in the changes of bath effective Al content with time shown in Fig. 4.27. This test used a stirring speed of 500 rpm. Soluble Fe content does not drop significantly until the effective Al level reaches 0.135%. The laboratory AAS results for total Al and total Fe are also shown in Fig. 4.27. The total Al does not differ from effective Al until the 0.135%Al level is reached, when the $Fe_2Al_{(5-x)}Zn_x$ dross begins to tie up significant amounts of Al. The difference between total and soluble Fe also does not change much until the 0.135%Al effective Al composition is reached; consistent with Fig. 4.8 phase diagram, adding Al does not significantly decrease iron solubility until this Al content is reached.

As shown in Fig. 4.28, dross particle sizes remain almost constant at about 12 μm, this was observed for rotational speeds of both 200 and 500 rpm. Aspect ratios do not vary greatly between 1.5 and 2.2 at 200 rpm and from 2 to 3 at 500 rpm, indicating the formation of $Fe_2Al_{(5-x)}Zn_x$ dross even at low Al contents, consistent with the isonucleation diagram of Fig. 4.20.

The addition of Al to a Fe-saturated bath to study dross growth was also carried out by Strutzenberger et al. [37] The bath in this case was unstirred and Al additions were made to a bath already containing 0.135%Al or higher. Similar to the results shown above, the $Fe_2Al_{(5-x)}Zn_x$ dross took 1 h to initially form; the diameter range was 16–64 μm with the main peak at 48 μm. The researchers also investigated the conversion from FeZn bottom dross to $Fe_2Al_{(5-x)}Zn_x$ top dross. The reactions occurring were much slower than the simple precipitation and dissolution reactions described above and involved conversion of the bottom dross to top dross took 1.5–2 days. This is consistent with the observations of Ajersch [28] where the top dross particles in the GA bath were observed to be in the process of slowly dissolving.

The effect of ingot preheating on coating bath productivity has been modeled and experimentally confirmed for aluminizing baths. It was shown that preheating a 10 kg Al-10%Si ingot to 500°C reduced melting time to 2 min and 10 s compared with 6 min and 40 s for a room temperature ingot after immersion in a 650°C aluminizing bath [38]. It has also been shown that increasing the galvanizing bath temperature reduces zinc ingot melting times [39], but this would be expected to affect overall bath equilibrium and dross production, overcoming any gains in bath productivity that might be achieved with a higher bath temperature.

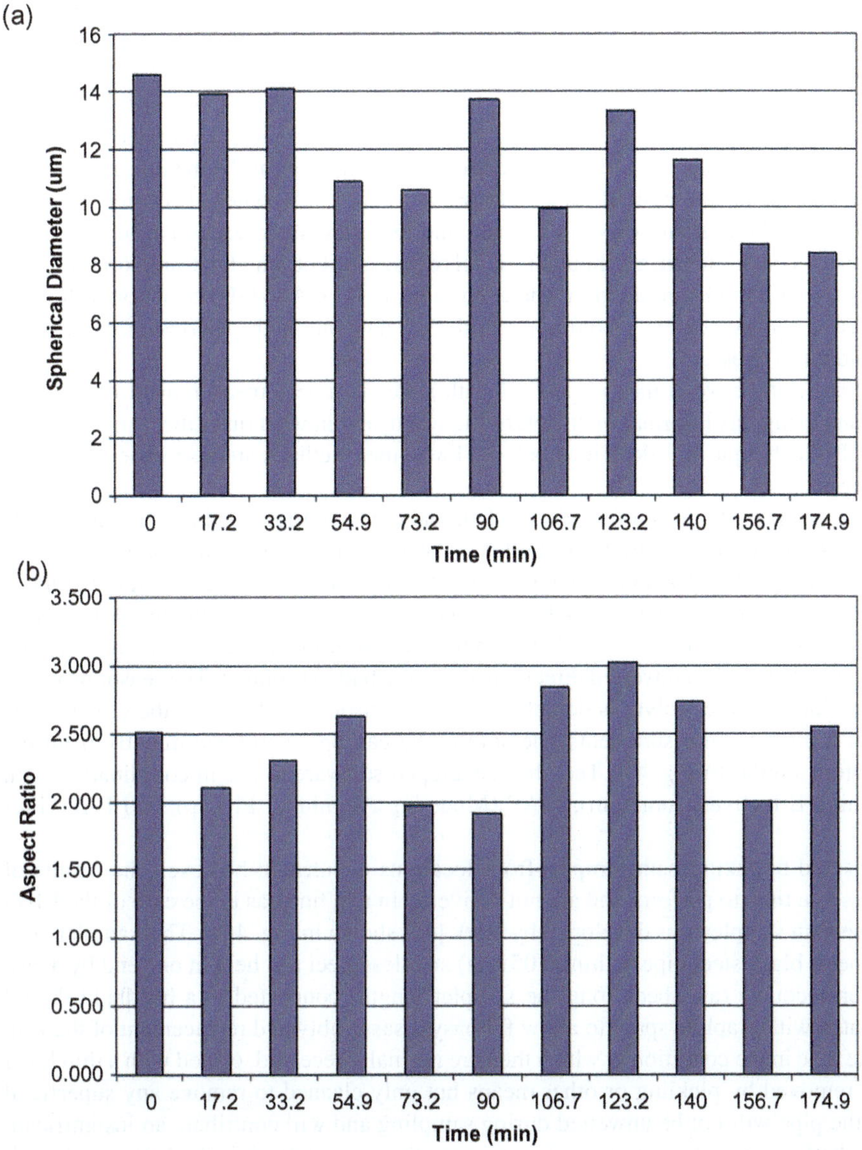

FIG. 4.28 Results of dross precipitation experiment for 460°C Fe-saturated 0%Al bath as Zn-5%Al brightener bar is added at the rate shown in Fig. 4.27. (A) Average spherical diameters for dross produced in a 460°C Fe-saturated 0%Al bath as Zn-5%Al brightener bar is added at the rate shown in Fig. 4.27. (B) Aspect ratios for dross produced in a 460°C Fe-saturated 0%Al bath as Zn-5%Al brightener bar is added at the rate shown in Fig. 4.27.

4.7 Bath management during transitions between galvanneal and galvanize (GA → GI), galvanize and galvanneal (GI → GA)

As described before, galvanizing baths used to make GA-coated steel sheet use galvanizing bath effective Al contents below the composition where the $Fe_2Al_{(5-x)}Zn_x$ is in equilibrium with the bath, 0.135% at 460°C, which enables FeZn phases to grow at the interface and eventually form the galvanneal coating. Baths for making GI-coated sheet steel operate at effective Al levels higher than this composition to permit formation of the thin $Fe_2Al_{(5-x)}Zn_x$ inhibition layer on the steel surface that is then coated with the zinc overlay coating as the sheet is withdrawn from the bath. This inhibition layer permits smooth, formable coatings to be produced. Therefore, GA baths operate in a range where FeZn dross compounds are formed that will eventually settle and form a layer on the floor of the zinc bath, and GI baths operate in a range where $Fe_2Al_{(5-x)}Zn_x$ dross is formed that eventually floats to the surface of the bath. Many lines operate GA baths near to the 0.135% "knee point" in which case both bottom and top dross are stable in the bath. As indicated before, the settling or rising velocities of these bottom or top drosses, respectively, are slow and are greatly influenced by local zinc bath flow patterns. It is therefore commonplace to find fine dross articles suspended in the bath, and such particles can be observed in bath samples unless the bath conditions are such that the bath is operating in an unsaturated (fully liquid) state, i.e., the portion of the phase diagram below the liquid phase boundaries in the phase diagram of Fig. 4.8. It is common practice to skim the surface of the bath to remove accumulated top dross during GI production, but much more difficult to remove accumulated bottom dross during GA production. Production must usually be stopped while bottom dross is removed. Details of dross and skimmings removal will be described later.

It is commonplace for CGLs producing GA-coated steel sheet to also produce GI-coated steel sheet. In North America in 2020, there were 44 operating lines that produced both GA and GI. Many of these lines producing GA and GI using two separate pots, one for each bath chemistry, that are exchanged for the product transition; this can be the quickest way to make this transition. But it is also beneficial for lines mainly dedicated to production of GA-coated sheet steel to occasionally transition to GI production to convert the bottom dross to top dross and therefore avoid excessive bottom dross accumulation. Bottom dross accumulation can be of such an extent that its level reaches the bottom of the sink roll [40]. The principal challenge in transitioning lines from GI to GA production, and even more so for GA to GI production, is how to change the bath chemistry from the effective bath Al content used for one product to the other in the shortest possible time, while producing the lowest possible amount of dross.

Key to maintaining control of the transition between the GA and GI bath states is timely measurement of the bath effective Al composition as it changes, and using this information to determine when, and in what quantities, zinc and zinc alloy additions need to be made. The LIBS technique was described before. Two other methods are also widely used: Al sensors based and manual bath sampling.

The Teck sensor for measuring soluble Al is in the form of an electrochemical concentration cell based on the Nernst Law. This consists of a quartz tube containing a proprietary salt that is molten at the galvanizing bath temperature and a high purity Al wire that serves the reference electrode [41]. The ions in the molten salt act as the electrolyte. The quartz tube is protected by a robust stainless steel tube that is also resistant to attack by the galvanizing bath. When the tip of the quartz tube is snapped off, the electrolyte establishes a potential difference between the Al in solution in the galvanizing bath and the pure Al wire, this can be read as a voltage that can be converted directly to effective bath Al content. The sensor provides a steady reading for about 6 days, after which the electrolyte is depleted, and the sensor is replaced. If the effective Al content and the temperature are known from these measurements, the soluble Fe can be determined directly from the ZnAlFe phase diagram for that temperature similar to Fig. 4.8. Teck has developed software to use in coordination with this sensor, including DEAL (Determining Effective Aluminum), MAP (Modeling Aluminum Pickup), and PAL (Predicting Effective Aluminum).

Bath compositions can also be measured by taking bath samples from locations of interest; however, the results of chemical analysis of these samples take some time to perform and are not delivered in real time, as in the case of the LIBS or Teck sensor readings. A very effective bath sampler was developed by Teck [42] shown in Fig. 4.29. This sampler consisted of a 1.5 or 3 m long 12.7 mm diameter black steel pipe, a thin (0.05 mm) stainless steel foil held in one end by a cast iron fitting, and a sheathed Type-K thermocouple (attached along the sampler length) connected to a handheld digital readout. The end of the sampler was coated with graphite spray to allow for easy disassembly and replacement of the foil. It is important to leave the surfaces of the pipe in the condition in which they are normally received, coated with a thin layer of hot rolling scale. If this scale is not removed by pickling or other means but only cleaned to remove any superficial contamination such as grease and dirt, the pipe will not be unwetted during sampling and will contribute an insignificant amount of Fe to the sample. To collect a bath sample, the sampler is immersed to the desired depth in the bath and the foil pierced using a pointed rod (or retractable weight), located within the sampler. After removing the pointed rod or weight, a 30-s fill time is sufficient to collect enough metal for an assay sample. The sampler is then quickly removed from the bath and the metal stream directed to sample disc molds. Typical sample molds used are a vertical split mold and a copper horizontal split mold, per ASTM Standard E-716 [43].

Fig. 4.30 shows the effective Al and soluble Fe contents for the 75 ton galvanizing bath at Sorevco, a subsidiary of ArcelorMittal in Coteau-du-Lac, Quebec, during a GA to GI transition. The starting Al content is just below the "knee point" in the ZnAlFe phase diagram of Fig. 4.8, 0.135%Al at 460°C, so this bath was producing a GA coating. There is excellent agreement in the readings reported by the LIBS and Al sensor techniques used to report real-time readings during this transition.

The transition from GA to GI coating production is accomplished by adding Al to the bath. The means by which this addition is made strongly affects the amount of dross that is produced. A variety of Zn-Al alloy ingots are available for feeding the bath. In Section 4.5, the amount of dross produced in steady-state GI operation using 1000 kg (jumbo) 0.5 or 1.0%Al content ingots was described. Performing the GA to GI transition requires the use of higher Al contents. The higher Al additions are typically made using 9 kg "brightener bars." If 10% Al bars are used, they will float on the bath surface, causing about 25% of the aluminum to be lost to oxidization and be removed with the top dross skimmings [45]. Moreover, the Al in these 10%Al bars that does not become oxidized and mixes poorly with the bulk of the bath because of its lower density. By contrast, solid 5% Al bars have a density greater than that of liquid Zn and therefore will

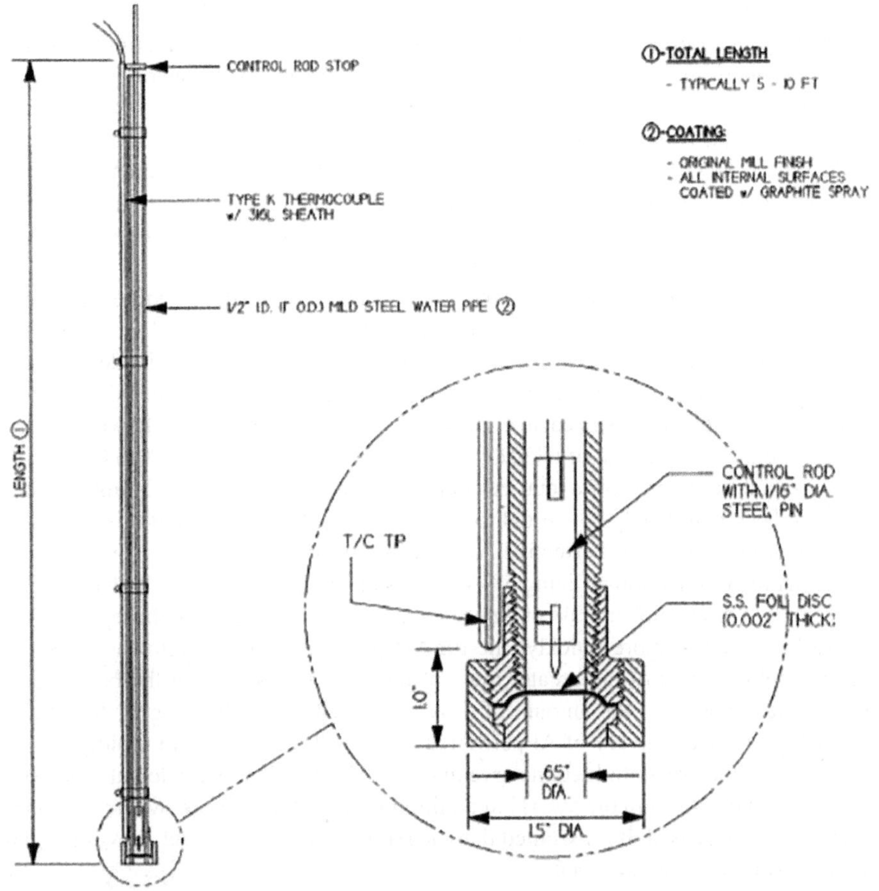

FIG. 4.29 Sectional view of bath sampler showing details at sample collection end [42].

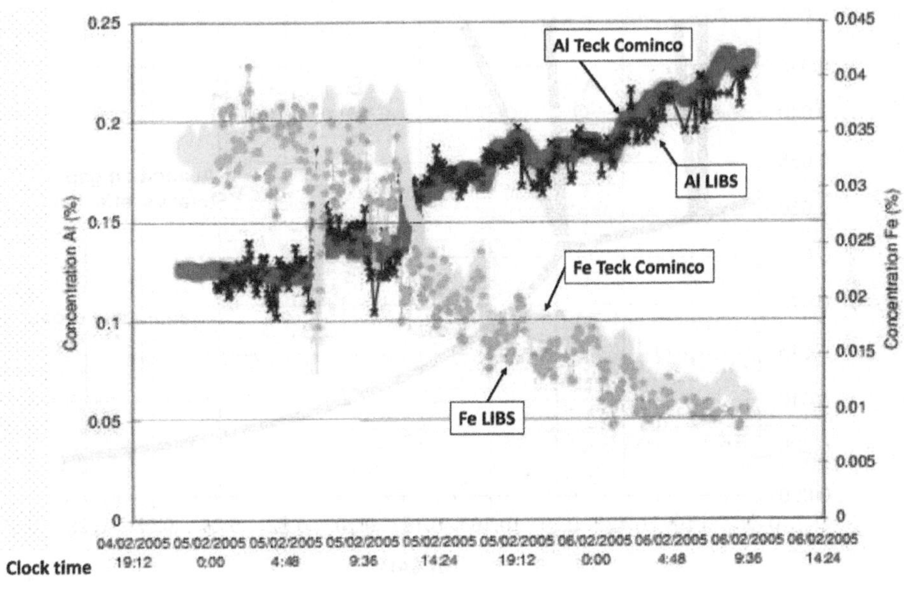

FIG. 4.30 Measured effective Al and soluble Fe contents in the Sorevco bath during a GA to GI transition using LIBS and the TeckCominco AlSensor probe [44].

sink in the bath, minimizing oxidation losses and greatly improving the mixing of the melting Al content into the bath. Additional Al is provided by the addition of 1% Al jumbos instead of the 0.45% aluminum jumbos customarily used in steady-state operation in this example. But because 10% Al bars are less costly overall than 5% Al bars, at least one producer has developed an effective and cost-saving practice using them with 1% Al jumbos as described below.

Fig. 4.31 shows the ZnAlFe phase diagram at 460°C, where the horizontal dashed line indicates the change in bath total Al content during a transition from GA to GI production. Al is added to increase bath content to above the "knee point" at 0.135% Al. As Al is increased, the amount of $Fe_2Al_{(5-x)}Zn_x$ top dross created is related to the increased length of the vertical/diagonal arrows as Al content increases. It can be seen by the length of the dashed red arrows that adding more Al (point C vs. B) results in more top dross being generated.

Fig. 4.32 depicts a transition from GA to GI using 10% Al bars. The initial bath effective Al composition was 0.136%. An initial jump in Al composition coincided with the addition at the start of 180 of these 9kg bars. The reduction in effective Al level after the transition started was caused by the feed of the 0.45% Al jumbos being inadequate to supply the Al needs of the strip, normal GI top dross generation, and the bottom to top dross conversion. To keep the effective Al from dropping below the target, two more additions of 30 bars each were needed after 5h and 19h, as shown in Fig. 4.32. The dross rate peaked at 15h into the 36-h transition. A total of 6350kg of dross was produced. The GA to GI transition is deemed to be completed when successive bath samples show a total Fe content of less than 0.02%, indicating that the bottom dross of the previous GA campaign had been completely converted to top dross. Fig. 4.31 shows that this Fe content is well below the solubility curve in the GA production range (effective Al < 0.135% at 460°C).

Fig. 4.33 illustrates a GA to GI transition utilizing 5%Al bars, starting from a 0.135% bath composition. Because of the lower Al content of these bars, a much higher mass of bars needed to be added. Because these bars sink to the bottom of the bath, allowing the Al to be mixed in more quickly, the time required for the transition was reduced from 38 to 33h. However, because of the cooling effect of the greater mass of ingots at the bottom of the bath, where the conversion of bottom dross to top dross was underway, the amount of dross produced was 450kg higher than the transition described in Fig. 4.32. However, this increased delivery of Al permitted the transition to be more rapidly achieved. This is evident by the more rapid rise and fall of the dross production curve in Fig. 4.33. Mass balance calculations showed that calculations would indicate that nearly all of the Al from the 5%Al bar addition was used in the bath, whereas approximately 75% of the Al from the 10%Al bar addition was used. It is assumed that the remainder of the unused Al was taken out with the dross in the form of oxidized Al or similar waste material.

Fig. 4.34 illustrates a GA to GI transition using an addition of 200 10%Al bars at the start, but this time the rest of the feed was 1% Al jumbos. Note that the effective Al is easily held at a desired higher level without adding more of the costly 10% Al bars. There is now sufficient Al supply to both convert bottom dross and to form the inhibition layer on the steel. Also, the transition time is about 6h less, the result of the increased rate of dross conversion due to the higher concentration of Al in the bath that was maintained throughout the transition.

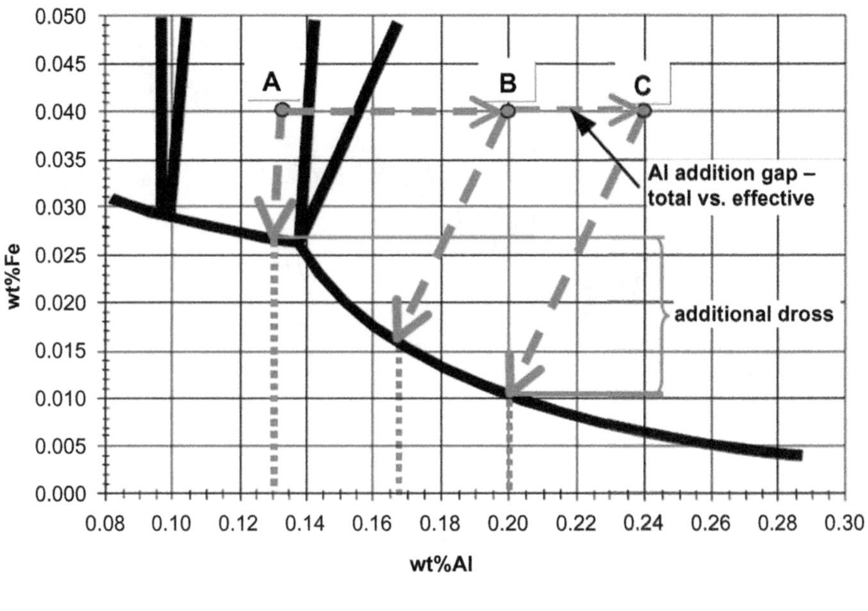

FIG. 4.31 Increasing Al content during the GA to GI transition reduced Fe solubility and therefore results in generation of $Fe_2Al_{(5-x)}Zn_x$ top dross. The endpoint C (0.24% total bath Al and 0.20% effective Al) results in more dross production than endpoint B (0.2% total bath Al, 0.175% effective Al) [46].

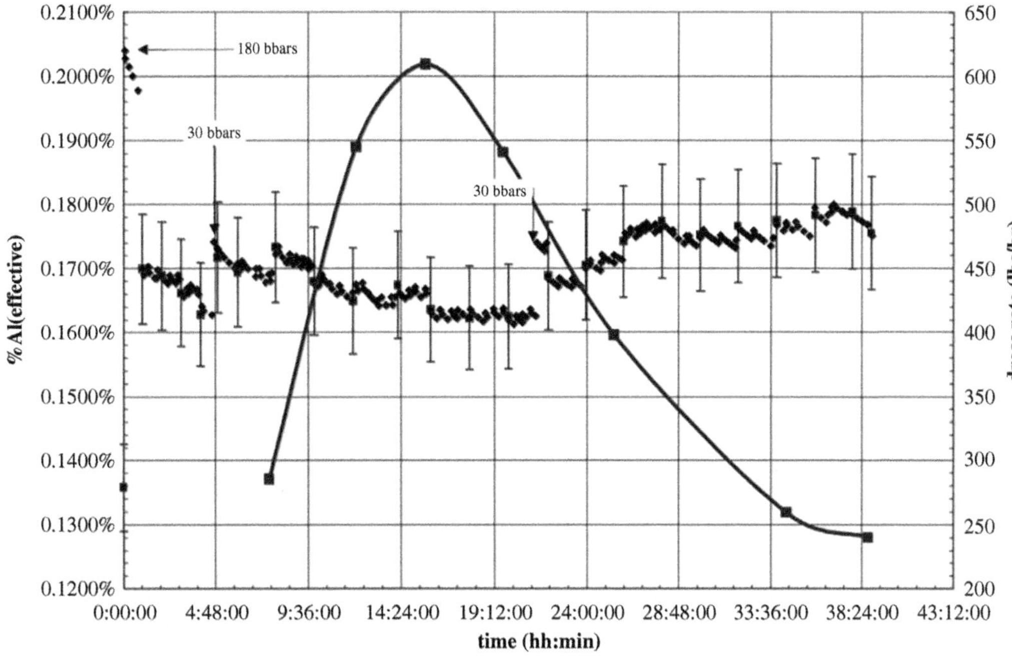

FIG. 4.32 Bath effective Al and dross rate for a GA to GI transition, plotting effective Al and dross production rate vs time. Brightener bars with 10%Al content were used with 0.45%Al jumbo ingots. Diamond symbols with 5% error bars represent manually taken bath samples, other diamond symbols represent mass balance prediction, square symbols give dross production rate [45].

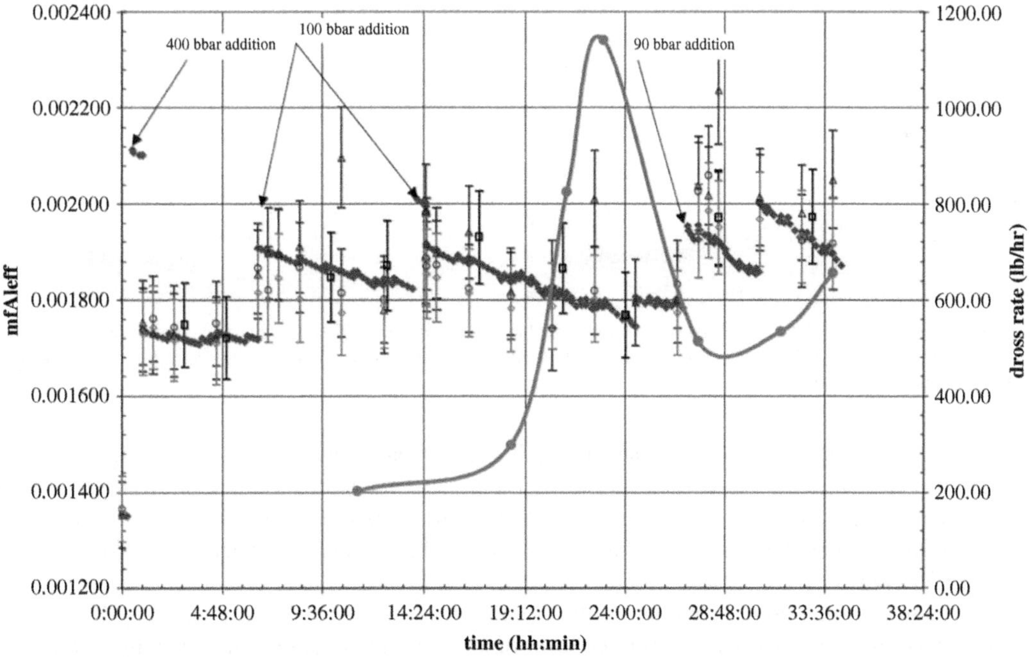

FIG. 4.33 Bath effective Al and dross rate for a GA to GI transition, plotting effective Al and dross production rate vs time. Brightener bars with 5%Al content were used. Bath Al predictive readings shown without error bars, bath samples with error bars, dross production given by solid line [45].

When changing bath chemistry, steady managed transitions are preferred [45,47]. Shocking the bath with sudden Al changes, large additions of cold Zn, bath, or strip temperature changes, or a change in bath level detracts from both line efficiency and product quality. Serious quality problems such as poor coating adherence and excessive dross generation that can become entrained in the coating causing appearance issues can appear, resulting in product downgrades or rejections.

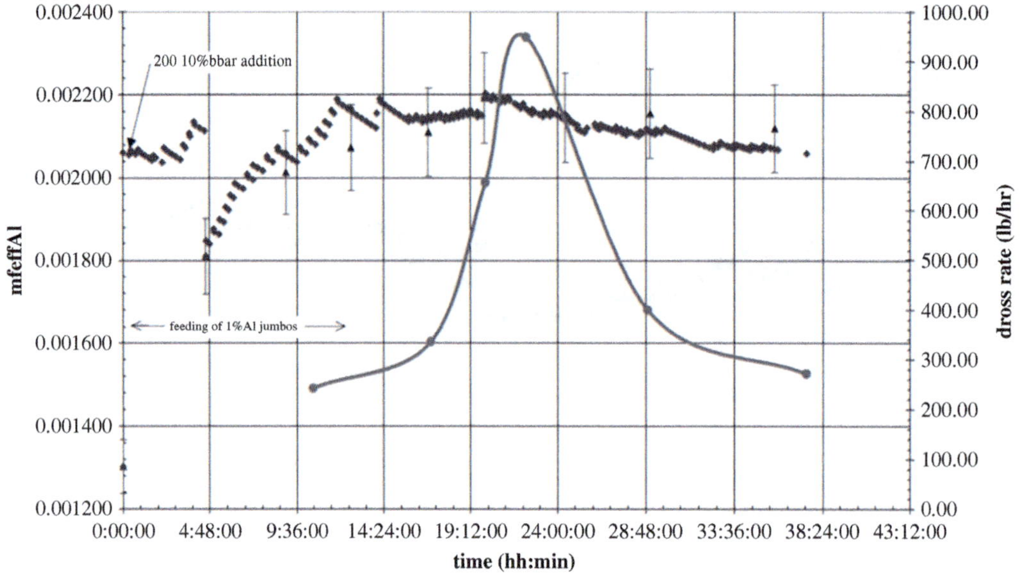

FIG. 4.34 Bath effective Al and dross rate for a GA to GI transition, plotting effective Al and dross production rate vs time. Brightener bars with 10%Al content were used with 1%Al jumbo ingots. Bath Al predictive readings shown without error bars, bath samples with error bars, dross production given by solid line [45].

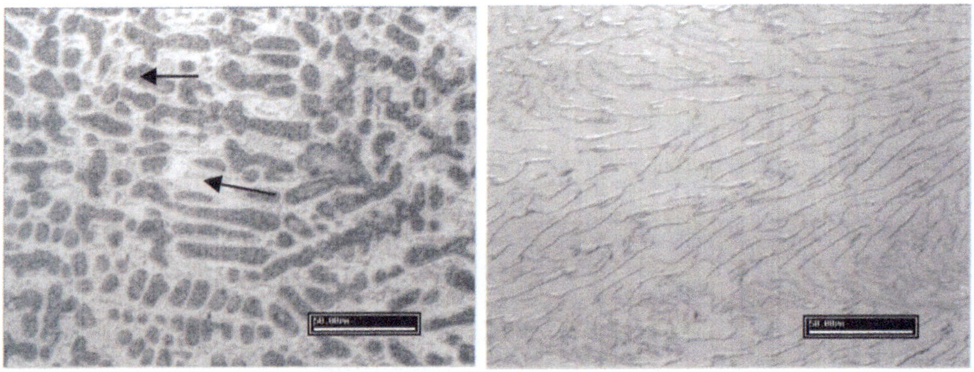

FIG. 4.35 Microstructures of 10%Al (left) and 5%Al (right) brightener bars. Note that the 10%Al bar microstructure comprises a high-melting-point, high Al content primary phase (darker color) and a low-melting-point 6%Al eutectic phase (lighter color). The 5%Al bar microstructure comprises the eutectic phase and a small amount of almost pure Zn [45].

A comparison of the microstructures of the 5 and 10%Al brightener bars provides additional insight into the reasons for the higher Al recovery seen with the 5%Al bars. As shown in Fig. 4.35, the microstructure of the 10%Al bar consists of a mixture of a 6%Al eutectic phase and a high Al content (22%Al) primary phase, the latter of which melts at a temperature higher (430°C) than that of pure Zn. The 5%Al bars consist of an almost entirely eutectic microstructure, with a single melting point of 385°C [45].

The nature of the transformation of dross during the GA to GI transition has been reported by Paik [22]. ZnFe delta particles were observed that were covered in a surface layer of an Al-rich amorphous phase having a composition similar to Fe_2Al_5. After thermodynamic stabilization, this transformed into the expected $Fe_2Al_{(5-x)}Zn_x$ crystalline phase.

The transition from GI to GA operation is accomplished by adding pure Zn ingots to the bath. In a reversal of the arrows shown in Fig. 4.31, the $Fe_2Al_{(5-x)}Zn_x$ top dross remains stable until the bath Al content is reduced to where knee point is reached. As bath Al content is lowered, Fe solubility is increased, raising the capacity of the bath for dissolved Fe. Upon crossing the knee point, slow dissolution of the top dross begins, together with nucleation of the FeZn delta bottom dross phase. The top dross dissolution process is very slow, therefore particles suspended in the bath do not reach equilibrium immediately. It was observed that during this stage, the $Fe_2Al_{(5-x)}Zn_x$ particles are in the process of dissolution, whereas the bottom dross particles are in a nucleation and growth stage. As always, the Fe supersaturation of the bath is the driving force for the precipitation and growth of FeZn particles, but because the Fe solubility in the bath increases with decreasing Al content, it can take many hours for the full transition, with complete dissolution of the $Fe_2Al_{(5-x)}Zn_x$ dross, to occur [28]. For this reason, zinc baths used for dual-purpose GI/GA production run their GI composition as close to the 0.135%Al "knee point" as possible; if a high Al (0.2%–0.3% soluble Al) bath is used, the time needed for transition to a GA bath composition becomes impracticably long.

Transitions between GI and GA bath chemistries can also be accomplished by pumping out most of the galvanizing bath. This requires the usage of two standby pots, one to receive the liquid from the in-line pot and another to hold the new bath composition that will later be pumped into the in-line pot. Channel induction pots cannot be pumped lower than the top of the channels, or else the heating capability of the bath will be lost, together with possible damage to the partially emptied inductors; the inductors are kept on low power to continue providing heat so that solidification does not occur. The composition of the remaining bath "heel" below the inductors level must be considered when preparing the composition of the new metal that will be pumped in to replace the old bath. As stated in Section 4.1, coreless induction furnaces can be pumped empty, as can the older pots heated through their walls by electric grids.

4.8 Entry snout phenomena and control of coating quality

The volume inside the entry snout includes the interface between the protective atmosphere of the pretreatment furnace and the zinc bath, through which the steel strip is passing over a wide range of speeds, typically between 0.5 and 3/s. Along with this interface, the snout walls can also interact with the bath, and as shown in Section 4.5, the snout dimensions influence the nature of flow, temperature, and composition in the portion of the zinc bath in and near the snout. Well-informed operation of the snout area is essential for the production of defect-free coatings.

A sketch of the relevant entry shout phenomena is shown in Fig. 4.36.

The surface of the zinc bath contained by the snout emits zinc vapor, the amount of vapor produced is related to the partial pressure of zinc vapor according to the formula [49]

$$P_{Zn}\ (\text{atm}) = 1.36 \times 10^{-8} * e^{(0.0236 * T)} \tag{4.3}$$

where T is the local zinc temperature in °C, for exposure to atmospheric air. This is for the ideal situation for a quiet surface, uncontaminated by oxides. The surface of the bath that is undisturbed by strip entry and exit, wiping knife effects and ingot entry and skimming operations, both discussed later, will very quickly be covered with a thin surface layer of Al oxide, even for the low Al contents used for GA and GI coatings production. This reduces the vaporization of zinc from these surfaces. However, inside the snout, the bath surface is continuously disturbed by strip entry, resulting in fresh liquid zinc being continuously exposed, from which vaporization occurs. Gas convection effects and the dryness of the furnace gas also affect zinc vaporization and removal of the zinc as it vaporizes from the liquid state [48]. The zinc temperature inside

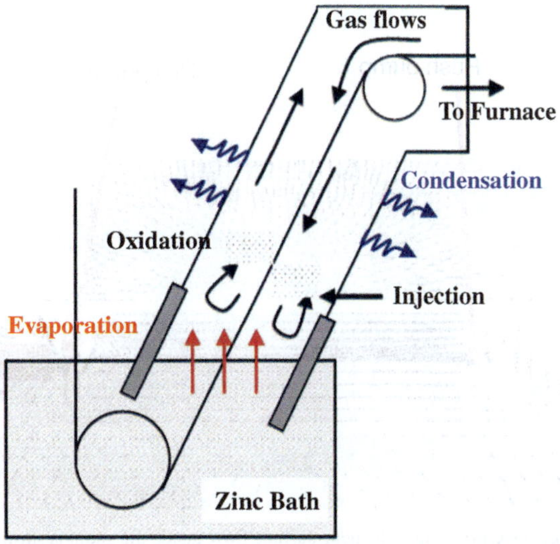

FIG. 4.36 Phenomena in the entry snout including injection of N_2H_2 gas atmosphere, evaporation of Zn from the bath surface, oxidation of both vapor and exposed bath metal, condensation of zinc vapor [48].

the snout may also differ from the average temperature of the bath if the steel strip is entering at a higher or lower temperature than the bath. As explained in Chapter 2, the vaporized zinc can condense in cooler parts of the snout and then deposit onto the strip resulting in coating defects. This effect can be reduced by insulating the snout walls. The use of wet snout practice to control zinc vaporization is described in Chapters 2 and 4. Another major cause of defects is entrainment of floating dross and other contaminants found in the bath at the point of strip entry. These can include oxides of Al and Mg and contaminants related to strip cleanliness as it enters the bath. Control of other defects depends upon control of the flow conditions in the snout, together with local temperature and bath composition.

For the Zn-55Al bath operating at 600°C, the partial pressure of zinc is about 10 times higher, the vapor pressure of Zn in this case is given by

$$\ln P_{Zn} = \left(\frac{-15644}{T}\right) - 1.05\ln(T) - 0.29 \times 10^{-3}T + 20.27 \tag{4.4}$$

where the pressure P is in units of kPa and T is in °K [50]. Zn55Al snout practice also focuses on creating and maintaining an oxide layer on the bath surface, but the injection of wet gas must be done so that it minimally disturbs the bath surface in the snout. The end of the snout for Zn55Al is usually narrower and better insulated than a snout used for GI or GA operations to prevent cold spots from developing. Snout heating may also be helpful. Frequent cleaning of the bath surface inside the snout also reduces the incidence of vapor-rated coating defects. The higher temperature of the Zn-55Al bath makes the use of hardware alloys more robust than the usual 316L stainless grade attractive; for this purpose, alloy 235MA (21Cr-11-Ni-1.6Si-0.17N-0.08Ce-balance Fe) was investigated. Although alloy 235MA provides excellent service in many aggressive gaseous environments at temperatures above 500°C because of formation of a protective chromium oxide scale, it was susceptible to Zn attack when used in an entry snout and did not perform any better than 316L. The best way to minimize degradation of snout interior surfaces was concluded to be minimizing zinc vapor condensation by control of local thermal conditions [50].

Bath flow in the vicinity of the GI-GA snout has been modeled using water analogs and also by computational fluid dynamics [51]. These models show that bath metal and entrained particles on the back side of the snout (on the side of the strip that does not contact the sink roll, more commonly termed the front or bottom side of the strip) exit the snout volume in this back side and do not reenter. However, there is some reentry of bath material on the front side of the snout because of the recirculating nature of flow in the "V region" above the sink roll. Kurobe [51] explains that this is how the snout volume is supplied with bath metal, and this reentry can contribute to coating defects. Flow patterns in the liquid volume contained in the snout have been found to be very sensitive to geometric factors, especially the symmetry of distance from the strip surface to the snout interior wall on the front and back sides [52].

Many remedies have been tried to improve bath metal conditions in the GI-GA snout. The two most common in practice are the use of snout pumps and the design of internal overflow channels to remove the top surface of the bath and with it floating materials such as top dross. [53,54]. An example showing the effect of a "push-pull" pump arrangement is shown in Fig. 4.37.

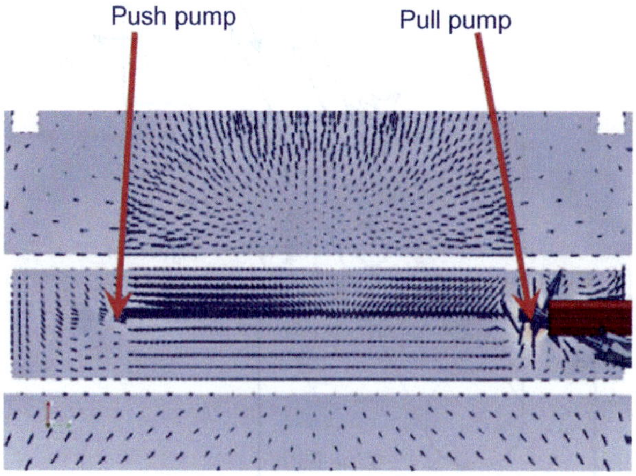

FIG. 4.37 Zinc flow patterns near and inside the entry snout. The near-surface flow inside the snout is shown in the middle section, the 2500 kg/min pull pump on the left side, and the 700 kg/min push pump on the right side. Note also the different bath flow patterns in front of the snout (in the "V region"), top, and behind the snout, bottom [55].

Like the walls of the bath and other items submerged in the bath, such as pot hardware support arms, the snout perimeter at the level of the bath attracts floating dross and oxides that build up. This buildup is driven by the reduction of surface energy, and the buildup is only loosely attached to the snout wall. Sudden changes in bath level can result in detachment of this buildup. Loosened buildup inside the snout is often drawn to the surface of the strip entering the bath, where it can cause coating defects. It is therefore desirable to keep the bath level as constant as possible.

As explained by George [53], comprehensive control of snout conditions to minimize coating defects involves:

- An accurate bath level control system to +/− 2mm.
- As described in Chapter 2, use of a wet H_2-N_2 atmosphere, together with an effective distribution system in the snout.
- Dross pumps, and a weir or similar collection system to collect dross and condensed zinc dust to keep it away from the steel strip. In this reference, a pull-pull pump system was used.
- Measurement devices in the snout for dew point and hydrogen.
- Cameras and lighting for viewing the surface of the bath inside the snout.
- Real-time continual monitoring of bath Al and Fe.

4.9 Galvanizing bath top skimmings and bottom dross removal

4.9.1 The nature of top skimmings

The surface of the zinc bath used for GA and GI coating that is not enclosed by the entry snout is covered by a thin layer of Al oxide that protects the underlying metal from further oxidation, except for those areas of the bath that are disturbed by strip exit, wiping knife effects that will be described later, and ingot entry and skimming operations. In each of these disturbed areas, exposure of fresh bath metal to the atmosphere results in formation of newly oxidized material. As described by Setargew [56], these oxides are then mixed back into the surrounding liquid, so that tangled oxide stringers are contained in a liquid metal matrix. With sufficient oxidation mixing, this mixture can become foamy or sponge-like. Thiounn [57] defined the various related bath contaminants as:

- Dross: intermetallic particles Fe-Zn (bottom dross) or $Fe_2Al_5Zn_{(5-x)}$ (top dross) as described in previous sections of this chapter, formed by supersaturation in Zn-Al-Fe phase relations within the bath. $Fe_2Al_5Zn_{(5-x)}$ dross particles are less dense than the galvanizing bath and consequently move up to the surface, Fe-Zn dross particles are heavier and sink to the bottom of the bath.
- Scum: the mixture of metallic zinc and oxides described above. Most of it is created around the area of strip exit from the bath. It floats on the bath surface.
- Skimmings is the global by-product, which is removed from bath surface during skimming operations. It contains scum, top dross, and liquid bath metal.

There are several means by which scum is formed. As the strip exits the bath, the thin layer of liquid coating adhering to either side is exposed either to the surrounding air or to a mixture of air and nitrogen if nitrogen gas is used for coating control, as described further in Chapter 7. This continuous exposure of fresh bath metal to the surrounding atmosphere results in oxidation of the coating surface, even in this short period of time. As observed by Thiounn [57], the portion of the coating withdrawn from the bath that is wiped off by the gas knives flows back downward into the zinc bath, flowing back down the strip surface in successive waves, this is termed "coating runback." Each wave on the bath surface overlaps the previous one; the oxides forming at the interface between air and zinc are mixed in the liquid metal. These waves overlap in a turbulent manner, resulting in the formation of gas pores, resulting in the foamy or spongy nature of this scum. The inner surfaces of the pores are observed to be coated with oxides. A photo of this flow is shown in Fig. 4.38.

The profile of the coating runback is required to determine the surface area and to determine the momentum and energy of this flow. This is shown in Fig. 4.39 for line speeds of 0.75, 1, and 1.5 m/s for a gas knife jet 0.3 m above the bath surface. Above 0.3 m the coating has been wiped to the required thickness; below 0.3 mm the excess zinc flows downward into the bath. The downward flow increases with strip speed. The turbulent nature of this flow is manifest in the undulating shape of zinc thickness below the gas knives.

The inertia of the coating runback and the wiping gas pushes the scum away from the exiting strip. This inertia carries the surface oxides on the bath surface that it flows into, mixing with it to create more scum. The upper (dashed) line in Fig. 4.40 shows the interface of the exiting steel strip, covered by downward flowing zinc. A semicircular surface of rapidly flowing metal is observed on the bath surrounding the exiting strip. Beyond this semicircle, indicated by the lower (dotted) line, the bath surface is covered with scum and can also be mixed with top dross and bath metal to form skimmings.

FIG. 4.38 Steel strip exiting the galvanizing bath, covered with downward-flowing zinc that has been wiped off by the gas knives [58].

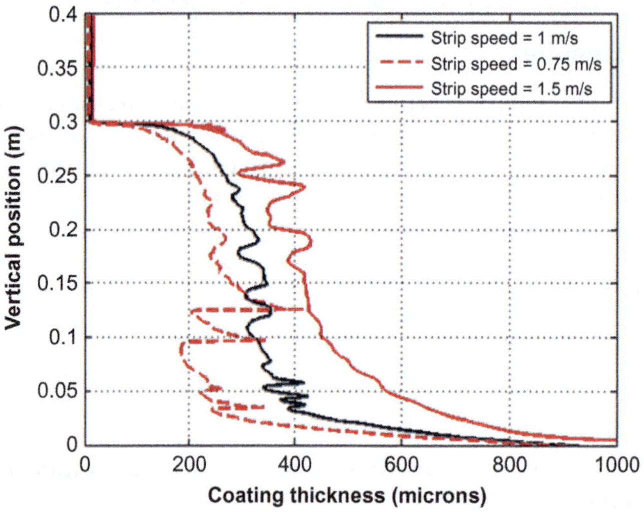

FIG. 4.39 Liquid zinc-free surface profiles for representative line speeds. Gas knife gap = 1 mm, strip-to-knife distance = 6 mm, wiping gas exit velocity from gas knives = 113 m/s (Re = 11,000) [59].

As production continues, the produced scum piles up, decreasing the size of his semicircle. The relationship between strip-to-scum distance and elapsed time was measured for 140 time periods by Strutzenberger [60]. Results for one set of line process variables are shown in Fig. 4.41. It is important to keep the strip-to-scum distance above a certain minimum value that depends upon process variables and line configuration that is specific for each line to minimize the chance of pick-up onto the exiting strip that can cause coating defects.

Removal of the scum by itself is not possible, any method of removing it from the bath surface will also remove top dross (in the case of GI coatings) and, more importantly, bath metal. For GA operation, top dross is not present unless the bath is operating with an Al and Fe content where both top dross and bottom dross are present. This results in a lower amount of skimmings. A relationship for skimmings production that is independent of configuration was derived by Yu [61].

$$\dot{s} = (14.84\,LS - 1.56)\frac{{O_l}^2}{21^2} f_B$$

where

$\dot{s}$ = skimmings generation rate, g/s/m^2 of strip coated
LS = line speed m/s
O_l = oxygen level in wiping gas, 10% for nitrogen, 21% for air
f_B = bath factor, GI bath = 1, GA bath = 0.67

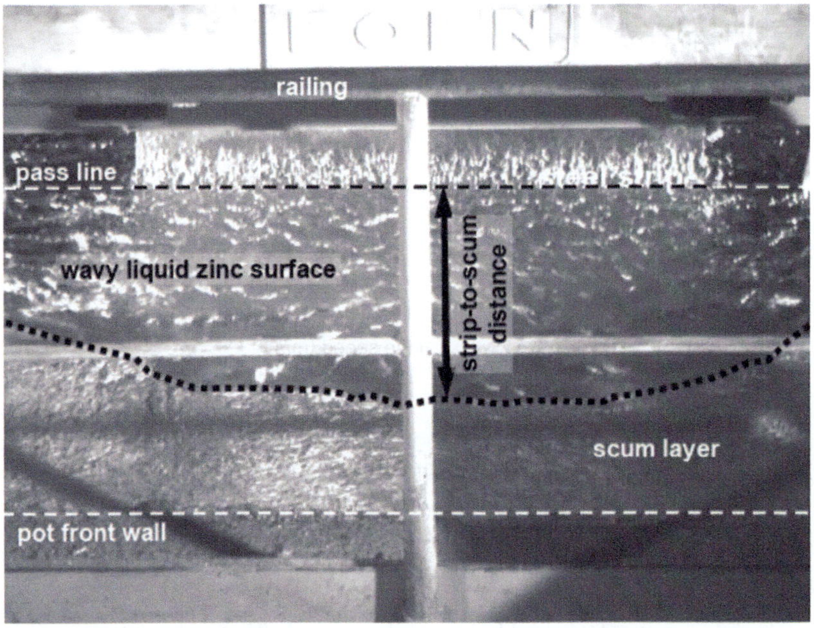

FIG. 4.40 View of the strip exiting the bath, showing coating runback (above dashed line) free bath surface (between dashed and dotted lines) and surface of bath covered with scum (below dotted line) [60].

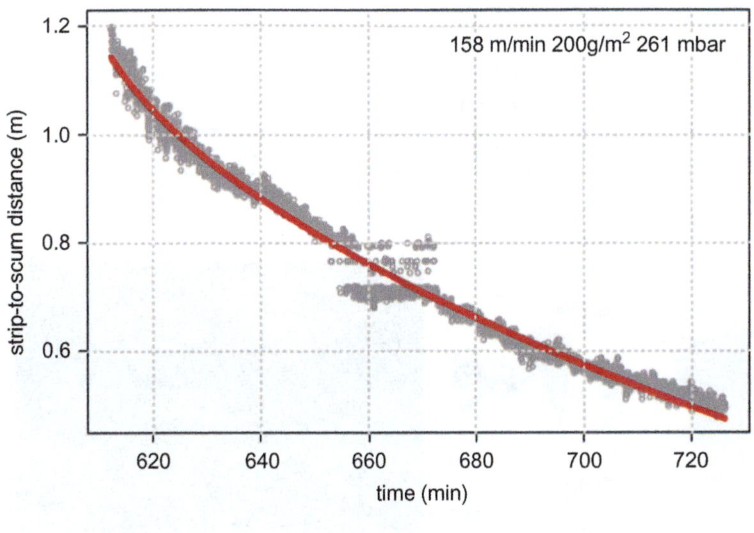

FIG. 4.41 Strip-to-scum distance vs. time for one selected growth period. Line speed = 158 m/min, two-side coating weight 200 g/m^2, wiping pressure 261 mbar [60].

The bath factor reflects the presence of top dross present in the GI bath.

This equation is plotted in Fig. 4.42 for a GI bath using air wiping.

These measurements are all in terms of grams per square meter of coated sheet, this is the only meaningful way of measuring skimmings production [62]. The sensitivity of skimmings formation rate to line operating variables is shown in Table 4.7. The reference conditions for this line are a wiping knife height of 0.3 m above the bath surface, a coil width of 1.3 m, a line speed of 2 m/s, and wiping using regular air, using a GI bath. The rates are not directly comparable to Fig. 4.42 but indicate the effects of the indicated variables. As indicated, the greatest sensitivity is to changes in line speed [29].

A low-magnification image of a skimmings cross section from a GI bath is shown in Fig. 4.43. It is normal for top skimmings to contain more than 90% bath metal. It is important to minimize this portion to minimize loss of zinc from the bath during skimming operations. The oxygen image in this figure indicates the location of oxides, the polygonal particles are rich in Al and Fe, indicating that they are top dross. The non-polygonal dark images in the zinc map are associated with porosity.

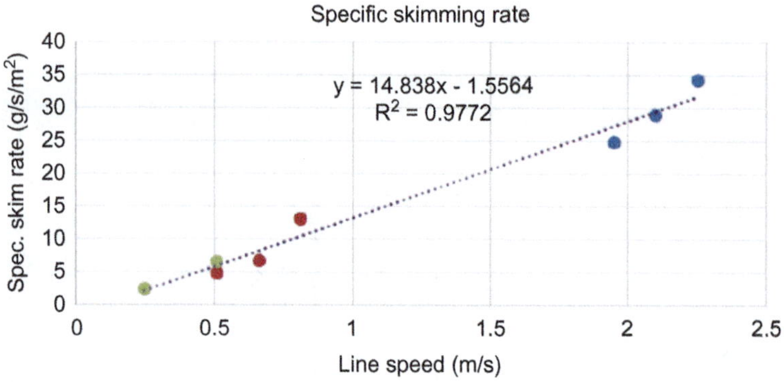

FIG. 4.42 Relationship between skimmings rate and line speed for a GI bath utilizing air wiping [61].

TABLE 4.7 Sensitivity of top skimmings formation rate to CGL operating variables [29].

Configuration	Skimming Rate (kg/min)	
	Air	N_2 (mixed)
Reference case	21.93	7.88
Lower air knife at 0.2 m	14.62	5.25
Higher air knife at 0.4 m	29.24	10.51
Slower line speed at 1 m/s	10.36	3.72
Faster line speed at 3 m/s	33.51	12.04
Narrower coil width at 1.2 m	20.25	7.27

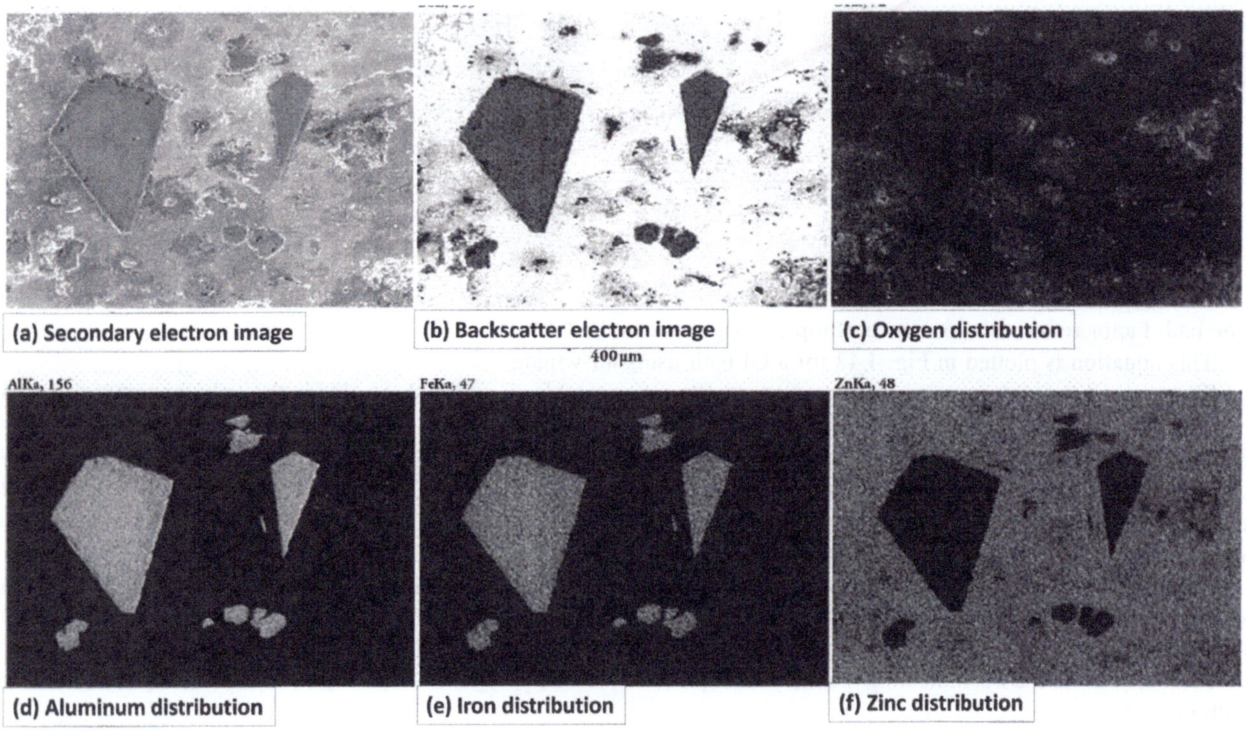

FIG. 4.43 Secondary electron and electron backscatter images together with X-ray mapping of a typical top skimmings sample from a GI bath [58].

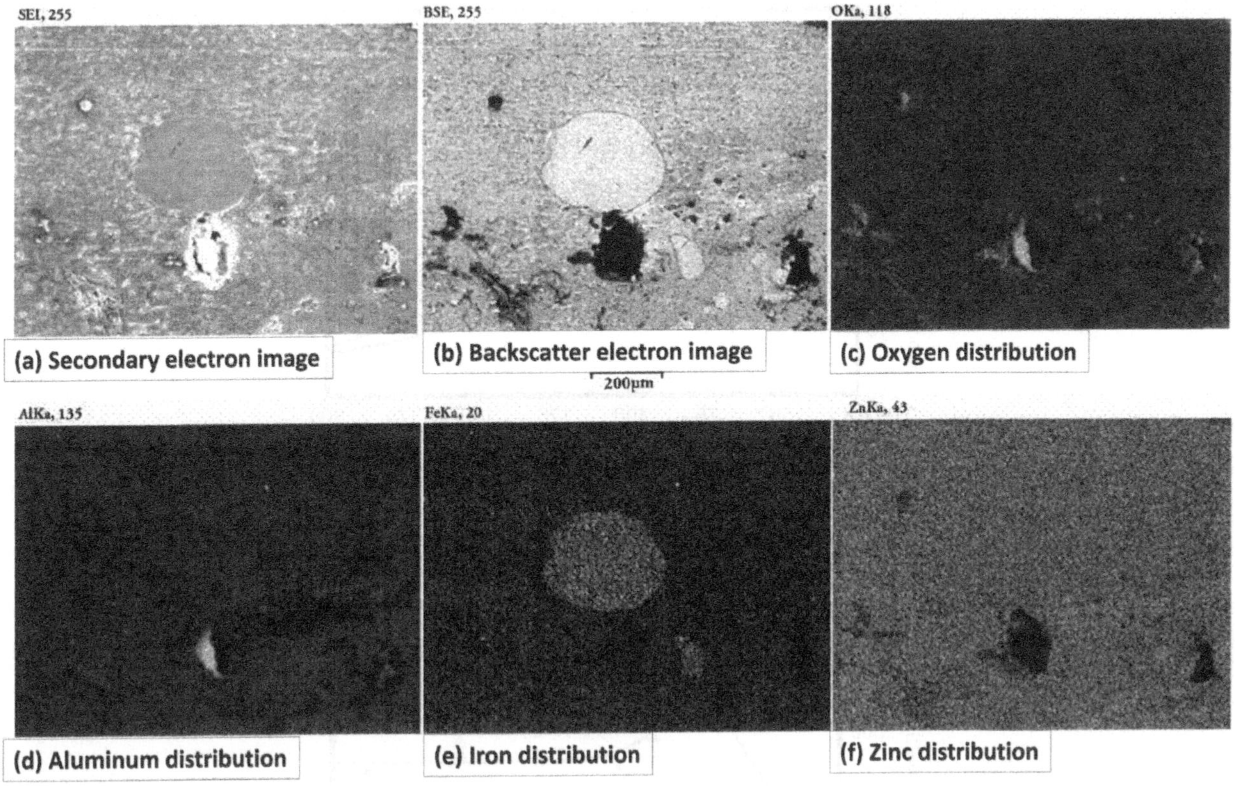

FIG. 4.44 Secondary electron and electron backscatter images together with X-ray mapping of a typical top skimmings sample from a GA bath [58].

A similar set of images for a top skimmings sample taken from a GA bath is shown in Fig. 4.44. Here, the only Al-rich area is associated with an oxide particle, the dross intermetallic particles are rich in Fe and Zn. As in the GI skimmings sample, much of the mass is bath metal.

4.9.2 Top skimmings removal

Dross removal techniques are either manual, using a rake-like or scoop-shaped tool, or robotic. The advantage of robotic skimming is the greatly reduced hazard of working over a bath of molten zinc and the associated moving machinery. Using the slowest possible traverse rate while skimming minimizes the amount of bath metal picked up with the top skimmings, but this prolongs worker exposure to the bath environment during manual skimming; robotic skimmers are able to traverse slowly without discomfort. When the skimmings are removed from the bath, they are customarily placed in a receptacle that provides for drainage of liquid metal from the bottom, so that it can be recovered for use in the bath. This can reduce the bath metal content in skimmings from 95% down to 85%.

The ability of bath metal to separate from collected skimmings was shown by Setargew [56], who conducted top skimmings measurements for both GI and a bath to which a low, unspecified amount of Mg had been added. Because of the more highly oxidizing nature of Mg, more top skimmings were produced. Fig. 4.45 shows the top skimmings rate, expressed as grams of top skimmings per square meter of coated steel strip, based on work with a laboratory apparatus. The GI bath (Bath 1) produced less top skimmings than the Mg-containing Bath 2.

The drainage or seepage rate from the collected top skimmings of the GI bath is shown in Fig. 4.46. In three experiments, initial quantities of 400, 650, and 1000 g of skimmings were collected, held at normal bath temperature, and the liquid metal was allowed to drain. It was observed that drainage was opposed by surface tension forces that hold the entrained metal entangled with the oxide. The drainage rate is observed to decrease with time, reaching a limit that depended upon the initial mass collected, after which no further drainage occurred, the remaining liquid was retained in the remaining scum morphology.

Typical rates of top skimmings produced on industrial lines running at 1.5 m/min are 72 kg/h for GA production and 102 kg/h for GA production [63]. This is many times higher than the "ideal" top and bottom dross production rates shown in Table 4.4, which do not include entrained bath metal and scum formation, nor departures in bath flow, chemistry, and temperature patterns from the data used for Table 4.4.

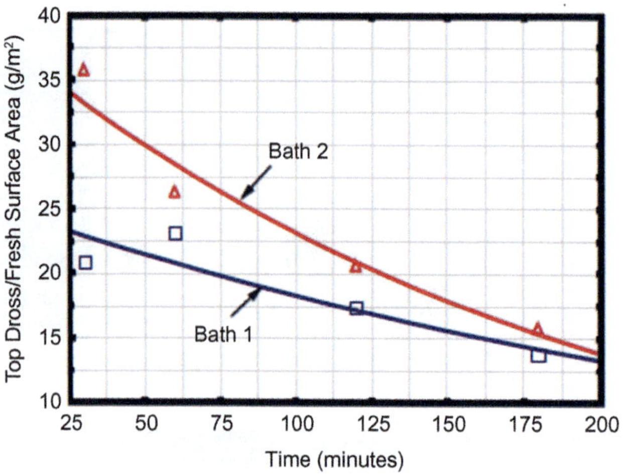

FIG. 4.45 Top skimmings rate, per square meter of steel coated, from a laboratory apparatus. The term "top dross" here refers to skimmings [56].

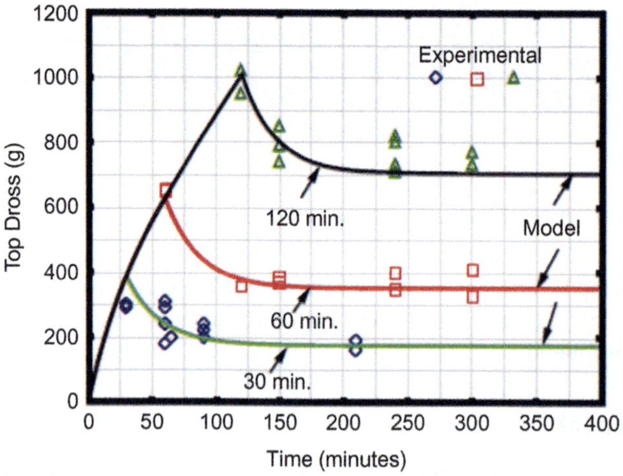

FIG. 4.46 Effect of drainage/seepage time on mass of collected top skimmings from a GI bath [56].

4.9.3 Bottom dross removal

Lines running extended GA campaigns will accumulate bottom dross simply because of the constant addition to iron to the bath coming from the steel strip being galvanized. It is normal practice to stop production for removal of bottom dross after about 20 days of production. There are several ways to remove bottom dross from a zinc bath, including excavation equipment and clamshell scoops, but these generally require removal of pot hardware, including the entry snout, to obtain sufficient working space. The most commonly used way of removing bottom dross is with liquid metal pumps. The pump discharge is fed into an ingot basket, such as illustrated in Fig. 4.47. The basket is kept at the 460°C bath temperature and liquid from the mixture of ZnFe bottom dross and bath metal allowed to drain.

X-ray analysis was then carried out using the scanning electron microscope used to examine samples. The compositions found were to be

- pump exit stream (0.13%Al, 0.06%Fe),
- basket drippings (0.12%Al, 0.03%Fe),
- semi-solid in basket (0.50%Al, 1.23%Fe),
- top of basket, drained (1.25%Al, 4.32%Fe)

The pump exit stream is a mixture of bath metal and bottom dross. The basket drippings can be seen from Fig. 4.8 phase diagram to be a saturated liquid solution of Al and Fe in Zn at 460°C, this includes a small quantity of FeZn intermetallics

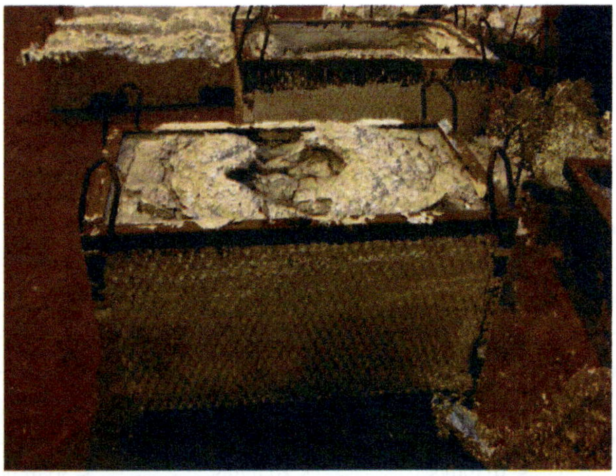

FIG. 4.47 Ingot basket used for bottom dross removal [64].

that can return to the zinc bath. The semisolid remaining in the basket is a mixture of bath metal with a higher amount of dross, while the top of the basket, with the lowest amount of bath metal, is a highly porous mixture of bottom dross and Al that oxidized during the draining period.

The effect of bath flow details on bottom dross accumulation in Zn55Al baths was modeled by Willis et al. [65] for two baths, large (27.4 m^3 or 82 ton) and small (13.9 m^3 or 46 ton). The larger bath had a curved floor, whereas the smaller bath has a flat floor. Moreover, the inductors in the small bath placed and directed so that there was more direct interaction with the strip. The larger bath has a bulkier pot roll arrangement located lower in the bath that interfered more with bath flow than the smaller bath rig. The difference between strip entry temperature and bath temperature is larger with Zn55Al than GI-GA production; strip entry temperature is between 520 and 580°C and bath temperature of 590–610°C. Lower strip entry temperatures are associated with smaller spangle sizes that are preferred by many customers. Because of this large difference, inductor power is significant and contributes more to bath flow patterns than in GI-GA baths. Also in contrast with GI-GA baths, most Zn55Al pots have four inductors, one each per side and on the front and back, whereas GA-GI pots usually have two inductors, one each per side. Therefore, a greater influence of inductor placement on bath flow patterns is expected with Zn-55Al. A diagram of the two pots that were compared is shown in Fig. 4.48.

A noticeable larger volume of bottom dross accumulation was noticed over time for the larger bath than the smaller one. This was attributed to the following:

- The flow from the inductors in the smaller bath causes much more stirring in the bath compared with the larger bath. The small bath inductors were also a jet flow design, contributing to greater influence on flow patterns. This flow interacted more directly with the steel strip. Exit velocities from the jet flow inductors were in the range of 0.72–1.140 m/s and 0.24–0.5 m/s for the larger pot.
- The lower position of the stabilizer roll and its greater distance from the pot wall resulted in a volume of recirculation about this roll in the larger pot. This was not observed in the smaller bath.

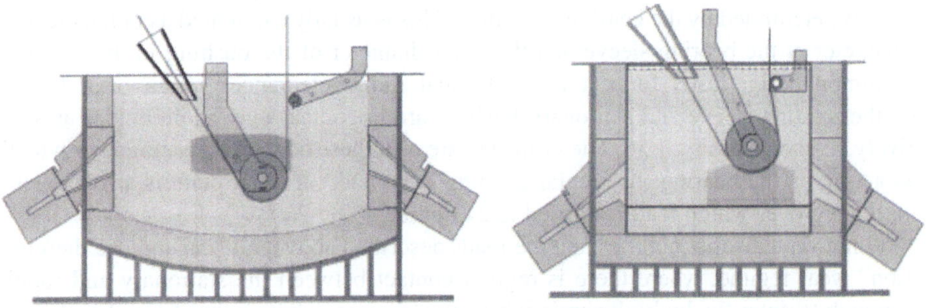

FIG. 4.48 Comparison of the 82 ton Zn55Al galvanizing pot arrangement (left) with the 46 ton pot. Note the difference in pot hardware placement and height of inductors on the front and back walls. Side inductors are not shown [65].

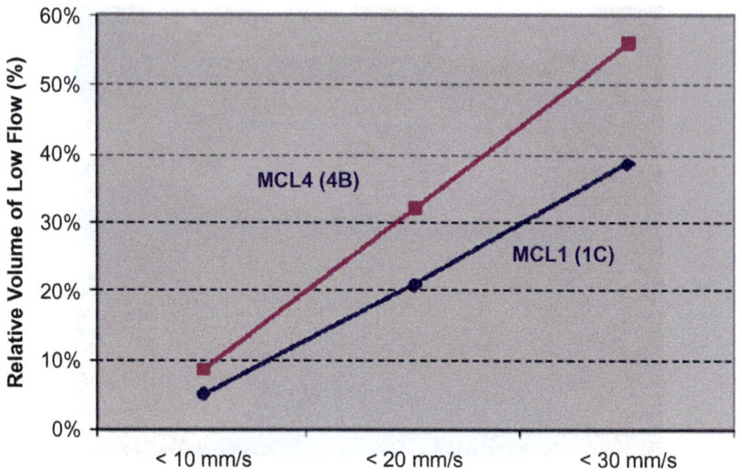

FIG. 4.49 Relative percentages of low flow for three different flow velocities for the 82 ton pot (MCL4) and the 46 ton pot (MCL1) producing Zn-55Al coatings [65].

- The volume fraction (and total volume) of the pot affected by low velocity is greater in the larger bath than the smaller one, as shown in Fig. 4.49. As a consequence, the temperature was much more uniform in the small bath than the larger one. Of particular note is a prominent hot zone behind the snout in the larger bath.

All these factors contributed to the much lower dross production rate in the smaller coating bath.

4.10 The metallurgy of galvanizing bath hardware

The engineering of submerged moving hardware used in the galvanizing bath encompasses physics, mechanical engineering, and other disciplines. This section is limited to a description of metallurgical phenomena encountered with bath hardware.

4.10.1 Pot roll bearings

Fig. 4.1 and associated text introduced the arrangement of the sink roll, corrector roll, and stabilizer roll in the bath. The strip passing around these rolls is subjected to a line tension that controls strip position and movement through the CGL. Line tension varies from one section of the line to another, with lowest tensions used in the higher temperature (annealing) sections of the line where deformation of the strip must be avoided, intermediate tensions though the coating section to provide sufficient turning torque to the submerged rolls to avoid skidding together with controlling vibration, and highest tensions in the temper mill and tension leveling sections. The highest tensions through the coating section are typically used when producing automotive products, a representative line tension is 66.7 kN (15,000 lb_f.) This must be handled by the two support bearings, one on each end of the sink roll. The resolved force acting on the bearings is related to the entry angle of strip into the bath and the position of the corrector roll. For a typical roll configuration such as that shown in Fig. 4.1 and an entry angle of 33 degrees, the resolved force on each bearing is 65.3 kN (14,674 lb_f.). The contact area of each bearing that is acted upon by this load is determined by the bearing clearance. This is usually calculated as radial clearance, the difference between the outer diameter of the bearing sleeve and the inner diameter of the bushing in the case of a journal bearing design, the most common design used for pot rolls. Typical radial clearances range from 0.64 to 4.6 mm. After 0.76 mm of wear-in, the rotating sleeve and stationary bushing are in contact over semicircular arcs of 125 degrees and 76 degrees, respectively, as shown in Fig. 4.50. The projected areas of these contact areas, together with the imposed loads, determine the stress acting on the bearing. A circular bushing in contact on a flat plate is also shown in Fig. 4.50; after 0.76 mm of wear-in, the arc over which load is carried is 24 degrees.

At the speeds of all hot-dip galvanizing lines and the loads described above, all bearings are therefore operating in the "boundary lubrication" wear regime, where there is regular contact between the stationary and rotating surfaces. Line speeds are not fast enough to enter the hydrodynamic lubrication regime, where a thin film of bath metal could be maintained in the bearing clearance. Roller and ball-bearing designs have been tried, until now without widespread success [67].

The most common metallic material used for bushings and sleeves is Stellite 6, a cobalt-based superalloy. This alloy derives its strength from carbide strengthening. Triballoy cobalt-based superalloys are also used; these are strengthened by Laves phases. Compositions of these alloys are shown in Table 4.8.

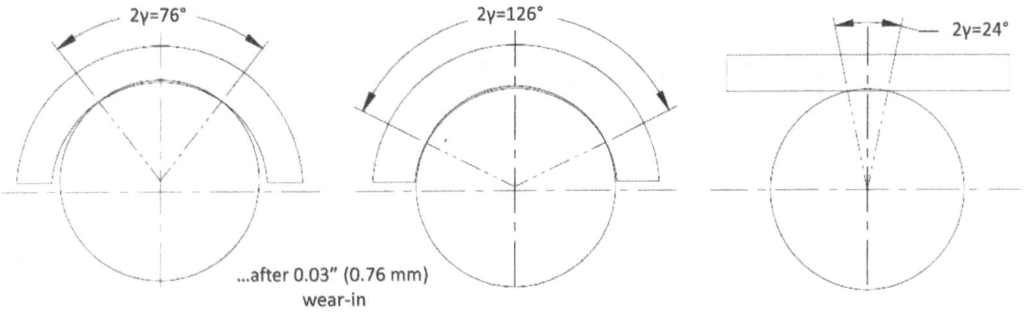

FIG. 4.50 Effect of radial clearance and initial wear on journal bearing contact area after 0.3 mm of running wear-in. Left: 4.6 mm initial radial clearance, Center: 0.64 mm, Right: zero (flat plate) [66].

TABLE 4.8 Compositions and room temperature Rockwell C hardnesses of Co-based superalloys used in pot roll-bearing applications [68].

Alloy	Co	Mo	W	Cr	Si	C	HRC
Tribaloy T-800	Balance	28.5	–	17.5	3.4	<0.1	58
Tribaloy T-400	Balance	28.5	–	8.5	2.6	<0.1	51
Tribaloy T-401	Balance	22.0	–	16.0	1.2	<0.3	48
Stellite 6	Balance	–	4.5	29.0	1.4	1.2	40

The yield stress of Stellite 6 at 460°C has been reported to be 621 MPa [69]. For perfectly smooth wear surfaces with a 0.5 mm radial clearance, a maximum stress of 346 MPa was computed. For a radial clearance of 10 mm, the maximum stress increases to 565 MPa. If the contacting surfaces are roughened so that in the unloaded condition only 10% of the surface is in contact, the maximum stress increases to 651 MPa for 1% radial clearance and 650 MPa for 10 mm radial clearance. [69]. Therefore, a small amount of plastic deformation can be expected under bearing loading conditions. A far greater influence on bearing performance originates from the attack of the liquid metal in the galvanizing bath on Stellite 6. The effective Al in the zinc bath has been observed to react with the Co matrix of the typically cast structure of Stellite 6, forming a cobalt aluminide intermetallic that is much harder and brittle than Stellite 6. This attack by Al has been observed in galvanizing baths with as low as 0.015% at 460°C [70]. An example of this reaction is shown in Fig. 4.51. for a surface that had experienced about 9000 m of sliding distance, equivalent to exposure of about 3 h. It was found that these hard particles could detach and damage the wear surfaces of the bearing. Dross from the galvanizing bath was also expected to contribute to this mode of wear [71]. Improvements in in performance can be achieved by milling grooves into the bearing surface. The hard

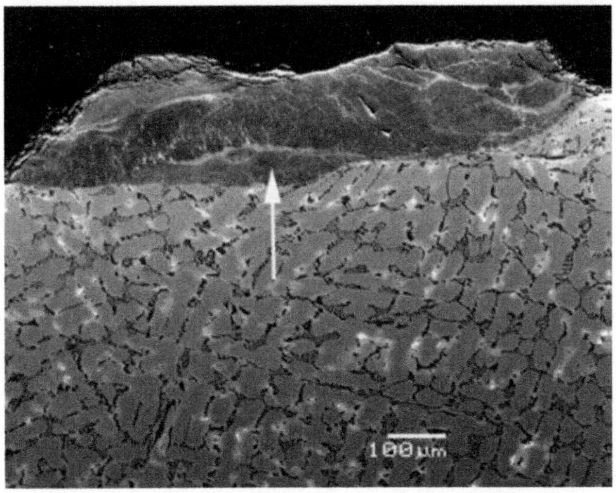

FIG. 4.51 Formation of cobalt aluminide (arrow) on as-cast microstructure of Stellite 6 after exposure to a zinc bath during bearing service [70].

FIG. 4.52 Use of chevron grooves to sequester wear particles in a Stellite 6 journal bearing set in zinc bath: stationary bearing sleeve shown [72].

intermetallics are found to collect in these grooves, rather than staying on the bearing surface, so that the rate and severity of wear are reduced. An example of a chevron groove milled into a Stellite 6 bushing surface is shown in Fig. 4.52. The wear particles are found to be contained in the liquid zinc that solidified after the bearing set was removed from the bath. The scoring lines seen on the surface of the bushing are typical of the observed abrasive wear. This wear is usually the rate-limiting factor for this bearing material and design, rather than overloading or general corrosion of Stellite 6 of the bearing or fatigue that would otherwise be associated with the high stresses associated with high line tensions. The corrosion rates of Stellite 6 and other alloys listed in Table 4.8 are shown in Fig. 4.53. The corroded layer thicknesses of 25–43 μm after 7 days of immersion are much less than the millimeter-scale mechanical wear observed in Fig. 4.52.

A variety of bearing materials and designs have been implemented to extend the life of pot roll bearings, which traditionally in a GI and GA bath was about 2 weeks. A Co-28Cr-4W-1.1C cast alloy, when heat treated for 2 h at 850°C, produced a Vickers hardness greater than 550, higher than the usual 400–490 hardness range of Stellite 6 of similar composition [73]. Other approaches include the coating of metallic substrates, usually 316L stainless steel, with a thermal sprayed or laser clad cermet WC-Co coating on the rotating sleeve and the use of ceramic materials, notably SiAlON, as the stationary bushing. The stationary bushing can take the form of two or three ceramic blocks placed to support the rotating sleeve, or a single V-shaped ceramic bushing as shown in Fig. 4.54, which has been found to greatly extend bearing life. This bushing, when used with a rotating Stellite sleeve, provides a service life in GA or GI baths of 4–7 weeks, and after changing the rotating Stellite sleeve, it can be placed back into service and used two or three more times. The Stellite sleeve also has a service life of 4–7 weeks but needs to be changed each time. In a Zn-55Al bath, the life is 10–14 days; the ceramic bushing can be used two or three times and the Stellite sleeve is replaced each time [75].

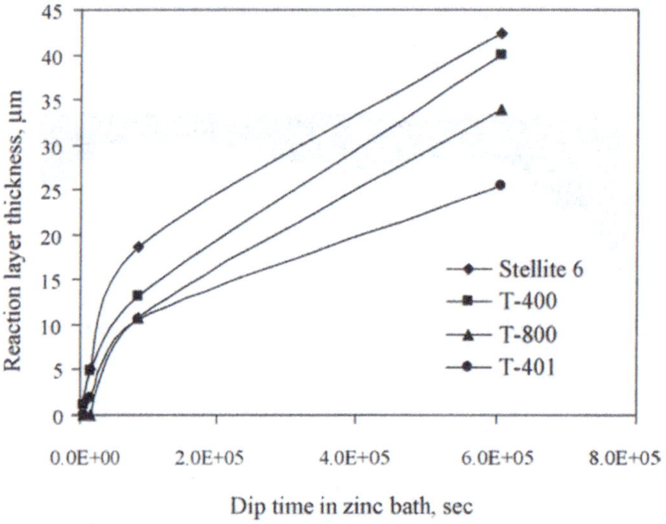

FIG. 4.53 Thickness of reactive layers on alloys listed in Table 4.7 in a 0.22% Fe-saturated Zn bath at 470°C for times up to 6 days. If the reaction layer is not disturbed by mechanical wear, parabolic reaction kinetics are observed [68].

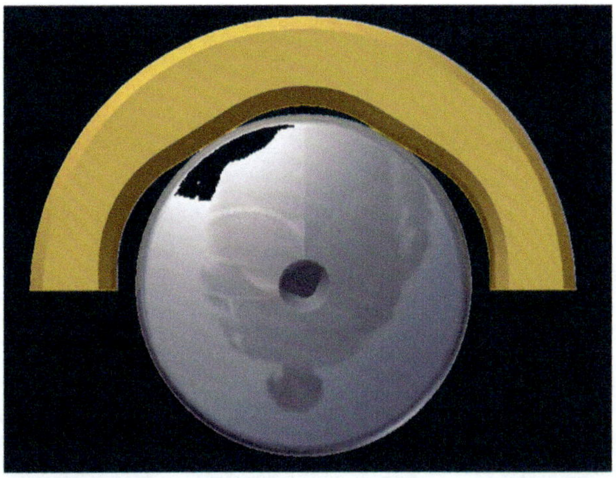

FIG. 4.54 V-shaped ceramic pot roll bushing of Duma Bandzink [74].

4.10.2 Pot roll buildup

4.10.2.1 GI-GA

The rotating surfaces of the pot rolls and their stationary support arms, together with other articles immersed in the galvanizing bath, are subject to attack by bath constituents that limits their service life and coating quality. Screening work by Brunnock on a wide variety of martensitic and austenitic stainless steels, both uncoated and protected with a variety of available coatings, found that all were attacked by a 0.135% Al bath saturated with Fe at 455 and 480°C. 316L stainless, both uncoated, performed well [76]. 316L test samples were attached to both stationary and rotating bath hardware of an industrial CGL by Zhang [77] using a bath at 460°C and an effective Al content that varied between 0.17% and 0.22%. The static samples were removed after various durations up to 336h, while the samples attached to rotating hardware were removed after the production campaign of 504h was completed. A two-component reaction layer was observed. The layer closest to the stainless steel was a compound similar to $Fe_2Al_5Zn_{(5-x)}$ with additions of Cr and Mo. This reaction layer grew continuously into the 316L substrate. In the stationary samples, this growth left behind a Zn-enriched boundary that further reacted to form $Fe_2Al_5Zn_{(5-x)}$ and Fe-Zn delta phase. Dross particles from the bath then attached to this layer, both tightly and loosely connected. Further growth of the reaction layer depends upon the Al supply through the built-up layer. The growth of the reaction layer was parabolic in nature reaching 40 μm after 336h, Fig. 4.55. This behavior is also descriptive of the growth of buildup on the outer surface of the entry snout.

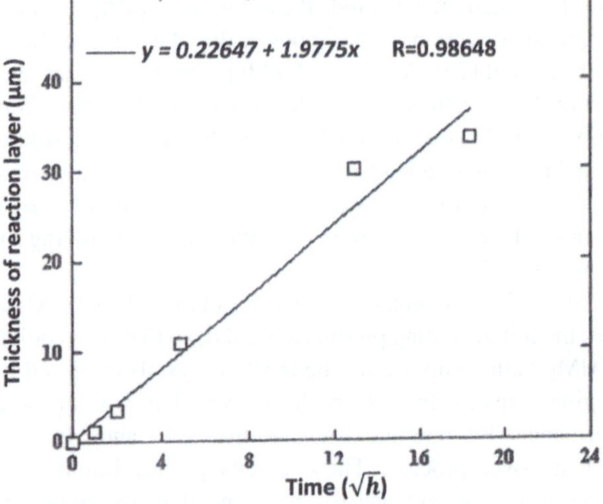

FIG. 4.55 Thickness of reaction layer on 316L stainless steel on stationary surfaces in a galvanizing bath with 0.17%–0.22% effective Al at 460°C [77].

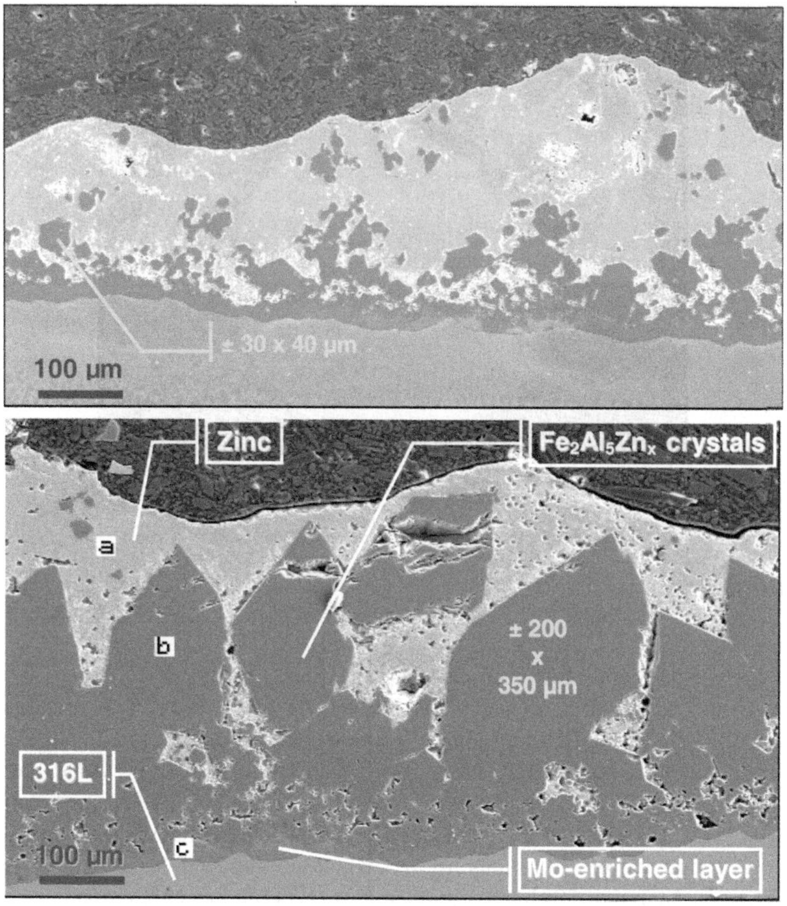

FIG. 4.56 Cross section of built-up layer on stationary (top) and moving (bottom) pot hardware surfaces during GI operation conditions [78].

By contrast, the growth of reaction layers on moving pot hardware surface is much more massive that that on stationary surfaces. Figs. 4.56 and 4.57 show the built-up layer on stationary and moving pot hardware surfaces exposed during typical GI operating conditions, respectively. In both cases, the crystal sizes and layer thickness are 400–600 μm compared with a maximum of 50 μm for the stationary surfaces. In the GA bath, occasional ZnFe zeta-delta outbursts are observed on the stationary samples, as shown in the top picture of Fig. 4.57. Also, the Zn content of the $Fe_2Al_5Zn_{(5-x)}$ reaction layer is 19%–22% in the GI bath vs. 15%–17% for the GA bath [78]. The mechanism for this rapid growth buildup on moving hardware surfaces is explained by two phenomena. First, these $Fe_2Al_{(5-x)}Zn_x$ crystals grow faster when they are in a moving liquid; the crystals grow into the direction of the imposed flow because of the reduced thickness of the diffusion boundary on the upstream side of the crystal [79]. Second, as buildup commences, the surface roughness of the moving pot roll surface increases, increasing turbulence in the boundary layer and therefore mass transfer. Flow is induced downward into the valleys of the rough surfaces. This downward flow has been shown to bring solute into these valleys and reduce the thickness of the local diffusion boundary. The roughess effect, and competitive growth resulting in larger crystals, becomes more important as buildup proceeds. [80]. Larger dross particle sizes, caused either by monotomic growth or agglomeration, have a higher probability than small particles to come in contact with the moving roll surface [81], and the adherence of these particles can also contribute to roll buildup.

Dross buildup on stationary 316L surfaces in a bath that produces both GI (0.2%Al) and Zn-1.6%Al-1.6%Mg coatings has been investigated. For this line, the initial coating produced was GI, so that dross buildup initially occurred as described above. Upon changing to the ZnAlMg bath composition, the faceted crystals observed in the GI bath transform to a more rounded crystals that crack and begin to separate from the built-up layer. This was attributed to the lower Fe solubility in the ZnAlMg bath but may also be influenced by the mutual insolubility of Fe and Mg. Cracks may also form as a result of thermal shock during the pot changeover process. These cracks permit liquid coating alloy to penetrate into the built-up layer. Fig. 4.58 shows the observed sequence for hardware that was exposed to two alternating sequences of GI followed by ZnAlMg production [82].

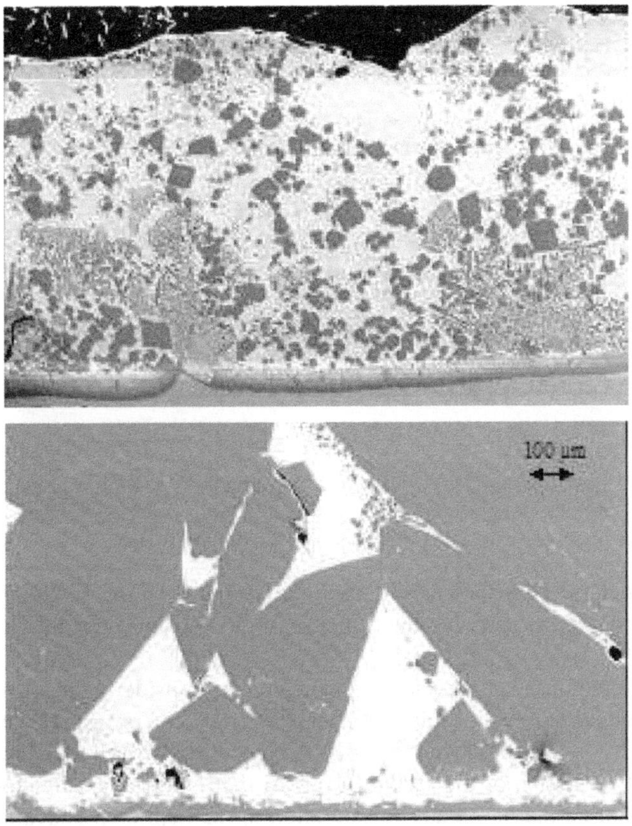

FIG. 4.57 Cross section of built-up layer on stationary (top) and moving (bottom) pot hardware surfaces during GA operation conditions [78].

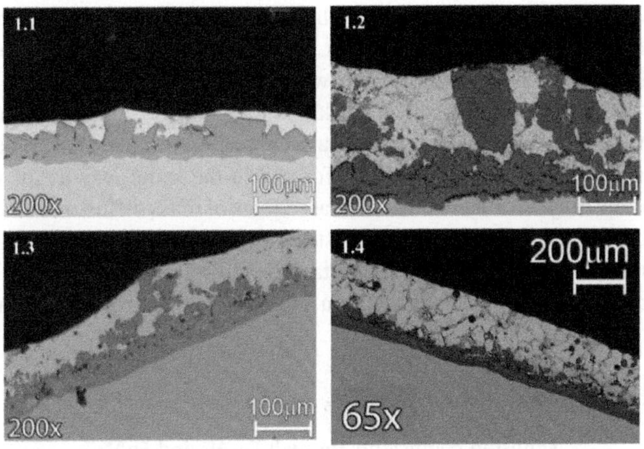

FIG. 4.58 Built-up layer formation on 316L stainless steel after (1.1) 1 week of GI production, (1.2) 1 week of ZnAlMg production immediately following (1.1), (1.3) 1 week of GI production immediately following (1.2), (1.4) 1 week of ZnAlMg production immediately following (1.3) [82].

4.10.3 Zn-55Al

Although the Zn-55Al galvanizing bath has much higher Al content together with an addition of 1.5%Si and an operating range of 550–600°C, there are similarities in the nature of attack of this bath on 316L stainless, which is, as in the cases of GI and GI baths, the principal material used for immersed bath hardware. As in the case of GI and GA, a two-layer buildup is formed, with the layer next to the 316L being in this case a compound based on the $FeAl_2$ phase, approximated by the formula (Fe, Cr, Ni)Al_2 with the composition (wt%): 46.4 Al, 8.3 Cr, 33.2 Fe, 5.1 Ni, and 5.2 Zn [83]. This is covered

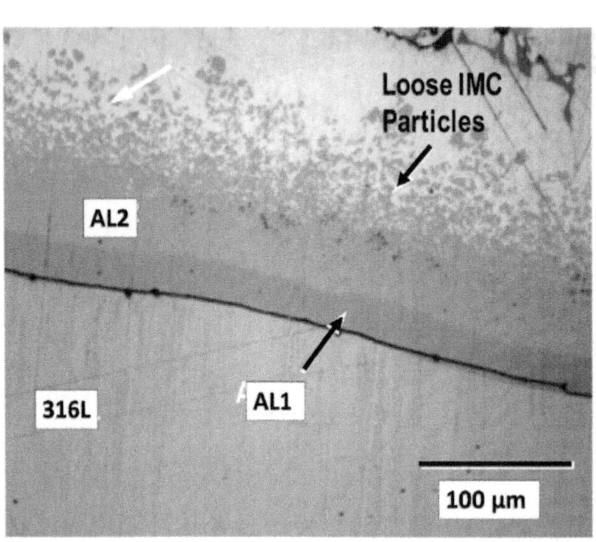

FIG. 4.59 Two-layer buildup structure on 316L stainless after 24 h immersion at 650°C AL1 is the (Fe, Cr, Ni)Al_2 phase and AL2 is the τ_{5c} phase [83].

by a layer base on the τ_{5c} intermetallic phase and has the approximate composition 57.2%Al, 7.2%Si, 6.7%Cr, 25.3 Fe, and 3.6 Zn. This τ_{5c} phase is identical to dross particles that form in a Zn-55Al bath, and in many cases is observed in a matrix of bath metal, such as in Fig. 4.59.

The role of Al in the formation of these phases can be appreciated by considering the diffusion coefficients of Fe and Al in the FeAl intermetallic compounds closest in nature to the two observed layers; as indicated in Table 4.9, Al is faster than Fe in both of these. Fe, Cr, and Ni on the surface of the 316L stainless initially react with liquid Al to form an initial (Fe,Cr, Ni)Al_2 layer. After subsequent formation of the second layer, Al that is left over after forming the τ_{5c} second layer by combining with local Zn, Si, and Fe that is usually in saturation in the bath then is available for continuing the growth of the (Fe,Cr,Ni)Al_2. The (Fe,Cr,Ni)Al_2 continues to grow by diffusion of the more rapid Al through this layer to its interface with the 316L substrate. It is not converted from the τ_{5c} outer layer because there is no evidence of Si in this intermetallic compound.

The rate of growth of the built-up layer in the Zn-55Al bath under stationary immersion conditions is shown in Fig. 4.60. This is thicker than the counterpart layers observed on 316L in GI and GA baths, for example, Fig. 4.55.

As in the case of GI and GA, the Zn-55Al built-up layer was much thicker for immersed samples that were attached to moving bath hardware. Liu [84] reported a built-up layer thickness of 9 mm for samples immersed for 1 week. The outer τ_{5c} layer was 1 mm thick; the inner layer on the 316L sample has a composition of 38%Al, 42%Fe, 1%Si, 5%Zn, 9%Cr, and 5% Ni. The presence of Ni and Cr indicates that this layer was formed in the same way as the counterpart layer in stationary samples, the lower Al and higher Fe may be related to the increased rates of mass diffusion in these moving immersed samples.

The surface of the pot rolls that are in contact with the steel strip during the galvanizing process has a much smaller amount of surface buildup than those surfaces that do not contact the strip. It is suspected, but has yet to be proven, that only the first layer of the two-layer buildup structure is present on roll surfaces in strip contact and that this first layer is much

TABLE 4.9 Diffusion coefficient of Fe and Al through Fe_2Al_5 and $FeAl_2$ at 600–650○C.

	Diffusion coefficients through Fe_2Al_5 and $FeAl_2$			
	D (Al), 10^{-11} cm^2/s		D (Fe), 10^{-11} cm^2/s	
T°C	Fe_2Al_5	$FeAl_2$	Fe_2Al_5	$FeAl_2$
600	7.97	8.4	1.95	0.103
615	10.2	11	2.64	0.144
650	17.7	19.5	5.2	0.305

Adapted from N. Setargew, W.Y.D. Yuen, R. Chen, The reactivity of 316L stainless steel in 55%Al-Zn bath, Rev. Métall. 110 (2013) 275–288. Originally derived by N. Setargew, W.Y.D. Yuen, R. Chen, The reactivity of 316L stainless steel in 55%Al-Zn bath, Rev. Métall. 110 (2013) 275–288 from data of R.W. Richards, R.D. Jones, P.D. Clements, H. Clarke, Metallurgy of continuous hot dip aluminizing, Int. Mater. Rev. 39 (5) (1994) 191–212.

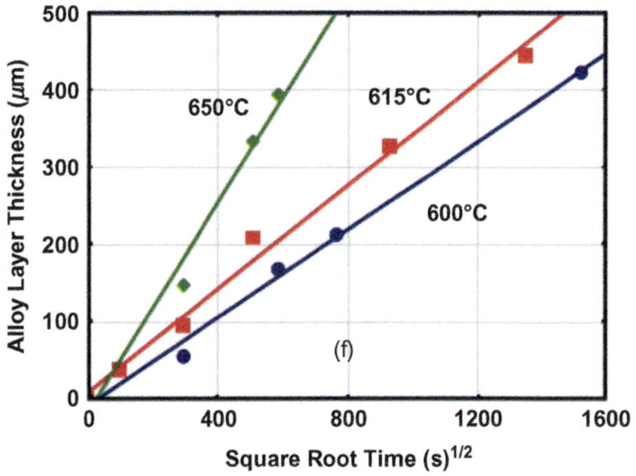

FIG. 4.60 Thickness of the total built-up layer on 316L stainless immersed in a Zn-55Al bath at 600–650°C for approximately 30 days [83].

thinner in these locations. The outer layer of buildup, which consists of a mixture of bath metal and intermetallics similar to the expected dross in the bath, is expected to be absent. The presence of the thicker and less adherent built-up layer on non-contacting roll surfaces makes it difficult to run wider strip after several tons of narrower strip have been coated, the edges of the wider strip contact the thicker buildup, resulting in adherence of the second layer to the strip, causing coating defects. For this reason, coils are generally scheduled from wider the narrower widths during a production campaign, so that the strip need not come into contact with the thicker built-up layer.

The interaction of the Zn-55Al bath with the entry snout, in areas of submerged contact, produces the same built-up layers as described before. Above the area of liquid metal contact, reaction of the 316L stainless snout material and zinc vapor can occur because of the high vapor pressure of zinc in the snout interior. This results in formation of two reaction products: NiZn with a composition of 44.9% Ni-49.8% Zn and $NiZn_3$ containing 20.8% Ni-75.7%Zn. These deplete Ni from the underlying stainless steel and result in the formation of alpha ferrite that is more susceptible to corrosion and the hard, brittle sigma phase $(Fe,Ni)_x(Cr,Mo)_y$ [85].

4.10.4 Coatings for roll surfaces

Pot hardware rolls coated with WC-Co cermets, usually with proprietary sealers, have been found to greatly reduce dross attachment and surface buildup. In one plant the coated sink rolls are used three or four times, while the bearings are replaced each time. Therefore, the roll needs less machining and can provide a longer life [86]. The lower rate of buildup of these coated rolls can be partly explained by the high work of adhesion between Fe_2Al_5 and WC, ranging between 3.62 and 5.55 J/m^2 depending upon the details of the attaching interfaces of these highly directional crystals [87]. By contrast, the work of adhesion between Zn and Fe_2Al_5 is 2.1 J/m^2 [88]. Other factors such as the increased hardness of these coatings and their resistance to pitting that influences coating quality are also important. As indicated before, WC-Co cermet coatings also provide improved performance to the journal bearings used to support pot rolls.

4.10.5 Role of pot roll grooves

It is customary for the sink roll to have grooves cut into it. Especially at higher line speeds, the purpose of these grooves is the same as for treads on an automobile tire, they prevent roll slippage that would be caused by an excessive layer of zinc between the moving strip and the turning roll. Three types of grooves in industrial use are shown in Fig. 4.61. The smaller-radius grooves of Geometries 1 and 3 are more common in GA production; the larger-radius groove of Geometry 2 is more typically used in GI production.

Built-up layers can develop inside these grooves and can grow to the extent that they cause defects on the zinc coating [90]. When the rotating sink roll initially makes contact with the steel strip, there is a high local fluid pressure as the liquid zinc is squeezed out of the gap, this ejected zinc then flows in a generally sideways direction, roughly parallel with the sink roll axis. If this local pressure is higher than the pressure imposed by local line tension, the roll will slip or stop. The effect of groove pitch (groove spacing along the roll surface) on fluid pressure as a function of line speed was calculated by Dubois [91]. The results are shown in Fig. 4.62. The pressure is higher for the smaller groove/pitch condition because the total flow

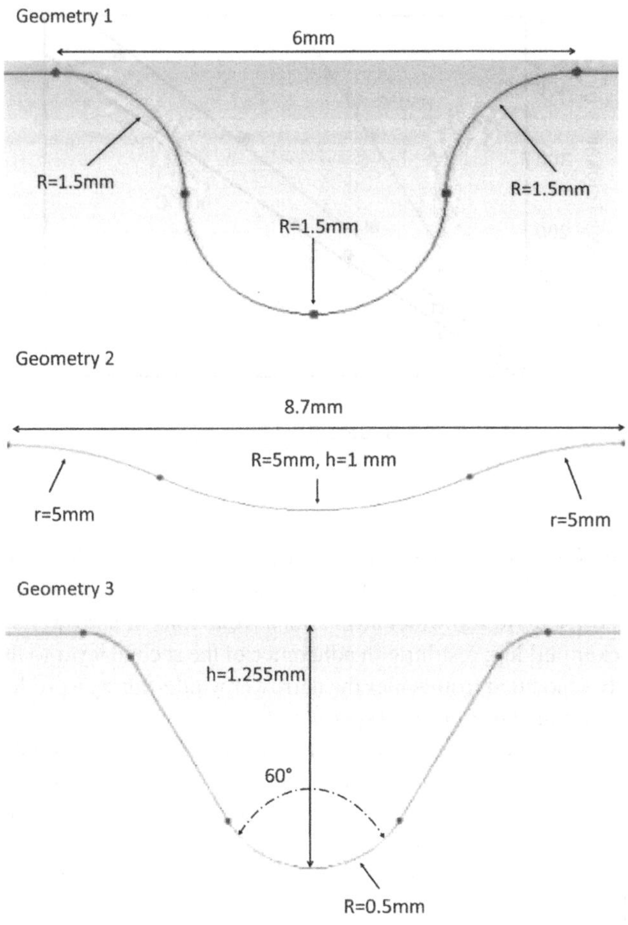

FIG. 4.61 Three types of sink roll grooves used in industrial production [89].

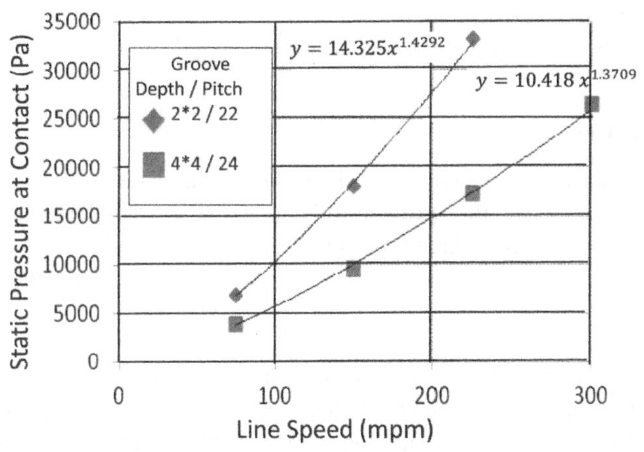

FIG. 4.62 Static liquid zinc pressure in the entry wedge as the steel strip contacts the rotating sink roll for two different sink roll sizes and pitches: a 2 × 2 mm groove with a 22 mm pitch and a 4.4 mm groove with a 24 mm pitch [91].

channel area of the grooved roll is comparatively lower. The average velocity of the liquid zinc trapped in these grooves is lower than the surface rotation speed of the sink roll. For a roll surface rotation speed of 2.5 m/s (equal to the line speed), the fluid velocity in the center of the groove is 0.5 m/s. It is expected that groove dimensions decrease as surface buildup progresses, resulting in higher fluid velocities in the groove and higher inlet pressures because of decrease channel size that increase the risk of roll slippage or stopping.

4.10.6 The nature of supermeniscus intermetallic climb

Pot hardware made from stainless steel that is immersed in zinc baths for long periods of time can exhibit attack just above the bath surface, both when exposed to air and inside the snout where a N_2-H_2 protective atmosphere is present. This attack, termed supermeniscus intermetallic climb (SMIC) by Brunnock [76], has the appearance of a mushy lump and consists of a mixture of intermetallic particles and liquid zinc. This attack is more severe in a GA bath than a GI bath. In contrast to stainless steel, carbon steel does not exhibit this kind of attack above the bath level, if it does occur it is very minor. A photo of SMIC on a 316L stainless coupon partially immersed in a GA bath is shown in Fig. 4.63. As explained by Liu et al. [92], a first reaction layer of $Fe_2Al_{(5-x)}Zn_x$ is formed at the liquid-solid interface. The Al is rapidly depleted, leading to formation of ZnFe delta or zeta intermetallics, depending upon local composition in the liquid. In GA baths, the reaction layer consists principally of FeZn phases mixed with liquid, enabling linear attack. In GI and Zn-5%Al baths, there is sufficient supply of Al to maintain the $Fe_2Al_{(5-x)}Zn_x$, $Fe_2Al_{(5-x)}Zn_x$ interface layer, resulting in parabolic reaction kinetics that slow attack in comparison with GA. The reason why the reaction can climb out of the bath is because the energy required for upward movement of Zn and Al at the liquid interface is much less than the reduction of chemical potential resulting from the reactions to form the intermetallics [92]. The dependence of the severity of SMIC on bath Al content is shown in Fig. 4.64 for baths over a range of Al contents between 0.12% and 12%.

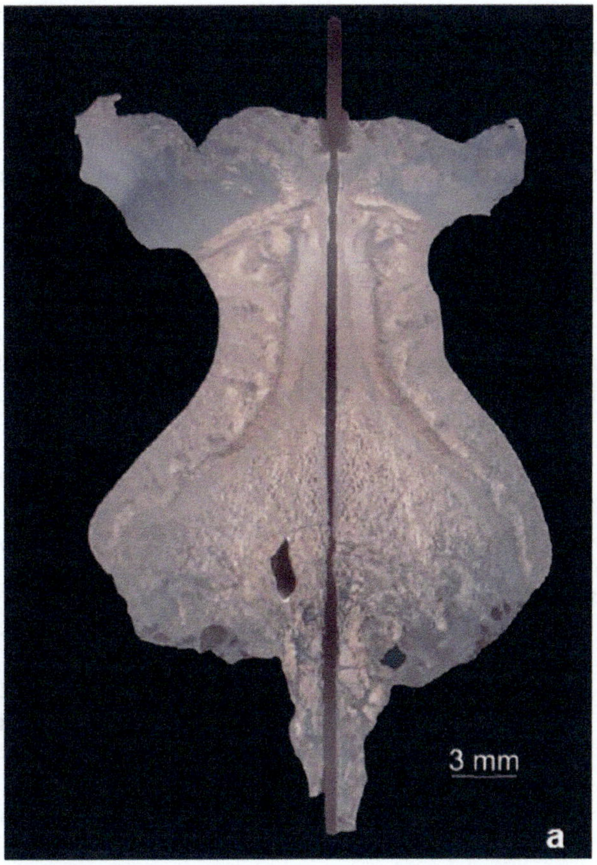

FIG. 4.63 Supermeniscus intermetallic climb on a 316L stainless coupon partially immersed in a GA bath after 336 h [92].

FIG. 4.64 Severity of SMIC after 600h for several bath compositions, at immersion temperature indicated for each type of bath. Together with GA and GI, GF (Zn-5Al) and Zn-12Al bath compositions were tested [92].

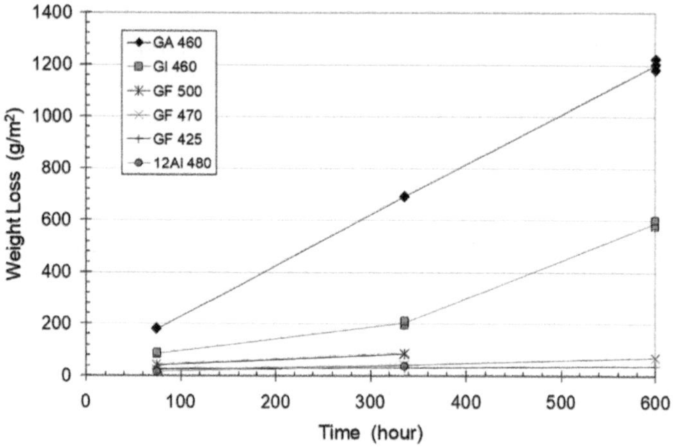

4.11 Production of exposed quality GI

The production of zinc coatings suitable for automotive exposed applications requires understanding and precise control over every aspect of production that can affect the quality of the steel surface to be galvanized and the coating of this surface with zinc. Key to a successful coating step is maintenance of a dross-free zinc bath through the period of exposed production. Referring to the ZnAlFe phase diagram in Fig. 4.8, a dross-free GI bath can be achieved by raising the Al level, which will lower iron solubility. The precipitated Fe resulting from this higher Al content is removed from the bath. Then the Al content is allowed to slowly decrease by controlling Al additions to the bath as production continues. As described by George [93], there can be no plateaus in this slow decrease in bath effective Al level during the exposed coating process, or dross will be created, the typical precision of bath control must be ±2°C. The bath temperature cannot be raised to recover the dross-free condition; the dross will not be re-dissolved during an acceptable period of time, and in the meantime, more Fe will be added to the bath as coating production continues. A plot of Al reduction with time in an exposed quality coating campaign is shown in Fig. 4.65. Here, effective Al content was raised from 0.24 to 0.38% during production of general-purpose coated steel. The bath was then stabilized at a content 0.38% Al level for 30h, during which time all dross was removed. Exposed quality production occurred during the following 8day period until the bath effective Al level reached 0.24%. General-purpose coating production then followed for a period of 25days before the cycle was repeated.

4.12 Summary of effects of bath Al composition

Table 4.10 summarizes the effects of bath Al composition over the GA and GI range on bath-related phenomena. Globally, most GI is produced with galvanizing baths in which 0.135–0.4 soluble Al is maintained. Exposed quality baths usually reach their lowest soluble Al level at around 0.25%Al, as indicated in the previous section. Baths in which both GI and GA production is alternated are run close to the 0.135% Al "knee point."

FIG. 4.65 Plot of 8000 on-line effective Al bath samples taken over an 8day period. Exposed quality coatings were made as Al content decreased from 0.38% to 0.25% [93].

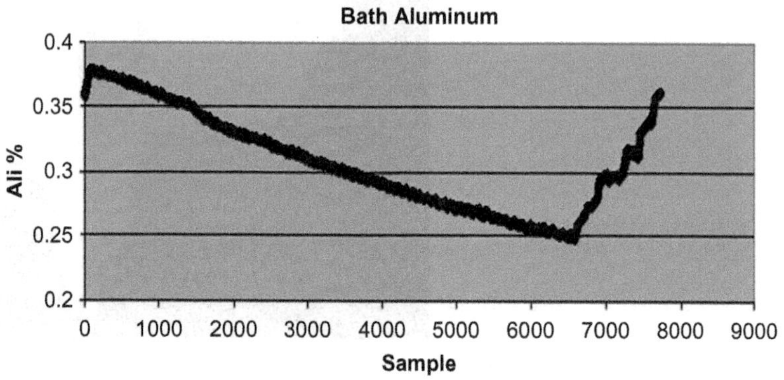

TABLE 4.10 Summary of bath Al effects on process conditions [94].

<table>
<tr><th>Bath effective Al, %</th><th>0.12</th><th>0.14</th><th>0.16</th><th>0.18</th><th>0.2</th><th>0.22</th><th>0.24</th><th>0.3</th></tr>
<tr><td>Al content in coating</td><td>Lower</td><td colspan="3">Low</td><td colspan="3">Limited</td><td>High</td></tr>
<tr><td>Intermetallic layer thickness</td><td colspan="2">High</td><td colspan="2">Intermediate</td><td colspan="4">Thin</td></tr>
<tr><td>Type of dross</td><td>Bottom</td><td colspan="7">Top</td></tr>
<tr><td>Dross production</td><td colspan="3">Higher</td><td colspan="3">Medium</td><td colspan="2">Lower</td></tr>
<tr><td>Dross defects</td><td colspan="3">More</td><td colspan="2">Acceptable</td><td colspan="3">Less</td></tr>
<tr><td>Pot roll dross buildup</td><td colspan="3">Higher</td><td colspan="3">Medium</td><td colspan="2">Low</td></tr>
<tr><td>Zn vapor sensitivity in snout</td><td colspan="3">Low</td><td colspan="3">Medium</td><td colspan="2">High</td></tr>
<tr><td>Thin coating possiblity</td><td colspan="3">Difficult</td><td>Acceptable</td><td colspan="4">Easier</td></tr>
<tr><td>Exposed quality GI</td><td>Impossible</td><td colspan="2">Very difficult</td><td>Difficult</td><td colspan="4">Possible</td></tr>
</table>

References

[1] J.F. Butler, The production of formable hot-dip galvanize sheet, in: Proceedings Galvanizers Association 83rd Meeting, October 7–10,1991, Lexington, KY, Galvanizers Association, Marine City, MI, 1991.

[2] Galvanized Autobody Partnership Project ZCO-8-7, Computer Aids for Galvanizing Bath Management—Aluminum Distribution in Bath, Final Report, International Zinc Assn., Durham, NC, October 3, 2002.

[3] G. Voss, G. Wallace, Crossbow control in hot dip galvanising lines, in: Proceedings Galvanizers Association 90th Meeting, November 1-4, 1998, Indianapolis, IN, Galvanizers Association, Marine City, MI, 1998.

[4] M. Dubois, Pot roll geometry and crossbow correction, in: Proc. 12th International Conference on Zinc and Zinc Alloy Coated Steel Sheet (Galvatech 2021), June 22-24, 2021, Virtual Conference, Austrian Society for Metallurgy and Materials, Vienna, 2021, pp. 695–705.

[5] M. Gagné, F. Ajersch, Galvanizing bath water model tests by monitoring pH changes from local acid additions, in: Proc. 3rd International Conference on Zinc and Zinc Alloy Coated Steel Sheet (Galvatech 1995), June 17–21, 1995, Chicago, IL, Iron & Steel Society, Warrendale, PA, 1995, pp. 687–694.

[6] F. Ajersch, F. Ilinca, M. Sabsabi, L.S. Onge, R. Heon, T.A.E. Cormode, N.-Y. Tang, C. Baril, M. Gagné, F.E. Goodwin, Monitoring of Al and Fe content and numerical simulation of the Sorevco galvanizing bath during GA to GI transition, in: Proceedings Galvanizers Association 97th Meeting, October 16–19, 2005, Lexington, KY, Galvanizers Association, Marine City, MI, 2005.

[7] Galvanized Autobody Partnership Project ZCO-71, Galvanizing Bath Management for Advanced High Strength sSteels, Final Report, International Zinc Assn., Durham, NC, May 31, 2017.

[8] F.E. Goodwin, E.A. Silva, North American developments and opportunities for galvanized steel sheet, in: Proc. 12th International Conference on Zinc and Zinc Alloy Coated Steel Sheet (Galvatech 2021), June 22–24, 2021, Virtual Conference, Austrian Society for Metallurgy and Materials, Vienna, 2021, pp. 24–38.

[9] Teck Metals, with permission.

[10] Galvanized Autobody Partnership Project ZCO-8-10, Dissolution and Growth of Dross Particles during Galvanize and Galvanneal Operations, Final Report, International Zinc Assn., Durham, NC, October 25, 2007.

[11] F. Ajersch, F. Ilinca, Review of modeling and simulation of galvanizing operations, Steel Res. Int. 89 (2018) 1700074.

[12] M. Gagné, M. Gaug, Investigating fluid flow in continuous galvanizing baths using numerical modeling, in: Proceedings Galvanizers Association 90th Meeting, November 1–4, 1998, Indianapolis, IN, Galvanizers Association, Marine City, MI, 1998.

[13] N.-Y. Tang, Dross management in continuous galvanizing, in: Proceedings Galvanizers Association 91st Meeting, October 24–27, 1999, Jackson, MS, Galvanizers Association, Marine City, MI, 1999.

[14] J. Kurobe, M. Iguchi, Size effect on dynamic behavior of dross in model hot dip plating Bath, Mater. Trans. JIM 44 (5) (2003) 885–892.

[15] A.K. Coker, Ludwig's Applied Process Design for Chemical and Petrochemical Plants, Chapter 6, Mechanical Separations, Gulf Professional Publishing, Elsevier, Amsterdam, 2007.

[16] H. Nishimura, Y. Kido, T. Nishimura, M. Nakazawa, A quantitative analysis of iron supersaturation: iron distribution in the zinc coating and in the zinc bath, in: Proc. 9th International Conference on Zinc and Zinc Alloy Coated Steel Sheet (Galvatech 2013), September 23–27, 2013, Beijing, Metallurgical Industry Press, Beijing, 2013, pp. 463–468.

[17] R.M. Mallens, C.J. Treadgold, M.J. Vlot, S.M. Meijers, Prevention of dross contamination in the Corus IJmuiden hot-dip galvanizing line, in: Proc. 5th International Conference on Zinc and Zinc Alloy Coated Steel Sheet (Galvatech 2001), June 26-28, 2001, Brussels, Verlag Stahleisen, Dusseldorf, 2001, pp. 255–261.

[18] P. Toussaint, L. Segers, R. Winand, M. Dubois, Short time investigation of hot dip galvanizing of steel under controlled hydrodynamics, in: Proc. 4th International Conference on Zinc and Zinc Alloy Coated Steel Sheet (Galvatech 1998), September 20–23, 1998, Chiba, Japan, Iron and Steel Institute of Japan, 1998, pp. 141–146.

[19] Galvanized Autobody Partnership Project ZCO-8-8, Computer Aids for Galvanizing Bath Management – Further Development of the Galvanizing Bath Model, Final Report, International Zinc Assn., Durham, NC, March 26, 2004.

[20] M. Dubois, Al content in intermetallic & free zinc on industrial galvanizing sheets, in: Proc. 6th International Conference on Zinc and Zinc Alloy Coated Steel Sheet (Galvatech 2004), April 4-7, 2004, Chicago, Association for Iron & Steel Technology, Warrendale, PA, 2004, pp. 1079–1088.

[21] Q. Yue, L. Zhang, J. Wang, H. Kong, L. Zhou, Flow field and dross behaviour in a hot-dip round coreless galvanizing bath, ISIJ Int. 57 (9) (2017) 1617–1624.

[22] D.-J. Paik, M.-H. Hong, Y. Huh, J.H. Park, H.-K. Chae, S.-H. Park, S. Choun, Metastable phases of dross particles formed in a molten zinc bath and prediction of soluble aluminum during galvannealing processes, Metall. Mater Trans. A 43A (2012) 1934–1943.

[23] J.H. Park, G.-H. Park, D.-J. Paik, Y. Huh, M.-H. Hong, Influence of aluminum on the formation behavior of Zn-Al-Fe intermetallic particles in a zinc bath, Metall. Mater. Trans. A 43A (2012) 195–207.

[24] T. Arioka, M. Hori, T. Toki, Growth mechanism of dross particles in molten Zn baths, in: Proc. 5th International Conference on Zinc and Zinc Alloy Coated Steel Sheet (Galvatech 2001), June 26-28, 2001, Brussels, Verlag Stahleisen, Dusseldorf, 2001, pp. 255–261.

[25] G.M. Michal, D.-J. Paik, M.-H. Hong, S.-H. Park, Dissolution of iron from automotive steel sheet in a molten zinc bath, in: Proc, 8th International Conference on Zinc and Zinc Alloy Coated Steel Sheet (Galvatech 2011), June 21-25, 2011, Genoa, Italy, Associazione Italiana di Metallurgia, Milan, 2011.

[26] Y.H. Liu, N.-Y. Tang, Computer modeling of aluminum uptake and iron dissolution in galvanizing and galvannealing, in: Proc. 6th International Conference on Zinc and Zinc Alloy Coated Steel Sheet (Galvatech 2004), April 4-7, 2004, Chicago, Association for Iron & Steel Technology, Warrendale, PA, 2004, pp. 1155–1164.

[27] S.M. Lee, S.K. Lee, D.-J. Paik, J.H. Park, Effect of initial iron content in a zinc bath on the dissolution rate of iron during a hot dip galvanizing process, Metall. Mater. Trans. A 48A (2017) 1788–1796.

[28] F. Ajersch, L. Trepannier, F.E. Goodwin, Particle size and composition of dross particles from galvanize to galvanneal operations, in: Proc. 44th Mechanical Working and Steel Processing Conference, September 9-11, 2002, Orlando, FL, Iron & Steel Society, Warrendale, PA, 2002.

[29] K.R. Yu, F. Ilinca, F.E. Goodwin, A modeling framework for top skimming formation in galvanizing bath, in: Proc. 12th International Conference on Zinc and Zinc Alloy Coated Steel Sheet (Galvatech 2021), June 22-24, 2021, Virtual Conference, Austrian Society for Metallurgy and Materials, 2021, pp. 653–661.

[30] T. Konishi, M. Shibat, J. Miki, H. Sawada, T. Doi, K. Ushioda, Structural and mechanical characterization of top dross in a molten zinc bath, in: Proc. 12th International Conference on Zinc and Zinc Alloy Coated Steel Sheet (Galvatech 2021), June 22-24, 2021, Virtual Conference, Austrian Society for Metallurgy and Materials, 2021, pp. 1345–1353.

[31] T. Arioka, M. Hori, T. Toki, A. Yagawa, M. Morikawa, CAMP, ISIJ 13 (2000) 1256 (in Japanese).

[32] T. Konishi, M. Shibata, J. Miki, Y. Nemoto, K. Ushioda, Structural and mechanical characterizations of top dross in a molten zinc bath, ISIJ Int. 61 (3) (2021) 937–944.

[33] J. Nakano, G.R. Purdy, D.V. Malakhov, Thermodynamic aspects of dross generation, in: Proc. 7th International Conference on Zinc and Zinc Alloy Coated Steel Sheet (Galvatech 2007), November 18–22, 2007, Osaka, Iron and Steel Institute of Japan, Tokyo, 2007, pp. 135–140.

[34] A. Einstein, On the movement of small particles suspended in stationary liquids required by the molecular-kinetic theory of heat, Ann. Phys. 17 (1905) 549–560.

[35] P. Perrot, J.-C. Tissier, J.-Y. Dauphin, Stable and metastable equilibria in the Fe - Zn - Al system at 450 °C, Z. Metallkunde 83 (1992) 786–790.

[36] M. Guttmann, Diffusive phase transformations in hot dip galvanizing, Mater. Sci. Forum 155–156 (1994) 527–548.

[37] J. Strutzenberger, R. Pree, W. Maschek, J. Faderl, Kinetic of dross formation and evolution of particle size distribution during bath transitions, in: Proc. 6th International Conference on Zinc and Zinc Alloy Coated Steel Sheet (Galvatech 2004), April 4-7, 2004, Chicago, Association for Iron & Steel Technology, Warrendale, PA, 2004, pp. 871–881.

[38] M. Didier, G. Ferrier, A. Settefrati, J. Benoit, Y. Binet, E. Dechassey, P. Vicente Hernandez, HDG pot productivity, in: Proc. 12th International Conference on Zinc and Zinc Alloy Coated Steel Sheet (Galvatech 2021), June 22-24, 2021, Virtual Conference, Austrian Society for Metallurgy and Materials, 2021, pp. 671–681.

[39] P. Toussaint, P. Vernin, L. Segers, R. Winand, M. Dubois, Experimental study and mathematical modeling of zinc ingot melting behavior in continuous hot dip galvanizing process, Ironmak. Steelmak. 22 (1995) 171–176.

[40] E.W. Fossen, J.P. Landriault, M.G. Lamb, Aluminum control on Stelco's Z-line, in: Proc. 3rd International Conference on Zinc and Zinc Alloy Coated Steel Sheet (Galvatech 1995), June 17–21, 1995, Chicago, IL, Iron & Steel Society, Warrendale, PA, 1995, pp. 795–800.

[41] Y.H. Liu, T.A.E. Cormode, A.B. Filc, N.-Y. Tang, M.S. Rhoads, Development of the Teck cominco new generation G2 aluminum sensor, in: Proceedings Galvanizers Association 96th Meeting, October 3-6, 2004, Charleston, SC, Galvanizers Association, Marine City, MI, 2004.

[42] G.N. Anderson, N.-Y. Tang, Aluminum distribution in a CGL bath, in: Proceedings Galvanizers Association 88th Meeting, October 20-23, 1996, Chicago, IL, Galvanizers Association, Marine City, MI, 1996.

[43] ASTM E716-16, Standard practices for sampling and sample preparation of aluminum and aluminum alloys for determination of chemical composition by spark atomic emission spectroscopy, in: ASTM Annual Book of Standards v. 03.05, ASTM International, Conshohocken, PA, 2016.

[44] Galvanized Autobody Partnership Project ZCO-8-9, Simulation of Al and Fe Distribution Using LIBS Analysis in the Sorevco Galvanizing Bath, Final Report, International Zinc Assn., Durham, NC, September 26, 2005.

[45] J.R. McDermid, C. Dewey, Optimizing the GA to GI transition at Pro-Tec CGL#1, in: Proceedings Galvanizers Association 92nd Meeting, October 29–November 1, 2000, Toronto, Galvanizers Association, Marine City, MI, 2000.

[46] J. McDermid, E. Baril, F.E. Goodwin, Galvanizing bath management during galvanize to galvanneal and galvanneal to galvanize product transitions, in: Proc. 6th International Conference on Zinc and Zinc Alloy Coated Steel Sheet (Galvatech 2004), April 4-7, 2004, Chicago, IL, Association for Iron & Steel Technology, Warrendale, PA, 2004, pp. 855–869.

[47] F. Ajersch, F. Ilinca, J.-F. Hetu, F.E. Goodwin, Numerical simulation of the rate of dross formation in continuous galvanizing baths, Iron Steel Technol. (2006) 93–101.

[48] H. Saint-Raymond, S. Gauthier, A. Lamandé, Modeling of zinc vaporization in snout, in: Proc. 8th International Conference on Zinc and Zinc Alloy Coated Steel Sheet (Galvatech 2011), June 21–25, 2011, Genoa, Italy, Associazione Italiana di Metallurgia, Milan, 2011.

[49] M. Dubois, A special equipment and area the snout & the liquid join, in: Online Workshop on Advanced Hot Dip Metallic Coatings, April 12-13, 2021, Jamshedpur, India, Organized by Indian Institute of Metals, Jamshedpur, 2021 (Chapter).

[50] N. Setargew, D. Parker, J. Hodges, Corrosion of 253MA stainless steel end plate in retractable snouts, in: Proc. 12th International Conference on Zinc and Zinc Alloy Coated Steel Sheet (Galvatech 2021), June 22-24, 2021, Virtual Conference, Austrian Society for Metallurgy and Materials, 2021, pp. 1256–1264.

[51] J. Kurobe, M. Iguchi, Effects of snout and support roll on transport phenomena in a hot dip plating bath, Mater. Trans. JIM 44 (7) (2003) 1429–1437.

[52] H. Unger, J. Strutzenberger, G. Angeli, G. Reiss, Flow Pattern Analysis Inside an Immersing Snout with a "Parametric Simulation Model", in: Proc. 12th International Conference on Zinc and Zinc Alloy Coated Steel Sheet (Galvatech 2021), June 22–24, 2021, Virtual Conference, Austrian Society for Metallurgy and Materials, 2021, pp. 1236–1246.

[53] S.J. George, Snout issues and minimising their impact on quality, in: AISTech 2018-Proceedings of the Iron & Steel Technology Conference, May 6–10 2018, Philadelphia, Association for Iron & Steel Technology, Warrendale, PA, 2018.

[54] D. Dauchelle, H. Baudin, P. Lucas, L. Gachet, Y. Prigent, Method and Installation for Dip Coating of a Metal Strip, U.S. Patent 6,936,307, August 30, 2005.

[55] F. Ilinca, F. Ajersch, F.E. Goodwin, Pumping flow and dross generation in galvanizing baths, in: Proceedings Galvanizers Association 106th Meeting, October 5–8, 2014, Jackson, MS, Galvanizers Association, Marine City, MI, 2014.

[56] N. Setargew, W.Y.D. Yuen, Quantitative analysis of top dross formation mechanism in metal coating baths, Rev. Métall. 110 (2013) 349–357.

[57] R. Thiounn, R. Zanfack, H. Saint-Raymond, P. Durighello, Understanding hot-dip galvanizing skimmings, in: Proceedings Galvanizers Association 100th Meeting, October 26–29, 2008, Baltimore, MD, Galvanizers Association, Marine City, MI, 2008.

[58] Galvanized Autobody Partnership Project ZCO-55, Dross Minimization in the Galvanizing Bat, in: Progress Presentation to Galvanized Autobody Partnership Program Review Meeting, October 21–22, International Zinc Assn., Durham, NC, 2009.

[59] K.R. Yu, F. Ilinca, F.E. Goodwin, Unsteady three-dimensional numerical model development for the investigation of the gas jet wiping process on the flow behaviors near the galvanizing bath surface, Iron Steel Technol. (2020) 80–87.

[60] J. Strutzenberger, R. Pree, Generation of scum due to wiping in hot-dip galvanizing, in: Proc. 8th International Conference on Zinc and Zinc Alloy Coated Steel Sheet (Galvatech 2011), June 21–25, 2011, Genoa, Italy, Associazione Italiana di Metallurgia, Milan, 2011.

[61] Galvanized Autobody Partnership Report ZCO-77, Dross Minimization in Galvanizing Bath, Final Report, International Zinc Assn., Durham, NC, August 2020.

[62] M. Dubois, Top drosses: the undocumented gold mine, in: Proceedings Galvanizers Association 95th Meeting, October 19-22, 2003, Monterrey, Mexico, Galvanizers Association, Marine City, MI, 2003.

[63] F. Ajersch, C. Koutsaris, S.G. Fountoulakis, M.A. Miller, F.D. Barrado, F.E. Goodwin, Characterizing top dross sampled from galvanizing lines using nitrogen and air wiping systems, in: Proceedings Galvanizers Association 101st Meeting, October 19-21, 2009, Louisville, KY, Galvanizers Association, Marine City, MI, 2009.

[64] Galvanized Autobody Partnership Program ZCO-55, Dross Minimization in the Galvanizing Bath, Final Report, International Zinc Assn., Durham, NC, May 7, 2014. Revised August 2014.

[65] D.J. Willis, F. Ilinca, F. Ajersch, Fluid flow modelling in a 55%Al-Zn pot, in: Proc. 6th International Conference on Zinc and Zinc Alloy Coated Steel Sheet (Galvatech 2004), April 4-7, 2004, Chicago, IL, Association for Iron & Steel Technology, Warrendale, PA, 2004, pp. 905–916.

[66] M. Bright, Dynamics of pot hardware bearings: understanding failure mechanisms, in: Presentation to AIST Galvanizing Technology Committee, March 10-11, 2010, Cleveland, OH, Association for Iron & Steel Technology, Warrendale, PA, 2010.

[67] P. de Kock, R. Jabs, G. Borrmann, A. Hissen, DUMAgaslock® sink rolls with rolling bearings for high performance galvanizing lines, in: Proceedings Galvanizers Association 103rd Meeting, October 16–19, 2011, Point Clear, AL, Galvanizers Association, Marine City, MI, 2011.

[68] M.X. Yao, J.B.C. Wu, R. Liu, Microstructural characteristics and corrosion resistance in molten Zn–Al bath of Co–Mo–Cr–Si alloys, Mater. Sci. Eng. A 407 (2005) 299–305.

[69] Galvanized Autobody Partnership Project ZCO-15-1, Improving performance of bearing submerged in galvanizing pots, in: Presentation to GAP Program Review Meeting, International Zinc Assn., Durham, NC, October 2004.

[70] K. Zhang, On the selection of materials for improved performance of pot bearings, in: AISTech 2005-Proceedings of the Iron & Steel Technology Conference, May 9–12, 2005, Charlotte, NC, Association for Iron & Steel Technology, Warrendale, PA, 2005.

[71] K. Zhang, Wear of cobalt-based alloys sliding in molten zinc, Wear 255 (2003) 545–555.

[72] Galvanized Autobody Partnership Program ZCO-15-1, Improving performance of bearings submerged in galvanizing pots, in: Presentation to Galvanized Autobody Partnership Program Review Meeting, May 8–9, International Zinc Assn., Durham, NC, 2003.

[73] L.J. Escott, D.J. Penney, A. Das, D. Thomas, Effects of heat treatments on the morphology and mechanical properties of a CoCrW alloy for hot dip galvanising applications, in: Proc. 12th International Conference on Zinc and Zinc Alloy Coated Steel Sheet (Galvatech 2021), June 22–24, 2021, Virtual Conference, Austrian Society for Metallurgy and Materials, 2021, pp. 1265–1274.

[74] Duma-Bandzink, Pot Hardware Features and On-Going Developments, Company Presentation of September 22, 2020. Pittsburgh, PA.

[75] D. Bollien, private communication, June 10, 2021.

[76] M.S. Brunnock, R.D. Jones, G.A. Jenkins, D.T. Llewellyn, Durability of bath hardware materials in continuous galvanizing, in: Proc. 88th Galvanizers Association Meeting, October 20-23, 1996, Chicago, IL. Proceedings Galvanizers Association 88th Meeting, October 20-23, 1996, Chicago, Galvanizers Association, Marine City, MI, 1996.

[77] K.E. Zhang, N.-Y. Tang, F.E. Goodwin, S. Sexton, Reaction of 316L stainless steel with a galvanizing bath, J. Mater. Sci. 42 (2007) 9736–9745.

[78] Y. Hardy, M. Dubois, J.-J. Bertrandie, H. Saint-Raymond, Mechanisms of dross build-up on hot-dipped hardware, in: Proc. 7th International Conference on Zinc and Zinc Alloy Coated Steel Sheet (Galvatech 2007), November 18-22, 2007, Osaka, Iron and Steel Institute of Japan, Tokyo, 2007, pp. 111–116.

[79] M. Stefan-Kharicha, A. Kharicha, K. Zaida, G. Reiss, W. Eßl, F. Goodwin, M. Wu, A. Ludwig, C. Mugrauer, Impact of hydrodynamics on growth and morphology of faceted crystals, J. Cryst. Growth 541 (2020), 125667.

[80] Galvanized Autobody Partnership Project ZCO-57, Improving galvanizing bath hardware, in: Presentation to GAP Program Review Meeting, International Zinc Assn., Durham, NC, October 2019.

[81] W. Eßl, G. Reiss, C. Mugrauer, R. Trasca, A. Kharicha, M. Stefan-Kharicha, H. Unger, J. Strutzenberger, G. Angeli, CFD-DPM modelling and analysis of dross particle dynamics in hot-dip galvanizing baths, in: Proc. 12th International Conference on Zinc and Zinc Alloy Coated Steel Sheet (Galvatech 2021), June 22-24, 2021 Virtual Conference, Austrian Society for Metallurgy and Materials, 2021, pp. 662–670.

[82] M. Bright, M. Kuperus, The modification process of the dross build-up structure on submerged hardware in Zn-Al and Zn-Al-Mg baths, in: Proc. 12th International Conference on Zinc and Zinc Alloy Coated Steel Sheet (Galvatech 2021), June 22-24, 2021, Virtual Conference, Austrian Society for Metallurgy and Materials, 2021, pp. 1275–1283.

[83] N. Setargew, W.Y.D. Yuen, R. Chen, The reactivity of 316L stainless steel in 55%Al-Zn bath, Rev. Métall. 110 (2013) 275–288.

[84] D. Liu, A. Filc, N.-Y. Tang, F.E. Goodwin, P. Giuffrida, P. Pendleton, T. Kitati, S. Sukumar, Dross buildup on various alloys in a galvalume bath, in: Proceedings Galvanizers Association 99th Meeting, October 21-24, 2007, Montreal, Galvanizers Association, Marine City, MI, 2007.

[85] N. Setargew, D.J. Parker, Zinc diffusion induced precipitation of σ-phase in austenitic stainless steel, Metall. Res. Technol. 116 (2019) 618.

[86] C.E. Dewey, K. Okamoto, E.E. Franks, E.A. Silva, Sink roll alloys and coatings, and dross print-through control, in: Proceedings Galvanizers Association 98th Meeting, October 22–25, 2006, Columbus, OH, Galvanizers Association, Marine City, MI, 2006.

[87] D. Scheiber, W. Eßl, J. Strutzenberger, F. Goodwin, J. Spitaler, G. Reis, Morphology of Fe_2Al_5 particles and the interface to WC coating in the context of hot-dip galvanizing: an Ab initio study, J. Alloys Compd. 824 (2020), 153854.

[88] G.-M. Song, W.G. Sloof, Effect of alloying element segregation on the work of adhesion of metallic coating on metallic substrate: application to zinc coatings on steel substrates, Surf. Coat. Technol. 205 (2011) 4632–4639.

[89] K. Neghab, A.N. Hrymak, F.E. Goodwin, Near-roll dross particle interactions in a galvanizing bath, Iron Steel Technol. (2015) 41–46.

[90] R. Plessinger, S. Seetharaman, E. DeMoor, F.E. Goodwin, Oxide stability analysis and characterization in a galvannealed dual-phase steel exhibiting coating defects, in: Proceedings Galvanizers Association 111th Meeting, October 6–9, 2019, Charleston, SC, Galvanizers Association, Marine City, MI, 2019.

[91] M. Dubois, Analysis of Zn flow inside the sink roll grooves, in: Proc. 11th International Conference on Zinc and Zinc Alloy Coated Steel Sheet (Galvatech 2017), November 12–16, 2017, Tokyo, Iron and Steel Institute of Japan, Tokyo, 2017, pp. 226–233.

[92] D.Y.H. Liu, N.-Y. Tang, F.N. Coady, Corrosion of 316L stainless steel in molten Zn-Al alloys, in: Proceedings Galvanizers Association 101st Meeting, October 19-21, 2009, Louisville, KY, Galvanizers Association, Marine City, MI, 2009.

[93] S. George, Sustainable productivity improvements in galvanised auto-body manufacture, in: AISTech 2017-Proceedings of the Iron & Steel Technology Conference, May 8-11 2017, Nashville, TN, Association for Iron & Steel Technology, Warrendale, PA, 2017.

[94] M. Dubois, private communication, March 2022.

Further reading

T.Q. Coddington, D.C. Cook, Mössbauer spectroscopy of aluminum substituted galvanneal steel, in: Proc. Int'l Conf on Applications of the Mössbauer Effect, September 10-16, 1995, Rimini, Italy, Paper #14-11, 1995.

I. Lee, K. Han, I. Ohnuma, R. Kainuma, Experimental determination of phase diagram at 450°C in the Zn-Fe-Al ternary system, J. Alloy Compd. 854 (2021) 157–163.

R.W. Richards, R.D. Jones, P.D. Clements, H. Clarke, Metallurgy of continuous hot dip aluminizing, Int. Mater. Rev. 39 (5) (1994) 191–212.

Chapter 5

Interface reactions

5.1 Introduction

Immersion of steel into a zinc or zinc alloy bath to produce a coated steel product with properties that are acceptable to industry is conceptually easy but is a difficult task in practice. For example, the automobile industry requires a combination of properties such as corrosion, formability, paintability, mechanical properties, wear, weldability, etc. In manufacturing an acceptable quality coating, an understanding of reactions that occur at the liquid zinc/steel interface is essential to the development of the coating type. Chapter 4, Zn Bath Phenomena, discussed in detail the many reactions that can occur in the Zn bath. As was pointed out, the Zn pot is where the steel is first reacted with the zinc bath and subsequently provides the necessary interface with a favorable microstructure for the formation of the overlay coating. Chapter 6 discusses overlay coating formation and the development of well-known coating types.

However, an understanding of the phase transformations that takes place at the liquid zinc/steel substrate interface is essential in order to predict and control the microstructure of the various galvanized coatings [1]. This understanding is further compounded by the use of zinc baths with additions of aluminum and other alloys, along with substrates that contain deliberate additions of solute elements and heat treatments to improve substrate properties. In 1995, Guttmann et al., enumerated the three main factors that contribute to the complexities in understanding and the subsequent underlying mechanisms [2]:

- Several reactions are occurring at the same time, including:
  1. wetting of the solid substrate by liquid zinc
  2. dissolution of the steel by zinc
  3. isothermal solidification of Fe-Al-Zn intermetallic compounds
  4. solid-state diffusional phase transformations
  5. solidification of the liquid zinc alloy
- The speed of the reactions is higher, and in some cases, the reactions take place in less than a second
- The transformation front often becomes unstable and therefore is not governed by simple equilibrium thermodynamics

These were the problems enumerated at Galvatech ’95 [2] and reviewed in 2000 [1]. Since that time, an explosion of research has been undertaken on the hot-dip process, particularly on the interface reaction, which has enabled numerous Zn-coated products to be offered to a range of industries including automotive, construction, and appliances.

5.2 The hot-dip process review

Hot-dip coating methods were reviewed in Chapter 2, Fe-Zn phase equilibria in Chapter 3 and Bath Phenomena in Chapter 4. The following is a summary of the various aspects of the process, best delineated in Ref. [3].

5.2.1 Substrate

The hot-dip process and the performance of the resultant coating are strongly influenced by the composition of the steel substrate. The composition of the steel can influence the rate and mode of attack at a given temperature of the molten zinc. Substrate chemical composition can influence the metallurgical properties of the hot-dip coating. Depending on concentration, the presence of Si, P, C, and Mn in the steel substrate can influence the Fe-Zn solidification mechanism and the presence of critical amounts of Si and P are necessary to control coating weight. C and P accelerate the growth of the alloy

The Metallurgy of Zinc Coated Steels. https://doi.org/10.1016/B978-0-323-99984-7.00017-8

layer, improving adherence. Various substrates are used in Zn hot-dip coating products from low carbon, nonkilled steel with low silicon, structural steels, deep-drawing steels, to dual-phase steels, and trip steels. Lately, other AHSS steels and third-generation steels have increasingly been used with hot-dip Zn coatings.

5.2.2 Surface preparation

A clean steel surface is required in the hot-dip process since Zn does not metallurgically react when the steel is not completely clean. Surface preparation consists of degreasing and pickling. For batch dipping, fluxing is required to dissolve any oxide films on the steel surface after pickling. In the case of sheet steel processing, a wet cleaning step is used prior to furnace cleaning for surfaces where a high-quality appearance is necessary, particularly in automotive coatings. This step involves cleaning with an alkali solution combined with abrasion with brushes. In many cases an electrolytic step is added to lift oil and iron residues from the surface (Chapter 2). Furnace cleaning is added after the wet cleaning step to burn off carbonaceous contaminants and reduce surface oxides prior to hot dipping. Another benefit of this step is to raise the temperature of the substrate to that of the Zn alloy bath temperature. As indicated in Chapter 2, annealing, cooling, and overaging steps can be included to produce the desired substrate microstructure and mechanical properties. These steps can influence the nature of the steel surface presented to the zinc bath.

5.2.3 Hot-dip variables

5.2.3.1 Bath temperature

Hot-dip bath temperatures depend upon the zinc alloy bath composition and range from 440°C to 460°C for Zn coatings with less than 1% Al to near 600°C for Zn-55%Al-X coatings. Strip temperature of the steel substrate and the addition of Zn ingots to replenish the bath are among the variables that can affect the bath temperature and the subsequent reaction between the zinc coating and the substrate, see Chapter 4.

5.2.3.2 Immersion time

Immersion time is critical to the formation of the interfacial layer and depends upon the type of coating required. In batch-type coating processes, where the coating is generally fully alloyed, immersion time can range up to 4–5 min duration to maintain an optimum alloying reaction. However, for sheet steel processing, the immersion time is dependent on the line speed through the furnace. Typical dipping times can range between 2 and 3 sec for line speeds more than 175 m/min. In sheet steel processing, the objective is to form an inhibition layer that prevents alloying of the overlay coating so that the required Zn coating can form.

5.2.3.3 Withdrawal speed

Withdrawal speed also depends upon line speed for sheet steel hot-dip processing. In batch dip processing, the steel substrate is taken out of the bath when the reaction between iron and zinc is complete. In this case, the thickness of the coating and surface smoothness and appearance depend upon withdrawal speed as described in Chapter 4.

5.2.3.4 Bath composition

Bath composition is an important variable that defines the type of coating required. Aluminum is added to the bath for several reasons. During galvanization, Al suppresses the growth Fe-Zn intermetallic layer, and for short times it is referred to as "inhibition." Additionally, Al is added to the bath to attain certain overlay compositions, e.g., Zn-55% Al, Galvalume, Zn-5% Al, Galfan, and Zn-0.2% Al, Galvanized. Mg is also added to the bath to obtain improved properties. The interfacial layer for all of these coatings will be discussed in detail later in the chapter, and the overlay of each of these alloy coatings will be reviewed in Chapter 6.

5.3 Interface reactions

5.3.1 Surface oxides

In the production of a Zn hot-dip coating, the steel strip undergoes a series of surface cleaning steps including wet cleaning and furnace cleaning. In the later step, the steel can also undergo a simultaneous annealing step. Before immersion into the liquid Zn bath, the strip is conventionally annealed between 780°C and 840°C, in a mixture of N_2 and H_2 (5–15 vol%) containing a low H_2O partial pressure (20–50 Pa) [4] with a dew point maintained under −40°C. The purpose of the heat treatment of the steel sheet is to:

- recrystallize the steel substrate after cold rolling
- reduce iron oxides on the surface of the sheet

Coincidentally, the less noble alloying elements of the steel such as Si, Mn, Al, Cr, Ti, and P can segregate to the surface where they can form both internal and external oxides. Prior to immersion in the Zn bath, the steel surface is composed of reduced metallic iron (wetted by liquid zinc) and oxide particles or films (not wetted by liquid zinc) [5]. Typical annealing treatments work well with low-level alloying additions because of the few oxides that are found on the surface as a result of the furnace cleaning step. However, when steels with higher levels of alloying additions (e.g., AHSS, 3rd Gen, etc.) are conventionally annealed, many more surface oxides are formed [6].

Similar results of surface oxide formation are found for dual-phase steels [7–9], TRIP steels [10], AHSS steels [11,12], and third-generation (3G) AHSS steels [13,14]. When the concentration of Mn and Si increases, as found in these higher-strength alloys, selective oxidation occurs and the type, morphology, and distribution of oxides (surface and internal) are then strongly dependent on processing parameters such as dew point, annealing temperature, and atmosphere composition [7]. An example of the amorphous manganese oxide formation at the substrate steel interface and at ferrite grain boundaries within the substrate steel is seen in Fig. 5.1 [8].

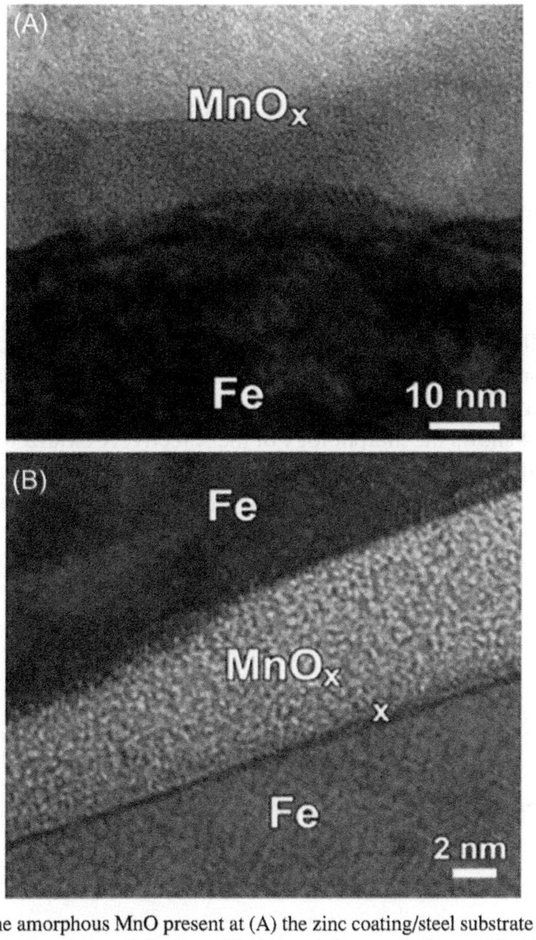

FIG. 5.1 High-resolution TEM image of the amorphous MnO present at (A) the zinc coating/steel substrate interface and (B) the ferrite grain boundary within the steel substrate just beneath the interface. *(From G.M. Song, T. Vystavel, N. van der Pers, J. Th.M. De Hosson, W.G. Sloof, Relation between microstructure and adhesion of hot dip galvanized zinc coatings on dual phase steel, Acta Mater. 60 (2012) 2973–2981.)*

5.3.2 Steel composition effects

Selected oxidation plays a critical role in the formation of the substrate surface because of the reaction of the less noble alloy elements with the available oxygen partial pressure during annealing. The selective oxidation behavior during recrystallization annealing of Si and Mn-added steels is an extremely complex phenomenon in which the stable oxide phases and the transition behavior of oxidation mode vary greatly with added Si and Mn contents in the steel, the oxygen potential, steel grain size, and other factors. Oxygen flow is the determining factor in these reactions because the diffusivity of oxygen is much faster in both ferrite and austenite than substitutional atoms in Fe such as Mn and Si. The transition from selective surface oxidation to internal oxidation can be explained by considering the oxygen flux in the oxidation film and the effect of the selective surface oxides on the inward diffusion behavior of oxygen. As a result, thermodynamic prediction of the selective oxidation behavior of Si and Mn-added steels can be determined [15]. Fig. 5.2 is a schematic of the possible phase regions found at the steel/coating interface [9].

The Si/Mn ratio plays a critical role in the development of surface oxides. For steels with low Si/Mn < 1, the stable phases normally produced in standard CGL processing are Mn_2SiO_4 or $MnSiO_3$ for low dew points and MnO+ Mn_2SiO_4 for higher dew points [16]. Avoiding the formation of a film-like amorphous SiO_2 is essential in creating a viable substrate surface enabling good wetting of the Zn liquid. At Si/Mn > 1, where the formation of SiO_2 is unavoidable, dew point becomes an important processing variable. Swaminathan et al. [17] studied oxidation behavior of model binary, ternary, and quaternary alloys. They found that in the binary alloys, Si tended to form sheet-like oxides on the surface, while Mn and Cr formed nodule-like oxides at grain boundaries and the steel surface. MnO was found in low dew point annealing, while Mn_3O_4 was produced when annealed at high dew points. As the substrate steel composition becomes more complex, the oxide coatings can be expected to also increase in complexity [12,13]. In high-strength dual-phase steels, additions of Mn, Si, P, Al tend to segregate to the steel surface on intercritical annealing and can be selectively oxidized prior to immersion in the zinc bath [18]. It was recently found that minor Sn additions significantly reduced both the external and internal oxidation kinetics [14]. From Fig. 5.3, it can be seen that the external oxide thickness decreased significantly with increasing process atmosphere pO_2 for both steel A (0.003 wt% Sn) and steel F (0.049 wt% Sn) (Fig. 5.3A). On the other hand, the depth of internal oxidation increased significantly with increasing process atmosphere pO_2 (Fig. 5.3B). as the process atmosphere pO_2 increased from 2.850×10^{-26} atm to 9.031×10^{-25} atm, an increase of an order of magnitude. As a result, internal oxidation increased owing to higher inward oxygen flow, which also suppressed the external oxidation kinetics. Fig. 5.3 also confirms that the minor Sn addition (~0.05 wt%) significantly reduced both the external and internal oxidation kinetics and therefore is quite surface-active, improving wettability in 3G steels.

5.3.3 Dew point

It has been observed that oxides will transform from external to internal oxides with increasing dew point [15,18]. External selective oxidation was observed for phosphorous on a steel annealed at 0°C dew point and for manganese, silicon, and aluminum at a lower dew point [19]. Variation in the dew point can also affect the morphology, composition, and distribution of external and internal oxides. Fig. 5.4 shows the morphology of the selectively oxidized surface obtained for the −30°C and +10°C dew points, respectively [20]. For the low dew point −30°C, Fig. 5.4A the surface oxides form a thin film. Most of the oxides are along the grain boundaries. XPS core-level spectra obtained for the sample annealed under −30°C dew point

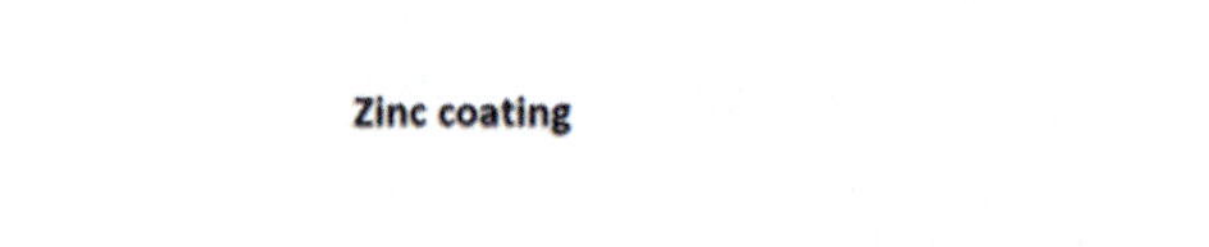

FIG. 5.2 A schematic of the interfacial region in galvanized DP steel. *(From I. Aslam, B. Li, R.L. Martens, J.R. Goodwin, H.J. Rhee, F. Goodwin, Transmission electron microscopy characterization of the interfacial structure of a galvanized dual-phase steel, Mater. Charact. 120 (2016) 63–68.)*

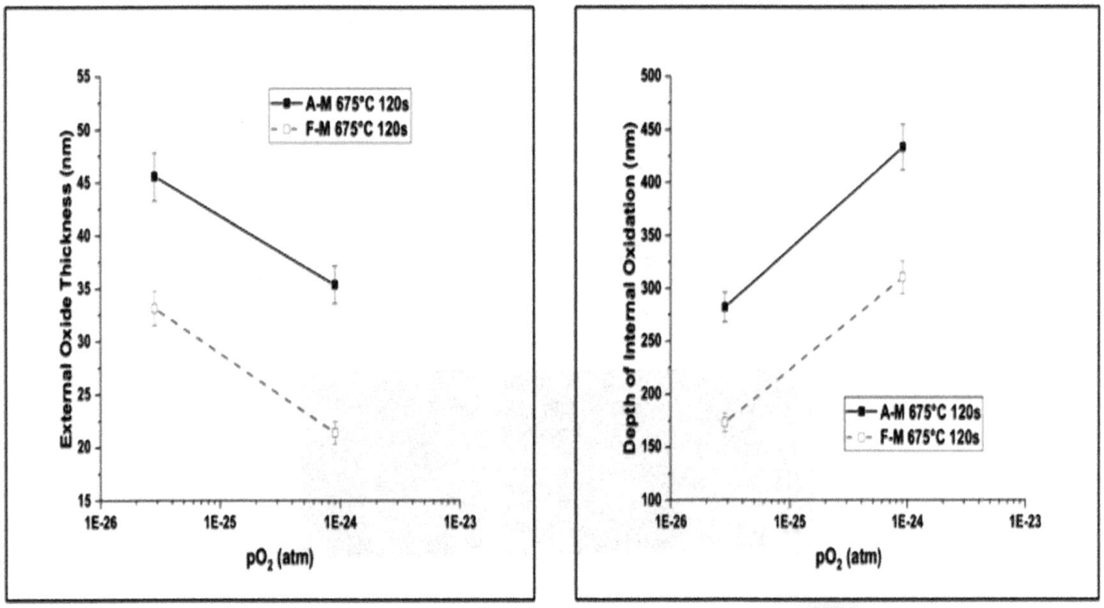

(A) (B)

FIG. 5.3 (A) External oxide thickness and (B) depth of internal oxidation as a function of process atmosphere pO_2. *(From K.M.H. Bhadhon, J.R. McDermid, F.E. Goodwin, Effect of process atmosphere dew point and minor Sn additions on the selective oxidation during intercritical annealing of a medium-Mn third generation advanced high strength steel, in: GALVATECH '21, 2021, pp. 404–412.)*

showed that the Si peak at 102.7 eV and Mn peak at 642.4 eV belonged to the Mn_2SiO_4/$MnSiO_3$. The Al surface segregation is very significant, the Al peak at 75.2 eV confirms the presence of Al_2O_3. It should be noted that there is no Fe and P detected at the surface for the sample annealed at low dew point. Increasing the dew point to +10°C results in different oxide morphology and chemical status. As shown in Fig. 5.4B, the oxides are distributed as discontinuous islands, EDS spectra show that the oxide contains Mn, P, Al, and Si. Thus, the surface oxides were found to be Mn_2SiO_4/$MnSiO_3$, $MnAl_2O_4$, and $Mn_3(PO_4)_2$. This agrees with the expected lower surface coverage of external oxidation at high dew point. It should also be noted that with the increase of the dew point from −30°C to +10°C, the high-temperature subsurface microstructure changed from austenite+ferrite to single ferrite. The change in microstructure at the surface was suggested to be due to the decarburization and consumption of alloy elements, particularly the oxidation of soluble Mn which, following the Fe-Mn-C phase diagram, results in local, isothermal transformation of ferrite to austenite as available C and Mn are oxidized. This can alter the properties of the surface of the substrate steel [20].

5.3.4 Annealing temperature

A review of the oxides observed on TRIP steels at various annealing temperatures is seen in Table 5.1. The complexity of the oxides formed are easily seen and that, as expected, with the addition of Al to the steel substrate, Al_2O_3 compounds became apparent.

A typical TEM cross section of the oxide morphology is shown in Fig. 5.5. In this study by Gong et al. [21] on the formation of oxides in TRIP steel, the annealing temperatures were 700°C, 870°C, and 930°C, corresponding to ferritic, intercritical, and austenitic annealing temperatures, respectively. The selection of temperatures made it possible to observe the effect of an increasing annealing temperature on the structure of the surface oxides, and an extended annealing time was chosen to allow for the precise determination of the equilibrium oxide phases in conditions of external oxidation. Fig. 5.6 is a detailed schematic comparing the observed evolution of the selective oxidation during the annealing of CMnSi TRIP steel. After annealing in the austenitic and intercritical temperature range, two distinct oxide layers are formed: an outer crystalline $2MnO \cdot SiO_2 + MnO \cdot SiO_2$ mixed oxide layer and an inner amorphous SiO_2 layer. During annealing in the ferrite stability range, a crystalline oxide film and amorphous lens-shaped oxide particles are formed at the surface. Both have a composition, which can be written as $xMnO \cdot SiO_2$: whereas the crystalline oxides have $1 < x < 2$, the amorphous oxide has $x < 0.5$. The latter was considered a solid solution of MnO in SiO_2. Internal oxide particles formed both in the grain matrix and at the grain boundaries were also detected after intercritical annealing and annealing in the austenite stability range, despite the pronounced external oxidation conditions. The oxide particles had a multilayer structure, consisting of an

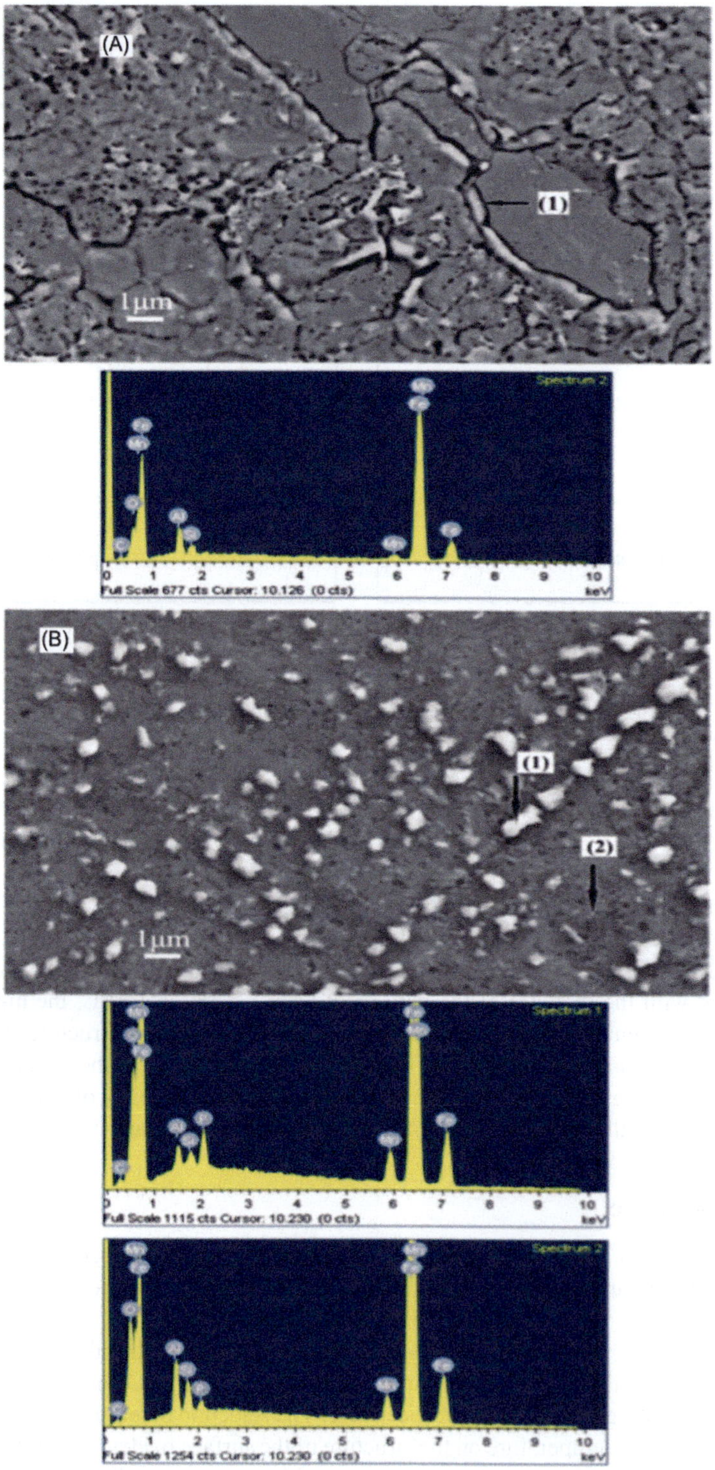

FIG. 5.4 Morphology and EDS analysis of surface oxides after annealing at different dew points. (A) Dew point −30°C, (B) dew point +10°C. *(From H. Liu, Y. He, S. Swaminathan, M. Rohwerder, Effect of dew point on the surface selective oxidation and subsurface microstructure of TRIP-aided steel, Surf. Coat. Technol. 206 (2011) 1237–1243.)*

TABLE 5.1 Composition of TRIP steels surface oxides observed after annealing in external oxidation conditions comparable with those in industrial continuous annealing furnace.

Type of TRIP steel	Dew point (°C)	Observed oxides						Annealing temperature (°C)	Analysis methods
CMnSi	−30	SiO_2 Amorphous	MnO	$MnO{\cdot}SiO_2$	$2MnO{\cdot}SiO_2$			700	HRSTEM
		SiO_2 Amorphous	MnO	$MnO{\cdot}SiO_2$	$2MnO{\cdot}SiO_2$	$MnO{\cdot}Al_2O_3$		870	
		SiO_2 Amorphous		$MnO{\cdot}SiO_2$	$2MnO{\cdot}SiO_2$	$MnO{\cdot}Al_2O_3$		930	
CMnSi	−35	Si-O oxides	MnO	$MnO{\cdot}SiO_2$				820	TEM
CMnSi	−30	SiO_2	MnO		$2MnO{\cdot}SiO_2$			800	XPS, EDS, SEM
CMnSiAl	−38	Si Complex oxides	MnO	Al_2O_3				780	XPS, SEM, GDOES
CMnSiAl	−53		MnO		$2MnO{\cdot}SiO_2$	$MnO{\cdot}Al_2O_3$	Al_2O_3	826	XPS, SEM
CMnSi	−30	SiO_2	MnO		$2MnO{\cdot}SiO_2$			780	XPS, EDS, SEM
CMnSiAl		SiO_2	MnO		$2MnO{\cdot}SiO_2$	$MnO{\cdot}Al_2O_3$	Al_2O_3	850	
CMnAl			MnO			$MnO{\cdot}Al_2O_3$	Al_2O_3	850	
CMnAl	−30		MnO			$MnO{\cdot}Al_2O_3$	Al_2O_3	825	XPS

From Y.F. Gong, H.S. Kim, B.C. De Cooman, Formation of surface and subsurface oxides during ferritic, intercritical and austenitic annealing of CMnSi TRIP steel, ISIJ Int. 48 (2008) 1745–1751.

FIG. 5.5 Cross-sectional TEM micrographs for CMnSi TRIP steel annealed at 700°C. (A) Overview with the lens-shaped xMnO · SiO$_2$ oxide particles on the surface and a larger subsurface MnO oxide. (B) Interface structure between the internal oxide and the matrix in zone A, and (C) corresponding diffraction pattern. *(From Y.F. Gong, H.S. Kim, B.C. De Cooman, Formation of surface and subsurface oxides during ferritic, intercritical and austenitic annealing of CMnSi TRIP steel, ISIJ Int. 48 (2008) 1745–1751.)*

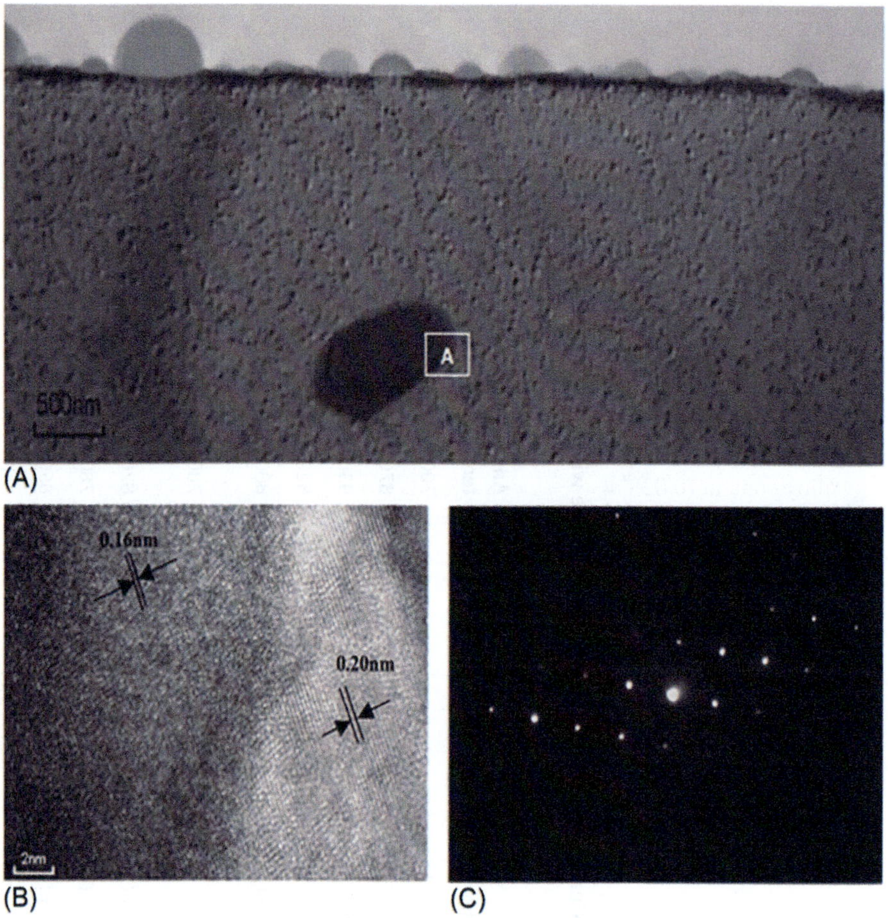

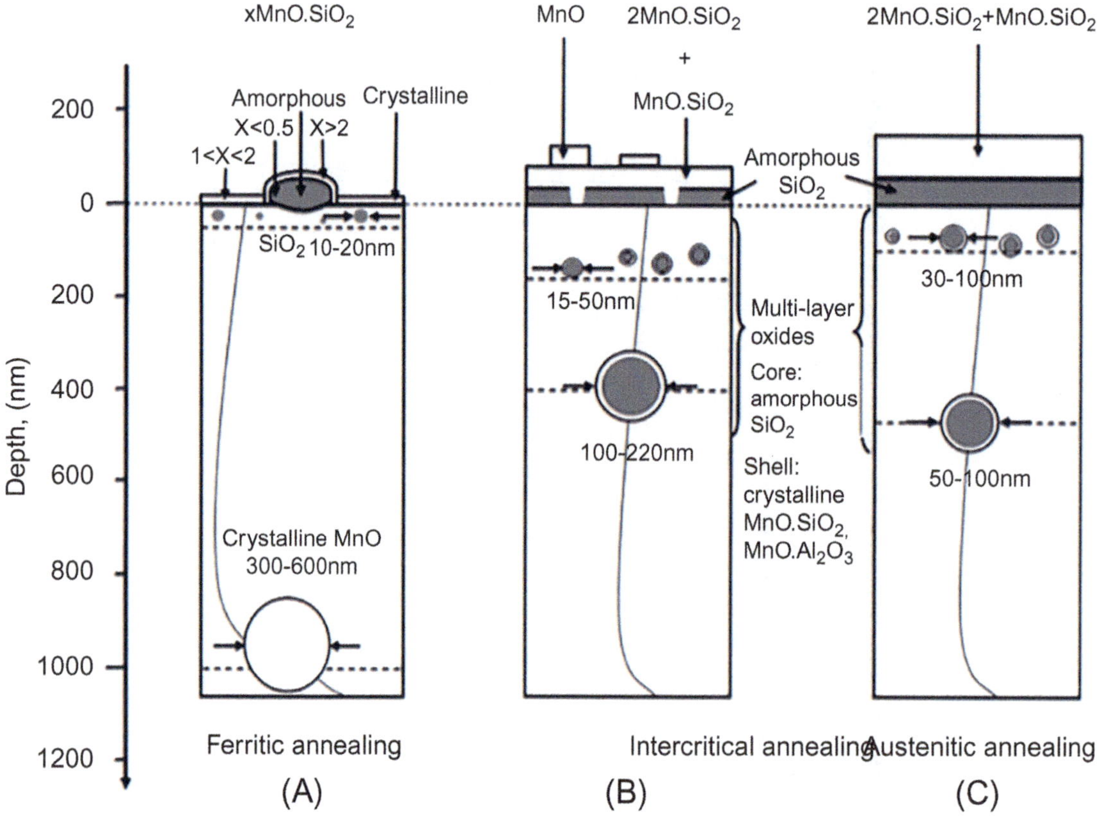

FIG. 5.6 Schematic of oxides formed at the surface region of MnSi TRIP steel during annealing at different temperatures in a gas atmosphere promoting external oxidation. *(From Y.F. Gong, H.S. Kim, B.C. De Cooman, Formation of surface and subsurface oxides during ferritic, intercritical and austenitic annealing of CMnSi TRIP steel, ISIJ Int. 48 (2008) 1745–1751.)*

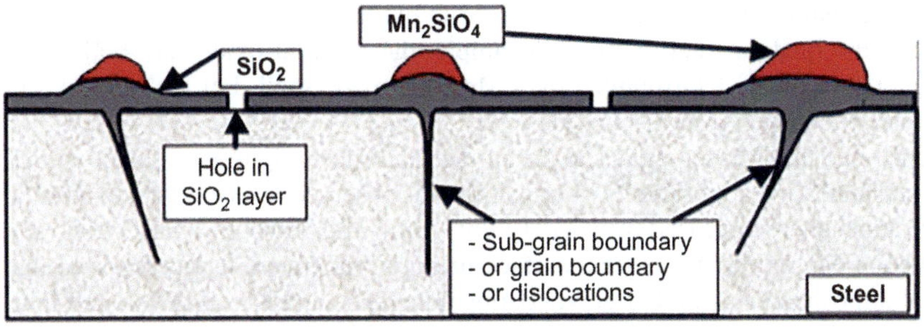

FIG. 5.7 Cross-sectional schematic view of the selective oxidation with high Si steels. *(From P. Drillet, Z. Zermout, D. Bouleau, J. Mataigne, S. Claessens, Selective oxidation of high Si, Mn and Al steel grade during recrystallization annealing, and steel/Zn reactivity, in: GALVATECH '04, 2004, pp. 1123–1134.)*

amorphous SiO_2 oxide core and a crystalline $MnO \cdot Al_2O_3 + MnO \cdot SiO_2$ mixed oxide layer. It was found that despite a very low alloy content (0.028 mass%), Al still showed a pronounced tendency to form internal oxide particles during annealing in the austenitic temperature range. The external $MnO \cdot SiO_2$-$2MnO \cdot SiO_2$ mixed oxide layer and the amorphous SiO_2 layer are expected to have a negative effect on galvanizing CMnSi TRIP steel as they are expected to act as effective barriers to the $Fe_2Al_{5-x}Zn_x$ inhibition layer formation, which requires the reaction of Fe atoms at the steel surface with solute Al atoms in the liquid Zn. The authors speculated that this may also explain why internal oxidation in higher dew point atmospheres has been shown to improve galvanizing of TRIP steels by causing internal rather than external oxidation.

Sagl et al. [15] reported that Mn oxides in the state of MnO form at 650°C, but change into a mixed state together with Si at higher temperature, forming Mn silicates or Mn-Si mixed oxides [22]. Another study stated that MnO starts to appear at 500°C but is transformed to Mn_2SiO_4 via the reduction of MnO with SiO_2 [23]. In contrast, Drillet et al. report that Mn_2SiO_4 grows on top of an amorphous SiO_2 layer, Fig. 5.7, which is assumed to be the first growing phase [24]. Earlier it was reported that at a temperature of 650°C, the first growing oxide has a higher Mn concentration, whereas for higher temperatures, diffusion of Si becomes more pronounced. Finally, at a temperature higher than 850°C, Mn_2SiO_4 starts to grow [24].

5.3.5 Annealing atmosphere

In a study of annealing furnace fuel-to-air ratio effects [7], a dual-phase steel was intercritically annealed in a direct-fired furnace to determine the effects on external and internal oxide formation and the interfacial structure resulting from galvanizing in a 0.2% Al bath [7]. It is expected that the oxygen partial pressure would decrease as the fuel-to-air ratio increases. For all the galvanized samples, independent of fuel-to-air ratio, the inhibition layer was continuous, due to the relatively high Al concentration in the bath. Also, both external and internal oxides were formed during annealing. However, the structure, distribution of the oxides, and the interfacial structure were different. Fig. 5.8 schematically shows the results of the structural analysis.

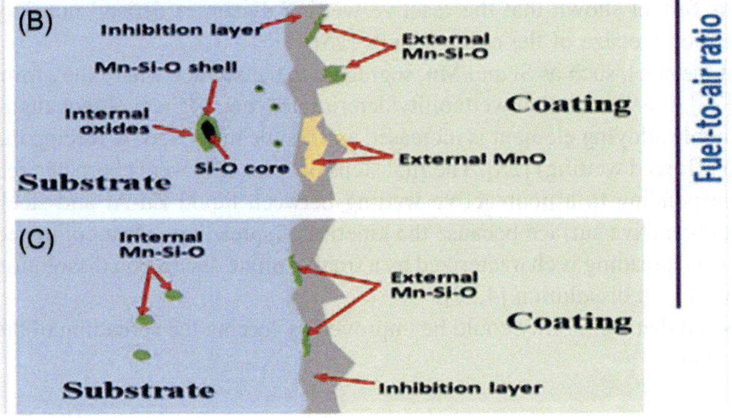

FIG. 5.8 Schematic illustration of oxides distributed after galvanizing with different fuel ratios (A) high fuel-to-air ratio, (B) medium fuel-to-air ratio, (C) low fuel-to-air ratio. *(From K. Chen, I. Aslam, B. Li, F.E. Goodwin, M. Horstemwyer, Effect of fuel-to-air ratio on oxidation and interfacial structure in galvanizing of a dual-phase steel, SAE Int. J. Mater. Manuf. 14 (2021), https://doi.org/10.4271/05-14-03-0021.)*

For the high fuel-to-air ratio sample, the external oxide is the complex Mn-Si-O oxide, and the internal oxide is the inhomogeneous Mn-Si-O complex oxide. For the medium fuel-to-air ratio sample, the external oxides are both MnO and Mn-Si-O oxides. The internal oxide had a core of Si-O surrounded by a Mn-Si-O oxide shell. For the low fuel-to-air ratio sample, both the internal and external oxides were Mn-Si-O oxides.

In another study, Khondker et al. [18] confirmed that Mn segregates to the surface and shows significant selective oxidation at the sample surface for all experimental annealing atmospheres. A transition was observed from external to internal oxidation when the oxidation potential of the atmosphere increased from a pH_2O/pH_2 ratio of 0.00844 to a pH_2O/pH_2 ratio of 0.03451. However, in spite of the presence of Mn oxides at the steel surface prior to immersion, it was reported that samples annealed under all the experimental atmospheres exhibited good coating adherence and a well-developed inhibition layer. It was proposed that an aluminothermic reduction of the Mn oxides by the Al present in the Zn bath allowed for the formation of a continuous inhibition layer.

All the many studies agree that the composition as well as the ratio of alloying elements (Si/Mn) and the resulting chemical composition of the oxides are the most critical factors for the final galvanizing performance [16]. It has also been shown that increasing (1) the dew point, as well as (2) increasing oxidation potential of the annealing atmosphere to encourage internal oxidation and increasing crystallinity of the oxides has a beneficial effect on the quality of the galvanized coating. However, increasing annealing temperature, especially from the two-phase intercritical region toward the single-phase austenite region produces a more complex surface oxide because of the increased diffusivities of both the surface-active species and oxygen at higher temperatures.

5.4 Wettability

Good adhesion of the zinc coating to the steel surface is one of the most important requirements of a hot-dip zinc coating product. Good adhesion directly depends on the formation of a defect-free Fe_2Al_5 inhibition layer at the steel substrate/Zn coating interface. The complex surface chemistry of the oxides, the overall oxide coverage, the island size, and the distribution of the oxides on the steel surface, which is a result of segregation and selective oxidation upon annealing, affect wettability and the interfacial reactions [25]. Wetting refers to the spreading behavior of a liquid deposited on a solid or liquid substrate. The main features of wetting are the surface energy, the triple line, the static and dynamic contact angles, and the contact angle hysteresis [26]. Fig. 5.9 schematically shows the various wetting possibilities for a liquid droplet on a solid substrate. Poor wetting results when the contact angle $\theta > 90$, good wetting occurs when $\theta < 90$ degrees, and excellent wetting happens when $\theta = 0$ degrees.

Frenznick et al. [25] modified the Cassie equation. According to the Cassie equation, zinc wetting is predominantly determined by the relative composition (i.e., surface coverage) of the oxide and metal phase,

$$\cos\theta = a_{\text{oxide}} \cos\theta_{\text{oxide}} + (1 - a_{\text{oxide}}) \cos\theta_{\text{metal}} \tag{5.1}$$

where a_{oxide} is the fraction of surface coverage of the oxide, θ_{oxide} and θ_{metal} are the contact angles in equilibrium state of the oxide and metal, respectively. Since reactive wetting in the hot-dip galvanizing process depends on the growth of stable Fe-Al interfacial layer, the modified Avrami growth law applied to Eq. (5.1), which leads to the new description of the dynamic Cassie equation as follows:

$$\cos\theta = \cos\theta_{\text{oxide}}\, a\, exp\,(1 - kt^{n}) + \cos\theta_{\text{metal}}[1 - a\, exp\,(1 - kt^{n})] \tag{5.2}$$

where a is the initial surface coverage of the oxide, k is the Avrami constant (defines kinetics), and n is the Avrami exponent. It was found that the new dynamic Cassie equation is a good approximation for the wetting kinetics and for the growth mechanism of the reaction layer. It was further shown that the reactive wetting dynamics depend not only on the overall oxide coverage of the surface but also on the size of the oxide islands [25].

As shown above, less noble alloying elements of the steel, such as Si and Mn, segregate to the surface where they form oxide particles that are poorly wetted by liquid Zn [27]. It is known that wettability deteriorates (i.e., higher contact angle, slower wetting rate) if the concentration of the less noble alloying element is increased and can be improved by forcing the spreading of the liquid on the solid by external action (forced wetting) [28]. The first steps of spreading were characterized by a high constant contact angle 120 degrees corresponding to a nonreactive wetting between liquid Zn-Al and steel. The liquid front at the triple line advances on a flat unreactive surface because the kinetics of spreading is fast compared with that of the substrate dissolution. The second step of spreading is characterized by a smaller angle due to iron dissolution as the kinetics of spreading became lower than that of iron dissolution [4,27].

In a series of papers, Giorgi et al. [4–6, 26] proposed that wettability could be improved by forcing the spreading of the liquid on the solid by external action or forced wetting.

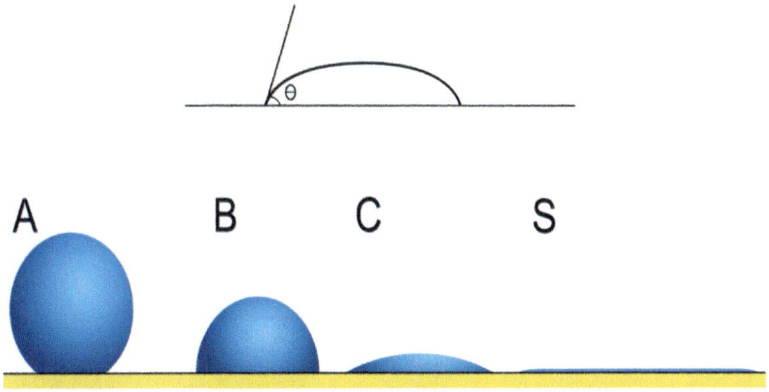

FIG. 5.9 Wetting. (A) Contact angle for a liquid droplet on a solid surface, (B) variable contact angle: Perfect wetting, S, $\theta = 0$ degrees; high wettability, C, $0 < \theta < 90$ degrees: low wettability, B, 90 degrees $\leq \theta \leq 180$ degrees: nonwetting, A, $\theta = 180$ degrees. *(From Wikipedia: Wetting. https://www.masterbond.com/techtips/surface-wetting.)*

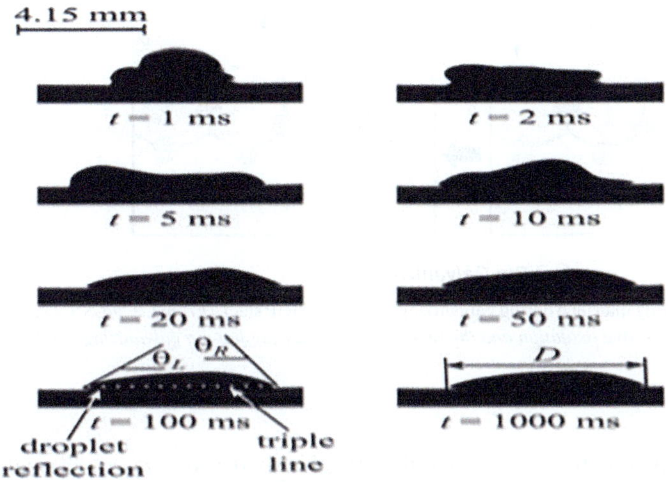

FIG. 5.10 Spreading of liquid droplets of Zn-0.2 wt% Al-0.015 wt% Fe on the steel substrate chosen at 450°C. *(From M.-L. Giorgi, J.-S. Diawara, H. Duval, Improvement of wettability between steel and liquid Zn-Al alloy by forced wetting, ISIJ Int. (2018), 147.)*

Fig. 5.10 shows the sequential time images of the Zn-Al-Fe liquid metal droplet on the solid substrate IF steel. These images are used to determine the evolution of the left and right contact angles, θ_L and θ_R, and the spreading diameter D as a function of time and are used to characterize wetting [5]. The first step of spreading is characterized by a high constant contact angle of 120 degrees corresponding to nonreactive wetting between liquid Zn and steel. The liquid front at the triple line advances on a flat unreacted surface because the kinetics of spreading is fast compared with that of the substrate dissolution. The second step of spreading is characterized by a smaller contact angle due to iron dissolution. As a result, they concluded that wettability of the steel substrate is improved by forced wetting. However, they questioned how forced wetting can be achieved in a commercial galvanizing process [5].

Reactive wetting can occur despite the presence of small amounts of Al in the Zn bath. Additional evidence has been provided to better describe the alumino-thermic reduction of surface compound oxides during hot dipping [29]. It was reported that the thickness of an MnO layer was significantly reduced for increasing reaction times and that Al_2O_3, the product of the alumino-thermic reaction, could be detected at the MnO/bath interface. Cho et al. [30] proposed a schematic of the reactive wetting mechanism in Fig. 5.11. In a low DP gas atmosphere, two types of surface oxides are formed on the surface: thick (30–40 nm) xMnO•SiO_2 layers and thin (5–10 nm) SiO_2 layers. When the steel is dipped in the liquid Zn bath, Al present in the stationary boundary layer in contact with the surface reduces the oxides. While the thicker parts of the oxide layer promote the localized depletion of Al, the thinner parts of the oxide layer are removed by the reduction reaction. This provides a location for the dissolution of Fe in the boundary layer with the subsequent formation of the $Fe_2Al_{5-x}Zn_x$ inhibition layer. Once the $Fe_2Al_{5-x}Zn_x$ nucleates at the steel/coating interface, the Al atoms close to the interface contribute to the growth of $Fe_2Al_{5-x}Zn_x$ rather than to the reduction of oxides. This process results in the formation of the observed larger and isolated $Fe_2Al_{5-x}Zn_x$ grains. When the rate of Al uptake decreases because of the exhaustion of Al in the stationary boundary layer, the Fe dissolution and the formation of ξ and δ-phase Fe-Zn intermetallics take place at the interface where the $Fe_2Al_{5-x}Zn_x$ layer has not formed. Some oxides also appear to be lifted off from their original position on the surface because of a rapid Fe dissolution during the formation of the Fe-Zn intermetallics.

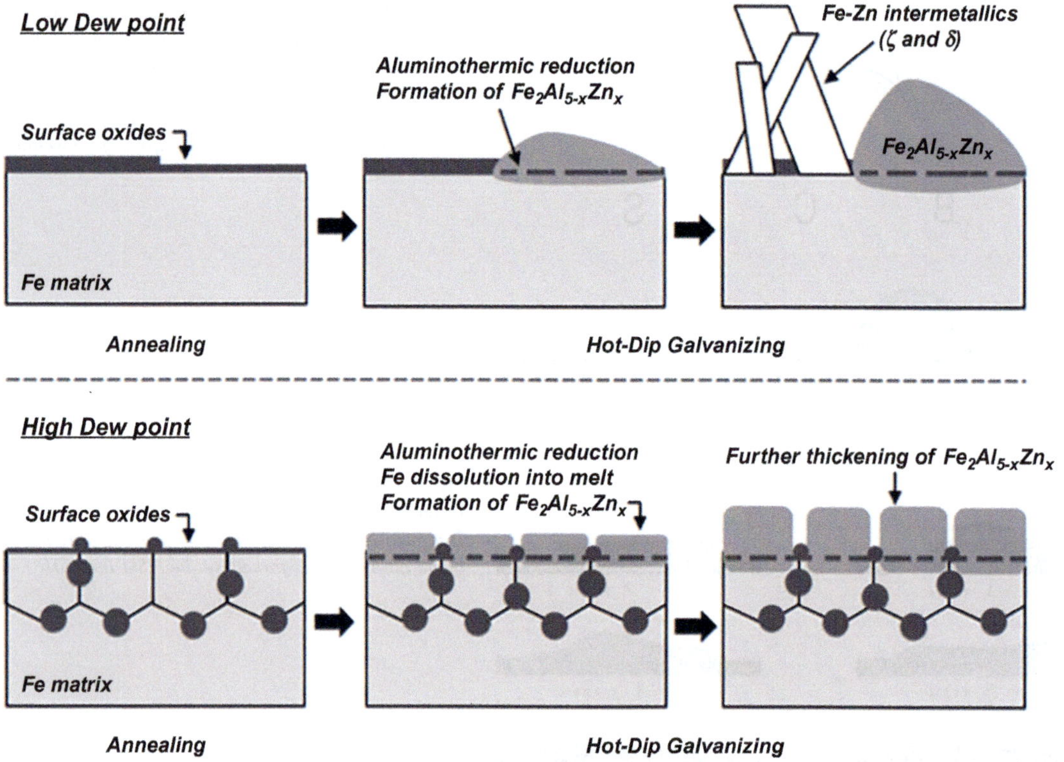

FIG. 5.11 Schematic of the reactive wetting mechanism at the steel/coating interface during galvanizing of CMnSi TRIP steel. *(From L. Cho, S.J. Lee, M. S. Kim, Y.H. Kim, et al., Influence of gas atmosphere dew point on the selective oxidation and the reactive wetting during hot dip galvanizing of CMnSi TRIP steel, Metall. Mater. Trans. A 44 (2013) 362–371.)*

In a high DP gas atmosphere, a thin oxide layer is formed on the surface, and the selective oxides are present as small oxide particles at grain boundaries. The subsurface grain boundaries are pinned by the internal oxides. This suppresses the grain growth during annealing and results in fine subsurface grains. During dipping in the Zn bath, Al in the stationary boundary layer partially reduces the oxides, whereas the oxides at the grain boundaries are not affected. Only a small amount of Fe diffusion is required to nucleate the $Fe_2Al_{5-x}Zn_x$ layer. As the oxide layer is partially reduced, the Fe dissolution and the Al diffusion occur simultaneously at the interface. As the $Fe_2Al_{5-x}Zn_x$ grains nucleate on the substrate ferrite grains, the size of the $Fe_2Al_{5-x}Zn_x$ grains are influenced by the small grain size in the subsurface region. When oxides are thicker, lateral growth of the $Fe_2Al_{5-x}Zn_x$ grains under the oxide layer also supports the formation of a continuous $Fe_2Al_{5-x}Zn_x$ inhibition layer. High DP gas atmosphere conditions are therefore effective in causing both internal oxidation and the reduction of the small amount of layer-type surface oxides initially present at the TRIP steel/coating interface. This results in the formation of a continuous fine-grained $Fe_2Al_{5-x}Zn_x$ layer on which a defect-free Zn coating can solidify.

5.5 Interface response

When the steel strip is immersed in a bath of molten zinc alloy with small amounts of Al (<1%), two reactions occur at the interface between the steel and the liquid zinc alloy: (1) dissolution of iron and (2) nucleation and growth of an intermetallic phase designated as $Fe_2Al_5Zn_x$ ($0<x<1$). The presence of aluminum inhibits the formation of Fe-Zn intermetallics by forming a thin layer of solid $Fe_2Al_5Zn_x$ at the substrate-coating interface. The interface layer (also known as a barrier or inhibition layer) is adherent to both the matrix and the coating, without affecting strip formability. The reaction between zinc and iron is delayed for a very short time, and the final coating is composed of a thin layer of solid $Fe_2Al_5Zn_x$ covered by a 10–15 μm layer of Zn [1,2]. The effectiveness of the aluminum in forming the inhibition η-$Fe_2Al_5Zn_x$ intermetallic at the substrate/bath interface is complex because of partitioning of aluminum between the coating and the intermetallic compounds formed within the galvanizing bath (dross) [25].

A schematic of the galvanizing reactions as a function of time is seen in Fig. 5.12. As soon as the steel strip enters the zinc alloy bath, two reactions occur at the interface between the steel substrate and the liquid zinc alloy: (1) dissolution of

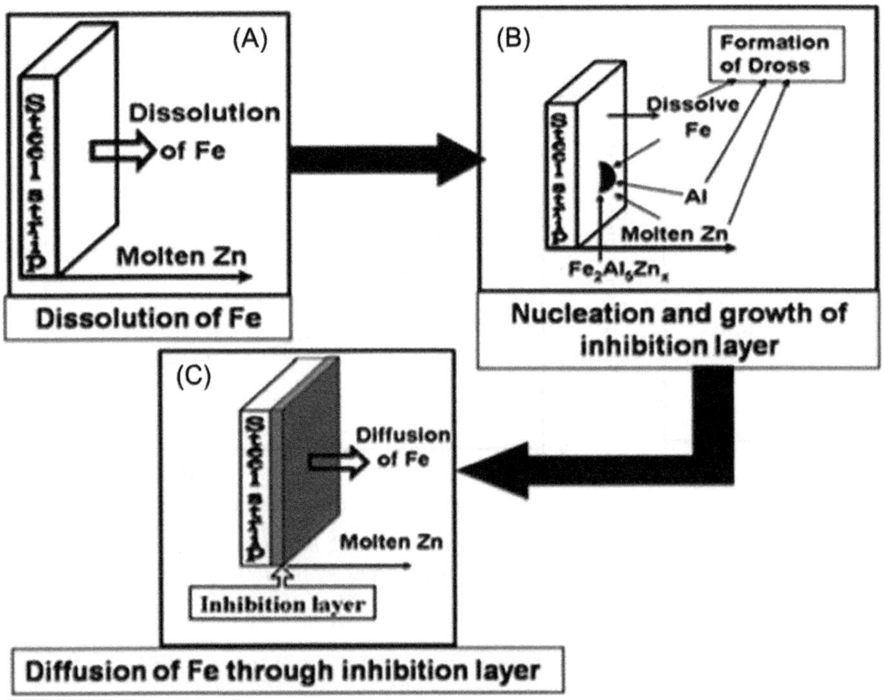

FIG. 5.12 Schematic representation of the galvanizing interfacial reactions. *(From G.K. Mandel, R. Balasubramaniam, S.P. Mehrotra, Theoretical investigation of the interfacial reactions between solid iron and liquid zinc, Metall. Mater. Trans. 40A (2009) 637–645.)*

iron (Fig. 5.12A) and (2) formation of an aluminum-rich inhibition layer (Fig. 5.12B). Iron dissolution from the strip occurs before (Fig.5.12A) and after (Fig. 5.12C) nucleation of the inhibition layer.

5.5.1 Dissolution of Fe

Part of the iron dissolved diffuses into the bath where it contributes to the formation of intermetallic particles about 20 μm in diameter, called dross. The zinc bath typically contains 0.025–0.06 wt% of iron from the steel. The dross is in thermodynamic equilibrium with the liquid zinc and depends on temperature and composition to form the dross (e.g., δ-$FeZn_9$ and η-$Fe_2Al_5Zn_x$ for 0.15 wt% Al, 0.04 wt% Fe at 450°C) (see Chapter 4). Dissolution of iron in liquid zinc containing only small amounts of aluminum (0.05 wt%) and iron (0.003 wt%) is controlled by the diffusion of iron in the liquid phase. Giorgi et al. [31,32] determine the diffusion coefficient of iron in liquid zinc to be:

$$D_{Fe}^{Zn/L} = (9.8 \pm 0.1) \times 10^{-10}\ \mathrm{m^2\ s^{-1}} \tag{5.3}$$

This diffusion coefficient is a key parameter for hot-dip galvanizing steels since it determines the flux of dissolved iron from the steel strip that diffuses into the zinc bath. They used the iron in liquid zinc diffusion coefficient to develop a model to describe the kinetics of galvanizing reactions that occur, i.e., dissolution of iron, heterogeneous nucleation, and growth of the intermetallic η-$Fe_2Al_5Zn_x$ phase. The model can calculate the following parameters as a function of time [31,32]:

- the concentration profiles for iron and aluminum in the boundary diffusion layer
- the nucleation rate
- the average critical embryo radius
- the establishment of supersaturation
- the time necessary to form a continuous layer

Because measurements are difficult to make in industrial plants, mathematical modeling can play an important role in improving the understanding of the process by arriving at the optimum setting of process parameters. A more recent iron-dissolution model was developed by Mandal et al. [33]. For example, the change in iron dissolution with immersion time is shown in Fig. 5.13A, and the increase in effective aluminum content in the bath led to a decrease in iron dissolution in the bath, Fig. 5.13B. Further, it was found that the theoretical predictions of inhibition layer thickness were in conformity with those estimated or measured by other researchers.

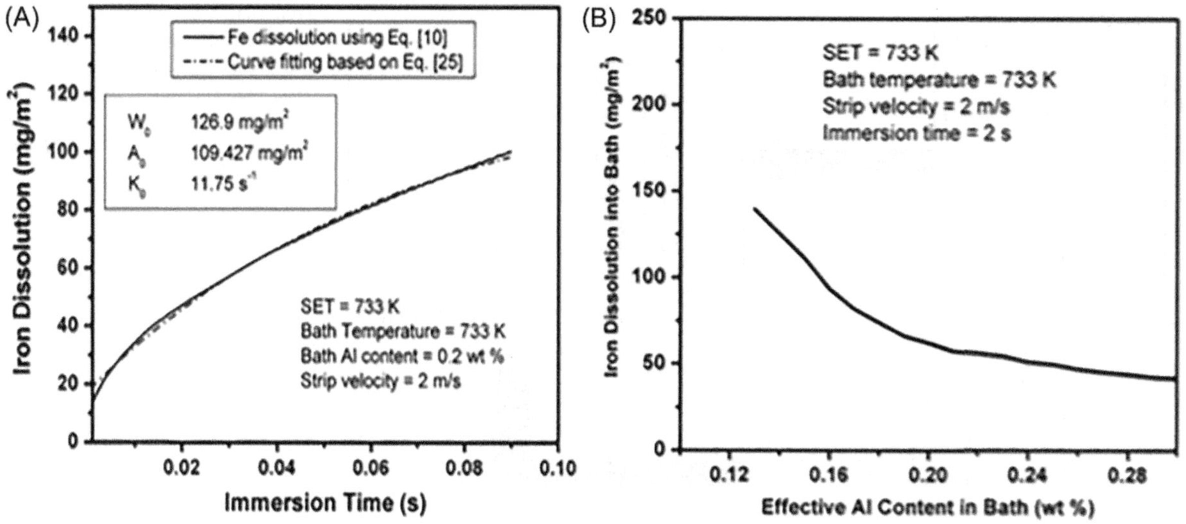

FIG. 5.13 Variation of iron dissolution with (A) immersion time and (B) Al content. *(From G.K. Mandel, R. Balasubramaniam, S.P. Mehrotra, Theoretical investigation of the interfacial reactions between solid iron and liquid zinc, Metall. Mater. Trans. 40A (2009) 637–645.)*

5.5.2 Inhibition layer formation

During continuous hot-dip galvanizing, the formation of a uniform Fe-Al inhibition layer at the steel interface with a proper thickness is an important issue for industrial production. The inhibition layer prevents the nucleation and growth of brittle Fe-Zn intermetallic compounds that deteriorate the adhesion of the galvanized coating and result in an inhomogeneous distribution of the coating. The inhibition layer was identified as Fe_2Al_5 with some dissolved Zn or $Fe_2Al_5Zn_x$ $(0<x<1)$ [1]. When the Al concentration in the zinc bath is lower than 0.14 wt%, a continuous inhibition layer does not form and some Fe-Zn intermetallic compounds develop; however, when the Al concentration is higher than this threshold, the Fe-Al inhibition compound forms and hinders the formation of Fe-Zn compound [34]. Experimental studies and industrial observations indicate that the inhibition layer at the coating-substrate interface becomes thinner when the bath aluminum content increases to over 0.2 wt% [35]. A typical inhibition layer microstructure is shown in Fig. 5.14. The microstructure shows the zinc layer, columnar $FeZn_{13}$ (ζ) phase, a continuous inhibition layer, and an underlayer in contact with the inhibition layer and steel matrix. The underlayer was found to be associated with Mn segregation and oxidation. A schematic of the microstructure of the region near the zinc/steel substrate interface is seen in Fig. 5.15.

The inhibition layer formation occurs quickly within only a few seconds and is affected by various factors including chemical constituent, temperature of molten zinc, substrate material, etc. [1,36,37]. Moreover, multiple processes are involved in the inhibition layer formation, including dissolution of Fe, diffusion of Fe and Al in liquid zinc and the solid inhibition layer, reaction between Fe and Al in competition with the Fe-Zn reaction [1,34,38,39]. The typical morphology of inhibition layer formed during galvanizing is a double-layer structure, with the lower layer next to substrate containing a continuous, compact, roughly equiaxed grains with sizes about tens of nanometers, and the upper layer adjacent to the liquid zinc comprising coarser, larger, elongated grains with a size about hundreds of nanometers [1,2,34,39–44]. Aslam et al. [9] showed a cross section of the zinc coating/substrate interface with a continuous inhibition layer, Fig. 5.16A. The thickness of the interfacial layer was measured at a higher magnification, Fig. 5.16B, showing an average thickness to be 93 ± 26 nm. An external MnO oxide film was captured on the surface of the steel substrate. The oxide film appears to be continuous and has a thickness about 20 nm. Internal oxide particles were observed as well, and the particles are about few tens of nanometers and are in a subsurface layer about 100 nm thick. However, no Si oxides were observed in this region. HRTEM results show that the oxide grains are crystalline. The Fe_2Al_5 or $Fe_2Al_{5-x}Zn_x$ inhibition layer was observed right on top of the external oxide film. The inhibition layer appears to be continuous in nature. Substructures inside the inhibition layer were resolved by HRTEM, which shows that grain boundaries or subgrain boundaries were present.

The fully developed inhibition layer, despite the presence of a continuous Mn oxide film on the surface of the steel substrate, may be attributed to diffusion of Fe through the porous MnO film, the short-circuit diffusion paths between the oxide grains, or the aluminothermic reduction of Mn oxides. On the other hand, Chen et al. [11] proposed that the

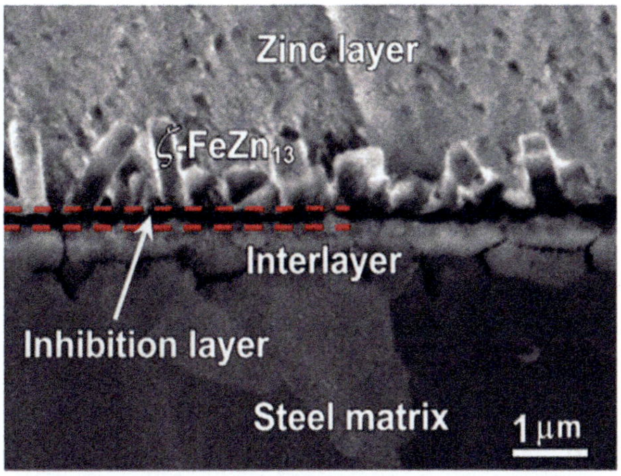

FIG. 5.14 SEM image of a cross section of the zinc-coated steel showing the zinc layer, columnar $FeZn_{13}$ (ζ) phase, continuous inhibition layer, and interlayer in contact with the inhibition layer. *(From G.M. Song, T. Vystavel, N. van der Pers, J. Th.M. De Hosson, W.G. Sloof, Relation between microstructure and adhesion of hot dip galvanized zinc coatings on dual phase steel, Acta Mater. 60 (2012) 2973–2981.)*

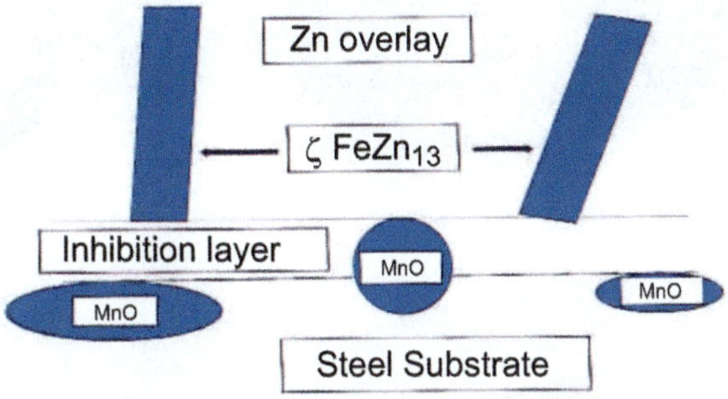

FIG. 5.15 Schematic representation of the microstructure near the zinc coating/steel interface.

surface MnO oxide was lifted off the surface of the substrate and subsequently entrapped in the Zn-Fe phase during galvanizing. A schematic of this microstructural result is shown in Fig. 5.17.

In a more recent study, Min et al. [45] confirmed that the interfacial layer was identified as a Zn-bearing Fe_2Al_5 phase and that the thickness increases with Al concentration in the bath. They also found that the particles of Fe_2Al_5 formed in a 0.8 wt% Al bath were smaller than those formed on a 0.2 wt% Al bath. On the surface of the inhibition layer generated in the 0.8 wt% Al bath, there are some tiny particles (approximately tens of nanometers) randomly dispersed on larger particles, indicating that the growth of the inhibition layer is so fast that the growth mode changes, and nucleation occurs on the inhibition layer/zinc interface.

5.5.3 Inhibition layer breakdown

Inhibition layer breakdown is most important in the galvanneal (GA) process (see Chapter 8). Different mechanisms have been proposed in the literature to explain inhibition layer breakdown [46]. The Zn-Fe outburst reaction is most commonly regarded as being responsible for inhibition layer breakdown. Several mechanisms can be found to explain the outburst formation [1]. One mechanism suggests that the diffusion of Fe toward liquid Zn causes the growth of the inhibition layer, which results in an Al depletion at the solid/liquid interface, leading to outbursts. A more accepted mechanism involves the diffusion of Zn toward the steel substrate. It was found that outbursts are formed at the steel grain boundaries. Further, it was shown that short circuit paths for Zn diffusion through the inhibition layer such as high angle grain boundaries, inhibition layer incorporated oxide interfaces, low angle boundaries, can allow Zn to react with the underlying steel substrate to cause outburst [1,2]. A schematic of the outburst mechanism through substrate grain boundaries, proposed by Hisamatsu, is shown in Fig. 5.18.

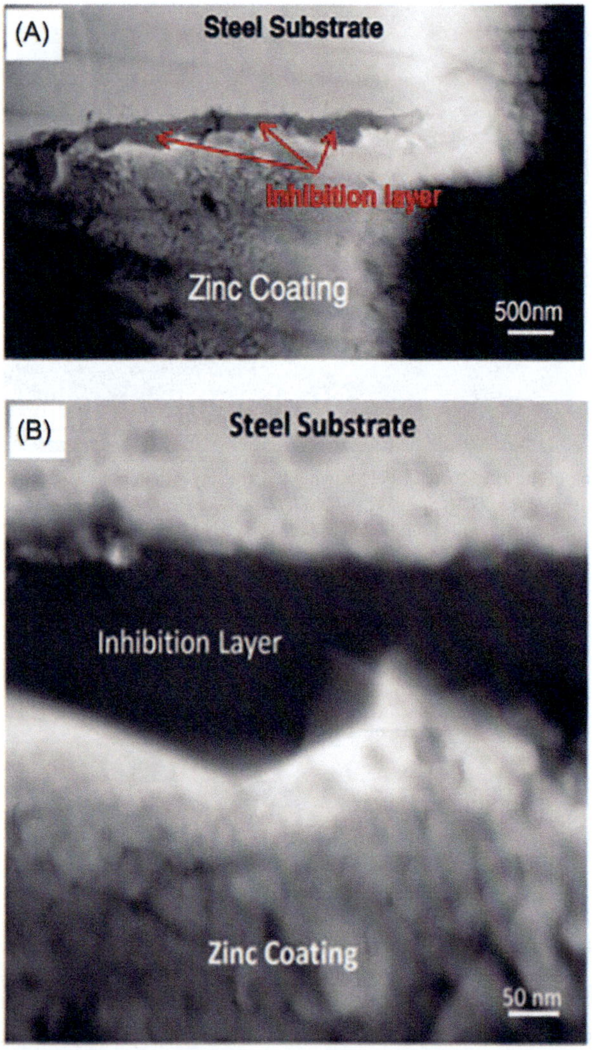

FIG. 5.16 The inhibition layer is captured in TEM using STEM mode. (A) A low magnification image that encompasses the entire inhibition layer present in the specimen. (B) A higher-resolution image detailing the morphology of the inhibition layer present in between the steel substrate and the zinc coating. *(From I. Aslam, B. Li, R.L. Martens, J.R. Goodwin, H.J. Rhee, F. Goodwin, Transmission electron microscopy characterization of the interfacial structure of a galvanized dual-phase steel, Mater. Charact. 120 (2016) 63–68.)*

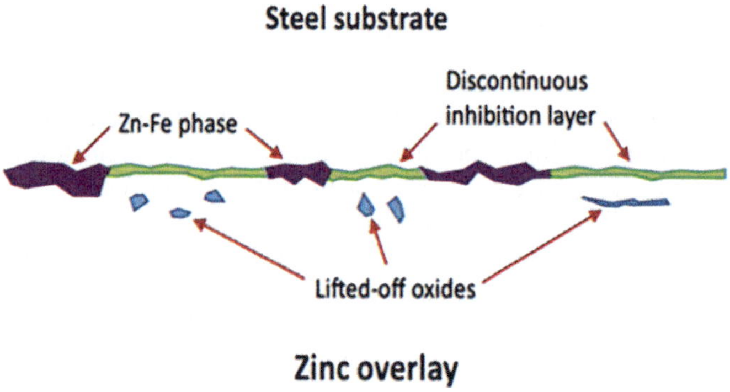

FIG. 5.17 Schematic of the interfacial structure of the hot-dipped DP steel in a Zn bath with 0.125% Al without annealing. The inhibition layer is discontinuous and the MnO oxides are lifted off the substrate surface due to interfacial reaction. This scenario is very different from the galvanized DP steel in a Zn bath with ~0.2% Al in which a fully developed inhibition layer directly grew on top of MnO oxides. *(From K.F. Chen, I. Aslam, B. Li, R.L. Martens, J.R. Goodwin, F. Goodwin, Lift-off of surface oxides during galvanizing of a dual-phase steel in a galvannealing bath, Metall. Mater. Trans. A 50A (2019) 3748–3757.)*

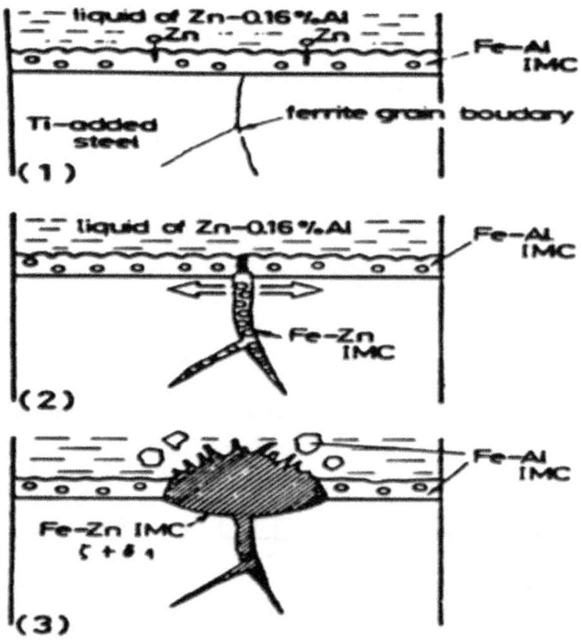

FIG. 5.18 Schematic diagram showing the Fe-Zn outburst growth behavior. *(From A.R. Marder, The metallurgy of zinc-coated steel, Prog. Mater. Sci. 45 (2000) 191–271.)*

5.6 Composition effects on interface reactions

5.6.1 Zn-<1.0% Al

The formation of an intermetallic interfacial layer, or inhibition layer, is required to prevent Fe-Zn phases from forming into the coating. In these alloys, the goal is to produce a relatively pure Zn overlay coating for enhanced corrosion resistance. In these coatings, Al is added to the bath to form the $Fe_2Al_5Zn_x$ inhibition layer. Table 5.2 summarizes the coating microstructure found in continuous galvanizing based on Al content of the bath and indicates that only $Fe_2Al_5Zn_x$ (η) phase is capable of complete inhibition of the formation of Fe-Zn phases. The well-known transition to complete inhibition at around 0.15 wt% Al bath composition was established by Tang [47].

The growth process for Fe-Al inhibition layer occurs with the $FeAl_3$ intermetallic nucleating and transforming to the $Fe_2Al_5Zn_x$ phase, Fig. 5.19. The microstructural development of the interfacial layer was strongly influenced by the zinc bath dissolved Al content, reaction time, and bath temperature. In a comparison study of the inhibition layer formation with two different Al bath compositions, 0.13 and 0.20 wt% Al, Chen et al. [34] found that the Al uptake in the interfacial layer increased with increasing reaction time for all experimental bath temperatures and for both bath compositions (0.20 and 0.13 wt% dissolved Al). For the 0.20 wt% dissolved Al bath, the interfacial layer comprised a two-layer structure, with the layer next to the substrate being composed of a finely crystalline, compact layer of $FeAl_3$, and the upper layer consisting of a course, noncompact layer of $Fe_2Al_5Zn_x$•$FeAl_3$ was the initial phase formed and subsequently transformed to $Fe_2Al_5Zn_x$ with longer reaction times. The final microstructure for the interfacial layer consisted of $Fe_2Al_5Zn_x$. Both the $FeAl_3$ and $Fe_2Al_5Zn_x$ intermetallics showed a preferred orientation with respect to the substrate surface [34,48,49], Fig. 5.19.

The evolution of the interfacial layer formed in a zinc bath with 0.13 wt% dissolved Al, including Fe-Al and Fe-Zn intermetallic compounds, was the result of competing processes. The Fe-Al phases formed initially and continued to grow by a diffusion mechanism. However, the Fe-Zn phases also nucleated and grew simultaneously via the mechanism of inhibition layer breakdown. The experimental results indicate that inhibition layer breakdown and formation of ζ $FeZn_{13}$ occur within immersion times associated with industrial galvanizing. For the 0.20 wt% dissolved Al bath, the incubation and nucleation events were completed, and the interfacial Fe-Al structure fully formed within a 2-s reaction time, after which the Fe-Al layer grew by a diffusive mechanism. The experimental results indicate that the Fe-Al interfacial layer structure is fully formed in immersion times also associated with industrial galvanizing [34]. In a study of strip entry temperature, Yang et al. [44] found that for a 0.20 wt% Al bath composition, the Fe-Al interfacial inhibition layer was fully formed at a strip entry temperature range of 440–480°C. For an Al bath composition of 0.12 wt%, the interfacial layer consisted of ζ $FeZn_{13}$ phase, and the size of the phase was increased with strip entry temperature. For the Al bath composition of 0.20 wt%, the absence of ζ $FeZn_{13}$ phase is notable, and the thickness of the Fe_2Al_5 layer increases with strip temperature, Fig. 5.20.

TABLE 5.2 Summary of coating microstructure in continuous galvanizing.

Al content (wt%)	Equilibrium compound	Intermetallics in coating	Alloy layer characteristics	Nucleation rate (s)	Growth rate (μm/s)
< 0.100	ζ	$\zeta/\delta/\Gamma'/\Gamma$	Continuous	$\zeta \approx 10^{18}$	Up to ≈ 10
0.100–0.135	δ	$\zeta/\delta/\Gamma'/\Gamma$	Gaps exist	$\zeta \approx 10^{18}$	≈ 1
0.135–0.140	η	Mostly ζ	Discontinuous	$\zeta \approx 10^{15}$	$\zeta \approx 0.5$, $\eta \approx 0.05$
0.140–0.145	η	ζ plus η	ζ dissolution	ζ, $\eta \approx 10^{12}$	$\zeta \approx 0.1$, $\eta \approx 0.05$
0.145–0.150	η	Mostly η	ζ dissolution	$\eta \approx 10^{12}$	$\eta \approx 0.05$
>0.150	η	η	Full inhibition	$\eta \approx 10^{12}$	$\eta \approx 0.05$

From N.-Y. Tang, Thermodynamics and kinetics of alloy formation in galvanized coatings, in: F.E. Goodwin (Ed.), Zinc-Based Steel Coating Systems: Production and Performance, TMS, Warrendale, PA, 1998. p. 3.

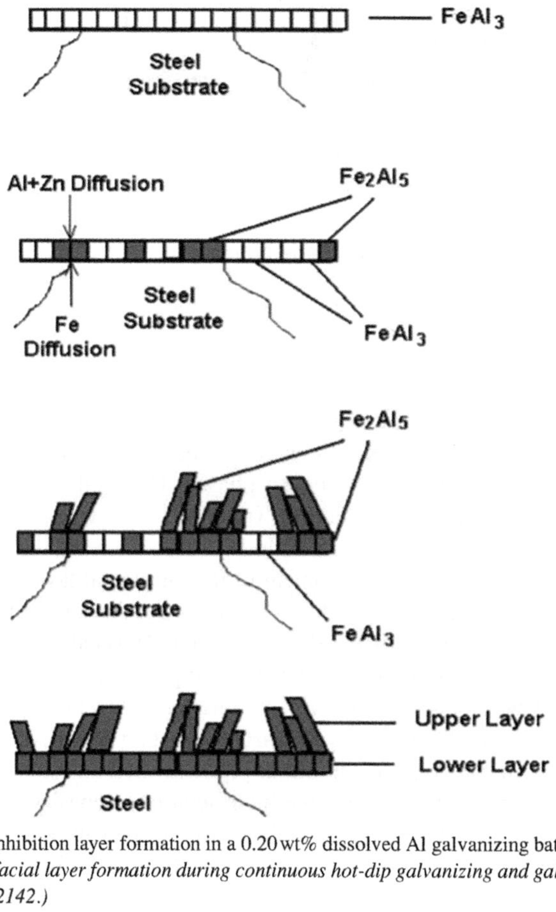

FIG. 5.19 Schematic representation of inhibition layer formation in a 0.20 wt% dissolved Al galvanizing bath. *(From L. Chen, R. Fourmentin, J.R. Mc Dermid, Morphology and kinetics of interfacial layer formation during continuous hot-dip galvanizing and galvannealing, Metall. Mater. Trans. A Phys. Metall. Mater. Sci. 39A (9) (2008) 2128–2142.)*

5.6.2 Zn-5%Al

Limited research has been conducted on the interfacial inhibition layer of Zn-5%Al coatings since Ghuman and Goldstein [50]. Most investigators agree that the interfacial layer is of the form $Fe_2Al_5Zn_x$ [51–54]. This result is not unexpected since it was reported in the previous section that increasing Al above approximately 0.15%Al enhances the stability of the $Fe_2Al_5Zn_x$ inhibition layer.

Wang et al. [55] recently reported that the interfacial layer was solely composed of $Fe_4Al_{13-x}Zn_x$ crystals with a monoclinic crystal structure, with a uniform thickness of 50–100 nm, and a well-defined orientation relationship with the underlying α iron substrate. According to Wang et al., Fig. 5.21 shows an SEM secondary electron image of the Zn-5 wt% Al

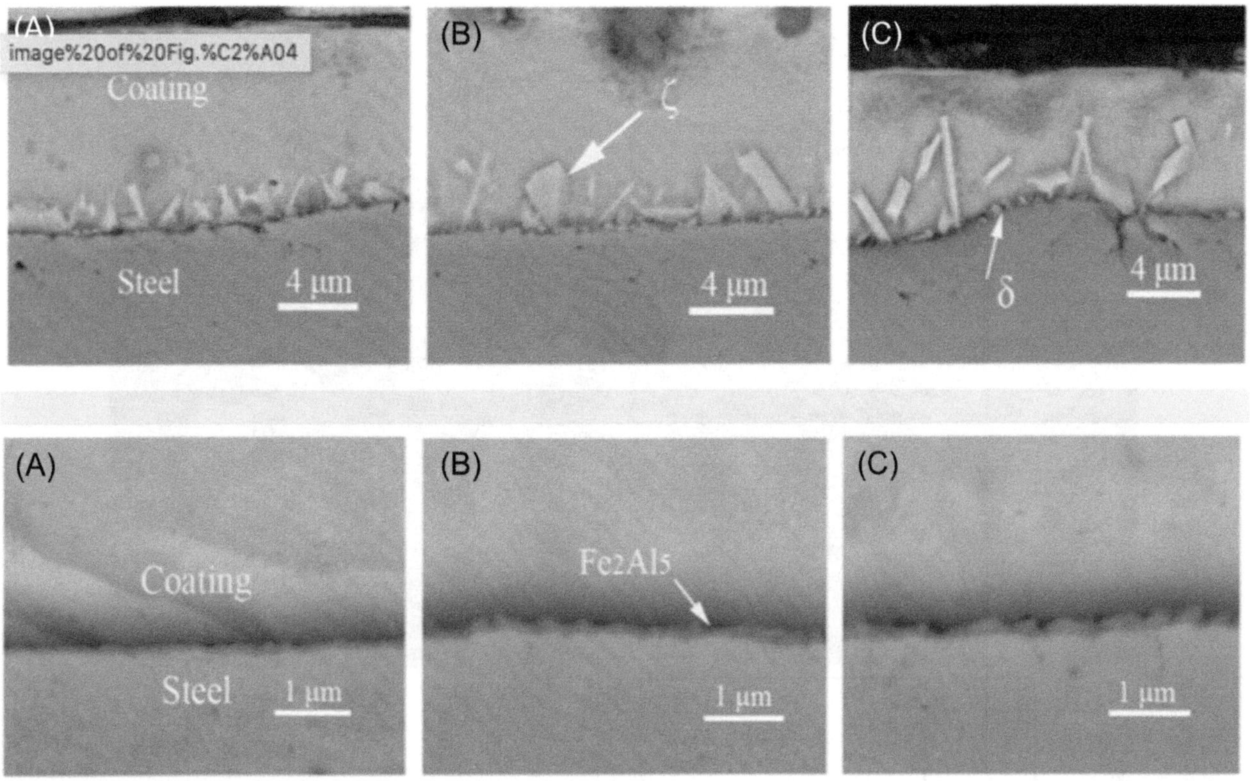

FIG. 5.20 SEM backscattered micrograph of cross sections of galvanized coatings in 0.12 wt% Al baths (upper row) and in 0.2 wt% Al baths (lower row). The coating was etched with 1% amyl nital for 3 s. (A) Strip entry temperature at 440°C, (B) strip entry temperature at 460°C, (C) strip entry temperature at 480°C. *(From H. Yang, S. Zhang, J. Liu, H. Wang, Effect of strip entry temperature on the formation of interfacial layer during hot-dip galvanizing of press-hardened steel, Surf. Coat. Technol. 240 (2014) 269–274.)*

coating, revealing a typical microstructure consisting of proeutectic Zn (η-Zn) islands surrounded by eutectic lamellae of β-Al and η-Zn phases. After dissolution of the overlying coating by acid, the morphology of the inhibition layer under the coating can be observed, as illustrated in the SEM micrograph in Fig. 5.21B. The steel substrate is covered with a dense layer of Fe-Al crystals. The layer can be divided into rough regions (labeled R) and smooth regions (labeled S), the locations of which do not correlate with the microstructure of the overlying Zn coating. In addition, high-magnification micrographs of the S and R regions are shown in Fig. 5.21C and D, respectively. The S region is covered with fine crystals without distinct facets, and the size of these crystals is 100 nm or less. By contrast, the R region is covered with crystals of larger sizes (~200 nm) of irregular shape. The crystals seem to pile up on each other to form an interfacial layer with nonuniform thickness. In addition, a few crystals coalesce to form a multiple leaved or flower-like morphology, as indicated by the white circles in Fig. 5.21D. A typical energy dispersive X-ray spectrum of the crystals, shown in Fig. 5.21E, demonstrates that these crystals contain Al, Zn, and Fe, with a 3:1 ratio of (Al + Zn) to Fe, which is consistent with the stoichiometric ratio of $FeAl_{3-x}Zn_x$ or $Fe_4Al_{13-x}Zn_x$, where x is approximately 0.6 for the latter.

Similar results are found for Zn-20% Al alloy coatings [52] when Si is added to a Zn-22.3 wt% Al coating. The results were found to be the same as that found when Si is added to a Zn-55% Al (Galvalume) bath [56]. Studies of the interfacial layer in coatings produced in a molten Zn-22.3% Al-0.4% Si alloy revealed a double layer with an Fe_2Al_8Si sublayer on the top and an Fe_2Al_5 sublayer on the bottom. This suggests that a diffusion path is established during the hot-dipping stage, starting from the liquid phase, and passing through the phase fields of the ternary inhibition compound of Fe_2Al_8Si and all binary Fe-Al intermetallic compounds available in the Al-Fe system. Thermodynamic calculations of the free energies of the relevant phases have confirmed the feasibility of such a diffusion path [56].

5.6.3 Zn-55%Al

Several important studies have been made on the intermetallic interfacial (inhibition) layers in 55 wt% Al-Zn-Si alloy baths. They include: (1) intermetallic layer development as a function of silicon bath content of 55wt%Al alloys [56]; (2) the role of bath temperature on intermetallic layer composition [57]; (3) the role of strip preheat temperature [58]; (4) study of the

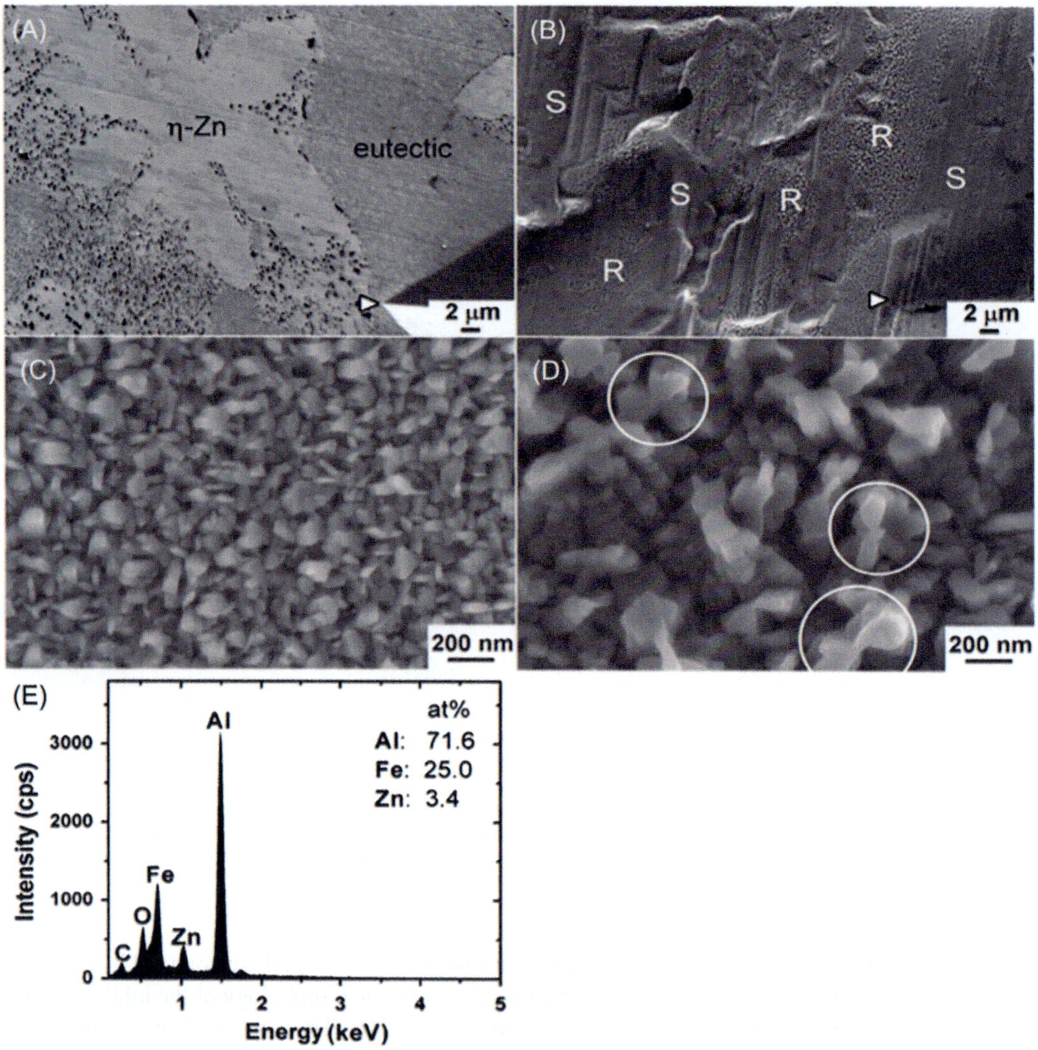

FIG. 5.21 SEM micrographs of the surfaces of (A) the Zn-5 wt% Al coating and (B) the Fe-Al interfacial layer of the same area, as well as high-magnification micrographs of the (C) S region and (D) R region, labeled in (B), of the interfacial layer. (E) Energy dispersive spectrum of the interfacial layer acquired at 5 kV. The *triangular marks* in (A) and (B) denote the position of an indent ensuring that (A) and (B) were acquired from the same area. *(From K.-K. Wang, C.-W. Hsu, L. Chang, W. J. Cheng, Characterization of the FeAl intermetallic layer formed at FeZn interface of a hot-dip galvanized coating containing 5 wt.% Al, Surf. Coat. Technol. 396 (2020) 125969.)*

reactions in Al-Si bath alloys [59]; and (5) the critical role silicon plays in the stabilization of the three primary intermetallic phases in the 55 wt% Al-Zn-Si hot-dip baths [60].

Selverian et al. [56] studied 55 wt% Al-Zn-Si baths with silicon contents of 0.7, 1.7, 3.0, 5.0, and 6.88 wt% that were coated at a temperature of 610°C. The bath was saturated with iron (approximately 2 wt%), and then steel coupons were cleaned and fluxed before being immersed for up to 3600 s. Fig. 5.22 shows that for a silicon bath content of 3.0 wt%, the intermetallic layer was the thinnest. For bath contents of 3.0 and 5 wt% Si, the growth was interface reaction controlled, and the growth of the intermetallic layer was found to be diffusion controlled for bath containing 0.7, 1.7, and 6.88 wt% silicon. It was concluded that the change to interface reaction control for 3.0 and 5.0 wt% Si baths was due to either the interface between Fe_2Al_5/τ_1, $\tau_1/FeAl_3$, or τ_2/τ_{5H}.

In a parallel study of Zn-Al baths without Si additions [61], the reaction that occurs between iron panels and Al-Zn baths with 45% Al, 55% Al, and 75% Al was very rapid and strongly exothermic. In all three of the baths, the iron panel was consumed within 2 min by the bath. The reaction between the iron panels and Al-Zn baths was controlled by the diffusion of Al through the liquid bath. The strong exothermic reaction, which caused the panels to become incandescent, can be explained in terms of the heat of formation of the $FeAl_3$ phase, which was the predominate phase that formed. When the panel was in the molten bath, the bath acts as a large heat sink for the heat evolved during the reaction. When the panel

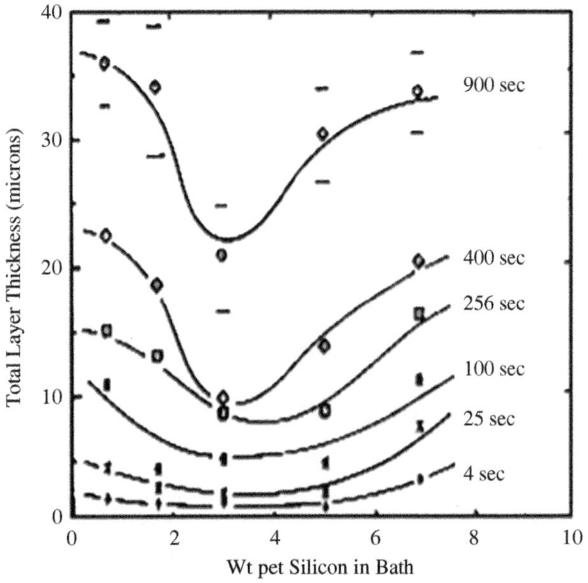

FIG. 5.22 The effect of silicon content in a Zn-55 wt% bath on total layer thickness. *(From J.H. Selverian, A.R. Marder, M.N. Notis, The effect of silicon on the reaction between solid iron and liquid 55 wt pct Al-Zn baths, Metall. Mater. Trans. A 20A (1989) 543–555.)*

is removed from the bath, the heat from the reaction is totally absorbed by the panel. Due to the small thermal mass of the panel, its temperature rose rapidly, becoming incandescent if not quickly water quenched. Thus, it was established that the Zn-55%Al baths require a Si addition to prevent the steel substrate from being destroyed.

In the work by Mercer [57], the temperature of hot dipping was increased from 605°C to 620°C and the dual intermetallic structure of $Fe_2Al_5/Fe_5Si_2Al_{20}$ changed to $Fe_2Al_5/FeAl_3$. This transition was used to explain the observation that the measured thickness of the intermetallic layer increased, as the bath temperature was raised from 590°C to 605°C, but no further increase in the thickness of the intermetallic phase was seen with a further increase in the bath temperature to 620°C. Although, no increase in intermetallic layer thickness was reported for an increase in bath temperature from 605°C to 620°C, samples presented in that study do show a clear difference in the appearance of the intermetallic layer.

The importance of strip preheat temperature on the growth of the intermetallic layer in 55 wt% Al baths was studied [58]. They used a bath composition of Al-43.4 wt% Zn-1.6 wt% Si, a bath temperature 610°C, and varied the strip preheat temperature between 480°C and 700°C, with immersion times between 0.02 and 0.3 s. The results showed that increasing strip preheat temperature produced a significant increase in the intermetallic layer thickness, and they proposed that $FeAl_3$ is the first phase that formed, followed by the epitaxial formation of Fe_2Al_5 on the substrate, and α-AlFeSi at the overlay interface. They also concluded that at low strip preheat, $FeAl_3$ was the only phase to form, and at higher strip preheat, Fe_2Al_5 formed adjacent to the substrate. Eggeler et al. [59] studied the influence of 2 wt% silicon on the intermetallic layer growth kinetics in the binary Al-Si system. They dipped steel coupons into either an Al bath or the Al-2 wt% bath at 780 or 791°C, for up to 900 s. It was determined that the intermetallic layer comprised two phases, Fe_2Al_5 next to the substrate and the $FeAl_3$ phase above. The addition of 2 wt% silicon to the bath decreased the thickness of the Fe_2Al_5 layer from 295 to 35 nm. They reasoned that silicon atoms occupy the structural vacancies of the Fe_2Al_5 phase.

Phelan et al. [60] in their study determined the formation of intermetallic phases on steel substrates immersed in Zn-55 wt% Al-Si hot-dip baths as a function of dipping time and bath Si content. Fig. 5.23 summarizes the generally accepted progress of intermetallic layer formation on steel strip immersed into a 55 wt% Al-Zn-Si bath. According to this process, the steel is immersed into the bath forming an interface between the liquid and steel, across which iron diffuses into the bath and aluminum diffuses into the steel. After some incubation period, thought to be of the order of milliseconds, either $FeAl_3$ or α-AlFeSi intermetallic phases form on the steel. After some further period of time Fe_2Al_5 forms between the steel and $FeAl_3$/α-AlFeSi phase (Fig. 5.23C). During solidification (Fig. 5.23D), facetted crystals of α-AlFeSi form in the interdendritic regions of the Al-Fe-Si coating or are entrained from the bath itself. A light optical micrograph of a sample dipped into Bath A is presented in Fig. 5.23E, which shows the various intermetallic phases as described schematically. They were able to determine from the experimental and modeling work that silicon plays a critical role in the stabilization of the three primary intermetallic phases in the 55 wt% Al-Zn-Si hot-dip baths. The three intermetallic

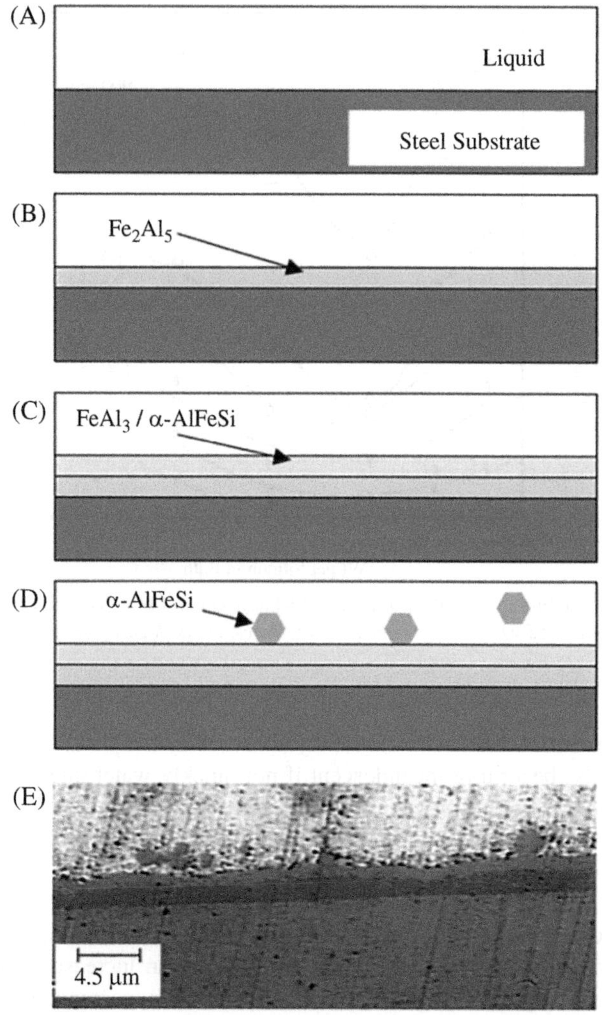

FIG. 5.23 Schematic representation of the development of intermetallic phases on hot-dipped Zn-55% Al-Si coatings. *(From D. Phelan, BJ. Xu, R. Dippenaar, Formation of intermetallic phases on 55 wt.% Al–Zn–Si hot dip strip, Mater. Sci. Eng. A 420 (2006) 144–149.)*

phases are $FeAl_3$, α-AlFeSi, and β-AlFeSi, which are stabilized in this sequence by an increase in silicon or decrease in temperature. The intermetallic phase stability in the bath plays a critical role in the development of the intermetallic reaction zone on hot-dipped steel substrates.

5.6.4 Zn-Al-Mg

Improvement in corrosion resistance due to the presence of Mg in Zn coatings has been well established. When small amounts of Mg are added to low Al Zn-Al baths (0.28 wt% Al, 0.28 wt% Mg or 1.0 wt% Al, 1.8 wt% Mg), it was found that the Fe-Al inhibition layer formed in the 0.28%Al-0.28%Mg bath was the $Fe_2Al_{5-x}Zn_x$ phase, which exhibited two morphologies [62]. One was characterized as a rough, granular layer composed of two or three variants having a common crystallographic direction parallel to the normal direction of the steel sheet. Another shows a smooth layer of single-crystal $Fe_2Al_{5-x}Zn_x$ grown epitaxially on the substrate. The epilayer has a (001) Fe_2Al_5//(110) Fe interface. On the other hand, numerous disk-shape Fe-Al crystals were found at the interface of the steel dipped in a 1.0%Al-1.8%Mg bath. The intermetallic compounds were thin decagon columns each of which consists of five variants of [010]-oriented Fe_4Al_{13} crystals containing high density of twins. No orientation relationship between the Fe_4Al_{13} crystals and α-Fe was found.

In a recent study of several steel substrates immersed in a bath of Zn-1.1 wt% Al-1.1 wt% Mg, Bordginon and Goodwin [63] evaluated wettability. They studied the effect of the Al signal in the inhibition layer as well as the morphology of the particles. In Fig. 5.24, the Al signal for the IF steel in contrast to the other substrates is high and tight, indicating a thin, compact inhibition layer as shown in Fig. 5.25. Moving from IF to DPs to TRIP steels, the Al peak gets broader and lower. This coincides with inhibition layers shown in Fig. 5.25.

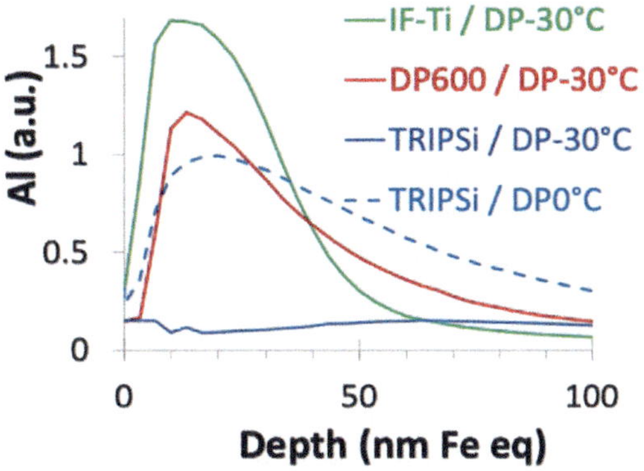

FIG. 5.24 GDOS profiles of the Zn11/11 inhibition layer after having removed the coating by electrochemical dissolution. *(From L. Bordignon, F. Goodwin, Galvanizing of TRIP-Si steels, in: GALVATECH '21, 2021, pp. 394–403.)*

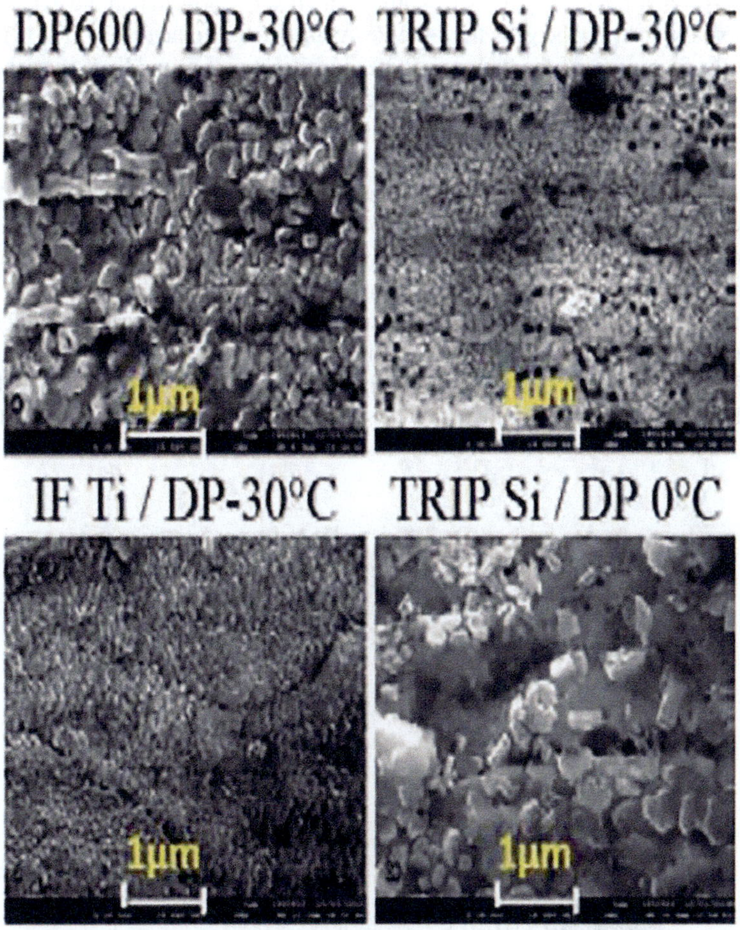

FIG. 5.25 SEM images of the inhibition layer. *(From L. Bordignon, F. Goodwin, Galvanizing of TRIP-Si steels, in: GALVATECH '21, 2021, pp. 394–403.)*

The SEM observation of the inhibition layer (Fig. 5.25) indeed indicates that fine Fe_2Al_5 crystals exist on the surface of all the steels. However, no inhibition layer is observed for the TRIP-Si. The Fe_2Al_5 crystals are bigger for the DP600 with numerous holes between them. The selective oxidation, which was in these holes, has been dissolved during the zinc etching process. The size of the Fe_2Al_5 crystals is larger than the one observed for the IF-Ti, and they do not cover the whole surface, which is characterized by a lower and large Al peak in the GDOS profile (Fig. 5.24). Therefore, the reactive wetting should be better for the IF-Ti grade and progressively deteriorated respectively for the DP600 and the TRIP-Si.

Studies on the effect of adding Mg up to 5% to a 5% Al bath shows the formation of two distinct layers. Mg tends to exist in the outer overlay coating layer while Al is found in the inner $Fe_2Al_5Zn_{0.4}$ intermetallic interface layer [64]. In a more recent study on a Zn-6 wt% Al-3 wt% Mg coating [65], it was found that the initial formation of a continuous θ-Fe_4Al_{13} phase layer was followed by the local growth of an η-Fe_2Al_5 phase toward both the steel and Zn alloy melt sides during the hot dipping process. Mg does not form any new phases with the Fe-Al intermetallic compounds [66]. However, it exists in two forms in the reaction layer: (1) during the nucleation and growth of Fe-Al intermetallic compounds, Mg is segregated to the grain boundaries of the Fe-Al phase resulting in grain refinement, and (2) Mg is in solid solution within the Fe-Al grains, which changes the lattice constants for the Fe-Al phases. This inhibits the diffusion of Al atoms into the Fe matrix and the growth of the $Fe_2Al_5Zn_{0.4}$ phase. The distribution of Al in a Zn-6Al-3Mg (ZAM) alloy for long-term exposure of 300 s at 450°C is seen in Fig. 5.26 [67]. This figure clearly shows the distribution of Al in the entire counting (red line). It demonstrates the existence of the large $Fe_2Al_5Zn_x$ interface phase adjacent to the substrate steel and the existence of $FeAl_3Zn_x$ between the overlay coating and the $Fe_2Al_5Zn_x$ interface phase and the Al distribution in these two phases.

Honda et al. [68] investigated the effect of Si additions to a Zn-10%Al-3%Mg bath. They found that at the interface between the coating and the substrate, Si additions resulted in the uniform formation very fine grains of 20–30 nm in diameter in the Fe_2Al_5 phase with dissolved Si and Zn in solid solution. Li et al. [69] studied the effect of Si on the growth of the Fe-Al intermetallic layer in a Zn-11%Al-3%Mg coating. The results showed that the existence of Si in the bath could affect the phase constitution and restrain the growth of the Fe-Al intermetallic layer, Fig. 5.27. The thickness of the Fe-Al layer is a minimum when the Si content of the bath was 0.2%. Further it was found that when the Si content of the bath is less than 0.1%, the phase constituents of the interfacial layer consist of $Fe_2(Al,Si)_5$ and $FeAl_3$ phase. However, when the Si content is 0.2%, the layer is only composed of $Fe_2(Al,Si)_5$ phase. Above 0.2% Si, the layer is composed of $Fe_2(Al,Si)_5$ phase and τ_4 phase.

Gao et al. [70] reported on the existence of a periodic layered structure that was developed in the overlay coating at long dipping times (up to 120 s) for Al contents from 15% to 55% Al and 0.5–3.0 wt% Mg. The alternating layers of the periodic layered structure consist of a $FeAl_3Zn_x$ and an (Al+Zn) layer, Fig. 5.28. In addition, the coating contains a thin $Fe_2Al_5Zn_x$ layer between the substrate and the periodic layered structure.

Further additions to Zn-Al-Mg coating baths have been studied. For example, Chen et al. [67] studied the effect of Ti on the growth of the Fe-Al layer in hot-dipped Zn-6Al-3Mg coatings. They found that the solubility of Ti in the $Fe_2Al_5Zn_x$ and $FeAl_3Zn_x$ phases is 0.052 wt% and 0.032 wt%, respectively. When the Ti content of the alloy bath was 0.05 wt%, the $FeAl_3Zn_x$ and $Fe_2Al_5Zn_x$ layers were found to be the thinnest. When the Ti content of the bath was less than 0.052 wt%, the thickness of the $FeAl_3Zn_x$ and $Fe_2Al_5Zn_x$ layers decreased with the increase in Ti content. The addition

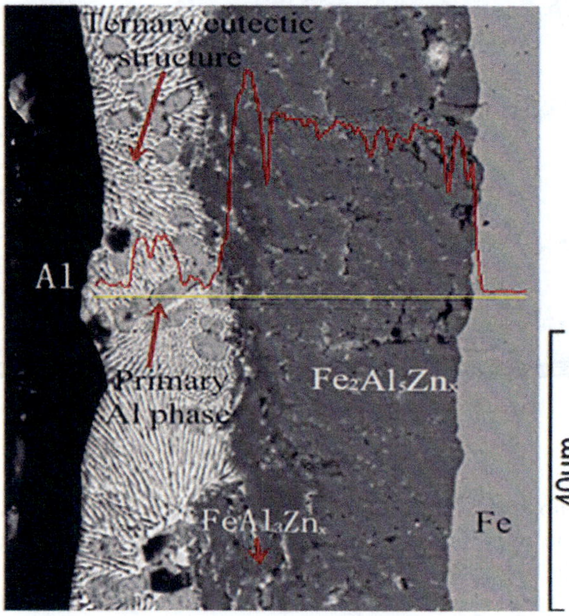

FIG. 5.26 The line scale curve of Al through the cross section of Zn-6Al-3Mg (ZAM) after 3300 s at 450°C. *(From Y. Chen, Y. Liu, H. Tu, C. Wu, X. Su, J. Wang, Effect of Ti on the growth of the Fe–Al layer in a hot dipped Zn–6Al–3Mg coating, Surf. Coat. Technol. 275 (2015) 90–97.)*

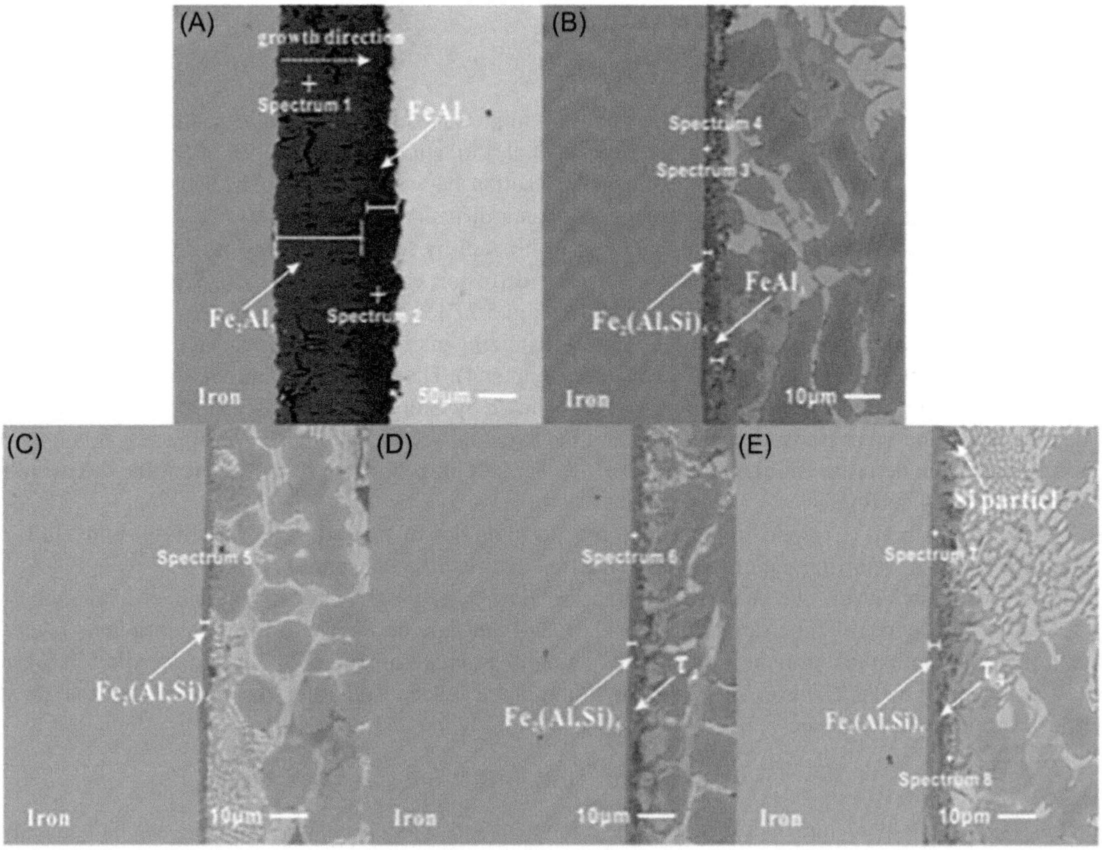

FIG. 5.27 Cross-sectional SEM images of Zn-11%Al-3%Mg-x%Si alloy coatings dipped 180 s at 510°C. (A) 0%Si; (B) 0.1%Si; (C) 0.2%Si; (D) 0.3%Si; (E) 0.6%Si. *(From K. Li, Y. Liu, H. Tu, X. Su, J. Wang, Effect of Si on the growth of Fe-Al intermetallic layer in Zn-11% Al-3% Mg coating, Surf. Coat. Technol. 306 (2016) 390–396.)*

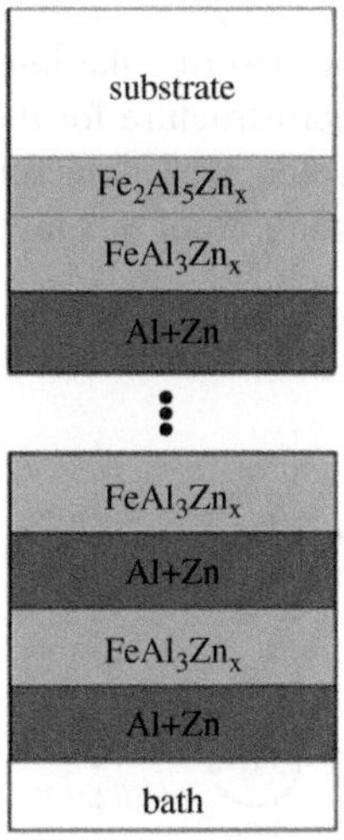

FIG. 5.28 Diagram of phases in the periodic layered structure. *(From L. Gao, Z. Li, X. Kuang, F. Yin, H. Ji, Formation of periodic layered structure during hot-dip galvanizing in Al-Zn-Mg bath, Surf. Coat. Technol. 304 (2016) 306–315.)*

of Ti in the bath inhibits the growth of the $Fe_2Al_5Zn_x$ layer in the Zn-6Al-3Mg coating. When the Ti content of the alloy bath was more than 0.052 wt%, the thickness of the $Fe_2Al_5Zn_x$ and $FeAl_3Zn_x$ layers significantly increased with the increase of Ti content. The addition of Ti in the alloy bath promotes the growth of the $Fe_2Al_5Zn_x$ layer in the Zn-6Al-3Mg coating. The growth of the $Fe_2Al_5Zn_x$ layer in the Zn-6Al-3Mg + 0.4Ti and +0.05Ti coatings is controlled by diffusion of Fe atoms, but the growth of the $FeAl_3Zn_x$ layer in the coatings is controlled by an interfacial reaction.

5.7 Kinetics of the interfacial reaction

5.7.1 Inhibition layer formation

Determination of the mechanisms of interface formation is an important step in developing models for the formation and breakdown of the inhibition layer. As an example, Selverian et al. [56] studying the kinetics of Si additions to 55% Al-Zn baths showed that silicon suppresses the rapid exothermic reaction between a binary Al-Zn bath and an iron panel by forming a solid reaction layer. The solid reaction layer acts as a diffusion barrier for the reactive species. The reaction layer then reduces the reaction rate between the bath and iron panel by several orders of magnitude, as compared with the binary Al-Zn baths. The growth of the reaction layer is diffusion controlled for the 0.7, 1.7, and 6.88 wt% baths and is interface controlled for the 3.0 and 5.0 wt% Si baths, Fig. 5.22 [56].

As schematically described earlier in Fig. 5.12 [25], the galvanizing interfacial reactions are a function of time and occur as soon as the steel strip enters the liquid zinc alloy bath (see Chapter 4). The process involves dissolution of iron in the bath, followed by nucleation and growth of the inhibition layer, as shown by Giorgi et al., Fig. 5.29 [32]. Afterward, diffusion of iron through the inhibition layer leads to breakdown and the eventual growth of other Fe-Zn phases. It is important that models be developed that describe inhibition layer development and breakdown in order to determine the overall process control of these Zn alloy coatings.

Several models have been developed to study the reaction kinetics of this process. The Giorgi et al. [32] model is capable of calculating the following parameters as a function of time: the concentration profiles for iron and aluminum in the boundary diffusion layer, the nucleation rate, the time t_{nucl} (=0.09 s), the average critical embryo radius (r^*_{mean} = 0.82 nm), the degree of supersaturation β (=2.3), and the time necessary to form a continuous layer (=0.2 s). They found that dissolution of iron in liquid zinc containing small amounts of aluminum (0.05 wt%) and iron (0.003 wt%) is controlled by the diffusion of iron in the liquid phase. As shown earlier, they determined that the diffusion coefficient of iron in liquid zinc was:

$$D_{Fe}^{Zn/L} = (9.8 \pm 0.1) \times 10^{-10}\ \text{m}^2\ \text{s}^{-1} \tag{5.3}$$

Chen et al. [34] studied the short-time microstructural development and kinetics of the galvanizing and galvannealing interfacial reaction layer. They determined that the incubation and nucleation events of the interfacial layer formation were completed by the 2-sec reaction time in all cases. For a 0.20 wt% dissolved Al bath, $FeAl_3$ nucleates and grows during the initial stages of interfacial layer formation followed by $Fe_2Al_5Zn_x$ formation by diffusion-controlled transformation and growth. The final microstructure of the interfacial layer consisted of $Fe_2Al_5Zn_x$ in a two-layer arrangement comprising a fine-grained, compact lower layer with a course, noncompact upper layer. The Al content of the interfacial layer increased with reaction time and reaction temperature. A model to describe the interfacial layer growth kinetics as a function of reaction time, bath temperature, and inhibition layer microstructure for the case of the 0.20 wt% dissolved Al bath was proposed by Chen et al. [34], which indicated that the development of microstructure of the interfacial layer had significant influence on the effective diffusion coefficient and growth rate of this layer.

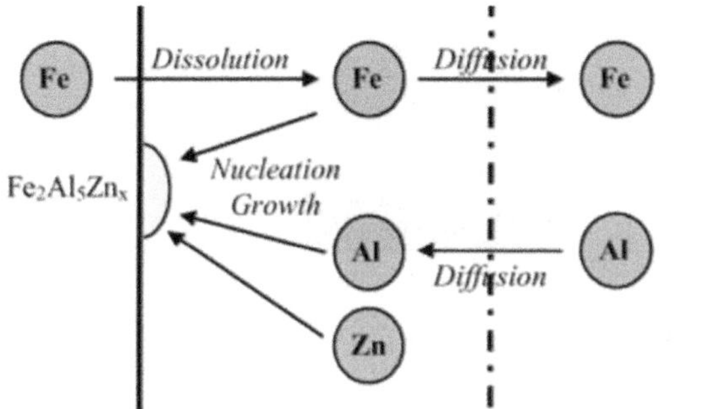

FIG. 5.29 Schematic representation for the galvanizing reactions. *(From M.-L. Giorgi, J.R. Guillot, R. Nicolle, Theoretical model of the interfacial reactions between solid iron and liquid zinc-aluminum alloy, J. Mater. Sci. 40 (2005) 2263–2268.)*

A power-law growth equation, as shown in Eq. (5.4), was used to fit the isothermal Al uptake data such that:

$$Y = K_1 t^n \tag{5.4}$$

where Y is the Al uptake of the interfacial layer in mg/m^2, K_1 the growth rate constant, t the reaction time in seconds, and n the time exponent. The value of n can be affected by the structure and growth mode of this layer. For example, volumetric diffusion-controlled growth is characterized by $n = 0.5$ (i.e., parabolic growth). However, liquid films between growing particles or microstructural coarsening of the growing layer can result in the value of n being less than 0.5. For example, growth of the ζ-$FeZn_{13}$ phase at a steel/pure liquid Zn interface was found to obey $t^{1/3}$ kinetics due to particle coarsening [71]. This finding was attributed to decreasing diffusivity, because the balance between grain boundary and bulk diffusion shifted to the latter during microstructural coarsening.

The Chen et al. model [34] utilized the growth of an oxide film as an analogy for the growth of the Fe-Al interfacial layer that resulted in an expression for parabolic growth controlled by volumetric (lattice) diffusion or an interfacial layer growth rate equation:

$$x^2 = 2\Omega D_{eff}\,\Delta c t = K_p(eff)t \tag{5.5}$$

where $K_p(eff)$ is the overall effective rate constant for mixed diffusion-controlled growth. $K_p(eff)$ is a function of the volume of $Fe_2Al_{5-x}Zn_x$ per Fe atom (Ω), the Fe concentration gradient across the interfacial layer (Δc), and the effective diffusion coefficient (D_{eff}). However, Ω is a constant parameter as $Fe_2Al_{5-x}Zn_x$ was the only phase observed in the interfacial layer for all experimental conditions. Because of local supersaturation of Fe in the vicinity of the steel substrate, Δc can be assumed to be constant. Thus, the only parameter that controls $K_p(eff)$ is D_{eff}, which is a function of the volume fraction of short-circuit diffusion paths in the structure. The model results are shown in Fig. 5.30 with reference to the 460°C and 480°C experimental data.

Alibeigi and McDermid [72] studied the short-time microstructural development and formation kinetics of the Fe-Al interfacial layer on Mn-containing steels as a function of bath dissolved Al (0.20 wt% and 0.30 wt% Al), substrate Mn content (0.2 wt%, 1.4 wt%, 2.5 wt% and 3.0 wt% Mn), and reaction time (2–8 s). They determined that the structure of the Fe-Al interfacial layer for the higher Mn steels (1.4Mn through 3.0Mn) was not homogeneous and consisted of a mixture of fine and coarse crystals. However, with increasing reaction time, the population of fine Fe-Al crystals decreased, and the interfacial layer comprised primarily larger, noncompact Fe-Al crystals. Also, for the higher Mn steels, Mn-rich grains remained at the Fe/Zn interface, resulting in a two-phase interfacial layer consisting of MnO and $Fe_2Al_{5-x}Zn_x$ where the thickness of the remaining Mn-rich layer increased with increasing steel Mn content. Also, fitting the data to a power-law growth equation determined that the interfacial layer Al uptake increased nonparabolically with increasing reaction time, as presented in Table 5.3. For the 0.20 wt% dissolved Al bath, the interfacial layer on the 1.4Mn steel showed

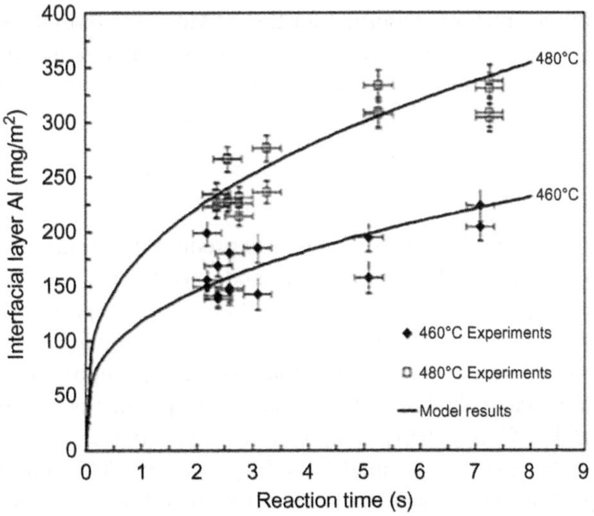

FIG. 5.30 Experimental data of Al uptake as a function of reaction time for a 0.20 wt% dissolved Al bath at 460°C and 480°C with adjustments to the model parameters (equation) such that $K_p(460°C) = 8.0 \times 10^2\,ms^{-1}$ and $K_p(480°C) = 2.0 \times 10\,ms^{-1}$. *(From L. Chen, R. Fourmentin, J.R. Mc Dermid, Morphology and kinetics of interfacial layer formation during continuous hot-dip galvanizing and galvannealing, Metall. Mater. Trans. A Phys. Metall. Mater. Sci. 39A (9) (2008) 2128–2142.)*

TABLE 5.3 Rate equation parameters for interfacial layer growth as a function of bath Al content.

Alloy name	$K_p(eff)$ (m^2/s)	R^2
0.2Mn-0.20 wt% Al	3.16×10^{-15}	0.96
1.4Mn-0.20 wt% Al	4.79×10^{-15}	0.89
2.5Mn-0.20 wt% Al	3.05×10^{-15}	0.92
3.0Mn-0.20 wt% Al	1.73×10^{-15}	0.92
0.2Mn-0.30 wt% Al	2.03×10^{-15}	0.91
1.4Mn-0.30 wt% Al	2.86×10^{-15}	0.87
2.5Mn-0.30 wt% Al	6.63×10^{-15}	0.95
3.0Mn-0.30 wt% Al	4.40×10^{-15}	0.93

From S. Alibeiga, J.R. McDermid, Modeling of interfacial growth kinetics on Mn-containing steels during continuous hot-dip galvanizing, in: Galvatech '15, 2015, pp. 334–341.

the highest Al uptake, with the 0.2Mn, 2.5Mn, and 3.0Mn substrates showing significantly lower Al uptake. It was found that increasing the dissolved bath Al to 0.30 wt% Al resulted in a significantly increased Al uptake being observed for the 2.5Mn and 3.0Mn steels for all reaction times. Further, in the case of the 2.5Mn and 3.0Mn substrates and 0.20 wt% Al bath, additional Al consumption through MnO reduction caused the interfacial layer growth to become Al limited, whereas the very open structure dominated growth in the case of the 0.30 wt% Al bath and resulted in changing the growth kinetics from mixed diffusion-controlled to a more interface-controlled growth mode, as characterized by $n > 0.50$ (Table 5.3) [73].

Additionally, Alibeigi and McDemid [73] used a kinetic model based on an oxide film growth analogy to describe the kinetics of the continuous galvanizing Fe-Al interfacial layer growth for Mn-containing steels as a function of substrate Mn content and bath dissolved Al. The results showed that the interfacial layer microstructural development, which was strongly affected by the presence of MnO at the interfacial layer/substrate interface, had a significant influence on the effective diffusion coefficient of Fe through the Fe-Al layer and the resulting interfacial layer growth rate. For the 0.20 wt% Al bath and 1.4Mn steel, it was determined that, despite additional dissolved bath Al consumption via aluminothermic reduction of the MnO layer, the presence of unreduced MnO in the Fe-Al interfacial layer and the consequently more open interfacial layer structure contributed to increasing the effective diffusion coefficient of Fe through interfacial layer and increased interfacial layer Al uptake versus the 0.2Mn base material. In the case of the 0.30 wt% dissolved Al bath, the more open interfacial layer microstructures observed for the 2.5Mn and 3.0Mn substrates, combined with the higher dissolved Al supply available for both Fe-Al layer growth and aluminothermic reduction of MnO, resulted in the growth kinetics shifting significantly toward being interface controlled from the mixed diffusion mechanism observed for all other alloys and bath Al contents and an increased interfacial layer Al pickup.

5.7.2 Inhibition layer breakdown

Control of the inhibition layer breakdown is important in order to produce the required Zn coating microstructure for galvanneal GA coatings. Zapico, Alvarez, et al. [46] proposed a model for the reaction mechanism for the inhibition layer breakdown kinetics in Ti IF steels, in particular for GA coatings. In review, a very thin layer of Al-rich intermetallic compounds rapidly nucleates (around 0.1–0.2 s) on the steel surface during galvanizing in Al-containing Fe-saturated Zn baths. As discussed earlier, this layer, known as an inhibition layer, inhibits at least temporarily, the alloying reactions between Fe and Zn. The inhibition layer that forms in typical baths for galvannealing production (Fe-saturated Zn baths with Al content ranging from 0.10 to 0.135 wt% at 460°C) is actually composed of a very thin layer of $Fe_2Al_5Zn_x$ (becoming discontinuous for low Al contents) and a thicker layer of Al-saturated δ ($FeZn_7$) on its top. This microstructure allows thermodynamic equilibrium to be reached along all interfaces and the system does therefore not require any other interfacial reaction. Increased temperature in the induction furnace, located at the exit of the galvanizing bath, activates Fe and Zn interdiffusion through the inhibition layer and, after a while, this layer is destabilized. Inhibition layer breakdown is a key step in the galvannealing process as it leads to the sudden development of Fe-Zn phases by consuming liquid Zn.

Different mechanisms have been proposed in the literature to explain the inhibition layer breakdown. It is generally agreed that an outburst reaction is responsible for the inhibition layer breakdown in the case of Ti IF steels [46].

Two mechanisms can be found to explain the outburst formation: (1) an Al depletion at the solid/liquid interface, leading to the outburst formation, and (2) that the triggering phenomenon in the mechanism of outburst formation is the diffusion of Zn toward the steel substrate. It had been previously determined that the steel grain boundaries are preferential locations for outburst formation. The kinetics of the inhibition layer breakdown is strongly influenced by the processing variables and the condition of the base steel.

A model describing the inhibition layer breakdown kinetics through outburst formation was suggested [46]. Because of the complexity of the reaction mechanisms involved in the inhibition layer breakdown, the following simplifications were made in order to facilitate modeling work:

- The Fe diffusion is neglected so that the thickness of the inhibition layer can be considered constant.
- The rate of Zn diffusion from the middle of the steel grains toward the grain boundaries is assumed to be infinite, so that the Zn concentration along the steel/inhibition layer interface can be assumed constant.
- The diffusion of Zn through the $Fe_2Al_5Zn_x$ and δ layers is described by a global diffusion coefficient considering the Zn diffusion in the entire inhibition layer.

Based on these assumptions, a simplified diffusion model can be proposed from Fick's law in steady state. The Zn diffusion flux through the inhibition layer $\phi_{Zn}^{i.\,l.}(\mathrm{mol\,m^{-2}\,s^{-1}})$ can be expressed as follows:

$$\phi_{zn}^{i.l.} = -D_{zn}^{i.l.}\nabla c_{zn}^{i.l.} = D_{zn}^{i.l.}\left|\nabla c_{zn}^{i.l.}\right| = D_o \exp\left(\frac{-E_a}{RT}\right)\left|\nabla c_{zn}^{i.l.}\right| \quad (5.6)$$

where $D_{zn}^{i.\,l.}$ ($\mathrm{m^2\,s^{-1}}$) is the diffusion coefficient of Zn in the inhibition layer, whose variation with temperature T has been expressed through an Arrhenius equation, D_o ($\mathrm{m^2\,s^{-1}}$) being the preexponential factor, E_a ($\mathrm{J\,mol^{-1}}$) the activation energy for diffusion, and R ($\mathrm{J\,mol^{-1}\,K^{-1}}$) the universal gas constant. As the Zn concentrations along all interfaces of the inhibition layer and its thickness are assumed to be constant, the Zn concentration gradient in the inhibition layer $\nabla c_{zn}^{i.\,l.}$ is also constant.

The model results in the following equation for isothermal annealing treatments. The activation energy E_a can be directly extracted from the slope of straight lines expected to be obtained when representing the results in the form of experimental holding time in logarithmic scale against the reciprocal of the holding temperature:

$$ln\, t_{hold.\,exp} = \ln\left(\frac{Q_{hold.\,exp}}{D_o\left|\nabla c_{zn}^{i.l.}\right|}\right) + \frac{E_a}{RT_{hold.\,exp}} \quad (5.7)$$

where $Q_{hold.exp}$ is the part of the critical quantity of Zn that diffuses through the inhibition layer to the steel grain boundaries during the isothermal treatment. Since Zn has already diffused during galvanizing in the pilot simulator, the real time for the inhibition layer breakdown t_c was underestimated when represented by $t_{hold.exp}$. When the value t_c is plotted on a logarithmic scale against the reciprocal of experimental holding temperature $T_{hold.exp}$ for the different Ti IF steels studied, straight lines were obtained, indicating that the time necessary for the inhibition layer breakdown varies with temperature according to an Arrhenius-type law. The values of E_a and $\left(\frac{Q_c}{D_o\left|\nabla c_{zn}^{i.l.}\right|}\right)$ for the different steels are shown in Table 5.4.

TABLE 5.4 Values of $x_{Ti_{excess}}$s, E_a, and $\left(\frac{Q_c}{D_o\left|\nabla c_{zn}^{i.l.}\right|}\right)$ for the Ti IF steels studied [46].

Steel	$x_{Ti_{excess}}$	E_a (J/mol)	$Q_c/\left(D_0\left|\nabla c_{Zn}^{i.l.}\right|\right)$ (s)
A	0.7	245500	3.2×10^{-17}
B	1.8	225500	6.6×10^{-16}
C	2.7	198300	5.7×10^{-14}
D	4.2	177600	1.5×10^{-12}

where $x_{Ti_{excess}}$ in these Ti IF steels is calculated from the weight fractions in the steel and reflects the cleanliness of the steel [46].

In the model proposed by Zapico, Alvarez, et al. [46], the diffusion through the inhibition layer only is considered. Therefore, the values obtained for E_a and D_o should not vary with the steel grade, and the differences in reactivity should only be linked to different values of Q_c, the quantity of Zn needed to cause the inhibition layer breakdown. However, the slopes of the straight lines E_a vary from one steel to another and cannot finally be considered as the activation energy of Zn diffusion through the inhibition layer. This energy was interpreted as the activation energy of Zn diffusion in the grain boundaries, the actual rate limiting step, which varies with $x_{Ti_{excess}}$. As seen in Table 5.4, the activation energy E_a decreases while the parameter $\left(\frac{Q_c}{D_o\left|\nabla c_{zn}^{I.L.}\right|}\right)$ increases when $x_{Ti_{excess}}$ increases. When the value of $x_{Ti_{excess}}$ is higher, the grain boundaries are cleaner. In this case, Zn diffusion in the grain boundaries is faster and the corresponding E_a is lower. However, a higher amount of Zn flux is needed to trigger the outburst reaction, because it requires a certain amount of Zn to be accumulated in a critical volume to stabilize the first nuclei of Γ phase.

The effect of the steel chemical composition on the inhibition layer breakdown kinetics depends upon the parameter $x_{Ti_{excess}}$, which describes the cleanliness of the steel grain boundaries and, consequently, the reactivity of the steel with respect to the inhibition layer breakdown. The experimental results for holding temperatures from 425°C to 460°C show that high values of $x_{Ti_{excess}}$, corresponding to the clean grain boundaries, accelerate the inhibition layer breakdown.

It was concluded by Zapico, Alvarez, et al. [46] that for the Ti IF steels studied, the inhibition layer breakdown by outburst formation occurs through Zn diffusion into the steel grain boundaries and where nucleation of a Fe-Zn phase occurs. The diffusion of Zn toward the steel enriches the steel/inhibition layer interface in Zn, and as a result of this enrichment, the disappearance of the $Fe_2Al_5Zn_x$ layer occurs first, stabilizing a steel/δ interface. Further enrichment of the steel grain boundaries in Zn leads to the nucleation of the Γ phase at these locations, which breaks the inhibition layer off locally. They further found that describing the inhibition layer breakdown kinetics by a simplified Zn diffusion model through the inhibiting layer is problematic because a diffusion model must consider Zn diffusion into the steel grain boundaries. The parameter $x_{Ti_{excess}}$ proposed in this study of Ti IF steels successfully describes the cleanliness of the steel grain boundaries and as a result the reactivity of the steel with respect to the inhibition layer breakdown.

Additional galvanneal kinetic studies have been conducted. Asgari et al. [74] investigated the influence of steel substrate grain size on the phase layer growth kinetics and thickness of zinc hot-dipped steel sheets. Oliveira and Rios [75] proposed a novel iron enrichment isothermal kinetic model for galvanneal coatings. The influence of substrate molybdenum on galvanealing kinetics in AHSS was studied by Krishnardula et al. [76]. A more detailed discussion of the galvannealing process coating formation and kinetics can be found in Chapter 8.

5.8 Summary

The immersion of steel alloys into zinc alloy baths to produce a coated steel product with excellent corrosion resistance and other properties required by industry depends upon the reactions that occur in the zinc bath in less than a few seconds. A critical step involves the interaction that takes place at the interface between the liquid Zn bath and the steel substrate. Fe dissolution and bath reactions at the interface depend upon an inhibition layer that initially forms. This inhibition layer appears to be of the form Fe_2Al_5 with solubility for Zn, in almost all types of baths studied from Zn-Al(<0.2 wt%) through Fe-Al-Mg-Si bath compositions. Short-lived intermediate phases like $ZnAl_3$ types do occur and along with Al_2O_3 in some baths, and sometimes remain after the formation of the inhibition layer $Fe_2Al_5Zn_x$. However, the dominant phase to form and inhibit the formation of Zn-Fe phases and allow the formation of the required overlay coating is the $Fe_2Al_5Zn_x$ inhibition phase.

References

[1] A.R. Marder, The metallurgy of zinc-coated steel, Prog. Mater. Sci. 45 (2000) 191–271.

[2] M. Guttmann, Y. Lepetre, A. Aubrey, M.-J. Roche, T. Moreau, P. Drillet, J.M. Montaine, H. Baudin, Mechanism of the galvanizing reaction. Influence of Ti and P contents in steel and of its surface microstructure after annealing, in: GALVATECH '95, Iron and Steel Society, 1995, pp. 295–307.

[3] S.M.A. Shibili, B.N. Meena, R. Remya, A review on recent approaches in the field of hot dip zinc galvanizing process, Surf. Coat. Technol. 262 (2015) 210–215.

[4] M.-L. Giorgi, J.-S. Diawara, H. Duval, Improvement of wettability between steel and liquid Zn-Al alloy by forced wetting, in: GALVATECH '21, 2021, pp. 445–452.

[5] M.-L. Giorgi, J.-S. Diawara, H. Duval, Improvement of wettability between steel and liquid Zn-Al alloy by forced wetting, ISIJ Int. (2018) 147.

[6] M.-L. Giorgi, J. Diawara, S. Chen, A. Koltsov, J.M. Mataigne, Influence of annealing treatment on wetting of steels by zinc alloys, J. Mater. Sci. 47 (2012) 8483–8495.

[7] K. Chen, I. Aslam, B. Li, F.E. Goodwin, M. Horstemwyer, Effect of fuel-to-air ratio on oxidation and interfacial structure in galvanizing of a dual-phase steel, SAE Int. J. Mater. Manuf. 14 (2021), https://doi.org/10.4271/05-14-03-0021.
[8] G.M. Song, T. Vystavel, N. van der Pers, J.T.M. De Hosson, W.G. Sloof, Relation between microstructure and adhesion of hot dip galvanized zinc coatings on dual phase steel, Acta Mater. 60 (2012) 2973–2981.
[9] I. Aslam, B. Li, R.L. Martens, J.R. Goodwin, H.J. Rhee, F. Goodwin, Transmission electron microscopy characterization of the interfacial structure of a galvanized dual-phase steel, Mater. Charact. 120 (2016) 63–68.
[10] G.M. Song, W.G. Sloof, T. Vystavel, J.T.M. De Hosson, Interfacial microstructure and adhesion of zinc coatings on TRIP steels, Mater. Sci. Forum 539–543 (2007) 1104–1109.
[11] K.F. Chen, I. Aslam, B. Li, R.L. Martens, J.R. Goodwin, F. Goodwin, Lift-off of surface oxides during galvanizing of a dual-phase steel in a galvannealing bath, Metall. Mater. Trans. A 50A (2019) 3748–3757.
[12] V.G. Krishnardula, C.C. Cheng, D.Y.H. Liu, AHSS surface oxide phase formation during annealing and predicting their type using Gibbs energy minimization (GEM) thermodynamic principle, in: GALVATECH '21, 2021, pp. 463–472.
[13] K. Ranganathan, J.R. McDermid, F.E. Goodwin, Continuous hot-dip galvanizing of a third generation (3G) advanced steel, in: GALVATECH '15, 2015, pp. 182–189.
[14] K.M.H. Bhadhon, J.R. McDermid, F.E. Goodwin, Effect of process atmosphere dew point and minor Sn additions on the selective oxidation during intercritical annealing of a medium-Mn third generation advanced high strength steel, in: GALVATECH '21, 2021, pp. 404–412.
[15] R. Sagl, A. Jarosik, G. Angeli, T. Haunschmied, et al., Tailoring of oxide morphology and crystallinity on advanced high-strength steel surfaces prior hot-dip galvanizing, Acta Mater. 72 (2014) 192–199.
[16] Y. Suzuki, T. Yamashita, Y. Sugimoto, S. Fujita, S. Yamaguchi, Thermodynamic analysis of selective oxidation behavior of Si and Mn-added steel during recrystallization annealing, ISIJ Int. 49 (2009) 564–573.
[17] S. Swaminathan, M. Spiegel, Thermodynamics and kinetic aspects on the selective surface oxidation of binary, ternary and quaternary model alloys, Appl. Surf. Sci. 253 (2007) 4607–4619.
[18] R. Khondker, A. Mertens, J.R. McDermid, Effect of annealing atmosphere on the galvanizing behavior of a dual-phase steel, Mater. Sci. Eng. A 463 (2007) 157–165.
[19] V.d.F.C. Lins, L. Madeira, J.M.C. Vilela, M.S. Andrade, V.T.L. Buono, J.P. Guimaraes, E. de Azvedo Alvarenga, Selective oxidation of dual phase steel after annealing at different dew points, Appl. Surf. Sci. 257 (2011) 5871–5878.
[20] H. Liu, Y. He, S. Swaminathan, M. Rohwerder, Effect of dew point on the surface selective oxidation and subsurface microstructure of TRIP-aided steel, Surf. Coat. Technol. 206 (2011) 1237–1243.
[21] Y.F. Gong, H.S. Kim, B.C. De Cooman, Formation of surface and subsurface oxides during ferritic, intercritical and austenitic annealing of CMnSi TRIP steel, ISIJ Int. 48 (2008) 1745–1751.
[22] A. Ollivier-Leduc, M.-L. Giorgi, D. Balloy, J.-B. Guillot, Nucleation and growth of selective oxide particles on ferritic steel, Corros. Sci. 52 (2010) 2498–2504.
[23] T. Van de Putte, Z. Zermout, D. Loison, S. Claessens, J. Penning, Selective oxidation during the austenitic annealing of a CMnSi steel, Adv. Mater. Res. 15–17 (2007) 129–134.
[24] P. Drillet, Z. Zermout, D. Bouleau, J. Mataigne, S. Claessens, Selective oxidation of high Si, Mn and Al steel grade during recrystallization annealing, and steel/Zn reactivity, in: GALVATECH '04, 2004, pp. 1123–1134.
[25] S. Frenznick, S. Swaminathan, M. Stratmann, M. Rohwerder, A novel approach to determine high temperature wettability and interface reactions in liquid metal/solid interface, J. Mater. Sci. 45 (2010) 2106–2111.
[26] M.-L. Giorgi, A. Koltsov, Wetting Assessment Using the dispensed drop method in the field of hot-dip galvanizing, in: GALVATECH '17, 2017, pp. 234–241.
[27] M.-L. Giorgi, J.-S. Diawara, H. Duval, A. Koltsov, Improvement of Wettability between steel and liquid Zn-Al alloy by forced wetting, in: GALVATECH '21, 2021, pp. 445–452.
[28] S.D. Aziz, S. Chandra, Impact, recoil and splashing of molten metal droplets, Int. J. Heat Mass Transf. 43 (2000) 2841–2857.
[29] R. Kavitha, J.R. McDermid, On the in-situ aluminothermic reduction of manganese oxides in continuous galvanizing baths, Surf. Coat. Technol. 212 (2012) 152–158.
[30] L. Cho, S.J. Lee, M.S. Kim, Y.H. Kim, et al., Influence of gas atmosphere dew point on the selective oxidation and the reactive wetting during hot dip galvanizing of CMnSi TRIP steel, Metall. Mater. Trans. A 44 (2013) 362–371.
[31] M.-L. Giorgi, P. Durighello, R. Nicolle, J.R. Guillot, Dissolution kinetics of iron in liquid zinc, J. Mater. Sci. 39 (2004) 5803–5808.
[32] M.-L. Giorgi, J.R. Guillot, R. Nicolle, Theoretical model of the interfacial reactions between solid iron and liquid zinc-aluminum alloy, J. Mater. Sci. 40 (2005) 2263–2268.
[33] G.K. Mandel, R. Balasubramaniam, S.P. Mehrotra, Theoretical investigation of the interfacial reactions between solid iron and liquid zinc, Metall. Mater. Trans. 40A (2009) 637–645.
[34] L. Chen, R. Fourmentin, J.R. Mc Dermid, Morphology and kinetics of interfacial layer formation during continuous hot-dip galvanizing and galvannealing, Metall. Mater. Trans. A Phys. Metall. Mater. Sci. 39A (9) (2008) 2128–2142.
[35] Y.H. Liu, N.Y. Tang, Computer modeling of aluminum uptake and iron dissolution in galvanizing and galvannealing, in: GALVATECH '04, 2004, pp. 1155–1164.
[36] M. Dutta, S.B. Singh, Effect of strip temperature on the formation of an Fe_2Al_5 inhibition layer during hot-dip galvanizing, Scripta Mater. 60 (2009) 643–646.

[37] G.K. Mandal, R. Balasubramaniam, S.P. Mehrotra, Theoretical investigation of the interfacial reactions during hot-dip galvanizing of steel, Metall. Mater. Trans. A 40 (2009) 637–645.

[38] N.Y. Tang, Modeling Al enrichment in galvanized coatings, Metall. Mater. Trans. A 26 (1995) 1699–1704.

[39] C.E. Jordan, A.R. Marder, Fe-Zn phase formation in interstitial-free steels hot-dip galvanized at 450°C: part II 0.20 wt% Al-Zn baths, J. Mater. Sci. 32 (1997) 5603–5610.

[40] P. Toussaint, L. Segers, R. Winand, M. Dubois, Mathematical modelling of Al take-up during the interfacial inhibiting layer formation in continuous galvanizing, ISIJ Int. 38 (1998) 985–990.

[41] C.E. Jordan, R. Zuhr, A.R. Marder, Effect of phosphorous surface segregation on iron-zinc reaction kinetics during hot-dip galvanizing, Metall. Mater. Trans. A28 (12) (1997) 2695–2703.

[42] E. McDevitt, Y. Morimoto, M. Meshii, Characterization of the Fe-Al interfacial layer in a commercial hot-dip galvanized coating, ISIJ Int. 37 (8) (1997) 776–782.

[43] Y. Morimoto, E. McDevitt, M. Meshii, Characterization of the Fe-Al inhibition layer formed in the initial stages of hot-dip galvannealing, ISIJ Int. 37 (9) (1997) 906–913.

[44] H. Yang, S. Zhang, J. Li, X. Liu, H. Wang, Effect of strip entry temperature on the formation of interfacial layer during hot-dip galvanizing of press-hardened steel, Surf. Coat. Technol. 240 (Supplement C) (2014) 269–274.

[45] T. Min, Y. Gao, X. Huang, Z. Gong, K. Li, S. Ma, Effect of aluminum concentration on the formation of inhibition layer during hot-dip galvanizing, Int. J. Heat Mass Transf. 127 (2018) 394–402.

[46] D. Zapico, F.B. Alverez, J.-M. Mataigne, M.-L. Giorgi, Mechanisms and kinetics of the inhibition layer breakdown in the case of Ti IF steel grades galvanized in a GA bath, in: GALVATECH '15, 2015.

[47] N.-Y. Tang, Thermodynamics and kinetics of alloy formation in galvanized coatings, in: F.E. Goodwin (Ed.), Zinc-Based Steel Coating Systems: Production and Performance, TMS, Warrendale, PA, 1998, p. 3.

[48] K.-K. Wang, L. Chang, D. Gan, H.-P. Wang, Heteroepitaxial growth of Fe2Al5 inhibition layer in hot-dip galvanizing of an interstitial-free steel, Thin Solid Films 518 (2010) 1935–1942.

[49] C.-W. Hsu, K.-K. Wang, L. Chang, D. Gan, Y.-L. Chang, H.-Y. Liang, H.-P. Wang, Formation of $Fe_2Al_{5-x}Zn_x$ intermetallic crystals at the Fe-Zn interface in hot-dip galvanizing, Mater. Charact. 137 (2018) 189–200.

[50] A.R. Ghuman, J.I. Goldstein, Reaction mechanisms for the coatings formed during the hot dipping of iron in 0 to 10 pct Al-Zn baths at 450 to 700 C, Metall. Trans. 2 (10) (1971) 2903–2914.

[51] Z.W. Chen, J.T. Gregory, R.M. Sharp, Intermetallic phases formed during hot dipping of zinc, Metall. Trans. A. 23 (9) (1992) 2393–2400.

[52] M. Manna, M. Dutta, A.N. Bhagat, Microstructure and electrochemical performance evaluation of Zn, Zn-5 wt.% Al and Zn-20 wt.% Al alloy coated steels, J. Mater. Eng. Perform. 30 (2021) 627–637.

[53] A. Du, R.N. Ma, Y.Z. Fan, X.M. Cao, J. Shi, F.H. Wang, Effects of dipping temperature on hot-dip Galfan coating of steel wire, Trans. Mater. Heat Treat. 31 (2010) 144–148.

[54] Y. Wang, A. Du, R. Ma, Y. Fan, X. Zhao, T. Liu, X. Cao, Effect of hot-dipping temperature on microstructure of interface layer of Galfan alloy coating by solvent method, Trans. Mater. Heat Treat. 04 (2017).

[55] K.-K. Wang, C.-W. Hsu, L. Chang, W.J. Cheng, Characterization of the FeAl intermetallic layer formed at FeZn interface of a hot-dip galvanized coating containing 5 wt.% Al, Surf. Coat. Technol. 396 (2020) 125969.

[56] J.H. Selverian, A.R. Marder, M.N. Notis, The effect of silicon on the reaction between solid iron and liquid 55 wt pct Al-Zn baths, Metall. Mater. Trans. A 20A (1989) 543–555.

[57] P.D. Mercer, The influence of temperature on the growth rate and composition of the interfacial alloy in 55% Al, 43.5% Zn, 1.5% Si hot-dipped coatings, in: A.R. Marder (Ed.), The Physical Metallurgy of Zinc Coated Steel, TMS, Warrendale, PA, 1994, p. 129.

[58] Y. Durandet, L. Strezov, N. Ebril, Formation of Al-Zn-Si coatings on low carbon steel substrates, in: 4th International Conference on Zinc and Zinc Alloy Coated Steel Sheet, Galvatech'98, 1998.

[59] G. Eggeler, A. Werner, H. Kaesche, Reactions between low alloyed steel and initially pure as well as iron-saturated aluminum melts between 670 and 800°C, Z. Met. 77 (1986) 239–244.

[60] D. Phelan, B.J. Xu, R. Dippenaar, Formation of intermetallic phases on 55 wt.% Al–Zn–Si hot dip strip, Mater. Sci. Eng. A 420 (2006) 144–149.

[61] J.H. Selverian, A.R. Marder, M.R. Notis, The reaction between solid iron and liquid Al-Zn baths, Metall. Trans. A. 19A (1994) 1193–1203.

[62] C.W. Hsu, G.L. You, K.K. Wang, L. Chang, D. Gan, L.J. Chiang, Characterization of the Fe-Al intermetallic phases formed in hot-dip Al-Zn-Mg coatings, in: Proceedings of the 10th International Conference on Zinc and Zinc Alloy Coated Steel Sheet (Galvatech 2015), 2015, pp. 845–851.

[63] L. Bordignon, F. Goodwin, Galvanizing of TRIP-Si steels, in: GALVATECH '21, 2021, pp. 394–403.

[64] C. Yao, H. Lu, T. Zhu, W. Zheng, X. Yuan, W. Gao, Effect of Mg content on microstructure and corrosion behavior of hot dipped Zn–Al–Mg coatings, J. Alloys Compd. 670 (2016) 239–248.

[65] H. Yokoi, N. Takata, A. Suzuki, M. Kobashi, Formation sequence of Fe–Al intermetallic phases at interface between solid Fe and liquid Zn–6Al–3Mg alloy, Intermetallics 109 (2019) 74–84.

[66] Y. Xie, A. Du, X. Zhao, R. Ma, Y. Fan, X. Cao, Effect of Mg on Fe–Al interface structure of hot–dip galvanized Zn–Al–Mg alloy coatings, Surf. Coat. Technol. 337 (2018) 313–320.

[67] Y. Chen, Y. Liu, H. Tu, C. Wu, X. Su, J. Wang, Effect of Ti on the growth of the Fe–Al layer in a hot dipped Zn–6Al–3Mg coating, Surf. Coat. Technol. 275 (2015) 90–97.

[68] K. Honda, K. Ushioda, W. Yamada, Influence of Si addition to the coating bath on the growth of the Al–Fe alloy layer in hot-dip Zn–Al–Mg alloy-coated steel sheets, ISIJ Int. 51 (2011) 1895–1902.
[69] K. Li, Y. Liu, H. Tu, X. Su, J. Wang, Effect of Si on the growth of Fe-Al intermetallic layer in Zn-11% Al-3% Mg coating, Surf. Coat. Technol. 306 (2016) 390–396.
[70] L. Gao, Z. Li, X. Kuang, F. Yin, H. Ji, Formation of periodic layered structure during hot-dip galvanizing in Al-Zn-Mg bath, Surf. Coat. Technol. 304 (2016) 306–315.
[71] C. Jordan, A.R. Marder, Fe-Zn formation in interstitial-free steels hot dip galvanized at 450oC, part I: 0.00 wt% Al-Zn baths, J. Mater. Sci. 23 (1997) 5593–5602.
[72] S. Alibeiga, J.R. McDermid, Effect of bath Al on interfacial layer formation for continuous galvanized Mn-containing steels, in: Galvatech '13, 2013, pp. 171–176.
[73] S. Alibeiga, J.R. McDermid, Modeling of interfacial growth kinetics on Mn-containing steels during continuous hot-dip galvanizing, in: Galvatech '15, 2015, pp. 334–341.
[74] H. Asgari, F. Ashrafizadeh, M.R. Toroghinejad, Influence of steel substrate grain size on phase layers growth kinetics and thickness of zinc hot dipped steel sheets, Ironmak. Steelmak. 36 (2) (2009) 145–148.
[75] A.C. de Oliveira, P.R. Rios, A novel iron enrichment isothermal kinetic model proposal for galvanneal coatings, J. Mater. Res. Technol. 2 (2013) 117–124.
[76] V.G. Krishnardula, C.C. Cheng, D. Bhattacharya, Influence of substrate molybdenum on galvannealing kinetics of an advanced high strength steel, ISIJ Int. 54 (8) (2014) 1890–1897.

Chapter 6

Overlay coating formation

6.1 Introduction

6.1.1 Zn-based coating systems

Zinc coatings for sheet steel applied by the immersion of steel into a liquid zinc bath can be classified into four categories:

1. Coatings with less than 1% Al and usually between about 0.1% and 0.5% Al in a solid solution overlay layer, covering a thin (<150 nm) interface layer of FeAlZn intermetallic, commonly termed (GI) coatings.
2. Coatings with higher Al content. These include the Zn-Al alloy coatings termed Galfan[a] (Zn-5%Al), called (GF) and Galvalume[b] (Zn-55%Al) referred to as (AZ).
3. Coatings with Zn-Al alloys to which Mg has been added. Coatings over the range 1% < Al < 11% and 1% < Mg < 3% have been separately developed. The Zn-55%Al-Mg coating is usually termed AM while the other Mg-containing coatings are termed ZM.
4. Heat-treated coatings with 10%–18% Fe formed by diffusion of liquid Zn into solid Fe after the exit of the steel from the hot-dip bath, termed galvanneal or GA coatings. These coatings will be discussed in Chapter 8. The initial Al content of these coatings before heat treatment is usually between 0.1% and 0.14%.
5. Coatings produced during the hot press forming process. The precursor coatings are either GI or GA coatings that are subjected to heating of the coated steel to the austenitic range, typically 650–900°C, followed by die quenching. These coatings normally have between 15 and 75% Zn, depending on steel, coating, and processing details. These coatings will be covered in Chapter 13.

Initially, after immersion in the alloy bath, an interfacial layer is formed (see Chapter 5 – Interface Reactions). The interfacial phase acts as a barrier layer for diffusion of the bath alloy into the substrate steel and the steel into the bath alloy coating.

The inhibition layer breaks down in a short amount of time and therefore is a limiting step in the formation of coatings 1, 2, and 3 above, whereas heat-treated coatings (Coatings 4 and 5 above) depend upon inhibition layer breakdown to form the desired coating.

6.1.2 Corrosion design

In considering the overlay structure of these Zn alloy coatings, the coatings morphology, i.e., the size, shape, and distribution of the various constituent phases, is designed for specific properties. For zinc coatings, corrosion resistance is a primary property requirement that has led to many innovations in the broad field of zinc coatings [1]. Zinc coatings add corrosion resistance to steel in several ways. As a barrier layer, a continuous zinc coating separates the steel from the corrosive environment and by galvanic protection, zinc acts as a sacrificial anode to protect the underlying steel at voids, scratches and cut edges of the coating. The sacrificial properties of zinc can be seen in a galvanic series where the potential of zinc is less noble than steel in most environments at ambient temperatures. In all coatings, the corrosion resistance is dependent on coating thickness. In Al added zinc coatings, in the first step of coating formation, a temporary protective aluminum oxide passive layer forms on the surface. This is followed by the formation of additional surface layers that depend upon the corrosive environment. For example, in marine environments, the surface layer will have chloride-containing reaction products, while in industrial environments, sulfate-containing reaction products are formed.

a. Galfan is a trademark of Galfan Technology Centre Inc. http//galfan.com/about.html.

b. GALVALUME is an internationally recognized trademark of BIEC International Inc. http//www.galvalume.com/.

The Metallurgy of Zinc Coated Steels. https://doi.org/10.1016/B978-0-323-99984-7.00012-9

These reaction products can form with constituent phases contained in zinc alloy coatings, the resulting surface strongly depends on the environment and coating microstructure.

In Galfan coatings, corrosion is slower than pure zinc galvanized coatings because the zinc-rich phase in both the proeutectic and eutectic lamella microstructure corrodes preferentially. Initially, the atmospheric corrosion of the Galvalume coating takes place in the zinc-rich interdendritic regions, enabling the coating to exhibit galvanic protection. As the coating continues to corrode, the zinc corrosion products become trapped in the interdendritic regions and act as a further passive barrier to corrosion. The aluminum dendrites, which also act as a barrier layer, add to the corrosion protection, as does the Fe-Al-Zn intermetallic alloy layer. Alloy additions to the coating serve to refine the coating structure as well as contributing new corrosion resistant phases. The complex microstructure of Galvalume coatings further improves corrosion resistance. Additional alloying elements to the coating composition, e.g., Mg, significantly enhance the corrosion resistance of Al-Zn-Si-xMg coatings due to the Mg_2Si and $MgZn_2$ phases formed in the interdendritic regions [2]. Additions of Mg (up to 0.8 wt%) in Galfan increase the amount of proeutectic Al phase, which improves the corrosion resistance. Thus, for the alloy coating designer, there are many ways to improve the corrosion properties of zinc coatings. A more detailed description of the corrosion resistance of zinc and zinc alloy coatings can be found in Chapter 11.

6.1.3 Thermal effects

When considering the development of these coatings and especially the formation of the overlay coating microstructure, it is first necessary to determine the thermal conditions encountered during solidification of the coating so that the nucleation and growth of the overlay may be determined. A solidification model for hot-dip galvanized zinc coatings was developed by Strutzenberger and Faderl [3]. The amount of undercooling and the temperature gradient in the liquid zinc film prior to solidification were evaluated in a hot-dip simulator using superimposed cooling rates from natural convective cooling in air (28 °K/s) to jet cooling in a gas flow of $10\,m^3/h$—5% H_2/N_2 atmosphere (15 °K/s). The measured undercooling and solidification period are given in Table 6.1. In their model, they used an undercooling of less than 1°C [4]. Since heat can only be removed by conduction through the liquid zinc film and considering a 25 μm zinc coating, the approximate difference in temperature between the steel/zinc interface and the zinc surface was estimated as:

$$\Delta T_{\text{min}} \quad 2.2 \times 10^{-3}\ \text{K at } 2\,\text{K/s} \tag{6.1}$$

$$\Delta T_{\text{max}} \quad 1 \times 6 \times 10^{-2}\ K \text{ at } 15K/s \tag{6.2}$$

These small temperature gradients and the slight undercooling suggest [4] that heterogeneous nucleation occurs at the rough steel/zinc interface, which offers preferred nucleation sites for zinc dendrites. It may be assumed that similar Zn alloy coatings will behave in the same manner in that nucleation of the overlay coating will occur at the zinc/substrate interface.

The importance of aluminum additions to the bath to control the inhibiting layer has been reviewed in Chapter 5. Aluminum additions are made to control both interfacial layer development to produce a specific as-cast microstructure in the overlay portion of the coating that influences certain properties in the coating such as corrosion resistance and formability. Interface reactions require a knowledge of ternary and quaternary equilibrium and metastability for the elements involved. Generally, binary and ternary equilibrium interactions are sufficient to explain the as-cast overlay structure, although further additions to the bath, e.g., Mg, can complicate the analysis.

TABLE 6.1 Correlation between cooling rate, solidification period, and undercooling.

Cooling rate (K/s)	Atmosphere	Solidification period[a] (s)	Undercooling[a] (K)
2	Air	10	<1
4	5% H_2/N_2	5	<1
15	5% H_2/N_2	2	<2

[a]*Results obtained from two different bath compositions (Zn + 0.2 wt% Al + 0.0025 wt% Pb and Zn + 0.2 wt% Al + 0.05 wt% Pb) did not vary significantly; thus, an average value is given.*

(From J. Strutzenberger, J. Faderl, Solidification and spangle formation of hot-dip-galvanized zinc coatings, Metall. Mater. Trans. 29A (1998) 631–646.)

6.2 Phase equilibria

Aluminum is added to the zinc bath to improve corrosion resistance by either allowing for the formation of a zinc-rich overlay for galvanic protection by inhibiting the formation of Fe-Zn phases or by introducing multiphase microstructures in the overlay coating. In each case, the resulting microstructures produce coatings that offer corrosion resistance by a combination of galvanic and barrier protection with corrosion products that in some cases are passive. The phase equilibria for Zn-Fe and Al-Fe are described in Chapter 3 as is the Zn-Al system. However, a slightly modified review of the Zn-Al equilibrium phase diagram [5] is repeated here in order to understand the overlay cast structure in the Zn-Al coatings, Fig. 6.1. *Caution: it should be noted that various investigators use different Greek letter designations for the same phase.* The solid phases present in the Zn-Al system include β^1, β, and η. β and β^1 represent the Al- and Zn-rich portions, respectively, of the aluminum fcc solid solution. The solubility of Zn in the Al solid solution increases with temperature to 32.4 wt% Zn (16.5 at.% Zn) at the eutectoid temperature (277°C) and 83.1 wt% Zn (67 at.% Zn) at the eutectic temperature (381°C). The Zn terminal solid solution is hexagonal close-packed with a maximum Al solubility of 1.2 wt% (2.8 at.%) at the eutectic temperature. This solubility decreases to 0.7 wt% Al (1.6 at.% Al) at the eutectoid temperature and 0.03 wt% Al (0.07 at.% Al) at room temperature. The transformations involving these phases, Table 6.2, are a eutectic at 95.0 wt% Zn (88.7 at.% Zn), a eutectoid at 77.7 wt% Zn (59.0 at.% Zn), and the existence of a miscibility gap with a critical temperature of 351.5°C at 61.3 wt% Zn (39.5 at.% Zn).

Reactions occurring between the steel substrate and the Zn alloy coating bath are described by equilibrium diagrams for the Zn-Fe binary and related multicomponent systems. The most accurate model of Zn-Fe phase equilibria is based on improved descriptions developed for all eight intermediate phases in the Zn-Fe system (intermetallic compounds

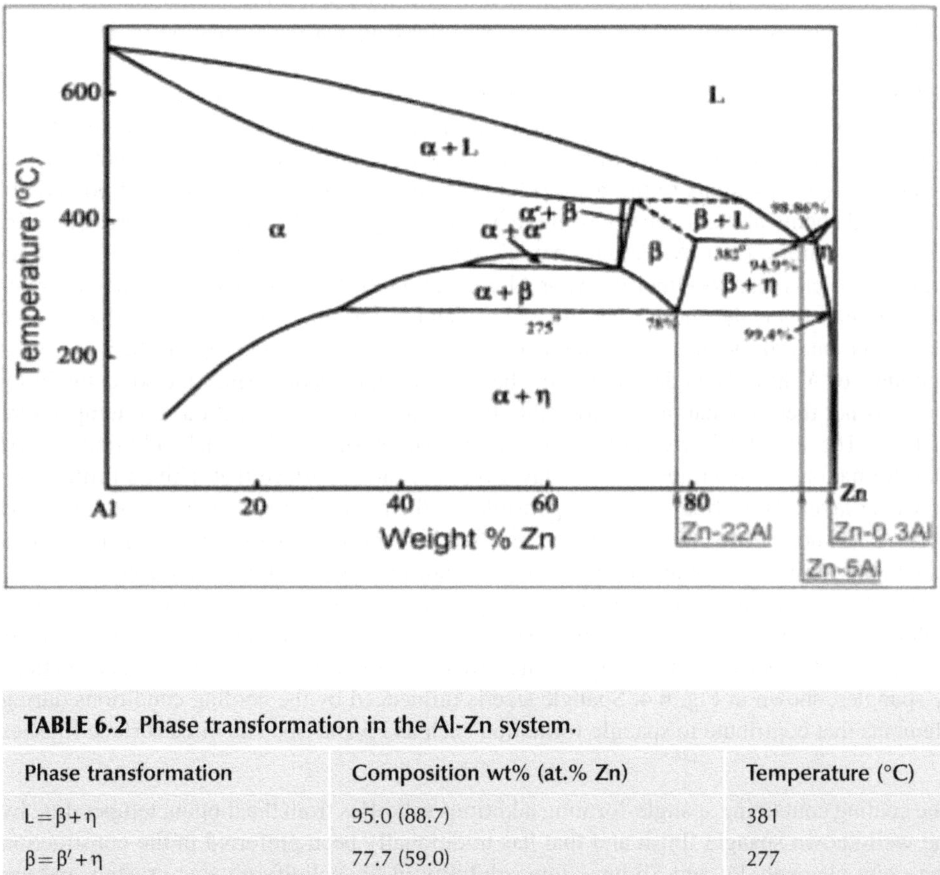

FIG. 6.1 The Al-Zn equilibrium binary diagram. *(From H. Chaieb, V. Maurel, K. Ammar, S. Forest, A. Koster, F. Nozahic, J. De Strycker, J.-M. Mataigne, Microstructure and texture analysis of a near-eutectic Zn-Al coating deposited on steel by hot dipping galvanizing, in: Galvatech, 2021, pp. 272–281.)*

TABLE 6.2 Phase transformation in the Al-Zn system.

Phase transformation	Composition wt% (at.% Zn)	Temperature (°C)	Transformation type
L = β + η	95.0 (88.7)	381	Eutectic
β = β′ + η	77.7 (59.0)	277	Eutectoid (monotectoid)
(Al) = β′ + β	61.3 (39.5)	351.5	Critical
L = (Al)	0 (0)	660.45	Congruent
L + η	100 (100)	419.58	Congruent

(From A.R. Marder, The metallurgy of zinc-coated steel, Prog. Mater. Sci. 45 (2000) 191–271.)

gamma-one, gamma, delta, and zeta together with FCC and BCC Fe-rich phases and the HCP Zn-rich phase and liquid). These were incorporated in the CALPHAD formalism and correlated with experimental results [5]. The sublattice models derived for the intermetallic compounds of the Zn-Fe diagram were examined to predict which sublattice(s) would be most capable of hosting Al. These were then combined with past assessments of Zn-Al-Fe equilibria, and Al activities introduced to the PARROT module of Thermo-Calc. Comparison of calculated quantities experimental results led to an improved thermodynamic model of the Zn-Al-Fe system within the 420–500°C range. The Zn-rich corner of the Zn-Al-Fe phase diagram, of the greatest interest for galvanizing phenomena, was then characterized in great experimental detail. Results were combined with thermodynamic modeling, permitting mapping of Al and Fe isoactivity contours on isothermal sections [6] and discussed in Chapter 3. Sections of this corner at three temperatures relevant to galvanizing are shown in more detail in Fig. 6.2. The important quaternary Zn-Al-Fe-Si system is also discussed in Chapter 3.

For the higher-Mn steels described in Chapter 2, it is also necessary to consider the Zn-Al-Fe-Mn system, because of the potential for dissolution of Mn from the steel into the Zn bath. Mn additions to the Zn bath favored the formation of δ-$FeZn_{10}$ and δ-$MnZn_9$ phases, shifting the invariant "knee" point of ô-Fe_2Al_5 in the Zn-Fe-Al ternary system to a higher Al level [7].

6.3 Galvanized (GI) coatings (<1 wt% Al)

Aluminum is the most common and important alloying element added to the continuous hot-dip galvanizing bath, with different levels required to produce particular properties in the bath [1]. Al levels of 0.005–0.02 wt% can be added to brighten the initial coating surface and small amounts of aluminum (0.1%–0.3%) are typically added in hot-dip galvanizing baths in order to improve the mechanical surface properties of zinc-coated steel. Aluminum has been reported to reduce the rate of oxidation of molten zinc, to reduce the spangle size, and to improve the uniformity of coating. Aluminum suppresses the growth of a brittle Fe-Zn intermetallic layer and reduces iron loss into the bath via the formation of Fe_2Al_5 and $FeAl_3$ intermetallic layers [7–13], see also Chapter 5.

Al exists in two forms in the bath: dissolved in the liquid and as intermetallic particles entrapped in the bath. The dissolved Al is the effective form needed to avoid the formation of brittle Fe-Zn phases. A galvanized zinc bath with 0.1%–0.3% Al may further contain Zn and Al oxides termed top skimmings, as well as Zn-Fe and Fe-Al intermetallic compounds. These segregated particles originate from Al additions to the bath, solution of iron from the steel strip, and potentially insufficient temperature uniformity and chemical homogeneity of the bath, see Chapter 4. Al has been reported to exist in four main forms in a solidified zinc coating: (i) in Al-Fe intermetallic compounds at the interface of steel and zinc, (ii) in segregated Al-containing dross particles that are either Fe-Al or Al-containing Zn-Fe compounds, (iii) in a thin continuous film of aluminum and oxygen that covers the entire zinc surface [14], and (iv) in solid solution within the metallic coating phases. The Al_2O_3 layer on the outermost surface is generated by diffusion of Al from the bulk because of its affinity to oxygen. Negligible amounts of Al have been detected in the bulk of the zinc coating, since the solid solubility of aluminum is very limited, and it is not thermodynamically favorable for Al to be dissolved in Zn as the temperature decreases during solidification [15]. The effect of brightening the initial coating surface by small additions of Al (0.005–0.02 wt%) is related to the formation of a continuous Al_2O_3 layer on the coating surface that inhibits further oxidation by acting as a protective barrier layer. This effect is also responsible for the reduced atmospheric oxidation of the zinc bath. As an example, using Tof-SIMS positive ion measurements (Fig. 6.3) [14] showed that the outermost surface of a temper rolled HDG sample was composed mostly of aluminum with minor contributions from zinc and iron.

Galvanized hot-dip coatings often have a structure consisting of very large grains called "spangles." In cross section, an Fe_2Al_5(Zn) inhibition layer develops first, preventing any Fe-Zn intermetallic phase formation. The overlay layer is made up of dendrites of pure zinc eta (η) phase and appears as a polycrystalline structure. The surfaces of as-galvanized coatings are commonly characterized by spangles, shown in Fig. 6.4. Spangle size is influenced by the cooling conditions during solidification and addition of elements that contribute to spangle formation such as Pb and Sb. The three surface finishes commonly produced are:

- Regular spangle, where the coating containing spangle-forming additions solidifies from the dipping temperature by air cooling, producing the well-known spangle finish and that has traditionally been preferred in the construction industry. The low-melting-point elements Pb and Sb have low solubility in the solidifying zinc crystals and are rejected to the grain boundaries. They also limit nucleation, resulting in larger crystal sizes and because of their limited solid solubility also promote dendritic crystal growth. All of these factors contribute to the development of the spangle appearance.
- Minimum spangle, where the coating is either produced without spangle-forming element additions or quenched using water, steam, chemical solutions, or by zinc powder spraying for the automotive and household applications.

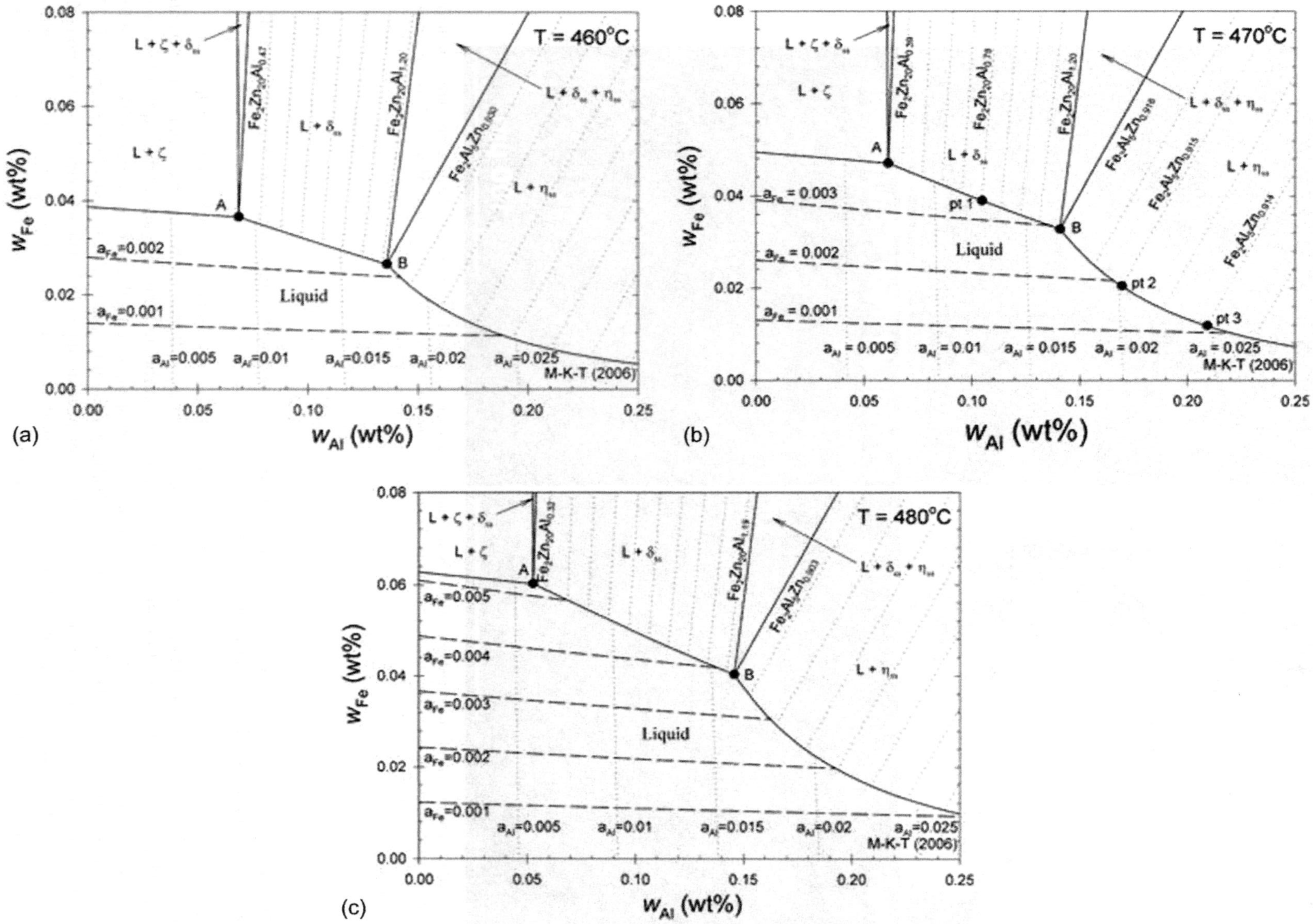

FIG. 6.2 Zn-rich corner of the Zn-Al-Fe equilibrium diagram at three temperatures relevant to galvanizing, (a) 460°C, (b) 470°C, (c) 480°C indicating isoactivity lines for Al and Fe. *(From J.R. McDermid, M.H. Kaye, W.T. Thompson, Fe solubility in the Zn-rich corner of the Zn-Al-Fe system for use in continuous galvanizing and galvannealing, Metall. Mater. Trans. B 38b (2007) 215–230.)*

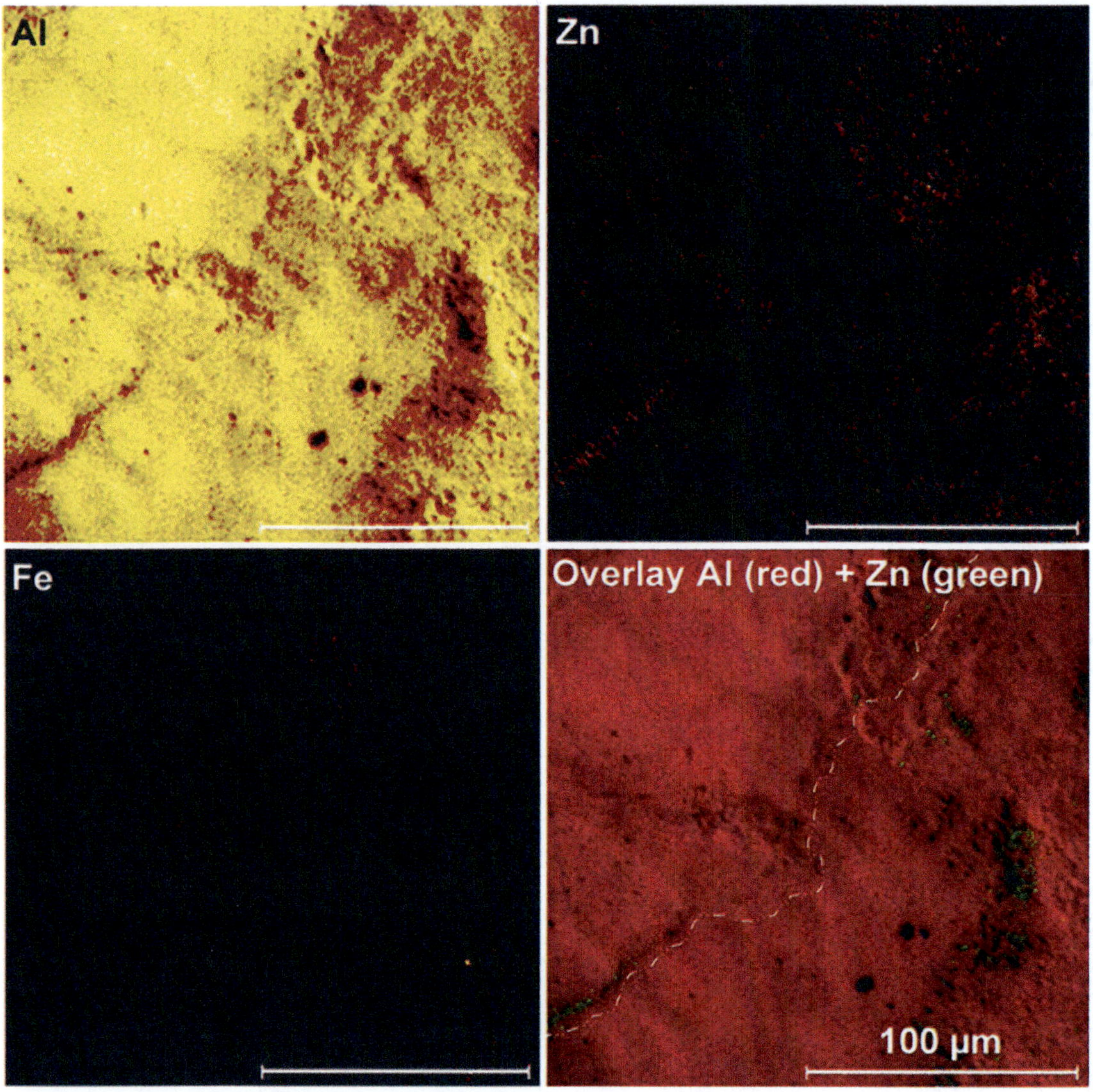

FIG. 6.3 Tof-SIMS element maps of Al, Zn, and Fe measured from a sample of temper rolled hot-dip galvanized steel without any surface treatment. A grain boundary is highlighted by the *dotted line* in the overlay map. *(From V. Saarimaa, A. Markkula, J. Juhanoja, B.-J. Skrifvars, Novel insight to aluminum compounds in the outermost layers of hot dip galvanized steel and how they affect the reactivity of the zinc surface, Surf. Coat. Technol. 306 (2016) 506–911.)*

- Extra-smooth temper roll finish carried out as an additional operation with regular and minimum spangle material, also for automotive and household industries.

Large-size spangles for construction are known for their aesthetic appearance and corrosion resistance, whereas the fine spangle and smooth surfaces for automotive and appliance applications are preferred for better image clarity after painting and a low friction coefficient for press formability [15]; therefore, controlling grain size and surface quality are major concerns for hot-dip galvanized coatings. One of the most common spangles on galvanized coatings is like a snowflake or 6-pointed star. The spangles can be classified as shiny, feathery, and dull, based on their optical appearance, Fig. 6.5.

Hot-dip galvanized coatings often show a strong (0001) basal plane texturing of the spangles (zinc crystals) upon solidification. The proportion of basal plane orientation increases with decreasing coating grain size [16] (i.e., minimized spangle and temper rolled coatings) and correlates well with improved paint adherence [17]. With preferred nucleation sites at the liquid Zn/substrate interface, growth of the Zn spangles occurs according to Strutzenberger and Faderl [3] in the following manner:

FIG. 6.4 Typical spangle structure of a galvanized coating. *(From A.R. Marder, The metallurgy of zinc-coated steel, Prog. Mater. Sci. 45 (2000) 191–271.)*

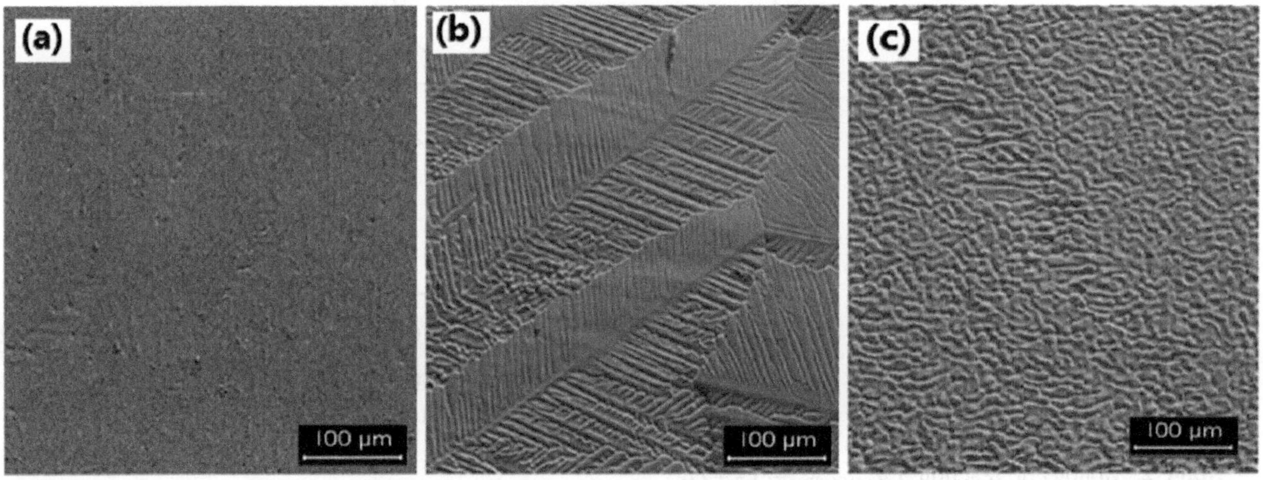

FIG. 6.5 SEM images of specimen with different morphologies: (a) I (shiny spangle), (b) II (feathery spangle), and (c) III (dull spangle). *(From S. Peng, S.K. Xie, J.-T. Lu, L.-C. Zhang, Surface characteristics and corrosion resistance of spangle on hot-dip galvanized coating, J. Alloys Compd. 728 (2017) 1002–1008.)*

- The initial growth stage (0.1 s) involves sideways growth of Zn basal planes (0001) parallel to the interface in the [1100] growth direction. The entire steel surface is covered by a solid layer of zinc dendrites, whereas the surface remains liquid.
- The second stage involves slow thickening of the solid zinc grain, which is dependent on grain orientation.
- The third stage, or continuous thickening of the zinc grain, causes solute enrichment of the remaining liquid and finally Pb precipitation between Zn dendrite arms.

Large spangles are generally associated with solute additions to the galvanizing bath, and Pb, at concentrations greater than 0.04 wt%, is the solute generally added to molten Zn to produce spangles, Fig. 6.6 [18]. Wang et al. [19] proposed that the behavior of the lead in the solidification of Zn-0.1Al-0.1Pb coating on hot-dipped steel occurred in the following sequence:

Stage 1: Forming the spangle framework, $L_{Zn\text{-}0.1\%Pb} \rightarrow Zn$
$T = 419°C–417.5°C$,
V_{liquid} (volume fraction of the liquid) from 100% to 11.1%, and c_{liquid} (lead concentration of the liquid): from 0.1% to 0.9%;

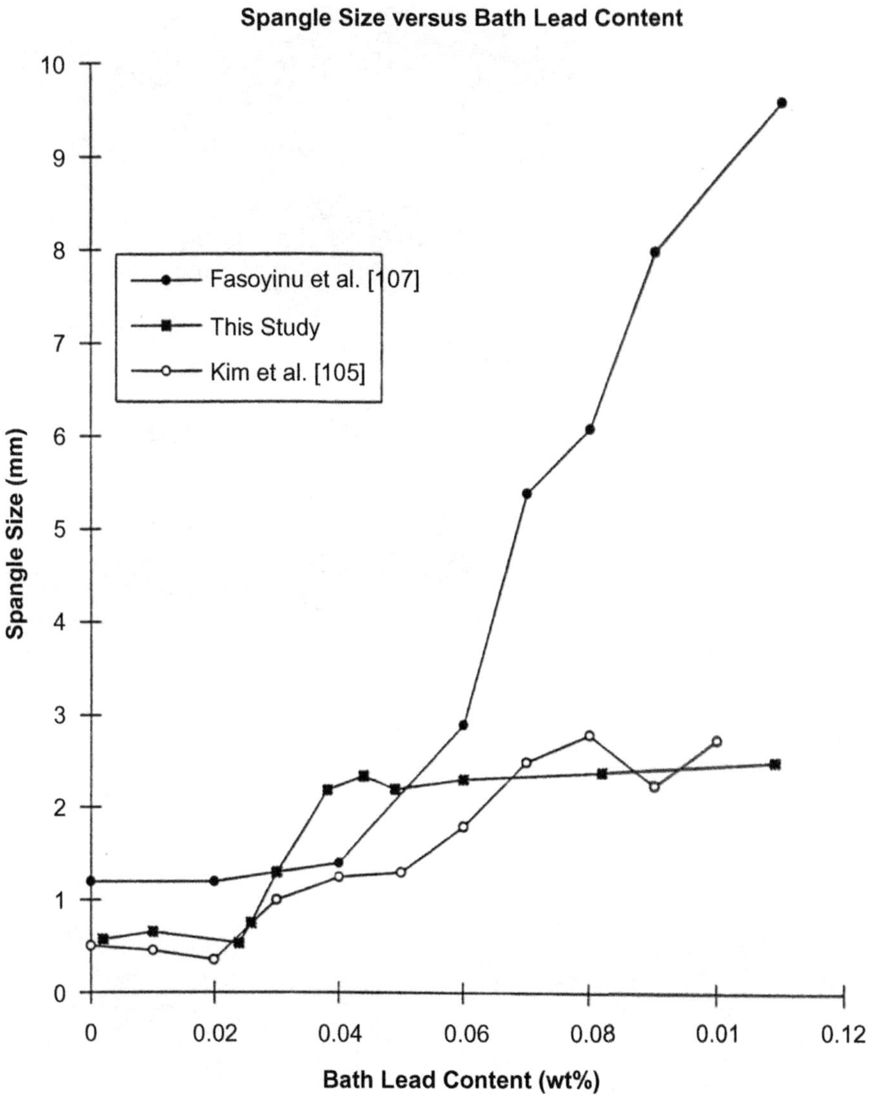

FIG. 6.6 Effect of lead content on spangle size. *(From R.E. Fraley, The Effect of Lead on the Solidification and Preferred Orientation of the Zinc Coating on Continuously Hot Dipped Galvanized Sheet Steel (M.S. thesis), Lehigh University, 1994.)*

Stage 2: Monotectic reaction, $L_{Zn\text{-}0.9\%Pb} \rightarrow L_2 + Zn$
$T = 417.5°C$,
V_{liquid}: from 11.1% to 0.1%,
c_{liquid}: from 0.9% to 98.3%;
Stage 3: Eutectic reaction, $L_{Zn\text{-}99.5Pb\%} \rightarrow \beta + \eta$
$T = 318.2°C$,
V_{liquid}: from 0.1% to 0%;
Stage 4: Forming equilibrium crystal shape of lead crystallites $T < 318.2°C$, in depressions of coating surface.

Al, Mg, Sn, and Cd additions do not influence spangle growth [4,20], but Sb and Bi additions produce spangles [21]. Fasoyinu and Weinberg [4] found that alloy additions such as Pb and Sb have lower surface tension and promote spangle formation. Strutzenberger and Faderl [3] proposed that these same elements segregate ahead of the growing dendrite, lowering surface tension and allowing the growth velocity of the zinc dendrite to increase, resulting in larger grains or spangles. Asagari et al. [22] confirmed the results that spangle size would be become larger by increasing the lead content of the zinc bath. However, they also found that the surface morphology would change, and the presence of basal planes decreased at the coating surface, due to an increase in Pb in the zinc bath.

In a recent study, Kim et al. [23] carried out 3D phase-field simulations to investigate the dendritic morphologies of HDG Zn-0.2 wt% Al coatings on a steel sheet as functions of the inclination angle and axis of the Zn basal plane with respect

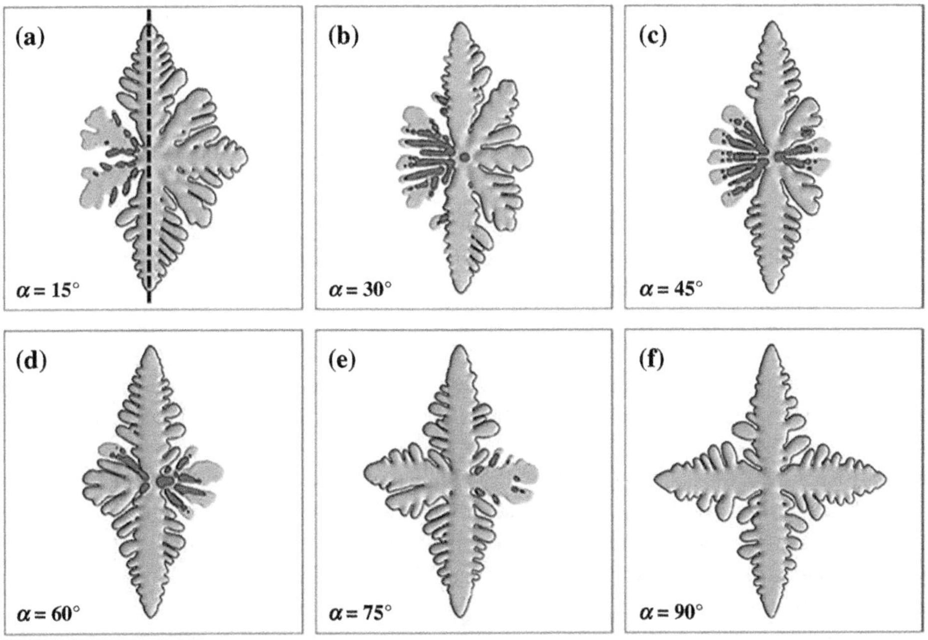

FIG. 6.7 Morphological change of dendrites as the inclination angle (α, degrees) of the Zn basal plane increases with the $\langle 1010 \rangle$ inclination axis: (a) $\alpha = 15$, (b) $\alpha = 30$, (c) $\alpha = 45$, (d) $\alpha = 60$, (e) $\alpha = 75$, and (f) $\alpha = 90$. *(From S.G. Kim, H.-S. Hwang, S.-K. Lee, J.-Y. Huh, Dendritic morphologies of hot-dip galvanized Zn-0.2wt % Al coatings: existence of third preferred growth directions, in: Galvatech, 2021, pp. 264–271.)*

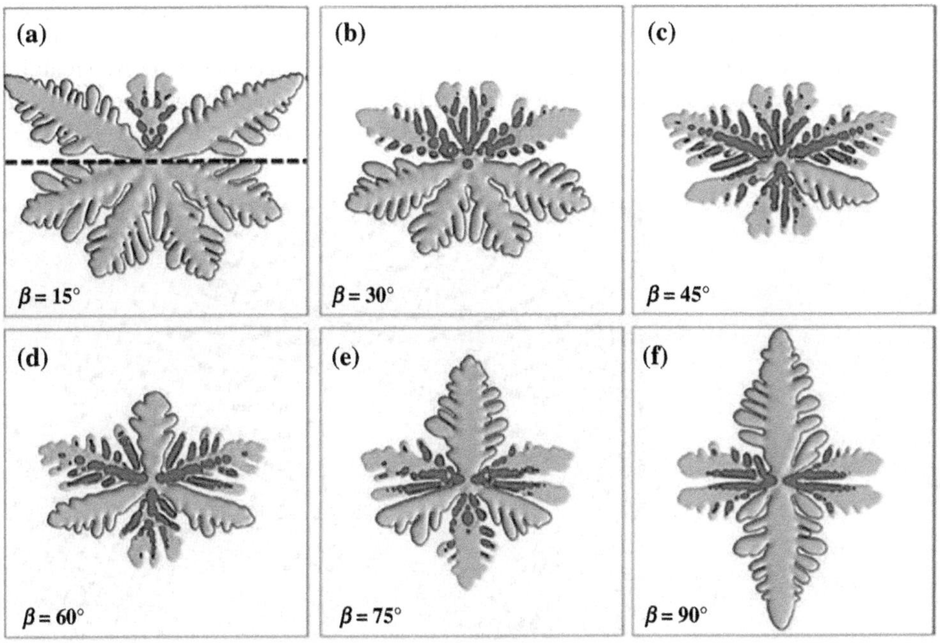

FIG. 6.8 Morphological change of dendrites as the inclination angle (β, degrees) of the Zn basal plane increases with the $\langle 1210 \rangle$ inclination axis: (a) $\beta = 15$, (b) $\beta = 30$, (c) $\beta = 45$, (d) $\beta = 60$, (e) $\beta = 75$, and (f) $\beta = 90$. *(From S.G. Kim, H.-S. Hwang, S.-K. Lee, J.-Y. Huh, Dendritic morphologies of hot-dip galvanized Zn-0.2wt % Al coatings: existence of third preferred growth directions, in: Galvatech, 2021, pp. 264–271.)*

to the sheet surface. The simulations considered that the growth directions of dendrites were controlled by the interface kinetic anisotropy rather than the interface energy anisotropy. The 3D simulations were performed for the dendritic growth of the primary Zn phase of which the basal plane is inclined to the sheet surface by various angles with the inclination axis of either $\langle 1010 \rangle$ or $\langle 1210 \rangle$. All the basic characteristics of the dendritic morphologies that were observed experimentally were reproduced. Fig. 6.7 shows the dendritic morphologies as the inclination angle (α) of the basal plane increases from 15 degrees to 90 degrees with the inclination axis of $\langle 1010 \rangle$. The inclination axis was denoted by the vertical dashed line in Fig. 6.7a. Fig. 6.8 shows the dendritic morphologies as the inclination angle (β) of the basal plane increases from 15 degrees to 90 degrees with the inclination axis of $\langle 1210 \rangle$. The inclination axis was denoted by the horizontal dashed line in Fig. 6.8a. The agreement of the simulated morphologies with the experimental observations strongly supports the existence of the third preferred growth direction, which are close to those normal to the {1211} planes, in Zn-rich alloys in addition to the two previously known directions of ⟨1010⟩ and ⟨0001⟩.

6.4 Coatings with Al (5–55 wt% Al)

6.4.1 Galfan (5 wt% Al-Zn)

Galfan is a Zn-5% Al alloy coating, which is near the eutectic point in the Al-Zn equilibrium phase diagram, Fig. 6.1. Two compositions have been reported based on additions to the eutectic composition: small mischmetal additions containing lanthanum and cerium contents up to about 0.5% [24] and additions of 0.5% magnesium [25]. These additions are made to improve the wettability and fluidity of the molten bath without affecting the corrosion resistance of the coating [26]. They also suppress bare spot formation [27] as well as producing a typical "minimized spangle" structure [28]. The microstructure of Galfan is characterized by a two-phase structure, a zinc-rich eta (η) proeutectoid phase surrounded by a eutectic-type phase consisting of beta (β) aluminum and eta (η) zinc lamellae, Fig. 6.9. However, the microstructure can be varied depending upon cooling rate [27].

Fig. 6.9a reveals the underlying microstructure of the coating (at half of the coating thickness): dendrites and eutectic phases are present. When solidification starts, the first phase to form is the proeutectic zinc, which is visible as dendrites. Then, the remaining liquid goes through the eutectic transformation. The resulting microstructure is a η-zinc matrix with lamellar/globular eutectic (β phase). This β phase undergoes a eutectoid transformation at 275°C, forming α-Al and η-zinc as shown in Fig. 6.1. So, inside the black dots and lamellae, a regular arrangement between zinc and aluminum is expected to be found. This is consistent with the findings of [29,30]. In Fig. 6.9b, two morphologies of eutectic are observed. The first one is a lamellar structure while the second one is a globular structure. (The globular structure could also be the lamellar structure on edge).

In the early stages of solidification, proeutectic zinc particles nucleate first and the solidification of the eutectic matrix proceeds from these particles [31]. The amount of the proeutectic was calculated to be 18% [32]. These results indicate nonreciprocal nucleating characteristics, i.e., one primary phase will act as an effective heterogeneous nucleation site for the other phase, but not vice versa, and the nucleation characteristics and microstructure in off-eutectic Al-Zn alloys have been detailed [33]. The Galfan overlay surface structure is shown in Figs. 6.10 and 6.11, and a schematic of the solidification sequence is found in Fig. 6.12. Although the nodules are on the order of 4 mm in diameter, each grain is divided

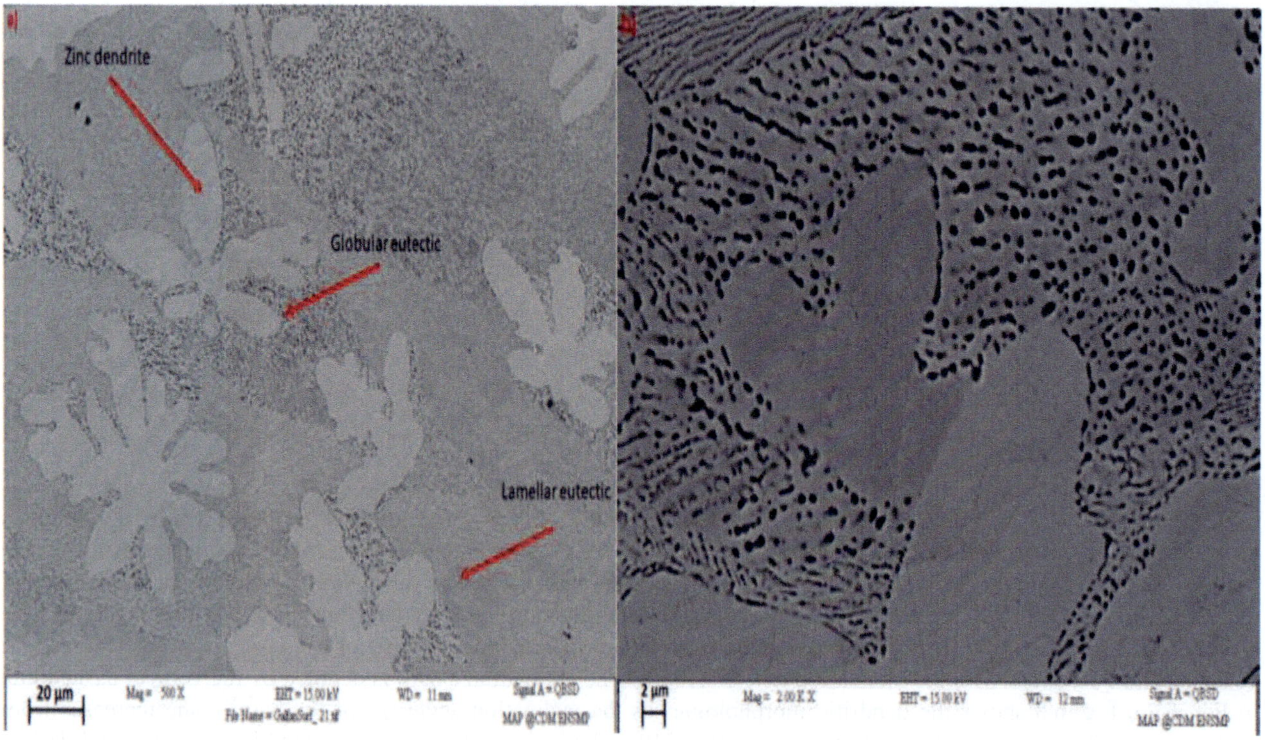

FIG. 6.9 Planar view of the lamellar microstructure of a Galfan coating. SEM image of the in-depth surface of the coating: (a) At a low magnification, (b) at a high magnification. *(From H. Chaieb, V. Maurel, K. Ammar, S. Forest, A. Koster, F. Nozahic, J. De Strycker, J.-M. Mataigne, Microstructure and texture analysis of a near-eutectic Zn-Al coating deposited on steel by hot dipping galvanizing, in: Galvatech, 2021, pp. 272–281.)*

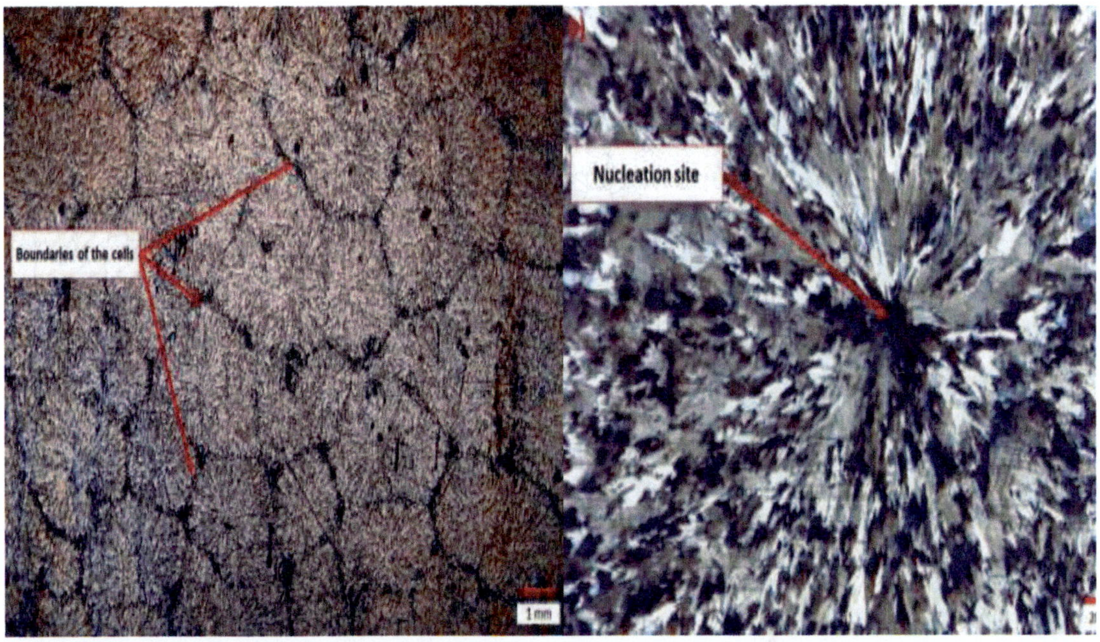

FIG. 6.10 Light optical coating surface images: (a) Macroscopic solidification cells, (b) the flake microstructure. *(From H. Chaieb, V. Maurel, K. Ammar, S. Forest, A. Koster, F. Nozahic, J. De Strycker, J.-M. Mataigne, Microstructure and texture analysis of a near-eutectic Zn-Al coating deposited on steel by hot dipping galvanizing, in: Galvatech, 2021, pp. 272–281.)*

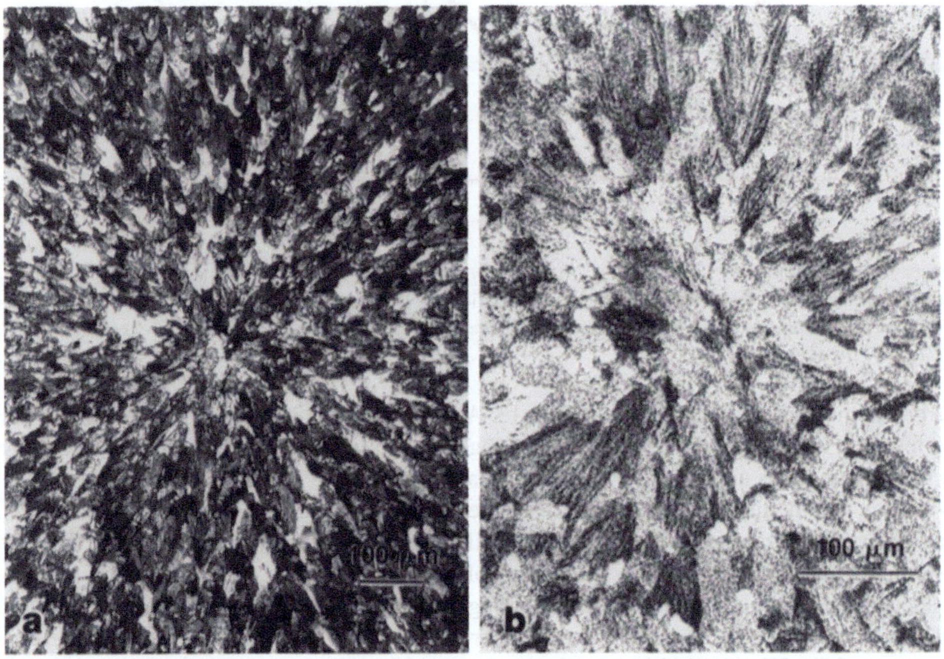

FIG. 6.11 Planar view showing (a) nodules and (b) colonies of a Galfan coating. *(From A.R. Marder, The metallurgy of zinc-coated steel, Prog. Mater. Sci. 45 (2000) 191–271.)*

into many smaller subgrains or colonies approximately 22.5 m in size. Each colony contains lamellae of alternate layers of beta (β) aluminum and eta (η) zinc, Fig. 6.10 (sometimes called flakes) [34]. In the range of normal bath temperatures, 420–440°C, it is reported that there is no visible intermetallic layer or at least an extremely thin layer (<0.5 m) at the interface between the steel substrate and the overlay coating [31]. This would indicate the presence of an inhibition effect in preventing Fe-Zn alloy formation [33,35]. The alloy layer in a Zn-5 wt% Al-0.1 wt% Mg coating was reported to be 50 nm and is composed of Al-Fe-Zn [31]. Thus, Galfan coatings have excellent formability and cut-edge corrosion protection.

The surface of Galfan coatings was shown to have solidification defects in association with nodule boundaries [36]. Surface depressions, termed dents, were found to occur at eutectic nodule boundaries and triple points and were typically

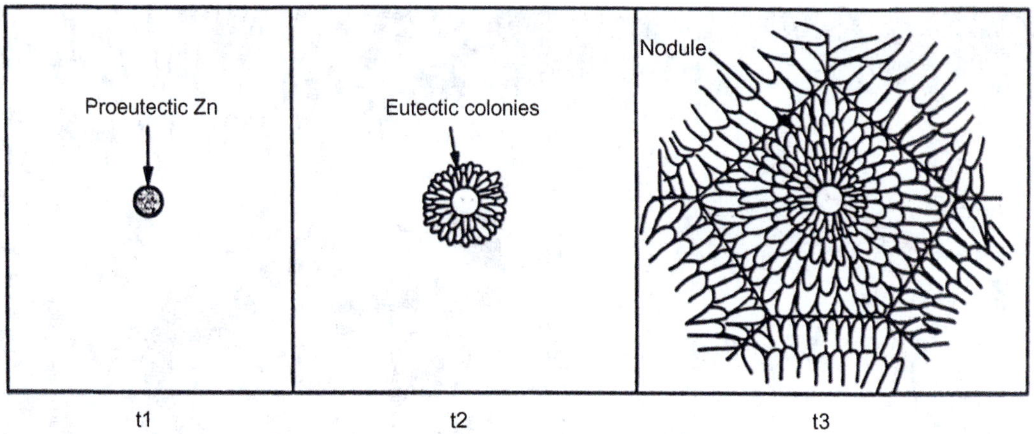

FIG. 6.12 Schematic of the solidification of a Galfan coating. *(From A.R. Marder, The metallurgy of zinc-coated steel, Prog. Mater. Sci. 45 (2000) 191–271.)*

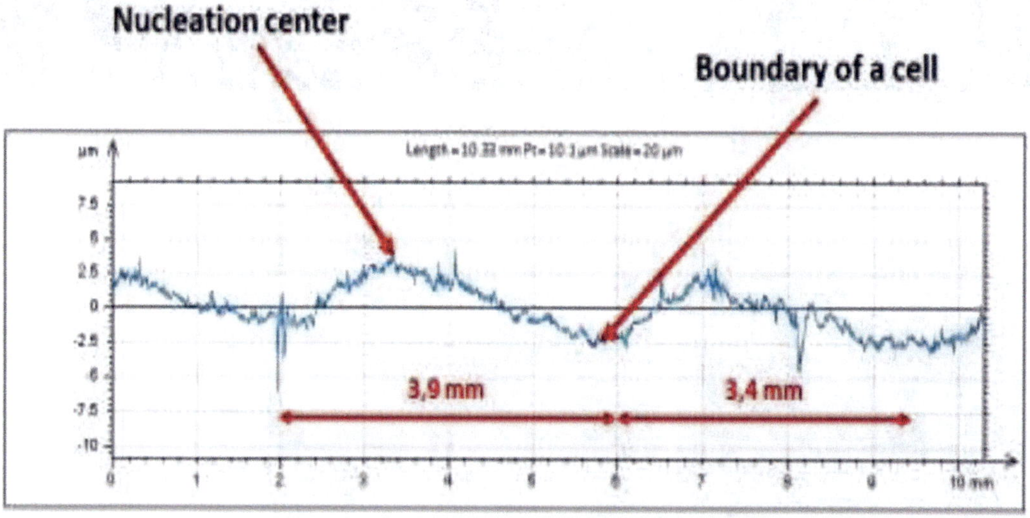

FIG. 6.13 Roughness profile of the surface of the coating showing the presence of a specific wavelength. *(From H. Chaieb, V. Maurel, K. Ammar, S. Forest, A. Koster, F. Nozahic, J. De Strycker, J.-M. Mataigne, Microstructure and texture analysis of a near-eutectic Zn-Al coating deposited on steel by hot dipping galvanizing, in: Galvatech, 2021, pp. 272–281.)*

10–15 µm deep. The results of a profilometry study are shown in Fig. 6.13. A wavelength of 3–4 mm corresponding to the size of the solidification cells is present. It is clear that the boundaries forming the solidification cells represent lower points of the roughness profile. Thus, those boundaries are expected to be the weakest within the coating. In fact, what can be concluded from such an observation is that the liquid starts to solidify at the center of the cells and the last solidification spot is located at the cell boundaries. Therefore, possible impurities will be rejected into these boundaries and the contraction of the last liquid will make the boundaries the weakest part of the microstructure [37]. This solidification shrinkage is the reason for the apparent boundaries and gives to the coating its "orange-peel" structure [38].

Substrate interactions were not expected to be the cause of these defects or shrinkage cavities, since these surface defects were reproduced in the solidification of Zn-5% Al-mischmetal alloy samples on an inert substrate. It was also found that impurities, particularly Pb, strongly segregate to eutectic nodules and triple points. Solidification cracking was found to be associated with both the shrinkage cavities (dents) and segregated impurities [36,39]. A mechanism for Galfan solidification and denting/cracking was proposed by Bluni and is schematically illustrated in Fig. 6.14. During the solidification process, the liquid metal is quickly consumed due to the relatively large volume changes associated with the solidification of Zn and Al. As a result, there is a shortage of liquid between the two or three adjacent growing eutectic nodules and, upon impingement of the nodules, the interface at the surface will be curved, creating a surface depression. Cracking can occur within weakened boundaries due to superimposed solidification stresses and the precipitation of Pb particles. Galfan coating surface appearance has been enhanced by decreasing the Pb content in the bath alloy and by an increase in skin-pass reduction [39].

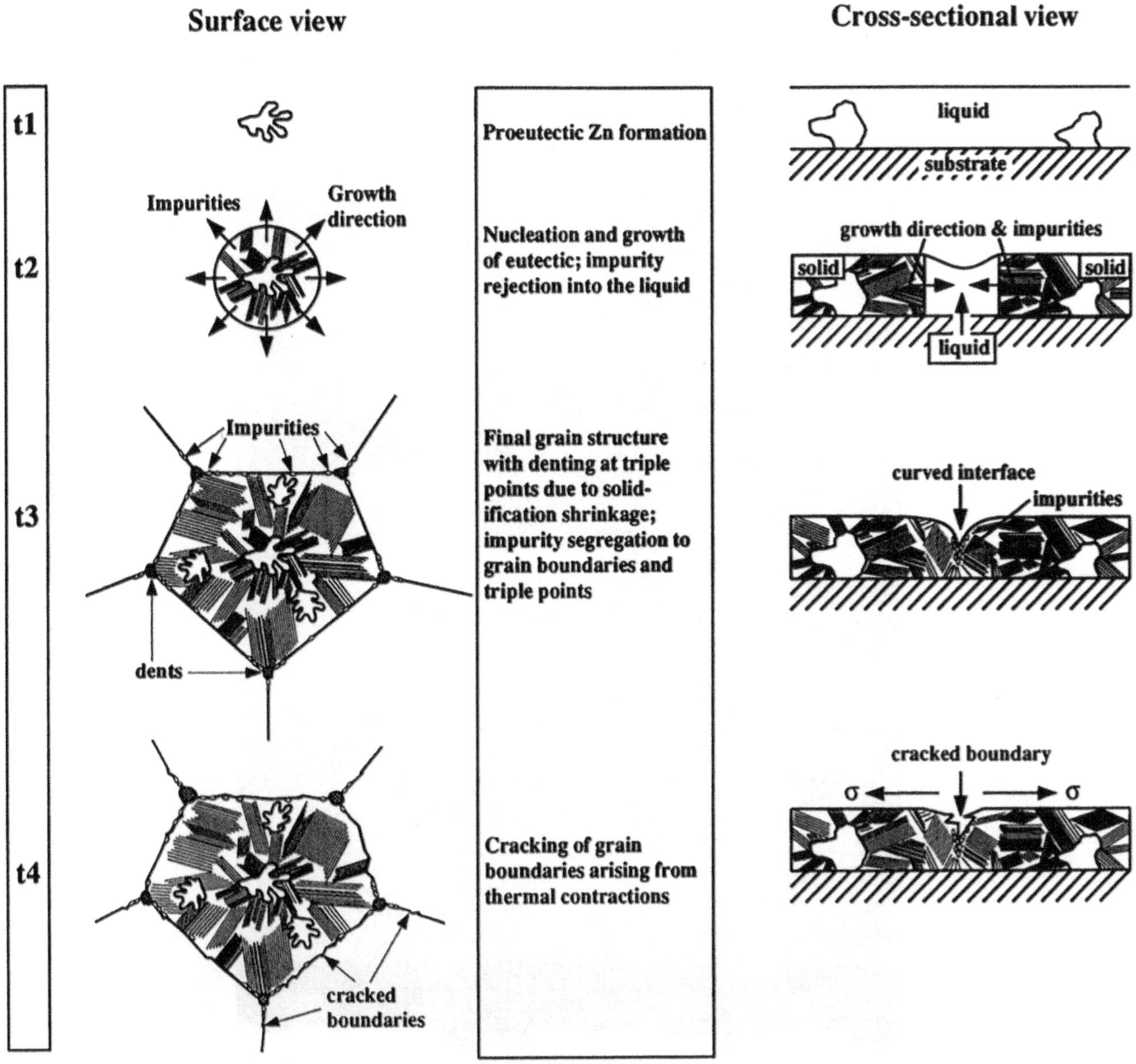

FIG. 6.14 Mechanism for Galfan formation and denting/cracking. *(From A.R. Marder, The metallurgy of zinc-coated steel, Prog. Mater. Sci. 45 (2000) 191–271.)*

6.4.2 Galvalume (55 wt% Al-Zn)

Galvalume is a 55% Al alloy coating containing about 1.5% Si added for the purpose of preventing an exothermic reaction at the coating overlay/substrate steel interface [40]. During the coating process, an interfacial Fe-Al-Zn intermetallic alloy layer forms at the interface between the steel substrate and the overlay coating [41]. The surface of the Galvalume coating contains characteristic spangles, Fig. 6.15a, which consist of aluminum dendrites with a clearly measurable dendrite arm spacing (DAS), Fig. 6.15b. In cross section, three features of the overlay (or solidified bath) are defined, Fig. 6.15c: the coating contains alpha (α) aluminum dendrites, Zn-rich interdendritic regions, and a fine dispersion of Si particles. The Al dendrites were reported to contain approximately 18 wt% Zn and up to 1.8 wt% Si, which is in good agreement with the Al-Zn phase diagram [42].

Cleary [43] has shown that increased cooling rates after dipping can significantly refine the DAS of the coating, Fig. 6.16, and Selverian et al. [42] have reported that spangle size, dendrite grain size, and DAS decrease as coating thickness decreases in a commercial Galvalume coating, 55 wt% Al, 43.4 wt% Zn, 1.6 wt% Si. The number of silicon particles in the overlay increased with cooling rate, Fig. 6.17 [43], and are primarily responsible for refining the microstructure by either constraining the growth of the Al dendrites or by providing numerous heterogeneous nucleation sites for dendrite formation. Si particles

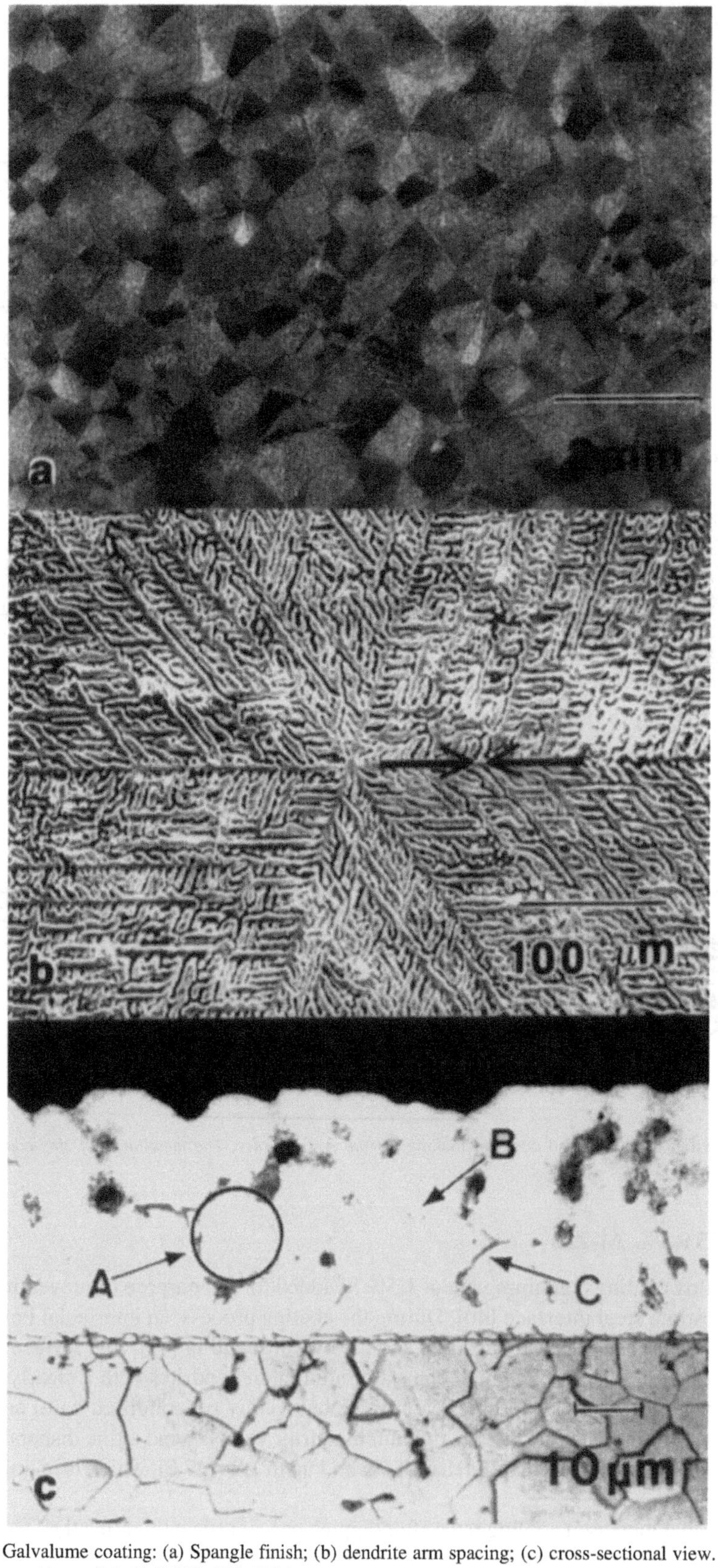

FIG. 6.15 Microstructure of Galvalume coating: (a) Spangle finish; (b) dendrite arm spacing; (c) cross-sectional view. *(From A.R. Marder, The metallurgy of zinc-coated steel, Prog. Mater. Sci. 45 (2000) 191–271.)*

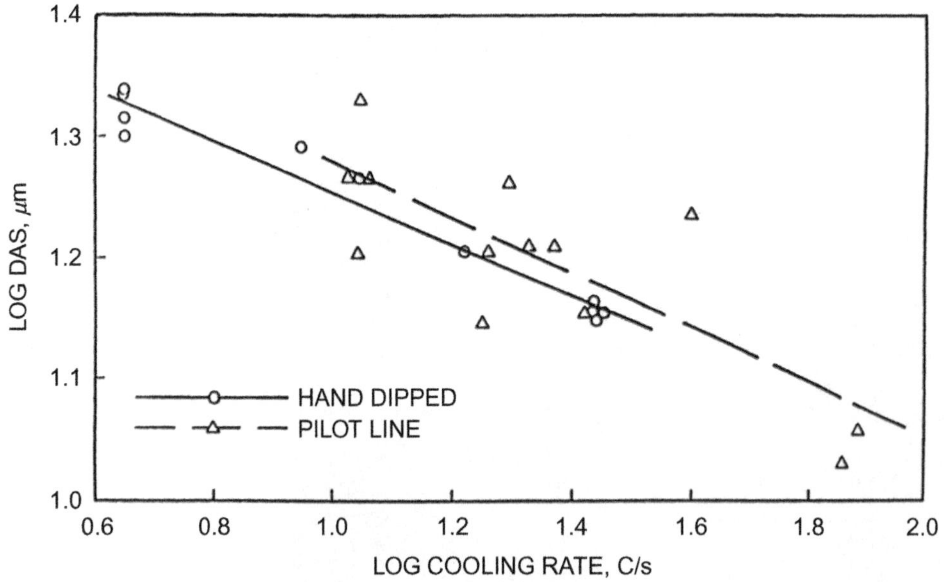

FIG. 6.16 The effect of cooling rate on DAS in Galvalume. *(From A.R. Marder, The metallurgy of zinc-coated steel, Prog. Mater. Sci. 45 (2000) 191–271.)*

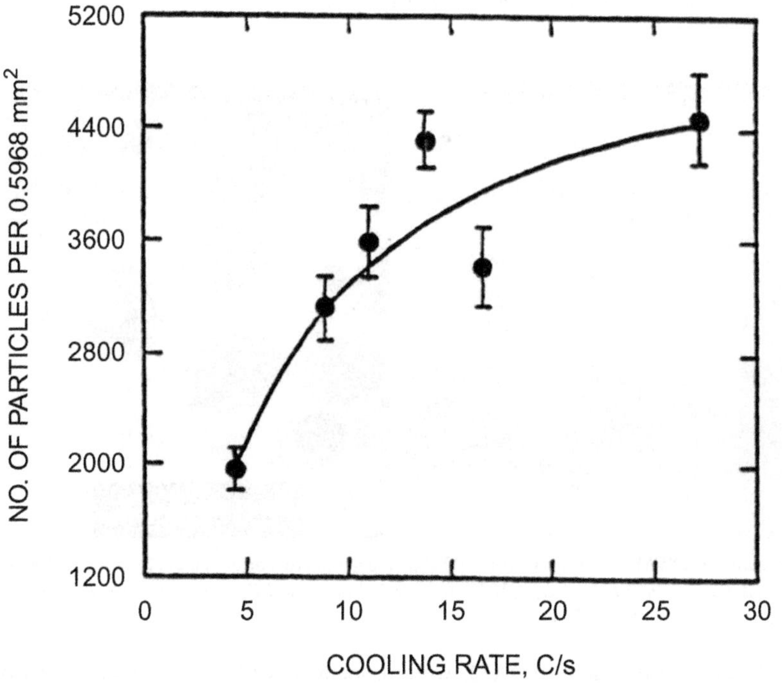

FIG. 6.17 The effect of cooling rate on Si particles in the Galvalume coating. *(From A.R. Marder, The metallurgy of zinc-coated steel, Prog. Mater. Sci. 45 (2000) 191–271.)*

appear to nucleate at the alloy layer/bath interface and grow into the bath. The zinc-rich interdendritic region is heavily populated by precipitates, Fig. 6.18, and it was proposed that the large precipitates could be the metastable R phase, whereas the small precipitates are possibly coarsened G.P. zones or spinodal phase precipitates [42]. More recent research confirmed the presence of these fine particles as aluminum-rich and zinc-rich precipitates, Fig. 6.19. The white regions were characterized as zinc-rich, and the dark globular regions were aluminum- rich. The presence of these two phases in the interdendritic region was explained by the use of binary Al-Zn equilibrium phase diagram [44].

Silicon suppresses the rapid exothermic reaction between the binary Al-Zn bath and an iron panel by forming a solid reaction layer [40,45]. Depending upon Si content, the growth of the reaction layer can be diffusion controlled (0.7, 1.7, and 6.88 wt% Si) [40]. Interface control occurs in 3.0 and 5.0 wt% Si [40] as well as in 0.0 wt% Si alloys [45], and silicon has been shown to affect the composition of the interface layers [40,45]. The presence of silicon in the coating is thought to play a role in the nucleation of alpha (α)-Al dendrites. Previous experimental studies found that the population of silicon particles observed in the coating increased with cooling rate or silicon content, e.g., Fig. 6.17 [43]. In both cases, finer

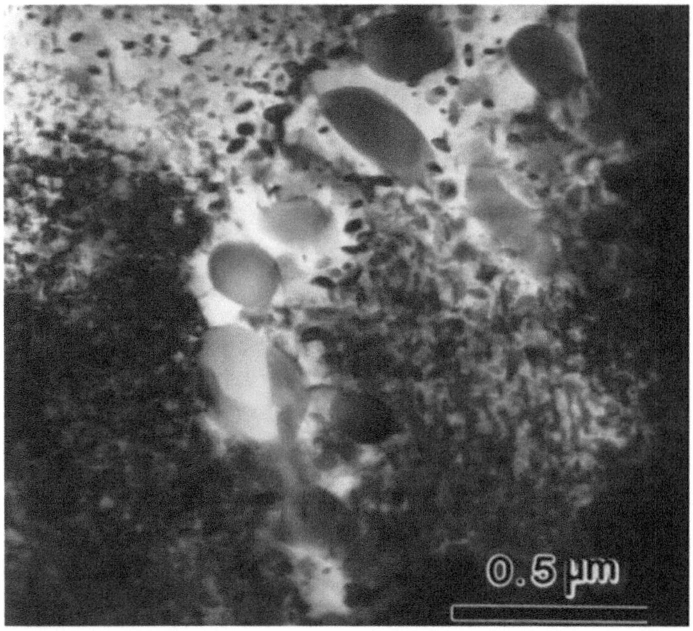

FIG. 6.18 Precipitates in the interdendritic boundaries of Galvalume. *(From A.R. Marder, The metallurgy of zinc-coated steel, Prog. Mater. Sci. 45 (2000) 191–271)*

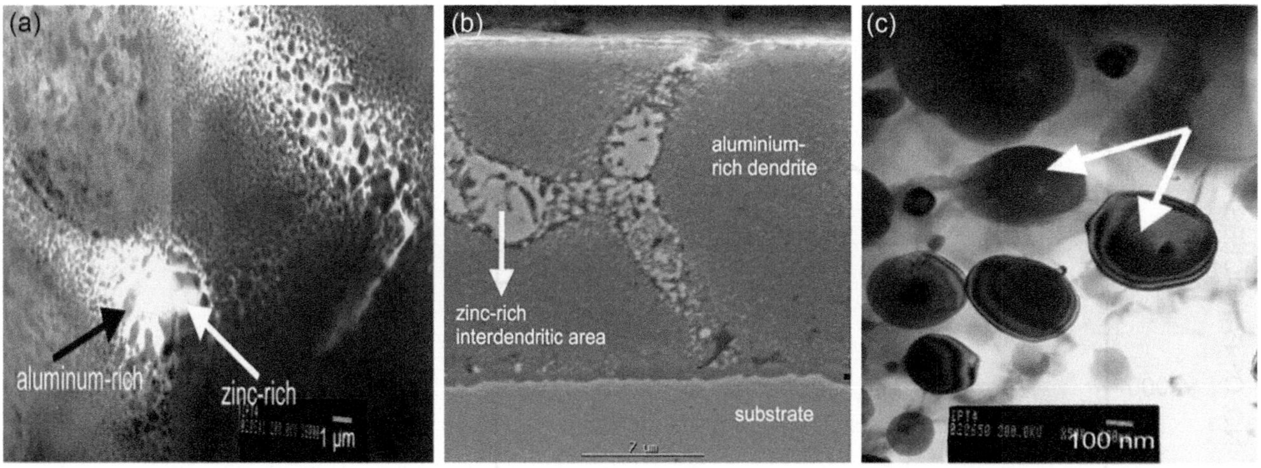

FIG. 6.19 The microstructure of 55% Al-Zn Galvalume coating. (a) SEM cross section, (b) TEM showing the interdendritic region, (c) Al-rich globular precipitates in the interdendritic region. *(From A.R. Moreira, Z. Panossian, P.L. Camargo, M.F. Moreira, I.C. da Silva, J.E. Ribeiro de Carvalho, Zn/55Al coating microstructure and corrosion mechanism, Corros. Sci. 48 (2006) 564–576.)*

alpha (α)-Al dendrite structures, i.e., smaller dendrite arm spacing (DAS), were associated with higher populations of silicon particles. Based on these observations, it was proposed [46] that silicon particles could form before the formation of alpha (α)-Al dendrites, by growing from the overlay/alloy layer interface. The silicon particles could then promote the nucleation of alpha (α)-Al by providing heterogeneous nucleation sites. Based on this assumption, the structural development of the coating was proposed to follow the following sequence [47]:

t1 (stage I): alloy layer formation at the coating/steel interface
t2 (stage II): silicon nucleation on the alloy layer and directly from the bath
t3 (stage III): alpha (α)-Al nucleation on existing silicon particles
t4 (stage IV): growth of alpha (α)-Al dendrites and further growth of silicon particles in the interdendritic region
t5 (stage V): solidification of zinc-rich phase in the interdendritic region.

Chen and Willis [48] noted that given the large population of the silicon particles observed, if silicon particles had acted as the heterogeneous nucleation centers for the alpha (α)-Al dendrites, the spangle size of the coating would have been much finer, and also related to the cooling rate and silicon level in the coating, i.e., finer spangle sizes at higher cooling rates or at higher silicon contents. However, it was reported [49,50] that the main effect of increasing the coating thickness is to decrease the cooling rate. Thus, as coating thickness decreases (increasing cooling rate), DAS, α-Al dendrite size, and spangle size decrease. Xu et al. [50] reported that $FeAl_3$ provides a higher heterogeneous nucleation density than τ_5 (α-AlFeSi), resulting in a smaller spangle size. Peng et al. [51] showed that spangle size was determined by the intermetallic layer, in disagreement with the proposal that the spangle size was determined by the grain size of the primary α-Al phase [48]. Further, Peng et al. [51] showed that Si influenced the formation of $FeAl_3$ over τ_5 (α-AlFeSi), with $FeAl_3$ having a better crystallographic fit with alpha (α)-Al. Therefore, additional Si can increase the heterogeneous nucleation of alpha (α)-Al dendrites on the intermetallic interface by promoting the formation of the $FeAl_3$ intermetallic, as well as the formation of Si particles that also nucleate at the interface.

Chen and Willis [48] also used a phase diagram to explain the observed microstructure in the overlay. Given that it takes a matter of seconds for the coating to solidify, equilibrium thermodynamics alone will have a limited role in explaining the overlay microstructure. The kinetics of the formation of the non- equilibrium microstructure is much more complex, and an understanding of the overlay microstructure and distribution of the phases formed (morphology) is required. Si particles morphology analysis is particularly important. In the study by Selverian [49], he found that, by selectively etching away the Al dendrites and Zn rich interdendritic regions using a 10% perchloric-ethanol mixture, the Si particles remaining were attached to the alloy layer and the number of particles increase as the Si content of the bath increases (Fig. 6.20), thus, it appears that the Si particles nucleate at the alloy/layer interface and grow into the bath. Furthermore, he found that as the dipping time increased, the Si crystals exhibit an increase in side branches (Fig. 6.21), indicating that the Si had time to grow into the liquid coating overlay, as opposed to quickly precipitating in the remaining Si-Al-rich liquid. As

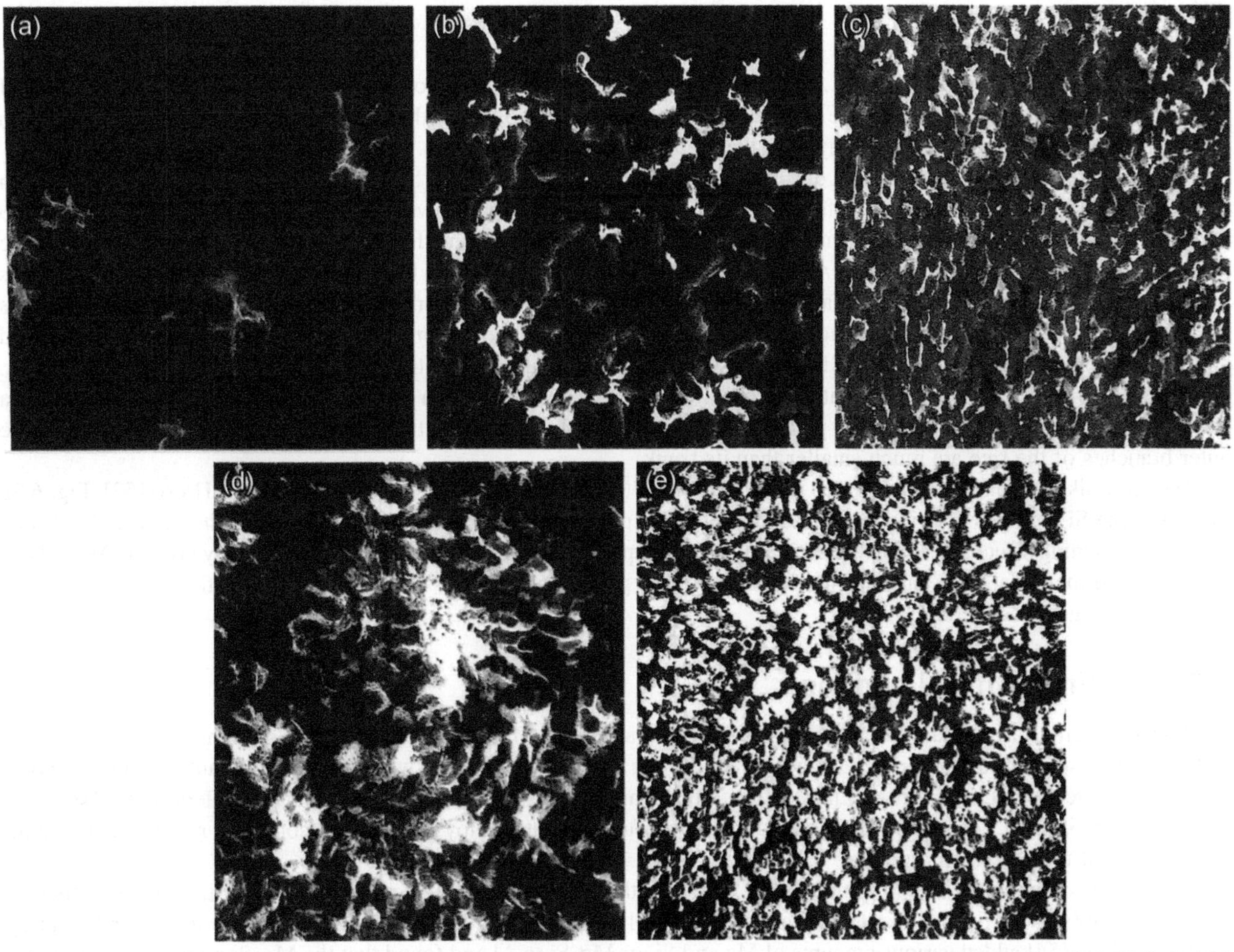

FIG. 6.20 The effect of Si content on the number of Si crystals after etching away of the alloy layer. All panels were dipped for 4 s. (a) 0.7% Si, (b) 1.7% Si, (c) 3.0% Si, (d) 5.0% Si, (e) 6.88% Si. Approximate magnification: 11mm = 25microns. *(From J.H. Selverian, An Electron Optical Study of the Effect of Silicon on Hot Dip 55 Wt% Al-Zn Coatings on Iron (Master's thesis), Lehigh University, 1985.)*

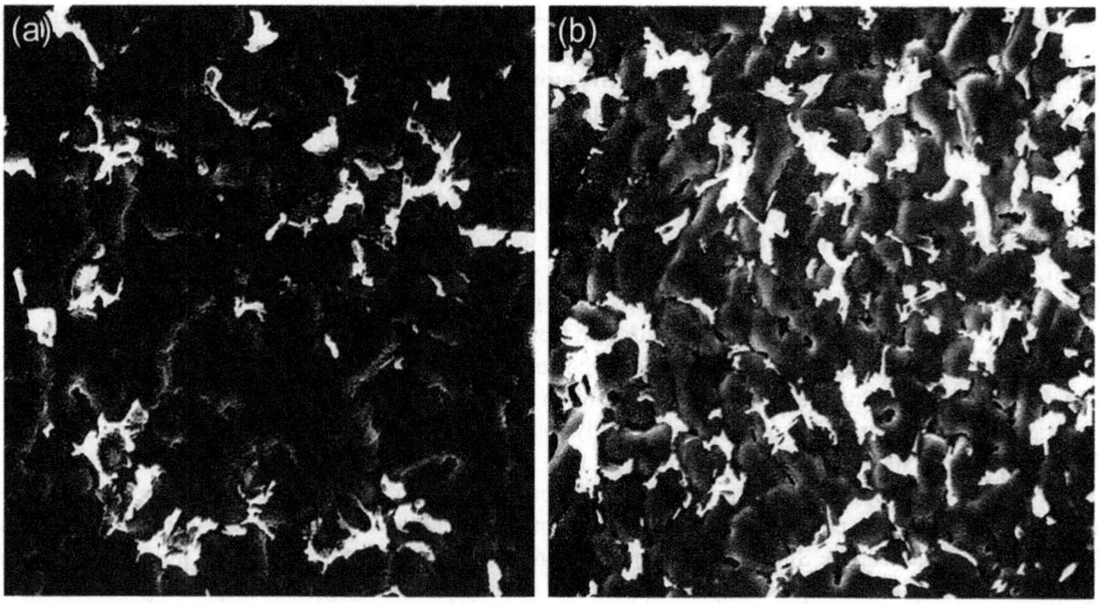

FIG. 6.21 The effect of increased time on the side branch growth of a 1.7% Si coating. (a) 4 s, (b) 900 s. Aproximate magnification 15mm = 25 microns. *(From J.H. Selverian, An Electron Optical Study of the Effect of Silicon on Hot Dip 55 Wt% Al-Zn Coatings on Iron (Master's thesis), Lehigh University, 1985.)*

an indicator, pure Si solidifies at 1414°C, well above the beginning of solidification of Al at 660°C. At a bath temperature of approximately 600°C, Si should be more stable and precipitate first. Also, due to their higher supercooling (1414°C–600°C) the Si crystals should grow more rapidly than the Al dendrites (660°C–600°C).

Therefore, it would be expected that during solidification of the overlay, the Si crystals nucleate at the alloy layer bath interface and grow toward the outer surface of the coating. Due to the high growth velocity, they grow unconstrained into the molten Al-Zn bath, until all of the Si in the overlay liquid is depleted. During this very short time, the outer surface of the coating has formed an Al_2O_3 "skin" and Al-rich dendrites can nucleate and begin to grow inward toward the Fe substrate as well as nucleating at the substrate interface and other heterogeneous sites. The amount, size, and spacing of the Si crystals (morphology) appear to be significant factors in determining the size of the Al dendrites. If the Si crystals were precipitated after the Al dendrites formed, then the Al dendrite size should not be affected by the Si content or the number of Si crystals. However, the dendrite size decreases with Si content (Fig. 6.22). In effect, it can be expected that the Si crystals grow as a "tree" and should become smaller in 2D planar analysis as the crystals approach the outer surface of the overlay, just as the outer branches of the tree are much smaller than its trunk.

The "tree-like" Si particle growth was confirmed by SEM photomicrographs reported by Li and Hreso [52]. Fig. 6.23 illustrates the Si particles growing out of the interface in a planar section. In a 3D view, Fig. 6.24, the Si particles "show root-like branching structures at the nucleation sites" of the particles attached to the intermetallic alloy layer. EDS analysis conducted on the isolated Si particles indicated that most particles contained 99% Si, confirming the previous Si morphology findings by Selverian [49].

6.5 Coatings with Al and Mg

Zn-Al-Mg systems may be classified into a high Mg system and a low Mg system. The high Mg systems includes 3 wt% Mg and is mainly used for building materials, while the low Mg systems include 1–2 wt% Mg is used in automobile and home appliance applications [53]. Further, Zn-Al-Mg coatings can be classified according to their Al content: 0.25–6 wt% Al; 11%–22% Al; 55% Al. Generally, Mg is added to the various Zn-coated alloys to improve corrosion resistance due to the stabilization of the corrosion products.

Some alloys contain deliberate amounts of Si across the range of Zn-Al-Mg compositions. Si, as in the other Zn coatings, notably 55% Al-Zn Galvalume, is added to control the interface interactions at the substrate steel/bath boundary. In 1962, Nikola [54] added various amounts of Mg and Si, or Mg_2Si to Al and found that the Mg_2Si retarded the growth of the interfacial layer normally found in Type 1 aluminized coatings. Magnesium was added to a galvanizing bath to reduce zinc oxides on the bath surface [55]. Several patents were granted at that time, for the addition of Mg to Zn-low Al Zn baths

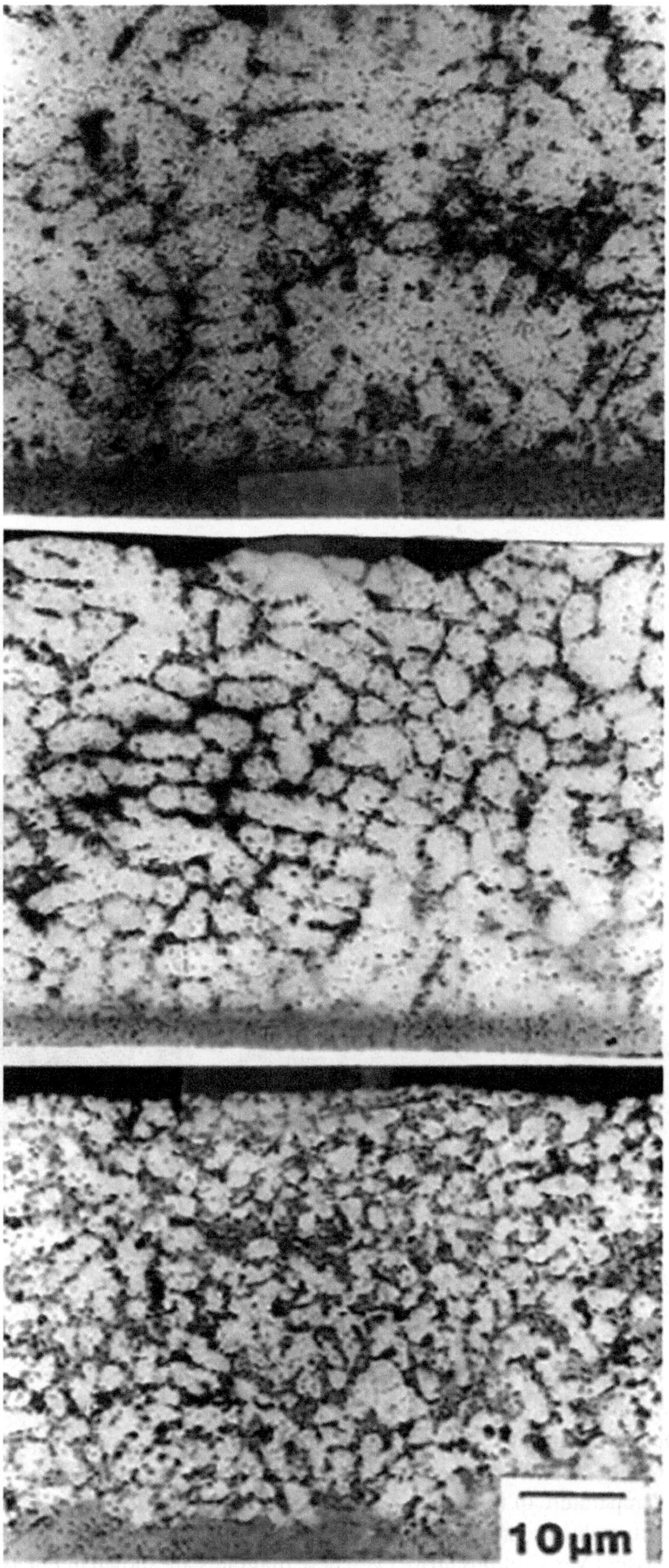

FIG. 6.22 Photomicrographs showing the effect of Si content of the bath on Al dendrite size. (A) 0.7% Si, (B) 3.0% Si, (C) 6.88% Si. *(From J.H. Selverian, An Electron Optical Study of the Effect of Silicon on Hot Dip 55 Wt% Al-Zn Coatings on Iron (Master's thesis), Lehigh University, 1985.)*

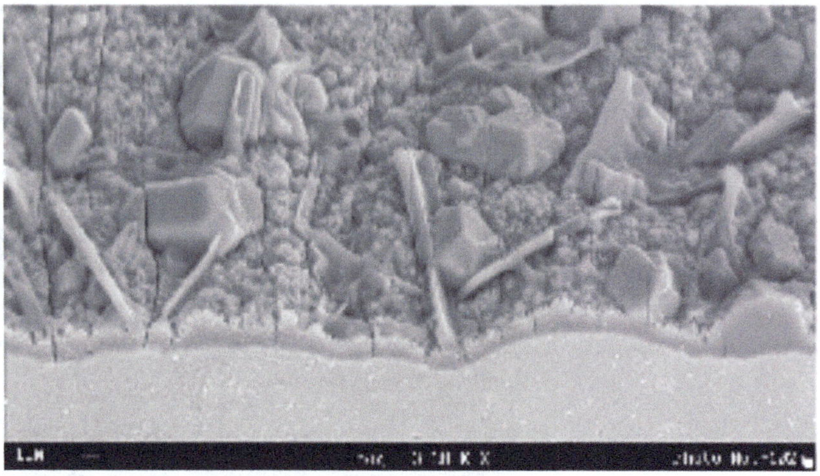

FIG. 6.23 SEM micrograph of the alloy layer. *(From J. Li, D. Hreso, Morphology of 55% Al-Zn coating and the Al-Zn coating with Mg additions, in: Galvatech, 2015, pp. 852–860.)*

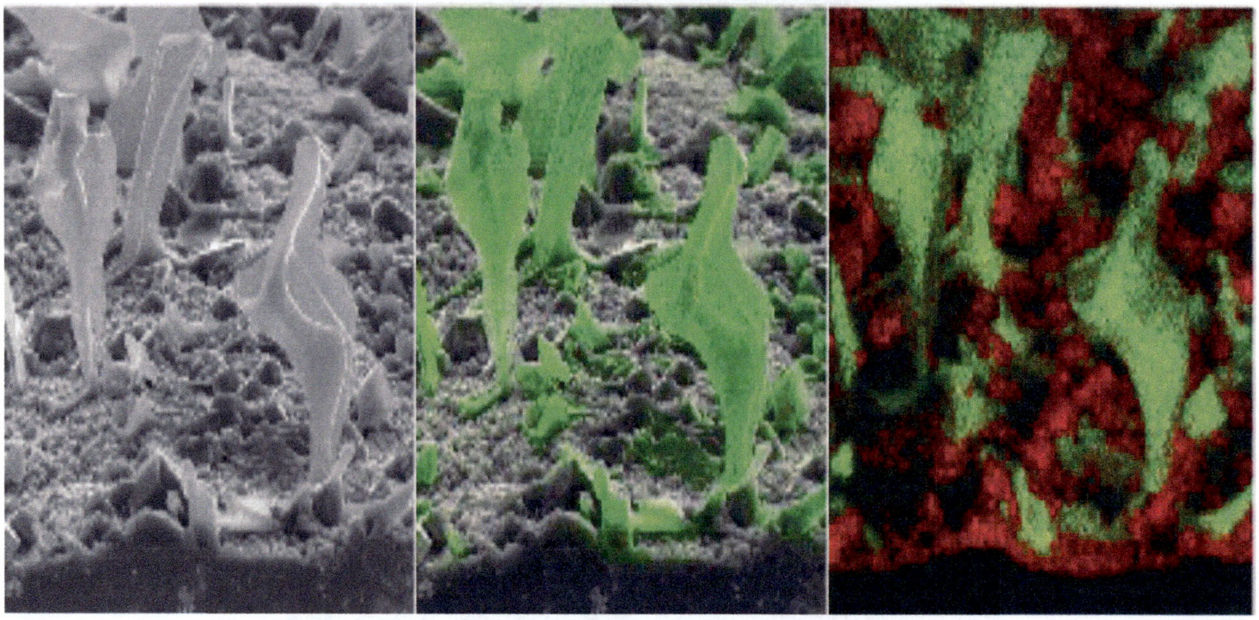

FIG. 6.24 SEM micrograph of tilted 3D view of Si particles in the 55%Al-Zn coating after the overlay was removed by etching with hydrochloric acid solution. *(From J. Li, D. Hreso, Morphology of 55% Al-Zn coating and the Al-Zn coating with Mg additions, in: Galvatech, 2015, pp. 852–860.)*

[56], as well as Zn–3%–17% Al baths [57]. A natural progression followed as Mg was added to 5% Al Galfan [58], Zn-10%–22% Al [59], Zn-20%–35%Al with 0.1%–0.2%Si [60], and 55%Al-Zn-1.5%Si Galvalume coatings [60]. Enhancement of corrosion resistance due to the presence of Mg in Zn-based coatings is well established and a good review can be found in [61].

The first work on ZnAlMg phase equilibria extended the Zn-rich corner of the ZnAlFe phase diagram by adding Mg up to 0.5% to a Zn-0.2%Al bath in a hot-dip galvanizing process simulator and comparing the phases observed with numerical analysis based on data that were, at the time, only available for the Al-rich corner [62]. Differential scanning calorimetry showed the ternary eutectic temperature of this system to be 344°C and the peritectic as 359°C. This was extended in the IZA Galvanized Autobody Partnership program (project ZCO-66) by Prof. A.R. Pelton, who derived phase diagrams for the same range of Al compositions and Mg additions up to 3%. These diagrams are shown in Fig. 6.25 [63]. It was observed that the Fe-Zn ζ phase is formed at low Al content over the entire temperature range only for 0% and 1% Mg. For 3% Mg, ζ phase forms up to 456°C, but no ζ phase for 5% Mg. When increasing Mg content $FeAl_2$ forms at Al compositions between those corresponding to δ and η (Fe_2Al_5) phases; Fe_2Al_5 forms at temperatures up to 462°C for 1% Mg, up to 472°C for 3%

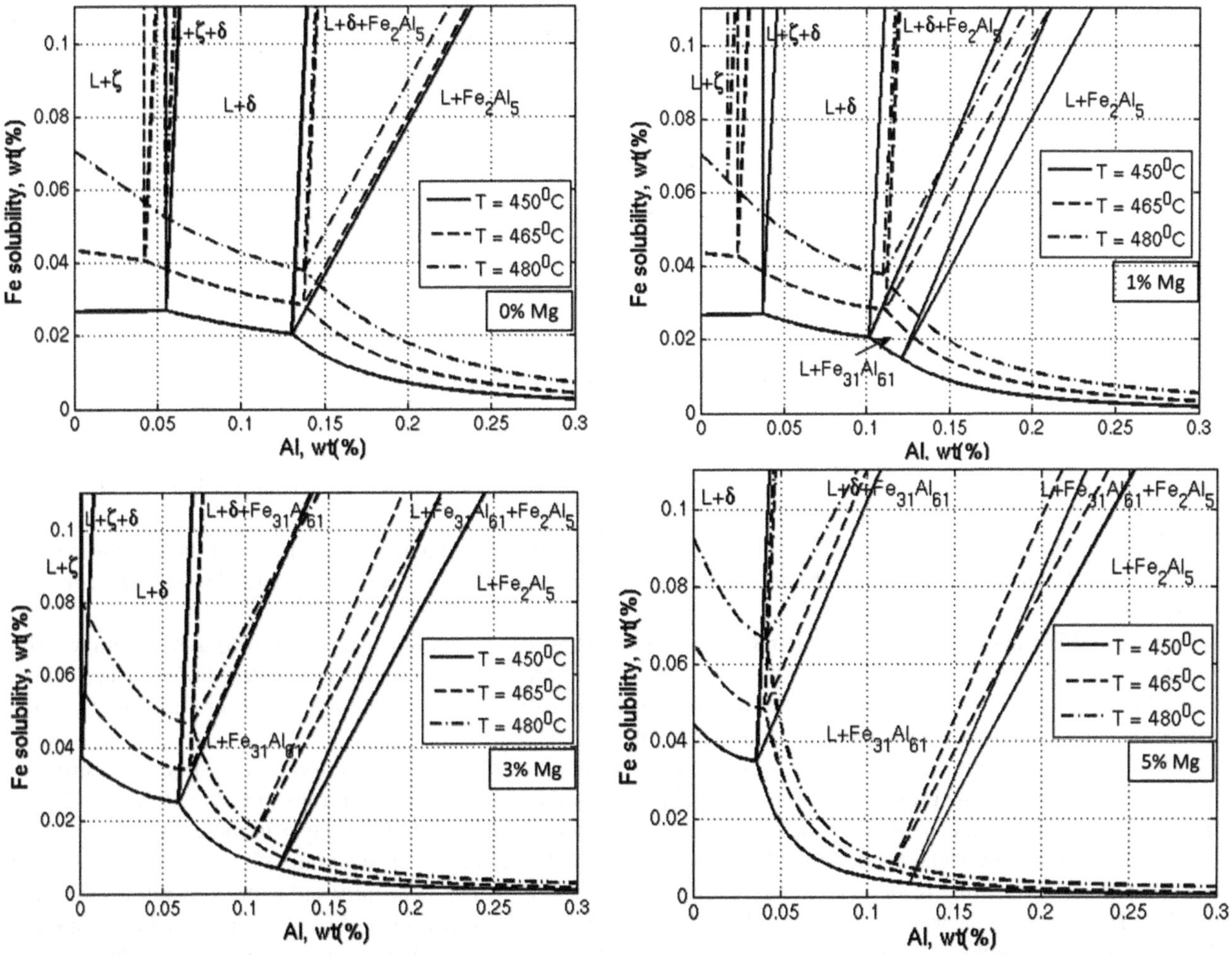

FIG. 6.25 Zn-Al-Fe-Mg phase diagrams for 0%, 1%, 3%, and 5% Mg. *(From A.D. Pelton, Thermodynamic Calculations of the Steps During Solidification of Certain Zn-Al-Mg Alloys, prepared for F. Ajersch, under contract with IZA, 2014. Galvanized Autobody Partnership Projects ZCO-55/66 Final Report, Issued May 7, 2014, revised August 2014, 135 pp. Available from www.zinc.org.)*

Mg, and up to 479.4°C for 5% Mg. The minimum Al content for which η phase forms is slightly lower for 1% Mg than for 0% Mg and then increases when increasing the Mg content (it remains in the range of 0.11%–0.14% Al). De Bruycker et al. [62] also examined the addition of Mg in compositions up to Zn-6Al-3Mg. They observed that, contrary to equilibrium predictions, the Mg_2Zn_{11} phase was never observed; the ternary eutectic consisted of Zn-hcp/Al-fcc/$MgZn_2$.

6.5.1 0.25%–6% Al

Results of additions of up to 5% Mg in Zn-Al coatings found that as the Mg increased, the corrosion loss decreased [53]. Most of the industrially used Zn-Al-Mg alloy coatings, particularly those used for automotive applications, have Al and Mg compositions in the range of 1%–3.7%. Because of this, their microstructures are formed in the same way, and they have similar physical and electrochemical properties [64].

Hot-dip Zn-Al-Mg coatings with 1%–2% Al and 1%–2% Mg were investigated [65]. These coatings consist of Zn grains, surrounded by the ternary eutectic and increased Al, and Mg increases the amount of eutectic. At a 3%Al-3% Mg coating, except for some Zn grains, the microstructure is almost fully ternary eutectic. The $MgZn_2$/Al/Zn ternary eutectic, which is not present in the equilibrium diagram, contains slightly more than 3.25% Al and 3.2% Mg. For comparison, the equilibrium ternary eutectic composition of the liquid for solid Mg_2Zn_{11}/Al/Zn formation is at 3.9%

TABLE 6.3 Phases and eutectics present in the Zn-Al-Mg coatings.

Name	Description
Zn	Zn-phase (α)
$MgZn_2$	Mg-Zn (Laves-C14) phase, no Mg_2Zn_{11} was found
Al	Zn-rich Al-phase, fee solid solution with 60%–80% Zn
Zn-Al	Binary eutectic of Zn and Al
Zn-Mg-Al	Ternary eutectic of Zn, $MgZn_2$ and Al
Si-Mg	Binary phase of Si and Mg (about $MgSi_2$)

(From M. Vlot, M. Zuijderwijk, M. Toose, L. Elliott, R. Bleeker, T. Maalman, Hot dip ZnAlMg coatings: microstructure and forming properties, in: Galvatech, 2007, pp. 574–579.)

Al and 2.4% Mg at 344°C. Examples of the phases found in these coatings are shown in Table 6.3. Oh et al. [66] found similar results for a 3 wt% Mg-2.5 wt% Al coating.

Surface morphology examinations of a Zn-0.5Mg-0.25 Al coating showed that the top surface of the coating revealed that Mg was preferentially segregated at the grain boundaries, with lower Mg content and finer lamellar structure compared with coatings without Al [66], Fig. 6.26.

Dutta et al. [67] reported ternary eutectic mixtures in Zn-0.5Mg-0.25Al coatings. Kim et al. [68] used in-situ real time synchrotron X-ray scattering and confirmed that the solidification proceeds in the order of Zn, Zn/$MgZn_2$ binary eutectic and Zn/$MgZn_2$/Al ternary eutectic phases. The cross-sectional microstructure is shown in Fig. 6.27 for these alloys, and the melting and crystallization temperatures are found in Table 6.4.

LeBozec et al. [69] found that increasing the cooling rate from 5 to 1000°C/s resulted in extremely fine microstructures with a large number of small primary zinc grains. The diameter of the zinc grains ranged from approximately 85 to 2.5 μm. The addition of small amounts of Al strongly influences the formation of an inhibition layer in these coatings.

The steel-coating interfacial layers formed with coatings made by dipping an IF steel into a Zn-1.0%Al-1.8%Mg bath were compared with coatings made in a Zn-0.28%Al-0.28%Mg bath [70]. The steel entry and bath temperatures were both 460°C, and the dipping time was 3 s. The lower alloyed bath formed coatings with an $Fe_2Al_{5-x}Zn_x$ interface layer consisting of two morphologies: either a rough, granular layer similar to that formed in GI baths with crystallographic direction parallel to the normal direction of the steel sheet, or a smooth layer of single crystal $Fe_2Al_{5-x}Zn_x$ grown epitaxially on the substrate. For the higher alloyed bath, the interfacial layer consisted of disk-shape [010]-oriented Fe_4Al_{13} crystals with diameters between 100 and 500 nm. No orientation relationship between the Fe_4Al_{13} crystals and α-Fe was found, and no Mg content was observed in any of the intermetallic compounds, a consequence of the mutual insolubility of Fe and Mg.

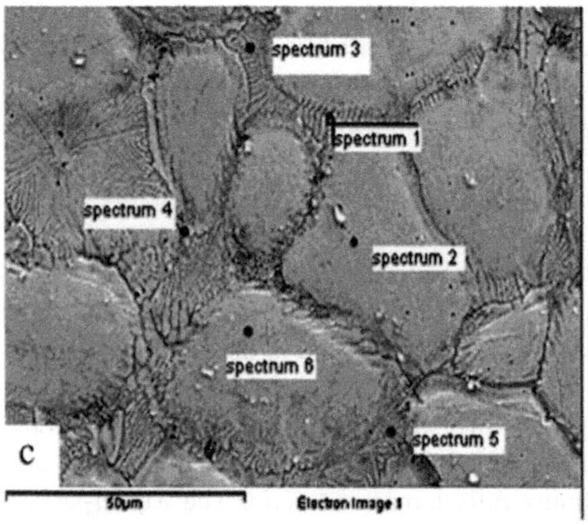

Spectrum	Mg	Al	Fe	Zn
1	2.03	0.99	0.43	96.55
2				100.00
3	6.44	2.74	0.14	90.68
4	3.82	6.73	0.18	89.27
5	5.31	8.59	0.22	85.88
6	0.37	1.59	0.69	97.44

FIG. 6.26 Typical SEM microstructure of the top surface of coating and EDS microanalysis (composition in wt%) of alloy 0.5% Mg and 0.25% Al. *(From M. Dutta, A.K. Halder, S.B. Singh, Morphology and properties of hot dip Zn–Mg and Zn–Mg–Al alloy coatings on steel sheet, Surf. Coat. Technol. 205 (2010) 2578–2584.)*

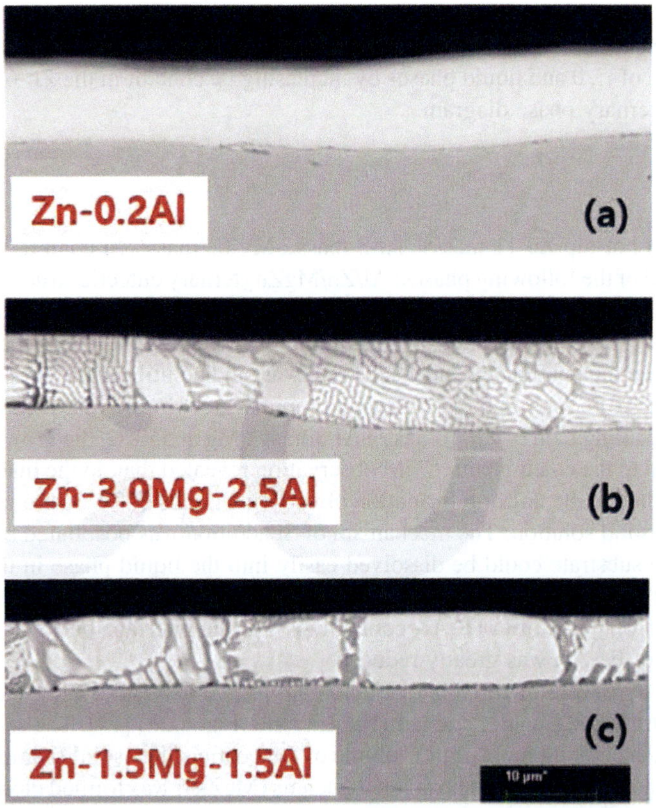

FIG. 6.27 The cross-section microstructure of hot-dip alloy coated steel sheet: (a) Zn-0.2%Al, (b) Zn-3.0%Mg-2.5%Al, (c) Zn-1.5%Mg-1.5%Al. *(From T.C. Kim, I.-R. Sohn, Y.H. You, M.S. Kim, J.S. Kim, Investigation of hot-dip ZnMgAl coatings for automotive applications, in: Galvatech, 2021, pp. 100–107.)*

TABLE 6.4 The melting and crystallization process of ZnMgSi coatings.

	Melting (phase transformation)		Crystallization	
Composition	Element	Temperature (°C)	Element	Temperature (°C)
Zn-3.0%Mg-2.5%Al	A(Al): disappear	329	Z(Zn): appear	336
	m($MgZn_2$): disappear	342	m($MgZn_2$): appear	336
	M(Mg_2Zn_{11}): appear		–	–
	Z(Zn): disappear	368	–	–
	M(Mg_2Zn_{11}): disappear	377	–	–
Zn-1.5%Mg-1.5%Al	m($MgZn_2$): disappear	360	Z(Zn): appear	393
	M(Mg_2Zn_{11}): appear	360	m(MgZn2): appear	Unclear
	M(Mg_2Zn_{11}): disappear	387	–	–
	Z(Zn): disappear	413	–	–

(From T.C. Kim, I.-R. Sohn, Y.H. You, M.S. Kim, J.S. Kim, Investigation of hot-dip ZnMgAl coatings for automotive applications, in: Galvatech, 2021, pp. 100–107.)

Takata et al. [71] studied the formation of Fe-Al intermetallic interface phases on an interstitial-free (IF) steel hot-dipped in the Zn-6Al-3 Mg (wt%) alloy melt at different temperatures for various times (2–3600 s). They found that the dissolution of Fe into the Zn alloy melt occurred even at 2 s in dipping. The microstructural characterization revealed that the continuous θ-Fe_4Al_{13} phase layer was initially formed, followed by the local growth of η-Fe_2Al_5 phase toward both solid α-Fe side and liquid Zn alloy side during the hot-dipping process. The further growth of η phase accompanied the dissolution of Fe into the Zn alloy melt, resulting in a significant thickness loss of IF steel sheets during the hot-dipping process. These experimental results were in good agreement with a phase transition from a two-phase region of θ and liquid phases to a three-phase region of η, θ and liquid phases by increasing Fe content in the Zn-6Al-3Mg alloy predicted by the calculated Zn-Al-Mg-Fe quaternary phase diagram.

6.5.2 11%–22% Al

The solidification structure of hot-dip Zn-11 mass% Al-3 mass%Mg-0.2mass%Si coating was studied [72]. The solidification structure was composed of the following phases: Al/Zn/$MgZn_2$ ternary eutectic structure, primary Al phase structure, and $MgZn_2$ phase structure. It was found that the results differed from the solidification structure estimated in the equilibrium state. In a further study, [73] the influence of Si addition to the coating bath on the growth of the Al-Fe alloy layer formed in the interface between the coating and steel substrate was studied utilizing a Zn-Al-Mg coating bath with 10–11 mass% Al content. While the Al-Fe intermetallic compound rapidly grew in the interface between the coating and steel substrate when dipped into a coating bath of Zn-10mass%Al-3mass%Mg at 550°C, the growth was remarkably suppressed by the addition of 0.2mass%Si to the coating bath. TEM observation revealed that, in the interface between the coating and the substrate, Si addition resulted in the uniform formation of very fine grains of 20–30 nm in diameter in the Fe_2Al_5 phase with dissolved Si and Zn in a solid solution. The mechanism of Si addition was postulated by the authors to be as follows: the large amount of Fe in the substrate could be dissolved easily into the liquid phase in the case of a ternary Fe-Al-Zn system, whereas in the case of a quaternary Fe-Al-Zn-Si system, the dissolution of Fe into the liquid phase was significantly suppressed by the presence of very fine, thin Fe_2Al_5 containing Si in the interface between the coating and the steel substrate. Thus, the growth rate of Fe_2Al_5 was greatly reduced.

Yu et al. [74] found that there were three typical structures in the Zn-11%Al-3%Mg-0.2%Si coating: snowflake-like dendrite; double hexagonal structure; granular and lamellar eutectic (Fig. 6.28) [74]. The snowflake-like dendrites are the main structure of the coating, occupying most of the volume of the coating. The solidification processes are: the dendrites structure was first nucleates and grows, then the hexagonal nucleus ($MgZn_2$) was formed close to the dendrites, and another

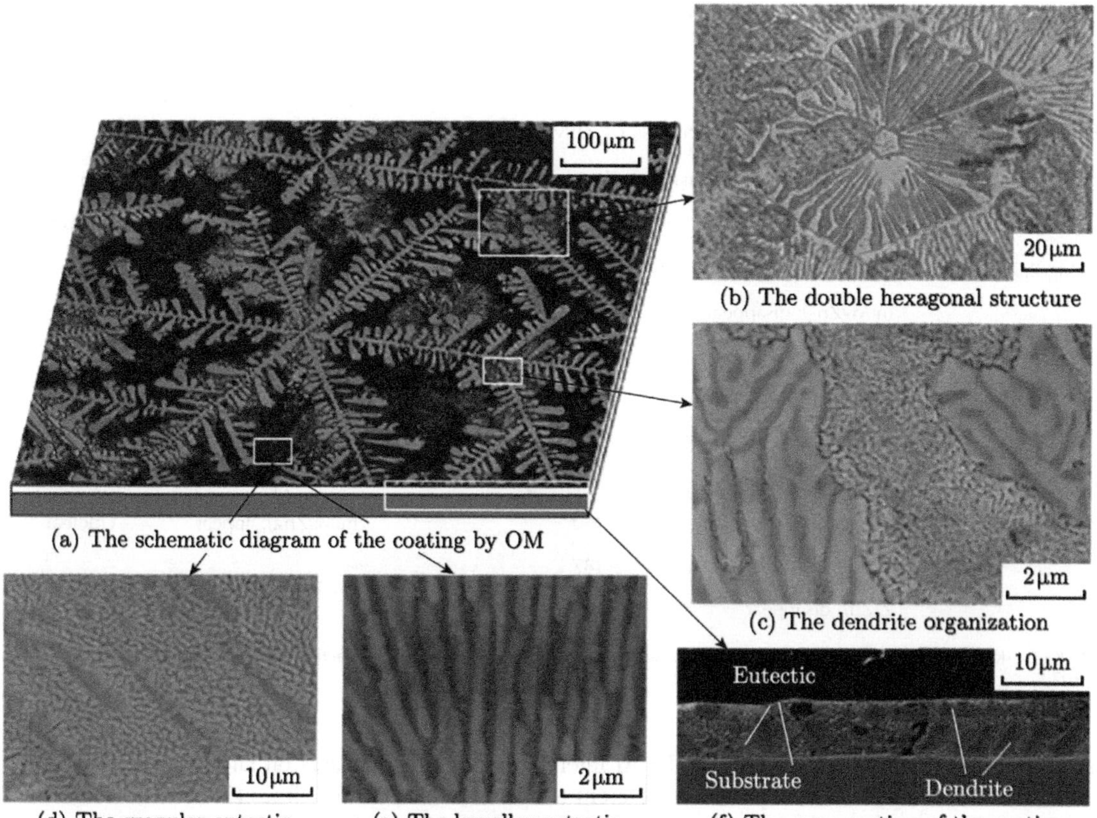

FIG. 6.28 The microstructure of the Zn-11%Al-3%Mg-0.2%Si. *(From K.-C. Yu, J. Li, X. Liu, J.-G. Li, X. Xue, Microstructure of hot dip galvanized Zn-Al-Mg alloy coating, J. Shanghai Jiaotong Univ. (Sci.) 17 (2012) 663–667.)*

phase of HCP grains grew relying on the nucleus, and afterward the ternary eutectics solidified. Lastly, the dendrites decomposed into granular eutectoid morphology. The coating was composed of Zn-rich, Al-rich, and $MgZn_2$ phases. The elemental composition of the dendrite is Zn and Al. The hexagonal nuclei are $MgZn_2$ phase, and the eutectic is Zn/Al/$MgZn_2$ ternary eutectic.

The addition of Mg in a Zn-22.3%Al-Si bath will affect morphology and structure of the surface microstructure and intermetallic layer. The amount of Mg is crucial to coating formation and growth. Tu et al. [75] found that five phases were found on the surface microstructure: η-Zn, -Al, (Si), Mg_2Zn_{11}, and $MgZn_2$.

The addition of Mg to the Zn-Al-Si bath improves the uniformity of the top surface. With increasing of Mg content, the quantity of Zn/Al/Mg_2Zn_{11} ternary eutectic increased. When the Mg content is increased up to 0.6 wt%, $MgZn_2$ appears on the top surface. Five phases were found in the alloy layers, i.e., Fe_2Al_5, $FeAl_3$, τ_{5C}, τ_{5H}, and τ_1. They also found that the addition of 0.2 wt% Mg delayed the emergence of $FeAl_3$ phase and the growth of total alloy layer was diffusion controlled for the various content Mg baths studied. When the Mg content was less than 0.6 wt%, the Zn/Al/Mg_2Zn_{11} ternary eutectic increased with more Mg additions, and the corrosion resistance of Mg-containing coatings gradually increased.

The absence of the Mg_2Zn_{11} phase was confirmed in a study of the Zn-11%Al-3%Mg-0.2%Si coating trademarked as SuperDyma®[c] [73]. Unlike the lower Al content coatings, in this coating precipitation of the Zn phase from the primary Al phase begins at 308°C. The metastable $MgZn_2$ forms as a Laves phase in the absence of Mg_2Zn_{11} and the associated peritectic reaction. The solidification path of hypoeutectic Zn-Al-Mg compositions, which are the most widely used Zn-Al-Mg coating compositions, was described in [63]. In the absence of the Mg_2Zn_{11} phase, the composition at the ternary eutectic point is reported as 3.8% Al and 2.2% Mg. In this case, solidification of 1.5% Al and 1.5% Mg alloy (point 1 in Fig. 6.29) begins at the liquidus temperature of 381.5°C. Precipitation of the primary Zn phase continues with the liquid composition following the indicated line from point 1 to point 2, which is at 350°C at a composition of 2.3% Al and 2.8% Mg. Thereafter, the liquid composition follows the univariant line from point 2 to the eutectic at point 5. During this stage, a binary eutectic constituent (Zn + $MgZn_2$) precipitates. At the eutectic point a ternary eutectic constituent (Zn + $MgZn_2$ + fcc) precipitates isothermally. The hcp Zn phase contains 0.9% Al and 0.1% Mg in solid solution, the $MgZn_2$ phase contains 0.7% Al in solution, and the composition of the fcc phase is 81.5% Zn-18.5% Al-0.06% Mg. For a 3.0% Al-3.0% Mg coating (point 3), solidification begins at the liquidus temperature of 361.5°C.

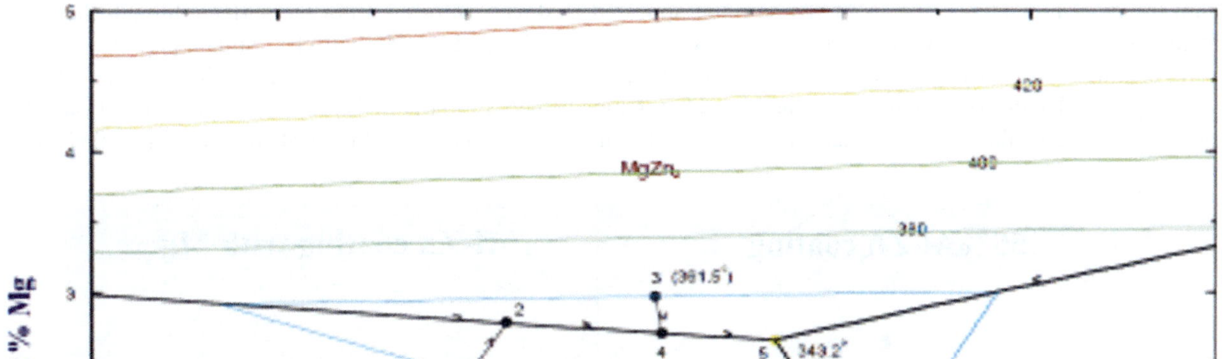

FIG. 6.29 FactSage liquidus projection of Zn-rich corner of ZnAlMg phase diagram, indicating solidification paths of Zn-1.1Al-1.1Mg alloy (beginning at point 1) and Zn-3%Al-3%Mg alloy (beginning at point 3). *(From A.D. Pelton, Thermodynamic Calculations of the Steps During Solidification of Certain Zn-Al-Mg Alloys, prepared for F. Ajersch, under contract with IZA, 2014.)*

c. Super Dyma® is a trademark of Nippon Steel and Sumitomo Metal Corporation. https://www-zam.nipponsteel.com/en/guideline/.

Precipitation of the primary $MgZn_2$ phase continues with the liquid composition following the line from point 3 to point 4, which is 346°C at a composition of 3.1%Al-2.7%Mg. Thereafter, the liquid composition follows the univariant line from point 4 to the eutectic at point 5. During this stage, a binary eutectic constituent ($Zn+MgZn_2$) precipitates. At the eutectic point a ternary eutectic constituent ($Zn+MgZn_2+fcc$) precipitates isothermally. The hcp Zn phase contains 0.9% Al and 0.1% Mg in solid solution, the $MgZn_2$ phase contains 0.7% Al in solution, and the composition of the fcc phase is 81.5% Zn-18.5%Al-0.06% Mg. Little difference was seen between equilibrium and Scheil-Gulliver solidification (no diffusion in the solid).

6.5.3 55% Al

Although Liu et al. [76] proposed that the addition of Mg significantly decreased spangle size, it was recently found that the DAS is not affected by the addition of Mg [77]. They also found that after Mg additions were made to 55%Al-Zn coating, the microstructure of the coating changes from Al-rich dendrite (α-Al), Al-Zn eutectic phase, and a small amount of Si-rich phase to an Al-rich dendrite (α-Al), Al-Zn-$MgZn_2$ phase, and a small amount of Si-rich phase.

Comparison of a Zn-55%Al-2%Mg-1.6%Si coating interface with that of a Mg-free version of the same coating showed the intermetallic layer of the Mg-added coating to consist of rounded particles with diameters 100–1000 nm, but to be largely free of the Si-containing intermetallic phases normally observed at the interface of the Mg-free Zn-55%Al-1.5%Si coating that has been industrially produced for many years [52]. The microstructure of the Zn-55%Al coating overlay also changed as Mg was added. The Zn-rich interdendritic areas were observed to be more connected in the Al-Zn coating with the Mg addition than in the conventional Zn-55%Al coating; the interdendritic areas of the Zn-55% Al coating normally contain isolated islands of Zn-rich particles in an Al-rich matrix. Si-Mg particles were also observed in the Mg-added coating, concentrated mostly in the bottom half of the coating, close to the alloy layer.

An SEM comparison of the Galvalume 55% Al-Zn-1.6% Si overlay structure with and without Mg was investigated [52]. It was found that the 2% Mg addition coating had a duller surface appearance, Fig. 6.30, relatively small spangles, Fig. 6.31, and finer dendrite arm spacing compared with the 55%Al-Zn product, Fig. 6.32. The dendrites have a similar surface morphology and the orientation changes through the thickness of the coating, but the microstructures in the interdendritic regions were different, exhibiting a nearly continuous zone of very fine, zinc-rich particles that are confined within the Al-rich dendrites, Fig. 6.33. These often appeared as direct pathways from the surface through the coating thickness to the alloy layer. Additionally, these areas contain large Zn and Zn-Mg particles. Silicon particles in the Al-Zn-Mg coating contain both Mg and Si and are concentrated mostly in the bottom half of the coating, closer to the alloy layer. It appears that the Mg_2Si particles in the Mg added coating behave similarly to the Si particles in standard Galvalume product. Both Si-containing particles nucleate at the interfacial alloy layer as seen in Fig. 6.24 and grow "tree-like" into the

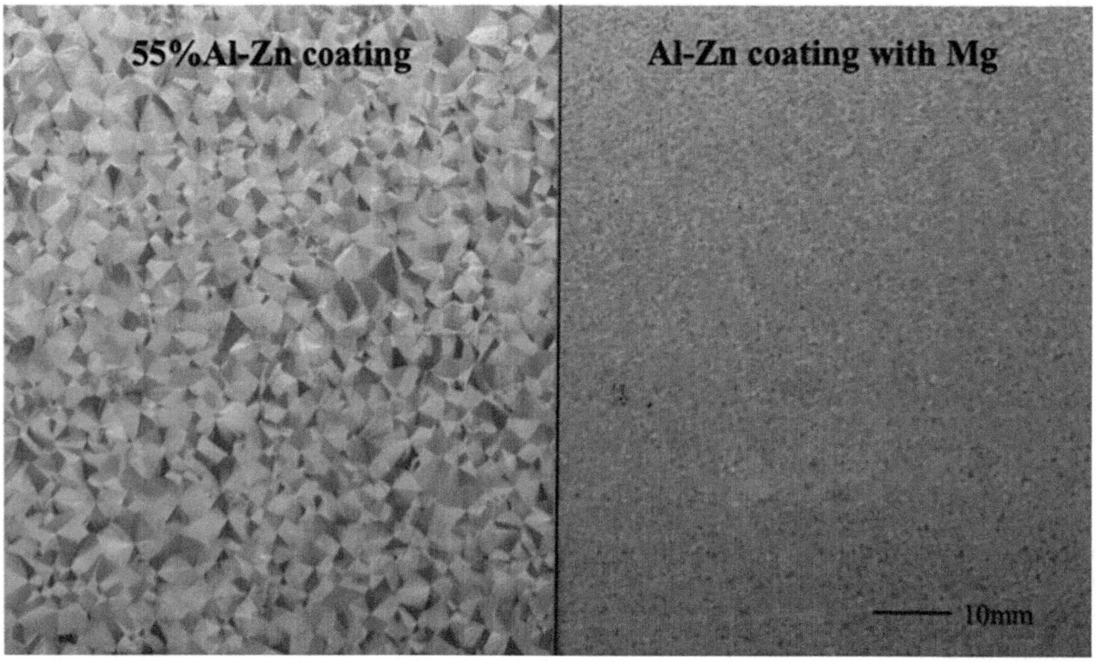

FIG. 6.30 Surface photographs of 55%Al-Zn coating without and with 2% Mg addition. *(From J. Li, D. Hreso, Morphology of 55% Al-Zn coating and the Al-Zn coating with Mg additions, in: Galvatech, 2015, pp. 852–860.)*

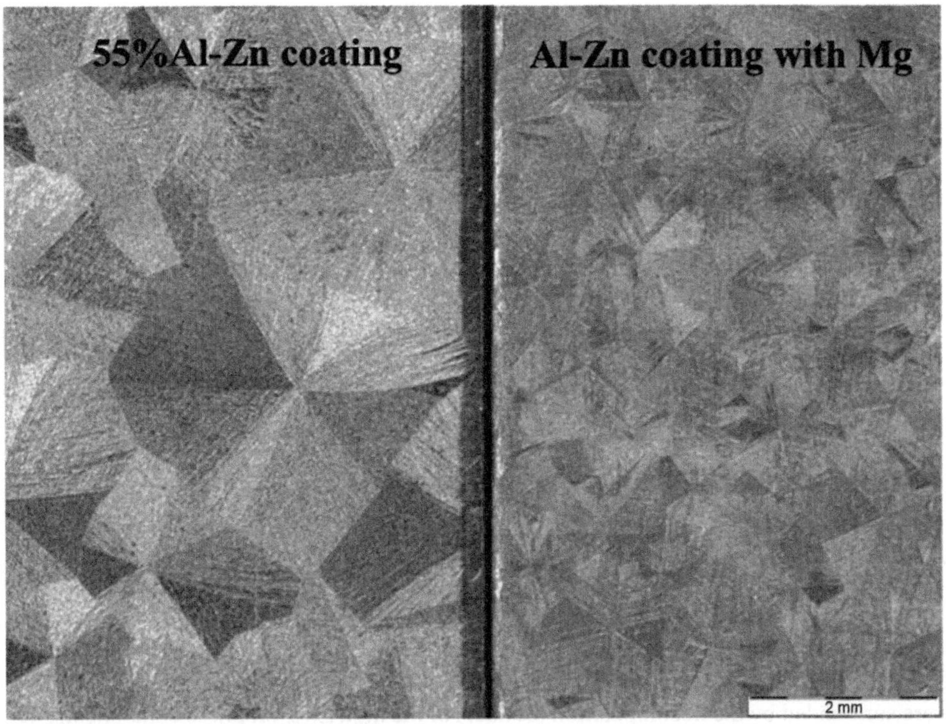

FIG. 6.31 Sample surfaces of Fig. 6.30 at higher magnification. *(From J. Li, D. Hreso, Morphology of 55% Al-Zn coating and the Al-Zn coating with Mg additions, in: Galvatech, 2015, pp. 852–860.)*

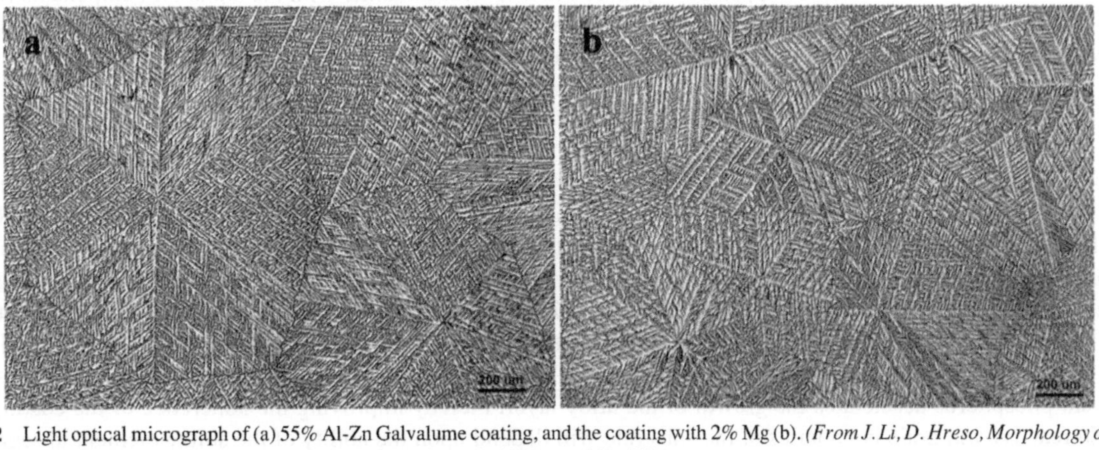

FIG. 6.32 Light optical micrograph of (a) 55% Al-Zn Galvalume coating, and the coating with 2% Mg (b). *(From J. Li, D. Hreso, Morphology of 55% Al-Zn coating and the Al-Zn coating with Mg additions, in: Galvatech, 2015, pp. 852–860.)*

liquid overlay. Thus, it would be expected that the Mg_2Si would be more apparent and concentrated at the interface as opposed to the surface of the coating.

Chen [78] recently published research on the details of the microstructure of the 2% Mg-55% Al-1.6% Si-Zn coating on steel and analyzed the formation mechanisms of different phases and constituents observed. He found that:

(1) The Al-Zn mixtures observed in the interdendritic region were the products of eutectoid decomposition of the -Al (high zinc-Al) components associated with the formation of $MgZn_2$ and Zn/$MgZn_2$ mixtures.

(2) Judging from the Al-Si-Mg liquidus projection, Mg_2Si should form first, followed by a eutectic reaction forming a mixture of a-Al + Mg_2Si + Si.

(3) It was also observed that when Si or Mg_2Si formed, they occupied the entire space of the interdendritic region at the location where they formed, suggesting that the α-Al phase formed as a divorced component with them had a lower Zn concentration.

(4) While thermodynamically, Mg_2Zn_{11} was a stable phase to form, it was not observed in the coating. The Mg-Zn phase observed was $MgZn_2$ instead, which was seen to develop into two different morphologies, a large blocky $MgZn_2$ phase and fine precipitates forming a lamellar structure with Zn.

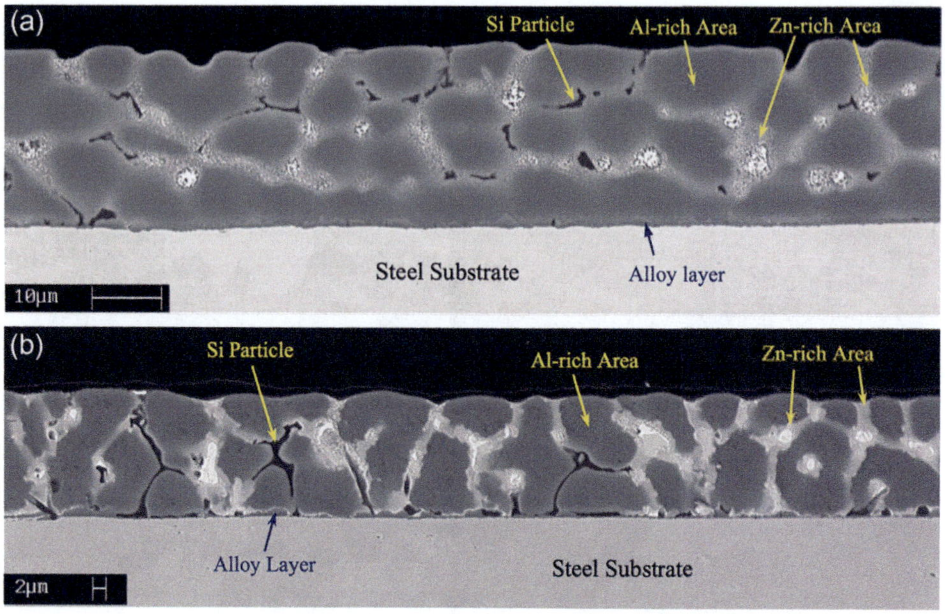

FIG. 6.33 SEM backscattered electron image of the cross section of (A) Galvalume 55%Al-Zn-Si coating, (B) Galvalume with Mg addition. *(From J. Li, D. Hreso, Morphology of 55% Al-Zn coating and the Al-Zn coating with Mg additions, in: Galvatech, 2015, pp. 852–860.)*

(5) Both blocky $MgZn_2$ and the lamellar Zn/$MgZn_2$ mixtures did not occupy the entire space of the interdendritic region, but rather, they were surrounded by the decomposition products of the high Zn α-Al phase components, i.e., Al-Zn mixtures.

(6) All eutectic reactions observed appeared to proceed in a divorced manner with the alpha-Al components formed at different locations from their eutectic counterparts, most likely via growth of the existing alpha-Al dendrite arms, with the exception that sometimes it formed as isolated islands embedded in the blocky $MgZn_2$ phase.

6.6 Substrate alloy additions

Steel substrate composition has a great influence on the hot-dipping process and the performance of the resultant hot-dip zinc coating. Compositional differences in the substrate affect the rate of attack of steel by molten zinc as well as the mode of attack at a given galvanizing temperature [79]. Compositional variation in the steel used for hot-dip zinc coating is an important factor in determining coating morphology and protection capacity of the coating. The chemical composition of the steel substrate also influences the metallurgical properties of the hot-dip zinc coatings. Si, P, C, and Mn in the steel substrate can influence the Fe-Zn solidification mechanism depending on their concentration. The presence of critical amounts of silicon and phosphorus in the steel substrate is necessary to control the coating weight and the presence of carbon and phosphorus accelerates the growth of the alloy layer, thereby improving the adherence of the coating.

Fe-Zn reactivity at grain boundaries of the substrate will depend upon the ability of the solute element to segregate to these sites [80]. These alloying additions can be separated into elements that segregate to the grain boundary (e.g., C, P), and those that will form compounds (e.g., Ti, Nb) that will precipitate throughout the grain, leaving the grain boundary "pure" or clean. Clean grain boundaries will have no barrier to Fe-Zn compound formation, while segregated boundaries will reduce the thermodynamic activity at these sites, decreasing outbursts and Fe_2Al_5 inhibition layer breakdown [1].

- Carbon—Carbon contents up to 0.3% have a relatively small impact on the reaction of iron in the zinc bath [81]. However, it is well known that outburst formation decreases with increased carbon content, since carbon segregates to the grain boundary [82].
- Phosphorous—In the batch-dip process, certain phosphorous contents can result in a Sandelin effect. To prevent this, a composition requirement is necessary:
  $Si + 2.5\ P < 0.09$.

In the CGL process, the growth of Fe-Zn phases is retarded by phosphorous additions [83]. Phosphorous was found to segregate to ferrite grain boundaries in rephosphorized low carbon steels, blocking the diffusion of Zn along grain boundaries and lowering the thermodynamic activity [84], and it has also been proposed that P segregates to the steel surface during recrystallization annealing. Jordan et al. [85] studied the ion implantation P on one surface of a large grain (10–20 mm) low-carbon steel sheet galvanized in a Zn-0.02 wt% Al bath and showed that Fe-Zn phase growth only occurred after extended reaction times on both the P ion implanted and nonimplanted surfaces. The outbursts were found to be related to a single grain boundary site, and a greater outburst reaction was associated with the P-free side where there was no P segregation.

- Titanium—Interstitial free (IF) alloy can contain additions of titanium and/or niobium at extremely low carbon levels. It is well known that Ti-IF steels are more sensitive to outburst formation than Al-killed or even Nb-Ti steels [80]. Because these steels are more likely to form carbide, nitride, sulfide, and phosphide precipitates in the grains, preventing segregation to the grain boundaries, the concept of excess titanium, Ti**, has been introduced [84]:

$$Ti^{**} = \text{Total Ti} - 3.99C - 1.49S - 3.42N - 1.55P$$

A positive Ti** indicates excess titanium and therefore a clean, carbide-free, and reactive grain boundary. A negative value of Ti** would indicate that not all of the solute carbon is tied up and zinc diffusion down the boundaries would be blocked. Many coating studies have been made using IF steels [34,74].

- Silicon—Silicon is one of several elements, e.g., Mn, Al, that are used to deoxidize the steel melt. Although silicon is a very efficient deoxidant, it can have a very unfavorable impact on the characteristics of the coating. Structural steels that have not been killed using silicon are usually recommended for hot-dip galvanizing. These steels are free of any abnormal effect on alloy layer growth, and control of the structure and coating thickness can proceed normally according to the requirement of the specific coating processing and properties.

For galvanized silicon killed steels, complications arise because silicon influences the reactivity of iron with zinc. Coatings in these steel types can be thick and brittle, prone to mechanical damage particularly in batch dip processes. This can result in the Sandelin Effect, Fig. 6.34, a breakaway of intermetallic phases that can cause an abnormally thick coating. However, if the silicon in the steel is kept at limited range, called the "Sebisty Range," between 0.15% and 0.25%, acceptable thicknesses can be achieved [86]. Again, this problem occurs almost exclusively in batch dip products.

- Sulfur can cause excessive dissolution of the steel in the zinc melt.
- Manganese—Levels of Mn above 1.25% will produce a more consistent product especially on coatings of Si-killed steel. Mn additions up to 2.5% have been used successfully in HSS steels.
- Other substrate steels—Many different substrate steels have been used with Zn coatings, including, e.g., 316SS and Triboloy [87]. AHSS steel substrates, such as TRIP, DP, CP, NexGen, coated with Zn alloys, have recently been used in the automotive industry and will be discussed in Chapter 13.

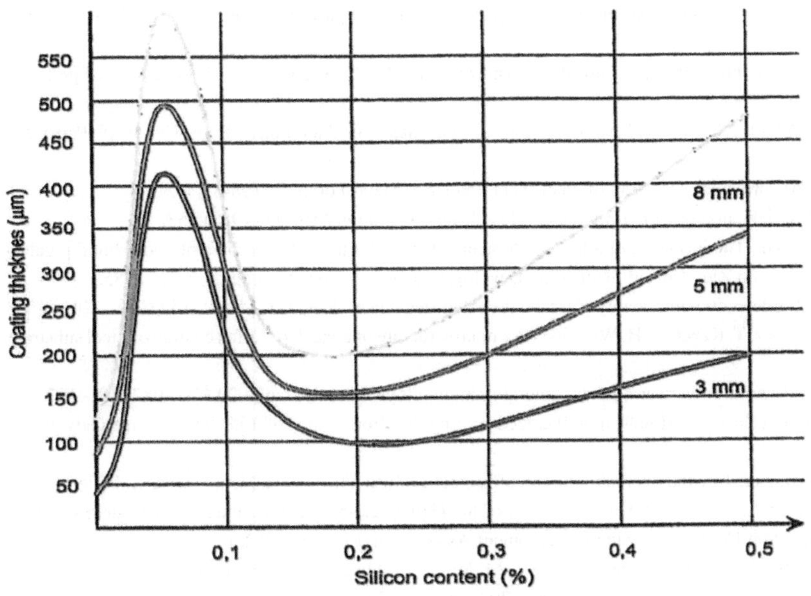

FIG. 6.34 The Sandelin effect. *(From V. Kuklik, J. Kudlacek, Hot Dip Galvanizing of Steel Structures, Butterworth-Heinemann, Cambridge, MA, 2016.)*

References

[1] A.R. Marder, The metallurgy of zinc-coated steel, Prog. Mater. Sci. 45 (2000) 191–271.

[2] W. Liu, M.-C. Li, Q. Luo, H.-Q. Fan, Y. Zhang, H.-S. Lu, K.-C. Chou, X.-L. Wang, Q. Li, Influence solution of alloyed magnesium on the microstructure and long-term corrosion behavior of hot-dip Al–Zn–Si coating in NaCl, Corros. Sci. 104 (2016) 217–226.

[3] J. Strutzenberger, J. Faderl, Solidification and spangle formation of hot-dip-galvanized zinc coatings, Metall. Mater. Trans. 29A (1998) 631–646.

[4] F.A. Fasoyinu, F. Weinberg, Spangle formation in galvanized sheet steel coatings, Metall. Trans. 21B (1990) 549–558.

[5] J. Nakano, D.V. Malakhov, S. Yamaguchi, G.R. Purdy, A full thermodynamic optimization of the Zn–Fe–Al system within the 420–500 °C temperature range, Comput. Coupling Phase Diagr. Thermochem. 31 (2007) 125–140.

[6] J.R. McDermid, M.H. Kaye, W.T. Thompson, Fe solubility in the Zn-rich corner of the Zn-Al-Fe system for use in continuous galvanizing and galvannealing, Metall. Mater. Trans. B 38b (2007) 215–230.

[7] N. Gao, Y.H. Liu, N.-Y. Tang, An experimental study of the liquid domain of the Zn-Fe-Al-Mn quaternary system at 460 °C, J. Phase Equilib. Diffus. 31 (6) (2010).

[8] F. Ilinca, F. Ajersch, C. Baril, F.E. Goodwin, Numerical simulation of the galvanizing process during GA to GI transition, Int. J. Numer. Meth. Fluids 63 (10) (2006) 1629–1646. John Wiley & Sons.

[9] J. Karlsson, S.E. Hörnström, H. Klang, J.-O. Nilsson, Sputter-induced effects on quantitative GD-OES depth profiles from hot-dip zinc-coated steel, Surf. Interface Anal. 21 (1994) 365–369.

[10] E. McDevitt, Y. Morimoto, M. Meshii, Characterization of the Fe–Al interfacial layer in a commercial hot-dip galvanized coating, ISIJ Int. 37 (1997) 776–7822.

[11] K. Bai, P. Wu, Assessment of the Zn–Fe–Al system for kinetic study of galvanizing, J. Alloy. Compd. 347 (2002) 156–164.

[12] S. Lazik, C. Esling, J. Wegria, Cracking in zinc layers on continuous galvanized sheets, Texture Microstruct. 23 (1995) 131–147.

[13] J.R. McDermid, F.E. Goodwin, Current issues in processing 3G steels for zinc coating, in: Proceedings AISTech 2016 Symposium on Advanced High Strength Steel for Automotive Applications, Pittsburgh, PA, May 16-19, 2016.

[14] V. Saarimaa, A. Markkula, J. Juhanoja, B.-J. Skrifvars, Novel insight to aluminum compounds in the outermost layers of hot dip galvanized steel and how they affect the reactivity of the zinc surface, Surf. Coat. Technol. 306 (2016) 506–911.

[15] J.D. Culcasi, C.I. Elsner, A.R. Di Sarli, Effect of zinc crystals size on galvanized steel deformation and electrochemical behavior, Mater. Res. 12 (3) (2009) 273–279.

[16] N.J. Wall, J.A. Spittle, R.D. Jones, The deformation characteristics of zinc coatings on hot-dip galvanized mild steel strip, in: Proceedings of 1st International Conference on Zinc Coated Steel, Zinc Development Association, Munich, 1985, p. C/1.

[17] H. Leidheiser, D.K. Kim, Crystallographic factors affecting the adherence of paint to deformed galvanized steel, J. Met. (JOM) 28 (1976) 19.

[18] R.E. Fraley, The Effect of Lead on the Solidification and Preferred Orientation of the Zinc Coating on Continuously Hot Dipped Galvanized Sheet Steel (M.S. thesis), Lehigh University, 1994.

[19] X. Wang, J. Lu, C. Che, G. Kong, The behavior of lead during the solidification of Zn-0.1Al-0.1Pb coating on batch hot-dipped steel, Appl. Surf. Sci. 254 (2008) 2466–2471.

[20] J.J. Sebisty, R.H. Palmer, Proceedings of 6th International Hot Dip Galvanization Conference, Interlaken, Zinc Development Association, London, 1962, p. 215.

[21] W. Radeker, W. Friehe, Proceedings of 7th International Hot Dip Galvanization Conference, Paris, Zinc Development Association, Oxford, 1964, p. 167.

[22] H. Asgari, M.R. Toroghinejad, M.A. Golozar, On texture, corrosion resistance and morphology of hot-dip galvanized zinc coatings, Appl. Surf. Sci. 253 (2007) 6769–6777.

[23] S.G. Kim, H.-S. Hwang, S.-K. Lee, J.-Y. Huh, Dendritic morphologies of hot-dip galvanized Zn-0.2wt % Al coatings: existence of third preferred growth directions, in: Galvatech, 2021, pp. 264–271.

[24] D.C. Herrschaft, S.F. Radke, D. Coutsouradis, J. Pelerin, Galfan—A New Zinc-Alloy Coated Steel for Automotive Body Use, SAE Paper No. 830517, SAE, Warrendale, PA, 1983.

[25] G.J. Harvey, R.N. Richards, Zinc-based coatings for corrosion protection of steel sheet and strip, Metal Forum 6 (1984) 234.

[26] S.F. Radtke, D.C. Herrschaft, Role of mischmetal in galvanizing with a Zn-5%Al alloy, J. Less Common Met. 93 (1983) 253.

[27] A.F. Skenazi, D. Davin, D. Coutsouradis, F.E. Goodwin, The influence of alloying elements in zinc-aluminum baths for continuous hot dip galvanizing, in: Proceedings of 1st International Conference on Zinc Coated Steel, Zinc Development Association, Munich, London, 1985.

[28] S.J. Makimattila, On the production possibilities of a deep drawing quality Zn-5% Al coated steel sheet, Scand. J. Metall. 15 (1986) 224.

[29] M. Zelechower, J. Klis, E. Augustyn, J. Grzonka, D. Stroz, T. Rzychon, H. Woznica, The microstructure of annealed Galfan coating on steel substrate, Arch. Metall. Mater. 57 (2012) 01.

[30] P. Porot, N. Ventham, R. Jones, J. Spittle, Structural variation in near-eutectic zinc-aluminum alloys, Metallography 20 (2) (1987) 181–197.

[31] S.M. Makimattila, E. Ristolainen, On the structure of a Zn - 5% Al hot-dip galvanized coating, in: Proceedings of 14th International Galvanizing Conference, Munich, Germany, June 9-14, 1985.

[32] S.T. Bluni, M.R. Notis, A.R. Marder, Nucleation characteristics and microstructure in off-eutectic Al-Zn alloys, Acta Metall. Mater. 43 (1995) 1775.

[33] J. Pelerin, J.P. Servais, D. Coutsouradis, D.C. Herrschaft, S.F. Radke, Characteristics of the Galfan. Zn-5% Al-mischmetal coated steel sheet, in: Proceedings of 13th International Hot Dip Galvanization Conference, Zinc Development Association, London, 1982, p. 49/I-9.

[34] H. Chaieb, V. Maurel, K. Ammar, S. Forest, A. Koster, F. Nozahic, J. De Strycker, J.-M. Mataigne, Microstructure and texture analysis of a near-eutectic Zn-Al coating deposited on steel by hot dipping galvanizing, in: Galvatech, 2021, pp. 272–281.

[35] A. Takahashi, T. Kato, S. Funaki, CAMP-ISIJ 2 (1989) 1662.

[36] S.T. Bluni, A.R. Marder, J.I. Goldstein, Surface characterization of hot-dip Galfan coatings, Mater. Charact. 33 (1994) 93.

[37] F. Ajersch, F. Ilinca, Review of modeling and simulation of galvanizing operations, Steel Res. Int. 89 (2018).

[38] S. Alibeigi, J.R. McDermid, Modeling of interfacial layer growth kinetics on Mn-containing steels during continuous hot-dip galvanizing, in: Proceedings of Galvatech 2015, Toronto, AIST, 2015.

[39] S.T. Bluni, A.R. Marder, Improvement of Galfan surface appearance for coil coating uses, in: 4th International Zinc Coated Sheet Conference, Zinc Development Association, Paris, 1994, p. SC6/1.

[40] J.H. Selverian, A.R. Marder, M.R. Notis, The effect of silicon on the reaction between solid iron and liquid 55 wt pct Al-Zn baths, Metall. Trans. 20A (1989) 543.

[41] J.H. Selverian, A.R. Marder, M.R. Notis, Preparation of cross sections of thin-coated sheet steel for analytical electron microscopy observation, J. Electron. Microsc. Technol. 5 (1987) 223.

[42] J.H. Selverian, A.R. Marder, M.R. Notis, The microstructure of 55 w/oAl-Zn-Si (galvalume) hot dip coatings, J. Mater. Eng. 9 (1987) 133.

[43] H.J. Cleary, in: D.O. Northwood, W.E. White, G.F. Van der Voort (Eds.), Microstructural Science, 12, ASM, Metals Park, OH, 1985, pp. 103–113.

[44] A.R. Moreira, Z. Panossian, P.L. Camargo, M.F. Moreira, I.C. da Silva, J.E. Ribeiro de Carvalho, Zn/55Al coating microstructure and corrosion mechanism, Corros. Sci. 48 (2006) 564–576.

[45] D. Wei, H. Tu, S. Zhou, Y. Liu, C. Wu, J. Wang, X. Su, Effect of silicon on the reaction between solid iron and liquid Zn-22.3 Wt.% Al bath, Surf. Coat. Technol. 305 (2016) 29–35.

[46] J.H. Selverian, M.R. Notis, A.R. Marder, The microstructure of 55 w/o Al—Zn—Si (Galvalume) hot dip coatings, J. Mater. Eng. 9 (1987) 133–140.

[47] A.R. Marder, Microstructural characterization of zinc coatings, in: G. Krauss, D.K. Matlock (Eds.), Zinc-Based Steel Coating Systems: Metallurgy and Performance, TMS, Warrendale, PA, 1990, pp. 55–82.

[48] R.Y. Chen, D.J. Willis, The behavior of silicon in the solidification of Zn-55Al-1.6Si coating on steel, Metall. Mater. Trans. A 36A (2005) 117–128.

[49] J.H. Selverian, An Electron Optical Study of the Effect of Silicon on Hot Dip 55 Wt% Al-Zn Coatings on Iron (Master's thesis), Lehigh University, 1985.

[50] B. Xu, D. Phelan, R. Dippenaar, Role of silicon in solidification microstructure in hot-dipped 55 wt% Al–Zn–Si coatings, Mater. Sci. Eng. A 473 (1–2) (2008) 76–80.

[51] H. Peng, G. Wu, W. Peng, B. Cao, J. Zhang, The effect of silicon on spangle size in hot-dipped 55 wt%Al-Zn coatings, Surf. Coat. Technol. 327 (2017) 110–117.

[52] J. Li, D. Hreso, Morphology of 55% Al-Zn coating and the Al-Zn coating with Mg additions, in: Galvatech, 2015, pp. 852–860.

[53] H.-Y. Kim, S.-J. Kim, M.-S. Kim, New ZnAlMg alloy coatings with ultra-high corrosion resistance, in: Galvatech, 2021, pp. 1201–1209.

[54] R.A. Nikola, U.S. Pat. No. 3,026,606, Mar. 27, 1963.

[55] J.A. Heath, U.S. Pat. No. 3,164,464, Jan. 5, 1965.

[56] W.P. Roe, W.J. Lantz, U.S. Pat. No. 3,320,040, May 16, 1967.

[57] H.H. Lee, J.W. Halley, U.S. Pat. No. 3,505,043, Apr. 7, 1970.

[58] M. Memmi, G. Giardetti, U. S. Pat. No. 5,091,150, Feb. 25, 1992.

[59] A. Komatsu, N. Yamaki, A. Ando, U.S. Pat. No. 6,709,770 B2, Mar. 23, 2004.

[60] N.S. Berke, H.E. Townsend, U.S. Pat. No. 4,401,727, Aug. 30, 1983.

[61] N.C. Hosking, M.A. Ström, P.H. Shipway, C.D. Rudd, Corrosion resistance of zinc–magnesium coated steel, Corros. Sci., 49, 3669–3695.

[62] E. De Bruycker, Z. Zermout, B.C. De Cooman, Zn-Al-Mg coatings: thermodynamic analysis and microstructure related properties, Mater. Sci. Forum 539–543 (2007) 1276–1281.

[63] A.D. Pelton, Thermodynamic Calculations of the Steps During Solidification of Certain Zn-Al-Mg Alloys, prepared for F. Ajersch, under contract with IZA, 2014.

[64] G. Angeli, R. Brisberger, M. Bülter, L. Diez, C. Filthaut, W. Fischer, T. Koll, T. Maiwald, C. Pesci, A. Richter, J. Schulz, U. Stellmacher, N. van Landschoot, Zinc-magnesium-aluminium for automotive industry, in: Galvatech, Beijing, 2013, pp. 632–636.

[65] M. Vlot, M. Zuijderwijk, M. Toose, L. Elliott, R. Bleeker, T. Maalman, Hot dip ZnAlMg coatings: microstructure and forming properties, in: Galvatech, 2007, pp. 574–579.

[66] M.-S. Oh, H.-J. Lee, J.-S. Kim, Investigation on the characteristics of hot-dip Zn-3%Mg-2.5%Al alloy coated sheets, in: Galvatech, 2015, pp. 822–828.

[67] M. Dutta, A.K. Halder, S.B. Singh, Morphology and properties of hot dip Zn–Mg and Zn–Mg–Al alloy coatings on steel sheet, Surf. Coat. Technol. 205 (2010) 2578–2584.

[68] T.C. Kim, I.-R. Sohn, Y.H. You, M.S. Kim, J.S. Kim, Investigation of hot-dip ZnMgAl coatings for automotive applications, in: Galvatech, 2021, pp. 100–107.

[69] N. LeBozec, D. Thierry, D. Persson, The effect of microstructure on the corrosion stability of zinc aluminum magnesium alloy coated steel in atmospheric conditions, in: Galvatech, 2021, pp. 220–228.

[70] C.-W. Hsu, G.-L. You, K.-K. Wang, L. Chang, D. Gan, L.-J. Chiang, Characterization of the Fe-Al intermetallic phases formed in hot-dip Al-Zn-Mg coatings, in: Galvatech, 2015, pp. 845–851.

[71] N. Takata, H. Yokoi, A. Suzuki, M. Kobashi, Formation sequence of FeAl intermetallic phases in hot-dip Zn-Al-Mg alloy galvanizing steels, in: Galvatech, 2021, pp. 1321–1327.

[72] K. Honda, W. Yamada, K. Ushioda, Solidification structure of the coating layer on hot-dip Zn-11%Al-3%Mg-0.2%Si-coated steel sheet, Mater. Trans. 49 (2008) 1395–1400.

[73] K. Honda, K. Ushioda, W. Yamada, Influence of Si addition to the coating bath on the growth of the Al-Fe alloy layer in hot-dip Zn-Al-Mg alloy-coated steel sheet, ISIJ Int. 51 (2011) 1895–1902.
[74] K.-C. Yu, J. Li, X. Liu, J.-G. Li, X. Xue, Microstructure of hot dip galvanized Zn-Al-Mg alloy coating, J. Shanghai Jiaotong Univ. (Sci.) 17 (2012) 663–667.
[75] H. Tu, D. Wei, S. Zhou, Y. Liu, C. Wu, J. Wang, X. Su, Effect of Mg on microstructure and growth kinetics of Zn-22.3 Al-1.1Si coating, J. Wuhan Univ. Technol. (2019) 373–382.
[76] Q.Y. Liu, J. Williams, D. Nolan, B. Shedden, R. Smith, A. Neufeld, Development of Mg-containing 55%Al-Zn coated steel for building appliance, in: Galvatech 10th International Conference on Zinc and Zinc Alloy Coated Steel Sheet, Toronto, 2015, pp. 208–215.
[77] Z.-L. Ding, B.-H. Yan, X.-Y. Jin, X.-J. Wang, J. Zhang, S.-N. Jiang, Q.-F. Zhang, Effect of magnesium on the microstructure and corrosion resistance of 55% Al-Zn coating, in: Galvatech, 2021, pp. 164–171.
[78] Y.R. Chen, Examination and analysis of 2 pct Mg-55 pct Al-1.6 pct Si-Zn coating on steel, Metall. Mater. Trans. A 51A (2020) 5758–5770.
[79] S.M.A. Shibli, B.N. Meena, R. Remya, A review of recent approaches in the field of hot dip zinc galvanizing process, Surf. Coat. Technol. 262 (2015) 210–215.
[80] Y. Hisamatsu, Science and technology of zinc and zinc alloy coated sheet steel, in: GALVATECH '89, The Iron and Steel Institute of Japan, Tokyo, 1989, p. 3.
[81] V. Kuklik, J. Kudlacek, Hot Dip Galvanizing of Steel Structures, Butterworth-Heinemann, Cambridge, MA, 2016.
[82] M. Saito, Y. Uchida, T. Kittaka, Y. Hirose, Y. Hisamatsu, Formation behavior of alloy layer in initial stages of galvanizing, Tetsu-to-Hagane 77 (1991) 947.
[83] M. Abe, S. Kanbara, T. Okuyama, Iron zinc alloy behavior in galvannealing P and P-Si containing steel, Trans. ISIJ 25 (1985) B15.
[84] T. Toki, K. Oshima, T. Nakamori, Y. Saito, T. Tsuda, Y. Hobo, Effect of P content in ultra-grain boundary diffusion during hot-dip galvanizing, in: A. R. Marder (Ed.), The Physical Metallurgy of Zinc Coated Steel, TMS, Warrendale, PA, 1994, p. 169.
[85] C.E. Jordan, R. Zuhr, A.R. Marder, Effect of phosphorous surface segregation on iron-zinc reaction kinetics during hot-dip galvanizing, Metall. Mater. Trans. 28A (1997) 2695.
[86] J.J. Sebisty, G.E. Ruddle, in: Hydrogen-Atmosphere Galvanizing of Iron-Base Alloys, Research Report No. R 255, DOE, Mines Branch, Ottawa, 1972.
[87] S. Yashchuk, R. Adigamov, O. Kazandzhiyan, Optimal content of aluminum in aluminum-zinc alloy in hot-dip coating on the steel strip, Galvatech, 2021 pp. 645–652.

Chapter 7

Hot-dip galvanized coating weight control

7.1 Introduction

Much of the value of the galvanized sheet depends upon proper control of coating thickness occurring after exit of the steel sheet from the bath up to the time when the coating is finally solidified. As shown in the simple schematic of Fig. 7.1, in this portion of the overall galvanizing process, the zinc alloy from the galvanizing bath is entrained on the surface of the sheet, it then passes between the gas jets produced by the wiping knives, and then either solidifies, or in the case of galvannealed product, enters the galvannealing furnace. The galvanneal process is described in detail in Chapter 8, but the successful and reliable operation of the galvannealing process is highly dependent on production of a predictably uniform coating as described below.

The many factors influencing the thickness or weight of the coating produced on the steel sheet will be described. These factors or input variables are subject to a great deal of variability on their own and therefore influence the variability of the coating thickness or weight. Because zinc alloys have different densities, the primary term that will be used in this chapter to describe the coating output will be coating thickness in microns.

Following ASTM Standard Specification A653 [2], the following equivalences are useful for converting between coating weights and thicknesses for galvanized (GI) and galvanneal (GA) coatings, using a coating alloy density of 7140 kg/m^3 (446 lb./ft^3):

$$1.00\,\text{oz/ft}^2 = 0.00168\,\text{in.} = 1.68\,\text{mils} = 305\,\text{g/m}^2 = 0.0427\,\text{mm} = 42.7\,\mu\text{m} \tag{7.1}$$

Care must be taken when using these conversions, because coating weight is often specified as two-side coating weight, following the standards described later in this chapter. Therefore, ideally, the 42.7 μm thickness calculated above would be divided in half, with half of the 42.7 μm coating thickness being present on each side of the sheet for a coating weight of 305 g/m^2. The same applies to the other measures of thickness shown, i.e., inches, mils, and mm.

For the Zn-55%Al coating, the equivalences using a coating alloy density of 3750 kg/m^3 (234 lb./ft^3), also following ASTM Standard Specification A792 [3], are:

$$1.00\,\text{oz/ft}^2 = 0.00320\,\text{in.} = 3.20\,\text{mils} = 305\,\text{g/m}^2 = 0.0813\,\text{mm} = 81.3\,\mu\text{m} \tag{7.2}$$

The reason why coating weight is usually reported as two-side is because of the historical use of the weigh-strip-weigh method specified in ASTM A90 [4], which gives the mass of the coating removed from both sides. Online measurements on modern continuous galvanizing lines use X-ray or radioisotope fluorescence techniques that continuously scan both sides of the sheet surface and report the results as coating weight. The gauges used in these on-line devices, and similar laboratory devices, are always calibrated using the weigh-strip-weigh method.

The stability and accuracy of coating weight control strongly affect the profitability of a galvanizing line. Customers can reject coated sheet containing less than the specified coating weight, mainly because thinner coatings will not provide as long of a lifetime of corrosion protection; therefore, operators apply excess coating to be sure the minimum is always exceeded. However, applying too much excess consumes more zinc than necessary. Zinc is usually the highest-cost consumable in a galvanizing line operation. Also, excessively thick zinc coatings can impair operations carried out by customers such as forming and welding. Like many other processes, coating weights produced during stable galvanizing operations are not exactly precise but fall into a range of coating weights defined by a frequency or distribution curve. A normal or Gaussian distribution is common; the width of the distribution curve is related to the scatter of actual results around the mean value and is characterized by the standard deviation, termed sigma. In this case of coating control, if the gas jet wiping device shown in Fig. 7.1 operates with a normal distribution where the upper and lower limits are 2.58 sigma from the mean, the device should be able to be held in a position so that 0.5% of the coils will be undercoated and 99.5% meet or exceed the coating thickness. However, coil width and thickness change, line speed changes, coating

The Metallurgy of Zinc Coated Steels. https://doi.org/10.1016/B978-0-323-99984-7.00006-3

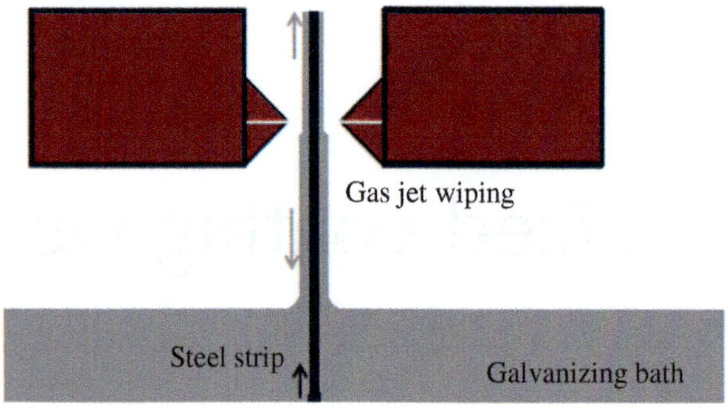

FIG. 7.1 Basic schematic of the gas jet coating control process. Liquid metal is entrained on the steel strip as it exits the galvanizing bath, excess coating metal is stripped by the gas jet pressure and returned to the bath [1]. *(Reproduced with permission from C. Pfeiler, M. Mataln, A. Kharicha, C.K. Riener, G. Angeli, Importance of the zinc film modeling for gas jet wiping simulations, in: Proc. 10th International Conference on Zinc and Zinc Alloy Coated Steel Sheet (Galvatech 2015), May 31–June 4, 2015, Toronto, Association for Iron & Steel Technology, Warrendale, PA, 2015, pp. 682–689.)*

thickness target changes, and other factors influence how closely the coating control equipment can be held to this position over time. In the best case of this example, a long-time steady-state condition was achieved, with no change in the coating target and less than a 10 m/min variation in line speed. In this case, the average coating thickness produced was 10.605 μm, minimally above the target 10.5 μm coating thickness. Even this small (0.98%) bias toward a higher coating thickness resulted in a consumption of 99 more tons of zinc than the ideal situation. At a zinc price of $2000 per ton, this excess zinc cost was $198,527. During another period of operation including numerous changes in target coating thicknesses, coil width and thickness changes, and line speed changes, the overcoat bias averaged 16.1%, resulting in an excess zinc consumption of 1632 tons compared with the ideal situation; this excess zinc costs $3,264,617 [5]. Upgrading the coating control equipment to the latest version running the same production schedule would have still resulted in excess zinc consumption of 612 tons with a cost of $1,223,884; this zinc reduction can in many cases justify improvements in coating control equipment. However, the costs of excess zinc resulting from process line condition variabilities and limitations of measurement and control equipment are still significant. Details on the nature of these variabilities and limitations are described in this chapter, with the objective of providing the reader an understanding of how these can be controlled to improve productivity and quality of a continuous hot-dip galvanizing line while minimizing both zinc consumption and product out-of-specification costs.

7.2 Liquid metal properties affecting coating control

The three bulk properties of liquid metal alloys influencing their behavior in the coating control section are density, viscosity, and surface tension. The equations relating these variables to resultant coating weight will be described in Section 7.4. The oxidation behavior of these alloys also influences behavior in the coating control section. The use of nitrogen wiping to help control oxidation is described in Section 7.11.

The density of liquid Zn varies with temperature, and it also varies with alloy content. Liu [6] showed that the density of ZnAl alloys could be expressed by dividing the molar weight of an alloy by its molar volume V of each of the constituent Zn and Al alloying elements, which vary linearly with temperature, to which a term including the excess Gibbs free energy ΔG^{ex} of the liquid metal mixture is added

$$V = X_{Al}V_{Al} + X_{Zn}V_{Zn} + C\Delta G^{ex} \tag{7.3}$$

where X is the molar fraction of either Zn or Al in the alloy, and C is a constant. When ΔG^{ex} is negative, it indicates affinity between the two alloying elements so that negative deviation of the molar volume from the weighted average is expected, whereas a positive ΔG^{ex} indicates repulsion between the two alloy constituents in this case and therefore a positive deviation in volume. Values of ΔG^{ex} are given in the literature [7,8]. Calculated values were compared with those found using pycnometry [9], which is suitable for use when measuring substances with high vapor pressures. Fig. 7.2 shows that Liu's calculated values agree well with these experimental values.

Liu extended this work to examine the effect of combined Al and Mg compositions on Zn alloy densities, and these were determined over a range of temperatures of interest to continuous galvanizing, Fig. 7.3 [10]. The measurements begin at the liquidus temperature of each alloy and decrease with increasing temperature. The addition of the lower density Al and Mg elements reduces density in a predictable way.

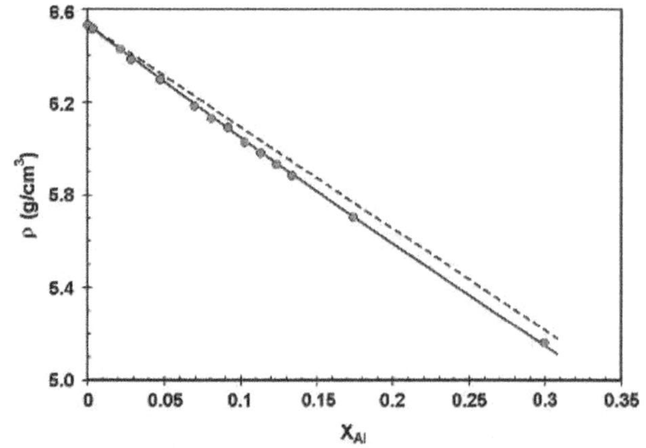

FIG. 7.2 Density of Zn-Al alloys at a bath temperature of 460°C. Data points shown as closed circles represent experimental measurements. The density predicted by using Eq. (7.3) is shown as a solid line, whereas that calculated from the simple weighted average of Al and Zn, without inclusion of the excess Gibbs free energy term, is shown as a dashed line [6]. *(Reproduced with permission from Y.H. Liu, Density and viscosity of zinc alloys, Metall. Mater. Trans. A 37A (2006) 2767–2771.)*

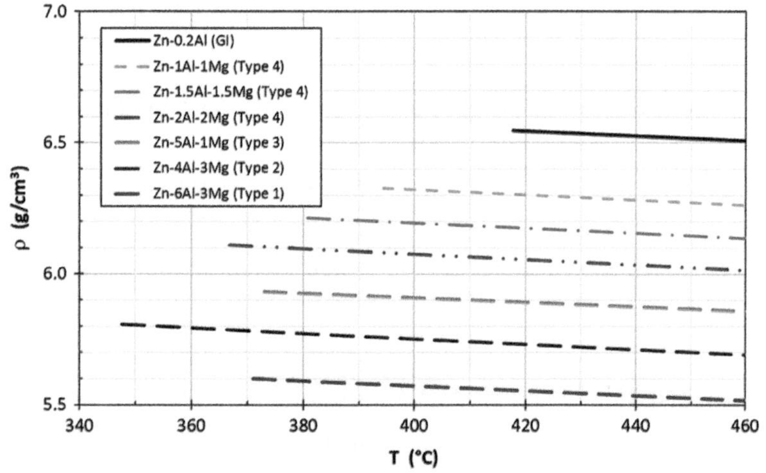

FIG. 7.3 Effect of alloy composition and temperature on densities of a range of compositions relevant to continuous galvanizing [10]. *(Reproduced with permission from D. Liu, N. Gao, A comparative study on different types of Zn-Al-Mg coatings, in: Proceedings Galvanizers Association 112th Meeting, March 27–30, 2022, Myrtle Beach, SC, Galvanizers Association, Marine City, MI, 2022.)*

To determine viscosity, Liu began with the Arrhenius equation that gives the basic relationship between viscosity and temperature.

$$\eta = \eta_0 e^{(E/RT)} \tag{7.4}$$

where η = viscosity, E = activation energy, R = universal gas constant, and T = absolute temperature. The activation energy term was replaced by a new term, the Gibbs energy of activation, which depends upon molar volumes and molar fractions of the Zn and Al alloy constituents and also the excess Gibbs free energy described above for the density calculations. This was refined using the work of several other researchers to give the relationship between viscosity and alloy composition shown in Fig. 7.4.

For convenience in using the data in Figs. 7.2 and 7.4, conversions from weight percent Al to molar fraction Al in binary Zn-Al alloys of interest to continuous sheet galvanizing are shown in Table 7.1.

The dependence of viscosity on temperature for binary Zn-Al alloys between 0 and 15 wt% is shown in Fig. 7.5. As an example of the decrease of viscosity with increasing Al content, it can be observed that Zn-5%Al at 400°C has about the same viscosity as the 0.1–0.3 Al alloys have at 480°C. Viscosity values for several Zn-Al compositions and temperatures from this figure are tabulated in Table 7.2; the 0.1% and 0.9% values are from plotted results while the 0.3% Al values are interpolations.

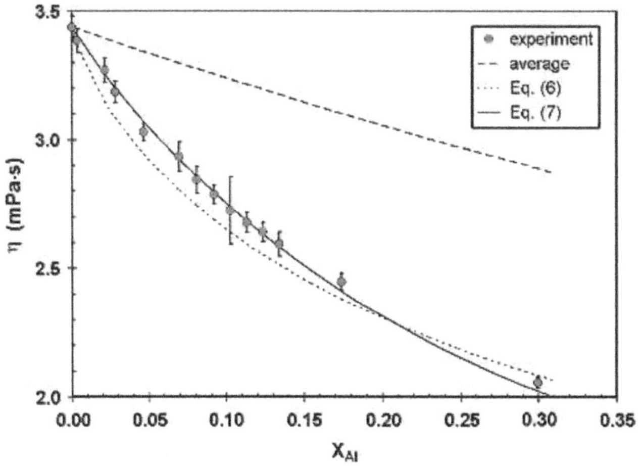

FIG. 7.4 Relationship between Al mole fraction X_{Al} and viscosity for binary Zn-Al alloys at 460°C, measured experimentally using an oscillating sphere viscometer (data points) [11]. Use of the Gibbs activation energy term provides good agreement between calculations and experimental results (solid line), compared with a prior model that did not include an entropy term (dotted line [12]), and a significant improvement over use of only weighted averages of Al and Zn properties viscosities. Each error bar shown with the data points in Fig. 7.4 represents $\pm 3\sigma$, the standard deviation of the experimental data [6]. *(Reproduced with permission from Y.H. Liu, Density and viscosity of zinc alloys, Metall. Mater. Trans. A 37A (2006) 2767–2771.)*

TABLE 7.1 Conversion table for weight percent Al to molar fraction Al for Zn-Al binary alloys.

Weight percent Al	Molar Fraction Al (X_{Al})
0.1	0.002
0.135	0.003
0.2	0.005
0.3	0.007
1.5	0.036
2	0.047
4.5	0.103
5	0.113
6	0.114
11	0.23
55	0.748

The viscosities of Zn-Al-Mg alloys over the range of compositions of interest for continuous galvanizing have been predicted using thermodynamic modeling and confirmed by limited experimental work [10]. Fig. 7.6 shows the effect of temperature on the viscosities of several alloys, starting from their liquidus temperature. The viscosity of the Zn-0.2% alloy is in this graph compares well with the data of Fig. 7.4. The Zn-5%Al-1%Mg alloy has a lower viscosity, but as its composition is raised to Zn-6%Al-3%Mg viscosity increases. The viscosity of Zn-6%Al-3%Mg is between that of Zn-1%Al-1%Mg and Zn-1.5%Al-Zn-1.5%Mg. Higher levels of Mg, with Al compositions up to 4%, further increase viscosity.

With regard to surface tension, several studies have concluded that for normal engineering analysis related to continuous galvanizing coating control, it can be neglected [13,14]. It is expected to influence the development of coating thickness at low line speeds, e.g., below 20 m/min. The role of surface tension in general galvanized coating behavior is described in Chapter 15. When wiping with air, surface tension is observed to decrease because of the presence of oxides formed on the surface the liquid metal; these oxides are also observed to increase viscosity [15]. Surface tension also becomes an important property when considering splashing phenomena discussed later in the chapter.

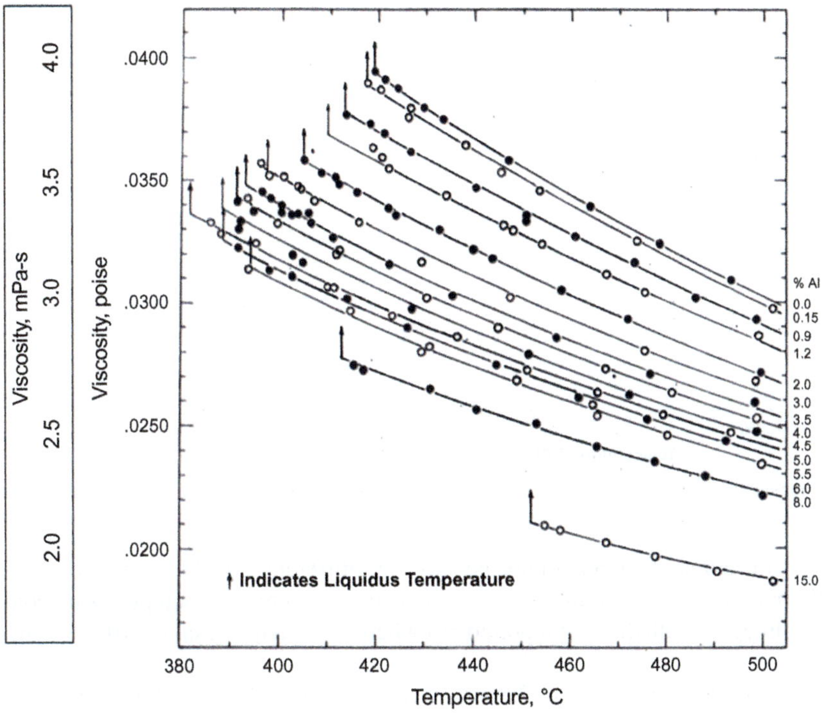

FIG. 7.5 Dependence of viscosity on temperature for binary Zn-Al alloys between 0 and 15 wt% [11]. *(Reproduced with permission from H.R. Thresh, D. W.G. White, J.O. Edwards, J.W. Meier, Properties of molten zinc and zinc alloys final report: Part II: Viscosity, International Lead Zinc Research Organization Project ZM-54, October 1, 1965, Available from International Zinc Assn., Durham, NC,1965.)*

TABLE 7.2 Representative values for Zn-Al galvanizing alloy viscosities at 453°C and 473°C.

wt% Al	Temperature (°C)	Density (g/cm^3)	Viscosity (mPa-s)
0.15	453	6.52	3.45
0.15	473	6.5	3.26
0.3	453	6.52	3.41
0.3	473	6.49	3.27
0.9	453	6.43	3.32
0.9	473	6.41	3.16

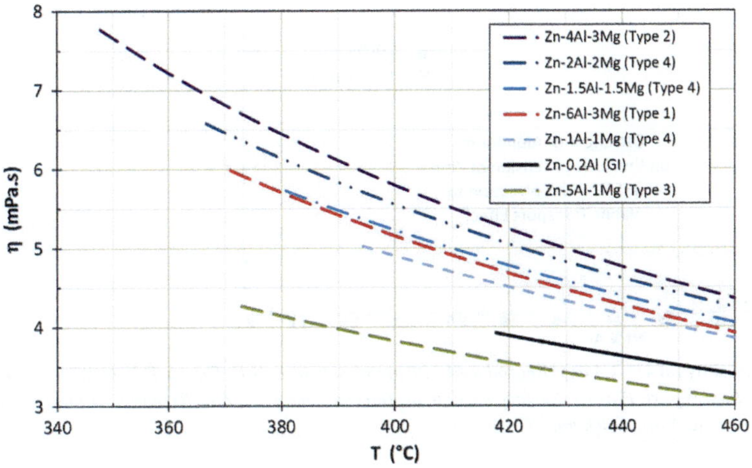

FIG. 7.6 Viscosities of Zn-Al and Zn-Al-Mg alloys over the range of temperatures relevant to continuous galvanizing, starting at the liquidus point of each alloy [10]. *(Reproduced with permission from D. Liu, N. Gao, A comparative study on different types of Zn-Al-Mg coatings, in: Proceedings Galvanizers Association 112th Meeting, March 27–30, 2022, Myrtle Beach, SC, Galvanizers Association, Marine City, MI, 2022.)*

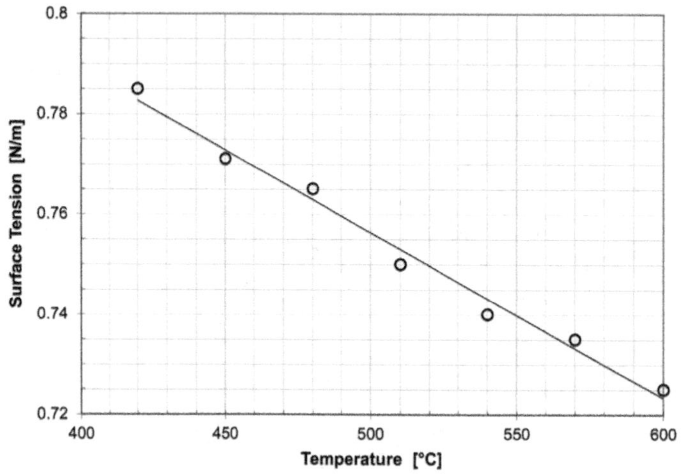

FIG. 7.7 Surface energy of pure Zn between its melting point and 600°C [17]. *(Reproduced with permission from Daniel Liu, private communication.)*

Surface tension is expected to decrease with increasing temperature because the difference of the energetic states between liquid and gas states decreases with increasing temperature [16]. This was confirmed by work on pure zinc conducted at Teck Metals, Fig. 7.7 [17]. These results agree with those of reported elsewhere, (777 dyn/cm, same as 0.777 N/m) at 460°C [18].

Surface tensions for a range of Zn-Al alloys for Al contents up to 2.8 wt% were determined by Thresh [19]. Fig. 7.8 shows the results; it was found that mass losses caused by vaporization of the liquid metal droplet during the sessile drop test reduced its surface area and therefore the measured surface tension. Therefore, surface tensions are shown in Fig. 7.8 for both uncorrected experimental results and after correction for vaporization losses that unavoidably occurred during the experiments. There is little effect of Al content on the surface energy of Zn over the range studied.

Water modeling has been widely used to simulate hot-dip galvanizing coating control [14]; however, the surface tension of water is much lower than that of galvanizing alloys. Therefore, surface tension effects wiping cannot be easily validated using water models.

The thermal conductivity of liquid pure zinc is quite stable with temperature; it decreases from 59 W/m × °K at 450°C to 57 W/m × °K at 700°C [20].

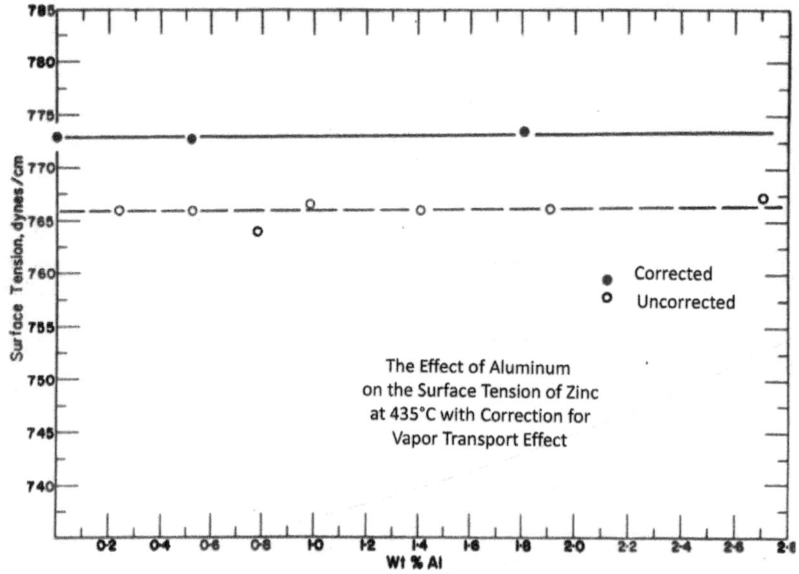

FIG. 7.8 Surface tensions measured for a range of binary Zn-Al alloys at 435°C [19]. *(Reproduced with permission from H.R. Thresh, D.W.G. White, J.O. Edwards, J.W. Meier, Properties of molten zinc and zinc alloys final report: Part III: Surface tension, in: International Lead Zinc Research Organization Project ZM-54, October 15, 1964, Available from International Zinc Assn., Durham, NC, 1964.)*

7.3 History of development of coating controls

From the perspective of the history of the steel industry, continuous sheet galvanizing is a relatively new process. In his 1962 article, C. Cone noted that as in industrial process, it was only about 25 years old, meaning it began about 1937. In his paper, it is noted that coating control was performed by use of coating rolls or asbestos pad wiping if thin coatings were desired. The speed limit of the line was in many cases limited by coating roll performance to 76–91 m/min [21]. According to Thornton and Graff [22], the minimum coating weight that could be achieved with coating rolls was 183 g/m^2 (corresponding to an equal-side coating thickness of 13 μm), and the minimum strip thickness that could be passed through the coating rolls was 0.389 mm. Excessive strip breaks at the coating rolls were experienced with thinner strip [23]. A thinner (5.5 μm per side) Zn-Fe coating was produced by asbestos pad wiping rather than using coating rolls [22]. It was noted that new coating control techniques were under development.

Air jets were used as early as the 1930s to regulate coating thickness on narrow strip typically used for electrical cable at Triangle Conduit and Cable Company [23]. In 1947, British patent GB588281A "Procedure and apparatus for continuously coating strip metal" was issued to inventor Warren Worthington of Pittsburgh, PA [24]. He described the operation of a hot-dip tinning line using a flux pretreatment and the use of "a blast of fluid under pressure to strip excess coating metal therefrom and spread the coating." It was also noted that "The thickness of the coating may be adjusted by changing the angle of incidence of the fluid from the nozzles, varying the fluid pressure delivered to the nozzles, varying the speed of the strip or varying the temperature of the fluid delivered to the nozzles, or a combination of these." Other means of control were not indicated. It was not until 1970 that the Canadian J.T. Mayhew received US Patent 3,499,418 "Continuous Metallic Strip Hot-Dip Metal Coating Apparatus," assigned to National Steel Corporation [25]. A drawing from this patent is included in Chapter 1. It is the first complete technical description of a coating control apparatus for zinc-coated strip using air knives. Both the Worthington and Mayhew patents teach the usefulness of wiping with hot gas or steam, which has since been shown to not confer any advantages in comparison with unheated gas, although hot gas may confer some benefits to very thin (<0.2 mm) strip [26]; Thornton and Graff noted that three Armco coating lines were converted from steam to air wiping to give improvements in coating quality and a 5 dB reduction in noise level in vicinity of the zinc pot [22]. At the same time, intensive development and implementation work was underway elsewhere. Beam, Butler, and Hawkins describe the installation and operation of air knives for coating control at Jones & Laughlin facilities in Hennepin, IL, beginning in 1967 after developmental work on their Pittsburgh, PA line [23]. Beam et al. reported that within 3 months of operation, zinc consumption was reduced by 10% compared with operating with coating rolls. The average coating variation for the same period was reduced to ±13.44 g/m^2 compared with ±27.45 g/m^2, and a 33% improvement in edge-center-edge variation was achieved with the use of air knives.

Beam et al. also identified the principal causes of coating weight variability known to them:

- Air flow rate
- Nozzle distance from the strip
- Angle of air impingement on the strip
- Nozzle height above the bath
- Nozzle slot opening
- Strip speed
- Strip temperature
- Zinc bath temperature
- Zinc bath composition
- Strip thickness
- Strip width
- Strip shape
- Strip surface characteristics

In the years since 1970, much work has been done to model, verify, and generalize these effects, and other causes of variability, such as jet instability, have been identified and modeled. This has led to the development of improved coating control equipment, coating weight or thickness measurement, and control programs and control schemes that could not be imagined in 1970, together with the use of alloy coatings. Much of the rest of this chapter is devoted to a description of these improvements that have led to today's state of the art.

7.4 Development of process models and operation correlations: Steady-state models

In their pioneering study, Thornton and Graff [22] considered the behavior of the liquid zinc coating as a boundary layer problem. They divided the problem into the six regions shown in Fig. 7.9:

- Region I: the transition of liquid zinc flow from a boundary layer formed by strip motion in the bath into, after emergence from the bath, a layer capable of being supported against gravity by the viscous shear forces at the strip.
- Region II: the distance from Region I to the point where interaction with the wiping jet begins. Thornton and Graff considered the flow in this region to be constant, with an interior layer of the coating supported by viscous shear and an outer layer, the runback layer, descending because of gravitational force. At some location between these two layers, a vertical boundary of zero upward velocity exists, indicated by the vertical dashed line in Region II in Fig. 7.9.
- Region III: The region where jet stripping occurs. This is caused by a gas pressure gradient created by the wiping jet that caused a pressure-induced flow. One point in this region important to understand for this analysis and those that follow is the jet stagnation point. This is the point in Region III where gas velocity is zero. For a stable, horizontal jet, this point is in line with the gas knife gap. The pressure at this point is the stagnation pressure. In the two-dimensional Fig. 7.9, gas from the air knife flows both upward and downward from the stagnation point. The liquid metal coating flow induced by the gas pressure gradient combines with the flow induced by movement of the steel strip; on the down-facing side of the air knife, the flows oppose each other, whereas on the up-facing side, the flow induced by the knife adds to the strip-induced flow, causing the local thinning indicated as δ_3 in Fig. 7.9.

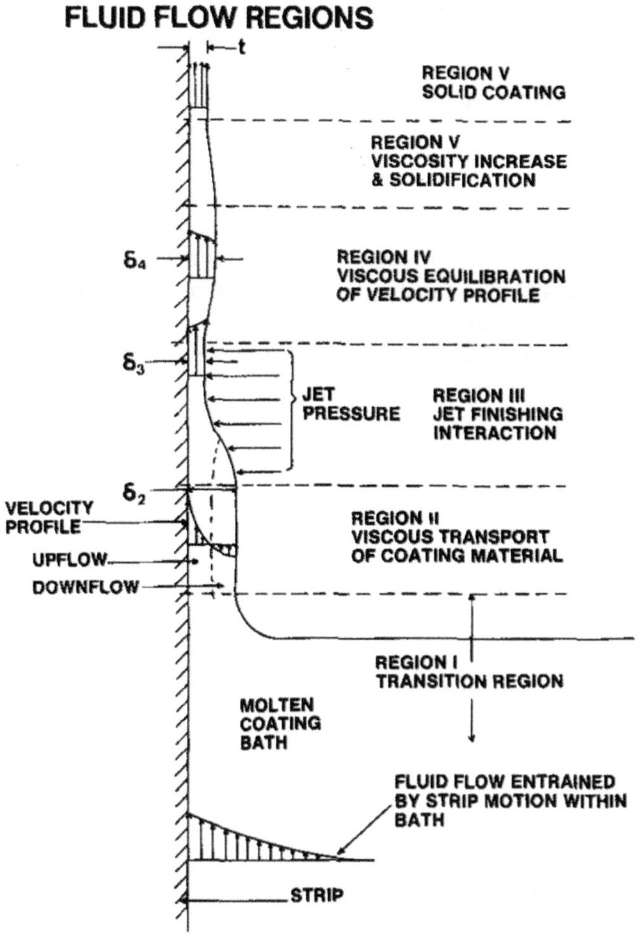

FIG. 7.9 Regions of coating flow considered in the Thornton and Graff analysis [22]. *(Reproduced with permission from J.A. Thornton, H.F. Graff, An analytical description of the jet finishing process for hot-dip metallic coatings on strip, Metall. Trans. B 7B (1976) 607–618.)*

- Region IV: In this region beyond the influence of the air knife, there were assumed to be no pressure-induced forces. The viscous entrainment forces were assumed to equilibrate with gravity forces although the outer portion of the liquid coating could still have a downward velocity under the influence of gravity.
- Region V: In this region, cooling becomes significant, increasing the viscosity of the liquid coating, a more constant velocity profile approximating strip velocity is developed. The coating begins to solidify by heterogeneous nucleation and growth.
- Region VI: The coating with its final thickness is fully solidified.

The nature of flow of the liquid zinc during its time outside the zinc bath was described as incompressible, constant viscosity, creep flow (with the nature of diffusional flow best described as a manifestation of Nabarro-Herring creep [27] because the viscous forces are larger than the inertial forces. The Reynolds number within the zinc layer is quite low, approximately 1300, even when high strip speeds are used. Thorn and Graff also neglected oxidation and surface tension effects. This allowed for development of the equation of motion for the liquid zinc:

$$\left(\frac{dp}{dx}\right)_{max} = \frac{4}{9}\alpha\,\eta_0\frac{U}{t^2} - \rho g \tag{7.5}$$

In the absence of air jet finishing, $dp/dx = 0$ and Eq. (7.5) can be solved to find the coating thickness when the strip is simply withdrawn from the bath without any regulation by air knives.

$$t_0 = \frac{\sqrt{2}}{3}\left(\frac{\alpha\eta_0 U_s}{\rho g}\right)^{\frac{1}{2}} \tag{7.6}$$

where t_0 is the final coating thickness, α is the ratio of effective viscosity in the system to reference viscosity (determined experimentally to be 2.22), η_0 is the reference viscosity (similar to values shown in Figs. 7.4–7.6), U_s is the strip velocity, ρ is the liquid zinc density (similar to values shown in Fig. 7.2), and g is the gravitational constant. The pressure characteristic that determines the coating thickness is not the peak pressure (usually equal to the stagnation pressure) but the maximum pressure gradient. This is located at the point of maximum slope in the pressure curve, moving in a vertical distance up or down from the stagnation point. After invoking further assumptions, including that the jet is subsonic and follows certain turbulent flow relationships, that the system is isothermal, and that the influence of gravity is much less than that of the jet stripping force and nozzle efficiency is 100%, the dependence of coating thickness on fluid and operating variables is obtained:

$$t = \frac{1}{10\sqrt{2}}\left(\beta\alpha^{\frac{1}{2}}\right)Z_0\left[\frac{\eta_0\,U_s\,(\gamma-1)}{p_A\gamma d}\right]^{\frac{1}{2}}\left[\left(\frac{p_0}{p_A}\right)^{(\gamma-1)/\gamma} - 1\right]^{-\frac{1}{2}} \tag{7.7}$$

where Z_0 = strip-to-knife distance; β = a scaling constant for stagnation pressure that was experimentally found to be equal to 1.5; γ = the ratio of gas specific heat at constant pressure C_p to the gas specific heat at constant volume C_v, approximately 1.4; p_A = atmospheric pressure, d = nozzle slot opening; p_0 = nozzle plenum pressure. With regard to the main line operating variables, coating thickness is observed to increase linearly with strip-to-knife distance, increase with the square root of line speed, decrease with the square root of nozzle slot opening, and decrease with approximately the square root of nozzle plenum pressure. Several experiments on laboratory pilot and industrial lines confirmed that $\beta\alpha^{\frac{1}{2}} = 2.2$,

Laboratory and mill experiments carried out to determine the accuracy of Eqs. (7.6) and (7.7) showed that in both cases, coating thickness varied with strip velocity in more of a linear than an inverse square root manner. Coating thicknesses obtained without stripping inferred viscosities that were about half of those indicated in Figs. 7.4–7.6. Industrial trials carried out with both air and steam confirmed the relationship between plenum pressure and coating thickness given in Eq. (7.7). In an anticipation of the findings of later researchers, the effect of reducing strip-to-knife distance on coating thickness was not as large as predicted, in the range of $Z/d < 10$. Overall, the coating weight model predicted the coating weights produced on industrial lines to an accuracy of 25%.

Thornton and Graff considered the use of both air and superheated steam in their analysis. It was shown that the momentum flux of steam is about the same (and even slightly less) than an ideal gas; therefore, its wiping behavior is very similar to air. The use of superheated steam vs. air also did not influence cooling of the coating in the jet area. The temperature difference between the zinc and the wiping medium is a much more important factor, there is much more rapid heat transfer between 460°C zinc and 37°C air than 460°C steam. It was also found that because of the lower density of superheated steam, a lower mass flux resulted for the same momentum flux as air. Under similar conditions, the superheated mass flux is about 80% of the equivalent air flow. Therefore, to obtain the same coating weight, the mass flow rate of steam through the wiping knives needs to be higher. Combined with the lower density of superheated steam, this results in noise levels near the coating rig that are 5 dB higher than the equivalent air knives. For this and other reasons, including reduction of top skimmings production, steam is no longer used as a wiping medium for coating control.

Thornton and Graff concluded by theoretically determining the maximum performance of a coating control unit. The minimum possible strip-to-knife distance was set at $Z/d = 8$, so that for the minimum conceivable nozzle gap distance of 1.27 mm (within today's state of the art), $Z = 10.2$ mm. The viscosity value used was about half of the values indicated in Figs. 7.4–7.6, so that the calculated minimum coating thickness for a line speed of 2 m/s of 4.23 µm per side is overestimated by a factor of the square root of 2, the corrected value is 6 µm, equally divided per side for a two-side coating weight of 43 g/m^2.

Eight years later, the above model was refined by Ellen and Tu [28] to include the effects of the shear stress of the wiping jet on the liquid coating. Together with making coating thickness prediction more accurate, the inclusion of shear stress also helped explain the occurrence of coating waviness caused by this shear stress, which can cause a defect termed "sag lines," as illustrated in Fig. 7.10.

Ellen and Tu related this waviness to the presence of "wind-driven waves" on the coating surface in Region IV of Fig. 7.9. Ellen and Tu began their analysis with the Navier-Stokes equation, ignoring surface tension and assuming variations on coating thickness only vary slowly

$$\eta \frac{d^2u}{dz} - \left(\rho g + \frac{dp}{dx}\right) = 0 \tag{7.8}$$

where, consistent with Eq. (7.5), x is the coordinate parallel to the direction of strip movement (normally vertical), z is the coordinate normal to the strip surface, $u(z)$ is the fluid velocity in the z direction. A constant viscosity η is assumed.

As with the analysis of Thornton and Graff, one boundary condition is $u_x = U_s$ at $z = 0$ (no slip). Additionally, the boundary condition

$$\eta \frac{du}{dz} = \tau \text{ at } z = t \tag{7.9}$$

where τ = the shear stress imposed by the gas-liquid metal interface at the coating thickness t was introduced. Shear stress had not been considered in any previous analysis.

Eq. (7.8) was then solved to yield

$$u/U_s = 1 - (GT^2/2)(z/t)(2 - z/t) + ST(z/t) \tag{7.10}$$

where T, S, and G are the nondimensional coating thickness, shear stress, and effective gravitational acceleration, respectively, defined as:

$$T = t(g/vU)^{\frac{1}{2}} \tag{7.11}$$

$$S = \tau(\eta\rho U)^{-\frac{1}{2}} \tag{7.12}$$

FIG. 7.10 Sag line defects on hot-dip galvanized sheet. Strip movement in vertical direction [29]. *(Reproduced with permission from M. Dubois, private communication, March 2022.)*

$$G = 1 + \left(\frac{dp}{dx}\right)/\rho g \tag{7.13}$$

where v = kinematic viscosity, η/ρ.

Eq. (7.10) can be integrated to give the mass flux Q of coating mass being withdrawn

$$Q = T\left(1 - GT^2/3\right) + ST^2/2 \tag{7.14}$$

$$Q = \left(q/U_s\right)\left(g/vU_s\right)^{\frac{1}{2}} \tag{7.15}$$

where $q = \int_0^t udz$, and Q is the dimensionless expression of mass flux.

The relationship between the local pressure at any point moving up or down vertically (in the x direction) from the stagnation point was determined by experiment. The location where the local pressure p was half the peak pressure p_s was termed b, so that the term $x/b = \xi$ is dimensionless. The relationship between local pressure and distance x was indicated as

$$p/p_s = f(\xi) = e^{-0.693\xi^2} \tag{7.16}$$

This relationship was accurate for $|\xi| < 1.6$. At locations on the strip not near the strip edges, Ellen and Tu showed that, for the strip-to-knife distances investigated, the pressure curve closely fit the Gaussian distribution described by Eq. (7.16).

The maximum shear stress was given as

$$\tau = \tau_{\max}\left[erf(0.833\xi) - 0.2\xi f(\xi)\right] \tag{7.17}$$

where τ_{max} depends upon jet angle and is described in the work of Beltaos on oblique jets [30]. Surprisingly, the location of τ_{max}, where maximum stripping occurs, was found to be $\xi = 0.849$, less than the predicted location where peak stripping should occur for either the pressure gradient or surface shear stress models $1.0 < \xi < 1.4$. This means that the location is nearer to the pressure peak location than expected.

After converting mass flux to coating thickness, Eq. (7.10) was plotted with and without the shear (S) term to produce the graphs shown in Fig. 7.11.

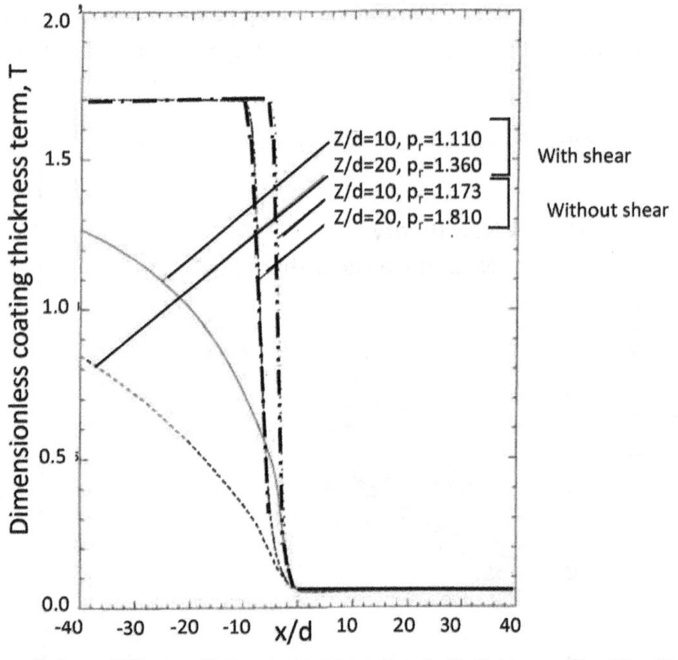

FIG. 7.11 Coating weight profile predictions of Ellen and Tu for near and far strip-to-knife distances, $z/D = 10$ and 20, result expressed as dimensionless term T, Eq. (7.11), moving up (+) and down (−) from the stagnation point (in the x direction) P_r = ratio of gas pressure to ambient. Wiping pressure was adjusted to give the same range of coating thickness. Line speed = 100 m/min., single-side coating mass 100 g/m^2, coating thickness 14 μm [28]. *(Reproduced with permission from C.H. Ellen, C.V. Tu, An analysis of jet stripping of liquid coatings, J. Fluids Eng. 106 (1984) 399–404.)*

It can be seen that the thickness transition zone occurs over a much longer vertical distance underneath the gas knives when shear stress is included, for both values of strip-to-knife distances. As expected, much thinner coatings are predicted when shear stress is included in the prediction equations. Also, the shear force was shown to act over a longer x-distance at higher strip-to-knife distances, so that to minimize jet effects on coating waviness, the strip-to-knife distance should be kept as close as possible.

Ellen and Tu also estimated heat transfer effects: For a 0.5 mm thick strip, it was estimated that the coating would cool as much as 3°C as it passed between closely spaced jets and three times as much at twice the jet distance. The higher rate of cooling would limit the time when waviness would influence the coating profile.

The improvement in predictive accuracy for the Ellen and Tu model is shown in Fig. 7.12. The points on this graph were taken from 35 line operating conditions and show that the prediction including shear is within 10% of the measured points, on average. The uncertainty in coating mass measurements was given as $\pm 5\,g/m^2$. The equivalent coating thickness range of Fig. 7.12 is 6–32 μm.

Twenty years later, the Ellen and Tu model was further refined in a Galvanized Autobody Partnership project [31] focusing on coating thicknesses relevant to automotive coated sheet, i.e., less than 10 μm per side. Such coatings are typically made at $Z/d \leq 8$, lower than the strip-to-knife distances considered before. The Ellen and Tu equations were solved and verified. When compared with results from an industrial line operation at $Z/d \leq 8$, they were found to underpredict coating thickness by 20%, that is, in Fig. 7.12, the calculated points would be 20% below the values for corresponding measured points. To improve prediction, laboratory studies were done to measure pressure profiles of a model air knife with $Z/d = 2,4,6,8$ using a plenum pressure of 137 mbar (13.7 kPa). When a jet is less than six jet diameters from the surface, the jet potential core impinges directly on the strip surface. The jet velocity profile is then not self-similar because of the uniform velocity of the potential core, so that the pressure profile becomes a top hat shape [32] rather than the Gaussian shape utilized by prior analyses. The pressure profiles for the range of investigated Z/d with $d = 1.52$ mm are shown in Fig. 7.13.

Further improvement of model accuracy depended upon accurate assessments of wall shear stress. Experimental work was carried out to assess this using oil film interferometry experiments to obtain wall surface friction measurements using the relationship

$$C_f = \tau_w / \frac{1}{2}\rho U^2{}_{\text{jet}} \tag{7.18}$$

where τ_w is the shear stress at the wall (strip surface), and $\frac{1}{2}\rho U^2{}_{\text{jet}}$ is the dynamic pressure of the jet. This produced a set of skin friction curves shown in Fig. 7.14. It is counterintuitive that skin friction should increase with Reynolds number because shear stress increases more rapidly with jet velocity (Reynolds number) than jet energy. The values in

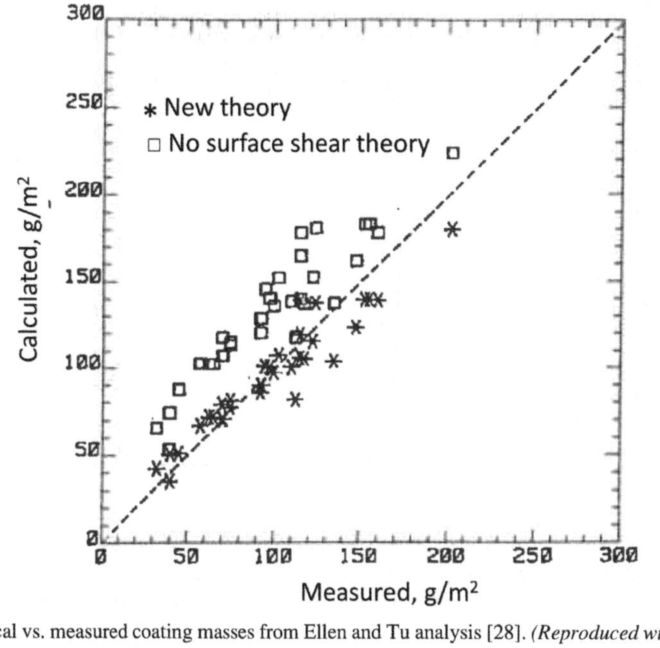

FIG. 7.12 Comparison of theoretical vs. measured coating masses from Ellen and Tu analysis [28]. *(Reproduced with permission from C.H. Ellen, C.V. Tu, An analysis of jet stripping of liquid coatings, J. Fluids Eng. 106 (1984) 399–404.)*

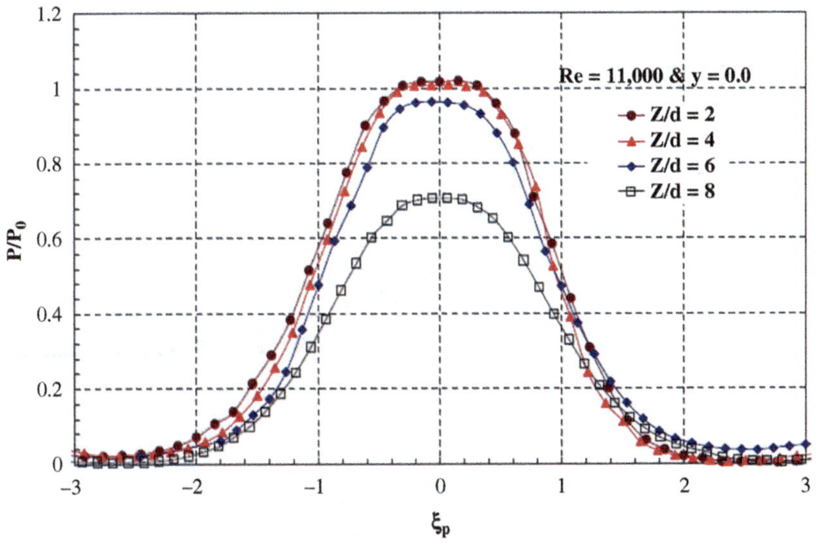

FIG. 7.13 Pressure profiles in the x (vertical) direction for air knives with indicated Z/d, far from the strip edge. The coordinate ξ is the same used by Ellen and Tu to normalize vertical distance to the half-width of the pressure peak in each case [31]. *(Reproduced with permission from Galvanized Autobody Partnership Project ZCO-2-3, Improving Coating Weight Consistency, Final Report, July 2004, International Zinc Assn., Durham, NC.)*

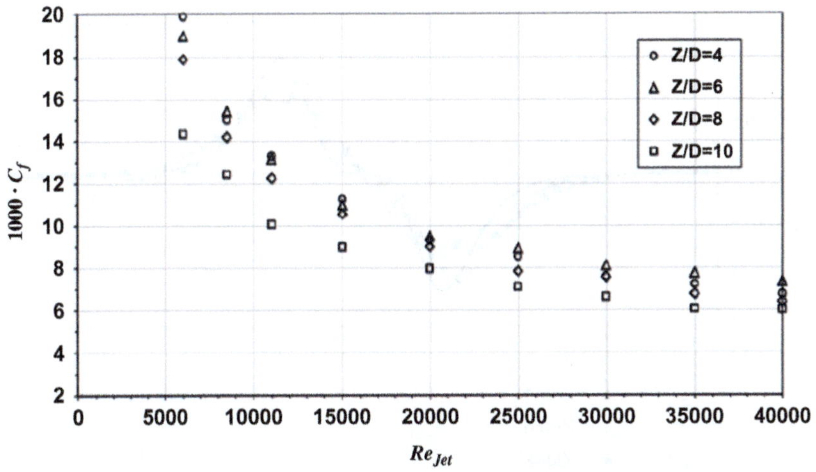

FIG. 7.14 Skin friction measured by oil film interferometry for small Z/d values [33]. *(Reproduced with permission from Galvanized Autobody Partnership Project ZCO-56, Improving Coating Weight Consistency, Final Report, March 2017, International Zinc Assn., Durham, NC.)*

Fig. 7.14 are judged to be more accurate at lower Reynolds numbers and agree with theoretical data for $Z/d = 6$ [34]. Development of the improved coating weight model using computational modeling better incorporated jet compressibility properties.

Wall shear stress was calculated using FLUENT 5.0, using the standard k-ε turbulence model and the nonequilibrium wall function. A large number of computational runs were required before a suitable turbulent flow solution was obtained. The empirical results were then fitted using regression techniques to obtain correlations for the half-width of the pressure distribution and the pressure profile as a function of the x-direction from the jet centerline

$$\frac{b_p}{d} = 0.0453\left(\frac{Z}{d}\right) + 0.7921, \quad \frac{Z}{d} \le 8 \tag{7.19}$$

$$\frac{P}{P_{\max}} = \left[1 + 0.6\xi^4{}_p\right]^{-1.5}, \quad \frac{Z}{d} \le 8 \tag{7.20}$$

Differentiation of Eq. (7.20) gives the pressure gradient

$$\frac{d\frac{P}{P_{\max}}}{d\xi_p} = -3.6\xi^3{}_p\left[1 + 0.6\xi^4{}_p\right]^{-2.5}, \quad \frac{Z}{d} \le 8 \tag{7.21}$$

Figs. 7.15 and 7.16 show the pressure distribution and pressure gradients based on these correlations together with the experimental results. The Gaussian distributions of Ellen and Tu are shown for comparison

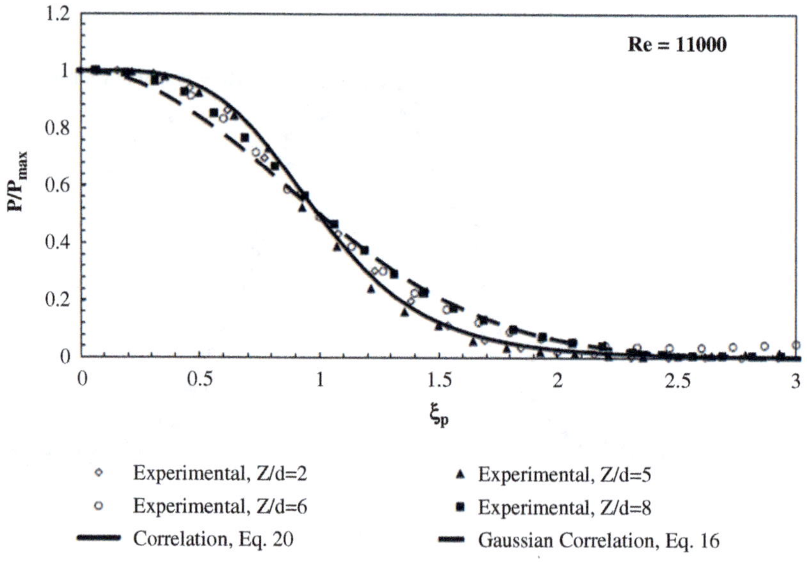

FIG. 7.15 Pressure correlations for use in small Z/d coating thickness prediction modeling [31]. *(Reproduced with permission from Galvanized Autobody Partnership Project ZCO-2-3, Improving Coating Weight Consistency, Final Report, July 2004, International Zinc Assn., Durham, NC.)*

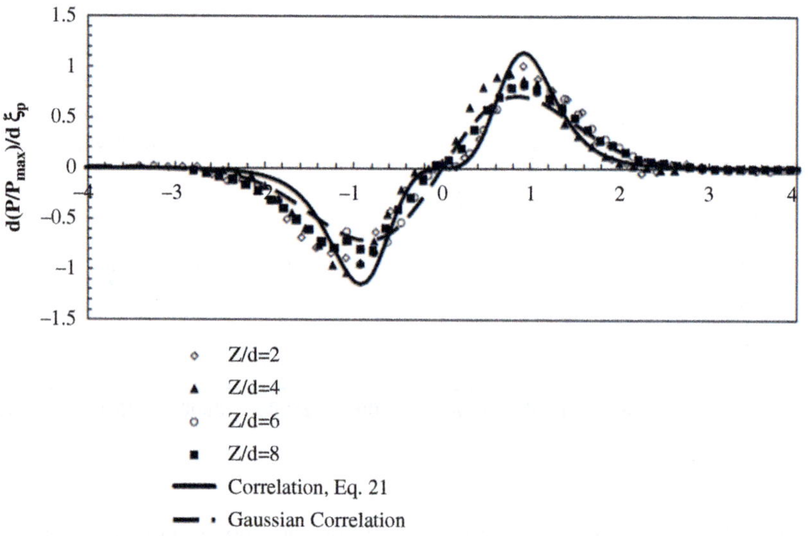

FIG. 7.16 Pressure gradient correlations for use in small Z/d coating thickness prediction modeling [31]. *(Reproduced with permission from Galvanized Autobody Partnership Project ZCO-2-3, Improving Coating Weight Consistency, Final Report, July 2004, International Zinc Assn., Durham, NC.)*

Similar regression fitting enabled correlations to be developed for the shear stress. The half-width of the shear stress distribution curved was computed to be

$$\frac{b_\tau}{d} = 0.0443\frac{Z}{d} + 1.1687, \quad \frac{Z}{d} \le 8 \tag{7.22}$$

And the maximum shear stress was computed to be

$$\frac{\tau_{\max}}{P_0} = -0.0001\left(\frac{Z}{d}\right) + 0.0035, \quad \frac{Z}{d} \le 8 \tag{7.23}$$

Figs. 7.17 and 7.18 show correlations developed by applying regression techniques to the FLUENT computational results. The results of this model are demonstrated in Figs. 7.19 and 7.20. In Fig. 7.19, the dependence of coating weight on Z/d is quite insensitive for low Z/d, when the jet potential core extents to, or nearly to, the strip surface. Increasing sensitivity is seen at higher Z/d where jet dispersion becomes important. This agrees with the work of both Phan and Gosset, who found that when $Z/d < 9$, the jet flow has not diffused much, and axial flow contains most of the flow energy [36,37]. In Fig. 7.20, the relationship between coating weight and line speed at high Z/d is linear, agreeing with the findings of Ellen and Tu, but becomes more of an inverse quadratic relationship at low Z/d.

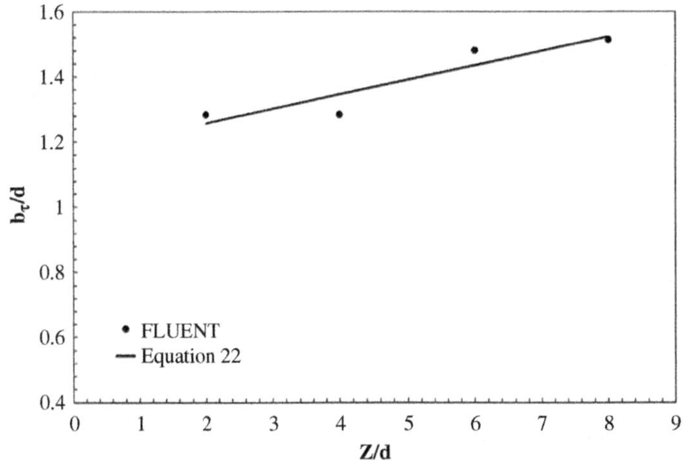

FIG. 7.17 Correlation of the location of the maximum wall shear stress profiles with dimensionless impingment distance (plot of Eq. 7.22). [31]. *(Reproduced with permission from Galvanized Autobody Partnership Project ZCO-2-3, Improving Coating Weight Consistency, Final Report, July 2004, International Zinc Assn., Durham, NC.)*

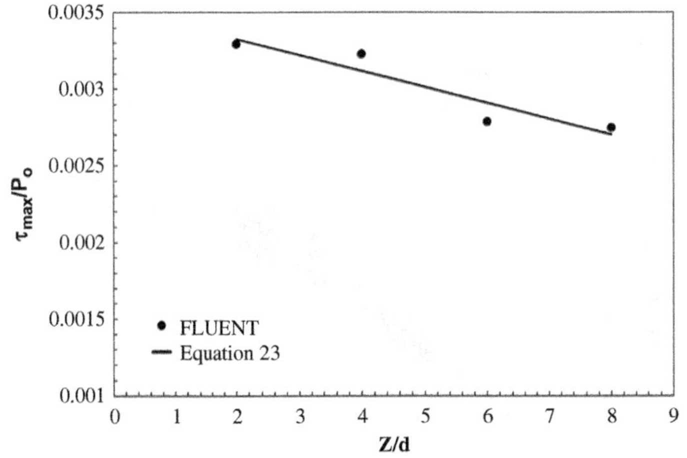

FIG. 7.18 Correlation of the maximum wall shear stress with dimensionless impingement distance (plot of Eq. 7.23) [31]. *(Reproduced with permission from Galvanized Autobody Partnership Project ZCO-2-3, Improving Coating Weight Consistency, Final Report, July 2004, International Zinc Assn., Durham, NC.)*

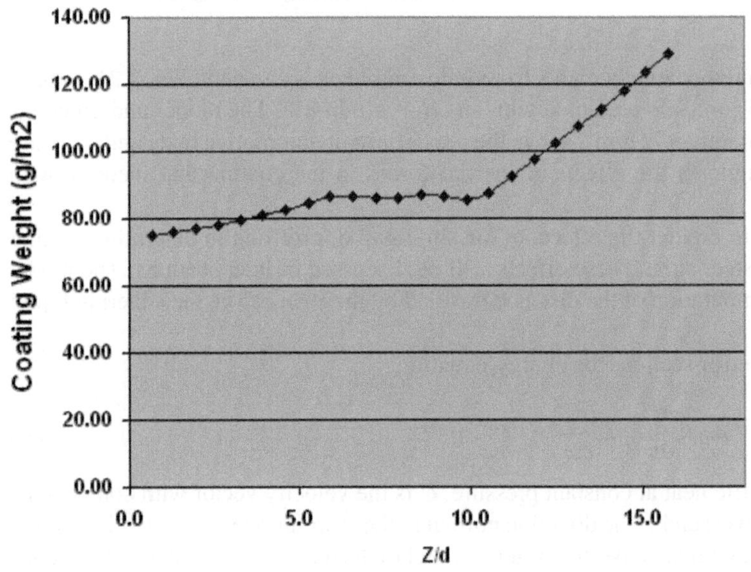

FIG. 7.19 Relationship between coating weight and Z/d from GAP model. $P_0 = 68.95$ mbar, $d = 1.52$ mm, $U_s = 2$ m/s [35]. *(Reproduced with permission from A.N. Hrymak, E.A. Elsaadawy, G. Hanumanth, J.R. McDermid, F.E. Goodwin, Air Knife Coating Control Models, in: AISTech 2005-Proceedings of the Iron & Steel Technology Conference, May 9–12, 2005, Charlotte, NC, Association for Iron & Steel Technology, Warrendale, PA, 2005.)*

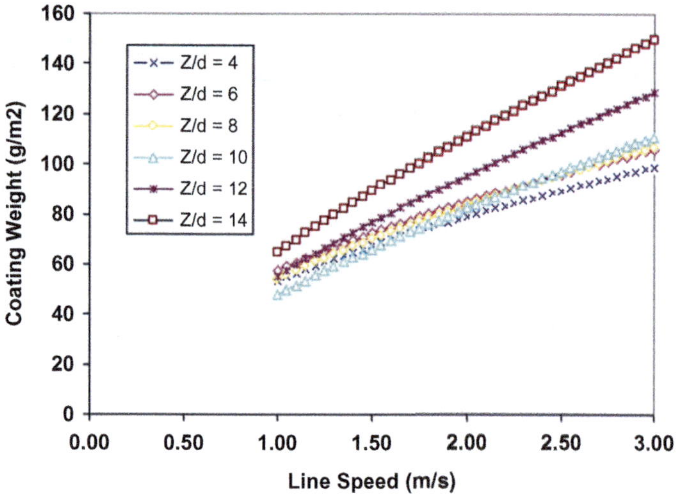

FIG. 7.20 Relationship between coating weight and line speed from GAP model. $P_0 = 20.684$ mbar, $d = 1.52$ mm [35]. *(Reproduced with permission from A.N. Hrymak, E.A. Elsaadawy, G. Hanumanth, J.R. McDermid, F.E. Goodwin, Air Knife Coating Control Models, in: AISTech 2005-Proceedings of the Iron & Steel Technology Conference, May 9–12, 2005, Charlotte, NC, Association for Iron & Steel Technology, Warrendale, PA, 2005.)*

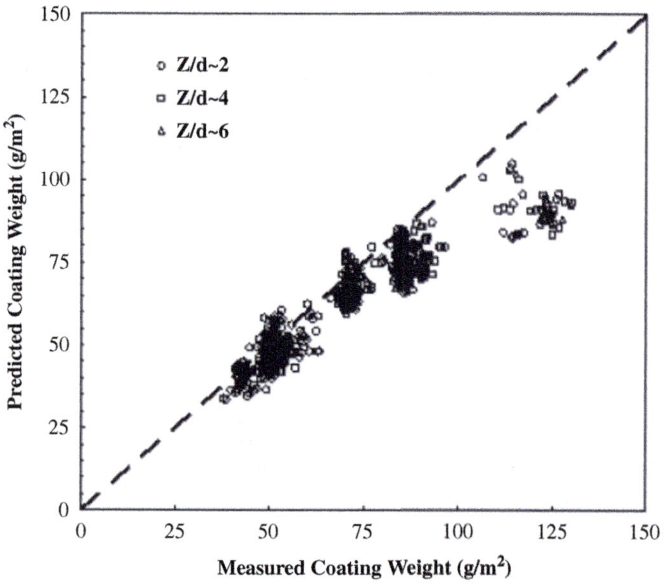

FIG. 7.21 Correlation of computed single-side coating weights from the GAP model with measured industrial coating weights [35]. *(Reproduced with permission from A.N. Hrymak, E.A. Elsaadawy, G. Hanumanth, J.R. McDermid, F.E. Goodwin, Air Knife Coating Control Models, in: AISTech 2005-Proceedings of the Iron & Steel Technology Conference, May 9–12, 2005, Charlotte, NC, Association for Iron & Steel Technology, Warrendale, PA, 2005.)*

The correlation of computed values with measured coating weights from industrial lines is shown in Fig. 7.21. Single-side coating weights, for coating weights up to 75 g/m^2, are predicted with an error within 8%. The model underpredicts coating weight for single-side coating weights greater than 75 g/m^2, above the upper limit of automotive body and structure galvanized coatings. This is because the model ignores the effects of inertial forces in the coating that increase with increasing coating weight [35].

The GAP model has the additional capability to predict the effects of air knife gap d, profiling to determine coating weight deviations caused by cross-bowing of the steel strip. These effects will be discussed in later sections. The model includes a heat transfer analysis that allows the computation of the distance traveled by the strip before the adhering liquid coating completely solidifies.

To characterize heat transfer in the coating control section, the energy equation

$$\rho c_p u \frac{\partial T}{\partial x} = k \frac{\partial^2 T}{\partial z^2} \tag{7.24}$$

applies, where ρ is the mass density, c_p is the specific heat at constant pressure, U is the velocity vector with components u and v along coordinates x (the direction of strip travel) and z (the direction normal to the strip surface) respectively, T is the temperature, and k is the heat conductivity. Flow is assumed to be steady and parallel to the direction of strip travel. Viscous dissipation of the coating liquid on the strip surface and heat generation are neglected. After application of relevant boundary conditions, the thermal gradient in the direction of strip travel becomes

$$\frac{dT}{dx} = \frac{-h_i(T - T_\infty) - \varepsilon\sigma\left(T^4 - T^4{}_\infty\right)}{A} \tag{7.25}$$

where

$$A = \left(a\rho_s c_{p_s} U_s + \rho_z c_{p_z} q\right) \tag{7.26}$$

h is the convection heat transfer coefficient, ε is strip emissivity, σ is the Stefan-Boltzmann constant 5×10^{-8} W/m$^{2\circ}$K^4, T_∞ is the ambient temperature, a is the half-thickness of the steel strip, U_s is the strip velocity, the subscripts s and z refer to the respective properties of the steel strip and the zinc coating, and q is the liquid coating volume flow rate. The latter term will be much higher below the air knife; for locations above the air knife, q is given by Eqs. (7.14) and (7.15) above, from the Ellen and Tu analysis. Convection heat transfer coefficients were computed from Nusselt-Prandtl-Reynolds number correlations available in the literature [38] to give values for $Z/d = 12$ at positions along the strip moving by the air knife for different plenum pressures shown in Fig. 7.22.

Knowing the heat transfer coefficient, the strip temperature profile in the region near the air knives can be computed, Fig. 7.23. The temperature drop is computed to be about 40°C, bringing coatings with GA or GI compositions very near to their liquidus temperature; below this, freezing begins. This is higher than the 9°C drop estimated by Ellen and Tu for widely spaced jets [28].

The melting point of pure zinc is 419.5°C; small additions of Al depress the melting point. The Hrymak model assumed that the jet boundary extended to the distance where the coating is cooled to its melting temperature, shown in Fig. 7.23 for three plenum pressures. Beyond that (Region IV in Fig. 7.9), a constant convective heat transfer coefficient of 5.05 W/m$^{2\circ}$C was determined from an analysis similar to this type of configuration [39]. This permitted the height above the air knives where the zinc coating begins to solidify (the "freeze line") to be calculated, Table 7.3.

Fig. 7.24 shows progression of solidification of the zinc coating from this initial appearance of solid in the coating film, plotted as fraction solidified versus strip distance above the air knives, for a coating thickness of 20 μm.

The results shown in Fig. 7.24 are for a conventional galvanized (GI) coating. For a Zn-55%Al-1.5% coating, solidification begins when the coating temperature is less than 567°C and then continues until the Al-Zn-Si ternary eutectic point is reached. It is reported that especially for thin steel strip, typically used in construction applications, a minimum strip velocity is required to prevent the solidification of the Al-rich phase in the coating from commencing before reaching the air knives [40].

All of the above models modeled wiping actions in the absence of a liquid film on the steel strip. Further refinements to include the presence of the liquid film on pressure and shear stress distributions were made by Gosset and Pfeiler [1,41]. Both used large eddy simulation-volume of fluid (LES-VOF) computational techniques. Considering the more recent of

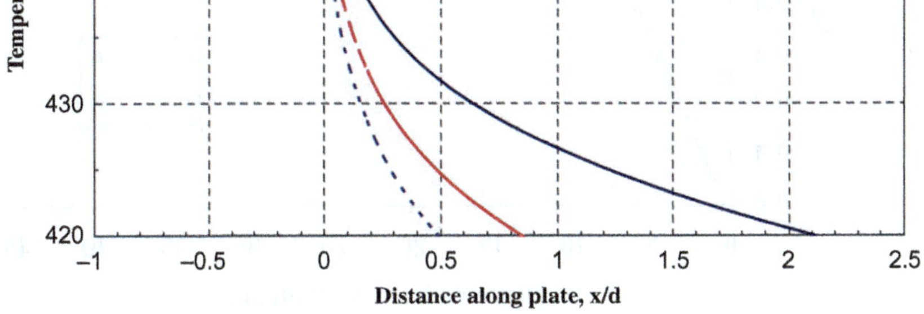

FIG. 7.23 Temperature distribution along the moving steel strip passing by the air knives ($Z/d = 12$, strip velocity = 2 m/s, strip half-thickness = 0.0175 mm, ε = 0.23 [31]. *(Reproduced with permission from Galvanized Autobody Partnership Project ZCO-2-3, Improving Coating Weight Consistency, Final Report, July 2004, International Zinc Assn., Durham, NC.)*

FIG. 7.22 Convection heat transfer coefficients along the steel strip moving past the air knife that is located at $x/d=0$ [31]. *(Reproduced with permission from Galvanized Autobody Partnership Project ZCO-2-3, Improving Coating Weight Consistency, Final Report, July 2004, International Zinc Assn., Durham, NC.)*

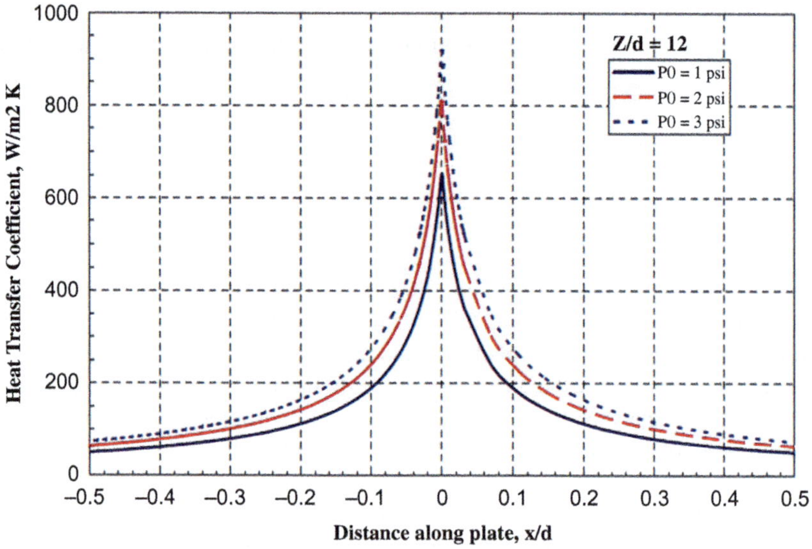

TABLE 7.3 Strip height above the air knives where solidification starts ("freeze line") ($Z/d=12$, strip speed = 2 m/s, strip half-thickness = 0.175 mm, and $\varepsilon=0.23$) [31].

Plenum pressure (psig, [mbarg])[a]	Coating weight (g/m^2)	Stagnation value of h (W/m^2 K)	Height of freezing line (m)
1 [68.9]	213	660	2.7
2 [137.9]	134	820	1.4
3 [206.8]	100	930	1

[a] *psig = pounds per square inch gauge—add ambient pressure to calculate absolute pressure, psia. mbarg = millibars gauge—add ambient pressure to calculate absolute millibars.*

FIG. 7.24 Solid fraction in the coating as a function of distance from the air knife. Plenum pressure = 207 mbar, strip-to-knife distance = 18.24 mm, nozzle gap = 1.52 mm, ($Z/d=12$), strip speed = 2 m/s, coating thickness = 20 μm (130 g/m^2), $\varepsilon=0.23$ Curve A: interface heat transfer coefficient of 5.05 W/m^{2}°C imposed at coating liquidus temperature, Curve B: interface heat transfer coefficient of 5.05 W/m^{2}°C imposed at height of 0.5 m over the stagnation point [35]. *(Reproduced with permission from Galvanized Autobody Partnership Project ZCO-2-3, Improving Coating Weight Consistency, Final Report, July 2004, International Zinc Assn., Durham, NC.)*

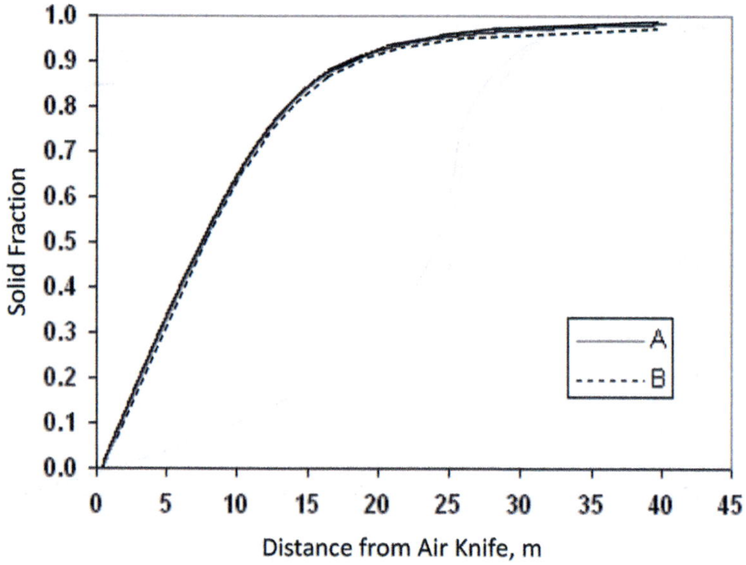

these studies, significant eddies were found to be formed between the gas jet and the surrounding air, moving from the edge of the wiping nozzle toward the moving strip. As they meet the strip, they are shed and move up or down the strip or can be caught in stagnant eddies near the stagnation point. This vortex shedding results in an undulating jet, moving erratically with time. Jet instability effects on coating thickness stability will be discussed in later sections. The shape of the gas flow underneath the wiping knife is greatly affected by the presence of the zinc coating; it is deflected outward from the strip and forms a large eddy below the wiping knife. The direction of downward gas flow is much more in the downward direction rather than being deflected back toward the wiping knife. The presence of the zinc coating has a significant effect on peak gas pressure, pressure gradient, and shear stress. Figs. 7.25–7.27 show these variables in both the presence and absence of the zinc coating. The peak pressure, Fig. 7.25, is slightly higher when the zinc coating is present and is shifted slightly upstream (downward) on the strip. At $x = -0.006$, the air flow is deflected from the zinc, resulting in a lower pressure at this position along the strip as it ascends to the stagnation point. In Fig. 7.26., the pressure gradient can be seen to be significantly reduced when the zinc coating is present, and like the peak pressure, it is shifted to the left. On the right side of the peak, above the wiping knives, the pressure gradients are more similar. In Fig. 7.27, the maximum mean shear

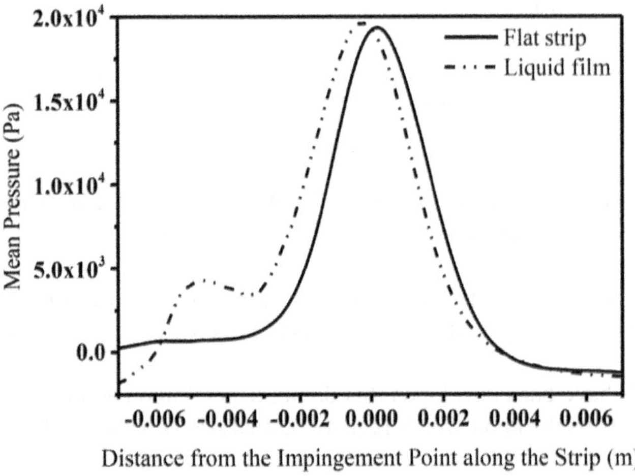

FIG. 7.25 Maximum pressure along the moving strip passing beneath the air knife at $x = 0$, showing the effect of the presence of a zinc coating [1]. *(Reproduced with permission from C. Pfeiler, M. Mataln, A. Kharicha, C.K. Riener, G. Angeli, Importance of the zinc film modeling for gas jet wiping simulations, in: Proc. 10th International Conference on Zinc and Zinc Alloy Coated Steel Sheet (Galvatech 2015), May 31–June 4, 2015, Toronto, Association for Iron & Steel Technology, Warrendale, PA, 2015, pp. 682–689.)*

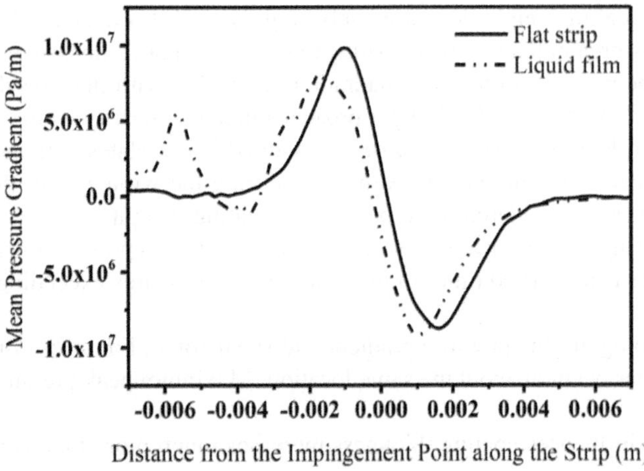

FIG. 7.26 Mean pressure gradient along the moving strip passing beneath the air knife at $x = 0$, showing the effect of the presence of a zinc coating [1]. *(Reproduced with permission from C. Pfeiler, M. Mataln, A. Kharicha, C.K. Riener, G. Angeli, Importance of the zinc film modeling for gas jet wiping simulations, in: Proc. 10th International Conference on Zinc and Zinc Alloy Coated Steel Sheet (Galvatech 2015), May 31–June 4, 2015, Toronto, Association for Iron & Steel Technology, Warrendale, PA, 2015, pp. 682–689.)*

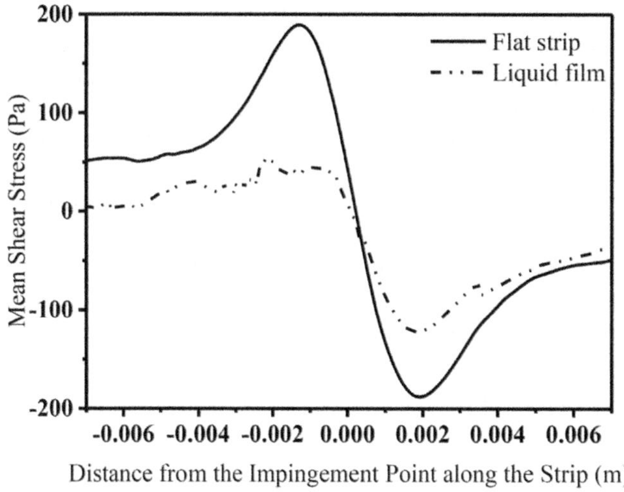

FIG. 7.27 Mean shear stress along the moving strip passing beneath the air knife at $x = 0$, showing the effect of the presence of a zinc coating [1]. *(Reproduced with permission from C. Pfeiler, M. Mataln, A. Kharicha, C.K. Riener, G. Angeli, Importance of the zinc film modeling for gas jet wiping simulations, in: Proc. 10th International Conference on Zinc and Zinc Alloy Coated Steel Sheet (Galvatech 2015), May 31–June 4, 2015, Toronto, Association for Iron & Steel Technology, Warrendale, PA, 2015, pp. 682–689.)*

stress along the zinc coating is much lower in value than in the absence of zinc. This is because the liquid film surface is accelerated by the gas jets, causing the shear stress to be lower compared with a strip surface where there is no zinc coating to be accelerated.

The effect of the presence of the coating on pressure and shear stress variables helps explain the some of the inaccuracies in prediction of measured coating weight by models that do not take it into account, for example, those shown in Figs. 7.12 and 7.21.

Other coating weight models have been developed that do not depend upon assumption of underlying functional relationships, or random sampling of results upon which regression analysis is performed [42,43].

7.4.1 Summary of coating control relationships

The mathematical relationships between process variables and coating thickness are complex and customarily solved by machine-based computational methods. However, several "rules of thumb" can be deduced from these relationships:

- For *Z*/*d* values greater than 7, there is an approximately linear relationship between increasing *Z*/*d* and increasing coating thickness (Fig. 7.19).
- For *Z*/*d* values less than 7, coating weight decreases only slightly as *Z*/*d* is decreased (Fig. 7.19).
- At high line speeds, typically greater than 1.5 m/s, coating thickness increases linearly with line speed, while at low line speeds, coating thickness increases in an inverse quadratic relationship with line speed. The transition between these relationships depends on *Z*/*d*, with high *Z*/*d* giving a broader range of linearity. (Fig. 7.20)
- The expression $\{\text{line speed}/(\text{plenum pressure} * \text{nozzle gap width})\}^{0.7}$ correlates with increasing coating thickness. For GA coatings, where some FeZn alloying begins before and during the wiping step, the total coating weight will be higher because the alloy layer is considered to be part of the coating weight.
- Coating thickness varies in a highly nonlinear way with plenum pressure. At pressures below 275 mbar, the sensitivity of coating thickness to pressure can be 1000 times higher than that at pressures exceeding 450 mbar, depending upon *Z*/*d* and line speed values.
- The main influences on wiping are the pressure gradient and shear force. For strip-to-knife distances less than about 15 mm, their maximum values occur at about the same location. Maximum peak pressure is not a primary influence on coating thickness.
- Coating viscosity increases influences coating thickness increases much more than coating density decreases.

7.5 Gas knife designs affecting coating control

In their 1976 paper, Thornton and Graff noted the importance of providing a uniform gas pressure to the nozzle across the full width of the strip. They describe the use of a multiplenum gas feed system, that is, one plenum for each side of the strip, to produce this uniform flow [22]. After exiting the plenum, the gas flows into the nozzle interior, is diffused to break up turbulence, and then exits to form the gas jet. The higher the efficiency of this system, the more wiping power will be available to achieve required coatings with minimum energy use. For one facility it was noted that pressure losses through a conventional single-slot knife system are 48 mbar [44]. Nozzle gaps typically vary from 1.5 to 3 mm, it was noted that excessive nozzle clogging by zinc droplets occurred for nozzle gaps below 1.27 mm [22]; however, many lines now use automatic nozzle gap cleaners [45]. Small nozzle gaps also have more difficulties producing uniform coatings at strip edges, where remedies described below are used.

Both interior and exterior nozzle shapes influence the nature of the gas jet regulating coating thickness. Ideally, the gas jet would be a stable, narrow core extending uniformly from the nozzle tip to the strip surface. This can be achieved by keeping the strip-to-knife distance shorter than the length of the jet core, but at such a short distance problems of zinc splashing, nozzle clogging and the risk of contact between the moving strip and the nozzle end become important [45]. At longer distances, jet instabilities observed in the above modeling work [1] become important, i.e., the jet is turbulent in nature and sheds flow vortices as it progresses from the nozzle tip. Minimizing this cause of coating variabilities in jet behavior by optimizing nozzle shape has been addressed by several investigators.

All knife designs have in common an expansion volume that receives air flow from the plenum that can include a diffuser partition. In this volume turbulent structures are broken up and flow equalized to deliver to the channel delivering the wiping gas to the nozzle exit. The design of the plenum itself has been analyzed, showing that if the pressurized wiping gas is delivered to the plenum from one side, both the plenum cross section and the exit channel to the diffuser section should be tapered to deliver uniform flow to the diffuser and the plenum inlet area should be more than twice that of its exit area, Fig. 7.28.

In other designs, the plenum is fed with pressurized wiping gas from both ends, and the interior geometry designed to minimize turbulence and other flow instabilities [47]. Several investigators have examined the effect of constant area, convergent and divergent geometries for the channel leading to the nozzle exit. Takeda showed the advantages of a convergent channel in creating more stable jet flow patterns compared with constant area channels and shows that the convergent channel helped reduce coating waviness related to this just instability [45]. In some cases, a linear reduction in area was used in the convergent channel [48], in another it was reduced following a reciprocal function [46], while in another the channel was reduced in area following an elliptical function [49]. Convergent flow channels always produced the highest peak pressures, as shown in Fig. 7.29.

Fig. 7.30 shows a drawing of the interior of a wiping knife used on industrial lines, including a plenum fed from both ends, feeding of the expansion volume through holes whose total area is much less than the total plenum inlet area, the expansion chamber to equalize flow and reduce turbulent structures, a diffuser to a second chamber, a second diffuser to the nozzle chamber, and the convergent area nozzle exit channel. This optimized design produces a higher, sharper pressure peak, and therefore a higher pressure gradient and shear stress, Fig. 7.31.

The effect of nozzle lip misalignment and shape was examined by Dubois [48]. The effect of lip misalignment, Fig. 7.32, was much less than the effect of flow channel convergence or divergence, but the shape and direction of the jet were strongly influenced by the sharpness of the lip at the nozzle exit. Gas flow deflected to the rounded lip side for both straight and convergent channels. The attachment of a cylinder to the lower lip, along the width of the knife, had a similar effect [50].

Knife design also influenced the uniformity of in the spanwise direction, across the width of the jet. Alibegi [49] showed the benefit of a convergent nozzle channel with an elliptical interior profile compared with a constant area exit channel for reducing peak pressure variability across the nozzle width for a laboratory-scale galvanizing process simulator, Fig. 7.33. The pressure variation across an industrial nozzle is shown in Fig. 7.34. Because of the finer scale of the vertical axis, the

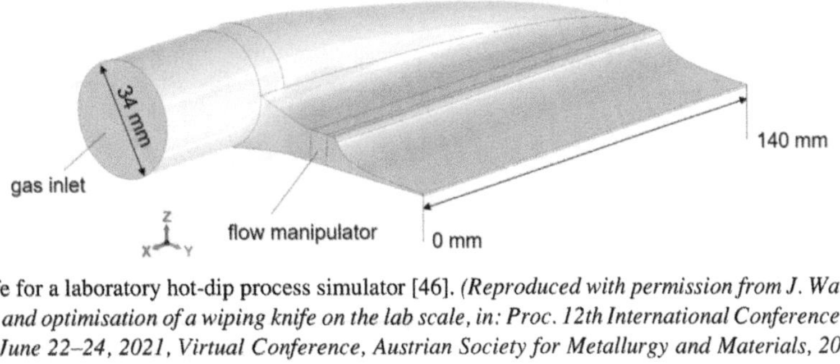

FIG. 7.28 Design of gas knife for a laboratory hot-dip process simulator [46]. *(Reproduced with permission from J. Wang, T. Klinkberg, M. Teschner, D. Weirich, T. Koll, Simulation and optimisation of a wiping knife on the lab scale, in: Proc. 12th International Conference on Zinc and Zinc Alloy Coated Steel Sheet (Galvatech 2021), June 22–24, 2021, Virtual Conference, Austrian Society for Metallurgy and Materials, 2021, pp. 1180–1191.)*

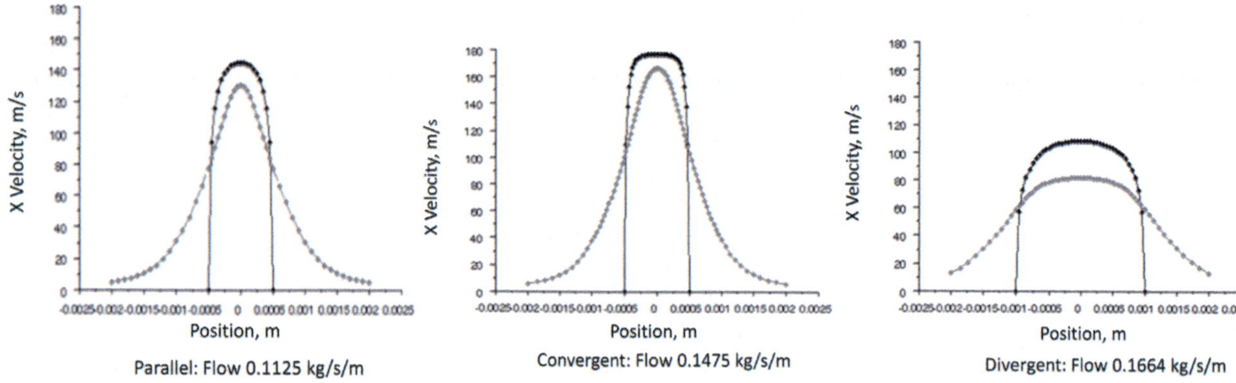

FIG. 7.29 Gas flow profile at nozzle channel exit (upper lines) and 8 mm from the gas knife (2 mm before impinging on moving strip) (lower lines) for: (left) a constant area channel 1 mm wide over the 20 mm channel length, (center) a channel converging from 2 mm to 1 mm over the channel length, (right) a channel diverging from 1 mm to 2 mm over the channel length [48]. *(Reproduced with permission from M. Dubois, F.E. Goodwin, Impact of lip adjustment on wiping efficiency, Iron Steel Technol. (2021) 50–57.)*

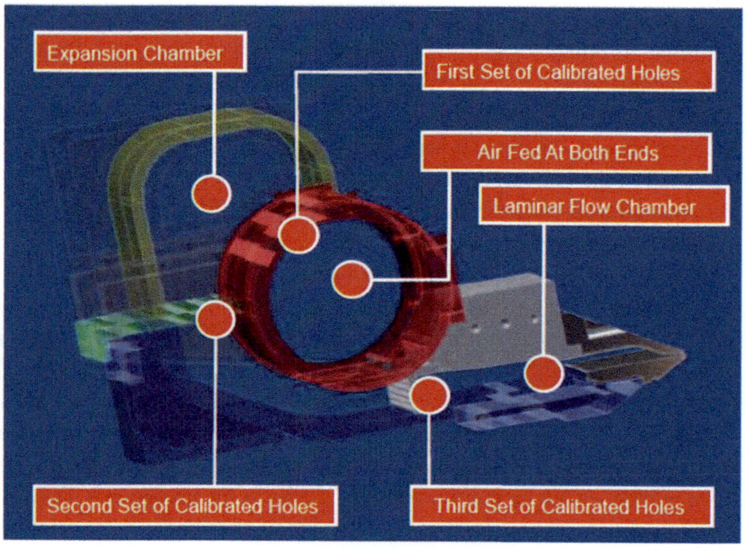

FIG. 7.30 Section view of the Danieli Kohler X-JET gas knife [47]. *(Reproduced with permission from G. Kohler, G. Caporal, Ancillary benefits of the high performance of Danieli X-JET, in: Proc. 10th International Conference on Zinc and Zinc Alloy Coated Steel Sheet (Galvatech 2015), May 31–June 4, 2015, Toronto, Association for Iron & Steel Technology, Warrendale, PA, 2015, pp. 714–718.)*

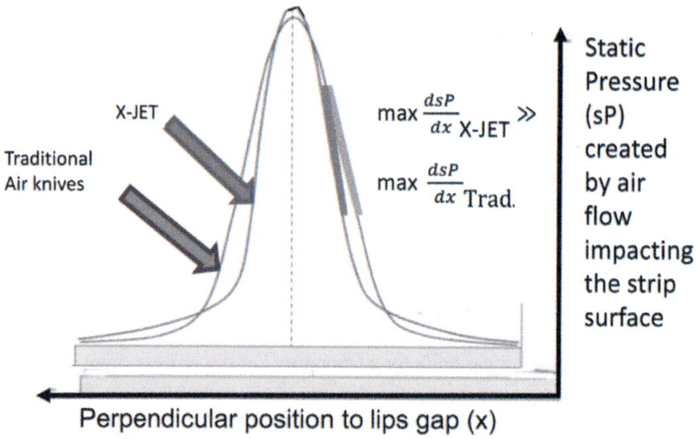

FIG. 7.31 Comparison of pressure peak and pressure gradient for optimized air knife design versus traditional design [47]. *(Reproduced with permission from G. Kohler, G. Caporal, Ancillary benefits of the high performance of Danieli X-JET, in: Proc. 10th International Conference on Zinc and Zinc Alloy Coated Steel Sheet (Galvatech 2015), May 31–June 4, 2015, Toronto, Association for Iron & Steel Technology, Warrendale, PA, 2015, pp. 714–718.)*

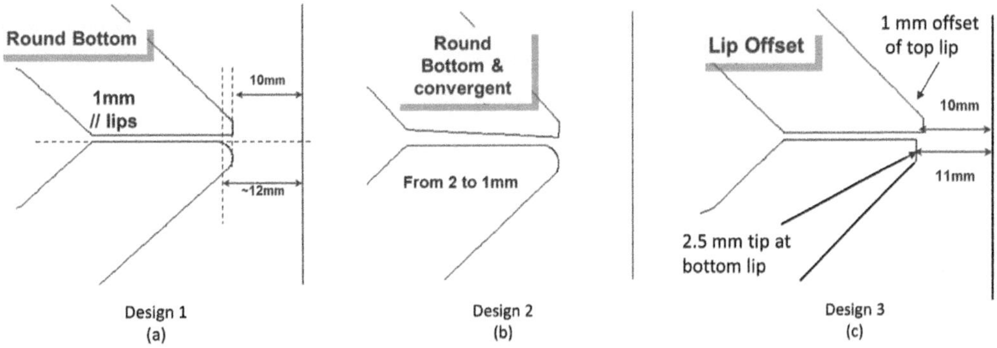

FIG. 7.32 Nozzle shape and misalignment geometries examined by Dubois [48]. *(Reproduced with permission from M. Dubois, F.E. Goodwin, Impact of lip adjustment on wiping efficiency, Iron Steel Technol. (2021) 50–57.)*

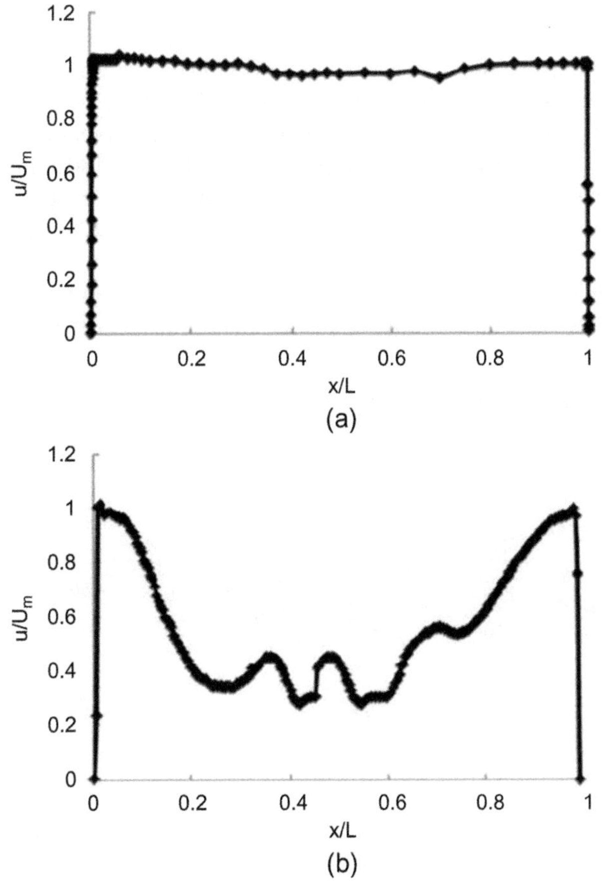

FIG. 7.33 Spanwise (across the half the air knife width, from center to edge) velocity at the jet exit for optimized laboratory process simulator air knife: (A) optimized design with elliptical shape nozzle exit channel (B) originally supplied air knife [49]. *(Reproduced with permission from S. Alibeigi, J.R. McDermid, J. Thomson, Characterization of a new air knife design for the McMaster University galvanizing simulator, in: Proc. 10th International Conference on Zinc and Zinc Alloy Coated Steel Sheet (Galvatech 2015), May 31–June 4, 2015, Toronto, Association for Iron & Steel Technology, Warrendale, PA, 2015, pp. 697–701.)*

variations in Fig. 7.34 appear higher than the velocity measurements (Fig. 7.33), and care must be taken in making comparisons. (Pressure and velocity are related, for example, by Ellen and Tu's solution to the Navier-Stokes equation, Eq. 7.10.) The pressure uniformity shown in Fig. 7.34 is better than ±0.5% across the shown strip width. It should be noted that similar pressures for laboratory equipment such as galvanizing process simulators under normal operating rarely exceed 80 mbar, while the values shown in Fig. 7.34 are in the range of 595 mbar.

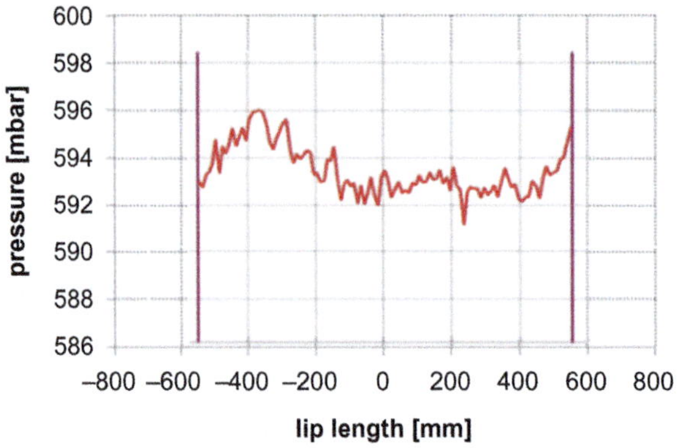

FIG. 7.34 Spanwise (full strip width) pressure variation for X-JET design shown in Fig. 7.30. Note the vertical axis scaling [48]. *(Reproduced with permission from G. Kohler, G. Caporal, Ancillary benefits of the high performance of Danieli X-JET, in: Proc. 10th International Conference on Zinc and Zinc Alloy Coated Steel Sheet (Galvatech 2015), May 31–June 4, 2015, Toronto, Association for Iron & Steel Technology, Warrendale, PA, 2015, pp. 714–718.)*

The exterior shape of the air knife also affects the character of wiping gas flow once it exits the nozzle. A tapered outside gives more room, reducing turbulence and vorticity in this area [46]. Using a well-designed air knife with a sharp-profiled gas jet reduced average coating weight was reduced by 9% for the same plenum pressure and Z/d, while the average process speed was increased by 15%, i.e., Z80 coating weight at process speeds up to 180 m/min.

A representative operating chart for an optimized industrial knife design, at constant line speed, is shown in Fig. 7.35. The shapes of these curves agree well with the analytical results of Ellen and Tu [28] and Hrymak [35] described earlier in this chapter.

It is useful to have a technique for comparing the performance of different air knives. The simplest equation relating gas flow to the pressure gradient is the isentropic nozzle equation, without the presence of an orifice [51].

$$V = c\sqrt{\left[\left(\frac{P + P_\infty}{P_\infty}\right)^{\frac{\gamma-1}{\gamma}} - 1\right]\frac{2}{\gamma - 1}} \tag{7.27}$$

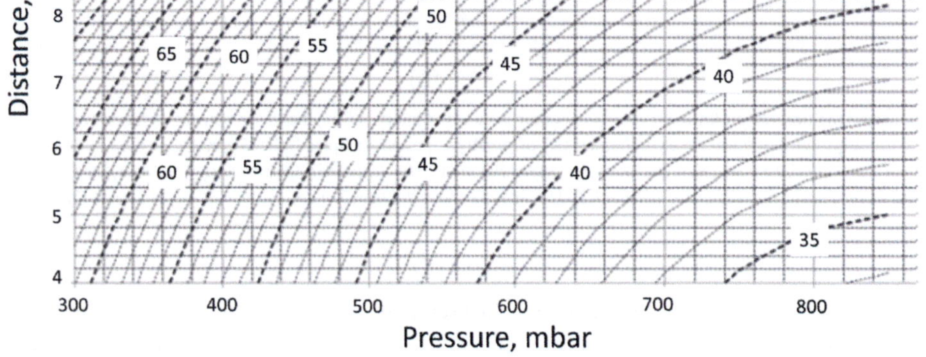

FIG. 7.35 Typical wiping model for air knife design shown in Fig. 7.30, showing predicted coating weight in g/m^2 per side [47]. *(Reproduced with permission from G. Kohler, G. Caporal, Ancillary benefits of the high performance of Danieli X-JET, in: Proc. 10th International Conference on Zinc and Zinc Alloy Coated Steel Sheet (Galvatech 2015), May 31–June 4, 2015, Toronto, Association for Iron & Steel Technology, Warrendale, PA, 2015, pp. 714–718.)*

where c = speed of sound; P = plenum pressure; P_∞ = ambient pressure; γ is the ratio of gas specific heat at constant pressure C_p to the gas specific heat at constant volume C_v, approximately 1.4. This velocity is the maximum velocity that can be achieved, with no losses in the system.

A more detailed description relevant to air knife comparisons was developed by van Eenennaam [52]. The measure of performance being evaluated was the stagnation pressure P_s that is related to jet velocity exit V as

$$P_s = P_a\left(1 + \frac{V^2}{2T_j c_p}\right)^{\frac{\gamma}{\gamma-1}} \tag{7.28}$$

where P_a = ambient pressure, c_p = specific heat of the wiping gas (air) at constant pressure, and T_j = the temperature of the gas jet at the point where it impinges upon the moving strip, found from the equation

$$T_j = \left(\frac{P_{in}^{\,1-\gamma}\, T_{in}^{\,\gamma}}{P_a^{\,1-\gamma}}\right)^{\frac{1}{\gamma}} \tag{7.29}$$

where P_{in} is the inlet (plenum absolute pressure), T_{in} is the inlet temperature in absolute temperature.

A plot of stagnation pressure versus inlet pressure for several nozzle gap widths is shown in Fig. 7.36. The upper limit of the stagnation pressure, at the highest achievable plenum pressure, corresponds to the speed of sound (Mach = 1) for this nozzle design, when the plenum pressure is 675 mbar on the gauge reading (to convert from gauge pressure to absolute pressure, add the ambient absolute pressure, nominally 1000 mbar, to the gauge pressure, i.e., 1675 mbar absolute pressure gives an impingement pressure of 1800 mbar by the accelerated gas jet). The indicated maximum stagnation pressure of 1800 mbar is higher than the indicated normal operating pressure shown in Fig. 7.36. Stagnation pressure is seen to be rather insensitive to nozzle gap width. Also in Fig. 7.36, the impingement pressure for an inlet gauge pressure of 1000 mbar is shown as the "ideal curve," it is only slightly higher than the other curves and indicates the limitations of this particular gas knife design.

The minimum practically achievable coating thickness per side for available air knives, plenum pressures, and strip-to-knife distances has been stated as 5 μm for a nozzle with a 0.8 mm width, 5.6 μm for a 1 mm gap width, and 7.1 μm for a 1.2 mm gap width, all for a line speed of 180 m/min. Because its lower density and viscosity, a Zn-10%Al coating would have a minimum coating thickness approximately 1 μm less for the same wiping process variables [53].

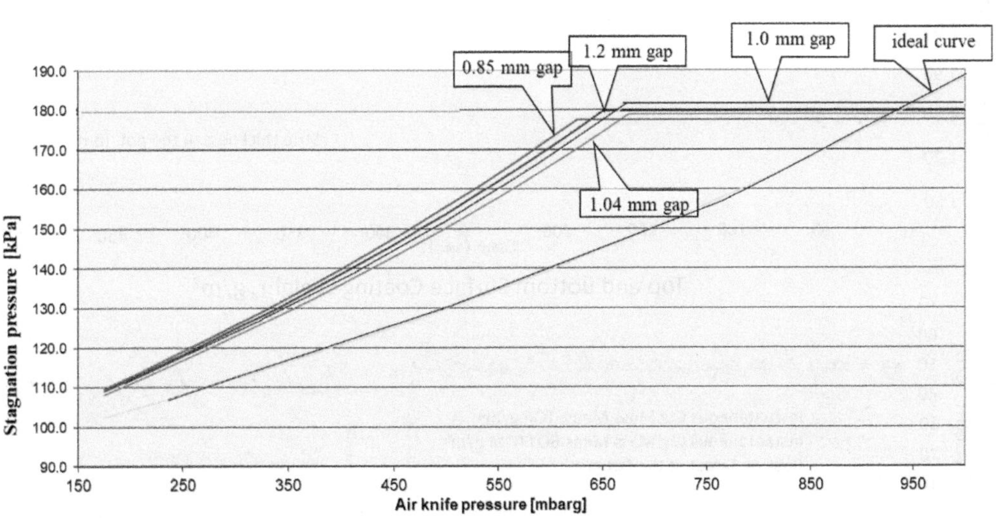

FIG. 7.36 Air knife performance evaluation graph. Stagnation pressure (absolute pressure) capable of being produced at a given plenum pressure (shown in gauge pressure) is shown for several nozzle gap widths. Also, the "ideal curve" for a plenum pressure of 1000 mbarg is shown. The impingement pressure limit is dictated by the point where the jet velocity reaches the speed of sound [52]. *(Reproduced with permission from J. van Eenennaam, J. Link, Easy analysis of air knife performance by stationary flow calculations, in: Proc. 12th International Conference on Zinc and Zinc Alloy Coated Steel Sheet (Galvatech 2021), June 22–24, 2021, Virtual Conference, Austrian Society for Metallurgy and Materials, 2021, pp. 713–724.)*

7.6 Factors affecting coating weight variability

7.6.1 Strip position

The position of the strip as it passes through the air knives is measured in terms of its mean distance from the theoretical passline, its skew from a position parallel to the plane strip width along this passline, and its transverse shape, usually termed crossbow. The first two of these are described in this section, while crossbow is covered in the next section.

Static deviations in strip position that are parallel to the theoretical passline position can be caused by improper positioning of the pot rolls, giving a fixed error in offset that can be compensated by moving the air knife position. These deviations are also influenced by strip thickness, yield strength of the strip at the local temperature, and line tension. An example showing the effect of strip thickness on strip position relative to this passline is shown in Fig. 7.37. Here, the strip thickness changes from 1.8 to 2.05 mm (middle chart), while the measured strip position changes from −6 to -4 mm relative to the theoretical passline. With thicker (4–6 mm) strip, the position changes can be even greater, as high as 20–30 mm [55]. The air knife control system responds by repositioning the knives during the period between 160 and 260 s of elapsed time (top chart). The resultant coating weights are shown in the bottom chart.

Static strip skew deviations can be caused by pot rolls or top roll misalignment, improper thickness profiles in the strip that can be related to hot or cold mill processing or nonuniform elongation of strip as it is pulled through the higher temperature sections of the galvanizing line including the degree of strip crowning and steering problems. Twisting about the strip center has been related to strip thickness profiles where the center is thicker than the edges (full center), whereas a cyclic back and forth bowing of the centerline has been related to profiles where the strip edges are thicker than the center (center buckling) [56]. These problems that are not related to vibration can be corrected by the adjusting the position of the air knives as shown in Fig. 7.39. Once these deviations are addressed, there can still be considerable dynamic deviation in

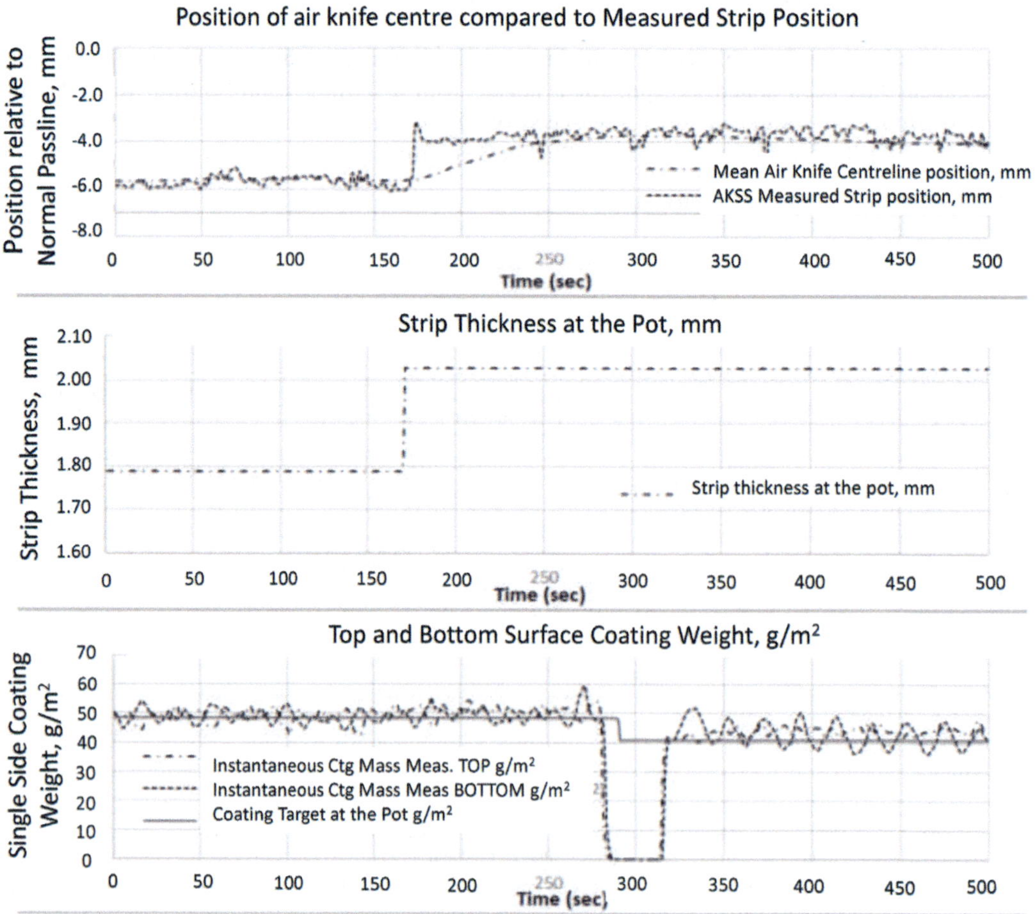

FIG. 7.37 Effect of strip thickness on strip position relative to the theoretical passline and effect on coating weight [54]. *(Reproduced with permission from O. Pearcey, H. Brown, I. Marincic, J. Tham, The evolution of coating control systems as the need for new coated products expands, Berg. Huetenmaenn Monastch. 166 (11) (2021) 525–534 (Springer).)*

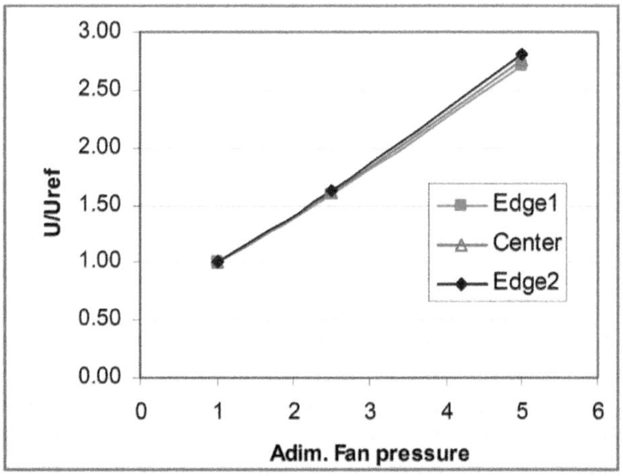

FIG. 7.38 Dependence on relative amplitude of vibration displacement in the wiping zone on fan pressure in the after-pot cooler, measured at the strip edges and center [58]. *(Reproduced with permission from M. Hamide, K. Beaujard, Predictive model of strip vibrations in gas jets cooling sections, in: Proc. 8th International Conference on Zinc and Zinc Alloy Coated Steel Sheet (Galvatech 2011), June 21–25, 2011, Genoa, Italy, Associazione Italiana di Metallurgia, Milan.)*

strip position caused by several sources of vibration, including those induced by worn pot roll bearings or roll surfaces, after-pot cooling equipment, and the natural frequency of the free span of strip supported between the top roll and the stabilizer roll. Strip vibration amplitude in the wiping zone was found to be highly sensitive to after-pot cooling fan pressure, Fig. 7.38. In this case the free span length between the pot and top rolls was 54 m, the strip tension was 3000 kg, and the strip cross section was 1550 × 0.65 mm. Other studies showed that the amplitude of strip vibration increased with blower pressure in a nonlinear manner [58,59]. The increasing demands for strip cooling related to the metallurgy of advanced high strength steels will require further attention to the design of cooling units without a concomitant increase in strip vibration. Example of a current design is shown in [60].

The frequency of strip vibration in the wiping zone was measured to be 1.13 Hz at another producer, this related to aerodynamic instabilities in after-pot cooling equipment. No influence of twisting that would have caused variations in skew or flutter related to variations in crossbow is included in this measurement [61]. This is in the range of 1–10 Hz measured for three galvanizing lines; it was found that the resonant frequency could only be shifted by changing either the line speed or strip tension by more than 25% [62]. A much lower vibration frequency, 0.0222 Hz, was coincident with a shift from the specified passline of 1.56 mm and a slow oscillation with a peak-to-peak amplitude of 2 mm. For the air knife settings being used, this corresponded to a coating weight variation of 0.7–1.4 μm per side. This correlated with strip temperature deviations in several of the pretreatment furnace zones that could have affected the phase makeup of the strip when it reached the wiping zone [63] and, as described above, consequent differences in volume expansion caused by transformation from austenite to ferrite and other lower-temperature phases; varying quantities of initial austenite related to varying furnace soaking temperatures would give variations in volume expansion after cooling and therefore variations in strip shape and also affect strip yield strength.

Strip position near the wiping zone can be monitored by several techniques, including those based on lasers [63] and eddy currents [64]. In the former, this position information has been used to dynamically adjust wiping variables to improve the accuracy of coating thickness control and also adjust the intermesh between the stabilizer and corrector rolls to control crossbow. In the latter, it is incorporated into an electromagnetic strip stabilization unit that can minimize vibration and, for thin strip, help correct crossbow. The electromagnetic strip stabilizer itself cannot fully correct crossbow and is integrated with automatic positioning of the corrector roll intermesh to assist this correction. The use of the integrated package including strip-to-knife distance sensing, closed-loop control of wiping variables, and the electromagnetic strip stabilizer permits a coating weight tolerance band of 1 g/m^2 to be reached in about 150 m of strip [63]. Several other electromagnetic strip stabilization techniques are available. [65–67]. These objectives are also met by using an air floatation system instead of stabilization by magnetic fields; such a system was found to reduce vibration-induced flutter on thinner gauge strips (<0.75 mm) by approximately 20%–35% [54]. Other air floatation systems are also available [68].

Reduction of strip vibration by increasing strip tension has been reported to be minimal. Indeed, the increase loading of pot hardware at high tensions will cause higher wear and can be expected to lead to vibrations caused by bearing wear sooner than during low-tension conditions. Therefore, typical current practice to maintain the stress on the strip through the pot at 20–30 MPa for strip thicknesses less than 1 mm, decreasing to 10–15 MPa for strip thicknesses between 4 and 6 mm thick. It has been reported that the pressure of coating control air jets can also help reduce vibration [56].

The stabilization of strip position by fluid motion has also been the subject of recent pilot line trials. In this technique, the liquid metal entrained with the moving strip as it exits the bath enters a convergent wedge serves to stabilize its position [69].

7.6.2 Strip shape

As indicated in the previous section, deviations in mean strip-to-knife distance and skew are accommodated by positioning air knives to obtain the desired strip-to knife distance. Two kinds of knife designs are used: Those that are always parallel, so that the skew for one side is the same as the other, and those that can be adjusted independently, Fig. 7.39. Crossbow, a deviation in strip shape, can also be accommodated by pot roll positioning and air knife practices.

As described in Chapter 4, Fig. 7.2, when a strip is wrapped around a roll, plastic stretch elongation can be realized on the outer surface that remains after contact with the roll. When the strip is then only in longitudinal tension, the strip assumes a bowed shape. More complex shapes, approximating W or S shapes, can also occur. It was shown that crossbow in typical continuous galvanizing line pot roll configurations can exceed 100mm, depending strongly on strip thickness and yield strength, the conditions of intermesh of the corrector and stabilizer rolls, the diameter of these rolls and their spacing, together with the amount of offset of the corrector roll from the sink roll and the sink roll diameter [70]. A typical pot roll arrangement and the dependence of crossbow values on intermesh are shown in Fig. 7.40.

Combinations of roll diameters and positions that minimize crossbow have been calculated using an elastic approximation, but there are conditions of strip thickness, yield strength, and pot roll configurations for which crossbow cannot be eliminated [54,71]. Several of the strip stabilization systems described later can also help minimize crossbow, especially for thin, low-strength strip but in other cases, it is necessary to use variable gap air knives to accommodate crossbow. The coating weight profile for a crossbowed strip where the centerline strip-to-knife distance was set at 4.6mm is shown in Fig. 7.41. As indicated before, the sensitivity of coating weight to strip-to-knife distance is relatively low at 4.6mm, increasing at higher values for a nozzle gap width of 1.52mm. Therefore, the coating weight increases as the crossbow shape produces higher strip to knife distances about 30cm way from the edges; edge effects discussed later become important for locations nearer to the edge [35].

To accommodate strip shape issues arising during campaigns, variable gap air knives have been developed. In one of these, the upper air knife lip is divided into 10 separately controllable segments along the span of the knife. Ten equally spaced servo motors adjust the nozzle width on each air knife local to an actuator. At the beginning of the campaign, the nozzle gap width is first selected for each product, this then dictates the range of strip-to-knife distance and plenum pressure that will be used. Slot openings at the edges are also adjusted to minimize edge buildup. The gap thicknesses are controlled by a constrained optimization program that first subtracts the coating weight deviations caused by mean and skew, this gives the residual deviation. From this an edge coating deviation is subtracted to give the deviation that is controlled by adjustment of the local nozzle gap. A Smith Predictor time delay compensation algorithm (discussed later) that accounts for both the transport delay from the knives to the scanning gauge measurement located 50–125m further down the line, after the cooling tower [72]. Another equipment producer notes that nozzle gap widths on such a configuration can vary from 0.1 to 2.5mm [73].

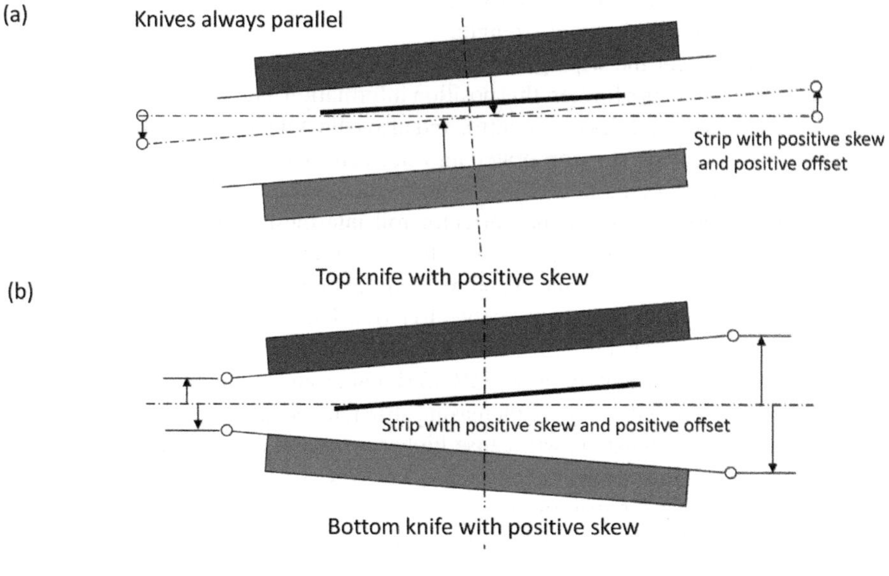

FIG. 7.39 Air knife positioning to accommodate mean strip distance and skew. (A) Fixed-parallel knives and (B) independently positioned knives [57]. *(Reproduced with permission from (A, B) R. Wilhelm, Coating Weight Control & Air Knives, presentation to GalvInfo Galvanizing Line Operators Course, June 24-26, 2007, Indianapolis, IN, International Zinc Assn., Durham, NC.)*

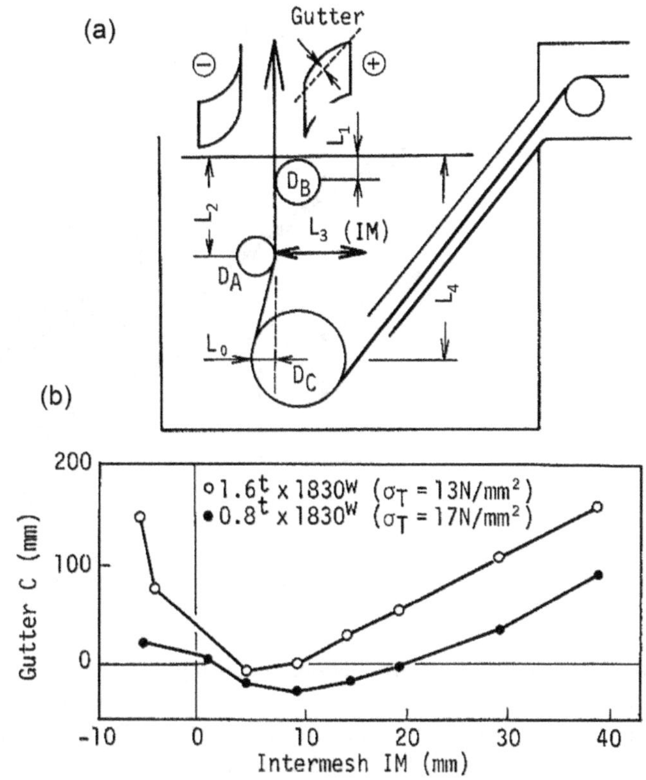

FIG. 7.40 (A) Laboratory pot roll arrangement used to determine strip crossbow values in (B) $\{D_A = 215\,\text{mm}$, $D_B = 270\,\text{mm}$, $D_C = 650\,\text{mm}$, $L_1 = 135\,\text{mm}$, $L_2 = 380\,\text{mm}$, L_3 (intermesh between corrector and stabilizer rolls) $= -10$ to 40, $L_4 = 850\,\text{mm}$, L_0 (offset between corrector and sink rolls) $= 10\,\text{mm}\}$. (B) Crossbow values for two strip thicknesses of 1.6 and 0.8 mm with respective yield strengths of 13 and 17 MPa. Strip width 1830 mm [70]. *(Reproduced with permission from (A, B) T. Masui, Y. Kaseda, K. Ando, Warp control in strip processing plant, ISIJ Int. 31 (3) (1991) 262–267.)*

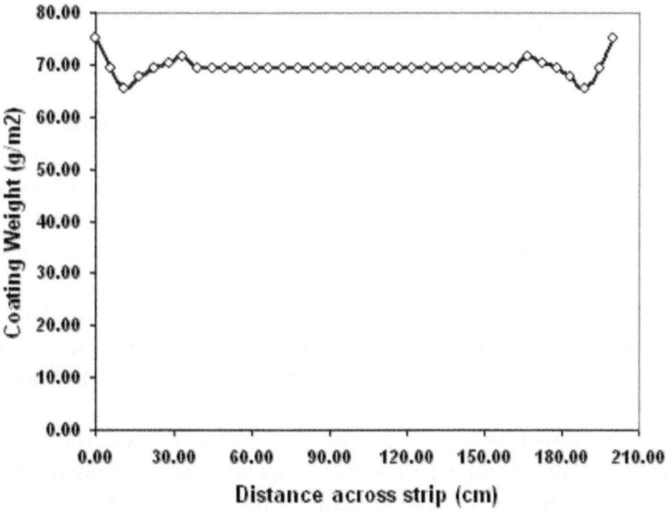

FIG. 7.41 Variation of coating weight across strip due to the cross-bow effect: Plenum pressure = 280 mbar, $Z = 4.6\,\text{mm}$, $U_s = 2\,\text{m/s}$ [35]. *(Reproduced with permission from A.N. Hrymak, E.A. Elsaadawy, G. Hanumanth, J.R. McDermid, F.E. Goodwin, Air knife coating control models, in: AISTech 2005-Proceedings of the Iron & Steel Technology Conference, May 9–12, 2005, Charlotte, NC, Association for Iron & Steel Technology, Warrendale, PA.)*

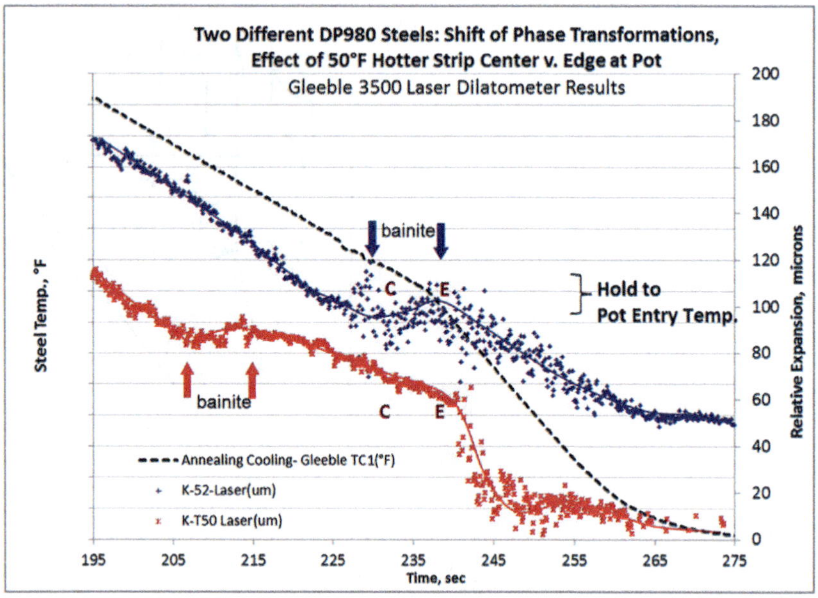

FIG. 7.42 Laser dilatometry and Gleeble process simulation for two different dual phase steels, showing effect of temperature history through the hot-dip coating and wiping zones on volume expansion. Steel K-52 shows significant expansion related to phase transformations through these zones, whereas Steel K-150 completed these transformations at higher temperatures [74]. *(Reproduced with permission from M.M. McCosby, E.A. Silva, Improved properties of galvannealed dual phase 980, in: Proceedings Galvanizers Association 104th Meeting, October 21–24, 2012, Columbus, OH, Galvanizers Association, Marine City, MI.)*

Strip shape issues can also be related to phase transformations occurring as the strip is passing through the wiping zone. This is especially observed in advanced high-strength steels whose M_f (martensite finish) temperature is below the customary zinc pot temperature range of 440–460°C. Control of metallurgical transformations as the strip passes into the galvanizing bath through the wiping zone is complicated by variabilities in strip cross-width temperature that can occur when the strip is carried through furnace zones prior to entry into the pot on rolls that have cooler outboard contact surfaces relative to the middle of the roll. Strip edges 10–16°C cooler than the center can cause phase transformations to begin at positions upstream in the galvanizing line relative to where they begin occur in the center of the strip [61]. Because most of these transformations, i.e., austenite to bainite or martensite produce expansions in the volume of the steel, changes in strip shape occur; this is especially manifest as waviness near the edges of the strip to accommodate the volume expansion. This waviness can be present through the wiping zone, but then disappears after the full width of the strip completes the intended phase transformations. Fig. 7.42 illustrates the responses of two different dual phase steels in the temperature range including the coating and wiping zones. The upper steel shows the dilatometry behavior for steel termed K-52, its transformations occur in the range of these steps, as illustrated by its dilatometry behavior. The transformations for Steel K-T50 occurred safely above the temperatures associated with coating and wiping zones.

7.6.3 Control of coating thickness at the strip edge

Edge overcoating is one of the largest tonnage reject causes for many galvanizing lines [75]. Without taking countermeasures, the zinc coating thickness of each side is 1.4–1.8 times thicker than at the strip centerline. This causes problems in coiling and poor flatness of the steel strip after uncoiling. Even when air knives are not used, a significant edge overcoat is observed. This has been attributed to temperature differences between the strip edge and center that cause changes in viscosity. Also, the thickest portion of the coating is about 10–15 mm from the strip edge, suggesting a surface tension effect. When wiping is used without countermeasures, gas flow in the direction transverse to the strip movement is indicated as a cause of transport of liquid zinc to the edges [76]. Complicating the assessment of edge overcoating is the customary reduction in thickness of the rolled strip, over a distance beginning 20 mm from each edge (edge drop), the strip edge is normally 50 μm (25 μm per side) thinner than the nominal thickness of the strip. Therefore, a strip of uniform total thickness near the edges would require a thicker coating to compensate for the edge drop of the strip.

One way of reducing edge overcoating is to use variable nozzle width air knives. The simplest of these are the "bow tie" designs shown in Fig. 7.43. A constant nozzle thickness is shown together with a symmetric bow tie and an asymmetric bow tie design that is simpler to construct because the bottom knife lip remains flat. These give the stagnation pressure distributions across the strip width shown in Fig. 7.44. For conditions where strip crossbow is a recurring condition, such designs where the bow tie taper can be kept constant can show benefits over a coating campaign period.

The use of tapered nozzles can result in higher gas flow at the strip edge, which can cause release of the liquid coating from the strip edge, a phenomenon known as edge splashing.

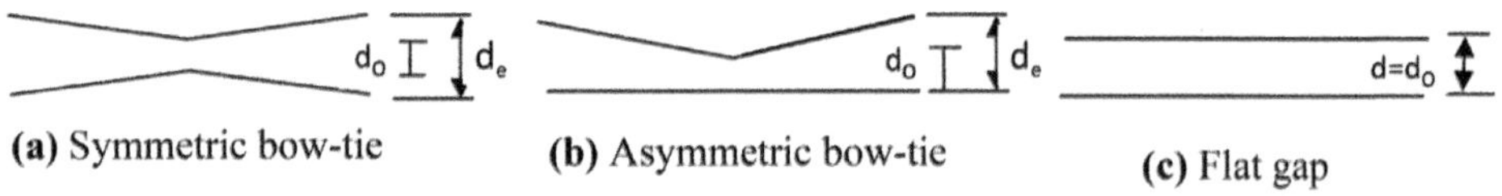

FIG. 7.43 Variable-gas air knife designs for accommodating crossbow. $d_e/d_0 = 1.3$ in (A) and (B), $d_0 = 1.52$ mm [31]. *(Reproduced with permission from Galvanized Autobody Partnership Project ZCO-2-3, Improving Coating Weight Consistency, Final Report, July 2004, International Zinc Assn., Durham, NC.)*

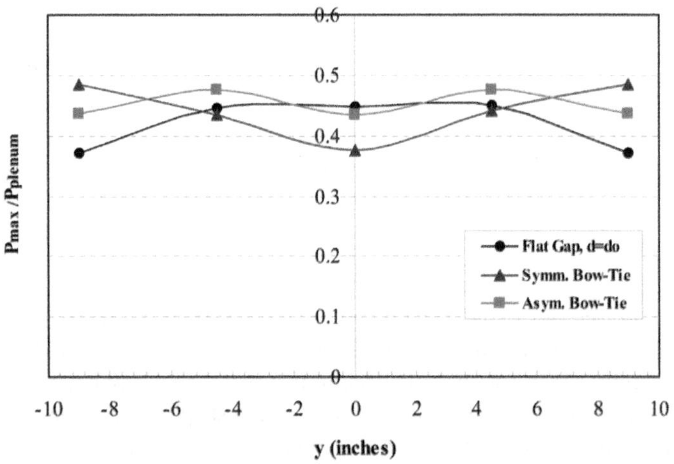

FIG. 7.44 Stagnation pressure profiles for the air knife profiles shown in Fig. 7.43. $P_{plenum} = 138$ mbarg [31]. *(Reproduced with permission from Galvanized Autobody Partnership Project ZCO-2-3, Improving Coating Weight Consistency, Final Report, July 2004, International Zinc Assn., Durham, NC.)*

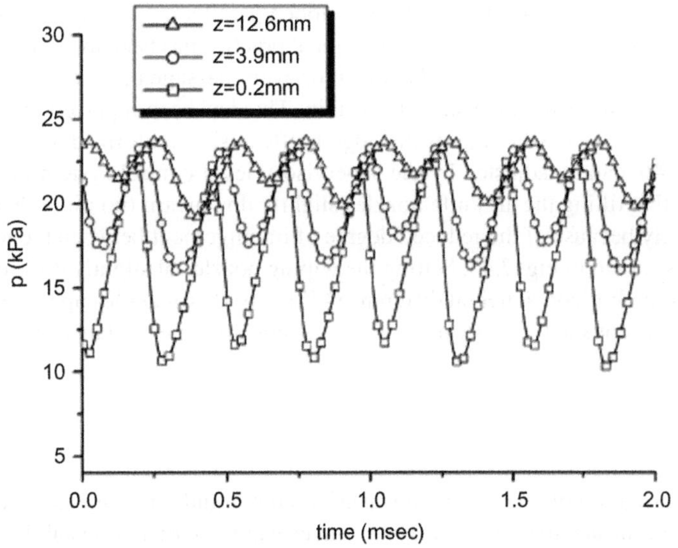

FIG. 7.45 Wiping jet pressure oscillations from two horizontally opposed and aligned jets. The ratio of nozzle-to-nozzle tip distance divided by nozzle gap width is 12. Z is the distance from the strip edge to the point of pressure measurement. N_2 gas nozzle exit velocity Mach 0.2 [77]. *(Reproduced with permission from S.J. Kim, J.W. Cho, K.-J. Ahn, M.K. Chung, Numerical analysis of edge overcoating in continuous hot-dip galvanizing, ISIJ Int. 43 (10) (2003) 1495–1501.)*

Another, and far more important, cause of edge overcoating is related to the impingement of the opposing gas jets of the air knives placed on either side of the strip. When they are separated by the strip, each jet acts on the strip in isolation to provide the pressure gradient and shear force required for wiping, as described above. At the strip edge, they interact. Because of the turbulent nature of these jets, they oscillate ("flap") near the strip edge as they alternate in their shedding of vortices; this gives an oscillating pressure distribution. These oscillations are shown in Fig. 7.45 for several distances z from the edge of the strip [77]. The oscillations are observed to be most severe at location nearest the strip edge. The loss of

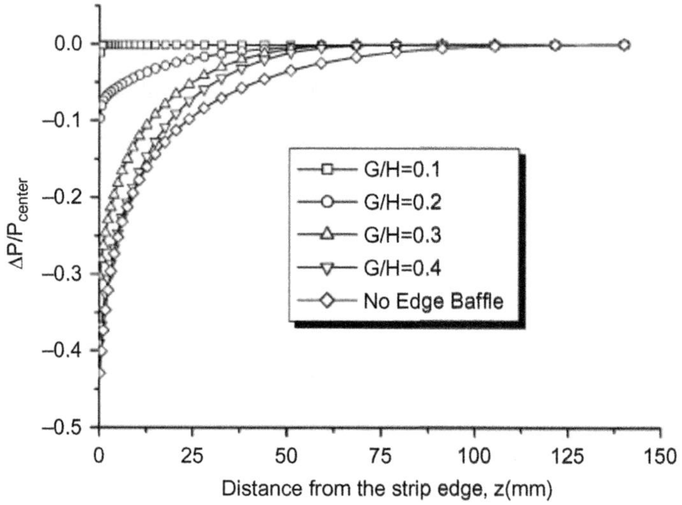

FIG. 7.46 Effect of distance from the edge of the strip outward on local peak pressure normalized by peak pressure for the gas flow impinging directly on the strip, showing dependence upon the distance between strip edge and adjacent parallel plate baffle edge *G* on pressure, expressed as ratio with *H*, the distance between the opposing jet nozzle tips [77]. *(Reproduced with permission from S.J. Kim, J. W. Cho, K.-J. Ahn, M.K. Chung, Numerical analysis of edge overcoating in continuous hot-dip galvanizing, ISIJ Int. 43 (10) (2003) 1495–1501.)*

pressure associated with these oscillations reduces the overall values of pressure gradient and shear force relative to strip locations away from the edge. More recent work has confirmed the findings of this model [78].

The control of edge overcoating is therefore usually accomplished by the use of edge baffles or edge masks, which are stationary plates placed between the air knives beyond the edge of the strip in each side. These plates can be in the same plane as the strip (parallel baffles) or less commonly, perpendicular to the strip edge or U-shaped, with the open ends of the U-shape overlapping the strip edge on each side [76]. For example, a parallel plate edge baffle was placed at distances *G* away from the strip edge, expressed in terms of ratio *G*/*H* where *H* is the distance between the two nozzle exits. Fig 7.46 shows effect of the baffle on local pressure, moving from the strip edge outward. The benefit of placing the baffle edge as close as possible to the strip edge, typically 20 mm or less, to maintain the desired pressure is clear [77,79,80].

It is an operational challenge to keep the baffle plates positioned parallel to the strip, given the changes in strip position and shape through the wiping zone described above. Conventional practice has been to use mechanical touch rolls that track the edge of the strip that must be moved outward and back in for every passage of a weld or change in strip width [81] and can leave tracking marks and possibly bend the strip if it is thin. This has been replaced by a noncontact system utilizing inductive sensors that enable precise positioning of the edge baffles using motorized actuators. The baffle-to-strip edge distance, termed *G* in Fig. 7.46, can be adjusted by the supervisory level control system of the galvanizing line [82].

It has also been reported that tilting the air knife nozzles slightly downward (no more than 5 degrees) can help reduce overcoated edges. This is likely because of the reduced degree of impingement between the jets beyond the strip edge that would reduce the oscillations shown in Fig. 7.45. Setting the wiping nozzles at slightly different heights is also reported as beneficial, for example, a 36 mm horizontal height difference [75]. Nozzle tilt angles up to 30 degrees have been explored; coating thickness is reduced at constant *Z*/*d* because of the greater effective distance between nozzle lip and the strip impingement point [14].

7.6.4 Jet instability

Ellen and Tu described the formation of "wind-driven waves" on the liquid zinc film as it exits the wiping zone that they associated with the jet-induced surface shear effects. These waves were regarded as instabilities in the liquid film behavior arising from the steady behavior of the jet with a steady shear force. The shapes of these waves were further described by So [13], who noted that close to the stagnation point the coating thickness is mainly determined by the pressure gradient, while further away from the wiping zone the shear stress is a more important influence. Variations in both pressure and shear stress caused by instabilities related to the turbulent nature of the wiping jet were determined using large eddy simulation (LES) techniques. The effects of turbulent jet flow on coating thickness, both in the (longitudinal) direction of travel and in the transverse direction, were determined using a perturbation model, but the effect of coating thickness on the subsequent character of the jet was not considered. The shape of the waves produced by the nonsteady-state jet is shown in Fig. 7.47. So concluded that periodically alternating sidewise movement of the peak pressure points along the transverse stagnation line with the upward moving steel strip at a constant speed results in the sag line formation on the zinc coating strip surface as shown in Fig. 7.10. So related the final appearance of surface waves to a relaxation time defined as

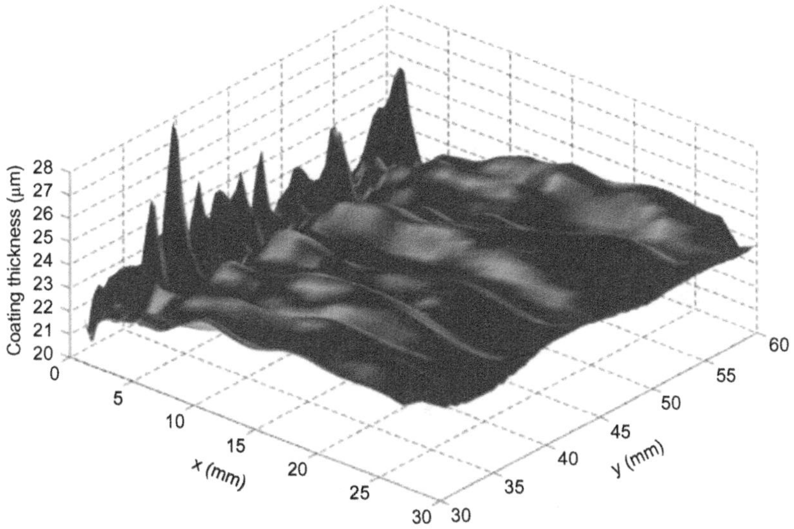

FIG. 7.47 Three-dimensional representation of waves in the liquid zinc coating. y is the transverse direction, and x is the direction of strip travel with $x=0$ at the lower lip of the nozzle. Strip and nitrogen knife width = 100 mm. *(Reproduced with permission from H. So, H.G. Yoon, M.K. Chung, CFD analysis of sag line formation on the zinc-coated steel strip after the gas-jet wiping in the continuous hot-dip galvanizing process, ISIJ Int. 51 (1) (2011) 115–123.)*

$$t_{\text{relax}} = \frac{\eta}{\tau_w} \tag{7.30}$$

where η = dynamic viscosity and τ_w = local wall shear stress. This relaxation time was estimated to be greater than 0.5 ms. Recalling Fig. 7.24, a typical coating is expected to be 10% solid by the time it reaches a height of 2.5 m above the wiping knives, so at a line speed of 2 m/s solidification would become significant after 1.25 s. For situations where the relaxation time is longer than the onset of solidification, waviness defects, sometimes termed "check mark" defects [83], could be expected. This was confirmed by a later multiphase (gaseous jet and liquid zinc) simulation using the VOF-LES technique indicated earlier to describe the waviness of the liquid coating for a distance of 100 mm after leaving the stagnation point because of the variations in pressure gradient and shear stress arising from the unsteady nature of turbulent jet flow. Resultant coating behavior was related to the waviness-related defects [84].

Further insights into the behavior of wiping jets can be gained from studies of made in the absence of the zinc coating. Phan [85] gave experimental results and accompanying mathematical descriptions of time-varying and time-averaged impingement pressure and pressure profiles for wiping jets over a range of plenum pressures between 18 and 84 mbarg for a 1.1 mm nozzle gap and between 8 and 32 mbarg for a 3.3 mm nozzle gap. This covered a Reynolds number range of 4090–16,050. Strip-to-knife distances between 5 and 24 mm were used. It was found that there is little effect of Reynolds number on mean impingement pressure profile over the range investigated in this study, agreeing with past investigations, and that they were all Gaussian profiles. Because of the flat, broad shape of the face of the air knife (absence of a tapered nose), the channel of flow between the nozzle and the strip was narrow for small strip-to-knife distances, making it impossible to obtain valid instantaneous impingement pressure readings for vertical distances above and below the jet greater than about twice the nozzle gap width. The variations in instantaneous pressure were used to determine the standard deviation from mean impingement pressure for various strip-to-knife and nozzle gap width settings. These are shown for a constant Reynolds number of 8000 in Fig. 7.48. The dimension y in the horizontal axis is the distance upward (in the direction of strip movement) from the wiping knife. The lowest standard deviations are all associated with closer strip-to-knife distances, indicating better jet stability. Also, the position of the highest standard deviation values is not at the jet centerline for the closer strip-to-knife distances, it is at $x/d=0.8$ for the 1.1 mm wide nozzle gap with a strip-to-knife distance of 6.6 mm and $x/d=1.8$ for the 3.3 mm wide nozzle gap with a strip-to-knife distance of 6.6 mm.

A prior study carried out by the same investigator [86] using the LES method showed that two shear layer vortices are formed immediately at the jet lip. These shear layers spread out and mix (merge) at the end of the potential core. The length of the potential core was found to be solely dependent on the nozzle gap width. In terms of Fig. 7.48, it was shown that for $Z/d \leq 4.8$, the two shear layer vortices are not mixed before they are redirected when the jet impinges on the strip. When mixing does not occur, the core of the efficiently transmits energy to the moving resulting in very similar mean pressure profiles for all $Z/d \leq 4.8$. This is consistent with past findings including Thornton and Graff, who observed that as the nozzle was moved progressively closer to the strip at these small distances, the pressure remained relatively constant [22]. The two conditions indicated above (1.1 mm wide nozzle gap with a strip-to-knife distance of 6.6 mm and $x/d=1.8$ for the 3.3 mm

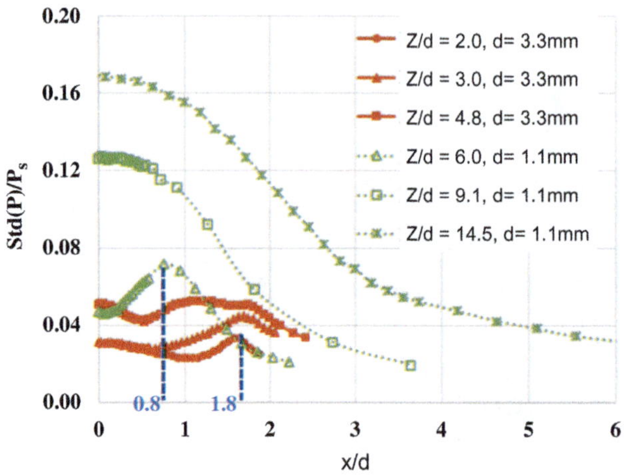

FIG. 7.48 Effect of strip-to-knife distance and nozzle gap width on pressure standard deviations moving up from the air knife centerline, $Re = 6000$ [85]. *(Reproduced with permission from L.Q. Phan, A.D. Johnstone, B. Kosasih, W. Renshaw, Wiping pressure fluctuation and its effect on galvanizing coating smoothness, in: Proc. 12th International Conference on Zinc and Zinc Alloy Coated Steel Sheet (Galvatech 2021), June 22–24, 2021, Virtual Conference, Austrian Society for Metallurgy and Materials, pp. 764–778.)*

wide nozzle gap with a strip-to-knife distance of 6.6 mm) were concluded to be between the conditions where the jet core is stable until it impinges on the strip and where significant vortex shedding occurs. The peaks in standard deviation away from the centerline correspond to the impact locations of vortices just shed from the jet core, whereas at higher strip-to-knife distances the vortices are mixed into the jet flow, coinciding with the observed larger standard deviation values.

The frequencies of the pressure variation were also measured and found to be highly dependent upon strip-to-knife distance. Fig. 7.49 shows measured frequency spectra for two representative conditions as a function of the Strouhal number

$$St = \frac{fZ}{U_j} \tag{7.31}$$

where d is the nozzle gap width, U_j is the mean jet exit velocity, and f is the frequency of the pressure fluctuations. The jets operating at closer strip-to-knife distance had much higher frequencies than those at higher distances.

There were concluded to be two scenarios: the close strip to knife distance, $Z/d \leq 4.8$, where the spectrum is dominated by high frequencies characterized by the jet, and (ii) $Z/d \geq 7.3$, where the spectrum is dominated by a wide range of low frequencies. As described above, jets with Z/d around 6.0 are in a transitional state between the nonmixing and full-mixing of the shear layer vortices.

Coupling a simplified version of these observations, in which the jet vibrations are assumed to be sinusoidal rather than exhibiting the patterns in Fig. 7.49 in a thin coating film model, it was shown that the variation in coating thickness is related to the dimensionless quantity df/U, where U is the coating line speed [87]. The physical meaning of this quantity is a traveling distance of the nozzle gap width d in one cycle of jet pressure fluctuation. The relationship between the variation in coating thickness with respect to the average coating thickness is defined as r, and df/U was determined and is shown in Fig. 7.50. From this it is observed that for a value of $r < 0.01$, then df/U must be greater than 1. Given the jet behavior described above, the $df/U > 1$ condition would be met by low Z/d jets because their unstable impingement pressure behavior is seen at occurring at high frequency. In contrast, wiping conditions where $Z/d \geq 7.3$ are dominated by low frequencies and would therefore be predicted to result in conditions where coating waviness is observed.

The wiping jet pressure fluctuation study was extended by conducting a proper orthogonal decomposition (POD) analysis, which identified the fluctuation modes [88]. Because there was negligible eddy separation from the core of the nonmixing jet for $Z/d \leq 4.8$, the spatial modes that contributed to jet fluctuation complement each other and synchronized to minimize fluctuations. For higher Z/d jets, dominant pressure modes were observed that had similar spatial distributions appear, at the same time that erratic fluctuations occurred in these instantaneous pressures. For example, for $Z/d = 7.3$, the first mode in pressure fluctuation has a flapping nature, with its dominant pressure variation appearing as a trough and peak that cyclically swings from one side of the jet centerline to the other. The time-averaged shear stress follows a similar pattern for this jet mode. The flapping from mode at $Z/d = 7.3$ was found to take up 19% of the total fluctuation energy. For higher Z/d values, i.e., 9.1 and 14.5, two other modes, pulsing and out-of-phase multipulsing, were

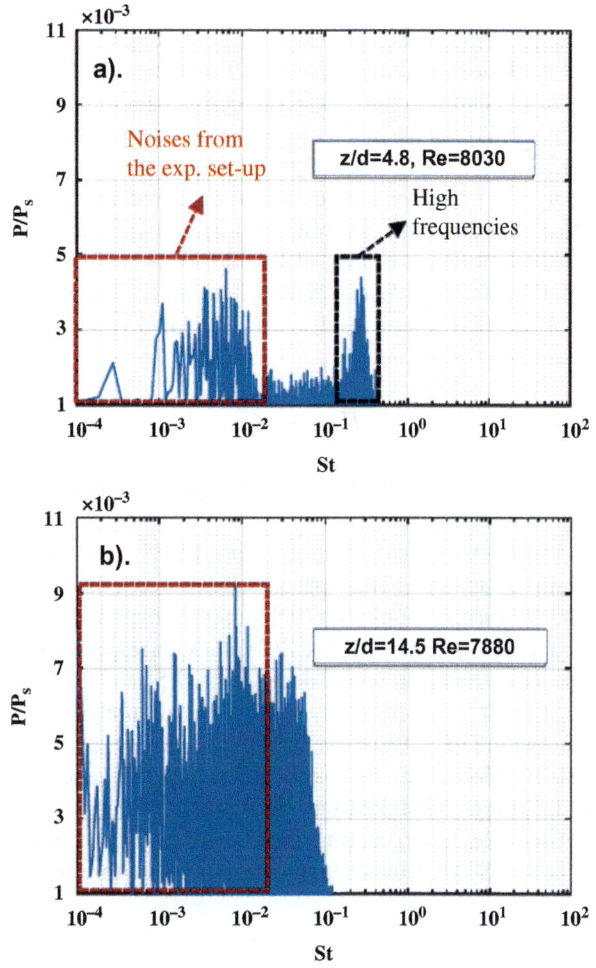

FIG. 7.49 (A) Spectra of the instantaneous impingement pressure measured at the jet centerline for different strip-to-knife distance normalized by nozzle gap width and $Re \sim 8000$. (A) $Z/d = 4.8$, (B) $z/d = 14.5$. *((A, B) Reproduced with permission from L.Q. Phan, A.D. Johnstone, B. Kosasih, W. Renshaw, Wiping pressure fluctuation and its effect on galvanizing coating smoothness, in: Proc. 12th International Conference on Zinc and Zinc Alloy Coated Steel Sheet (Galvatech 2021), June 22–24, 2021, Virtual Conference, Austrian Society for Metallurgy and Materials, 2021, pp. 764–778.)*

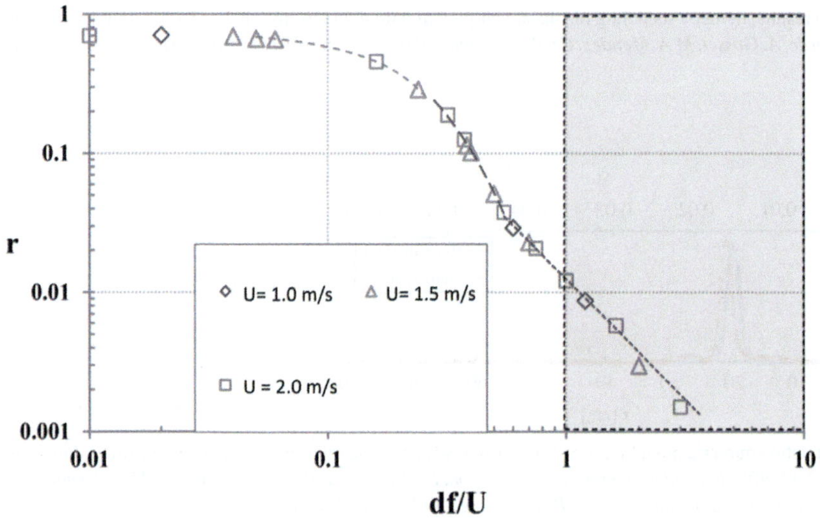

FIG. 7.50 Effect of quantity df/U (nozzle gap width d, pressure variation frequency f, and coating line speed U) on r, the magnitude of the surface disturbance relative to the averaged thickness of the coating [85]. *(Reproduced with permission from L.Q. Phan, A.D. Johnstone, B. Kosasih, W. Renshaw, Wiping pressure fluctuation and its effect on galvanizing coating smoothness, in: Proc. 12th International Conference on Zinc and Zinc Alloy Coated Steel Sheet (Galvatech 2021), June 22–24, 2021, Virtual Conference, Austrian Society for Metallurgy and Materials, 2021, pp. 764–778.)*

observed. The position of the pressure fluctuation during the pulsing mode does not appreciably change, but it alters the height of the Gaussian-shaped pressure curve and therefore also the pressure gradients in the curve; the greatest variation is occurring at the jet centerline. The third mode features a pressure fluctuation profile that alternates between two troughs with a central peak and two peaks with a central trough. This fluctuation pattern influences both the peak and the width of the overall pressure profile. These modes do not always occur in the same order, they can be of equal importance and any one of them can be the most energetic.

Linking wiping gas fluctuations to their effects on the liquid coating is mathematically and computationally challenging and a subject of current investigation. Moreover, the simplifying assumption that the gas jet-liquid coating coupling is a one-way interaction has received recent attention; effects of liquid film behavior on gas jet dynamics have begun to be described. This was related to experimental work using dipropylene glycol as the coating liquid and air as the wiping gas to help confirm numerical analyses [89], these results can be related to cases involving liquid zinc coatings, for example, by the wiping number

$$\Pi_g = \frac{\Delta P_N d}{\rho_l g Z^2} \tag{7.32}$$

where ΔP_N is the stagnation pressure, d is the nozzle slot gap width, ρ_l is the density of the liquid coating, g is the gravitational constant, and Z is the strip-to-knife distance [89].

As in other work described above, mathematical approaches used LES-VOF that was then further interpreted using multiscale POD together with correlation analysis. This approach permitted a description of fluctuation behavior in both the gas jet and the liquid coating and their interactions with each other. Remarkably, the frequencies of gas undulation and liquid film undulation were both found the be near 20 Hz for Z/d values corresponding to gas conditions where significant vortex shedding and mixing with the surrounding air would occur. Figs. 7.51 and 7.52 show the power spectra for the undulation frequencies of the liquid film and air jet, respectively. The Strouhal number on Fig. 7.52 top horizontal axis can be directly compared with that of the air jet in Fig. 7.49, derived by another investigator, they can be seen to coincide. For the results in these two figures, $\Pi_g = 0.16$. Moreover, the coupling between frequencies was observed for coating runback below the gas jet and for the wiped coating as it exited the wiping zone.

This indicates a relationship between the liquid coating undulations, on both sides of the wiping zone, and the gas jet undulations. As in the waves described by So [13], the coating undulations are three-dimensional in character and therefore

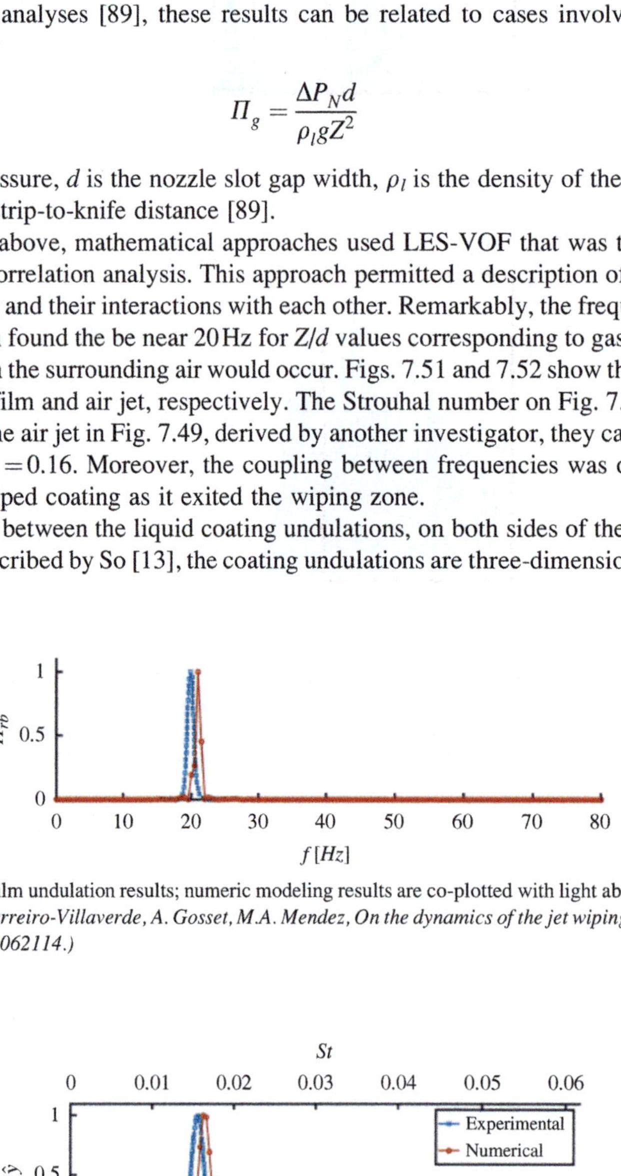

FIG. 7.51 Power spectrum for liquid film undulation results; numeric modeling results are co-plotted with light absorption data from reference [89,90]. *(Reproduced with permission from D. Barreiro-Villaverde, A. Gosset, M.A. Mendez, On the dynamics of the jet wiping process: numerical simulations and modal analysis, Phys. Fluids 33 (2021) 062114.)*

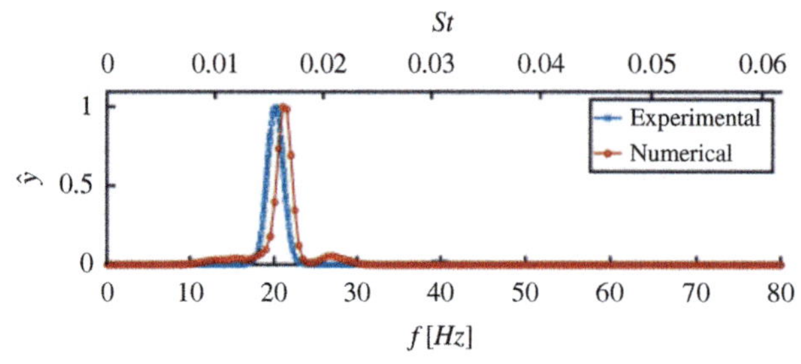

FIG. 7.52 Power spectrum from gas jet pressure undulation numeric modeling results compared with particle interference velocimetry (PIV) data from [89]. Compare Strouhal number correlations with Fig. 7.49 [90]. *(Reproduced with permission from D. Barreiro-Villaverde, A. Gosset, M.A. Mendez, On the dynamics of the jet wiping process: numerical simulations and modal analysis, Phys. Fluids 33 (2021) 062114.)*

complex to mathematically model. The coupling between the wiping gas and coating undulation frequencies was observed to persist as the strip departed the immediate wiping zone, when viscosity and surface tension begin to modulate the coating undulations in the transverse direction [90].

Further investigation into the starting location of the coating waviness revealed that it is always at a plane just below the plane of the air knife centerline. It is remarkable that both the waves in the runback flow and those in wiped coating originate from this location. The origin point of the waves did not change appreciably with time, meaning that the coating undulations are caused by fluctuations in the intensity of the wiping pressure rather than displacement fluctuations that would be anticipated to be a consequence of turbulent gas flow. The sequence of gas jet undulation and formation of runback coating waves is illustrated in Fig. 7.53 for three time steps of 30, 46, and 63 ms. Two counter-rotating vortices being shed from the lower side of the jet are indicated as w1 and w2. The slight deflection of the impinging jet is indicated as "*d*." The sequence of vortex shedding is exactly coordinated with the formation of waves on both the wiped coating and coating runback side. In the top image, at 30 ms, a runback wave is far from the impingement point, and the pressure gradient is fairly symmetric around this point, as shown in the accompanying pressure and shear profiles. The shear stress is relatively high in the runback flow direction because of the higher difference in velocity between the liquid and solid phases, considering the contribution of the velocity of the moving strip.

The action of the vortices results in an upward flow direction of the liquid coating. This upward flow results in growth of a wave, which influences the pressure gradient. Whether the pressure gradient is principally influenced by the simple passage of the wave or by wave-induced deflection of the gas flow has still yet to be understood. Regardless, wave passage past the stagnation point at the 46 ms time step causes a pulsation in the magnitude of the pressure gradient and a shift in the point of maximum shear as seen in the accompanying profiles. The point of highest downward acceleration corresponds to the point of highest shear stress. The rising wave lifts the boundary layer of the gas jet, but the falling waves cannot be accelerated to the local velocity of the gas jet; therefore, there is an interaction between the flow of the coating runback

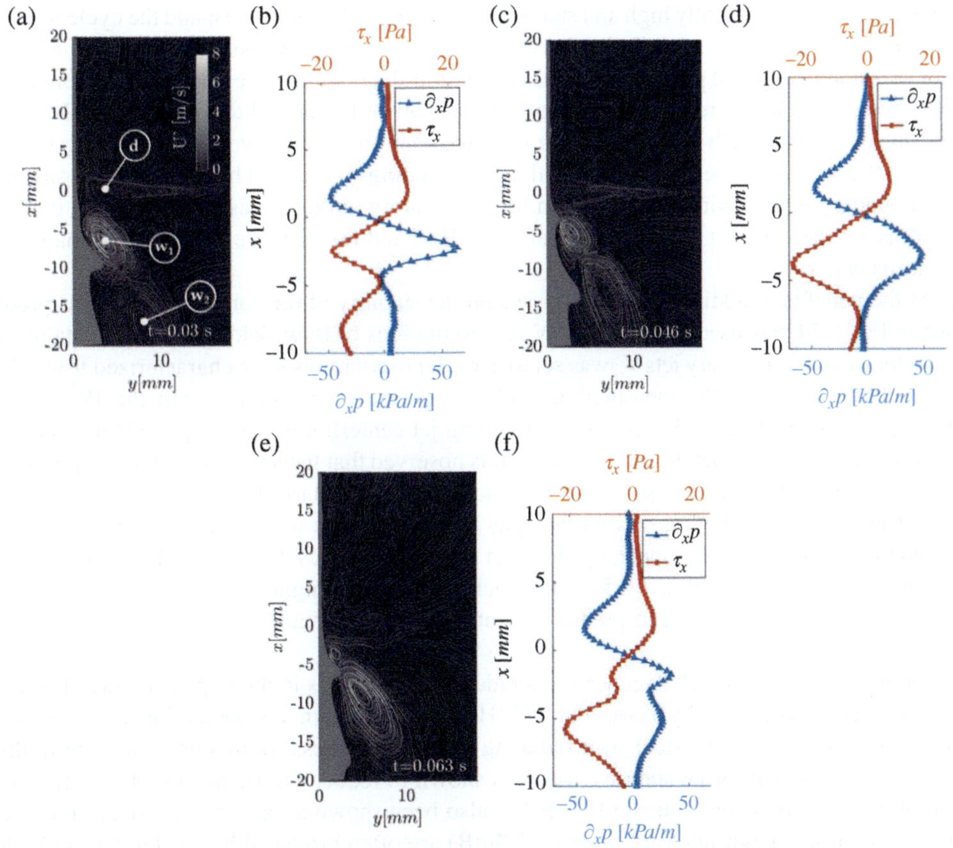

FIG. 7.53 Sequence of coupling of wiping gas flow and coating wave formation for a representative gas wiping configuration, $\Pi_g = 0.16$. Instantaneous pressure gradient and shear force profiles are shown for each time step of 30, 46, and 63 ms [90]. *(Reproduced with permission from D. Barreiro-Villaverde, A. Gosset, M.A. Mendez, On the dynamics of the jet wiping process: numerical simulations and modal analysis, Phys. Fluids 33 (2021) 062114.)*

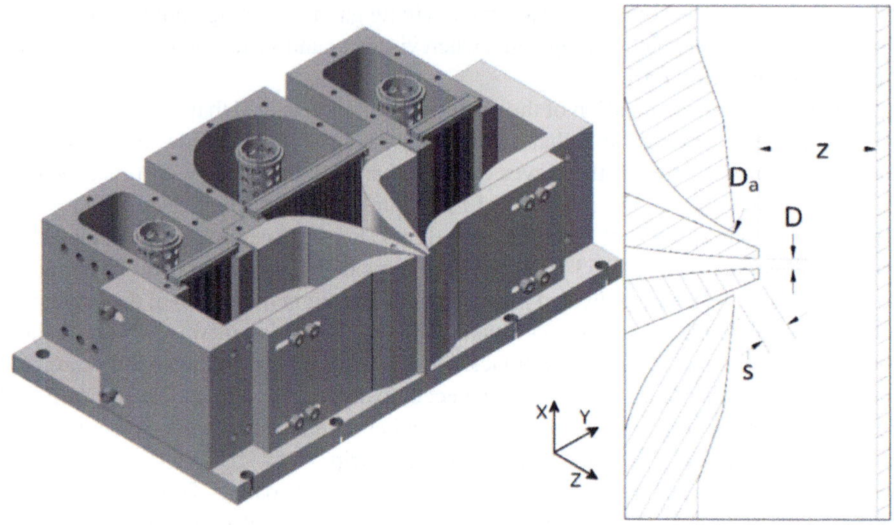

FIG. 7.54 Design of a multislot wiping knife. Each jet is fed from a flow distribution tube feeding a surrounding settling plenum. Gas then passes through stainless steel mesh screen that acts as a flow conditioner into a nozzle chamber that feeds the jet nozzle [91]. *(Reproduced with permission from A.Y. Soufiani, D. Finnerty, J.R. McDermid, S. Ziada, A.N. Hrymak, F.E. Goodwin, A novel multi-slot air-knife design for lighter more consistent coating weights in the gas-jet wiping process, in: Proc. 12th International Conference on Zinc and Zinc Alloy Coated Steel Sheet (Galvatech 2021), June 22–24, 2021, Virtual Conference, Austrian Society for Metallurgy and Materials, 2021, pp. 746–755.)*

surface and the impinging gas jet. In the third time step, at 63 ms, a vortex is observed to be attached to the back side of the falling wave, when the wave is sufficiently high and steep, the vortex detaches at its crest and the cycle repeats. Of particular interest in these time step images is the correlation between the pattern of evolution of the pressure gradient and the variation in film thickness, indicating that pressure rather than position fluctuation is the dominant influence. This agrees well with the work of Pfeiler, who considered the direct application of an earlier model to the hot-dip galvanizing process [1].

Auxiliary jets, one or more on each side of the main jet, have been shown by several investigators to stabilize the main jet. The magnitude of the shear layer between the jet and its surroundings is reduced by introducing a stream of moving gas that surrounds the boundary of the main jet. One design for a laboratory multislot gas knife is shown in Fig. 7.54, showing the nozzle gap widths for the main jet "D" and auxiliary jets "D_s" and the distance "s" from the main knife body to the nozzle tip (setback distance).

Fig. 7.55 shows the benefits of adding the auxiliary jets on the stability of the combined jet that is produced. The multislot knife shown in Fig. 7.54 was used to produce PIV representations of flow fields. To obtain data on single-slot knife performance, the velocity of the auxiliary jets V_a was set to zero. Jet oscillations were characterized using POD of a number of randomly selected PIV images. The measured jet velocities were reconstructed with the POD-derived velocities, removing the turbulent vortices. Fig. 7.55 shows the impinging jet centerline for $0 \leq V_a \leq 100$ m/s and main jet velocity $V_m = 250$ m/s at the same phase in the jet oscillation cycle. It is observed that the amplitude of the impinging jet oscillations of the impinging jet decreases as auxiliary jet velocity is increased, particularly for $V_a/V_m \geq 0.32$.

The reduction of jet oscillations by use of auxiliary jets has been shown to permit the achievement of lower coating weights than possible with single-slot jet, especially at higher line speeds, Fig. 7.56. This is obtained by keeping the relative auxiliary jet velocity low such that $Re_a/Re_m \leq 0.5$. Under such conditions, a higher pressure gradient and higher wall shear stress are obtained in the wiping region. The predicted coating weights for all geometries agreed with the approach of Hrymak et al. [35].

Reduction of jet oscillations should also reduce the incidence of waviness in the wiped coating. The use of a multislot knife for this purpose was demonstrated by Yoon et al. [92] Here, it was shown that the auxiliary jet prevents the formation of alternating impinging vortices on the steel strip reducing coating waviness. Both single-slot and multislot trials used nitrogen wiping. The improvement of jet stability was also shown to reduce the frequency of nozzle clogging [93].

The reduction of jet oscillations by multislot knives has also been shown to be very beneficial in reducing noise originating from single-slot knives. High-intensity tones (~100 dB) are often produced by gas knives, and it has been shown that these tones are caused by the same mechanisms described above that produce jet instability. Fig. 7.57 shows the reduction in acoustic amplitudes that were achieved with multislot jets using three slot widths, over a range of auxiliar jet velocities. The reduction in noise was attributed to the effect of the auxiliary jets on increasing the width of the overall

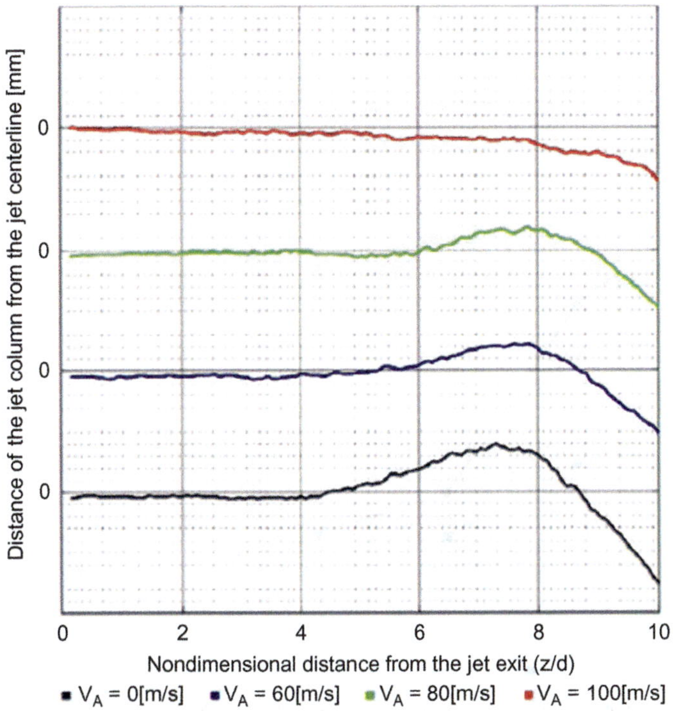

FIG. 7.55 Benefit of auxiliary jets on improving stability of a single jet. All measurements taken at the same phase of oscillation, 270 degrees. Highest oscillation observed with single jet operation ($V_a = 0$), bottom curve, oscillations decrease as V_a increases (ascending order of curves) [91]. *(Reproduced with permission from A.Y. Soufiani, D. Finnerty, J.R. McDermid, S. Ziada, A.N. Hrymak, F. E. Goodwin, A novel multi-slot air-knife design for lighter more consistent coating weights in the gas-jet wiping process, in: Proc. 12th International Conference on Zinc and Zinc Alloy Coated Steel Sheet (Galvatech 2021), June 22–24, 2021, Virtual Conference, Austrian Society for Metallurgy and Materials, 2021, pp. 746–755.)*

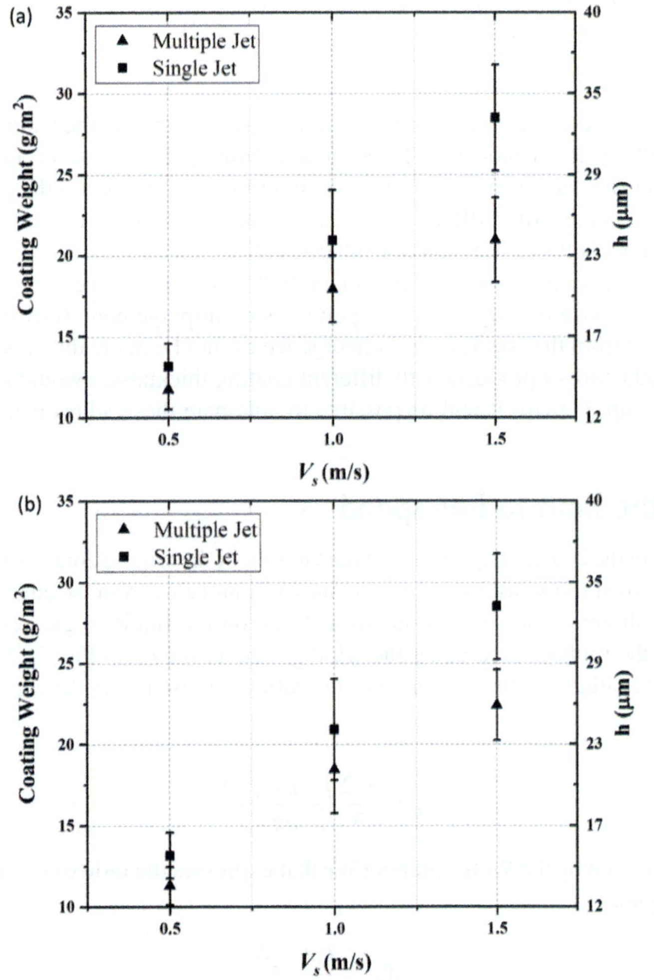

FIG. 7.56 Comparison of experimentally measured coating weights for the multislot jet with $0 \leq V_s \leq 1.5$ m/s, $Re_m = 20{,}000$, $z/D = 15$, $D = 1.5$ mm, $D_a = 1.5$ mm, and $s = 10$ mm for (A) $Re_a = 5000$ and (B) $Re_a = 10{,}000$, with the single slot jet at $Re_m = 20{,}000$, $D = 1.5$ mm, and $Z/d = 15$ [91]. *(Reproduced with permission from A.Y. Soufiani, D. Finnerty, J.R. McDermid, S. Ziada, A.N. Hrymak, F.E. Goodwin, A novel multi-slot air-knife design for lighter more consistent coating weights in the gas-jet wiping process, in: Proc. 12th International Conference on Zinc and Zinc Alloy Coated Steel Sheet (Galvatech 2021), June 22–24, 2021, Virtual Conference, Austrian Society for Metallurgy and Materials, 2021, pp. 746–755.)*

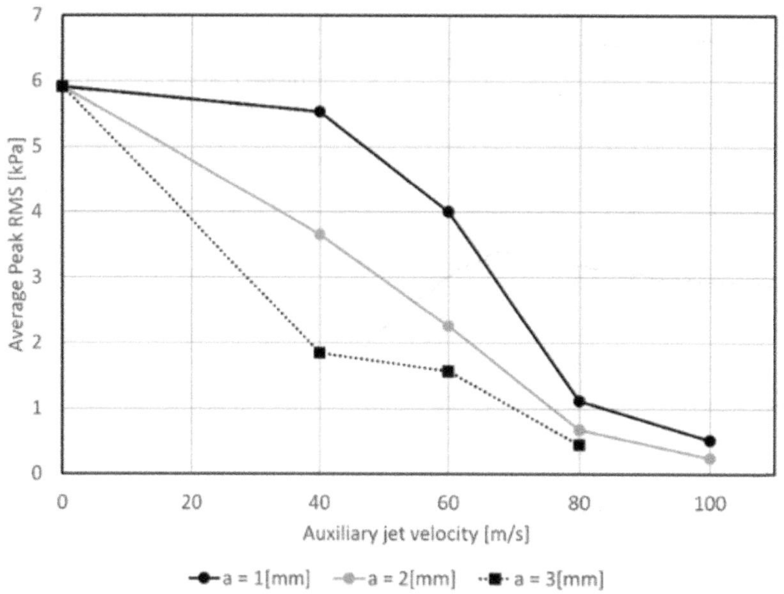

FIG. 7.57 Averaged peak RMS values of the maximum acoustic amplitude produced by the multislot impinging jet for $6 \leq z/D \leq 16$, $1 \leq D_a \leq 3$ mm, and $0 \leq V_a \leq 100$ m/s. $V_m = 250$ m/s, $D = 2$ mm, $s = 5$ mm. Geometric nomenclature from Fig. 7.54 [91]. *(Reproduced with permission from A.Y. Soufiani, D. Finnerty, J.R. McDermid, S. Ziada, A.N. Hrymak, F.E. Goodwin, A novel multi-slot air-knife design for lighter more consistent coating weights in the gas-jet wiping process, in: Proc. 12th International Conference on Zinc and Zinc Alloy Coated Steel Sheet (Galvatech 2021), June 22–24, 2021, Virtual Conference, Austrian Society for Metallurgy and Materials, 2021, pp. 746–755.)*

jet and decreasing the velocity gradient between the surrounding environment and the main jet, consequently reducing the shear layer velocity gradient [91]. It was noted by Thornton and Graff [22] that use of air instead of steam as a wiping medium reduced noise level in zinc bath area by 5 dB, the use of multislot knives gives the potential for a further reduction of the same magnitude. Noise reduction of 10 dB has also been accomplished by use of edge baffles spaced 20 mm from the strip edge with plenum pressures between 150 and 300 mbar [79].

From a practical viewpoint, the improvements seen with multislot air knives dependent upon maintaining convergence of the auxiliary and main jets at a position very near to the point of strip impingement. If strip position cannot be maintained at a relatively constant distance from the nozzles, this convergence cannot be maintained. Also, adjustment of the strip-to-knife distance to permit manufacture of products with different coating thicknesses would require adjustment of auxiliary nozzle geometries for each setup. Perhaps it will be possible to automate these adjustments in the future.

7.7 Splashing and the limit to line speed

The ejection of liquid zinc from the coated strip surface in the vicinity of the wiping zone, especially adjacent to the runback flow below the gas knife, is termed full splashing, while ejection of liquid zinc from the edge of the strip, mentioned before, is termed edge splashing. Both are induced when the shear force on the liquid is greater than the surface tension. An example of full splashing in the runback area under the wiping zone is shown in Fig. 7.58.

The thickness of the liquid coating withdrawn from the zinc bath and presented to the wiping zone was shown previously as Eq. (7.6):

$$t_0 = \frac{\sqrt{2}}{3}\left(\frac{\alpha\eta_0 U_s}{\rho g}\right)^{\frac{1}{2}} \tag{7.6}$$

Splashing behavior is characterized by the Weber number We that expresses the ratio of the dynamic pressure of the jet and the surface tension of the liquid.

$$We = \frac{\rho_g {U^2}_g \delta_0}{\sigma} \tag{7.33}$$

FIG. 7.58 Full splashing in the runback zone. Strip speed 170m/min, strip-to-nozzle distance 10mm, nozzle gap width 0.85mm, final coating thickness 20μm. Splashing in this example is seen only on one side because of strip shape or pressure fluctuations [94]. *(Reproduced with permission from M. Dubois, J.M. Buchlin, A. Gosset, V. Perrot, Effect of nozzle tilting on splashing in jet wiping, in: Proc. 6th International Conference on Zinc and Zinc Alloy Coated Steel Sheet (Galvatech 2004), April 4–7, 2004, Chicago, IL, Association for Iron & Steel Technology, Warrendale, PA, 2004, pp. 197–206.)*

where ρ_g is the density of the wiping gas, U_g is the impingement velocity of gas on the strip, and σ is the surface tension of the liquid. The critical Weber number We_c, above which value splashing initiates, is correlated with the Reynolds number of the initial zinc coating thickness presented to the knives for wiping [95].

$$We_c = 42.4\left(\frac{\delta}{\delta_0} Re\right)^{-0.369} \tag{7.34}$$

where δ is the thickness of the wiped coating, and Re is the Reynolds number in the unwiped coating

$$Re = \frac{\rho_l V \delta_0}{\eta_l} \tag{7.35}$$

where ρ_l is the density of the liquid coating, V is the strip velocity, and η_l is the dynamic viscosity of the liquid coating. These equations use time-averaged values of velocity and thickness and are therefore based on steady flow assumptions.

In practice, splashing in the runback flow area such as that shown in Fig. 7.58 is observed when higher gas velocities are used to obtain thinner coatings, or when increasing strip speed, when jet velocity must be increased to maintain the same coating thickness, consistent with Fig. 7.20. This can be understood by only considering steady gas flow. Splashing from the edges also caused by the inability of liquid shear forces to sustain shear forces in the liquid film, but in this situation it is exacerbated by the undulations in jet movement (flapping) beyond the edge of the strip, giving the pressure patterns shown in Fig. 7.45. Droplets of zinc detach and are splashed at times of peak pressure and resultant shear force during these fluctuations. Higher gas velocities at strip edges are commonly caused by the use of wider nozzle gaps near the edge (bow tie gap geometry). The usefulness of edge baffles in operating under conditions where edge splashing was observed absent the edge baffles was confirmed by experiment [95].

As described by Oke [96], splashing is preceded by the coating waviness behavior described before. At the point where surface tension can no longer keep the waves formed on the runback film intact because of increasing shear forces, splashing by droplet ejection from the tips of runback waves such as those illustrated in Fig. 7.53 commences. As either strip velocity or gas pressure increases, the surface of the runback waves is separated, resulting in ejection of a continuous sheet of liquid metal shown in Fig. 7.58. If strip speed is reduced in order to halt the splashing, it is observed that splashing ceases at a strip speed 7% slower than where it initiated, indicating a hysteresis effect [14]. This effect may be related to the nature of vortex attachment and separation from the liquid film in the wiping zone.

It is interesting that downward tilting of the wiping nozzle, into the upstream direction, is observed to delay the onset of splashing. This may be due to the stronger shearing effect of the downward wall jet flow, which helps to keep the runback flow attached. Also, jet at the impingement point, which is lowered by increasing the jet angle, is at a point where the jet is less confined, permitting modifying the curvature of the streamlines in the impingement region [89]. Fig. 7.59 shows that downward tilt angles up to 20 degrees increase the splashing limit to higher Weber-Reynolds combination, while upward tilting (in the direction of the wiped coating) lowers the splashing limit below that of horizontal nozzles. The latter effect is likely caused by the additive effect of the wiping gas velocity and the strip velocity.

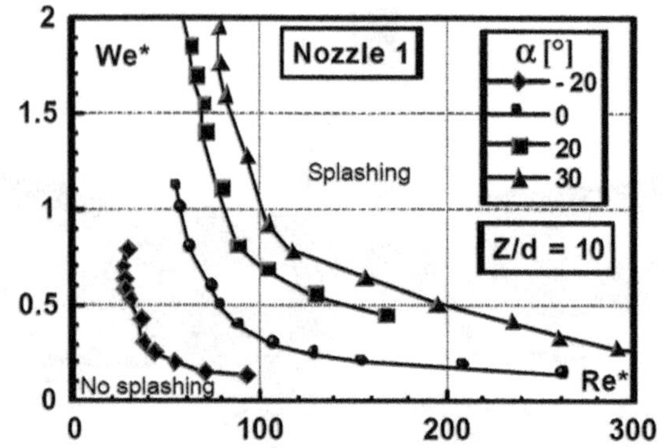

FIG. 7.59 Splashing limit operating curves, expressed as Reynold-Weber number combinations, for several knife tilt angles. Positive angles are those directed counterflow to strip travel [94]. *(Reproduced with permission from M. Dubois, J.M. Buchlin, A. Gosset, V. Perrot, Effect of nozzle tilting on splashing in jet wiping, in: Proc. 6th International Conference on Zinc and Zinc Alloy Coated Steel Sheet (Galvatech 2004), April 4–7, 2004, Chicago, IL, Association for Iron & Steel Technology, Warrendale, PA, 2004, pp. 197–206.)*

In practical terms, this suggests achievable line speed limits using state-of-the-art equipment for specified coating weights. The achievable limit for a nozzle gap width of 1 mm, a strip-to-knife distance of 7 mm, and a coating thickness of 6.6 μm is 180 m/min. Because of its lower density and viscosity, the same conditions produce a 5.3 μm thick Zn-10%Al alloy coating [53]. However, advanced nozzle designs recently placed in production indicate that speeds of 240 m/min are within reach [64].

However, tilted nozzles produce higher coating thicknesses because of the higher effective strip-to-knife distance resulting from the tilting angle [97,98]. It was reported that a 5 degrees tilt reduced the maximum impingement pressure by 18%, which therefore reduced the onset of splashing, but at the same time it increased coating thickness by 11%. For this reason, even though splashing limits are raised when using tilted nozzles, tilt angles no more than 5 degrees are recommended [97].

7.8 Production of thick coating weights on heavy gauge cold and hot rolled steels

Galvanizing of thicker (>1.3 mm) steels, which are usually hot rolled but also available in the cold rolled condition, with coating thicknesses greater than 30 μm, is a growing portion of zinc-coated steel production. These are used for a wide variety of construction and agricultural applications, including support frames for solar panels.

The heating and cooling capacities of galvanizing line pretreatment furnaces and after-pot cooling equipment, even in facilities that only heat the strip to about 550°C to reduce surface iron oxides without providing any other heat treatments, generally limit strip speed to 30–60 m/min. Typical strip-to-knife distances are in the range 8–13 mm, and the plenum pressure is typically in the range 10–100 mbar [98]. At the low end of this pressure range, it can be difficult to achieve a uniform spanwise pressure condition, requiring attention to air knife designs that assure uniform flow at low pressures. As explained below, strip surface roughness can also affect coating control behavior; typical surface roughnesses for hot rolled grades are in the range $0.8 < R_a < 2\,\mu m$. For these conditions zinc splashing in the wiping zone is not expected and the wiping knifes are run in the horizontal (untilted) position. Lower line speeds can permit the development of thicker intermetallic layers at the interface, which, together with the wiped metallic overlay coating, sum to give the total coating thickness. As explained in Chapter 4, lower soluble Al contents in the galvanizing bath permit earlier inhibition layer breakdown and development of outbursts than higher Al contents; it is therefore common to use bath effective Al contents between 0.15% and 0.18%.

Recalling Eq. (7.6), the calculation of coating thickness on the withdrawn strip in the absence of wiping knives, it can be determined for a pure zinc layer that exiting at a strip speed of 40 m/min is about 65 μm. However, the thickness of the withdrawn layer changes as it approaches the wiping zone for two reasons: first, a velocity profile develops across the coating thickness as gravity exerts a force that counteracts viscous drag. This is manifest as a surface velocity on the coating that is calculated as 87% of the strip velocity. Second, the runback flow contributes to the total thickness of the withdrawn layer as it approaches the wiping zone. At the low wiping pressures use for these thick coatings, there is little effect of jet shear on this surface layer. Also, the portion of the withdrawn layer that is wiped and directed into runback flow is lower than in the case of thinner coatings. This runback layer induces a washing effect on the layer withdrawn from the bath; the total coating layer thickness has been calculated to be about 260 μm for a 40 m/min strip speed [99]. This layer thickness is much more sensitive to strip speed than the final wiped coating weight; it doubles from 200 to 400 μm as strip speed increases from 20 to 75 m/min but only increases by about 20 μm as the wiped coating thickness decreases from 20 to 42 μm, Fig. 7.60.

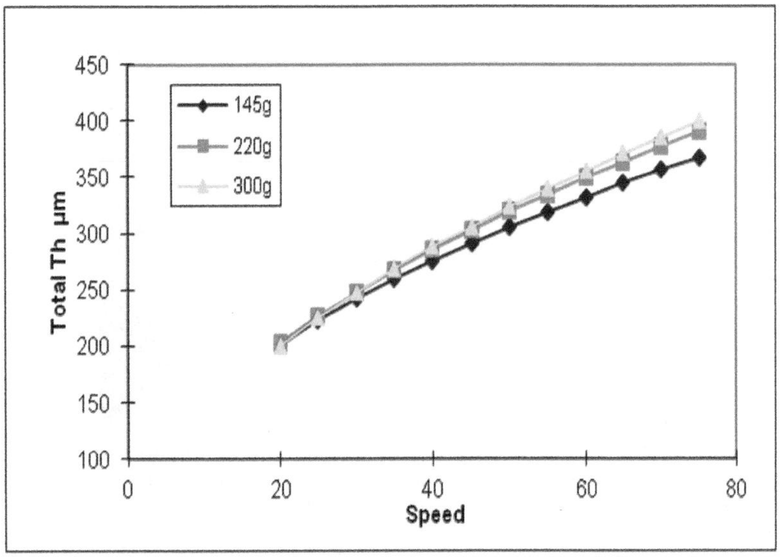

FIG. 7.60 Effect of line speed and wiped one-side coating weight on the thickness of the withdrawn layer, including runback flow at a distance 100mm below the impingement point [99]. *(Reproduced with permission from M. Dubois, Production of heavy coatings at low line speed, in: Proc. 12th International Conference on Zinc and Zinc Alloy Coated Steel Sheet (Galvatech 2021), June 22–24, 2021, Virtual Conference, Austrian Society for Metallurgy and Materials, 2021, pp. 733–745.)*

The ability to effectively wipe the zinc coating to achieve any coating thickness depends on sufficient supply of coating on the strip as it enters the wiping zone. Generally, a runback flow of 100g/s per meter of strip width has been judged to be the minimum required; as line speed decreases and wiped coating thickness increases, this becomes increasingly difficult to achieve [99]. To compensate for this, the strip speed can be increased if furnace capacity and other constraints permit. Or, the height of the impingement zone over the surface of the bath is lowered, typically in the range 75–200mm [98]. This is below the typical 250–500mm knife height used for thin strip operating at speeds between 120 and 180m/min [57]. Adding to these difficulties is the longer time that the liquid metal is exposed to air, increasing the chance of oxidation of the exposed liquid metal.

A coating flow effect unique to thick coating production on thicker steels is observed at the strip edges beneath the gas knives. Because of the low downward surface shear force, the surface tension of the liquid coating at the strip edge is observed to become dominant and causing the coating to shrink away from the edge. A view of this situation, showing the difference in appearance of the strip, is shown in Fig. 7.61, where this liquid metal is transported toward a position on the strip 30–50mm from the strip edge. An example of this vicinity on the finished strip is shown in Fig. 7.62.

FIG. 7.61 Appearance of hot rolled strip being withdrawn from the galvanizing bath at as line speed of 40m/min, showing withdrawal of liquid coating from the edge [99]. *(Reproduced with permission from M. Dubois, Production of heavy coatings at low line speed, in: Proc. 12th International Conference on Zinc and Zinc Alloy Coated Steel Sheet (Galvatech 2021), June 22–24, 2021, Virtual Conference, Austrian Society for Metallurgy and Materials, 2021, pp. 733–745.)*

FIG. 7.62 Edge coating defect on hot rolled strip with a thick zinc coating related to surface tension effects at strip edge [99]. *(Reproduced with permission from M. Dubois, Production of heavy coatings at low line speed, in: Proc. 12th International Conference on Zinc and Zinc Alloy Coated Steel Sheet (Galvatech 2021), June 22–24, 2021, Virtual Conference, Austrian Society for Metallurgy and Materials, 2021, pp. 733–745.)*

As the wiped coating exits the impingement zone, it is acted upon by shear forces from the coating jet, as in the case of thin coatings on lighter strip. In the case of heavy coatings, the strip-to-knife distance is typically larger, permitting the shear forces to influence coating waviness, as observed by Ellen and Tu [28]. At the same time, gravity is exerting a downward force on the wiped coating. Viscous and surface tension effects result in the downflow of the surface of the wiped coating to be insignificant for thin coatings, but the surface velocity of a 40 μm thick coating can reach 15 mm/s, which is five times higher than that of a 20 μm coating, which in turn is six times higher than that of a 10 μm coating. Also, the greater mass of the coating and strip result in longer period before solidification begins. Solidification times for thick coatings on hot rolled strip have been reported to be 20–25 s for a GI coating [100]. These times can be contrasted with the freeze line locations of 1–3 m shown in Table 7.3 for thin strip, which at a line speed of 120 m/m would correspond to 0.5–1.5 s. These factors, together with the increased time available for oxide formation on the wiped coating and oxides withdrawn from the bath under these slow strip speed and low runback conditions, contribute to opportunities for coating defects to be formed. Use of a strip cooling unit can help minimize the duration of time for which the coating needs to remain liquid.

Because the moment of inertia of a strip increases with third power of strip thickness [55], strip vibration issues in the wiping zone are usually not of concern. However, correction of strip shape by using the conventional intermesh positioning of the corrector and stabilizer rolls can present challenges, particularly for "full hard" (not annealed) strip because of its higher yield strength. Such shape issues can lead to coating uniformity issues discussed earlier; the addition of longer times before solidification contributes to oxidation of over-thick areas.

7.9 Strip roughness effects

It seems intuitive that a rougher strip would entrain more zinc from the galvanizing bath and therefore result in a thicker coating. However, the origin of the actual substrate roughness to which the withdrawn coating is adhering should not only be dependent on the roughness of the uncoated coil, but upon the interfacial reactions and consequent interface layers that develop by the time of coating withdrawal from the bath. To determine the relative importance of initial substrate roughness and the interfacial layers, samples of bright cold rolled, conventional cold rolled and holt rolled strips were characterized before and after galvanizing (GI) and galvannealing (GA). The zinc overlay of the GI-coated samples was stripped using fuming nitric acid to reveal the interface layer, while the mixed η-$Fe_2Al_{5-x}Zn_x$ + ζ-$FeZn_{13}$ layer was revealed with an inhibited sulfuric acid solution. Roughness was characterized by white light interferometry. Images of the roughnesses of these substrates before and after GI and GA processing, with the top surface stripped as described, are shown in Figs. 7.63 and 7.64, respectively. of the surface of the GI interface layer showed that there was little effect on the as-received surface roughness for any of the three strip conditions, the as-rolled roughness pattern "printed through" the thin inhibition layer. For the hot rolled and bright cold rolled substrates, the influence of short-wavelength features such as rolling marks or sharp corners on surface roughness was reduced by the GI interface, less so with the conventional cold rolled condition. This was attributed to the higher rate of iron dissolution at these features. Therefore, the as-received roughness was concluded to be well representative of the strip roughness in the GI condition as it enters the wiping zone.

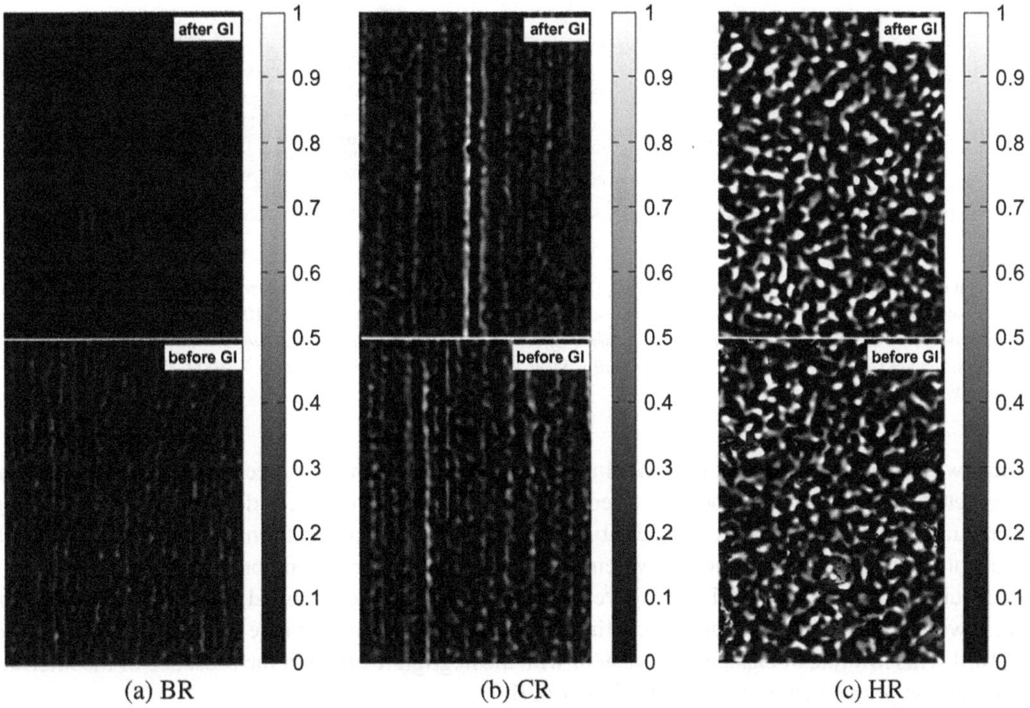

FIG. 7.63 White light interferometry roughness images of (A) bright cold rolled, (B) conventional cold rolled, and (C) hot rolled substrates before and after coating in a galvanizing (GI) bath, overlay removed by fuming nitric acid. The indicated vertical length scale is 0.7mm long [101]. *(Reproduced with permission from J.R. McDermid, A.N. Hrymak, R. Fourmentin, S. Salari, F.E. Goodwin, Effect of steel substrate roughness on GA and GI coating roughness during the gas-jet wiping process, in: AISTech 2009-Proceedings of the Iron & Steel Technology Conference, May 4–7, 2009, St. Louis, MO, Association for Iron & Steel Technology, Warrendale, PA.)*

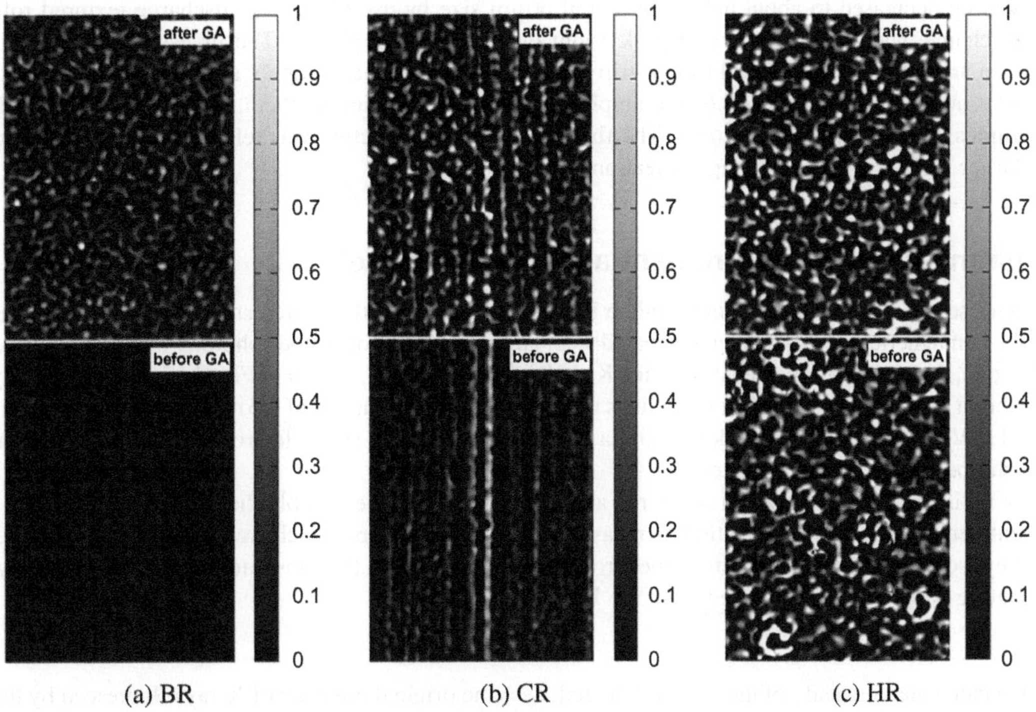

FIG. 7.64 White light interferometry roughness images of (A) bright cold rolled, (B) conventional cold rolled, and (C) hot rolled substrates before and after coating in a galvannealing (GA) bath, overlay removed by inhibited sulfuric acid solution. The indicated vertical length scale is 0.7mm long [101]. *(Reproduced with permission from J.R. McDermid, A.N. Hrymak, R. Fourmentin, S. Salari, F.E. Goodwin, Effect of steel substrate roughness on GA and GI coating roughness during the gas-jet wiping process, in: AISTech 2009-Proceedings of the Iron & Steel Technology Conference, May 4–7, 2009, St. Louis, MO, Association for Iron & Steel Technology, Warrendale, PA.)*

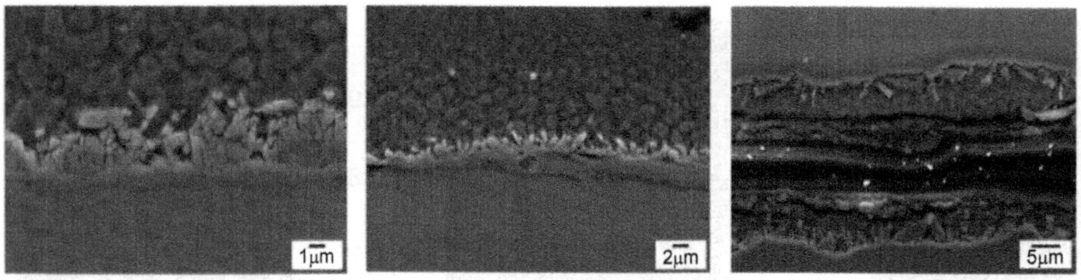

FIG. 7.65 Interfacial microstructures for (A) bright cold rolled, (B) conventional cold rolled, and (C) hot rolled substrates before and after coating in a galvannealing (GA) bath showing the ζ-$FeZn_{13}$ phase outbursts superimposed over the original surface [101]. *(Reproduced with permission from J.R. McDermid, A.N. Hrymak, R. Fourmentin, S. Salari, F.E. Goodwin, Effect of steel substrate roughness on GA and GI coating roughness during the gas-jet wiping process, in: AISTech 2009-Proceedings of the Iron & Steel Technology Conference, May 4–7, 2009, St. Louis, MO, Association for Iron & Steel Technology, Warrendale, PA.)*

In contrast, there was a significant effect of the GA interface layer on roughness, as see in Fig. 7.65. The roughness of this layer is also expected to be dependent upon bath effective Al content, temperature, and immersion time in the galvanizing bath. It was concluded that for all of the examined substrates, the condition upon entering the wiping done could be approximated as a hot rolled substrate with a short wavelength (1–10 μm) roughness component superimposed.

The effect of substrate roughness on GI coating weight for a group of 865 cold rolled coils was examined in an industrial sampling trial. It was concluded that galvanized coating thickness changed in the range of 0.06–0.08 μm for every 0.25 μm change in R_a [102]. This low rate of change is consistent with the negligible effect of the presence of the interfacial layer on GI substrate roughness prior to entering the bath, Fig. 7.63B.

The effects of substrate roughness on surface appearance are more significant. Different surface conditions were produced on steel strip that was first processed on the final stand of a tandem cold rolling mill using shot blasted or electric discharge textured work rolls. The samples were then galvanized in a 0.2% effective Al bath at 100 m/min to produce coating thicknesses of 10 μm per side. The cold rolled surface texture had a significant effect on spangle size, Fig. 7.66. The top row shows the roughness profiles, average roughness, and peak counts for the shot blasted condition. Under this, surface views of the as-produced coating are shown. The third row shows counterpart information for the electric discharge condition, followed by the coating appearance micrographs. It can be seen that the spangle, or grain, size of the coating decreased to about half of its initial 60 μm size by use of electric discharge textured rolls, and that the grain size change was more sensitive to peak count than average roughness. The investigators also correlated the exposed surface area of the textured cold rolled strip and found that it was also highly sensitive to peak count. A surface area of 5 m^2 was measured for the 0.8 × 0.6 mm samples that had a peak count of 50/cm, while a surface area of 15 m^2 was found for samples with a 150/cm peak count. The finer coating microstructure was related to the larger peak count and consequent larger surface area for coating nucleation and growth [103].

7.10 Coating weight measurement and online control

Coating thicknesses or weights are measured online by gauges that excite the coated steel using X-ray, gamma, or other sources of electromagnetic waves, causing the coated steel target to emit fluorescence photons within a known spectrum of energies. For zinc, the target for this radiation is the K shell of the zinc atom, shown in Fig. 7.67. A typical X-ray tube used for coating weight measurement in galvanizing lines runs at 43 kV with a current of 0.5 mA, giving radiation in the energy range of 1–11 keV. This is sufficient to knock out a K shell electron; a L shell electron quickly takes its place with an emission of photoelectric radiation (photons).

The X-ray radiation leaving the zinc is centered at a peak of 8.6 keV, measurably distinct from the energy leaving the steel strip, centered at 6.4 keV. This radiation then passes through Ross filters, which are copper and nickel foils of precise thickness. They isolate the zinc fluorescence lines from those of the iron in the substrate by creating an energy band-pass filter. The coating weight is then obtained from the Beer-Lambert law,

$$I_T = I_0 e^{-\mu \rho x} \tag{7.36}$$

where I_T = the calculated intensity of the X-rays detected, I_O = the original intensity of X-rays fluoresced by the substrate, μ = the mass absorption coefficient, ρ = the density of the material, and x = the thickness of the coating [105].

This permits calibration curves similar to Fig. 7.68 to be developed.

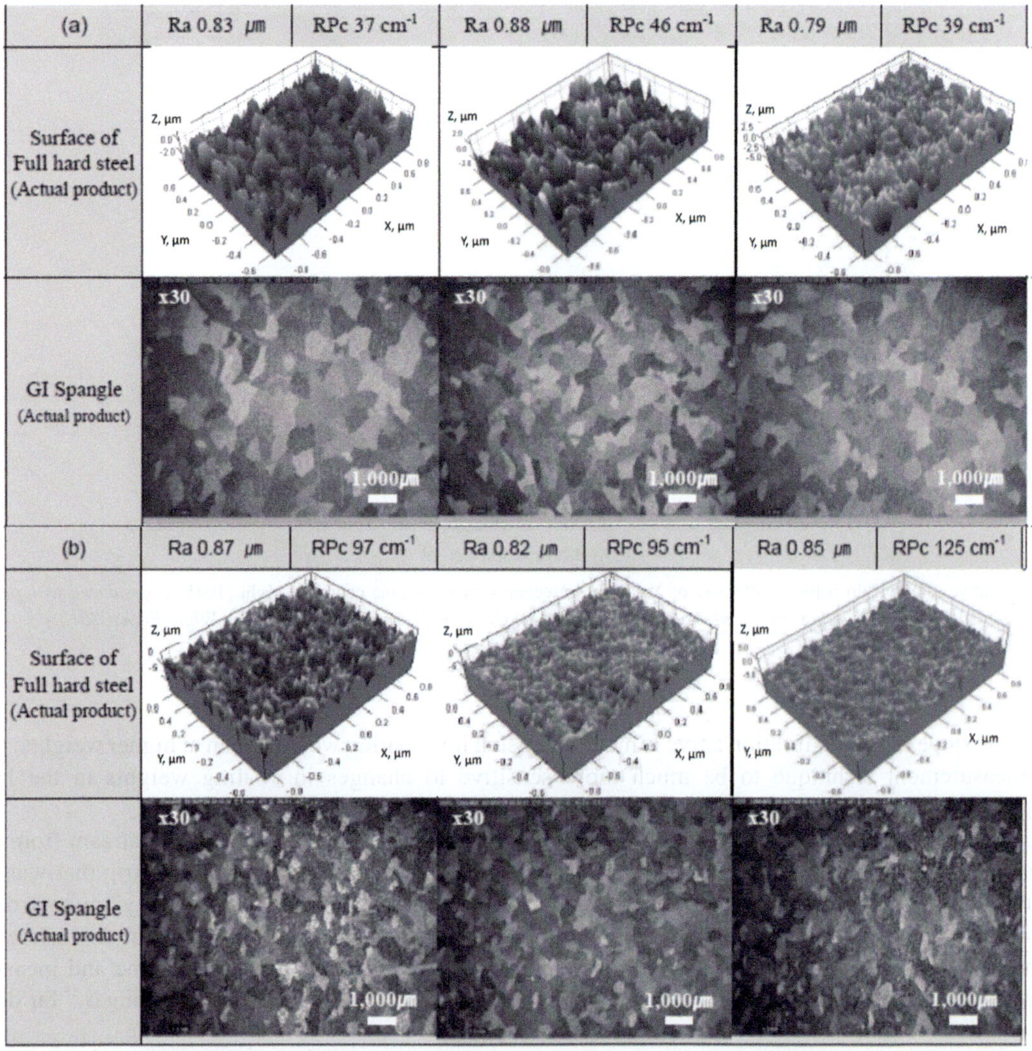

FIG. 7.66 Comparison of full hard steel surface morphologies and spangle size reduction changes of galvanized steel coated after conditioning with shot blasted or electric discharge texturing rolls. (A) Strip surface morphology prepared using shot blasted roll (B) appearance of coating after galvanizing shot blasted prepared strip. (C) Strip surface morphology prepared using electric discharge textured roll, (D) appearance of coating after galvanizing electric discharge prepared strip [103]. *(Reproduced with permission from T.J. Park, S.Y. Choi, Y.H. Kim, Influence of surface roughness of full hard steel on spangle of galvanized steel and actual application results, in: Proc. 12th International Conference on Zinc and Zinc Alloy Coated Steel Sheet (Galvatech 2021), June 22–24, 2021, Virtual Conference, Austrian Society for Metallurgy and Materials, 2021, pp. 1284–1291.)*

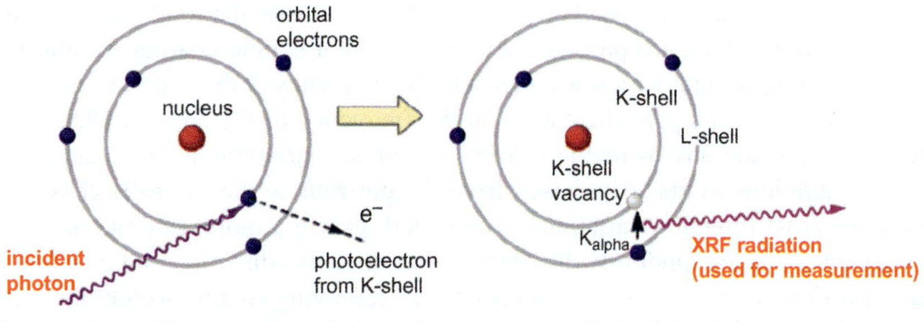

FIG. 7.67 Principle of X-ray coating weight measurement. Incident electromagnetic radiation (photon) knocks an electron out of the zinc K shell (left). An L shell electron fills the K shell vacancy, emitting radiation in the X-ray frequency range usable for measurement [104]. *(Reproduced with permission from C. Burnett, The Benefits of 'Hot' Coating Weight Measurements to Coating Control, Thermo Fisher Scientific White Paper, 2012.)*

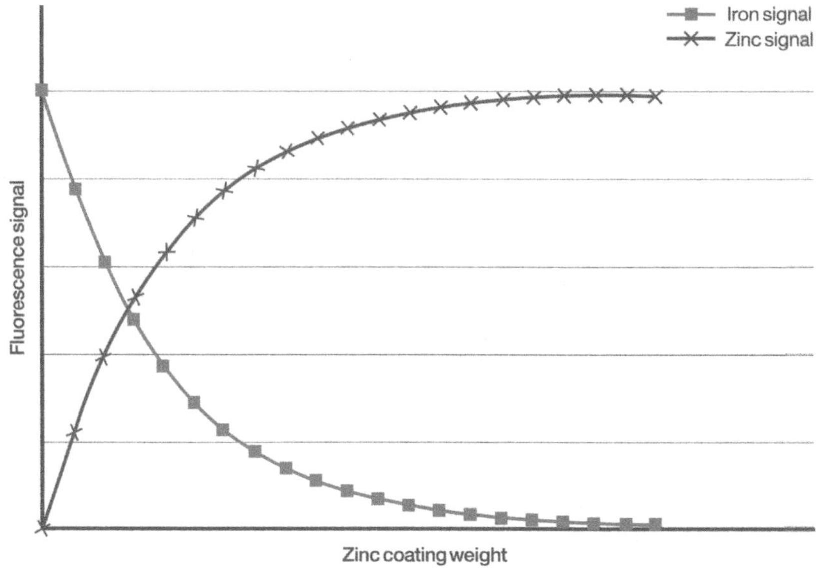

FIG. 7.68 Qualitative relationship between intensity of X-ray fluorescence signal and zinc coating weight [104]. *(Reproduced with permission from C. Burnett, The Benefits of 'Hot' Coating Weight Measurements to Coating Control, Thermo Fisher Scientific White Paper, 2012.)*

Note that the slope of this calibration curve is much steeper at low coating weights than at higher weights; this permits the X-ray measurement technique to be much more sensitive to changes in coating weights in the low coating weight range.

Such detectors are customarily located after the strip has cooled, which can be 100m downstream from the coating control knives. At a strip velocity of 100m/min, this means that the obtained reading is for strip that was wiped 60s before. Various feedback control strategies can be used to compensate for this "transport delay" in control of the gas knives; however, it is becoming increasingly common to use a "hot gauge" that is located 2m downstream from the wiping zone, thereby reducing the delay between the time of passing through the wiping zone and measurement of the coating weight to 2s. However, the majority of process lines still depend upon the "cold gauges," far downstream, to measure coating weight.

There are no affordable coating weight gauges that can simultaneously report the coating weight across the width of the strip. Therefore, detectors that scan back and forth across the width of the strip are used. The path of coating weight measurement is therefore a zig-zag pattern, from one edge of the strip to the other, in the direction of strip travel. This technique measures only a small portion of the coating, for example, if it takes 10s to traverse a 1500mm wide strip moving at 120m/min, 20m of strip will have passed in the direction of strip travel (the "machine direction') versus 1.5m in the "cross direction." Because of this scanning pattern, it is impossible to determine by simple observation if any detected variation in coating weight was caused by a heavy edge or other variations across the strip width such as a crossbow shape, or because of a wiping zone adjustment that caused a variation along the length of the strip. Because cross-direction variations are usually periodic for reasons of strip shape, knife geometry, or presence of a built-up edge, they can be used to determine the cross-direction component of coating weight variation. This can be subtracted from the total variation. In practice, this is performed by moving average filters, which are programmed to compute a simple coating weight average value over one complete scan. The periodic coating weight variations related to the frequency of the edge-to-edge scan are then subtracted, removing the fluctuations that occur in a periodic pattern at the frequency of the edge-to-edge scan.

It may be possible to also include any available information about variations in the machine direction (caused, for example, by string-shaped vibrations as the strip passes from the pot rolls to the top roll), these can be subtracted from the total variation before the cross-direction variation is subtracted, giving a more accurate measure of the residual variations, which are the variations that are randomly distributed after the cross-direction and machine-direction variabilities are known. It is may also possible to program a "virtual edge trim," removing coating weight measurements within 50mm of each edge, to eliminate variations related to built-up edges and shape considerations related to strip edge drop. Whether or not it is possible to disregard the residual variation when setting the coating weight target for the control system depends on the magnitude of these variations. If they are large, the target may need to be increased to be sure than minimum coating weight requirements are met [106]. As discussed in the specifications section at the end of this chapter, the usual criteria for coating weights are either single spot minimum and maximum coating weights, and triple spot minimum and maximum coating weights, per side and total, therefore the target set in the control system will likely include these to be acted upon by the control system algorithm.

Measurements from these coating weight gauges are then fed into a control system, of which a typical example is shown in Fig. 7.69.

At top left, the physical properties of the liquid coating, e.g., density and viscosity are entered. Below this on left side, the process variables for the coating control system, e.g., strip-to-knife distance and plenum pressure are entered. These all feed the coating weight control model, based on the relationships between these input variables and the resultant coating weight described before. The model can be manually tuned to improve accuracy. The model then sets the reference conditions under the heading "CMC (Coating Mass Control) Functions" for strip-to-knife distance, plenum pressure, and knife height over the bath; information on stabilizer roll position to assist with control of strip shape is also entered for manual action. To the right of this column, under the heading "Level 1 Functions" are the variables governed by the closed loop controller. This receives the coating weight feedback from the right side column under the heading "Plant." Closed-loop correction of wiping variables is achieved in the Level 1 Function column. This closed-loop controller needs to take into account the transportation lag described before, techniques such as Smith Controller [107] are used to reduce the lag in feedback control [108]. Provision is also made for manual adjustment of wiping variables without leaving automatic model. Feed-forward control is shown in the CMC Function column, this algorithm is invoked when changing line speed, coating weight, or other conditions to effect changes in wiping zone variables with minimal production of out-of-specification material. For this example, control of the strip-to-knife distance is used as the main control variable for continuous and

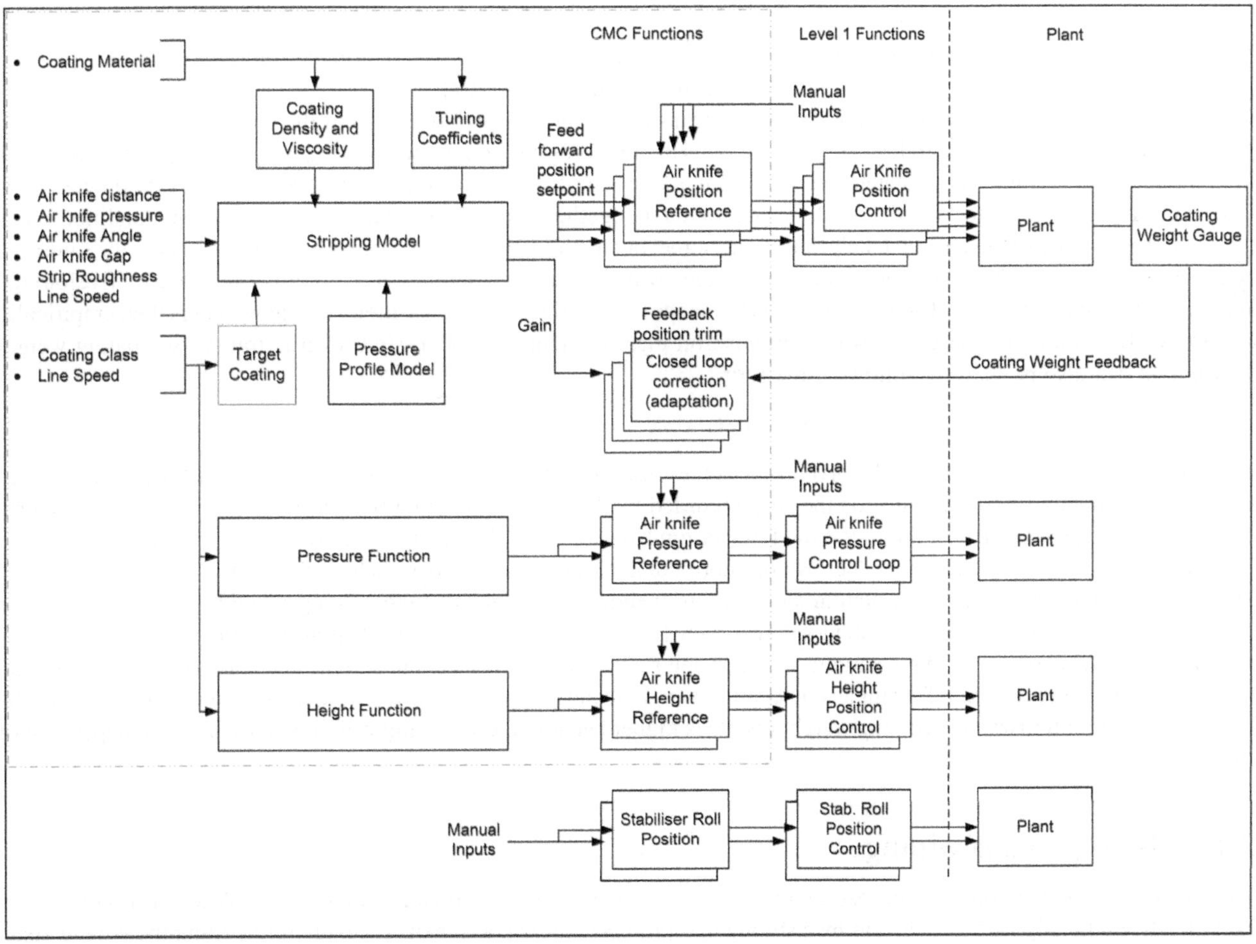

FIG. 7.69 Block diagram for a coating weight control system [54]. *(Reproduced with permission from O. Pearcey, H. Brown, I. Marincic, J. Tham, The evolution of coating control systems as the need for new coated products expands, Berg. Huetenmaenn Monastch. 166 (11) (2021) 525–534 (Springer).)*

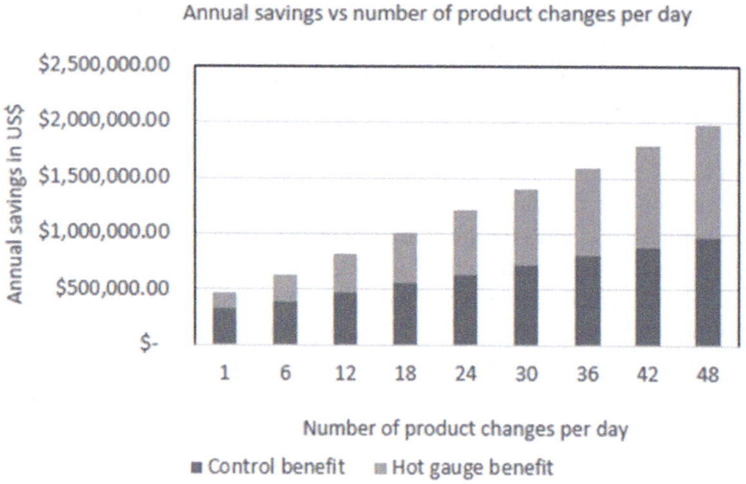

FIG. 7.70 Economic benefits of automatic coating weight control and added benefit of hot coating weight gauge installation. Based on zinc price of $2400/ton [104]. *(Reproduced with permission from C. Burnett, The Benefits of 'Hot' Coating Weight Measurements to Coating Control, Thermo Fisher Scientific White Paper, 2012.)*

automatic adjustment. Plenum pressure control is achieved using a variable speed blower, with finer pressure control adjustments provided through pressure control valves located upstream of each air knife [54].

For the measurement of coating weights of Zn-Al coatings with appreciable amounts of Al such as Zn-55% Al, two X-Ray detectors are required: one for Zn and one for Fe. An Al signal is not directly measured. A Ross filter for each detector isolates the Zn and Fe signals, these are then converted, and the Al portion of the coating weight obtained by fitting the data to polynomial or similar functions [109].

The value of implementation of automatic coating control and hot gauges is shown in Fig. 7.70. The values here are based on a zinc price of $2400 per metric ton and are based only on zinc savings. Reduction of scrap, shorter throughput, and reduction of product changeover time would result in increased savings [104].

Analysis of batches of coating weight data to determine the standard deviation (1σ) of the coating weight distribution and its relationship to strip shape has been described. This involves recording both coating weight and position readings from the scanning coating weight gauge every 0.5 s for a duration of 5–10 min. At the same time, the relevant process variables are sampled every 1 to 5 s. Edge readings are rejected, together with readings taken during transient conditions. The 0.5 s sampling rate can be expected to help determine some, but not all variations in coating weight; increasing the sampling rate to a frequency of 0.1 s would introduce inaccuracies with available techniques and moreover increase by fivefold the amount of data to be analyzed. The relationship between process variables and measured coating weight is then empirically developed. By combining these data with strip position measurements, an expression for true transverse coating weight distribution can also be derived, following the form

$$W = \mathrm{AVG} + aX + bX^2 + cX^3 + dX^4 \tag{7.37}$$

where CW is the coating weight at the given transverse location, AVG is the average coating weight for the duration of sampling, x is the transverse distance on the strip recorded by the scanning coating weight gauge, a is the skewness term in units of g/m^2 per meter, b is the crossbow term in g/m^2 per m^2, c is the symmetry term in g/m^2 per m^3, and d is the crossbow term away from the center (describing more complex crossbow shapes such as S and W shapes) g/m^2 per m^4. This can be done for both top and bottom side coating weights. From plots of the coating weight results, in particular the differences in top and bottom coating weights over the measured length of the strip at each position, the variations coating weight related to strip shape variations can be determined. Using this technique, it has been found that on industrial lines the 1σ standard deviation for single side coating weight is never less than 2.5% of the measured coating weight. The standard deviation for two-side coating weight results is higher because these coating weight values are not independent of each other [110].

7.11 Nitrogen gas wiping

The advantages of using nitrogen instead of air as a wiping gas have been known for many years. When air is completely excluded from the wiping zone, such as by a shroud that protects the strip as it exits the bath until exiting the wiping zone, oxide formation on the liquid zinc coating is greatly reduced. The benefit of reduced coating oxidation has reported to begin with oxygen contents in the shrouded volume as high as 1000 ppm, with optimal results obtained below 100 ppm [111].

Reduction of the thickness of built-up edges was reported to be related to prevention of development of an oxide envelope that surrounds the edge if no wiping force is directed at the edge surface; the oxides was said to form a "container" permitting zinc wraparound that led to formation of the built-up edge. The reduction of ripples was indicated as being related to the prevention of formation of a thin oxide film that can initially be restrained by jet forces, but ultimately results in surface ripples as the oxide fractures and is entrained with the rippled surface. During production of thick coatings at low line speeds, the nitrogen shrouding was reported to help prevent edge defects termed "feathered oxides," which are discontinuous patches of oxide on the strip below the wiping zone that are pulled through the gas knives, with the appearance of feathers pointing toward the strip centerline.

It is expected that the oxidation film formed on the surface of the liquid coating is disproportionately high in Al, and when present, Mg because of their greater reactivity with air; one measure of this is the standard reduction potentials for Zn, Al, and Mg, which are −0.76 V, −1.676 V, and −2.536 V, respectively. Oxidation of liquid metals is known to significantly reduce their surface tension [112]. Also, if formed oxides are mixed into the liquid coating as a result of jet impingement, the viscosity of the liquid film will decrease. Dubois [15] defined a factor related to the initial wave amplitude of a wavy coating after it exits the wiping zone as

$$\beta \approx 520\frac{\sigma h^3{}_0}{\eta \lambda^4} \tag{7.38}$$

where σ is the surface tension, h is the coating thickness, η is the dynamic viscosity, and λ is the wavelength of the coating wave. As surface tension decreases and viscosity increases, the amplitude of the wavy surface decreases; because this term appears as a negative exponential in Dubois's estimation of relaxation time, the relaxation time increases with viscosity (consistent with Eq. 7.30) and decreases, as expected, as surface tension increases. Therefore, lowering surface tension by oxidation of the surface increasing the risk that the leveling effect driven by surface energy minimization will not be complete before solidification commences.

If nitrogen is used as the wiping medium in knives without the use of a shroud, dilution of the nitrogen gas jets will begin after exit from the nozzle because of mixing with the surrounding air. Calculations by Dubois [15] used a nitrogen wiping gas mass flow rate of 0.345 kg/s per meter of nozzle span width, at 25°C this corresponds to a nozzle exit velocity of 316 m/s for a 1 mm nozzle gap width. By comparison, the speed of sound under these conditions is 343 m/s. Fig. 7.71 shows the effect of strip-to-knife distance, using a 1 mm gap, for this gas flow rate. According to this graph, there is significant dilution of jet at strip-to-knife distance of $Z = 10$ mm using a 1 mm nozzle. At the impingement point, the oxygen content expressed in mole fraction O_2 is about 0.025 or 25,000 ppm; this further dilutes as the jet flow is diverted to flow up and down the strip. Use of a shorter strip-to-knife distance of 6 mm brings the O_2 content at the impingement point down to 3000 ppm. This is still above the 1000 ppm threshold where the benefits reported by Pierson [111] were observed.

The benefits of using nitrogen wiping gas with a low O_2 content on coating quality of a Zn-6%Al-3%Mg were reported by Nagata [113] where the incidence of areas of heavy coating and other defects as greatly reduced when local oxygen content near the strip surface just below the wiping zone was less than 2 vol% (approximately 20,000 ppm), Fig. 7.72.

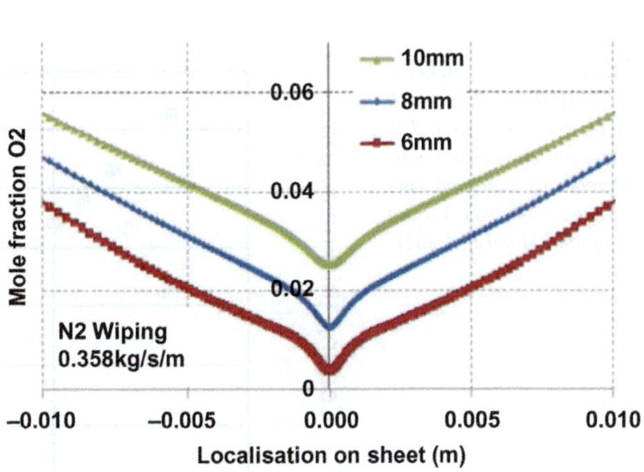

FIG. 7.71 Effect of strip to knife distance on dilution of nitrogen wiping gas for a nozzle gap width of 1 mm, the horizontal axis is the direction of strip movement, centered on the jet impingement point [15]. *(Reproduced with permission from M. Dubois, Minimization of the N2 dilution when wiping in air, in: AISTech 2019-Proceedings of the Iron & Steel Technology Conference, May 6–9, 2019, Pittsburgh, PA, Association for Iron & Steel Technology, Warrendale, PA.)*

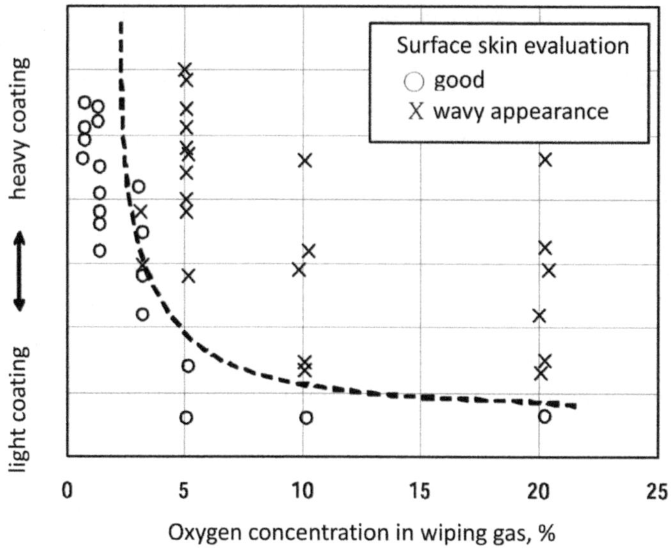

FIG. 7.72 Effect of nitrogen wiping gas dilution by oxygen on coating quality for a Zn-6%Al-3%Mg coating [113]. *(Reproduced with permission from Nagata, S. Masui, R. Yamashita, Nisshin-Steel's Toyo works starts up a state-of-the-art galvanizing line, in: Proc. 6th International Conference on Zinc and Zinc Alloy Coated Steel Sheet (Galvatech 2004), April 4–7, 2004, Chicago, IL, Association for Iron & Steel Technology, Warrendale, PA, 2004, pp. 185–196.)*

Coating quality was further improved by covering the entire zinc bath and the gas knife equipment with a shroud, similar to Pierson [111].

In Section 7.2, it was noted that Mg additions increase liquid metal viscosity and decrease density. The overall effect of composition on obtained coating thickness is shown in Fig. 7.73 for ZnAlMg coatings with up to 3% Mg wiped with nitrogen. These results are for several coatings made on industrial lines while holding all other conditions constant.

As described in Chapter 4, the use of nitrogen wiping was also found to reduce the volume of top skimmings produced on the surface of a galvanizing bath [115]. In Eq. (7.39). the oxygen content in the diluted nitrogen jet is stated as 10%, compared with the 21% oxygen content in normal air.

$$\dot{s} = (14.84\,LS - 1.56)\frac{O^2{}_l}{21^2} f_B \tag{7.39}$$

where $\dot{s}$ =skimmings generation rate, g/s/m^2 of strip coated; LS =line speed m/s; O_l =oxygen level in wiping gas, 10% for nitrogen, 21% for air; f_B =bath factor, GI bath=1, GA bath=0.67.

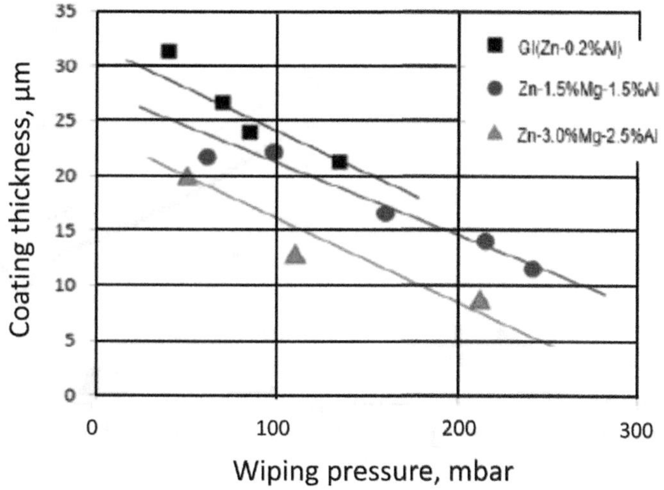

FIG. 7.73 Effect of bath alloy composition on coating thickness achieved with a range of wiping pressures [114]. *(Reproduced with permission from T.C. Kim, I.-R. Sohn, Y.H. You, M.S. Kim, J.S. Kim, Investigation of hot-dip ZnMgAl coatings for automotive applications, in: Proc. 12th International Conference on Zinc and Zinc Alloy Coated Steel Sheet (Galvatech 2021), June 22–24, 2021, Virtual Conference, Austrian Society for Metallurgy and Materials, 2021, pp. 100–107.)*

7.12 Other technical considerations

7.12.1 Supersonic wiping

The possibilities for use of supersonic wiping to increase productivity of lower coating thickness material have been discussed for many years. Thornton and Graff noted that it is the pressure gradient and related shear force of the wiping jet that determines coating weight rather than the magnitude of the maximum jet impact pressure [22]. The shock waves produced by supersonic flow would produce an uneven surface on the coating [52]; the magnitude of these irregularities has yet to be determined. Modeling work using the same convergent nozzle design typical of that used for subsonic jets has determined wiping gas exit velocities for four mass flow rates using a 1 mm nozzle gap width, Table 7.4 [116]. The same modeling work predicted that after exit from the nozzle a supersonic jet will have the same core length before turbulent mixing broadens its width and lowers the pressure gradient near the stagnation point; this means that no gains in stability are gained with longer strip-to-knife distances with supersonic flow. In fact, if longer strip-to-knife distances were required to be used, increased waviness from higher shear forces, as predicted by Ellen and Tu [28] for subsonic jets, would be predicted. No benefits for reduction of dilution of nitrogen gas jets are predicted either. In theory, the same modeling work predicted that use of a conventional convergent flow nozzle with a 1 mm gap width with a gas flow rate of 0.672 g/s per meter of knife span width, a strip velocity of 180 m/min, and a strip-to-knife distance of 8 mm should produce supersonic flow and a resultant coating weight of about 3 μm, but this has yet to be verified by experiment or practice.

7.12.2 Hot gas wiping

The possibilities of wiping with air that exits the nozzle gap at temperatures between 50°C and 250°C were modeled by Dubois [26]. As gas temperature increases, its density decreases. Therefore, for the same mass flow rate of gas delivered to the wiping knife plenum, the higher gas temperature will result in a higher nozzle exit velocity. Conversely, to obtain gas at a specified pressure, the mass flow rate is reduced in comparison with room temperature gas; this offers a possibility of lowering costs when wiping with nitrogen. The mass consumption of gas in kg/min at 250°C is 85% of that of gas at 100°C. However, the lower density at higher temperature results in lower momentum transfer to the strip and therefore there is no effect on coating thickness; coating thickness was shown to be independent of wiping gas temperature. After the hot wiping gas exits the nozzle, it first cools by adiabatic expansion and later by mixing with the surrounding air. The effect of hot wiping gas on the temperature of the zinc coating is therefore reduced, both for this reason and the greater influence of the strip temperature. It was shown both by modeling and by in-plant trials that the liquid zinc temperature in the wiping zone increased by less than 25°C when wiping gas temperature was increased from 60°C to 250°C for a 0.4 mm strip thickness. Higher strip thicknesses reduced the effect of wiping gas temperature on coating temperature; a 0.7 mm thickness gave less than a 15°C temperature change. Another consideration is the effect of wiping gas temperature on the nozzle gas thickness; this would be expected to increase as a result of thermal expansion as wiping gas temperature increased. Coating thickness is known to be extremely sensitive to nozzle gap thickness if all other variables are held constant, for example, with room temperature wiping gas, the effect of a variation of only 0.05 mm on a nozzle gap thickness of 1.1 mm is about 0.336 μm on a

TABLE 7.4 Computed mass flow rates, required plenum pressure, and mach numbers using conventional convergent flow nozzle with 1 mm gap.

Wiping gas mass flow rate at nozzle exit (kg/s per meter of nozzle span)	Wiping gas volume flow rate (Nm^3/h per meter of nozzle span)	Plenum pressure (mbar)	Mach number
0.287	800	590	0.75
0.358	1000	885	0.95
0.45	1250	1320	1.15
0.672	1875	2400	1.55

Derived from data in M. Dubois, Supersonic wiping: new window or wrong trail, in: Proc. 9th International Conference on Zinc and Zinc Alloy Coated Steel Sheet (Galvatech 2013), September 23–27, 2013, Beijing, Chinese Society for Metals, Metallurgical Industry Press, Beijing, pp. 427–432.

10.5 μm coating. For comparison, a change of only 0.5 mm of the strip-to-knife distance results in a 0.52 μm. It was concluded that the main benefit for use of hot gas wiping would be for production of thin coatings, around 5 μm, with strip speeds of 150 m/min or greater.

7.12.3 Magnetic wiping

The possibility of regulating coating thickness by electromagnetic rather than pneumatic wiping was investigated by CRM [117]. Three different configurations were investigated: longitudinal magnetic flux, where one or more turns of an induction coil around the width of the strip were made, and two kinds of transverse flux, where either two loops of a magnetic or several slots of conductive material fed by three-phase alternating current were placed in front of the face of each side of the strip. The longitudinal flux technique was found to give the most uniform magnetic field pattern that is useful in supplying a wiping force. Use of magnetic wiping alone always resulted in an increase of strip temperature, sometimes as high as 300°C. To moderate this increase, the possibility of combined pneumatic and magnetic wiping was investigated. It was found that this combined technique worked best if the two kinds of wiping were applied simultaneously. The inclusion of a single turn of the induction coil in the upper lip of the wiping knife was found to avoid ripples in the coating. The results from a trial on the CRM pilot line are shown in Fig. 7.74. The frequency of the voltage in the induction coil was 200 kHz and the nozzles were held in a horizontal (0 degrees tilt) position, the strip thickness was 0.7 mm, and strip width was 250 mm. For a coating weight reduction of 25%, a strip temperature increase of 74°C was measured. Coating iron content is also shown on the right-hand vertical axis, reaching as high as 8% for an inductor current of 650A. This suggests that magnetic wiping might best be used in conjunction with an adjacent furnace for production of galvannealed coatings.

7.13 Coating weight standards

Coating weight standards originated in the general galvanizing industry, where zinc coatings cut pieces of sheet were permitted to "drip dry" rather than being subjected to coating weight control. Galvanized sheet in North America was sold by weight per square foot using the Galvanized Sheet Gauge (GSG) system [118]. The decimal thicknesses in the GSG system were equal to the steel sheet Manufacturers Standard Gauge (MSG), but with a 0.0037″ (94 μm) allowance for a 47 μm thick zinc coating to be present on each side. The gauge numbers did not refer to only one thickness, but instead represented a thickness range within the allowable weight tolerance for each gauge number. With the development of coating controls, first with coating rolls and later with gas wiping knives, thinner coatings entered the market. The most common of the

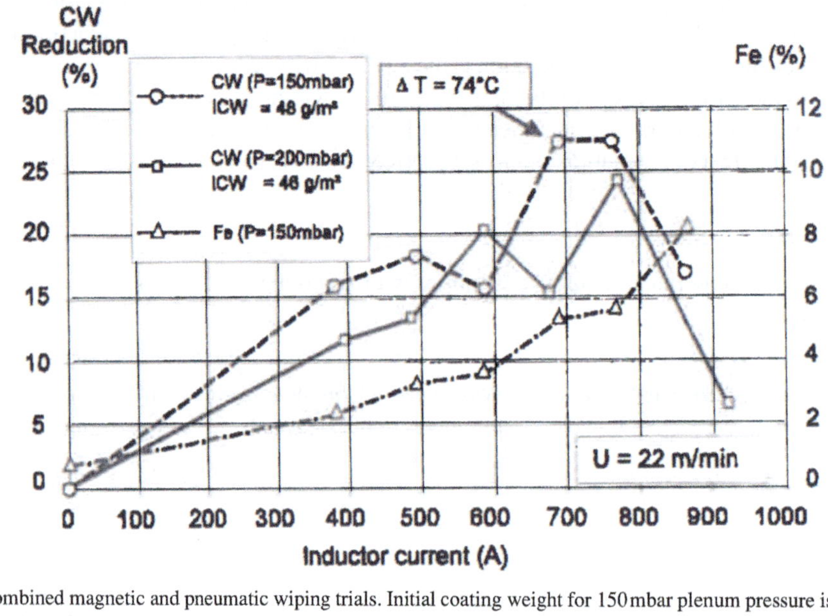

FIG. 7.74 Results of combined magnetic and pneumatic wiping trials. Initial coating weight for 150 mbar plenum pressure is 48 g/m^2, for 200 mbar it is 46 g/m^2. Line speed 22 m/min [117]. *(Reproduced with permission from M. Malmendier, J. Crahay, J.-F. Noville, S. Wilmotte, Wiping by electromagnetic techniques in hot dip galvanizing, in: Proc. 4th International Conference on Zinc and Zinc Alloy Coated Steel Sheet (Galvatech 1998), September 20–23, 1998, Makuhari, Chiba, Japan, Iron and Steel Institute of Japan, Tokyo, 1998, pp. 751–756.)*

TABLE 7.5 Example G90 coating weight results—illustrating passing and failing the requirements of A653/A653M (Table 1)—values in oz/ft^2.

Example	Test	E1	E2	E3	TST	Comments
1	TBS	0.92	0.93	0.96	0.94	Passes min TST
2	TBS	0.85	0.93	0.96	0.91	Passes min TST
3	TBS	0.85	0.87	0.96	0.89	Fails on min TST
4	TBS	0.78	0.95	0.96	0.9	Fails on min SST
5	OS-top OS-bottom TBS	*0.30*[a] 0.55 0.85	0.32 0.62 0.94	0.40 0.55 0.95	0.91	Passes on min TST-TBS Review low OS-top value

Reference tables in ASTM 653 Standard:
Requirements of A653 (Table 1):
Minimum triple spot test average (TST), total both sides (TBS)—0.90 oz/ft^2.
Minimum single spot test (SST), total both sides (TBS)—0.80 oz/ft^2.
Single spot test (SST) on one side (OS) "normally not less than" 0.32 oz/ft^2 on any one side (per Footnote A of A653 Table 1).
[a]*This value is marginally below the aim of "normally not less than" 40% (0.32 oz/ft^2) of the SST-TBS value of 0.80 oz/ft^2 on any one side of a G90 SST. Production coating weight [mass] data should be reviewed to determine if there is a need for retesting before the product can be approved.*

original thinner coatings was labeled "1.25 Commercial," which aimed at minimum 0.90 oz./ft^2 (275 g/m^2 or a two-side coating thickness of 38.4 μm), split between both sides. In 1964, such coatings were formally standardized in ASTM specification A525 "General Requirements for Steel Sheet, Zinc-Coated (Galvanized) by the Hot-Dip Process" [119]. Coating Class "1.25 Commercial" became Coating Designation "G90," with a specified minimum weight of 0.90 oz./ft^2, total both sides. The coating class known as "Light Commercial" became Coating Designation "G60." Coating Class "2.50" (equivalent to the original 1.25 per side coating) became Coating Designation "G210."

There are three separate coating weight designations used in ASTM standards, any of which can be specified by a customer according to their preferences. These designations are:

- Minimum triple spot test (TST) value. There are two types of TSTs.
- the average of three total coating weight measurements, made on samples from near each edge of the sheet and the strip center, analyzed by the weigh-strip-weigh procedure described in ASTM Standard A90/A90M for galvanized (GI) and galvannealed (GA) sheet, zinc-55% aluminum alloy-coated sheet, zinc-5% aluminum alloy-coated sheet, and zinc-aluminum-magnesium alloy-coated sheet. There are special procedures required for other types of alloy coatings such as aluminized and zinc-nickel alloy electroplated sheet.
- A total-both-sides (TBS) requirement (See ASTM A924/A924M [120]) for testing details).
- Minimum single spot test (SST) value. This is the same as a single spot TBS requirement.
- Minimum single side requirement. This is normally not less than 40% of the minimum SST, TBS requirement.

The TST test only applies to the original, full-width (as-coated) sheet. Narrow sheet cut from full-width sheet is subject only to a minimum SST, TBS requirement. With the exception of Supplementary Table S2.1 in A653/A653M, ASTM specifications state that the minimum weight on one side of a single spot is normally not less than 40% of the minimum TBS requirement. Examples of full-width pass/fail test results for A653 are given in Table 7.5.

7.13.1 Designation system for galvanized and galvannealed sheet products

Galvanized (GI) coatings: For galvanized sheet, common inch-pound coating weight designations (ordered using ASTM Standards A653 or A1063) are, in oz/ft^2: G30, G40, G60, G90, G 115.

These designations specify the minimum average TST, total-both-sides, tested per ASTM A924/A924M, e.g., G90 requires a minimum average TST of 0.90 oz/ft^2 total-both-sides. The specification stipulates that TST samples shall be taken from defined positions at the edge-center-edge of the as-coated sheet.

There are designations for heavier coatings, such as G165 and G210, but these products are used for very specialized applications and are generally not available on thinner gauge sheet.

In SI units (ordered as A653M or A1063M), the comparable coating mass designations for galvanized sheet are, in g/m^2: Z90, Z120, Z180, Z275, Z350.

These designations specify the minimum average TST, total-both-sides, tested per A924/A924M, e.g., Z275 requires a minimum average TST of 275 g/m^2 total-both-sides.

In 2007, ASTM added the option of ordering single side, single spot test (SST) coating designations to A653/A653M. These are SI designations only (ordered to A653M) and specify the minimum and maximum allowable coating mass per side for any SST. They take the familiar form of automotive coating designations (numeric characters first—signifying a per side requirement). No inch-pound designations are used since single side coatings are traditionally ordered in SI units only. Examples are 60G, 70G, 90G. These designations specify the minimum and maximum SST value on each surface, e.g., 60G requires a minimum of 60 g/m^2 and a maximum of 110 g/m^2 of zinc on each surface for any SST.

When specifying single side single spot coatings, the designation for each surface must be shown, e.g., 60G60G.

Zinc-Iron (Galvanneal): The common inch-pound coating weight designations (ordered as A653) for galvannealed sheet (zinc-iron alloy-coated) are, in oz/ft^2:

A25, A40, A60. As with galvanized product designators, A40, for example, requires a minimum average TST coating weight of 0.40 oz/ft^2, total-both-sides. While the coating contains approximately 8%–10% iron, resulting in the density being slightly higher than a zinc coating and the coating thickness being slightly less than a G40 galvanized coating, the difference is too small to be of concern.

The SI equivalent coating mass designations (ordered as A653M) for galvannealed sheet are, in g/m^2: ZF75, ZF120, ZF180. ZF120, for example, requires a minimum average TST of 120 g/m^2 total-both-sides.

As with galvanizing, the option of ordering zinc-iron coatings to single side, SST coating designations has been added to A653/A653M. Again, these are SI designations only (ordered to A653M), specifying the minimum and maximum allowable coating mass per side for any single spot, and taking the familiar form of automotive coating designations (numeric characters first—signifying a per side requirement). No inch-pound designations are used since single side coatings are traditionally ordered in SI units only. Examples are: 45A, 50A. These designations relate to the minimum and maximum SST value on each surface, e.g., 45A requires a minimum of 45 g/m^2 and a maximum of 75 g/m^2 of zinc-iron alloy on each surface for any SST. When specifying single side, SST coatings, the designation for each surface must be shown, e.g., 45A45A.

7.13.2 Keeping ASTM designation systems straight

As hot-dip galvanized and galvannealed coatings saw more use by the automotive industry, it became the practice to manufacture these products to conform to single side, SST g/m^2 values; that being the requirement of automotive manufacturers. Products ordered for construction and other general end uses continue to be ordered to total-both-sides, TST, inch-pound designations. For hot-dip galvanize, as we have seen, ASTM uses "G" (prior to the numerals) in the designator for inch-pound coatings and "Z" for SI coatings—total-both-sides in each case. On the other hand, for electrogalvanized "G" (after the numerals) means SI units and "Z" means inch-pound units—single spot, single side in each case.

The use of both dimensional units, and the reversal of "G" and "Z" between TST hot-dip, and single side, SST EG in ASTM specifications, certainly can lead to confusion in the marketplace. Table 7.6 below summarizes what the various designations mean in terms of single spot and triple spot requirements.

TABLE 7.6 Galvanized sheet coating designations explained.

Product type and coating requirements							
				Coating tests required			
				Single side		Total both sides	
Coating designation format	Specification	Coating	Units	SST	TST	SST	TST
G*nn*	A653-Table 1	Zinc-HD	oz/ft^2	None	Min	Min	Min
Z*nn*	A653M-Table 1	Zinc-HD	g/m^2	None	Min	Min	Min
A*nn*	A653-Table 1	oz/ft^2	oz/ft^2	None	Min	Min	Min
ZF*nn*	A653M-Table 1	gm/m^2	g/m^2	None	Min	Min	Min

TABLE 7.6 Galvanized sheet coating designations explained—cont'd

Product type and coating requirements							
				Coating tests required			
				Single side		Total both sides	
Coating designation format	Specification	Coating	Units	SST	TST	SST	TST
*nn*Z*nn*Z	A879	Zinc-EG	oz/ft^2	Min and max	None	None	None
*nn*G*nn*G	A879M	Zinc-EG	g/m^2	Min and max	None	None	None
*nn*G*nn*G	A653M-Table S2.1	Zinc-HD	g/m^2	Min and max	None	None	None
*nn*A*nn*A	A653M-Table S2.1	Zinc-iron-HD	g/m^2	Min and max	None	None	None
*nn*A*nn*A	Auto (typical) 1[a]	Zinc-HD and EG	g/m^2	Min and max	None	None	None
*nn*A*nn*A	Auto (typical) 2[a]	Zinc-iron-HD and EG	g/m^2	Min and max	None	None	None

Notes: nn = numerals (2 or 3) specific to coating weight (mass); HD = hot-dip; EG = electrogalvanize; SST = single spot test; TST = triple spot test.
[a]*Some auto manufacturers require a minimum TST.*

For additional clarification, see Table 7.7 below, which provides the requirements of selected coating weight (mass) examples for galvanized sheet made to ASTM specifications. It is not easy to keep the terminology straight. Users should be aware that both units are in common use today and are advised to pay close attention when ordering, knowing precisely what is meant by the terminology being used. See Table 7.8 at the end of this section, which summarizes the designations used for most hot-dip products and may be useful in keeping terminology clear.

TABLE 7.7 Selected ASTM galvanized sheet designations—requirements.

Product type	Example designation	Requirement
Hot-dip galvanize ASTM A653/A653M	G90 (A653, Table 1, in-lb)	TST[a] average 0.90 oz/ft^2 min—total both sides SST[b] 0.8 oz/ft^2 min—total both sides and normally not less than 40% on any one side
	Z275 (A653M, Table 1, SI)	TST[a] average 275 g/m^2 min—total both sides SST[b] 235 g/m^2 min—total both sides and normally not less than 40% on any one side
	60G60G (A653M, Table S2.1, SI)[c]	SST 60 g/m^2 min, 110 g/m^2 max—each side
Hot-dip galvanneal ASTM A653/A653M	A40 (A653, Table 1, in-lb)	TST[a] average 0.40 oz/ft^2 min—total both sides SST[b] 0.3 oz/ft^2 min—total both sides and normally not less than 40% on any one side
	ZF120 (A653M, Table 1, SI)	TST[a] average 120 g/m^2 min—total both sides SST[b] 90 g/m^2 min—total both sides and normally not less than 40% on any one side
	45A45A (A653M, Table S2.1, SI)	SST 45 g/m^2 min, 75 g/m^2 max—each side

Continued

TABLE 7.7 Selected ASTM galvanized sheet designations—requirements—cont'd

Product type	Example designation	Requirement
Electrogalvanize ASTM A879/A879M	13Z13Z (A879, Table 1, in-lb)	SST 0.13 oz/ft² min, 0.23 oz/ft² max—each side
	40G40G (A879M, Table 1, SI)	SST 40 g/m² min, 70 g/m² max—each side

[a]*Triple spot test.*
[b]*Single spot test.*
[c]*For information purposes, Table S2.1 in A 653M shows inch-pound values for the SI coating designations.*

TABLE 7.8 Summary of designators for zinc-based coatings on steel sheet.

	Example of common coating designations			
Product	inch-pound	SI	Coating weight inch-pound (oz/ft²)	Coating mass SI (g/m²)
Total both sides—Minimum triple spot average				
ASTM A653/A653M galvanize	G90	Z275	0.9	275
ASTM A653/A653M galvanneal	A40	ZF120	0.4	120
ASTM A792/A792M 55%Al-Zn alloy-coated	AZ55	AZM165	0.55	165
ASTM A875/A875M Zn-5%Al alloy-coated	GF75	ZGF225	0.75	225
ASTM A1046/A1046M Zn-Al-Mg alloy-coated	ZM90	ZMM275	0.9	275
Single side[a]—Minimum single spot				
ASTM A653M galvanize (Table S2.1)	N/A[b]	60G	0.2	60
ASTM A653M galvanneal (Table S2.1)	N/A[b]	45A	0.15	45
ASTM A879/A879M electrogalvanize	13Z	40G	0.13	40
Automotive specified electrogalvanize	N/A[b]	100G	N/A[b]	100
Automotive specified galvanneal	N/A[b]	45A	N/A[b]	45

[a]*Single side designators are used to specify the coating mass for each side and are written, for example, 60G60G, or in the case of differential coating masses, 90G60G.*
[b]*Not applicable.*

7.13.3 Total both side TST versus single side SST coatings

Because the ASTM total-both-sides, TST designators allow an uneven split of the coating (one side SST values are normally has at least 40% of the specified minimum SST total-both-sides coating weight), it is not possible to precisely convert them to SST, single side designators since the latter specify exact minimums per surface. It is sometimes useful, however, to provide an approximate conversion based on the total coating thickness on both surfaces.

Fig. 7.75 is a chart that allows this to be done, both in terms of coating designators and thickness of the total coating. For instance, it can easily be seen that a G60 coating has a minimum total-both-sides thickness of about 1.0 mil, which is the

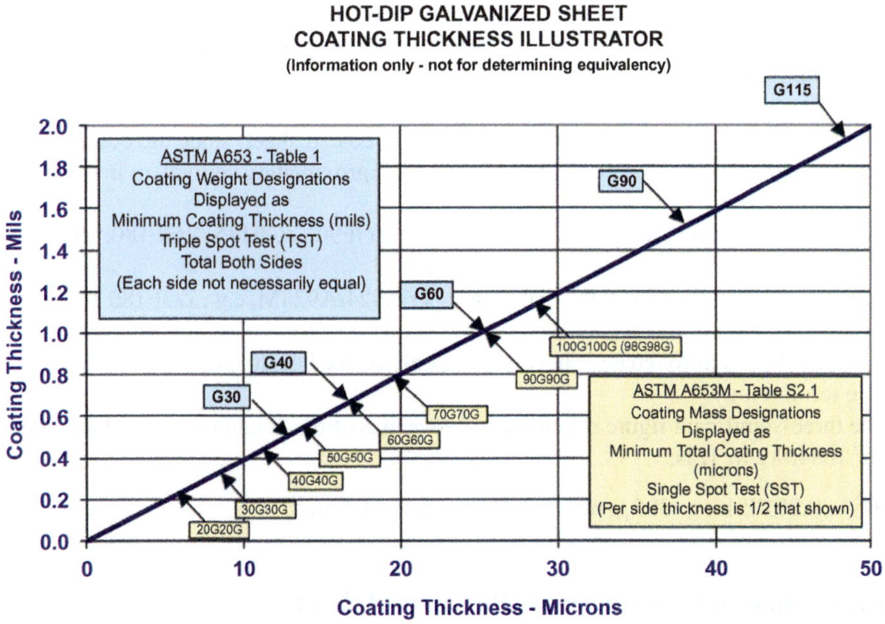

FIG. 7.75 Guide for estimating coating thickness in terms of single spot and triple spot tests. *(Reproduced with permission from GalvInfo Note 1.1, Understanding Coating Weight Designations for Zinc-Based Coatings on Steel Sheet, Rev. 4, December 2020, GalvInfo Center of International Zinc Assn. of International Zinc Assn., Durham, NC.)*

very close to twice the minimum per side thickness ($2 \times 12.5\,\mu m$) of a 90G/90G coating. Remember, however, that a G60 coating is an average of 3 readings (TST) and can have an uneven split of the total coating thickness, while a 90G90G coating must have a minimum of $12.5\,\mu m$ on each side for any single spot. Fig. 7.75 is a guideline only for estimating coating thickness in terms of the two systems and is not meant to suggest equivalency. Also, the values shown are specified minimums. Actual coatings are always about 5% thicker in order to guarantee the minimums.

7.13.4 Designation system for zinc-55% aluminum alloy-coated sheet

Steel sheet with a zinc-55% aluminum coating is in common use today throughout the construction and appliance industries. It, too, has very specific coating designators. Fortunately, there are only a few designators, but that doesn't mean there is no confusion about the meaning. The designation systems for coating weight and coating mass are given in ASTM Standard A792/A792M. The four inch-pound coating weight designations (ordered as A792) are, in oz./ft^2: AZ50, AZ55, AZ60, AZ70.

These designations specify the minimum average of a TST, total-both-sides, tested per A924/A924M, e.g., AZ50 requires a minimum average TST of 0.50 oz./ft^2 total-both-sides. These designators are comparable to those used for galvanized sheet in that the dimensions are oz./ft^2. Be aware, however, that the designation AZ60 is not equivalent to a G60 coating with respect to the thickness of the coating. Here is where the issue of density comes into play. The coating on 55% Al-Zn alloy coated sheet has 55% aluminum and about 45% zinc. Actually, the coating has a small addition of silicon, but for purposes of this discussion, the silicon is not important. Since aluminum is less dense than zinc (a given volume weighs less than the same volume of zinc), an AZ60 coating is thicker than a G60 galvanize coating. The equivalencies between coating mass and thickness are shown in Eq. (7.2).

For 55% Al-Zn alloy-coated sheet, there is also a SI coating mass designator system (ordered as A792M). The SI equivalents to AZ50, AZ55, AZ60, and AZ70 are, in g/m^2: AZM150, AZM165, AZM180, AZM210.

These designations specify the minimum average of a TST, total-both-sides, per A924/A924M, e.g., AZM150 requires a minimum TST of 150 g/m^2 total-both-sides.

Because 55% Al-Zn alloy-coated sheet is produced only by the hot-dip process, there is no additional terminology or specification related to the manufacture of an electroplated product. Also, there are no SST, single side designations for this product.

7.13.5 Designation system for zinc-5% aluminum alloy-coated sheet

Steel sheet with a zinc-5% aluminum alloy coating (Zn-5% Al alloy-coated) is used by the construction, automotive, and appliance industries. Zn-5% Al alloy-coated sheet has a coating that consists of 95% zinc and 5% aluminum, and small amounts of other elements to improve processing and product characteristics. The designation systems for coating weight and coating mass are given in ASTM Standard A875/A875M. The common inch-pound coating weight designations (ordered as A875) are, in oz/ft2: GF30, GF45, GF60, GF75, GF90.

These designations specify the minimum average of a TST, total-both-sides, tested per A924/A924M, e.g., GF60 requires a minimum average TST of 0.60 oz/ft^2 total-both-sides. For Zn-5% Al alloy-coated sheet, since the coating contains about 95% zinc, and thus has nearly the same density as zinc, a GF90 coating is approximately equivalent in thickness to a G90 galvanized coating.

The equivalent SI coating mass designations (ordered as A875M) are, in g/m^2: ZGF90, ZGF135, ZGF180, ZGF225, ZGF275.

These designations specify the minimum average of a TST, total-both-sides, per A924/A924M, e.g., ZGF180 requires a minimum TST of 180 g/m^2 total-both-sides.

As with 55% Al-Zn alloy-coated sheet, Zn-5% Al alloy-coated sheet is made only by the hot-dip process so there are no designator systems that involve per side terminology.

For Zn-5% Al alloy-coated sheet, the three-significant figure relationship between coating weight (mass) and thickness (based on an alloy density of 427 lb./ft^3 [6840 kg/m^3]) is:

$$1.00\,\mathrm{oz/ft^2} = 0.00175\,\mathrm{in.} = 305\,\mathrm{g/m^2} = 0.0446\,\mathrm{mm} = 44.6\,\mu\mathrm{m} \tag{7.40}$$

7.13.6 Designation system for zinc-aluminum-magnesium alloy-coated sheet

A fourth zinc-based coating that is used where superior corrosion resistance is needed, is zinc-aluminum-magnesium alloy-coated (Zn-Al-Mg alloy-coated) sheet. This alloy-coated sheet has five coating types, each with different amounts of aluminum and magnesium, as shown below. The designation systems for coating weight and coating mass are given in ASTM Standard A1046/A1046M.

The common inch-pound coating weight designations (ordered as A1046) are, in oz./ft^2: ZM30, ZM40, ZM60, ZM75, ZM90.

These designations specify the minimum average of a TST, total-both-sides, tested per A924/A924M, e.g., ZM60 requires a minimum average TST of 0.60 oz/ft^2 total-both-sides.

The equivalent SI coating mass designations (ordered as A1046M) are, in g/m2: ZMM90, ZMM120, ZMM180, ZMM220, ZMM275.

These designations specify the minimum average of a TST, total-both-sides, per A924/A924M, e.g., ZMM180 requires a minimum TST of 180 g/m^2 total-both-sides.

As with 55% Al-Zn alloy-coated sheet, Zn-Al-Mg alloy-coated sheet is made only by the hot-dip process so there are no designator systems that involve per side terminology.

For Zn-Al-Mg alloy-coated sheet, the three-significant figure relationship between coating weight (mass) and thickness, based on a density of 375 lb./ft^3 (6001 kg/m^3) for Type 1, 5%–9% Al; 356 lb./ft^3 (5700 kg/m^3) for Type 1, 9%–13% Al; 385 lb./ft^3 (6170 kg/m^3) for Type 2; 401 lb./ft^3 (6430 kg/m^3) for Type 3; 415 lb./ft3 (6650 kg/m^3) for Type 4; and 388 lb./ft^3 (6210 kg/m^3) for Type 5 is:

Type 1–5–9% Al; 2–4% Mg:

$$1.00\,\mathrm{oz/ft^2} = 0.00200\,\mathrm{in.} = 305\,\mathrm{g/m^2} = 0.0508\,\mathrm{mm} \tag{7.41}$$

Type 1: 9%–13% Al; 2%–4% Mg:

$$1.00\,\mathrm{oz/ft^2} = 0.00210\,\mathrm{in.} = 305\,\mathrm{g/m^2} = 0.0533\,\mathrm{mm} \tag{7.42}$$

Type 2: 3%–<5% Al; 2%–4% Mg:

$$1.00\,\mathrm{oz/ft^2} = 0.00195\,\mathrm{in.} = 305\,\mathrm{g/m^2} = 0.0495\,\mathrm{mm} \tag{7.43}$$

Type 3: 3%–6% Al; 0.4%–<2% Mg:

$$1.00\,\mathrm{oz/ft^2} = 0.00187\,\mathrm{in.} = 305\,\mathrm{g/m^2} = 0.0475\,\mathrm{mm} \tag{7.44}$$

Type 4: 0.5%–<3% Al; 0.4%–2.6% Mg:

$$1.00\,\mathrm{oz/ft^2} = 0.00181\,\mathrm{in.} = 305\,\mathrm{g/m^2} = 0.0460\,\mathrm{mm} \tag{7.45}$$

Type 5: 0.5%–3% Al; 2.6%–4% Mg:

$$1.00\,\text{oz/ft}^2 = 0.00193\,\text{in.} = 305\,\text{g/m}^2 = 0.0490\,\text{mm} \tag{7.46}$$

Table 7.8 gives examples of some of the designators discussed above.

A Appendix

A.1 Definitions of variables and constants used in Chapter 7 equations

a, half-thickness of steel strip.
b, location where local pressure p is half the peak pressure p_s.
c, speed of sound.
C_f, coefficient of friction.
d, nozzle gap width.
d_e, nozzle gap width at strip edge.
d_o, nozzle gap width at strip center.
E, activation energy (Eq. 7.4).
f, frequency of the pressure fluctuations.
g, gravitational constant.
G, nondimensional effective gravitational acceleration.
h, convection heat transfer coefficient.
I, X-ray intensity.
k, heat conductivity.
P, p pressure.
P_∞, P_a ambient pressure.
P_{in}, inlet pressure (plenum absolute pressure).
p_r, ratio of plenum pressure to ambient pressure.
q, liquid coating volume flow rate.
Q, nondimensional mass flux.
r, relationship between variation in coating thickness with respect average coating thickness.
R, universal gas constant.
Re_a, Reynolds number at auxiliary jet exit.
Re_m, Reynolds number at main jet exit.
$\dot{s}$, skimmings generation rate.
S, nondimensional shear stress.
St, Strouhal number.
t, coating thickness.
T, temperature.
T_{in}, inlet temperature.
T_j, temperature of the gas jet at the point where it impinges upon the moving strip.
t_{relax}, local relaxation time.
U_j, mean jet exit velocity.
U_s, strip velocity.
V molar volume.
V_a, gas velocity exiting auxiliary jet.
V_m, gas velocity exiting main jet.
We, Weber number.
x, distance in direction of strip travel.
X, mole fraction.
y, distance transverse to strip travel.
z, distance normal to plane of the strip surface.
Z, strip-to-knife distance.
α, ratio of effective viscosity in the system to reference viscosity (Fig. 7.5).
γ, ratio of gas specific heat at constant pressure C_p to the gas specific heat at constant volume C_v.

ΔG^{ex}, excess Gibbs free energy.
ΔP_N, stagnation pressure.
ε, strip emissivity.
η, dynamic viscosity.
λ, length of coating wave.
μ, X-ray mass absorption coefficient.
ν, kinematic viscosity.
ξ, x/b (nondimensional).
Π_g, wiping number.
ρ density.
σ, Stefan-Boltzmann constant 5×10^{-8} W/m$^{2\circ}$K^4.
σ, surface tension.
τ, shear stress.
τ_w, local wall shear stress

References

[1] C. Pfeiler, M. Mataln, A. Kharicha, C.K. Riener, G. Angeli, Importance of the zinc film modeling for gas jet wiping simulations, in: Proc. 10th International Conference on Zinc and Zinc Alloy Coated Steel Sheet (Galvatech 2015), May 31-June 4, Association for Iron & Steel Technology, Warrendale, PA, 2015, pp. 682–689. Toronto.

[2] ASTM A653/A653M-20, Standard Specification for Steel Sheet, Zinc-Coated (Galvanized) or Zinc-Iron Alloy-Coated (Galvannealed) by the Hot-Dip Process, ASTM International, West Conshohocken, PA, 2020.

[3] ASTM A792/A792M-21a, Standard Specification for Steel Sheet, 55% Aluminum-Zinc Alloy-Coated by the Hot-Dip Process, ASTM International, West Conshohocken, PA, 2021.

[4] ASTM A90/A90M-21, Standard Test Method for Weight [Mass] of Coating on Iron and Steel Articles with Zinc or Zinc-Alloy Coatings, ASTM International, West Conshohocken, PA, 2021.

[5] O. Pearcey, Performance results and future developments for pot equipment, in: Proceedings Galvanizers Association 105th Meeting, October 6-9, Galvanizers Association, Marine City, MI, 2013. Newport Beach, CA.

[6] Y.H. Liu, Density and viscosity of zinc alloys, Metall. Mater. Trans. A 37A (2006) 2767–2771.

[7] S.L. Chen, Y.A. Chang, A thermodynamic analysis of the Al-Zn system and phase diagram calculation, Calphad 17 (1993) 113–124.

[8] H. Liang, Y.A. Chang, A thermodynamic description for the Al-Cu-Zn system, J. Phase Equilib. 19 (1998) 25–37.

[9] H.R. Thresh, D.W.G. White, J.O. Edwards, J.W. Meier, Properties of molten zinc and zinc alloys final report: preface and part I: density, in: International Lead Zinc Research Organization Project ZM-54, September 15, International Zinc Assn., Durham, NC, 1964. Available from.

[10] D. Liu, N. Gao, A comparative study on different types of Zn-Al-Mg coatings, in: Proceedings Galvanizers Association 112th Meeting, March 27–30, Galvanizers Association, Marine City, MI, 2022. Myrtle Beach, SC.

[11] H.R. Thresh, D.W.G. White, J.O. Edwards, J.W. Meier, Properties of molten zinc and zinc alloys final report: part II: viscosity, in: International Lead Zinc Research Organization Project ZM-54, October 1, International Zinc Assn., Durham, NC, 1965. Available from.

[12] D. Sichen, J. Bygden, S. Seetharaman, A model for estimation of viscosities of complex metallic and ionic melts, Metall. Mater. Trans. B Process Metall. Mater. Process. Sci. 25B (1994) 519–525.

[13] H. So, H.G. Yoon, M.K. Chung, CFD analysis of sag line formation on the zinc-coated steel strip after the gas-jet wiping in the continuous hot-dip galvanizing process, ISIJ Int. 51 (1) (2011) 115–123.

[14] A. Gosset, J.M. Buchlin, Jet wiping in hot-dip galvanization, J. Fluids Eng. 129 (2007) 466.

[15] M. Dubois, Minimization of the N_2 dilution when wiping in air, in: AISTech 2019-Proceedings of the Iron & Steel Technology Conference, May 6–9, Association for Iron & Steel Technology, Warrendale, PA, 2019. Pittsburgh.

[16] S.J. Palmer, The effect of temperature on surface tension, Phys. Educ. 11 (2) (1976) 119.

[17] D. Liu, Private Communication, March 8, 2022.

[18] T. Iida, R.I.L. Guthrie, The Physical Properties of Liquid Metals, Oxford University Press, New York, NY, 1993, p. 134.

[19] H.R. Thresh, D.W.G. White, J.O. Edwards, J.W. Meier, Properties of molten zinc and zinc alloys final report: part III: surface tension, in: International Lead Zinc Research Organization Project ZM-54, October 15, International Zinc Assn., Durham, NC, 1964. Available from.

[20] Y.S. Touloukian, Thermophysical Properties Research Center Data Book, Purdue Univ., W. Lafayette, IN, 1964.

[21] C. Cone, Continuous strip galvanizing Developments, in: Iron and Steel Engineer, vol. 39, March, 1962, pp. 80–87.

[22] J.A. Thornton, H.F. Graff, An analytical description of the jet finishing process for hot-dip metallic coatings on strip, Metall. Trans B 7B (1976) 607–618.

[23] D.J. Beam, J.J. Butler Jr., J.C. Hawkins, The development for air coating control for continuous strip galvanizing, in: Proceedings Galvanizers Association 61st Meeting, April 8–9, Galvanizers Association, Marine City, MI, 1970. St. Louis, MO.

[24] British Patent GB588281A W. Worthington, Procedure and Apparatus for Continuously Coating Strip Metal, May 19, 1947.

[25] US Patent 3,499,418 J.T. Mayhew, Continuous Metallic Strip Hot-Dip Metal Coating Apparatus, March 10, 1970.

[26] M. Dubois, Hot gas wiping: a review of the effects on the process, in: Proceedings Galvanizers Association 98th Meeting, October 22-25, Galvanizers Association, Marine City, MI, 2006. Columbus, OH.

[27] F.R.N. Nabarro, F. de Villiers, The Physics of Creep, CRC Press, 1995.

[28] C.H. Ellen, C.V. Tu, An analysis of jet stripping of liquid coatings, J. Fluids Eng. 106 (1984) 399–404.

[29] M. Dubois, Private Communication, March 2022. with permission.

[30] S. Beltaos, Oblique impingement of plane turbulent jets, in: T.M. Shih (Ed.), Proc. Second National Symp. On Numerical Properties and Methodologies in Heat Transfer, Sept. 1981, Univ. of Maryland, Hemisphere Publ. Corp., 1983, pp. 405–417.

[31] Galvanized Autobody Partnership Project ZCO-2-3, Improving Coating Weight Consistency, Final Report, International Zinc Assn., Durham, NC, July 2004.

[32] D.J. Phares, G.T. Smedley, R.C. Flagan, The inviscid impingement of a jet with arbitrary velocity profile, Phys. Fluids 12 (8) (2000) 2046–2055.

[33] Galvanized Autobody Partnership Project ZCO-56, Improving Coating Weight Consistency, Final Report, International Zinc Assn., Durham, NC, March 2017.

[34] A. Ritcey, J.R. McDermid, S. Ziada, The maximum skin friction and flow field of a planar impinging gas jet, J. Fluids Eng. 135 (2017). 101204-1-13.

[35] A.N. Hrymak, E.A. Elsaadawy, G. Hanumanth, J.R. McDermid, F.E. Goodwin, Air knife coating control models, in: AISTech 2005-Proceedings of the Iron & Steel Technology Conference, May 9–12, Association for Iron & Steel Technology, Warrendale, PA, 2005. Charlotte, NC.

[36] L.Q. Phan, A.D. Johnstone, B. Kosasih, W. Renshaw, Numerical investigation of flow characteristics of galvanizing line air knife, in: Proc. 11th International Conference on Zinc and Zinc Alloy Coated Steel Sheet (Galvatech 2017), November 12–16, Iron and Steel Institute of Japan, Tokyo, 2017, pp. 222–306.

[37] A. Gosset, P. Rambaud, L. Castellano, M. Dubois, J.M. Buchlin, Modeling of gas-jet wiping at small standoff distances, in: 6th European Coating Symposium, Bradford, Great Britain, 2005.

[38] H. Martin, Impinging jet flow heat and mass transfer, Adv. Heat Tran. 13 (1977) 1–60.

[39] H.T. Lin, H.-L. Hoh, Mixed convection from an isothermal vertical flat plate moving in parallel or reversely to a free stream, Heat Mass Transf. 32 (1997) 441–445.

[40] D. Delaunay, 55% Al/Zn coating: which furnace, for which product? in: Proceedings Galvanizers Association 100th Meeting, October 26–29, Galvanizers Association, Marine City, MI, 2008. Baltimore, MD.

[41] A. Gosset, D. Lacanette, S. Vincent, E. Arquis, P. Gardin, LES-VOF simulations of gas-jet wiping: confrontation to experiments, in: 6th European Coating Symposium, Bradford, Great Britain, 2005.

[42] A. Jarošik, G. Angeli, Statistical modelling of gas-jet wiping in hot-dip galvanizing, in: Proc. 8th International Conference on Zinc and Zinc Alloy Coated Steel Sheet (Galvatech 2011), June 21–25, Associazione Italiana di Metallurgia, Milan, 2011. Genoa, Italy.

[43] F. Vecchiet, M. Zorzut, M. Pavlicevic, A. Poloni, L. Vignolo, M. Turchetto, G. Kohler, Development of air knives using fluid dynamics modeling, in: Proceedings Galvanizers Association 98th Meeting, October 22–25, Galvanizers Association, Marine City, MI, 2006. Columbus, OH.

[44] M. Volpato, D. Garassino, A. Alessandrini, Efficient and effective operation of a hot dip galvanizing line, in: Proc. 12th International Conference on Zinc and Zinc Alloy Coated Steel Sheet (Galvatech 2021), June 22–24, Austrian Society for Metallurgy and Materials, 2021, pp. 806–817. Virtual Conference.

[45] G. Takeda, Y. Ito, K. Yamashiro, H. Takahashi, H. Uoya, N. Ninomiya, Effect of initial velocity distributions in transition region impingement of slot jet, in: Proc. 12th International Conference on Zinc and Zinc Alloy Coated Steel Sheet (Galvatech 2021), June 22–24, Austrian Society for Metallurgy and Materials, 2021, pp. 756–763. Virtual Conference.

[46] J. Wang, T. Klinkberg, M. Teschner, D. Weirich, T. Koll, Simulation and optimisation of a wiping knife on the lab scale, in: Proc. 12th International Conference on Zinc and Zinc Alloy Coated Steel Sheet (Galvatech 2021), June 22–24, Austrian Society for Metallurgy and Materials, 2021, pp. 1180–1191. Virtual Conference.

[47] G. Kohler, G. Caporal, Ancillary benefits of the high performance of danieli X-JET, in: Proc. 10th International Conference on Zinc and Zinc Alloy Coated Steel Sheet (Galvatech 2015), May 31–June 4, Association for Iron & Steel Technology, Warrendale, PA, 2015, pp. 714–718. Toronto.

[48] M. Dubois, F.E. Goodwin, Impact of lip adjustment on wiping efficiency, Iron Steel Technol. (2021) 50–57.

[49] S. Alibeigi, J.R. McDermid, J. Thomson, Characterization of a new air knife design for the McMaster University Galvanizing simulator, in: Proc. 10th International Conference on Zinc and Zinc Alloy Coated Steel Sheet (Galvatech 2015), May 31–June 4, Association for Iron & Steel Technology, Warrendale, PA, 2015, pp. 697–701. Toronto.

[50] K.J. Ahn, M.K. Chung, A Noble Gas wiping system to prevent the edge overcoating in continuous hot-dip galvanizing, ISIJ Int. 46) (4 (2006) 573–578.

[51] S. Alibeigi, J.R. McDermid, S. Ziada, F.E. Goodwin, Effect of air knife geometry on coating weight control in the continuous galvanizing line, in: Proc. 9th International Conference on Zinc and Zinc Alloy Coated Steel Sheet (Galvatech 2013), September 23–27, Chinese Society for Metals, Metallurgical Industry Press, Beijing, 2013, pp. 437–440. Beijing.

[52] J. van Eenennaam, J. Link, Easy analysis of air knife performance by stationary flow calculations, in: Proc. 12th International Conference on Zinc and Zinc Alloy Coated Steel Sheet (Galvatech 2021), June 22-24, Austrian Society for Metallurgy and Materials, 2021, pp. 713–724. Virtual Conference.

[53] F.E. Goodwin, M. Dubois, Review on wiping, a key process limiting CGL productivity, in: AISTech 2012-Proceedings of the Iron & Steel Technology Conference, May 7–12, Association for Iron & Steel Technology, Warrendale, PA, 2012. Atlanta, GA.

[54] O. Pearcey, H. Brown, I. Marincic, J. Tham, The evolution of coating control systems as the need for new coated products expands, Berg- Huettenmaenn. Monatsh. 166 (11) (2021) 525–534.

[55] M. Dubois, F. Pizzuto, Thick strip: specific defects and pot equipment, in: AIST Galvanizing Technology Committee Virtual Meeting, September 22–23, Association for Iron & Steel Technology, Warrendale, PA, 2021.
[56] M. Dubois, Strip vibration at the pot: causes and investigations, in: Proceedings Galvanizers Association 104th Meeting, October 21–24, Galvanizers Association, Marine City, MI, 2012. Detroit, MI.
[57] R. Wilhelm, Coating weight control & air knives, in: Presentation to GalvInfo Galvanizing Line Operators Course, June 24–26, International Zinc Assn., Durham, NC, 2007. Indianapolis, IN.
[58] M. Hamide, K. Beaujard, Predictive model of strip vibrations in gas jets cooling sections, in: Proc. 8th International Conference on Zinc and Zinc Alloy Coated Steel Sheet (Galvatech 2011), June 21–25, Associazione Italiana di Metallurgia, Milan, 2011. Genoa, Italy.
[59] M. Renard, J. Muller, D. Van de Vyver, F. Gurniki, K. Beaujard, Control of strip vibrations in gas jets cooling areas, in: Proc. 7th International Conference on Zinc and Zinc Alloy Coated Steel Sheet (Galvatech 2007), November 18–22, Iron and Steel Institute of Japan, Tokyo, 2007, pp. 57–62. Osaka.
[60] M. Boyer, Latest choices for new or modernized automotive CGLs after pot cooling (APC) sections, in: Presented at AISTech 2012—Iron & Steel Technology Conference, May 8–11, Association for Iron & Steel Technology, Warrendale, PA, 2017. Nashville, TN.
[61] M.M. McCosby, High speed video as an industrial diagnostic tool, in: Proceedings Galvanizers Association 105th Meeting, October 6–9, Galvanizers Association, Marine City, MI, 2013. Newport Beach, CA.
[62] S. Gaignard, M. Dubois, Characterization of strip vibration at the wiping nozzles, in: Proc. 6th International Conference on Zinc and Zinc Alloy Coated Steel Sheet (Galvatech 2004), April 4–7, Association for Iron & Steel Technology, Warrendale, PA, 2004, pp. 207–216. Chicago, IL.
[63] O. Pearcey, Air knife wiping distance and strip stability measurement, in: Proceedings Galvanizers Association 107th Meeting, October 18-21, Galvanizers Association, Marine City, MI, 2015. Toledo, OH.
[64] L. Luciano, G. Caporal, F. Vecchiet, F. Guastini, A. De Luca, M. Minen, Danieli wiping process evolution: from new 3rd air knives generation to one integrated technological package able to produce tight coating weight tolerances at higher processing speed and reduced operating costs, in: Proc. 9th International Conference on Zinc and Zinc Alloy Coated Steel Sheet (Galvatech 2013), September 23–27, Chinese Society for Metals, Metallurgical Industry Press, Beijing, 2013, pp. 441–447.
[65] P. Rybing, O. Sjöden, J. Kim, ABB Integrated EM stabilizer with state-of-the-art coating knife POSJET, in: Proceedings Galvanizers Association 110th Meeting, October 14–17, Galvanizers Association, Marine City, MI, 2018. Memphis, TN.
[66] T. Daube, P. Fontaine, C. Sasse, Integrated strip stabilization system FOEN DEMCOJET®, in: Proceedings Galvanizers Association 109th Meeting, October 15–18, Galvanizers Association, Marine City, MI, 2017. Troy, MI.
[67] S. Weber, S. Dombrowski, J. Koch, M. Black, E. Owsianny, D. Plätzer, Integrated solution of electromagnetic strip stabilization and air knives: first installations, experiences and results, in: Proc. 11th International Conference on Zinc and Zinc Alloy Coated Steel Sheet (Galvatech 2017), November 12–16, Iron and Steel Institute of Japan, Tokyo, 2017, pp. 465–472. Tokyo.
[68] P. Henderson, J. O'Dwyer, Installation and operating experience with a spooner after-pot strip cooler and stabilizer at DJ galvanizing, in: Proc. 10th International Conference on Zinc and Zinc Alloy Coated Steel Sheet (Galvatech 2015), May 31–June 4, Association for Iron & Steel Technology, Warrendale, PA, 2015, pp. 708–713. Toronto.
[69] U.S. Patent 10,550,459 M. Larnicol, Y. Hardy, O. Bregand, P. Gerkens, Device for Hydrodynamic Stabilization of a Continuously Travelling Metal Strip, February 4, 2020.
[70] T. Masui, Y. Kaseda, K. Ando, Warp control in strip processing plant, ISIJ Int. 31 (3) (1991) 262–267.
[71] M. Dubois, Pot roll geometry and crossbow correction, in: Proc. 12th International Conference on Zinc and Zinc Alloy Coated Steel Sheet (Galvatech 2021), June 22–24, Austrian Society for Metallurgy and Materials, 2021, pp. 695–705. Virtual Conference.
[72] K.M. Hochstetter, S.D. Domanti, G.A. Wallace, E. Garcia, J. Salinas, A. Treviño, Variable geometry air knife control utilizing constrained model predictive control, in: Proceedings Galvanizers Association 98th Meeting, October 22–25, Galvanizers Association, Marine City, MI, 2006. Columbus, OH.
[73] J.-J. Hardy, B. Grenier, S. Özer, M. Baras, Industrial benefits of dynamic air knives implementation on continuous galvanizing lines, in: Proceedings Galvanizers Association 104th Meeting, October 21–24, Galvanizers Association, 2012. Columbus, OH. Marine City, MI.
[74] M.M. McCosby, E.A. Silva, Improved properties of galvannealed dual phase 980, in: Proceedings Galvanizers Association 104th Meeting, October 21–24, Galvanizers Association, Marine City, MI, 2012. Columbus, OH.
[75] J. Sandqvist, Edge buildup detection system, in: Proceedings Galvanizers Association 107th Meeting, October 18–21, Galvanizers Association, Marine City, MI, 2015. Toledo, OH.
[76] Y. Takeishi, H. Morino, Mechanism and prevention of edge over coating in continuous hot-dip galvanizing, ISIJ Int. 40 (11) (2000) 1127–1133.
[77] S.J. Kim, J.W. Cho, K.-J. Ahn, M.K. Chung, Numerical analysis of edge overcoating in continuous hot-dip galvanizing, ISIJ Int. 43 (10) (2003) 1495–1501.
[78] M. Beck, C. Contreras, P. Durighello, R. Valentin, Wiping pilot use for baffle designs and setup improvements, in: Proc. 12th International Conference on Zinc and Zinc Alloy Coated Steel Sheet (Galvatech 2021), June 22–24, Austrian Society for Metallurgy and Materials, 2021, pp. 779–806. Virtual Conference.
[79] J.-R. Park, Noise, edge overcoating, and edge buildup phenomena associated with gas wiping in continuous hot dip galvanizing and role of edge baffles, Ironmak. Steelmak. 28 (1) (2001) 53–57.
[80] C. Bao, Y. Kang, Y. Li, Numerical analysis of edge over coating and baffle effect on hot-dip galvanizing, Surf. Rev. Lett. 24 (1) (2017) 1750010.
[81] J.P. Gimigliano, G.A.E. Kohler, Experience with the use of baffles to improve the coating uniformity at strip edges, in: Proceedings Galvanizers Association 85th Meeting, October 10–13, Galvanizers Association, Marine City, MI, 1993. Baltimore, MD.

[82] S. Devorovich, A. Jordan, S. Dombrowski, J. Koch, Optimizing the galvanizing process—electromagnetic strip stabilization and complementary solutions: eMASS, eWIPE, eBACS, in: Proceedings Galvanizers Association 102nd Meeting, October 24–27, Galvanizers Association, Marine City, MI, 2010. Huntsville, AL.

[83] H.G. Yoon, G.J. Ahn, S.J. Kim, H.K. Chung, Aerodynamic investigation about the cause of check-mark stain on the galvanized steel surface, ISIJ Int. 49 (11) (2009) 1755–1761.

[84] K. Myrillas, P. Rambaud, J.-M. Mataigne, P. Gardin, S. Vincent, J.-M. Buchlin, Numerical modeling of gas-jet wiping process, Chem. Eng. Process. 68 (2013) 26–31.

[85] L.Q. Phan, A.D. Johnstone, B. Kosasih, W. Renshaw, Wiping pressure fluctuation and its effect on galvanizing coating smoothness, in: Proc. 12th International Conference on Zinc and Zinc Alloy Coated Steel Sheet (Galvatech 2021), June 22–24, Austrian Society for Metallurgy and Materials, 2021, pp. 764–778. Virtual Conference.

[86] L.Q. Phan, A.D. Johnstone, B. Kosasih, W. Renshaw, Vortex dynamics and fluctuations of impinging planar jet, ISIJ Int. 60 (5) (2020) 1030–1039.

[87] A.D. Johnstone, B. Kosasih, L.Q. Phan, A. Dixon, W. Renshaw, Coating film evolution subject to time dependent pressure and shear stress profiles, in: Proc. 11th International Conference on Zinc and Zinc Alloy Coated Steel Sheet (Galvatech 2017), November 12–16, Iron and Steel Institute of Japan, Tokyo, 2017, pp. 314–321.

[88] L.Q. Phan, A.D. Johnstone, B. Kosasih, W. Renshaw, Patterns and reduced-order reconstruction of impinging wiping jet pressure profile fluctuation using proper orthogonal decomposition, J. Fluids Eng. 144 (2) (February 2022), 021502 (20 pages).

[89] A. Gosset, M.A. Mendez, J.-M. Buchlin, An experimental analysis of the stability of the jet wiping process: part I—characterization of the coating uniformity, Exp. Thermal Fluid Sci. 103 (2019) 51–65.

[90] D. Barreiro-Villaverde, A. Gosset, M.A. Mendez, On the dynamics of the jet wiping process: numerical simulations and modal analysis, Phys. Fluids 33 (2021), 062114, https://doi.org/10.1063/5.0051451.

[91] A.Y. Soufiani, D. Finnerty, J.R. McDermid, S. Ziada, A.N. Hrymak, F.E. Goodwin, A novel multi-slot air-knife design for lighter more consistent coating weights in the gas-jet wiping process, in: Proc. 12th International Conference on Zinc and Zinc Alloy Coated Steel Sheet (Galvatech 2021), June 22–24, Austrian Society for Metallurgy and Materials, 2021, pp. 746–755. Virtual Conference.

[92] H.G. Yoon, M.K. Chung, Development of novel air-knife system to prevent check-mark stain on galvanized strip surface, ISIJ Int. 50 (5) (2010) 752–759.

[93] G. Takeda, H. Takahashi, Method for Manufacturing Molten Metal Plated Steel Strip, U.S. Patent 8,529,998, September 10, 2013.

[94] M. Dubois, J.M. Buchlin, A. Gosset, V. Perrot, Effect of nozzle tilting on splashing in jet wiping, in: Proc. 6th International Conference on Zinc and Zinc Alloy Coated Steel Sheet (Galvatech 2004), April 4–7, Association for Iron & Steel Technology, Warrendale, PA, 2004, pp. 197–206. Chicago, IL.

[95] G. Takeda, H. Takahashi, Quality assessment of gas wiping process by simple simulator, in: Proc. 9th International Conference on Zinc and Zinc Alloy Coated Steel Sheet (Galvatech 2013), September 23–27, Chinese Society for Metals, Metallurgical Industry Press, Beijing, 2013, pp. 433–436. Beijing.

[96] T. Oke, D. Okamoto, The behavior of liquid zinc flow in full splashing, in: Proc. 12th International Conference on Zinc and Zinc Alloy Coated Steel Sheet (Galvatech 2021), June 22–24, Austrian Society for Metallurgy and Materials, 2021, pp. 706–712. Virtual Conference.

[97] T.-S. Cho, Y.-D. Kwon, S.-B. Kwon, A study of the influence of air-knife tilting on coating thickness in hot-dip galvanizing, J. Therm. Sci. 18 (3) (2009) 262–267.

[98] A. Alessandrini, A. Russo, Heavy coatings on heavy gauges: air knife operational settings for effective wiping, in: AIST Galvanizing Technology Committee Virtual Meeting, September 22–23, Association for Iron & Steel Technology, Warrendale, PA, 2021.

[99] M. Dubois, Production of heavy coatings at low line speed, in: Proc. 12th International Conference on Zinc and Zinc Alloy Coated Steel Sheet (Galvatech 2021), June 22–24, Austrian Society for Metallurgy and Materials, 2021, pp. 733–745. Virtual Conference.

[100] M. Dubois, Heavy coatings and zinc flows, in: Proceedings Galvanizers Association 111th Meeting, October 6–9, Galvanizers Association, Marine City, MI, 2019. Charleston, SC.

[101] J.R. McDermid, A.N. Hrymak, R. Fourmentin, S. Salari, F.E. Goodwin, Effect of steel substrate roughness on GA and GI coating roughness during the gas-jet wiping process, in: AISTech 2009-Proceedings of the Iron & Steel Technology Conference, May 4–7, Association for Iron & Steel Technology, Warrendale, PA, 2009. St. Louis, MO.

[102] R.G. Wilhelm Jr., D.L. Maas, Evaluating on-line surface roughness measurement for coating weight control, in: Proceedings Galvanizers Association 96th Meeting, October 3–6, Galvanizers Association, Marine City, MI, 2004. Charleston, SC.

[103] T.J. Park, S.Y. Choi, Y.H. Kim, Influence of surface roughness of full hard steel on spangle of galvanized steel and actual application results, in: Proc. 12th International Conference on Zinc and Zinc Alloy Coated Steel Sheet (Galvatech 2021), June 22–24, Austrian Society for Metallurgy and Materials, 2021, pp. 1284–1291. Virtual Conference.

[104] C. Burnett, The benefits of 'hot' coating weight measurements to coating control, in: Thermo Fisher Scientific White Paper, 2012. http://www.thermoscientific.com/content/dam/tfs/ATG/CAD/CAD%20Documents/Application%20&%20Technical%20Notes/Flat%20Sheet%20Gauging%20for%20Metals%20and%20Web%20Applications/Coating%20Weight%20Gauges%20for%20Metals/The-Benefits-of-Hot-Coating-Weight-Measurements-to-Coating-Control.pdf.

[105] A. Thurston, Thickness control during the galvanizing process, in: Presented at AISTech 2018—Iron & Steel Technology Conference, May 7–10, Association for Iron & Steel Technology, Warrendale, PA, 2018. Philadelphia, PA.

[106] R. Wilhelm, What does it mean to control coating weight? in: Proceedings Galvanizers Association 98th Meeting, October 22–25, Galvanizers Association, Marine City, MI, 2006. Columbus, OH.

[107] T.G. Moldar, D. Hajdu, T. Insperger, Chapter 10—The smith predictor, the modified smith predictor and the finite spectrum assignment: a comparative study, in: Q. Gao, H.R. Karimi (Eds.), Stability, Control and Application of Time-Delay Systems, Elsevier, 2019, pp. 209–226, https://doi.org/10.1016/C2017-0-02175-4.

[108] G. Romero, L. García, I. Pérez, R. Domínguez, M.A. Panduro, A. Méndez, Robust stability analysis of the hot-dip galvanizing control system, Int. J. Autom. Control. 1 (2/3) (2007) 220–232.

[109] R. Gouel, New high performance X-ray coating weight gauge for metals, in: Presented at AISTech 2021—Iron & Steel Technology Conference, June 27–July 1, Association for Iron & Steel Technology, Warrendale, PA, 2021. Nashville, TN.

[110] M. Dubois, J. Callegari, Methodology to quantify objectively coating weight uniformity, Iron Steel Technol. (2017) 68–74.

[111] M.B. Pierson, C. Flinchum, Finishing Method for Conventional Hot Dip Coating of a Ferrous Base Metal Strip with a Molten Coating Metal, U.S. Patent 4,330,574, May 18, 1982.

[112] I. Egry, E. Ricci, R. Novakovic, S. Ozawa, Surface tension of liquid metals and alloys—recent developments, Adv. Colloid Interf. Sci. 159 (2010) 198–212.

[113] K. Nagata, S. Masui, R. Yamashita, Nisshin-steel's Toyo works starts up a state-of-the-art galvanizing line, in: Proc. 6th International Conference on Zinc and Zinc Alloy Coated Steel Sheet (Galvatech 2004), April 4–7, Association for Iron & Steel Technology, Warrendale, PA, 2004, pp. 185–196. Chicago, IL.

[114] T.C. Kim, I.-R. Sohn, Y.H. You, M.S. Kim, J.S. Kim, Investigation of hot-dip ZnMgAl coatings for automotive applications, in: Proc. 12th International Conference on Zinc and Zinc Alloy Coated Steel Sheet (Galvatech 2021), June 22–24, Austrian Society for Metallurgy and Materials, 2021, pp. 100–107. Virtual Conference.

[115] Galvanized Autobody Partnership Report ZCO-77, Dross Minimization in Galvanizing Bath, Final Report, International Zinc Assn., Durham, NC, August 2020.

[116] M. Dubois, Supersonic wiping: new window or wrong trail, in: Proc. 9th International Conference on Zinc and Zinc Alloy Coated Steel Sheet (Galvatech 2013), September 23–27, Chinese Society for Metals, Metallurgical Industry Press, Beijing, 2013, pp. 427–432.

[117] M. Malmendier, J. Crahay, J.-F. Noville, S. Wilmotte, Wiping by electromagnetic techniques in hot dip galvanizing, in: Proc. 4th International Conference on Zinc and Zinc Alloy Coated Steel Sheet (Galvatech 1998), September 20–23, Iron and Steel Institute of Japan, Tokyo, 1998, pp. 751–756. Makuhari, Chiba, Japan.

[118] Steel Sheet Facts 10, Steel Gauges and Thicknesses, Canadian Steel Sheet Building Institute, September 2012. http://cssbi.ca/assets/resources/Fact_Sheets/CSSBI-Facts10-12.pdf.

[119] ASTM A525, Specification for General Requirements for Steel Sheet, Zinc-Coated (Galvanized) by the Hot-Dip Process, (Withdrawn 1994), ASTM International, West Conshohocken, PA, 1994.

[120] ASTM A924/A924M-20, Standard Specification for General Requirements for Steel Sheet, Metallic-Coated by the Hot-Dip Process, ASTM International, West Conshohocken, PA, 2020.

Chapter 8

Galvannealing

8.1 Introduction—History

Galvannealing is a hot-dip galvanize post heat treatment process to produce Zn-Fe intermetallic phases in the coating that have improved properties over the original hot-dip product. The heat treatment of galvanized Zn coatings to produce Fe-Zn intermetallic phases on the surface of a coated article was proposed nearly 100 years ago. Herman in 1922 [1] received a patent on "heat treating the coated material at a relatively high temperature as compared with the temperature at which such materials are coated." In 1925, Cornell [2] proposed heating an electrodeposited or hot-dip coated part. Heat treating Zn coatings in the 1930s utilized several heating methods: impinging flames [3], Pb bath [4], and molten salt bath [5]. Several sheet steel processing patents soon followed, e.g., [6], and in 1940, Sendzimir received his famous patent for continuous processing of steel sheet in a reducing environment without fluxing, which also included a final heat-treating step after hot dipping [7]. A patent granted to Inland Steel in 1962, adapting the Sendzimir configuration, appeared to show the more modern general arrangement for continuous process of galvanneal sheet, Fig. 8.1 [8], and galvanneal processing was detailed by Inland in 1979 [9]. Concurrently, heat treatment of electrodeposited zinc alloy pipe, e.g., [10], and Zn-Fe galvanneal sheet, e.g., [11] was also reported.

8.2 Galvanneal processing

The galvanneal (GA) coating on sheet steel typically has an iron content of 8%–11% Fe. It can be either produced by co-deposition on an electrogalvanizing line or by diffusion alloying of the hot-dip zinc coating with the steel by additional heating in a tower above the coating bath. During the 1980s, 10 electrogalvanizing lines capable of producing co-deposited Zn-Fe coatings were constructed in Japan, and several lines were also constructed in North America and Europe [12]. Since then, continual improvement in the development of the hot-dip process has led to rapid expansion of capacity for production of hot-dip GA, which has displaced much of the more costly electrogalvanized product. Presently, almost all production of the GA coating in the world is by the hot-dip process. The attractions of the GA coating for automotive body applications are its ease in spot welding and painting, compared with the normal galvanized product. GA generally exhibits less undercutting corrosion beneath the paint at exposed edges, scratches, or other defects in the paint than the conventional galvanized product. It is also characterized by a high coating hardness and sometimes brittle behavior during forming, and this has led to most of the processing challenges in recent years.

In 1976, Thornton and Graff [13] discussed a very thin (5.5 μm per side) GA coating that can be produced by intensive wiping of the coating after it emerges from a low-Al Zn bath that permits the ZnFe outburst layer to grow on the strip, removing almost all of the liquid Zn overlay. However, this coating is used for construction products only, such as painted metal doors, because its thickness and surface uniformity tolerances do not meet auto quality requirements.

Hot-dip galvannealed coatings are produced in a continuous galvanizing line, Fig. 8.2, in which the steel strip is fed through a bath of molten zinc. The steel substrate can be selected from several cold-rolled steel products. During the continuous process, the steel is suitably cleaned, prior to immersion in a zinc bath. After leaving the zinc bath, the strip is subjected to heat treatment. In this process, the strip is further heated by passing it through a furnace directly above the coating bath. By heating to approximately 1000–1050°F (538–565°C) and holding the strip at this temperature for a specific amount of time, the zinc coating is alloyed with iron by diffusion between the molten zinc and iron from the steel strip. The result is that the final product has a coating that is approximately 90% zinc and 10% iron. The final iron concentration depends on the heating cycle since the total amount of diffusion is a function of the time/temperature cycle. A schematic representation of the galvannealing process is shown in Fig. 8.3. The resulting coating is shown in Fig. 8.4.

Galvanized coatings are essentially pure Zn, with between 0.2% and 0.5% total Al, and most of the Al concentrated in the thin inhibition layer next to the steel. In contrast, the GA coating contains 9%–12% bulk Fe, together with 0.11%–0.14%

The Metallurgy of Zinc Coated Steels. https://doi.org/10.1016/B978-0-323-99984-7.00008-7

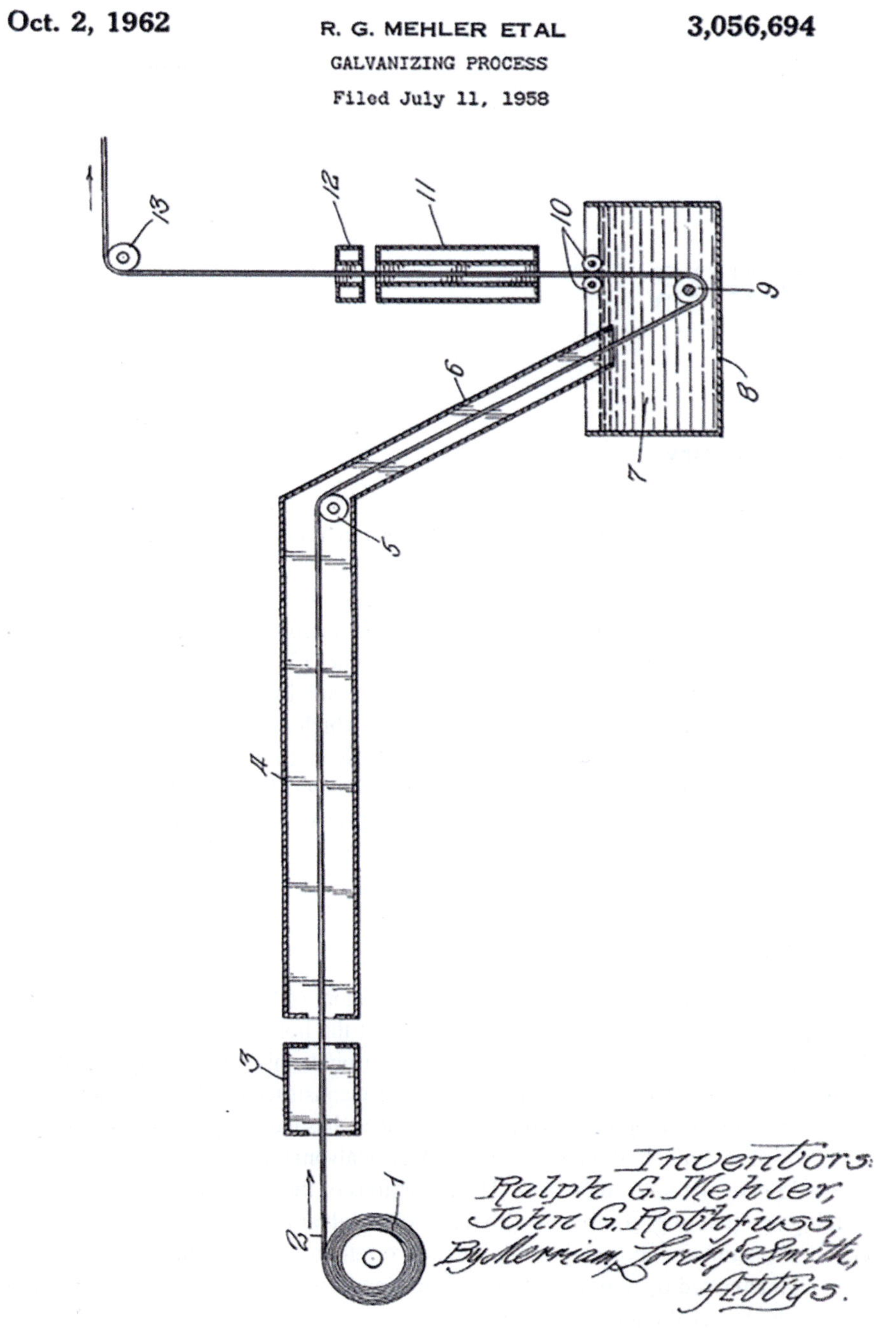

FIG. 8.1 An Inland Steel galvannealing process: US patent 3,056,694. *(From R.G. Mehler, J.G. Rothfuss, Galvanizing process, US Pat. 3,056,694, Oct. 2, 1962.)*

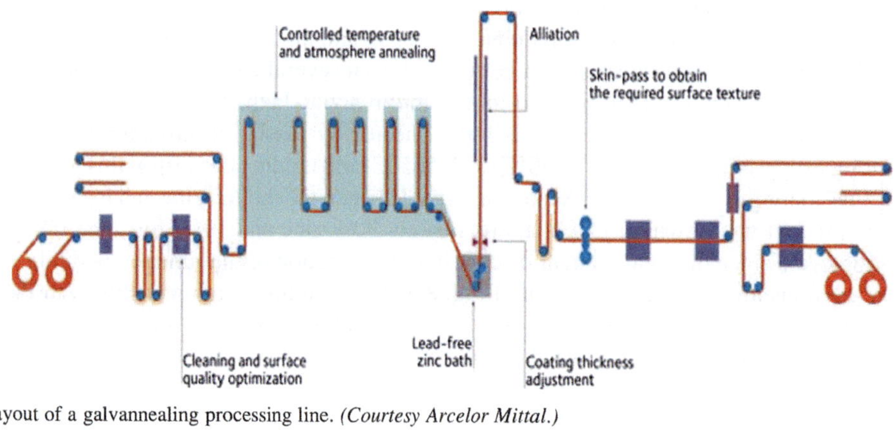

FIG. 8.2 Typical layout of a galvannealing processing line. *(Courtesy Arcelor Mittal.)*

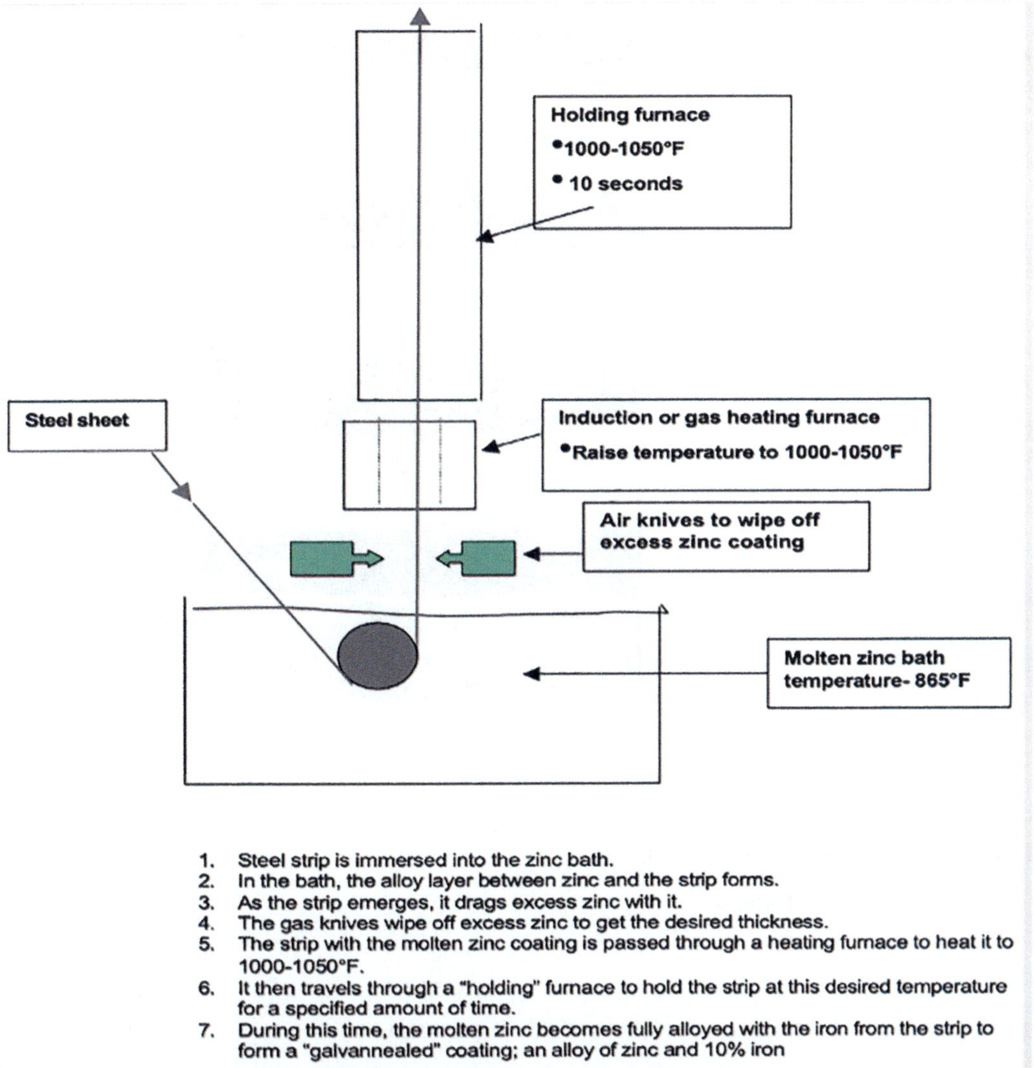

FIG. 8.3 The galvannealing process. *(From GalvInfoNote 1.3, Galvanneal—Differences From Galvanized? IZA, Rev. 1.4, Sept. 2018, www.galvinfo.com.)*

total Al that is more uniformly spread through the coating thickness than in the case of the normal galvanized product. The Fe in the GA coatings is, however, not uniformly distributed, being combined with Zn in three distinct iron phases. These phases are shown in Table 8.1, along with their Al and Fe contents. The phases in Table 8.1 are shown in the order that they occur in the coating, with the high iron gamma layer next to the steel substrate. It is also important for good appearance and formability that the top zeta layer contains no less than about 5% Fe, or there will be a risk of free zinc on the surface. The Al content shown for the three alloy layers is a result of its diffusion outward from the inhibition layer next to the steel, where it was concentrated prior to the reheating cycle. Fig. 8.5 shows a micrograph of a cross section of the GA coating showing the three alloy phase layers. GA coatings are hard and brittle; therefore, during bending and press forming, coating cracks and powdering are always present to some degree. The as-produced GA coating usually exhibits the presence of cracks, as shown in Fig. 8.6. Powdering is the detachment of FeZn particles from the coating under compressive strain and will be discussed in Section 8.6.1 (see Fig. 8.27). Each coating line must develop practices to produce the optimum coating properties for the particular end use in order to balance performance in forming versus coating line productivity.

In continuous galvannealing, heating, holding, and cooling sections are added to the upleg of the hot-dip coating line, Fig. 8.7, above the jet finishing nozzles that are used for coating thickness control. A gas-fired furnace or an induction coil is used to heat the strip, and in some designs a holding zone is utilized prior to cooling. Besides the variables associated with hot-dip galvanizing, e.g., bath temperature, substrate alloy content, and bath composition, a major variable of the

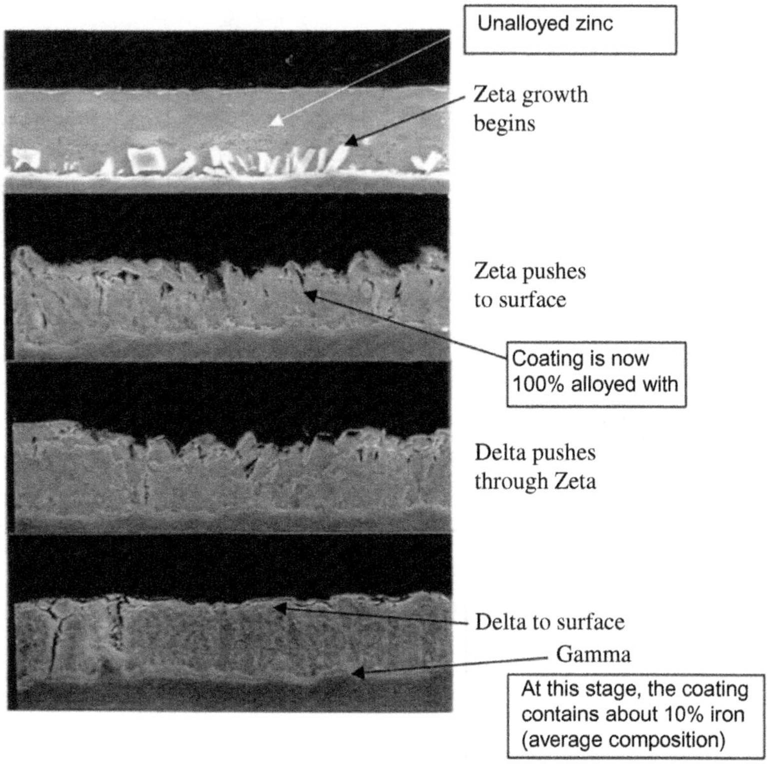

FIG. 8.4 The stages of alloying between the steel strip and the molten zinc coating to produce "galvanneal." *(From GalvInfoNote 1.3, Galvanneal—Differences From Galvanized? IZA, Rev. 1.4, Sept. 2018, www.galvinfo.com.)*

TABLE 8.1 Galvanneal phases and compositions.

Phase	Formula	wt% Fe	wt% Al
Zeta (ζ)	$FeZn_{13}$	5.2–6.1	0.7
Delta (δ)	$FeZn_{10}$	7.0–11.5	3.7
Gamma (Γ)	Fe_4Zn_9	23.5–28.0	1.4

Adapted from F.E. Goodwin, Developments in the production of galvannealed steel for automotive, in: International Symposium of NMD-ATM-2012, Nov. 2012, pp. 1–21. Trans. India Inst. Metals. N. Okamamoto, K. Tanaka, A. Yasuhara, H. Inui, Structure of the δ1p phase in the Fe-Zn system by single crystal X-ray diffraction combined with scanning transmission electron microscopy, Acta Crystallogr. B70 (2014) 275–282.

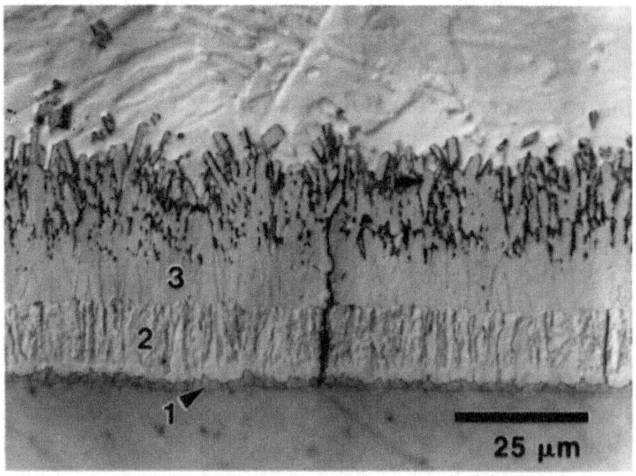

FIG. 8.5 Microstructure of Zn coating formed after 300s immersion in a 450°C, 0.00wt% Al bath on a ULC steel substrate. (1) Gamma (Γ) phase, (2) delta (δ) phase (3), zeta (ζ) phase. *(From A.R. Marder, The metallurgy of Zn-coated steel, Prog. Mater. Sci. 45 (2000) 191–271.)*

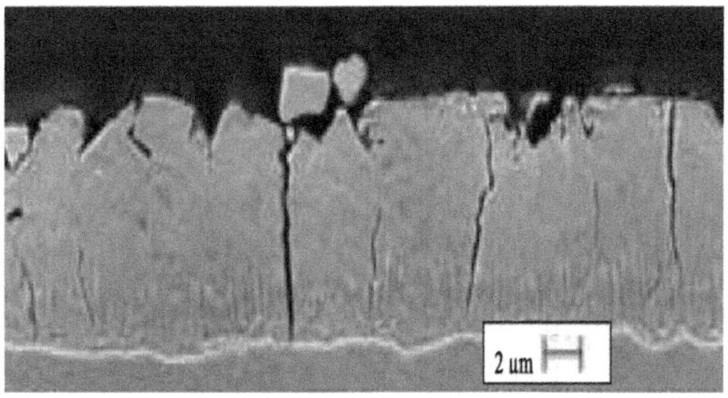

FIG. 8.6 Typical appearance of cracks in the fully galvannealed coating. *(From F.E. Goodwin, Developments in the production of galvannealed steel for automotive, in: International Symposium of NMD-ATM-2012, Nov. 2012, pp. 1–21. Trans. India Inst. Metals. Galvanneal Products and Processing, Presentation to GalvInfo Operations Course, 7-9 Nov, 2012)*

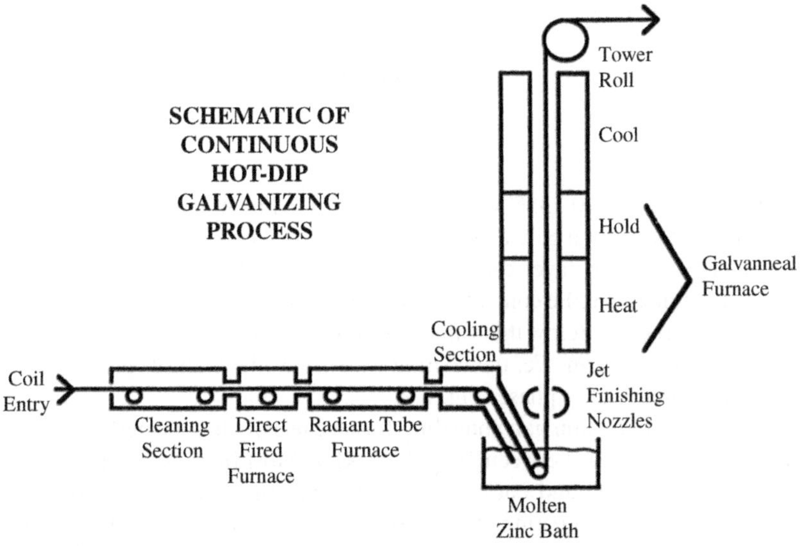

FIG. 8.7 Schematic of a typical continuous hot-dip galvanizing line with furnace section for galvannealing. *(From A.R. Marder, The metallurgy of Zn-coated steel, Prog. Mater. Sci. 45 (2000) 191–271.)*

galvannealing process is the line speed, which will control the dependent galvannealing variables such as (1) heating rate, (2) peak or holding temperature and time, and (3) cooling rate.

Characterizing the temperature profile of the entire cooling tower is extremely important. A typical calculated temperature profile of the strip moving through the cooling tower portion of a modern hot-dip galvannealing line is seen in Fig. 8.8 [14]. It is apparent from this profile that the temperature exposure of the strip is never isothermal. The galvannealing furnace takes the strip from the air knives at 450°C to a peak temperature of about 505°C at a heating rate of approximately 15°C/s. The strip is held in the soaking furnace at the peak temperature of around 500°C for 8–9 s, where some slight cooling occurs (2°C/s), and in the air coolers the strip cools down to below 350°C at a rate of over 17°C/s. Kanamura and Nakayama [15] showed that the heating rate during galvannealing should be as high as possible to limit the growth of both zeta (ζ) and gamma (Γ) phases. Hayes [16] modeled the galvanneal temperature profile and studied the effect of heating rate, peak temperature, and cooling rate on morphology development in a series of alloy substrates and bath Al compositions. He found that during galvannealing it is entirely possible for alloying to occur on heating and cooling, as well as during isothermal holding. For example, heating rates necessary to prevent alloying ranged between 25 and 50°C/s depending on the substrate composition and peak temperature. These results were based on the transition in coating morphology and Fe pick-up in the coating. Likewise, cooling rates less than 50°C/s resulted in coatings that were more alloyed than the peak temperature microstructure.

Typical target variables for successful galvannealing include heating after the wiping process to a temperature between 480 and 550°C, depending upon the line speed and the type of steel. Soaking is greater than 10 s, usually in a range of 15–20 s. The coating must then be cooled to 300–320°C to allow it to pass over the top roll without any defects. The

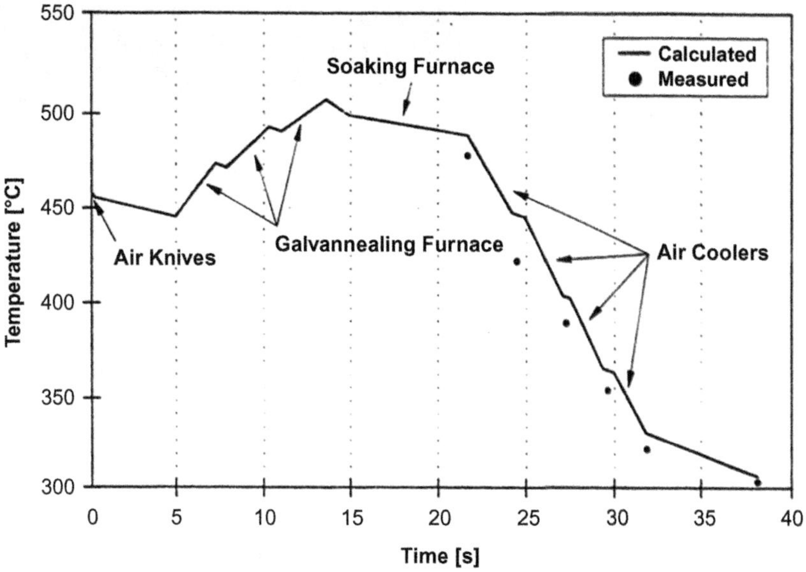

FIG. 8.8 Typical temperature profile of the cooling tower. *(From A.R. Marder, The metallurgy of Zn-coated steel, Prog. Mater. Sci. 45 (2000) 191–271.)*

effective Al content in the galvanizing bath is usually held around 0.135% or lower. At 460°C a higher bath Al content will result in a thick inhibition layer, typical of that seen with the normal galvanizing process, but still on the order of a fraction of a micrometer. Below 0.135% effective Al, a patchy inhibition layer can form, interspersed with zeta outbursts.

Claus et al. [17] measured the phase distribution as a function of Fe content, Fig. 8.9. They found that below 5 g/m^2 Fe or 10% Fe was a critical composition for the galvannealing process. Below this value, the coating consists mainly of the η and ζ phases and to a lesser extent of the δ and Γ phases. Above 5 g/m^2 Fe, the coating consists of the δ and Γ phases and a residual surface ζ phase, possibly mixed with a small amount of η phase, Fig. 8.9.

Regarding typical equipment used on modern lines, the induction galvannealing furnace usually has a length of 2–6 m, with a power of 1000–4000 kW. The induction frequency range is between 10 and 100 kHz. Most lines are dual-purpose; therefore, rails are installed to allow the furnace to be placed in and out of position for galvannealing production. The furnace is removed from the pass line to produce the galvanized coating. The holding section is also a movable section in many cases. In this case, a short holding section is used. In the case of a fixed holding section, a length of 20–30 m is usually used, with electrical heating.

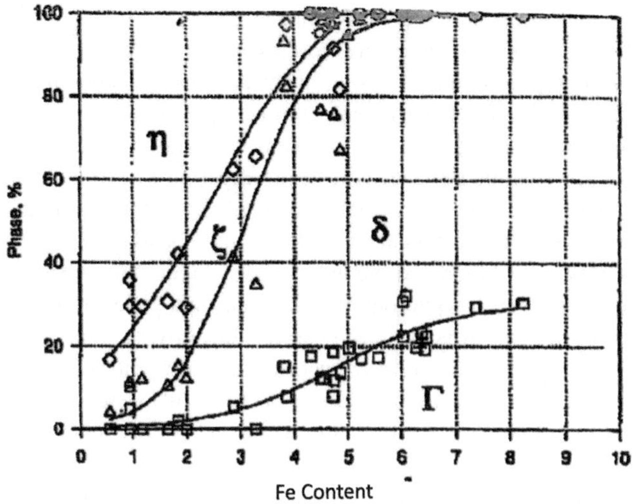

FIG. 8.9 Phase distribution as a function of the Fe content. *(From G. Claus, J. Dilewijns, B.C. De Cooman, U. Meers, Determination of the process window for optimal galvannealing, in: Galvatech '95, 1995, pp. 107–113.)*

The concept of processing windows for optimum microstructure was introduced to determine the kinetics of coating microstructure development during galvannealing [18]. Time-temperature maps were able to illustrate the isothermal temperature and times necessary to produce various morphologies, as well as iron content and powdering properties. A Gleeble thermomechanical simulator was used to replicate the galvannealing temperature and time, while the heating rate was set at 500°C/s and the cooling was fixed at 25°C/s [19,20]. Similar test conditions were used to determine the effect of the galvannealing isothermal temperature and time on coating microstructure, galvanneal TTT diagrams [21], and Fe content [17]. Still others produced processing windows for powdering [18,22]. Powdering is also influenced by coating weight, and powdering increases rapidly with coating thickness.

8.3 Phase equilibria

The Fe-Zn equilibrium phase diagram is shown in Fig. 8.10, and a detailed description of the zinc rich portion of the diagram is seen in Fig. 8.11 [23]. A brief description of each iron-zinc intermetallic phase found in hot-dip galvanized coatings appears below in order of increasing iron content [24]. A detailed description of the crystallography of each phase is found in reference [25]. Table 8.2 details the crystallography of each Zn-Fe phase [26].

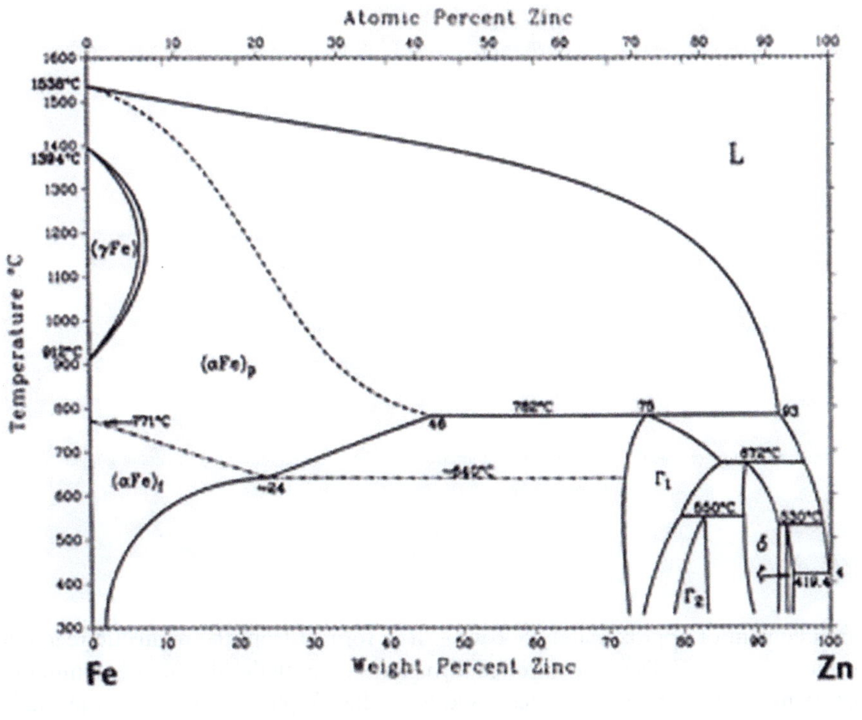

FIG. 8.10 Fe-Zn equilibrium phase diagram. *(From A.R. Marder, The metallurgy of Zn-coated steel, Prog. Mater. Sci. 45 (2000) 191–271.)*

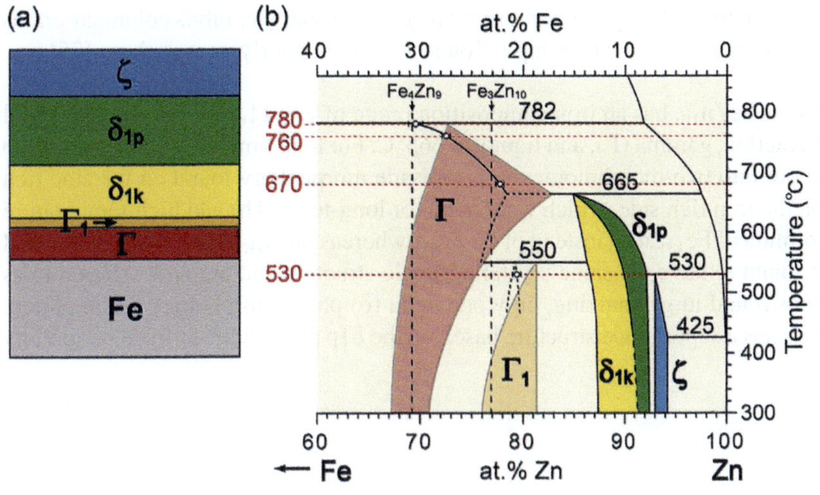

FIG. 8.11 (a) Cross-sectional schematic of GA steel sheet. A series of Fe-Zn intermetallic compounds, Γ, Γ_1, δ_{1k}, δ_{1p}, and ζ, form thin layers in the coat according to the corresponding phase diagram. (b) Zn-rich side of the phase diagram of the Fe-Zn system. The chemical compositions of the Γ- and Γ_1-phase compounds obtained by EDS analysis in the present study are plotted. *(From N. L. Okamoto, M. Inomoto, H. Takebayashi, H. Inui, Crystal structure of the Γ- and Γ1 phases in the coating layer of galvannealed steel, J. Alloys Compd. 732 (2018) 52–63.)*

TABLE 8.2 Crystal structure parameters, Burgers vectors, and interplanar distances of slip planes for the Fe-Zn intermetallic compounds and pure iron [26].

	Formula	Pearson symbol	Space group	Lattice constant (nm)	Burgers vector, *b*	Interplanar distance, *h*	*h*/*b*	Plastic deformation in polycrystalline micropillars
Γ^a	Fe_4Zn_9	*cI*52	$I\bar{4}3m$	a=0.8976	1/2 ⟨111⟩[b] 0.777 nm	{110}[b] 0.135 nm	0.174	Yes
$\Gamma_1{}^a$	$Fe_{21.2}Zn_{80.8}$	*cF*408	$F\bar{4}3m$	a=1.7989	1/2 ⟨110⟩[c] 1.272 nm	{111}[c] 0.137 nm	0.108	No
$\delta_{1k}{}^d$	$FeZn_7$	–	(higher-ordered structure based on δ_{1p})	a=2.2148	1/3 ⟨1120⟩[b] 2.222 nm	(0001)[b] 0.1298 nm	0.058	No
$\delta_{1p}{}^e$	$Fe_{13}Zn_{126}$	*hP*556	$P6_3/mmc$	a=1.2830 c=5.7286	1/3 ⟨1120⟩[b] 1.283 nm	(0001)[b] 0.1298 nm	0.101	No
ζ^f	$FeZn_{13}$	*mC*28	*C*2/*m*	a=1.0862 b=0.7608 c=0.5061 β=100.32 degrees	1/2 ⟨112⟩[b,g] 0.770 nm	{110}[b,g] 0.1412 nm	0.183	Yes
Fe^h	–	*cI*2	$Im\bar{3}m$	a=0.2867	1/2 ⟨111⟩ 0.2483 nm	{110} 0.2027 nm	0.817	Yes

[a] *N.L. Okamoto et al.*
[b] *Slip systems were determined via compression tests of single-crystal micropillars.*
[c] *The shortest Burgers vector and the plane with the largest interplanar distance were selected for the anticipated operative slip system.*
[d] *N.L. Okamoto et al.*
[e] *N.L. Okamoto et al.*
[f] *R. Belin et al.*
[g] *N.L. Okamoto et al.*
[h] *R. Kohlhaas et al.*
From H. Inui, N.L. Okamoto, S. Yamaguchi, Crystal structure and mechanical properties of Fe-Zn intermetallic compounds formed in the coating layer of galvannealed steels, ISIJ Int. 58 (2018) 1550–1561.

Zeta (ζ) phase:

The zeta (ζ) phase, $FeZn_{13}$, has an iron content of approximately 5–6 wt%. It is formed from the peritectic reaction between the delta (δ) phase and liquid zinc at 530 ± 10°C. During controlled diffusion studies in the absence of aluminum, the zeta (ζ) phase was found to form between the free zinc eta (η) phase and the delta (δ) phase. The zeta (ζ) phase is isomorphous with a base-centered monoclinic unit cell and an atomic structure that contains an iron atom and a zinc atom surrounded by 12 zinc atoms at the vertices of a slightly distorted icosahedron. The icosahedra link together to form chains and the linked chains pack together in a hexagonal array. The ζ phase in the coating layer usually exhibits columnar grains elongated along the [0001] directions, while no orientation relationship is found with the underlying δ_{1k} phase [25].

Delta (δ) phase:

The delta (δ) phase, sometimes referred to as $FeZn_{10}$, has an iron composition range of 7.0–11.5 wt% and a hexagonal unit cell. It is formed from another peritectic reaction, gamma (Γ), and liquid, at 665°C. For long time exposures, or close to equilibrium, the delta (δ) phase is separated out into two morphologies, δ_{1P}, palisade morphology found on the zinc rich side, and δ_{1K}, compact morphology found on the iron rich side, which was found for long-term (4 h) and high-temperature (553°C) immersions. δ_{1K} has the chemical formula of $Fe_{13}Zn_{126}$ instead of Fe Zn_{10}, whereas the δ_{1K} phase has the chemical formula of $FeZn_7$. Both morphologies were found to have the same crystallographic structure and are now referred to as delta (δ) phase. For the short-term immersions found in galvanizing, only one delta (δ) phase morphology was reported. Okamoto et al. [25] reported that the δ1k phase has a superlattice structure based on the δ1p phase with a tripled periodicity along the [1 1 -2 0]p direction accompanied by one-dimensional stacking disorder of structural blocks along the c-axis direction and the structure can thus be crystallographically described in terms of the order-disorder (OD) theory.

Gamma$_1$ (Γ_1) phase:

The gamma$_1$ (Γ_1) phase, $Fe_{21.2}Zn_{80.8}$, has a face-centered cubic lattice structure with an iron composition of 17–19.5 wt % at 450°C. It forms as a result of a peritectoid reaction between the gamma (Γ) phase and delta (δ) phase at 550 ± 10°C. The gamma$_1$ (Γ_1) phase appears as an uninterrupted layer between the gamma (Γ) and delta (δ) layers and can be produced upon heating at low temperatures over long periods of time.

Gamma (Γ) phase:

The gamma (Γ) phase, Fe_4Zn_9, is body-centered cubic and has an iron composition range of 23.5–28.0 wt% at 450°C. It forms as a result of a peritectic reaction at 782°C between iron and liquid zinc and exhibits a maximum Zn solubility of Zn in Fe at the delta (δ) phase peritectic temperature of 665°C.

8.4 Diffusion

Various chemical reactions or microstructural changes in solids can occur by solid state diffusion, i.e., the movement and transport of atoms in a solid phase or by liquid/solid interactions in which the liquid and solid interdiffuse in the diffusion couple. The diffusion may take place because of the presence of defects in solids, e.g., point defects (vacancies and interstitial ions) are responsible for lattice diffusion. Diffusion also takes place along line and surface defects, which include grain boundaries, dislocations, inner and outer surfaces, etc., termed high diffusivity or easy diffusion paths or short-circuit diffusion.

Diffusion is a process driven by temperature and described by statistical thermodynamics. It occurs in various processes, and it is responsible for many important physical transport effects in all fields of science and engineering. In the field of coatings, e.g., liquid zinc on a steel substrate, diffusion is the process by which the zinc diffuses into the substrate steel and similarly the iron diffuses into the liquid. The existence of intermetallic phases between the liquid zinc, steel substrate, and other additives to the liquid zinc such as aluminum, as predicted by phase diagrams, can limit diffusivity as these phases form and eventually breakdown and reform new intermetallic phases over time. Because of the very short times involved in the coating process, the metallurgy is very complex. However, interface reactions and overlay microstructures can sometimes be characterized by diffusion equations and their derivatives.

Fick's Laws are universally applied to diffusion processes. Fick's First Law states that the diffusive flux is proportional to the concentration gradient. Steady-state diffusion is determined by the amount of diffusion flux that passes through the cross-sectional area of the mating surfaces. Fick's first law of diffusion states:

$$J_x = -D\frac{dC}{dx}J$$

where J is the diffusion flux, D is a diffusion coefficient, and dC/dx is the concentration gradient through the materials in question. The negative sign is a product of the gradient.

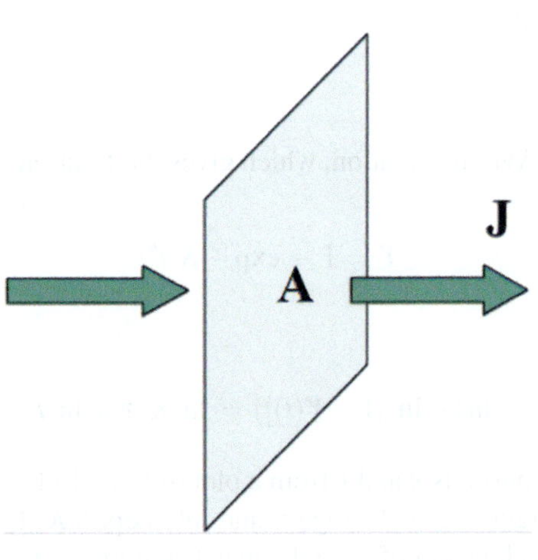

Another form of Fick's law states:

$$J = M/At$$

where M is defined as either the mass or number of atoms being diffused, A is the cross-sectional area, and t is the time required. Equating the two equations and rearranging, the following result for time, t:

$$t = (1/D)(M/A)(\mathrm{d}C/\mathrm{d}x)^{-1}$$

The diffusion coefficient or "diffusivity" usually follows an Arrhenius relationship with Q_d, the activation energy. The diffusion coefficient is determined by the equation:

$$D = D_o e^{-Q_d/RT}$$

where:

- Q_d is the activation energy for diffusion,
- R is the universal gas constant,
- T is the thermodynamic temperature experienced during the process, and
- D_o is a temperature-independent preexponential factor that depends on the materials being diffused.

The activation energy is the slope of the line on a plot of log(D) versus $1/RT$. Typical activation energies for solid-state diffusion are ~3.3–4.4 eV. As mass and area are constant for a given couple, the time required is largely dependent on the concentration gradient, which changes by only incremental amounts through the couple, and the diffusion coefficient.

Fick's second law predicts how diffusion causes the concentration to change with respect to time. It is a partial differential equation, which in one dimension reads:

$$\frac{\mathrm{d}C}{\mathrm{d}t} = D\left(\mathrm{d}^2c/\mathrm{d}x^2\right)$$

- C is the concentration in dimensions of [(amount of substance) length^{-3}], example mol/m^3
- t is time,
- D is the diffusion coefficient in dimensions of [length^2 time^{-1}], example m^2/s,
- x is the position [length]

The differential equation can be solved for any distance within the diffusion couple

$$\frac{C(x,t) - Co}{Cs - Co} = 1 - \operatorname{erf} \frac{x}{2\sqrt{Dt}}$$

- Co = initial concentration in material,
- Cs = initial concentration at surface,
- Cx = concentration at position x

Utilizing the more familiar form of the Avrami equation, which gives the fraction of transformed material after a hold time at a given temperature:

$$Y = 1 - \exp\left[-K\,t^n\right]$$

where: $K = \pi NG^3/3$ and $n = 4$

This can be rewritten as

$$\ln\left(-\ln\left[1 - Y(t)\right]\right) = \ln K + n \ln t$$

which allows the determination of the constants n and k from a plot of ln ln(1/(1 − Y)) vs ln t. If the transformation follows the Avrami equation, this yields a straight line with slope n and intercept ln$K$, Fig. 8.12.

In order to understand the actual development of the galvanneal coating, a review of the diffusion couple mechanism is required. Fig. 8.13 shows a schematic of a diffusion couple, and Fig. 8.14 shows an example of a phase diagram with three intermetallic phases, α, β, and γ along with the compositional changes with time at a given temperature for a diffusion couple of A+50% X_B (in the single phase β region). At equilibrium, t_∞, it would be expected that the diffusion couple would consist of the single β phase. Initially, after a short time, t_1, a γ phase appears along with the α phase of the A alloy

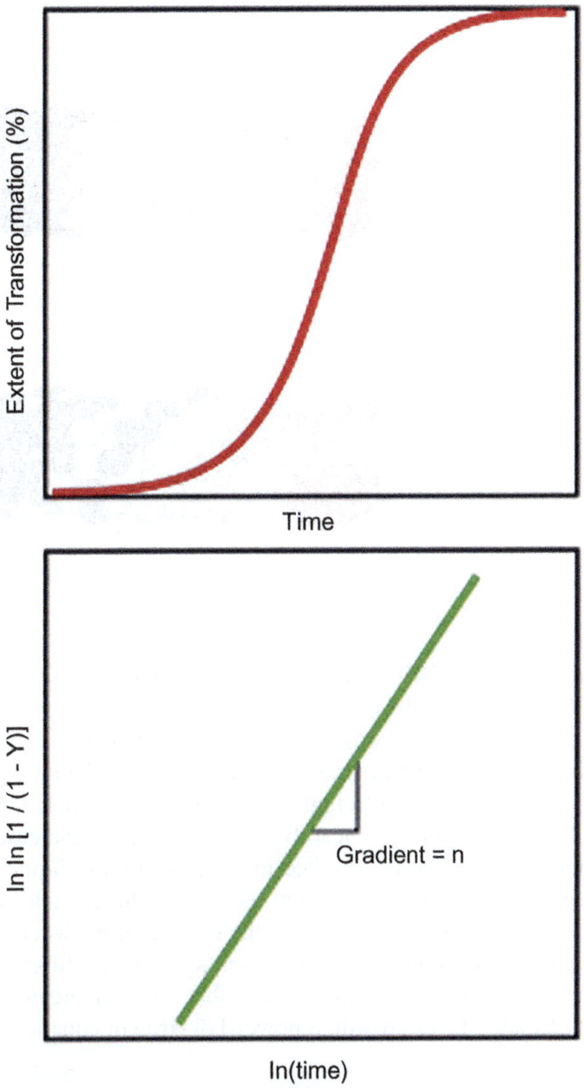

FIG. 8.12 Typical isothermal transformation plot (top). The transformation can be described using the Avrami equation as a plot of ln ln(1/(1 − *Y*)) vs ln *t*, yielding a straight line.

(0%X_B). With additional time, t_2, the β phase also appears. With more time, t_3, the β region increases at the expense of both the α phase and the γ phase. Eventually, as equilibrium is approached, t_∞, both the α phase and the γ phase disappear. Thus, it can be seen how multiple phases nucleate and grow or diminish with time.

8.5 Kinetics

Each phase layer in the zinc coating exhibits different growth kinetics, depending upon the immersion temperature, which together affect the total layer kinetics. For example, in short-time immersions up to 300s at 450°C, Fig. 8.15, the zeta (ζ) phase layer grows rapidly at first, then slows down, while the delta δ phase layer grows slowly and after time its thickness increases more rapidly. The gamma ($\Gamma+\Gamma_1$) phase layer forms only after long periods of time and seems to reach a maximum thickness of about 1 μm [27]. Similar effects were reported at 457°C for times up to 6h [28]. Horstmann [29,30] reported that there is an overall inward movement of the gamma ($\Gamma+\Gamma_1$) phase layer toward the iron, whereas the zeta (ζ) phase layer is displaced toward the zinc melt. The delta (δ) phase layer expands in both directions, but generally toward the zinc melt. Thus, as the gamma ($\Gamma+\Gamma_1$) phase layer grows into the iron, it is also consumed by the growing delta (δ) phase layer. Similarly, the delta (δ) phase layer expands into the growing zeta (ζ) phase layer that is advancing into the zinc melt. All of these transformations are governed by the diffusion of Zn into the iron substrate based on inert marker experiments in both liquid zinc [31] and solid-state diffusion studies [32]. However, Fe probably also diffuses outward through the alloy phases into the zinc melt, but at a much slower rate [31]

The effect of temperature on the kinetics of the reaction between liquid zinc and pure iron has been investigated. Verma [33] compared the coatings formed in the temperature range of 520–550°C with that formed at 450°C and found that

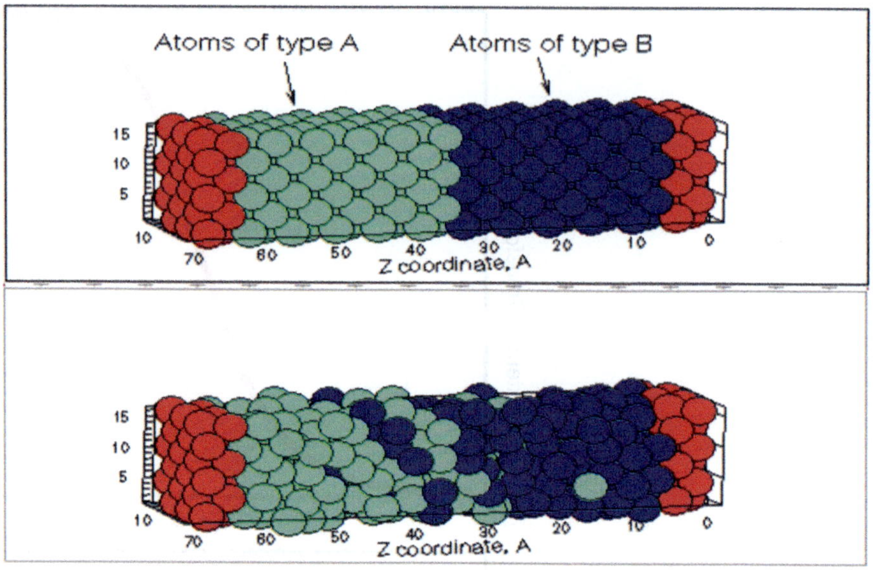

FIG. 8.13 Schematic of a typical diffusion couple. *(From N.V. Priezjev lecture notes.)*

the coating thickness reaches the maximum at 530°C when low carbon steel (0.042 wt% C, 0.2 wt% Mn) was galvanized in a zinc bath. In contrast, Bicao et al. [34] galvanized pure iron sheets in the entire temperature range of 450–530°C with immersion times of 1, 2, 3, 5, and 8 min. Microstructural analysis was conducted, and coating thickness was measured and plotted in Fig. 8.16. According to the microstructural and morphological changes, the entire temperature range was divided into three partial ranges, which represent different coating growth kinetics: (1) 450–470°C; (2) around 480°C, and (3) 490–530°C. It was found that the layer significantly changes with the increase in temperature. Segregation of the alloying elements in the boundary of the ζ phase resulted in the formation of liquid paths that prevent the formation of coherent and compact ζ layer. As a result, liquid zinc contacts the δ phase directly and erodes the δ phase. Rapid interface reactions accelerate the formation of the ζ phase, which makes the coating grow faster. It was concluded that the coating grows fastest at 480°C in hot-dip galvanizing at the same immersion time, which differs from previous studies [33]. The ζ phase formed at 480°C is incoherent and loose; thus they concluded that steels should not be galvanized at this temperature.

To evaluate the kinetics of Fe-Zn alloy layer growth, a power-law growth equation is generally used to interpret the growth rate data [14], as follows:

$$Y = Kt^n$$

where:

- Y = growth layer thickness,
- K = growth rate constant,
- t = reaction time, and
- n = growth-rate time constant.

The growth-rate time constant, n, is an indication of the type of kinetics controlling the growth of the layer under study. An n value of 0.5 is indicative of parabolic diffusion-controlled growth, while an n value of 1.0 is representative of linear kinetics in which growth is interface controlled. Table 8.3 [14] is taken from a compilation by Mackowiak and Short [28] of the many determinations made in the past. Unfortunately, many of the studies were made on steel substrates with significant additions of carbon and other alloying elements. Most of the studies were made for long immersion times in excess of 1 h, which is not indicative of the continuous hot-dip process. Nevertheless, it was concluded from these results that in the lower parabolic range, values of n tend to be around 0.5 for delta (δ) phase layer growth and total layer growth. Values for the zeta (ζ) phase layer and gamma ($\Gamma + \Gamma_1$) phase layer growth were lower, around 0.35 and 0.25, respectively. In a further experiment [27] with short time immersions of up to 300 s at a temperature of 450°C, Table 8.4, the short-time immersion results confirmed the previous work [14]. The only significant difference is found with the total layer n value. However, the total layer should reflect the dominant alloy layer phase in the total coating. For short times, less the 300 s, the zeta (ζ) phase layer dominates the coating morphology [27], whereas for long times, the delta (δ) phase layer dominates the coating structure.

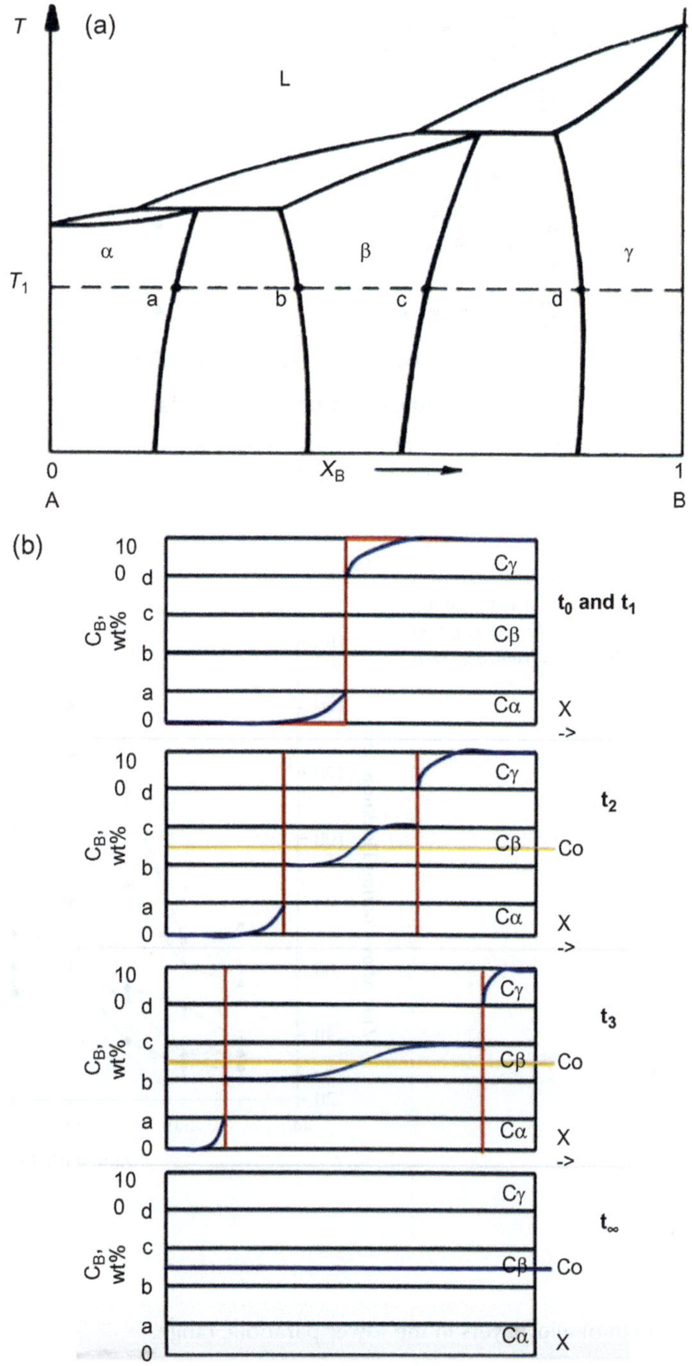

FIG. 8.14 Diffusion couple composition changes with time for a composition of 50% X_B at a temperature of T_1. Time is a variable with $t_\infty > t_3 > t_2 > t_1$. *(From J.N. Dupont lecture notes.)*

8.5.1 Inhibition layer

The inhibition layer in a galvanneal coating (GA) bath acts as an unstable transient barrier that is disrupted during the galvannealing heat treatment as Fe and Zn interdiffusion is increased by higher temperatures. This phenomenon, commonly known as *inhibition breakdown*, is a critical key step in the GA process. [14,35]. A precise understanding of the structure and composition of the inhibition layer is critical to ensure or improve the quality of GA coatings for at least two reasons according to Zapico-Alverez et al. [36]: (1) As the kinetics for the inhibition rupture is controlled by diffusion mechanisms in the inhibition layer, its nature is expected to play a direct role on the quality of the final coating, and (2) the production of high-quality GA coatings demands accurate bath management in order to keep the bath composition at the targeted values.

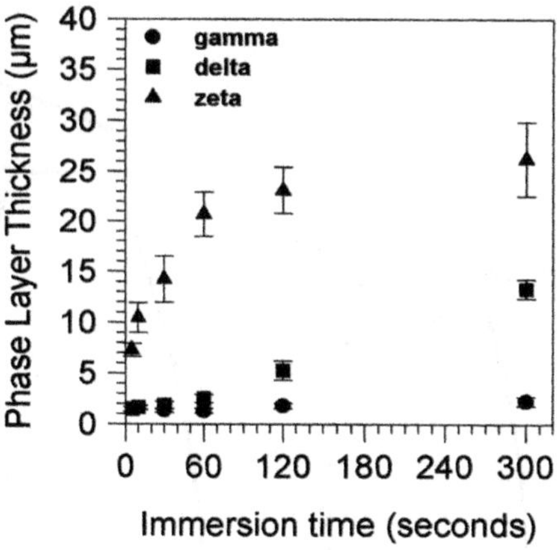

FIG. 8.15 Individual Fe ± Zn gamma (Γ) phase, delta (δ) phase, and zeta (ζ) phase layer growth for a ULC steel substrate hot dipped at 450°C in a 0.00 wt % Al ± Zn bath. *(From A.R. Marder, The metallurgy of Zn-coated steel, Prog. Mater. Sci. 45 (2000) 191–271.)*

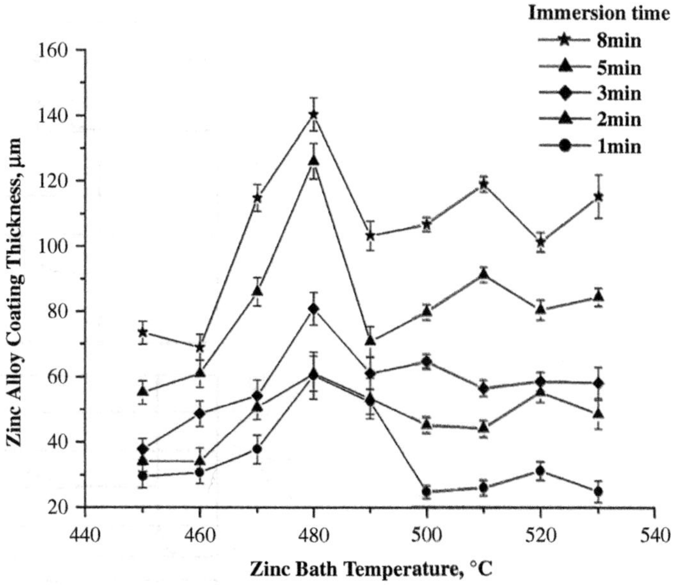

FIG. 8.16 Plot of coating thickness as a function of temperature. *(From P. Bicao, W. Jianhua, S. Xuping, L. Zhi, Y. Fucheng, Effect of zinc bath temperature on the coatings of hot-dip galvanizing, Surf. Coat. Technol. 202 (2008) 1785–1788.)*

TABLE 8.3 Values of *n* for growth of alloy layers in the lower parabolic range.

Source	Γ	δ	ζ	Total layer
Allen	0.25	0.65	0.35	0.55
Rowland	0.13	0.53	0.31	–
Blickwede	0.10	0.60	0.16	–
Horstmann and Peters	0.50	0.50	–	–
Sjoukes	0.23	0.58	0.26	–
Onishi et al.	0.23	0.49	0.36	0.43

From A.R. Marder, The metallurgy of Zn-coated steel, Prog. Mater. Sci. 45 (2000) 191–271.

TABLE 8.4 Individual phase layer growth at 450°C up to 300 s on an iron substrate (0.003 wt% C) in a pure zinc iron saturated (0.03 wt% Fe) bath.

Alloy layer	*n* value
Gamma (Γ)	0.24 ± 0.06
Delta (δ)	0.51 ± 0.11
Zeta (ζ)	0.32 ± 0.03
Total	0.35 ± 0.02

From A.R. Marder, The metallurgy of Zn-coated steel, Prog. Mater. Sci. 45 (2000) 191–271.

Once the galvanizing bath is saturated in iron, the main parameter to be controlled is the bath Al content, since the inhibition layer is highly sensitive to Al uptake.

The GA baths are composed of a ternary Al-Fe-Zn liquid phase in thermodynamic equilibrium with an intermetallic compound (δ phase) rich in iron and zinc (2.8 wt% Al, 9.0 wt% Fe) [37–39]. Earlier, Paunoiu et al. [40] provided evidence that the inhibition layer formed in a GA bath (0.122 wt% Al) had a biphasic structure consisting of a thick layer of δ on a thin discontinuous layer of $Fe_2Al_5Zn_x$. Some authors have reported the transient presence of other metastable phases, such as $FeAl_3$ [41–44] and $FeAl_2$ [42] in coexistence with $Fe_2Al_5Zn_x$. In 2001, a biphasic interfacial layer composed of $Fe_2Al_5Zn_x$ with δ on its top was also reported by Dionne et al. [45]. Metastable epitaxial ς,$FeZn_{13}$, containing less than 1 wt% Al rod crystals may also appear, together with the inhibition layer [46–48]. Recently, it was reported that the inhibition layer is composed of a very thin layer of $Fe_2A_{15}Zn_x$ and a thicker layer of δ on top of it [36]. For two Al baths (0.112 and 0.128 wt% Al), the main difference observed in the structure of both inhibition layers is the discontinuity of $Fe_2Al_5Zn_x$ in the case of the lowest bath Al content. Other intermetallic compounds were also found.

It was suggested that knowledge of the exact nature of the inhibition layer would lead to robust models to improve the management of industrial galvanizing/galvannealing lines: one model describes the kinetics of reactions occurring in the GA galvanizing bath, and a second model describes the growth of Fe-Zn phases in the galvannealing furnace located at the exit of the liquid metal bath, which begins with the breakdown of the inhibition layer. This breakdown, for example, the time required for breakdown, is highly dependent on the continuity or discontinuity of the $Fe_2Al_5Zn_x$ layer.

Although GA coatings consisting of Fe-Zn intermetallic phases are predicted from the equilibrium diagrams shown in Fig. 8.11, the first steel-coating interface phase to form is $Fe_2Al_{(5-x)}Zn_x$ occurs because of the stronger chemical affinity of Fe for Al rather than Zn and the rapid lateral growth of this thin layer from FeAl nuclei [35]. The extremely rapid rate of growth of the $Fe_2Al_{(5-x)}Zn_x$ interface layer was demonstrated by Toussaint et al. [49] on a fully reactive steel substrate, full coverage by Fe_2Al_5 is realized after 100 ms after which it continues to grow; two-thirds of the Al pick-up after 2 s is already realized at 100 ms. After this short reaction time, a thicker layer of Al-saturated δ ($FeZn_7$) is formed on top of the $Fe_2Al_{(5-x)}Zn_x$.

Wang et al. [50] confirmed that the δ instead of the ζ phase was the first Fe-Zn compound to form at the $Fe_2Al_{5-x}Zn_x$/steel interface followed by the Γ phase at the δ-steel substrate interface for the Al-containing coating in the range of Zn-Fe phase stability. Transmission electron microscopy indicated no epitaxial growth of the δ phase (reported as $FeZn_7$ with an HCP crystal structure) on the steel substrate. However, the Γ phase (Fe_3Zn_{10} with a BCC crystal structure), has a cube-on-cube orientation relationship with the underlying steel substrate. Occasionally Γ_1 ($Fe_{11}Zn_{40}$ with an FCC crystal structure) was observed between the Γ and δ layers, and only in samples galvannealed at temperatures at the high end of the galvannealing temperature range, but never above 550°C, which is the highest temperature where Γ_1 is stable. No Al-containing phase was ever found in the vicinity of ζ outbursts, confirming the understanding of the outburst mechanism. Wang confirmed observations described before on Mn-Si dual phase steels. Surface oxide coverage results in a patchy inhibition layer and strongly inhibits galvannealing kinetics. But as long as a continuous inhibition layer is present, the galvannealing behavior of Mn-Si steels was found to be similar to low carbon and IF steels. The surface oxides never interfered with galvanneal coating growth under this condition. Wang also observed Mn-Si oxide particles in the Γ layer. The steel grain boundaries near the interface were also decorated with the oxide films and particles. Interfacial oxides were observed to be detached and pushed into the liquid Zn where they were later surrounded by the δ phase or remained on the steel surface where they were surrounded by the Γ phase.

GA bath management is described in detail in Chapter 4. The effective, or soluble, Al content in the galvanizing bath is usually held around the 0.135% "knee point" described in Fig. 4.8 of Chapter 4 or lower, where Zn-Fe phases are stable. In all cases, the Zn-Al-Fe inhibition layer that is kinetically favorable to form, although it is not thermodynamically stable below the 0.135% Al composition, must be broken down to permit formation of the stable Zn-Fe phases. At 460°C a higher bath Al content will result in a thick inhibition layer, typical of that seen with the normal galvanizing process, but still on the order of a fraction of a micrometer. Production rates are typically 60% of normal when a 0.14%Al bath composition is used, especially for rephosphorized IF and some DP steels that have lower than usual reactivity during the GA process. Steel reactivity effects are described further in Section 8.7. Bath compositions between 0.11% and 0.13% Al produce partial or patchy inhibition layers with the uncovered areas quickly becoming the sites of zeta outbursts. This results in a rapid GA reaction, especially for IF steels stabilized with Nb or Ti, DP, and TRIP steels. However, the use of Al compositions in this range is associated with higher production of bottom dross that must be periodically removed during a line stop. The 0.135% Al "knee point" at 460°C provides constant Al activity and transformation of Zn-Fe to Zn-Al-Fe dross as production proceeds; however, precise Al control is required.

The initial formation of the coating in the bath is also influenced by bath temperature. A high strip entry temperature, more than 5°C higher than the bath temperature, can result in a higher Fe dissolution rate, especially at low bath Al contents where, following the Zn-Al-Fe phase diagram, Fe solubility is higher. Excessive Zn-Fe alloying can then occur in the bath rather than in the galvannealing furnace, increasing thickness and reducing phase control. This results in making it impossible to produce coatings in the 80 g/m^2/side range typically specified for automotive applications. The coating will already be excessively thick by the time it reaches the wiping knives.

Zinc bath Al content also strongly influences the growth behavior of the GA coating and consequent microstructure. Dissolved Al contents of less than 0.12% are associated with partial coverage by the inhibition layer leading to nonuniform growth of the galvanneal coating, together with excessive Zn diffusion into the grain boundaries of reactive steels such as Ti-IF that can cause embrittlement [51]. Dissolved bath Al contents greater than 0.145% are associated with thicker Γ layers that also compromise the formability of the GA coating and result in powdering during press forming [52]. The thicker gamma layer is a consequence of the higher galvannealing temperature required to fully react in the coating within the required galvannealing time, typically 20 s. The surface of galvanneal coatings on a normally well-behaved IF steel is much rougher and nonuniform because of the thicker inhibition layer produced at higher bath dissolved Al levels that causes nonuniform growth of the GA coating, as shown in Fig. 8.17.

In production, the moving steel sheet is immersed in the galvanizing bath, forming a thin, Al-bearing inhibition alloy layer at the zinc-steel interface. The higher the strip entry temperature or bath temperature, the higher the Fe dissolution

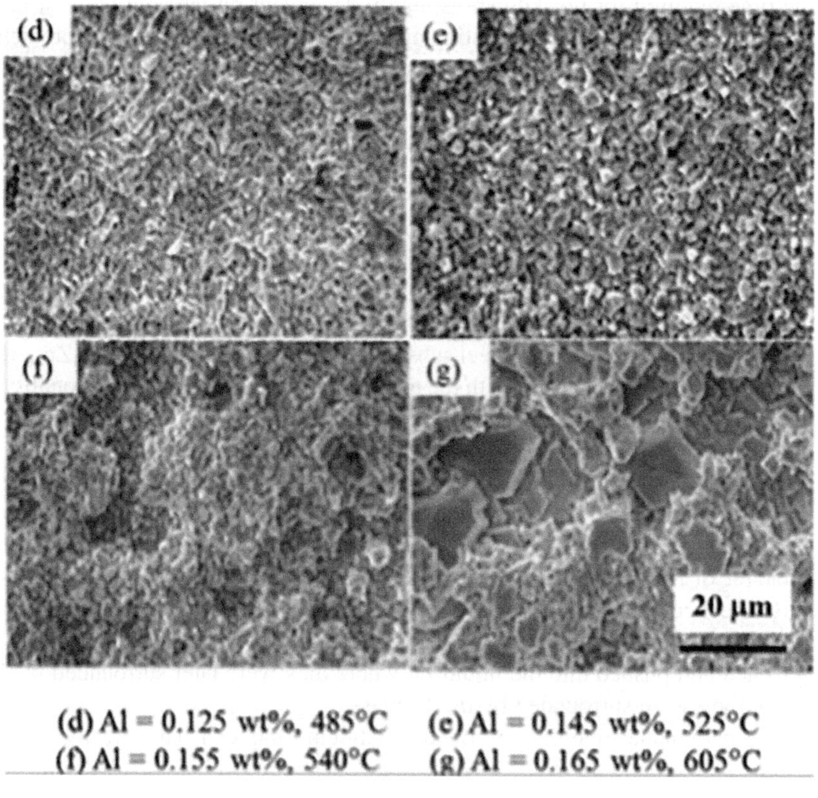

FIG. 8.17 Effect of Zn bath dissolved Al content and galvannealing temperature on surface appearance of a Ti-IF steel grade. *(From D.-J. Paik, M.-H. Hong, Sustainable production and technology challenges for galvanized sheet steel, in: Galvatech '17, 2017, pp. 11–17.)*

with the zinc will be, influencing the amount of dross that is formed. However, Fe dissolution is temporarily halted when the inhibition layer is formed. Usually, the effective, or soluble, Al content in the galvanizing bath that is used for the production of GA is between 0.12 and 0.135% Al. At effective Al contents below 0.135% at 460°C, the typical bath temperature, continued exposure of the zinc overlay, and inhibition layer on the steel substrate will lead to Zn enrichment of the inhibition layer and its transformation to a layer of delta phase. Also, local colonies of zeta crystals can be formed that have a columnar appearance, particularly at effective Al contents below 0.135% when discontinuous inhibition layer coverage can occur because of lower Al availability. These are termed "outbursts." The control of outbursts depends upon the correct control of effective aluminum in the galvanizing bath so that a complete inhibition layer is formed. However, the inhibition layer cannot be so thick as to impede the further development of the GA coating after the hot-dip process.

8.5.2 Inhibition layer breakdown

A conventional galvannealed (GA) coating undergoes different and complex reactions: (1) inhibition layer formation, (2) inhibition layer breakdown, (3) liquid Zn consumption, and (4) enrichment in Fe of the coating by solid state diffusion. A very thin layer of Al-rich intermetallic compounds nucleates very rapidly (around 0.1–0.2 s) [53] on the steel surface during galvanizing in Al-containing Fe-saturated Zn baths. Called the *inhibition layer*, because it temporarily inhibits the alloying reactions between Fe and Zn. Previously [54], it was shown that the inhibition layer, which forms in typical baths for galvannealing production, was composed of a very thin layer of $Fe_2Al_5Zn_x$ (becoming discontinuous for low Al contents) and a thicker layer of Al-saturated δ ($FeZn_7$) on its top. This microstructure allows thermodynamic equilibrium to be reached along all interfaces, and the system does therefore not require any other interfacial reaction. As the process continues, increased temperature in the galvannealing induction furnace, located at the exit of the galvanizing bath, activates Fe and Zn interdiffusion through the inhibition layer eventually destabilizing this layer. This processing step, called *inhibition layer breakdown*, is a key step in the galvannealing process as it leads to the sudden development of Fe-Zn phases by consuming liquid Zn.

Different mechanisms have been proposed in the literature to explain the inhibition layer breakdown. It is generally agreed that an outburst reaction is responsible for the inhibition layer breakdown [35,42,44,46,55–63]. Two mechanisms have been proposed to explain the outburst formation. One group [42,44,55] proposed that the diffusion of Fe toward liquid Zn causes the growth of the inhibition layer and, as a result, an Al depletion at the solid/liquid interface, leading to the outburst formation. A second and larger group [40,46,56–62] proposed that the diffusion of Zn toward the steel substrate is the critical step in outburst formation. Early investigators [53] suggested that the steel grain boundaries were preferential sites for outburst formation. Nishimoto et al. [55] showed that the outbursts are formed at the steel grain boundaries for long immersion times in GA baths. Hisamatsu [56] confirmed afterward that the steel grain boundaries are preferential locations for outburst formation. Later, other authors [57–59] provided experimental evidence of the formation of outbursts at the emergence of the steel grain boundaries.

The inhibition of Fe-Zn reactions is always transient. Al delays the Fe-Zn reaction rather than suppressing it completely and eventually Fe-Zn "outbursts" form, Fig. 8.18 [14]. Outbursts appear to occur in association with the inhibition layer [42], since it has been shown that normal linear interface growth occurs in the absence of Al in the Zn bath [27]. For example, outbursts were not seen for a low carbon steel (0.006 wt% C) dipped in a 0.12 wt% Al bath at 460°C after immersion times of up to 10 s; however, the appearance of outbursts was observed after 3 s in a 0.16 wt% Al bath [63].

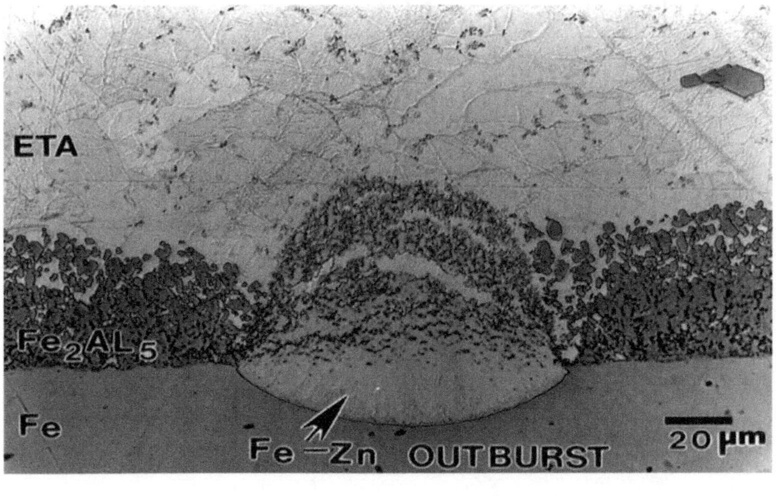

FIG. 8.18 Typical outburst formation in a 0.20 wt% Al-Zn bath immersed at 450°C for 3600 s. *(From A.R. Marder, The metallurgy of Zn-coated steel, Prog. Mater. Sci. 45 (2000) 191–271.)*

Even thick layers of Fe_2Al_5 are destroyed after sufficiently long holding times in high aluminum baths [55,64]. According to the ternary equilibrium phase diagram, liquid Zn with more than 0.12 wt% Al is in principle saturated with respect to Fe_2Al_5, and this phase should not redissolve in the liquid after it is formed. Thus, the transient character of inhibition cannot be explained by thermodynamic equilibrium between Fe_2Al_5 and the liquid. In fact, lumps of Fe_2Al_5 inhibition layer have been found scattered throughout an Fe-Zn outburst as if the outburst had caused the Fe_2Al_5 to break apart by nucleating beneath the inhibition layer [55,64].

The competing mechanisms of inhibition layer breakdown required for growth of the galvanneal coating to continue, including the dissolution of these interface layers in liquid Zn and the diffusion of solid Fe through the layers to the zinc bath, all of which are thermodynamically possible [14]. The dominant mechanism for inhibition layer breakdown, first deduced from the nucleation and rapid growth of ζ outbursts in more reactive steels, has been conclusively shown to be intergranular diffusion of Zn through these interface layers to the steel substrate, as first reported by Hisamatsu [56]. This mechanism is observed in a wide variety of ferritic and multiphase steels.

Diffusion of zinc through the inhibition layer so that as Zn atoms reach the substrate, they cause nucleation and growth of Fe-Zn intermetallic compounds at the Fe_2Al_5/substrate interface. Guttmann [35,65] has proposed that inhibition layer breakdown occurs by Zn diffusion down short-circuit paths in the Fe_2Al_5 inhibition layer, as originally suggested by Hisamatsu [56]. In reported studies on thick Fe_2Al_5, Guttmann [65] estimated that the average Zn diffusion coefficient would be $5 \times 10\,cm^2/s$. This implies that Zn could diffuse across a 70 nm thick Fe_2Al_5 layer in 1 s and a 250 nm layer in 3 s at 460°C. It was found that appreciable intergranular diffusion took place, thus real inhibition layers in galvanized steel can be traversed under usual galvanizing conditions. These results support the hypothesis that Fe_2Al_5 destabilization results from Zn diffusion through the Fe_2Al_5 layer along short-circuit paths [35]. When Zn reaches the substrate, it reacts with Fe, nucleating Fe-Zn intermetallic phases at the Fe_2Al_5 inhibition layer/substrate interface that bursts the layer apart into the surrounding bath. Short-circuit paths through the Fe_2Al_5 are expected to be the grain boundaries of the pancake structure layer or interfaces between the surface particles and the layer, Fig. 8.19.

Iron oxide was used as a simulated inhibition layer to test the Guttmann hypothesis [35,65]. In these experiments, a pure Zn bath (0.00 wt% Al) was used in order to prevent the confounding effect of the formation of an additional Fe_2Al_5 inhibiting layer. The iron oxide initially acted as a physical barrier for Zn diffusion, but eventually outbursts form beneath the oxide layer broke through the layer, causing sections to break away and become incorporated into the coating, Fig. 8.20. Increased immersion time for a constant oxide layer thickness led to an increase in the number of outbursts. Cracks and other microdefects in the oxide acted as fast diffusion paths for Zn, simulating the short-circuit diffusion mechanisms for Fe_2Al_5 inhibition layer breakdown in Al-containing Zn baths [35,65].

A comprehensive description of the mechanisms involved in inhibition layer breakdown, together with a model describing inhibition layer breakdown kinetics through outburst formation, was reported by Zapico-Alvarez et al. [53]. This was described using a FeZnAl ternary section at the zinc bath temperature of 460°C, Fig. 8.21. The blue line, diffusion path 1, indicates the equilibrium of all phases immediately after dipping in the zinc bath. As galvannealing proceeds, the inhibition layer can initially grow by interdiffusion of Fe toward the liquid Zn and Zn toward the substrate. Following the black line, in equilibrium the $Fe_2Al_{(5-x)}Zn_x$ layer becomes enriched in Zn and disappears, becoming δ phase (red diffusion path 2).

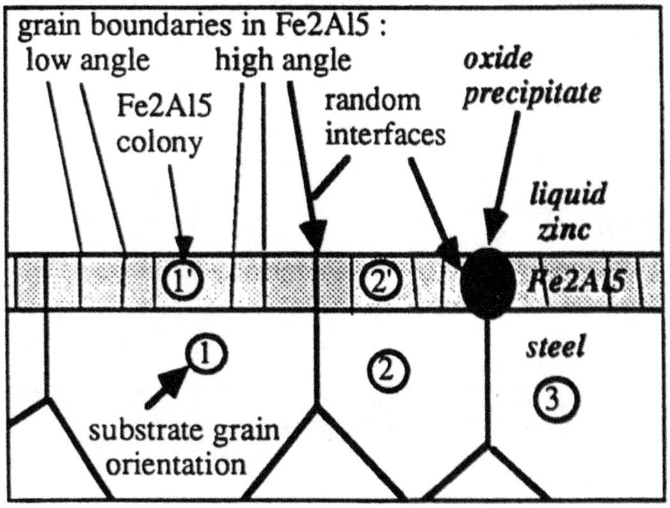

FIG. 8.19 Short-circuit Zn diffusion path through Fe_2Al_5. *(From A. R. Marder, The metallurgy of Zn-coated steel, Prog. Mater. Sci. 45 (2000) 191–271.)*

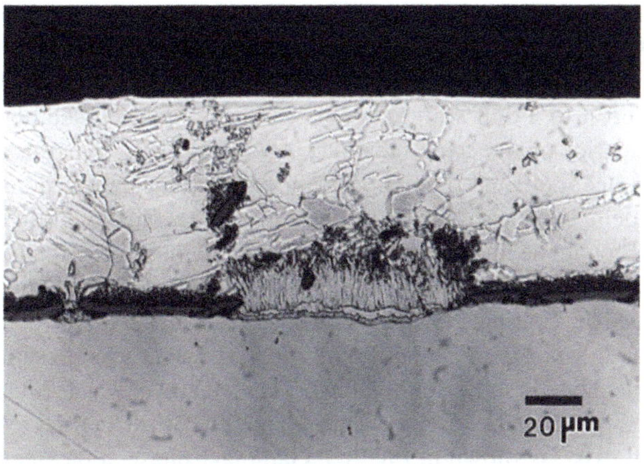

FIG. 8.20 Outburst formation in a 0.00 wt% Al bath with a simulated iron oxide inhibition layer. *(From A.R. Marder, The metallurgy of Zn-coated steel, Prog. Mater. Sci. 45 (2000) 191–271.)*

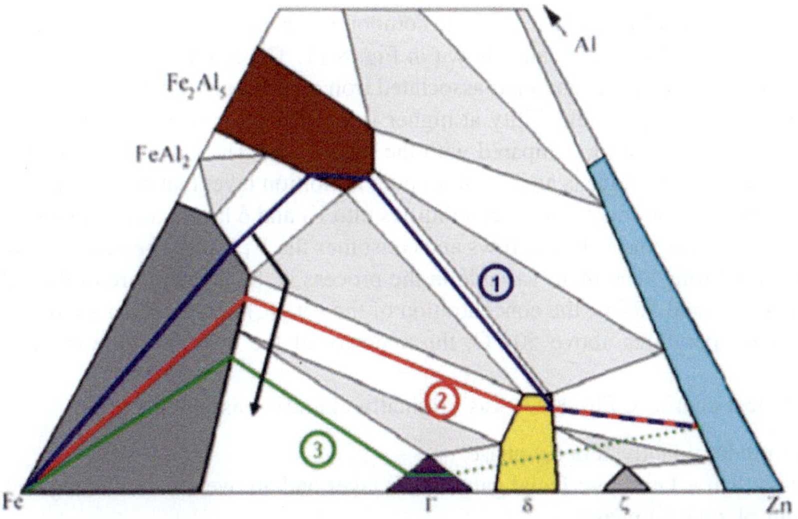

FIG. 8.21 Isothermal section of FeZnAl phase diagram indicating key diffusion paths followed during galvannealing. *(From D. Zapico Álvarez, F. Bertrand, J.-M. Mataigne, M.-L. Giorgi, Mechanisms and kinetics of the inhibition layer breakdown in the case of Ti IF steel grades galvanized in GA baths, in: Galvatech '15, 2015, pp. 307–315.)*

As noted earlier, there are two morphologies of δ phase: a compact morphology that is richer in Fe (δ_{1k}, $FeZn_7$) and a palisade morphology (δ_{1p}, $Fe_{13}Zn_{126}$). Still following an equilibrium hypothesis, the steel grain boundaries act as fast diffusion paths for Zn and become Zn-enriched. At a sufficient Zn content the δ is destabilized and Γ, with its lower density, begins to grow. Its volume expansion then peels away the δ phase, following the green path 3. Because there is no equilibrium between Γ and liquid Zn, a boundary of other ZnFe phases must exist between them. Following the dotted green line, ζ ($FeZn_{13}$) is then the dominant phase observed in outbursts that nucleate from the freshly exposed steel surface as the Γ layer peels away because of the higher local Zn content in the adjacent liquid (the local soluble Fe having been consumed in growth of Γ phase), represented in the two-phase ζ + Zn(*l*) region in the lower right corner of the phase diagram section. These observations and experimental data were then used to derive a model for inhibition layer breakdown that followed Fick's law. The time required for breakdown varied with temperature following an Arrhenius-type law. The quantity of Zn required for interface breakdown decreased as excess Ti content increased in the studied Ti-IF grades; a measure of reactivity that was found to depend on the rate of Zn diffusion in the grain boundaries. Separately, the Zn diffusion coefficient through the inhibition layer was reported as about 5.0×10^{-11} cm²/s, which implies a time of 1 s to cross a layer of 70 nm at 460°C [66].

The hot-dip galvanized sheet, with its formed inhibition layer and perhaps a scattering of outbursts, emerges from the galvanizing bath through the coating control apparatus, usually consisting of gas knives. The coating control is regulated to produce the coating weight required by the customer. In the United States, the ASTM A653 standard is adhered to and prescribes a range of two-side coating weights between 150 g/m^2 maximum down and no minimum for the lowest designated value. The strip, with the still-liquid coating, enters the GA furnace about 3–5 m above the gas wiping knives. Before the zinc can solidify, reheating of the strip begins. As the strip temperature rises, the zinc-iron diffusion action restarts and breaks down the Al-Zn-Fe inhibition layer. After 5 s of heating and up to about 10 s of soaking, enough Fe diffuses into the coating to convert it to a dull matte gray appearance. As discussed above, the development of the GA coating begins at the grain boundaries of the steel substrate. At these points, Zn first diffuses through the delta layer, resulting in zinc enrichment at the steel-delta interface. Local zinc enrichment at these points results in nucleation of gamma and gamma-1 phases at the steel-delta interface. Nucleation continues to be rapid at grain boundaries because intergranular diffusion is more rapid than that through the bulk of the ferrite grains. The rupturing of the delta phase occurs because of the volumetric expansion of gamma and gamma-1 phase. After this rupture, gamma and zinc are in direct contact, resulting in dissolution and diffusion to produce the coating shown in Fig. 8.5 [14]. In all cases, the delta layer that initially forms on the surface is the product of the breakdown of the Fe-Al-Zn inhibition layer that was present upon emergence from the bath. The sequence of rupture of the delta layer is shown in Fig. 8.22. The delta layer was formed by the breakdown of the Fe-Al-Zn inhibition layer formed during hot-dipping of the steel sheet in the Zn bath. The final result is a gradient coating in which gamma, delta, and zeta phases all exist, as shown in Fig. 8.5. Gamma-1 phase, a face-centered cubic phase, is sometimes observed growing ahead of the gamma phase and has been reported to result in improved formability.

8.6 Galvanneal microstructure

Galvanneal coatings are mainly composed of four layers of Fe-Zn intermetallic compounds, namely gamma (Γ), gamma1 (Γ_1), delta (δ), and zeta (ζ), actually seen in Fig. 8.5 and schematically shown in Fig. 8.11. The required property of the coating depends upon the amount and distribution of these layers and their associated iron content, e.g., spot weldability—high iron, powdering—lower iron. The improvement in spot weldability at higher iron contents is related to the higher electrical resistivity and higher melting point of the GA coating compared with the GI coating. The galvanneal structure develops over four stages. First to form at the zinc-steel interface is an iron-aluminum inhibition layer that quickly decomposes as the zinc coating becomes solid. As the process continues, the layer solidifies into Γ_1 and δ layers, with a thin layer of Γ forming against the steel. Continuing heat treatment, the Γ-layer grows and consumes the Γ_1 layer. The zeta (ζ) layer continues to be pushed toward the surface of the coating. The main variable in the process is the temperature of the galvannealing furnace. At temperatures between 481°C and 504°C, the concentration of the zeta (ζ) phase decreases to zero, while that of the delta (δ) phase increases at temperatures above 504°C; the majority of the coating is the delta (δ) phase [67].

The cross-section microstructure of galvanneal coatings, Fig. 8.23, was originally classified as follows [19]:

- Type 0—underalloyed coating containing predominantly zeta (ζ) phase.
- Type 1—optimum-alloyed coating with less than a 1 mm interfacial gamma (Γ) layer and an overlay containing delta (δ) phase interspersed with a small amount of zeta (ζ) phase.
- Type 2—overalloyed coating with a gamma layer >1 mm and an overlay containing a delta (δ) phase with basal plane cracks perpendicular to the coating/substrate interface and an occasional top layer of zeta (ζ) phase.

After an outburst of Fe-Zn intermetallic compounds occurs at the substrate grain boundary, rapid growth of these solid phases occur into the liquid. The liquid Zn will remain present between the crystals, which enables very fast transport of iron from the substrate [56]. Capillary effects of the pores and channels between the crystals drain the nearby regions

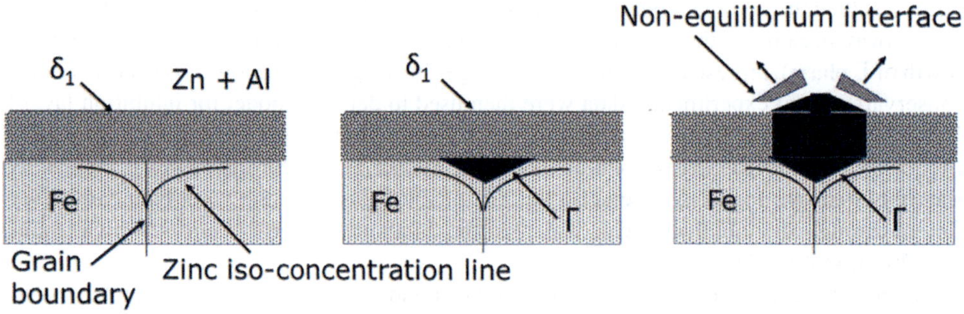

FIG. 8.22 Sequence of delayed rupture of the Fe-Zn delta layer by growing gamma phase in the formation and growth of the galvanneal coating. *(From F.E. Goodwin, Developments in the production of galvannealed steel for automotive, in: International Symposium of NMD-ATM-2012, Nov. 2012, pp. 1–21. Trans. India Inst. Metals.)*

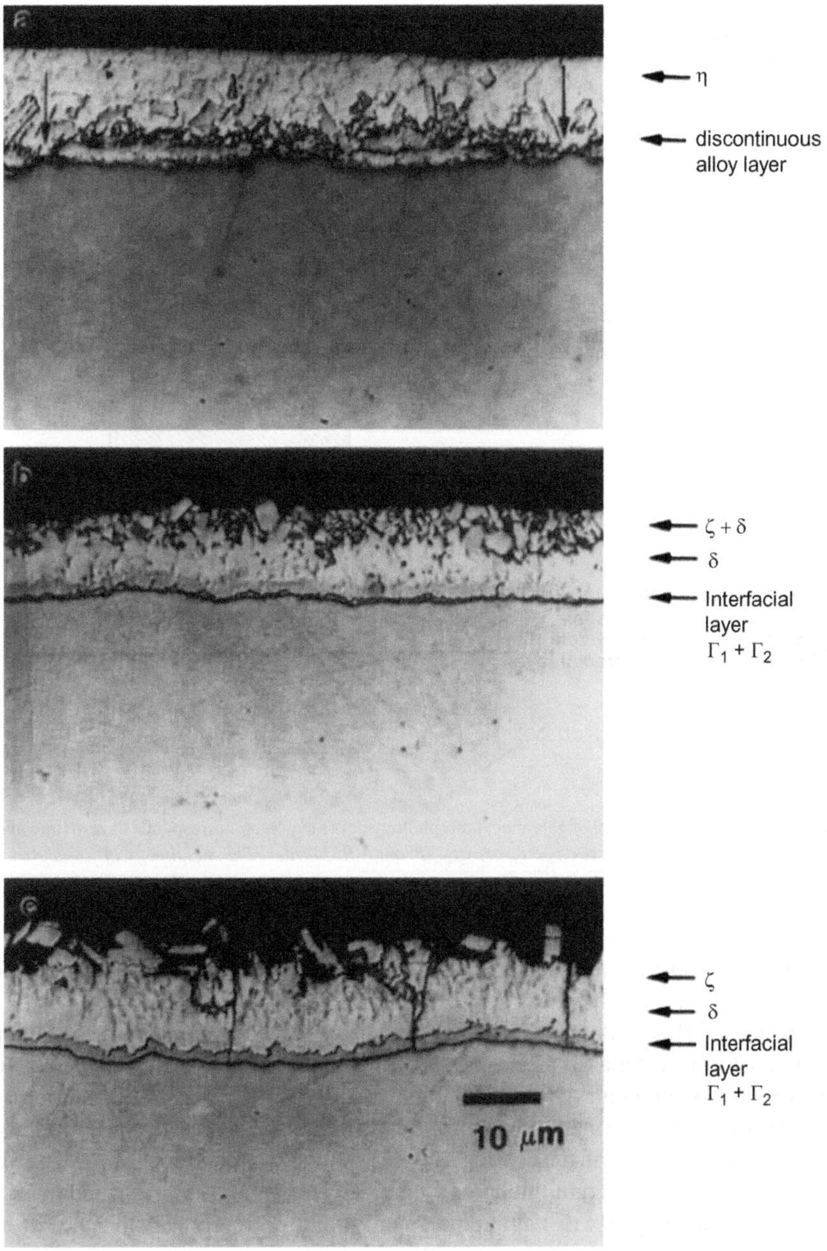

FIG. 8.23 Morphology of different galvanneal coatings: (A) Type 0, (B) type 1, and (C) type 2. *(From A.R. Marder, The metallurgy of Zn-coated steel, Prog. Mater. Sci. 45 (2000) 191–271.)*

where no alloy reaction takes place, whereby "dry spots" or craters are formed. A similar phenomenon has been used to describe denting in Galfan coatings [68]. These results have led to a phenomenological model that describes the sequence of events that occur during the microstructure development of a fully alloyed galvanneal coating, Fig. 8.24 [14].

The kinetics of formation will greatly depend on the processing parameters and composition of the bath and substrate steel; however, it was proposed that the time sequence was as follows:

t_0—the steel develops an Fe-Al interfacial layer during hot-dip galvanizing which, depending on Al content in the Zn bath, inhibits the formation of Fe-Zn phases [69,70]. The effectiveness of the inhibition will depend on the Al content in the bath [71], as well as time of immersion and bath temperature [15].

t_1—Fe-Al inhibition layer breakdown during annealing causing nucleation and growth of delta (δ) phase at the coating/steel interface [44]. Outburst formation can also occur with accelerated growth of Fe-Zn phases [56]. Simulated annealing of a galvanized coating in an environmental SEM shows the formation of outbursts at the interface, Fig. 8.25 [72]. The rate

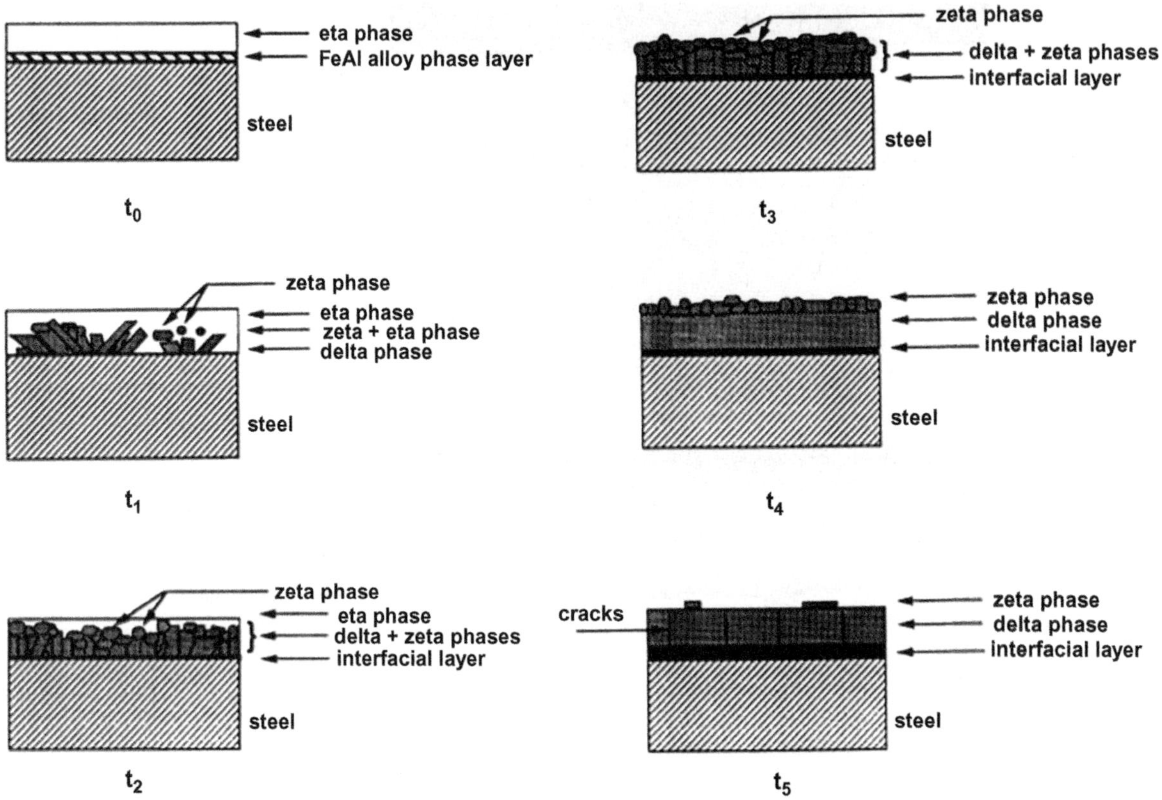

FIG. 8.24 Schematic of the phenomenological model of galvanneal morphology development. t_0 corresponds to zero time of the as-galvanized structure, and development occurs according to time such that $t_1 < t_2 < t_3 < t_4 < t_5$. *(From A.R. Marder, The metallurgy of Zn-coated steel, Prog. Mater. Sci. 45 (2000) 191–271.)*

of Zn attack will depend on the substrate alloy addition as well as the temperature profile in the annealing process. This microstructure is referred to as Type 0 galvanneal [20].

t_2—As annealing continues, diffusional growth of delta (δ) phase in a columnar growth morphology occurs. Previously nucleated zeta (ζ) phase transforms to delta (δ) phase. Additional zeta (ζ) phase may nucleate due to oversaturation of iron in the liquid eta (η) phase or upon cooling. An interfacial gamma (Γ) phase forms at the coating/steel interface [20].

t_3—Zn depletion occurs with longer temperature exposure resulting in the complete consumption of eta (η) phase at the coating surface. Fe concentration increases in the coating as the delta (δ) phase continues to form, pushing the zeta (ζ) phase to the surface and maintaining a constant 1 mm thickness gamma (Γ) phase [73]. This has led many to rely on an Fe concentration measurement (approximately 10 wt%) as a requirement for the optimum galvanneal coating, see, e.g., Ref. [16]. This microstructure is referred to as Type 1 galvanneal [20].

t_4—With longer times at temperature, delta (δ) phase diffusional growth continues toward the surface of the coating consuming the zeta (ζ) phase, while maintaining a constant 1 mm thickness gamma (Γ) phase [73].

t_5—Once the delta (δ) phase reaches the surface, it serves as the Zn rich side of the delta (δ) phase Fe-Zn/steel diffusion couple, allowing for the continued growth of gamma (Γ) phase at the expense of the delta (δ) phase. Cracking occurs along delta (δ) phase basal planes parallel to the coating/steel substrate [74]. This microstructure is referred to as Type 2 galvanneal [20].

Recently, gas-fired furnaces have given way to induction furnaces, and the galvannealing sheet steel heating rate has increased from <10°C/s for the gas-fired furnace to near 100°C/s for the latest induction furnaces. Interface outburst and inhibition layer dissolution are two parallel mechanisms in the galvannealing process. However, the interface outburst involves solid state diffusion, which is relatively slow, whereas the inhibition layer dissolution relies on the high temperature. With induction furnaces, the operating condition changes significantly (shorter times and higher temperature), thereby favoring the inhibition layer dissolution mechanism, which yields smooth coatings. Higher temperature increases the Fe solubility in molten zinc and also expands the δ phase regime at the expense of ζ and Fe_2Al_5 (pushing the knee point higher), thus promoting the dissolution of Fe_2Al_5 particularly for normal to low bath Al contents (moving from L

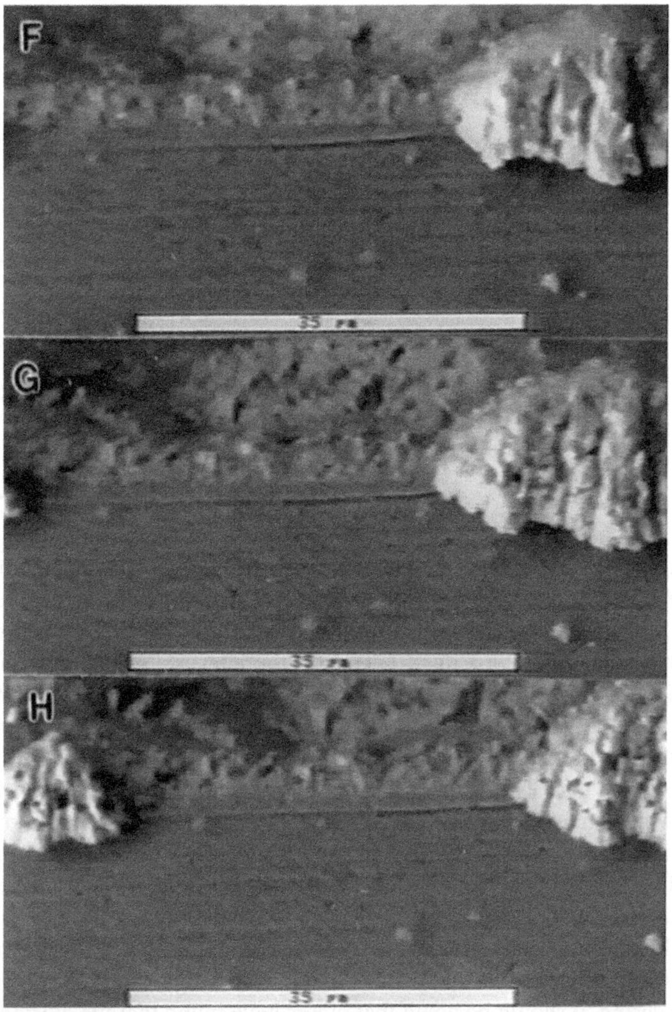

FIG. 8.25 Sequence of outburst formation during simulated in-situ galvannealing. *(From A.R. Marder, The metallurgy of Zn-coated steel, Prog. Mater. Sci. 45 (2000) 191–271.)*

+ δ + Fe_2Al_5 or even L + Fe_2Al_5 into L or L + δ). GA coatings with low bath Al are typically smooth, whereas those with high bath Al can be rough.

For the growth controlled by solid-state diffusion, the inward diffusion of Zn is much more significant than the outward diffusion of Fe. However, before the Zn-Fe phases form, the Fe dissolution from solid to liquid is significant and plays an important role in defining the final coating structure. Cheng and Krishnardula [75] have emphasized the importance of rapid heating. Fig. 8.26 is a schematic drawing of the Zn-Fe binary phase diagram at the Zn corner to illustrate the effects of the heating rate during galvannealing of IF steels. The galvannealing heating rate in Fig. 8.26 was shown as an atypical heating rate of Δtemperature/ΔFe%, instead of typical heating rate of Δtemperature/Δtime, and it actually is a measure of the competition between the rise of temperature (ΔT) and the diffusion of Fe into liquid Zn (ΔFe%). Assuming the inhibition layer is thick (complete coverage of the steel surface) and stable enough from the knife exit temperature up to the δ + L zone, the authors believe that the strip temperature should follow Path 1 along the liquidus line in Fig. 8.26. Under this condition, following Path 1, ζ phase will not precipitate out from liquid Zn and the final GA coating will have no ζ phase. Path 1 never enters the ζ + L zone (follows along the liquidus line) and no solidification takes place until it reaches the δ + L zone. They suggest that this is an extreme case and would produce the ideal GA coating structure, with only the δ phase having the minimum Fe composition profile in the GA coating.

Chen and Krishnardula [75] reasoned that for IF steels, including Ti-IF and Nb-Ti-IF steels, due to the incomplete coverage of the inhibition layer, it is very difficult, if not impossible, to follow Path 1 to avoid the precipitation of ζ phase out from liquid Zn. Instead, they tend to follow Path 2 of Fig. 8.26. Path 2 is an estimate path, based on the GA coating structure,

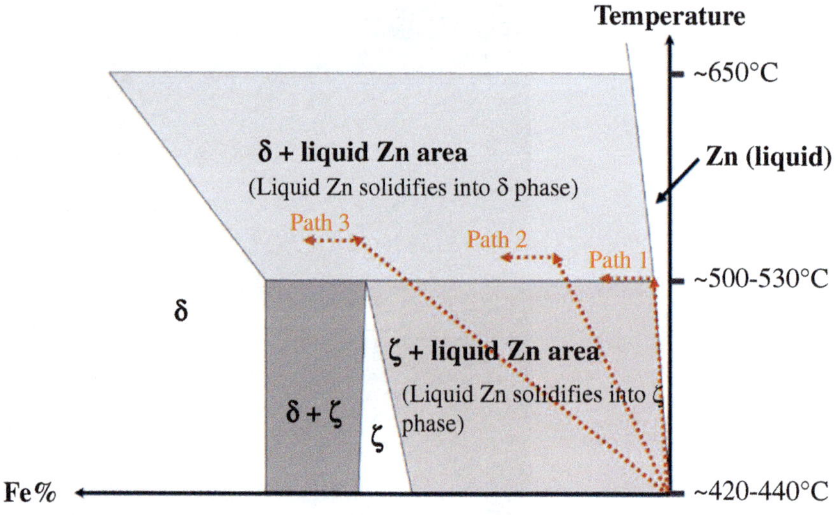

FIG. 8.26 Schematic drawing of the Zn-Fe binary phase diagram at the zinc corner. *(From C. Cheng, V. Krishnardula, Coating property comparisons of commercial GA Nb-Ti-IF and GA Ti-IF products produced under a similar hot-dip galvannealing condition, in: Galvatech'17, 2017, pp. 705–712.)*

for the galvannealing of IF steels with a near knee point Zn bath and induction heating. In order to get rid of ζ phase at the coating surface for the final minimized ζ products, one has to rely on a high holding zone temperature to convert excessive ζ phase into delta phase. Thus, the Γ phase grows thicker during this holding process. Based on this observation, the final coating structure of GA IF steels tend to have a thick Γ layer at the interface rather than too much ζ at the surface.

For the case of further slowing down the galvannealing heating rate, Path 3 in Fig. 8.26, e.g., simulates a gas-fired furnace or a very slow galvannealing rate in an induction furnace. The time the strip stays inside the ζ +L zone is considerably long such that a great deal of ζ phase will precipitate out from the liquid Zn. This high ζ phase structure is more pronounced for GA Ti-IF steel than that of GA Nb-Ti-IF steel when using the slow galvannealing heating rate of the gas-fired furnace (path 3). Meanwhile, an even higher holding zone temperature yet again promotes an even thicker Γ layer at the interface. In considering this analysis the authors wrote that, "under the conditions of a rapid heating and high-volume outburst of Fe that diffuses into liquid Zn, this is not a thermodynamic equilibrium condition, rather it is a dynamic situation," and therefore, the analysis may not exactly reproduce the actual processing conditions. In their analysis, they found that Ti-IF GA coatings had slightly more ζ phase than Nb-Ti-IF steel, but the Nb-Ti-IF steel had slightly better powdering resistance. From these results the authors have assumed that better powdering resistance occurs when ζ phase is minimized. As discussed later under powdering, it should be noted that others have found that the existence of ζ phase increases powdering resistance [73,76].

This work does point out the importance and drawbacks to the various simulation techniques. Most lab hot-dip simulators utilize heating rates <10°C/s, close to a gas-fired furnace [51]. Therefore, the structure before galvannealing begins, has an inhibition layer and an overlay that is probably predominantly liquid Zn. Others utilize a Gleeble with heating rates around 500°C/s, in excess of an induction heating furnace [20]. However, in these Gleeble experiments the sample was pregalvanized, having an existing inhibition layer and a solidified relatively pure Zn η phase. Nevertheless, the galvannealing temperatures studied (450–550°C) are above the melting point of Zn (419°C). Although it may be assumed that the η phase has been completely melted, it should be noted that the process is kinetic rather than being governed by thermodynamic equilibrium and may not be all liquid at the very short reaction times (1–5 s). As with most simulation experiments, it is important to understand the shortcomings of each experimental method before interpreting the results.

8.6.1 Galvanneal coating defects

Galvanneal coatings have received most of the attention in autobody applications because of its excellent spot weldability, better paintability, and superior corrosion resistance when used with paint. However, poor incoming surface quality, improper substrate cleaning, inadequate process control during heat treatment, coating and surface conditioning, surface damage in handling, and customer fabrication are the main sources of coating defects [77]. It is well known that zinc does not hide any substrate defects. Zinc, by reacting with the substrate, can even magnify the defects that are not visible to the naked eye in the substrate cold-rolled product. Chapter 16 discusses these defects in greater detail.

The topography of the galvanneal coatings is important. Van der Heiden et al. [18] studied the surface of typical production galvanneal coatings by SEM stereophotography and found the existence of "craters." Crater formation was

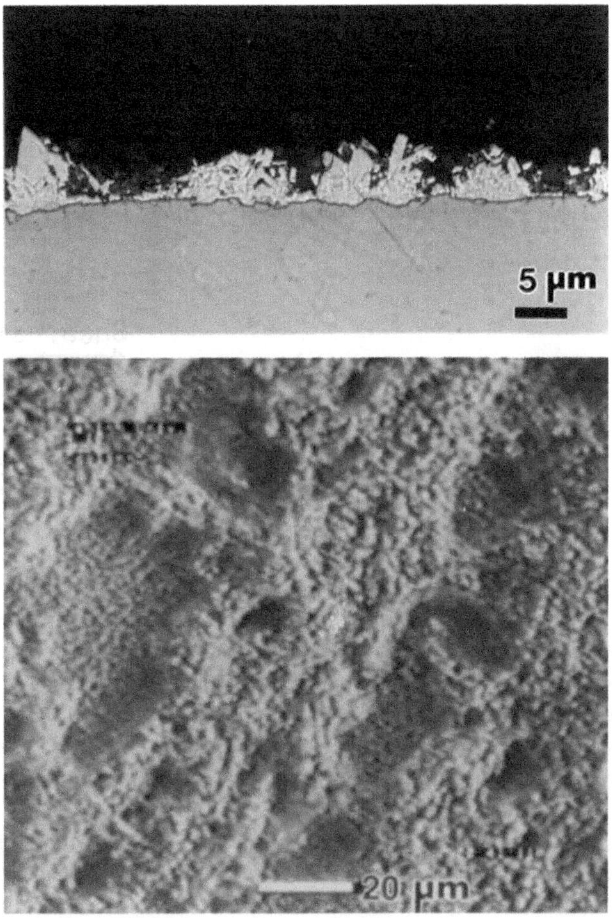

FIG. 8.27 Crater formation in galvanneal coatings: (A) Cross section, (B) planar surface. *(From A.R. Marder, The metallurgy of Zn-coated steel, Prog. Mater. Sci. 45 (2000) 191–271.)*

observed in situ in a hot-stage environmental scanning electron microscope [78]. Craters, Fig. 8.27 [14], were found to be inherent to the galvanneal coating layer, causing major fluctuations in coating thickness [18], and appear to be associated with outburst formation [78,79]. A mechanism was proposed that can explain the relationship between outburst formation and the appearance of craters [18]. Nevertheless, von Koesveld et al. [22] reported that "craters have no negative affect on GA user properties, including corrosion." Actually, it was found that the presence of craters had a positive effect on properties such as coating adhesion. The presence of "crater" within the galvanneal coating has been reported to reduce the amount of powdering [80,81]. It was proposed by O'Dell et al. [80] that the "craters" act as stress relievers for the galvanneal coating when subjected to compressive strains. Similarly, "bare spots" were investigated for C-Mn steels. Kawano et al. [82] pointed out that "bare spots" were not always related to oxides produced by thermal treatment of the steel substrate and resulted in outburst structures at the edge of the bare spots by galvannealing. Norden et al. [83] working on AHSS steels confirmed Kawano's research that zeta (ζ) phase was found at the edge of the bare spots on GA Ti-IF steels. In contrast, zeta (ζ) could not be found on GA in an Al-TRIP steel. Instead, complex oxides formed on the surface of this TRIP grade prior to the galvanizing step were found. Mataigne [84] confirmed the presence of this liquid bath oxide film on the surface of the steel substrates intended for galvannealing.

During the forming process, it was found that the major failure mode for galvanneal coatings on IF steels was "flaking," whereas the major failure mode for the galvanneal high-level P bearing steel was powdering [85]. A schematic of powdering and flaking is shown in Fig. 8.28 [14]. The major microstructural differences of these galvannealed coatings were reported as the Γ phase at the interface regions of the IF steel and a Γ_1 phase in the high-level P-bearing steel. A more detailed discussion of flaking and powdering is presented in reference [14]. Powdering was found to increase as the Γ layer increases above 1.0 μm [78]. Other coating characteristics that may be deleterious to powdering resistance are high iron content, in excess of 11.0 wt%, and cracking of the δ phase [78]. It was reported that liquid Zn penetration of grain boundaries and relatively smooth coating/steel interface can contribute significantly to the poor flaking resistance of

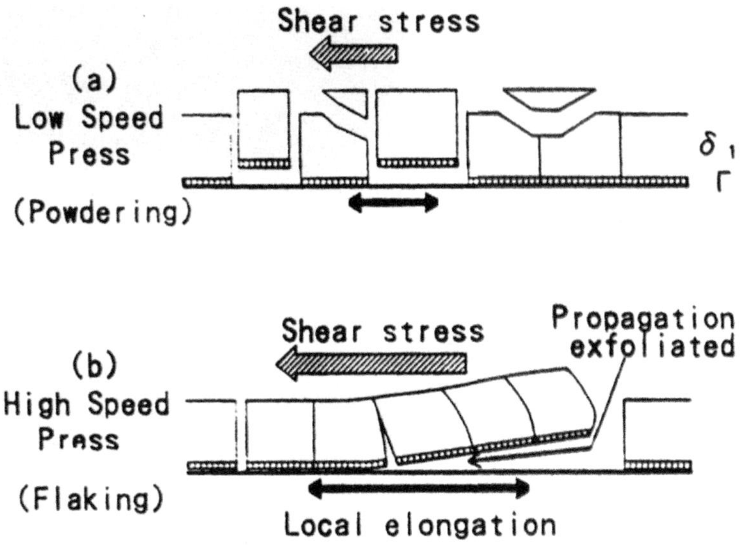

FIG. 8.28 Schematic model for the coating exfoliation process. *(From A.R. Marder, The metallurgy of Zn-coated steel, Prog. Mater. Sci. 45 (2000) 191–271.)*

GA IF steels [86,87]. Fast heating rates associated with induction heating, as discussed previously, were expected to minimize ζ formation and improve powdering resistance [75]. Yet, it has been reported that the plastic deformation of ζ phase is a major mechanism by which powdering can be prevented [76]. Jordan et al. [73] proposed that a mixed overlay structure of ζ and δ phases, while maintaining a 1 μm interfacial layer and 11.0 wt% Fe content, will exhibit good powdering resistance properties [73]. However, it should also be noted that the softer ζ phase on the surface of the coating may also exfoliate and adhere to the press forming dies, causing potential damage to the pressed part as well as the dies themselves.

Jagannathan [88] studied the origin of "striations" on the surface of iron-zinc-coated steel sheet produced by hot-dip galvannealing. He found that the formation of these dark striations was a result of contact with the sink roll in the molten Zn alloy pot because of the formation of a nonuniform iron-aluminum intermetallic layer during the galvannealing process. The uniformity in the Fe-Al diffusion barrier layer mirrored the machined grooves on the sink roll, with a greater amount of the barrier layer being formed in the areas of the sheet corresponding to the grooves on the sink roll. More recently, Plessinger et al. [89] observed "streaking defects" and found the presence of globular particles on the surface of DP steel that was thought to be Mn-rich Si-containing oxides. Due to the periodicity of the streaking defects, and no observed bare spots, it was thought that there are two mechanisms that could be influencing the creation of these defects: (1) the Zn pot sink role micro grooved surface degrading the inhibition layer formed in the Zn bath, and (2) Mn-rich Si-containing oxides influencing the formation of the inhibition layer, as well as intermetallic growth, influencing coating thickness.

It is well known that the intermetallic coating layer on GA steels is brittle, and therefore, it exhibits multiple cracking under thermally induced residual stresses and externally applied stress. Ochiai et al. [90] reported that the "multiple-cracking phenomenon" of the Fe-Zn intermetallic coating layer on hot-dip galvanneal steels (GA) was apparent in as-supplied samples, was found to be due to a thermally induced residual stress, and that further multiple cracking with increasing applied strain was a result of increased stress in the coating. It was also shown that the thinner the coating layer and the higher the flow stress of the substrate, the higher the stress of the coating layer becomes and, therefore, the smaller the crack spacing becomes [90].

Shanker et al. [91] found a peculiar type of "pin prick spots /dents" defect on the surface of some galvanneal sheets. These defects, which caused significant appearance problems in the finished product, were found to align in one direction. The investigators found the defect exhibited three distinct regions on the surface apart from the matrix: (A) particle, (B) dense layer surrounding the particle, and (C) a crater region surrounding the particle, and (D) is the parent matrix, Fig. 8.29. They proposed that dross/zinc dust falling from the snout wall was intermetallic phase or reaction product particles that embedded in the craters on the strip surface. During subsequent skin-pass rolling, these particles peel out the zinc coatings in the craters, thus densifying and flattening the region surrounding the particle.

An "uncommon coating defect" on industrial galvannealed high-strength interstitial free steel was reported by Chakraborty et al. [77]. A macroscopic view of two typical defects is shown in Fig. 8.30. The defects are described as whitish gray in color and easily visible to the naked eye. It was found that the root cause of the defect is the formation of a sticky oxide layer of Fe_2O_3 on the hot-rolled sheet, which survives even after cold rolling. Subsequent galvannealing operation does not produce the desired Fe-Zn reaction at these spots. The defect occurs as a mixture of galvanized layer coupled with

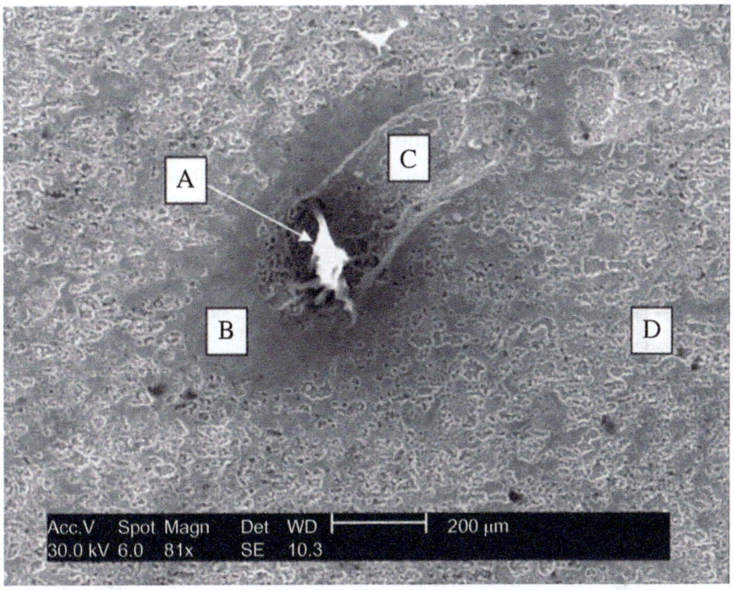

FIG. 8.29 SEM micrograph of surface defect in galvannealed sheet showing different regions. *(From A.R. Shankar, U.K. Mudali, B. Raj, Failure analysis of pin prick defects in galvannealed sheet—a case study, Eng. Fail. Anal. 16 (2009) 2485–2492.)*

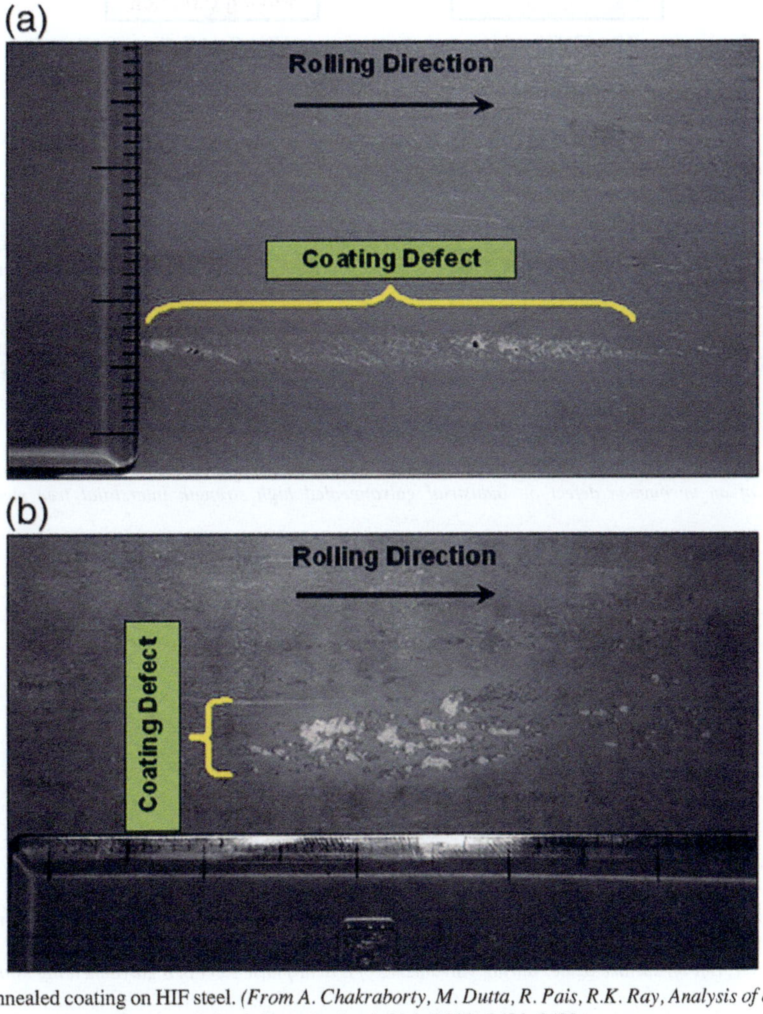

FIG. 8.30 Defects on galvannealed coating on HIF steel. *(From A. Chakraborty, M. Dutta, R. Pais, R.K. Ray, Analysis of an uncommon defect on industrial galvannealed high strength interstitial free steel, Surf. Coat. Technol. 204 (2010) 3481–3489.)*

an underalloyed delta (δ) phase that forms over and above the pickling-resistant oxide layer during the galvannealing process. A schematic representation of the formation of these coating defects is shown in Fig. 8.31. Another similar defect, "lift-off," also involves oxide formation in AHSS steels [92]. In this case a surface MnO oxide, a product of heat treating prior to galvanizing, was lifted off the surface of the substrate during dipping. The "lift-off" of the MnO oxide was attributed to the fast dissolution of Fe from the substrate into the GA bath when the concentration of Al in the bath is low. As a result, the MnO oxides are pushed away from the substrate and then subsequently entrapped in the Fe-Zn phase, which grows directly on the steel substrate where the inhibition layer is absent, Fig. 8.32.

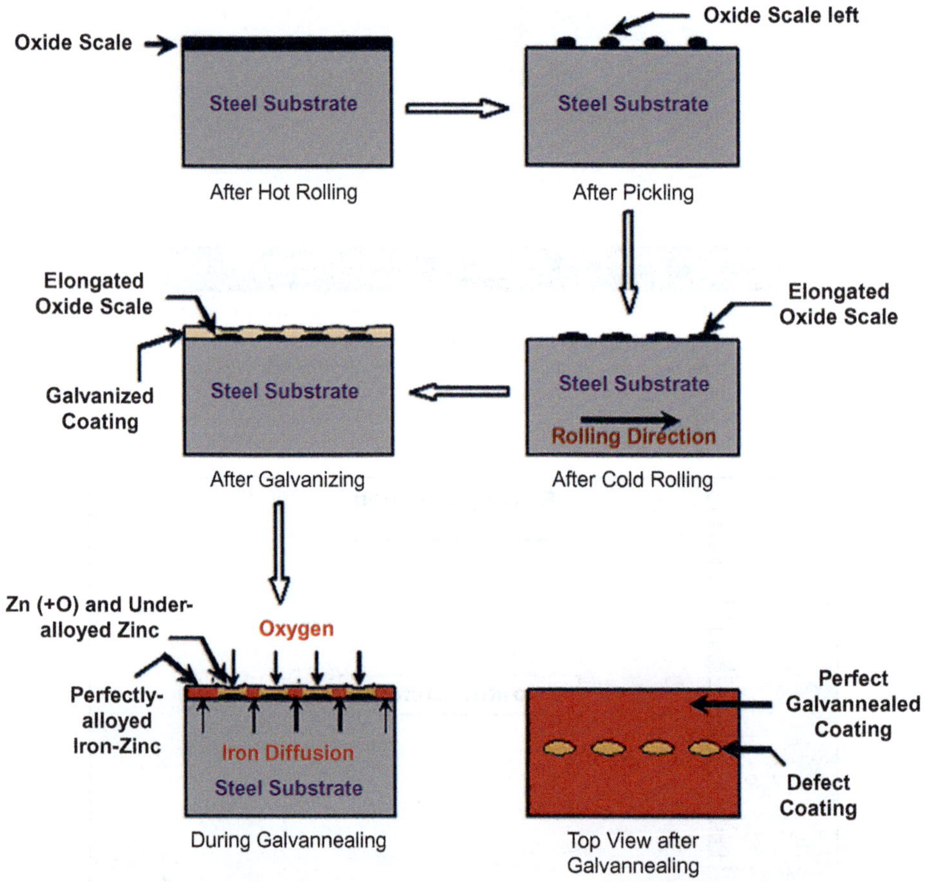

FIG. 8.31 Schematic representation of sequential steps of formation of galvannealed coating defect on HIF steel. *(From A. Chakraborty, M. Dutta, R. Pais, R.K. Ray, Analysis of an uncommon defect on industrial galvannealed high strength interstitial free steel, Surf. Coat. Technol. 204 (2010) 3481–3489.)*

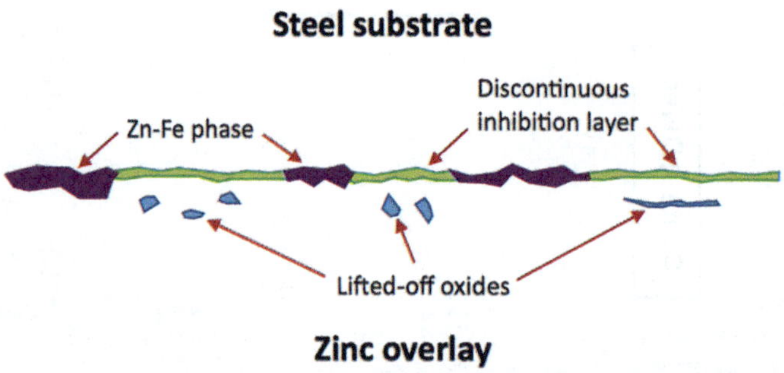

FIG. 8.32 Schematic of the interfacial structure of the hot-dipped DP steel in a Zn bath with 0.125% Al without annealing. The inhibition layer is discontinuous and the MnO oxides are lifted off the substrate surface due to interfacial reaction. *(From K.F. Chen, I. Aslam, B. Li, R.L. Martens, J.R. Goodwin, F.E. Goodwin, M.F. Horstemeyer, Lift-off surface oxides during galvanizing of a dual-phase steel in a galvannealing bath, Metall. Mater. Trans. A, 50 (2019) 3748.)*

8.7 Substrate steel chemistry

The galvannealing post dipping heat treatment temperature and heating profile are critical process parameters that control the iron zinc reaction kinetics. In addition, steel chemistry can also affect galvanneal kinetics. Alloying elements such as manganese, silicon, chromium, molybdenum, titanium, etc., are added to the steel to obtain desired mechanical properties. Additions of 0.5 wt% Al to the steel can alter surface segregation of alloying elements such as manganese, chromium, and silicon due to subsurface internal oxidation that enhances galvannealing kinetics [93]. Carbon and phosphorous have a retarding effect on the alloying reaction between Fe and Zn during galvannealing. It has been found that these elements segregate to grain boundaries and limit short-circuit diffusion of Zn atoms into the ferrite substrate [59,94]. Ti additions have been found to promote the galvanneal reaction by forming carbides in the ferrite matrix and reducing the concentration of free carbon in the grain boundaries [95,96].

Substrate solute additions have been reported to affect the rate of attack of zinc on steel during galvanizing. However, most of the research has been conducted on steel substrates containing a multiplicity of solute additions that can confound the determination of any systematic solute effect. Few studies have been concerned with pure Zn baths without the addition of Al. In general, small additions of carbon are found to segregate to grain boundaries, impeding the reaction rate [56], whereas Mn additions of up to 1–2 wt% have very little effect on weight loss and coating microstructure [97]. Interstitial-free (IF) steel [98] produced by the addition of titanium and/or niobium to an extra low-carbon grade to precipitate interstitial carbon and nitrogen atoms is being utilized as a galvanized coated sheet product [99]. Phosphorous is also sometimes added to these steels to improve mechanical properties. The effect these solute additions on the formation and kinetics of galvanizing has been studied in Al-containing Zn baths [30,76,96].

The effect of Fe-Zn alloy formation in IF steels galvanized in a pure Zn (Al free) bath has also been investigated [27]. Table 8.5 shows the total alloy layer growth-rate time constant, n, for a series of IF steels hot-dip galvanized at 450°C for up to 300 s, and Table 8.6 shows the growth-rate time constant, n, for the individual Fe-Zn phases formed. It was reported [27] that phosphorous solute additions retard the kinetics of gamma (Γ) phase layer growth but did not affect the growth kinetics of any other Fe-Zn phases. On the other hand, titanium and titanium + niobium solute additions had no effect on the growth kinetics on any of the Fe-Zn phase layers found in the coating. In a 0.003 wt% C substrate, the IF substrates exhibited a parabolic growth-rate time constant, $n = 0.5$, for the delta (δ) phase layer, indicative of bulk diffusion and the solid + liquid reaction at the interface. The gamma (Γ) phase layer for the 0.003 wt% C substrate and the other IF steels (except for the phosphorous containing alloys) had a growth-rate time constant of $n = 0.25$, indicative of grain boundary diffusion-controlled growth. The phosphorous added alloys reduced the gamma (Γ) phase layer n to 0.035 ± 0.13, most likely by blocking Zn diffusion down solute grain boundaries and preventing gamma (Γ) phase formation [21,100]. In general, it was found the structure of the galvanized coatings was not strongly influenced by their substrate composition. However, it was found that a rephosphorized grade had a somewhat thicker and more Al-rich inhibition layer and displayed slower galvannealing kinetics than a Ti-stabilized steel [60,101].

The growth of Fe-Zn phases is retarded by phosphorous additions [102]. Phosphorous was found to segregate to ferrite grain boundaries in rephosphorized low-carbon steels, blocking the diffusion of Zn along grain boundaries and lowering the thermodynamic activity [100,103]. It has also been proposed that P segregates to the steel surface during recrystallization annealing, stabilizing the inhibition layer and retarding the rate of Fe-Zn phase growth reactions during galvanizing

TABLE 8.5 Total Fe-Zn alloy layer growth-rate time-constant, *n*, values for steels hot-dip galvanize in a 0.00 wt% Al-Zn bath.

Sample	Growth-rate time constant, *n*
ULC	0.35 ± 0.02
ULC-P	0.35 ± 0.03
Ti-IF	0.31 ± 0.02
Ti-P IF	0.33 ± 0.02
Ti-Nb-IF	0.37 ± 0.03
Ti-Nb-P IF	0.34 ± 0.03

From C.E. Jordan, A.R. Marder, Fe-Zn phase formation in interstitial-free steels hot-dip galvanized at 450°C, Part I 0.00 wt% Al-Zn baths, J. Mater. Sci. 32 (1997) 5593.

TABLE 8.6 Individual Fe-Zn phase layer growth-rate time constant, *n*, values for steels hot-dip galvanized in a 0.00 wt% Al-Zn bath.

Sample/layer	Growth-rate time constant, *n*
Gamma-phase layer	
ULC	0.24 ± 0.06
ULC-P	0.10 ± 0.08
Ti IF	0.22 ± 0.11
Ti-P IF	0.035 ± 0.080
Ti-Nb IF	0.23 ± 0.05
Ti-Nb-P IF	0.13 ± 0.28
Delta-phase layer	
ULC	0.51 ± 0.11
ULC-P	0.44 ± 0.12
Ti IF	0.34 ± 0.12
TI-P IF	0.37 ± 0.13
Ti-Nb IF	0.44 ± 0.15
Ti-Nb-P IF	0.39 ± 0.15
Zeta-phase layer	
ULC	0.32 ± 0.03
ULC-P	0.32 ± 0.03
Ti IF	0.28 ± 0.05
Ti-P IF	0.29 ± 0.03
Ti-Nb IF	0.33 ± 0.05
Ti-Nb-P IF	0.33 ± 0.03

From C.E. Jordan, A.R. Marder, Fe-Zn phase formation in interstitial-free steels hot-dip galvanized at 450°C, Part I 0.00 wt% Al-Zn baths, J. Mater. Sci. 32 (1997) 5593.

[94,104]. A study of ion implanted P on one surface of a large grain (10–20 mm) low-carbon steel sheet galvanized in a Zn-0.02 wt% Al bath showed that Fe-Zn phase growth only occurred after extended reaction times on both the P ion implanted and nonimplanted surfaces [105]. The Fe-Zn outbursts were found to directly correspond to the location of a single grain boundary site, Fig. 8.33, and that a greater Fe-Zn reaction occurred on the P-free side (NP), since there was no P segregation in the grain boundary. It was shown that substrate grain boundary cleanliness is the dominant steel substrate structural feature that controls the kinetics of Fe-Zn alloy phase formation in aluminum containing zinc baths. These results were confirmed with a study of phosphorous additions in IF-P steels when outbursts were first found after 100 h in the 0.105 [106].

The details of growth of the GA coating vary with the composition and condition of the steel substrate. Makimizu [107] studied dual-phase steel compositions, each with 1.5% Si and with Mn levels of 1.4, 1.9, and 2.7% Mn. At lower Mn levels, significant film-like SiO_2 was found on the surface after preparation in a N_2-10%H_2 atmosphere with a −35°C dew point. At higher Mn levels, the dominant surface oxide was Mn_2SiO_4. A patchy inhibition layer formed on the low-Mn substrates after dipping in a 0.13% Al bath. Growth of the GA coating using soaking temperatures between 540°C and 560°C showed them to have a lower coating Fe content on the low-Mn substrates, and they were significantly rougher. The fraction of SiO_2 on the steel surface was observed to be a barrier to galvanneal growth.

The addition of Al to a dual-phase steel with 0.025% Si and a Mn + Cr total composition of 2% was shown to increase subsurface oxidation of Mn and Cr even when annealed with a dewpoint of −35°C [108]. The Al-added and Al-free variants

FIG. 8.33 Outbursts at the same grain boundary in a phosphorous ion-implanted (P) side and a nonimplanted side (NP). *(From A.R. Marder, The metallurgy of Zn-coated steel, Prog. Mater. Sci. 45 (2000) 191–271.)*

were then galvanized in a 0.126% dissolved Al-containing Zn bath and galvannealed at temperatures between 493°C and 527°C. Typical of all galvanneal coatings, the lowest galvannealing temperature produced coatings with a thin ζ phase on the surface, covering a thick δ layer comprising most of the coating, and a thin Γ layer at the coating-steel interface. Higher temperatures led to continued Fe enrichment, causing the ζ phase to disappear and the Γ layer to grow thicker. In all cases the Al-added steel exhibited a thicker Γ layer than the Al-free steel, so that galvannealing kinetics were more rapid for the Al-added steel at every temperature. Consistent with this, a higher coating Fe content was measured in each instance for the Al-added steel, at least 2% higher at each galvannealing temperature, e.g., 12% vs 10% respectively at 515°C. Curiously, the coating Al content of the Al-added steel was higher, 0.3% vs 0.2% for the Al-free steel, and this was attributed to dissolution of the substrate Al into the coating as the galvanneal coating grew, rather than a contribution from any Al-containing oxide. GDOES depth profiling of the steel substrates concluded that Al encouraged subsurface oxidation of the steel as it was prepared for galvanizing, greatly limiting the growth of surface oxides containing Mn and Si and therefore increasing galvanneal growth kinetics of the Al-added steel.

References

[1] J.L. Herman, Process of coating and treating materials having an iron base, US Pat. 1,430, 648, Oct. 3, 1922.
[2] E. Cornell, Jr., Method of treating coated material, US Pat. 1,565,420, Dec. 15, 1925.
[3] C.K. Lytle, Nonspangled galvanized sheet, US Pat. 2,034,348, March 17, 1936.
[4] A.H. Waltman, P.W. Dillon, Galvanizing process, US Pat. 2,111,826, Mar. 22, 1938.
[5] J.R. Daesen, A.F. Bradley, Galvanizing process, US Pat. 2,172,933, Sept. 12, 1939.
[6] C.A. Davis, Method and apparatus for coating, US Pat. 1,732,504, Oct. 22, 1929.
[7] T. Sendzimir, Process for galvanizing sheet metal, US Pat. 2,197,622, April 16, 1940.
[8] R.G. Mehler, J.G. Rothfuss, Galvanizing process, US Pat. 3,056,694, Oct. 2, 1962.
[9] W.C. Sievert, J.B. Cundiff, P.A. Klobuchar, L.H. Lindberg, J.A. Kargol, Process of producing one-side alloyed galvanized steel strip, US Pat. 4,171,392, Oct.16, 1979.
[10] F.J. Kennedy, Ferrous metal body with alloyed zinc coating, US Pat. 2,994,126, Aug. 1, 1961.
[11] A.R. Marder, M. Gu, Method of producing a Zn-Fe galvanneal on a steel substrate, US Pat. 4,913,746, Apr. 3, 1990.
[12] F.E. Goodwin, Developments in the production of galvannealed steel for automotive, in: International Symposium of NMD-ATM-2012, Nov. 2012, pp. 1–21. Trans. India Inst. Metals.
[13] J.A. Thornton, H.F. Graff, An analytical description of the jet finishing process for hot-dip metallic coatings on strip, Metall. Trans. B 7B (1976) 607–618.
[14] A.R. Marder, The metallurgy of Zn-coated steel, Prog. Mater. Sci. 45 (2000) 191–271.
[15] T. Kanamaru, M. Nakayama, Alloying reaction control in production of galvannealed steel, Mater. Sci. Res. Int. 1 (1995) 150.
[16] S. Hayes, Morphology Development of Galvannealed Coatings Based on Steel Substrate Chemistry and Processing Parameters (M.S. thesis), Lehigh University, 1994.
[17] G. Claus, J. Dilewijns, B.C. De Cooman, U. Meers, Determination of the process window for optimal galvannealing, in: Galvatech '95, 1995, pp. 107–113.

[18] A. van der Heiden, A.J.C. Burghardt, W. van Koesveld, E.B. van Perlstein, M.G.J. Spanjers, Galvanneal microstructure and anti-powdering process windows, in: A.R. Marder (Ed.), The Physical Metallurgy of Zinc Coated Steel, TMS, 1993, pp. 251–263.

[19] C.E. Jordan, K.M. Goggins, A.R. Marder, Kinetics and formability of hot-dip galvanneal coatings, in: GALVATECH '92, Stahl and Eisen, Amsterdam, 1992, p. 137.

[20] C.E. Jordan, A.R. Marder, Morphology development in hot-dip galvanneal coatings, Metall. Mater. Trans. 25a (1994) 937.

[21] C. Coffin, S.W. Thompson, The effect of 0.07 weight percent phosphorous on the galvannealing behavior of a niobium/titanium-stabilized interstitial-free sheet steel, in: GALVATECH '95, 1995, p. 121.

[22] W. van Koesveld, M. Lamberigts, A. van der Heiden, L. Bordignon, Coating microstructure assessment and control for advanced product properties of galvannealed IF steels, in: GALVATECH '95, 1995, pp. 343–355.

[23] N.L. Okamoto, M. Inomoto, H. Takebayashi, H. Inui, Crystal structure of the Γ- and Γ_1 phases in the coating layer of galvannealed steel, J. Alloys Compd. 732 (2018) 52–63.

[24] GalvInfoNote 1.3, Galvanneal—Differences From Galvanized? Sept. 2018, IZA, Rev. 1.4 www.galvinfo.com.

[25] N. Okamamoto, K. Tanaka, A. Yasuhara, H. Inui, Structure of the δ_{1p} phase in the Fe-Zn system by single crystal X-ray diffraction combined with scanning transmission electron microscopy, Acta Crystallogr. B70 (2014) 275–282.

[26] H. Inui, N.L. Okamoto, S. Yamaguchi, Crystal structure and mechanical properties of Fe-Zn intermetallic compounds formed in the coating layer of galvannealed steels, ISIJ Int. 58 (2018) 1550–1561.

[27] C.E. Jordan, A.R. Marder, Fe-Zn phase formation in interstitial-free steels hot-dip galvanized at 450°C, Part I 0.00 wt% Al-Zn baths, J. Mater. Sci. 32 (1997) 5593.

[28] J. Mackowiak, N.R. Short, Metallurgy of galvanized coatings, Int. Metals Rev. (1979) 1.

[29] D. Horstmann, F.K. Peters, The reactions between iron and zinc, in: Proceedings of 9th International Galvanizing Conference (Intergalva 1970), Dusseldorf, 1970, pp. 75–106.

[30] D. Horstmann, Formation and growth of iron-zinc alloy layers, in: Proceedings of 14th International Galvanizing Conference, (Intergalva 1985), Munich, 1985, pp. 6/1–6/5.

[31] C. Allen, J. Mackowiak, Application of inert-marker technique to solid/solid and solid/liquid iron/zinc couples, J. Inst. Met. 91 (1962–3) 369.

[32] M. Onishi, Y. Wakamatsu, H. Miura, Formation and growth kinetics of intermediate phases in Fe-Zn diffusion couples, Trans. JIM 15 (1974) 331.

[33] A.R.B. Verma, W.J. van Ooij, High-temperature batch hot-dip galvanizing. Part 2. Comparison of coatings formed in the temperature range 520–555 °C, Surf. Coat. Technol. 89 (1997) 143–150.

[34] P. Bicao, W. Jianhua, S. Xuping, L. Zhi, Y. Fucheng, Effect of zinc bath temperature on the coatings of hot-dip galvanizing, Surf. Coat. Technol. 202 (2008) 1785–1788.

[35] M. Guttmann, Diffusive phase transformations in hot dip galvanizing, Mater. Sci. Forum 155 (1994) 527–548.

[36] D. Zapico-Álvarez, P. Barges, C. Musik, F. Bertrand, J.-M. Mataigne, M. Descoins, D. Mangelinck, M.-L. Giorgi, Further insight into interfacial interactions in iron/liquid Zn-Al system, Metall. Mater. Trans. A 51 (5) (2020) 2391–2403. ASM International.

[37] N.Y. Tang, Refined 450°C isotherm of Zn–Fe–Al phase diagram, Mater. Sci. Technol. 11 (1995) 870–873.

[38] N.Y. Tang, Determination of liquid-phase boundaries in Zn-Fe-Mx systems, J. Phase Equilib. 21 (2000) 70–77.

[39] J.R. McDermid, M.H. Kaye, W.T. Thompson, Fe solubility in the Zn-rich corner of the Zn-Al-Fe system for use in continuous galvanizing and galvannealing, Metall. Mater. Trans. B 38B (2007) 215–230.

[40] A. Paunoiu, J.-M. Mataigne, F. Bertrand, D. Zapico-Alverez, J. Staudte, M.-L. Giorgi, Study of the inhibition layer formed in a GA bath on a TRIP Mn-Al steel, in: Galvatech '17, 2017.

[41] L. Chen, R. Fourmentin, J.R. McDermid, Morphology and kinetics of interfacial layer formation during continuous hot-dip galvanizing and galvannealing, Metall. Mater. Trans. A 39A (2008) 2128–2142.

[42] M. Urednıcek, J.S. Kirkaldy, An investigation of the phase constitution of iron-zinc-aluminum at 450°C, Z. Met. 64 (1973) 419–427.

[43] M. Sugiyama, M. Okada, Y. Takada, J. Jpn. Inst. Metals 65 (2001) 397–403 (in Japanese).

[44] H. Yamaguchi, Y. Hisamatsu, Reaction mechanism of the sheet galvanizing, Trans. Iron Steel Inst. Jpn. 19 (1979) 649–658.

[45] S. Dionne, G. Botton, M. Charest, F.E. Goodwin, A study of interrupted galvannealing of interstitial-free steels with different substrate compositions, in: Proceedings of Symposium on Materials in the automotive industry International Symposium on Materials in the Automotive Industry, Metallurgical Society of CIM 2001 Conference, Toronto, August 27-29, 2001, pp. 351–365.

[46] J.M. Mataigne, Key mechanisms in galvanization of steel sheet, Rev. Metall. 106 (2009) 27–33.

[47] Y. Adachi, T. Nakamori, K. Kamei, J. Jpn. Inst. Metals 56 (1992) 1235–1241 (in Japanese).

[48] E.T. McDevitt, M. Meshii, Electron microscopy study of the microstructure evolution in the early stages of galvannealing, in: F.E. Goodwin (Ed.), Zinc-Based Steel Coating Systems: Production and Performance, TMS, Warrendale, PA, 1998, pp. 127–136.

[49] P. Toussaint, L. Segers, R. Winand, M. Dubois, Short time investigation of hot dip galvanizing of steel under controlled hydrodynamics, in: Galvatech'98, 1998, pp. 141–146.

[50] K.-K. Wang, C.-W. Hsu, L. Chang, D. Gan, L.-J. Chiang, Formation of Fe-Zn intermetallic phases in galvannealed Mn-Si TRIP steels, in: Galvatech '15, 2015, pp. 81–87.

[51] V. Krishnardula, C. Cheng, Influence of steel chemistry and bath aluminum content on the integrity of the inhibition layer, in: Galvatech '15, 2015, pp. 324–333.

[52] D.-J. Paik, M.-H. Hong, Sustainable production and technology challenges for galvanized sheet steel, in: Galvatech '17, 2017, pp. 11–17.

[53] D. Zapico Álvarez, F. Bertrand, J.-M. Mataigne, M.-L. Giorgi, Mechanisms and kinetics of the inhibition layer breakdown in the case of Ti IF steel grades galvanized in GA baths, in: Galvatech '15, 2015, pp. 307–315.

[54] D. Zapico Álvarez, F. Bertrand, J.-M. Mataigne, M.-L. Giorgi, in: Nature of the Inhibition Layer in GI Baths, Proceedings of the 9th International Conference on Zinc and Zinc Alloy Coated Steel Sheet, Beijing, 2013, p. 743.
[55] A. Nishimoto, J. Inagaki, K. Nakaoka, Effect of surface microstructure and chemical composition of steels on formation of Fe-Zn compounds during continuous galvanizing, Trans. ISIJ 26 (1986) 807–813.
[56] Y. Hisamatsu, Science and technology of zinc and zinc alloy coated steel sheet, in: Proceedings of the 1st International Conference on Zinc and Zinc Alloy Coated Steel Sheet, Tokyo, 1989, pp. 3–12.
[57] Y. Leprêtre, J.-M. Mataigne, M. Guttmann, J. Philibert, Reactive interdiffusion in the Fe-Al-Zn system: reaction mechanism during hot-dip galvanizing, in: F.E. Goodwin (Ed.), Zinc-Based Steel Coating Systems: Production and Performance, Proceedings of the International Symposium, TMS, Warrendale, PA, 1998, p. 95.
[58] C.E. Jordan, A.R. Marder, Effect of substrate grain size on iron-zinc reaction kinetics during hot dip galvanizing, Metall. Mater. Trans. 28A (1997) 2683.
[59] C.E. Jordan, A.R. Marder, Inhibition layer breakdown and outburst Fe-Zn alloy formation during galvanizing, in: F.E. Goodwin (Ed.), Zinc Based Steel Coating Systems: Production and Performance, TMS, 1998, p. 115.
[60] M. Guttmann, in: Y. Leprêtre, A. Aubry, M.J. Roch, T. Moreau, P. Drillet, J.-M. Mataigne, H. Baudin (Eds.), Mechanisms of the Galvanizing Reaction. Influence of Ti and P Contents in Steel and the Surface Microstructure after Annealing, Proceedings of the 3rd International Conference on Zinc and Zinc Alloy Coated Steel Sheet, Chicago, 1995, pp. 295–307.
[61] M. Saito, Y. Uchida, T. Kittaka, Y. Hirose, Y. Hisamatsu, Tetsu-to-Hagane 77 (1991) 947 (in Japanese).
[62] J.-S. Kim, J.H. Chung, Alloy behavior of hot-dip galvannealed sheet steel containing silicon, in: F.E. Goodwin (Ed.), Zinc-Based Steel Coating Systems: Production and Performance, Proceedings of the International Symposium, TMS, Warrendale, PA, 1998, pp. 157–164.
[63] R. Kiusalaas, G. Engberg, H. Hlang, E. Schedin, L. Schon, Control of texture and formation intermetallic phases in continuously hot-dip galvanized coatings, in: Galvatech '89, 1989, p. 485.
[64] M. Saito, Y. Uchida, T. Kittaka, Y. Hirose, Y. Hisamatsu, Tetsu-to-Hagane 77 (1991) 947–954 (in Japanese).
[65] M. Guttmann, Y. Leprêtre, A. Aubry, M.-J. Roche, T. Moreau, P. Drillet, J.M. Mataigne, H. Baudin, Mechanism of the galvanizing reaction. Influence of Ti and P contents in steel and of its surface microstructure after annealing, in: Galvatech '95, 1995, p. 295.
[66] S. Goulart-Santos, A.H.A. Barbosa, Study of the Fe-Zn phases formation during the galvannealing treatment of coatings with different aluminum contents, in: Galvatech'15, 2015, pp. 342–347.
[67] J.M. Long, D.A. Haynes, P.D. Hodgson, Characterization of galvanneal coatings on strip steel, Mater. Forum 27 (2004) 62–67.
[68] S.T. Bluni, A.R. Marder, Improvement of galfan surface appearance for coil coating uses, in: 4th International Zinc Coated Sheet Conference, Zinc Development Association, Paris, 1994, p. SC6/1.
[69] A.R.P. Ghuman, J.I. Goldstein, Reaction mechanisms for the coatings formed during hot dipping of iron in 0 to 10 Pct Al-Zn baths at 450-700°C, Metall. Trans. 2A (1971) 2903.
[70] J.S. Kikaldy, M. Urednicek, Mechanism of iron attack inhibition arising from additions of aluminum to liquid zinc(Fe) during galvanizing at 450°C, Z. Met. 64 (1973) 2903.
[71] N.Y. Tang, Thermodynamics and kinetics of alloy formation in galvanized coatings, in: F.E. Goodwin (Ed.), Zinc-Based Steel Coating Systems: Production and Performance, TMS, Warrendale, PA, 1998, p. 3.
[72] W. Maschek, S.P. Hayes, A.R. Marder, Cross sectional studies of zinc iron phase growth in an environmental scanning electron microscope, in: Galvatech '95, 309, 1995.
[73] C.E. Jordan, K.M. Goggins, A.R. Marder, Interfacial layer development in hot-dip galvanneal coatings on interstitial free (IF) steel, Metall. Mater. Trans. 25A (1994) 2101.
[74] K.M. Goggins, A.R. Marder, Crack initiation and propagation in hot-dip galvanneal steel sheet during bending, in: 3rd International Conference on Zinc Coated Sheet, Barcelona, S4I, 1991, pp. 1–11.
[75] C. Cheng, V. Krishnardula, Coating property comparisons of commercial GA Nb-Ti-IF and GA Ti-IF products produced under a similar hot-dip galvannealing condition, in: Galvatech'17, 2017, pp. 705–712.
[76] L. Garza, C.J. Van Tyne, Intermetallic phase distribution effect on the friction behavior of galvannealed I-F sheet steels, in: Galvatech'11, 2011.
[77] A. Chakraborty, M. Dutta, R. Pais, R.K. Ray, Analysis of an uncommon defect on industrial galvannealed high strength interstitial free steel, Surf. Coat. Technol. 204 (2010) 3481–3489.
[78] C. Coffin, S.W. Thompson, Galvannealing of interstitial-free sheet steels strengthened by manganese, silicon, or phosphorous: an initial study, in: A. R. Marder (Ed.), The Physical Metallurgy of Zinc Coated Steel, TMS, Warrendale, PA, 1994, p. 181.
[79] C.E. Jordan, A.R. Marder, A model for galvanneal morphology development, in: A.R. Marder (Ed.), The Physical Metallurgy of Zinc Coated Steel, TMS, Warrendale, PA, 1994, pp. 197–219.
[80] S.P. O'Dell, J. Charles, The formation of craters within the galvanneal coating and their effects on processing properties, in: 44th MWSP Conference Proceedings, XL, 2002, pp. 1299–1307.
[81] P.M. Hale, R.N. Wright, F.E. Goodwin, The role of "cratering" in the processing and properties of galvannealed steel sheet, 2001-01-0084, SAE Trans. (2001) 1–5.
[82] T. Kawano, H. Noro, K. Sato, Y. Suzuki, M. Sakurai, Y. Tobiyama, Influence of thin oxide films on hot-dip galvanize and galvanneal behaviors of steel sheet, in: Galvatech '07, 2007, pp. 466–471.
[83] M. Norden, F. Macherey, F. Friedel, M. Peters, Influence of thin zinc-melt oxide films on galvanizing of IF and advanced high strength steels, in: Galvatech '11, 2011.

[84] J.-M. Mataigne, The role of furnace atmosphere in hot dip galvanizing, in: Galvetech "11, 2011.

[85] C. Cheng, L. Franks, Forming characteristics of galvanneal coatings on IF and P-bearing steels, in: Galvatech '95, 1995, pp. 723–732.

[86] C. Cheng, V. Krishnardula, Flaking behavior of GA IF and GA AHSS and its correlation with the adhesion strength and interfacial residual stress, in: Galvatech'11, 2011.

[87] C. Cheng, V. Krishnardula, H. Hahn, The effect of Al content in the coating on the flaking resistance of GA IF steels, in: Galvatech'15, 2015, pp. 96–103.

[88] V. Jagannathan, The origin of striations on the steel surface of iron-zinc coated steel sheet produced by hot-dip galvannealing, J. Mater. Sci. 32 (1997) 1093–1097.

[89] R. Plessinger, S. Seetharaman, E. De Moor, Oxide stability analysis and characterization in a dual-phase steel exhibiting coating defects, AIST Trans. 18 (2021).

[90] S. Ochiai, S. Iwamoto, T. Tomida, T. Nakamura, H. Okuda, M. Tanaka, M. Hojo, Multiple-cracking phenomenon of the galvannealed coating layer on steels under thermal and tensile stresses, Metall. Mater. Trans. A 36A (2005) 1807–1816.

[91] A.R. Shankar, U.K. Mudali, B. Raj, Failure analysis of pin prick defects in galvannealed sheet—a case study, Eng. Fail. Anal. 16 (2009) 2485–2492.

[92] K.F. Chen, I. Aslam, B. Li, R.L. Martens, J.R. Goodwin, F.E. Goodwin, M.F. Horstemeyer, Lift-off surface oxides during galvanizing of a dual-phase steel in a galvannealing bath, Metall. Mater. Trans. A 50 (2019) 3748.

[93] V.G. Krishnadula, C.C. Cheng, Effect of steel substrate aluminum on galvannealing kinetics of dual phase steels, in: Galvatech '17, 2017, pp. 182–188.

[94] C.S. Lin, M. Meshii, The effect of steel chemistry on the formation of Fe-Zn intermetallic compounds of galvanneal-coated steel sheet, Metall. Mater. Trans. B 25B (1994) 721–730.

[95] C.E. Jordan, A.R. Marder, Fe-Zn phase formation in interstitial-free steels hot-dip galvanized at 450°C, J. Mater. Sci. 32 (1997) 5603–5610.

[96] E.T. McDevitt, L. Pelayo, Y. Morimoto, M. Meshii, Investigation of the Fe-Al inhibition layer formed during hot-dip galvannealing, in: 39th MWSP Conference Proceedings, ISS, XXXV, 1998, pp. 127–136.

[97] J.J. Sebisty, G.E. Ruddle, Hydrogen-Atmosphere Galvanizing of Iron-Base Alloys, Research Report No. R 255, DOE, Mines Branch, Ottawa, 1972.

[98] J.A. Elias, R.E. Hook, Interstitial Free Sheet Steel Applications and Performance, SAE Paper No. 720018, 1972.

[99] A.R. Marder, Zinc coating of interstitial-free steel sheet, in: Proceedings of Conference on Interstitial-Free Steel Sheet: Processing, Fabrication and Properties, Metallurgical Society of CIM, Ottawa, 1991, p. 157.

[100] L. Allegra, R.G. Hart, H.E. Townsend, Intergranular zinc embrittlement and its inhibition by phosphorous in 55 pct Al-Zn coated sheet, Metall. Trans. A 14A (1983) 401.

[101] S. Dionne, M. Charest, G. Botton, F. Goodwin, Influence of substrate characteristics on the inhibition layer breakdown and galvannealing kinetics, in: 44th MWSP Conference Proceedings, XL, 2002, pp. 405–415.

[102] M. Abe, S. Kanbara, T. Okuyama, Iron zinc alloy behavior in galvannealing P and P+Si containing steel, Trans. ISIJ 25 (1985) B15.

[103] P.D. Mercer, Factors which affect the alloy growth rate in galvannealed coatings, in: GALVATECH '92, 1992, p. 204.

[104] C.S. Lin, W.A. Chiou, M. Meshi, Effect of phosphorous content in base-steel on the formation of alloy layer of hot-dip coated steel sheets, in: A.R. Marder (Ed.), The Physical Metallurgy of Zinc Coated Steel, TMS, 1994, p. 31.

[105] C.E. Jordan, R. Zuhr, A.R. Marder, Effect of phosphorous surface segregation on iron-zinc reaction kinetics during hot-dip galvanizing, Metall. Mater. Trans. A 28A (1997) 2695.

[106] Y. Leprêtre, J.M. Mataigne, Reaction mechanisms during hot dip galvanizing: effect of phosphorous on coating development, in: Galvatech '98, 1998, p. 133.

[107] Y. Makimizu, Y. Suzuki, W. Tanimoyo, T. Aoyama, N. Yoshimi, Influence of Mn content on Fe-Zn galvannealing reaction of Si-added steel sheet, Tetsu-to-Hagane 102 (2016) 583–590.

[108] M. Arimura, M. Urai, J. Iwaya, M. Iwai, Effect of press forming factors and flash plating on coating exfoliation of galvannealed steel sheet, Galvatech '95, 1995, p. 733.

Chapter 9

Electrogalvanizing processes (EG coatings)

9.1 Introduction

One important method of forming a protective zinc coating on steel is to use an electrogalvanizing process. This process is possible because Zn is more electronegative than Fe. Therefore, a plating cell setup with zinc, in some form as the anode and steel as the cathode will, under the imposition of an electrical potential in a suitable electrolyte, result in plating a steel substrate with a zinc coating. The process of electrogalvanizing was invented in the 1800s with the initial efforts of Sorel, Davey, and Faraday (see Chapter 1 for an historical review of the development of electrodeposited coatings). Since that time, development of the electrogalvanized processing of sheet steel product and fasteners has continued, although recently electrodeposition of flat-rolled products has largely been replaced by hot-dip coatings. Electrogalvanizing involves producing a homogeneous layer of zinc on a steel substrate that exhibits improved corrosion resistance, good spot weldability, as well as excellent surface appearance and painting adhesion properties [1]. Excellent reviews of electrodeposition of Zinc and Zinc alloys have been reported by R. Winand [2] and S. Fountoulakis [3]. According to Winand [2], the main trends in zinc and zinc alloy plating are as follows:

- Continued replacement of cyanide baths
- Development of high-current-density plating of zinc and zinc alloy on continuous wire and steel sheet
- Environmental constraints

Further, Winand [2] recognized the advantage in generating nanocrystalline zinc deposits is to understand the role of organic additives to control the morphology, properties, texture and surface morphology development in these coatings.

9.2 Processing methods

The process of electroplating occurs by causing a current of electricity to run through an electrically conductive solution. The anode can be inert, in which case zinc in solution in the electrolyte is attracted to the steel cathode, or the anode can be zinc that dissolves into the electrolyte before it is attracted to the steel cathode. Electrogalvanizing is industrially important in the production of small items such as fasteners, where the coating must follow the thread shape, although it remains viable for certain sheet applications.

9.2.1 Classification of continuous electrodeposition processes

Strip plating processes are generally classified by three process characteristics: anode type, electrolyte chemistry, and plating cell geometry [3].

Anode type is either soluble or insoluble. Soluble anodes are consumable, dissolving anodically to replenish the metal or metals, which eventually deposit onto the steel strip as the coating. Insoluble anodes generally oxidize in water and have oxygen evolution as the anodic reaction within each plating cell.

Electrolyte chemistry is generally characterized by the primary anionic species, such as sulfate or chloride. The key factor is obviously the capability of the particular chemistry to produce a high-quality, smooth, and compact coating. Zinc and zinc alloy electrodeposition of steel strip takes place from either a sulfate- or chloride-based electrolyte. These electrolytes are aqueous solutions of zinc ions and ions of the alloying nickel or iron metal with sulfate or chloride ions, acidified with sulfuric or hydrochloric acid, respectively. Alkali or ammonium salts may also be used to increase the conductivity of these solutions. Chloride electrolytes are used exclusively with soluble anodes. The most significant advantage of the chloride process is that the conductivity of the bath is many times higher than that of the sulfate bath, and so less electric power is consumed to deposit the same amount of coating as in the sulfate process. The most widely

The Metallurgy of Zinc Coated Steels. https://doi.org/10.1016/B978-0-323-99984-7.00021-X

used zinc and zinc alloy steel plating process is the sulfate type. The main attribute of sulfate electrolytes is that they are compatible with insoluble anodes. Selecting the anode type is the first step in deciding on the plating process to use. Sulfate-based processes are also simple in chemistry. The main advantages of sulfate electrolytes are that they deposit finer-grain coatings without organic additives and operate over a wide range of current densities, temperatures, and acidity [3].

Plating cell geometry. The many different plating cells can be classified into three main groups based on cell geometry: vertical, horizontal, and radial. All three cell types can be used with either soluble or insoluble anodes [3].

Vertical cell design. Fig. 9.1 shows several major variations of vertical cell design. In electrogalvanizing, which is done at about a third of the speed used for tin and chromium, the vertical cell is normally modified to become a forced-flow or counter-current-flow cell in order to operate at higher current densities [3]. The forced flow is provided by pumping electrolyte through nozzles, as shown in Fig. 9.1a. Generally, these electrogalvanizing cells are used with insoluble anodes, lead alloy, or noble-metal-coated titanium. However, they have also been used with soluble (zinc) anodes, with the anodes in the form of solid bars suspended from busbars (Fig. 9.1b) or pellets contained in a special anode basket (Fig. 9.1c) [3]. A variation of the insoluble anode vertical cell is called Gravitel (Fig. 9.2). In this cell design, only the anode-to-strip gap is filled with electrolyte, and although the cell is housed in a similar tank, the tank is not filled with solution but simply contains the splashing of the solution, sending it down under the cell and into a large recirculation tank. The solution is pumped inside each hollow anode box, overflows the top of the box through a special V-shaped weir, falls into the gap between the anode and the strip, fills the gap, falls out the bottom of the gap, and returns by gravity to the recirculation tank.

Radial cell design. Fig. 9.3 shows three major variations of radial cell design. The first radial cell evolved into the CAROSEL (Consumable-Anode Radial One-Side Electroplating) technology. The distinguishing characteristics of this cell are a single but very-large-diameter conductor roll (about 2.4 m, or 8 ft., in diameter) in the center of the cell and large cylindrical consumable anode sections under this roll. Like all radial cells, this cell plates only one of the surfaces of the steel strip. However, by arranging multiple cells within a line into two tiers or elevations, and by inverting the strip as it goes from one tier to the next, the second surface can also be plated.

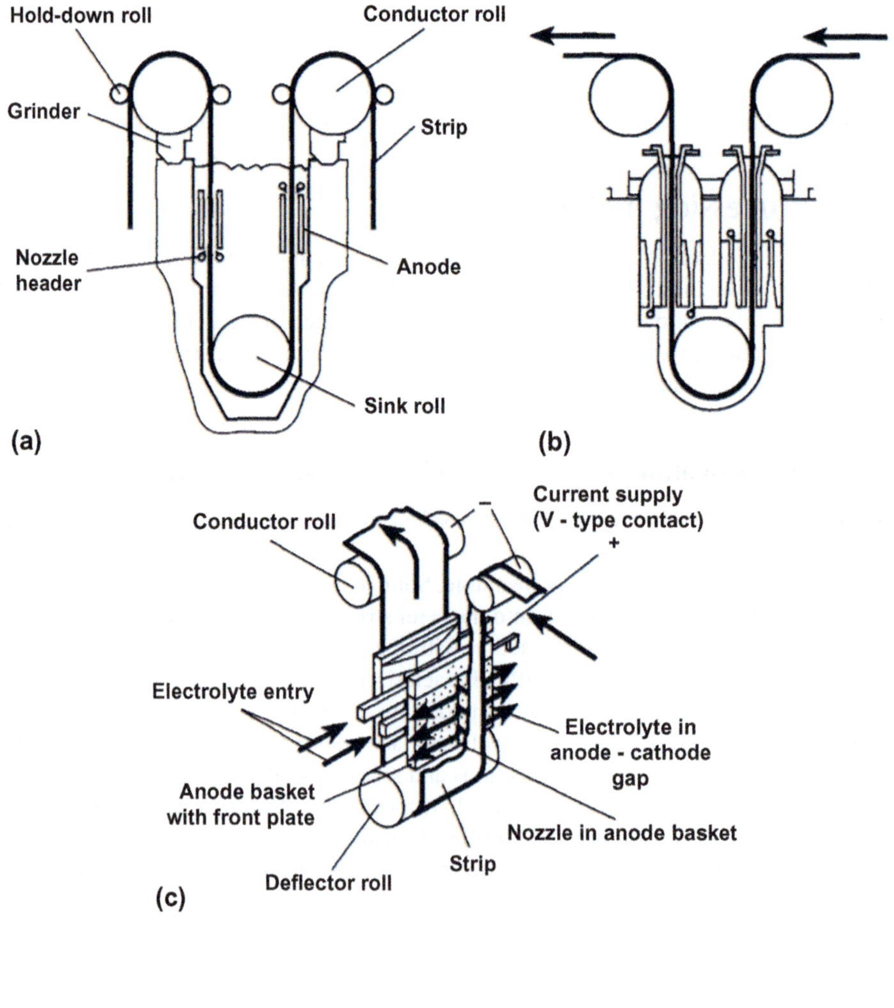

FIG. 9.1 Three major variations of vertical plating cell design. (a) Flooded and counter-current-flow plating cell with insoluble anodes. (b) Flooded and counter-current-flow cell with soluble anode bars. (c) Flooded cell with anode baskets containing metal granules [3].

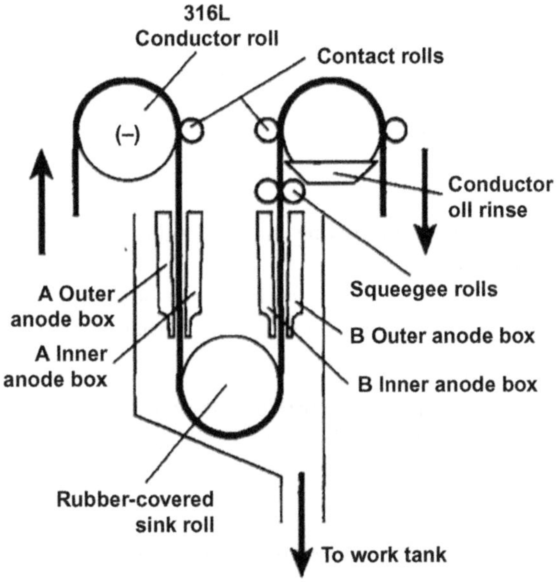

FIG. 9.2 Gravitel plating cell with insoluble anodes [3].

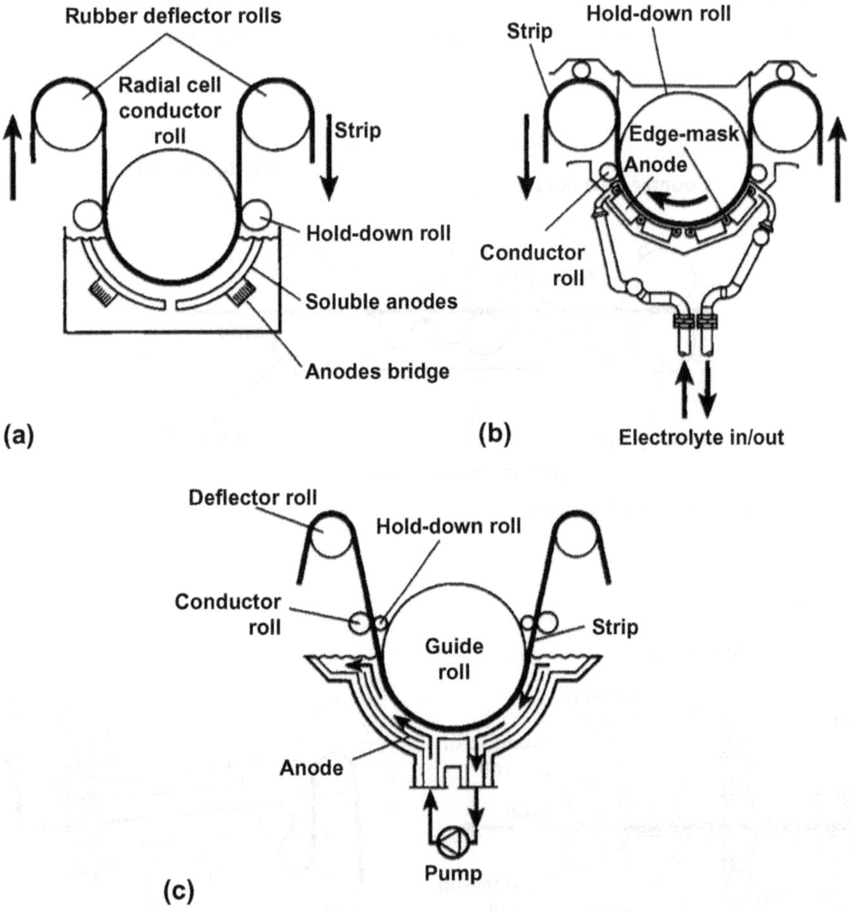

FIG. 9.3 Three major variations of radial plating cell design. (a) CAROSEL radial cell with soluble anode, a large 8 ft. diameter conductor roll, and the more recent addition of hold-down rolls. (b) Soluble-anode radial cell with modifications made in the way new anodes are inserted into the cell. (c) A radial cell with insoluble anodes and small-diameter conductor rolls on either side of the guide roll [3].

Horizontal cell design with top and bottom anodes (initially soluble anodes) was later developed for electrogalvanizing (Fig. 9.4). Then, in the 1980s, high-current-density (high-productivity) horizontal cells with insoluble anodes were introduced (Fig. 9.5) to meet the high productivity requirements for electrogalvanized sheet production.

9.2.2 Barrel and rack plating

For barrel and rack plating, pure zinc is the main coating material although zinc-alloy plating has been more recently developed. The following processes are used:

Rack plating—Rack plating is a method that involves fixturing individual parts into an array that is mounted on a solid frame called the plating rack. The plating rack serves the following purposes:

- To physically hold the parts in a secure manner
- To maintain electrical contact with and to pass DC current to the parts
- To orient the parts appropriately in accordance with the thickness specifications, part geometry, drainage, gas elimination, and quality requirements
- To optimize throughput

The anodes in Fig. 9.6 are the sources of new metal ions to replace the ions being plated onto the part. This process involves the physical movement of ions through the plating solution at a rate on the order of 0.01 cm per second. The rate of ion flow can be manipulated by the current density, e.g., the ratio of current supplied by the DC power supply to the surface area of the cathode, the parts being plated. Current density is especially important in rack plating. Current density manipulation is often a principal reason for developing a custom rack. For a given array of parts on a plating rack, the current density is not distributed evenly across the parts; almost without exception, there will be a gradient.

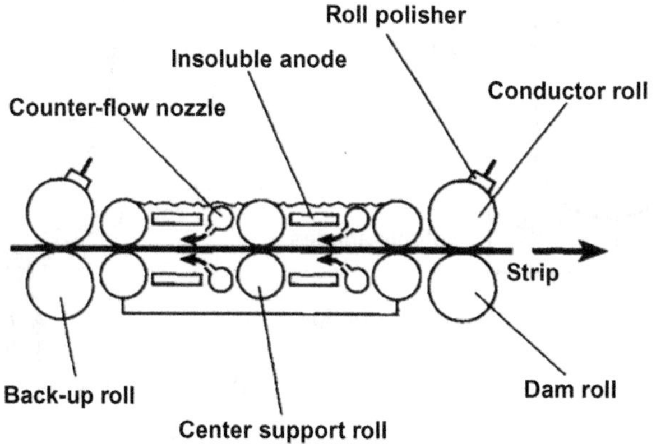

FIG. 9.4 Conventional horizontal cell with insoluble anodes [3].

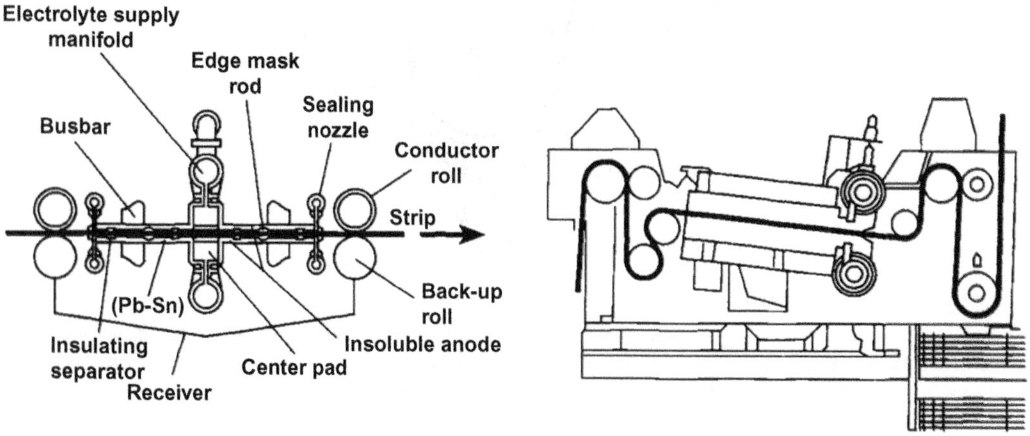

FIG. 9.5 Additional horizontal cells: (a) LCC-H, a liquid-cushioned horizontal cell. (b) Horizontal cell for high-current-density tin-free steel plating [3].

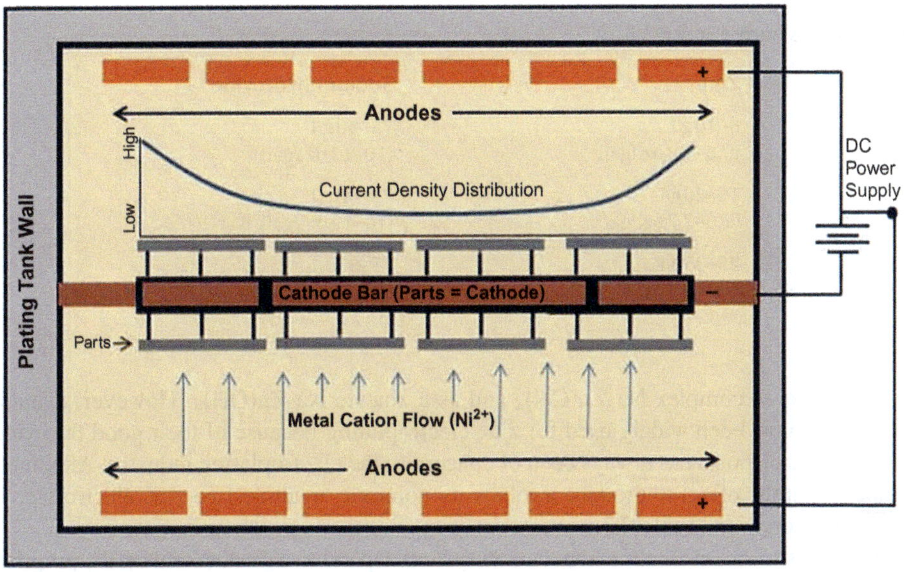

FIG. 9.6 Schematic of a rack plating configuration [4].

Barrel plating [5]—Most fasteners are barrel plated. Barrel plating typically involves a rotating vessel that tumbles a contained, bulk workload, Fig. 9.7 [6].

The barrel is immersed, sequentially, in a series of chemical process tanks, including plating baths, while tumbling the workload. Utilizing interior cathode electrical contacts to polarize the workload, dissolved metals are attracted out of solution onto the individual workpieces. The workload then becomes part of the plating equipment during processing because the individual pieces function as bipolar electrical contacts to the other pieces in the workload. This bipolar contact is a significant contributor to the high efficiencies of barrel plating because the entire surface of the workload, in the current path at any time, is in cathode contact. Barrel plating has distinct advantages; the ability to finish a larger variety of work and producing a greater volume of work for a specified period over a rack-type finishing line.

9.2.3 Electrolytes

The economic development of the electrogalvanizing process depends greatly upon bath solutions and composition, as well as the effect and type of addition agents. Modern electrolytes are both alkaline and acidic. The electrolyte compositions presented below are typical for continuous plating. For details on the differences in electrolyte composition between continuous plating and barrel-and-rack plating, see ref. [2].

9.2.3.1 Alkaline electrolytes

Alkaline electrolytes can be of the cyanide or noncyanide variety. Cyanide electrolytes contain sodium sulfate and sodium hydroxide, along with proprietary brightening agents. The various compositions of the cyanide baths are shown in Table 9.1.

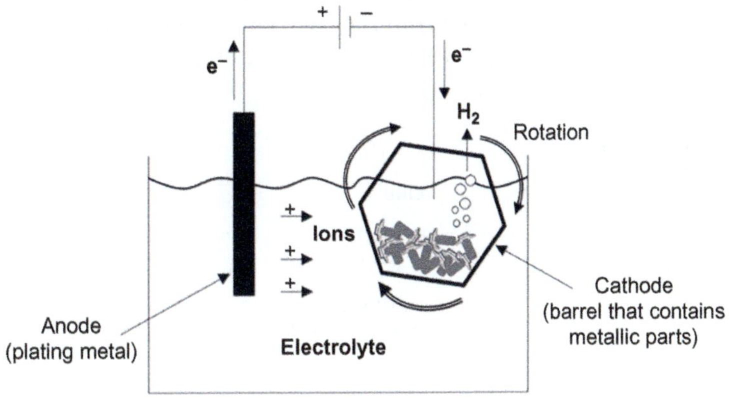

FIG. 9.7 Schematic of a simple barrel plating process [6].

TABLE 9.1 Typical cyanide bath compositions.

	Zinc	Sodium hydroxide	Sodium cyanide
Low cyanide	6–10 g/L (0.8–1.3 oz./gal)	75–90 g/L (10–12 oz./gal)	10–20 g/L (1.3–2.7 oz./gal)
Mid cyanide	15–20 g/L (2.0–2.7 oz./gal)	75–90 g/L (10–12 oz./gal)	25–45 g/L (3.4–6.0 oz./gal)
High cyanide	25–35 g/L (3.4–4.7 oz./gal)	75–90 g/L (10–12 oz./gal)	80–100 g/L (10.7–13.4 oz./gal)

Zinc is soluble as a cyanide complex $Na_2Zn(CN)_4$ and as a zincate $Na_2Zn(OH)_4$. However, cyanide baths are well-known toxic solutions that have been widely used for zinc electroplating because of their good throwing power and high deposition rates. Thus, waste disposal has always been of concern to the electroplating industry. Alkaline noncyanide baths are a logical development in the effort to produce a relatively nontoxic, cyanide-free zinc electrolyte. These electrolytes contain zinc and sodium hydroxide, Table 9.2.

The operating characteristics of an alkaline noncyanide system depend to a great extent on the proprietary additives and brightening agents used in the bath. Alkaline noncyanide baths are inexpensive to prepare and maintain. They also produce bright deposits and cyanide-free effluents (see Ref. [7] for a complete review of these plating baths).

9.2.3.2 *Acidic electrolyte*

Acidic electrolyte zinc plating solutions can be divided into two types, mildly acidic solutions (using chlorides or sulfate anions) and alkaline-zincate solutions. The mild baths contain zinc chloride dissolved in a solution of excess ammonium chloride or potassium chloride that are much less corrosive. Zincate baths are made up solely of a small concentration of zinc metal dissolved in a sodium hydroxide solution. The chloride zinc solution operates with improved bath efficiency and exceptional brightness while eliminating cyanide in the bath, Table 9.3.

Acid zinc baths are used when a high plating rate and low cost is desirable. With all the advantages of a chloride bath over a cyanide bath, there are some disadvantages. Acid zinc sulfate baths have relatively lower costs, improved safety features and pollution control, but it exhibits poor throwing power and insufficient brightness. Acid chloride zinc baths are either based on ammonium chloride or based on potassium chloride. The ammonium chloride baths can be operated at higher current densities than the potassium baths. Additional agents for acid zinc chloride baths are mainly proprietary but consist of aldehydes or unsaturated ketones. Bright zinc electrodeposits have been produced in acid zinc sulfate baths with the use of organic additional agents, such as glycine and thiourea. (See ref. [8] for a complete review of these plating baths.)

9.2.4 Hydrogen embrittlement

A major problem in the electrodeposition process is the evolution of hydrogen at the cathode [6]. In the initial stages of zinc electroplating, hydrogen evolution can occur. Although the evolved hydrogen is dissipated into the atmosphere as gas bubbles at a constant rate, some hydrogen is distributed to the electrolyte while others diffuse into the substrate [9]. Hydrogen may also be trapped into the deposits prior to diffusing into the substrate iron [10]. As a result, hydrogen embrittlement occurs, when hydrogen diffuses and is trapped within the steel substrate, detrimentally affecting the properties, e.g.,

TABLE 9.2 Typical noncyanide baths.

	Zinc	Sodium hydroxide
Ideal metal distribution	6–14 g/L (0.8–1.9 oz./gal)	120 g/L (16 oz./gal)
High productivity	14–25 g/L (1.9–3.4 oz./gal)	120 g/L (16 oz./gal)

TABLE 9.3 Chloride bath types [2].

Constituent	All ammonium chloride	Low ammonium: Potassium chloride	Low ammonium: Sodium chloride	No ammonium, potassium chloride
Zn	15–30	15–30	15–30	22–38
NH_4Cl	120–180	30–45	30–45	–
KCl	–	120–150	–	185–225
NaCl	–	–	120	–
H_3BO_3	–	–	–	22–38
Carrier brightener	4vol%	4vol%	4vol%	4vol%
Primary brightener	0.25%	0.25%	0.25%	0.25%

ductility and mechanical strength, especially in high-strength martensite containing steels [11]. Fig. 9.8 shows the effect of various electrodeposited coatings on the embrittlement index EI [12]. The EI was calculated from slow strain rate tests as follows:

$$EI = 1 - t_c/t_a$$

where (t_c) was the time to failure of a coated specimen, and (t_a) was the mean failure time of uncoated specimens tested in air.

Additives such as sodium benzoate and polyethylene glycol are often added to the plating baths to alter the hydrogen evolution and permeation process [13]. Post baking at 190–220°C at least 4–22h should be performed within 3h immediately after electroplating, depending upon the tensile strength of the substrate [14].

Hydrogen in the bcc iron lattice is found in interstitial sites that weaken the metal-metal bond and is responsible for hydrogen embrittlement. In addition to interstitial sites, hydrogen is strongly attracted to defect sites in the steel, called traps. Trap sites include vacancies, solute atoms, dislocations, grain boundaries, voids, and nonmetallic inclusions. The most important trap sites in steel appear to be phase boundaries, dislocations, voids and inclusions. High-strength steels,

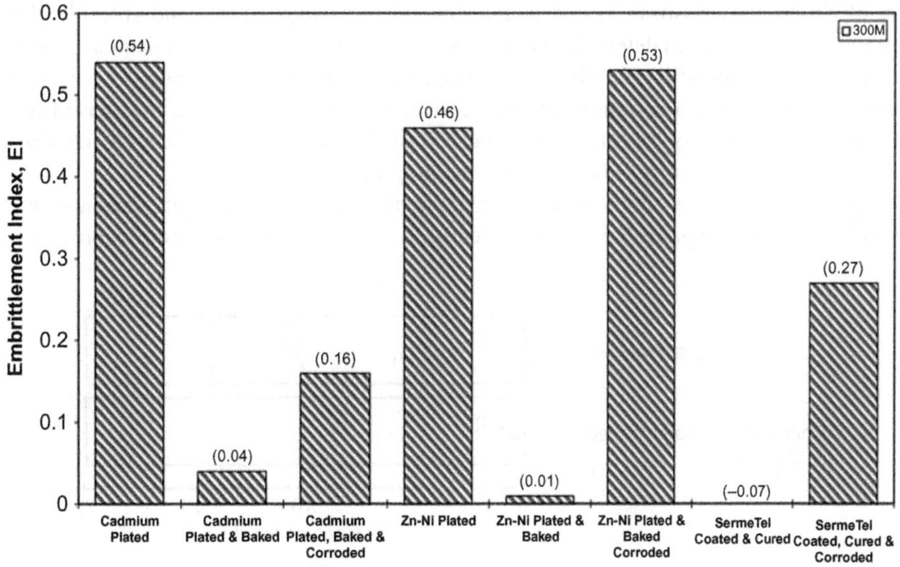

FIG. 9.8 Comparison of embrittlement indices for 300M steel with different coatings [12].

especially containing metastable microstructures such as martensite and bainite, are particularly susceptible to hydrogen embrittlement [14] because they exhibit a much higher defect structure than annealed ferritic microstructures due to their mode of transformation on cooling from the high-temperature austenite phase. Thus, electrogalvanized high-strength fasteners and bolts are particularly susceptible to hydrogen embrittlement.

9.3 Zn-M coatings and phase formation

9.3.1 Introduction

As a sacrificial coating under normal atmospheric conditions, zinc will provide adequate protection to the underlying ferrous substrate. However, electrodeposited binary zinc alloys have resulted in improved properties over electrodeposited pure zinc coatings. In pure zinc coatings, dissolution of the zinc occurs rapidly because of the relatively large difference in electronegativity of zinc and the steel substrate, e.g., 4340, $\Delta E_o = 0.42$ V. Thus, alloying with an iron-group metal will bring the E_o of the coating closer to that of the substrate steel, therefore increasing the life of the coating [15]. The zinc alloy systems find application in three areas [16]:

- As a more corrosion-resistant replacement for electrodeposited zinc coatings on small components
- As a replacement for electrodeposited cadmium coatings on small components
- As a more corrosion-resistant replacement for electrodeposited zinc coatings on steel strip and wire

Electrodeposition of alloys requires careful control of the process operating variables, such as current density, convection, temperature, and bath composition to produce the required properties of the coating. The electrodeposition of Fe-M alloys presents an additional challenge, anomalous co-deposition [17]. By this process, a less noble metal deposits preferentially to the more noble metal. In the case of Zn-M coatings, the less noble metal zinc deposits preferentially at a ratio that is higher in the deposit than in the electrolyte. Anomalous co-deposition can lead to the simultaneous deposition of several phases [18].

9.3.2 Zn binary alloy coatings

9.3.2.1 Zn-Ni

Electrodeposited Zn-Ni alloys exhibited improved corrosion resistance [19], mechanical properties and thermal stability than the corresponding bare zinc and other zinc alloy coatings [20–22]. The Zn-Ni alloys with 15%–22% Ni exhibit low hydrogen embrittleness and find application in the aeronautical industry [23], whereas 10%–15% Ni exhibits maximum corrosion protection to steel and is used as a substitute for cadmium coatings in the automotive industry [24]. These coatings are extensively used in the production of automobile sheet, nuts, bolts, etc.

The mechanism of Zn-Ni co-deposition has been investigated in various sulfate and chloride electrolytes, with and without ammonium ions, and it has been determined that anticorrosive and physical–mechanical properties of Zn-Ni alloy coatings depend on the phase composition [25]. It has long been known that in a large range of compositions from 0.1% to 0.8% zinc, the cathode deposits have a complex structure corresponding to the gamma [γ] phase [2]. These results have been distinguished from the normal equilibrium phase diagram by Winand [26], Fig. 9.9, clearly demonstrating that the electrodeposition process results in nonequilibrium phase development.

It is well known that electrodeposits often contain metastable phases or phases that are deficient or oversaturated compared with the equilibrium phase diagram [15]. More recently, it was reported that electrodeposits of Zn-Ni alloys may

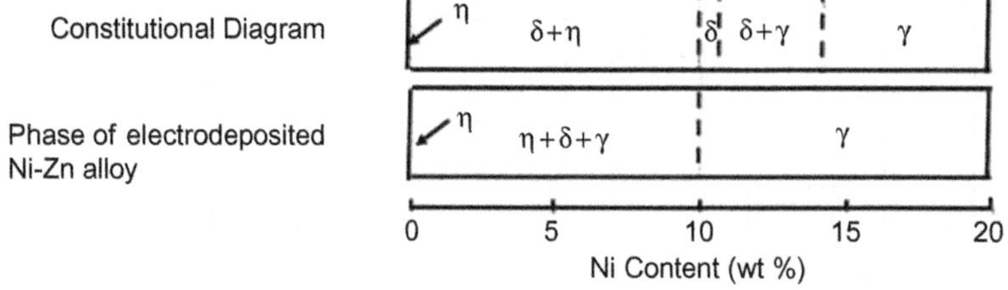

FIG. 9.9 Relationship between nickel content and phase composition of Zn-Ni coatings [26].

contain a solid solution of Ni in Zn (η phase), an intermetallic compound Ni_5Zn_{21} (γ phase), a solid solution of Zn in Ni (α phase), and on occasion an intermetallic compound Ni_3Zn_{22} (δ phase) [25]. Again, the zinc-nickel alloy at about 13% Ni with a metastable γ phase structure exhibited the best corrosion resistance, excellent adhesion and weldability. Although current efficiencies were lower than for zinc deposition due to hydrogen evolution and phosphating and paintability were poorer, it was easy to obtain an almost constant composition of the alloy over a wide range of electrode positions and plating conditions [2].

Zn-Ni deposits from sulfate baths were studied at different deposition variables of current density and the concentration of sodium sulfate and boric acid. Different proportions of the two metals were obtained by using different deposition parameters, but in all cases, it was found that preferential deposition of zinc occurs and anomalous co-deposition occurred [27]. The authors postulated the following sequence during deposition: (1) Ni^{2+} is adsorbed, followed by (2) adsorption of Zn^{2+} onto the freshly adsorbed and deposited nickel. The adsorption of the zinc ions inhibits subsequent deposition of nickel. At low deposition current densities (1 mA cm^{-2}), the deposit consists of a high nickel-rich phase. The deposition process first step is the deposition of a Ni rich ZnNi α phase and is then followed by an increase in Zn content and forms a new ZnNi phase, which is probably γ phase [27].

Further, it has been shown that by changing deposition conditions, the nickel content of the Zn-Ni alloy can be effectively controlled in a wide range (from ≤2.0 to 89.2 at.%) [24]. Table 9.4 shows the effect of the current density on the phase composition of deposited Zn-Ni alloys. The Zn-Ni alloys grown in different baths at the same current density have similar phase compositions. At low current densities, the coating is enriched with Ni and consists of a solid solution of zinc in nickel, α phase. Increasing the current density results in the formation of Ni_5Zn_{21}, γ phase. Further increase in current density produces, depending upon plating bath composition, a single-phase solid solution of Ni in the γ phase with an addition of Ni crystallites for the ammoniacal diphosphate bath.

Table 9.5 shows the effect of deposition potential on the structure of zinc-nickel coatings. Under low polarization, the solid solution of zinc in nickel, α phase, is deposited from both chloride and ammoniacal diphosphate electrolytes. As the electrode potential shifts more negatively, γ phase forms in the chloride bath and a solid solution of zinc in nickel, α phase, tends to form in the ammoniacal diphosphate electrolyte.

9.3.2.2 Zn-Fe

Although more difficult to electroplate than Zn-Ni, Zn-Fe alloys have been studied with the aim of improving phosphatability and paintability, the two weaknesses of Zn-Ni alloys [26]. Winand [2] reviewed the research of Adaniya et al. [28,29] and showed the excellent weldability and workability obtained between 15% and 25% Fe, Fig. 9.10, whereas Fig. 9.11 from Adaniya et al. [28,29] shows excellent corrosion resistance in a similar iron range. Good paintability is obtained when the iron content is higher than 50%.

In his review, Winand [26] showed a comparison of the electrodeposited phases with Zn-Fe thermal equilibrium phases that was adapted from Adaniya et al. [30], Fig. 9.12. A more complete tabulation of the phases that exist in electrodeposited

TABLE 9.4 Effect of current density on the structure of zinc-nickel coatings [24].

Plating bath	[Zn(II)]/[Ni(II)] molar ratio	i_c (A dm^{-2})	C_{Ni} (at.%)	Phase composition
Weak acid chloride	0.5	0.05	66.1	α-phase
		0.15–0.5	18.5–15.1	γ-phase
		1	20.7	γ-phase + polycrystalline Ni
		2–4	25.2–28.0	Solid solution of Ni in γ-phase + polycrystalline Ni
Ammoniacal diphosphate	3	0.05	48.8	α-phase
		0.15–1	19.2–19.0	γ-phase
		2–4	20.2–22.3	Solid solution of Ni in γ-phase

TABLE 9.5 Effect of the deposition potential on the structure of zinc-nickel coatings [24].

Plating bath	[Zn(II)]/[Ni(II)] molar ratio	*E* (V)	C_{Ni} (at.%)	Phase composition
Weak acid chloride	0.5	−0.75	75.3	α-phase+traces of polycrystalline Zn
		−0.95	50.0	α-phase
		−1.15	17.1	γ-phase
		−1.3	14.6	γ-phase+traces of polycrystalline Zn
Ammoniacal diphosphate	3	−0.95	89.2	α-phase+traces of polycrystalline Zn
		−1.15	76.6	Partly amorphizated α-phase+traces of polycrystalline Zn
		−1.3	21.2	Partly amorphizated α-phase

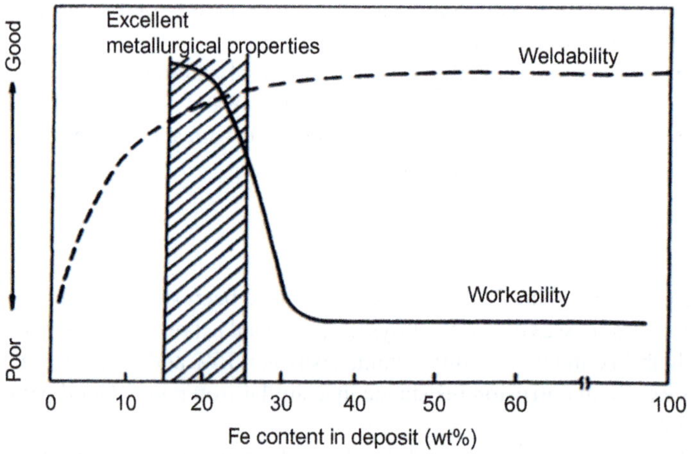

FIG. 9.10 Effect of Fe content in Zn-Fe coatings on weldability and workability, from [2].

FIG. 9.11 Effect of Fe content in Zn-Fe coatings on corrosion resistance and printabilty. *(From R. Winand, Electrodeposition of zinc and zinc alloys, in: M. Schlesinger, M. Paunovic (Eds.), Modern Electroplating, 5th ed., J.W. Wiley, 2010 (Ch. 10).)*

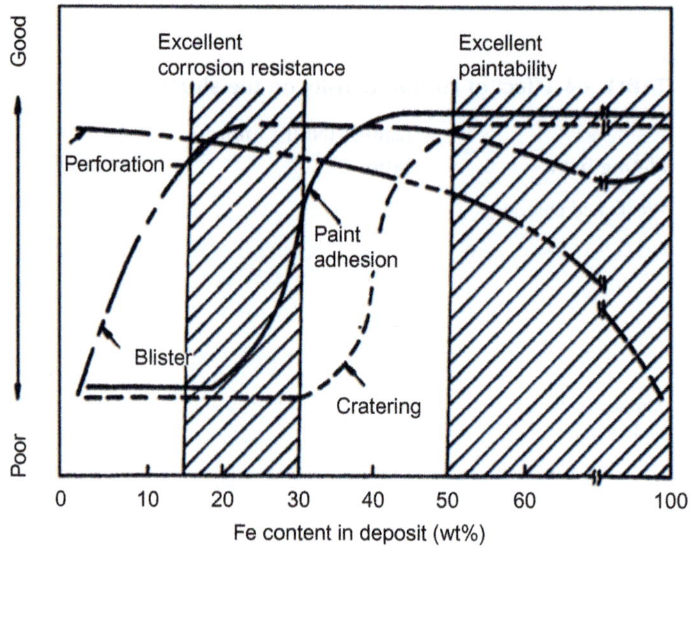

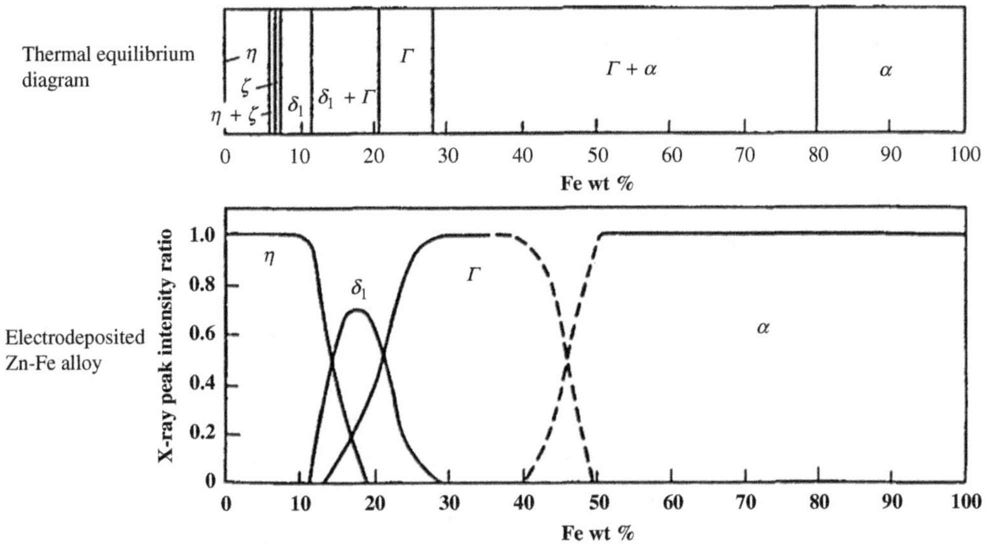

FIG. 9.12 Comparison of electrodeposited phases with thermal equilibrium phases for Zn-Fe alloys [26].

Zn-Fe coatings are found in Table 9.6. Obviously, the phases observed in the electrodeposited coatings are metastable nonequilibrium phases.

9.3.2.3 Zn-Mn

Zn-Mn alloys have a lower corrosion rate than pure zinc with the highest corrosion rate reported for alloys containing 30–40 at.% Mn [32]. It was reported [33] that Zn-Mn coatings have received attention for the following reasons:

- Extremely high corrosion resistance without the aid of painting
- For the formation of a dense corrosion product, γ phase Mn_2O_3
- Excellent corrosion resistance after painting with thin alloy coating layers
- Excellent paintability properties

Thus, allowing the Zn-Mn alloys to meet the requirements for both interior and exterior surfaces of autobody panels. In contrast, Boshkov [34,35] suggested that the high corrosion resistance of Zn-Mn alloys was due to Mn being more electronegative, first dissolves and then promotes the slower formation of $Zn_5(OH)_8Cl_2.H_2O$ (ZHC) compound, with lower solubility.

Zn-Mn alloy coatings can be electrodeposited from several baths including citrate-based electrolytic baths; however, these later baths are unstable and have low efficiency in terms of current density [36]. Further, sulfate-citrate electrolytic baths are the most widely used baths for electrodeposition of high Mn content (>40%). Earlier, Bonzini et al. [37] found that Zn-Mn (0.3%–2.5%) electrodeposited alloys showed a marked improvement for automobile applications over Zn and Zn-Mn (20%–30%) alloy coatings. He found that it was possible to plate these alloys from simple salt baths with easy maintenance and higher cathodic efficiency. Although the corrosion resistance of these low Mn coatings is weaker than the higher Mn coatings, the low Mn coatings showed improved plastic behavior. Recently, it was shown that the problems, e.g., efficiency, control, and the production of hydrogen in electroplating high-strength steels from aqueous solutions, can be overcome by using low pulse current deposition [38]. For details on the Zn-Mn electrodeposition coatings, a recent literature review has been published [39].

Tsuchiya et al. [40] investigated the structure of the electrodeposited Zn-Mn alloys using the SEM observation, the SIMS depth profile, and X-ray diffraction analysis, and observed that alloy phases of the coating are the η phase, the Γ phase, the ε phase, the γ-Mn phase, and the α-Mn phase. The electrodeposited Zn-Mn alloy coatings are composed of either a single phase or two phases from these phases. Single ε phase is observed at around 20 mass% or less Mn, the ε phase and the γ-Mn phase coexist in the range from 20 to 50 mass% of Mn, and single γ-Mn phase exists over 50 mass % Mn. The phases that exist are the thermally nonequilibrium alloy phase, which is seen in the high temperature region in the binary alloy system, Fig. 9.13. The ε phase and the γ-Mn phase in the coating are stable when heating at 170°C. The γ-Mn phase is transformed to the β-Mn phase and the α-Mn phase by the heating beyond 250°C while the ε phase is unchanged.

TABLE 9.6 Existing composition ranges of various phases in EZF coatings [31].

Phase designation	Crystal structure	Equilibrium composition range (at 400°C)[13]	Existing composition range	Diagnostic technique
η	hcp	0 pct Fe	0–20 wt pct Fe	XRD
η			0–27 at. pct Fe	XRD
η			0.9–17 wt pct Fe	XRD, TEM, MS
η			0–20 at. pct Fe	XRD
δ	hcp	8–12 wt pct Fe	10–30 wt pct Fe	XRD
δ			10–28 wt pct Fe	XRD
δ			6–50 at. pct Fe	XRD
Γ_1	fcc	17–20 wt pct Fe	22–38 at. pct Fe	XRD
Γ	bcc	24–28 wt pct Fe	10–50 wt pct Fe	XRD
Γ			13–52 at. pct Fe	XRD
Γ			18.5–40 wt pct Fe	XRD, TEM, MS
Γ			20–50 at. pct Fe	XRD
α	bcc	97–100 wt pct Fe	40–100 wt pct Fe	XRD
α			38–100 at. pct Fe	XRD
α			18 wt pct Fe	TEM
α			>50 wt pct Fe	XRD, TEM, MS
α			20–100 at. pct Fe	XRD

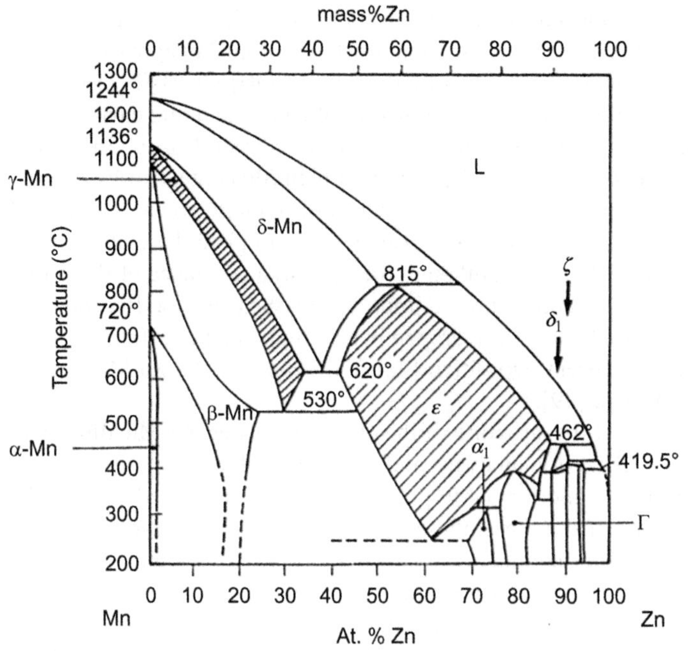

FIG. 9.13 The Zn-Mn equilibrium phase diagram [40].

9.3.2.4 Zn-Co

Zinc-cobalt electrodeposition is normally carried out from two classes of baths: mildly acid chloride and alkaline cyanide-free. Acid-chloride baths are used to produce deposits with less than 1% Co in the electrodeposit. This level of alloying is achieved with the aid of organic addition agents that ensure consistent cobalt concentrations over a wide range of conditions [16]. Prasad et al. [41] reported that triethanolamine and gelatin were added to alkaline baths to improve the deposit quality and boric acid was used to obtain a fine-grained zinc-rich deposit. In addition, XRD studies of alkaline solution deposits with Co <12% revealed a single γ phase, the substitutional solid solution of Co in Zn. In all cases, the corrosion performance of Zn-Co deposits was found to be superior to that of plain Zn coatings. It was found that all Zn-Co electrodeposits showed greater corrosion resistance than the Zn electrodeposits [42]. The Zn-18%Co layer showed the best corrosion resistance with a lifetime of about three times longer than that of the Zn layer. For a 2.8% Co concentration, it was found that the deposits contained $CoZn_{13}$, pure Zn phase, and hexagonal η phase [43]. In order to improve the properties and structure of the Zn-Co coatings, additional processing variables were introduced. The effect of a magnetic field on crystal structure and morphology was studied by Kermoune et al. [44], the effect of a pulsed current by Thangaraj et al. [44], and ultrasound influence by Ridosic et al. [45].

9.3.2.5 Zn-Cr

Several patents claim better corrosion protection of electrodeposited Zn-Cr alloy coatings on steel in comparison to Zn coatings [46–49], and the electrodeposition of Zn-Cr has been optimized for high-speed plating on steel sheet for car body applications in the automotive industry [50,51]. One benefit of Cr in Zn-Cr coatings is the fact that it behaves almost like a noble metal because of surface passivation from the formation of dense and largely insoluble surface passivation layers of Cr-oxides that could inhibit the dissolution of the sacrificial Zn coatings [52]. Zn-Cr alloy electrodeposited coatings prepared from chromium (III) salt solutions have become an alternative to the chromating process in an attempt to find non-toxic alternative to the highly toxic and carcinogenic chromium (VI) [53].

The properties of the coatings are determined not only by composition but by the surface morphology and structure of Zn-Cr coatings. It is well known that the electrodeposition process produces nonequilibrium phases in comparison to the binary equilibrium diagrams. Several unidentified nonequilibrium Zn-Cr phases have been found, e.g., ref. [54]. With increasing Cr content, the phase composition changes in the sequence:

$$\begin{aligned}\text{Hexagonal } \eta \text{ (Zn, Cr)} &\rightarrow \text{hexagonal } \delta\text{(Zn, Cr) (supersaturated solid solution)}\\ &\rightarrow \text{cubic } \Gamma \text{ (Zn, Cr) (disordered } Cu_5Zn_8 \text{ g} - \text{b brass type)}\end{aligned}$$

including concentration intervals in which two or all three phases coexist [54]. The nonequilibrium diagram has not been studied extensively, but the available data indicate that δ - and Γ-phases may be regarded as nonequilibrium, super saturated solid solutions. By heating the coatings, the equilibrium structures form and δ-phase transforms to a mixture of phases ($\xi+\delta$) or a single phase (ξ), depending on temperature and Cr content, while the Γ-phase is stable up to 275°C [54]. In the first stages of electrodeposition, the Zn-Cr alloy deposits are mostly composed of Cr-based Γ-phase containing about 33 at.% Cr. With advancing deposition, the relative Cr content in the Γ-phase gradually decreases to a critical average value of 27 at.% and the relative share of the η-Zn phase increases. As a result, the average Cr content in the alloy gradually decreases [55].

9.3.3 Zinc multiple component coatings

In order to meet industry's high demand for corrosion resistance, the corrosion behavior of electrodeposited binary alloy coatings has been further enhanced by the use of a third metal or ternary alloy as proposed by a number of investigators (e.g., [56,57]).

9.3.3.1 Zn-Ni-Mn

Zn-Ni electrodeposited coatings have greater corrosion protection, improved mechanical properties, and better thermal stability than the corresponding bare zinc and other zinc alloy coatings [20], and Zn-Mn electrodeposited coatings were found to have lower corrosion rate than pure zinc [32]. It has been proposed that combining the two coatings as a Zn-Ni-Mn electrodeposited coating would improve the properties further, especially for automotive applications, e.g. [56]. These coatings were produced using both chloride baths [56] and sulfate baths [57]. It was discovered that current density significantly affected the composition, structure, and corrosion resistance of this ternary coating. It was found that the presence of higher Mn content in the Zn-Ni-Mn coatings could lead to the formation of a good passive layer [58,59].

9.3.3.2 Zn-Ni-Co

It has been observed that the ternary Zn-Ni-Co is characterized by enhanced corrosion resistance compared with binary Zn-Ni and Zn-Co alloys, in particular the best corrosion resistance was observed for ternary deposits containing 11.24% Ni and 6.52% Co [60]. It was shown that the electrodeposition of Zn-Ni-Co alloy exhibited an anomalous codeposition phenomena and the improved properties were due to the phase structure and the formation of high nickel γ-alloy phase [61]. The improved corrosion resistance was attributed to a change in the phase structure of the coatings [62]. The corrosion resistance was further attributed to its unique surface chemistry, phase content and orientation, homogeneity, surface morphology and reduced roughness, as well as to the change in the intrinsic electrical properties of the surface film [63]. Surface chemistry and topography were found to be responsible for the better corrosion resistance in a 14.36% Ni and 6.39% Co ternary Zn coating [64].

9.3.3.3 Zn-Ni-M

Abe et al. [57] studied the addition of Cr to Zn-Ni electrodeposited coatings and found that these alloys had excellent corrosion resistance and plating adhesion. It was found that the addition of 4% magnesium to Zn-Ni electrodeposited coatings increased the tribocorrosion, i.e., wear and corrosion, performance of a Zn-Ni-Mo coating in a chloride environment [65]. The corrosion protection of ternary Zn-Ni-Mo coatings (1.2 wt% Ni) was found to be worse than the binary Zn-Ni coating with the optimum Ni content (12 wt% Ni), which may have been caused by the two-phase composition of the ternary alloy. The coating contained both the η phase and the γ phase [66].

9.3.3.4 Zn-Ni-Fe

Electrodeposited Zn-Fe-Ni alloy coatings were reported to have good adhesive and anticorrosion properties [67]. All deposited coatings were found to be of the same crystallographic structure as zinc, but with different crystallographic orientations. Electrodeposition of Zn-Ni-Fe alloy coatings in acidic chloride baths resulted in the lowest corrosion rates for this alloy in comparison to a Zn-Ni and Zn-Fe coating [15]. Their results showed that the rate of Zn deposition was heavily influenced by mass-transport limitations at high applied current densities in these baths, while the rates of Ni and Fe deposition were not. The effect of current density and pH values was studied using sulfate baths [68]. The results of the study produced bright and uniform surface appearance deposits of Zn-Ni-Fe that were obtained at a pH range of 2–5, with high corrosion resistance.

9.3.3.5 Zn-Fe-Mo

Known for their catalytic application in desulfurization processes, Zn-Fe-Mo ternary alloy coatings produced by electrodeposition have been studied. Electrodeposited zinc-iron-molybdenum ternary alloys result in coatings with increased resistance to corrosion and wear in comparison to binary alloy coatings without molybdenum [69]. Also, co-deposition of molybdenum with nickel or iron leads to the improvement in microhardness. It was reported that ternary Zn-Fe-Co alloy coatings containing as little as 4 wt% Mo and 6 wt% Co have much better mechanical properties and higher resistance to corrosion in a chloride medium than zinc coatings [70]. The relatively high corrosion resistance of the coatings is reported to be associated with a different chemical composition of the passive layer as compared with Zn-Fe coatings. The passive layer on the Zn-Fe-Mo coatings shows in the entire thickness (approximately 7 nm) the clear predominance of the oxide form (Mo and Zn oxides) over the hydroxide form (Mo and Zn hydroxides). Further, the layer is enriched in molybdenum. The effect of pH [71] and current density [72] on the production of Zn-Fe-Mo alloy coatings electrodeposited in a sulfate bath was studied. It was found that at a pH of 5.7–6.0, bright, homogeneous coatings are developed and are characterized by a significantly lower corrosion rate [71]. Also, it was shown that increasing cathodic current density from 5 to 30 mA/cm^2 had no effect on the molybdenum content in the coating which remained constant at close to 1.5 at.% [72]. In addition, the effect of deposition duration was studied [73].

9.3.3.6 Zn-Fe-Co

Electrodeposition of Zn-Co alloys exhibits a significantly higher corrosion resistance and better surface morphology than pure zinc coatings. Coatings with high Co content are nobler than steel and provide a barrier-type protection. On the other hand, Zn-Fe alloy coatings have high corrosion resistance, excellent paintability, and good weldability in comparison to pure zinc. The co-deposition of Fe into the binary Zn-Co alloy or Co into the binary Zn-Fe alloy leads to improved performance of the ternary Zn-Fe-Co coating [74]. The amount of Co introduced into the Zn-Fe coating is critical since high Co content (18–40 wt%) showed superior corrosion resistance, whereas low Co content (0.7–5.5 wt%) exhibited poor corrosion performance [75] using a chloride bath. However, it was also reported that for an 8.9% Co coating, good corrosion resistance was found in comparison to zinc alone, Zn-Co, and Zn-Fe binary alloys [74] using a sulfate bath. XRD showed that the Zn-Co-Fe coating consisted of Zn solid solution and an iron cobalt intermetallic phase, Fe-Co, [74]. In a later study [76] utilizing a sulfate bath, it was shown that increased corrosion protection of steel can be achieved by electrochemical deposition of ternary Zn-Fe-Co alloys with higher amounts of Co and Fe. The composition of the ternary alloy can be optimized by proper choice of the current density. The influence of deposition temperature in a sulfate bath was studied [77], and it was found that Co and Fe content abruptly increases once the plating temperature reaches 60°C. Therefore, it was reported that the most suitable electroplating temperature was between 30°C and 40°C since the coating achieved a compact morphology and no cracks.

9.3.4 Thermal treatment of coatings

It has been clearly demonstrated that electrodeposited Zn and Zn alloy coatings as deposited are made up of nonequilibrium metastable phases. As examples, see Zn-Cr coatings [54]; Zn-Ni coatings [78]; Zn-Fe coatings [31]. Gu et al. [31] categorized the effect of continuous heating on structural changes in Zn-Fe alloy coatings into two temperature ranges. In the low temperature range, the average iron content in the coating did not change and the nonequilibrium structure transformed gradually to the equilibrium structure. For example, in an 18 wt% Fe coating, the η phase transforms to δ phase and the δ phase transforms to Γ_1 phase, Fig. 9.14 [79]. In the high temperature range, the coating and the steel substrate form a diffusion couple of Fe-Zn coating and the steel substrate, and the predicted equilibrium phases form between the couple. In the case of Zn-Fe coatings, δ-phase is expected to have the best uniform properties.

For the Zn-Ni coating [80], it was found that the phases formed by electrodeposited Zn-Ni alloys are different than the phases found on the thermodynamic equilibrium diagram. The δ phase is replaced by distorted η_d and γ_d. It was shown by thermal analysis that the electrodeposited alloys containing up to 13 wt% nickel were out of equilibrium. The first transformation, at 200°C, corresponds to the formation of δ and η phases from the supersaturated hexagonal phase η_d than the second transformation around 250°C, corresponds to the precipitation of δ and γ phases from the unsaturated cubic phase γ_d phase.

For Fe-Cr electrodeposited coatings, nonequilibrium δ- and Γ-(Zn,Cr) phases are the major constituents. However, on annealing the coatings at 260°C equilibrium ζ-$CrZn_{13}$ phase precipitates from δ- and Γ-supersaturated solid solutions [54,81,82]. In an annealed 5.6% Cr alloy coating containing only ζ-$CrZn_{13}$ phase, electrochemical and corrosion studies showed a more positive corrosion potential in 0.5 M Na_2SO_4 solution at pH 5.9 and less weight loss at pH 1.3 of the thermally treated coatings as more regular corrosion occurred compared with the as-deposited layer [82]. In all the alloyed coatings, improvement in uniform properties can be obtained by annealing.

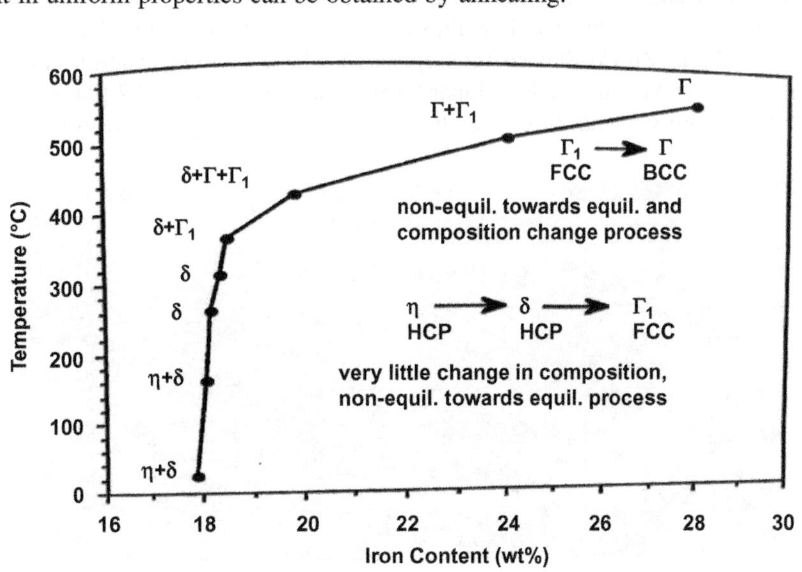

FIG. 9.14 Variations of chemical composition of the 18 wt% Fe electrodeposited coating during continuous heating [79].

9.3.5 Morphology and microstructure

The characterization of the cross section and surface morphology of electrodeposited coatings is essential in understanding the properties of these coated alloys. The surface morphology is very important to the coating deformation characteristics, paint adhesion, and corrosion resistance [83]. The cross section of the coating that may be made of bands and fine grains is important in determining corrosion resistance, particularly compositional and phase homogeneity. In addition to alloy content, impurities, current density, variations of potential, additives, hydrodynamics of the catholyte flow, bath composition, bath temperature, plating time, pH, substrate, etc., all have significant effects on the morphology of the coating [84]. For a detailed discussion of "Electrocrystallization," see the review by Winand [85]. In addition, the general development of morphology in electrodeposition, "Morphogenesis," is described in ref. [86].

9.3.5.1 Surface morphology

Typical surface morphologies of Zn electrodeposited coatings are shown in Fig. 9.15. The Zn-18%Fe coating has a nodular surface morphology, whereas the pure zinc EG coating is plate-like. This study showed that the nodular surface morphology was also present in an 11 wt% Fe coating. Current density is well known as an important variable in producing superior coatings [87]. Although different additives can affect the surface morphology of pure zinc coatings, it was found that the critical current density for Zn deposition was higher than that without additives [64]. For electrodeposited zinc coatings, results show that as the current density increases from 3 to 50 A/dm^2, the zinc coating surface evolves from the morphology of hexagonal platelets and ridges to the pyramidal morphology [88]. A strong basal texture had better corrosion resistance than deposits with a weak basal texture with a pyramidal fiber component. Filho and Lins [89] found that zinc coating morphology tends to become more homogeneous with smaller crystals as the current density is increased up to 150 A/dm^2 and was more significant than the effect of electrolyte rate. Three different crystallization forms were observed:

- Type 1: large crystals showing hexagonal ridges parallel to each other (CD – 15 A/dm^2)
- Type 2: smaller crystals than those in Type 1 perpendicular to each other (CD – >60 A/dm^2)
- Type 3: hexagonal crystals parallel to each other showed more surface area than Types 1 and 2 (CD – >60 A/dm^2)

The effect of temperature, pH, and current density on the morphology and texture of electrodeposited zinc was studied [90], and it was also found that the electrodeposited zinc had mostly basal and low angle planes parallel to the surface. At pH 2, increasing overvoltage (increasing current density or decreasing temperature) reduced the percentage of basal planes, increased the percentage of low angle planes parallel to the substrate surface, and decreased the zinc crystal size. At pH 4, increasing current density increased the number of basal and low angle planes parallel to the surface, but increased the zinc crystal size. These observations were related to nucleation and growth mechanisms at the two different pH environments.

For Zn-Mn electrodeposited coatings, a transition from a hexagonal plate structure similar to pure zinc deposits (0.28 at.% Mn) to a nodule surface morphology (at >30 at.% Mn) [91]. The surface morphology of Zn-Ni electrodeposited coatings shows a similar transition from hexagonal plate structure at 2% Ni to a nodule surface morphology at Ni contents more than 6% [92]. It was found that for sulfate baths, nodule grain size increased with increasing Ni content and porosity is introduced [93]. It was reported [42] that Zn-Co alloy coatings appeared to have a two-layer surface morphology, an outer layer of nodules (18 at.% Co), and a cracked morphology inner layer (17 at.% Co). Table 9.7 [94] details the surface morphologically characteristics for Zn, Zn-Co, and Zn-Co-Fe alloys. The effect of alloy composition is strongly apparent.

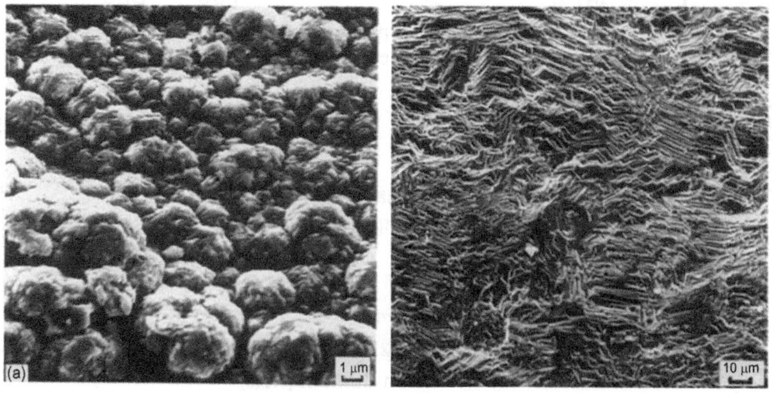

FIG. 9.15 Scanning electron microscopy showing the surface morphology of (a) 18 wt% Fe and (b) EG Zn coatings [79].

TABLE 9.7 The variation in surface morphology for Zn-Ni electrodeposited Zn-Co alloys [79].

Morphologies and structures of various coatings shown in the results				
Coating	GDOES composition analysis	Morphology	XRD analysis	TEM analysis
Zinc	100% Zn	Hexagonal	hcp [4,8,10,17]	Crystalline
Zn-Co alloys				
Alloy 1	0.7–1% Co	Angular grains	η [6,13,14]	–
Alloy 2	2% Co	Angular/platelet grains	η+γ [6,13,14]	–
Alloy 3	2.5% Co	Angular/platelet grains	//	–
Alloy 4	3% Co	Nodular dense grains	Unidentified structure	–
Alloy 5	5.5% Co	Polyhedral and not dense grain	//	Crystalline
Alloy 6	9% Co	Two types of grains but same composition: (1) small polyhedral grains; (2) big globular grain	//	–
Alloy 7	15% Co	Two types of grains; but different compositions: (1) Cauliflower-like; (2) matrix	//	–
Alloy 8	32% Co	Homogeneous featureless	Nanocrystalline or amorphous	Nanocrystalline
Zn-Co-Fe				
Alloy 9	3.5% Co+0.5% Fe	Angular/platelet grains	Unidentified structure	Crystalline
Alloy 10	4% Co+0.5% Fe	Two types of grains; (1) polyhedral; (2) unidentifiable and textured	//	–
Alloy 11	18% Co+0.7% Fe	Two types of grains: (1) cauliflower-like; (2) matrix	//	–
Alloy 12	32% Co+1% Fe	Homogeneous featureless	Nanocrystalline or amorphous	Nanocrystalline

9.3.5.2 Cross-section morphology

Winand [87] has developed a scheme of how the cross-section morphology changes with current density, Fig. 9.16. Using results for Zn-Fe coatings, the current density vs. % Fe space is divided into regions characterized as:

- field-oriented isolated texture type (FT)
- basis-oriented reproduction type (BR)
- nodular type (N)
- rhythmic lamellar type (RL)

The various cross-section microstructures found in Fe-Zn alloy coatings are shown in Fig. 9.17. The nodular structure (N) is found in a pure zinc electroplated alloy [95] and forms by epitaxial growth at low overpotentials by 2D nucleation. The nonepitaxial growth of this structure occurs at a higher overpotential and proceeds by 3D nucleation. The rhythmic lamellar structure, RL, of a Zn-18 wt% Fe electrodeposited coating is seen in Fig. 9.18. The bandwidth increases with deposition temperature and current density [79]. As plated the average band width was 0.2–0.3 μm, and it was found that the average grain size within each band was about 30 nm. For Fe-Ni coatings, it was found that the grain size in Zn-Ni coatings ranged between 13 and 55 nm, depending on Ni content ([96]. The banded structure may result from small variations in deposition parameters, such as the current density in the deposition process. Variations in the current density can produce small compositional gradients between adjacent bands or rows of grains.

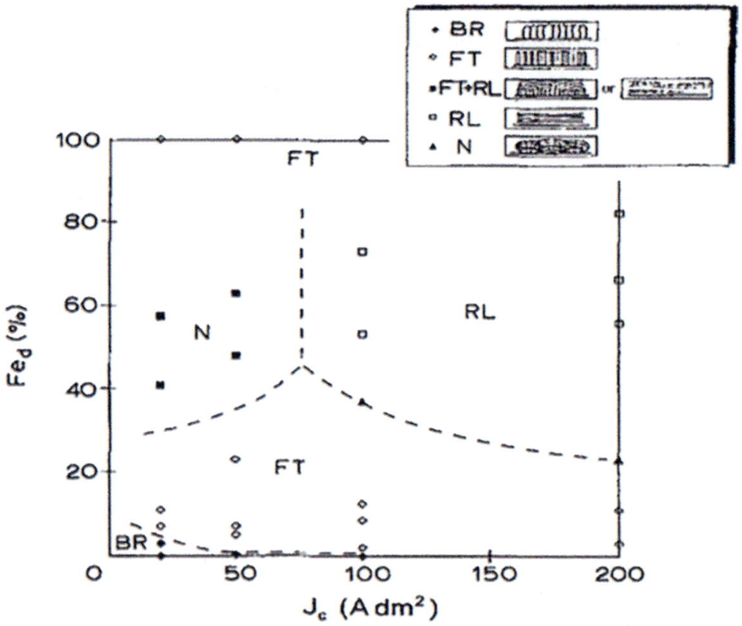

FIG. 9.16 Cross-section metallographic structures of alloy deposits as a function of current density and composition [87].

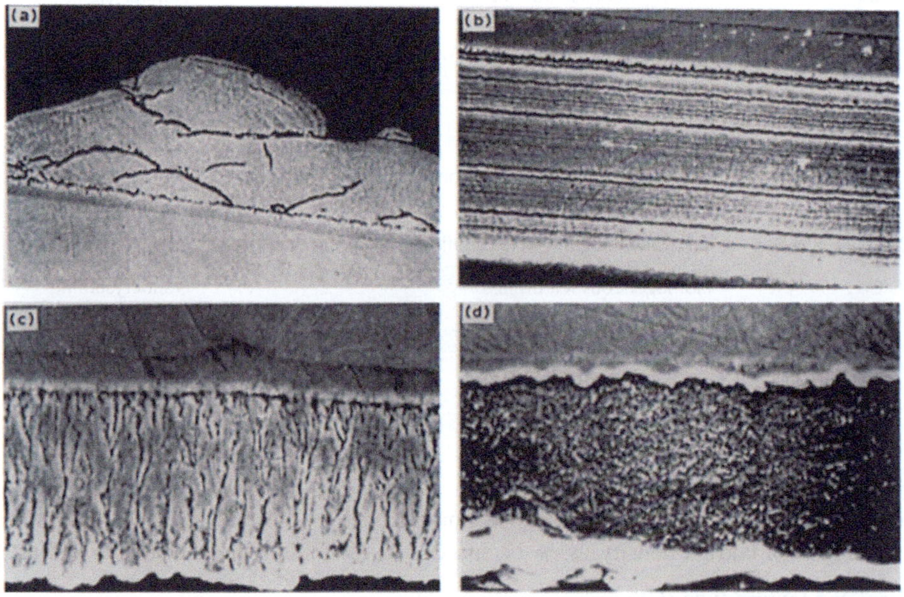

FIG. 9.17 Various metallographic structures observed in the FeZn system. (A) Nodular (N), (B) rhythmic lamellar (RL), (C) field-oriented (FT), (D) unoriented dispersed (UD) [87].

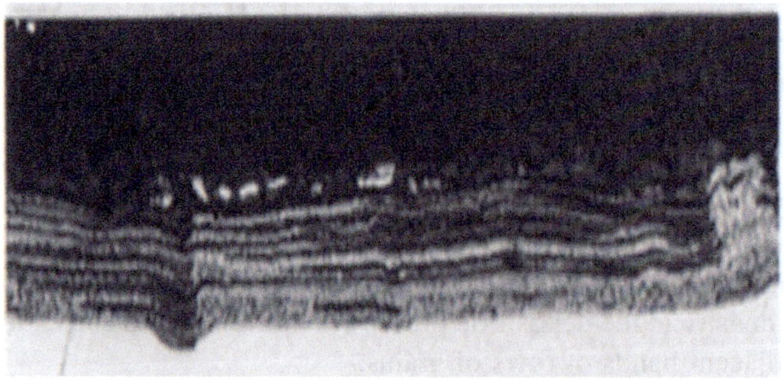

FIG. 9.18 The rhythmic lamellar (RL) cross-section microstructure of a Zn-18 wt% Fe alloy [79].

9.4 Electrodeposited nanocrystalline Zn coatings

9.4.1 Introduction

The utilization of zinc coatings has continuously increased with the need for new global requirements for steel structures applications such as storage cells and energy harvesting [97]. Standard electrodeposited Zn coatings offer excellent galvanic protection to the underlying steel substrate; however, because of the required thicker coatings for corrosion resistance, poor mechanical properties of the conventional coarse grain Zn coating can result and prevent its application in many fields. Nanocrystalline electrodeposit research has led to the development of thinner electrodeposited coatings with improved properties. The unique nanocrystalline coatings contain a small grain size below 100 nm and a high-volume fraction of grain boundaries of up to as much as 50% of the total crystal volume [98]. The resulting coating has a smoother and brighter surface and exhibits many unusual mechanical, physical, chemical, electrochemical properties. As a result, these coatings exhibit excellent wear resistance, corrosion resistance, ductility, and hardness, compared with conventional coarse-grained zinc coatings [97].

9.4.2 Zn processing parameters

Pulse electrodeposition is a useful process for producing nanocrystalline materials [99–102]. For pulse electrodeposition (PED) [103–105], the potential or current is alternated swiftly between two different values, which results in a series of pulses of equal amplitude, duration, and polarity, separated by zero current. Each pulse consists of an on-time (T_{on}) during which potential and current are applied, and an off-time (T_{off}) during which zero current is applied as shown in Fig. 9.19. It is possible to control the deposited film composition and thickness in an atomic order by regulating the pulse amplitude and width [107–110], which favors the initiation of grain nuclei and greatly increased the number of grains per unit area producing a finer-grained deposit with better properties than conventionally plated coatings. Process and circuitry for generating current pulses for electrolytic metal deposition are found in the literature [111].

Chandraseker and Pushpavanam [106] have outlined the concepts of pulse current (PC) and pulse reverse current (PRC) techniques as follows:

(a) In electroplating, a negatively charged layer is formed around the cathode as the process continues. When using direct current (DC), this layer charges to a defined thickness and obstructs the ions from reaching the part. In pulse electrodeposition (PED), the output is periodically turned off to cause this layer to discharge somewhat. This allows easier passage of the ions through the layer and onto the part.

(b) High-current-density areas in the bath become more depleted of ions than low-current-density areas. During T_{off}, ions migrate to the depleted areas in the bath. When pulse T_{on} occurs, more evenly distributed ions are available for deposition onto the part. The effects of pulse current (PC) and pulse reverse current (PRC) variables on electrodeposition are summarized in literature [112].

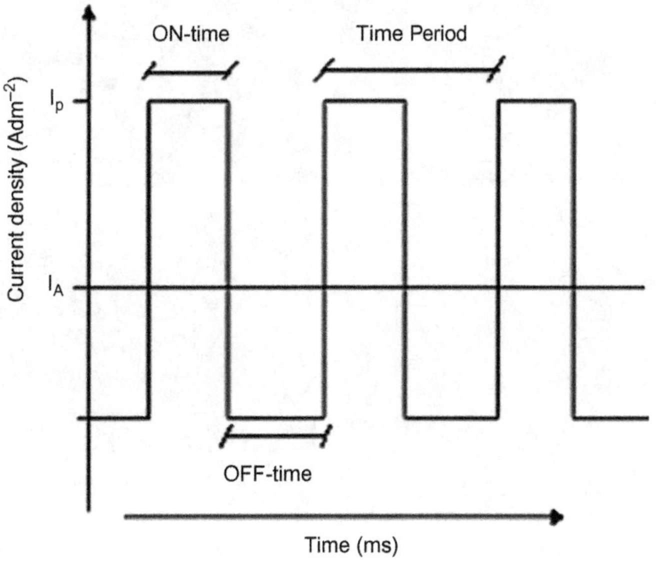

FIG. 9.19 Typical pulse-current waveform [106].

The effect of various pulse parameters such as current density, duty cycle, and other parameters such as additive concentration, alloying addition on corrosion resistance is very well documented in literature [20,60,88,113–115]. Pulse current electrodeposition (PC), in which current is imposed in a periodic manner with a rectangular wave form, Fig. 9.19), has been used to produce unique structures, such as coatings with properties not obtainable by direct current (DC) plating [116]. In pulse current electrodeposition (PC), three parameters influence the deposit properties: peak current density (J_p), current on-time (T_{on}), and current off-time (T_{off}). It is well known that pulse current density has a beneficial effect on the properties of the deposits [117]. Pulse current electrodeposition (PC) yields a finer-grained and more homogeneous surface appearance of the deposit because a higher instantaneous current density is possible during deposition by using pulse control than that by direct current (DC) plating. As a result, an increased nucleation rate leads to the formation of finer grains [118]. Paatsch [119] investigated the influence of pulse current electrodeposition (PC) on the deposition of bright zinc layers on a steel substrate using commercial high cyanide concentration ($100\,g\,L^{-1}$ NaCN) as well as low-acid content, ammonium chloride-containing solutions. Paatsch found that a finer grain size zinc deposit was obtained by a high-pulse current density ($J_p = 0.44\,A\,cm^{-2}$) and the resulting high overpotential in comparison to zinc produced by direct current (DC) plating. According to Gurovich and Krivtsov [120], the most appreciable reduction in grain size of zinc deposited from sulfate-based electrolyte was obtained at a pulse peak current density of $0.03\,A\,cm^{-2}$.

Youssef et al. [121] studied the influence of pulse plating parameters on the synthesis and preferred orientation of nanocrystalline zinc. The results of their systematic study were as follows:

9.4.2.1 Current on-time (T_{on})

Figs. 9.20 and 9.21 demonstrate the influence of pulse duration current on-time (T_{on}) on the morphology and grain size of zinc deposits, with constant pulse current density and current off-time. The photomicrographs in Fig. 9.20 show that an increase in current on-time (T_{on}) results in a decrease in grain size of the zinc deposits.

Zinc platelets with an average grain size of 195 nm are formed at $T_{on} = 1$ ms, Figs. 9.20a and 9.21, while increasing T_{on} to 3 ms, Fig. 9.20b results in disappearance of the sharp edges of zinc platelets with a slight reduction of the average grain size to 157 nm, Fig. 9.21. Further increase of current on-time to 5 and 7 ms drastically reduces the average grain size of zinc deposits to 75 and 51 nm, respectively, Figs. 9.20c, d and 9.21. The effect of decreasing grain size with increasing current on-time (T_{on}) is explained by an increased number of nucleation sites caused by the higher over-potential at longer current on-times. Similar behavior of increasing overpotential with current on-time was also found during zinc deposition from a chloride-based electrolyte [100]. A decrease in grain size was also found for gold [122] and palladium [123] electrodeposits with increasing current on-time and was also related to an increased nucleation rate resulting from the higher overpotential.

FIG. 9.20 FESEM micrographs of the surface morphology of zinc deposits obtained at $T_{off} = 9$ ms, $J_p = 800$ mA/cm^2 and different current on-times: (a) 1 ms, (b) 3 ms, (c) 5 ms, and (d) 7 ms [121].

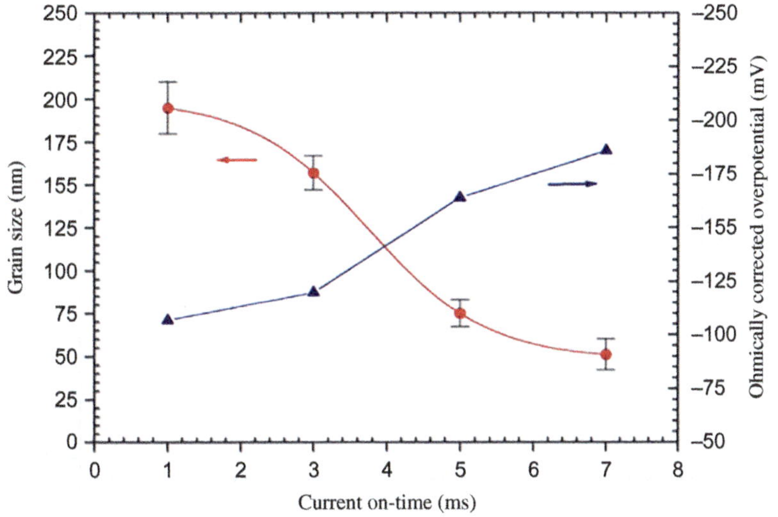

FIG. 9.21 Variation of grain size and ohmically corrected overpotential of zinc deposits [121].

9.4.2.2 Current off-time (T_{off})

Fig. 9.22 shows FESEM photomicrographs of the effect of current off-time (T_{off}) on the grain size and surface morphology of zinc deposits. T_{off} was varied from 9 to 50ms at an "optimized" T_{on} value (7ms) and pulse current density of 800mA/cm^2. The "optimized" parameter is defined by Youseff et al. [121] as the value that yields deposits with the smallest grain size. The average grain size of the deposits increased gradually from 51 to 117nm with increasing current off-time (T_{off}) from 9 to 50ms, respectively, Figs. 9.22 and 9.23. The formation of large grains at longer current off-times (T_{off}) was explained by desorption of some species (most likely additives and/or hydrogen) during the longer current off-times (T_{off}), which activates the growth centers and results in grain growth. In addition, zinc adatoms at longer current off-times (T_{off}) have sufficient time to migrate over the crystal surface and enhance the grain growth process. Puippe [116] obtained grain coarsening during pulse plating of copper and gold deposits at long current off-times (T_{off}) while the current on-time (T_{on}) and pulse current density (J_p) were held constant. They explained this trend in terms of a recrystallization process and

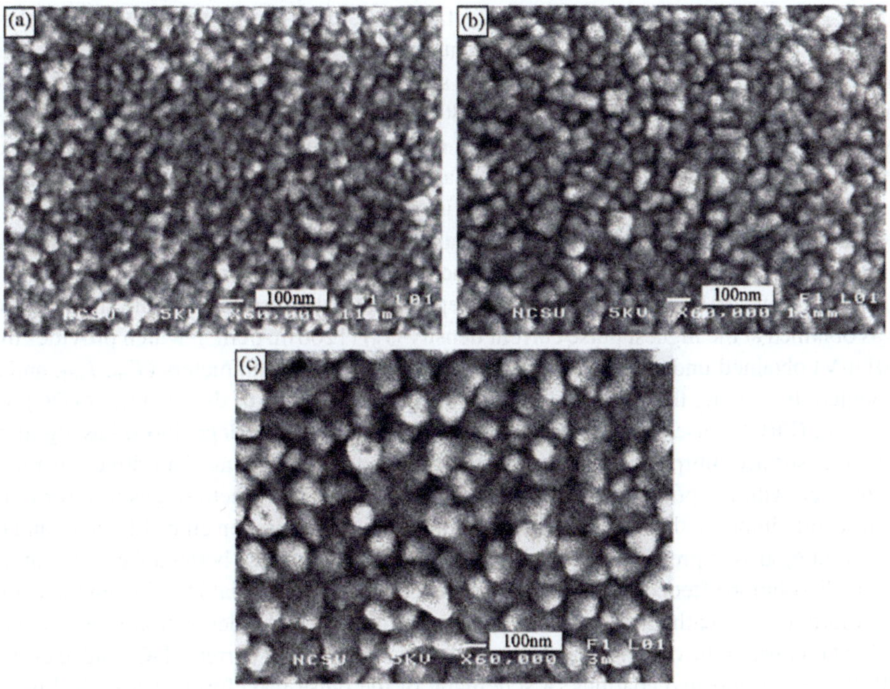

FIG. 9.22 FESEM micrographs of the surface morphology of zinc deposits obtained at T_{on} = 7ms, J_p = 800mA/cm^2 and different current off-times (a) 9ms, (b) 25ms, and (c) 50ms [121].

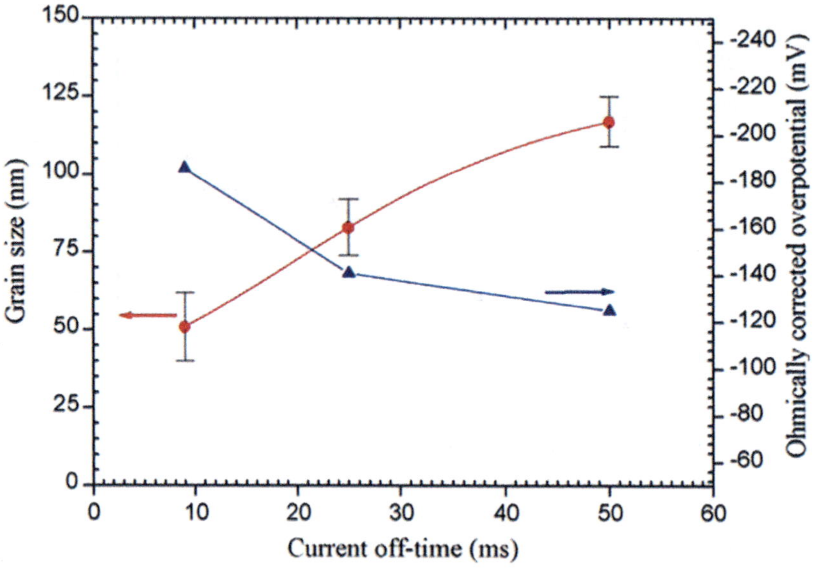

FIG. 9.23 Variation of grain size and ohmically corrected overpotential of zinc deposits with current off-time at $T_{on}=7$ ms, and $J_p=800$ mA/cm^2 [121].

argued that coarse grains are thermodynamically more stable and longer current off-time (T_{off}) provides enough time for the system to stabilize and therefore grain growth takes place. Fig. 9.23 shows the variation of grain size and ohmically corrected overpotential of zinc deposits with current off-time, (T_{off}). The cathodic overpotential decreases gradually with increasing current off-time, (T_{off}). It was explained that this result may also be plausibly used to justify the grain coarsening at longer off-times, (T_{off}).

9.4.2.3 *Pulse current density* (J_p)

Saber et al. [124] in Fig. 9.24 showed the SEM surface morphology images of zinc deposited at pulse current densities (J_p) of 0.4, 0.8, 1.6, and 2 A cm^{-2} from a chloride-based electrolyte without additives. It can be seen that the grain size of zinc deposits decreases with increasing J_p. The smallest grain size (4–5 μm) was produced at $J_p=2$ A cm^{-2}. Similar results were found by Youssef et al. [121], and Fig. 9.25 shows the effect of pulse (peak) current density (J_p) on the grain size of zinc deposits obtained at 400, 800, and 1200 mA/cm^2. The current on-time (T_{on}) and off-time (T_{off}) times were kept at an optimized value 7 and 9 ms, respectively. An average zinc grain size of 300 nm was obtained at a pulse current density (J_p) of 400 mA/cm^2 and increasing the pulse current density (J_p) to 800 mA/cm^2 resulted in a progressive decrease in the grain size of the deposits down to 51 nm. Several studies [99–102,116] observed grain-size refinement with increasing pulse current density (J_p), which was attributable to the high overpotential associated with increasing pulse current density (J_p). The high overpotential was reasoned to increase nucleation rate, by increasing the free energy for formation of new nuclei, and therefore refine the grain size. Puippe [116] attained finer grains of gold, copper, palladium, and cadmium electrodeposits at higher pulse current density (J_p), and it was reported that increasing the pulse current density (J_p) resulted in a decrease in the grain size of nickel deposits [101]. Youssef et al. [121] reported that the formation of the smallest nanocrystalline zinc deposits (38 nm) was obtained at the highest pulse current density (J_p) (1200 mA/cm^2), which provides the highest cathodic overpotential (−201 mV) obtained under the studied ranges of pulse plating parameters (T_{on}, T_{off}, and J_p) (see Fig. 9.25.

It is well documented that among the various electrodeposition techniques, of direct current (DC), pulse current (PC), and pulse reverse current (PRC), pulse electrodeposition (or single pulse electrodeposition) has significant advantages in the control of grain size, surface morphology, and preferred orientation of coating than direct current electrodeposition [121]. However, compared with the pulse electrodeposition, the pulse reverse electrodeposition (or double-pulse electrodeposition) can significantly improve the thickness distribution, eliminate hydrogen embrittlement, and keep the surface of coating in an activated state, so as to prepare smoother, denser, lower porosity, and better adhesion coatings [106,125–127]. In addition, the anodic dissolution effect of pulse reverse current (PRC) is beneficial in reducing the actual thickness of the diffusion layer and improving the cathodic current efficiency and speed up the deposition rate of coating.

Waseker et al. [128] compared the coating morphology of the various direct current (DC), pulse current (PC), and pulse reverse current (PRC), electrodeposited coatings. A schematic of the pulse waveforms is seen in Fig. 9.26. They characterized the coating surface morphology of these electrodeposited Zn coatings as well-developed colonies and measured the size of the colonies, Table 9.8. Measurement of grain sizes was made from cross-section samples that can be seen in Fig. 9.27. Ruan and Schuh [129] previously described the microstructure of electroplated coatings as having "grain

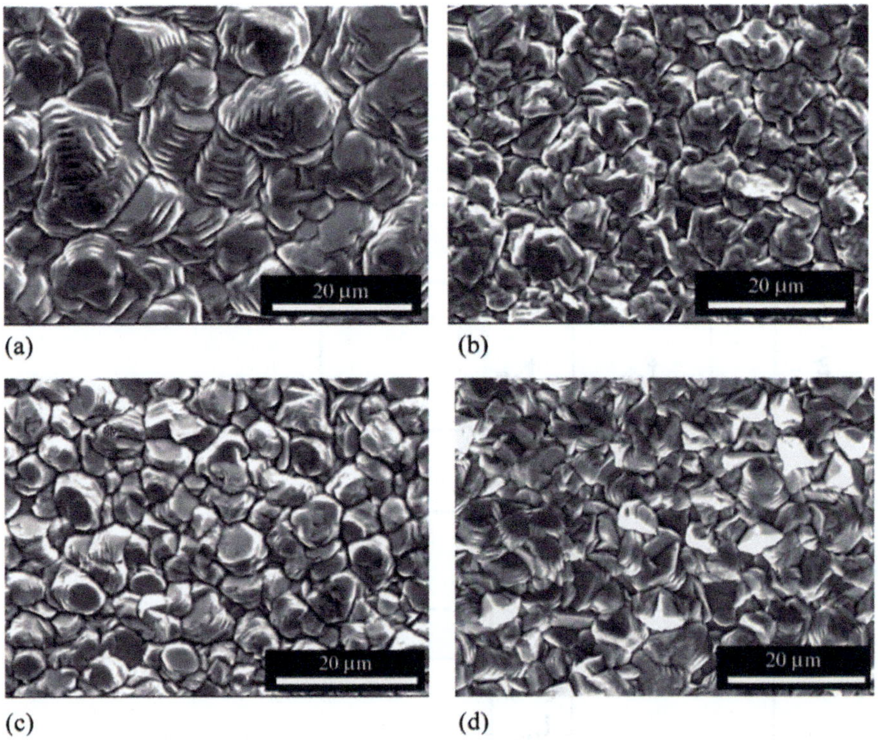

FIG. 9.24 Effect of peak current density on the surface morphology of zinc deposited from chloride-based electrolyte without additives: (a) $0.4\,A\,cm^{-2}$, (b) $0.8\,A\,cm^{-2}$, (c) $1.6\,A\,cm^{-2}$, and (d) $2\,A\,cm^{-2}$ [124].

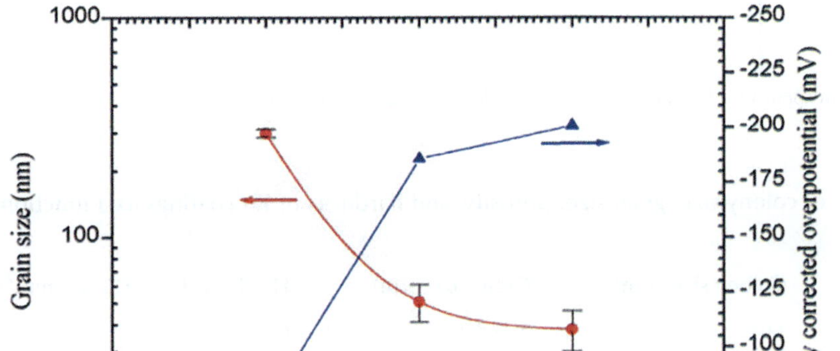

FIG. 9.25 Variation of grain size and ohmically corrected overpotential of zinc deposits with pulse current density at $T_{on}=7\,ms$ and $T_{off}=9\,ms$ [121].

colonies" that are orders of magnitude larger than the nano-scale grain size. They further inferred that intercolony boundaries have a different character or composition from the colony interior regions, and furthermore, that the colony boundary structure may depend upon the current waveform used in electrodeposition. They proposed that although properties such as yield strength and hardness of electrodeposited nanocrystalline alloys may be primarily dictated by grain size, it is clear that many other properties, such as corrosion resistance, toughness or ductility, coefficient of friction, and wear resistance, cannot be explained by grain size alone [130]. For these properties, mesoscale structures or composition fluctuations may have a dominant effect. In agreement with the previous work, Wasekar et al. [128] found that in these electroplated Zn coatings, generally each colony contains many grains. Although colony size was nearly the same for all three processes, the PRC coating process had a significantly finer grain size and that the PRC Zn coating had the best properties in terms of the highest hardness, lowest porosity, and fine grain size.

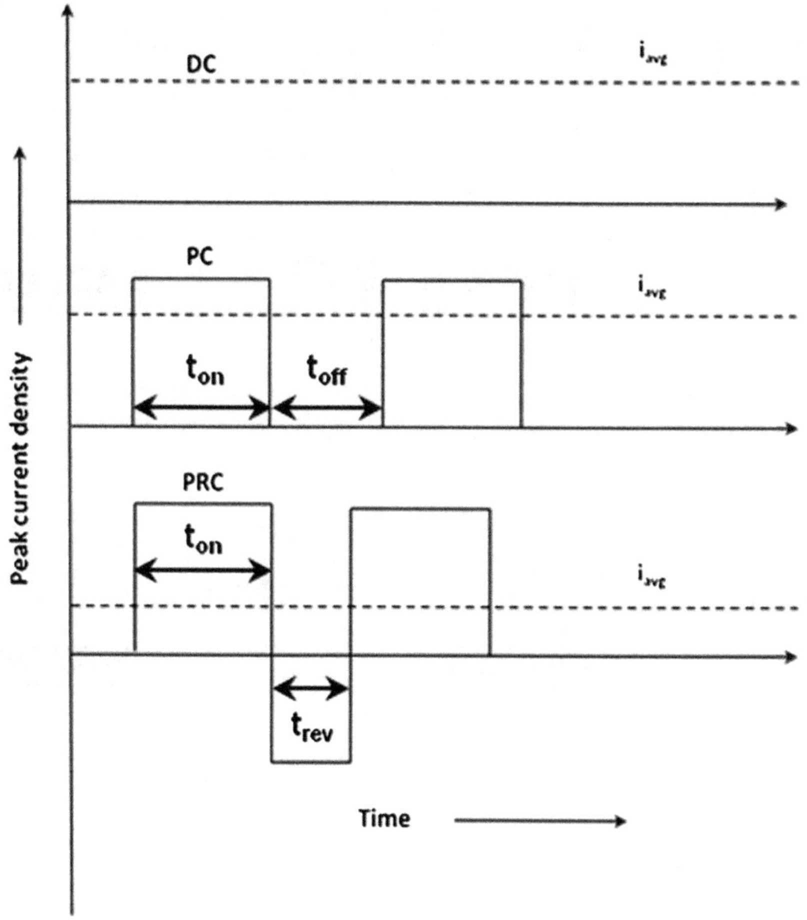

FIG. 9.26 Pulse waveform applied for DC, PC, and PRC deposition (i_{avg}—average current density) [128].

TABLE 9.8 Variation of colony size, grain size, porosity, and hardness of Zn coatings as a function of deposition mode [128].

Deposition mode	Colony size (μm)	Grain size (μm)	Hardness HV0.025 (kg/mm^2)	Porosity (%)
DC	11±2	5.37±1.0	63±3	12.00
PC	15±3	4.36±1.1	63±3	4.86
PRC	11±2	0.7±0.2	71±3	0.50

9.4.3 Additives

A major objective in Zn electrodeposition is to avoid dendrite growth, which occurs mainly at high current densities and depends on the concentration of soluble zinc ions. The morphology of Zn deposits has been characterized by Wang et al. [131], see Table 9.9. Several methodologies have been developed to avoid dendrite growth. Suppression of the dendritic growth is possible under the conditions supporting transport of zinc ions such as increased concentration in the electrolyte and low viscosity of the electrolyte or forced electrolyte convection [131]. Another way to avoid dendrite formation is a controlled zinc deposition by special current profiles as discussed above. Still another approach for inhibition of dendrite formation is utilizing special organic additives (surfactants) that should prevent a nonuniform zinc deposition and thus keep up the smooth and bright surfaces without dendrites [126,132]. The surfactants have the ability to form a barrier for the reactants (zincate) on the electrode surface, thereby balancing their uneven deposition on the surface of the electrode, called a blocking effect [133]. This at higher current densities leads to suppression of the preferred growth of dendrites. Not only

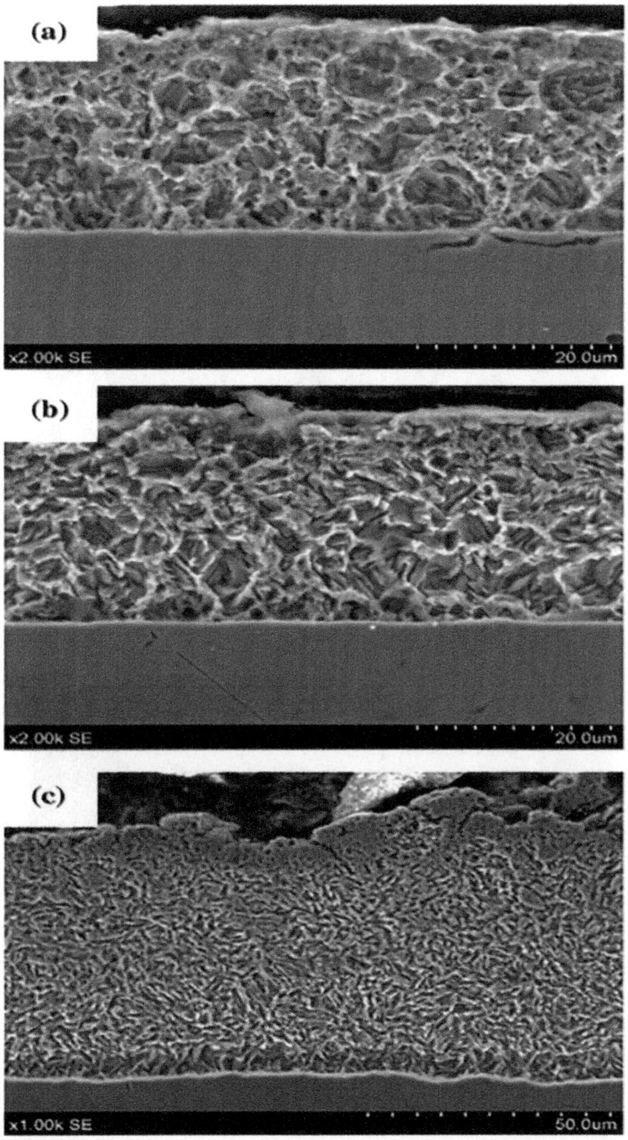

FIG. 9.27 SEM micrographs illustrating coating cross section after etching for (a) DC, (b) PC, and (c) PRC Zn-coated mild steel [128].

TABLE 9.9 Categorization of Zn deposits [131].

	Heavy spongy	Dendritic	Boulder	Layerlike	Mossy
Appearance (unaided eye)	Black powder	Metallic crystal	Gray metal	Shiny metal	Black powder
Microstructure (under microscope)	Agglomerate large boulder	Fern, leaf, hexagon	Granular boulder	Ridge, layerlike	Filament, whisker
Adherence	Nonadherent	Nonadherent	Adherent	Adherent	Nonadherent
Porosity	Dispersed	Dispersed	Compact nonporous	Compact nonporous	Highly porous
Crystallinity	Anisotropically oriented	Isotropically oriented	Anisotropically oriented	Epitaxial oriented	Isotropically oriented
Growth cd	Highest	Very high	Moderate	Low	Lowest
Nucleation site selectivity	Nonselective	Nonselective	Selective Selective	Nonselective	Highly selective
Category	Type II	Type I	Type II	Type I	Type I

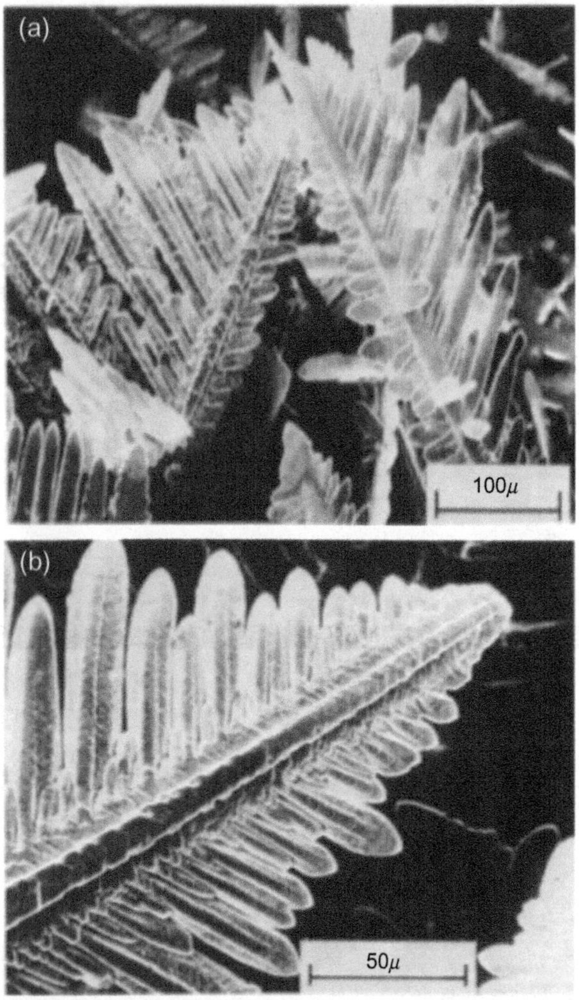

FIG. 9.28 Typical dendritic growth of electrodeposited Zn [134].

surfactants, but also other metal ions, e.g., Pb and Sn, and organic and inorganic compounds, are important additives that can affect the shape and size and thus change the morphology of the deposit and avoid the presence of dendrites [134,135].

The effect of Pb ions and Sn added to the electrolyte was studied by Mansfield and Gilman in 1970 [134,135]. For example, they were able to show that classical dendrites, Fig. 9.28, were converted by additions of Pb ions to compact cylinders with rounded tops consisting of many small crystallites, Fig. 9.29. It was postulated that lead blocks most of the sites active for dissolution and deposition of Zn. Growth is nucleated preferentially at sites of macroscopic defects in the crystal, where impurity segregation is also expected. Similarly, Sn showed the same results [134].

Additives in aqueous electrodepositing solutions are also known to control the surface morphology [136] and refine the grain size of zinc deposits [119,137,138]. Organic additives and pulse electrodeposition are utilized most frequently to produce homogeneous and ultrafine deposits. Organic additives, added commonly as leveler and brightener to electroplating baths, can increase the overpotential and nucleation rate and inhibit the surface diffusion of adions. These conditions also can be fulfilled by using pulse electroplating, because a higher peak current density is possible in comparison with direct current plating, i.e., much higher than the limiting current density attained during direct current plating in the same electrolyte [117,139,140]. Thus, organic additives and pulse control are powerful means to achieve grain refinement down to the nanosized dimension of metallic deposits. In acetate-based bath without additives, several nanocrystalline zinc deposits were obtained with the smallest grain size about 68 nm [141]. In chloride-based bath with a mixture of thiourea and polyacrylamide, a zinc deposit with an average grain size of 50 nm was produced by pulse current control [99,124]. In a sulfate-based bath in the presence of surfactants cetyltrimethylammonium bromide CTAB, sodium dodecyl sulfate SDS, and Triton X-100 and N_2 bubbling, zinc coatings with different crystal shape and size in the range 40–20 nm were prepared on stainless steel substrates by pulse current electrolysis [142]. Li et al. [143] also found that organic compounds PEG,

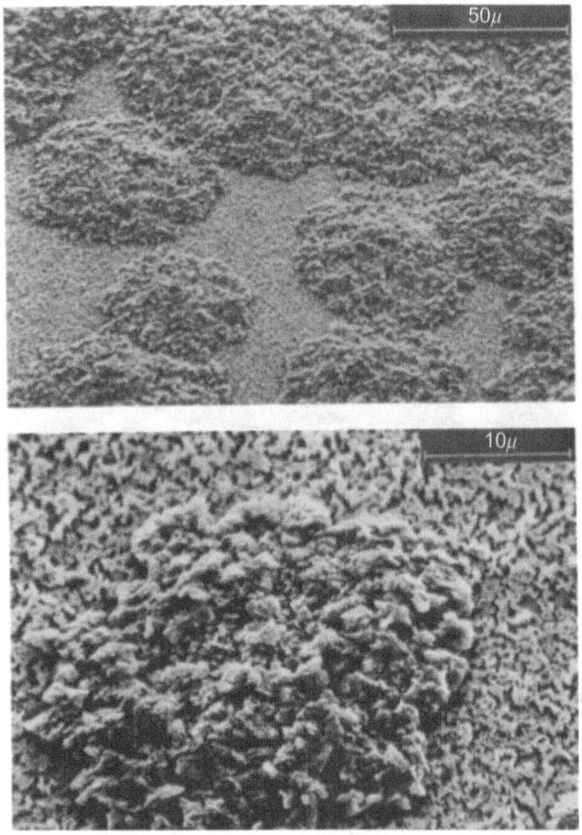

FIG. 9.29 The effect of Pb additions on the deposition characteristics of electrodeposited Zn [135].

CTAB, BA, and TU, used separately or in combination as additives, have a great effect on the morphology and structure of zinc electrodeposits obtained in concentrated acidic sulfate baths by a high pulse current control.

It was reported that mixtures of polyacrylamide and thiourea (TU) were used as additives in zinc chloride electrolytes for electrodepositing nanocrystalline zinc coatings [100,124]. The presence of additives in chloride-based baths, such as o-vanillin, polyethyleneglycol, benzylideneacetone, etc., also results in smoother surface and smaller grain size of the electrodeposited zinc layers [144–146]. The surfactants including cetyl trimethyl ammonium bromide, sodium dodecyl sulfate, and octylphenolpoly (ethylenegly-colether) were presented in zinc sulfate solutions for electrodepositing nanocrystalline zinc [142]. Thiourea, gelatin, glycine, benzalacetone (BA), etc., are traditional additives in sulfate baths, usually as grain refiners or brighteners for zinc electrodeposition [147–153]. Li et al. [102] showed that thiourea reduced the grain size to a value of 0.3–1 μm, Fig. 9.30. However, they found that nanocrystalline zinc deposits can be obtained from an acid sulfate bath containing mixtures of thiourea and benzalacetone by using pulse electrodeposition, with average grain sizes in the range of 60–77 nm and a $(10\bar{1}1)$ random orientation.

Youssef et al. [100] reported that nanocrystalline zinc deposits with average grain sizes ranging from 50 to 95 nm were produced from a zinc-chloride-based electrolyte with additives (PAA and thiourea). Their systematic studies showed:

Effect of polyacrylamide (PAA). Fig. 9.31 shows the effect of PAA concentration on the ohmically corrected deposition overpotential. The asymptotic grain size with increasing PAA concentration is likely due to the relative leveling off the overpotential and/or saturation of adsorption sites at the cathode surface. The ability of PAA to refine the grain size may be due to a combined effect of: (i) increasing the overpotential with increasing PAA concentration and (ii) retardation of growth and increase in the effective nucleation rate.

Effect of thiourea. The effect of thiourea on the surface morphology and grain size of zinc deposits was examined at the optimized PAA concentration (0.7 g/L).

Thiourea is known to have a brightening and leveling effect on deposits [139] The brightness of zinc deposits produced increases with thiourea content and reached a specular reflection "mirror-like surface" at a thiourea content of 0.05 g/L. In addition to this brightness effect, increasing thiourea content in the bath reduces the grain size, Fig. 9.32. The asymptote in

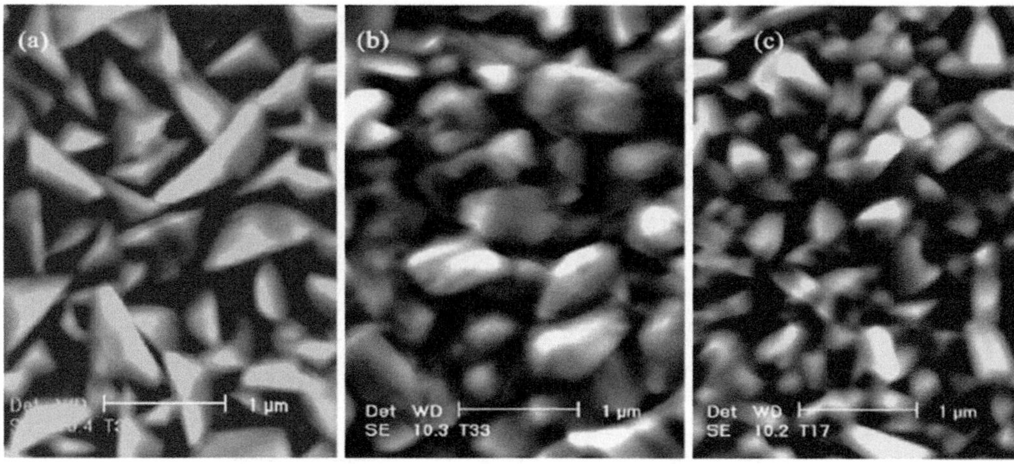

FIG. 9.30 SEM morphologies of zinc electrodeposited at $i_p = 2\,A/cm^2$ from sulfate-bath containing different TU concentrations: (a) at 0.1 g/L, (b) at 1 g/L, and (c) at $i_p = 4\,A/cm^2$ from the same solution as in (B) [102].

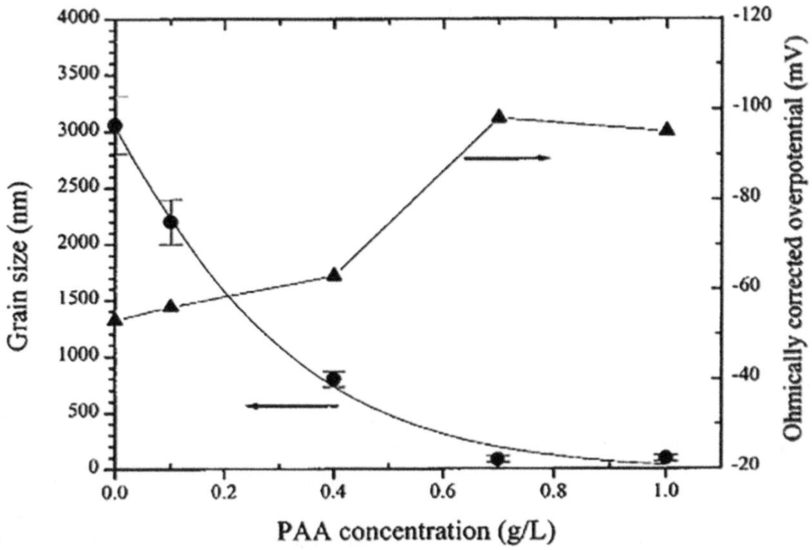

FIG. 9.31 Variations of grain size and ohmically corrected overpotential of zinc deposits with increasing PAA concentration in the bath at thiourea contents of 0.05 g/L, $T_{on} = 5\,ms$, $T_{off} = 9\,ms$, and $J_p = 800\,mA/cm^2$ [100].

the grain size of zinc deposits at the thiourea content of 0.05 g/L and higher may be due to the plateau behavior of the cathodic overpotential at these concentrations.

Li et al. [97] obtained the nanocrystalline zinc coatings from sulfate baths containing mixtures of thiourea and benzalacetone by pulse electrodeposition, with average grain sizes in the range of 60–77 nm. The electrochemistry behaviors of electrodeposition of nanocrystalline zinc coatings in a sulfate bath with cetyl trimethyl ammonium bromide (CTAB), sodium dodecyl sulfate (SDS), and octylphenolpoly(ethyleneglycolether) n, $n = 10$, (Triton X-100) as surfactants and obtained the needle shape zinc coating with average grain sizes in the range of 20–40 nm. It is also proved that the additive plays a crucial role in the process of grain refinement.

Li et al. [97] investigated pulse reverse current for nanocrystalline zinc electrodeposition from a sulfate bath containing polyacrylamide (PAM) as the only additive, which resulted in a more homogeneous, finer, and denser distribution of grains, in comparison to single pulse electrodeposition samples. Their investigation found that:

(1) in additive-free electrolytes, no nanocrystalline zinc coating was obtained just through changing the pulse current density.
(2) PAM can effectively reduce the grain size from micro to nano-scale mainly due to the increases in the overpotential of the cathode by absorption on the electrode surface.
(3) the grain size of nanocrystalline zinc coatings decreases gradually with increased pulse current density.

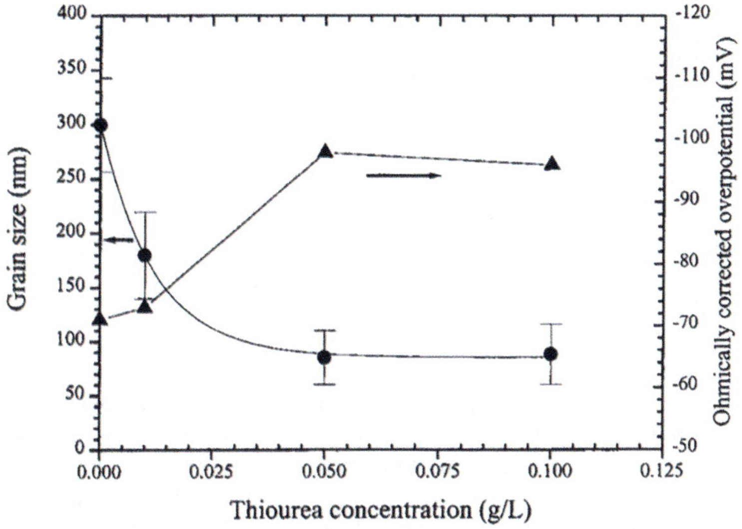

FIG. 9.32 Variations of grain size and ohmically corrected overpotential of zinc deposits with increasing thiourea concentration in the bath at PAA content of 0.7 g/L, $T_{on} = 5\,ms$, $T_{off} = 9\,ms$, and $J_p = 800\,mA/cm^2$ [100].

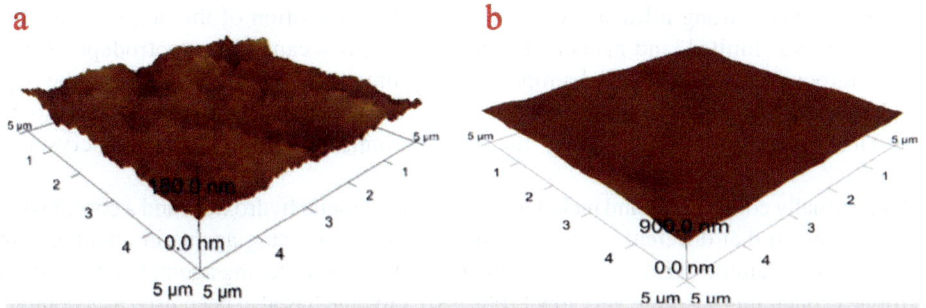

FIG. 9.33 The effect of pulse reverse current density, J_r, on surface morphology. AFM images of nanocrystalline zinc coatings obtained at (a) $J_r = 0\,A/dm^2$, $J_f = 3\,A/dm^2$, and (b) $J_r = 0.3\,A/dm^2$, $J_f = 3\,A/dm^2$, from sulfate bath with a PAM additive (1 g/L) [97].

(4) the pulse reverse current has the effect of altering the preferred orientation and leveling off coatings.
(5) the surface morphology of the coatings electrodeposited at different pulse reverse current densities measured by AFM shows that although the pulse reverse current does not have an obvious effect on the grain size, but achieves the effect of leveling off the coating surface while keeping the pulse forward current density, J_f, constant. Fig. 9.33 shows the effect of pulse reverse current density, J_r, on the surface topography.

Martin [138] used proprietary organic additives in a zinc sulfate electrolyte and was able to alter the electro crystallization processes, resulting in a smoother surface with a finer grain size. Martin also demonstrated the improved formability of zinc deposits in a standard draw bead test. Loto and Olefjord [154] used thiourea, dextrin, a mixture of the two, and glycine as additives with acid chloride solution. Although, they report no reduction in the grain size, nonporous zinc electrodeposits were obtained.

9.4.4 Zn alloy coatings

In order to achieve the desired levels of corrosion protection, electrodeposited zinc coating must be thick enough (typically up to 25 μm) to endure the corrosive environment. The major disadvantages of using thick coatings are the poor formability and weldability and a thick coating makes it difficult to achieve a specular finish after painting. This has led to the development of thinner electrodeposited coatings with improved properties compared with zinc. Alloys of zinc with the metals of the 8th-group (iron, nickel, cobalt, etc.) have been considered as prime candidates for this purpose. In addition, pulse plating techniques have enabled the development of nanocrystalline coatings as outlined above.

9.4.4.1 Zn-Ni

Nanocrystalline pulse electrodeposited Zn-Ni alloys have received much attention in the literature. Hansen and Jessen [155] were the first group of researchers who obtained nanocrystalline zinc-nickel coatings from baths containing ammonium chloride. Alfantezi and Erb [156] have obtained nanocrystalline zinc-nickel coatings from a chloride-based electrolyte in the presence of inorganic additives by pulse electrodeposition with an average grain size of less than 20 nm. Brooks and Erb [157] have studied the hardness of nanocrystalline zinc-nickel coatings. These coatings were obtained from acid or weak acid baths. Li et al. [158] produced nanocrystalline zinc-nickel coatings from alkaline baths containing a grain refiner using the heliotropin, *N*-[1-(2-pyrrolidonyl) ethyl] methyacrylamide and a laboratory-made additive in an alkaline bath. Their results produced a grain size of nanocrystalline Zn-Ni alloys that ranged from 14 to 33 nm. Muresan et al. [159] synthesized nanocrystalline Zn-Ni alloys (9.6 nm) from an alkaline bath containing the mixtures of polyethyleneglicol, coumarin, piperonal, and vanillin as the additives. Rao et al. [160] deposited nanocrystalline Zn-Ni alloys (51 nm) using gelatin and glycerol as the additives and found a synergistic effect of gelatin and glycerol on the electrocrystalline process. There were also other studies on the electrodeposition of nanocrystalline Zn-Ni alloys with additives [161,162].

Based on the above research, Feng et al. [163] concluded that additives play a very important role in the nanocrystallization of Zn-Ni alloys. They observed that almost all of the above nano-electrodeposition was obtained from baths with two or more additives. Feng et al. [163] noted that it was important to consider the concentrations of zinc ions, nickel ions, and complexing agents, as well as the current density, pH, and temperature of the electrolyte that can also affect the electrocrystallization of Zn-Ni alloy and refine the grain of Zn-Ni alloys.

Alfantazi et al. [164] systematically studied the effect of $NiCl_2{\cdot}6H_2O$ concentration in the bath, bath temperature, peak current density, the on-time (T_{on}), and off time (T_{off}) on the Ni content of the coating, Fig. 9.34. The bath temperature appears to have a very strong influence on the chemical composition of the deposit. However, reducing the grain size of Zn-Ni alloys was limited, and nanocrystalline Zn-Ni alloys cannot be electrodeposited in the absence of the additives. Also, heterocyclic compounds and amines are the conventional additives that can be used in Zn-Ni alloy electrodeposition [158,165]. 2,2-Bipyridine not only has the heterocyclic ring structure but also has the nitrogen-containing group. Thus, it is considered among the most promising additives in nanocrystalline Zn-Ni alloy electrodeposition.

Basic alkaline baths usually contain zinc and nickel ions, an alkaline metal hydroxide, and a complexing agent for nickel [166,167]. Studies have shown that different amines can be used as complexing agents in alkaline baths [22,167–169]. Among these amines, triethanolamine (TEA) is conventionally used as a complexing agent, but deposition rate and current efficiency of electrolytes containing TEA are very low [168–170]. Glycine has also been used as a complexing agent in the electrodeposition of Zn-Ni [169,171]. These studies have shown that an alkaline bath containing glycine offers high-quality deposits, high deposition rate, and current efficiency, but glycine is expensive. In addition to complexing properties, glycine acts as a buffer and maintains the bath pH. Organic additives are often present in zinc and zinc alloy electroplating baths to improve the properties of the coating and refine the grain size of deposits [102,172,173]. The mixtures of polyacrylamide and thiourea were used as additives in zinc chloride electrolytes for electrodepositing nanocrystalline zinc coatings [100,124]. Saccharin is a common additive, usually as a grain refiner or a brightener for nanocrystalline Ni electrodeposition [174,175]. The effect of saccharin on the crystallite size is seen in Fig. 9.35.

According to early Zn-Ni phase diagram studies, e.g. [176], electrodeposited Zn-Ni alloys have four main phases: α-phase, a solid solution of zinc in nickel with an equilibrium solubility of about 30% Zn; γ-phase, an intermediate phase with a composition Ni_5Zn_{21}; δ-phase, an intermediate phase with a composition Ni_3Zn_{22}; and η-phase, a solid solution of nickel in zinc with less than 1% nickel. For most nanocrystalline Zn-Ni alloy coatings, Mosavat et al. [165] showed that X-ray analysis of all coatings studied had only a single cubic γ-phase structure (Ni_5Zn_{21}). According to the phase diagram, Zn-Ni alloys with 9–17 wt% Ni have a γ-phase structure. Fig. 9.36 shows a typical X-ray diffractogram for the Zn-Ni coating. In the X-ray diffraction patterns, the peaks related to the steel substrate are marked Fe.

The microhardness of electrodeposited Zn-Ni alloy coatings was measured by Mosavat et al. [165] and plotted versus grain size in Fig. 9.37. Only the coatings with single γ-phase were used for this study to eliminate the interference of the phase structure. Similar results were found by Feng et al. [177]. The well-known Hall-Petch relationship was used to show that the microhardness of the Zn-Ni coating is influenced by the crystalline size.

$$H_v = H_o + kd^{-1/2}$$

where H_V is the measured hardness, H_o and k are constants, and d is the average grain size. The Hall-Petch equation is one of the main reasons why the materials with ultrafine crystallite grain sizes are considered for enhanced strength and hardness. In this way, nanocrystalline materials will have the best properties. As in this study, the results of Feng et al. [177] showed that there are some points far from the fitting line, indicating that there may other factors of concern. One such variable for these studies is the Ni content of the coatings. Fig. 9.38 shows that Ni content strongly affects the

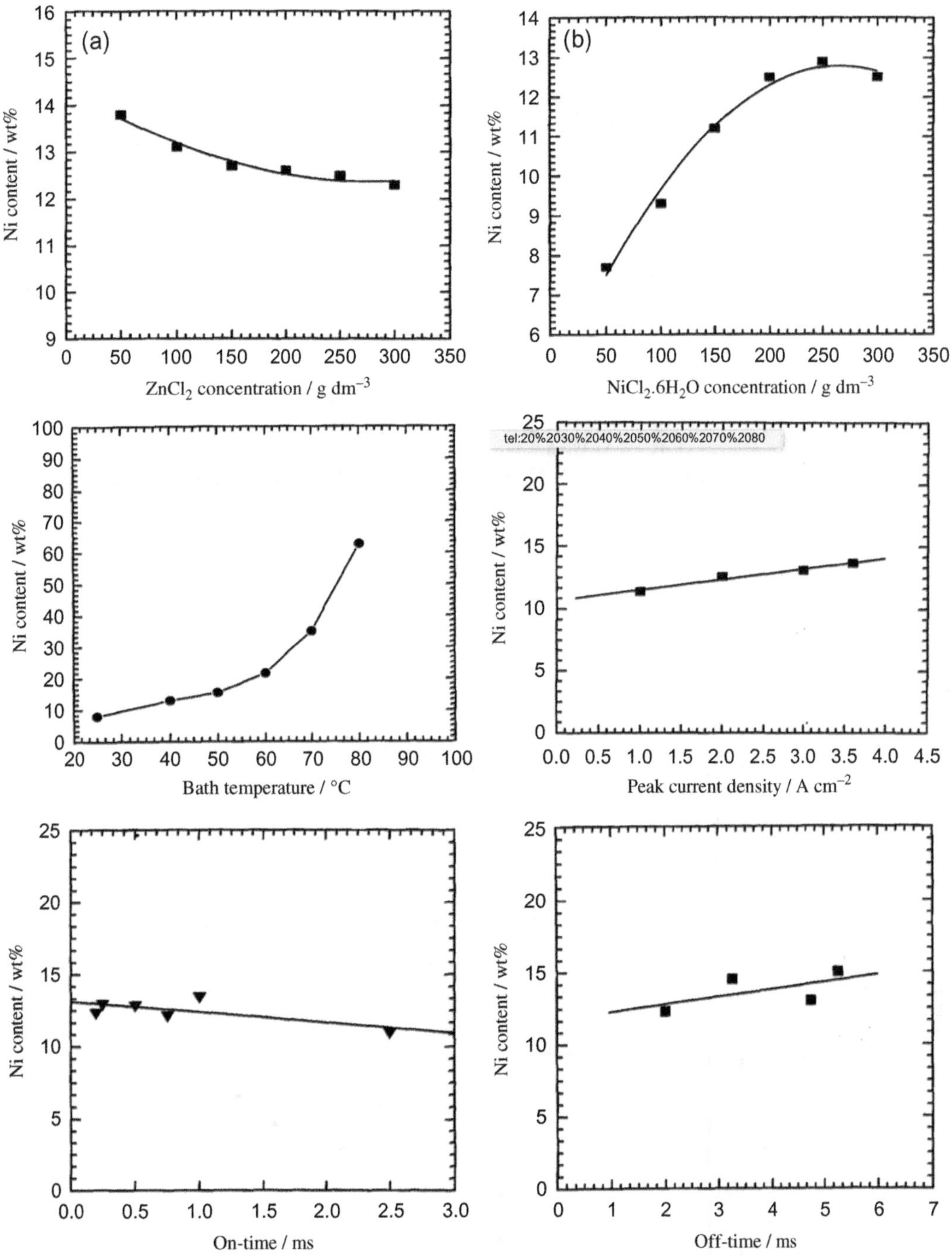

FIG. 9.34 The change in Ni content in the coating as a function of $NiCl_2{\cdot}6H_2O$, temperature, peak current density, T_{on}, and T_{off} [164].

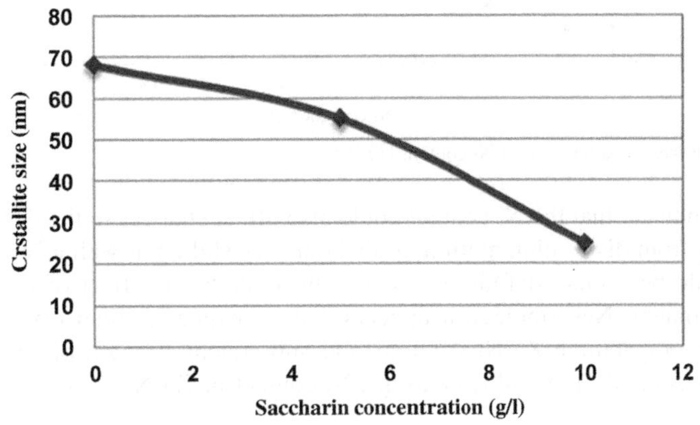

FIG. 9.35 Crystallite size of the electrodeposited nanocrystalline Zn-Ni alloy coatings as a function of saccharin concentration ($i = 2\,A\,dm^{-2}$) [165].

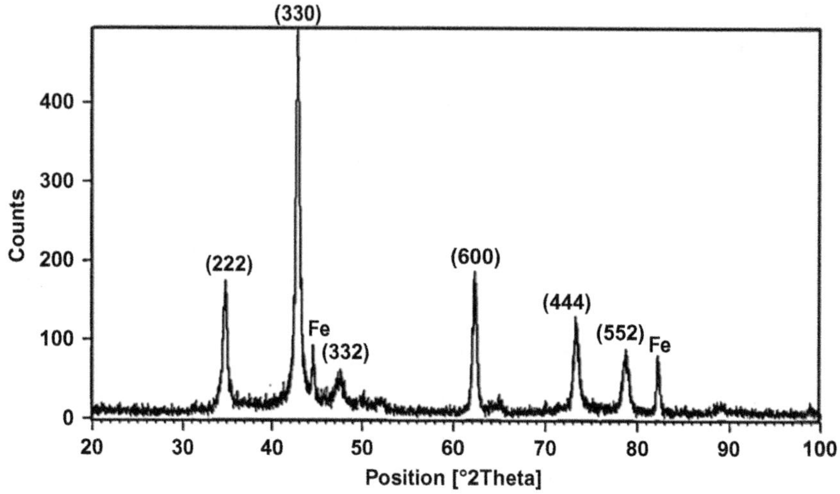

FIG. 9.36 X-ray diffractogram for the Zn-Ni coating obtained at $0.5\,A\,dm^{-2}$ and $5\,g\,L^{-1}$ saccharin concentration [165].

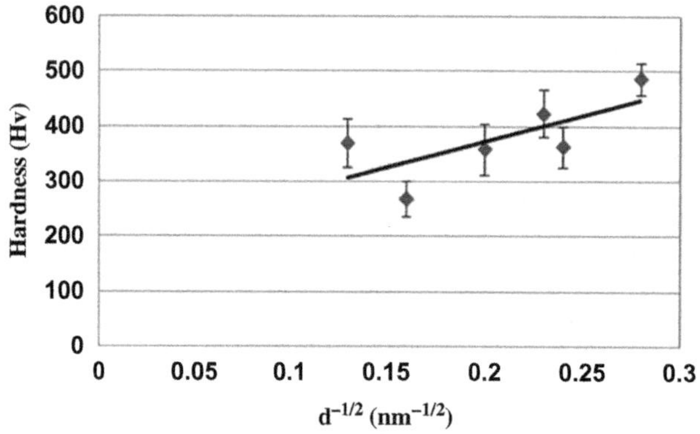

FIG. 9.37 Hardness (H_V) of nanocrystalline Zn-Ni alloy coatings as a function of inverse square root of crystallite size ($d^{-1/2}$) [165].

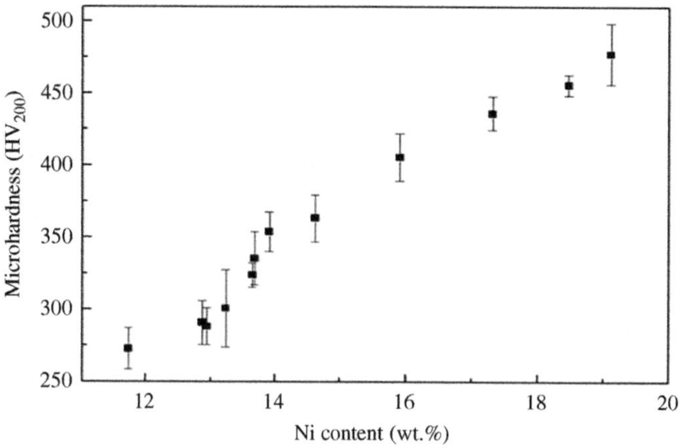

FIG. 9.38 The change in microhardness as a function of Ni content [177].

microhardness. This could indicate that the Ni content could also affect changes in the deformation mechanism undergoing microhardness testing from dislocation motion, grain boundary sliding as well as grain texture. Because of these confounding effects, it would be a more definitive experiment to study the effect of nanocrystalline size on microhardness at a constant Ni content. Nevertheless, it appears that, as expected, finer crystallite size contributes to the increase in the hardness nanocrystalline Zn-Ni coatings and subsequent strength of Zn-Ni coatings. In addition, it has been found that for pulse plated Zn-Ni alloy coatings, Ni content of Zn-Ni alloy coatings plays an important role in corrosion resistance behavior. The deposits with Ni content in the range of about 13–16 wt% show better corrosion resistance, which agrees with the study of Mosavat et al. [165].

9.4.4.2 Zn-Fe

Many other binary and ternary electrodeposited coatings have been evaluated, e.g., Zn-Fe, Zn-Co, Zn-Sn, Zn-Co-Fe. Among zinc alloy coatings, zinc-iron (Zn-Fe) alloys have attracted considerable attention due to their excellent corrosion resistance, good paintability, formability, and weldability. Zn-Fe alloys exist in various phases and their structures affect their corrosion resistance significantly [178–182]. It was shown that pulse reverse coatings exhibit excellent resistance to corrosion in comparison with normal pulse and direct current conditions [183]. All pulse plated Zn-Fe coatings, in spite of the differences in their composition, consisted of a single γ Zn-Fe phase. The authors showed that increasing current density substantially refined the deposited crystals. By decreasing pulse frequency, the deposit structure became more uniform and compact, and that pulse reverse coatings showed higher resistance to corrosion in comparison with normal pulse and DC conditions [183]. According to the equilibrium phase diagram [184], the solubility limit of iron in the HCP η-Zn phase is lower than 0.03 wt%. Different phases can be obtained according to the Fe content: ζ-$FeZn_{13}$ (5–6 wt% Fe), δ-$FeZn_{10}$ (7–11.5 wt% Fe), Γ_1-Fe_5Zn_{21} (17–19.5 wt% Fe), Γ-Fe_3Zn_{10} (23.5–28 wt% Fe), and α-Fe solid solution. Based on the literature, a link can be established between the structure of electrodeposits obtained in acidic electrolytes and their iron content [15,177,179,184–191]. For low iron contents, η is the main crystallographic phase, with small amounts of δ. The amount of η decreases with the increase of the iron content, which is related to the increase of δ and/or Γ_1/Γ. Early on it was claimed that only a small amount of iron or cobalt was required to improve the corrosion resistance of zinc electrodeposits and that higher iron contents could lead to a performance decrease due to the presence of brittle zinc-iron phases, and it was mentioned that a significant improvement of corrosion properties was attained with low iron contents, compared with zinc [192,193]. The impact of higher iron contents (up to about 30 wt%) on the corrosion resistance of Zn-Fe electrodeposits has been investigated. However, contradictory results were obtained, probably due to differences in metallurgical states of the Zn-Fe coatings deposited from acidic or alkaline baths with or without additives. Lan et al. [194] and Park et al. [88] noticed an ennoblement of the corrosion potential with the increase of the iron content while Nayana et al. [187] observed a reverse tendency. On the other hand, Bhat et al. [195] highlighted an increase followed by a decrease of the corrosion potential with the increase of the iron content (from 2.1 to 6.8 wt% Fe).

Arrighi et al. [196] found that the modification of plating variables influenced the morphology of the deposits, by reducing the number of pores and forming finer grains, thus reducing the roughness of the coating. Fig. 9.39 compares the evolution of the iron content with the applied current density. As iron content increased, the crystallographic phases evolved from η Zn, δ for $ZnFe_7$, and Γ_1 for the highest iron content alloy (14.5%) and increased iron led to an increase of the coating microhardness. Pulse electroplating of Zn alloys in comparison with direct current deposition resulted in the grain size reduction from 61 to 36 nm and from 72 to 50 nm for Zn-Co (<1%) and Zn-Fe (<!%) coatings, respectively [197]. However, the corresponding reduction in the corrosion potential for these Zn-1%Fe and 1%Co added alloys was not observed [197]. The high plastic deformation of the low iron content alloy and pure zinc limited the coating abrasion, suggesting a very good adhesion of the coatings. The more brittle behavior of the 14.5 wt% Zn-Fe alloy reduced its adhesion strength onto the steel substrate [196]. Polarization curves in 3.5 wt% NaCl highlighted the sacrificial behavior of all

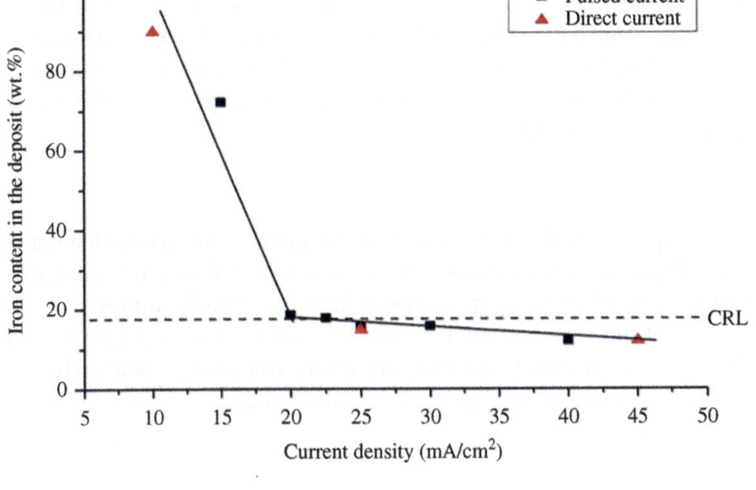

FIG. 9.39 Evolution of the iron content in DC and PC deposits in function of the applied current density (j_m in the case of PC). The dash line represents the compositional reference line (CRL), which is the iron content in the electrolyte [196].

coatings. Adding 7 wt% of iron to zinc was not enough to increase its corrosion potential. This observation was associated to the similarities of microstructure between Zn-7%Fe and Zn. For the Zn-14.5%Fe, the corrosion potential was significantly increased, and this deposit seemed promising regarding corrosion resistance.

9.4.4.3 Zn-Co

While several studies have shown that nanocrystalline structures in Zn-Co electrodeposited coatings can be obtained without pulse plating, see e.g., Lodhi et al. [94], Zn-Co alloy coatings deposited by pulsed current (PC) typically exhibit higher adhesion, lower porosity, and smaller grain size than those deposited by direct current (DC). All of these features contribute to the improved anticorrosive performance observed in PC Zn-Co coatings. Additionally, an increase in cobalt content in Zn-Co coatings tends to decrease the material's grain size, which may influence the anticorrosive performance of the coating [198]. As previously noted by Ramanauskas et al. [197], pulse electroplating of Zn alloys in comparison with direct current deposition resulted in the grain size reduction from 61 to 36 nm for Zn-Co. In a study by Garcia et al. [198], the best anticorrosive performances were observed for coatings containing 7.4–13 wt% Co, presenting nanocrystalline grain size and one predominant crystalline phase (γ-$Zn_{21}Co_5$). These coatings showed a high stability in an aggressive environment for 24 h of exposure, thus indicating that they could be useful as long-term corrosion protection coatings.

9.4.4.4 Zn-Sn

Marino in 1915 [199] first introduced the co-electrodeposition of zinc and tin, but the application of zinc-tin alloy coating (containing >50 wt% of tin) was only established commercially in the 1940s [200]. Zinc-tin alloy coatings offer sacrificial properties of zinc and barrier protection of tin to the steel substrate. In addition to corrosion protection, these coatings provide anti-frictional properties, as well as ductility and solderability [200,201]. Zinc-tin alloys with high content of tin (tin content ≥70 wt%) have been produced from alkaline bath such as stannate/zincate [202] and acidic baths such as sulfate-tartrate and sulfate-gluconate [203,204]. Electrodeposition of Zn-Sn alloy coatings with low content of Sn (<30 wt%) has been reported from acidic electrolytes such as citrate and gluconate baths [205,206]. Esfahani et al. [207] studied pulse electrodeposition of Zn-Sn alloys utilizing CTAB [$CH_3(CH_2)_{15}N\ (CH_3)_3Br$] and polyethylene glycol (PEG) [$(C_2H_4O)_nH_2O$] with a mean molecular weight of 400 g $mole^{-1}$ and 2050 g $mole^{-1}$ at various concentrations (1, 2, 4 g/L) were used as additives. In the presence of CTAB and equal concentrations of PEG, the surface morphology of deposits exhibited a fine and uniform structure. Nanocrystalline Zn-Sn alloy coating was produced from the bath containing average nanocrystalline grain sizes from 423 nm (UFG) down to 78 nm (NC) depending upon the electrolyte used. All zinc-tin alloy deposits exhibited a low tin content (< 35 wt%).

Additional research on the mechanical properties of these coatings was conducted [208]. Fig. 9.40 shows the surface morphology, cross section, and EDS spectra of the 78 nm (NC) and the 423 nm (UFG) Zn-Sn coatings. Nanocrystalline materials have grains smaller than 100 nm in which a great number of atoms belong to the grain boundaries or interfacial boundaries. Surface morphology of the NC coating, Fig. 9.40a, exhibits a uniform structure of nano grains packed together tightly with some protruding areas. The average grain size of the NC coatings was 78 ± 18 nm. Surface morphology of the UFG coatings, Fig. 9.40d, composed of closely packed grains with an average grain size of the UFG coatings of 423 ± 96 nm. Fig. 9.40b, e show the difference in the fine structure of NC and UFG. The average zinc and tin contents of the alloys are presented in the inset of Figs. 9.40c, f.

The data show that the two coatings have very close chemical compositions. EDS spectra of the coatings exhibit major peaks of tin, zinc, and minor peaks of oxygen and iron. Iron peaks come from the steel substrate. The results of their investigation showed that the NC coatings exhibited superior mechanical properties such as higher hardness, lower coefficient of friction, and smaller wear volume than the UFG coatings. Both NC and UFG have either rich tin or rich zinc areas on their structure, which leads to nonhomogeneous distribution of hardness on the surface. Modulus of elasticity was not affected by grain refinement and remained almost constant after grain refinement.

9.4.4.5 Multilayer coatings

Nanocrystalline materials have considerably different properties from conventional coarse-grained polycrystalline materials. One class of such nanocrystalline materials are compositionally modulated alloys (CMAs). In CMA materials, successive layers of alloys with alternating composition are created by bringing modulation in the deposition process itself (usually by a complex pulse current sequence). Because of the very fine grain sizes, CMA alloys exhibit properties not attainable in any of the metallurgically prepared alloys. These include better morphological/structural properties, higher electrical resistivity, increased strength/hardness, enhanced diffusivity, improved ductility/toughness, reduced density,

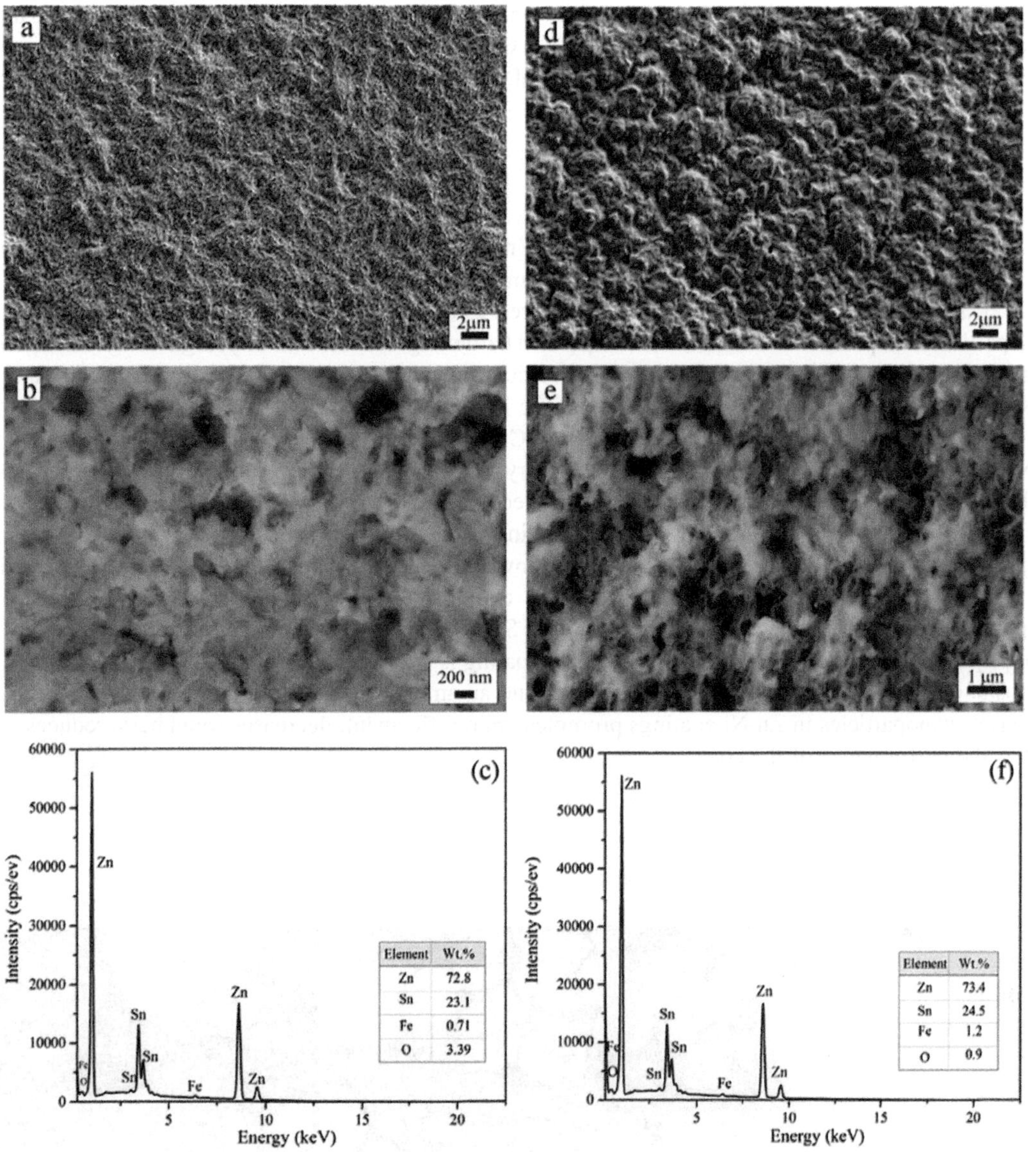

FIG. 9.40 Characteristics of the NC (a, b, and c) and the UFG (d, e, and f) Zn-Sn coatings. (a and d) SEM images of the surface morphology, (b and e) SEM images of the cross section and (c and f) corresponding EDS spectra [208].

reduced elastic modulus, increased specific heat, higher thermal expansion coefficient, lower thermal conductivity, enhanced corrosion and wear resistance, superior reflectance, and soft magnetic properties, in comparison with those of conventional coarse-grained materials. The literature reports results for Zn/Ni [114,209–217], Zn/Co [218–222], and Zn/Fe [222–227] Compositionally Modulated Alloy (CMA) coatings.

Electrochemical deposition method of CMA can be broadly classified as (A) electrodeposition from dual baths [210,211,214], (B) electrodeposition from single bath using switching currents/potentials [114,209–213,215,216], and (C) electrodeposition from single bath using pulsing flow [228–230]. The better performance of multilayers, for example, in corrosive medium is the result of improved barrier properties imparted to the coating by layering. During multilayer deposition, defects of each layer, due to higher energy, would be the preferential nucleation site for the nucleation of the subsequent layer. Consequently, the pores, cracks, and continuous defects from surface to substrate are filled, and this effect is enhanced as the thickness of the layers is decreased [213,216]. Increasing the number of layers of Zn-Ni CMA coatings resulted in a finer surface morphology with the surface roughness reduced by a factor of approximately 2.

An increase in the number of layers improved the hardness and corrosion resistance up to an optimum value and afterward both of them decreased. Similar results were found for Zn-Co CMA coatings [231]. To maximize the corrosion resistance of Zn-Ni CMA multilayers, it was found that the number of layers were optimized around 200 in a 20 μm thickness coating [232].

9.4.4.6 Composite coatings

As various nanoparticles become increasing more available, nano-composite coatings based on a Zn-Ni matrix have been developed [233]. Due to the unique physical and chemical properties, nanoparticles, such as CeO_2 [234,235], TiO_2 [236,237], Al_2O_3 [238], and SiC [239], have been used as a second strengthening phase for the composites. Use of these nanoparticles can provide self-lubrication, dispersion hardening, thermal stability, high hardness, superior antiwear and anticorrosion properties [238]. It is known that the incorporation of Si_3N_4 into the metal matrix significantly affects the structural, surface, and physical properties of the composite [240].

Pulse electrodeposition compared with direct current (DC) could enhance the quality of the composite coatings [241]. As for Zn-Ni coatings, pulse electrodeposition could refine crystal grains and further enhance the corrosion resistance compared with DC electrodeposition [238,242]. The results of a recent study ([243] show that the Zn-Ni alloy contains 7–8 wt% Ni and exhibits a cauliflower-like morphology containing a nanocrystalline phase, Fig. 9.41. The inclusion of Si_3N_4 nanoparticles in the Zn-Ni matrix changed the cauliflower-like Zn-Ni to plate-like Zn-Ni/Si_3N_4 covered by a nanocrystalline layer, Fig. 9.42. The plated-like Zn-Ni/Si_3N_4 composite coating comprises 4–7 wt% Ni and 1.5–3 wt% Si_3N_4 nanoparticles, covered by a nanocrystalline layer. The Si_3N_4 nanoparticles fill crevices and micron holes, enhancing the effect of physical barrier. The Zn-Ni/Si_3N_4 composite coating is smoother than the Zn-Ni coating. The Zn-Ni, Zn-Ni/Si_3N_4 composite coating consists of metallic zinc, nickel, and an amount of oxidized/hydroxide zinc species. The incorporation of Si_3N_4 nanoparticles in Zn-Ni coatings promotes grain refinement, decreases roughness, reduces porosity, and thus improves the performance of the coating.

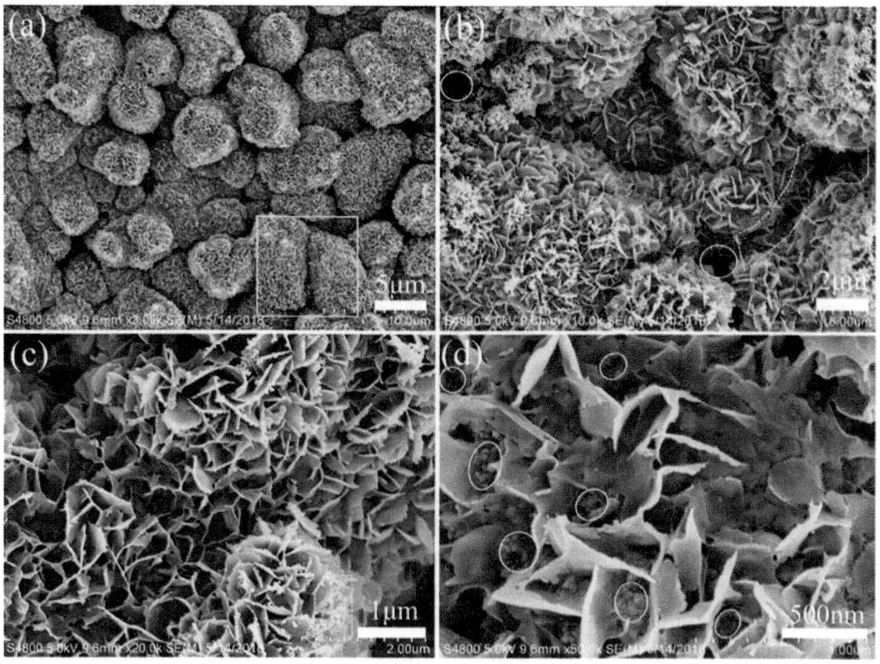

FIG. 9.41 SEM micrographs of the cauliflower-like Zn-Ni alloy at different magnifications [243].

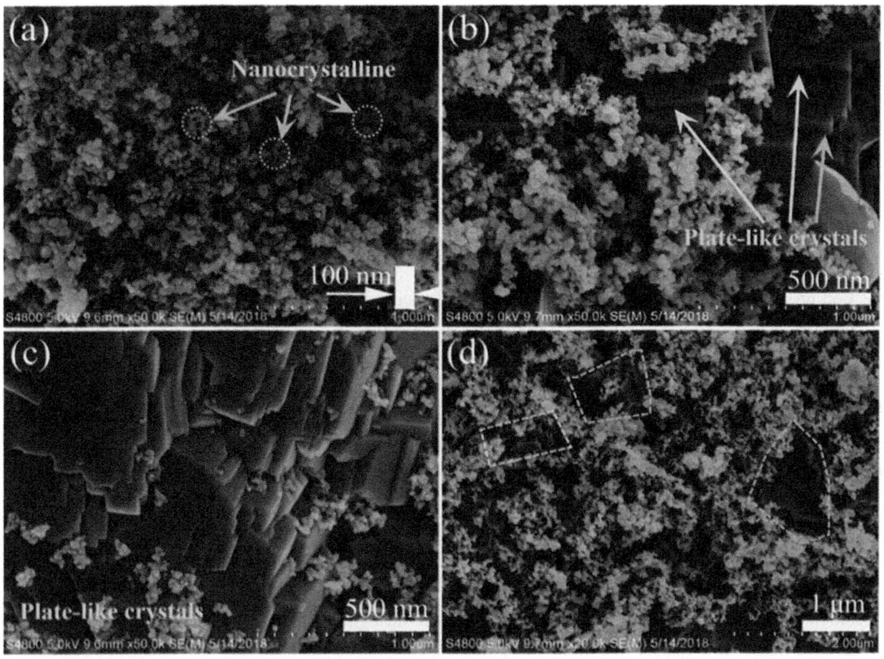

FIG. 9.42 SEM micrographs of plate-like Zn-Ni/Si_3N_4 composite coating (a) nanocrystalline structure with average diameter 25 nm, 50 kx; (b) and (c) plate-like nanostructure, 50 kx; (d) nanocrystalline covered plate-like Zn-Ni/Si_3N_4 composite coating, 20 kx [243].

9.5 Summary of ED processing

Electrodeposited zinc coatings exemplify the classic materials science and engineering paradigm:

Processing =====> Structure =====> Properties

An understanding of all the exceedingly large electrodeposition process variables:

- bath composition
- bath chemistry
- bath temperature
- bath time
- electroplating parameters
- pH

will determine the structure of the coating, for example:

- surface morphology
- cross-section microstructure
- chemistry of the coating
- composition of the coating
- phase stability

In turn, the structure of the coating will determine the properties of the electroplated component, such as:

- corrosion resistance
- adhesion
- wear
- weldability
- machineability
- printability

Or, given a set of properties, the processing of the coating could be determined to produce the required structure.

References

[1] M. Debeaux, M. Witte, T. Koll, Influence of electrolyte composition and temperature on energy demand and zinc structure during electrogalvanizing, Galvatech (2021) 292–301.
[2] R. Winand, Electrodeposition of zinc and zinc alloys, in: M. Schlesinger, M. Paunovic (Eds.), Modern Electroplating, fifth ed., J.W. Wiley, 2010. Ch. 10.
[3] S.G. Fountoulakis, Continuous electrodeposited coatings for steel strip, in: C.M. Cotell, J.A. Sprague, F.A. Smidt Jr. (Eds.), ASM Handbook, vol. 5, Surface Engineering, 1994, pp. 349–359.
[4] J. Lindstedt, Plating Methods & Tooling Design Considerations for Rack Plating, Posted in Metal Finishing Resources, APT Blog, March 22, 2016.
[5] R. Singleton, E. Singleton, Fundamentals of barrel plating, Metal Finish 109 (2011) 14–22.
[6] P.P. Chung, J. Wang, Y. Durandet, Deposition processes and properties of coating on steel fasteners—a review, Friction 7 (5) (2019) 389–416.
[7] A. Sato, Zinc plating, in: C.M. Cotell, J.A. Sprague, F.A. Smith Jr. (Eds.), ASM Handbook, vol. 5—Surface Engineering, ASM International, 1994, pp. 227–235.
[8] C.A. Loto, Electrodeposition of zinc from acid base solutions: A review and experimental study, Asian J. Appl. Sci. 5 (6) (2012) 314–326.
[9] M. Monev, L. Mirkova, I. Krastev, H. Tsvetkova, S. Rashkov, W. Richtering, Effect of brighteners on hydrogen evolution during zinc electroplating from zincate electrolytes, J. Appl. Electrochem. 28 (10) (1998) 1107–1112.
[10] T. Casanova, F. Soto, M. Eyraud, J. Crousier, Hydrogen evolution and permeation into steel during zinc plating on steel, Corros. Sci. 39 (3) (1997) 529–537.
[11] R.A. Cottis, Hydrogen embrittlement, Shrier's Corros. 2 (2010) 902–922.
[12] D. Figueroa, M.J. Robinson, The effect of sacrificial coatings on hydrogen embrittlement and re-embrittlement of ultra-high strength steels, Corros. Sci. 50 (2008) 1066–1079.
[13] L. Mirkova, G. Maurin, I. Kratsev, C. Tsvetkova, Hydrogen evolution and permeation into steel during zinc electroplating; effect of organic additives, J. Appl. Electrochem. 31 (6) (2001) 647–654.
[14] ASTM B850-98, Standard Guide for Post-Coating Treatments of Steel for Reducing the Risk of Hydrogen Embrittlement, ASTM, 1998.
[15] A. Chitharanjan Hegde, K. Venkatakrishna, N. Eliaz, Electrodeposition of Zn-Ni, Zn-Fe, Zn-Ni-Fe alloys, Surf. Coat. Technol. 205 (2010) 2031–2041.
[16] G.D. Wilcox, D.R. Gabe, Electrodeposited zinc alloy coatings, Corros. Sci. 35 (1993) 1251–1258.
[17] A. Brenner, Electrodeposition of Alloys, Vol. 1,2, Academic Press, 1963.
[18] S. Swathirian, Electrodeposition of zinc+nickel alloy phases and electrochemical stripping studies of the anomalous codeposition of zinc, J. Electroanal. Chem. 221 (1987) 211–228.
[19] J.Y. Fei, G.Z. Liang, W.L. Xin, W.K. Wang, Surface modification with zinc and Zn-Ni alloy compositionally modulated multilayered coatings, J. Iron Steel Res. Int. 13 (2006) 61–67.
[20] C. Muller, M. Sarret, M. Benballa, Some peculiarities in the codeposition of zinc–nickel alloys, Electrochem. Acta 46 (2001) 2811–2817.
[21] D.H. Coleman, B.N. Popov, R.E. White, Hydrogen permeation inhibition by thin layer Zn-Ni alloy electrodeposition, J. Appl. Electrochem. 28 (9) (1998) 889–894.
[22] M.G. Hosseini, H. Ashaassi-Sorkhabi, H. Ghiasvand, Electrochemical studies of Zn-Ni ally coatings from non-cyanide alkaline bath containing titrate as complexing agent, Surf. Coat. Technol. 202 (13) (2008) 2897–2904.
[23] M.E. Soares, C.A.C. Souza, S.E. Kuri, Characteristics of a Zn-Ni electrodepositerd alloy obtained from controlled electrolyte flux with gelatin, Mater. Sci. Eng. 402 (1–7) (2005) 16–21.
[24] T.V. Byk, T.V. Gaevsaya, L.S. Tsybuskaya, Effect of electrodeposition conditions on the composition, microstructure, and corrosion resistance of Zn-Ni alloy coatings, Surf. Coat. Technol. 202 (24) (2008) 5817–5823.
[25] A. Petrauskas, L. Grinceviciene, A. Cesuniene, R. Juskenas, Studies of phase composition of Zn-Ni alloy obtained in acetate-chloride electrolyte by using XRD and potentiodynamic stripping, Electrochem. Acta 50 (2005) 1189–1196.
[26] R. Winand, Contimuous strip coating: new zinc alloy tailored coatings, Surf. Coat. Technol. 37 (1989) 65–87.
[27] M.M. Abou-Krisha, A.M. Zaky, A.A. Toghan, Morphology, composition and corrosion properties of electrodeposited Zn-Ni alloys from sulphate electrolytes, Asian J. Biochem. 1 (1) (2006) 84–97.
[28] T. Adaniya, T. Hara, M. Sagiyama, T. Honma, T. Watanabi, Zinc-iron alloy electroplating on strip steel, Plating Surf. Finish. 72 (1985) 52–56.
[29] T. Adaniya, T. Hara, M. Sagiyama, T. Honma, T. Watanabi, Proc. 4th AES Continuous Strip Plating Symp., Chicago, 1984, pp. H1–H15.
[30] T. Adaniya, M. Sagiyama, T. Honma, Phase composition and properties of electrodeposited Zn-Fe alloy, Nippon Kokan Tech. Rep. Overseas 43 (1985) 33–39.
[31] M. Gu, M.R. Notis, A.R. Marder, The effect of continuous heating on the phase transformations in zinc-iron electrodeposited coatings, Met. Trans. A 22A (1991) 1737–1743.
[32] S. John, S.M. Silaimani, V. Anand, T. Vasudivan, Zinc-manganese alloy plating—a critical review, Bull. Electrochem. 18 (2002) 407–412.
[33] T. Hara, M. Sagiyama, T. Urakawa, T. Adanyia, Y. Fukuda, Y. Tsuchiya, Properties of Zn-Mn alloy electroplated steel, Nippon Kokan Tech. Rep. Overseas 48 (1987) 29–37.
[34] N. Boshkov, Galvanic Zn-Mn alloys -electrodeposition, phase composition, corrosion behavior and protective ability, Surf. Coat. Technol. 172 (2003) 217–226.
[35] N. Boshkov, K. Petrov, D. Kovacheva, S. Vikova, S. Nemska, Influence of the alloying component on the protective ability of some zinc galvanic coatings, Electrochim. Acta 51 (2005) 77–84..

[36] P. Diaz-Arista, Z.I. Ortiz, H. Ruiz, R. Ortega, Y. Meas, G. Trejo, Electrodeposition and characterization of Zn-Mn alloy coatings obtained from a chloride-based acidic bath containing ammonium thiocyanate as an additive, Surf. Coat. Technol. 203 (2009) 1167–1175.
[37] B. Borzini, V. Accardi, P.L. Cavallotti, F. Pavan, Electrodeposition and plastic behavior of low-manganese zinc-manganese alloy coatings for automotive applications, Autom. Finish. (1999) 33–42.
[38] M. Marin-Sanchez, P. Ocon, A. Conde, I. Garcia, Electrodeposition of Zn-Mn coatings on steel from 1-ethyl 3-methylimidazolium bis (trifluoromethanesulfonyl) imide ionic liquid, Surf. Coat. Technol. 258 (2014) 871–877.
[39] N. Loukil, M. Feki, Review—Zn-Mn electrodeposition: a literature review, J. Electrochem. Soc. 167 (2020) 022503.
[40] Y. Tsuchiya, S. Hashimoto, Y. Shibashi, T. Urakawa, M. Sagiyama, Y. Fukuda, Structure of electrodeposited Zn-Mn alloy coatings, ISIJ Int. 40 (2000) 1024–1028.
[41] K.A. Prasad, P. Giridhar, V. Ravindran, Zinc-cobalt alloy: electrodeposition and characterization, J. Solid State Electrochem. (2001) 63–68.
[42] P. De Lima-Neto, A.N. Correia, R.P. Colares, W.S. Araujo, Corrosion study of electrodeposited Zn and Zn-Co coatings in chloride medium, J. Braz. Chem. Soc. 18 (2007).
[43] H. Kermoune, A. Levesque, J. Douglad, R. Rehamnia, J.P. Chopart, Magnetic field effect on Zn-Co alloy electroplating, Ionics 23 (2017) 3565–3570.
[44] V. Thangaraj, N.K. Udayashankar, A.C. Hegde, Development of Zn-Co alloy coatings by pulsed current from chloride bath, Indian J. Chem. Technol. 15 (2008) 581–587.
[45] M. Ridosic, E. Garcia-Lecina, A. Salicio-Paz, J. Bajat, The advantage of ultrasound during electrodeposition on morphology and corrosion stability of Zn-Co alloy coatings, trans IMF, Int. J. Surface Eng. Coat. 98 (2020) 114–120.
[46] T. Kanamaru, M. Nakayama, K. Araj, S. Suzuki, R. Naka, Corrosion resistant plated steel strip and method of producing same, U.S. Patent, 4,877,494, 1989.
[47] K. Hasagawa, H. Nakamura, R. Mochizuki, T. Katagiri, N. Morito, S. Kurokawa, Method of producing zinc-chromium alloy plated steel sheet with excellent plating adhesiveness, U.S. Patent 5,273,643, 1993.
[48] H. Nakamura, T. Fugimura, H. Ohnuma, K. Mochizuki, N. Morito, M. Kanamura, Corrosion resistant steel sheets improved in corrosion resistance and other characteristics, U.S. Patent 5,510,196, 1996.
[49] M. Nakazawa, A. Takahashi, K. Matsumura, Process for producing zinc-chromium alloy-electroplated steel plate, U.S. Patent 5,616,232, 1997.
[50] T. Steck, J. Gerdenitsch, A. Tomandl, W. Achtliener, J. Faderl, T. Lavric, T. Boiadjieva-Scherzer, H. Kronberger, Zinc chromium coated steel sheet—properties and production, in: Galvatech, 2011.
[51] G. Luckeneder, M. Fleischanderl, T. Steck, K.H. Stellinberger, J. Faderl, S. Schuerz, G. Mori, Corrosion mechanisms and cosmetic corrosion aspects of zinc-aluminum-magnesium and zinc chromium alloy coated strip, in: Galvatech, 2011.
[52] V. Chakarova, T. Boiadjieva-Scherzer, D. Kovacheva, H. Kronberger, M. Monev, Corrosion behavior of electrodeposited Zn-Cr alloy coatings, Corros. Sci. 140 (2018) 73–78.
[53] E. Osuchowska, Z. Buczko, K. Olkowicz, Zn-Cr alloy coatings: electrodeposition and properties, J. KONBIN 51 (2021) 29–42.
[54] T. Boiadjieva, K. Petrov, H. Kronberger, A. Tomandl, G. Avdeev, W. Artner, T. Lavric, M. Monev, Composition of electrodeposited Zn-Cr alloy coatings and phase transformations induced by thermal treatment, J. Alloy Compd. 480 (2009) 259–264.
[55] T. Boiadjieva, D. Kovacheva, K. Petrov, S. Hardcastle, A. Sklyarov, M. Monev, Electrodeposition, composition and structure of Zn-Cr alloys, J. Appl. Electrochem. 34 (2004) 315–321.
[56] M.M. Younan, T. Oki, Electrodeposition of Zn-Ni-Fe alloy in acidic chloride bath with separate anodes, J. Appl. Electrochem. 26 (1996) 537.
[57] M. Abe, S. Ando, M. Sagiyama, Zinc-nickel chromium alloy electroplated steel sheet excellent in corrosion resistance, plating adhesion, chemical conversion treating property and adhesion of coating film, 1991. Japanese Patent JPH0673591A.
[58] F.H. Assaf, A.M.A. Seidy, M.M. Abou-Krishna, A.A. Eissa, Electrodeposition and characterization of Zn-Ni-Mn alloy from sulfate bath: influence of current density, Int. J. Electrochem. Soc. 10 (2015) 5465–5478.
[59] B. Abedini, N. Parvini, Ahmad, S. Yazdani, L. Magagnin, Electrodeposition and corrosion behavior of Zn-Ni-Mn alloy deposited from alkaline solution, Trans. Nonferrous Metals Soc. China 30 (2020) 548–558.
[60] M.M. Younan, Surface microstructure and corrosion resistance of electrodeposited Zn-Ni-Co alloy, J. Appl. Electrochem. 30 (2000) 55–60.
[61] M.M. Abou-Krisha, Influence of Ni2+ concentration and deposition potential on the characterization of thin electrodeposited Zn-Ni-Co coatings, Mater. Chem. Phys. 125 (2011) 621–627.
[62] R.S. Bhat, B.U. Bhat, A.C. Hegde, Corrosion behavior of electrodeposited Zn-Ni, Zn-Co and Zn-Ni-Co alloys, Anal. Bioanal. Electrochem. 3 (2011) 302–315.
[63] N. Eliaz, K. Venkatakrishna, A.C. Hegde, Electroplating and characterization of Zn-Ni, Zn-Co and Zn-Ni-Co alloys, Surf. Coat. Technol. 205 (2010) 1969–1978.
[64] R.S. Bhat, K.B. Manjunatha, R.P. Shankara, K. Venkatakrishna, A.C. Hegde, Electrochemical studies on the corrosion resistance of Zn-Ni-Co coating from acid chloride bath, Appl. Phys. A Mater. Sci. Process. 126 (772) (2020) 1–9.
[65] M. Kowalski, A. Stachowiak, Tribocorrosion performance of Zn, Zn-Ni, Zn-Ni with magnesium electroplated on medium carbon steel in chloride environment, Surf. Coat. Technol. 366 (2019) 75–85.
[66] B. Szczygiel, A. Laszczynska, W. Tylus, Influence of molybdenum on properties of Zn-Ni and Zn-Co alloy coatings, Surf. Coat. Technol. 204 (2010) 1438–1444.
[67] I.H. Karahan, Electrodeposition and properties of ZnFeNI alloys, Chin. J. Phys. 46 (2008) 105–112.

[68] M.M. Abou-Krisha, Effect of pH and current density on the electrodeposition of Zn-Ni-Fe alloys from sulfate bath, J. Coat. Technol. 9 (2012) 775–783.

[69] J.M. Costa, M.S. Hori, A.F. de Almeida Neto, Effect of forced convection and current density on the electrodeposition of Zn-Fe-Mo alloys, Chem. Phys. 527 (2019) 110502.

[70] J. Winiarski, W. Tylus, M.S. Krawczyk, B. Szczygiel, The influence of molybdenum on the electrodeposition and properties of ternary Zn-Fe-Mo alloy coatings, Electrochim. Acta 196 (2016) 708–726.

[71] J. Winiarski, A. Lesniewicz, P. Pohl, B. Szczygiel, The effect of plating bath on electrodeposition and properties of protective ternary Zn-Fe-Mo alloy coatings, Surf. Coat. Technol. 299 (2016) 81–89.

[72] J. Winiarski, The effect of current density on the structure and mechanical properties of protective Zn-Fe-Mo alloy coatings electrodeposited on a mild steel, Mater. Chem. Phys. 239 (2020), 122258.

[73] R. Solmaz, B. Deniz Karahan, Corrosion behavior of Ni-Fe-Mo deposits obtained under different electrodeposited conditions, J. Mater. Eng. Perform. 30 (8) (2021) 5593–5602.

[74] M. Abou-Krisha, F. Assaf, O. Alduaij, A. Alshammari, F. El-Sheref, Corrosion behavior of electrodeposited Zn-Co-Fe alloy, Indian J. Chem. Technol. 23 (2016) 271–278.

[75] Z.F. Lodhi, J.M.C. Mol, A. Hovestad, L. 't Hoen-Velterop, H. Terryn, J.H.W. de Wit, Corrosion resistance of Zn-Co-Fe alloy coatings on high strength steel, Surf. Coat. Technol. 203 (2009) 1415–1422.

[76] F. Assaf, M. Abou-Krisha, T.A. Yousef, A. Abushoffa, F. El-Sheref, A. Toghan, Influence of current density on the mechanism of electrodeposition and dissolution of Zn-Fe-Co alloys, Coll. Chem. Electrochem. Russ. J. Phys. Chem. 94 (2020) 1708–1715.

[77] O.K. Duaij, M.M. Abou Krisha, M.I. Attia, Influence of the deposition temperature on the electrodeposition mechanism of Zn-Co-Fe alloy, Int. J. Electrochem. 12 (2017) 11972–11986.

[78] B. Fedi, M.P. Gigander, J.-Y. Hihn, S. Mierzejewski, Structure determination of electrodeposited zinc-nickel alloys: thermal stability and quantification using XRD and potentiodynamic dissolution, Electrochim. Acta 215 (2016) 652–666.

[79] M. Gu, The Transformation from Non-equilibrium to Equilibrium of Electrodeposited Zinc-iron Alloy Coatings (PhD dissertation), Lehigh University, December 1989.

[80] C. Bories, J.P. Bonino, A. Rousset, Structure and thermal stability of zinc-nickel electrodeposits, J. Appl. Electrochem. 29 (1999) 1045–1051.

[81] T. Boiadjieva-Scherzer, G. Avdeev, T. Vassilev, V. Chakarova, H. Kronberger, M. Monev, Influence of annealing temperature on z-$CrZn_{13}$ formation in electrodeposited Zn-Cr coatings, Surf. Eng. 35 (2019) 1055–1060.

[82] V. Chakarova, T. Boiadjieva-Scherzer, D. Kovacheva, H. Kronberger, M. Monev, Corrosion behavior of the phase obtained by annealing an electrodeposited Zn-Cr coating, Electrochem. Commun. 122 (2021) 106904.

[83] J.H. Lindsley, R.F. Paluch, H.D. Nine, V.R. Miller, T.J. Okeefe, The interactive between electrogalvanized zinc deposit structure and the forming properties of sheet steel, Plat. Surf. Finish. 76 (1989) 62–69.

[84] V. Jiriciny, H. Choi, J.W. Evans, Zinc electrodeposit morphology under conditions of fluctuating current density, J. Appl. Elecrtrochem. 17 (1987) 91–103.

[85] R. Winand, Electrocrystallization—theory and applications, Hydrometallurgy 29 (1992) 567–598.

[86] B. Bozzini, D. Lacitignola, C. Mele, I. Sgura, Morphogenesis in metal electrodeposition, Note de Matematica 32 (2012) 7–46.

[87] R. Winand, Electrocrystallization: fundamental considerations and application to high current density continuous steel sheet plating, J. Appl. Electrochem. 21 (1991) 377–385.

[88] H. Park, J.A. Szpunar, The role of texture and morphology in optimizing the corrosion resistance of zinc-based electrogalvanized coatings, Corr. Sci. 40 (1998) 525–545.

[89] J.F. Silva Filho, V.F.C. Lins, Crystallographic texture and morphology of an electrodeposited zinc layer, Surf. Coat. Technol. 200 (2006) 2892–2899.

[90] K. Raeissi, A. Saatchi, M.A. Golozar, Effect of nucleation mode on the morphology and texture of electrodeposited zinc, J. Appl. Electrochem. 33 (2003) 635–642.

[91] M. Eyraud, A. Garnier, F. Mazeron, J. Crousier, Morphology and composition of electrodeposited zinc-manganese alloys, Plat. Surf. Finish. (1995) 63–70.

[92] S. Ghaziof, W. Gao, Electrodeposition of single gamma phased Zn-Ni alloy coatings from additive-free acidic bath, Appl. Surf. Sci. 311 (2014) 635–642.

[93] M.J. Rahman, S.R. Sen, M. Moniruzzaman, K.M. Shorowordi, Morphology and properties of electrodeposited Zn-Ni alloy coatings on mild steel, J. Mech. Eng. ME40 (2009) 9–14.

[94] Z.F. Lodhi, F.D. Tichelaar, C. Kwakernaak, J.M.C. Mol, H. Terryn, J.H.W. de Wit, A combined composition and morphology study of electrodeposited Zn-Co and Zn-Co-Fe alloy coatings, Surf. Coat. Technol. 202 (2008) 2755–2764.

[95] K. Raeissi, A. Saatchi, M.A. Golozar, J.A. Szpunar, Texture and surface morphology in zinc electrodeposits, J. Appl. Electrochem. 34 (2004) 1249–1258.

[96] S.H. Mosavat, M.H. Shariat, M.E. Bahrololoom, Study of corrosion of electrodeposited nanocrystalline Zn-Ni alloy coatings, Corros. Sci. 59 (2012) 81–87.

[97] Q. Li, Z. Feng, J. Zhang, P. Yang, F. Li, M. An, Pulse reverse electrodeposition and characterization of nanocrystalline zinc coatings, RSC Adv. 4 (94) (2014) 52562–52570.

[98] K.M. Youssef, C.C. Koch, P.S. Fedkiw, Improved corrosion behavior of nanocrystalline zinc produced by pulse-current electrodeposition, Corros. Sci. 46 (1) (2004) 51–64.

[99] K.M. Youssef, C.C. Koch, P.S. Fedkiw, Mater. Sci. Eng. A 341 (2003) 174.

[100] K.M. Youssef, P.S. Fedkiw, C.C. Koch, J. Electrochem. Soc. 151 (2004) C103.
[101] A.M. El-Sherik, U. Erb, J. Page, Surf. Coat. Technol. 88 (1996) 70.
[102] M.C. Li, L.L. Jiang, W.Q. Zhang, Y.H. Qian, S.Z. Luo, J.N. Shen, J. Solid State Electrochem. 11 (2007) 549.
[103] M. Ghaemi, L. Binder, J. Power Sources 111 (2) (2002) 248.
[104] A. Marlot, P. Kern, D. Landolt, Electrochim. Acta 48 (1) (2002) 29.
[105] K.M. Yin, S.L. Jan, C.C. Lee, Surf. Coat. Technol. 88 (1997) 219.
[106] M.S. Chandrasekar, M. Pushpavanam, Pulse and pulse reverse plating—conceptual, advantages and applications, Electrochim. Acta 53 (8) (2008) 3313–3322.
[107] S. Kainuma, S. Ishikura, K. Hisatake, J. Magn. Soc. Jpn. 21 (1995) 889.
[108] T. Houga, A. Yamada, Y. Ueda, J. Jpn. Inst. Metals 64 (2000) 739.
[109] Y. Ueda, N. Hataya, H. Zaman, J. Magn. Magn. Mater. 156 (1996) 350.
[110] M. Alper, K. Attenborough, R. Hart, S.J. Lance, D.S. Lashmore, C. Younes, W. Schwarzacher, Appl. Phys. Lett. 63 (1993) 2144.
[111] N.V. Mandich, Met. Finish. 101 (2002) 8–10.
[112] J.J. Kelly, P.E. Braddey, D. Landolt, J. Electrochem. 147 (2000) 2975.
[113] M. Mouanga, L. Ricq, J. Douglade, P. Bercot, Corros. Sci. 51 (2009) 690–698.
[114] P. Ganesan, S.P. Kumaraguru, B.N. Popov, Surf. Coat. Technol. 201 (2007) 7896–7904.
[115] K. Boto, Electrodepos. Surf. Treat. 3 (1975) 77–95.
[116] J.C. Puippe, Influence of pulse plating on crystallization, in: J.C. Puippe, F. Leaman (Eds.), Theory and Practice of Pulse Plating, AESF, Orlando, Florida, 1986, pp. 17–39.
[117] G. Devaraj, S. Guruviah, S.K. Seshadri, Mater. Chem. Phys. 25 (1990) 439–461.
[118] J.C. Puippe, N. Ibl, Plat. Surf. Finish. 67 (1980) 68–72.
[119] W. Paatsch, First AES International Pulse Plating Symposium, Abstract, Boston, 1979.
[120] R.I. Gurovich, A.K. Krivtsov, J. Appl. Chem. USSR 41 (1968) 1227.
[121] K.M. Youssef, C.C. Koch, P.S. Fedkiw, Influence of pulse plating parameters on the synthesis and preferred orientation of nanocrystalline zinc from zinc sulfate electrolytes, Electrochim. Acta 54 (2) (2008) 677–683.
[122] D.I. Rehrig, H. Leidheiser Jr., M.R. Notis, Plat. Surf. Finish. 64 (1977) 40.
[123] S. Yoshimura, S. Chida, E. Sato, Metal Finish. 84 (1986) 39.
[124] K. Saber, C.C. Koch, P.S. Fedkiw, Pulse current electrodeposition of nanocrystalline zinc, Mater. Sci. Eng. A 341 (1–2) (2003) 174–181.
[125] L. Chang, J. Alloys Compd. 466 (2008) L19.
[126] K.M. Yin, Surf. Coat. Technol. 88 (1997) 162.
[127] N.V. Mandich, Met. Finish. 98 (2000) 375.
[128] N.P. Wasekar, A. Jyothirmayi, N. Hebalkar, G. Sundararajan, Influence of pulsed current on the aqueous corrosion resistance of electrodeposited zinc, Surf. Coat. Technol. 25 (272) (2015) 373–379.
[129] S. Ruan, C.A. Schuh, Mesoscale structure and segregation in electrodeposited nanocrystalline alloys, Scr. Mater. 59 (11) (2008) 1218–1221.
[130] U. Erb, Nanostruct. Mater. 6 (1995) 533.
[131] R.Y. Wang, D.W. Kirk, G.X. Zhang, Effects of deposition conditions on the morphology of zinc deposits from alkaline zincate solutions, J. Electrochem. Soc. 153 (5) (2006) C357.
[132] J. Zhu, Y. Zhou, C. Gao, Influence of surfactants on electrochemical behavior of zinc electrodes in alkaline solution, J. Power Sources 72 (2) (1998) 231–235.
[133] C. Lan, C. Lee, T. Chin, Tetra-alkyl ammonium hydroxides as inhibitors of Zn dendrite in Zn-based secondary batteries, Electrochim. Acta 52 (17) (2007) 5407–5416.
[134] F. Mansfeld, S. Gilman, The effect of tin and tetraethylammonium ions on the characteristics of zinc deposition on a zinc single crystal in aqueous KOH, J. Electrochem. Soc. 117 (9) (1970) 1154.
[135] F. Mansfeld, S. Gilman, The effect of lead ions on the dissolution and deposition characteristics of a zinc single crystal in 6N KOH, J. Electrochem. Soc. 117 (5) (1970) 588.
[136] B.K. Thomas, D.J. Fray, J. Appl. Electrochem. 11 (1981) 677–683.
[137] R.J. Lash, Proceedings of the 53rd Annual Convention of the Wire Association International, November 14, vol. 18, 1983, pp. 211–217.
[138] S. Martin, Proceedings of the AES Symposium on Continuous Strip Plating, Chicago, May, 1984, pp. 1–17.
[139] L. Oniciu, L. Muresan, J. Appl. Electrochem. 21 (1991) 565.
[140] R.T.C. Choo, A.M. El-Sherik, J. Toguri, U. Erb, J. Appl. Electrochem. 25 (1995) 384.
[141] B.H. Juarez, C. Alonso, J. Appl. Electrochem. 36 (2006) 499.
[142] Gomes, M.I. da Silva Pereira, Electrochim. Acta 51 (2006) 1342.
[143] M. Li, S. Luo, Y. Qian, W. Zhang, L. Jiang, J. Shen, Effect of additives on electrodeposition of nanocrystalline zinc from acidic sulfate solutions, J. Electrochem. Soc. 154 (11) (2007) D567.
[144] S.J. Kim, H.T. Kim, S.M. Park, J. Electrochem. Soc. 151 (2004) C850.
[145] G. Trejo, H. Ruiz, R.O. Borges, Y. Meas, J. Appl. Electrochem. 31 (2001) 685.
[146] P. Diaz-Arista, Y. Meas, R. Ortega, G. Trejo, J. Appl. Electrochem. 35 (2005) 217.
[147] T.V. Venkatesha, J. Balachandra, S.M. Mayanna, R.P. Dambal, Plat. Surf. Finish. 74 (1987) 77.
[148] S.E. Afifi, A.R. Ebaid, M.M. Hegazy, A.K. Barakat, JOM 44 (1992) 32.
[149] K.D. Song, K.B. Kim, S.H. Han, H.K. Lee, Electrochem. Solid-State Lett. 7 (2004) C20.

[150] D.J. Mackinnon, J.M. Brannen, R.M. Morrison, J. Appl. Electrochem. 18 (1988) 252.

[151] A. Stankeviciute, K. Leinartas, G. Bikulcius, D. Virbalyte, A. Sudavicius, E. Juzeliunas, J. Appl. Electrochem. 28 (1998) 89.

[152] D. Mockute, G. Bernotiene, J. Appl. Electrochem. 27 (1997) 691.

[153] P.F. Mendez, J.R. Lopez, Y. Meas, R. Ortega, L. Salgado, G. Trejo, Electrochim. Acta 50 (2005) 2815.

[154] C.A. Loto, I. Olefjord, Corrosion Prevent. Control (1990) 158–163.

[155] P.L. Hansen, C.Q. Jessen, Scripta Metall. 23 (1989) 1387.

[156] A.M. Alfantazi, U. Erb, J. Mater. Sci. Lett. 15 (1996) 1361.

[157] I. Brooks, U. Erb, Scripta Mater. 44 (2001) 853.

[158] G.Y. Li, J.S. Lian, L.Y. Niu, Z.H. Jiang, Surf. Coat. Technol. 191 (2005) 59.

[159] L.M. Muresan, J. Eymard, D. Blejan, E. Indrea, Zn-Ni alloy coatings from alkaline bath containing triethanolamine. Influence of additives, Stud. Univ. Babes-Bolyai Chem. 1 (2010) 37–44.

[160] V.R. Rao, A.C. Hegde, Synergistic effect of gelatin and glycerol on electrodeposition of Zn-Ni alloy, Metall. Mater. Trans. B Process Metall. Mater. Process. Sci. 44B (2013) 1236–1242.

[161] C.M. PraveenKumar, T.V. Venkatesha, K. Vathsala, K.O. Nayana, Electrodeposition and corrosion behavior of Zn-Ni and Zn-Ni-Fe_2O_3 coatings, J. Coating Technol. Res. 9 (2012) 71–77.

[162] S. Fashu, C.D. Gu, J.L. Zhang, M.L. Huang, X.L. Wang, J.P. Tu, Effect of EDTA and NH_4Cl additives on electrodeposition of Zn-Ni films from choline chloride-based ionic liquid, T. Nonferr. Metal. Soc. 25 (2015) 2054–2064.

[163] Z. Feng, D. Li, Q. Sun, L. Wang, P. Xing, M. An, Insight into the role and mechanism of 2, 2-bipyridine as a novel additive for nano-electrodeposition of Zn-Ni alloy, J. Alloys Compd. 765 (2018) 1026–1034.

[164] A.M. Alfantazi, J. Page, U. Erb, Pulse plating of Zn-Ni alloy coatings, J. Appl. Electrochem. 26 (12) (1996) 1225–1234.

[165] S.H. Mosavat, M.E. Bahrololoom, M.H. Shariat, Electrodeposition of nanocrystalline Zn-Ni alloy from alkaline glycinate bath containing saccharin as additive, Appl. Surf. Sci. 257 (2011) 8311–8316.

[166] M. Schlesinger, M. Paunovic, Modern Electroplating, fourth ed., John Willy and Sons, New York, 2000.

[167] C. Muller, M. Sarret, M. Benballa, J. Electroanal. Chem. 519 (2002) 85.

[168] H.Y. Lee, S.G. Kim, Surf. Coat. Technol. 135 (2000) 69.

[169] L.S. Tsybulskaya, T.V. Gaevskaya, O.G. Purovskaya, T.V. Byk, Surf. Coat. Technol. 203 (2008) 234.

[170] M.R. Kalantary, G.D. Wilcox, D.R. Gabe, Electrochem. Acta 40 (1995) 1609.

[171] I. Rodriguez-Torres, G. Valentin, S. Chanel, F. Lapicque, Electrochim. Acta 46 (2000) 279.

[172] R. Albalta, E. Gomez, C. Muller, M. Sarret, E. Valles, J. Pregonas, J. Appl. Electrochem. 20 (1990) 635.

[173] R. Albalta, E. Gomez, C. Muller, M. Sarret, E. Valles, J. Pregonas, J. Appl. Electrochem. 21 (1990) 44.

[174] E. Moti, M.H. Shariat, M.E. Bahrololoom, Mater. Chem. Phys. 111 (2008) 469.

[175] P. Najafi Sayar, M.E. Bahrololoom, J. Appl. Electrochem. 39 (2009) 2489.

[176] D.E. Hall, Plat. Surf. Finish. 70 (1983) 59.

[177] Z. Feng, Q. Li, J. Zhang, P. Yang, H. Song, M. An, Electrodeposition of nanocrystalline Zn–Ni coatings with single gamma phase from an alkaline bath, Surf. Coat. Technol. 270 (2015) 47–56.

[178] Z.N. Yanga, Z. Zhanga, J.Q. Zhang, Electrodeposition of decorative and protective Zn–Fe coating on to low-carbon steel substrate, Surf. Coat. Technol. 200 (2006) 4810–4815.

[179] E. Gomez, E. Pelaez, E. Valle, Electrodeposition of zinc_iron alloys I. Analysis of the initial stages of the anomalous codeposition, J. Electroanal. Chem. 469 (1999) 139–149.

[180] E. Gomez, X. Alcobe, E. Vallés, Electrodeposition of zinc_iron alloys II. Relation between the stripping results and ex-situ characterization, J. Electroanal. Chem. 475 (1999) 66–72.

[181] J.B. Bajat, V.B. Mišković-Stanković, Z. Kačarević-Popović, The influence of steel surface modification by electrodeposited Zn–Fe alloys on the protective behaviour of an epoxy coating, Prog. Org. Coat. 47 (2003) 49–54.

[182] Z. Zhang, W.H. Leng, H.B. Shao, J.Q. Zhang, J.M. Wang, C.N. Cao, Study on the Behavior of Zn–Fe Alloy Electroplating, J. Electroanal. Chem. 516 (2001) 127–130.

[183] A. Roshanghias, M.H. Sohi, The effects of pulse plating variables on morphology and corrosion behavior of Zn–Fe alloy coatings, J. Coat. Technol. Res. 9 (2) (2012) 215–218.

[184] A.R. Marder, The metallurgy of zinc-coated steel, Prog. Mater. Sci. 45 (2000) 191–271.

[185] K. De Wit, A. De Boeck, B.C. De Cooman, Study of the influence of phase composition and iron content on the formability characteristics of zinc-iron electroplated sheet steel, J. Mater. Eng. Perform. 8 (1999) 531–537.

[186] C.N. Panagopoulos, E.P. Georgiou, P.E. Agathocleous, K.I. Giannakopoulos, Mechanical behaviour of Zn-Fe alloy coated mild steel, Mater. Des. 30 (2009) 4267–4272.

[187] K.O. Nayana, T.V. Venkatesha, K.G. Chandrappa, Influence of additive on nanocrystalline, bright Zn-Fe alloy electrodeposition and its properties, Surf. Coat. Technol. 235 (2013) 461–468.

[188] C.A. Drewien, J.I. Goldstein, A.R. Marder, Structure of as-deposited iron-zinc coatings from chloride bath, Metall, Mater. Trans. 25 (1994) 249–255.

[189] I.H. Karahan, H.A. Çetinkara, H.S. Güder, Electrodeposition and characterization of Zn, Zn-Fe and Zn-Fe-Ni coatings in presence of gelatin as additive, Trans. Inst. Met. Finish. 86 (2008) 157–161.

[190] Z.N. Yang, Z. Zhang, J.Q. Zhang, Electrodeposition of decorative and protective Zn-Fe coating onto low-carbon steel substrate, Surf. Coat. Technol. 200 (2006) 4810–4815.

[191] G.U. Mingyuan, M.R. Notis, A.R. Marder, The effect of continuous heating on the phase transformations in zinc-iron electrodeposited coatings, Metall. Trans. A. 22 (1991) 1737–1743.

[192] R. Ramanauskas, P. Quintana, L. Maldonado, R. Pomés, M.A. Pech-Canul, Corrosion resistance and microstructure of electrodeposited Zn and Zn alloy coatings, Surf. Coat. Technol. 92 (1997) 16–21.

[193] R. Ramanauskas, L. Gudaviciute, R. Juskenas, O. Scit, Effect of pulse plating on the composition and corrosion properties of Zn–Co and Zn–Fe alloy coatings, Trans. Inst. Met. Finish. 19 (2008) 7–13.

[194] C.J. Lan, W.Y. Liu, T.S. Chin, Electrodeposition of Zn–Fe alloys using electrolytes of the system ZnO–(ferrous gluconate)–KOH, J. Electrochem. Soc. 154 (2007) D30.

[195] R.S. Bhat, A.C. Hegde, Electroplating and corrosion study of Zn-Co, Zn-Fe and Zn-Co-Fe alloys, Anal. Bioanal. Electrochem. 4 (2012) 593–609.

[196] C. Arrighi, C. Savall, S. Cohendoz, J.L. Grosseau-Poussard, L. Baissac, M.G. Olivier, J. Creus, Optimization of the morphology, structure and properties of high iron content Zn–Fe coatings by pulse electrodeposition, Mater. Chem. Phys. 15 (263) (2021), 124366.

[197] R. Ramanauskas, L. Gudavičiūtė, R. Juškėnas, Effect of pulse plating on the composition and corrosion properties of Zn–Co and Zn–Fe alloy coatings, Chem. Aust. 19 (1) (2008) 7–13.

[198] J.R. Garcia, do Lago DC, Cesar DV, Senna LF., Pulsed cobalt-rich Zn–Co alloy coatings produced from citrate baths, Surf. Coat. Technol. 306 (2016) 462–472.

[199] P. Marino, "British Patent", 10, 133, 1915.

[200] J.W. Cuthbertson, R.M. Angles, J. Electrochem. Soc. 94 (1948).

[201] E. Budman, M. McCoy, Tin-zinc plating, Met. Finish. 93 (10–11) (1995) 14–15.

[202] O.A. Ashiru, J. Shirokoff, Electrodeposition and characterization of tin-zinc alloy coatings, Appl. Surf. Sci. 103 (1996) 159–169.

[203] E. Guaus, J. Torrent-Burgués, Tin-zinc electrodeposition from sulphate-gluconate baths, J. Electroanal. Chem. 549 (2003) 25–36.

[204] E. Guaus, J. Torrent-Burgués, Tin-zinc electrodeposition from sulphate-tartrate baths, J. Electroanal. Chem. 575 (2005) 301–309.

[205] M. Arici, H. Nazir, M.L. Aksu, Investigation of Sn-Zn electrodeposition from acidic bath on EQCM, J. Alloys Compd. 509 (2011) 1534–1537.

[206] S. Vitkova, V. Ivanova, G. Raichevsky, Electrodeposition of low tin content zinc-tin alloys, Surf. Coat. Technol. 82 (1996) 226–231.

[207] M. Esfahani, J. Zhang, Y.C. Wong, Y. Durandet, J. Wang, Electrodeposition of nanocrystalline zinc-tin alloy from aqueous electrolyte containing gluconate in the presence of polyethylene glycol and hexadecyltrimethylammonium bromide, J. Electroanal. Chem. 813 (2018) 143–151.

[208] M. Esfahani, K.S. Munir, C. Wen, J. Zhang, Y. Durandet, J. Wang, Y.C. Wong, Mechanical properties of electrodeposited nanocrystalline and ultrafine-grained Zn-Sn coatings, Surf. Coat. Technol. 333 (2018) 71–80.

[209] I. Ivanov, I. Kirilov, Corrosion resistance of compositionally modulated multilayered Zn–Ni alloys deposited from a single bath, J. Appl. Electrochem. 33 (2003) 239–244.

[210] T. Ivanov, Corrosion resistance of compositionally modulated Zn–Ni multilayers electrodeposited from dual baths, J. Appl. Electrochem. 32 (2002) 85–89.

[211] M. Rahsepar, M.E. Bahrololoom, Corrosion study of Ni/Zn compositionally modulated multilayer coatings using electrochemical impedance spectroscopy, Corros. Sci. 51 (2009) 2537–2543.

[212] S. Yogesha, A.C. Hegde, Development of composition modulated multilayer alloy coatings and their corrosion behavior, J. Met. Mater. Miner. 21 (2011) 83–92.

[213] R.S. Bhat, A.C. Hegde, Development of nano-structured cyclic multilayer Zn–Ni alloy coatings using triangular current pulses, Surf. Eng. Appl. Electrochem. 47 (2011) 112–119.

[214] J. Fei, G.D. Wilcox, Electrodeposition of zinc–nickel compositionally modulated multilayer coatings and their corrosion behaviours, Surf. Coat. Technol. 200 (2006) 3533–3539.

[215] K. Venkatakrishna, A.C. Hegde, Electrolytic preparation of cyclic multilayer Zn–Ni alloy coating using switching cathode current densities, J. Appl. Electrochem. 40 (2010) 2051–2059.

[216] A. Maciej, G. Nawrat, W. Simka, J. Piotrowski, Formation of compositionally modulated Zn–Ni alloy coatings on steel, Mater. Chem. Phys. 132 (2012) 1095–1102.

[217] F. Jing-yin, L. Guo-zheng, X. Wen-li, W. Wei-kang, Surface modification with zinc and Zn-Ni alloy compositionally modulated multilayer coatings, J. Iron Steel Res. 13 (2006) 61–67.

[218] I. Kirilova, I. Ivanov, S. Rashkov, Anodic behaviour of composition modulated Zn-Co multilayers electrodeposited from single and dual baths, J. Appl. Electrochem. 28 (1998) 1359–1363.

[219] V. Thangaraj, N. Eliaz, A. Chitharanjan Hegde, Corrosion behavior of composition modulated multilayer Zn–Co electrodeposits produced using a single-bath technique, J. Appl. Electrochem. 39 (2009) 339–345.

[220] F. Jingyin, L. Guozheng, X. Wenli, L.I.U. Jianghong, Corrosion performance of zinc and zinc-cobalt alloy compositionally modulated multilayer (CMM) coatings, J. Wuhan Univ. Technol. Mater. Sci. Ed. 21 (2006) 40–44.

[221] M.E. Bahrololoom, D.R. Gabe, G.D. Wilcox, Development of a bath for electrodeposition of zinc-cobalt compositionally modulated alloy multilayered coatings, J. Electrochem. Soc. 150 (2003) C144–C151.

[222] F. Jingyin, L. Guozheng, X. Wenli, Electrodeposition of compositionally modulated alloy zinc-cobalt multilayer coatings, Chin. J. Chem. Eng. 13 (2005) 259–265.

[223] M. Rahsepar, M.E. Bahrololoom, Study of surface roughness and corrosion performance of Ni/Zn–Fe and Zn–Fe/Ni compositionally modulated multilayer coatings, Surf. Coat. Technol. 204 (2009) 580–585.

[224] K. Venkatakrishna, A. Chitharanjan Hegde, Composition modulated multilayer Zn–Fe alloy coatings on mild steel for better corrosion resistance, Mater. Manuf. Process. 26 (2011) 29–36.

[225] M. Rahsepar, M.E. Bahrololoom, Corrosion resistance of Ni/Zn–Fe/Zn and Ni/Zn/Zn–Fe compositionally modulated multilayer coatings, Corrosion Eng. Sci. Technol. 46 (2010) 70–75.

[226] R.S. Bhat, R. Udupa K, A.C. Hegde, Development of layered Zn-Fe coatings for better corrosion protection, J. Met. Mater. Miner. 20 (2011) 43–51.

[227] S. Yogesha, A.C. Hegde, Optimization of deposition conditions for development of high corrosion resistant Zn–Fe multilayer coatings, J. Mater. Process. Technol. 211 (2011) 1409–1415.

[228] D.T. Schwartz, P. Stroeve, B.G. Higgins, Multilayered alloys induced by fluctuating flow, J. Electrochem. Soc. 138 (1989) 53C–56C.

[229] S.D. Leith, D.T. Schwartz, Flow-induced composition modulated Ni-Fe thin films with nanometer-scale wavelengths, J. Electrochem. Soc. 146 (1996).

[230] D.T. Schwartz, P. Stroeve, B.G. Higgins, Electrodeposition of composition-modulated alloys in fluctuating flow field, J. Am. Inst. Chem. Eng. 35 (1989) 1315–1327.

[231] S. Yogesha, A.C. Hegde, Electrodeposition of high performance multilayer coatings of Zn–co using triangular current pulses, Trans. IMF 88 (6) (2010) 317–323.

[232] B. Bahadormanesh, M. Ghorbani, N.L. Kordkolaei, Electrodeposition of nanocrystalline Zn/Ni multilayer coatings from single bath: influences of deposition current densities and number of layers on characteristics of deposits, Appl. Surf. Sci. 404 (2017) 101–109.

[233] H. Zheng, M. An, J. Lu, Surface characterization of the Zn–Ni–Al_2O_3 nanocomposite coating fabricated under ultrasound condition, Appl. Surf. Sci. 254 (2008) 1644–1650.

[234] S. Ranganatha, T.V. Venkatesha, K. Vathsala, M.K.P. Kumar, Electrochemical studies on Zn/nano-CeO_2 electrodeposited composite coatings, Surf. Coat. Technol. 208 (2012) 64–72.

[235] T. Xiang, M. Zhang, C. Li, C. Dong, L. Yang, W. Chan, CeO_2 modified SiO_2 acted as additive in electrodeposition of Zn–Ni alloy coating with enhanced corrosion resistance, J. Alloys Compd. 736 (2018) 62–70.

[236] A. Shamsolhodaei, H. Rahmani, S. Rastegari, Effects of electrodeposition parameters on morphology and properties of Zn–TiO_2 composite coating, Surf. Eng. 29 (2013) 695–699.

[237] O.S.I. Fayomi, A.P.I. Popoola, L.R. Kanyane, T. Monyai, Development of reinforced in-situ anti-corrosion and wear Zn–TiO_2/$ZnTiB_2$ coatings on mild steel, Results Phys. 7 (2017) 644–650.

[238] S. Ghaziof, W. Gao, The effect of pulse electroplating on Zn–Ni alloy and Zn–Ni–Al_2O_3 composite coatings, J. Alloys Compd. 622 (2015) 918–924.

[239] S. Pouladi, M.H. Shariat, M.E. Bahrololoom, Electrodeposition and characterization of Ni–Zn–P and Ni–Zn–P/nano-SiC coatings, Surf. Coat. Technol. 213 (2012) 33–40.

[240] S. Ranganatha, T.V. Venkatesha, K. Vathsala, Development of electroless Ni–Zn–P/nano-TiO_2 composite coatings and their properties, Appl. Surf. Sci. 256 (2010) 7377–7383.

[241] N.P. Wasekar, S.M. Latha, M. Ramakrishna, D.S. Rao, G. Sundararajan, Pulsed electrodeposition and mechanical properties of Ni–W/SiC nano-composite coatings, Mater. Des. 112 (2016) 140–150.

[242] Y. Boonyongmaneerat, S. Saenapitak, K. Saengkiettiyut, Reverse pulse electro-deposition of Zn–Ni alloys from a chloride bath, J. Alloys Compd. 487 (2009) 479–482.

[243] B. Li, D. Li, W. Xia, W. Zhang, Synthesis and characterization of a novel Zn-Ni and Zn-Ni/Si3N4 composite coating by pulse electrodeposition, Appl. Surf. Sci. 458 (2018) 665–677.

Chapter 10

Physical vapor deposited coatings

10.1 Introduction

Physical vapor deposition (PVD) has been used for decades to coat a wide variety of metallic and nonmetallic substrates with a thin film of inorganic or organic material. Common applications include CdTe deposited with other layers onto glass in the manufacture of solar cells [1], the coating of polyethylene terephthalate (PET) with a very thin layer of Al for use in food packaging and balloons [2] and hard coatings for cutting tools [3]. Such layers are much thinner than normal galvanized coatings, usually in the range of nanometers. During the PVD process the material to be coated is transformed from a condensed (liquid or solid) phase to vapor, in which state it is deposited on the substrate, after which it condenses to the solid state. To both increase free path lengths of the vaporized atoms and prevent undesired reactions such as oxidation of the vapor, the process is carried out under a moderately high vacuum, for example the mean free path of atoms in a Zn vapor is on the order of several decimeters at 0.001 mbar [4]. Ionized inert gas can be useful in enhance deposition rates, increasing coating density adherence. Ref. [5] is a useful general reference on PVD processing [5].

10.2 PVD properties of zinc

Because of its relatively low boiling point in both ambient and relevant partial vacuum conditions, zinc is an attractive metal to coat using PVD. The vapor pressure of zinc in solid and liquid form is calculated from Eqs. (10.1) and (10.2) [6]:

Solid (from 298 to 692°K (melting point at 1 atm)

$$\log_{10}P = 6.102 - \frac{6776}{T} \tag{10.1}$$

Liquid (from 692 to 750°K)

$$\log_{10}P = 5.378 - \frac{6286}{T} \tag{10.2}$$

where T is the temperature in °K and P is the vapor pressure in atm.

For consistency, pressures in this chapter will be given in mbar (1 atm = 1013.25 mbar) As an example, the vapor pressure of solid zinc at its melting point of 692°K is 0.207 mbar. The vapor pressure of liquid zinc at 750°K is 1.01 mbar. Pressures in the literature are also often reported in Pa (1 Pa = 0.01 mbar).

Other elements alloyed with Zn for PVD coating include Mg (melting point 923°K, boiling point at 1 atm 1364°K). Al (melting point 933°K, boiling point at 1 atm 2743°K) cannot be thermally evaporated from a Zn alloy melt for practical application because at the boiling point of Zn the vapor pressure of Al is very low.

PVD offers new possibilities for alloy coatings that are not possible with other steel coating processes. In the hot-dip process, zinc alloy possibilities are limited by solubilities and melting point compatibilities; in electrogalvanizing, elements with a reduction potential lower than water, such as Al and Mg, cannot be electroplated using a water-based electrolyte [7].

A schematic of the temperature-entropy diagram for pure Zn is shown in Fig. 10.1 [4]. The triple point, where vapor, solid and liquid exist simultaneously occurs at a temperature of 692.5°K and a pressure of 0.65 mbar. Starting at point A, the solid zinc is melted, either outside of the vacuum deposition facility or inside. The transformation from liquid to vapor, between points D and E, always occurs under a partial vacuum. The zinc must be kept in a vapor state for successful deposition; if adiabatic cooling occurs during vapor transport, re-condensation can occur before the vapor reaches the substrate, for example the path between points E and G. Temperature and pressure conditions indicated by the path E′–G′ are more favorable for efficient deposition of zinc on the substrate.

The Metallurgy of Zinc Coated Steels. https://doi.org/10.1016/B978-0-323-99984-7.00003-8

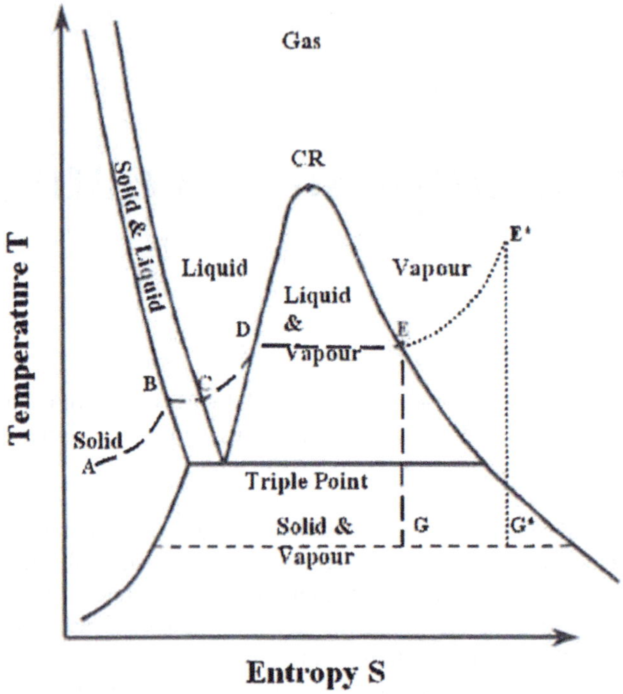

FIG. 10.1 Schematic temperature-entropy diagram for zinc, indicating relevant points in PVD processing [4]. *(From J.P. Zijp, J.A.F.M. Schade van Westrum, Physical vapour deposition for metal strip coating, in: Proc. 5th International Conference on Zinc and Zinc Alloy Coated Steel Sheet (Galvatech 2001), June 26–28, 2001, Brussels, Verlag Stahleisen, Duesseldorf, 2001, pp. 131–136.)*

10.3 PVD process basics

Prior to coating, the steel substrate must be cleaned; the initial cleaning steps can include alkaline cleaning similar to other galvanizing processes. An early application of PVD coating technology used an H_2-N_2 furnace similar to that used for hot-dip CGL processing to pretreat the steel surface, but this was abandoned for reasons described later in this chapter. It is customary to provide a final cleaning to the surface with a plasma or other technique under partial vacuum. It has been reported that bombardment of the substrate with 10^{16} Ar^+ ions/cm^2 provides sufficient cleaning [8]. Ar^+ plasma cleaning has been reported to be very effective in removal of steel substrate surface oxides prior to PVD coating, using a background pressure (partial vacuum) of 1×10^{-4} mbar [9].

Vaporization of the liquid metal in the coating section can be achieved by either bombarding the zinc target with high energy electron beams or by evaporating in a crucible. For the high production rates required for steel coil coating lines, evaporative techniques are more common. Each of these will be described further. An electric arc can used to vaporize the coating material, however it must be operated below the vapor breakdown voltage to avoid disturbance of the vapors. For Zn and Mg vapors these voltages have been reported to be 350 and 110V respectively, with the product of local pressures and electrode-to-electrode distances of 0.04–0.07 mbar-m and 0.015 mbar-m respectively [10].

Evaporation of zinc occurs at much lower temperatures than other industrial metals, a comparison is shown in Fig. 10.2. The next lowest temperature metal shown on the graph is Mg; other metals of interest for alloying with Zn for coatings on this graph include Al and Ti [4].

10.4 Electron beam vaporization and deposition

Early work on PVD of steel substrates with zinc alloys used beams of high energy electrons to vaporize metal from one or more targets made of metals used for coating. One example of a batch-type coater is shown in Fig. 10.3, built by equipment supplier Andritz-Ruthner in cooperation with von Ardenne Anlagentechnik that has been used in several research laboratories [11]. In this apparatus the metal to be evaporated is contained in a 3kW resistance heated crucible (1). A drum rotating at 10rpm (2) holds a steel sheet of dimensions 300mm × 1800mm × 0.7mm. The 80kW electron beam (3) vaporizes coating metal from the crucible. A shutter (4) controls the cross section area through which the vapor flows to the substrate. The drum rotation speed corresponds to a line speed of 15m/min. Using this apparatus, several

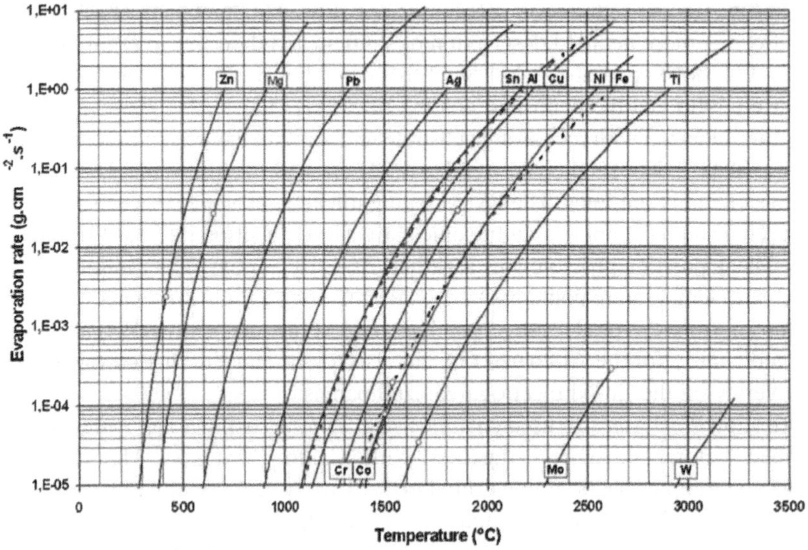

FIG. 10.2 Evaporation rates for selected pure metals [4]. *(J.P. Zijp, J.A.F.M. Schade van Westrum, Physical vapour deposition for metal strip coating, in: Proc. 5th International Conference on Zinc and Zinc Alloy Coated Steel Sheet (Galvatech 2001), June 26–28, 2001, Brussels, Verlag Stahleisen, Duesseldorf, 2001, pp. 131–136.)*

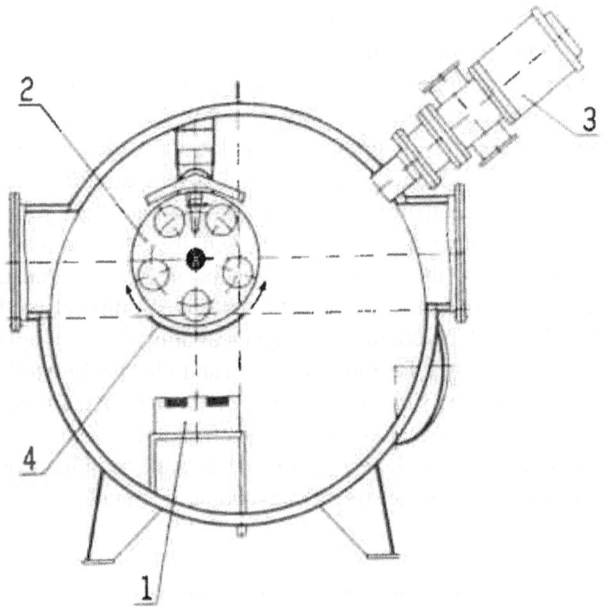

FIG. 10.3 A drum PVD coater using an electron beam to vaporize the coating metal [11]. *(G. Goldschmied, W. Korner, G. Josh, New methods of coating steel strip, in: Proc. 2nd International Conference on Zinc and Zinc Alloy Coated Steel Sheet (Galvatech 1992), September 8–10, 1992, Amsterdam, Verlag Stahleisen, Duesseldorf, 1992, pp. 396–401. ISBN: 3-514-00458-7.)*

combinations of coatings were deposited: 2 μm of Al, 2 μm Al on top of 0.1 μm Ti, 7.5 μm Zn, 2 μm Zn on top of 2 μm Al and 2 μm Al on top of 2 μm Zn. The temperature of the substrate never exceeded 100°C, as described later, coating adhesion improves with increasing substrate temperature. An unspecified heat treatment was used to improve adhesion, likely by reacting the coating with the substrate to form a thin intermetallic interface. The salt spray corrosion resistance of the Zn-Al double layer was equivalent to an Zn-Ni alloy electrogalvanized coating produced on an industrial line; the pure Zn PVD coating had equivalent corrosion resistance to a Zn electrogalvanized coating when evaluated by both salt spray and a VDA cycle scab blister test.

The shutter, item 4 in Fig. 10.3, is required to regulate vapor flow. This mechanical shutter can be moved quickly to permit rapid changes in flow. A more detailed description of shutters and their use in regulating PVD metal vapor flow is given in US Patent 4,587,134, developed at Nisshin Steel that describes the basic relationships between power input for evaporation, shutter aperture ratio and the deposited coating thickness [12].

Improvements in coating adhesion with increasing substrate temperature have been described for electron beam PVD by several researchers. Shin et al. [13] deposited layers of Al by PVD on steel substrates that had been prepared with GI, GA or electrogalvanized coatings on industrial lines. The PVD coating thicknesses were between 1 and 6 μm. Samples were prepared by cleaning with acetone and alcohol in an ultrasonic bath. The deposition rate of 0.08 g/m^2-s (0.012 μm/s) is typical for this coating process. The deposition pressure was less than 2.6×10^{-5} mbar. The minimum strip temperature for good adhesion was 170°C, but strip temperature could not exceed 250°C or else the Al re-evaporated off the substrate, together with some of the metallic Zn from the predeposited coatings. Salt spray tests showed great improvement for these PVD overlay-coated panels, particularly for the Al-coated GA but also for GI and Zn-Ni electrogalvanized substrates; the best results seen for the Al overlay on the Zn-Ni electrogalvanized panels.

Electron beam vaporization PVD was also used coat bare steel samples using a von Ardenne Anlagentechnik apparatus similar to that described before, developed together with Centro Sviluppo Materiali, coating the same size 300 mm × 1800 mm × 0.7 mm samples on a rotating drum [14]. A variety of pretreatments were used. Pure Zn, Zn-Al alloy layers and double layers of Zn and Al were deposited. The minimum deposition chamber pressure was 5×10^{-5} mbar. Coating adherence was checked using bend and Scotch tape pull tests according to ASTM A525 that was active until 1994, superseded by ASTM A653. Adhesion was always poor for the samples cleaned only with thermal pretreatments. Use of the electron beam removed only physically absorbed contaminants but was not able to remove much of the oxide film present on the steel substrate. Substrates cleaned under vacuum with mechanical brushing or Ar^+ ion plasma cleaning showed greatly improved adhesion, especially those where the substrate was above 100°C when coating deposition commenced. However, brushing was detrimental to surface finish. Ar^+ plasma cleaning removed surface oxides without affecting appearance. A range of Zn coating microstructures were produced as the substrate temperature ranged from 70°C to 260°C with a coating drum rotation speed running between equivalent strip speeds of 0.2–20 m/min.

Coatings with discontinuous crystallization near the steel-coating interface had poor adhesion while those with continuous columnar structures had improved adhesion. This was related to the kinetics of the absorption of the condensed metal on the surface and subsequent nucleation of groups of zinc atoms with critical dimensions for further growth. Too high a temperature reduced absorption kinetics while too low a temperature reduced nucleation kinetics. Zn-Al coatings exhibited greatly reduced porosity compared to the pure Zn coating, these were made either by overlaying the Zn layer with a thin layer of Al or simultaneously depositing Al and Zn; the Al content of the layered coating ranged from 1.6% to 3% down to a depth of 0.6 μm, whereas the simultaneously deposited coating had a composition of Zn-2% Al. The Zn-Al coatings had improved salt spray corrosion resistance and adhesion compared to the pure Zn coating.

A coil coating line termed "MAXI", in which crucibles of metals or alloys to be evaporated were heated with electron beams, was constructed at the Fraunhofer Institute for Electron Beam and Plasma Technology in Dresden [15]. As show in Fig. 10.4, this line consisted of 8 vacuum chambers including pretreatment chambers for plasma cleaning or sputter etching, deposition chambers containing several crucibles including plasma-activated deposition and posttreatment. The purpose of the plasma-activated deposition was to increase the deposition rate of the vapor coming from the crucibles. Two techniques were used. For both, a spotless arc is burned within the stream of evaporating metal, generating a pure metal plasma, with no

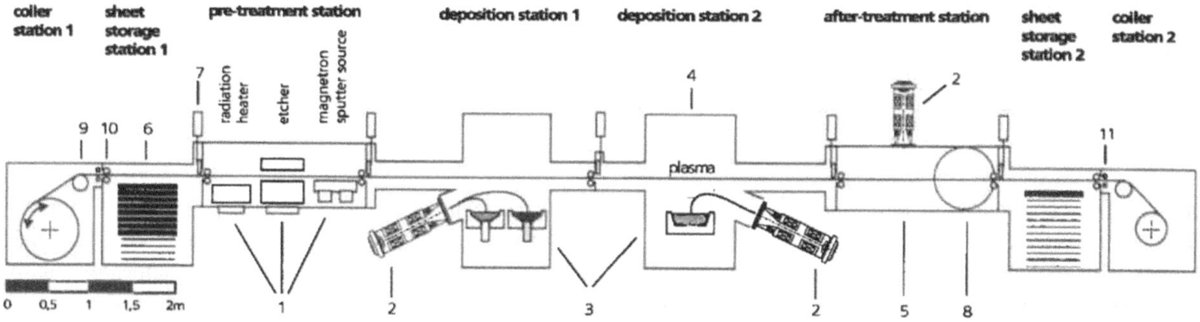

FIG. 10.4 The "MAXI" coil coating line at Fraunhofer Institute, Dresden, Germany [15]. *(B. Scheffel, C. Metzler, O. Zywitzki, J. Faber, High-rate physical vapour deposition in combination with plasma processes for metal strip coating, in: Proc. 5th International Conference on Zinc and Zinc Alloy Coated Steel Sheet (Galvatech 2001), June 26–28, 2001, Brussels, Verlag Stahleisen, Duesseldorf, 2001, pp. 35–41.)*

gas addition is needed. The arc burns on a hot cathode with a current density of 10–100 A/cm^2. In one case, termed Spotless Arc Activated Deposition (SAD), the hot evaporating surface is used as the cathode of the arc discharge, with a water cooled electrode serving as the anode. In the other, termed Rod Cathode Activated Deposition (RAD), a tungsten rod protrudes into the vapor stream, it is heated as a secondary effect of the electron beam and serves as a cathode. In the RAD technique the surface of the metal being evaporated in the crucible serves as the anode with a current in the range of 100–500 A, therefore this process is well-suited for low melting point metals. Using these techniques, a deposition rate of up to 14 g/m^2-s (2 μm/s) was possible at a maximum line speed of 1 m/s.

To enhance coating adhesion, this line includes sputter etching from a magnetron discharge to plasma clean the surface of the substrate. This was applied to the surface of galvanized sheet, and it was found that energy input of 2–7 W s/cm^2 provided satisfactory cleaning.

The "MAXI" line was used to coat GI-coated steel sheet with a Mg coating [16]. The GI coating thickness was 5 μm per side. It was first cleaned ultrasonically with a wet cleaner after which the substrate was admitted to the vacuum line where it was further cleaned by magnetron sputtering under a pressure of 2×10^{-3} mbar, vapor coated with 0.65 μm of Mg under a pressure of 1×10^{-5} mbar and held at a maximum temperature of 300°C for about 7 min under a pressure of 750 mbar with an Ar backfill. During the deposition process the Mg coating rate was 0.1336 g/m^2-s (0.167 μm/s), and the sample travel speed was 10 m/min through the deposition zone. The heat treatment resulted in the formation of a top layer of $MgZn_2$ between 1.5 and 2 μm thick. Further heat treatments up to 352°C transformed the top layer to Mg_2Zn_{11}. This phase is not stable above the Mg-Zn peritectic temperature of 381°C, so that only $MgZn_2$ was found in samples heat treated above this temperature, up to a maximum of 500°C.

Improvement of both coating density and vapor coating adhesion can also be achieved by a multistage Ar^+ ion bombardment process that delivers the atoms in the vapor to the substrate with high kinetic energy. The coating section of the University of Heidelberg coil coating mode PVD "Alligator" apparatus operating in a 1.2 m diameter vacuum chamber is shown in Fig. 10.5. The Ar^+ ions come from a ECWR (electron cyclotron wave resonance) source and pass through an extraction slit of dimensions 1 × 30 cm shown in Fig. 10.6 with an extraction voltage of 1500 V. The current density of the ion source in the extraction region is 1 mA/cm^2. Substrates to be coated had widths equal to the extraction slit width. Also visible in Fig. 10.6 is one of the crucibles in its evaporators. The substrate being coated is above the view shown in Fig. 10.6. The substrate is in the line of sight of the evaporators and therefore a mask is used to control the area of the substrate being contacted by vapor flow. The extraction slit holder can be angled so that incident ions strike the substrate either before the coating section or in the coating section [18]. The reasons for this are detailed below.

Three different modes of ion beam assisted deposition are used with this apparatus:

1. Cleaning and surface activation. Before insertion into the vacuum chamber, the substrate surface must be cleaned. This can involve initial cleaning with acetone/2-propanol followed by an alkaline degreasing solution, then rinsing in deionized water, each step being done in an ultrasonic bath. After drying the substrate can then be transferred to the vacuum

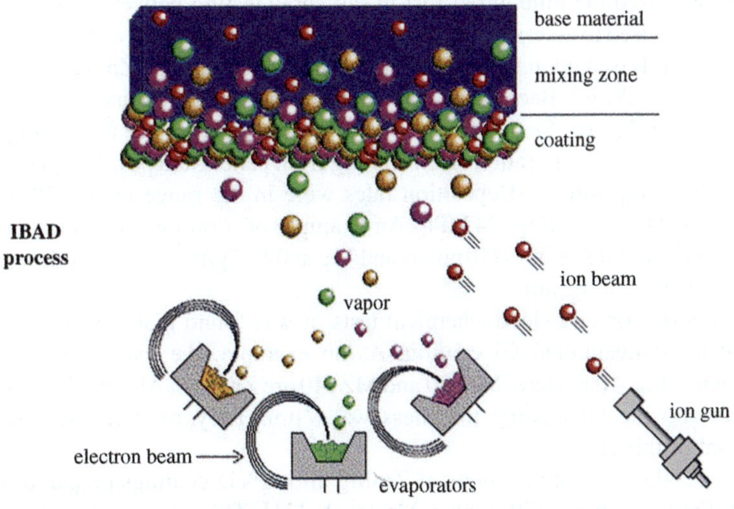

FIG. 10.5 PVD coater of University of Heidelberg showing three electron beam evaporators and Ar^+ gun for ion beam assisted deposition (IBAD) [17]. *(ZCO-45-3 New Zinc Based Automotive Steel Protection Technologies, Progress Presentation to Galvanized Autobody Partnership Program Review Meeting, November 7–8, 2006, International Zinc Assn., Durham, NC, 2006.)*

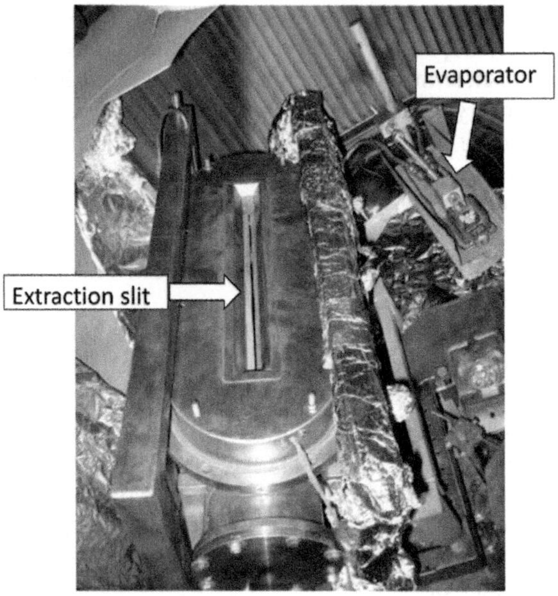

FIG. 10.6 Ar^+ ion extraction slit (*left*) and evaporator (*right*) [17]. *(ZCO-45-3 New Zinc Based Automotive Steel Protection Technologies, Progress Presentation to Galvanized Autobody Partnership Program Review Meeting, November 7–8, 2006, International Zinc Assn., Durham, NC, 2006.)*

chamber and dust-cleaned with a jet of nitrogen gas. This is followed with Ar^+ ion bombardment of the surface from the 1500 V source; 10^{16} ions/cm^2 cleans the surface sufficiently to permit good coating adhesion.

2. Development of intermediate or barrier coating: the vapor flowing from the evaporative sources mixes with high energy Ar^+ ions, producing a surface layer thickness of 10–20 nm. The Ar^+ ion to vapor atom arrival ratio at the surface of the substrate is 0.01. This intermediate layer can improve coating adhesion or produce a diffusion barrier.
3. Bulk coating of the substrate. The ion to vapor ratio remains 0.01. Given the capacity of this apparatus, the deposition of a 3 μm coating would require a line speed not greater than 6 cm/minute.

To increase line speed to 1 m/min it would be required to use ion sources with much higher current densities, or the same current density over a much larger area [18].

This apparatus was then used to assess the performance of Zn alloy coatings. Initially, a set of Zn-Cr, Zn-Ti and Zn-Mn coatings were deposited on substrates with an 8 μm thick zinc electrogalvanized coating. Variables include the presence or absence of interface layers, gradient coatings and multilayered coatings. With the exception of the Zn-Mn multilayer coating, the other coatings increased the time to red rust in salt spray testing between two and tenfold [19].

This same apparatus was then used to assess a wide variety of Zn coatings alloyed with Li, Al, Zr, Ce, Mn, Mg, Ti and Ca on uncoated steel substrates [20]. Based upon the outstanding results obtained with Zn-Mg, ternary coatings of Zn-Mg-Al, Zn-Mg-Zr and Zn-Mg-Ti were produced. Background pressure was between 0.65 and 1.5×10^{-5} mbar, and the pressure during deposition was around 10^{-4} mbar. Several techniques were used to produce the coatings: (1) simultaneous evaporation from different crucibles, (2) co-evaporation of Zn and Mg from pellets and powders placed in the same crucible, and (3) layer coatings of different compositions. Deposition rates were in the range of 0.4–20 nm/s. The highest adhesion strength was seen with the Zn-Mg-Ti coating, 24 MPa. An example of a produced coating is shown in Fig. 10.7. This was produced by evaporating Zn at 0.069 g/m^2 s (10 nm/s) and Mg at 0.093 g/m^{2}s (50 nm/s) from two different evaporators, producing a total coating thickness of 4.5 μm.

Using a wide variety of corrosion and electrochemical tests, it was found that the Zn-Mg and Zn-Mg + Al, Ti or Zn consistently outperformed the conventional GI coating. As an example, the mass gain of samples after 8 weeks of ECC1 cyclic testing is shown in Fig. 10.8. Here, MZ100 and MZ140 are Zn-1.6%Al-1.6%Mg hot-dip coatings with coating thicknesses of 7 or 10 μm and the GI coating thickness was 7 μm. They overlay each other and the PVD sample "E-ZnMgAl2" at the bottom of this figure.

More recent work has extended the compositions of Zn-Mg alloy PVD coatings prepared with electron beam evaporation and shown the benefits of coatings with higher Mg levels [21]. This was combined with a sputtering process to produce Zn-Mg coatings with Mg levels up to 22%. In these experiments the background pressure was 1.4×10^{-5} mbar, and the Ar pressure was held at 5×10^{-3} mbar during sputter cleaning and coating deposition. Changes in crystallography

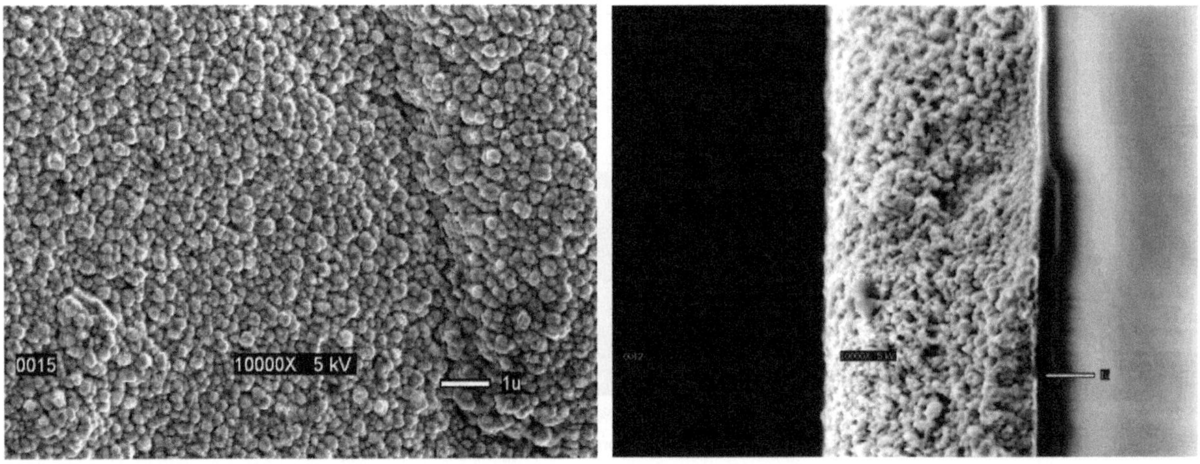

FIG. 10.7 A Zn-Mg coating produced by simultaneous evaporation of Zn and from separate evaporators: (*left*) top view, (*right*) side view. Total coating thickness 4.5 μm [20]. *(Galvanized Autobody Partnership Project ZCO-45-5, Investigation of Advanced Zinc-Based Coatings, Final Report, International Zinc Assn., Durham, NC, July 2011.)*

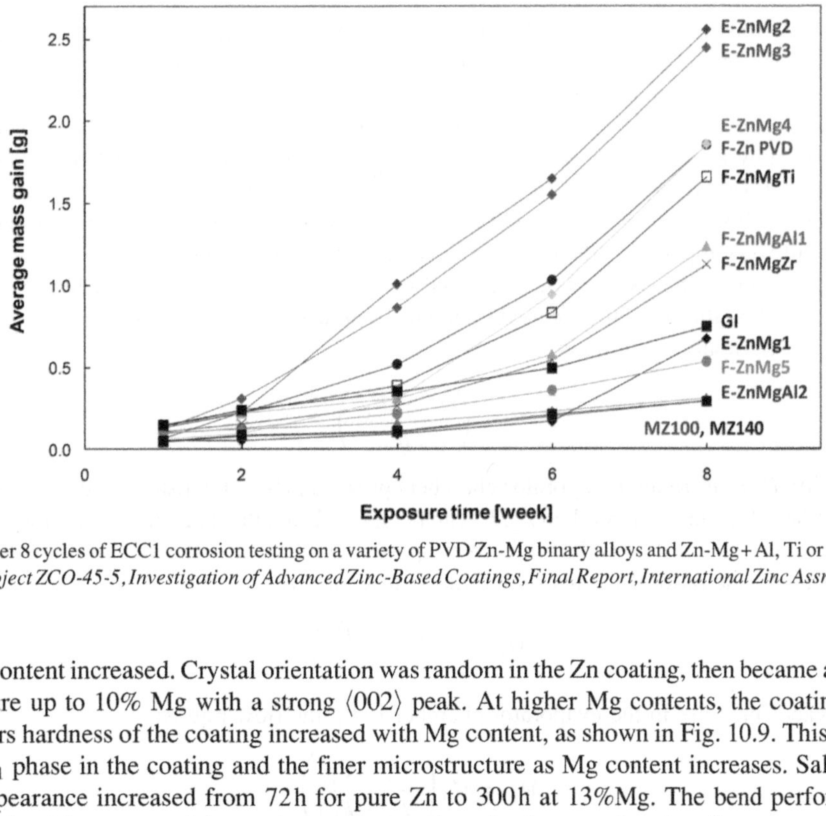

FIG. 10.8 Mass gain after 8 cycles of ECC1 corrosion testing on a variety of PVD Zn-Mg binary alloys and Zn-Mg + Al, Ti or Zr alloys [20]. *(Galvanized Autobody Partnership Project ZCO-45-5, Investigation of Advanced Zinc-Based Coatings, Final Report, International Zinc Assn., Durham, NC, July 2011.)*

observed as the Mg content increased. Crystal orientation was random in the Zn coating, then became a columnar hexagonal close-packed structure up to 10% Mg with a strong ⟨002⟩ peak. At higher Mg contents, the coating had an amorphous structure. The Vickers hardness of the coating increased with Mg content, as shown in Fig. 10.9. This a result of increasing amounts of Mg_2Zn_{11} phase in the coating and the finer microstructure as Mg content increases. Salt spray corrosion test hours to red rust appearance increased from 72 h for pure Zn to 300 h at 13%Mg. The bend performance was found to decrease with increasing Mg content. A layered structure was found to improve bend performance, using a 3%Mg bottom layer and 10%Mg top layer. A (1:2) layer thickness ratio showed the best combination of adherence and corrosion resistance.

10.5 Jet vapor deposition (JVD)

Heating of zinc under partial vacuum to the temperature where its vapor pressure exceeds local pressure will result in boiling and effusion of vapor. If the velocity and direction of this vapor is not controlled, it will condense onto cool surfaces at a low rate, including interior chamber walls and other surfaces in the coating chamber that have not been heated to

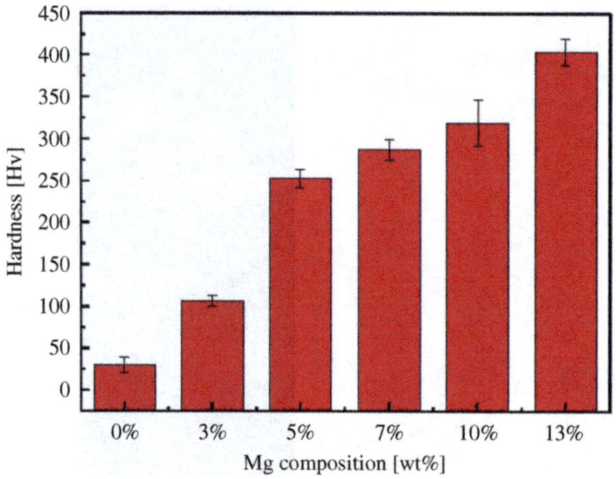

FIG. 10.9 Effect of Mg composition in a Zn-Mg binary PVD coating on Vickers hardness [21]. *(H.K. Kim, S.H. Lee, S.Y. Lee, Compositional and structural optimization of the Zn-Mg alloy coating for industrial applications, in: Proc. 12th International Conference on Zinc and Zinc Alloy Coated Steel Sheet (Galvatech 2021), June 22–24, 2021, Virtual Conference, Austrian Society for Metallurgy and Materials, 2021, pp. 108–115.)*

approximately 700 °C [22]. Examples of typical substrate coating rates in uncontrolled flow are in a batch coating apparatus where zinc coatings between 25 and 50 μm thick were produced in 1 h with background pressures between 0.1 and 10 mbar [23] and an overlay coating of 100 nm of Mg on a GI coating [24].

Producing industrially relevant coating application rates in PVD is based on vapor flow through an orifice, which in its simplest form is described by the Bernoulli equation. Shimozato [12] has applied this to JVD coating, noting that if the pressure on the upstream (evaporator) side of the orifice is very large, then the flow velocity at the orifice will reach the speed of sound, and the flow rate will then be determined only by the upstream pressure. For an ideal gas, the speed of sound is dependent only on temperature; the pressure and density factors balance each other. This condition is generally termed throttled flow, it becomes the special case of choked flow when the gas velocity reaches the speed of sound at the orifice. Further background on this and other related phenomena in near-vacuum conditions is found in Suurmeijer [25]. The simplest form of the Bernoulli equation is

$$V = \left(\frac{(P_1 - P_2)}{\rho}\right)^{0.5} \tag{10.3}$$

where V = flow velocity, P_1 = upstream (evaporator chamber) pressure, P_2 = downstream (deposition chamber) pressure and ρ = gas density. When P_1 is much larger than P_2, P_2 can be ignored, and this becomes the relationship for choked flow. The mass flow rate M through the orifice is then directly proportional to P_1 times the area of the orifice. If the orifice area A is changed from A_1 to A_2, the ratio of the respective mass flow rates is

$$M_1 = M_0\left(\frac{A_1 - P_1}{A_2 - P_2}\right) \tag{10.4}$$

and if the saturation vapor pressure in the evaporator chamber is P_s, the flow rate is:

$$M_1 = M_0\left(\frac{(P_s - \mathrm{P}_1)}{(P_s - P_0)}\right) \tag{10.5}$$

because the evaporation rate is proportional to the difference between the saturation vapor pressure and the pressure in the evaporation chamber. In the case of choked flow, P_1 is eliminated from these two expressions for M_1 giving

$$\frac{M_1}{M_0} = \left(\frac{A_1}{A_0}\right) / \left\{\left(\frac{P_0}{P_s}\right) + \frac{A_1}{A_0}\left(1 - \left(\frac{P_0}{P_s}\right)\right)\right\} \tag{10.6}$$

According to this equation, the flow rate after a change in the area of the orifice is determined by the orifice ratio (A_1/A_0) and the ratio of the pressure in the evaporation chamber P_0 before the orifice area change to the saturation vapor pressure for the evaporation bath at its given temperature P_s.

Shimozato also described the condition under which choking flow occurred

$$\frac{P_{ch}}{P_v} < \left\{\frac{2}{(\gamma - 1)}\right\}^{\{\gamma/(\gamma-1)\}} \tag{10.7}$$

where P_{ch} is the pressure in the channel downstream from the orifice at the onset of choking flow, P_v is the pressure in the evaporation chamber and γ is the ratio of the gas specific heat at constant pressure C_p to the gas specific heat at constant volume C_v. For zinc vapor $\gamma = 1.4$; near to the value of $\gamma = 1.667$ for a monoatomic gas [4].

The mass flow rate of the zinc vapor through the orifice at the speed of sound is given by Shimozato as

$$M = AP\ \{2/(\gamma + 1)\}^{\left\{\frac{(\gamma+1)}{2(\gamma+1)}\right\}} \left(\frac{\gamma M}{RT}\right) \tag{10.8}$$

where A is the orifice area, P is the pressure in the evaporation chamber, γ is the specific heat ratio described above, M is the atomic weight of zinc, R is the universal gas constant and T is the vapor temperature in the evaporation chamber. This is for the simplified situation where the vapor flow is considered to be adiabatic and frictionless, the flow is isentropic and the initial flow velocity at the evaporation source is near zero [26]. To produce different coating thicknesses, the orifice area is changed in coordination with the evaporator temperature. An example of the dependence of the deposited coating thickness on evaporator temperature is shown Fig. 10.10 for several line speeds [4].

Deposition chamber pressures between 0.06 and 0.2 mbar have been indicated. To obtain flow at the speed of sound, the ratio between the deposition chamber and the vapor ejector (Fig. 10.11) pressures with a JVD coating is recommended to be kept between 0.002 and 0.055 for deposition of Zn-Mg coatings [28]. Within the limits of these conditions, an evaporation rate of 200 g/s has been reported [27], which means that, before consideration of coating efficiencies discussed below, 720 kg/h of vapor must be generated, passed through the orifice and deposited on the substrate. For this example, a separate evaporator is used to coat each side of the strip, as shown in Fig. 10.11. It is desirable to use a diffuser immediately before the entrance of the orifice, this permits turbulent flow structures to be broken up and the orifice can be shaped as a slot, of the same width as the substrate to be coated, to improve coating speed and efficiency [29].

Evaporation temperature also strongly influences evaporation rate and consequent line productivity. Working at deposition pressure of 10^{-4} mbar, it was found that when the evaporation temperature was raised from 660°C to 722°C at constant line speed of 2 m/min, the thickness of a Zn-14.5%Mg coating increased from 1.3 to 6.8 μm for a constant orifice size, Fig. 10.12. Also, defects in the coating were observed when the evaporation temperature was less 700°C. These defects were related to both the kinetic and thermal energies of the arriving vapor and substrate temperature, which was kept at

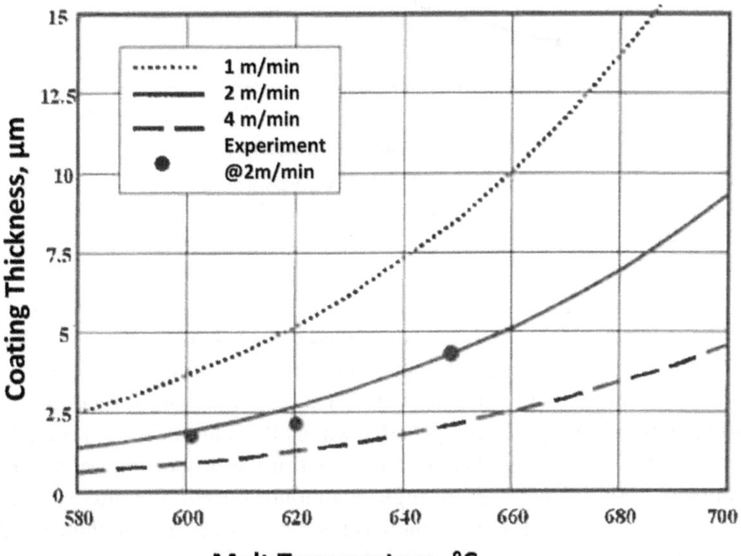

FIG. 10.10 Measured (solid circles) and predicted (lines) coating thicknesses for a jet vapor deposition configuration [4]. *(J.P. Zijp, J.A.F.M. Schade van Westrum, Physical vapour deposition for metal strip coating, in: Proc. 5th International Conference on Zinc and Zinc Alloy Coated Steel Sheet (Galvatech 2001), June 26–28, 2001, Brussels, Verlag Stahleisen, Duesseldorf, 2001, pp. 131–136.)*

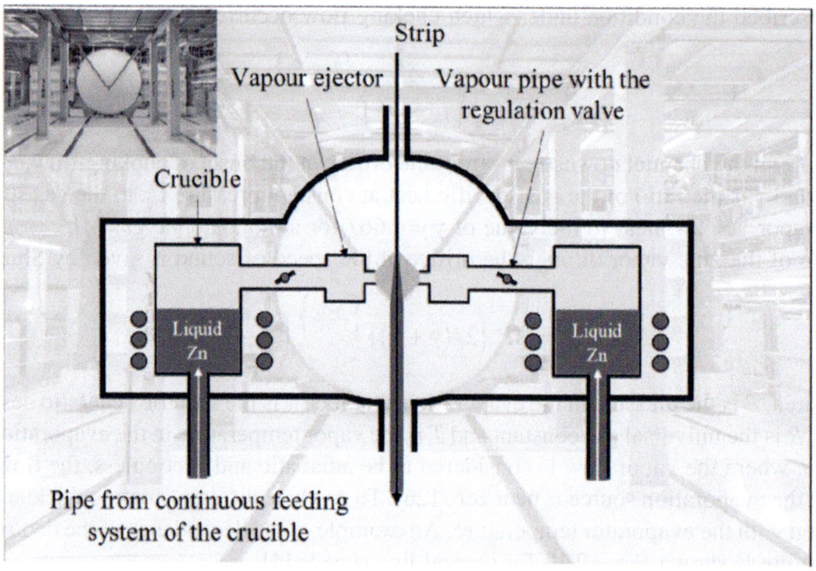

FIG. 10.11 Vacuum section of ArcelorMittal JVD line, Kessales, Belgium [27]. *(D. Chaleix, E. Jacqueson, C. Pesci, N. Amimi, S. Pace, E. Silberberg, Jet vapour deposition: a technical economic alternative to electro-galvanizing for Zn coatings of future steels, in: Proc. 12th International Conference on Zinc and Zinc Alloy Coated Steel Sheet (Galvatech 2021), June 22–24, 2021, Virtual Conference, Austrian Society for Metallurgy and Materials, 2021, pp. 1134–1143.)*

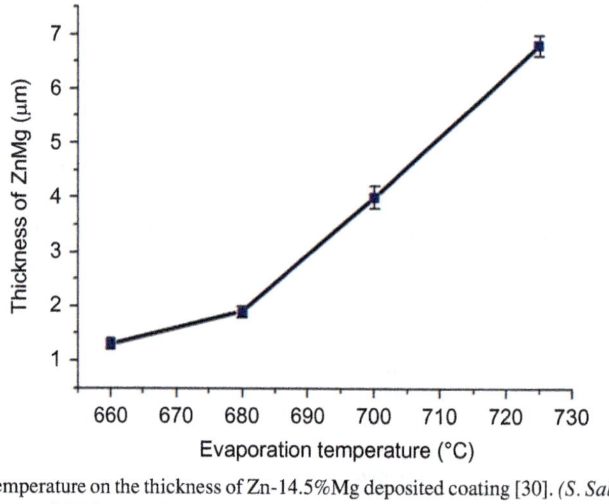

FIG. 10.12 Effect of the evaporation temperature on the thickness of Zn-14.5%Mg deposited coating [30]. *(S. Sabooni, M. Ahmad, E. Galinmoghaddam, R.J. Westerwaal, C. Boelsma, E. Zoestbergen, G.M. Song, Y.T. Pei, Fundamentals of the adhesion of physical vapor deposited ZnMg-Zn bilayer coatings to steel substrates, Mater. Des. 190 (2020) 108560.)*

150°C to avoid re-evaporation of the deposited coating as described before. This limitation on the substrate temperature led to the conclusion that 700°C was the minimum recommended evaporation temperature [30].

Several means of heating the evaporation crucible are noted. These include ohmic heating [22], levitation melting in an induction furnace [31] and induction heating [29]. Especially when alloys are being vaporized in the same crucible, it is important to provide stirring to the liquid being evaporated to control the proportions of vapor being produced [31].

For the jet vapor deposition of pure zinc coatings, a single evaporator is used. For alloys such as Zn-Mg, it is possible to use a separate evaporator each for Zn and Mg, joining their vapor flows together via channels that can be regulated with slide, butterfly, or other kinds of compatible valves [12,22,26]. The temperatures of each evaporator can be separately regulated, depending upon the vaporization properties of each pure metal. This makes it possible to create alloy coatings with a wide variety of alloying elements including Ni, Co, Mn, Fe, Cr, Mo and Cu [22]. However, as shown in Fig. 10.2, the low partial pressures for some of these elements, even at high temperatures, would result in impractically low evaporation rates.

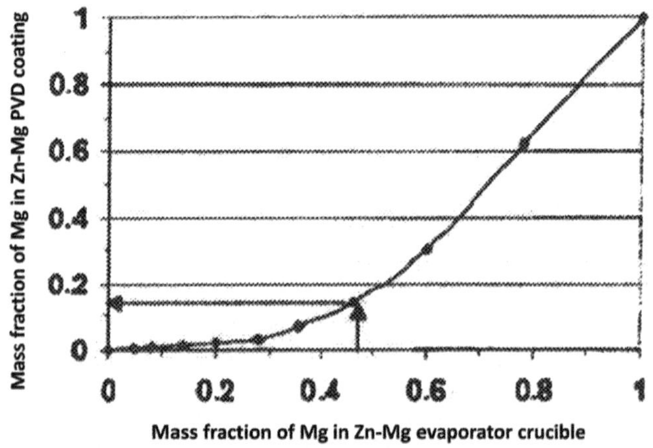

FIG. 10.13 Relationship between mass fraction of Mg in a Zn-Mg evaporator and the produced Zn-Mg coating. For the example shown, a Zn-48% Mg melt in the evaporator results in a Zn-16%Mg coating [32]. *(P. Choquet, E. Silberberg, B. Schmitz, D. Chaleix, Method for Coating a Substrate and Metal Alloy Vacuum Deposition Facility, U.S. Patent 8,481,120 issued July 9, 2013.)*

When a single evaporator is used to deposit a Zn-Mg coating, the proportions present in the crucible will be different than that achieved in the coating because of the different vaporization properties of Zn and Mg. Fig. 10.13 shows the relationship between the Mg composition of a coating prepared by JVD on the vertical axis and the Mg content of a Zn-Mg alloy in an evaporator on the horizonal axis. For the example shown in this figure, it is necessary to provide a 48%Mg alloy in the Zn bath to obtain a Zn-16% Mg coating [32].

After an extended period of time of feeding and evaporating the metals to be vapor coated, it can be anticipated that the metal in the evaporator will become enriched with elements of a lower vapor pressure than the evaporating species. To determine if this was a concern for JVD, a zinc melt was contaminated separately with aluminum and lead. These elements are commonly found in various grades of zinc alloys. Aluminum is completely soluble in zinc, while lead is only sparingly soluble, forming two liquids that co-exist with each other up to 798°C over much of the composition range. Addition of Al to the zinc melt greatly reduced zinc evaporation, the evaporation rate was found to be inversely proportional to the aluminum content when measure in atomic percent. Conversely, the addition of lead had little effect on evaporation rate, at 40 at.% lead (68 wt% lead) the evaporation rate is still more than 95% of that of pure zinc [33].

The merits of different feeding techniques of the evaporators during extended continuous processing have been considered [7]. If a single side of steel sheet is to be coated at 200 g/s, then a two-sided coating, before consideration of coating efficiency, will require 242 tons of metal to be delivered each week to the coating line. The merits of delivering solid zinc and zinc alloy ingots to the vacuum coating section were considered. Although it is much simpler to handle solid metal, heat must be provided for raising the temperature of the ingots to the correct temperature and also for the heat of fusion (112 kJ/kg for Zn, 368 kJ/kg for Mg) although the heat of vaporization (1764 kJ/kg for Zn, 5242 kJ/kg for Mg) is much higher. The solid ingots will also be covered with an oxide skin that would need to be removed from the vacuum system. Considering liquid feed, the hazards that need to be dealt with include, the handling of liquid metal, the flammability of liquid Mg and problems in precision measurement of melt constituents. However, the ability to drain the system during an unscheduled shutdown is of great advantage and improves the safety of the operation. Therefore, the feeding of liquid metal is preferred. Liquid metal feeding arrangements have been described for many years, these include both gravity-fed and pumped arrangements [7,34,35]. Pumps with no moving parts, based on magnetohydrodynamic designs, are in industrial use, capable of delivering 250 g/s of liquid metal [7].

The first industrial-scale JVD line for continuous zinc coating of steel strip was located at Nisshin Steel's Sakai Works near Osaka. This line consisted of a conventional annealing furnace utilizing a N_2-5%H_2 atmosphere to reduce the surface of the steel strip before it was introduced into the vacuum section where the coating was deposited. The vapor emitted from the electric resistance heated evaporator passed through an orifice as described above to accelerate flow to the speed of sound. This permitted zinc coating rates of 40 g/m^2-s (6 μm/s) for each side of the strip, which was exposed to the vapor jet as it passed around a heated deflector roll located in the path of the vapor stream. Line speeds up to 200 m/min were reported. Indeed, one of the objectives of building this line was to produce very thin coatings, around 30 g/m^2 that could not be produced on the company's hot-dip galvanizing lines were possible. It was also claimed that such coatings could be produced at lower cost than equivalent electrogalvanized coatings [36]. A major drawback in the use of this design was the impracticality of overaging with the N_2-5%H_2 furnace atmosphere because of the risk of explosion. The auto-

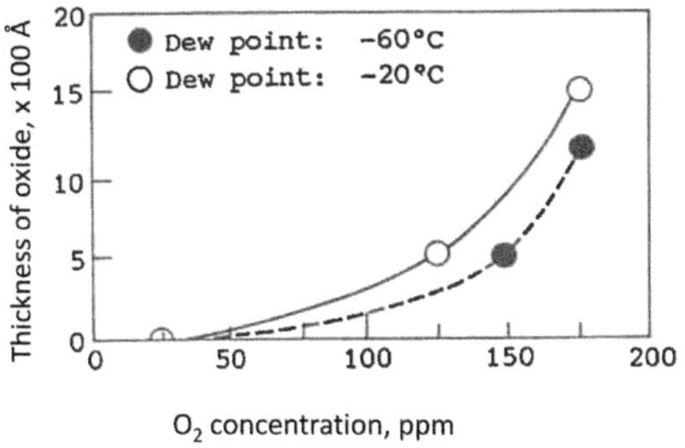

FIG. 10.14 Effect of furnace atmosphere O_2 concentration on oxide thickness on steel substrates for dewpoints of −20°C and −60°C in conditions simulating operation of the Nisshin Steel vapor coating line [38]. *(K. Chohata, M. Saito, Y. Hirose, A. Morita, Effect of surface activation conditions in continuous annealing process on adhesion of vapor zinc coated steel sheets, in: Proc. 1st International Conference on Zinc and Zinc Alloy Coated Steel Sheet (Galvatech 1989), September 5–7, 1989, Tokyo, Iron and Steel Institute of Japan, Tokyo, 1989, pp. 37–44.)*

ignition temperature of pure H_2 is 585°C [37], below the reported 700°C annealing temperature but above reported 400°C overaging temperature [38].

It was found the quality of the vapor deposited coating was much more sensitive to the risk of oxygen intrusion into the furnace and its consequent effects on dewpoint than comparable conditions for hot-dip galvanized coatings, especially with regard to coating adhesion. The inability to overage greatly limited the steel grades that could be processed, as described in more detail in Chapter 2. Therefore, the use of N_2-2%H_2 gas, which is below the flammability limit of H_2 [39] was tried in a series of laboratory experiments. The effects of dewpoint and furnace oxygen content on coating adhesion was assessed. It was found that slow cooling of 3°C/min between the annealing and overaging temperatures using a dewpoint of +10°C and higher resulted in sufficient oxide film surface coverage to prevent achievement of good coating adhesion, regardless of the temperature of the strip in the vapor coating section. The nature of this surface oxide was believed to be wüstite (FeO) from the annealing temperature down to 575°C; at that temperature, following the Fe-O equilibrium diagram, the FeO transformed to magnetite (Fe_3O_4) with a granular morphology during further slow cooling and overageing. Thinner oxide layers were observed with −60°C and −20°C dewpoint treatments that increased as the ppm of O_2 increased in the furnace atmosphere, Fig. 10.14. When a much lower furnace gas dewpoint of −60°C was used, corresponding to furnace gas oxygen content in the range of 0–25 ppm, good adhesion was realized when the strip was vapor coated at temperatures above 150°C. At this same dewpoint, furnace gas oxygen contents between 50 and 150 ppm permitted good coating adhesion to be realized at substrate temperatures greater than 200°C, Fig. 10.15. For a furnace gas dewpoint of −20°C, an oxygen content of 0–25 ppm permitted good adhesion with substrate temperature higher 150C, Fig. 10.16. It was concluded that when using the H_2-2%H_2 for vapor coating pretreatment, the oxygen content needed to be kept below 25 ppm O and the substrate temperature maintained about 150°C.

Deposition efficiency is greatly improved for the JVD process compared to the electron beam methods described in the previous section. Part of this is related to the strong directionality of flow provided by the high-speed jet, this means a much smaller area of the substrate received the deposited vapor in comparison with the line of sight mode of deposition provided

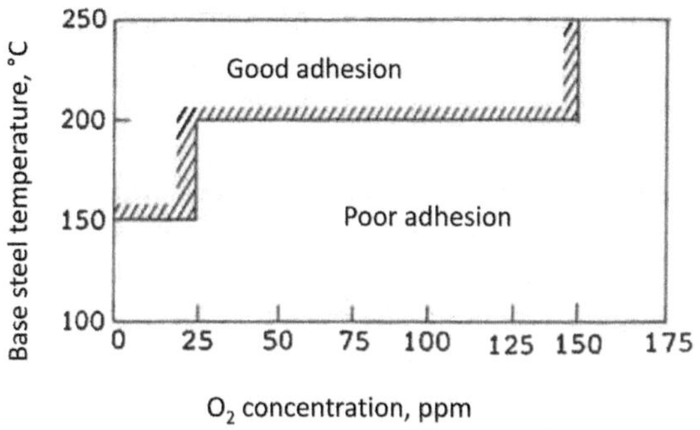

FIG. 10.15 Operating window for production of coatings on the Nisshin PVD line, −60°C furnace atmosphere dewpoint [38]. *(K. Chohata, M. Saito, Y. Hirose, A. Morita, Effect of surface activation conditions in continuous annealing process on adhesion of vapor zinc coated steel sheets, in: Proc. 1st International Conference on Zinc and Zinc Alloy Coated Steel Sheet (Galvatech 1989), September 5–7, 1989, Tokyo, Iron and Steel Institute of Japan, Tokyo, 1989, pp. 37–44.)*

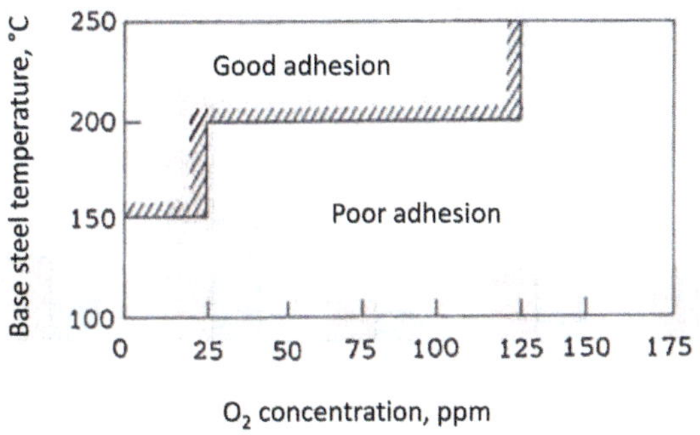

FIG. 10.16 Operating window for production of coatings on the Nisshin PVD line, −20°C furnace atmosphere dewpoint [38]. *(K. Chohata, M. Saito, Y. Hirose, A. Morita, Effect of surface activation conditions in continuous annealing process on adhesion of vapor zinc coated steel sheets, in: Proc. 1st International Conference on Zinc and Zinc Alloy Coated Steel Sheet (Galvatech 1989), September 5–7, 1989, Tokyo, Iron and Steel Institute of Japan, Tokyo, 1989, pp. 37–44.)*

by the electron beam methods. As with the electron beam melting methods described before, substrate temperature during vapor deposition affects both deposition efficiency and coating adherence. It was reported that maintaining the substrate temperature at 150°C limited re-evaporation of the JVD coating to significantly less than 0.1 nm/s, allowing a 98% deposition efficiency to be achieved [4]. Control of the substrate temperature must consider the heat of sublimation of the zinc vapor, 1995 kJ/kg. [40] as it condenses on the substrate and increases substrate temperature. It has been calculated that the transformation of zinc from vapor to solid contributes about 100°C to the substrate when a 5 μm thick layer is deposited on one side of 0.2 mm thick strip [4]. The strip temperature from the preceding cleaning section, which for modern lines is based on sputter or plasma cleaning techniques that also add heat to the strip, must be controlled to give the desired strip temperature in the deposition section. A lower temperature rise occurs with a thicker strip.

As in the case of the electron beam techniques, the temperature of the substrate during deposition also strongly effects the microstructure of JVD coatings when sputter or plasma cleaning techniques are used. Below a 120°C temperature, adhesion was found to be poor. Below this temperature a fine columnar or needle-like coating structure was observed [4]. This is consistent with observations made on sputter-deposited coatings [41]; as substrate temperature increases, both surface and bulk diffusion become more important, resulting in the microstructures that are expected to become more equiaxed and free of grain boundary voids [42].

Both alloyed and multilayer coatings can be produced using JVD [32,43]. The layered coatings can be either used as-deposited as in the case of the Sabooni work [43] or heat treated after completion of deposition to make an alloyed coating. When producing multilayered coatings, a Zn layer is usually deposited first for two reasons: to provide the maximum galvanic protection to the adjoining steel, and to maximize work of adhesion. It has been determined that the work of adhesion, defined as the amount of work required to separate the coating from the substrate per unit of the interface area, at the ZnMg/Zn interface is ~1.6 J/m^2, notably lower than that of the steel/Zn interface, ~3 J/m^2. This implies that the interface of ZnMg/Zn is inherently weaker than that of steel/Zn interface [30]. Depositing a Zn-11 Mg alloy layer on top of a pure Zn layer, both from single-source evaporators, then heat treating this layered structure at 250°C for 3 min, produced the optimized phase mixture; the coating cross-sections are shown in Fig. 10.17. The initially deposited Zn-11%Mg layer is made up of nanocrystalline $MgZn_2$ and

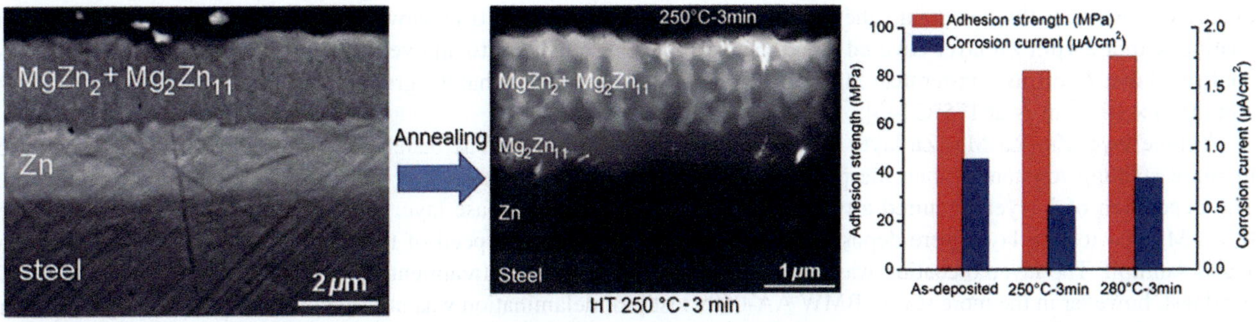

FIG. 10.17 Initially deposited PVD layers of Zn and Zn-11%Mg (*left*), heat treated coating (*center*), adhesion strengths and corrosion currents measured during electrochemical testing (*right*) [43]. *((left, center, right) S. Sabooni, E. Galinmoghaddam, A.A. Turkin, L.Q. Lu, R.J. Westerwaal, C. Boelsma, E. Zoestbergen, Y.T. Pei, Tailoring vapor-deposited ZnMg–Zn bilayer coating for steels by diffusion-driven phase transformation, J. Alloys Compd. 836 (2020) 155448.)*

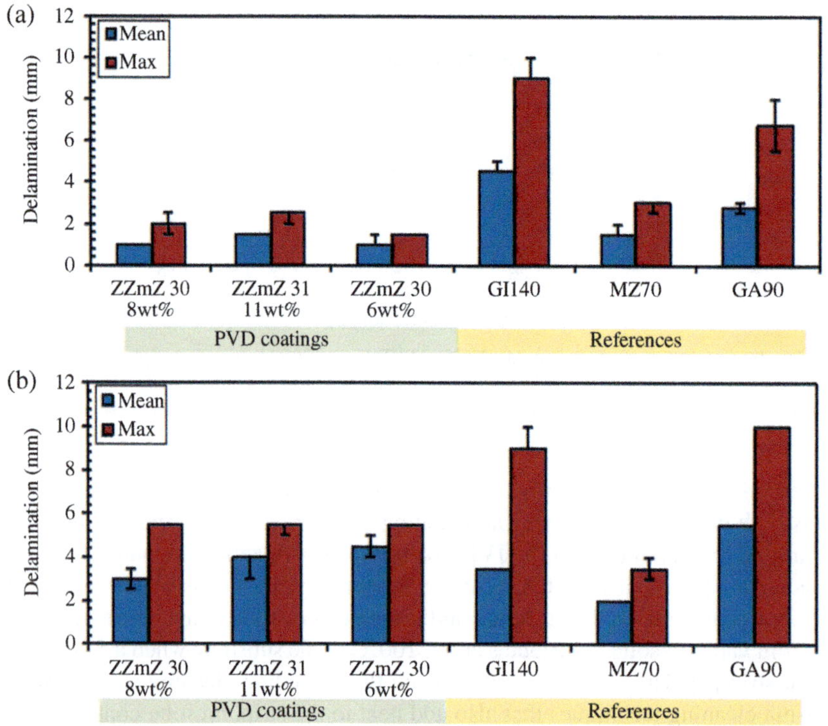

FIG. 10.18 (A) Comparison of the triple layer coatings to reference hot-dip coatings for delamination after 20 weeks of ECC1 cyclic corrosion testing on panels scribed through a standard electrophoretic primer to the metallic coating. The results are median values of the average (mean) and highest performing sample (max). Three different intermediate layer compositions are shown: 8%, 11%, and 6% Mg [44]. (B) Comparison of the triple layer coatings to reference hot-dip coatings for delamination after 20 weeks of ECC1 cyclic corrosion testing on panels scribed through a standard electrophoretic primer to the steel substrate. The results are median values of the average (mean) and highest performing sample (max). Three different intermediate layer compositions are shown: 8.11% and 6% Mg [44]. *((A, B) R. Westerwaal, A. de Vooys, J. van de Langkruis, P. Beentjes, E. Zoestbergen, Performance of multilayer PVD coatings containing Zn and Zn-Mg for automotive applications, in: Proc. 11th International Conference on Zinc and Zinc Alloy Coated Steel Sheet (Galvatech 2017), November 12–16, 2017, Tokyo, Iron and Steel Institute of Japan, Tokyo, 2017, pp. 520–527.)*

Mg_2Zn_{11} phases, with no amorphous phases are observed. Also shown in this Figure is the increased adhesion strength from 65 to 82 MPa that was developed in the optimized structure, and the decrease in corrosion current measured by potentiodynamic polarization in a 3.5% NaCl solution, indicating a more corrosion-resistant coating [43].

Multilayer PVD coatings were also made by depositing pure Mg on top of pure Zn. The bottom layer of zinc varied from 2.4 to 3.1 μm in thickness, while the top layer of Mg varied from 0.5 to 5.1 μm. These were then heat treated at the customary paint bake cycle temperature of 175°C for times between 1 and 500 h. In the lower Mg content coatings, the $MgZn_2$ phase was found and after extended times this consumed all the Mg in the coating. At coating Mg compositions where there was excess Mg relative to this compound, the Mg_2Zn_{11} phase also formed. Fig. 10.18 shows the evolution of microstructure for a coating with a 2.4 μm Zn layer covered by a 0.5 μm Mg layer, equivalent to an overall Mg composition of 5 wt%. It was confirmed that Zn diffuses faster than Mg as these phases are formed, and that the growth constant of the $MgZn_2$ phase was $1.8(\pm0.5) \times 10^{-15}\ m^2/s$ at 175°C [45].

A three-layer Zn-(Zn-Mg)-Zn layer system has also been developed. The benefit of adding the outer zinc layer is to increase staining resistance when the product is shipped with only a protective oil coating for temporary protection. The deposition of 3 layers required a two-run sequence, where the Zn base layer was deposited in the first run, while to Zn-Mg and top Zn layer were deposited in the second run. The line speed of this von Ardenne coater was between 1 and 4 m/min. The layered coating was not given a postdeposition heat treatment [44]. All test samples passed the 0T bend test, however in the more severe BMW AA-0509 test [46] delamination was observed between the Zn-Mg and lower Zn layers. It was difficult to compare the corrosion performance of these PVD coatings to conventional hot-dip coatings. The PVD coatings had single-side coating weights of 26.5–39 g/m^2 while the single-side coating weights of the GI, GA and Zn-1.6%Al-0.6%Mg hot-dip coatings were 66, 45, and 37 g/m^2 respectively. However, after receiving a conventional phosphating and PPG electrophoretic primer coating, edge creep from scribes down to the metallic coating was observed to be much less than the GI and GA coatings, and slightly less than the Zn-1.6%Al-1.6%Mg hot-dip coating, Fig. 10.18A. Results for samples scribed down to the substrate showed the 3-layer coatings to have mean edge creep values lower than GA but higher than the other hot-dip coatings tested, Fig. 10.18B. These panels were exposed to the ECC1 D17 2028 cyclic corrosion test for 14 weeks. There are also small differences in edge creep performance between the three-layer coatings deposited with different intermediate layer compositions.

10.6 Performance of vapor deposited coated steels

10.6.1 Pure Zn PVD coatings

Because of the nature of the PVD process, the roughness profile of the steel substrate is precisely followed by the PVD coating, there is no influence of intermetallic reactions and coating solidification as with hot-dip coatings, or the interaction of substrate features with electrodeposition details. Fig. 10.19A shows the steel-coating interface for a martensitic 1500 MPa steel; Fig. 10.19B shows the correlation between substrate roughness and coating roughness.

As shown in Fig. 10.20 the Zn PVD coating has a much finer grain or surface crystal size than the coatings produced by the hot-dip and electrogalvanizing processes. These coatings, of course, have very different origins. The fine zinc granules on the JVD surface develop from condensation. The larger prismatic crystals on the electrogalvanized coating develop by deposition of charged ions where both electric field, electrolyte composition and electrolyte flow parameters control crystal nucleation, growth and texture. The much larger crystal sizes on the hot-dip coating are influenced by thermal and constitutional effects on coating nucleation and growth. The finer-scale PVD coating features would not be expected to influence the appearance of high-gloss paints and other critical surfaces as much as the other coatings.

In contrast to electrogalvanizing, there is no risk of hydrogen pickup in the PVD process when cleaning in the vacuum chamber is used. Steel sheet pretreated on continuous annealing lines (CAL) to achieve targeted microstructures and properties before PVD coating has the chance to pick up diffusible hydrogen in the CAL furnace where hydrogen solubility and

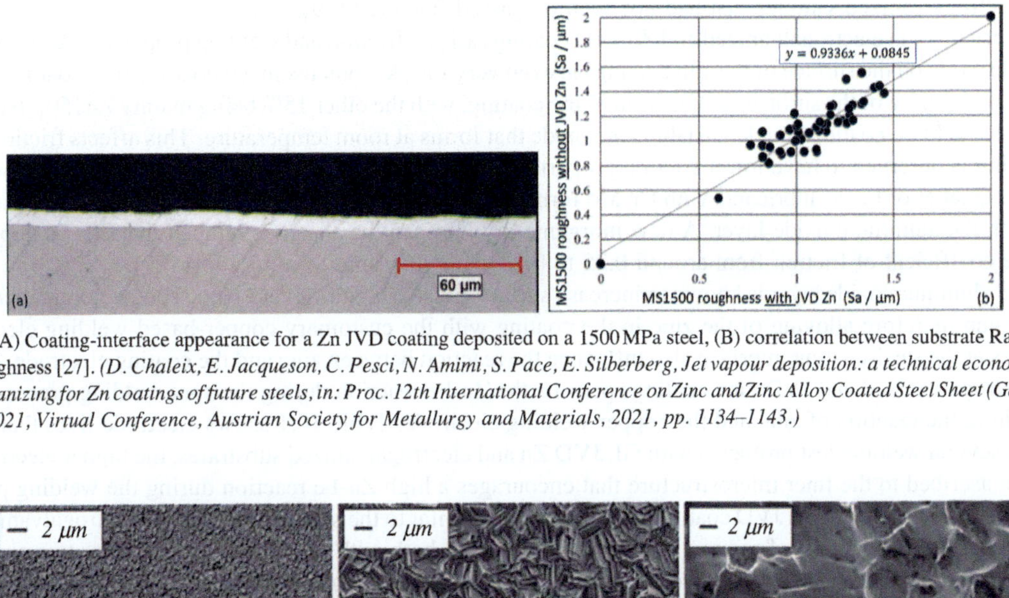

FIG. 10.19 (A) Coating-interface appearance for a Zn JVD coating deposited on a 1500 MPa steel, (B) correlation between substrate Ra roughness and coating Ra roughness [27]. *(D. Chaleix, E. Jacqueson, C. Pesci, N. Amimi, S. Pace, E. Silberberg, Jet vapour deposition: a technical economic alternative to electro-galvanizing for Zn coatings of future steels, in: Proc. 12th International Conference on Zinc and Zinc Alloy Coated Steel Sheet (Galvatech 2021), June 22–24, 2021, Virtual Conference, Austrian Society for Metallurgy and Materials, 2021, pp. 1134–1143.)*

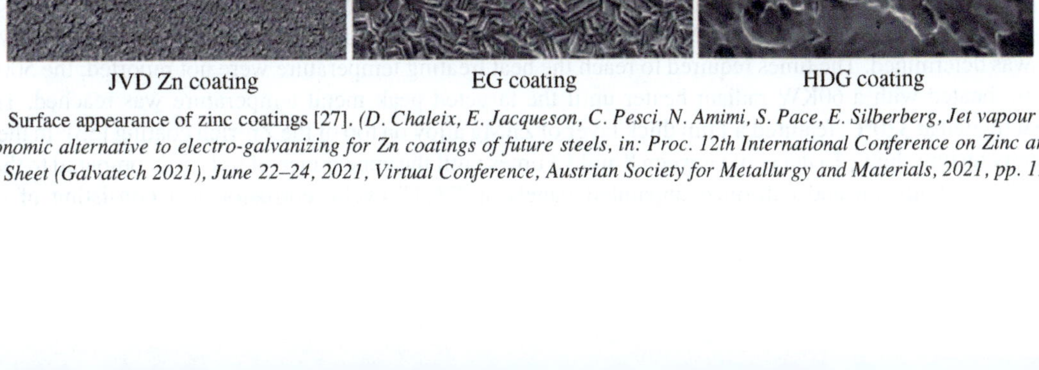

FIG. 10.20 Surface appearance of zinc coatings [27]. *(D. Chaleix, E. Jacqueson, C. Pesci, N. Amimi, S. Pace, E. Silberberg, Jet vapour deposition: a technical economic alternative to electro-galvanizing for Zn coatings of future steels, in: Proc. 12th International Conference on Zinc and Zinc Alloy Coated Steel Sheet (Galvatech 2021), June 22–24, 2021, Virtual Conference, Austrian Society for Metallurgy and Materials, 2021, pp. 1134–1143.)*

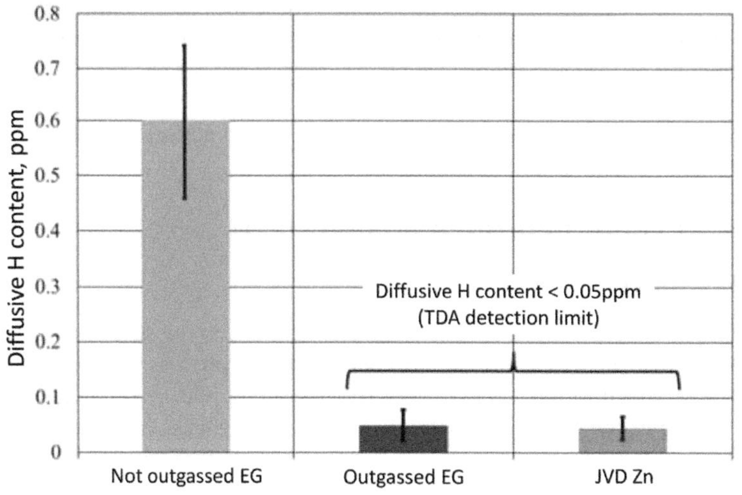

FIG. 10.21 Hydrogen pickup and outgassing behavior of a Zn electrogalvanized DP 1180 steel compared to the same grade coated by Zn PVD without the outgassing treatment of 10h at 170°C [27]. *(D. Chaleix, E. Jacqueson, C. Pesci, N. Amimi, S. Pace, E. Silberberg, Jet vapour deposition: a technical economic alternative to electro-galvanizing for Zn coatings of future steels, in: Proc. 12th International Conference on Zinc and Zinc Alloy Coated Steel Sheet (Galvatech 2021), June 22–24, 2021, Virtual Conference, Austrian Society for Metallurgy and Materials, 2021, pp. 1134–1143.)*

diffusivity are high, but this hydrogen quickly exits the steel during overaging treatments in the range of 400–500°C or even after a few days of storage at room temperature [47,48]. Fig. 10.21 shows that an intermediate degassing temperature of 170°C allowed for complete removal of hydrogen from an Zn electrogalvanized steel after 10h. U-bend samples of Zn PVD coated ultra-high strength steels showed no delayed cracking or fracture after immersion in 0.1 N HCl solution for 300h, while the electrogalvanized samples that had not been outgassed fractured [49].

The absence of Al in electrogalvanized and Zn JVD coatings affects friction and welding properties. Al is known to form a hard, friable oxide on the surface of hot-dip coatings at even very low Al contents in the coating after coating at 460°C; Al typically makes up 85% of the surface composition of the coating, with the other 15% being mainly Zn [50]. By contrast the surface of the other two coatings is a less friable zinc oxide that forms at room temperature. This affects friction properties. The GI coating is observed to have lower friction properties than Zn electrogalvanized or Zn PVD coatings when using a conventional Quaker or Fuchs lubricants with Cr-Mo forming tools. This is believed to be related to the interaction of this lubricant with the aluminum oxide layer. A new inorganic dry lubricant has been developed that can be applied on-line, reducing the coefficient of friction from around 0.25 to 0.10 using the same forming tools [27].

The same aluminum oxide layer is known to increase surface resistivity during resistance spot welding, increasing local temperature and therefore alloying of the zinc in the coating with the customary copper-based welding electrodes [51]. Intermetallics at the steel-coating interface also influence the reaction between zinc and the coating electrode. The absence of Al in electrogalvanized and JVD zinc coatings permits the Zn-Fe reaction during the spot welding process to be uninhibited, limiting the reaction of zinc and the copper welding electrode. Table 10.1 shows electrode lives that were determined using several welding test protocols with GI, JVD Zn and electrogalvanized substrates, the higher electrode tip lives of JVD were ascribed to the finer microstructure that encourages a high Zn-Fe reaction during the welding process [27].

The corrosion resistance of Zn JVD coating was found to be similar to the same thickness Zn electrogalvanized coating. Fig. 10.22 shows the appearance of red rust on these two coatings, both with a 7.5 μm thickness after two exposure durations. The evolution of red rust surface coverage with time is graphed in Fig. 10.23 [52].

10.6.2 Zn-Mg PVD coatings

The performance of a Zn-Mg coating prepared by overlaying a 6 μm thick GI coating with a 300–700 nm layer of electron beam evaporated Mg using the Fraunhofer MAXI coater described above and followed by heat treatment between 300°C and 400°C was determined. The times required to reach the heat treating temperature were not reported; the 500 × 500 mm samples were heated with a 60KW radiant heater until the targeted peak metal temperature was reached. The optimal coating, heat treated at 320°C, featured a 1 μm thick layer of ZnMg alloy on top of the Zn-rich coating [53]. In the unpainted state, salt spray testing showed a delay of between 8 and 13 times until the appearance of red rust, compared to the reference GI sample. Using both flat and deformed unpainted panels, a "3CT" cyclic corrosion test consisting of one day of

TABLE 10.1 Comparison of resistance spot welding parameters and electrode tip lives for GI Zn electrogalvanized and Zn JVD coatings [27].

Steel grade	Coating	Welding criterion (mini weld diameter)	Welding range/Mini welding current	Electrode lifetime: (mini weld number)
Welding standard: ISO18278-2				
DC06 0.7mm	HDG	4mm mini	1.2kA/8.5–9kA	500 welds
	EG	4mm mini	1kA/9–9.5kA	800 welds
	JVD Zn	4mm mini	1kA/9–9.5kA	1200 welds
MS1500 1.5mm	EG	4.9mm mini	2.4kA/6.5–7kA	800 welds
	JVD Zn	4.9mm mini	2.3kA/6.5–7kA	1200 welds
Welding standard: SEP 1220-2AC				
MS1500 1.5mm	EG	4.9mm mini	1.6kA/7–7.5kA	1200 welds
	JVD Zn	4.9mm mini	1.4kA/7–7.5kA	1800 welds

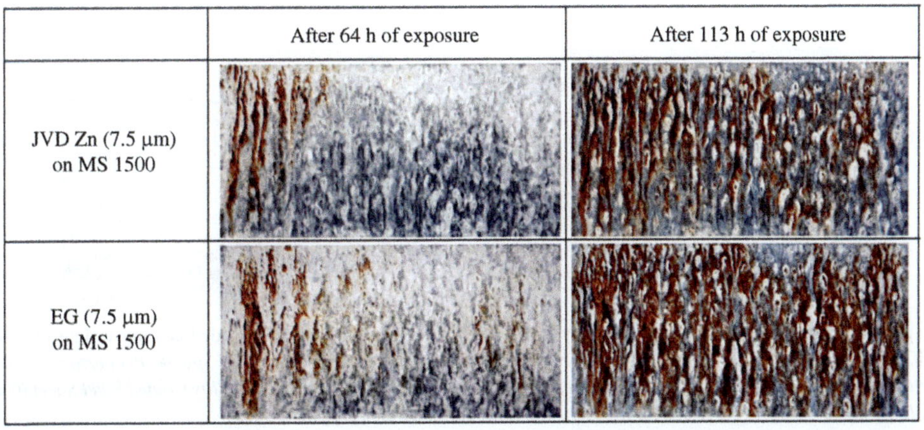

FIG. 10.22 Appearance of *red* and *white* rust on Zn JVD and electrogalvanized coatings in a neutral salt spray test after 64 and 113h of exposure [52]. *(C. Pesci, L. Diez, D. Chaleix, B. Chatelain, E. Silberberg, Jet vapour deposition: coatings steels of the future, in: Proc. 11th International Conference on Zinc and Zinc Alloy Coated Steel Sheet (Galvatech 2017), November 12–16, 2017, Tokyo, Iron and Steel Institute of Japan, Tokyo, 2017, pp. 275–282.)*

immersion in 5% NaCl at 35°C, 4days of cycling between 8h at 95% relative humidity at 40°C and 16h at 70% relative humidity at 20°C and 2days at 65% humidity at 20°C, showed a delay of 8–12 times to red rust compared to the reference GI sample. Using the VDA 621-415 cyclic corrosion test on glass-flange samples, the samples with an initial top layer of 600–700nm shows a delay of 2–3 times compared to the reference GI samples.

The Zn-Mg and 10μm thick (Z70) GI coatings were then prepared by phosphating and electrophoretic painting. The painted panels were then scribed down to the metallic coating and also down to the steel substrate, then exposed to 20weeks of the 3CT cyclic corrosion test described above. Fig. 10.24 shows the paint delamination or undercreeping observed. The coatings showed equivalent performance for creepage from the scribe down to the coating, while the Zn-Mg coating showed better performance for creepage from the scribe down to the steel substrate, as summarized in Table 10.2.

Regarding tribological properties, both the ZnMg and Z70 reference had the same surface roughness, $R_a = 0.2\,\mu m$. The flat die test was used to measure the coefficient of friction. This was 0.03–0.07 lower for the ZnMg samples, the decrease was attributed to the presence of the relatively hard $MgZn_2$ compounds on the surface of the ZnMg coating.

Adhesive bonding test samples were prepared using Dow Betamate 1496 with an adhesive layer thickness of 3mm on samples with a range of initial Mg coating thicknesses between 350 and 700nm. These were then lap shear tested both as-is

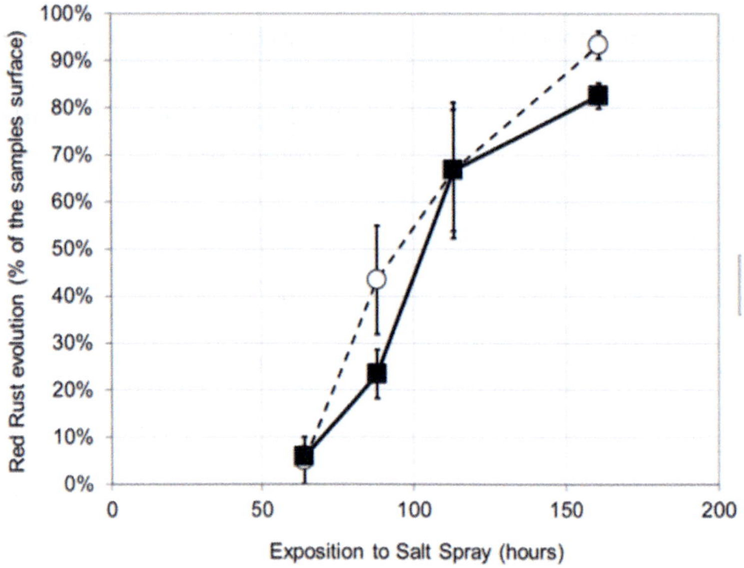

FIG. 10.23 Percentage of *red* rust coverage of the surfaces of test panels coated with 7.5 μm thick Zn JVD and electrogalvanized coatings [52]. *(C. Pesci, L. Diez, D. Chaleix, B. Chatelain, E. Silberberg, Jet vapour deposition: coatings steels of the future, in: Proc. 11th International Conference on Zinc and Zinc Alloy Coated Steel Sheet (Galvatech 2017), November 12–16, 2017, Tokyo, Iron and Steel Institute of Japan, Tokyo, 2017, pp. 275–282.)*

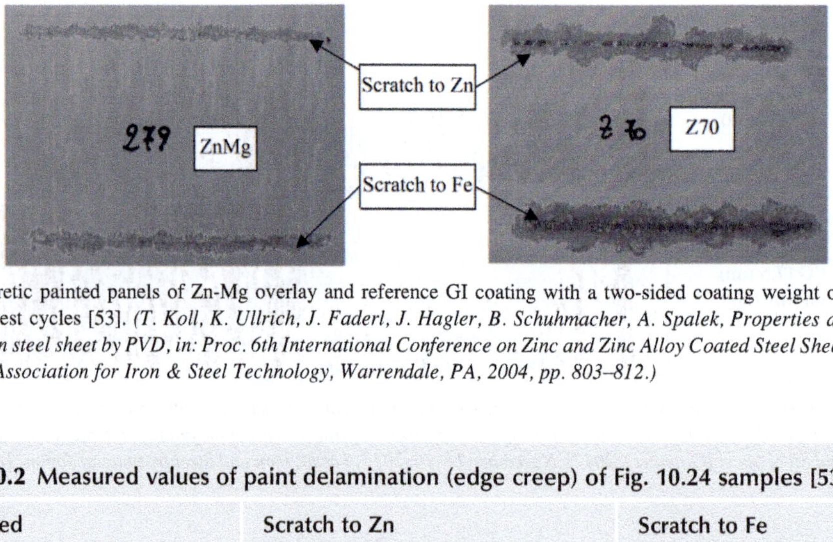

FIG. 10.24 Electrophoretic painted panels of Zn-Mg overlay and reference GI coating with a two-sided coating weight of 70 g/m^2 (10 μm coating thickness) after 20 3CT test cycles [53]. *(T. Koll, K. Ullrich, J. Faderl, J. Hagler, B. Schuhmacher, A. Spalek, Properties and potential applications of ZnMg-alloy-coatings on steel sheet by PVD, in: Proc. 6th International Conference on Zinc and Zinc Alloy Coated Steel Sheet (Galvatech 2004), April 4–7, 2004, Chicago, IL, Association for Iron & Steel Technology, Warrendale, PA, 2004, pp. 803–812.)*

TABLE 10.2 Measured values of paint delamination (edge creep) of Fig. 10.24 samples [53].

ED-painted	Scratch to Zn	Scratch to Fe
ZnMg	1.2 mm	1.2 mm
Z70	3.3 mm	4.8 mm

and after 5 cycles of VDA 621-415 corrosion testing. Reference samples of GI coated samples with either a two-sided coating weight of 70 g/m^2 (10 μm coating thickness) or 100 g/m^2 (14 μm coating thickness) were also tested. Fig. 10.25 shows that in the as-is state, all samples had comparable tensile strengths and working loads (work of adhesion). There was little effect of initial Mg thickness or heat treatment temperature. After corrosion testing, there was a higher tendency to an adhesive fracture failure mode in the ZnMg samples.

A similar overlay coating was produced using a 0.5 mm thick steel substrate coating with a 20 g/m^2 (3 μm thick) electrogalvanized coating. This was sputter-cleaned with Ar^+ ions before coating with a Zn-Mg alloy levitated in an induction-heated evaporator [54]. A drawing of the coating section is shown in Fig. 10.26. This apparatus held a Zn-10%Mg alloy in

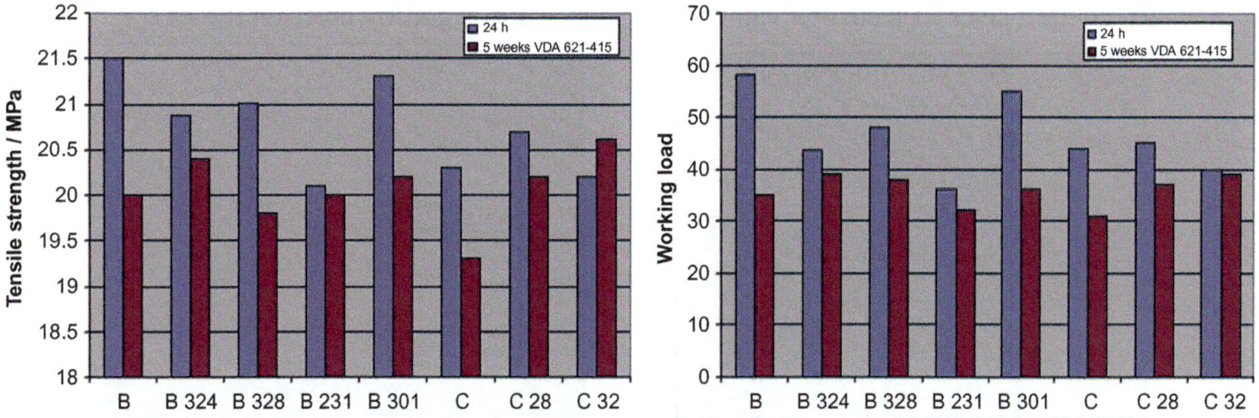

FIG. 10.25 Adhesive bonding testing test results: tensile strength (left) and working load (work of adhesion, right) of hot-dip galvanized reference samples B and C (Z70, Z100) and ZnMg-coated samples in the initial state (lighter bars) and after aging (darker bars) in the cyclic corrosion test (5 cycles acc. to VDA 621-415). The samples B 324, B 301 and C 28 received a 350 nm Mg vapor coat on top of the reference GI layer. The samples B 328, B 231 and C 32 received a 700 nm Mg vapor coat on top of the reference GI layer. The samples B 324, B 328, C 32 and C 28 received a 320°C a heat treatment after vapor coating. Samples B 231 and B 301 received a 380°C heat treatment [53]. *(T. Koll, K. Ullrich, J. Faderl, J. Hagler, B. Schuhmacher, A. Spalek, Properties and potential applications of ZnMg-alloy-coatings on steel sheet by PVD, in: Proc. 6th International Conference on Zinc and Zinc Alloy Coated Steel Sheet (Galvatech 2004), April 4–7, 2004, Chicago, IL, Association for Iron & Steel Technology, Warrendale, PA, 2004, pp. 803–812.)*

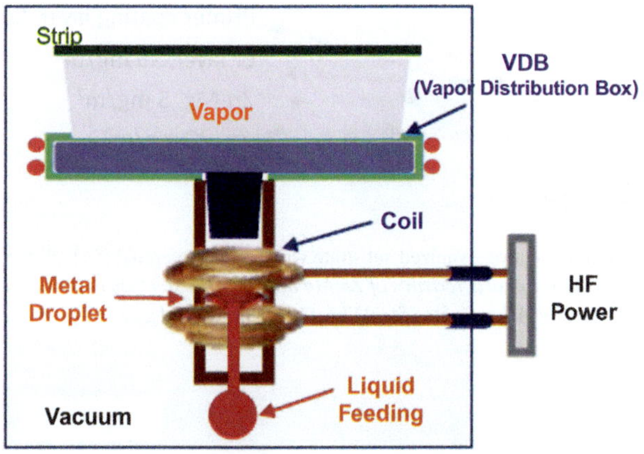

FIG. 10.26 Electromagnetic Levitation PVD coating apparatus [54]. *(Y.J. Kwak, T.Y. Kim, D.Y. Lee, K.H. Nam, Y.H. Jung, W.S. Jung, M.J. Eom, S.J. Hong, A study of the characteristics of Zn-Mg alloy coated steel sheet for constructional appliances by PVD process, in: Proc. 8th International Conference on Zinc and Zinc Alloy Coated Steel Sheet (Galvatech 2011), June 21–25, 2011, Genoa, Italy, Associazione Italiana di Metallurgia, Milan, 2011.)*

the evaporator with a background pressure of 3×10^{-5} mbar and produced a 1 μm thick Zn-Mg coating layer on the electrogalvanized substrate.

As in the case of the Zn JVD coating, the surface of the ZnMg coating was composed of smaller crystals (in the range of 1–2 μm) compared to the 3–5 μm range in the initial electrogalvanized coating, Fig. 10.27.

Test panels from this coater were then given a standard architectural/appliance paint system shown in Fig. 10.28. The topcoat was a high molecular weight polyester paint.

The hardness of the Zn-Mg top layer was found to be almost two times higher than that of the electrogalvanized coating, this is likely due to the presence of $MgZn_2$ compounds on the surface. Painted panels were then subjected to crosscut, cross Erichsen (6 mm) and ball impact tests (1/2 in. ball × 50 cm × 500 g) No powdering was observed between the paint system, the Zn-Mg/Zn metallic coating and the steel substrate, Fig. 10.29. As in the case of the reference Zn-55%Al and GI samples with heavier coating weights, the ZnMg coating therefore appears to combine good coating adherence with good formability.

Fig. 10.30 shows the results of edge creep tests for Zn-55Al, GI and the Zn-Mg layer coating after 500 h of neutral salt spray testing. The edge creep from the scribe down to the metal coating is much less for the Zn-Mg coating than the

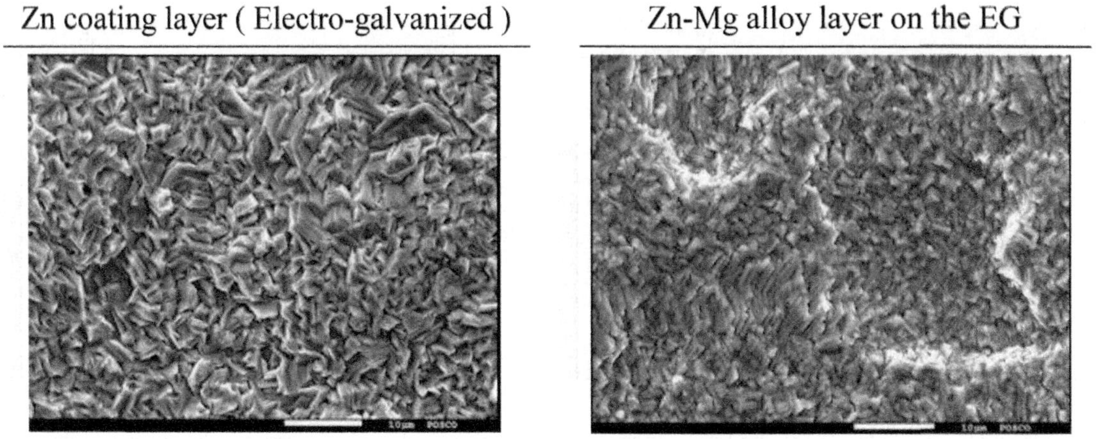

FIG. 10.27 Surface appearance of Zn electrogalvanized coating (left) and Zn-Mg overlay coating on an electrogalvanized substrate (right) The bottom white bar indicates a length of 10 µm [54]. *(Y.J. Kwak, T.Y. Kim, D.Y. Lee, K.H. Nam, Y.H. Jung, W.S. Jung, M.J. Eom, S.J. Hong, A study of the characteristics of Zn-Mg alloy coated steel sheet for constructional appliances by PVD process, in: Proc. 8th International Conference on Zinc and Zinc Alloy Coated Steel Sheet (Galvatech 2011), June 21–25, 2011, Genoa, Italy, Associazione Italiana di Metallurgia, Milan, 2011.)*

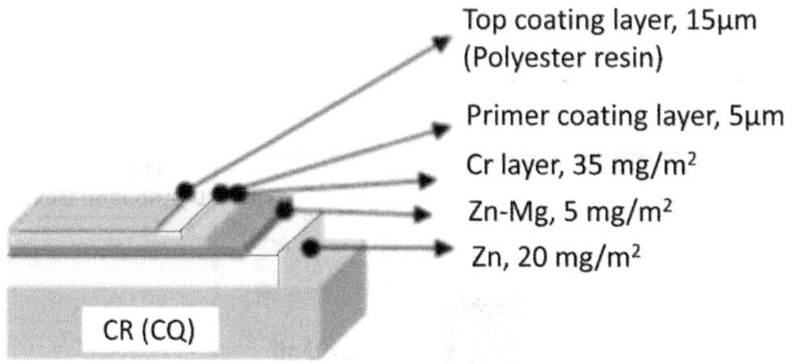

FIG. 10.28 Paint system used for evaluation of electrogalvanized substrate with Zn-Mg overcoat [54]. *(Y.J. Kwak, T.Y. Kim, D.Y. Lee, K.H. Nam, Y.H. Jung, W.S. Jung, M.J. Eom, S.J. Hong, A study of the characteristics of Zn-Mg alloy coated steel sheet for constructional appliances by PVD process, in: Proc. 8th International Conference on Zinc and Zinc Alloy Coated Steel Sheet (Galvatech 2011), June 21–25, 2011, Genoa, Italy, Associazione Italiana di Metallurgia, Milan, 2011.)*

reference coating. This was attributed to the oxidation and formation of Mg-containing simonkolleite, $Zn_5Cl_2(OH)_8 \cdot H_2O$ in the scribe that slowed the progression of corrosion [55]. This is consistent with the blister width results shown in Fig. 10.31.

Close examination of the interface layer in a coating comprising a base layer of PVD zinc onto which a Zn-11%Mg layer had been vapor deposited followed by heat treatment at 250°C for 3 min, showed a thin 200 nm layer of Mg_2Zn_{11} between the zinc base layer and $MgZn_2$ top layer. From the Zn-Mg binary phase diagram shown in Fig. 10.32 calculated from FactSage software [56], it can be seen that the reaction between the two vapor deposited layers begins with formation of a Mg_2Zn_{11} layer followed by simultaneous nucleation and growth of $MgZn_2$ (indicated in Fig. 10.32 as Laves C14 phase) and further growth of Mg_2Zn_{11} if sufficient Mg is present [30]. The 250°C heat treatment was below the Mg-Zn peritectic temperature of 381°C, so that only the initially formed Mg_2Zn_{11} phase did not decompose, it remained stable through this heat treatment.

The adhesion strength of a co-deposited ZnMg coating was found to be consistently weaker than a coating consisting of a base Zn layer and a Zn-Mg top layer. Tensile tests showed that significant portions of the co-deposited layer have detached by the time the tensile specimen fractures. In contrast, for the layered coating, both the ZnMg top layer and Zn base layer are completely adherent to the steel substrate, the fracture in the steel and coating occurs at the same transverse location. The adhesion strength has been found to be dependent upon both coating Mg content and the amount of defects in these developmental coatings. Fig. 10.33 shows adhesion strength decreasing from 171 to 66 MPa as coating Mg content increases from zero to 16.5%, with the defect-free coatings all falling on or above the pass-fail line. Table 10.3 indicates that adhesion strength is independent of the thickness of the zinc base layer for a series of coatings

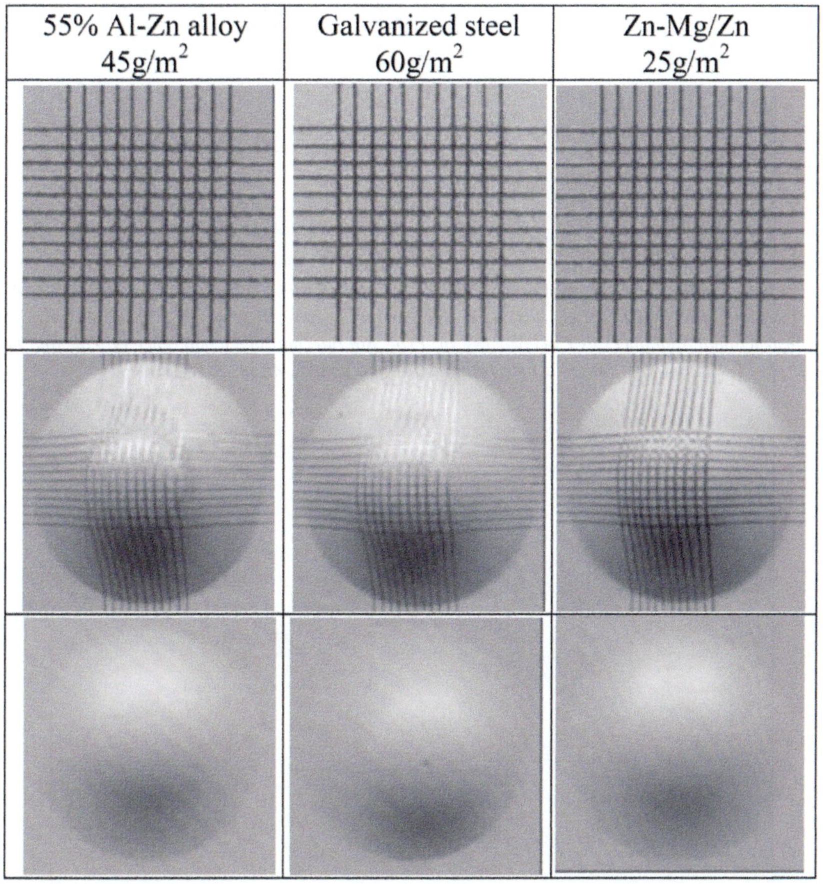

FIG. 10.29 Appearance of Zn-55Al, GI and Zn-Mg layer coating after cross-cut, cross-cut Erichsen bulge and ball impact tests [54]. *(Y.J. Kwak, T.Y. Kim, D.Y. Lee, K.H. Nam, Y.H. Jung, W.S. Jung, M.J. Eom, S.J. Hong, A study of the characteristics of Zn-Mg alloy coated steel sheet for constructional appliances by PVD process, in: Proc. 8th International Conference on Zinc and Zinc Alloy Coated Steel Sheet (Galvatech 2011), June 21–25, 2011, Genoa, Italy, Associazione Italiana di Metallurgia, Milan, 2011.)*

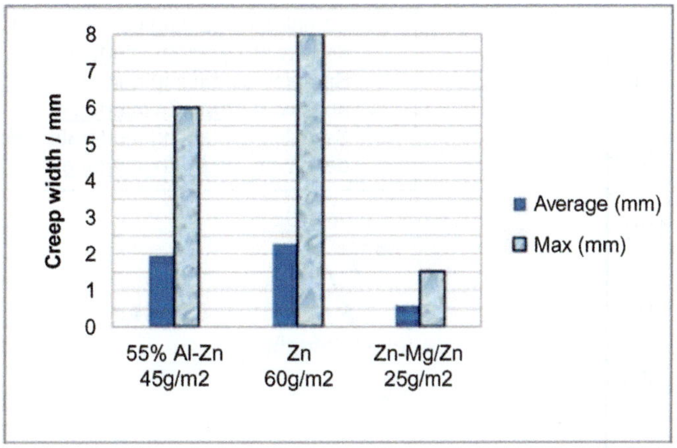

FIG. 10.30 Undercutting corrosion from scribes down to the metallic coating after 500h salt spray testing. (A) Zn-55Al, (B) GI, (C) Zn-Mg layer coating [54]. *(Y.J. Kwak, T.Y. Kim, D.Y. Lee, K.H. Nam, Y.H. Jung, W.S. Jung, M.J. Eom, S.J. Hong, A study of the characteristics of Zn-Mg alloy coated steel sheet for constructional appliances by PVD process, in: Proc. 8th International Conference on Zinc and Zinc Alloy Coated Steel Sheet (Galvatech 2011), June 21–25, 2011, Genoa, Italy, Associazione Italiana di Metallurgia, Milan, 2011.)*

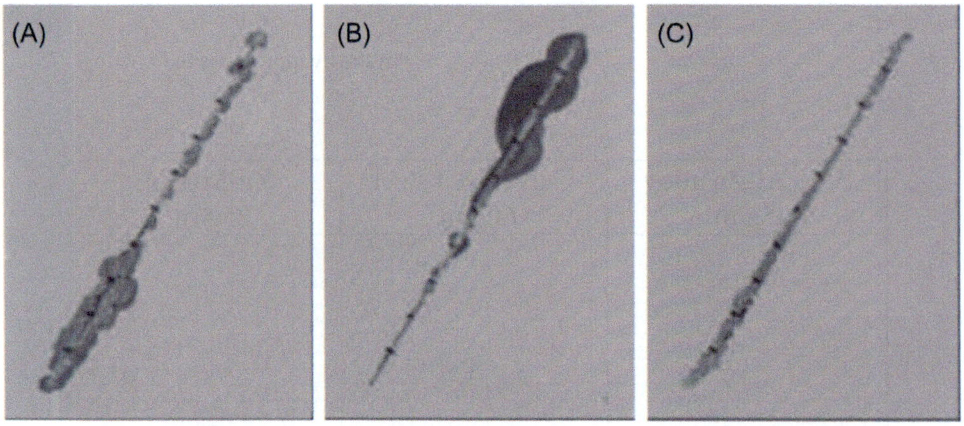

FIG. 10.31 Appearance of blister width from the scribe of prepainted sheets after 500 h salt spray testing (A) 55%Al-Zn 45 g/m^2, (B) Zn 60 g/m^2, and (C) Zn-Mg/Zn 25 g/m^2 [54]. *(Y.J. Kwak, T.Y. Kim, D.Y. Lee, K.H. Nam, Y.H. Jung, W.S. Jung, M.J. Eom, S.J. Hong, A study of the characteristics of Zn-Mg alloy coated steel sheet for constructional appliances by PVD process, in: Proc. 8th International Conference on Zinc and Zinc Alloy Coated Steel Sheet (Galvatech 2011), June 21–25, 2011, Genoa, Italy, Associazione Italiana di Metallurgia, Milan, 2011.)*

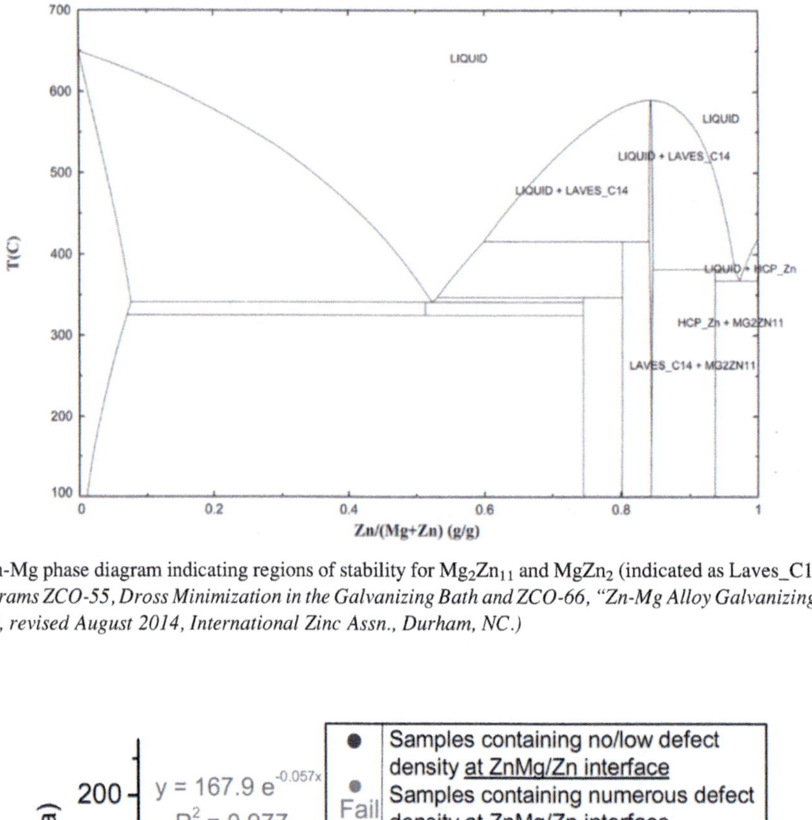

FIG. 10.32 Calculated Zn-Mg phase diagram indicating regions of stability for Mg_2Zn_{11} and $MgZn_2$ (indicated as Laves_C14 phase) [56]. *(Galvanized Autobody Partnership Programs ZCO-55, Dross Minimization in the Galvanizing Bath and ZCO-66, "Zn-Mg Alloy Galvanizing Bath Management" Final Report Issued May 7, 2014, revised August 2014, International Zinc Assn., Durham, NC.)*

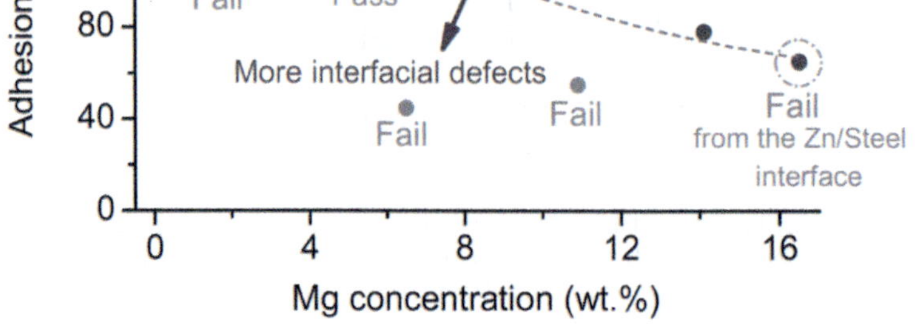

FIG. 10.33 Adhesion strength of a two-layer coating, Zn base layer and Zn-Mg co-deposited top later. The indicated Mg concentration is the initial concentration of the top layer [30].

TABLE 10.3 Summary of the effect of the thickness of Zn interlayer (t_{Zn}) on the adhesion of ZnMg-Zn bi-layered coatings [30].

	t_{Zn} (μm)	t_{ZnMg} (μm)	Critical load of delamination (N)	Adhesion strength (MPa)	Results of BMW adhesion test
ZnMg6.5-Zn[a]	0.2	3	11.6±0.4	108±1.8	Fail
ZnMg6.5-Zn[a]	0.7	3	13.1±0.5	111±2.5	Pass
ZnMg6.5-Zn[a]	1.3	3.1	13.3±0.2	105±1.2	Pass

[a]*The thickness of the Zn Interlayer is different for these samples.*
Reproduced with permission from S. Sabooni, M. Ahmad, E. Galinmoghaddam, R.J. Westerwaal, C. Boelsma, E. Zoestbergen, G.M. Song, Y.T. Pei, Fundamentals of the adhesion of physical vapor deposited ZnMg-Zn bilayer coatings to steel substrates, Mater. Des. 190 (2020) 108560.

with a 6.5%Mg top layer for the 0.2–1.3 μm Zn thickness range. Adhesion strength has been found to decrease with the zinc base layer thickness decreases below 500 nm [57].

Sabooni et al. conclude by noting that adhesion performance of the ZnMg-Zn bilayer coating during the bending test is a complex function of different parameters such as the thickness of Zn and ZnMg layers, interfacial adhesion strength and the interfacial defect density.

Although Fig. 10.33 indicates lower Mg concentrations may lead to a higher interfacial adhesion strength, they do not provide the superior corrosion resistance achieved at higher Mg contents. A desirable range of Mg contents including layer thicknesses that provide acceptable mechanical performance while considering the economics of industrial production was suggested. The appropriate thickness range for the Zn intermediate layer and the ZnMg top layer is suggested as 1–2 and 2–3.5 μm, respectively. Thickness control is governed by the evaporation temperature, strip speed and plasma cleaning parameters [30].

10.7 Summary

Although PVD Zn and Zn alloy coatings are relatively new, their production and use are growing. Because processing temperatures do not exceed the range of 200°C, Zn PVD coatings can be applied without the use of heat treatments that would, for example, temper the high strengths desired in martensitic steels. The ability to prepare and coat such steels in the absence of hydrogen exposures, either from H_2 containing gases used in hot-dip pretreatments or from electrogalvanizing cells, means the risk of hydrogen absorption is prevented. The development of jet deposition has opened this process to industrial-scale production that has now been realized, making it possible to the JVD process to compete with the established coating processes, especially for ultra-high strength (UTS > 1 GPa) grades. The JVD process unit has been reported to have a lower environmental impact than competing processes [27], aiding the steel industry's progress toward sustainable production.

References

[1] N. Selvakumar, H.C. Barshilia, Review of physical vapor deposited (PVD) spectrally selective coatings for mid- and high-temperature solar thermal applications, Sol. Energy Mater. Sol. Cells 98 (2012) 1–23, https://doi.org/10.1016/j.solmat.2011.10.028.

[2] J.F. Hanlon, R.J. Kelsey, H. Forcinio, Coatings and Laminations, in: Handbook of Package Engineering, third ed., CRC Press, Boca Raton, FL, 1998, ISBN: 978-1566763066. Ch. 4, April 23.

[3] F. Cai, Tribological and Electrochemical Corrosion Behaviours of Titanium Nitride and Chromium Nitride Based PVD Coating Systems, Ph.D. thesis, Carleton Univ., Ottawa, April 2011. https://curve.carleton.ca/4567f5b6-ee3d-47d8-abd4-de5aa6ff4042.

[4] J.P. Zijp, J.A.F.M. Schade van Westrum, Physical vapour deposition for metal strip coating, in: Proc. 5th International Conference on Zinc and Zinc Alloy Coated Steel Sheet (Galvatech 2001), June 26–28, Verlag Stahleisen, Duesseldorf, 2001, pp. 131–136. Brussels.

[5] D.M. Mattox, Handbook of Physical Vapor Deposition (PVD) Processing, second ed., Elsevier, Amsterdam, March 31, 2000. ISBN hardcover 978-0-8155-2037-5, eBook 978-0-8155-2038-2.

[6] C.B. Alcock, V.P. Itkin, M.K. Horrigan, Vapour pressure equations for the metallic elements: 298–2500K, Can. Metall. Q. 23 (1984) 309.

[7] E. Zoestbergen, C. Commandeur, R.J. Snijders, E.P.M. Bakker, P.W. Hazelett, D.A. Hamilton, S.J. Widdis, T.D. Kaiser, Feeding technology for a continuous PVD coating line, in: Proc. 12th International Conference on Zinc and Zinc Alloy Coated Steel Sheet (Galvatech 2021), June 22–24, Austrian Society for Metallurgy and Materials, 2021, pp. 1144–1153. Virtual Conference.

[8] G.K. Wolf, G. Preiss, R. Munz, L. Guzman, A. Miotello, M. Adami, Vapour deposition of corrosion resistant Zn-alloy coatings in the coil coating mode, in: Proc. 5th International Conference on Zinc and Zinc Alloy Coated Steel Sheet (Galvatech 2001), June 26–28, Verlag Stahleisen, Duesseldorf, 2001, pp. 27–34. Brussels.

[9] J. Rolls, G. West, M. Auinger, R.J. Westerwaal, E. Zoestbergen, B. Shollock, Systematic comparison of zinc coatings on a dual phase steel, using various sample preparation methods, in: Proc. 12th International Conference on Zinc and Zinc Alloy Coated Steel Sheet (Galvatech 2021), June 22–24, Austrian Society for Metallurgy and Materials, 2021, pp. 413–425. Virtual Conference.

[10] E. Zoestbergen, C. Commandeur, T. Maalman, Breakdown voltage of zinc and magnesium vapours, J. Phys. D 47 (24) (2014), 245501, https://doi.org/10.1088/0022-3727/47/24/245501.

[11] G. Goldschmied, W. Korner, G. Josh, New methods of coating steel strip, in: Proc. 2nd International Conference on Zinc and Zinc Alloy Coated Steel Sheet (Galvatech 1992), September 8–10, Verlag Stahleisen, Duesseldorf, 1992, pp. 396–401, ISBN: 3-514-00458-7. Amsterdam.

[12] Y. Shimozato, T. Wada, K. Yanagi, M. Kato, H. Furukawa, K. Wake, A. Morita, N. Tsukiji, T. Aiko, T. Kittaka, Y. Nakanishi, Method of Rapidly Changing Deposition Amount in a Continuous Vacuum Deposition Process, U.S. Patent 4,587,134 issued May 6, 1986.

[13] J.C. Shin, Y.P. Lee, K.D. Choi, J.H. Jun, Corrosion resistance of zinc and zinc alloy coated sheets with Al deposition in the upper layers, in: Proc. 2nd International Conference on Zinc and Zinc Alloy Coated Steel Sheet (Galvatech 1992), September 8-10, Verlag Stahleisen, Duesseldorf, 1992, pp. 341–345, ISBN: 3-514-00458-7. Amsterdam.

[14] F. Arezzo, P. Gimondo, V. Ferrari, Advanced Zn-based coatings by vapor deposition, in: Proc. 3rd International Conference on Zinc and Zinc Alloy Coated Steel Sheet (Galvatech 1995), September 17–21, Chicago, Iron & Steel Society, Inc., Warrendale, PA, 1995, pp. 235–241.

[15] B. Scheffel, C. Metzler, O. Zywitzki, J. Faber, High-rate physical vapour deposition in combination with plasma processes for metal strip coating, in: Proc. 5th International Conference on Zinc and Zinc Alloy Coated Steel Sheet (Galvatech 2001), June 26–28, Verlag Stahleisen, Duesseldorf, 2001, pp. 35–41. Brussels.

[16] B. Scheffel, O. Zywitzki, C. Metzner, Effect of annealing procedure on phase formation in zinc magnesium alloy coatings on steel sheets, in: Proc. 7th International Conference on Zinc and Zinc Alloy Coated Steel Sheet (Galvatech 2007), November 18–22, Iron and Steel Institute of Japan, Tokyo, 2007, pp. 556–561. Osaka.

[17] ZCO-45-3 New Zinc Based Automotive Steel Protection Technologies, Progress Presentation to Galvanized Autobody Partnership Program Review Meeting, November 7–8, International Zinc Assn., Durham, NC, 2006.

[18] G.K. Wolf, G. Preiss, L. Guzman, Ion beam assisted deposition of corrosion resistant coatings in the coil coating mode, Surf. Coat. Technol. 128–129 (2000) 28–35.

[19] G.K. Wolf, G. Preiss, R. Münz, L. Guzman, IBAD-multilayers on steel prepared by coil coating, Surf. Coat. Technol. 158–159 (2002) 125–130.

[20] Galvanized Autobody Partnership Project ZCO-45-5, Investigation of Advanced Zinc-Based Coatings, Final Report, International Zinc Assn., Durham, NC, July 2011.

[21] H.K. Kim, S.H. Lee, S.Y. Lee, Compositional and structural optimization of the Zn-Mg alloy coating for industrial applications, in: Proc. 12th International Conference on Zinc and Zinc Alloy Coated Steel Sheet (Galvatech 2021), June 22–24, Austrian Society for Metallurgy and Materials, 2021, pp. 108–115. Virtual Conference.

[22] K. Shimigori, H. Satoh, M. Toyama, H. Shimoto, K. Ikeda, J. Kawahuku, S. Miyake, S. Nomura, H. Sakai, Zn-Mg Alloy Vapor Deposition Plated Metals of High Corrosion Resistance, As Well As Method of Producing Them, U.S. Patent 5002837 issued March 26, 1991.

[23] E.J. Petit, B. Gay, Bearing general vapor phase galvanizing to the industrial scale, in: Proc. 12th International Conference on Zinc and Zinc Alloy Coated Steel Sheet (Galvatech 2021), June 22–24, Austrian Society for Metallurgy and Materials, 2021, pp. 1154–1162. Virtual Conference.

[24] B. Schuhmacher, D. Wolfhard, Combined processes—novel approaches for alloy coating on steel sheet, in: Proc. 5th International Conference on Zinc and Zinc Alloy Coated Steel Sheet (Galvatech 2001), June 26–28, Verlag Stahleisen, Duesseldorf, 2001, pp. 3–10. Brussels.

[25] B. Suurmeijer, T. Mulder, J. Verhoeven, Vacuum Science and Technology, The High Tech Institute/Settels Savenije Van Amelsvoort, Eindhoven, Netherlands, 2016. ISBN 978-90-9029137-6.

[26] J.A.F.M. Schade van Westrum, J.L. Linden, J.F.M. Velthuis, Process for Coating Substrates Using Vapor Deposition, U.S. Patent 7220450, issued May 22, 2007.

[27] D. Chaleix, E. Jacqueson, C. Pesci, N. Amimi, S. Pace, E. Silberberg, Jet vapour deposition: a technical economic alternative to electro-galvanizing for Zn coatings of future steels, in: Proc. 12th International Conference on Zinc and Zinc Alloy Coated Steel Sheet (Galvatech 2021), June 22–24, Austrian Society for Metallurgy and Materials, 2021, pp. 1134–1143. Virtual Conference.

[28] D. Chaleix, D. Jacques, S. Pace, E. Silberberg, B. Schmitz, X. vanden Eynde, Painted Steel Sheet Provided with a Zinc Coating, US. Patent 10400326 issued September 3, 2019.

[29] P. Banaszak, D. Marneffe, E. Silberberg, L. Vanhee, Industrial Vapor Generator for the Deposition of an Alloy Coating onto a Metal Strip, U.S. Patent Application 12/681, 969 published January 6, 2011.

[30] S. Sabooni, M. Ahmad, E. Galinmoghaddam, R.J. Westerwaal, C. Boelsma, E. Zoestbergen, G.M. Song, Y.T. Pei, Fundamentals of the adhesion of physical vapor deposited ZnMg-Zn bilayer coatings to steel substrates, Mater. Des. 190 (2020) 108560.

[31] T.-Y. Kim, D.-Y. Lee, M.-J. Eum, S.-W. Cho, K.-H. Nam, Y.-J. Kwak, S.-J. Hong, Y.-H. Jung, W.-S. Chung, D.-G. Kim, Y.-K. Park, J.-D. Lee, J.-H. Shin, Installation of a *wide* pvd pilot plant in POSCO, in: Proc. 9th International Conference on Zinc and Zinc Alloy Coated Steel Sheet (Galvatech 2013), September 23–27, Metallurgical Industry Press, Beijing, 2013, pp. 521–526. Beijing, Chinese Society for Metals.

[32] P. Choquet, E. Silberberg, B. Schmitz, D. Chaleix, Method for Coating a Substrate and Metal Alloy Vacuum Deposition Facility, U.S. Patent 8481120 issued July 9, 2013.

[33] E. Zoestbergen, T.F.J. Maalman, C. Commandeur, M. Goodenough, Influence of contamination on the thermal evaporation of a zinc melt, Surf. Coat. Technol. 218 (1) (2013) 108–113, https://doi.org/10.1016/j.surfcoat.2012.12.036.

[34] J.P. Gimigliano, J.F. Butler, Supplying Liquid to a Vacuum Chamber, U.S. Patent 3581766 issued June 1, 1971.

[35] P. Banaszak, D. Marneffe, B. Schmitz, E. Silberberg, L. Vanhee, Automatic Supply Device for an Industrial Metal Vapor Generator, U.S. Patent 10011905 issued July 3, 2018.

[36] M. Maeda, T. Ito, T. Aiko, K. Hashimoto, H. Furukawa, K. Yanagi, Installation of a continuous Zinc vapor deposition line, in: Proc. SAE Int'l Congress and Exposition, February 24–28, SAE International, Warrendale, PA, 2006. Detroit, MI. SAE Paper Number 860272.

[37] J. Gummer, D. Hawksworth, "Spontaneous Ignition of Hydrogen" U. K. Health and Safety Executive Research Report RR615, HSE Books, Norwich, U.K, 2018.

[38] K. Chohata, M. Saito, Y. Hirose, A. Morita, Effect of surface activation conditions in continuous annealing process on adhesion of vapor Zinc coated steel sheets, in: Proc. 1st International Conference on Zinc and Zinc Alloy Coated Steel Sheet (Galvatech 1989), September 5–7, Iron and Steel Institute of Japan, Tokyo, 1989, pp. 37–44. Tokyo.

[39] M.G. Zabetakis, Flammability Characteristics of Combustible Gases and Vapors, U.S. Bureau of Mines Bulletin 627, U.S. Department of the Interior, Washington, DC, 1965.

[40] R.F. Barrow, P.G. Dodsworth, A.R. Downie, E.A.N.S. Jeffries, A.C.P. Pugh, F.J. Smith, J.M. Swinstead, The heats of sublimation of inorganic substances. Part 1. The heat of sublimation of zinc, Trans. Faraday Soc. 51 (1955) 1354–1362.

[41] J.A. Thornton, The microstructure of sputter-deposited coatings, J. Vac. Sci. Technol. A 4 (1986) 3059, https://doi.org/10.1116/1.573628.

[42] S. Minshell, J.P. Carlos, W. Newton, E. Zoestbergen, C. Commandeur, D. Penney, N.P. Lavery, Simulating physical vapour deposition on steel substrate using the direct simulation Monte Carlo (DSMC) method, in: Proc. 11th International Conference on Zinc and Zinc Alloy Coated Steel Sheet (Galvatech 2017), November 12–16, Iron and Steel Institute of Japan, Tokyo, 2017, pp. 393–400. Tokyo.

[43] S. Sabooni, E. Galinmoghaddam, A.A. Turkin, L.Q. Lu, R.J. Westerwaal, C. Boelsma, E. Zoestbergen, Y.T. Pei, Tailoring vapor-deposited ZnMg–Zn bilayer coating for steels by diffusion-driven phase transformation, J. Alloys Compd. 836 (2020) 155448, https://doi.org/10.1016/j.jallcom.2020.155448.

[44] R. Westerwaal, A. de Vooys, J. van de Langkruis, P. Beentjes, E. Zoestbergen, Performance of multi-layer PVD coatings containing Zn and Zn-Mg for automotive applications, in: Proc. 11th International Conference on Zinc and Zinc Alloy Coated Steel Sheet (Galvatech 2017), November 12–16, Iron and Steel Institute of Japan, Tokyo, 2017, pp. 520–527. Tokyo.

[45] E. Zoestbergen, J. van de Langkruis, T. Maalman, E. Batyrev, S. Melzer, M. Zuijderwijk, Diffusion of Zinc and Magnesium in physical vapour deposited thin films at 175°C, in: Proc. 10th International Conference on Zinc and Zinc Alloy Coated Steel Sheet (Galvatech 2015), May 31–June 4, Association for Iron & Steel Technology, Warrendale, PA, 2015, pp. 36–43. Toronto.

[46] BMW Group Labortechnik, Prüfung der Zinkhaftung (Bend test to determine the adhesion of the zinc-coating), AA-0509, BMW Group Labortechnik, January 2010.

[47] C. Georges, X. Vanden Eynde, F. Goodwin, Hydrogen solubility effects in galvanized advanced high strength steels, SAE Int. J. Mater. Manuf. 9 (2) (2016) 494–500, https://doi.org/10.4271/2016-01-0360.

[48] M. Mandy, B. Nabi, M. Larnicol, X. vanden Eynde, C. Georges, F. Goodwin, Influence of Zn-based coating alloys on hydrogen diffusion and corrosion resistance in a DP steel, in: Proc. 12th International Conference on Zinc and Zinc Alloy Coated Steel Sheet (Galvatech 2021), June 22–24, Austrian Society for Metallurgy and Materials, 2021, pp. 985–993. Virtual Conference.

[49] J.-S. Kim, T.-C. Kim, S.-K. Lee, Recent progress of Zinc and Zinc alloy coated steel sheets in Korea, in: Proc. 12th International Conference on Zinc and Zinc Alloy Coated Steel Sheet (Galvatech 2021), June 22–24, Austrian Society for Metallurgy and Materials, 2021, pp. 9–15. Virtual Conference.

[50] F.M. Kilbane, Relationship between surface composition, steel mill processing and paint pretreatment of galvanized strip, in: Proceedings Galvanizers Association 74th Meeting, April 14–16, Galvanizers Association, Marine City, MI, 1982. Monterrey, Mexico.

[51] J.A. Wist, C.L. White, M.D. Gugel, F.E. Goodwin, Influence of resistance spot welding electrode alloy type on the nature of electrode wear and tip life, in: Proceedings of 4th International Zinc-Coated Sheet Conference, June 8–10, European General Galvanizers Assn, London, 1994. Paris, France.

[52] C. Pesci, L. Diez, D. Chaleix, B. Chatelain, E. Silberberg, Jet vapour deposition: coatings steels of the future, in: Proc. 11th International Conference on Zinc and Zinc Alloy Coated Steel Sheet (Galvatech 2017), November 12–16, Iron and Steel Institute of Japan, Tokyo, 2017, pp. 275–282. Tokyo.

[53] T. Koll, K. Ullrich, J. Faderl, J. Hagler, B. Schuhmacher, A. Spalek, Properties and potential applications of ZnMg-Alloy-coatings on steel sheet by PVD, in: Proc. 6th International Conference on Zinc and Zinc Alloy Coated Steel Sheet (Galvatech 2004), April 4–7, Association for Iron & Steel Technology, Warrendale, PA, 2004, pp. 803–812. Chicago, IL.

[54] Y.J. Kwak, T.Y. Kim, D.Y. Lee, K.H. Nam, Y.H. Jung, W.S. Jung, M.J. Eom, S.J. Hong, A study of the characteristics of Zn-Mg alloy coated steel sheet for constructional appliances by PVD process, in: Proc. 8th International Conference on Zinc and Zinc Alloy Coated Steel Sheet (Galvatech 2011), June 21–25, Associazione Italiana di Metallurgia, Milan, 2011. Genoa, Italy.

[55] E. Diler, B. Lescop, S. Rioual, G. Nguyen Vien, D. Thierry, B. Rouvellou, Initial formation of corrosion products on pure zinc and $MgZn_2$ examined by XPS, Corros. Sci. 79 (2014) 83–88.

[56] Galvanized Autobody Partnership Programs ZCO-55, "Dross Minimization in the Galvanizing Bath" and ZCO-66, "Zn-Mg Alloy Galvanizing Bath Management" Final Report Issued May 7, 2014. revised August 2014, International Zinc Assn., Durham, NC.

[57] S. Sabooni, E. Galinmoghaddam, H.T. Cao, R.J. Westerwaal, E. Zoestbergen, J.T.M. DeHosson, Y.T. Pei, New insight into the loss of adhesion of ZnMg-Zn bi-layered coatings, Surf. Coat. Technol. 370 (2019) 35–43.

Chapter 11

Corrosion behavior

11.1 Introduction

Although the versatility of steel has led to its usage in many applications, it is susceptible to corrosion. Zinc and zinc alloy coatings provide both barrier and galvanic corrosion to steel at low cost; because of their hardness and strength, these coatings also resist mechanical damage during installation and service better than softer organic coatings. This chapter focuses on atmospheric corrosion behavior of zinc-coated steel and includes vehicle applications that are an important use of zinc-coated steel.

In outdoor exposures, the freshly exposed zinc surface reacts quickly to form corrosion products that are quite protective in many environments, as further explained in Sections 11.2, 11.3, and 11.5. In rural and industrial environments, these protective layers can include basic carbonates, oxides, or hydrated sulfates. In marine environments, the corrosion products formed on pure zinc surfaces are less protective and include basic zinc chloride ($ZnCl_2 \cdot 4Zn(OH)_2$) and simonkolleite ($Zn_5(OH)_8Cl_2 \cdot H_2O$). In Zn coatings with Al, or Al and Mg, the corrosion products especially in marine environments are altered to provide more protective surface layers, as detailed in Section 11.3. These protective layers contribute greatly to providing a barrier to corrosion of both the zinc coating and steel substrate.

If a cut edge or scribe is present in the coating that exposes the underlying steel, the zinc coating can provide galvanic protection to the exposed steel. The extent of this protection depends upon the nature of the environment and the composition of the coating. Galvanic protection is provided because zinc is more electronegative (corrodes preferentially) to steel; steel in these damaged areas will not corrode until the supply of available zinc is consumed. The electrochemical potentials for zinc, iron, and each of their intermetallic compounds are shown in Fig. 11.57. As described in Section 11.4, self-healing of damaged areas occurs by migration of zinc ions into the damaged portion, where they inhibit oxygen reduction by the steel that would otherwise produce rust products. In alloy coatings, the zinc oxide and chloride corrosion products that participate in the self-healing reaction in purer zinc coatings are later replaced by more protective Al- and Mg-corrosion products, even though the zinc-rich phase in these coatings corrodes first to provide galvanic protection to the steel. The amount of visible red rust at the cut edge of a coated steel sheet decreases with time as a result of this self-healing effect, as shown in Fig. 11.28. Because of this self-healing capability, crevice corrosion is rarely observed in zinc-coated steels.

Great care must be taken when using the results of accelerated tests to predict long-term corrosion performance of zinc-coated steels. Temperature, time of wetness, and the presence of pollutants, especially sulfates and chlorides, are all important influences on corrosion rates in outdoor exposures; also the extent to which the coated steel is directly exposed to weathering as opposed to sheltered exposures that limit rainfall washing of the coating surface can be of importance especially in more aggressive environments. Carlsson [3] provided an excellent review of 20 kinds of accelerated corrosion tests for zinc-coated steel and a comparison of results with actual outdoor exposures. The most widely used accelerated test is the neutral salt spray test [4], which is most valuable for checking for the presence of discontinuities in the coating. It is not recommended for comparing coatings to each other to assess corrosion resistance, although in many of the references cited in this chapter, it is used for that purpose. More accurate assessments of service life are made using cyclic corrosion testing. A comparison of corrosion rates from long-term outdoor exposures with the two of the accelerated tests cited in this chapter, the ISO 9227 salt spray test [5] and the automotive cyclic test ISO 11997 described in Section 11.6, is shown in Table 11.1.

Although the salt spray test is rapid, it never permits the surface of the sample to dry, preventing the formation of many of the protective surface films that can only be formed during dry intervals. Also, the 5% salt content of the spray, applied at 35°C, is much higher than natural environments. Cyclic corrosion tests include a drying step, also cycles of high and low humidity, and sometimes freezing, which have been found to better predict actual performance. Details of several cyclic tests are given in Section 11.6.

The Metallurgy of Zinc Coated Steels. https://doi.org/10.1016/B978-0-323-99984-7.00005-1

TABLE 11.1 Comparison of zinc corrosion losses in ISO C3 (marine) environment, neutral salt spray, and automotive cyclic tests.

Test method	Exposure time	Metallic zinc loss (g/m^2)
Outdoor ISO C3 site	10 years	40–140
ISO 9227 salt spray	19 days	240–710
ISO 11997-1:C2 cyclic test	32 days	241

Adapted from B. Carlsson, G. Engström, A.-L. Hög Lejre, M. Johansson, R. Johnsen, R. Lahtinen, M. Ström, Guideline for selection of accelerated corrosion test for product qualification, NORDTEST Technical Report TR 597, Nordic Innovation Center, Oslo, 2006. ISSN 0283-7234.

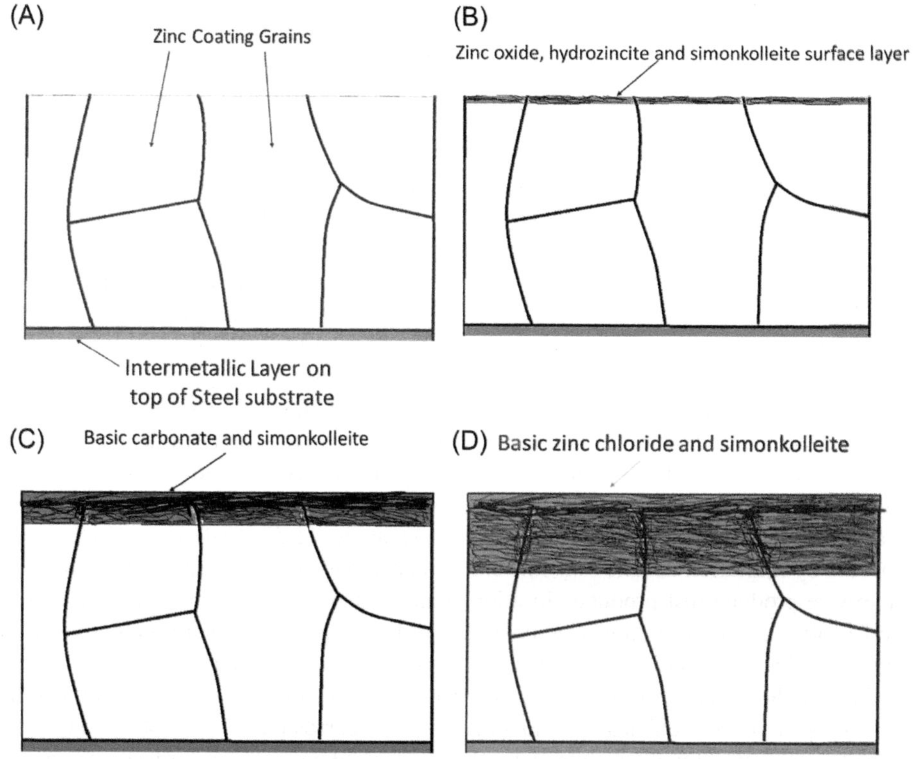

FIG. 11.1 Progress of corrosion in a Zn-0.2%Al (GI) coating in a moderate marine-industrial environment. (A) Initial microstructure, (B–D) Progression in microstructural corrosion and development of corrosion products until reaching about 1 year of outdoor exposure.

The details of coating corrosion depend on the composition and microstructure of the coating, together with the effects of the corrosive environment. Much of this chapter is devoted to describing these details. The initial steps of corrosion in a chloride-containing environment are described in Section 11.2. For a Zn-0.2%Al coating, the first corrosion products formed are zinc oxide, hydrozincite, and simonkolleite, as shown in Fig. 11.6 and described in Table 11.6. A schematic of the nature of corrosion of this coating is shown in Fig. 11.1. After exposure and drying, which expose the surface to atmospheric carbon dioxide, the bulk of the surface layer is transformed to basic zinc carbonate. Repeated exposure to chloride-coating conditions again reacts with the surface to form the chloride-containing corrosion products. Local anode and cathode locations on the coating surface, resulting from local differences in moisture and contaminant levels, result in further corrosion as exposure proceeds. For the Zn-5%Al coating, primary zinc crystals are present, surrounded by Zn-Al eutectic lamellae, as shown in Fig. 11.2.

The more anodic zinc phases corrode first, forming the same corrosion products as the Zn-0.2%Al coating, but with a lower proportion of zinc oxide being present. But shortly after this, a layered double hydroxide, also described in Table 11.6, is formed. This indicates that the Al-rich portion of the eutectic begins to corrode shortly after exposure, in the portion of the coating microstructure that was originally cathodic [8]. In the case of the Zn-55%Al-1.5%Si coating, a representative microstructure, consisting of Al-rich dendrites surrounded by Zn-Al eutectic, is shown schematically in Fig. 11.3, also in Fig. 11.20, with more details shown in Fig. 11.20 of Chapter 6. During the initial stage of corrosion in

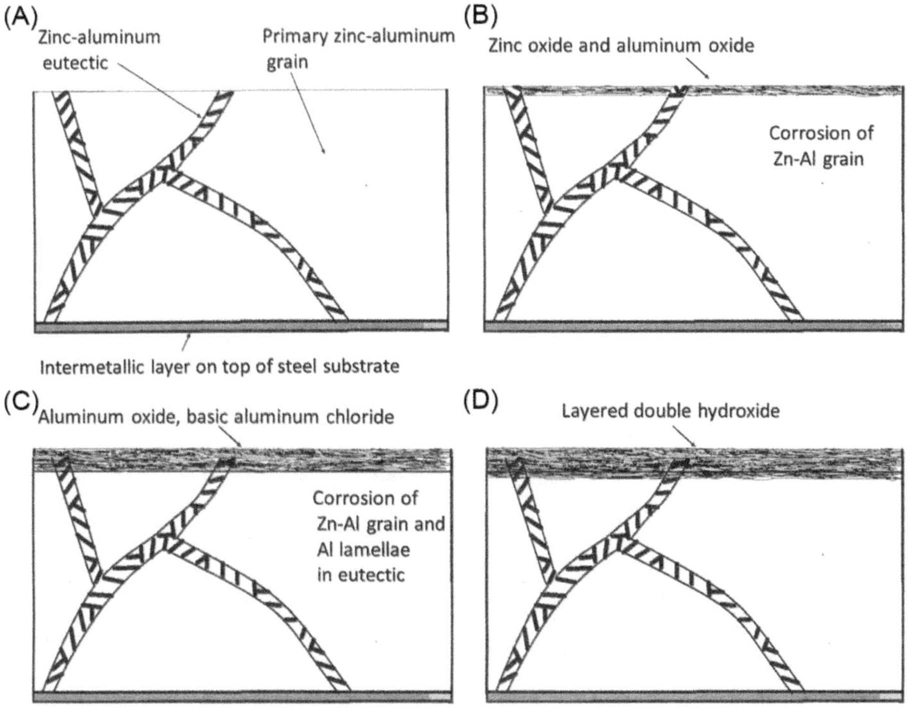

FIG. 11.2 Progress of corrosion in a Zn-5%Al (GF) coating in a moderate marine-industrial environment. (A) Initial microstructure, (B–D) Progression in microstructural corrosion and development of corrosion products until reaching about 1 year of outdoor exposure.

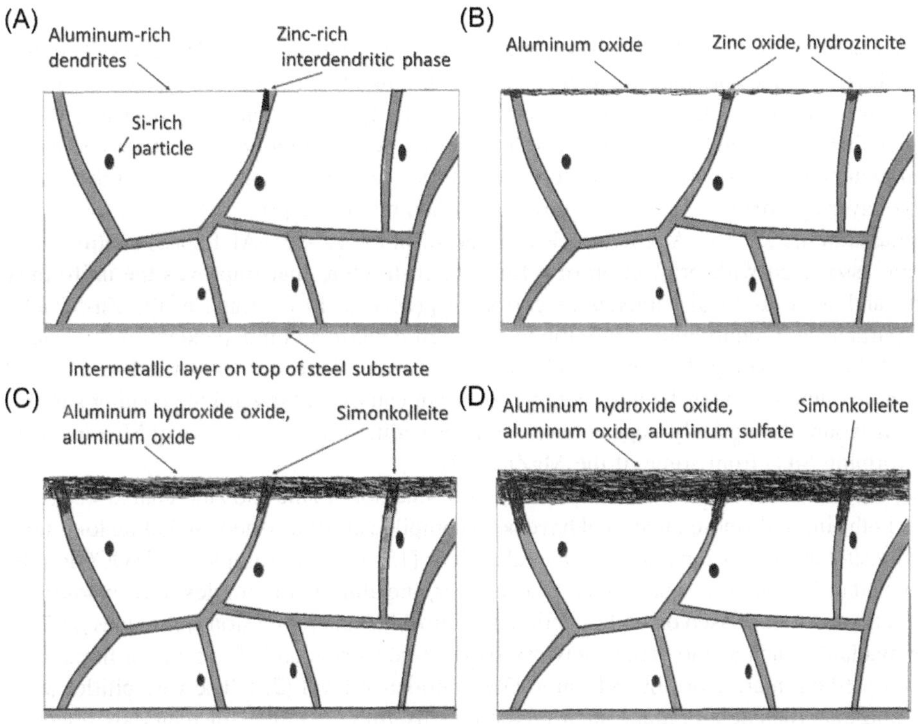

FIG. 11.3 Progress of corrosion in a Zn-55%Al-1.5%Si (AZ) coating in a moderate marine-industrial environment. (A) Initial microstructure, (B–D) Progression in microstructural corrosion and development of corrosion products until reaching about 1 year of outdoor exposure.

a chloride-containing environment, the zinc-rich phases corrode preferentially, forming the same zinc oxide, hydrozincite, and simonkolleite corrosion products indicated for the higher zinc content coatings. These phases have a higher solubility than the aluminum-rich corrosion products AlOOH, Al_2O_3, and aluminum sulfate that are the result of later corrosion of the Al-rich phases, so that the Zn-rich phases become a lower proportion of the corrosion product mass as corrosion proceeds. The less soluble Al-rich corrosion products Al_2O_3, AlOOH, and eventually $Al(OH)_3$ form highly protective surface layers for much of

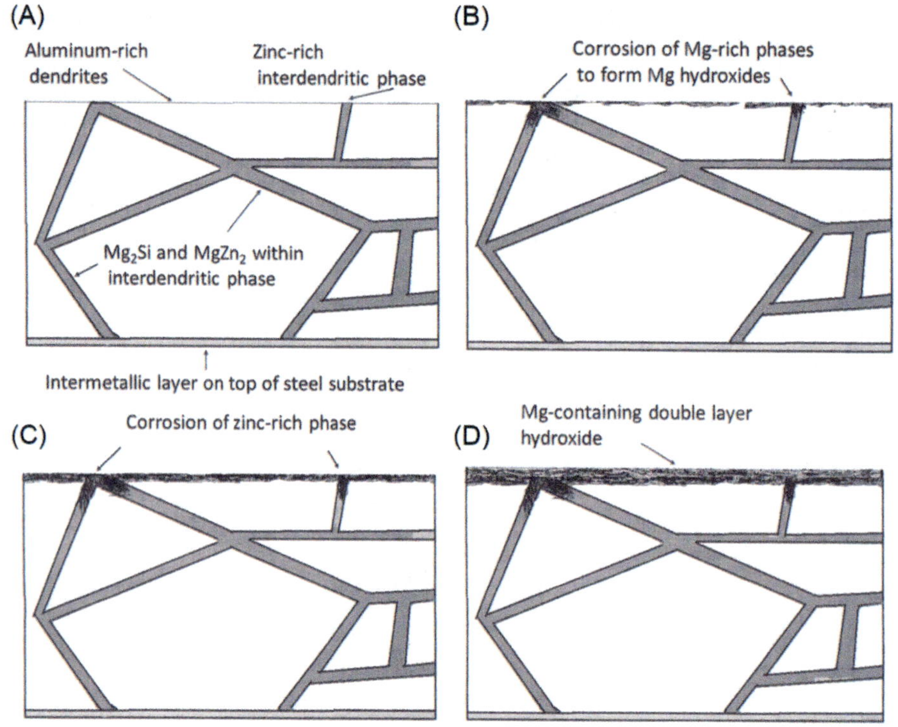

FIG. 11.4 Progress of corrosion in a Zn-55%Al-1.5%Si-2%Mg (AM) coating in a moderate marine-industrial environment. (A) Initial microstructure, (B–D) progression in microstructural corrosion and development of corrosion products until reaching about 1 year of outdoor exposure.

the lifetime of the Zn-55%Al-1.5%Si coating [10]. The addition of Mg to Zn-Al coatings with at least 1.5%Al results in formation of a ZnMg eutectic that corrodes first in the undamaged coating, resulting in formation of double-layered hydroxides described in Section 11.3, as shown schematically in Fig. 11.4. However, when a cut edge or scratch in the steel substrate is exposed, the zinc-rich phase corrodes first, as detailed in Section 11.4. In all of the Mg-coating coatings described in the chapter, Mg ions migrate to oxygen reduction sites, controlling the corrosion of more cathodic areas such as exposed steel, forming the highly protective double-layer hydroxide layer that improves corrosion resistance [8].

The role of the Si-containing phases in the Zn-55%Al coatings is of interest. In the Zn-55%Al-1.5%Si coating, Si is found in discrete particles. They are associated with production of a finer microstructure that improves the uniformity of corrosion in the overall coating and serve as local cathodes as corrosion proceeds. In contrast, in the Zn-55%Al-1.5%Si-2%Mg coating, the Mg_2Si particles are found both within the interdendritic phase and independent of this phase, consistent with the solidification path for this coating described in Chapter 6. As noted there and shown in Fig. 11.33 of Chapter 6, the Mg_2Si particles are more numerous near the intermetallic layer but can extend toward the coating surface. Both Mg_2Si and $MgZn_2$ serve as local anodes during early stages of corrosion, providing for quick release of Mg ions, with the Si of the Mg_2Si particles then forming SiO_2 from some of the $MgZn_2$ [12].

There are many publications that describe the electrochemical and corrosion behavior of zinc and zinc-coated steels, for example [13,14]. Corrosion data for both zinc and zinc-coated steel have been compiled and published, including long-term corrosion rates from outdoor exposure sites and corrosion maps for several regions [15–18] For example, the ISOCORAGG study included 53 exposure sites located in 13 countries where steel, zinc, copper, and aluminum samples were exposed for 1, 2, 4, and 8 years [19]. Corrosion prediction tools, based on both functional relationship and regression techniques [20] and neural network inference [21], are available, such as the neural-network-based Zinc Corrosion Life Predictor http://zclp.galvinfo.com:8080/zclp/ that gives a prediction error of 36% + 13 at a 95% confidence level [22]. The variabilities associated with exposure, application, and materials details make such predictions useful as a guide for materials selection rather than guaranteeing materials performance. Accelerated corrosion tests have been used to help qualify materials for specific applications, but care must be taken in interpretation of results; a review of the main issues involved in inferring long-term performance is a useful guide [23].

Because most applications of zinc-coated steel sheet involve atmospheric exposure, most of this chapter will be devoted to discussing how the metallurgy of zinc and zinc alloy coated steel influences atmospheric corrosion behavior. Corrosion performance in automotive and appliance applications is also discussed, although most such applications involve painted product that is discussed in Chapter 12.

The nature and severity of atmospheric exposure environments are described in ISO Standard 9223 "Corrosion of Metals and Alloys – Corrosivity of Atmospheres – Classification, Determination and Estimation" [24]. Five corrosion severity categories are set forth, from C1 = Very Low to CX = Extreme. Factors influencing corrosion severity are (1) time of wetness that ranges from less than 10 h per year for internal climate-controlled environments to less than 5500 h per year (a year has 8760 h) for damp climates or unventilated sheds, (2) sulfur dioxide deposition rates of less than 4 mg/m^2 per day for rural atmospheres to between 80 and 200 mg/m^2 per day for highly polluted industrial atmospheres (3). Chloride deposition rates of less than 3 mg/m^2 per day for mild environments to between 300 and 1500 mg/m^2 per day for severe marine environments. (4) Temperature. Extreme marine exposures that include frequent splash and spray are not included in this characterization. These variables have been used to develop a polynomial correlation between first-year zinc corrosion rate and environmental variables shown in this standard:

$$r_{corr} = (0.0129)\, P_d^{\,0.44} e^{(0.0046\, RH + f_{Zn})} + (0.0175) S_d^{\,0.57} e^{(0.008\, RH + 0.085T)} \tag{11.1}$$

where r_{corr} = first-year corrosion rate μm/yr, T = average annual temperature °C, RH = average annual relative humidity %, P_d = average annual sulfur dioxide deposition rate mg/m^2 per day, S_d = average annual chloride deposition rate mg/m^2 per day, $f_{Zn} = 0.038(T - 10)$ when $T < 10$°C, otherwise $f_{Zn} = -0.071(T - 10)$. P_d can be replaced by $0.8P_c$, where P_c is the average annual sulfur dioxide content in the air, μg/m^3. The length of time when relative humidity is greater than 80% for temperatures greater than 0°C is defined as the time of wetness. This corrosion rate correlation is valid for temperatures between −17.1 and 28.7°C, relative humidities between 34% and 93%, sulfur dioxide deposition rates between 0.7 and 150.4 mg/m^2 per day, and chloride deposition rates between 0.4 and 750.5 mg/m^2 per day. The level of uncertainty for this correlation is −33% to +50%. Uncertainty is related to both the validity of the dose-response relationships shown in Eq. (11.1) and those relating to environmental parameters. This correlation is only for the first year of exposure; as explained later, passivation of the zinc surface with time as corrosion products are formed is expected to result in lower rates of corrosion for succeeding years. Section 11.6 gives several examples of long-term exposures illustrating this; when protective corrosion product layers can be formed, corrosion kinetics changes from reaction-controlled (linear) to diffusion-controlled (parabolic), for example, Figs. 11.41 and 11.45.

Indications of time to first maintenance for galvanized steel sheet in corrosion severity categories C3 through CX are given in ISO 14713 "Zinc coatings — Guidelines and recommendations for the protection against corrosion of iron and steel in structures — Part 1: General principles of design and corrosion resistance" [27]. Two single-side coating thicknesses are shown: 20 and 42 μm. For the thinner coating, the range of lives to first maintenance in a C3 environment is 10–29 years (High durability class), while for the thicker coating it is 20–60 years (Very High durability class). For categories C5 and CX, the durability classes range from Medium to Very Low; under these conditions, the protective function of the galvanized coating would be expected to be augmented by a paint system as described in Chapter 12.

11.2 The nature of the surface of zinc and zinc alloy coated sheet steel and initial reactions with the environment

Because of its higher reactivity compared with zinc, Al is surface-selective for oxidation after the hot-dip coating exits from the galvanizing bath. Al typically makes up 85% of the surface composition of the coating, with the other 15% being mainly Zn. A high portion of the Al on the surface is in the form of aluminum oxide [28]. The total amount of Al on the surface of a GI coating with a bulk Al content of 0.2% in the overlay (not including the $Fe_2Al_{(5\text{-}x)}Zn_x$ interface layer described in Chapter 5) just after solidification was measured as about 2.5 mg/m^2, and this increased with exposure to air, reaching 3 mg/m^2 after 2000 h [29]. Temper rolling, described in Chapter 2, exposes fresh coating surface because of the texture imparted by the work roll surface and the slight elongation of the strip length. Al surface enrichment does not immediately increase from the as-coated level of 2.5 mg/m^2 but increases gradually with exposure time to room temperature air. Whereas in the as-coated samples, the Al surface enrichment was related to segregation of Al oxides at coating grain boundaries, the Al segregation in the temper rolled samples included precipitates of metallic Al that grew from the supersaturated solid solution near the coating surface together with Al oxides. Al precipitation occurred at a faster rate than in the as-coated samples because of the mixture of recrystallized and unrecrystallized grains that made up the coating after temper rolling; this finer-grained structure provided more grain boundaries for metallic Al nucleation and growth, together with a high dislocation density. Al enrichment was observed in both the surface areas that contracted the textured features of the temper roll and on grain boundaries. Fig. 11.5 shows that Al surface enrichment increased with the square root of time, suggesting diffusion-driven growth.

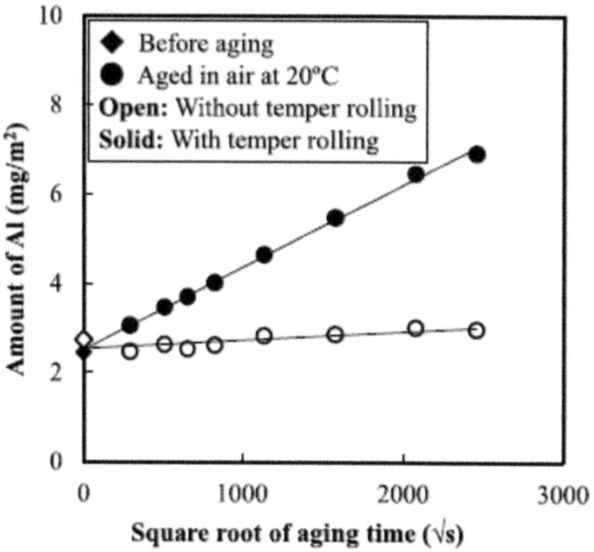

FIG. 11.5 Aluminum present on the surface of a 0.2% GI coating surface, with and without temper rolling as function of square root of aging time in room temperature air [29].

The amount of surface Al has been found to increase with coating Al content, ranging from 5 to 8 mg/m^2 to 15–20 mg/m^2 for thin gauge material with coating overlay compositions of 0.15% and 0.29%Al, respectively [30].

The reaction products formed on the surface of galvanized coatings during production, transport, and storage set the stage for the initiation of corrosion during actual use. The range of Al coating weights given in the previous paragraph, 5–20 mg/m^2 per side, correspond to surface layer thicknesses of 1–4 nm. This very thin layer can initially corrode by pitting or dissolution, depending upon levels of wetness and surface contamination. Depending upon the environment and local coating chemistry, portions of this initial layer may remain, or other compounds may be formed as corrosion proceeds. Very early stage corrosion has been related to zinc coating and microstructural details using several techniques. Corrosion in the early stages of exposure was investigated by depositing NaCl solutions on test surfaces to give NaCl coatings weights of 0.01, 0.1, and 1 g/m^2 [6]. Samples in this work comprised hot-dip Zn coated sheets with containing 0.2, 5, and 55%Al together with pure Zn and Al sheet references. Samples were diamond polished at a 0.5° tilt to the surface. After NaCl deposition, samples were exposed to saturated water vapor at 50°C for 16 h, then dried in air at 60°C for 4 h, then stored in vacuum until they were analyzed using photoemission spectroscopy using synchrotron radiation (SR-PES) and Fourier transform infrared (FT-IR) spectroscopy. The composition of the corrosion products at a depth of 1 nm below the surface film varied with the strength of the NaCl solution. The schematic in Fig. 11.6 shows that for samples exposed to 0.01 g/m^2 NaCl, the corrosion products at a depth of 1 nm on Zn-0.2%Al consisted mainly of zinc oxide together with some zinc chloride, while those on Zn-5%Al consisted of aluminum oxide or aluminum chloride. Deeper in the corrosion product layer, down to at least 100 nm, the composition was observed to be mainly basic zinc carbonate {$2ZnCO_3 \cdot Zn(OH)_2$} with some basic zinc chloride {$ZnCl_2 \cdot 4Zn(OH)_2$} at deeper levels, with more basic zinc chloride present in the Zn-5%Al coating. For samples exposed to 1 g/m^2 NaCl, the Zn-0.2%Al had a layer composition 1 nm below the surface that was mainly zinc chloride with some zinc oxide, while the Zn-5%Al 1 nm depth consisted mainly of aluminum chloride together with some aluminum oxide. For the top 100 nm of layer thickness after the 1% NaCl exposure, the main corrosion product for the Zn-0.2%Al coating continued to be basic zinc carbonate with a presence of some basic zinc chloride {$ZnCl_2 \cdot 4Zn(OH)_2$}. For this NaCl concentration, the top 100 nm thickness of Zn-5%Al was mainly basic zinc chloride with a lesser amount of basic zinc carbonate. Aluminum oxide and chloride possess high electrical resistivity, whereas zinc oxide in this situation has been reported to be an n-type semiconductor [31]. Therefore, one would expect the higher Al coating to have a higher resistivity to current flow. It was indicated that from the results of the experimental results of the study that the corrosion product structure for the Zn-55%Al should be the same as Zn-5%Al. The high corrosion resistance of both Zn-5%Al and Zn-55%Al would then be attributed to the formation of aluminum chloride or aluminum oxide at the surface, which restricts the reduction of oxygen owing to their low electric conductivities, therefore slowing anodic dissolution of these two coatings relative to Zn-0.2%Al. Raman spectroscopy of a Zn-12%Ni electrogalvanized exposed to 3%NaCl droplets showed that the surface NiO coating delayed corrosion product formation, but the principal corrosion eventually formed was basic zinc chloride [32].

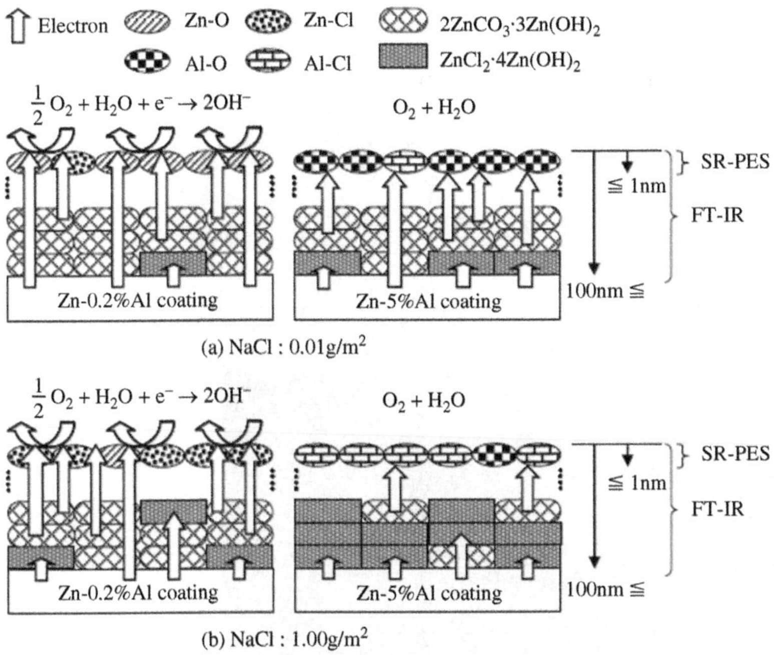

FIG. 11.6 Corrosion products at 1 nm and up to 100 nm from the surface for Zn-0.2%Al and 5%Al coating in the early stages of corrosion by 0.01 and 1.00 g/m^2 NaCl solutions [6].

Time lapse microscopy has also been used to observe early stage corrosion. In this technique, the corrosive solution is placed in a small glass or plastic container; the lens of the microscope is protected from immersion in this solution, the rest of the surface is covered to prevent evaporation. The test sample with a polished surface is exposed with or without connection to a polarization voltage and the surface observed as corrosion proceeds. A coated steel containing between 1% and 2% of both Al and Mg in the 10 μm coating was exposed to a 0.1%NaCl solution and its surface examined without polarization in this manner [33]. The initial microstructure, as described in Chapter 6, consists of primary zinc-rich dendrites surrounded by binary and ternary eutectics. Between 10 and 80 min, corrosion initiated and progressed at the Zn-$MgZn_2$ constituent of the eutectic lamellae, leading to dealloying of both the binary and ternary Zn-Al-$MgZn_2$ eutectic. The appearance of the microstructure before exposure to the corrosive solution and after initial corrosion is shown in Fig. 11.7.A–D. In Fig. 11.7.B, dealloying of the binary eutectic is observed; after this reaction is locally exhausted, attack of the zinc-rich dendrites proceeds. During the corrosion of the eutectic phases, they become local anodes in the microstructure. Chloride ions build up around the anodes together with development of a pH gradient as $(OH)^-$ ion concentration increases at local cathodes; this was confirmed by adding 1% phenolphthalein to show the pH gradient related to the anode and cathode. Fig. 11.8A and B show low and high magnification images of the nature of corrosive attack, including the preferential attack on the eutectic phases in Fig. 11.8A and the ring of chloride ions in Fig. 11.8B.

A similar experiment was performed using a binary Zn-4.8%Al coating with a 20 μm coating thickness [34]. As in the ZnAlMg alloy work described before, the top 2 μm of coating thickness was removed; in this case the removed layer was a smaller proportion of the initial coating thickness. The 1% NaCl electrolyte was 10 times stronger for this experiment compared with the above ZnAlMg work. As in the case of the ZnAlMg alloy, corrosion occurred first in the eutectic phase that for this near-eutectic composition comprised a much higher proportion of the microstructure than the primary zinc phase whose dendrite arms were much more rounded compared with the ZnAlMg samples. As corrosion proceeded across the eutectic portion of the coating surface, the primary zinc phase was left intact. The portions of the coating surface that were not being attacked were observed to brighten in appearance as the anodic portions darkened; the brightening of the cathodic portions was related to the electrochemical potential change or removal of surface oxides. The pattern of corrosion attack was related to the surface oxides present at the initiation of exposure. Aluminum oxide formed a thin layer on the surface of the eutectic phase exposed to the surface. Zinc oxide was present on the primary zinc-rich phase that penetrated the eutectic matrix to the surface. As indicated before, these oxides have very different electrical resistivities, so that there is high electron mobility in the zinc-rich primary phase in comparison to the aluminum-rich eutectic phase. Electron mobility is essential for the cathodic reaction to occur, making the zinc oxide the likely site for oxygen reduction and the eutectic phase the site for metal dissolution, consistent with the ZnAlMg results described before.

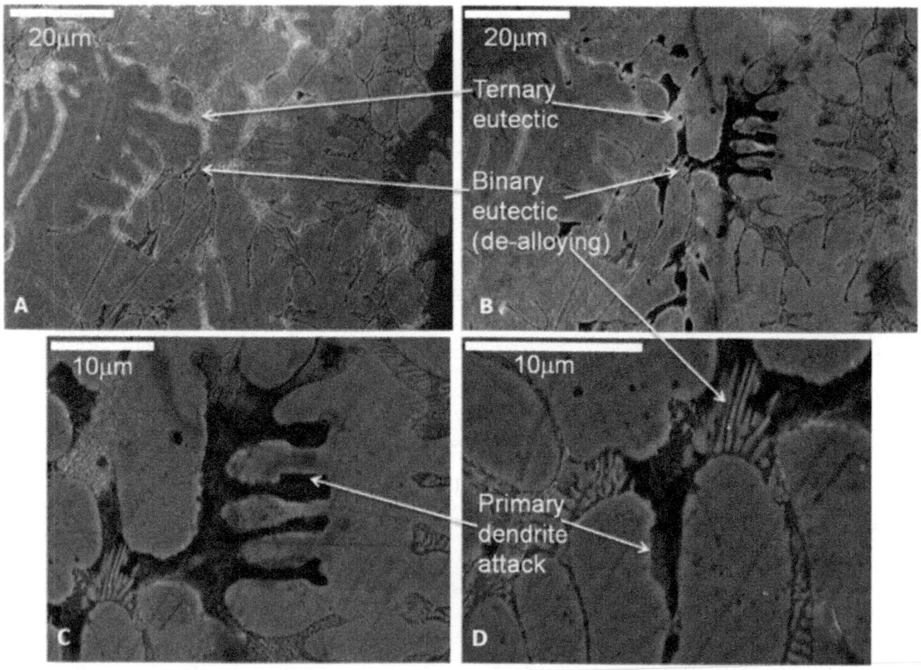

FIG. 11.7 (A) Polished surface of Zn-Al-Mg coating before corrosion test. (B) Preferential attack on eutectic phases dealloying and dissolution of lamella by 0.1%NaCl solution. (C, D) Attack of zinc-rich dendrites after eutectic corrosion has progressed [33].

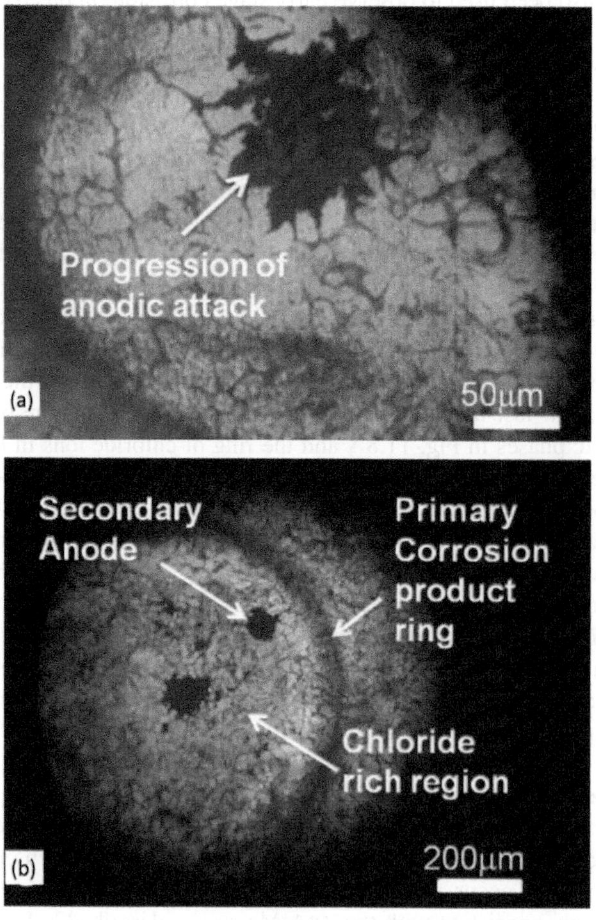

FIG. 11.8 Development of (A). local anodes within the eutectic phase and (B). local chloride corrosion product rings in a Zn coating containing between 1% and 2% of both Al and Mg [33].

A similar experimental arrangement was used to study the initial corrosion behavior of the Zn-55%Al alloy [35]. This utilized a driving voltage applied in an in-situ galvanostatic anodic polarization test. The corrosion cell was set up under the microscope using naturally aerated 0.1 M NaCl at 25°C. An anodic current density of 150 A/m^2 was applied. With this arrangement, anodic dissolution and corrosion were concentrated in the Zn-rich regions during anodic polarization. The Zn-rich phases were observed to corrode at lower (more negative) potentials than the dissolution potential region of the Al-rich matrix phases.

The protective ability of the surface oxide film in initial stages of atmospheric corrosion can also be assessed by a droplet spreading test. For this test, a 25-μL droplet of water solution of 0.5 M NaCl is applied to preheated (40°C) samples of the surface to be tested. Because of its higher temperature relative to ambient air, the drop dries rapidly, leaving a surface deposit. The specimen with its surface deposit were exposed to humid air at 20°C and at 95% relative humidity for 24 h. The droplet area was measured with an optical microscope before and after the exposure. The difference between the final and initial area of NaCl deposits, the spreading, is an indication of the corrosion stability of the metal surface [36]. In atmospheric conditions, it is usually the cathodic reaction of oxygen reduction that prevails around the original droplet [37]. The larger the area, the lower the corrosion stability in atmospheric conditions. Table 11.2 gives the details of the samples tested and the magnitude of droplet spreading after 24 h. Low rates of spreading are observed for any of the Mg-containing coatings and Zn-55%Al. This is attributed to the effectiveness of the Mg component of the surface oxide in slowing oxygen reduction; the higher Al content of the Zn-55%Al coating would be expected to give a thicker later of aluminum oxide, also improving the surface electrical resistivity of the surface layer in comparison to the lower Al content Zn-Al alloys that did not contain Mg [38].

The relative protectiveness of the surface oxide film can also be assessed with a scanning Kelvin probe (SKP). Zn-0.2% Al and Zn-2.5%Al-1.5%Mg coatings were scanned after corrosion was initiated by depositing NaCl on the surface of the samples: 2 μL of a NaCl solution 0.06 M in methanol was applied with a micropipette and spread over 2 cm^2 to obtain a layer of NaCl crystals with density close to 10 μg/cm^2. The SKP was used to determine the potential distribution in humid air (65% RH) at the NaCl contaminated/noncontaminated areas boundary. Under the NaCl surface deposit, the attack on and deterioration of the surface oxides and consequent activation of the metallic coating surface led to a decrease in potential. The reference electrode was made of Ni-Cr alloy. A surface contour (topography) measurement was made simultaneously with potential measurements. The air gap between the probe and the sample was kept 50 μm. The lateral resolution of the probe was in the range of 100 μm, and the scanned area was 3 × 3 mm. The distributions of potential for the two coating surfaces are shown in Fig. 11.9. Both the maximum and minimum potentials for the ZnAlMg are more negative than the Zn-0.2%Al coating, Table 11.3. The difference between the maximum and minimum voltages of the ZnAlMg coating is also less than the Zn-0.2%Al coating, so that there is less driving force for corrosion between the anodic and cathodic portions of the coating surface. Both of these behaviors are related to the higher electrical resistivity of the ZnAlMg oxide film [39].

TABLE 11.2 Droplet spreading test—sample characteristics and droplet spreading behavior [38].

Coating	Coating weight sum of both sides (g/m^2)	Equiv. one-side coating thickness (μm)	Steel thickness (mm)	Droplet spreading, % after 24 h[b]
Zn-0.2Al	270	20	0.5	176±6
Zn-5Al-MM	130	10	0.95	68±9
Zn-1.5Al-1.5Mg	185	14	[a]	10±1
Zn-2.4Al-2.8Mg	140	11	[a]	4±1
Zn-11Al-3Mg	100	9	[a]	21±2
Zn-55Al-1.5Si	150	20	0.55	5±3

[a]*Not known, assumed to be construction grade steel.*
[b]*Average of three samples.*

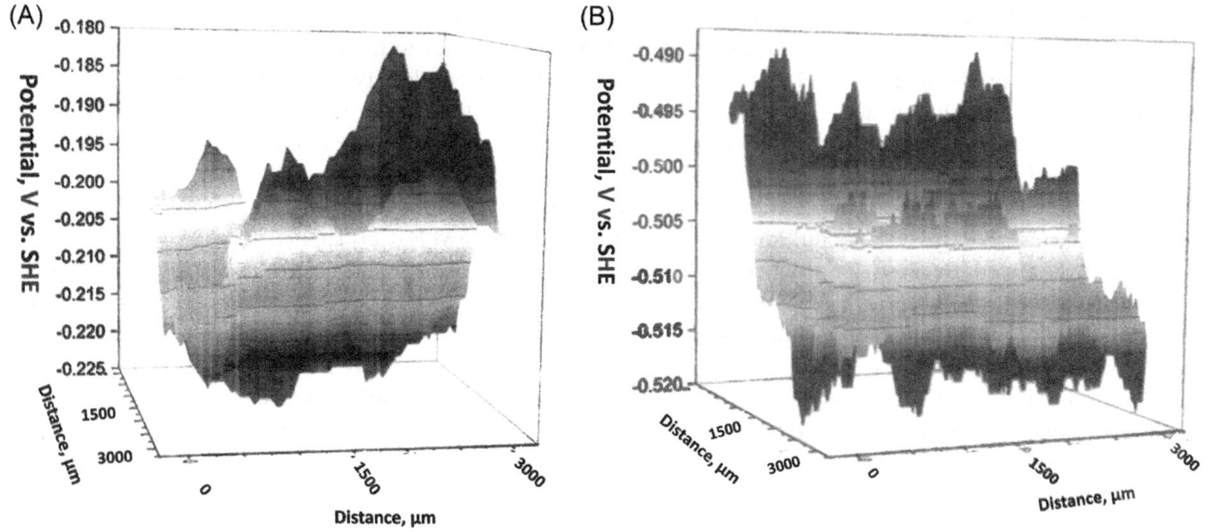

FIG. 11.9 Distribution of potentials measured by scanning Kelvin probe in 65% relative humidity under a NaCl film solution for (A) Zn-0.2%Al and (B) Zn-2.5%Al-1.5%Mg coatings [39].

TABLE 11.3 Maximum and minimum potentials for scans in Fig. 11.9, 65% relative humidity [39].

Coating	Minimum potential (V)	Maximum potential (V)	Potential difference (mV)
Zn-0.2Al	−0.22	−0.18	40
Zn-2.5Al-1.5Mg	−0.52	−0.49	30

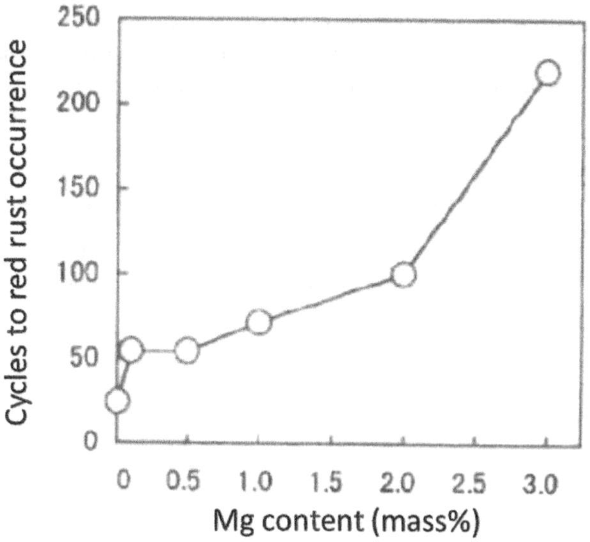

FIG. 11.10 Effect of Mg content in the Zn-6%Al coating on the number of cycles of the JASO M609 cyclic corrosion test before red rust is observed [40].

Early work on the Zn-6%Al-Mg coating used a range of Mg compositions between 0% and 3% to determine their effect on corrosion performance. Fig. 11.10 shows the time to appearance of red rust using the cyclic corrosion test JASO M609-91 of the Society of Automotive Engineers of Japan, where a cycle includes 2h of salt spraying at 35°C, 4h of drying at 60°C, and 30% relative humidity, concluding with 2h exposure to 95% humidity at 50°C [41]. An improvement with even small Mg additions is observed, with a rapid increase occurring at about 2%Mg.

TABLE 11.4 Variation in SKP potential from inside the NaCl droplet contaminated area to outside when the surrounding atmosphere is changed from air to nitrogen, constant 95% relative humidity [39].

	Potential variation (V)	
Coating	Inside the NaCl contaminated area	Outside the NaCl contaminated area
Zn-0.2Al	0.4	0.14
Zn-2.5Al-1.5Mg	0.13	0.11

An additional insight can be gained by changing the atmosphere surrounding the SKP from air to nitrogen. When this was done, maintaining a 95% relative humidity, the probe showed that the Zn2.5Al1.5 Mg alloy showed a smaller change in potential outside or inside of the sodium chloride deposited droplet, indicating the best stability, Table 11.4. This behavior is related to the more rapid formation of protective surface oxides on the ZnAlMg coating as corrosion continues [39]. These measurements agree with observations made by Matsumoto, who observed that additions of Al to a Zn coating decreased the cathodic reaction rate on the coating, namely the reduction of oxygen, and that Zn-5%Al and Zn-55%Al are more resistant to corrosion than Zn-0.2%Al [42].

11.3 Corrosion mechanisms and coating microstructure

The progress of atmospheric corrosion of the surface of zinc-coated steel from its initial stages was described by Zhang [43]. At first, the corrosion products are loosely attached to the surface, but as the surface experiences cyclic wetting and drying during weathering, these products gradually become more adherent and denser. After the formation of the dense corrosion product layer, further corrosion can proceed only via locations on the surface where the zinc surface is not sealed by the corrosion products, while the rest of the surface area, which is sealed by the corrosion products, is protected from corrosion. This is a dynamic process that forms local anodes and cathodes. With time, some pores become sealed by newly formed corrosion products while some pores are opened due to dissolution of the corrosion products.

Yoshizaki [44] confirmed Zhang's observations for pure Zn and extended them to hot-dip Zn-0.2%Al, Zn-4.5%Al-0.1% Mg, and Zn-6%A1-3%Mg coatings. Samples were exposed for 5 years at a severe marine (Okinawa, 30 m from the sea) and rural (Kiryuu, 100 km from the sea) sites. Coating cross sections, including the appearance of corrosion products, are shown in Fig. 11.11, and the surface appearance of the corrosion products are shown in Fig. 11.12. Fig. 11.11 shows that in the Okinawa exposures, there are locations where the Zn-0.2%Al and Zn-4.5%Al-0.1%Mg coatings have corroded through to the steel substrate, whereas the corrosion depth of the Zn-6%A1-3%Mg coating is 5 μm or less. Corrosion at the rural site was less severe, but there are still locations where the Zn-0.2%Al and Zn-4.5%Al-0.1%Mg coatings have corroded to, or nearly to the steel substrate, while the corrosion depth of the Zn-6%A1-3%Mg coating is 1 μm or less.

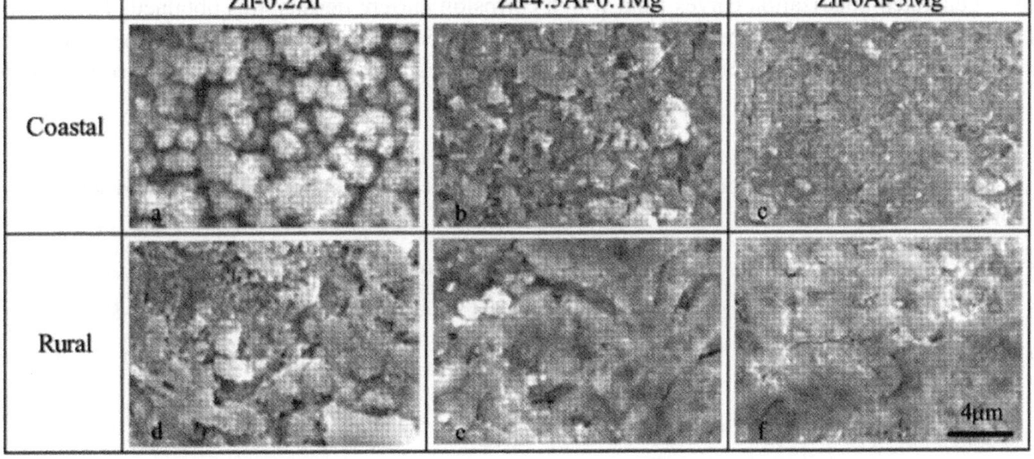

FIG. 11.11 Cross sections of coated steel samples exposed for 5 years at Okinawa severe marine site and Kiryuu rural site [44].

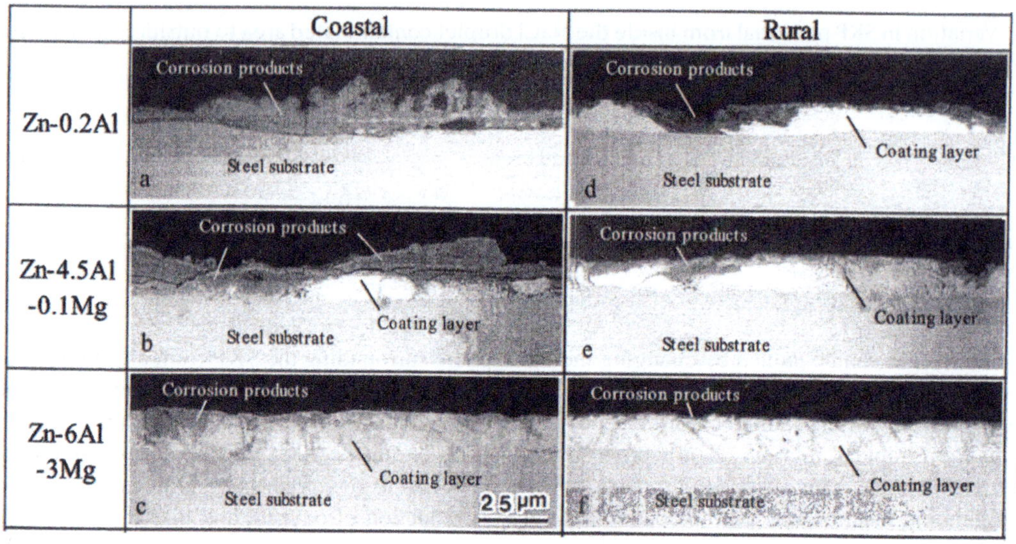

FIG. 11.12 Appearance of corrosion products on surfaces of Fig. 11.11 samples [44].

After 5 years exposure at both the Okinawa and Kiryuu sites, Fig. 11.11 shows that the entire surface of the Zn-0.2%Al coating is covered with coarse and porous corrosion products, consistent with Zhang [43]. On the Zn-4.5%Al-0.1%Mg surface, granular corrosion products are observed on the Okinawa surface while a mixture of granular and fine corrosion products is seen on the Kiryuu surface. In both environments, the surface of the Zn-6Al-3Mg coating is covered with fine and smooth corrosion products. The Okinawa exposure confirmed the NaCl droplet studies described in the previous section; the observed corrosion products on the Zn-0.2%Al surface were basic zinc carbonate and, as expected with the longer exposure time compared with the laboratory droplet studies, basic zinc chloride. In the rural environment, the Zn-0.2%Al corrosion products were basic zinc carbonate together with a small amount of zinc oxide. For the Zn-4.5% Al-0.1%Mg and Zn-6%A1-3%Mg coatings, the Okinawa samples were mostly covered with basic zinc aluminum carbonate ($5Zn(OH)_2{\cdot}2Al(OH)_3{\cdot}ZnCO_3{\cdot}4H_2O$); additionally minor amounts of zinc oxide and basic zinc carbonate were detected on the Zn-4.5%Al-0.1%Mg, explaining the minor differences in surface appearance between these two coatings in Fig. 11.11. The Zn-4.5%Al-0.1%Mg surface exposed at the rural site was covered with basic zinc carbonate, similar to the Zn-0.2%Al site. Similar to the Okinawa site, the rural Zn-6%A1-3%Mg surface was covered with basic zinc aluminum carbonate. In contrast with ZnO, which is bulky and porous, the other corrosion products observed are not only more compact but are more electrically insulating. Therefore, they are better at inhibiting the oxygen reduction reaction on cathodic surfaces, regardless of the atmospheric corrosion environment. The protectiveness of these mixed oxide layers was assessed by making potentiodynamic polarization measurements before and after the 5-year exposures at the two sites. An artificial rainwater solution was used containing both chloride and sulfate ions at a pH of 5.8. A potential sweep of 1 mV/s at 25°C generated polarization curves from which corrosion current density was obtained. The results are shown in Table 11.5. Higher corrosion rates are measured on the samples before exposure because of the absence of corrosion products, corresponding to a higher chemical activity of the surfaces. The relative performance of the samples in each

TABLE 11.5 Corrosion current densities of samples before and after 5-year exposure to rural and severe marine environments [44].

	Current density (mA/m^2)		
Condition	Zn-0.2%Al	Zn-4.5%Al-0.1%Mg	Zn-6%Al-3%Mg
Before exposure	80	70	100
5y rural exposure	12	7	3.5
5y marine exposure	30	24	11

environment was related to the amount of the more porous and crystalline ZnO product; corrosion performance improved as the corrosion products became more amorphous in character rather than crystalline. Al and Mg, but especially Mg, were observed to suppress the transformation from amorphous to crystalline corrosion product forms and suppress the growth of these crystals when they did form. This influence on corrosion product morphology, together with the higher electrical resistivity of compounds that suppress the oxidation reduction reaction at the cathode, improves the corrosion resistance of the alloy coatings.

For ZnAlMg alloys, the corrosion products formed are also influenced by microstructural fineness, which is in turn influenced at a fixed composition by cooling rate. After 24 months of exposure in the Brest, France marine atmosphere, six corrosion products were identified during study of a reference Zn-0.36%Al (GI) coating together with coatings containing between 2.7% and 2.9%Al and 1.5%–1.6%Mg [7]. The cooling rate of the ZnAlMg coatings was varied between 5 and 1000°C/s to give a range of primary Zn grain sizes of 2.5 and 81 μm; a typical industrial coating produced with a cooling rate of 30°C/s and a grain size of 30 μm was also included. The GI coating thickness was 20.7 μm while the ZnAlMg coating weights ranged from 10.4 to 11 μm. The corrosion products found on each of the coatings are shown in Table 11.6. After the 24-month exposure, the GI coating showed 11.7 g/m^2 of coating mass loss, about double the 1-year loss so that little additional protection from buildup of protective corrosion product layers occurred. In contrast, several of the ZnAlMg coatings showed greatly slowed corrosion rates between the first and second year of exposure. Consistent with Fig. 11.6, the principal corrosion product for the ZnAl coating was the basic carbonate product termed hydrozincite in Table 11.6, this was present in a much larger quantity than the ZnAlMg corrosion products because of its higher corrosion rate. Gordaite was found on the two thickest ZnAlMg coatings after 2 years of exposure, but the corrosion product common to all of the ZnAlMg coatings was the layer double hydroxide. This compound was most dominant on the ZnAlMg coating with highest amount of eutectic near the surface, the 20 μm thick Zn-2.7Al-1.5Mg coating. This rapidly formed the layered double hydroxide coating on its surface which limited the formation of hydrozincite, resulting in a more proactive corrosion product layer and consequently the lowest corrosion rate after both 1 and 2 years. The relationship between the percent of the coating surface area that was attacked after 2 years of marine exposure and fraction eutectic is shown in Fig. 11.13. Because the eutectic is more electronegative than the primary zinc phase, it corrodes first, releasing Mg that then reacts to form the double hydroxide layer described in Table 11.6.

The corrosion mechanism for ZnAlMg coatings is shown in Fig. 11.14. This shows the preferential attack of the surface eutectic phases Zn-$MgZn_2$ and Zn-$MgZn_2$-Al. The cathodic primary zinc phase provides sites for the oxygen reduction

TABLE 11.6 Corrosion products observed after 2 years of marine exposure on ZnAl and ZnAlMg coatings with different primary zinc grain sizes [7].

		Coating type and grain size (μm)				
Name	Chemical formula	Zn-0.36% Al	Zn-2.9Al-1.6Mg	Zn-2.7Al-1.5Mg	Zn-2.7Al-1.5Mg	Zn-2.7Al-1.5Mg
		81	50	85	20	2.5
Hydrozincite	$Zn_5(OH)_6(CO_3)_2$	•	•	•	•	•
Simonkolleite	$Zn_5(OH)_8Cl_2 \cdot H_2O$	•				
Gordaite	$NaZn_4(SO_4)(OH)6Cl \cdot 6H_2O$	•	•	•		
Zinc Hydroxy Sulfate	$Zn_4(OH)_6\,SO_4 \cdot nH_2O$, $n=3, 5$	•				
Zincite	ZnO					
Layered Double Hydroxide	$M(II)_xM(III)_y(A)_m(OH)_n \cdot zH_2O$ where $M(II)=Zn^{2+}$, (Mg^{2+}) $M(III)=Al^{3+}$ $A=CO_3^{2-}$, Cl^-, SO_4^{2-} $Zn_2Al(OH)_6(CO_3)_{1/2} \cdot zH_2O$		•	•	•	•

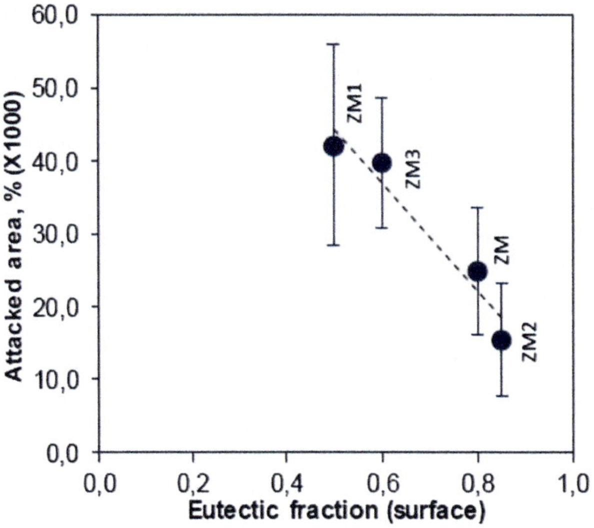

FIG. 11.13 Relationship between surface eutectic fraction and percent of surface area attacked by corrosion after 2-year marine exposure [7].

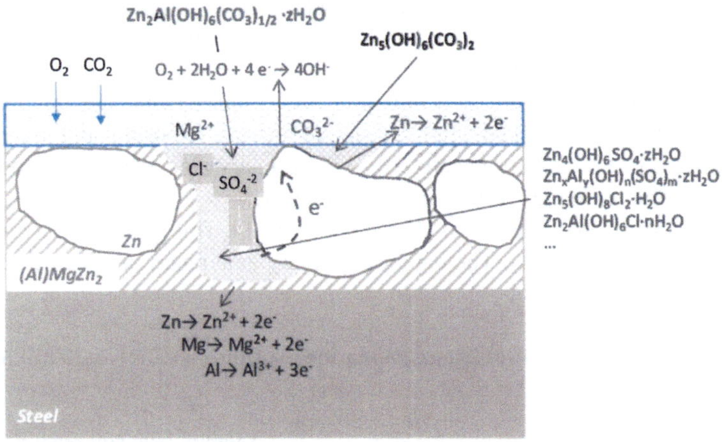

FIG. 11.14 Schematic of the mechanisms of corrosion and buildup of corrosion products on a Zn-2.7Al-1.5Mg coating. Corrosion compounds are further described in Table 11.6 [7].

reaction, resulting in zinc dissolution that produces hydrozincite, contributing to corrosion mass loss. The ternary eutectic has been observed to experience a greater level of attack on initial exposure than the binary eutectic. Two reasons for this have been postulated: First, it could be related to the initial dissolution of Al in the ternary microstructure, lowering the local pH as hydrolysis proceeds and thereby increasing the corrosivity of the local environment. Second, the larger-scale structure of the binary eutectic can maintain its integrity for a longer time than the fine structured ternary eutectic as corrosion proceeds. In both the binary and ternary eutectics, the more anodic $MgZn_2$ phase is attacked first, resulting in de-alloying of the eutectics as corrosion commences. The finer-scale Zn lamellae and Al nodules remaining in the ternary eutectic structure after $MgZn_2$ phase is attacked will be more rapidly dissolved than the coarser Zn lamellae in the binary eutectic and thus the binary eutectic structure. Therefore, the binary eutectic remains visible for longer even though the $MgZn_2$ lamellae have been de-alloyed [33]. Recent work suggests that for a given corrosive environment, there may be a critical zinc crystal size, below which the local electrolyte (anolyte containing chloride ions) can reach a critical concentration on both sides of the crystal, resulting in propagation of corrosion across the surface of the crystal, resulting in more uniform corrosion. At higher crystal sizes, the local anolyte drops below the critical level for oxygen reduction, resulting in corrosion around, rather than across the dendrite, resulting in nonuniform corrosion [45].

Another industrial coating with a composition of Zn-3.5%Al-3%Mg was analyzed after exposures in 5% neutral salt spray cyclic corrosion tests, both involving chloride corrosion for various periods of time [46]. Corrosion product analysis confirmed that for the reference Zn-0.2%Al coating the corrosion products were ZnO, zinc carbonate, and hydrozincite, as reported in the above studies. On the Zn-3.5%Al-3%Mg corroded surface, a mixture of simonkolleite ($Zn_5(OH)_8Cl_2{\cdot}H_2O$)

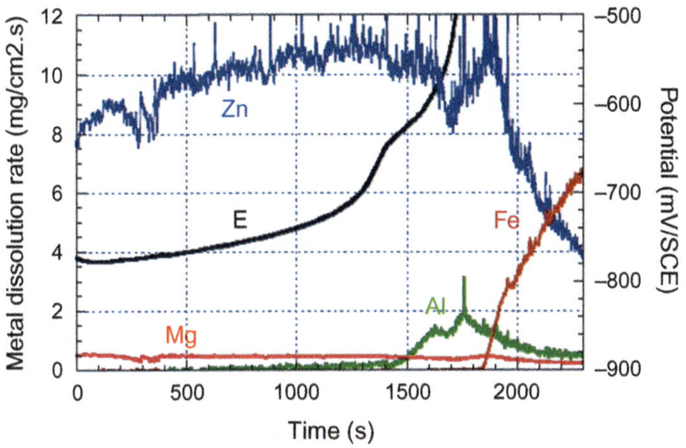

FIG. 11.15 Selective dissolution rates of Zn, Mg, Al, and Fe together with increase of potential (E) during dissolution of the Zn-3.5%Al-3%Mg coating at a current density of 4 mA/cm^2 using the in-situ atomic emission spectroelectrochemical technique [46].

layered double hydroxides that included AlMg and/or AlZn was observed; as before, these double layer hydroxides suppressed the appearance of hydrozincite. Again, the layered double hydroxides were highly protective, this was attributed to their compact morphology that inhibited ion transfer and thus created a strong barrier for hydrogen transport toward the underlying coating. From thermodynamic modeling of the corrosion process, it was suggested that simonkolleite is stabilized in the presence of Mg ions released from the corroding Mg-rich part of the eutectics by the reaction of these ions with carbonate ions ($CO_3{}^{-2}$), preventing the transformation of the crystalline form of simonkolleite into smithsonite (zinc carbonate), $ZnCO_3$, amorphous hydrozincite, and zincite (ZnO) during wet-dry exposure cycles in the presence of CO_2. In the early stages of corrosion, the metallic Al, primarily associated with the ternary eutectic as a discrete phase, is not dissolved; there is an absence of Al-rich corrosion products on the surface of the corroded samples. Further evidence for this mechanism comes from in-situ atomic emission spectroelectrochemical experiments (AESEC) that revealed a strong preferential dissolution of Zn and Mg in the early stages of exposure to the electrolyte, Fig. 11.15.

The AESEC experiments showed that when the electrochemical potential or the surface pH in the 3.5%Al-3%Mg coating exposed to a NaCl solution increases sufficiently for Al dissolution to occur, the presence of Al ions in the solution inhibits corrosion by two different mechanisms: first, by buffering the pH of the electrolyte because of precipitation of $Al(OH)_3$ that prevents the dissolution of zinc based corrosion products into soluble hydroxide complexes and promoting the formation of simonkolleite in place of zincite; second, by promoting the formation of double based metallic salts (layered double hydroxides, LDH), which as described before are quite protective against further corrosion of the underlying metallic coating. The barrier properties of the LDH corrosion products were observed to be similar to simonkolleite including their plate-like morphologies, low electron densities, and layered structures with difficult ion transfer through the layer. These corrosion products therefore inhibit oxygen transport toward the metallic surface, suppressing the oxygen reduction reaction of cathodic regions. The relative electronegativities of several coatings were evaluated by coating steel bolts and exposing them to a 5%NaCl solution. The Zn-3.7%Al-3%Mg coating had a corrosion potential more negative than −1050 mV referenced to an Ag/AgCl/KCl saturated electrode. All other zinc coatings including those on general galvanized, zinc mechanical plated, zinc electroplated bolts were more positive than −1000 mV and therefore provided less driving force for the galvanic reaction [47].

Except for the rural Japanese site (Kiryuu), corrosion conditions described to this point have all included chlorides. Indeed, chlorides dominate corrosion phenomena in marine environments and in locations where deicing salts are used, i.e., many transportation environments. However, sulfur dioxide is the dominant pollutant in many industrial areas, which can coincide with many urban areas where there is intensive use of galvanized steel in construction and automobile applications. Therefore, even though sulfur dioxide emission has been greatly reduced since the 1970s in Europe and North America, they still can play an important role in corrosion. Pure Zn and Zn-2%Al-2%Mg coated steel samples were exposed to an air regulated at 82% relative humidity containing 80 ppb SO_2 and 350 ppm CO_2 (below the average atmospheric content of 400 ppm CO_2) [48]. Like the chloride exposures, the bulk of the ZnAlMg coating first corroded by dissolution of the anodic $MgZn_2$ phase, while the Zn lamellae in the eutectics served as cathodes, the site of the oxygen reduction reaction, with some cathodic activity also occurring on the interfaces between the primary Zn phase and the eutectics. The progress of corrosion is shown in Fig. 11.16, where the microstructures before and after 16 h exposure are shown. The corrosion products on the corroded sample were stripped off with a saturated glycine solution. Analyzing the corrosion products, no products were detected by X-ray diffraction and other analysis tools, the corrosion product was determined to be a noncrystalline (amorphous) sulfate salt that for shorter exposure durations contained Mg; at more prolonged exposures

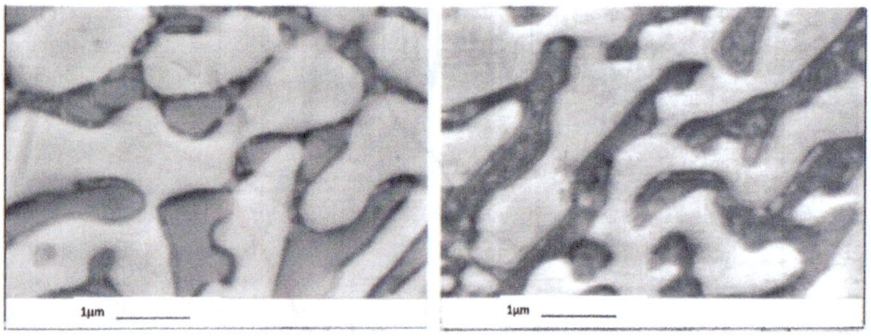

FIG. 11.16 SEM micrographs of pickled (in saturated glycine) surfaces of (left) unexposed $ZnAl_2Mg_2$ coated steel and (right) $ZnAl_2Mg_2$ coated steel exposed 16h to humid air at 82% relative humidity and 80ppb SO_2 [48].

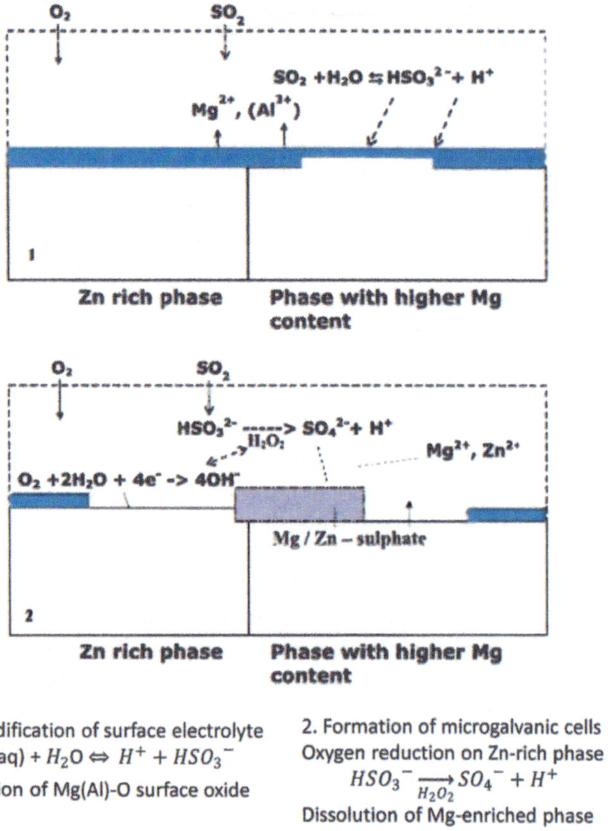

FIG. 11.17 Schematic illustration of the mechanisms of initial atmospheric corrosion for a Zn-2%Al-2%Mg coating in humid air containing an 80ppb concentration of SO_2 [48].

it also includes some Zn. There was a longer delay time until corrosion of the bulk of the coating began; during this time sulfur dioxide deposits on the oxidized surface of the ZnAlMg coating react with water to form bisulfite (HSO_3^-) ions and H^+, Fig. 11.17. As the hydrogen ion concentration increases, the pH is lowered, acidifying the surface and leading to dissolution of the surface oxide. Corrosion of the bulk of the ZnAlMg coating begins more quickly when chloride ions are added to the solution because of the more rapid attack of the surface oxide layer on the coating. When Cl contamination is present, a surface layer of gordaite (Table 11.6) is produced, whereas when only sulfur-base corrosion occurs, a magnesium-containing sulfate is produced.

A similar condensing humidity test conducted both with constant humidity and alternating with periods of drying in accordance with ISO 6270-2 [49] was conducted, during which no other corrosive agents such as chlorides or sulfates were added [50]. A passive layer of $Zn_6Al_2(CO_3)(OH)_{16}{\cdot}4H_2O$ was formed, this is in the family of layered double hydroxides described in Table 11.6 with the presence of only Zn and Al cations; apparently this environment was insufficiently aggressive to attack and then mobilize Mg ions to participate in the corrosion reactions, although, per Fig. 11.17, Mg would also be expected to be present in the initial surface oxide layer. A cross section of this layered double hydroxide after 400h of neutral salt spray testing is shown in Fig. 11.18. In this case, the Mg was mobilized by the presence of chlorides in agreement with above descriptions so that a Mg signal is detected in the EDS spectrum included in this figure.

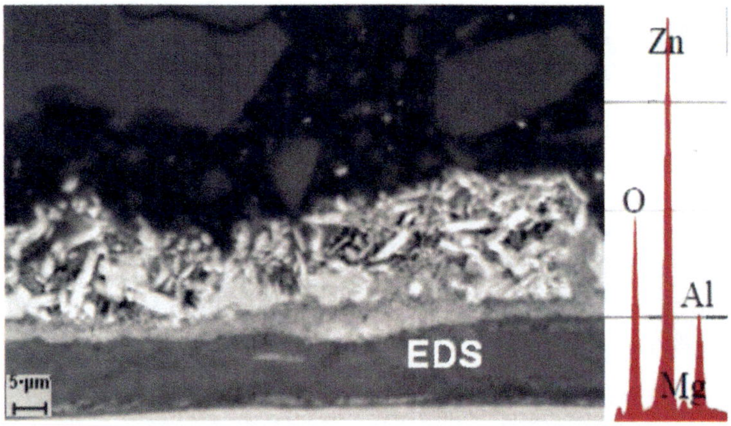

FIG. 11.18 Cross section of a Zn-2%Al-2%Mg coated steel sample after 400h of ISO 9227 neutral salt spray testing test with an Energy Dispersive Spectroscope analysis of the observed double layer hydroxide [50].

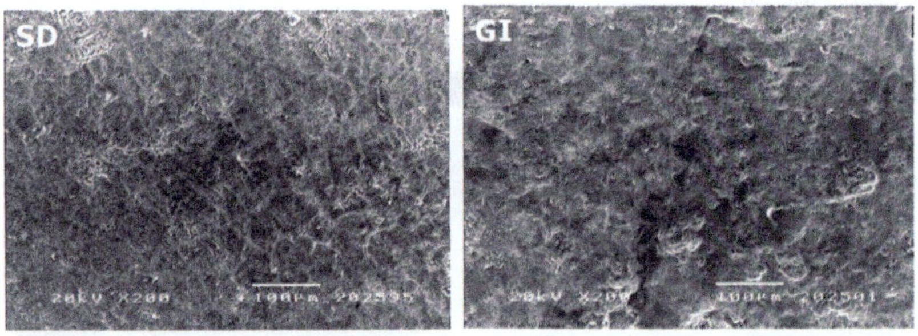

FIG. 11.19 Surface appearance of Zn-11%Al-3%Mg-0.2%Si (left) and Zn-0.2%Al (right) coatings after 5-year outdoor exposure at Sodegaura, Chiba, Japan [9].

The development of corrosion products on a Zn-11%Al-3%Mg-0.2%Si coated steel with a two-side coating weight of 270 g/m^2 was reported by Shimoda [9]. Panels were exposed for 5 years at an urban industrial site 1.6 km from Tokyo Bay. Also exposed were samples of Zn-0.2%Al with the same coating weight and Zn-4.5%Al-0.1%Mg and Zn-55%Al-1.6%Si samples with coating weights of 180 and 150 g/m^2 respectively. The backs and edges of the panels were sealed, and they faced southward at a 30° inclination. The Zn-0.2%Al corrosion products found agreed with those observed in other results described above and indicative of the presence of nearby industry included zinc sulfate hydroxide ($Zn_7(OH)_{12}SO_4{\cdot}H_2O$) with the zinc oxide, hydrozincite, and simonkolleite reported for chloride-containing exposures above. Only the double-layer hydroxide $Zn_6Al_2(OH)_6CO_3{\cdot}4H_2O$) was found on the surface of the Zn-11%Al-3%Mg-0.2%Si coating after 5 years. Photos of the surface appearance of these two coatings are shown in Fig. 11.19. Localized attack, resulting in craters and voids, can be seen on the Zn-0.2%Al coating, while the surface of the Zn-11%Al-3%Mg-0.2%Si coating is relatively smooth. Cross sections of all four tested coatings after 5-year exposure are shown in Fig. 11.20. The Zn-11%Al-3%Mg-0.2%Si and Zn-55%Al-1.6%Si coatings exhibit the greatest thickness. As in the case of the coatings with ZnAlMg contents with Al contents between 1.5% and 3%, the Zn-11%Al-3%Mg-0.2%Si showed comparatively low coating thickness loss. The Zn-55%Al-1.6%Si coating, which had a much lower initial coating weight, showed some attack on the Zn-rich crystals at the surface but still showed a relatively low loss of coating thickness. In both the Zn-11%Al-3%Mg-0.2%Si and Zn-55%Al-1.6%Si coatings, the first phase to solidify is on the aluminum side of the Zn-Al-(Mg) eutectic, further described in Chapter 6 Sections 6.5.1-2; the zinc-rich crystals are constituents of the interdendritic matrix in both coatings and make up a much smaller volume fraction of the entire coating than they do in the other two coatings tested. The Zn-0.2%Al and Zn-4.5%Al-0.1%Mg coatings exhibited more severe attack, with some corrosion progressing down to the steel substrate.

In all of the above coatings, the Al composition was equal to or higher than the Mg composition. Examination of a Zn-3%Mg-2.5%Al coating showed that the fraction of binary eutectic was larger and coarser than either of the Zn-2.7%Al-1.5%Mg or Zn-6%A1-3%Mg described above; a representative micrograph is shown in Fig. 11.21. In the binary eutectic, the Zn-Mg lamellae consisted of Zn and the $MgZn_2$ phases whose composition was found to be 14.6 wt% Mg-1.7 wt% Al [51].

The ternary eutectic consisted of lamellae of Zn and $MgZn_2$ together with Al particles with a diameter smaller than the interlamellar spacing that adhered to the $MgZn_2$ phase. As in the case of the ZnAlMg alloys described before, the structure of the binary lamellae is finer and denser in the ternary eutectic than the binary. Details of these microstructures are shown in Fig. 11.22.

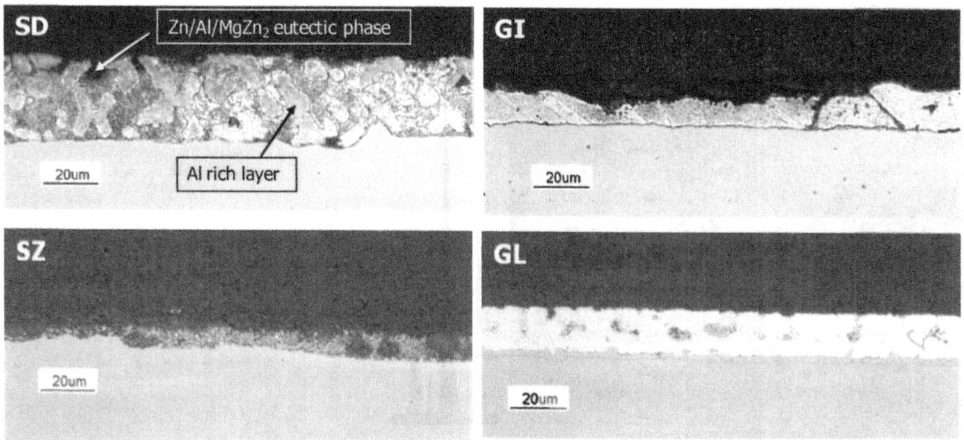

FIG. 11.20 Cross-sectional micrographs of (SD) Zn-11%Al-3%Mg-0.2%Si, (GI) Zn-0.2%Al, (SZ) Zn-4.5%Al-0.1%Mg, and (GL) Zn-55%Al-1.6%Si coatings after 5 years of outdoor exposure at Sodegaura, Chiba, Japan [9].

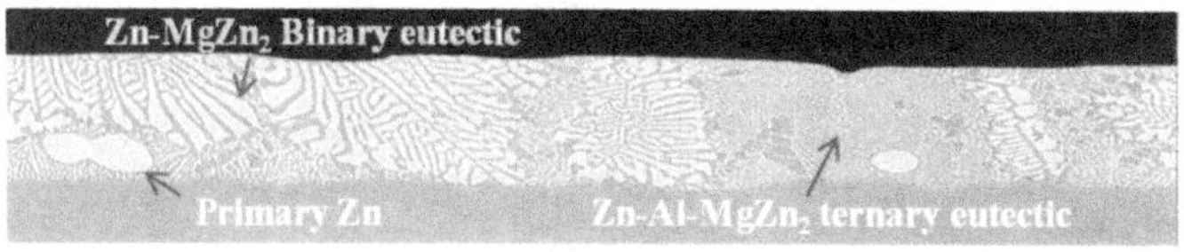

FIG. 11.21 Microstructure of a Zn-3%Mg-2.5%Al coating [51].

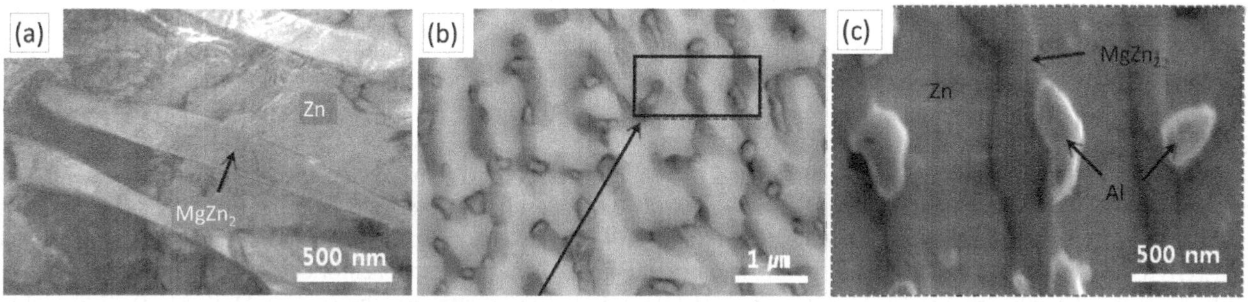

FIG. 11.22 High-resolution transmission electron microscopy images of microstructures in Zn-3%Mg-2.5%Al coated steel sheet. (A) Zn/$MgZn_2$ binary eutectic (B) and (C) Zn/Al/$MgZn_2$ ternary eutectic, magnified area from (B) [51].

After 500 h of neutral salt spray testing, the corrosion products of the Zn-3%Mg-2.5%Al coating, shown in Fig. 11.23, had the same relatively smooth, continuous appearance as the Zn-6%A1-3%Mg coating shown in Fig. 11.12.

To better understand the nature of corrosion behavior in the ZnAlMg eutectics, single crystals of Zn and $MgZn_2$ in several chloride electrolytes were studied. $MgZn_2$ released cations into the electrolyte. The Mg in solution formed surface layers on zinc and iron substrates, decreasing the corrosion of zinc and the formation of red rust on iron, in each case by inhibiting the oxygen reduction reaction (ORR) [52]. Confirming the SKP work on the very early stages of corrosion described in the previous section, potentiodynamic polarization tests were conducted in a 0.5%NaCl solution and showed the open-circuit potential of $MgZn_2$ to be −1.5V (SCE), lower than that of Zn, −1V (SCE). It was found that the $MgZn_2$ phase dissolves first in the mixture of binary and ternary eutectics and primary Zn in the Zn, confirming that Mg ions become available in the electrolyte as the early stage of corrosion proceeds. This early stage $MgZn_2$ corrosion was confirmed by observations of ZnAlMg coating microstructures after up to 72 h of 5% neutral salt spray testing [53].

The addition of 2%Mg to the Zn-55%Al-1.5%Si coating has been observed to significantly improve its corrosion performance [54]. The nature of solidification is little changed, with Al dendrites first formed; however, the final solidification includes both Mg_2Si and $MgZn_2$ compounds rather than the fine Si particles observed in the Mg-free composition, described in Chapter 6 Section 6.4.2, compared with Chapter 6 Fig. 11.30A and B. The corrosion behavior of both Zn-55%Al coatings was characterized and compared using electrochemical impedance spectroscopy in a 3.5%NaCl solution where the test sample was first stabilized for 10 min to give the measured open-circuit potential. Polarization curves gave the results shown in Table 11.7 [55].

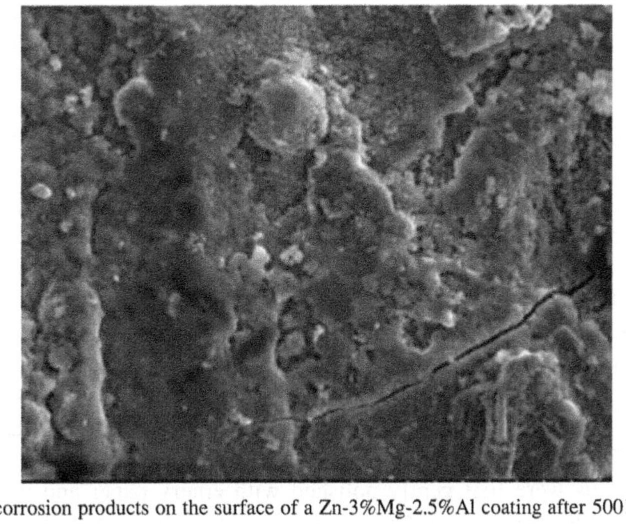

FIG. 11.23 Surface appearance of corrosion products on the surface of a Zn-3%Mg-2.5%Al coating after 500h of 5% neutral salt spray testing [51].

TABLE 11.7 Electrochemical impedance spectroscopy polarization resistance values measured for Zn-55%Al-1.5%Si and Zn-55%Al-1.5%Si-2%Mg coatings.

Self-corrosion potential/current	Zn-55%Al-1.5%Si	Zn-55%Al-1.5%Si-2%Mg
E_{corr} (mV)	−1.02	−1.03
I_{corr} (μA/cm^2)	0.74	0.27

Adapted from Z.-L. Ding, B.-H. Yan, X.-Y. Jin, X.-J. Wang, J. Zhang, S.-M. Jiang, Q.-F. Zhang, Effect of magnesium on the microstructure and corrosion resistance of 55% Al-Zn coating, in: Proc. 12th International Conference on Zinc and Zinc Alloy Coated Steel Sheet (Galvatech 2021), June 22–24, 2021, Virtual Conference, Austrian Society for Metallurgy and Materials, pp. 164–171.

The corrosion current i_{corr} of the 55%Al-Zn coating in 3.5% NaCl aqueous solution is 0.74 μA/cm^2, about three times higher than the 0.27 μA/cm^2 observed for the 55%Al-Zn-Mg coating. Moreover, potentiodynamic polarization testing showed that the impedance of the Zn-55%Al-Mg coating is about a third of the Zn-55%Al coating at low frequencies, less than 1 Hz. Table 11.7 also indicates that the open-circuit potential E_{corr} is about the same for both coatings. The Mg_2Si and $MgZn_2$ compounds are expected to contribute Mg ions to the formation of corrosion products in the same manner as previously described for the Zn-11%Al-3%Mg-0.2%Si coating.

An observation of corrosion mechanisms in a Zn-2%Al-2%Mg coating that produced results rather contrary to the above descriptions was reported by Schuerz [56]. After 5% neutral salt spray testing at 34.5°C, where the pH of the salt solution was regulated to 6.71 by the addition of NaOH, coated steel samples that had been precleaned with customary alkaline treatments showed patchy light and dark areas on the surface after 100h, which remained for the 600h total salt spray test duration. The dark portion was compact and included simonkolleite together with another corrosion product highly enriched in Al; this appeared to overlay a coarse crystal of Zn-rich primary phase that had only begun to corrode. The light portion was highly porous and consisted of products containing Zn, O, and small amounts of Mg overlaying corroding $MgZn_2$, the oxide layer was much thicker in the light portion. The corrosion products were found to be stable, adherent, highly protective oxides dominated by Al, contrary to other observations, Mg did not play a role in the development of these corrosion products. It is speculated that the high volume of electrolyte used in salt spray testing never permitted Mg ion concentrations to rise to levels where they could influence the development of the corrosion products observed in the NaCl film tests described above, either from runoff or dilution effects. Still, the corrosion resistance of this coating was greatly superior to a Zn-0.2%Al coating that underwent the same tests.

An examination of binary Zn-Mg cast alloys, rather than coatings of these alloys on steel substrates, permitted the use of a wide range of cooling rates, with times to solidification between 1 and 8 min for the 1%–3%Mg compositions investigated [57]. At slower cooling rates, the intermetallic compound Mg_2Zn_{11} was observed. In industrial coatings the cooling rate is too fast to form Mg_2Zn_{11}; $MgZn_2$ is always detected in Zn-Mg coatings on steel in this range of compositions, there was $MgZn_2$ present in the more rapidly cooled samples. This study also investigated the nature of corrosion of these alloys in the absence of CO_2. This was expected to prevent the formation of the various carbonate compounds such as the layered double hydroxides described in Table 11.6. The lowering of pH during initial corrosion is also favorable for the formation of simonkolleite that does not include carbon in its chemistry. No Mg-containing corrosion products were found after exposure to 0.1 M NaCl solution; this was attributed to the absence of atmospheric CO_2. Still, increases in Mg content were associated with significant improvements in corrosion resistance, so that the benefit of Mg in the coating may arise from effects other than its influence on development of certain corrosion products. This was related to the denser distribution of the Mg_2Zn_{11} phase, which has a finer microstructure than $MgZn_2$; this in turn was associated with a much finer-scale distribution of anode and cathode regions. Because the pH in the absence of CO_2 was neutral to slightly alkaline, the formation of denser, better adhering films of zinc-based corrosion products could be expected. Mg under these conditions would form $Mg(OH)_2$, also buffering the pH. The moderation of pH and the widespread availability of Mg cations in the fine microstructures produced were attributed with the observed improvements in corrosion resistance.

The distribution of species in the corrosion products of several hot-dip coatings for steel sheet was investigated by Prosek [58]. Details of specimens used in this study are shown in Table 11.8. They can be seen to include temper rolling, oiled, and annealed treatments. All are from industrial production except the Zn-16%Mg coating that was produced by coating an Al-free Zn coating with 0.7 μm Mg, then annealed at 320°C to produce the indicated coating composition and coating weight. All specimens were first gently polished with emery paper and then preexposed to 50% relative humidity air at 20°C for 24 h. A solution of 7.21 g/L NaCl in methanol was then gradually applied to give a chloride surface deposition of 1.4 g/m^2; it was noted that deposition rates in the range of 0.01–4 g/m^2 per day were typically experienced in coastal atmospheres. The samples were then placed in a chamber whose air was maintained at 80% relative humidity (RH) and at 20°C for 1, 2, 5, 12, and 28 days. Soluble corrosion products were extracted with deionized water in an ultrasonic bath, while the insoluble products were stripped with glycine following ISO 8407 [59].

Fig. 11.24 shows the relative proportion of corroded coating and corrosion products measured in these coatings after 28 days, together with the compositions of each constituent. Here, wet mass gain was measured within 15 min after removal of the panels from the climate chamber and dry mass gain after drying for 7 days in a desiccator with relative humidity less than 15%. Soluble corrosion products were extracted with deionized water in an ultrasonic bath, while the insoluble products were stripped with glycine following ISO 8407 [59] to determine the mass loss during exposure. The "Corr. Metal" on the left side of each diagram is the mass of corroded metal that remains on top of the underlying uncorroded substrate after stripping. For the Zn-0.2%Al coating, the highest mass of corroded metal, 14.7 g/m^2, is observed. Of the 908 mg/m^2 dose of Na cations applied prior to exposure of the Zn-0.2%Al panels, 69% were bonded in corrosion products on the surface of the Zn-0.2%Al coating. This ratio is observed to be much lower, in the range of 2–7% in all the alloyed coatings. In total, 99.6% of the corrosion products on the Zn-0.2%Al coating were observed to be insoluble corrosion products, meaning that most of the chloride ions were incorporated in stable corrosion products, whereas most chloride ions were soluble on the alloyed coatings. The introduction of Na and Cl into stable corrosion products was inhibited most efficiently on coatings containing magnesium.

Only a small portion of Zn was present in a water-soluble form. On the Zn-0.2%Al coating, it was below 1%. The most soluble amount of Zn, 7%, was found on Zn-0.5Al (100% minus the indicated 93% of insoluble Zn). Except first 2 days of exposure in the humidity cabinet, the absolute quantity of water-soluble Zn in corrosion products was found to be kept almost constant. As expected, the amounts of both Al and Mg in corrosion products increased with their respective contents

TABLE 11.8 Sample details for study of soluble and insoluble corrosion products.

Coating composition	Zn-1.5Al-1.5Mg	Zn-11Al-1.5Mg-0.2Si	Zn-Mg	Zn-5Al	Zn-0.2Al
Al (wt.%)	1.5	11	–	4.5	0.2
Mg (wt.%)	1.5	3	16^(3)^	–	–
Coating weight^(1)^ (g/m^2)	140	100	70	265	275
Coating thickness^(2)^ (μm)	10.7±2.5	9.6±2	4.8±1	20.0±2.5	16.9±2.7
Treatment	TR	TR, 0	A	TR	TR

TR = temper rolled, O = oiled, A = annealed, (1) two sides, (2) measured by metallographic microscope at 20 random points, (3) Mg content in the outer coating layer (bulk coating average 3.4%).
From T. Prosek, D. Thierry, D. Persson, J. Stoulil, Corrosion products formed on Zn-Mg, Zn-Al and Zn-Al-Mg coatings in model atmospheric conditions, in: Proc. 8th International Conference on Zinc and Zinc Alloy Coated Steel Sheet (Galvatech 2011), June 21–25, 2011, Genoa, Italy, Associazione Italiana di Metallurgia, Milan.

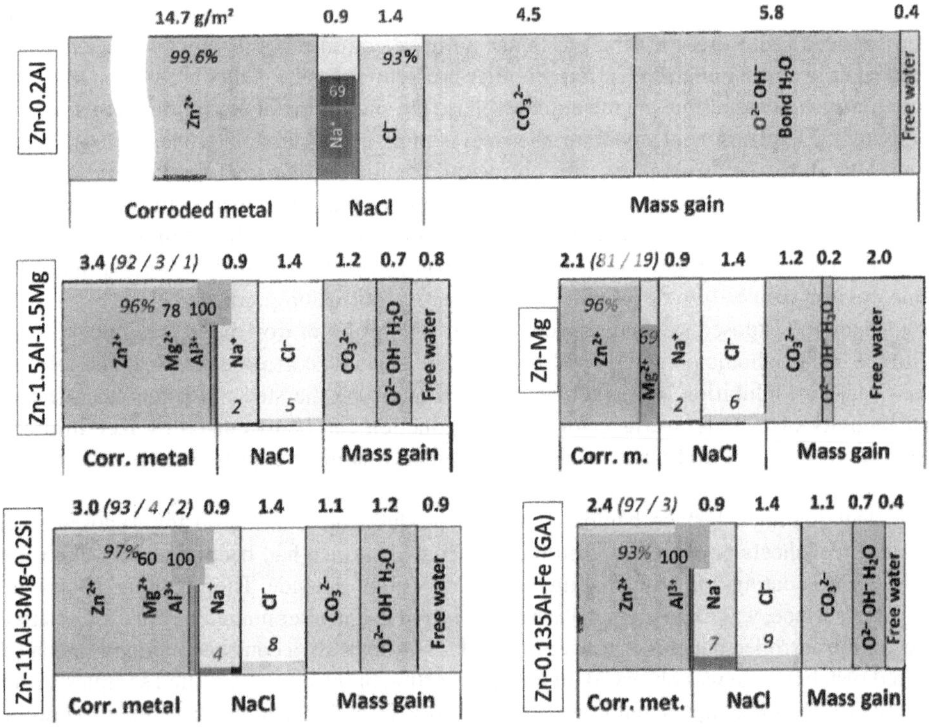

FIG. 11.24 Distribution of species in corrosion products formed on Zn-0.2Al, Zn-1.5Al-1.5Mg, Zn-11Al-3Mg-0.2Si, ZnMg, and Zn-5Al coatings after dosing with 1.4 g/m^2 Cl and exposure to 80% relative humidity at 20°C for 28 days. "Corr. Metal" indicates the corrosion products remaining after stripping in glycine, the percent numbers inside the boxes give the percentage of species in water-insoluble corrosion products, the other numbers in the boxes give the ratios of corroded Zn, Al, and Mg in the corroded metal [58].

in the coatings. From 22% to 40% of corroded Mg was water-soluble (100% minus the indicated 60%–78% of insoluble Mg). No Al was detected in water extracts of the corrosion products, it was all insoluble, nor was any water-soluble Al detected on the coating surface. It is instructive to compare ratios of the alloying metals in the coatings and in the corrosion products. In all cases, Mg dissolved preferentially, whereas Al corroded less readily than Zn. There were 114%, 38%, and 21% more Mg in corrosion products compared with the bulk compositions of the ZnAlMg, ZnAlMgSi, and ZnMg coatings, respectively. The opposite effect is seen for Al: the corrosion products in the ZnAlMg, ZnAlMgSi, and Zn5Al coatings had 19, 79, and 40% less Al than the bulk compositions, respectively. However, the ratio of Mg in the corrosion products of the ZnAlMg coating decreased in time as testing progressed from 2 to 28 days. These corrosion products contained 7.5% Mg after 2 days of exposure and 3.2% after 28 days. Similarly, the Mg content in the corrosion products of the ZnMg decreased from 89% after 1 day to 19% after 28-day exposure. However, the behavior of the ZnMg coating is likely related to its fabrication that involved vacuum deposition of a Mg layer on a Zn coating followed by interdiffusion; depletion of Mg in the outer layer of the coating prior to corrosion testing had been measured.

11.4 Galvanic corrosion and cut edge protection

Corrosion protection of bare cut edges by metallic coating involves two different phenomena: cathodic protection that is provided by sacrificial protection by the coating (galvanic corrosion of the coating) and possible subsequent self-healing of the corroding coating. Self-healing depends upon the corrosive environment and the composition of the coating. In all zinc-base coatings, galvanic corrosion usually initiates with corrosion activity on the highest zinc content phase, whether it is the primary zinc crystal phase or the Zn-rich secondary phases when the primary phase is Al-rich. This is in contrast to general surface corrosion of ZnAl and ZnAlMg alloy coatings, where initiation and propagation of corrosion occur in the binary eutectic (Zn-Al, Zn-Mg) portions of the microstructure. In contrast, during cut-edge corrosion where the coating galvanically protects exposed steel the site of initiation is in the primary zinc dendritic phase. The efficiency of galvanic protection of the zinc-rich crystals in hypoeutectic coating compositions was observed to be related to the near-perfect coherency of the crystallographic lattices of these crystals from the substrate to the surface, so, all crystals that extend from the substrate to the surface of the coating would be electrically connected with a minimal number of grain boundaries to overcome, in comparison to the lamellar grains, which have a multitude of grain boundaries for the path of the electron to overcome. The preferential corrosion of zinc rather than the ZnAl in a Zn-4.5%Al coating was also related to the pattern of oxide formation on the coating surface. Aluminum oxide, an electrical insulator, was observed to form a very fine layer on portions of the coating surface where the Zn-Al eutectic protrude to the surface. Zinc oxide, a semiconductor, was observed to form on the zinc-rich crystals protruding to the surface. Therefore, the zinc phase has a high electron mobility in comparison to the eutectic phase. In the case of general coating corrosion, this electron mobility is essential for the cathodic reaction to occur, making the zinc oxide the likely site for oxygen reduction and the eutectic phase the site for metal dissolution. However, in the case of galvanic corrosion involving the exposed steel substrate, the steel is more noble (less electronegative) than the alloy coating on the galvanic series. This shifts the driving force for corrosion, resulting in the steel becoming the site of cathodic activity; the anodic attack then initiates in the primary zinc crystals rather than the eutectic because the zinc crystals are covered with semiconductive zinc oxide rather than the insulating aluminum oxides [34]. Uninhibited galvanic corrosion was found to run for 3-month exposure on a relatively heavy gauge, 3 mm steel sheet coated with a 25 μm Zn-3% Al-3.7%Mg coating after in marine environment before development of protective corrosion products [2].

At the early stages of corrosion, where the exposed steel edge is covered by a thin film of iron oxide, a high pH can be present at the cut edge together with the local cathodic potential resulting from the galvanic corrosion couple. Zinc cations, migrating to the cut edge, can then cause an inhibition of oxygen reduction directly on the steel. This has been shown directly by Kelvin probe [60]. The activity of the edge surface for oxygen reduction can be evaluated by Kelvin probe measurements during change from air to nitrogen and back to air, as shown in a similar experiment summarized in Table 11.4.

After galvanic corrosion at the steel-coating interface begins, corrosion products are formed on the exposed steel surface. Uranaka [61] exposed coated steel sheets on which a portion of the coated surface had been removed, exposing the bare steel. GA, GI, and Zn-6%Al-3%Mg coatings, all with coating weights of 50 g/m^2 per side. The edges and back were sealed. The steel-coating interface, a flat surface, was exposed to a 5% NaCl solution in chamber maintained at 98% relative humidity and 25°C. The appearance of the corroded panels is shown in Fig. 11.25. After 48 h, a control coupon of uncoated steel was completely covered with red rust, but none was observed on the steel portion of the partially coated samples. After 144 h, some red rust was noted on the GA and GI samples, but none on the Zn-6%Al-3%Mg samples. Examination of the

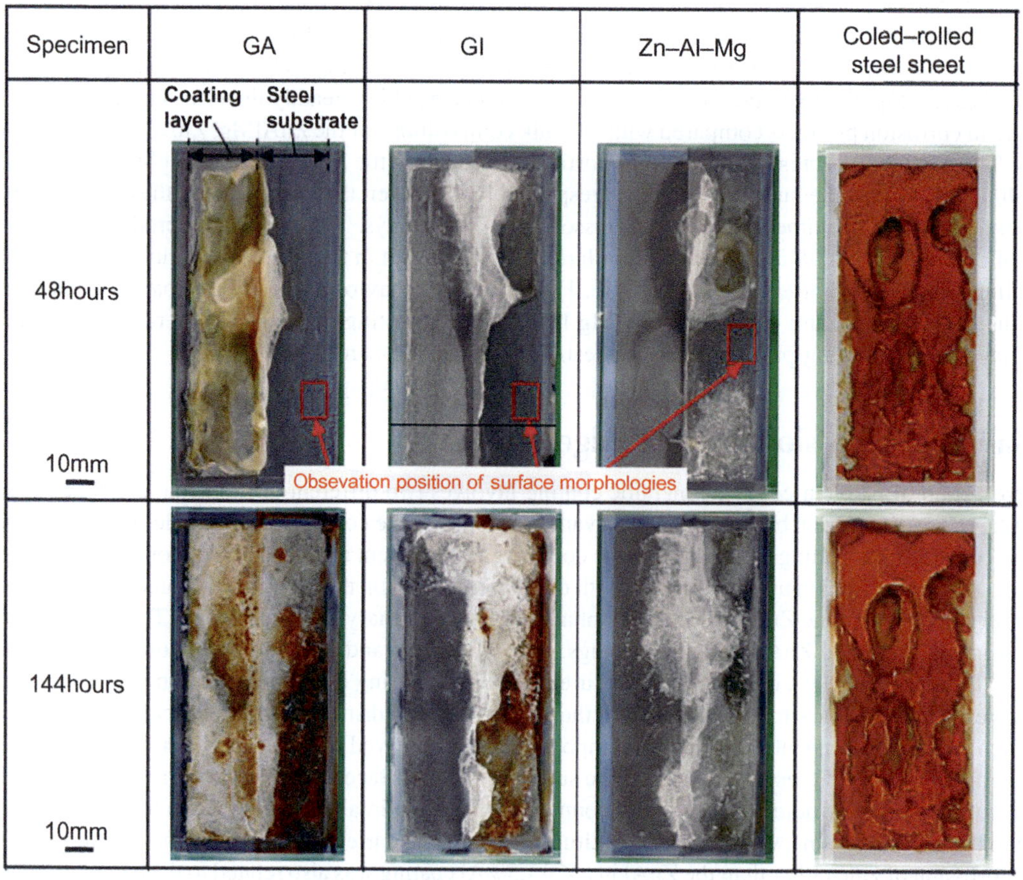

FIG. 11.25 Surface appearance of GA, GI, and Zn-6%Al-3%Mg partially coated panels after exposure to 5%NaCl film for 48 and 144 h [61].

surface morphologies and chemistries of the corrosion products formed after 48 h of corrosion testing on the GA coating showed the presence of hydrated hematite $Fe_2O_3{\cdot}H_2O$ and zinc oxide in a porous morphology. These corrosion products that included a reddish tinge are formed on the GA coating adjacent to the bare steel surface, but only white-colored corrosion products are seen on the steel immediately adjacent to the steel-coating surface boundary. This indicated the early mobility of dissolved zinc cations that migrated to oxygen reduction sites at the substrate, the Fe in the GA coating was also undergoing early stage corrosive attack, but the Fe ions resulting from this did not migrate to the steel-coating boundary. After 48 h of corrosion testing, simonkolleite and zinc hydroxide corrosion products were observed on the GI and Zn-6%Al-3%Mg samples. After 144 h, the simonkolleite and zinc hydroxide corrosion products previously observed on the GI sample were joined by the hematite and zinc oxide corrosion products previously observed on the GA sample; these products also covered about half of the exposed steel surface on which patches of red rust also appeared on the GI sample, whereas the GA sample was fully covered with red and white corrosion products. Cross-sectional views indicated that both the GA and GI coatings were largely consumed after 144 h. After this duration of testing, the Zn-6%Al-3%Mg samples continued to show dense white corrosion simonkolleite on the surface. As in the case of general corrosion, it was concluded that Mg ion migration from corrosion of the Zn-6%Al-3%Mg suppressed the formation of the less protective ZnO. From observations described above, it is expected that the Zn-rich crystals in the coating participated first in the galvanic reaction; the Mg reaction enabling the suppression of ZnO and encouraging the growth of simonkolleite occurring after the initial stage of galvanic corrosion. This does not happen with the GA coating and only occurs to a lesser extent in the GI coating.

A similar effect of Mg on the development of corrosion products in corrosion of exposed steel was also observed by Oh [51] for the Zn-3%Mg-2.5%Al coating described before. After 500 h of salt spray testing, red rust formation in the exposed steel cut-edge areas of this alloy coating was retarded by white corrosion products, while edges of the GI and Zn-55%Al-1.6%Si coatings without Mg were severely corroded and fully covered by red rust, Fig. 11.26. It was concluded that $Zn/MgZn_2$ binary eutectic lamellar structures formed in this relatively Mg-rich coating alloy were coarser than that formed in coatings with lower proportions of Mg to Al, resulting in an increased amount of Mg dissolution.

Significant differences in cut edge behavior are observed in Zn-55%Al-1.6%Si coated sheets after prolonged exposure to river water or seawater compared with salt spray testing. After river or seawater exposure of cut edges, $CaCO_3$ and $Mg(OH)_2$ reaction product films, respectively, are formed, as pH increases by a value of about 10 in the electrolyte adjacent to the edge because of the cathodic current in the water medium that serves as the electrolyte. These films form as a consequence of galvanic corrosion at the edge of coated steel sheets. The films were shown to control the diffusion of oxygen, resulting in suppression of the cathodic reaction on the cut edge. Corrosion of the coating near the shear cut edge is significantly less than observed in salt spray testing [62].

The "self-healing" behavior of the corroded cut edge resulting from the formation of protective films has been observed with other coatings and in a range of corrosive environments. This was investigated by using a salt spray exposure cabinet, but using a sprayed electrolyte that was formulated to represent the mean composition of rainfalls sampled worldwide, therefore including SO_4^{-2}, Cl^-, NO_3^-, HCO_3^-, Na^+, Ca^{+2}, and $NH4^+$ ions [2]. The ratio between the different ions was kept constant and their concentrations were increased to increase electrolyte corrosivity. The corrosion ratio calculated between Zn-3.7%Al-3%Mg and the standard GI coating using this test gave much closer approximation to results obtained in actual environments. Fig. 11.27 shows the percent of cut edge area covered by red rust amount observed on samples of these two coatings, each with a coating thickness of 25 μm coating per side and substrate thicknesses ranging from 2 to 5 mm after 8000 h exposure in the rainfall spray cabinet. For similar substrate and coating thickness, the ZnMg3Al3.7 coating shows less red rust than standard HDG-Zn. Self-healing is observed for both coating types. The evolution of red rust coverage with time is shown in Fig. 11.28 for the 3 mm thick samples.

FIG. 11.26 Red rust appearance on the cut edges of coated steel sheets after 500 h of 5% neutral salt spray testing (A) GI, (B) Zn-3%Mg-2.5%Al, and (C) Zn-55%Al-1.6%Si [51].

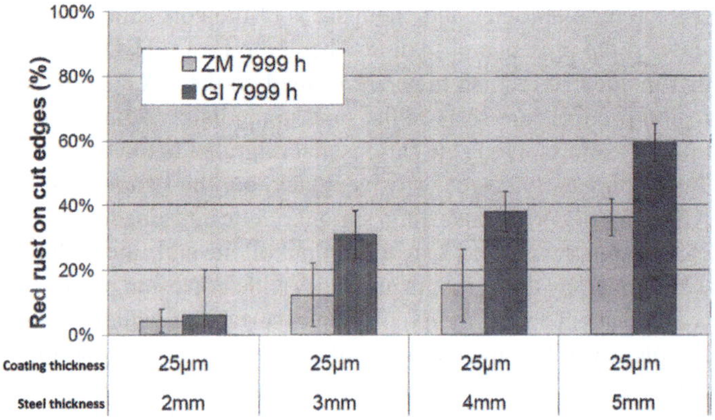

FIG. 11.27 Percent red rust coverage at edges after 8000h of rainfall spray testing for several steel thicknesses coated with either Zn-3.7%Al-3%Mg or standard GI, coating thickness of 25 µm per side [2].

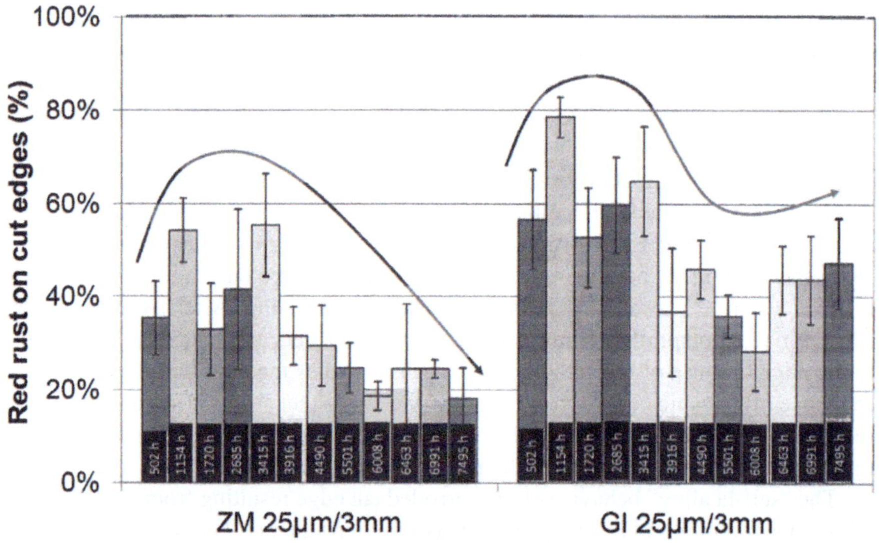

FIG. 11.28 Percent red rust coverage at edges after up to 7500h of rainfall spray testing for 3mm thick steel panel coated with either Zn-3.7%Al-3%Mg or standard GI, coating thickness of 25 µm per side. The solid line in the graph illustrates data trends only [2].

The percent red rust coverage reached almost 80% for the GI samples after 1154h of exposure, while coverage reached about 55% for the Zn-3.7%Al-3%Mg after the same time, after which the increase in red rust coverage ceased and then decreased until the end of the test. Self-healing in the GI coating proceeded until 6008h, after which red rust coverage again increased. These laboratory test results were confirmed by 5y exposure in a marine/industrial (ISO 9223 Class C4) environment [24]. The percent red rust coverage on several steel thicknesses coated with the Zn-3.7%Al-3%Mg coating with a thickness of 25 µm per side is shown in Fig. 11.29. The self-healing phenomenon is observed for all substrate thickness. For the thickest panels, 4 and 5mm, red rust coverage at the edges increases quickly; self-healing begins after 5-month exposure, then decreasing until 15-month exposure. The 1–3mm panels exhibit a slower progression of red rust that was attributed to more effective cathodic protection of uncoated steel area; the percent red rust coverage on the edge reached a maximum after 9-month exposure, after which self-healing commenced. The percent red rust level coverage after 15 months of exposure was observed to be nearly the same for all substrate thicknesses. Beyond 15 months of exposure, red rust increases on cut edges of the 3 and 4mm thick panels that was attributed to coating consumption in the vicinity of cut edges, so that the self-healing effect that depends upon the supply of metal anions from the coating could no longer be operative.

One application where a large proportion of the sample area consists of cut edges is perforated plate. Fig. 11.30 shows the evolution of the percent red rust coverage at all edges in a 2mm thick sample of perforated plate after 5 years of marine (ISO 9223 C3) environment exposure. The general appearance of the test panels after this exposure duration is shown in Fig. 11.31. Here, the proportion of cathode to anode area is much higher than with customary test panels. Together with the corrosivity of the environment, this results in higher loss of zinc near the exposed edges as it produced cations that progress to the edge to participate in the oxygen reduction reaction. In contrast, the Zn-3.7%Al-3%Mg shows noticeable white

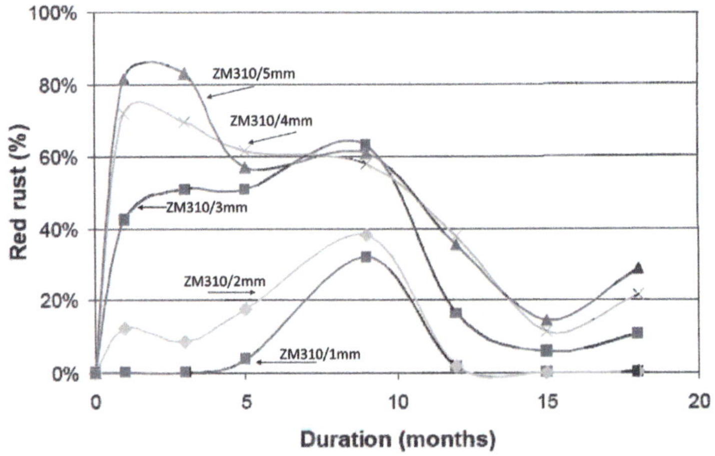

FIG. 11.29 Percent red rust coverage of edges for up to 18 months of exposure in an ISO 9223 Class C4 marine/industrial on Zn-3.7%Al-3%Mg coated samples, 25 μm coating thickness per side) for steel panel thickness of 1–5 mm [2].

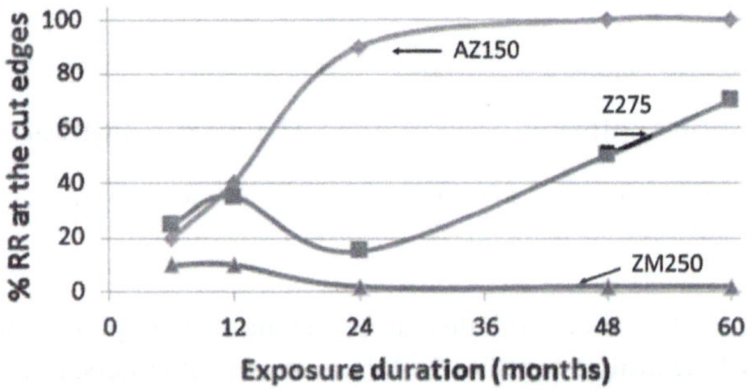

FIG. 11.30 Percent red rust coverage of edges for up to 5y of exposure in an ISO 9223 Class C3 marine environment on 2 mm perforated panels shown in Fig. 11.27 for Zn-3.7%Al-3%Mg (ZM250), standard GI (Z275), and Zn-55%Al-1.5%Si (AZ150) coated samples, each with 20 μm coating thickness per side [2].

FIG. 11.31 Appearance of perforated exposure panels after 5y of exposure in an ISO 9223 Class C3 marine environment, percent red rust coverage on edges shown in Fig. 11.30 [2].

rusting on the coated surface, but little red rust on the edges. This again is attributed to the migration of Mg to the cut edge to participate in the creation of stable protective films that are more durable than those produced by the Mg-free coatings [2].

Further insights on the nature of galvanic corrosion and the influence of Mg were obtained by Nishihara [11], who exposed couple of both pure Zn and pure Fe and Zn-55%Al-1.5%Si ingot (AZ) and pure Fe to NaCl or $MgCl_2$ solutions and ISO 11130 [63] artificial seawater, each over a range of dilutions. For NaCl, the electrolyte compositions were 0.85, 0.085, and 0.0085 M, for $MgCl_2$, the electrolyte compositions were 0.425 M, 0.0425 M, and 0.00425 M and for artificial seawater the solution strengths were full strength, 10% and 1% of the standard solution concentration. Fourier transform infrared spectroscopy (FTIR) was then used to analyze corrosion products. Basic zinc carbonate or simonkolleite was found to precipitate on Fe near Fe/Zn boundary after exposure of the couple containing pure zinc to the NaCl aqueous solutions. In contrast, exposure to the $MgCl_2$ and artificial sea water solutions resulted in the dominant component of the formed corrosion products being changed from Zn to $Mg(OH)_2$ with increase of the distance from the galvanic couple boundary. The distribution of OH and $Mg(OH)_2$ in corrosion products on galvanic couples agreed with past modeling work. The progression of the formation of corrosion products is shown in Fig. 11.32. Here, the Fe/Zn junction is shown at the middle of the lower border of the figure. The buildup of these compounds with time is shown in the sequence from top to bottom. In all cases, galvanic corrosion results on transport of zinc hydroxide anions being transported to the

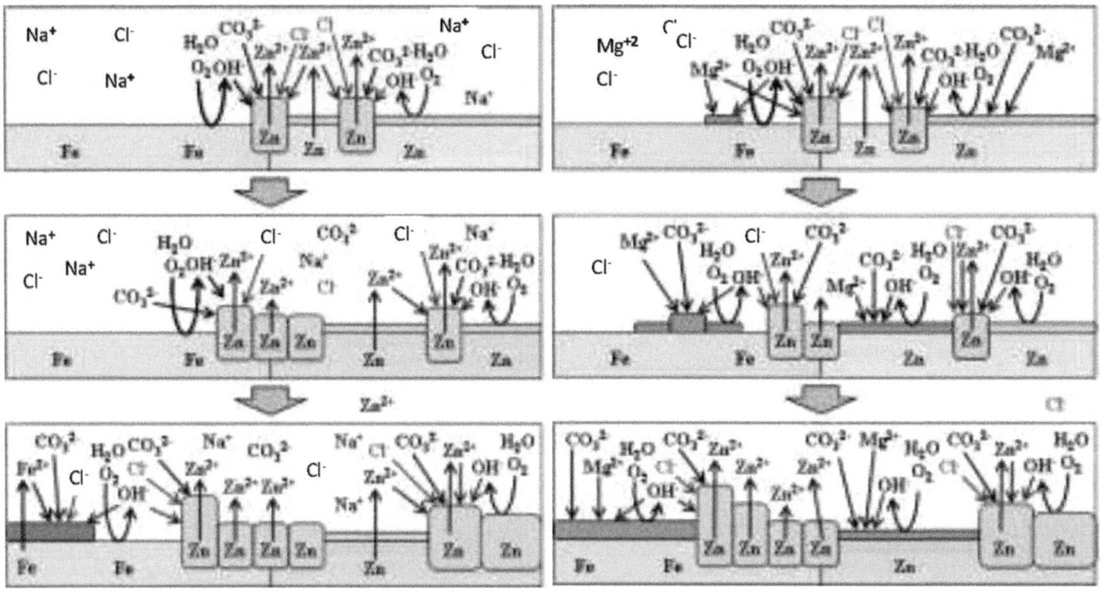

FIG. 11.32 The progression of corrosion product formation, from top to bottom for pure zinc/iron galvanic couples in (A) NaCl aqueous solutions and (B) solutions of $MgCl_2$ and artificial seawater [11].

iron side where they are precipitated. Both simonkolleite (Zn-Cl containing precipitates) and basic zinc carbonate (Zn-CO_3) precipitates are shown in the NaCl electrolyte together with Fe corrosion products. For the $MgCl_2$ and artificial seawater electrolytes, Mg-containing compounds are shown, including hydroxides and carbonates. Some precipitated Zn compounds redissolve as corrosion progresses and re-precipitate further from the Fe/Zn boundary. No Fe corrosion products were observed in the Mg-containing electrolytes, indicating that the Fe had not been significantly attacked.

For the AZ-Fe galvanic couple, the Zn and Al-containing corrosion products that are precipitated onto the Fe are not redissolved into aqueous solution; therefore, the area of corrosion products resulting from galvanic corrosion quite localized on pure iron near the AZ-Fe boundary. In the NaCl aqueous solution, the anodic dissolution of Fe occurs on pure Fe from the boundary, where some Fe corrosion products are precipitated; other Fe corrosion products also precipitate further away. In the $MgCl_2$ solution and artificial sea water solutions, Mg hydroxides and carbonates are precipitated at the surface on pure iron and progress outward from the couple, as shown in Fig. 11.33.

More details on the nature of the corrosion products and solution conditions in a galvanic corrosion cell similar to Nishihara's were given in the in-situ Raman spectroscopy observations of Ohtsuka [64]. In these tests, only a 3% NaCl electrolyte was used. As the start of corrosion, both NaCl and $ZnCl_2$ were dissolved in the solution covering the coated steel side of the galvanic couple. After saturation by $ZnCl_2$, the solution precipitated simonkolleite; additionally basic zinc carbonate was formed on the steel size of the couple in the 25°C, 84% relative humidity atmosphere, agreeing with Nishihara's results, this was also favored by the local alkaline pH. The Fe corrosion products formed near the couple were identified as amorphous ferric oxyhydroxide or hydrated zinc ferrite. Five different coatings were tested in this way, the time to form the simonkolleite film was found increase as corrosion rate decreased. The pure zinc coupon joined to the galvanic couple corroded the fastest. Coatings of Zn-12%Ni and Zn-5%Al corroded slower than Zn and at roughly the same rate, and the Zn-55%Al-1.6%Si coating exhibited the slowest corrosion rate. These same basic mechanisms were also confirmed in the cyclic corrosion tests results of Lee [65], where during the portion of the cycle when the galvanic couples were dry, there was no electrolyte to provide a galvanic reaction. Lee experimented with both Zn-Al and Zn-Al-Mg coatings and found, agreeing with other work described above, that the Mg-containing corrosion products were more stable. For the Zn-55%Al-1.6%Si coating, this behavior was confirmed by salt spray testing, where the time to appearance of red rust at a scribe in the coating was delayed by 20%–40% when 2%Mg was added [55]. The galvanic protection benefits of Mg additions during wet-dry cycling were further confirmed by Thébault [66], who observed that Mg ions deposited over the cut edges and adjacent coating on sheet steel during the wet part of the cycle also give benefits to corrosion resistance during the dry portion of the cycle, limiting the amplitude between local electrode potentials over the steel surface and consequently preventing the creation of local corrosion cells and the appearance of red rust on the less noble areas.

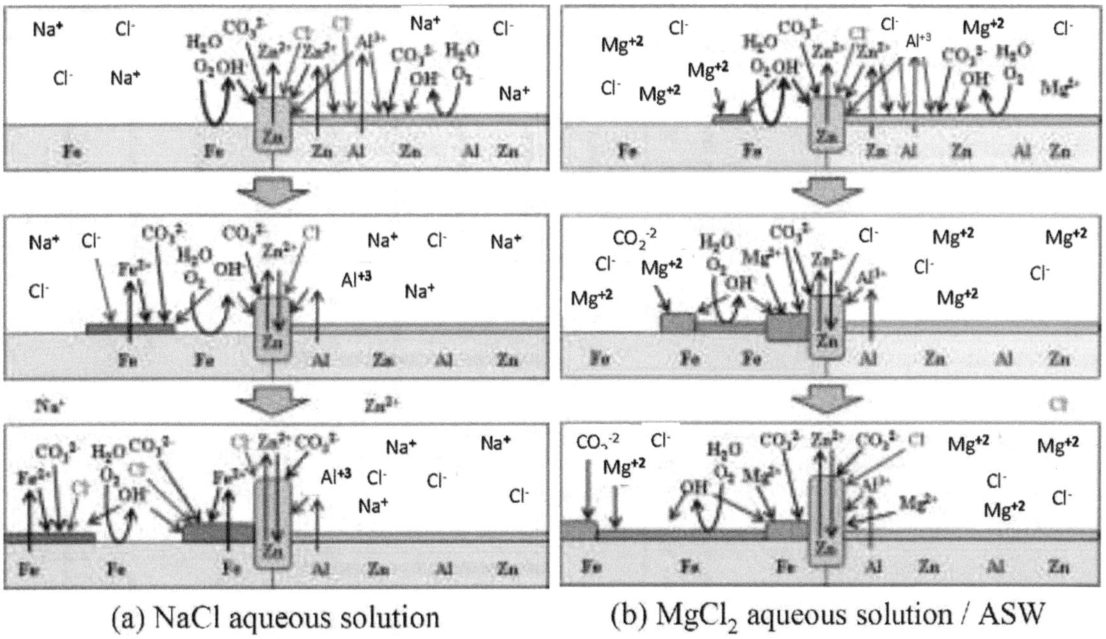

FIG. 11.33 The progression of corrosion product formation, from top to bottom for Zn-55%Al-1.6%Si/iron galvanic couples in (A) NaCl aqueous solutions and (B) solutions of $MgCl_2$ and artificial seawater [11].

The beneficial effect of Si in improving corrosion resistance of ZnAlMgSi coatings was examined by Šefl, who used galvanic ZRA (zero-resistance ammeter) and OCP (open-circuit potential) measurements on cast rather than coated steel samples. The phases present in the alloy compositions tested were η-Zn (Zn-1%Al), α-Al (Al-32%Zn), $MgZn_2$, Si, and Mg_2Si. The η and α phases are expected to be near the compositions of primary dendrites formed during solidification of hypoeutectic and hypereutectic alloys, respectively. None of the compositions contained Fe. While, as expected, the $MgZn_2$ phase was anodic for the entire 3-day experiment, the Mg_2Si was not stable and shifted from anodic to cathodic behavior. Elemental Si was inert and showed high polarization. With time, Mg was dissolved from the Mg_2Si phase, so that the cathodic potentials of this and the elemental Si became closer in value, the release of Mg from these Mg-containing compounds was expected to be prolonged by the presence of the Mg_2Si phase [12].

11.5 Corrosion behavior in atmospheric exposures

As shown in the previous section, laboratory and accelerated tests are very useful in the determination of corrosion mechanisms and resulting corrosion products in zinc and zinc alloy-coated steel. To provide data related to anticipated customer use, long-term exposures are required to confirm these mechanisms and corrosion products, also to determine phenomena that only appear a longer exposure times. These can include the continuing evolution of interactions between local corrosion cells on the surface of the coating that change as the coating is consumed or chemically transformed, long-term environmental effects related to the deposition and reaction of chlorides and sulfates, times of wetness, and temperature variations. The character of galvanic corrosion would be expected to change as additional corrosion products, both on and in the steel and zinc components of the galvanic couple, evolve with time, and both the coating and substrate are consumed. A recent example of a 6-year study is described by Thierry [67] that exposed steel panels with Zn-0.2% coatings having two coating weights of 275 and 600 g/m^2 and a Zn-3.7%Al-3%Mg coating with a two-side coating weight of 250 g/m^2. The heavier coating weight Zn-0.2%Al coating was given a trivalent passivation treatment that applied 25 g/m^2 per side, while the other coatings were untreated. The substrate thicknesses were in the range of 1–1.4 mm. Masking tape was applied to the backs and edges of all panels. The panels were placed on 11 sites from which data were reported. One of these, Bangkok, was a relatively noncorrosive ISO 9223 C2 [24] site with low chloride and sulfate deposition rates and moderate humidity. Seven sites were classified as more severe C3 sites, with predominately chloride depositions in Brest (France), the two Bohus-Malmö (B.-M., Sweden) and Ijmuiden (Netherlands) while the Qingdao (PR China) and Singapore sites had a mix of sulfate and chloride deposition, and the Cadiz site had a low sulfate deposition rate. Three sites were of C4 severity, with strong chloride deposition at Daytona Beach (Florida, USA) while the Wanning (PR China) and Dubai (U.A.E.) sites have significant chloride deposition together with lower levels of sulfate deposition.

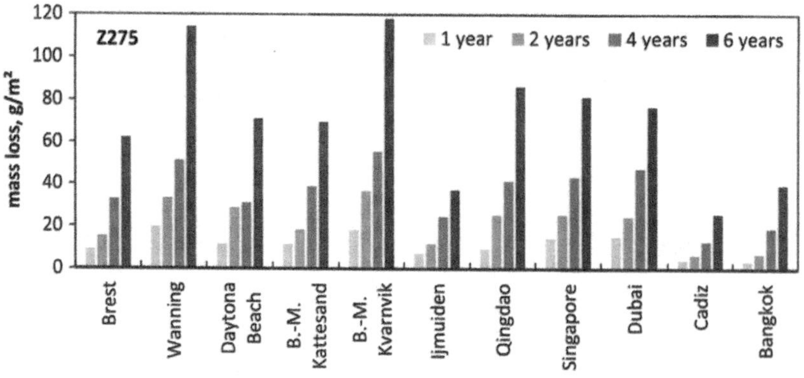

FIG. 11.34 Mass loss of Zn-0.2%Al 275 g/m^2 (two-sided coating weight) panels after 1, 2, 4, and 6 years of exposures at 11 different sites [67].

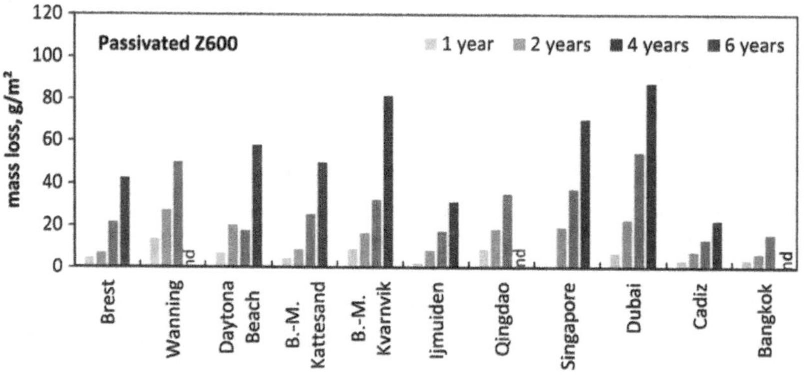

FIG. 11.35 Mass loss of passivated Zn-0.2%Al 600 g/m^2 (two-sided coating weight) panels 1, 2, 4, and 6 years of exposures at 11 different sites [67].

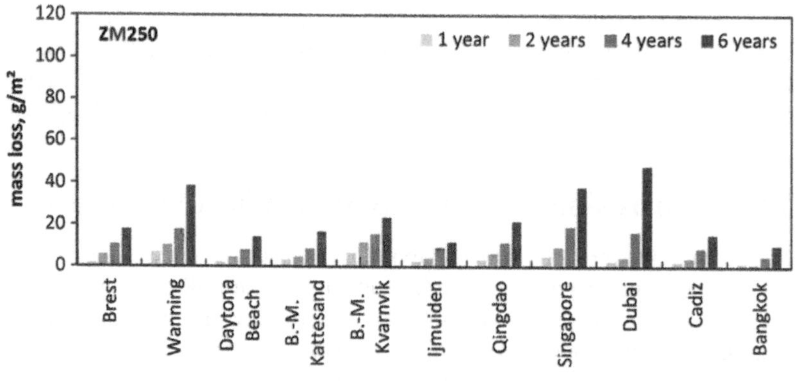

FIG. 11.36 Mass loss of Zn-3.7%Al-3%Mg 250 g/m^2 (two-sided coating weight) panels after 1, 2, 4, and 6 years of exposure at 11 different sites [67].

The mass losses from each of the sites for each set of samples for 1, 2, 4, and 6 y are shown in Figs. 11.34, 11.35, and 11.36. None of the coatings at any of the sites showed any red rust on their surfaces, meaning that none of the coatings were completely consumed after 6 y exposure. The passivated coating, which also had a heavier coating weight, performed better in all cases than the lighter unpassivated coating; the beneficial effects of passivation are observed during the first 2 years of exposure.

When the data for the European sites and Qingdao are analyzed together, Fig. 11.37, it is found that the ratio of corrosion rate of the ZnAlMg coating and the lighter weight Zn-0.2%Al coating is slightly greater than 1 in the least-corrosive environments (where the lowest coating weight losses are observed) after both 4 and 6 years; the ratio becomes larger, rising to approximately 4, at the most severe of the exposure sites, indicating an improved relative corrosion resistance of the ZnAlMg coating as corrosivity becomes more severe.

The corrosion products present on the Zn-3.7%Al-3%Mg and Zn-0.2%Al 275 g/m^2 coated steel sheet panels from the Brest and Qingdao sites, together with those from an industrial site in Jiangjin, PR China, were analyzed using visual and

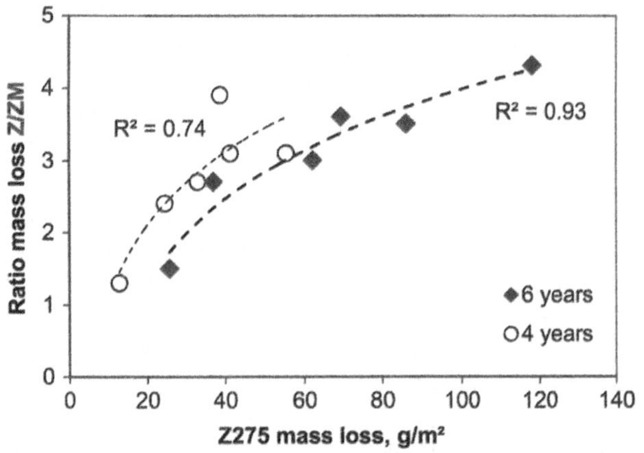

FIG. 11.37 Ratio of mass losses of Zn-3.7%Al-3%Mg 250 g/m^2 to Zn-0.2%Al 275 g/m^2 coated steel sheet panels from aggregated data of European and Qingdao sites for the 4 and 6 years of exposure [67].

TABLE 11.9 Average environmental conditions for three exposure sites where corrosion products were analyzed [67].

Exposure site	Average temperature (°C)	Average relative humidity (%)	Average Cl^- deposition ($mg/m^2/day$)	Average SO_2 deposition (μm/day)
Brest, France	12	83	1064	<1
Qingdao, PR China	14	70	16	32
Jiangjin, PR China	20	71	–	218

with FTIR spectroscopy and X-Ray diffraction. The Jiangjin site is a C4 severity site, the environmental conditions measured for these three sites are shown in Table 11.9. Quantitative corrosion loss data were not published for the Jiangjin site.

At the Brest site, the compositions of the corrosion products on the Zn-0.2%Al coatings were similar after 2 and 6 years of exposure. After 2 years, in order of highest to lowest proportion of observed corrosion products, hydrozincite, simonkolleite, gordaite, and $Zn_4(OH)_6SO_4{\cdot}xH_2O$ (zinc hydroxysulfate) where $x =$ either 3 or 5. The order changed after 6 years to simonkolleite, hydrozincite, and zinc hydroxysulfate. For the ZnAlMg coating at the Brest site, the order after 2 years was aluminum hxdroxyl sulfate present as two compositions: $Al_2(OH)_4(SO_4){\cdot}7H_2O$ and $Al_4(OH)_{10}(SO_4){\cdot}4H_2O$ followed by zinc hydroxysulfate. After 6 years, the layered double hydroxide becomes dominant followed by hydrozincite. It is of interest that the chloride-containing corrosion products for the ZnAlMg coating only became a significant proportion of the product layer after several years of exposure, being contained within the layered double hydroxide. Consistent with the weight loss data, there was a higher volume of corrosion products present for the Zn-0.2%Al coatings, whereas the ZnAlMg coating had a duller appearance with more compact, less massive corrosion products. In the ZnAlMg coatings, the sulfate and carbonate products within the layered double hydroxide were concentrated in areas that are believed to be the locations of local attack of the eutectic phases. These observations are consistent with laboratory and accelerated tests described before, of interest is the role of sulfur-containing corrosion products in these extended exposures rather than the dominance of chlorides seen in the shorter exposures.

At the Qingdao site, the corrosion products Zn-0.2%Al coating after 2 years included, in decreasing order of prevalence, hydrozincite and zinc hydroxysulfate. After 6 years, the order changed to simonkolleite, zinc hydroxysulfate, and gordaite. For the ZnAlMg coating after 2 years, the corrosion products in decreasing order of prevalence included aluminum hxdroxyl sulfate, zinc hydroxysulfate, and a layered double hydroxide. After 6 years, the order for ZnAlMg changed to aluminum hxdroxylsulfate, zinc hydroxysulfate, a layered double hydroxide, and simonkolleite. Chloride-containing layered double hydroxides and aluminum hydroxychlorides were also detected on the ZnAlMg coating after 6 years. Some carbonates are also expected to be present after 6 years. Two different scenarios for carbonate formation were described: first, sulfates and chlorides would be expected to participate in localized corrosive attack at the reaction front on the bottom of the corrosion product layer, while carbonates are formed in the outer layer. Second, the zinc hydroxysulfate and aluminum hydroxylsulfate compounds, together with the layered double hydroxides, can transform to carbonates in the presence of carbon dioxide that can be absorbed in the electrolyte film at local cathodic areas on the outer part of the coating. However, these carbonates are not a significant part of the corrosion product layer.

At the Jiangjin site, a heavily polluted inland location near the Yangtze River, the Zn-0.2Al panels were dominated by zinc hydroxysulfate, with a trace of simonkolleite after 6 years. After both 2 and 6 years, the ZnAlMg corrosion products in decreasing order of prevalence were zinc hydroxysulfate, aluminum hxdroxylsulfate, and a small amount of layered double hydroxides. At this site, as well as at the other two sites, no Mg-containing corrosion products were found; this contrasts significantly with accelerated test results described earlier such as those of Prosek [58] where the proportion of Mg in corrosion products exceeded those within the Mg-containing test samples.

Zinc-coated steel sheet used in outdoor applications is commonly subjected to tension bending. A 7-year outdoor corrosion study compared the behavior of panels with flat sections, of which a portion was bent around a radius of twice the sheet thickness (2T bend) [68]. The marine site in Okinawa, Japan, has slightly less average chloride deposition, 51 mg $NaCl/m^2/day$, than the Wanning site utilized in the Thierry study described above, but without the average 12 μg/day of SO_2 deposition of Wanning. The rural site in Gunma, Japan, was similar to the Cadiz, Spain site in the Thierry study. Coatings included were Zn-0.2%Al (GI), Zn-4.5%Al-0.1%Mg (GF), Zn-6%Al-3%Mg (Zn-Al-Mg), and Zn-55%Al-1.6%Si (AZ). The first three had coating weights on each side of 150–160 g/m^2 and were chromated after galvanizing while the AZ coating weight was 80 g/m^2 per side and had a thin organic resin coating applied after galvanizing. All substrate thicknesses were 1.6 mm. These coatings responded differently to bending: the GI and GF coatings developed few cracks during bending, instead the coating was thinned in response to deformation. The Zn-Al-Mg and AZ coatings exhibited more cracks, and their coatings did not noticeably thin during bending deformation.

After 5 years of exposure, none of the coatings from either site exhibited any red rust except for the AZ panels at the bends, so the other three coatings were still present to provide protection to the steel in all locations. The GI coating was covered with a thick layer of white rust; in contrast, the GF and Zn-Al-Mg coatings were gray-black in color with isolated patches of white rust. Fig. 11.38 shows that in all cases, corrosion was much more advanced after 5 years for the bent panels, the results from the Okinawa site are shown in this figure. This was related the thinner coating layer in the case of GI and GF and the presence of cracks in the Zn-Al-Mg and AZ panels in the deformed areas. For the flat panels, the AZ coating showed the thickest remaining coating; however, bending opened cracks that permitted eventual steel attack.

Observations on corrosion products on the Zn-Al-Mg coating at the Okinawa site, shown in scanning electron microscope and corresponding X-Ray images of bent coating cross sections in Fig. 11.39, are similar to that Thierry described above. Corrosion products were found on this together with the GI and GF coatings after the first year of exposure were identified as basic zinc carbonate and simonkolleite; after 5 years, an additional layer that is likely to be a chloride-containing double-layer hydroxide added a third component to the corrosion products. Unlike Thierry, Shimizu observed Zn, Mg, and Al in the initially formed corrosion products; Zn and Mg were in the third layer that formed later. This may be related to the higher alloy content (6%Al, 3%Mg) of this coating compared with the (3.7%Al, 3%Mg) coating investigated by Thierry. Fig. 11.39 shows taken from samples after 1 and 5 years of exposure at Okinawa. Concentrations of chloride and sulfur are observed near the bottom of the coating after 5 years, supporting Thierry's hypothesis on the relationship of corrosion products containing these species and those nearer to the surface containing carbonates that would be in the double-

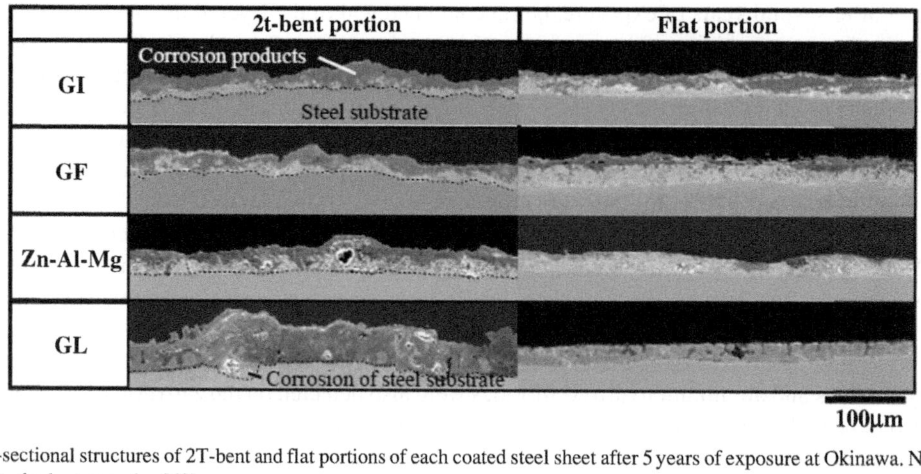

FIG. 11.38 Cross-sectional structures of 2T-bent and flat portions of each coated steel sheet after 5 years of exposure at Okinawa. Note the much thinner remaining coating on the bent samples [68].

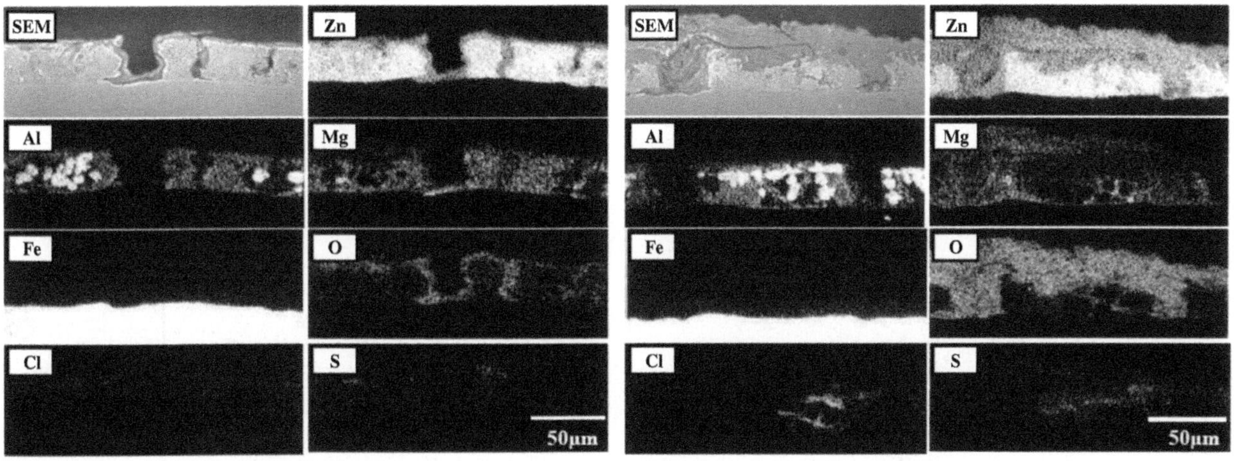

a) 1 year b) 5 years

FIG. 11.39 X-ray images of 2T-bent portions of Zn-Al-Mg after 1 year and 5 years of exposure at Okinawa [68].

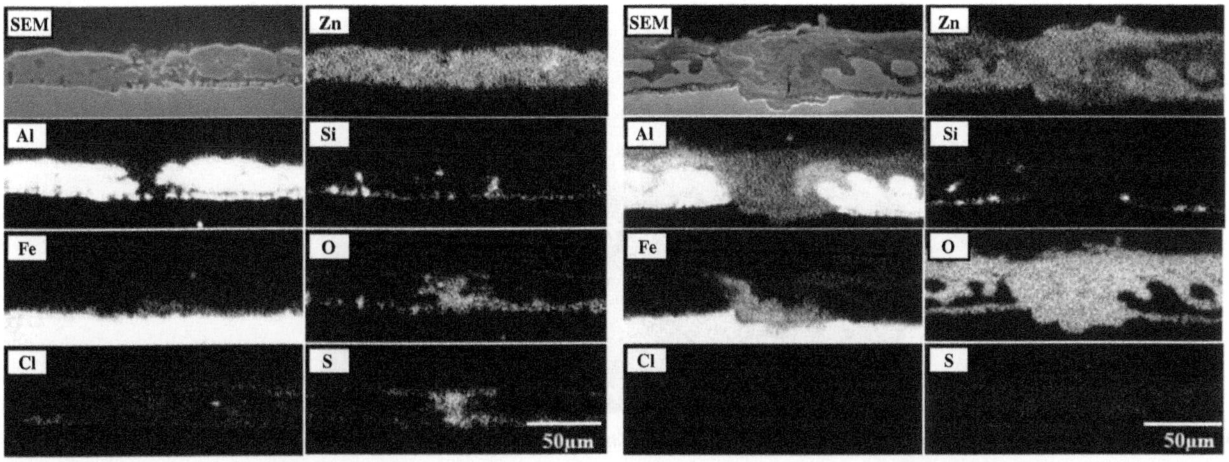

a) 1 year b) 5 years

FIG. 11.40 X-ray images of 2T-bent portions of Zn-55%Al-1.5%Si coating after 1 year and 5 years of exposure at Okinawa [68].

layer hydroxide. Because outdoor exposure sites are subject to alternate wetting and drying, the cracks in the ZnAlMg coating corrosion product layer observed by Keppert may also play a role in the formation of chloride and sulfur-rich corrosion products at the bottom of the coating by permitting localized attack; this localized attack was found to be further accentuated at lower chloride concentrations [69], such as in actual atmospheric tests versus acerated laboratory tests.

Fig. 11.40 shows corrosion products on the Zn-55%Al-1.5%Si coating at the Okinawa site, in scanning electron microscope and corresponding X-Ray images of bent coating cross sections. After 1 year, aluminum-containing zinc corrosion products form in the locations of the cracks. The sequence of corrosion for the AZ coating was set forth as: (i) corrosion of the aluminum-rich primary crystals and passivation of their surface, (ii) selective corrosion of the Zn-rich intercrystalline phase and consequent formation of zinc corrosion products, and (iii) secondary corrosion of the Al-rich phase related to depletion of the Zn-rich phase and formation of Al corrosion products. However, in the cracked portion of the AZ coating, once the Zn-rich phase is depleted, the steel substrate, whose potential is anodic to than the passive Al-rich phase, is corroded instead. In atmospheric exposures, the iron corrosion products (red rust) are then cathodic to the aluminum-rich crystals. Therefore, the rusted iron accelerates local corrosion of the Al-rich phase.

Quantitative analysis of corrosion mass losses from the 7-year exposures at the Okinawa and Gunma sites for the GI, GF, and Zn-Al-Mg coatings is shown in Fig. 11.41. The corrosion rate (the slope in the graph of Fig. 11.41) of the Zn-Al-Mg and GF coatings gradually decreased with exposure time at a higher rate, whereas the slope of the GI coating, so it is expected that the difference in corrosion loss between the coatings increases with the exposure time [25]. Agreeing with other results from accelerated and outdoor exposure tests, the corrosion products formed on the Zn-Al-Mg coating are a mix

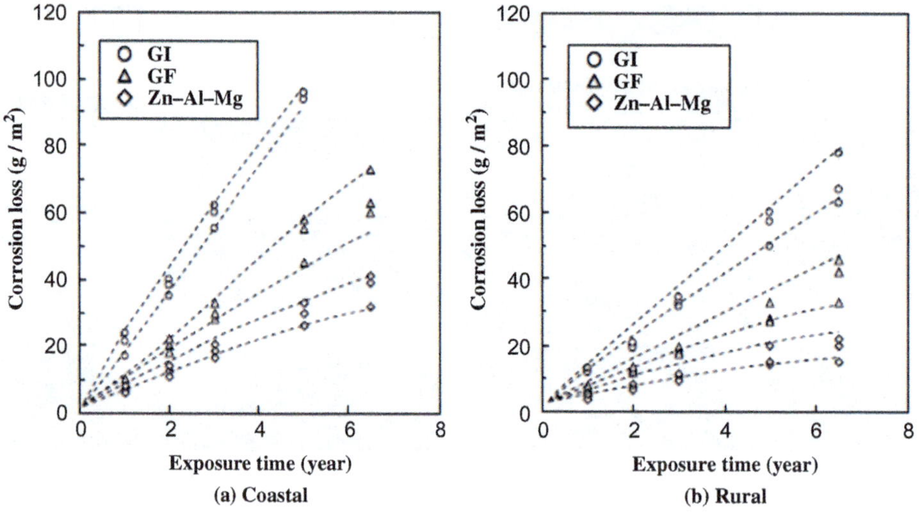

FIG. 11.41 Results of 7-year outdoor exposure tests at (A) Okinawa and (B) Gunma, Japan, for GI, GF, and Zn-6%Al-3%Mg coatings [25].

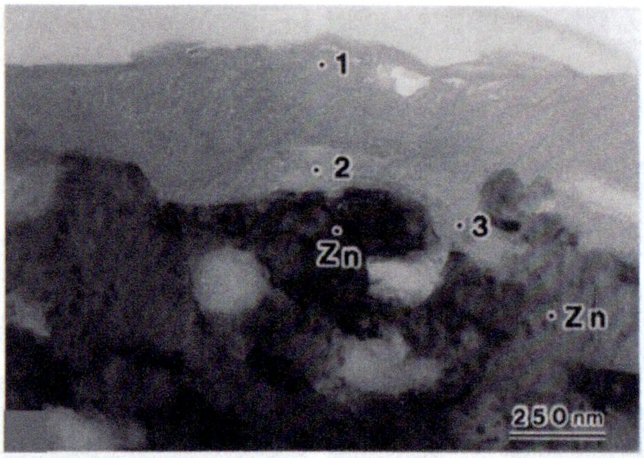

FIG. 11.42 Transmission electron micrograph of corrosion protects on Zn-6%Al-3%Mg coating after 5 years of exposure at Okinawa [25].

of fine crystalline and amorphous corrosion products [25]. A transmission electron micrograph showing these products after 5 years of exposure at Okinawa is shown in Fig. 11.42, showing the crystalline zinc corrosion products (points 1, 2, and 3) under the noncrystalline corrosion products containing Al and Mg.

Differences in the corrosion products formed on cut edges were observed for long-term outdoor tests versus the laboratory tests described before. For these 7-year exposure tests, 0.8 mm thick steel panels were coated with either Zn-0.2%Al or Zn-11%Al-3%Mg, with coating weights on each side of 63 and 35 g/m^2, respectively. Outdoor panels at the semi-tropical Miyakojima island in Japan were placed at a 45 degrees angle, while sheltered panels were placed in a horizontal orientation. The edges were left unprotected, but the panels were given an inorganic chromate-free passivation treatment after applying the zinc coatings. After 7 years of exposure, the corrosion products on the edges of both the outdoor and sheltered Zn-0.2%Al panels were ZnO, hydrozincite, and simonkolleite, agreeing with the laboratory galvanic tests described before. On the edges of the outdoor Zn-11%Al-3%Mg panels, simonkolleite, hydrozincite, and the double-layered hydroxide (Zn, Al) hydrotalcite ($Zn_6Al_2(OH)_{16}CO_3{\cdot}4H_2O$, see Table 11.6) were formed together with amorphous Mg-containing corrosion products, also agreeing with the accelerated tests. On the sheltered Zn-11%Al-3%Mg panels that did not experience the periodic washing by rain that the outdoor panels were exposed to, the amorphous corrosion product on the cut edge was much thicker than on the outdoor panels and was rich in Mg and O. In contrast, low Zn and Cl levels were found [70]. This product was likely a Mg-containing double-layer hydroxide of the type described in Table 11.6. A similar set of tests for 3 years at the same location showed the progress of edge creep in Zn-11%Al-3%Mg-0.2%Si, Zn-5%Al-0.1%Mg, and Zn-0.2%Al coated samples. All were tested with coating weights of 20, 60, 90, 120, and 140 g/m^2 per side with a steel sheet thickness of 0.8 mm and received a chromate passivation treatment after galvanizing. Fig. 11.43A. shows the effect of initial sample coating weight on the width of corrosion from the edge at the outdoor exposed site, while Fig. 11.43B shows the results for the sheltered site. The corrosion from the cut edge is observed to be more severe at the sheltered site.

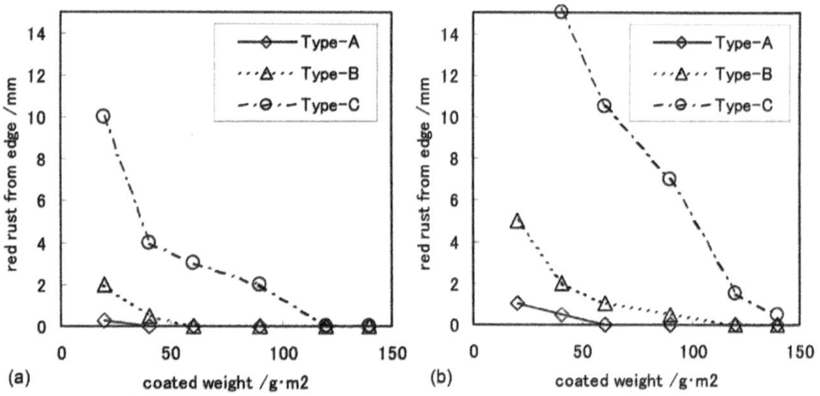

FIG. 11.43 Effect of initial coating weight on corrosion from the edge of cut edge samples for 11%Al-3%Mg-0.2%Si (Type A), Zn-5%Al-0.1%Mg (Type B), and Zn-0.2%Al (Type C) coatings at (A) outdoor and (B) sheltered exposure sites at Miyakojima, Japan [71].

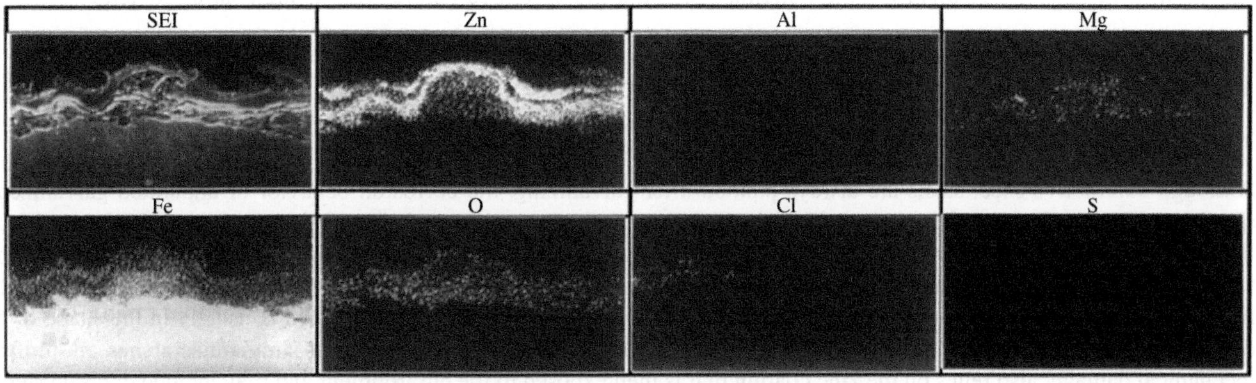

FIG. 11.44 X-ray signals from electron probe microanalysis analysis of corrosion products formed on the cut edge of Zn-6%Al-3%Mg alloy coated steel sheet subjected to 3-year exposure test. Sheet thickness: 3.2 mm, coating weight: 120 g/m^2 (each side) [25].

Consistent with corrosion observations, the alloy coatings show less edge corrosion, and alloying with Mg increases corrosion resistance. The higher severity of the sheltered site was again related to the absence of rainfall washing that resulted in more chloride deposition [71].

Similar results for corrosion products found on the cut edges at Miyakojima were reported for the Zn-6%Al-3%Mg coating exposed at Okinawa for the 7-year period described before. A scanning electron micrograph of a sample removed after 3 years is shown in Fig. 11.44 together with X-Ray signals from electron probe microanalysis.

Here, agreeing with the accelerated test results, Zn and Mg have been transported to the edge to protect the exposed steel. The edge had a blackish-brown appearance without any appearance of red rust [25]. Chlorides appear to be a minor part of these products; from the accelerated tests, oxides, hydroxides, and carbonates can be expected to be present.

Heavier gauge (1.5 mm thick) steel panels were coated with Zn-Al alloy coatings on a flux-based (Cook-Norteman, see Chapter 2 for details) galvanizing line with galvanizing bath Al compositions of 0.2%, 4.5%, 7%, 10%, 12%, and 15% at a severe industrial-marine site at Belmont Beach (Newcastle, NSW, Australia) [26]. Weight changes in the samples before and after stripping with 10% NH_4Cl solution (for the 0.2%Al samples only) or in a 20% CrO_3 solution (for the other coating compositions for 2–3 min. at 70°C) for intervals during the 1851 days (~5 years) of exposure are shown in Fig. 11.45. Corrosion weight changes do not change monotonically with Al content; after 1851 days, the 15%Al samples had higher changes, both before and after stripping, than the other higher-Al content coatings. After 1851 days, the Zn-0.2%Al coating was completely corroded, therefore no data points are shown for this time period for this composition.

After 497 days, gordaite ($NaZn_4(OH)_6(SO_4)Cl{\cdot}6H_2O$) and zinc hydroxysulfate ($Zn_4(SO_4){\cdot}(OH)_6{\cdot}4H_2O$) were identified as main corrosion products for the 0.2Al, 7Al, and 15Al coatings, simonkolleite was also detected in the 0.2%Al coating. After 1851 days, when the Zn-0.2%Al coating was consumed, the corrosion products found on the panel in an area free of red rust were mainly hydrozincite and ZnO, with very little gordaite being detected. The importance of reaction with atmospheric carbon dioxide was therefore observed, even when the carbonation reaction was competing with the high content of sulfur species evident after 497 days of exposure. For the 7%Al coating, gordaite formed a large portion of the corrosion

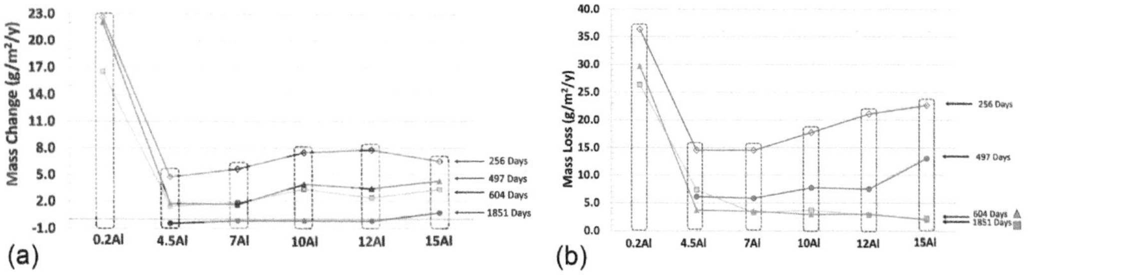

FIG. 11.45 Mass changes (A) before and (B) after the removal of corrosion products for 0.2%–15%Al coatings on 1.5 mm steel sheet for intervals up to 1851 days [26].

products, together with a small portion of zinc hydroxysulfate after 497 days, but this changed after 1851 days so that simonkolleite became the principal corrosion product. For the 15%Al coating, zinc hydroxysulfate and gordaite together made up the main portion of the corrosion products, after 497 and 1851 days. This evolution of corrosion products that include sulfur-containing species is different from other sites with lower sulfur pollution; the Zn-0.2%Al corrosion product evolution is consistent with Thierry's results from the Jiangjin, PR China site described before with its high sulfur content.

11.6 Corrosion in automobile environments

11.6.1 Hem flange corrosion

Although automotive sheet steels are always painted after galvanizing, the corrosion behavior of unpainted galvanized sheet provides important insights on perforation corrosion performance. Corrosion behavior of painted product, which is related to cosmetic corrosion, will be described in Chapter 12. Corrosion at hem flanged joints is of particular concern for perforation corrosion, because the paint system often does not fully protect the entire galvanized surface and therefore remediation such as application of a protective wax is often applied, the area being corroded is usually in the shape of a crevice. Corrosion at damaged areas, such as locations of stone chipping and scribes that simulate scratches or similar mechanical damage, also relies on the zinc coating that is then exposed to the environment and may include exposed steel in the damaged area. Both hem flanges and damaged areas can be located in the lower portion of the vehicle that is more severely exposed to liquid splashing from the roadway; the splashed liquids can contain high concentrations of deicing salts in cold climates, deposited sea salts in marine climates, and other corrosive constituents such as alkaline salts in desert climates or sulfates in industrial climates. Recently, multimaterial approaches in automobile bodies and structures have become common, especially for closure applications such as doors, hoods, and trunk lids. This results in the possibility of galvanic corrosion, especially between zinc and aluminum or magnesium when alloys based on these metals are included.

The corrosion behavior of zinc-coated steels has been characterized using laboratory accelerated testing and in proving ground tests [72]. An example of a sample configuration used to assess hem flange corrosion is shown in Fig. 11.46. Panels of galvanized (GI), galvanneal (GA), and a Zn-Al-Mg alloy coating in which the Al and Mg contents summed to 3% with respective coating weights of 60, 45, and 35 g/m^2 on each side, together with uncoated cold rolled sheet, were given a standard automotive phosphating plus electrophoretic paint primer on one side only. They were assembled into the

FIG. 11.46 Hem flange corrosion simulation test sample. The outside surfaces are given a standard automotive primer, while the inside surfaces are in the as-galvanized condition. The upper panel dimensions are 20 × 30 mm, the lower panel is 50 × 100 mm [72].

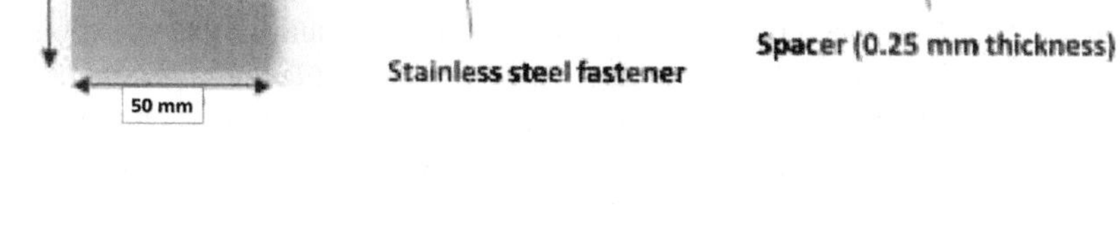

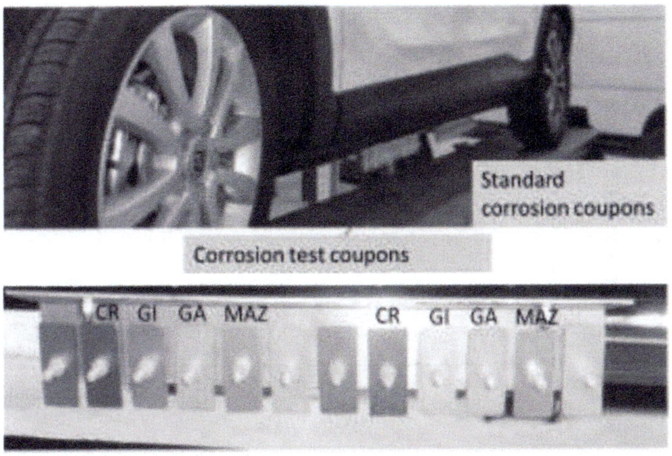

FIG. 11.47 (upper) Location of corrosion test samples on proving grounds vehicle (lower) detail of sample mounting using a single nylon fastener for each sample [72].

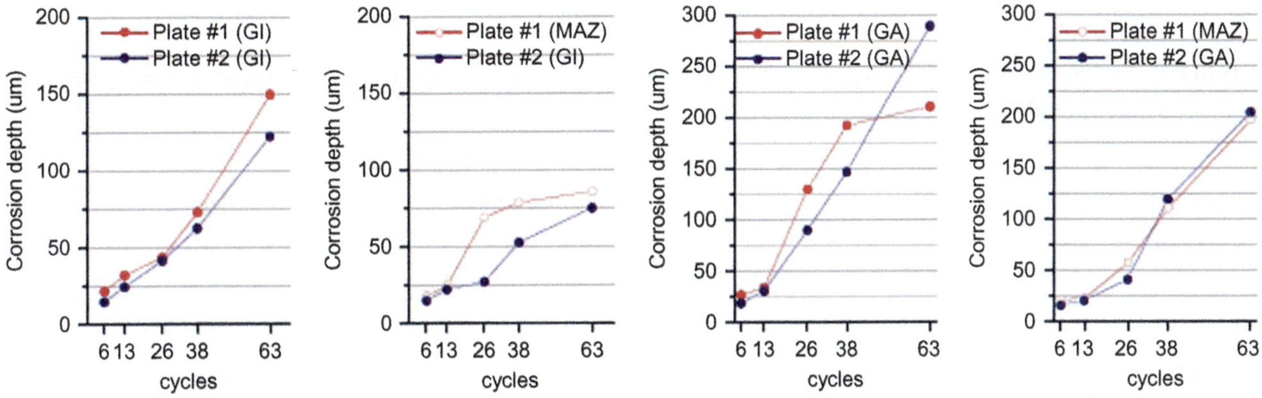

FIG. 11.48 Hem flange test sample configuration, coating thickness loss for samples on each side of the gap for GI/GI, ZnAlMg(MAZ)/GI, GA/GA, and ZnAlMg(MAZ)/GA pairs of test coupons [72].

configuration shown in Fig. 11.46, where the untreated zinc-coated surfaces face inward to a 0.25 mm thickness maintained by a nylon spacer and joined by stainless steel fasteners. The dimensions of each panel facing the gap was 20 × 30 mm, simulating a hem flange. Several mixed pairs of coating types were also tested in this way. Another set of coated samples were exposed prepared as single panels for accelerated corrosion testing, these were compared with single panels connected with nylon fasteners to a mounting bar that was located below the door of a vehicle, shown in Fig. 11.47.

The hem flange samples were then subjected to a test intended to simulate underbody corrosion, GMW14872, a General Motors worldwide cyclic corrosion laboratory test [73] Each 24 h cycle consists of three stages: a dry stage for 8 h at 25°C and 45% relative humidity followed by a wet stage for 8 h at 49°C and 100% relative humidity, then a drying stage 8 h at 60°C at 30% relative humidity. Every 2 h within the first 8 h stage, a neutral salt solution consisting of 0.9% NaCl, 0.1% $CaCl_2$, and 0.075% $NaHCO_3$ is sprayed for 5 min duration at 2 h intervals. The proving ground cycle for the test vehicle included periods of driving in salt water, on gravel roads, and soaking the vehicle in a high temperature and humidity chamber. The condition of the samples after 10 cycles of the proving ground test was reported to simulate 10 years of typical vehicle service [72].

Fig. 11.48 shows the results of the hem flange corrosion test after up to 63 GMW14872 corrosion test cycles. Two matched-pair results are shown, for GI/GI and GA/GA. Two mixed-pair results, for Zn-Al-Mg (MAZ)/GI and Zn-Al-Mg (MAZ)/GA, are also shown. The results are termed as corrosion depth, which is the same as coating thickness reduction; results are shown for the coatings on each side of the 0.25 mm gap. The severity of corrosion decreases for both the GI and GA coatings when they are used together with the Zn-Al-Mg coating, this was attributed to the formation of the dense corrosion products simonkolleite and layered double hydroxide that formed in the 0.25 mm gap when the Zn-Al-Mg coating was present; only porous, poorly-protecting corrosion products that could not prevent the spread of corrosion were formed in the gap of the GI and GA pairs.

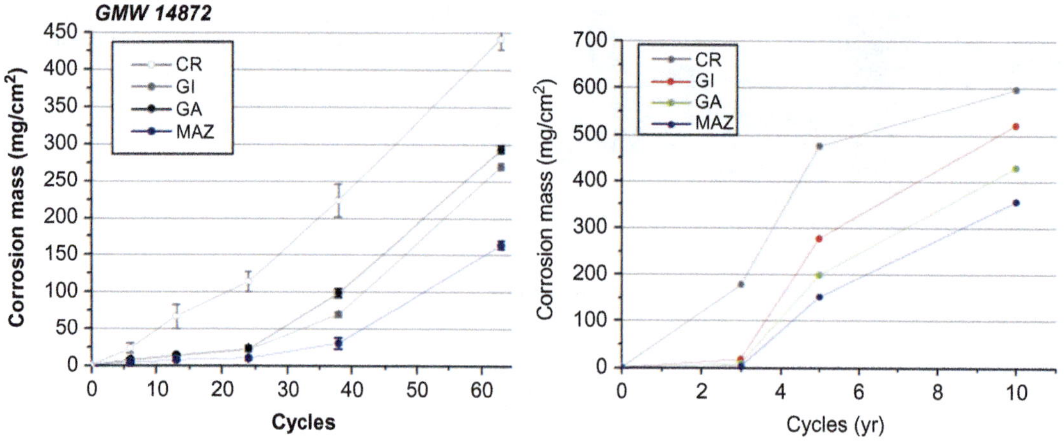

FIG. 11.49 Results of corrosion panel test result in GMW14872 cyclic tests (left) and vehicle proving ground tests (right). Each cycle of the proving ground test was reported to be equal to 1 year of typical vehicle service [72].

TABLE 11.10 Corrosion rates for samples in the GMW14872 and proving ground tests, and ratio of proving ground to GMW14872 corrosion rates.

	Corrosion rate ($mg/cm^2/cycle$)		
Coating	GMW14872	Proving ground	Ratio
Uncoated	0.27	59.64	221
GI	0.04	6.25	156
GA	0.04	3.26	78
ZnAlMg	0.02	1.4	67

Derived from Y.H. Yoo, T.C. Kim, S.J. Yoo, Anti-corrosion performance of ZnMgAl-coated steels for automotive applications, in: Proc. 12th International Conference on Zinc and Zinc Alloy Coated Steel Sheet (Galvatech 2021), June 22–24, 2021, Virtual Conference, Austrian Society for Metallurgy and Materials, pp. 116–121.

The corrosion rates of the coated steel samples after various cycles of the GMW14872 and proving ground tests are shown in Fig. 11.49. After removal from the corrosion test in both cases, the coupons were blasted with alumina grit that had a 75 μm average particle size. For the GMW14872 test, the weight loss due to corrosion of the coating is observed for about the first 20 cycles; after that corrosion of the steel substrate begins and the slope for the coated samples becomes similar to the slope shown for the uncoated sample. For the proving ground tests in Fig. 11.49, a steep increase in corrosion rate between cycles 3 and 5 is observed; this is due to adjustment of the severity of the test in response to the lower than expected weight losses during the first three cycles. After the first five cycles, the slopes of the coated steels become similar to the uncoated steel, indicating that the coating layers were completely consumed and the corrosion of steel was responsible for the measured weight corrosion loss. The weight loss of GI is observed to be higher than GA for the proving ground samples, conflicting with the laboratory tests, the reason for this is unknown. The average corrosion rate per cycle for the laboratory and proving ground tests for each of the coatings is shown in Table 11.10; for the proving ground tests, the slopes up to three cycles are used. Comparing the corrosion rates in Table 11.10, the proving ground rates for ZnAlMg and GA are 67–78 times that of the accelerated test; the GI ratio is double this because of the unexplained high rate of corrosion in the proving ground test. The uncoated samples, with their higher rate of corrosion in both instances, have a much higher ratio, indicating that the benefit of zinc coating protection increases as corrosion severity increases.

Yoo also estimated the expected service life of the ZnAlMg coating, based on assumptions made from the proving ground results where this 35 g/m^2/side coating took about 2.5 years to completely corrode, indicating a corrosion rate of 13.97 g/m^2/year. After depletion of this coating, the uncoated steel was then calculated to corrode at a rate of

59.637 mg/cm^2/year. Converting this to thickness by using the density of Fe, 7.87 g/cm^3, the steel substrate would then corrode at a rate of 0.076 mm/year. The corrosion life was then estimated by the equation

$$\text{Corrosion life (years)} = \frac{\text{Coating wt.}(\text{g/m}^2)}{13.97\ (\text{g/m}^2/\text{year})} + \frac{\text{Thickness of sheet (mm)}}{0.076\ (\text{mm/year}) \times (\text{safety factor})} \tag{11.2}$$

If a sheet thickness of 1 mm is used with a coating weight of 35 g/m^2 per side and the safety factor is 1.3, a corrosion life of 12.6 years is calculated. It must be emphasized that this is only in the case where the coated steel receives not further protective treatment such as application of an automotive paint system, or in the case of hem flanges, application of protective waxes or undercoatings.

The effect of the thickness of the gap in the hem flange sample was investigated by Mizuno [74]. One kind of sample had a 0.3 mm gap, slightly larger than Yoo's sample described above. The other had no gap. The areas of the test coupons facing each other in this test were zinc coatings without further surface treatments, as in Yoo's experiments. These were subjected to 60 days of SAE J2334 cyclic corrosion testing, where the daily cycle consists of 6 h at 50°C and 100% relative humidity, then salt application by dip, fog or spray for 15 min, then drying for 17 h 45 min. at 60°C and 50% relative humidity. For the 5-day cycle, the samples are then left for a 2-day weekend at 60°C and 50% relative humidity before resuming, whereas for the 7-day cycle used in fully automatic cabinets, the cycle used for each of the 5 days continues for each of the 7 days of the week [75]. Three types of coatings were tested, a 35 g/m^2/side Zn-Ni electrogalvanized coating, a 40 g/m^2/side GA coating, and a 100 g/m^2/side GI coating. A sketch of the sample configuration and appearance of the corroded surfaces from inside the gap are shown in Fig. 11.50 after sample disassembly on samples exposed for 60 days of corrosion test cycling. For the samples with no crevice gap, corrosion tends to be more suppressed as coating thickness increased and coating composition did not have a significant impact on the severity of observed corrosion. With the 0.3 mm gap, the ZnNi coated steel showed better corrosion resistance than GA and GI. It was concluded that coating composition could be more crucial than coating thickness when specimen has wide crevice, agreeing with Yoo's work. However, in a sufficiently wide gap, it is expected that the zinc coating will be protected by the usual phosphating and electrophoretic primers. Therefore, only narrow or zero-gap hem flange specimens were concluded to give accurate simulations of actual vehicle service.

Mizuno then compared the results of disassembled hem flanges from door side-sills taken from actual vehicles that had seen 5–9 years use in Europe (Austria, Denmark, Sweden) or 10–20 years use in Thailand with results from standard automotive cyclic tests. The selected vehicles utilized zinc-coated steels in their initial construction together with customary full automotive paint systems but without further use of waxes and undercoatings. Corrosion products formed in the disassembled hem flanges were removed with an inhibited HCl solution in accordance with ISO 8407C.3.1 [59]. Corrosion depth was measured with a needle micrometer within the lapped portions of the parts. Fig. 11.51 shows corrosion depth in the hem flange parts of the doors of the test vehicles as a function of the use period of these vehicles. The numbers in parentheses represent the one-side coating thickness. In Europe, the zinc coating of 70–80 g/m^2 suppressed corrosion of the steel for approximately 7 years, after which the steel started to corrode. In contrast, no corrosion was observed on the steels until nearly 10 years of vehicle service in Thailand.

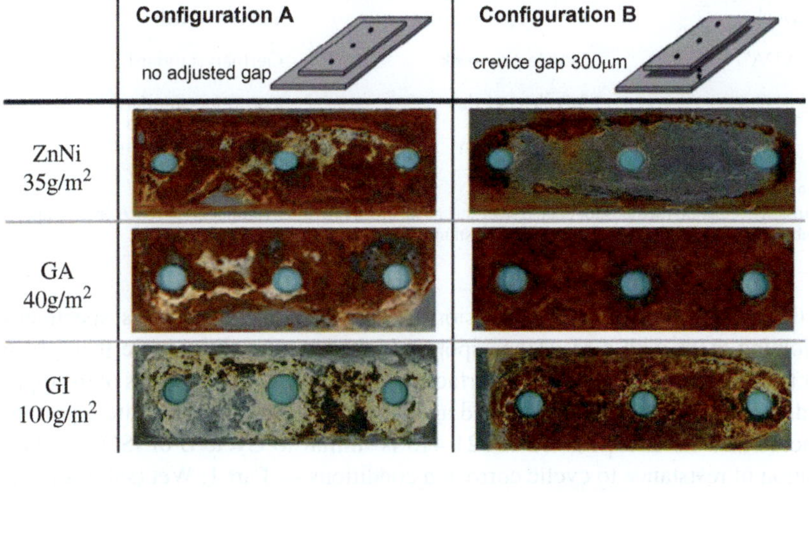

FIG. 11.50 Sketch of hem flange corrosion configuration and appearance of disassembled facing surfaces after 60 SAE days of J-2334 corrosion test cycles [74].

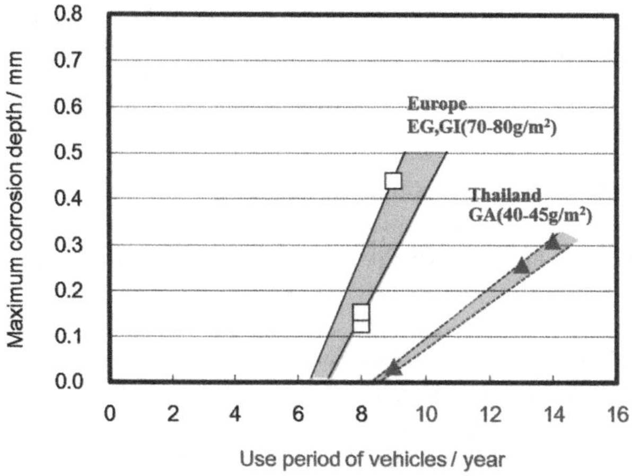

FIG. 11.51 Depth of corrosive attack from disassembled hem flanges of vehicles after various years of service in Europe (Austria, Denmark, Sweden) and Thailand [74].

TABLE 11.11 Coated steel types used together with the cyclic corrosion tests shown in Table 11.12, the coating weights shown are the coating weights on each side of the coated sheets with equivalent coating thickness.

Sheet sample type	Sample symbol	Coating thickness (g/m^2)	Equiv. coating thickness (μm)
Cold rolled	CRS	0	0
Zn electrogalvanized	EG	18–68	2.5–9.5
Zn-Ni electrogalvanized (11%–13%Ni)	ZnNi	27–34	3.7–4.7
Galvanneal (10%–11%Fe)	GA	43–51	6.0–7.1
Galvanized	GI	64–100	9.0–14.0

Adapted from D. Mizuno, Automotive corrosion and corrosion evaluation methods for coated steel sheets, in: Proc. 11th International Conference on Zinc and Zinc Alloy Coated Steel Sheet (Galvatech 2017), November 12–16, 2017, Tokyo, Iron and Steel Institute of Japan, Tokyo, pp. 553–558.

TABLE 11.12 Cyclic tests used to compare performance of the coatings shown in Table 11.11 with on-vehicle tests [74].

Test number	Corrosion test	Cycle frequency	Standard type
T1	JASO M609-91	3 cycles per day	Japanese standard
T2	SAE J2334 5 cycles/week	5 cycles per week	USA standard
T3	SAE J2334 7 cycles/week	7 cycles per week	USA standard
T4	VDA 621-415 ("old VDA")	1 cycle per week	German standard
T5	CCT-1	3 cycles per day	Corporate standard
T6	CCT-2	1 cycle per day	Corporate standard

VDA 621-415 is similar to Cycle B of ISO Standard 11997-1.
Adapted from D. Mizuno, Automotive corrosion and corrosion evaluation methods for coated steel sheets, in: Proc. 11th International Conference on Zinc and Zinc Alloy Coated Steel Sheet (Galvatech 2017), November 12–16, 2017, Tokyo, Iron and Steel Institute of Japan, Tokyo, pp. 553–558.

The results of these on-vehicle tests were then compared with cyclic corrosion testing of simulated flange test specimens made by overlapping two panels and then joining them with a resistance spot weld. These assemblies were then given phosphating and electrophoretic paint primer treatments typical of vehicle surface treatments. Duplicate panels of the types of samples shown in Table 11.11 were tested. The cyclic corrosion tests used are shown in Table 11.12 and are representative of tests used in global practice. It is noted that the European VDA 621-415 is similar to Cycle B of ISO Standard 11997-1 "Paints and varnishes — Determination of resistance to cyclic corrosion conditions — Part 1: Wet (salt fog)/dry/

humid" [76]. Two JFE corporate standard tests, CCT-1 and CCT-2, were also used. The corrosion products and paint were stripped using the same procedure utilized for the on-vehicle tests above. After measuring the corrosion depth of the test coupons with a needle micrometer, maximum corrosion depth was determined by the extreme value analysis of Gumbel distribution [77].

The appearance of the interior surfaces of the lapped coupons of the hot-dip coated steels after disassembly is shown in Fig. 11.52. As expected, the results difference because of the differences in overall wetness periods and severity of the tests. The appearance of these samples after exposure to test T1 is similar, even though they had different coatings and coating thicknesses. On the other hand, the amount of red rust formed on the three types coated steels was different in the test T3. The entire surface of the GA coating was covered with red rust, whereas white rust was still dominant in the heavier coating weight (100 g/m^2) GI. It was concluded that the corrosion performance of the tested coated steels strongly depends on the details of the cyclic test conditions.

Mizuno then compared the results of all the cyclic tests by computing the corrosion rates of the coating and the corrosion rate of the steel substrate that begins after the coating is consumed. A perforation corrosion index (PCI) was also calculated, which is the ratio of the steel corrosion rate to the zinc coating corrosion rate. The PCI was also computed for the results taken from analysis of the hem flanges of the on-vehicle tests. The values are shown in Table 11.13. Lower PCI values indicate that, for those tests, the zinc coating is not being very protective; high values, near 100, more accurately simulate actual vehicle service conditions for these flanged samples.

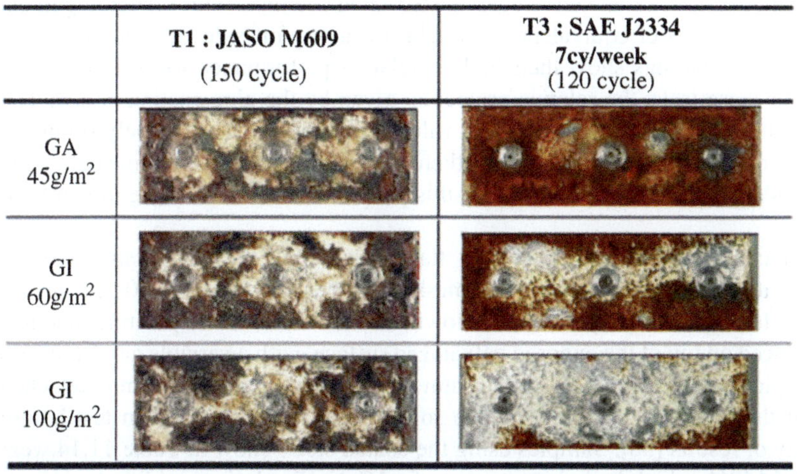

FIG. 11.52 Appearance of the interior surfaces of lap coupons of hot-dip coating panels after 150 cycles of JASO M609 or 120 cycles of SAE J-224 corrosion testing (Tests T1 and T3 in Table 11.12, respectively) [74].

TABLE 11.13 Corrosion rates and perforation corrosion index (PCI) for coatings tested using the results of the cyclic tests of Table 11.12 and the on-vehicle test result of Fig. 11.51 [74] together with results from reported USA tests.

Test number from Table 11.12	Steel corrosion rate per cycle of test (μm/cycle)	Zinc corrosion rate per cycle of test (μm/cycle)	PCI (perforation corrosion index)
T1	3.4	0.18	19
T2	6.6	0.23	29
T3	7.2	0.13	56
T4	20	0.53	38
T5	7.7	0.47	16
T6	7.3	0.07	98
PCI computed from results from vehicles in service in North America [78], Europe and Thailand			75–120

Adapted from L.A. Roudabush, T.E. Dorsett, A review of perforation corrosion testing 1980–1990, in: Proc. 5th Automotive Corrosion & Protection Conference, October 21–23, 1991, Detroit, MI, Warrendale, PA, Society of Automotive Engineers, Publication P-250, Paper 912285, p. 221.

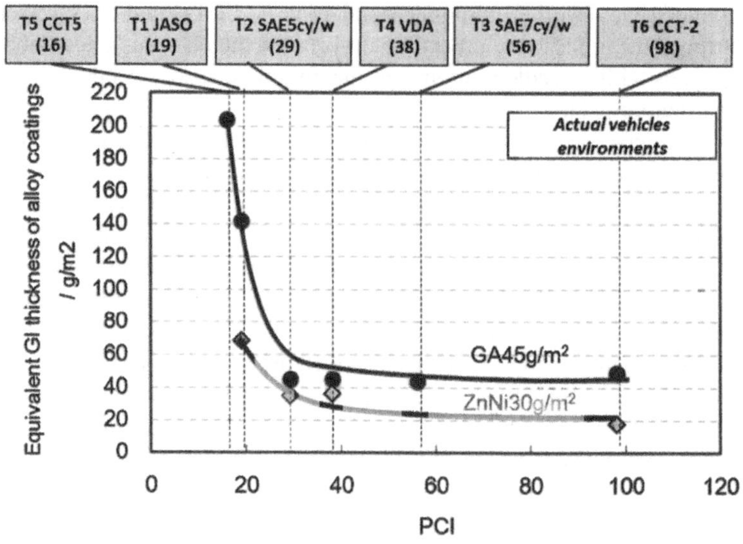

FIG. 11.53 Performance in GA and Zn-Ni coatings expressed in terms of equivalent GI thickness for various cyclic tests and actual vehicle service conditions [74].

Fig. 11.53 compares the corrosion performance of the GA and electrogalvanized ZnNi coatings to the GI coating in the various tests by expressing their corrosion rates in terms of equivalent GI weight, for a 45 g/m^2/side GA coating, and a 30 g/m^2/side Zn-Ni electrogalvanized coating. The coatings show their highest relative performance for tests with low PCI values, but care must be taken because for these tests, the relative protection given by the zinc coating is low. The relative performance of these two coatings is much closer to GI when the PCI value is high, which more closely simulates corrosion performance in actual service environments. In these environments, there is a significant period of time before the initiation of coating corrosion that is not included in Mizuno's calculations; this period of time can add significantly to vehicle service life.

Another test sample configuration for simulating the corrosion at hem flanges is called the glass flange crevice specimen. In this test, the galvanized sample with no extra pretreatments is masked with tape over a 66 × 20 mm surface area, the overall sample is then phosphate and electrophoretically primed, following normal automotive painting practice. The mask is then removed, and a glass panel placed on top of the exposed galvanized surface, with a variable gap from zero (the thickness of the electrophoretic primer, typically 0.025 mm [79]) to 0.25 mm over the glass length of 75 mm, attached with plastic clips [80]. Some researchers fill the crevice with the corrosive solution used in the corrosion test before beginning, claiming this reduces the variability of results [38]. Samples using the coated steels show in Table 11.14 were subjected to Volvo cyclic standard test 432-0014 [81], which includes spraying of a 1% NaCl solution, acidified with sulfuric acid to a pH of 4.2. The main cycle begins with a wet stage at 95% relative humidity at 35°C for 4 h, followed by 2 h transition during which temperature is increased from 35°C to 45°C while relative humidity is lowered from 95% to 50%. The higher temperature, dryer stage is then held for 4 h. A 2 h transition from 45°C to 35°C while simultaneously increasing humidity from 50% to 95% then completes each 12 h cycle. On Mondays and Fridays, the first 15 min of the initial wet stage

TABLE 11.14 Coatings used for glass flange crevice tests.

	Coating weight evaluated side (g/m^2)	Coating thickness (μm)	Al%	Mg%	Fe%
GI100	57	8	0.5	<0.01	0.38
GI140	82	12	0.4	<0.01	0.32
GA90	41	6	0.3	<0.01	10.2
MZ70	35	5	1.8	1.8	0.17
MZ100	53	8	1.7	1.6	0.16
MZ140	68	10	1.6	1.6	0.13

The designator in the left-hand column indicates nominal two-side coating weights, g/m^2 [80].

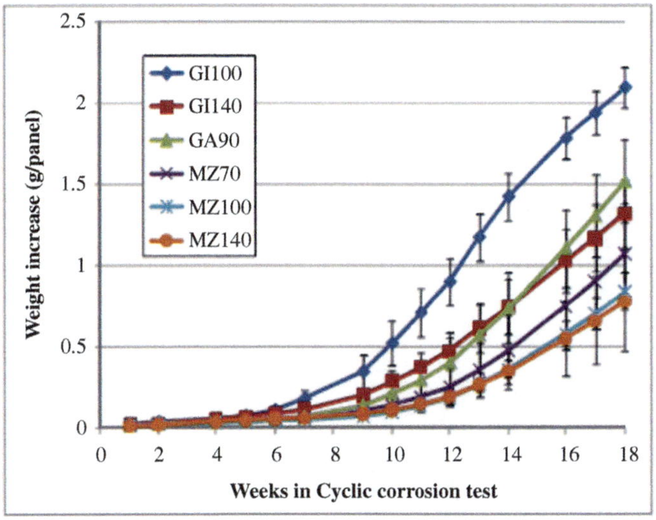

FIG. 11.54 Weight gain of unstripped Table 11.14 coatings in glass hem flange configurations for up to 18 weeks of Volvo 432-0014 cyclic corrosion testing [80].

is replaced by spraying with the acidified salt solution followed by holding for 1 h 45 min at 35°C and 95%–99% relative humidity. This spraying and humid exposure cycle is repeated twice more, for a total of 6 h. This is followed by increasing the temperature from 35°C to 45°C while relative humidity is lowered from 95% to 50% during 2 h. The higher temperature, dryer stage is then held for 2 h before resuming with the lower temperature, higher humidity stage.

Weight gain results after up to 18 weeks of Volvo cyclic testing are shown in Fig. 11.54. In this case, corrosion products were not removed. The lighter coating weight (57/g/m^2 per side) GI coating increases in mass gain the earliest and reaches a higher level than the other specimens. The first appearance of red rust on the GA sample was seen after 1 week; this is expected because of the iron content of this coating. For the GI coating red rust appeared between 2 and 4 weeks, while for the MZ coatings, red rust did not appear until 4–6 weeks, time until the appearance of red rust in both the GI and MZ sample groups increased with increasing coating weight. As in the cyclic test results described before, the slope of the sample weight gain curves shown in Fig. 11.54 is related to the type of corrosion occurring at a particular time; the low slope during the first few weeks is related to corrosion of the zinc coating, with a relatively small weight gain from the formed corrosion products. In each case, the slope increases as the steel begins to corrode while some zinc coating is still present, finally the coating is consumed inside the hem flange sample gap and only steel is corroding. All samples except for MZ100 and MZ140 reached the point where only steel is corroding after 18 weeks in this test; after this time these two coatings were still providing some cathodic protection to the steel substrate. Despite the early appearance of red rust, the curves in Fig. 11.54 showed that GA90 had a weight gain similar to GI140, within the limits of the shown error bars. Fig. 11.55 shows the depth of attack on the steel substrate, which was measured after removal of the corrosion products; the deepest points of attack are shown in this figure. All MZ samples show less attack of the steel. As in the weight gain results, GA90 and GI140 performed similarly and both better than GI100. The MZ-coated samples showed the smallest depth of attack. The highest attack points for the MZ-coated samples were in the thinnest gap area of the glass flange crevice. This is different from the result of Yoo above who found that corrosion for the Zn and Zn-Ni coatings was suppressed in the zero-gap flange specimens for the Zn and Zn-Ni coatings. The difference in these results may be related to the types of corrosion products formed in each case: for the Zn and Zn-Ni, the products are more porous and more electrically conductive than the formed on the ZnAlMg coatings. Therefore, the narrowest gap may prevent the early formation of double-layer hydroxide and other compact, electrically insulating corrosion products, resulting in improved comparative performance for the ZnAlMg coating with hem flange gaps greater than zero. The narrowest gap is very difficult to dry in the Volvo cyclic test, so that it would have a longer time of wetness. Also, O_2 and CO_2 transport into a very narrow gap would be more difficult, slowing the development of protective corrosion products and the extent of the oxygen reduction reaction on cathodic sites. As the gap widens, the ZnAlMg coatings would then be comparatively more corrosion-resistant; overall the depth of attack shown in Fig. 11.55 is lower for these coatings because much of the glass hem flange crevice is sufficiently large to permit the formation of the stable corrosion products associated with these coatings.

Hem flange corrosion test results from another laboratory, using the VDA 621-415 cyclic test, gave the results shown in Fig. 11.56. Here, a GI 275 coating shows a longer time to red rust appearance than a ZnAlMg 100 coating [50]. The difference in the results of Figs. 11.55 and 11.56 is related to the heavier coating weight of the GI coating in Fig. 11.56 vs. the heaviest GI weight in Fig. 11.55, and also the fact that the VDA 621-415 cyclic corrosion test is drier than the Volvo

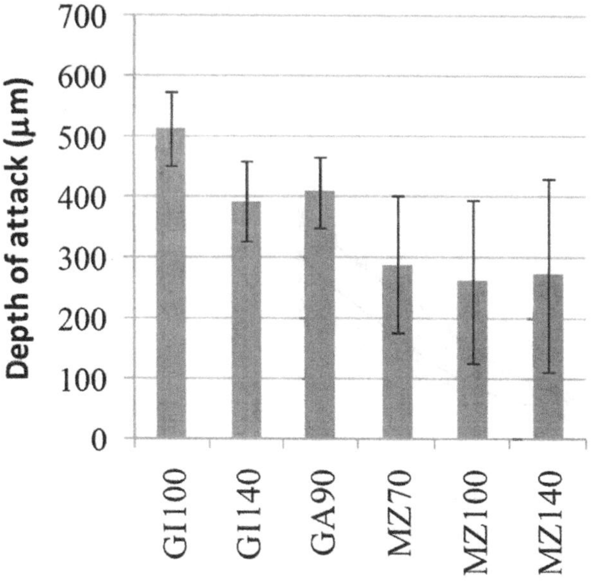

FIG. 11.55 Depth of attack on steel substrates used with Table 11.14 coatings on stripped samples after 18 weeks of Volvo 432-0014 cyclic corrosion testing [80].

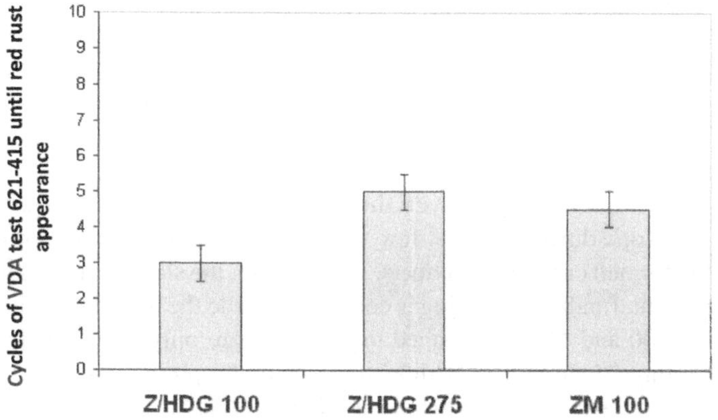

FIG. 11.56 Cycles of VDA 621-415 testing until appearance of red rust for two coating weights of a GI coating and a Zn-2%Al-2%Mg coating. All coating weights are nominal two-side coating weights [50].

432-0014 cyclic corrosion test. The Cycle B of ISO Standard 11997-1 (which as noted above, is similar to VDA 621-415) is also drier than the Volvo test and therefore expected to be advantageous to coatings with less protective corrosion products.

The hem flange test was also used to evaluate the corrosion performance of a Zn-5%Cr electrogalvanized coating. When this coating with a coating weight of 35 g/m^2 per side was evaluated using the VDA 621-415 test, it was found to give a sixfold longer time to the appearance of 5% red rust in the hem flange area compared with a pure Zn electrogalvanized coating with a coating weight of 75 g/m^2 per side [50].

11.6.2 Corrosion potentials in the galvanneal coating

As described in Chapter 8, the galvanneal coating can include Zn-Fe phases that are produced as the galvanneal heat treatment step proceeds, with the phases becoming richer in Fe as the coating develops, with the Fe content generally increasing from the surface to the substrate; the interfaces between phases can be diffuse rather than sharp; features such as outbursts can be present, disrupting stratified Fe-Zn phase layers. The Fe content affects the electronegativity of the coating, i.e., the capability of the coating to provide sacrificial protection for the steel substrate, this is measured by the characteristic potential in volts relative to a standard potential. Chakraborty [1] measured these potentials relative to a standard silver/silver chloride electrode using a galvanostat containing a solution of 250 g/L NaCl and 50 g/L $ZnSO_4$, giving a pH of 5. The voltage was varied to give a constant current density of 0.3 mA/cm^2 on 15 × 15 mm coupons of a

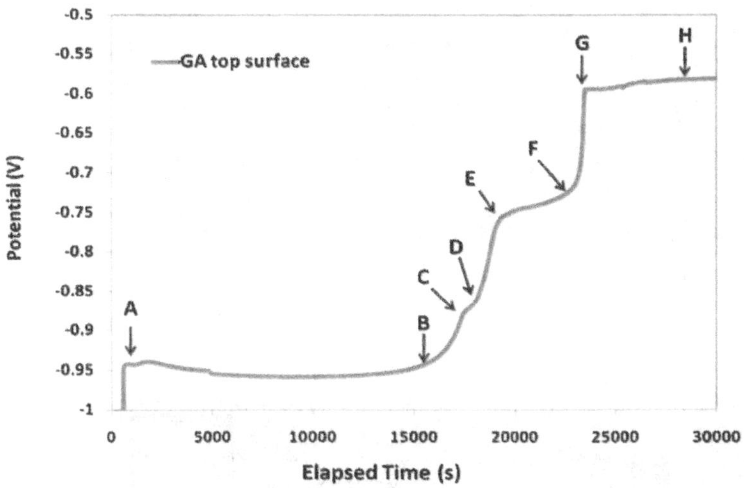

FIG. 11.57 Progressive dissolution of a GA coating, indicating the characteristic voltages for each Zn-Fe phase measured with a galvanostat [1].

0.8 mm thick IF steel galvanized in a bath with a 0.132%–0.135% effective Al composition that was galvannealing at 550°C followed by a 0.8% skin pass elongation after galvannealing. Fig. 11.57 shows that voltage increased (became less negative) with time as successive layers of the GA phases were dissolved. The location of the top surface is also indicated. This experiment was run in triplicate to account for variabilities in the GA coating morphology. The initial increase in voltage, marked as point A, marks the interface between the ζ and δ phases described in Chapter 8. This interface appeared to be of negligible width and could not be observed using microscopy examinations. The voltage remains at a basically constant level of −0.95 V until about 15,000 s, point B, this long duration corresponds to dissolution of the δ phase, which is the thickest constituent of this coating. Voltage then rises from −0.950 to −0.875 V (point C) and plateaus at this volage for about 1000 s, until point D is reached. The trace between points B and C is related to the δ-Γ_1 interface, so that the plateau between points C and D corresponds to dissolution of the thin Γ_1 phase. Voltage then increases between points D and E to −0.75 V, corresponding to the $\Gamma_1 + \Gamma$ interface; the plateau between points E and F corresponds to dissolution of the Γ plateau, which persists for 4000 s. After dissolution of the Γ phase, voltage rises to 0.6 V, reaching the steel potential at point G. These readings were correlated with scanning electron microscope examinations of the surfaces of the GA samples from which corrosion products had been stripped after removal from the galvanostat. Sample (i) in Fig. 11.58 was taken just after the voltage of 0.95 V was read, 1000s after beginning the experiment shown in Fig. 11.57. This shows the zeta-delta interface. Sample (ii), taken after 10,000 s, shows the delta phase surface. Sample (iii), after 19,000 s of testing, shows the gamma-1-gamma interface; some cracking of the coating is evident; this was not observed in the delta phase. Sample (iv) was taken after 22,000 s in the gamma phase plateau, a higher surface crack density is observed. Sample (v) was taken after 24,000 s in the transition from gamma to the steel substrate; sample (vi) is the appearance of the steel surface from which all coating has been dissolved after 28,000 s. The compositions measured by EDS are also shown for each sample in Fig. 11.58 and can be compared with the Fig. 11.10 phase diagram in Chapter 8 to confirm phase identification.

11.6.3 Exposed underbody corrosion

The beneficial effect of galvanic and barrier protection of a GI coating with a nominal two-side very light coating weight of 50 g/m^2 GI coating for steel in several auto underbody applications was demonstrated by Suzuki, who instrumented a typical passenger automobile with sensors that measured corrosion current; at the same time temperature and humidity readings were recorded [82]. For this test, the driver's side of the car (the right-hand side in Japan, facing forward) was sprayed once a week with 10 L of 0.5% NaCl solution while the passenger side was sprayed with tap water. The spray was applied to the outer surface of the side body panels and the underside of the floor pan. The car was periodically driven. Whereas wet/dry cycles for the side of the car sprayed with pure water, and also the salt-sprayed surfaces of the car that were exposed to washing by rainfall correlated with incidences of spray application and rainfall events, the times of wetness on the outer surface of the floor pan and the driver's side door inner panel were higher, indicating the difficulty of washing away the deposit salt. Unpainted test panels were fastened to several locations and correlated with sensor output: the back sides and edges of the panels, both cold-rolled and GI-coated, were covered with a protective film. Fig. 11.59 shows the electrical current, measured in coulombs per day, measured by sensors adjacent to the samples mounted in four locations.

Spectrum	Al	Fe	Zn
1	1.31	12.46	86.24
2	1.45	12.32	86.23
3	1.24	12.16	86.60

Spectrum	Al	Fe	Zn
1	1.18	11.72	87.11
2	1.08	12.11	86.81
3	-	12.39	87.61

Spectrum	Al	Fe	Zn
1	-	18.84	81.16
2	1.77	25.27	72.46
3	1.67	29.19	69.14

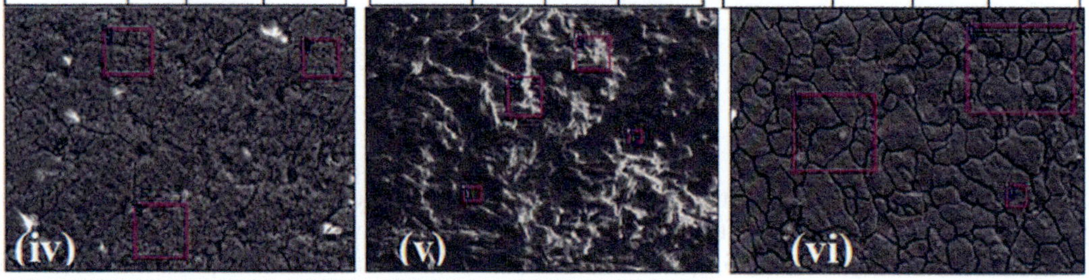

Spectrum	Al	Fe	Zn
1	2.01	42.15	55.83
2	1.57	41.54	56.88
3	2.26	41.41	56.33

Spectrum	O	Al	Fe	Zn
1		0.96	96.76	2.28
2	33.32	1.08	13.78	40.91
3	33.89	1.39	12.73	41.29
4		0.70	96.86	2.44

Spectrum	Fe	Zn
1	99.23	0.77
2	100.00	-
3	100.00	-

FIG. 11.58 Appearance of stripped GA coating surfaces after removal from galvanostatic testing after intervals of immersion (i) −0.95 V, 1000 s (ζ-δ phase interface); (ii) −0.95 V, 10,000 s (δ phase); (iii) −0.8 V, 19,000 s (Γ_1-Γ gamma interface); (iv) −0.73 V, 22,000 s (Γ phase); (v) −0.65 V, 24,000 s (Γ-steel interface); (vi) −0.6 V 28,000 s IF (steel surface) [1].

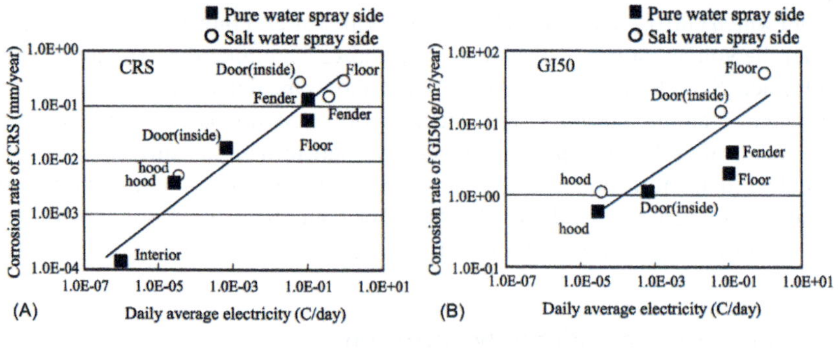

FIG. 11.59 Correlation of corrosion rate and current flow, measured in coulombs per day from corrosion sensor output, for a test vehicle sprayed on one side with tap water and the other side with a 0.5%NaCl solution (A) cold-rolled steel (B) steel with GI coating with a two-side coating weight of 50 g/m^2 [82].

The corrosion thickness loss of the cold rolled steel is reported in mm/y, whereas the GI coating weight loss is reported in g/m^2/y; conversion of the latter shows that a 1 g/m^2/y loss is equivalent to a GI thickness loss of 0.37 μm/y for these samples where the corrosion behavior is measured only on one side of the panel. The higher corrosion rates for the salt sprayed side are evident, together with the magnitudes of differences in corrosion rates between the cold-rolled and zinc-coated samples. These data were used to calculate the predicted corrosion rates for the investigated locations, Fig. 11.60. The differences in the corrosion rates shown were the difficulty of washing away the corrosive NaCl solution in the locations that had the highest corrosion rates. For the location with the highest GI corrosion rate, 16 g/m^2 per year corresponds to a thickness loss of 5.9 μm/y. The time to initiation of red rust was then calculated for the different vehicle locations, Fig. 11.61, for this coating with a nominal one-side coating thickness of 9.3 μm. It is observed that for the salt-sprayed underside of the floor, the time to red rust is very short; it is clear that to obtain satisfactory performance in this location, the customary protective organic coatings would need to be applied; a heavier coating weight would also assist in providing a longer duration until red rust is observed.

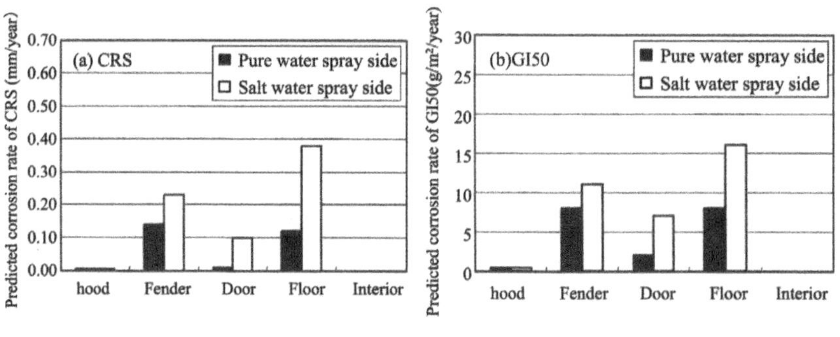

FIG. 11.60 Calculated annual corrosion rates for several vehicle locations exposed to tap water or 0.5% NaCl solution for (A) cold-rolled steel (B) steel with GI coating with a two-side coating weight of $50\,g/m^2$ [82].

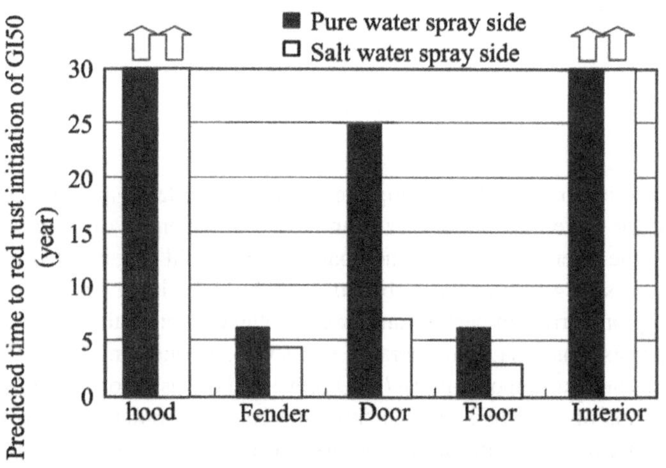

FIG. 11.61 Calculated time to appearance of red rust for a GI coating with a two-side coating weight of $50\,g/m^2$ for several vehicle locations exposed to tap water or 0.5% NaCl solution [82].

11.7 Corrosion of galvanized press-hardened steels

Processing conditions and phase equilibria associated with Zn-coated press hardened steels are reviewed in Chapter 13. For the conventional hot press forming grades such as 22MnB5 that contains 0.25%C, 0.4%Si, 1.35%Mn, 0.25%Cr, and 0.004%B, soaking temperatures between 890°C and 930°C to achieve full austenization before beginning the hot press forming process are typical. This range is well above the Fe-Zn peritectic temperature of 782°C depicted in the Fig. 11.37 Fe-Zn phase diagram in Chapter 13. Above this temperature, Γ (Fe_3Zn_{10}) phase is not stable, transforming to a mixture of Zn-containing ferrite, termed α-ferrite(Zn) and liquid. Therefore, Zn or Zn-Fe coated steel samples held above this peritectic can be expected to experience loss of the volume fraction of Γ phase as soaking time proceeds. Fig. 11.62 shows the relationship between soaking time in air at 900°C and the phases found to be present in quenched samples taken at various times [83]. The 1.5 mm thick 22MnB5 steel sample was initially coated with a GI composition with $70\,g/m^2$ per side. The Zn coating soaked at 900°C transforms rapidly to Fe-Zn phases as its temperature increases through the conventional galvannealing temperature range of 500–550°C; beyond this range the coating is in the Γ stability range until it reaches the peritectic at 782°C. The start of the annealing time axis in 58 is when the sample temperature reached 890°C, by which time the coating mainly consisted of Γ phase. The volume fraction of Γ is reduced to about 15% after 180 s of soaking time; a minor fraction of ZnO is observed with the rest of the coating consisting of α-ferrite (Zn). Because of the small thickness of these coating layers, it was necessary to use low angle X-ray diffraction to obtain phase fraction measurements. ZnO is found as a surface layer and is initially adherent; this adherence decreases as its thickness increases. ZnO plays an important role in slowing evaporation of Zn from the coating into the process atmosphere during soaking, therefore controlling its thickness helps control the development of the hot press formed coating. The corrosion potential, or capability of the coating to protect the steel substrate from corrosion, was assessed by potentiostatic polarization measurements. It was found that samples annealed at 900°C for 300 s or less had at least some Γ phase present, permitting the coating to achieve a corrosion potential of −0.85 V referenced to a standard calomel electrode (SCE). Note that this is a different reference electrode than that used in the Chakraborty work described above.

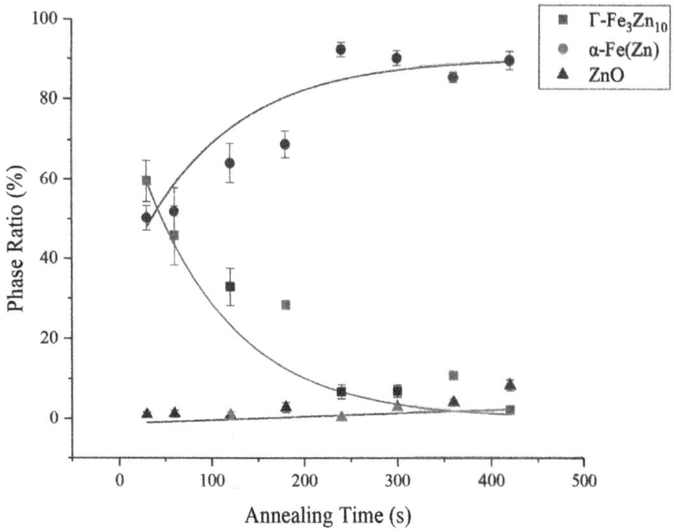

FIG. 11.62 Relationship between annealing time at 900°C and phase fractions present in a 22MnB5 steel initially coated with a 70 g/m² per side GI coating [83].

Increasing the soaking time increased the corrosion potential to −0.72 V, making the coating much less galvanically protective of the steel substrate. Coatings that were soaked for more than 300 s were found to have a quantity of Γ phase insufficient to provide galvanic protection. This is despite the overall growth of the coating that would be expected to improve its barrier protection capabilities, from 20 μm at 30 s to 40 μm at 420 s, because of the diffusional growth of α-Fe(Zn). Further insights into the effects of coating makeup on corrosion performance were obtained from galvanostatic corrosion tests. The corrosion test solution used for these tests was a naturally aerated salt solution consisting of 100 g $ZnSO_4 \cdot 7H_2O$, 200 g NaCl, and 1000 mL of deionized water. The addition of $ZnSO_4 \cdot 7H_2O$ prevented diffusion of zinc ions from the coating into the solution. Fig. 11.63 shows the results of these tests, which show the corrosion potential at two different currents. A zero current corresponds to the open-circuit potential, that is, the voltage is not decreased by connecting the steel substrate to the coating. The open-circuit potential is observed to remain at the highly protective value of −0.85 V for Γ phase contents greater than 15%, rising rapidly to less negative (less protective) values as the Γ content in the coating decreases below 30%. The other potential shown is the potential arrest, which is the potential at which the Γ stops dissolving in order to protect the substrate, in favor of the α-ferrite(Zn) phase. The current density at which this arresting dissolution potential was observed was 10 mA/cm². A similar pattern of increasing potential with decreasing Γ content is observed, the coating becomes less electronegative at Γ contents less that 15%. Interrupted galvanostatic

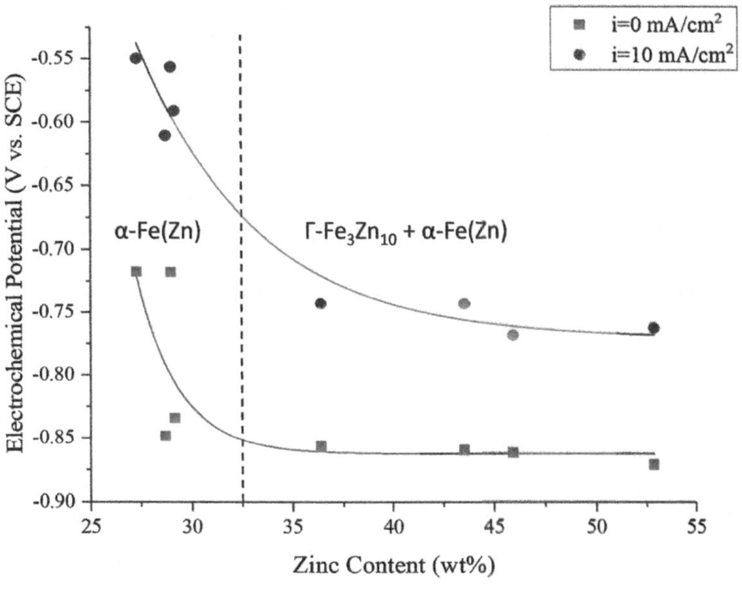

FIG. 11.63 Effect of zinc content on galvanic protection capability of hot press formed coatings. The lower curve shows the open-circuit potential (zero current), the upper curve shows values of potential for the current corresponding to the potential where the Γ ceases to dissolve in favor of the α-ferrite(Zn) phase [83].

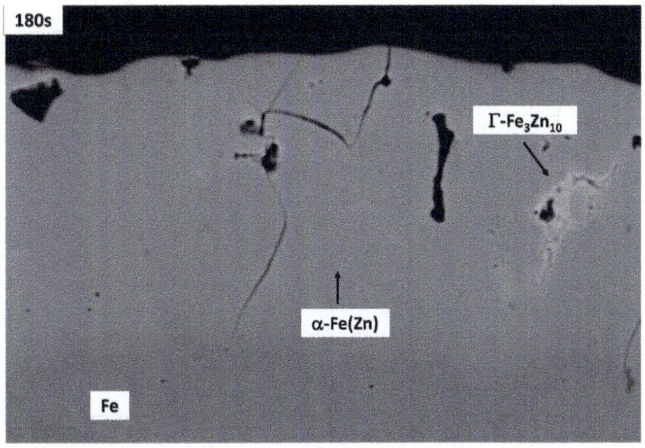

FIG. 11.64 Cross section of the microstructure of a GI coating with an initial coating weight of 70 g/m^2 per side soaked at 900°C for 180 s. The Γ Fe_3Zn_{10} phase has a lighter appearance than the surrounding α-ferrite(Zn) phase. The observed cracks in the coating occurred during rapid cooling and do not extend into the undeformed 22MnB5 substrate [83].

polarization tests together with X-Ray diffraction were used to identify corrosion products that formed as a result of the anodic dissolution of the Γ phase, also to assess how well coatings with different Γ phase fractions protected the steel substrate. The Γ phase corroded first, sacrificing itself to protect both the substrate and the α-ferrite(Zn). An example of the morphology of the coating after 180 s of soaking at 900°C, quenched without deformation, is shown in Fig. 11.64. The Γ phase exists as islands in the surrounding α-ferrite(Zn). Increasing soaking times resulted in both a decrease in the amount of α-ferrite(Zn) remaining in the coating and an increase in the corrosion products formed. Three main corrosion products were formed: akageneite (beta iron hydroxide, $Fe^{+3}O(OH)$), hydrozincite, and simonkolleite. In addition, trace amounts of β-$ZnCl_2$ from the corrosion test solution were detected. Unlike the corrosion products in conventionally produced coated steels, the hydrozincite and simonkolleite layers formed in the galvanostatic tests flaked off easily, their adherence was poor. For samples produced with 180 s of soaking time, the total fraction of corrosion products was 10%, whereas it was 25% for 240 s. Moreover, akageneite made up a larger proportion of the corrosion products at longer soaking times [84].

The corrosion performance of zinc-containing hot press formed coatings was investigated by Hassel [85], who used the VDA 621-415 cyclic corrosion test. A 1-week cycle of this test, shown graphically in Fig. 11.65, consists of a 24 h salt spray test followed by a 96 h in a condensing humidity, then exposure to the ambient climate for 48 h. After 10 weeks of this cycle, the visual appearance of the coating shows reddish-brown corrosion products, but as in the case of the early stage GA corrosion, this was not associated with corrosion of the steel substrate, it was the Fe contained in the coating that corroded to

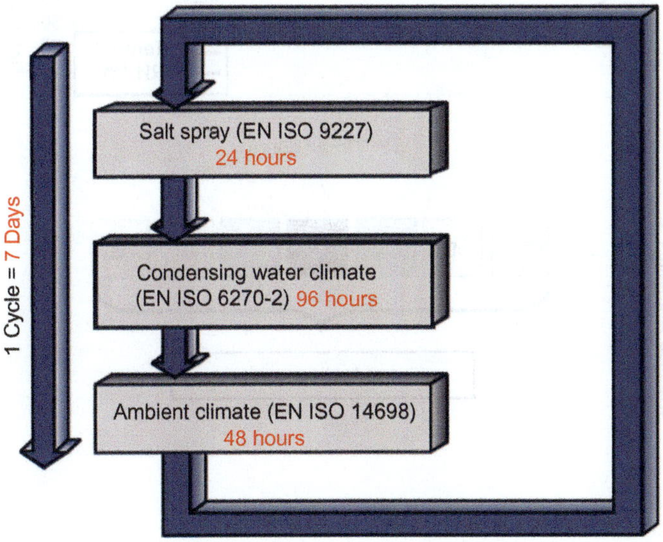

FIG. 11.65 Diagram showing the steps in a complete 1 week cycle of the VDA 621-415 cyclic corrosion test [86].

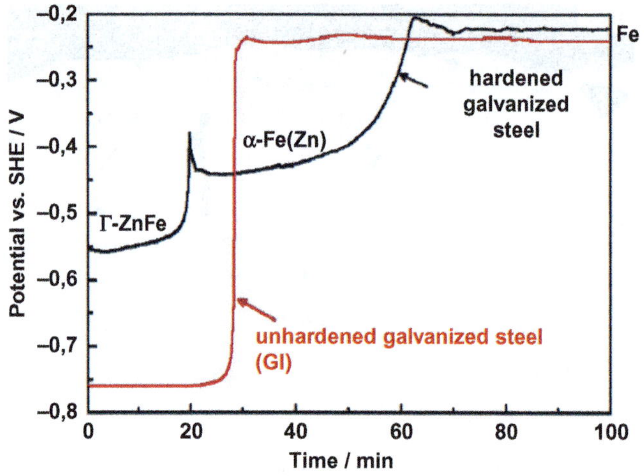

FIG. 11.66 Potential profiles for hot press formed coated steel and the galvanized steel blank prior to press hardening, measured by a galvanostat [86].

form these corrosion products. As in the galvanostatic tests described above, the corrosion products were akaganeite, hydrozincite, and simonkolleite. The electrochemical potentials of the different phases in the hot press formed coatings relative to a standard hydrogen electrode for the steel substrate, α-ferrite(Zn) and Γ (Fe_3Zn_{10}) the respective values measured were −0.22, −0.4, and −0.54. These are shown graphically in Fig. 11.66 from work in the same research group [86], together with the potentials for the starting GI coating. For the latter, the potential of the GI coating is indicated as −0.76 V, which then undergoes a step change to the steel potential of −0.22 V at the coating/steel interface. These values, after converting, are consistent with the voltages measured in the work described previously in galvanostatic testing that used either silver/silver chloride [1] or standard calomel reference electrodes [84] and confirm that the Γ phase is more galvanically protective of the steel substrate than the α-ferrite(Zn) phase. The electrolyte for the test results shown in Fig. 11.66 was 100 g/L $ZnSO_4.7H_2O$ + 200 g/L NaCl, the applied constant current density was 11.76 mA/cm^2, and the test was conducted in a ventilated cell at 35°C [86]. A second, newer cyclic test intended to simulate automotive service conditions is the VDA 233-102 test, sometimes termed the "new VDA test" [87]. As shown graphically in Fig. 11.67, this test is composed of three different subcycles forming a weekly cycle by the sequence of A-B-A-C-A-B-B. The VDA 233-102 test includes a freezing

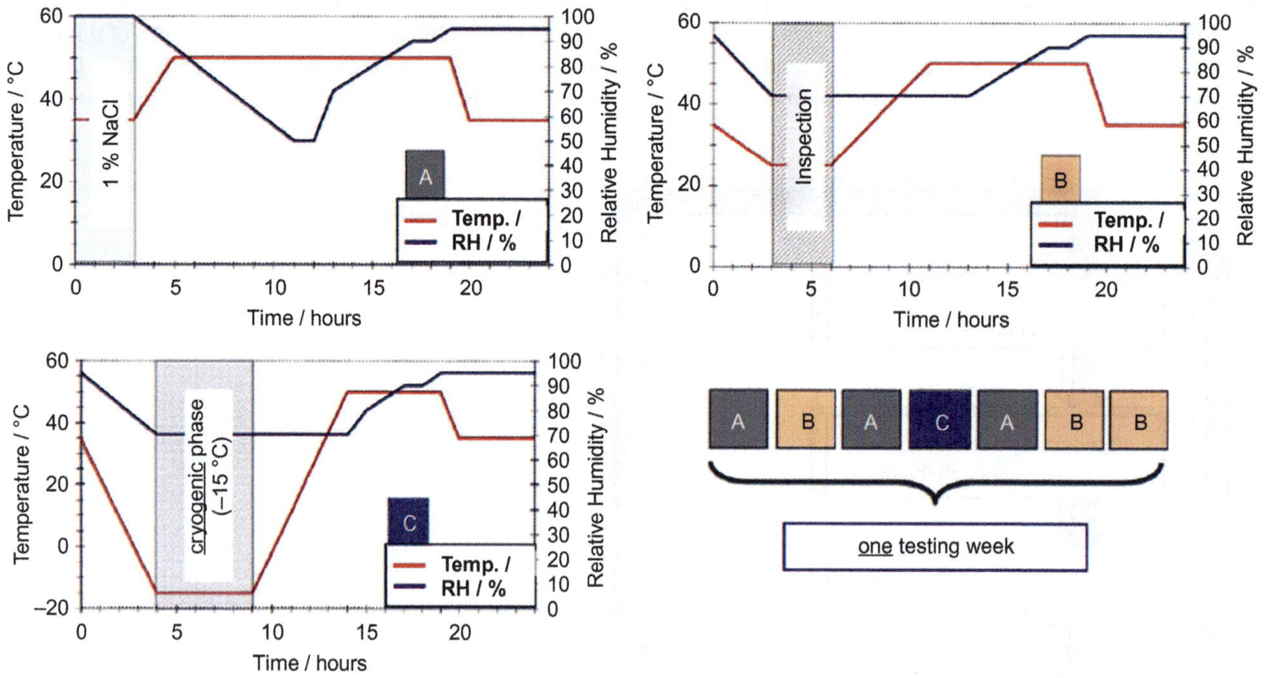

FIG. 11.67 Diagram showing the steps and subcycles of the VDA 233-102 corrosion test [88].

phase at −15°C (cycle C), which alternates with wet and dry cycles at temperatures up to 50°C and salt spray phases at a lower NaCl concentration (1 wt%) than the VDA 621-415 (5 wt%). In these experiments, the salt was introduced as a fog rather than a spray. Cross sections of coated steels subjected to either the VDA 621-415 or VDA 233-102 cyclic corrosion tests taken at 1-week intervals are shown in Fig. 11.68. The tests were run for either 10 weeks of VDA 621-415 cycles or 6 weeks of VDA 233-102 cycles; comparing the appearance of the samples after the longest durations shows that VDA 233-102 is a more severe test. In both cases, reddish-brown rust was observed after 1 week from corrosion of the iron component of the coating. The coating has not been completely consumed at the endpoint of either test and no steel attack is observed. Examination of the progress in corrosion of the samples in Fig. 11.68 shows that the Γ (Fe_3Zn_{10}) phase, with a lighter appearance in these cross sections, is the first to corrode, consistent with its higher electronegativity than the α-ferrite(Zn) phase. After the Γ phase is consumed, the α-ferrite(Zn) phase corrodes, providing sacrificial protection for the steel substrate, consistent with the electronegativity values determined from the above galvanostatic tests [86].

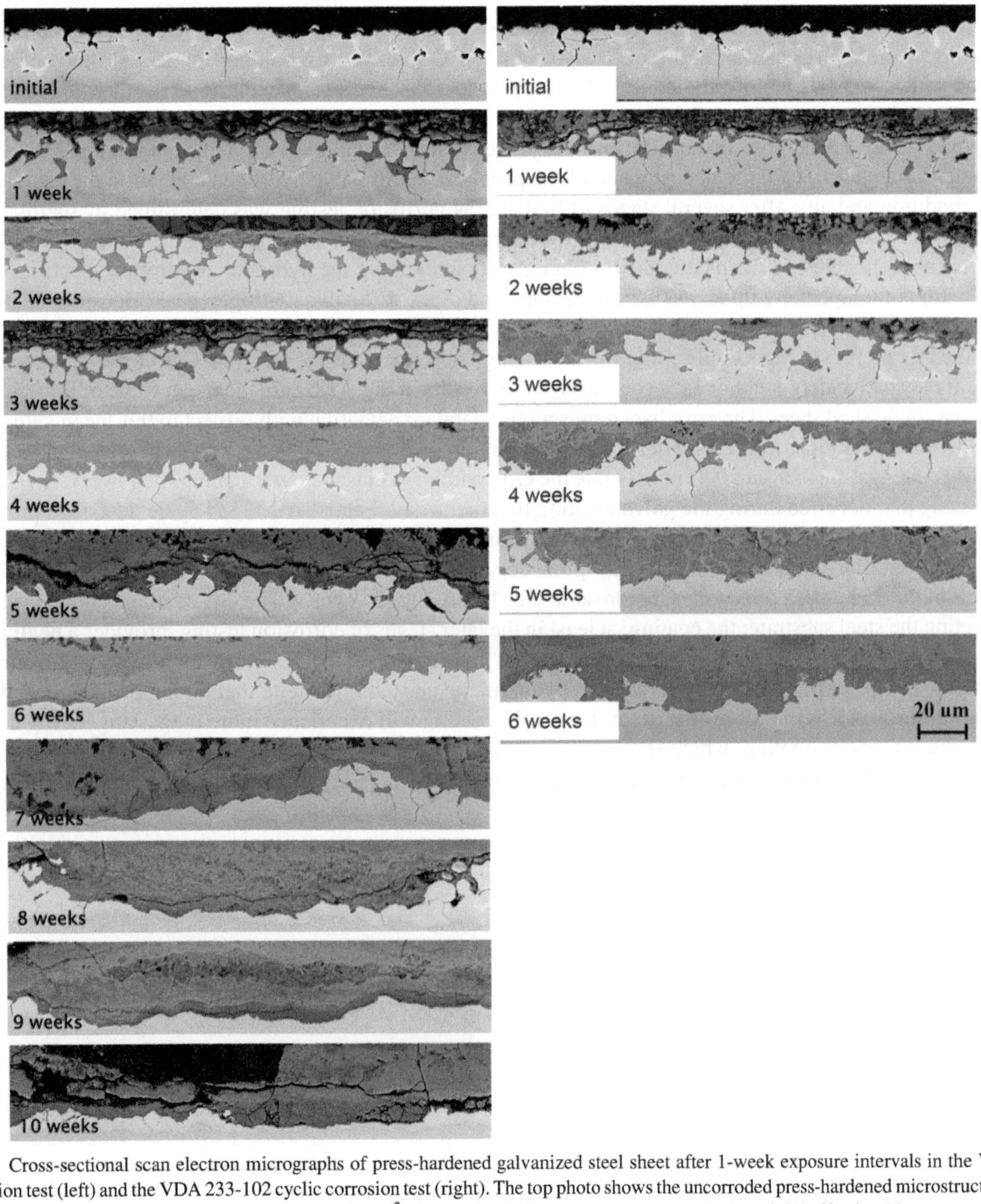

FIG. 11.68 Cross-sectional scan electron micrographs of press-hardened galvanized steel sheet after 1-week exposure intervals in the VDA 621-415 cyclic corrosion test (left) and the VDA 233-102 cyclic corrosion test (right). The top photo shows the uncorroded press-hardened microstructure, similar to Fig. 11.64. The starting microstructure was a GI coating, 70 g/m^2 per side, produced in a galvanizing bath with a 0.5% effective Al content and then soaked at 910°C to produce the shown "initial state" microstructures [86].

11.8 Corrosion and hydrogen uptake in the steel

It is well known that the presence of hydrogen can compromise the mechanical behavior of steels, especially those with strengths above 700 MPa when an appreciable quantity of hard phases such as martensite or bainite is found. Corrosion of a zinc coating results in creation of zinc ions by shedding of electrons by the zinc atoms. These electrons react with water in the corrosive electrolyte to form hydroxide and hydrogen atoms that can be adsorbed on the surface of the steel according to the reaction

$$H_2O + e^- \rightarrow H_{ad} + OH^- \tag{11.3}$$

The adsorbed hydrogen can then be absorbed into to bulk of the steel. The role of corrosion in hydrogen absorption in a 980 MPa dual-phase steel was examined by Mandy using GI (Zn-0.23Al), GA (Fe-Zn based on Zn-0.12Al), and Zn-1.2Al-1.2 Mg hot-dip coatings [89]. Steel panels were coated in a hot-dip process simulator by first annealing at 800°C for 60s, then overaging for 15 s at 480°C for the GI and GA coatings or 435°C for the Zn-Al-Mg coatings. The panels were then galvanized in the coating bath; the GI and Zn-Al-Mg coatings were then immediately quenched to room temperature while the GA coatings were then heated at a rate of 5°C/s up to 570°C followed by holding for 15 s at this temperature. The galvannealed panels were then slow cooled at −2°C/s down to 455°C before quenching to room temperature. Samples were then stored in liquid N_2 to prevent the hydrogen from exiting the samples. The samples were completely coated except for the top edges of the panel dipped in the bath; this was judged to have a negligible effect on outgassing, and all outgassing measurements were made toward the opposite end of the panels. The diffusible hydrogen content in the samples in the as-galvanized condition and also after several stages of SAE J2334 corrosion testing was determined by thermal desorption analysis (TDA). The setup was previously described by Georges [90]. This included heating the samples to 400°C at a rate of 20°C/min. The diffusible hydrogen amount was obtained by integrating the hydrogen desorption spectrum (in wt.ppm/min) as a function of temperature up to 350°C, which is the end of the diffusible hydrogen peak. Because most of the SAE J2334 test is performed at 50°C or 60°C (23 h 45 min out of the total 24 h cycle), samples were degassed in air at 60°C for 10 days for the GI and Zn-Al-Mg samples; this was not though necessary for the GA panels because of the galvannealing heat treatment they had undergone. For each coating type, the results in Fig. 11.69 are the mean value of four different TDA performed on a single steel sheet. The error bars represent the standard deviation with respect to this mean value. Fig. 11.69 shows that the initial hydrogen contents, after the galvanizing process and 60°C degassing (where applicable), for the GI and Zn-Al-Mg coatings, are 0.28 and 0.31 ppm, while the GA coating was just below 0.1 ppm. The lower GA value is related to the outgassing that occurred during the galvannealing treatment. The 7-day SAE J2334 cycle described previously was then used, and samples taken after 3 and 10 cycles (days) for TDA. It is observed that some degassing of the GI and GA occurred over the duration of this test, whereas negligible degassing occurred with the Zn-Al-Mg samples. Fig. 11.70 shows cross sections of the three types of coatings, before and after 10 SAE J2334 cycles. The GA coating is severely attacked but is still protecting the steel substrate; the coating, at least in the later stages of corrosion testing, provides a negligible barrier to the hydrogen outgassing observed in Fig. 11.69. The GI coating experiences significant attack, and from the results described before can be expected to be covered with hydrozincite and simonkolleite that do not provide sealed protection to the corroding surface. It is therefore consistent that the GI coating will experience more outgassing during the corrosion process than the Zn-Al-Mg coating, which shows minimal attack of the coating in Fig. 11.70, and from the results above would be expected to be mainly coated with a more protective double-layer hydroxide corrosion product. Some pits in the

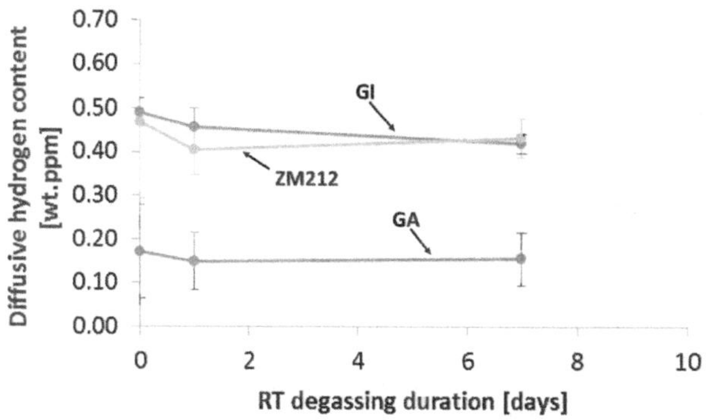

FIG. 11.69 Effect of SAE J2334 cyclic corrosion test exposure of GI, GA, and Zn-Al-Mg coated steel panels on diffusible hydrogen in the steel [89].

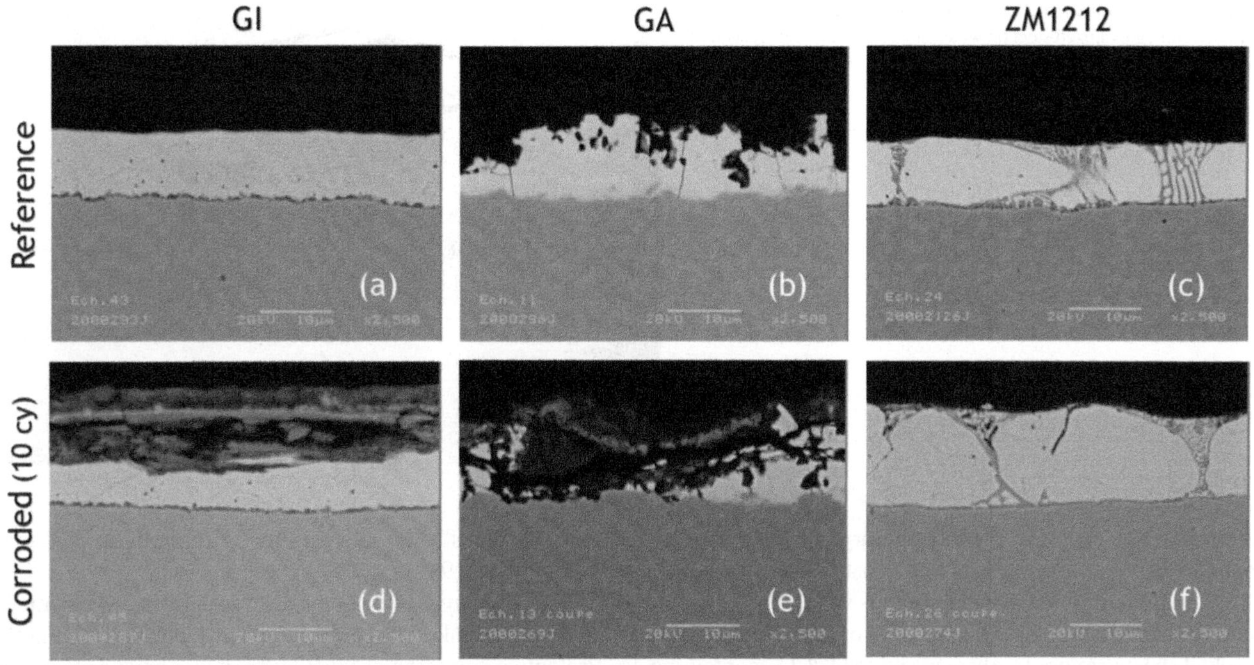

FIG. 11.70 Scanning electron micrographs of cross sections of GI, GA, and Zn-Al-Mg coatings after application of the coatings after 10 SAE J2334 corrosion cycles [89].

area of the eutectic phases are observed, as indicated in Section 11.3, these are the first to be attacked in a chloride electrolyte and would therefore offer pathways for hydrogen to outgas from the coated steel.

While the samples of Mandy were precharged with hydrogen during the continuous galvanizing process and appeared to reach an equilibrium value after corrosion testing, other investigators have measured the amount of hydrogen that can be charged into the steel only through corrosion. Uncoated steel plates with a composition of Fe-0.25%C-0.3%Si-1.5Mn-0.3V were exposed at 10 sites throughout Japan [91]. There was no influence on hydrogen pick-up of the exposure environment (inland, marine, industrial) or whether the exposure environment was open or sheltered. The higher-strength, 1500 MPa condition produced by quenching and tempering this composition reached a diffusible hydrogen content as high as 0.285, within the range of values shown in Fig. 11.69. The reduction in hydrogen flux decreases after corrosion products form, at which point the hydrogen reaction (cathode reduction) occurs on the corrosion products instead of on the steel substrate. Laboratory accelerated tests with scratched galvanized coupons showed the same trend; zinc corrosion products were concluded to play an important role in regulating the flux of hydrogen into the steel after they were formed [92]. The rapid mobility of diffused hydrogen out of the steel after charging shown by Mandy, even as corrosion of the coated samples occurred, indicates that an equilibrium diffusible hydrogen concentration is expected to be reached as hydrogen charging and outgassing reach a balance. Whether or not the equilibrium hydrogen concentration is low enough so that it does not influence mechanical properties also depends on the sensitivity of the steel microstructure and the level of applied stress. For the quench and tempered steels investigated by Kushida, a threshold diffusible hydrogen content between 1 and 2 ppm was indicated, well above the measured levels in the investigated steels.

11.9 White rust (storage stain)

Storage stain is a corrosion stain that is typically white, but which can also take the form of a gray or black deposit on the surface. Because the most common form of discoloration is white in appearance, storage stain is often called white rust. It can occur when sheets of galvanized steel that are in close contact (in a coil or stacked in lifts/bundles) get wet, either by direct water intrusion or condensation between the surfaces. The discoloration is caused by the corrosion products that form after zinc reacts with moisture in the absence of free air circulation. The typical appearance of storage stain is shown in Fig. 11.71; this shows a building made from galvanized sheets that had been stacked without ventilation between the sheets; the interfaces between the stacked sheets had become wet. In normal weathering, as described in earlier sections, the corrosion products formed on galvanized steel surfaces form protective barrier layers that slow the rate of further corrosion and

FIG. 11.71 Appearance of a metal building whose panels were constructed from galvanized steel on which storage stain (white rust) was formed [93].

are formed by cycles of wetting and drying. When galvanized sheet becomes wet while still in coil form, or stacked into bundles at a roll-forming plant or construction site, storage stain can result [94].

Storage stain (usually white rust) is simply the chemical compound, zinc hydroxide (ZnOH), which forms when zinc is in contact with moisture. It does not convert to basic zinc carbonate or simonkolleite protective films because the tightly packed sheets are not freely exposed to O_2/CO_2-containing air. Protective corrosion product layers never have a chance to form. Because the corrosion reaction continues to proceed as long as the surfaces are wet and starved for oxygen and carbon dioxide, a large accumulation of zinc hydroxide can form. When white rust does occur, there is loss of zinc coating thickness to create the zinc hydroxide corrosion product. The severity of this corrosive attack is primarily dependent on (A) the exposure time to moisture, (b) the temperature that is experienced during storage, and (c) the presence of accelerating corrosive agents, such as chloride-containing salts. Often, the amount of white rust appears to be quite heavy when, in fact, the amount of corroded zinc is small. This occurs because the formed zinc hydroxide is porous and nonadherent, building up in the area of the wetness. When the zinc hydroxide dries in the open air, it converts to zinc oxide. If the application is not aesthetically critical, the galvanized coating should perform very well and meet the requirements and expectations of the end user. In most instances involving outdoor exposure, the white rust will disappear over time, as it is either washed off by rainfall or is converted to zinc oxide and then to the customary longer-term passive layers of basic zinc carbonate or simonkolleite. In severe cases, the storage stain may take on a dark gray or black appearance. When the stains on galvanize turn black, it usually means that iron has become part of the corrosion product, i.e., enough zinc has been consumed that iron from the steel substrate is being attacked, with the formation of iron-rich corrosion products. Darker forms of storage stain are also seen on temper rolled material because of the higher concentration of aluminum near the surface as described in Section 11.2. The darker color is related to formation of an amorphous layer covering a sublayer of the mixed oxides of zinc and aluminum [95]; this becomes more prominent with higher levels of surface aluminum i.e., 6 rather than $2\,mg/m^2$ per Fig. 11.5.

References

[1] A. Chakraborty, A. Mondal, A. Pathak, A. Pandey, N. Rani, M. Dutta, A study on electrochemical impedance spectroscopy of galvannealed phases, in: Proc. 10th International Conference on Zinc and Zinc Alloy Coated Steel Sheet (Galvatech 2015), May 31–June 4, Association for Iron & Steel Technology, Toronto, Warrendale, PA, 2015, pp. 247–254.

[2] M. Monnoyer, B. Corlu, T.M. Amorim, L. Dosdat, C. Dieu, Interest of new generation Zn-Al3.7-Mg3.0 coatings for the industry and construction market, in: Proc. 10th International Conference on Zinc and Zinc Alloy Coated Steel Sheet (Galvatech 2015), May 31–June 4, Association for Iron & Steel Technology, Toronto, Warrendale, PA, 2015, pp. 231–238.

[3] B. Carlsson, G. Engström, A.-L. Hög Lejre, M. Johansson, R. Johnsen, R. Lahtinen, M. Ström, Guideline for Selection of Accelerated Corrosion Test for Product Qualification, NORDTEST Technical Report TR 597, 0283-7234, Nordic Innovation Center, Oslo, 2006.

[4] ASTM B117-19, Standard Practice for Operating Salt Spray (Fog) Apparatus, ASTM International, West Conshohocken, PA, 2019.

[5] ISO 9227, Corrosion Tests in Artificial Atmospheres—Salt Spray Tests, International Organization for Standardization, Geneva, 2017.

[6] K. Nishihara, M. Matsumoto, M. Kimoto, T. Kudo, H. Uchida, Y. Haruyama, K. Kanda, S. Matsui, Effect of Al content on structure of corrosion product films in the early stages on Hot-Dip Zn-Al coatings, in: Proc. 7th International Conference on Zinc and Zinc Alloy Coated Steel Sheet (Galvatech 2007), November 18–22, Iron and Steel Institute of Japan, Osaka, Tokyo, 2007, pp. 665–670.

[7] N. LeBozec, D. Thierry, D. Persson, The effect of microstructure on the corrosion stability of zinc aluminium magnesium alloy coated steel in atmospheric conditions, in: Proc. 12th International Conference on Zinc and Zinc Alloy Coated Steel Sheet (Galvatech 2021), June 22–24, 2021, Virtual Conference, Austrian Society for Metallurgy and Materials, 2021, pp. 220–228.

[8] T. Prosek, J. Hagström, D. Persson, N. Fuertes, F. Lindberg, O. Chocholatý, C. Taxén, J. Šerák, D. Thierry, Effect of the microstructure of Zn-Al and Zn-Al-Mg model alloys on corrosion stability, Corros. Sci. 110 (2016) 71–81.

[9] N. Shimoda, M. Nakazawa, H. Nomura, Y. Morimoto, Atmospheric corrosion resistance Of Zn-11%Al-3%Mg-0.2%Si coated steel, in: Proc. 8th International Conference on Zinc and Zinc Alloy Coated Steel Sheet (Galvatech 2011), June 21–25, Associazione Italiana di Metallurgia, Genoa, Italy, Milan, 2011.

[10] P. Qiu, C. Leygraf, I. Odnevall Wallinder, Evolution of corrosion products and metal release from Galvalume coatings on steel during short and long-term atmospheric exposures, Mater. Chem. Phys. 133 (2012) 419–428.

[11] K. Nishihara, N. Okada, M. Matsumoto, T. Kudo, Application of infrared microspectroscopy to structure analysis of corrosion products precipitated on galvanic couples in 1 mm thickness layer of aqueous solution, in: Proc. 8th International Conference on Zinc and Zinc Alloy Coated Steel Sheet (Galvatech 2011), June 21–25, Associazione Italiana di Metallurgia, Genoa, Italy, Milan, 2011.

[12] V. Šefl, T. Prošek, J. Švadlena, V. Kytka, Electrochemical behaviour of Si phases in novel ZnAlMgSi coatings, in: Proc. 12th International Conference on Zinc and Zinc Alloy Coated Steel Sheet (Galvatech 2021), June 22–24, 2021, Virtual Conference, Austrian Society for Metallurgy and Materials, 2021, pp. 209–219.

[13] X.G. Zhang, Corrosion and Electrochemistry of Zinc, Plenum, New York, 1996, https://doi.org/10.1007/978-1-4757-9877-7.

[14] F.C. Porter, Corrosion Resistance of Zinc and Zinc Alloys, CRC Press, Boca Raton, FL, 1994, https://doi.org/10.1201/9781482293524.

[15] L.T.H. Lien, P.T. San, H.L. Hong, Results of studying atmospheric corrosion in Vietnam 1995–2005, Sci. Technol. Adv. Mater. 8 (2007) 552–558.

[16] M. Natesan, G. Venkatachari, N. Palaniswamy, Corrosion and durability maps of India, Corros. Prev. Control 52 (2) (2005) 43–55.

[17] B. Chico, D. de la Fuente, J.M. Vega, M. Morcillo, Corrosivity maps of Spain for zinc in rural atmospheres, Rev. Metal. 46 (2010) 485–492.

[18] Y.M. Panchenko, A.I. Marshakov, L.A. Nikolaeva, V.V. Kovtanyuk, T.N. Igonin, T.A. Andryushchenko, Comparative estimation of long-term predictions of corrosion losses for carbon steel and zinc using various models for the Russian territory, Corros. Eng. Sci. Technol. 52 (2017) 149–157.

[19] D. Knotkova, K. Kreislova, S.W. Dean Jr., ISOCORRAG International Atmospheric Exposure Program: Summary of Results, ASTM Stock No. DS71, ASTM International, West Conshohocken, PA, 2010.

[20] S. Feliu, M. Morcillo, S. Feliu Jr., The prediction of atmospheric corrosion from meteorological and pollution parameters—II. Long-term forecasts, Corros. Sci. 34 (3) (1993) 415–422.

[21] M. Omoda, K. Nakatsuji, D. Mizuno, Application of data science to atmospheric corrosion prediction, in: Proc. 12th International Conference on Zinc and Zinc Alloy Coated Steel Sheet (Galvatech 2021), June 22–24, 2021, Virtual Conference, Austrian Society for Metallurgy and Materials, 2021, pp. 340–347.

[22] X.G. Zhang, F.E. Goodwin, A software for predicting coating life, in: Proc. 5th International Conference on Zinc and Zinc Alloy Coated Steel Sheet (Galvatech 2001), June 26–28, Verlag Stahleisen, Brussels, Duesseldorf, 2001, pp. 311–316.

[23] B. Carlsson, Accelerated corrosion testing for product qualification, in: CEEES First International Symposium on Environmental Testing Engineering, November 22–24, Royal Military Academy, Confederation of European Environmental Engineering Societies, Brussels, Belgium, Haasrode, Belgium, 2007.

[24] ISO 9223:2012, Corrosion of Metals and Alloys—Corrosivity of atmospheres—Classification, Determination and Estimation, International Organization for Standardization, Geneva, 2017.

[25] T. Shimizu, Recent progress of zinc and zinc-alloy coated steel sheets for construction applications, in: Proc. 7th International Conference on Zinc and Zinc Alloy Coated Steel Sheet, (Galvatech 2007), November 18–22, Iron and Steel Institute of Japan, Osaka, Tokyo, 2007, pp. 568–573.

[26] N. Gao, D. Harrison, Y.H. Liu, Five-year atmospheric corrosion test of various Zn-Al coatings in a severe marine environment, in: Proc. 10th International Conference on Zinc and Zinc Alloy Coated Steel Sheet (Galvatech 2015), May 31–June 4, Association for Iron & Steel Technology, Toronto, Warrendale, PA, 2015, pp. 294–301.

[27] ISO 14713-1:2017, Zinc Coatings—Guidelines and Recommendations for the Protection against Corrosion of Iron and Steel in Structures—Part 1: General Principles of Design and Corrosion Resistance, International Organization for Standardization, Geneva, 2017.

[28] F.M. Kilbane, Relationship between surface composition, steel mill processing and paint pretreatment of galvanized strip, in: Proceedings Galvanizers Association 74th Meeting, April 14–16, Galvanizers Association, Monterrey, Mexico, Marine City, MI, 1982.

[29] K. Hoshino, K. Oikawa, W. Tanimoto, M. Nagoshi, M. Koba, Segregation mechanism of Al-based oxides on surface of Zn-0.2mass%Al hot-dip galvanized steel sheets, ISIJ Int. 60 (8) (2020) 1765–1773.

[30] M. Dubois, Modifications in the skin Al weight with aging on hot dip galvanized sheet, in: Proc. 5th International Conference on Zinc and Zinc Alloy Coated Steel Sheet (Galvatech 2001), June 26–28, Verlag Stahleisen, Brussels, Duesseldorf, 2001, pp. 409–416.

[31] S. Fujimoto, D. Ohzaike, Electrochemical properties of corrosion products on Zn, CAMP-ISIJ 15 (2002) 1189 (in Japanese).

[32] M. Ueda, M. Matsuda, T. Ohtsuka, Corrosion products on zinc-coated steels detected by In-situ Raman spectroscopy, in: Proc. 6th International Conference on Zinc and Zinc Alloy Coated Steel Sheet (Galvatech 2004), April 4–7, Association for Iron & Steel Technology, Chicago, IL, Warrendale, PA, 2004, pp. 845–852.

[33] J. Sullivan, S. Mehraban, J. Elvins, In situ monitoring of the microstructural corrosion mechanisms of zinc-magnesium-aluminium galvanising alloys using time lapse microscopy, in: Proc. 8th International Conference on Zinc and Zinc Alloy Coated Steel Sheet (Galvatech 2011), June 21–25, Associazione Italiana di Metallurgia, Genoa, Italy, Milan, 2011.

[34] C. Gallagher, N. Cooze, J. Sullivan, P. Dodds, P. Barker, In-situ time-lapse microscopy to elucidate the corrosion mechanism of Zn-4.8wt.%Al galvanising metallic coating; surface and cut-edge, in: Proc. 11th International Conference on Zinc and Zinc Alloy Coated Steel Sheet (Galvatech 2017), November 12–16, Iron and Steel Institute of Japan, Tokyo, 2017, pp. 72–79. Tokyo.

[35] I. Muto, Y. Sugawara, N. Hara, Real-time microelectrochemical observations of initiation and very early stage of corrosion on steels, in: Proc. 11th International Conference on Zinc and Zinc Alloy Coated Steel Sheet (Galvatech 2017), November 12–16, Iron and Steel Institute of Japan, Tokyo, 2017, pp. 141–147. Tokyo.

[36] T. Prosek, A. Nazarov, U. Bexell, D. Thierry, J. Serak, Corrosion mechanism of model zinc-magnesium alloys in atmospheric conditions, Corros. Sci. 50 (8) (2008) 2216–2231.

[37] A. Nazarov, D. Thierry, Probing the atmospheric corrosion of metals. Zinc, Prot. Met. 42 (2006) 437–451.

[38] Galvanized Autobody Partnership Project ZCO-45-6, New Zinc-Based Automotive Steel Protection Technologies, Final Report, International Zinc Assn., Durham, NC, June 30, 2014.

[39] Galvanized Autobody Partnership Project ZCO-45-7, New Zinc-Based Automotive Steel Protection Technologies, Final Report, International Zinc Assn., Durham, NC, October 15, 2018.

[40] T. Tsujimura, A. Komatsu, A. Andoh, Influence of Mg content in coating layer and coating structure on corrosion resistance of hot-dip Zn-Al-Mg alloy coated steel sheet, in: Proc. 5th International Conference on Zinc and Zinc Alloy Coated Steel Sheet (Galvatech 2001), June 26–28, Verlag Stahleisen, Brussels, Duesseldorf, 2001, pp. 145–152.

[41] JASO M609, Corrosion Test Method for Automotive Materials, Japan Society of Automotive Engineers (JSAE), Tokyo, January 1, 1991.

[42] M. Matsumoto, M. Kimoto, A. Sakoda, T. Kudo, Corrosion behavior of 55% Al-Zn alloy coated steel sheet near its shear cut edge in salt spray test, Tetsu-to-Hagané 91 (2005) 16 (in Japanese).

[43] X.G. Zhang, Corrosion performance of zinc-coated steels—mechanism, life prediction and corrosivity mapping, in: Proc. 7th International Conference on Zinc and Zinc Alloy Coated Steel Sheet, (Galvatech 2007), November 18–22, Iron and Steel Institute of Japan, Osaka, Tokyo, 2007, pp. 694–698.

[44] F. Yoshizaki, T. Shimizu, Y. Miyoshi, A. Andoh, Atmospheric corrosion resistance of hot-dip Zn-6%AI-3%Mg alloy coated steel sheets, in: Proc. 6th International Conference on Zinc and Zinc Alloy Coated Steel Sheet (Galvatech 2004), April 4–7, Association for Iron & Steel Technology, Chicago, IL, Warrendale, PA, 2004, pp. 1025–1034.

[45] Galvanized Autobody Partnership Project ZCO-86 "Zinc Alloy Coatings Microstructure and Coating Thickness Effects on Performance" Presentation to GAP Program Review Meeting, International Zinc Assn., Durham, NC, November 3, 2021.

[46] P. Schouller-Guinet, C. Allely, P. Volovitch, ZnAlMg: an innovative metallic coating that offers protection in the harshest environments, in: Proc. 8th International Conference on Zinc and Zinc Alloy Coated Steel Sheet (Galvatech 2011), June 21–25, Associazione Italiana di Metallurgia, Genoa, Italy, Milan, 2011.

[47] B. Corlu, C. Dieu, C. Allely, T. Machado, L. Diez, J. De Strycker, On the anti-corrosion performance of Zn-Al-Mg substrates under natural exposure, in: Proc. 11th International Conference on Zinc and Zinc Alloy Coated Steel Sheet (Galvatech 2017), November 12–16, Iron and Steel Institute of Japan, Tokyo, 2017, pp. 673–678. Tokyo.

[48] D. Persson, T. Prosek, N. LeBozec, D. Thierry, G. Luckeneder, Initial SO_2-induced atmospheric corrosion of ZnAlMg coated steel studied with in situ infrared reflection absorption spectroscopy, Corros. Sci. 90 (2015) 276–283.

[49] ISO 6270-2:2017, Paints and Varnishes—Determination of Resistance to Humidity—Part 2: Condensation (in-Cabinet Exposure with Heated Water Reservoir), International Organization for Standardization, Geneva, 2017.

[50] G. Luckeneder, M. Fleischanderl, T. Steck, K.-H. Stellnberger, J. Faderl, S. Schuerz, G. Mori, Corrosion mechanisms and cosmetic corrosion aspects of zinc-aluminium-magnesium and zinc-chromium alloy coated steel strip, in: Proc. 8th International Conference on Zinc and Zinc Alloy Coated Steel Sheet (Galvatech 2011), June 21–25, Associazione Italiana di Metallurgia, Genoa, Italy, Milan, 2011.

[51] M.-S. Oh, H.-J. Lee, J.-S. Kim, Investigation on the characteristics of hot-dip Zn-3%Mg-2.5%Al alloy coated steel sheets, in: Proc. 10th International Conference on Zinc and Zinc Alloy Coated Steel Sheet (Galvatech 2015), May 31–June 4, Association for Iron & Steel Technology, Toronto, Warrendale, PA, 2015, pp. 822–828.

[52] R. Hausbrand, M. Rohwerder, M. Stratmann, C. Schwerdt, B. Schuhmacher, G. Grundmeier, Model study on the corrosion of magnesium-containing zinc coatings on steel sheet, in: Proc. 5th International Conference on Zinc and Zinc Alloy Coated Steel Sheet (Galvatech 2001), June 26–28, Verlag Stahleisen, Brussels, Duesseldorf, 2001, pp. 161–167.

[53] H.-K. Sohn, J.-W. Lee, Y. Yoo, J. Min, M.-S. Oh, S.-H. Kim, Y.-S. Jin, K.Y. Kim, Corrosion behaviors of Zn-$MgZn_2$ eutectic structure in Zn-Al-Mg coatings, in: Proc. 8th International Conference on Zinc and Zinc Alloy Coated Steel Sheet (Galvatech 2011), June 21–25, Associazione Italiana di Metallurgia, Genoa, Italy, Milan, 2011.

[54] Q.-Y. Liu, S.-M. Jiang, Z.-G. Xi, Q.-F. Zhang, X. Liu, C.-F. Yue, Microstructure and corrosion resistance of Zn-Mg alloy coatings, in: Proc. 10th International Conference on Zinc and Zinc Alloy Coated Steel Sheet (Galvatech 2015), May 31–June 4, Association for Iron & Steel Technology, Toronto, Warrendale, PA, 2015, pp. 834–844.

[55] Z.-L. Ding, B.-H. Yan, X.-Y. Jin, X.-J. Wang, J. Zhang, S.-M. Jiang, Q.-F. Zhang, Effect of magnesium on the microstructure and corrosion resistance of 55% Al-Zn coating, in: Proc. 12th International Conference on Zinc and Zinc Alloy Coated Steel Sheet (Galvatech 2021), June 22-24, 2021, Virtual Conference, Austrian Society for Metallurgy and Materials, 2021, pp. 164–171.

[56] S. Schuerz, M. Fleischanderl, G.H. Luckeneder, K. Preis, T. Haunschmied, G. Mori, A.C. Kneissl, Corrosion behaviour of Zn–Al–Mg coated steel sheet in sodium chloride-containing environment, Corros. Sci. 51 (2009) 2355–2363.

[57] R. Krieg, A. Vimalanandan, M. Rohwerder, Corrosion of zinc and Zn-Mg alloys with varying microstructures and magnesium contents, J. Electrochem. Soc. 161 (3) (2014) C156–C161.

[58] T. Prosek, D. Thierry, D. Persson, J. Stoulil, Corrosion products formed on Zn-Mg, Zn-Al and Zn-Al-Mg coatings in model atmospheric conditions, in: Proc. 8th International Conference on Zinc and Zinc Alloy Coated Steel Sheet (Galvatech 2011), June 21–25, Associazione Italiana di Metallurgia, Genoa, Italy, Milan, 2011.

[59] ISO 8407:2021, Corrosion of Metals and Alloys—Removal of Corrosion Products from Corrosion Test Specimens, International Organization for Standardization, Geneva, 2021.

[60] M. Rohwerder, Scanning Kelvin Probe techniques for the investigation of corrosion, hydrogen uptake and permeation properties of zinc alloy coatings on steel, in: Proc. 12th International Conference on Zinc and Zinc Alloy Coated Steel Sheet (Galvatech 2021), June 22–24, 2021, Virtual Conference, Austrian Society for Metallurgy and Materials, 2021, pp. 244–253.

[61] M. Uranaka, Y. Hattori, Corrosion resistance mechanism of hot-dip Zn-6%Al-3%Mg alloy coated steel sheet in corrosion test in water film formation state, in: Proc. 11th International Conference on Zinc and Zinc Alloy Coated Steel Sheet (Galvatech 2017), November 12–16, Iron and Steel Institute of Japan, Tokyo, 2017, pp. 583–588. Tokyo.

[62] M. Matsumoto, Edge corrosion behaviour of coated steel sheet, in: Proc. 11th International Conference on Zinc and Zinc Alloy Coated Steel Sheet (Galvatech 2017), November 12–16, Iron and Steel Institute of Japan, Tokyo, 2017, pp. 589–596. Tokyo.

[63] ISO 11130:2017, Corrosion of Metals and Alloys—Alternate Immersion Test in Salt Solution, International Organization for Standardization, Geneva, 2017.

[64] T. Ohtsuka, Y. Abe, Corrosion products on cut edge surface of zinc-coated steels in humidified air in the presence of NaCl precipitate by Raman spectroscopy, in: Proc. 7th International Conference on Zinc and Zinc Alloy Coated Steel Sheet, (Galvatech 2007), November 18–22, Iron and Steel Institute of Japan, Osaka, Tokyo, 2007, pp. 648–652.

[65] J.-W. Lee, Y.R. Yoo, H.-K. Sohn, M.-S. Oh, S.-H. Kim, Y.-S. Jin, K.Y. Kim, Investigation of the cut-edge protection behavior of Zn-Al-Mg coated steel sheets, in: Proc. 9th International Conference on Zinc and Zinc Alloy Coated Steel Sheet (Galvatech 2013), September 23–27, Chinese Society for Metals, Metallurgical Industry Press, Beijing, 2013, pp. 673–675. Beijing.

[66] F. Thébault, B. Vuillemin, R. Oltra, C. Allely, K. Ogle, O. Heintz, Influence of magnesium content on the corrosion resistance of the cut-edges of Zn–Mg-coated steel, Corros. Sci. 97 (2015) 100–106.

[67] D. Thierry, N. LeBozec, A. Le Gac, D. Persson, Long-term atmospheric corrosion rates of hot dip galvanised steel and zinc-aluminium-magnesium coated steel, Mater. Corros. 70 (2) (2019) 2220–2227.

[68] T. Shimizu, F. Yoshizaki, Atmospheric corrosion behavior of hot-dip zinc alloy coated steel sheet subjected to forming, in: Proc. 8th International Conference on Zinc and Zinc Alloy Coated Steel Sheet (Galvatech 2011), June 21–25, Associazione Italiana di Metallurgia, Genoa, Italy, Milan, 2011.

[69] T.A. Keppert, G. Luckeneder, K.-H. Stellnberger, G. Mori, H. Antrekowitsch, Investigation of the corrosion behavior of Zn-Al-Mg hot-dip galvanized steel in alternating climate tests, Corrosion 70 (12) (2014) 1238–1248.

[70] K. Ueda, Corrosion resistance of Zn-11%Al-3%Mg-0.2%Si alloy coated steel sheet in the marine atmospheric after long term outdoor exposure, in: Proc. 9th International Conference on Zinc and Zinc Alloy Coated Steel Sheet (Galvatech 2013), September 23–27, Chinese Society for Metals, Metallurgical Industry Press, Beijing, 2013, pp. 659–662. Beijing.

[71] Y. Kimata, A. Takahashi, K. Asai, Corrosion resistance Of Zn-11%Al-3%Mg-0.2%Si coated steel sheet after sheltered outdoor exposure test In Miyakojima, in: Proc. 7th International Conference on Zinc and Zinc Alloy Coated Steel Sheet, (Galvatech 2007), November 18–22, Iron and Steel Institute of Japan, Osaka, Tokyo, 2007, pp. 586–591.

[72] Y.H. Yoo, T.C. Kim, S.J. Yoo, Anti-corrosion performance of ZnMgAl-coated steels for automotive applications, in: Proc. 12th International Conference on Zinc and Zinc Alloy Coated Steel Sheet (Galvatech 2021), June 22–24, 2021, Virtual Conference, Austrian Society for Metallurgy and Materials, 2021, pp. 116–121.

[73] GMW14872, Cyclic Corrosion Laboratory Test, General Motors Worldwide Standard, fifth ed., General Motors Corp., Detroit, MI, February 2021.

[74] D. Mizuno, Automotive corrosion and corrosion evaluation methods for coated steel sheets, in: Proc. 11th International Conference on Zinc and Zinc Alloy Coated Steel Sheet (Galvatech 2017), November 12–16, Iron and Steel Institute of Japan, Tokyo, 2017, pp. 553–558. Tokyo.

[75] SAE Surface Vehicle Standard J-2334, Laboratory Cyclic Corrosion Test, Revised December, Society of Automotive Engineers, Inc., Warrendale, PA, 2003 (historical).

[76] ISO 11997–1, Paints and Varnishes—Determination of Resistance to Cyclic Corrosion Conditions—Part 1: Wet (Salt Fog)/Dry/Humid, International Organization for Standardization, Geneva, 2017.

[77] K.-H. Chang, Section 10.3.3.3, Extreme value distributions, in: e-Design: Computer-Aided Engineering Design, Academic Press (Elsevier), Amsterdam, 2015, https://doi.org/10.1016/C2009-0-63076-2.

[78] L.A. Roudabush, T.E. Dorsett, A review of perforation corrosion testing 1980–1990, in: Proc. 5th Automotive Corrosion & Protection Conference, October 21–23, Society of Automotive Engineers, Detroit, MI, Warrendale, PA, 1991, p. 221. Publication P-250, Paper 912285.

[79] D. Santos, H. Raminhos, M. Costa, T. Diamantino, F. Goodwin, Performance of conductive pre-primers applied on galvanized steel sheets for automotive bodies, in: SAE Congress and Exposition, April 2–6, Society of Automotive Engineers, Detroit, MI, Warrendale, PA, 2007. SAE Paper 2007-01-1750.

[80] M. Vlot, N. Van Landschoot, R. Bleeker, ZnAlMg coatings versus galvannealed, in: Proc. 8th International Conference on Zinc and Zinc Alloy Coated Steel Sheet (Galvatech 2011), June 21–25, Associazione Italiana di Metallurgia, Genoa, Italy, Milan, 2011.

[81] Volvo Standard 432-0014, Accelerated Corrosion Test- Atmospheric Corrosion, Issue 6, Volvo Personvagnar AB, Gothenburg, Sweden, January 2015.

[82] S. Suzuki, D. Mizuno, S. Fujita, Evaluation of automobile corrosion environments by ACM type corrosion sensor, in: Proc. 7th International Conference on Zinc and Zinc Alloy Coated Steel Sheet, (Galvatech 2007), November 18–22, Iron and Steel Institute of Japan, Osaka, Tokyo, 2007, pp. 705–709.

[83] Galvanized Autobody Partnership Project ZCO-70, Nature of Coating Development and Substrate Microcracking During Hot Press Forming, First Progress Report, International Zinc Assn., Durham, NC, May 2017.

[84] C. Dever, J. Kish, J. McDermid, Corrosion properties of hot dip zinc galvanized coatings on 22MnB5 press hardened steels, in: Proc. 11th International Conference on Zinc and Zinc Alloy Coated Steel Sheet (Galvatech 2017), November 12–16, Iron and Steel Institute of Japan, Tokyo, 2017, pp. 385–392. Tokyo.

[85] A.W. Hassel, Fundamental aspects of corrosion of press hardened galvanized steel, in: Proc. 11th International Conference on Zinc and Zinc Alloy Coated Steel Sheet (Galvatech 2017), November 12–16, Iron and Steel Institute of Japan, Tokyo, 2017, pp. 130–137. Tokyo.

[86] G. Luckeneder, R. Autengruber, K.H. Stellnberger, J. Faderl, T. Kurz, Corrosion protection of galvanized press-hardening steel: main influencing factors and mechanisms, in: Proc. 10th International Conference on Zinc and Zinc Alloy Coated Steel Sheet (Galvatech 2015), May 31–June 4, Association for Iron & Steel Technology, Toronto, Warrendale, PA, 2015, pp. 868–875.

[87] VDA 233-102, Cyclic Corrosion Testing of Materials and Components in Automotive Construction, Version 06/2013, VDA (Verband der Automobilindustrie), Berlin, 2013.

[88] R. Autengruber, G. Luckeneder, S. Kolnberger, J. Faderl, A.W. Hassel, Corrosion behavior of press-hardened hot-dipped galvanized steel sheet, in: Proc. 8th International Conference on Zinc and Zinc Alloy Coated Steel Sheet (Galvatech 2011), June 21–25, Associazione Italiana di Metallurgia, Genoa, Italy, Milan, 2011.

[89] M. Mandy, B. Nabi, M. Larnicol, X. Vanden Eynde, C. Georges, F. Goodwin, Influence of Zn-based coating alloys on hydrogen diffusion and corrosion resistance in a DP steel, Berg- Huettenmaenn. Monatsh. 166 (2021) 554–559.

[90] C. Georges, T. Sturel, P. Drillet, J.-M. Mataigne, Absorption/desorption of diffusible hydrogen in aluminized boron steel, ISIJ Int. 53 (8) (2013) 1295–1304.

[91] T. Kushida, Hydrogen entry into steel by atmospheric corrosion, ISIJ Int. 43 (4) (2003) 470–474.

[92] M. Sakairi, T. Hashimoto, Hydrogen permeation behavior of size controlled scratch formed coated steels during wet-dry cycle corrosion tests, in: Proc. 11th International Conference on Zinc and Zinc Alloy Coated Steel Sheet (Galvatech 2017), November 12–16, Iron and Steel Institute of Japan, Tokyo, 2017, pp. 687–693. Tokyo.

[93] G.W. Dallin, Control and treatment of hot-dip galvanize surfaces, in: Proceedings Galvanizers Association 97th Meeting, October 16–19, Galvanizers Association, Lexington, KY, Marine City, MI, 2005.

[94] GalvInfo Note 3.2, Protecting Galvanized Steel Sheet Products from Storage Stain, Rev. 3, GalvInfo Center of International Zinc Assn., Durham, NC, October 2020. https://www.galvinfo.com/wp-content/uploads/sites/8/2020/10/GalvInfoNote-3-2-1.pdf.

[95] Y. Uchida, M. Koda, Y. Fukui, Y. Hirose, Black patina developed on Zn-Al alloy coated steel sheet, Tetsu-to-Hagané 72 (8) (1986) 1013–1020 (in Japanese).

Chapter 12

Zn-coated steel product properties

12.1 Corrosion performance of painted Zn and Zn alloy coated sheet

12.1.1 Introduction

Often, painting of an object is performed to improve its appearance. Galvanized sheet in appliance, automotive, construction, and other applications is often painted for this reason, but another important reason is to further improve its corrosion resistance and lengthen service life. Not only does the paint system provide an additional barrier to corrosion, the synergy produced by the paint and the galvanized coating creates a system whose service life is greater than the sum of the service lives of the individual metallic and paint coatings. The corrosion protection ability of a paint system partly depends on the permeability of the paint system that provides a barrier effect. Also, pigments added to the paint contribute inhibitors for zinc anodic dissolution when they are added to primers or topcoats [1]. The traditional and still very prevalent inhibitor addition is hexavalent chromium, although this is being phased out in favor of trivalent chromium or chromium-free inhibitors for occupational and environmental reasons. When the coil-coated surface containing this inhibitor is wetted, hexavalent chromium is transported to the zinc surface where it is reduced to trivalent chromium and precipitated, forming a passive layer. This mechanism enables long-term protection, after the permeability of the paint has increased from weathering, providing a less protective barrier to corrosion [1].

12.1.2 Corrosion tests for coil-coated galvanized panels

The corrosion tests used to assess performance of painted galvanized steel are in many cases the same as those used for unpainted product, as described in Chapter 11, the choice of test depends on up the conditions expected in the intended application. In addition, the "prohesion test" standardized as ASTM G85 [2] was developed specifically for assessing the corrosion performance of painted galvanized sheet. More common laboratory corrosion tests such as the ASTM B117 salt spray test, the ASTM D2247 humidity test, and the ASTM G87 moist sulfur dioxide [3–5] are used to assess the response of painted galvanized steel coating systems to moisture and aggressive pollutants but give no resemblance to natural weathering. The most widely used version of the prohesion test uses a sprayed solution of 0.35% ammonium sulphate ($(NH_4)_2SO_4$) and 0.05% NaCl rather than the 5% NaCl of the ASTM B117 salt spray test, and an ambient test temperature rather than the 35°C used in the B117 test. Also, the spray cycle is 1 h wet/1 h dry, rather than the continuous wetness utilized in the other accelerated tests. This simulates urban-industrial environments rather than marine environments. De Souza [6] compared prohesion and salt spray test results for GI panels pretreated with zinc phosphate, primed with a flexible polyester layer and topcoated with silicon-modified polyester. Fig. 12.1 compares the appearance of panels after 400 h of prohesion or 500 h of salt spray testing. Paint blistering is observed to be much more severe on the salt sprayed panel and is related to cathodic delamination of the paint in areas of local alkalinity, especially along the drip edge, which is where the most severe corrosion is expected to occur. After a number of years of actual outdoor exposure in southern Ontario, the test panels exhibit less blistering than the salt spray panel, similar to the prohesion test panel of Fig. 12.1, both the prohesion and long-term exposure panels also show red rust at the scribe and the drip edge of the panel, which is absent in the salt spray panel. The corrosion mechanism for both the prohesion and long-term exposure panels was found to be anodic dissolution of the zinc coating. In contrast, the salt spray panel exhibits no red rust in the scribe or at the drip edge. The same visual conclusions are seen when the same paint systems and tests were used with the Zn-55%Al-1.5%Si coating, Fig. 12.1.

The Metallurgy of Zinc Coated Steels. https://doi.org/10.1016/B978-0-323-99984-7.00014-2

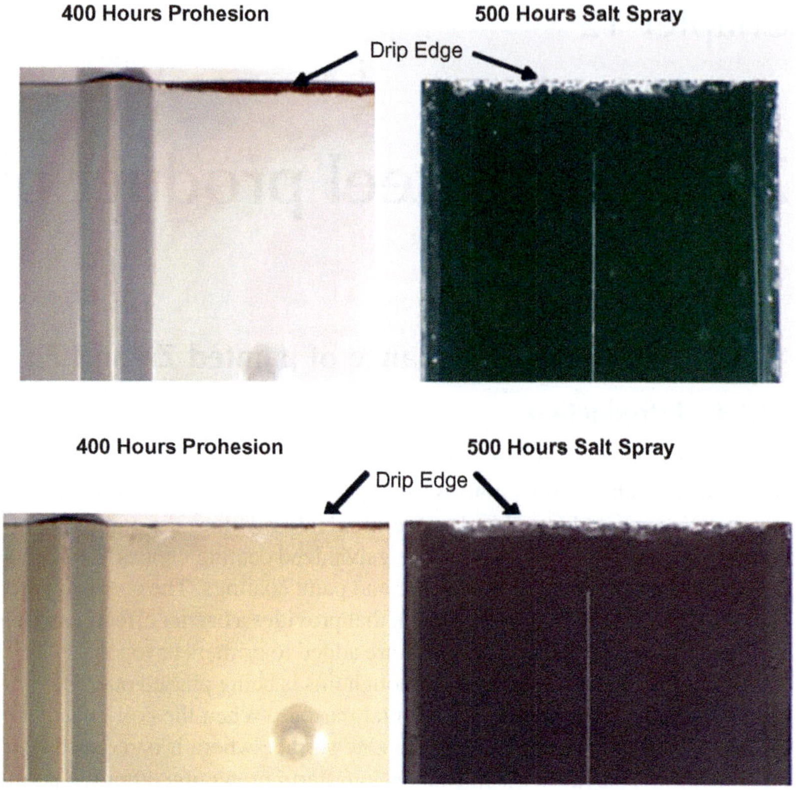

FIG. 12.1 Appearance of painted GI (top) and Zn-55%Al-1.5%Si (bottom) panels after laboratory testing using ASTM G85 prohesion and ASTM B117 salt spray laboratory tests. Both received a zinc phosphate pretreatment, a flexible polyester primer, and a silicon-modified polyester topcoat. The cathodic delamination of the paint on the drip edge of the salt sprayed panels does not correlate with long-term exposure observations; the appearance of red rust on the drip edge seen on the prohesion samples from anodic dissolution of the coating, and the relatively lower amount of red rust on the Zn-55%Al-1.5%Si panel, agrees with long-term exposure results [6].

12.1.3 Preparation methods for coil coating

Zinc-coated sheet is customarily temper rolled (skin passed) after coating to reduce yield point elongation and improve strip shape, as described in Section 2.15 of Chapter 2. As noted in that section, temper rolls can also be textured to roughen the surface of the zinc coating and break up the thermal oxide film resulting from hot-dip coating. Chapter 11 described the segregation of Al to the surface of both temper-rolled and nontemper-rolled galvanized surface with aging; to pretreat the surface of the zinc coating for painting, this oxide film is usually removed by alkaline cleaning. As described by Pélissier [7], for a GI coating the hydroxides in the alkaline cleaner react with the surface oxides to dissolve them and a thin portion of the underlying zinc coating following the reactions

$$Al_2O_3 + 2OH^- + 3H_2O \rightarrow 2Al(OH)_4^- \tag{12.1}$$

$$Zn + 2OH^- + 2H_2O \rightarrow Zn(OH)_4^{2-} + H_2 \tag{12.2}$$

Cleaning of the surface improves the uniformity of the chromate or phosphate chemical conversion coating treatment, reducing the incidence of paint blistering [7]. In the case of GI coatings, it was shown that the use of a KOH-based cleaning treatment with a pH greater than 13, after first solvent cleaning in heptane, reduced the surface Al/Zn atomic ratio from 3 to about 0.2. Samples pretreated and painted after this alkaline treatment exhibited better coating adherence during tension bending and less blistering after corrosion testing than samples that were only solvent cleaned. The effect of pH in the alkaline cleaner on the surface Al/Zn atomic ratio, compared with solvent cleaning only, is shown in Fig. 12.2. In practical terms, it is desired to produce a surface free of "water breaks," or discontinuities in a water film, on the cleaned surface. This is difficult to achieve when only solvent cleaning is used, but easier to achieve when alkaline cleaning is added.

Two kinds of chemical conversion treatments are used to improve the bond strength of paint systems to zinc-coated steel, based on either phosphate or chromate chemistries [8]. Additionally, chromate-free chemistries have been developed as substitutes for chromates because of environmental concerns. Traditional hexavalent chromate conversion treatments change the zinc surface to a complex oxide layer about 0.5–3 mm thick and contain chromium hydroxide, zinc hydroxy-chromate, and zinc chromate [9].

When used as paint pretreatments on coil coating lines, these coatings are usually heavier than when used as passivation treatments on unpainted sheet, and thus have a greenish/yellow-iridescent, brown, or drab appearance. The color varies with

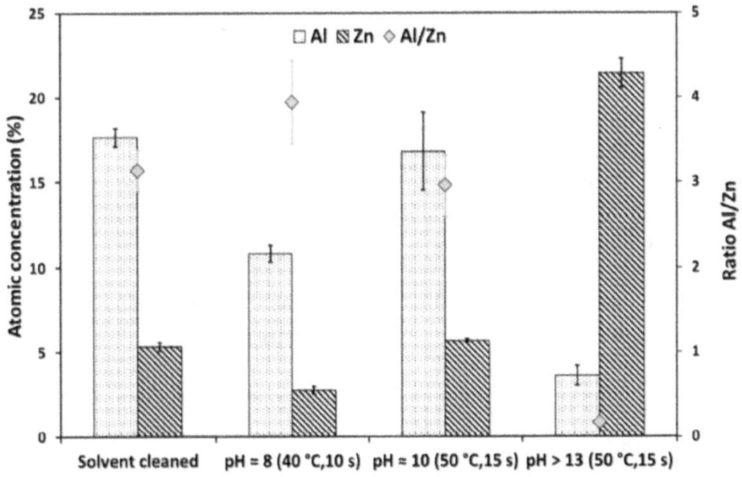

FIG. 12.2 Atomic concentrations of surface Al and Zn measured by X-ray photoelectron spectroscopy (XPS) on GI-coated steel sheet after solvent cleaning and subsequent alkaline cleaning with KOH at several levels of pH. The ratio of surface Al/Zn is plotted on the right side vertical axis [7].

bath formulation, process parameters, film thickness, and substrate. Coating weight is measured in terms of Cr coverage, typical coating weights are 25–50 mg/m^2. These treatments are used on both zinc and aluminum-zinc-coated steel sheet to improve paint adhesion by increasing surface polarity that strengthens covalent bonding between the conversion coating and the paint primer [10], enhancing paint adhesion and the corrosion resistance of the final prepainted product. On prepainting lines, these treatments can be applied with the traditional tank-spray process or by a dry-in-place (DIP) method using roll coaters. Galvanized sheet intended for prepainting is usually produced without application of a passivation treatment. The surface is only given a coating of oil that includes corrosion-inhibiting chemicals that can easily be removed before cleaning and painting. On the other hand, passivated Zn-55%Al-1.5%Si is routinely pretreated on prepaint lines to remove some of the surface treatment, afterward a fresh pretreatment is deposited on top of the remaining passivation treatment to give excellent corrosion resistance and paint adhesion. Chemical treatment suppliers should be consulted for specific products to be used for this application. Environmental pressures to cease using hexavalent chromium have resulted in these treatments beginning to be phased out and replaced with the less environmentally sensitive trivalent chromium [11] or chrome-free treatment. Hexavalent chromate coating imparts a higher corrosion resistance because hexavalent chromium is leached out of the conversion in a corrosive environment, resulting in a self-healing effect as hexavalent chromium is reduced to trivalent chromium by the reduction reaction [12]. This mechanism is not present in other treatments. Trivalent chromium also participates in a reduction reaction. The means by which hexavalent and trivalent chromium conversion treatment protects steel is shown in Fig. 12.3.

Many chromium-free systems have also been developed, including those based on self-assembling molecules [13] and alternative chemicals [12].

While a well-applied chromate conversion coating does afford substantial added corrosion resistance to many prepainted systems used on zinc-coated steel sheet, phosphate treatments have gained favor because of their superior resistance to under-film corrosion. Both zinc and iron phosphate treatments are used, with iron phosphating mainly used in postfabrication painting rather than in continuous coil coating. After cleaning on the coil coating line, a titanium phosphate conditioning or activating step is usually added before zinc phosphating, producing a finer zinc phosphate crystal structure. Coating weights of 1.6–3.2 g/m^2 are typical and are higher with GA coatings because of the higher reactivity of the

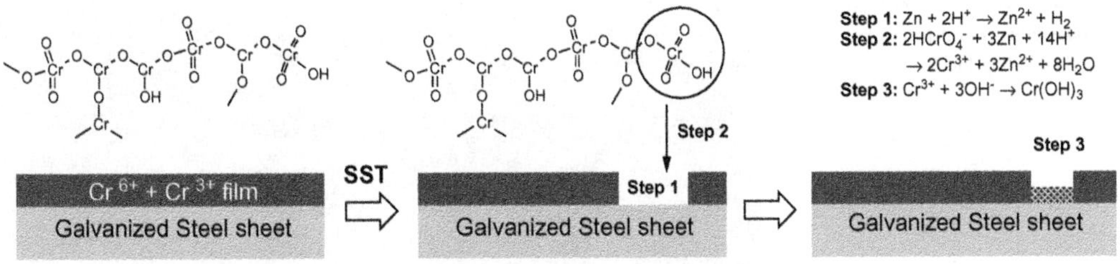

FIG. 12.3 Reactions of hexavalent and trivalent chromium. Initial application state (left), reduction of hexavalent chromium (center), reduction of trivalent chromium (right) [12].

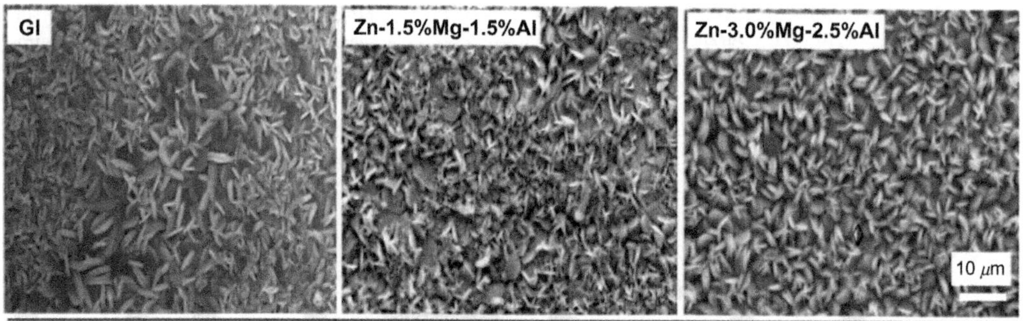

Sample	Coating weight g/m^2	Crystal size μm
GI	3.64	5.1
Zn-1.5%Mg-1.5%Al	3.69	3.7
Zn-3.0%Mg-2.5%Al	3.75	4.6

FIG. 12.4 Appearance of zinc phosphate chemical conversion treatments on GI and two ZnAlMg hot-dip coatings [14].

Fe-containing coating and the greater surface roughness of the GA coating that exposes more surface area to the phosphating reaction. The surface appearance and phosphate coating weights and crystal sizes of several hot-dip coatings are shown in Fig. 12.4.

Phosphate treatments are particularly effective at reducing paint undercutting corrosion at scratches and cut edges. Normally, the corrosion reaction would undercut the paint and move laterally along the sheet surface by corroding and dissolving the galvanized coating resulting in peeling off of the paint layer. A zinc phosphate pretreatment slows the rate of undercutting corrosion, resulting in a longer product life.

Adding Mg to the zinc phosphate treatment at levels of up to 5% resulted in spherical crystal morphologies rather than the needle-like morphologies seen on Fig. 12.4. When applied to an electrogalvanized coating, both friction and corrosion resistance were improved. The appearance of the surfaces with zero and 4%Mg is shown in Fig. 12.5.

12.1.4 Inhibition action from pretreatments and paints

In most cases of application, the cut edge of galvanized sheet painted on a coil coating line is left unprotected after cutting. Further, cut edges are often found at the drip edges of roofs and other locations where prolonged periods of wetness occur. These locations are usually the first locations where the coating fails. The delamination of paint from growth of corrosion inward from the sheet edge under the paint coating is termed "edge creep." The skyward-facing side of the painted sheet faces the weather and usually receives a full paint system that includes the chemical pretreatment, a primer, and one or more layers of topcoat. The groundward-facing side is more sheltered and usually receives only the chemical pretreatment and a backing coating that can be the primer only or a topcoat designed to be used without a primer. The inhibitors in the pigments of the paints are released as weathering occurs and help protect the cut edge of the painted sheet together with scratches, holes, or other damage to the barrier protection provided by the painted galvanized steel. An example of this was shown by Takata [16], who exposed coil-coated Zn-55%Al-1.5%Si and Zn-55%Al-1.5%Si-2%Mg alloy-coated sheet for 2.5 years at

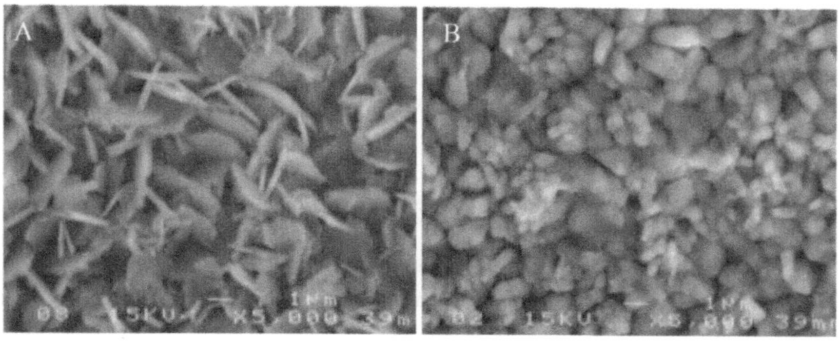

FIG. 12.5 Surface appearance of an electrogalvanized coating (30 g/m^2/side pure Zn) covered with a zinc phosphate conversion treatment containing (A) %Mg, (B) 4% Mg [15].

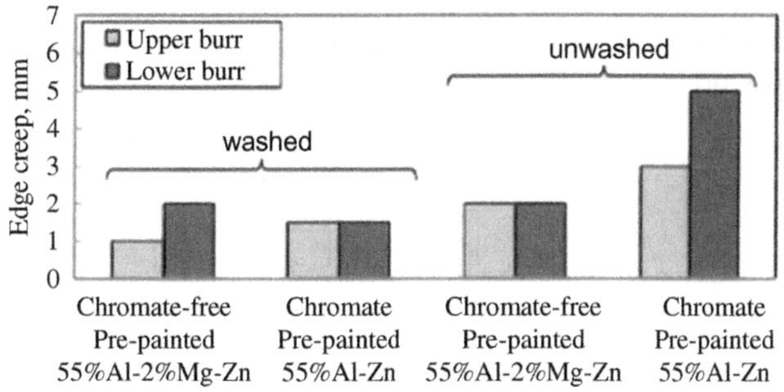

FIG. 12.6 Effects of edge shear burr orientation, metallic coating type, and pretreatment type on edge creep of coil-coated panels [16].

Noto, Ichikawa, Japan, an urban-industrial site 7 km from Tokyo Bay. Panels were sheared so that the trailing burr after shearing faced either up or down on the exposed panels. This gives slight differences in panel shape and how the coating system is wiped onto the sheared edge. More details about this will be discussed later in this chapter. Edge creep for both the skyward and groundward-facing burr orientations of the test panels is shown in Fig. 12.6.

Edge creep for the skyward-facing burrs never exceeded the groundward-facing burrs, this is likely related to the higher time of wetness experienced by the groundward-facing burr. In this case, the chromate-free pretreatment included an elutable (leachable) component that helped protect the cut edge from corrosion. The chromate-containing system provided inhibition as explained earlier. Fig 12.6 shows that the chromate-free system provided better protection, especially for the groundward-facing burr edge because the higher time of wetness allowed for more elution of its inhibitor. As explained in Chapter 11, sheltered panels do not receive periodic washing from rainfall to remove atmospheric contaminants such as chlorides and sulfates that contribute to the severity of corrosion. An example of the differences in sea salt deposited on sheltered and exposed surfaces of air conditioning units was given by Kajiyama [17] in Fig. 12.7 for four different locations: Okinawa in the Pacific Ocean, Choshi and Niigata on the Sea of Japan coast, and Tsukuba, 45 km from the seacoast. The effect of rainfall washing that periodically removes deposited sea salt is clearly shown in this figure.

As a result of environmental regulations, many paint coatings free of chromates have been developed; there are several examples of these in this chapter. For both the appliance and construction industries, resins containing corrosion inhibitors have been developed to replace chromate conversion treatments. For example, Suzuki [18] added phosphoric acid to a polyurethane resin water dispersion solution. The nonvolatile component concentration in the treatment solution other than the phosphate addition was 13.5 wt%. The phosphoric acid content was adjusted to vary from 0 to 30 wt% of the nonvolatile components, the details are shown in Table 12.1. This was then applied to an electrogalvanized steel 0.8 mm thick with a pure zinc coating of 20 g/m^2 per side. After curing, an X-shaped scribe, comprising two 55 mm long scratches cut with a utility knife down to the steel substrate, was made, after which panel back side and edges were covered with protective tape. Samples were then exposed to ISO 9227 salt spray testing. The width of paint creepage from the scribes is shown in Fig. 12.8 after 48 and 72 h of exposure, the trends show the effectiveness of phosphoric acid additions in reducing paint

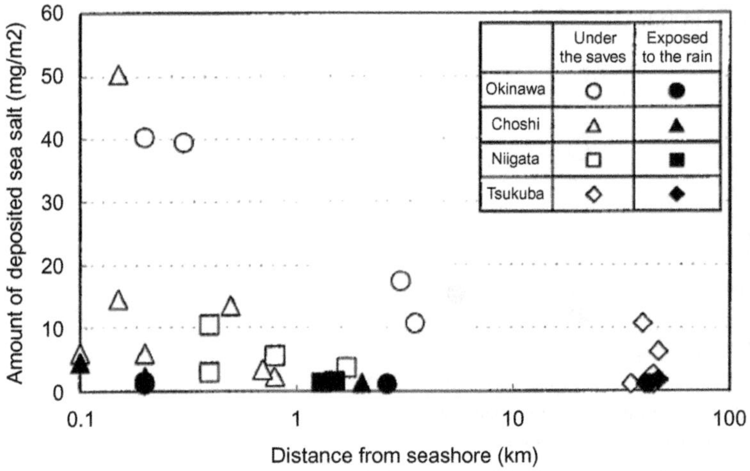

FIG. 12.7 Sea salt deposition on skyward surfaces of air conditioning units in exposed and sheltered conditions for four Japanese locations [17].

TABLE 12.1 Levels of phosphoric acid added to water-based polyurethane coating solution [18].

Specimen	Phosphoric acid content (wt%)
A	0
B	1.1
C	3.3
D	13.2
E	20.0

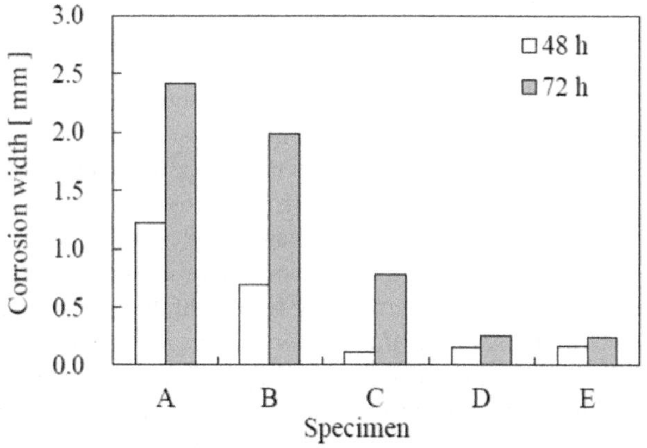

FIG. 12.8 Paint creepage from scribes after 48 and 72h of ISO 9227 salt spray testing for formulations shown in Table 12.1 [18].

corrosion creepage from the scribe. To determine the corrosion protection mechanism, the amount of phosphorus eluted from the coating was calculated by comparing the uncorroded area of the panel with the corroded area. During the first hours of salt spray testing, the steel was protected by the zinc coating, which preferentially corroded. As the phosphate coating eluted, Zn^{+2} and PO_4^{-3} reacted to form a compound in the scribe, protecting the steel from oxygen and the salt solution. Fig. 12.9 shows that the elution rate occurred most rapidly in the initial 24h of salt spray testing, permitting the protective zinc phosphate product to be formed.

The use of phosphates in a paint topcoat, rather than a water-based resin, was demonstrated by Kawamura [19], who prepared a topcoat with an elutable phosphate component, applied it with the rest of a paint system to a GI-coated steel,

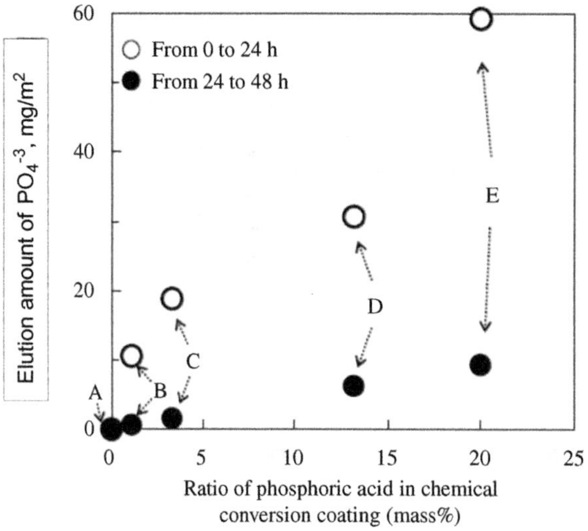

FIG. 12.9 Comparison of phosphate amount eluted after salt spray testing with the amount initially in the coating [18].

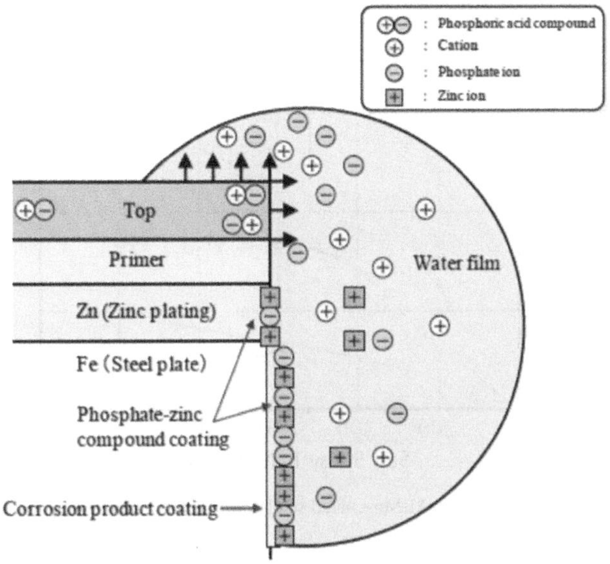

FIG. 12.10 Mechanisms of cut edge protection of painted galvanized sheet with a phosphate-containing paint [19].

sheared the edge to expose the steel, and exposed it to rain conditions. The mechanism of protecting the edge is shown in Fig. 12.10. Incoming rainwater, with its usual low electrical conductivity, first covers the cut edge of the steel sheet and the corrosion resistance pigments, which have high dissolution properties, begin to be eluted from the coating film. The inhibitors in the eluted corrosion resistance pigments then become ionic, and the electrical conductivity of the water is increased. As the electric conductivity of the water rises, the "throwing power" of the galvanic protection distance of the GI coating for the exposed steel edge becomes greater, permitting the coating to protect a greater thickness of the exposed edge. Zn from the GI coating then begins to elute, promoting sacrificial corrosion protection of the steel sheet. Additionally, the phosphate ion in the water film suppresses the cathode reaction of Fe by forming a barrier on the steel sheet.

Further insight into the differences between skyward and groundward performance of painted galvanized roof panels used in metal building construction was provided by Zapponi [1]. It was observed from long-term exposures of panels that the skyward-facing panels also accumulated dirt and insoluble corrosion products at a higher rate than the groundward-facing surfaces, giving additional barrier protection. While the time of wetness on the groundward side was higher than on the skyward side, the higher time of wetness permitted elution of chromate from the paint to occur for longer periods of time, contributing to improved inhibition of corrosion and formation of a more protective passive layer. The corrosion mechanisms in the areas of edge creep on the skyward and groundward sides were observed to be different at the Argentine coastal sites. In the samples of GI-coated and painted panels, the corrosion products in the edge creep areas on the skyward side early in the exposure period were hydroxycarbonates (hydrozincite), whose chemical formulas are shown in Chapter 11. After some time of exposure, depending upon distance from the sea, up to 1 km distant, chloride-including corrosion products replaced the carbonate and the hydrozincite corrosion products, transforming the surface into simonkolleite. On the groundward side, only hydroxycarbonates were detected. For Zn-55%Al-1.5%Si coated and painted panels in the skyward edge creep area, the zinc-rich phase corroded first, and the corrosion products were zinc hydroxycarbonates. On the back side, corrosion was more severe and aluminum zinc hydroxycarbonates were also detected.

12.1.5 Corrosion performance of coil coated panels—Edge corrosion and paint creepage

A comparison of edge creep performance of Zn-0.2Al, Zn-4Al-0.1Mg, and Zn-6Al-3Mg-coated steels was made by Yamamoto [20] using the standard salt spray test, 5% NaCl solution at 35°C. The coatings were all about the same coating weight, between 121 and 141 g/m^2 per side. They were given the traditional hexavalent chromium passivation treatment; the 5 μm epoxy primer contained a strontium chromate inhibitor and was covered with a 11 μm polyester topcoat. The

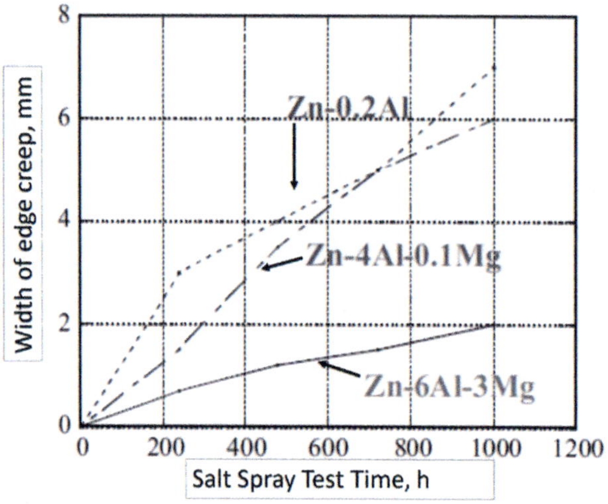

FIG. 12.11 Progression of edge creep for GI and two Zn-Al-Mg-coated steels during neutral salt spray testing [20].

150 × 60 mm panels had the edges on the short sides fully taped, together with the long sides, except for a portion in the middle of the long side. Also, the panels were sheared so that the burr faced up on one of the long sides and down on the other. Fig. 12.11 shows the progression of edge creep from the unprotected portion of the edge on the side with the downward-facing burr, measuring the largest distance found. Photographs of the panels showed little difference in edge creep for the upward versus downward-facing burr sides.

The reason for the better edge creep performance of the ZnAlMg-coated steels was attributed to the finer and denser corrosion products formed in the area of edge creep. Representative morphologies are shown in Fig. 12.12. The corrosion products on the Mg-containing panels were said to more effectively block oxygen from reaching the steel substrate than the Zn-0.2Al coating, greatly decreasing the oxygen reduction reaction that would have oxidized the iron.

Further insights into the corrosion mechanisms for hot-dip coatings containing Mg were provided by Schulz [21], who conducted various corrosion tests on GI and Zn-1%Mg coatings. These mechanisms are shown in Fig. 12.13. Ignoring any effects of inhibitors being released from the pretreatment or paint systems, hydroxyl (OH^-) ions in Fig. 12.13 are observed to react with Zn and Mg ions, resulting in formation of protective layers on the exposed edge. As in the case of the exposure of zinc-coated steels without paint systems, described in detail in Chapter 11, the protective layer on the ZnMg-coated steel is much denser and more ordered than the zinc hydroxide layer formed on the GI coating. Agreeing with Yamamoto, the reaction layer containing Zn and Mg results in a slowed cathodic oxygen reduction at the iron exposed at the cut edge. This in turn slows zinc dissolution at the surface of the metallic ZnMg coating. Agreeing with other investigators who examined the behavior of a Zn-1.5Al-1.5%Mg coating, a lower coating weight was found to provide the same edge creep performance for Mg-alloyed hot-dip coatings as a higher-coating-weight GI coating [22].

The addition of Al to the Zn-Mg coating was found to improve paint creep performance and protection of cut edges after 14 months of severe marine exposure in New Zealand and supporting laboratory tests [23] utilizing Zn-0.2Al, Zn-55Al-1.5Si, Zn-5Al-0.1Mg, Zn-6Al-3Mg, and Zn-11Al-3Mg-0.2Si coatings. Both traditional hexavalent chromium-phosphate and a chromium-free fluorotitanate conversion coating systems were used to pretreat panels. The chromium pretreatment was used with a chromium-containing primer, while the other was used with a chromium-free primer; both received polyester topcoats. The conversion coatings formed with the more Zn-rich coatings, Zn-0.2Al and Zn-5%Al-0.1Mg, were found to form conversion layer products rich in Zn, while the other more Al-rich coatings formed conversion layer products richer in Al that were more clustered and compact than those of the Zn-rich coatings, the Al-rich coatings were therefore found to have better adhesion prior to exposure. The chromium-free system did not perform as well as the traditional chromate system in helping to control paint creepage because it was not as efficient as eluting inhibitors, at least after 14 months of exposure. The higher Al hot-dip coatings were found to perform the best, and this was attributed to the very low solubility product of aluminum hydroxide (1.3×10^{-33} M^4 at 25°C) compared with that of zinc (3×10^{-17} M^3 at 25°C) and magnesium (5.61×10^{-12} M^3 at 25°C). Also, Al-rich alloys were found to passivate easily under severely corrosive conditions compared with pure zinc coatings. Adding Mg to the Al-containing coatings was found to provide two benefits when an exposed cut edge was present: first, the Mg-containing intermetallics are more anodic than other constituents in their coatings and corrode first to provide Mg ions that contribute to protective corrosion products; second, they provide more durable corrosion products.

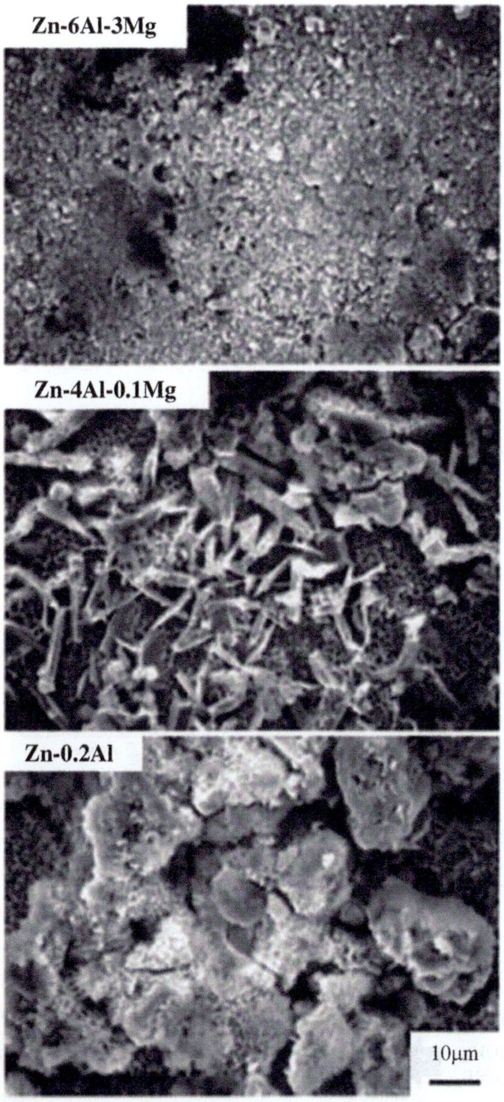

FIG. 12.12 Appearance of corrosion products in the edge creep areas of panels subjected to 96h of neutral salt spray testing [20].

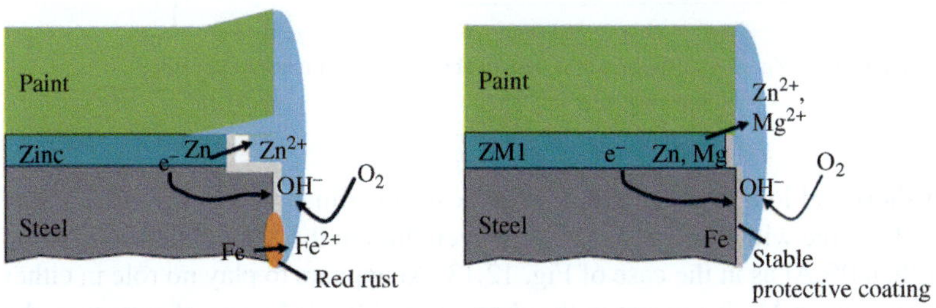

FIG. 12.13 Corrosion mechanisms in the edge creep area for a GI coating (left) and a Zn-1%Mg coating (right) derived from various coil coating paint systems and corrosion tests [21].

Further insights into the nature of corrosion of coil-coated Zn-0.2Al and Zn-11Al-3Mg-0.2Si coatings at cut edges were provided from 9.5-year outdoor exposure results at an Okinawa semitropical marine site [24]. Both types of metallic-coated panels, each with coating weights of 50 g/m^2 per side, received chromate-free paint systems including, on the skyward side, a 5 μm polyester primer with an inhibitor and a 15 μm polyester topcoat, the groundward side received a 5 μm polyester coating. After 9.5 years, edge creepage was seen on all panels, but red rust was only seen on the Zn-0.2Al panels. Fig. 12.14 shows the progress of edge creep during the time of exposure. The difference between edge creepage arose after the appearance of red rust on the Zn-0.2Al panels; minimal red rust was observed on the Zn-11Al-3Mg-0.2Si panels.

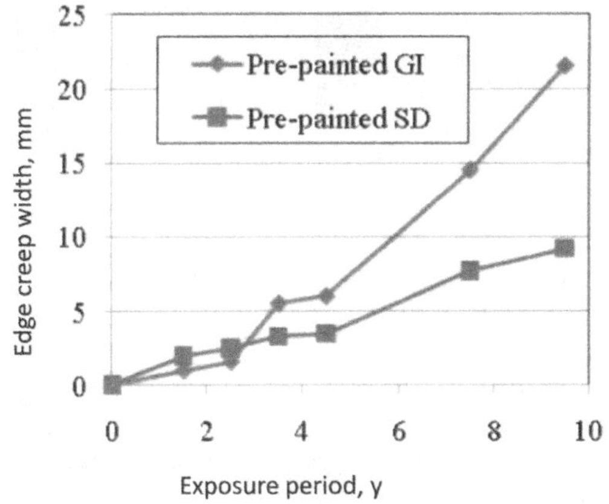

FIG. 12.14 Edge creep on Zn-0.2Al (GI) and Zn-11Al-3Mg-0.2Si (SD) panels during 9.5 years of exposure at Okinawa [24].

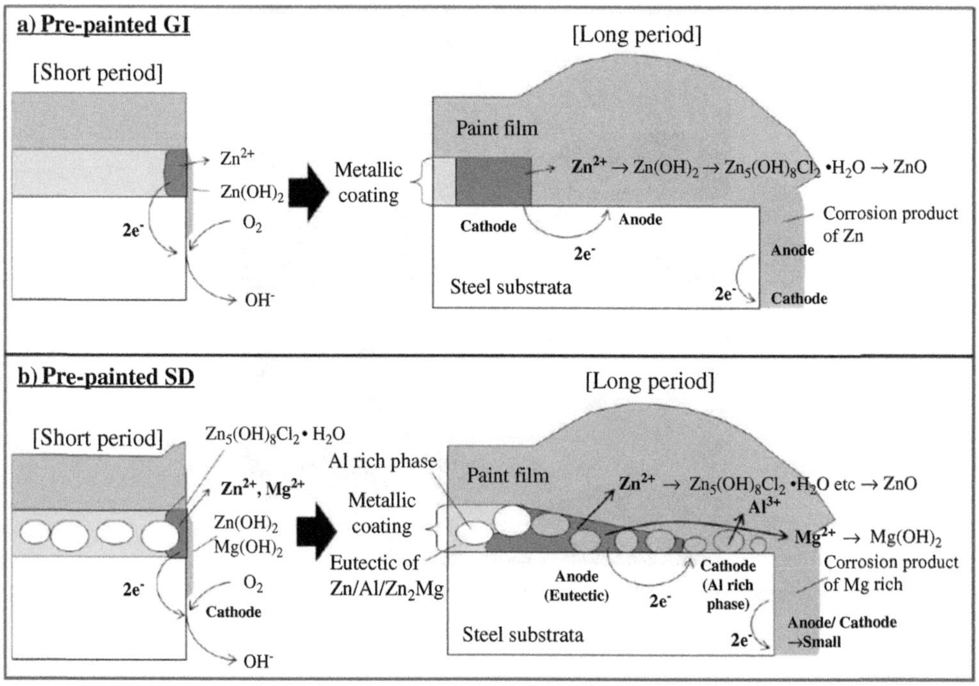

FIG. 12.15 Cross-section schematic of corrosion of Zn-0.2Al (GI) and Zn-11Al-3Mg-0.2Si (SD) up to about 2 years of marine exposure (left) and after 9.5 years of exposure (right) [24].

The sequence of reactions that occur after both short and longer time periods of exposure is illustrated in Fig. 12.15. The short-term illustrations on the left side of Fig. 12.15 agree with those of Fig. 12.13, even though the Mg-containing coating of Figs. 12.14 and 12.15 contains 11%Al rather than 0%Al as in the case of Fig. 12.13. Al appears to play no role in either cut edge protection or paint edge creep mechanisms when Mg is present in the short term. After 9.5 years of exposure, the right side of Fig. 12.15 shows the progression of Zn-0.2%Al corrosion products, changing from $Zn(OH)_2$ on the exposed cut edge in early stage of corrosion to the more stable and protective corrosion product simonkolleite with the progress of corrosion on the cut edge. For the Zn-11Al-3Mg-0.2Si panels, early stage dissolution of first Mg, then Zn from the eutectic portion of the coating forms hydroxides, as explained in Chapter 11. During the early stages of exposure, these are transformed to the more protective simonkolleite and gordaite under the paint layer, which are produced and deposited on the corroded metallic coating in the early period. Because of the formation of these corrosion products, the progression of edge creep for both metallic coatings is similar for about the first 2 years of exposure, as corrosion products are very effective in slowing as shown in Fig. 12.14. With further exposure of the ZnAlMg coating, a dual-protective layer including the corroding metallic coating and its corrosion products develops under the paint film, also there is significant Mg dissolution from the coating at the cut edge. This produces a Mg-rich deposit on the cut edge, so that red rust is much less observable than in the 0.2% Al system after 9.5 years of exposure. This dual-corrosion product layer under the corroded edge of the paint system, together with the Mg-rich layer on the exposed steel cut edge, provides better corrosion performance than in the Zn-0.2%Al system.

There are several ways of preparing cut edges on galvanized and coil-coated sheet; these include shearing, cutting with a saw, and nibbling. The latter technique makes a series of consecutive punches using a rapidly oscillating punch and die. For this example, a 1.2 mm diameter punch was used, and a guillotine shear was used to simulate the action of an industrial shear [25]. Zn-4.8%Al hot-dip galvanized panels were coil coated with a conventional system including a 200 μm thick PVC plastisol topcoat over a pretreatment and primer on the top side, while the back side received a pretreatment, primer, and a 15 μm polyester coating. Examination of the cut edges showed that the shear, cutting from the topcoat down through the panel, transferred zinc and some organic coating over the steel edge, Fig. 12.16. When using the nibbler or a saw, the zinc was not wiped across the steel substrate in the same way. The Scanning Vibration Electrode Technique (SVET) was used with a 0.1%NaCl solution to simulate corrosion exposures. A polished edge was included as a reference condition. After 24 h of SVET exposure, the amount of zinc loss from the sample, which was localized near the cut edge, was measured and is shown in Fig. 12.17.

Sawing significantly damaged the condition of the paint near the edge. This resulted in a high number of cathodic (exposed steel) sites being present, under the surface of the damaged paint and on the edge. The nibbling technique produces a very rough edge, resulting in peaks and valleys on the exposure surface that expose more steel to the corrosive environment. These contributed to the higher zinc losses shown in Fig. 12.17. The sheared (guillotined) samples had a

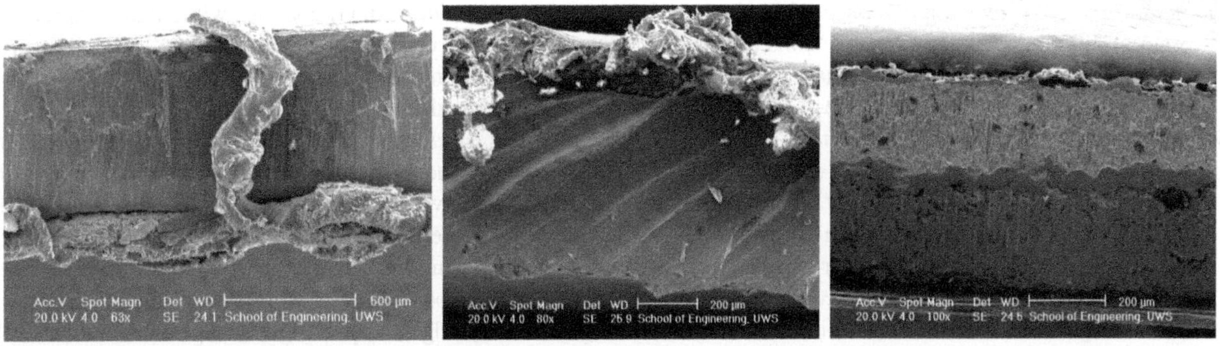

FIG. 12.16 Appearance of cut edges of coil-coated Zn-4.8%Al-coated steel sheet, scanning electron micrographs [25]. *Left:* nibbled, *center:* sawn, *right:* guillotined edge.

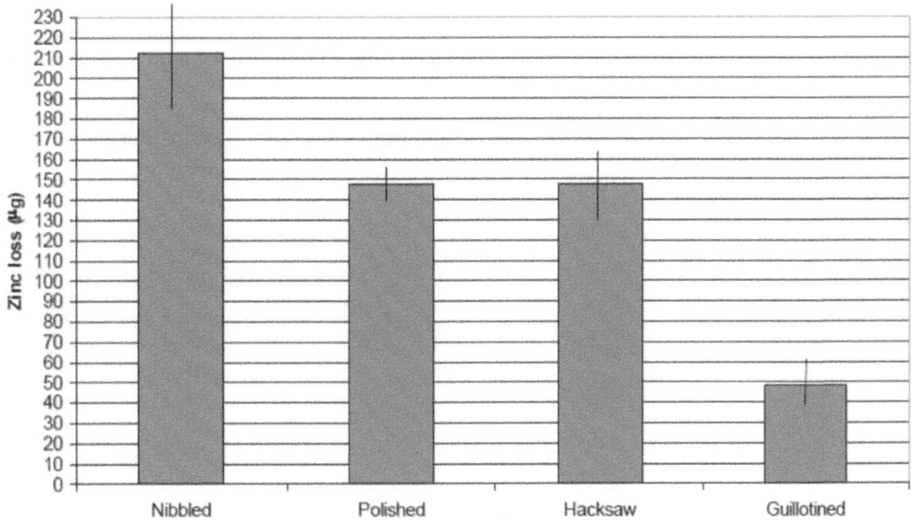

FIG. 12.17 Zinc loss after 24 h SVET testing in 0.1%Al for different edge cutting methods [25].

much higher fraction of anodic sites, not only near the boundary of the paint system with the sheet edge, but also on the surface of the cut edge where zinc was wiped as a result of the shearing action, providing more protection to the steel. Shearing was always performed downward from the topcoated side, so that portions of the topcoat could also contribute to the protection of the shared edge. The reason that polishing produced a relatively high zinc loss was that polished samples were freshly prepared. They did not have time to develop any oxides or other short-term corrosion products that would be present on the other samples.

The effect of both steel substrate thickness and coating thickness on long-term performance of coil-coated sheet was studied by Prosek [26]. A variety of GI-coated steel sheets were exposure to the marine site at Brest, France, for up to 5 years. The details of the coating weights, thicknesses, and steel substrate thicknesses are shown in Table 12.2.

All samples received a chromate-free system consisting of alkali cleaning, conversion coating utilizing using a Ti-based pretreatment solution with the target coating weight of 5 mg/m^2 Ti. The top side received a Cr-free polyester primer and topcoat. After curing at 224°C, the paint dry film thickness (DFT) of the primer and topcoat was 5 and 20 μm, respectively. The bottom side was painted with the primer paint at DFT of 7 μm. Both sheltered and fully exposed conditions were utilized.

Prosek described four factors related to steel and zinc coating thickness that influence the performance of coil-coated galvanized sheet:

(1) *Increase of size of steel area to be protected with time.* As the zinc coating dissolves to provide corrosion protection, the area of steel surface to be galvanically protected increases. The oxygen reduction reaction that results in oxidation (rust) of the steel surface can take place over a larger area, further increasing the driving force for further zinc coating dissolution and paint delamination. Therefore, an increase in zinc coating thickness provides an extended time of steel protection.

(2) *Inhibition of oxygen reduction on steel by deposited zinc corrosion products.* In GI coatings without any appreciable level of alloying elements, the Zn^{+2} ions released from corroding zinc coating diffuse and migrate to steel, forming low-solubility corrosion products that serve to lower the rate of the oxygen reduction reaction. This will lead to a lower demand of galvanic current and rate of zinc coating dissolution. Thicker zinc coatings increase the supply of zinc ions.

(3) *Limited "throwing power" (distance) of galvanic protection due to electrolyte conductivity.* Under atmospheric exposure conditions, the thickness of the electrolyte layer varies, depending upon climatic conditions, in particular precipitation and relative humidity. Electrolyte (surface film) conductivity is affected by the thickness of the film and the concentration of ions such chlorides and sulfates and the presence of the formed corrosion products. For cut edges on thick steel substrates or corroded cut edges after long exposure, a significant portion of the steel thickness does not have galvanic (electrical) contact with the zinc coating for long periods of time. Therefore, further increase in the cut edge area under these conditions does not lead to higher rates of edge creep. For given exposure conditions, the maximum edge creep rate can be defined based on the typical duration of the available galvanic action. Thicker steels reduce the amount of steel receiving galvanic protection.

TABLE 12.2 Combinations of steel substrate thicknesses and GI coating weights for 5-year exposure of coil-coated sheet steels [26].

	Coating weight, both sides (g/m^2)|coating thickness, each side (μm)		
Substrate thickness (mm)	100|7	140|10	275|20
0.2	200/7	–	–
0.35	–	–	350/20
0.5	500/7	500/10	500/20
0.7	700/7	700/10	700/20
1	1000/7	1000/10	1000/20
1.5	1500/7	1500/10	1500/20
2.5	2500/7	2500/10	2500/20

(4) *Formation of red rust, therefore increasing the cathodic activity.* Red rust increases the efficiency of the oxygen reduction reaction that oxidizes iron, thereby increasing the demand for galvanic current from zinc coating, consuming more zinc.

The first two factors were indicated as having strong and opposite effect on cut edge corrosion. Depending on steel substrate and zinc coating thicknesses, the released Zn^{+2} ions either would or would not be able to passivate efficiently the entire steel surface. If the amount of Zn^{+2} is insufficient for steel passivation, the first factor would dominate, and the rate of edge creep will increase in time. In the opposite case, i.e., for lower steel gauges or thicker zinc coatings, the second mechanism would dominate, and the delamination rate will decrease in time as a more efficient protective layer over steel is built. The chloride load is of importance too; openly exposed edges with a lower chloride load can passivate even at high steel to zinc thickness ratios, whereas the first factor will dominate at lower steel to zinc thickness ratios for sheltered edges where more chloride can accumulate.

The effect of steel and coating thickness on the corrosion of coil-coated sheet from the 5-year exposure results is shown in Figs. 12.18 and 12.19.

Figs. 12.18 and 12.19 are divided into three zones that depend on exposure time. Considering Fig. 12.18, in Zone 1, the zinc corrosion products that are formed on the cut edge inhibit the oxygen reduction reaction that oxidizes steel. Therefore, the rate of edge creep is nearly independent of steel thickness. Edge creep progresses slowly in time with a constant or even decreasing delamination rate. In Zone 2, the amount of released zinc is insufficient; paint creepage is seen to increase with

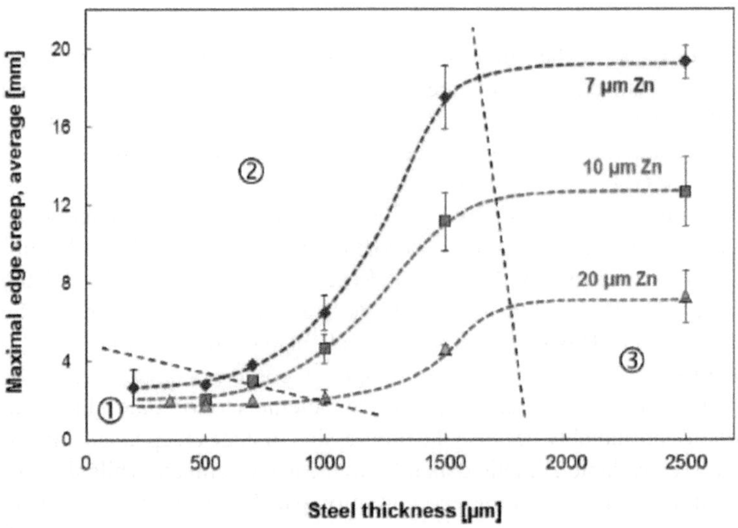

FIG. 12.18 Effect of steel and GI coating thicknesses on maximum paint delamination from sheltered cut edges after 60 months of marine exposure [26]. *Circled numbers* indicate the zones discussed in the text.

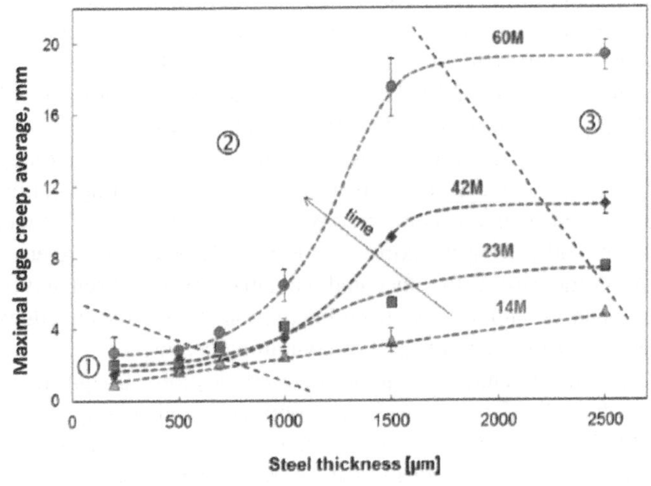

FIG. 12.19 Effect of steel thickness on maximum paint delamination from sheltered cut edges for materials with 7 µm GI coatings after exposure for 14–60 months [26]. *Circled numbers* indicate the zones discussed in the text.

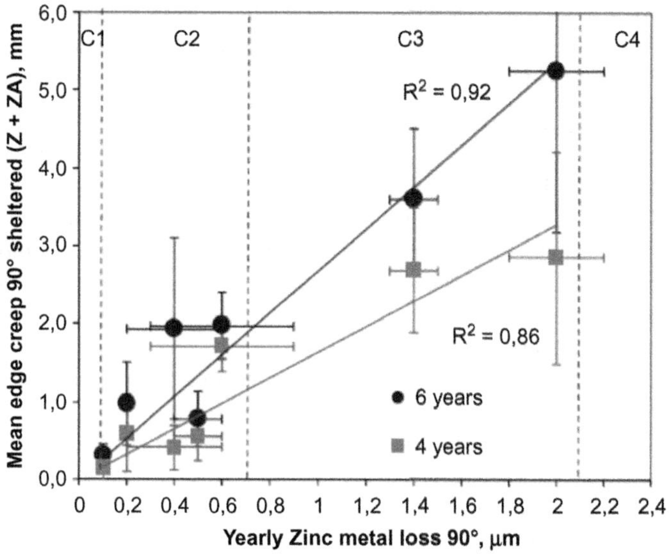

FIG. 12.20 Edge creep measured on coil-coated GI and Zn-5%Al panels after 4 and 6 years of exposure in sheltered conditions at 90 degrees as a function of the yearly zinc metal loss for corrosion sites with ISO 9223 corrosion severity classifications of C2 and C3 [27].

increasing steel thickness and increases at much higher rates for the thinnest coatings. Approaching Zone 3, as the Zn coating is increasingly consumed, the steel cathode is large, and a maximum creep rate is reached at the boundary between Zones 2 and 3. No further effect of steel thickness is observed in Zone 3. Boundaries between the three zones move in time. As exposure time proceeds, the zinc coating is consumed and the size of steel to be protected increases; therefore, Zone 3 enlarges and Zone 1 shrinks; this is shown for 7 μm thick GI coatings in Fig. 12.19.

Another set of long-term exposures from marine and industrial sites in Sweden, France, Australia, Vietnam, and P. R. China, using a variety of traditional chromate and chromium-free paint systems on GI, Zn-5%Al, and Zn-55%Al-Si substrates, was able to correlate zinc corrosion rates from unpainted samples, described in Reference 67 in Chapter 11, with the performance of coil-coated GI and Zn-5%Al steel sheets [27]. This is shown in Fig. 12.20. Also shown in this figure are the ISO 9223 corrosion severity factors described in Chapter 11. This correlation was made using samples exposed at 90 degrees (vertical) in sheltered conditions, whereas the ISO 9223 standard refers to samples exposed at a 45 degrees angle, and was found to be valid for ISO corrosion severity classifications of C2 and C3. However, attempts to correlate edge creep with the climatic data of the field sites were unsuccessful, presumably because the progression of edge creep is sensitive to various parameters that are interconnected. Moreover, the available climatic data refer to the macro-climate while the corrosion occurs at a microscale, for which no data (such as local temperature and humidity, panel temperature, etc.) were recorded, the only available microscale data that are possible to correlate are the metallic coating corrosion rates. Such a correlation was not possible for samples from higher corrosion severity classifications (C4 and C5), nor were any correlations between edge creep and metallic coating corrosion rates possible to make for the Zn-55%Al-1.5%Si coating.

However, there is a general correlation between the corrosion rates of Zn-55%Al-1.5%Si coating and the same coating with a 2%Mg addition [28]. The improved corrosion resistance of the Mg-added version of this coating was described from work by the same author in Chapter 11. As in the case of work described above, there is also a correlation between edge creep performance and coating weight for the same coating chemistries, Fig. 12.21. The CCT-1 cyclic test [29] consisting of repeated cycles of 4 h salt fog (5%NaCl solution) at 35°C, 2 h drying at 60°C and 35% RH, and 2 h humidity at 50°C and >95% RH was used to assess edge creep performance. Consistent with the correlations of coating weight and edge creep described above, the sensitivity of edge creep to coating corrosion resistance is more sensitive for the lower-coating-weight samples, because there is less available coating to provide protection. This was confirmed by the appearance of test panels that were exposed for 3 years at an Australian severe marine site, 60 m from breaking surf, Fig. 12.22. More severe edge creep and blistering are observed on the Mg-free coating. As in the case of the uncoated galvanized panels discussed in Chapter 11, the presence and distribution of $MgZn_2$ and Mg_2Si phases, and their contribution to protection of the cut edges during the corrosion process, are related to the improvement in performance of the Mg-containing coating.

Although GA (galvanneal) is normally thought of as a product intended for automotive applications, a significant amount is also used in prepainted construction applications such as metal doors. As in the continuous coil coating sector, strong efforts have been made to reduce the environmental burden of automotive paint systems. This includes the development of chromium-free paint systems and water-based paints. An example of an alternative pretreatment system that has

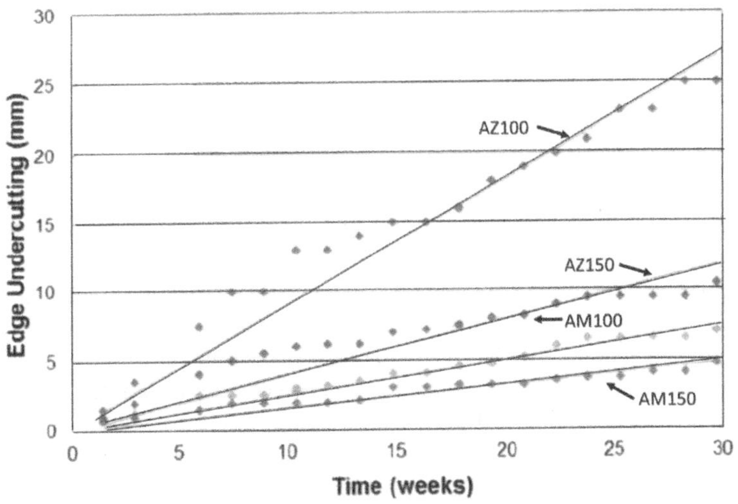

FIG. 12.21 Edge creep measured on coil-coated Zn-55% Al-1.5% (AZ) and Zn-55%Al-1.5%Si-2%Mg (AM)-coated panels after various periods in the CCT-1 cyclic corrosion test [28].

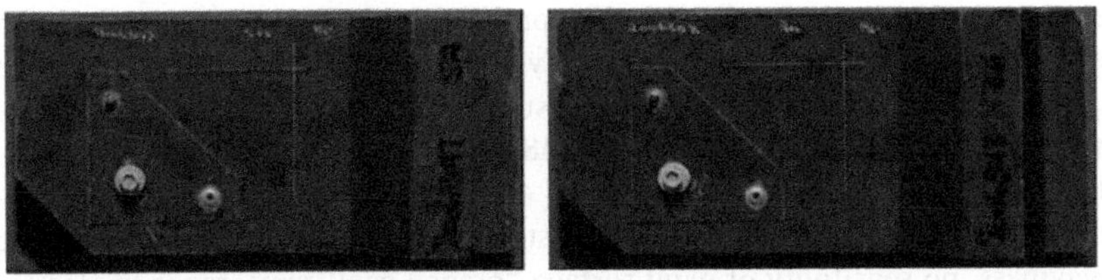

FIG. 12.22 Appearance of coil-coated Zn-55%Al-1.5%Si (left) and Zn-55%Al-1.5%Si-2%Mg (right) after 3 years of exposure at an Australian severe marine site [28].

been used with the galvanneal (GA) coating is γ-mercaptopropyltrimethoxysilane (MTMO), a silane-based pretreatment [30]. This was used with a three-component water-based polyurethane paint system. The pretreatment solution was prepared by mixing 3.6 mL MTMO with 5.4 mL of a solution of 60 vol% methanol and 40 vol% distilled water. Edge creep measurements for GI and GA panels using this pretreatment with the water-based paint system were compared with panel that received the paint system without any pretreatment by conducting salt spray testing for up to 42 days, the results are shown in Fig. 12.23. In these tests, the GA coating has higher paint creepage than the GI coating, this was attributed to the presence of maghemite (γ-Fe_2O_3) corrosion products in the creepage area that were formed from the Fe within the GA coating and formed a corrosion product more voluminous than the simonkolleite that was also found in the GA samples but that accounted for all of the corrosion product formed in edge creep area of the GI samples.

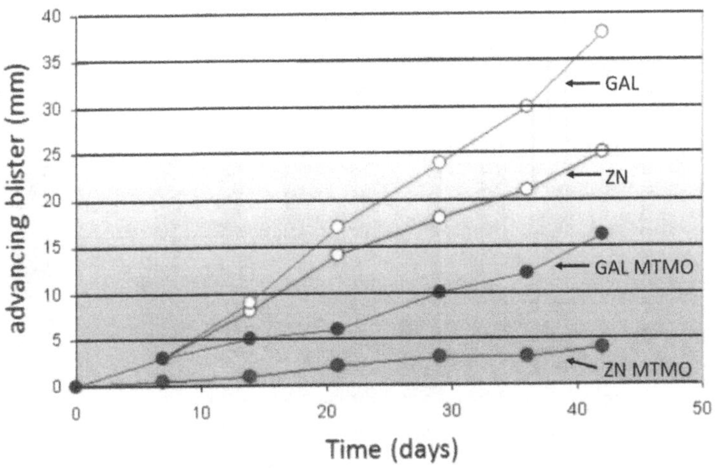

FIG. 12.23 Paint creepage from the unprotected edges of GI and GA panels given either no pretreatment or the silane-based MTMO treatment and a water-based polyurethane three-part paint system after up to 42 days of salt spray testing [30].

12.1.6 Performance of painted automotive galvanized steel

12.1.6.1 Substrate considerations

Paint appearance standards for automotive applications are customarily more stringent than those for construction and appliance applications; the objective is to supply the customer with a vehicle that has a body surface without any imperfections. The two technical measures of painted automotive surface quality were described by Steck [31] as being brilliance and smoothness, measured as "distinctness of image" (DOI) and "orange peel." ASTM E 284 [32] defines DOI gloss as "the aspect of gloss characterized by the sharpness of images of objects produced by reflection at a surface." DOI is mainly determined by structures below 0.1 mm often referred to as "dullness" and wavy patterns in the range of 0.1–1.2 mm. The DOI measure gives an indication for paint appearance at a viewing distance below 50 cm. It is also called the "short-wave" part of the spectrum. Structures between 1.2 and 30 mm typically influence the appearance at a viewing distance up to 3 m and are called "long-wave" structures or "orange peel." ASTM E 284 defines orange peel as "the appearance of irregularity of a surface resembling the skin of an orange." A range of electrogalvanized coatings with a Zn coating thickness of 7.5 μm were plated onto a bake hardenable substrate that had been cold-rolled and skin passed to provide an average roughness (R_a) of 1.1 μm and a peak count (RP_c) of 120/cm, typical of the surface quality used for exposed body panels. Electrogalvanizing process variables were varied to give a range of average coating roughnesses between 0.45 and 1.5 μm, peak counts between 47 and 139/cm, and longer-range waviness (W_{sa}), measured in the band of 1–5 mm, between 0.108 and 0.235 μm. The values of all three of these measurements increased in the same order for each of the seven sample conditions. The panels were then sent to three automotive painting lines for application of a full automotive paint system comprising a phosphate pretreatment, an electrophoretically applied primer, and a basecoat/clearcoat finish system. Figs 12.24 and 12.25 show the effects of substrate peak count and waviness on the DOI measurements made on the fully painted panels. The total paint system peak count for each of the three suppliers does not increase as substrate peak count increases from 47 to 139/cm, staying essentially in the range of 47–48/cm. Also, as longer-range waviness increases for the substrate, peak count for the full paint system stays in the same 47–48/cm range. Increasing the measurement window for waviness measurements to 1.2–30 mm for the full paint system showed the insensitivity of these measurements to substrate peak count and waviness, no sensitivity of paint surface "orange peel" appearance to electrogalvanized substrate measurements was observed. The automotive paint system therefore leveled any roughness and waviness effects of the substrate over the range of variables measured.

12.1.6.2 Corrosion performance of painted automotive galvanized sheet steel

An example of performance of state-of-the-art automotive steel corrosion protection systems is the 2-year study carried out in Japan and Indonesia by Sato et al. [33] This includes uncoated cold-rolled sheet and cold-rolled electrogalvanized zinc and zinc-nickel-coated sheet together with GI and GA hot-dip coated steel sheet, sample details are shown in Table 12.3.

A 3 g/m^2 zinc phosphate pretreatment was applied to all substrates followed by a 15 μm electrophoretic primer and spray-applied finish coats. Panels were then given an X-shaped scribe down to the steel substrate; the edges were protected from corrosion. Four exposure sites were used: A is an industrial site near Tokyo that experiences use of deicing salts, B

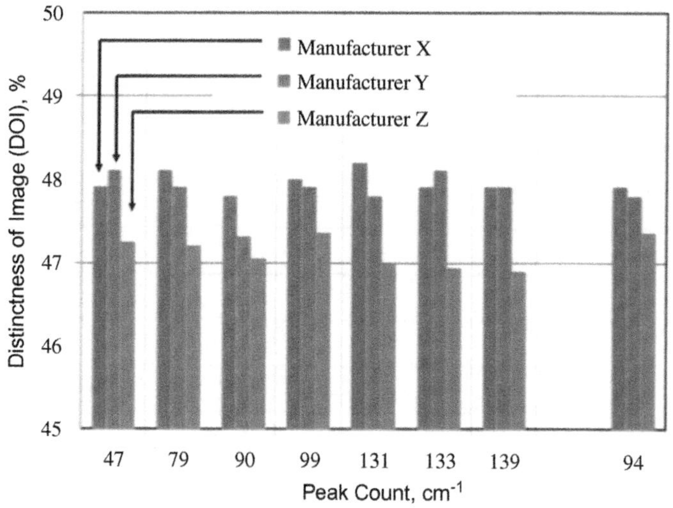

FIG. 12.24 Relationship between roughness peak counts made on Zn electrogalvanized substrates and the distinct optical image (DOI) measurements made on fully painted panels supplied by three automotive suppliers [31].

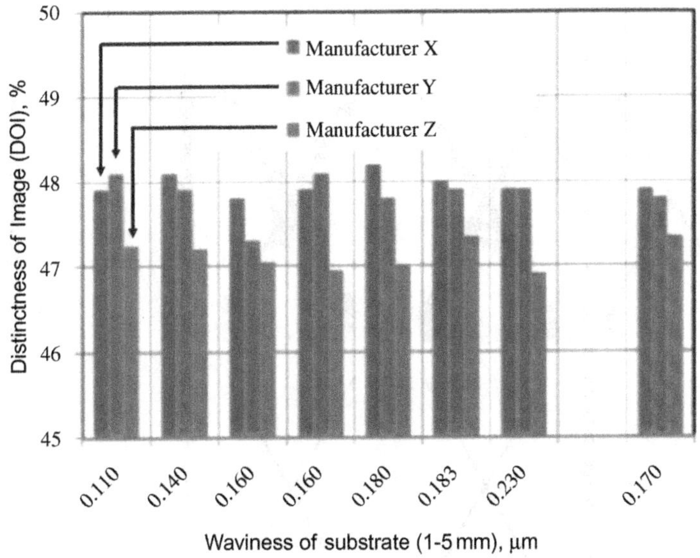

FIG. 12.25 Relationship between long-wave waviness measurements made on Zn electrogalvanized substrates and distinct optical image [31].

TABLE 12.3 Coating weights and substrate thicknesses used for 2-year exposure study in Japan and Indonesia [33].

Type and composition of coating	Coating weight (g/m^2)	Thickness of steel sheet (mm)	Code
Bare	0	0.8	SP
Electrogalvanized pure Zn	18	0.7	EG20
Electrogalvanized Zn-13 mass%Ni alloy	37	0.8	ZN30
Galvannealed Zn-10 mass%Fe	45	0.8	GA45
Hot-dip galvanized Zn-0.2 mass%Al	60	0.8	GI60

and C are subtropical marine sites in the Ryukyu Islands that include Okinawa, and D is an industrial site near Jakarta. The appearance of the ungalvanized, zinc electrogalvanized, and GI panels is shown in Fig. 12.26. Significant red rust is observed in the scribes of the painted ungalvanized panels, the corrosion was observed to be filiform in nature. Paint blisters with diameters greater than 1 mm were observed nearby the scribe in the ungalvanized panel specimens exposed at sites B and C. The painted zinc electrogalvanized panels showed slightly better performance after exposure at sites B and C, and the morphology of corrosion under the paint next to the scribe at sites B and C was in the form of blisters rather than filiform corrosion. The painted GI panels showed the lowest width of creepage from the scribe for sites B and C. However, at the A and D industrial sites, corrosion adjacent to the scribe is seen for the painted GI panels, including the presence of red rust. Still the G60 panels provided the best overall protection to the paint; although iron corrosion from the scribe is evident at site D, the paint creepage from the scribe was quite small. The relationship between coating weight and maximum scribe creep width for panels of all of the six sample conditions from the four exposure sites is shown in Fig. 12.27. The results from the marine sites B and C show the same benefit of increasing coating weight on minimizing paint creepage as the results from 6 years of coil-coated panel exposures at a French marine site described earlier, even though the paint systems, sample details, and coating types except for the GI coating are different.

As described in Chapter 8, the Zn-Fe crystals that make up the GA coating provide an intrinsically rougher surface than the zinc-rich eta phase that solidifies in place in the case of the GI coating, for example. Also, because the GA coating is less electronegative, less corrosion creep-back of this coating occurs at cut edges and scribes when the steel substrate is exposed in a corrosive atmosphere. This results in less blistering of painted GA at these locations than coatings that have an eta phase on their surface. This behavior was demonstrated in a laboratory test [34] by coating electrogalvanized (EG), GI, and GA-coated steel panels with a conventional electrophoretically applied primer system. The thickness of the EG, GI, and GA coatings was 8, 11, and 10 μm, respectively. The primer coating thickness was 4 μm for EG and GI and 2 μm for GA. The

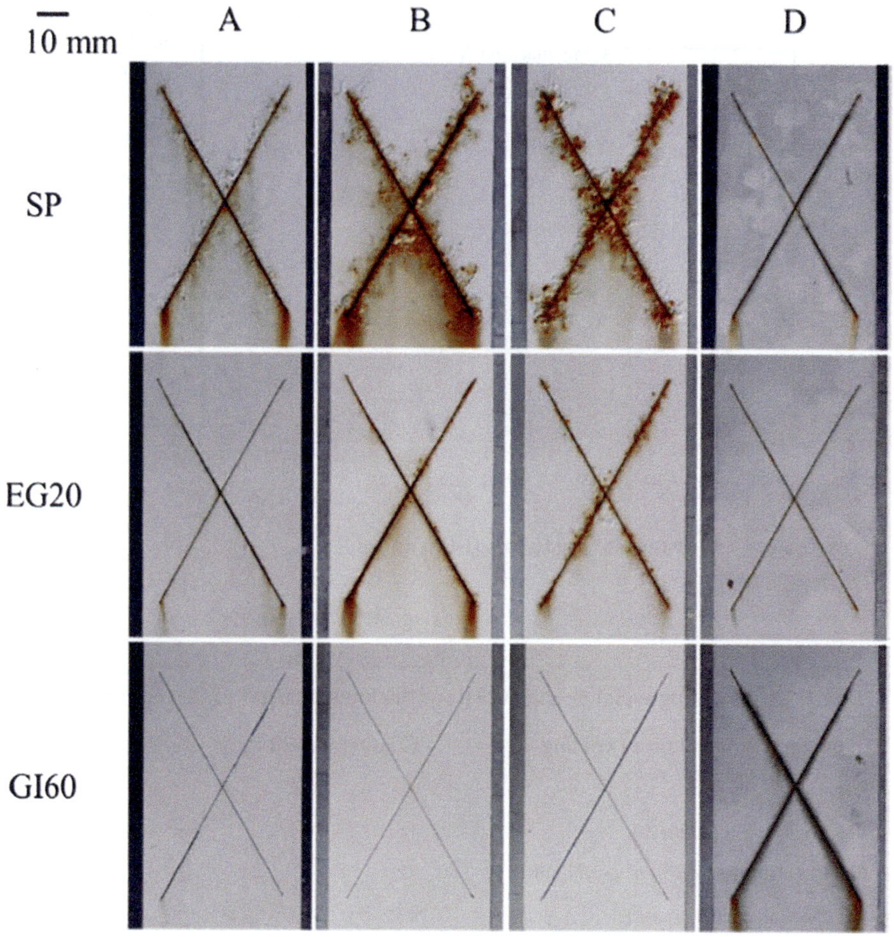

FIG. 12.26 Appearance of the ungalvanized (SP), zinc electrogalvanized (EG), and GI (GI60) panels after 2 years of exposure in Japanese and Indonesian site [33].

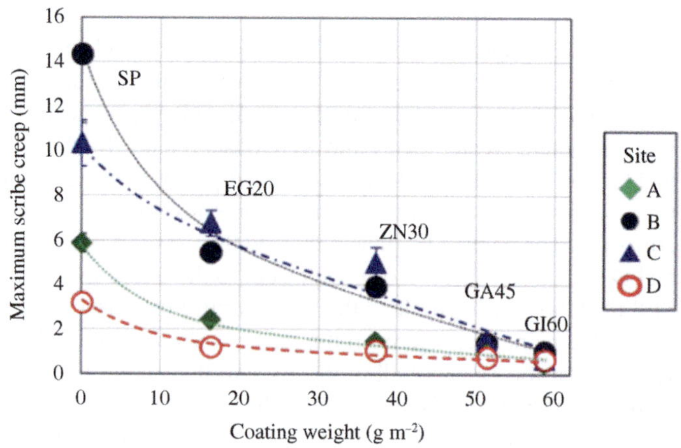

FIG. 12.27 Relationship between coating weight and maximum scribe creep width of painted test pieces described in Table 12.3 exposed at Japanese and Indonesian sites for 2 years [33].

primed panels were then subjected to a stone chipping test according to ASTM D3170 [35]. Following the standard practice, panels were chilled to 0°C before stone chipping commenced. The test lasted for about 15 s, using about 550 mL of gravel, with a maximum projection pressure of about 3.8 bar. The primed and stone-chipped panels were submitted to the VDA 621-415 cyclic corrosion test that is similar to Cycle B of ISO Standard 11997-1 [36]. Table 12.4 and Fig. 12.28 show that despite the reduced primer layer thickness, the GA coating provided better protection to the steel panel than the EG or GI coating, in both the undamaged and damaged part of the stone-chipped panels.

TABLE 12.4 Performance of EG, GI, and GA coatings with electrophoretic primer after ASTM D3170 stone chipping followed by exposure to 10 VDA 621-415 corrosion test cycles [34].

Area	EG + primer	GI + primer	GA + primer
General overview	Moderate to considerable Zn corrosion	Moderate Zn corrosion	Slight Zn corrosion
Stone chipping area	Severe Zn corrosion; moderate Fe corrosion	Severe Zn corrosion; moderate Fe corrosion	Slight Zn corrosion; slight Fe corrosion

Panels' appearance is shown in Fig. 12.28 [34].

FIG. 12.28 Appearance of EG (left), GI (center), and GA (right) coatings with electrophoretic primer after ASTM D3170 stone chipping followed by exposure to 10 VDA 621-415 corrosion test cycles [34].

The performance of EG, GI, and GA-coated panels with a full automotive finish system was also investigated by Santos [37]. The experimental design included two pretreatments: a conventional zinc phosphate (PT1) and an experimental treatment (PT2). A conventional electrophoretic primer was then applied, followed by one of two finish coats: finish coat 1 (FC1) was a system used at the Windsor, Canada assembly plant of Daimler Chrysler (now part of Stellantis), a PPG system. It included a waterborne base and clear system, being composed of PCV70118 Powder, stone white basecoat HWBS83542, and clearcoat DCT 555. Finish coat 2 (FC2) was a system used at the Warren, Michigan truck plant of Daimler Chrysler, a PPG system, composed of PVC 70118 Powder, white basecoat NHU90394E, and clearcoat CSRC 8002E. The thickness of the complete protective coating system, including the metallic coatings, ranged from 169 to 199 μm for PT1 and 154–169 μm for PT2. Samples were then corrosion tested using VDA 621-415, which is similar to Cycle B of ISO Standard 11997-1 "Paints and varnishes—Determination of resistance to cyclic corrosion conditions—Part 1: Wet (salt fog)/dry/humid" [36]. The test cycle consisted of:

- 1 day salt spray test according to ASTM B117 [3]
- 4 days cyclic humidity, being each daily cycle composed of 3 h at 40°C and 98% relative humidity and 16 h at 23°C and 50% relative humidity
- 2 days ambient climate according to DIN 50014 [38]

Before testing, a scribe 1 mm wide and 50 mm long, penetrating through to the steel substrate, was made with an increasing pressure from zero to a maximum of 4 bar, using the procedure given in ISO 12944-6, Annex A [39]. Ten cycles of VDA 621-415 corrosion testing were then performed. After testing, blisters near the scribe were evaluated using ISO 4628 Standard part 2 [40]. The appearance of the panels is shown in Fig. 12.29. No significant effects of corrosion testing were observed on any of the areas away from the scribes on any of the panels. White and red-colored corrosion products and severe blistering or delamination were observed on the scribe and growing from the scribe, respectively, on all samples. The highest proportion of white corrosion products were observed on the GI-coated panels; 100% red corrosion products were observed on the GA-coated panels because of the higher Fe content of the GA coating. This study also examined several weldable primers that were applied by continuous coil coating rather than the electrophoretic painting process. Equivalent

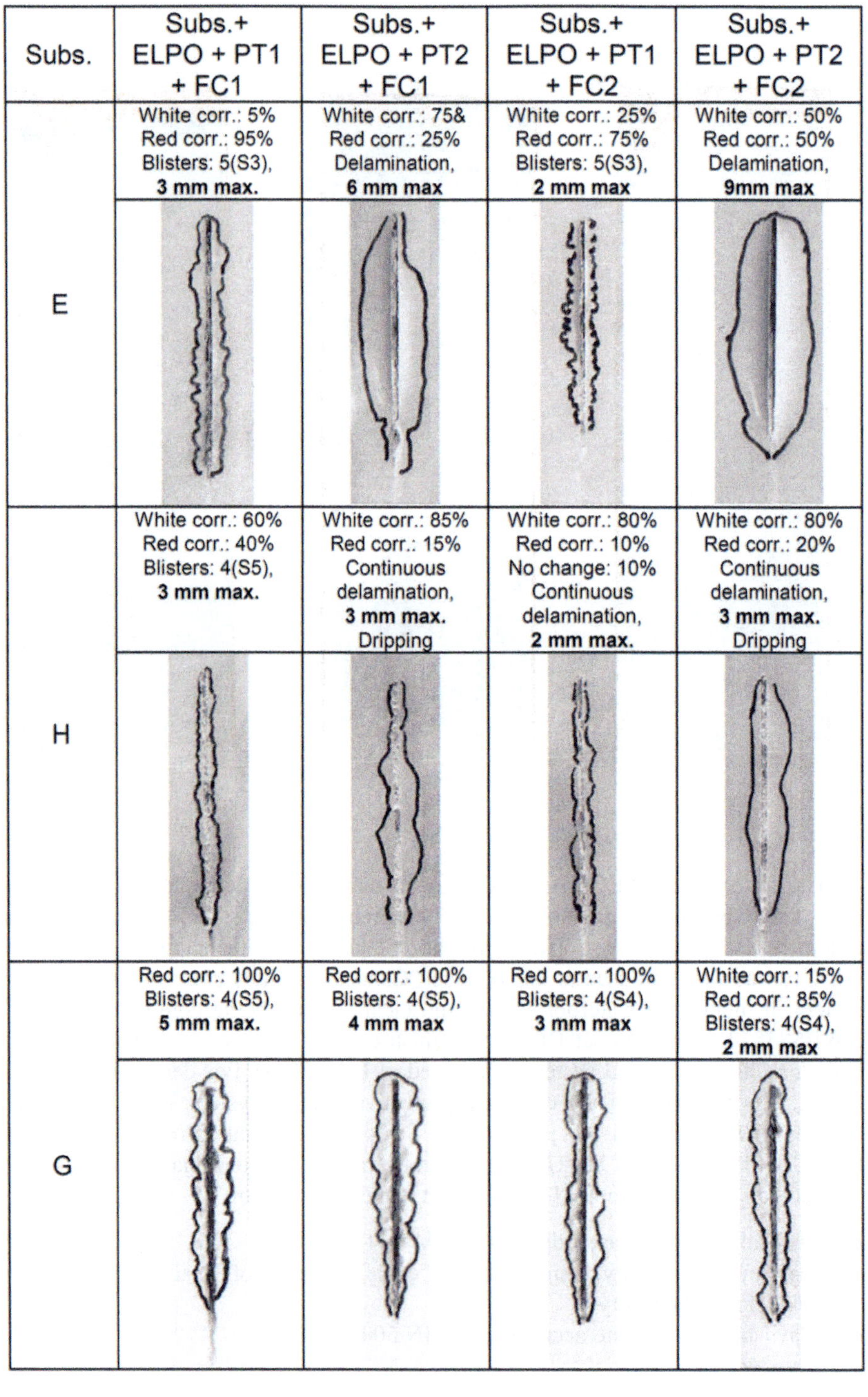

FIG. 12.29 Appearance of scribed full finished panels with a 50 × 1 mm long scribe after 10 cycles of VDA 621-415 corrosion testing. Zn electrogalvanized (E, top), GI (H, middle), and GA (G, bottom) metallic-coated substrates [37].

corrosion performance was observed for GA-coated panels for these panels in comparison with the results shown in Fig. 12.29; the EG and GI-coated panels were not as durable when the weldable primers were used. Several other studies have also examined the performance of weldable primers as replacements for the traditional electrophoretic primer application method. These have especially shown benefits in improving flange corrosion, where the penetration of electrophoretic primers to provide corrosion protection can be difficult [41,42].

12.1.6.3 *Painted ZnAlMg automotive coatings*

Zinc-aluminum-magnesium-coated steels have also been characterized for corrosion-resistant automotive applications. coatings with compositions in the range of Zn-1.5Al-1.5Mg were compared with GI and GA coatings, their details are shown in Table 12.5 [43]. These were then given a phosphate pretreatment and electrophoretically coated to produce a 20 μm thick primer. One set of panels, 90 × 190 mm in size with two vertical scribes of 100 × 0.5 mm down to the steel substrate, were exposed to the Renault ECC1 D172028 cyclic corrosion test [44] for 42 cycles (days). Another set of panels, 200 × 100 mm, was given two crossing scribes, one which penetrated down to the zinc coating and the other all the way to the steel substrate. Above the scribes, an area of about 90 × 90 mm was stone-chipped. These panels were exposed to the VDA 612–415 cyclic corrosion test for 20 cycles (weeks). The appearance of both sets of panels is shown in Fig. 12.30. The

TABLE 12.5 Details of GI, GA, and ZnAlMg-coated panels used for the results given in Figs. 12.30 and 12.31 [43].

	Coating weight evaluated side (g/m^2)	Coating thickness (μm)	Al%	Mg%	Fe%
GI100	57	8	0.5	<0.01	0.38
GI140	82	12	0.4	<0.01	0.32
GA90	41	6	0.3	<0.01	10.2
MZ70	35	5	1.8	1.8	0.17
MZ100	53	8	1.7	1.6	0.16
MZ140	68	10	1.6	1.6	0.13

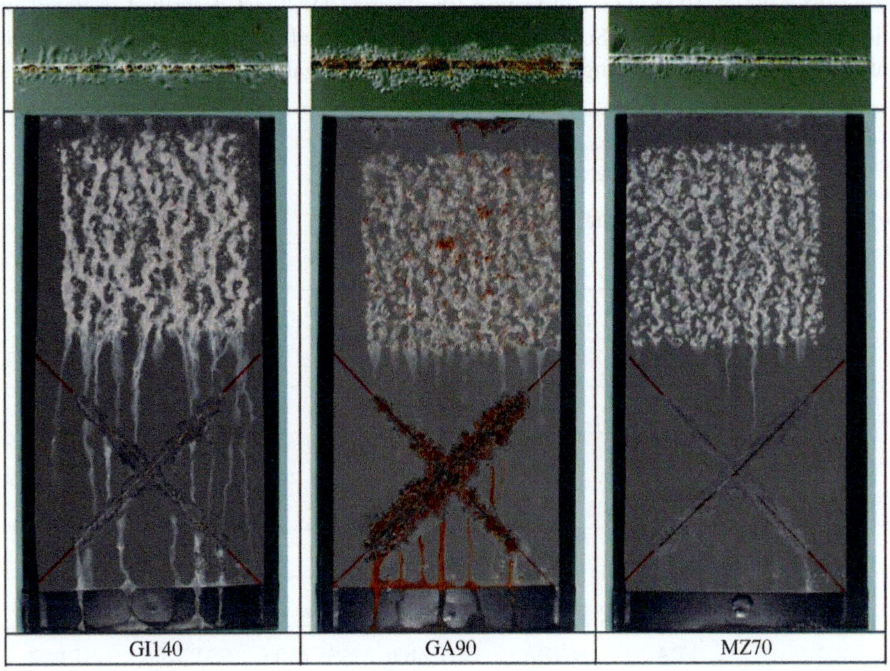

FIG. 12.30 (top) 2 cm long sections from scribed panels subjected to 42 days of Renault ECC1 cyclic corrosion testing (bottom) stone-chipped and scribed sections of panels subjected to 20 cycles of VDA 621-415 corrosion testing. The left top to bottom right scribe was aimed down to the zinc coating and the other scribe down to the steel substrate. The top of the panels was stone-chipped, and the top cut edge was unprotected [43].

FIG. 12.31 Average and maximum scribe delamination measurements in cyclic corrosion tests after 21 and 42 cycles in the Renault ECC1 test [43].

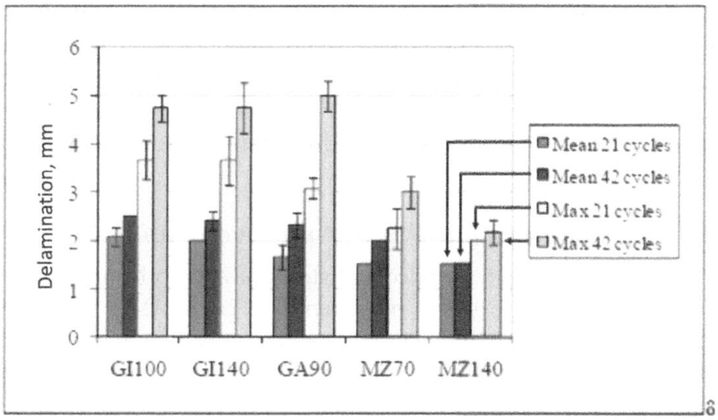

top of this figure shows about 2 cm length of the scribe from the first set of panels, subjected to the Renault ECC1 test for 42 days (6 weeks), the bottom part shows the stone-chipped and scribed panels after 20 cycles (20 weeks) of VDA 621–415 corrosion testing. As expected, the severity of edge creep is less after 6 weeks of ECC1 testing than 20 weeks of VDA 621-415 testing. Reflecting the higher Fe content of the GA coating, more red rust is observed in the stone-chipped and scribed areas of the painted GA panels. Less corrosion is seen on the ZnAlMg panels. Scribe creep measurements for all samples listed in Table 12.5 after ECC1 testing are summarized in Fig. 12.31. Both mean and maximum scribe creep values are shown; consistent with results described before, increased coating weight for each coating type is associated with better scribe creep performance. Also, the MZ70 performance is better than the GI140 performance. The GA90 performance is similar to that of the two GI coatings.

12.1.6.4 Corrosion performance of painted hot press-formed steels

The metallurgy of hot press-formed steels is described in Chapter 13; the corrosion performance of these steels in unpainted condition is documented in Chapter 11. To determine the corrosion performance of painted hot press-formed steels, samples were cleaned, activated, and zinc phosphated before electrophoretic coating with an automotive primer [45]. The initial 10 μm Zn-0.5%Al coating more than doubled in thickness during the hot press forming pretreatment during the heat treatment at 910°C for 6 min. Portions of the painted panels were then scribed down to the coating and subjected to a stone chipping treatment. For reference, GA samples with a coating weight of 50 g/m^2 per side and precursor samples for hot press forming with a 70 g/m^2 per side coating weight were also painted. Corrosion testing compared performance after 10 weeks of VDA 621-415 testing, 6 weeks of VDA 233-102 testing (the "new VDA" test described in detail in Chapter 11), and after various durations of salt spray testing. Scribe creep measurements after these tests are shown in Fig. 12.32. These results show that the much thicker coating of the press-hardened galvanized steel sheet shows

FIG. 12.32 Paint creepage from scribes of precursor GI coatings, hot press-formed coatings, and GA coatings after 10 weeks VDA 621-415; 6 weeks VDA 233-102 ("new VDA"); and 240, 480, 720, and 840 h of salt spray testing [45].

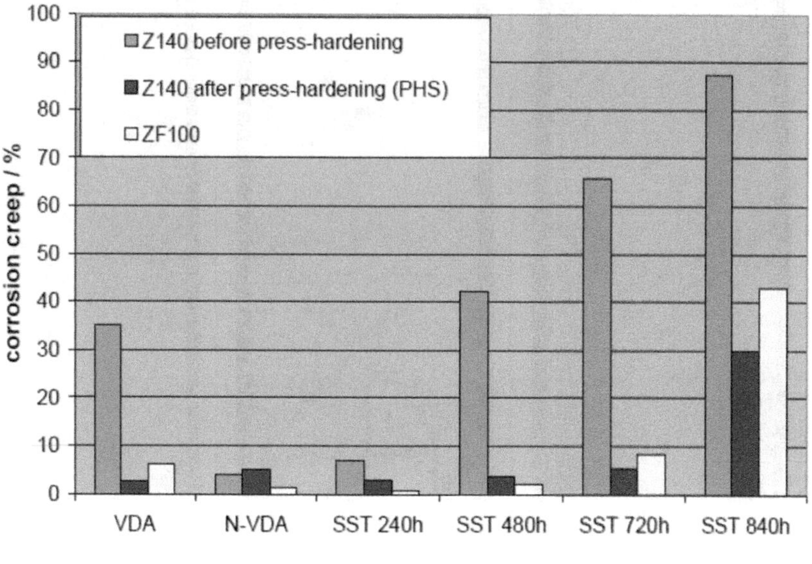

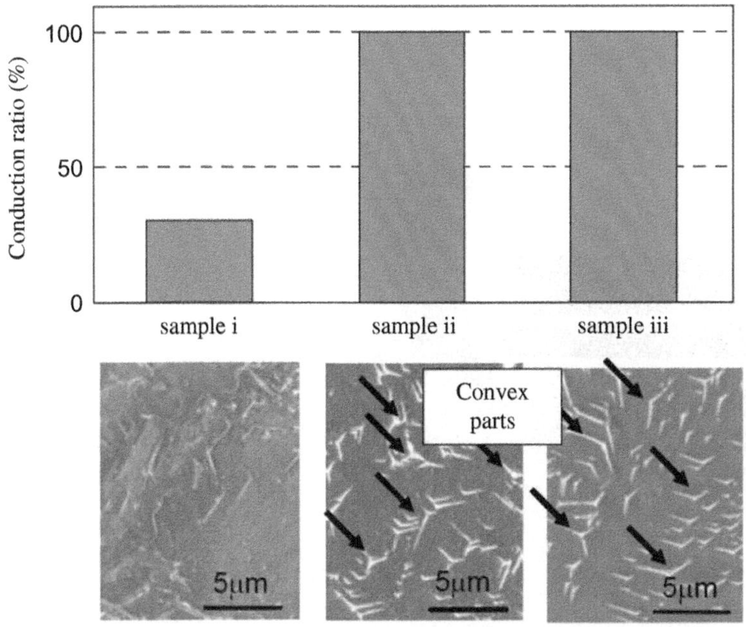

FIG. 12.33 Effect of surface morphology of a zinc electrogalvanized coating on relative electrical conductivity of samples treated with a resin containing phosphoric acid (left) and a resin containing a silicate binder and insoluble metal salts at a neutral pH (center and right) [46].

significantly more resistance to cosmetic corrosion than unhardened galvanized coatings in all but the "new VDA" test and are comparable to the galvannealed steel sheet samples in most of the tests. The similarity between the press-hardened and GA coatings is expected because of their similar electrochemical activities compared with the precursor GI coating.

12.1.7 Coil coating for appliance applications

Electrical appliance applications require high corrosion resistance from the coil-coated sheet steel used for wrappers, bases, and other applications. In many countries, appliances can be located in outdoor locations (on balconies, etc.) so that not only is resistance to detergents and bleaches required, but the coil-coated steel must also be suitable for outdoor exposure. Anti-fingerprinting resistance is also required. With the increased functionality of electronic devices, electrical conductivity of coil-coated galvanized steel is also becoming more important; of course, the entire unit must be grounded for safety reasons. Many appliance applications use zinc electrogalvanized sheet. As explained in Chapter 9, zinc electrogalvanized coatings normally exhibit a basal plane morphology. The electroplated zinc layer is not a perfect plane but is made of stacked layers, each of which has an edge, as shown in Chapter 9, Fig. 9.15B. When this surface is either chromate or phosphate pretreated for coil coating, the edges are attacked by the acidic pretreatment, resulting in a smoothing of the surface. As shown by Tsuchimoto [46], this smoothing effect reduces electrical conductivity by reducing conduction points on the coating surface. With traditional acid pretreatments, it has the opposite effect on corrosion resistance, improving service life. Fig. 12.33 shows how the smoothing of the ledges reduces electrical conductivity; if the convex ledges on the surface of the zinc electrogalvanized coating are not attached by a pretreatment with neutral pH, conductivity is higher. Optimization of the neutral permitted the goals of good corrosion resistance, anti-fingerprinting, sufficient lubricity for forming, and electrical conductivity to be met.

12.2 Formability of zinc-coated sheet steels

12.2.1 Overview of formability issues

Krupitzer [47] summarized the four principal forming issues involved in implementing advanced high-strength steels (AHSS) in automobile applications. These include:

Springback: This is the dimensional deviation of the shape of stamped part from the shape of the punches and dies and is caused by elastic recovery of the part. In general, springback is proportional to the flow stress achieved in a formed part before unloading (removing from the die), and so it is proportional to the flow stress of the part just before unloading the stamping die. AHSS, with their higher flow stresses, are associated with higher springback than lower-strength steels unless

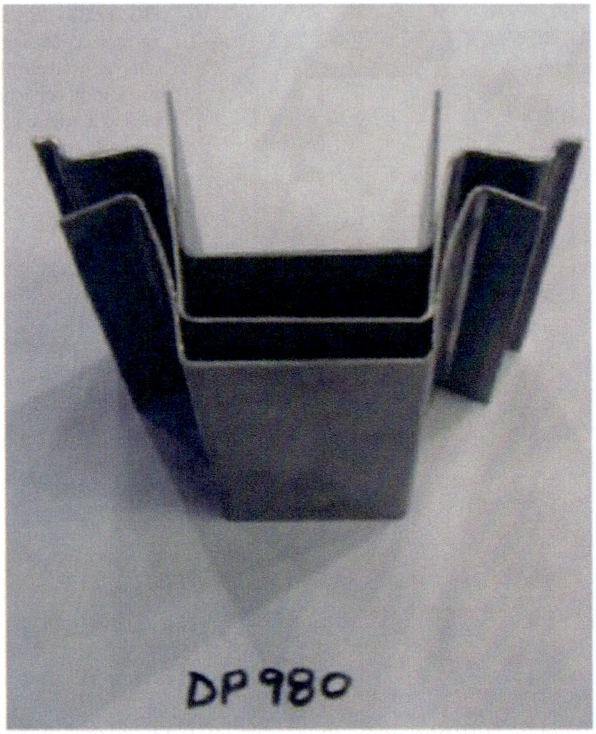

FIG. 12.34 Control of springback in DP980 using lock-flanging to control sidewall stretch [47].

FIG. 12.35 Control of springback by adding stiffening ribs on laser welded hat section. (A) 1.2 mm DP800/1.5 mm DP600; (B) 1.2 mm DP800/1.6 mm DQ; (C) 1.2 mm DP600/1.6 mm DQ [47].

compensating steps are taken. Such steps include overbending, by use of lock flanges, Fig. 12.34, or during the die design by adding ribs or other geometrical stiffeners, as shown in Fig. 12.35. The coating must be capable of being overbent and then partially unbent as the formed part recovers to its final shape after being stamped. The required lock flange forces and the behavior of zinc-coated steel during the deformation process depend in part on the coefficient of friction of the zinc or zinc alloy coating against the stamping die surface; friction properties will be addressed later in this section.

Press loading forces: Stamping press forces are influenced by the yield strength, work hardening (n-value), and ultimate tensile strength of the steel. The energy to deform a material corresponds to the area under the true stress-true strain curve. As shown in Fig. 12.36, the high work hardening of the DP600 leads to a higher tensile strength than for HSLA340, which causes a high press load. As the sheet is deformed, its thickness is reduced that reduces the required press load. Fig. 12.37

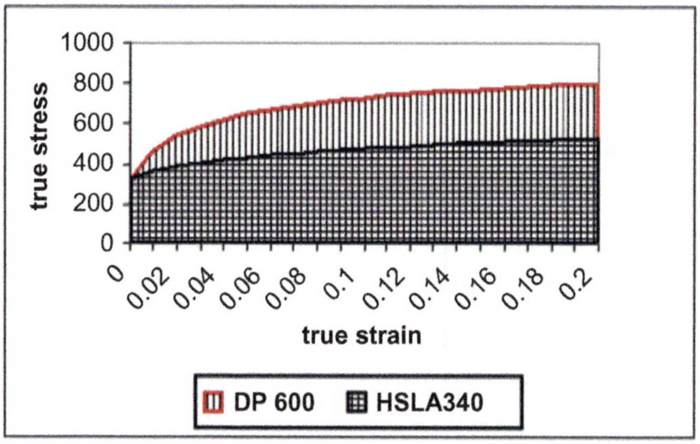

FIG. 12.36 True stress-true strain curves comparing the strain hardening behaviors of DP600 and HSLA340 [47].

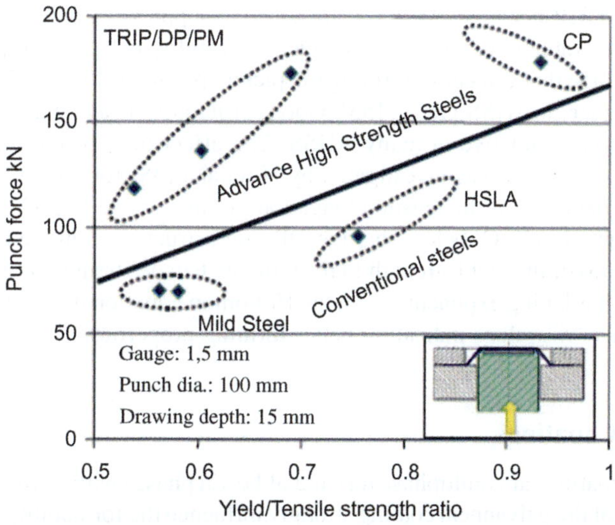

FIG. 12.37 Punch forces required in the Marciniak test to form various grades of AHSS and other steels, schematic of Marciniak test [47].

shows that the Marciniak "pie pan" test, used to determine stretch formability, requires higher punch forces for the same yield/tensile stress ratio, a schematic of this test is also shown in Fig. 12.38. AHSS usually require blankholder forces that are 1.5–3 times that required for conventional steels.

Tool wear: The higher forces required to hold the blank and to move the punch through the die with AHSS result in increased tool wear. Tool surfaces are therefore hardened by heat treatment or application of surface coatings such as chromium plating, laser-deposited alloy coatings, or nitriding. Each of these surfaces, when used to form zinc-coated steels, gives a characteristic coefficient of friction; frictional heating contributes to die wear and possible damage to the coating that can transfer to the tool surface.

Stretch flanging: For some structural and suspension parts, AHSS must be capable of stretch flanging and hole expansions of 70% or greater. Stretch-flangeable AHSS are characterized by high yield strength to tensile strength ratios. The most common approaches are to create a dual phase structure with bainite as the primary secondary phase or to strengthen the ferrite matrix in a ferrite/martensite dual phase structure.

The ability of the metallurgist to manipulate steel microstructures and mechanical properties to meet formability goals is more limited on a continuous galvanizing line, especially when a galvannealing heat treatment cycle is included, compared with continuous annealing line processing, because of line speed limitations, the holding time required for hot-dip coating and, when galvannealing, adding the additional heating and holding times. To increase hardenability, for example, permitting bainite or martensite to be formed at lower cooling rates, higher amounts of alloying elements such as Mn, B, Mo, and Cr can be added that suppress ferrite precipitation in the soaking and cooling sections of the annealing furnace, maintaining a higher austenite fraction. It is important to balance the levels of these elements with C to obtain the desired

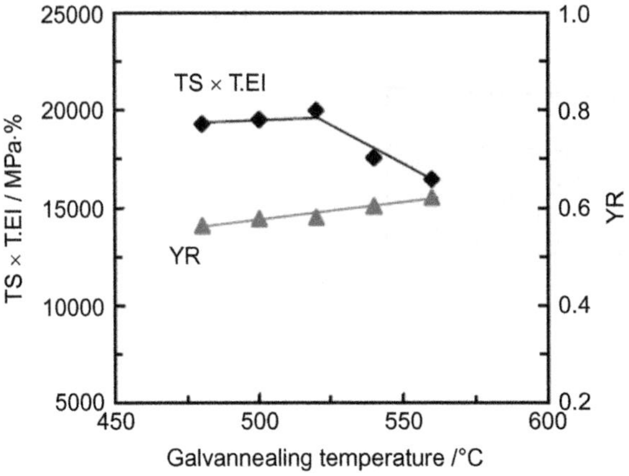

FIG. 12.38 Effect of galvannealing temperature on mechanical properties of 0.15%C-1.2%Si-2.2%Mn steel [48].

formability properties [49]. For TRIP steels, Si, Al, and P are added to suppress carbide formation [50]. Of these elements, only Cr and Mo additions are favorable to maintaining good galvanizability behavior [51]. High Si levels in IF steels can compromise zinc coating adherence that can cause forming die pickup problems [52], with the same surface-selective oxidation issues being found in higher-Si AHSS grades [53]. Good formability is associated with high elongations and low yield ratio, properties that are very attractive in many AHSS. The effect of galvannealing soaking temperature on the tensile-elongation product and yield ratio was investigated by Hironaka [48] for a 0.15%C-1.2%Si-2.2%Mn dual-phase steel. Fig. 12.38 shows a rapid decrease in the tensile-elongation product above 525°C; this was related to the presence of pearlite as a second phase above 525°C, below this the DP microstructure consisted only of ferrite and martensite. The yield ratio, defined as the ratio of the yield strength to the ultimate tensile strength, is a measure of strain hardenability (directly indicated by the strain hardening exponent "n" in the Hollomon equation $\sigma = k\varepsilon^n$) and the formability of the steel, with lower yield ratio and higher "*n*" values indicating better forming properties.

12.2.2 Formability of GA coatings

As described in Chapter 8, GA coatings are multiphase mixture of Fe-Zn phases with an overall Fe content of 9%–12%. The phase makeup and morphology of the galvanneal coating strongly influence the formability of galvannealed steel sheet. The five types of phases observed in GA have very different hardnesses, Table 12.6. These are much harder than pure metallic η zinc that typically has a Vickers hardness of 55. Hardness influences the coefficient of friction that the GA surface will exhibit when formed and depends upon the predominant Fe-Zn phase that is exposed to the surface during forming. The relationship between coating Fe content and coefficient of friction was determined by Garza [55] for a range of GA coatings made up of mainly delta and zeta phase, Fig. 12.39. The dependence of the coefficient of friction on the fraction of the surface occupied by zeta phase, rather than delta, is shown in Fig. 12.40. The hardness and related coefficient of friction influence two modes of failure in GA coatings, described further below in this section. Flaking occurrences are

TABLE 12.6 Vickers hardnesses and Zn compositions (at.%) of Fe-Zn phases prepared by equilibrating at 450°C [54].

Phase	Equilibrium composition (at.%)	Hardness (N/mm^2)
Γ	73.2	425
Γ_1	77.5	586
δ_{1k}	85.1	516
δ_{1p}	90.7	394
ζ	92	242

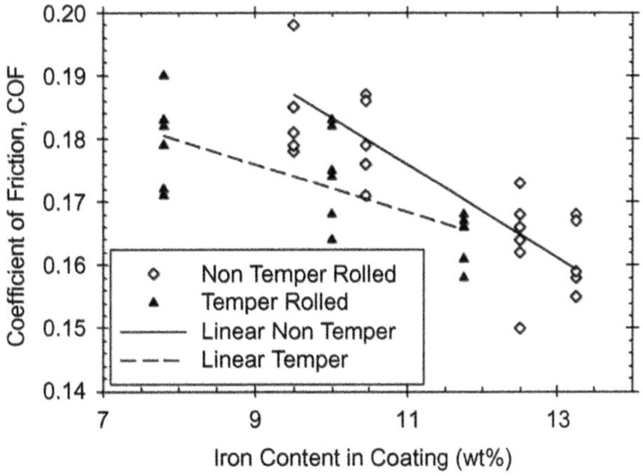

FIG. 12.39 Coefficient of friction as a function of GA coating Fe content [55].

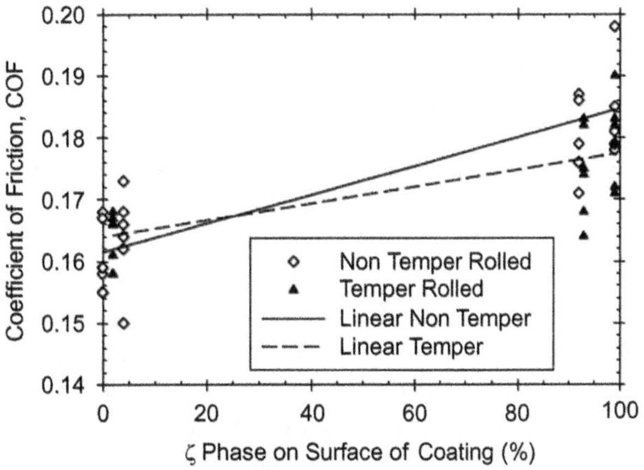

FIG. 12.40 Coefficient of friction as a function of ζ-phase on the surface of the coating [55].

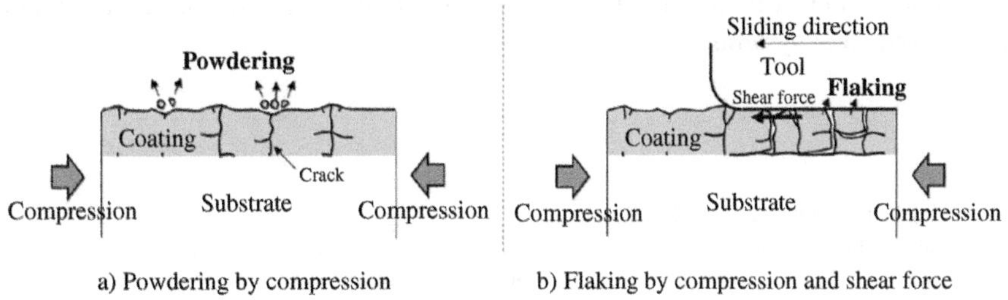

FIG. 12.41 Schematic figure of typical failure behaviors on GA [56].

related to a high friction coefficient between GA coating surface and the stamping die surface and increase as the difference of hardness values between coating and die increases. Powdering is associated with excessively galvannealed coatings and the presence of Γ and Γ_1 phases [54].

The principal modes of failure of GA coatings, Fig. 12.41, and EG and GI coatings, Fig. 12.42, were described by Yamasaki [56]. He categorized coating failures into three types: (1) destruction of the coating itself, (2) destruction of the interface between the coating and substrate, and (3) adhesion abrasion. These types of failure can occur simultaneously. Coating failures are affected by the deformation mode of the coatings and the nature of sliding friction between the zinc coating and the die. The deformation mode and the sliding conditions depend on the shape of the formed part, the forming method, and the forming conditions.

Many investigations of coating failures have focused on steel mechanical properties and stamping die-coating interactions. In GA coatings, formability is also strongly related to the interaction between the GA coating and the steel substrate

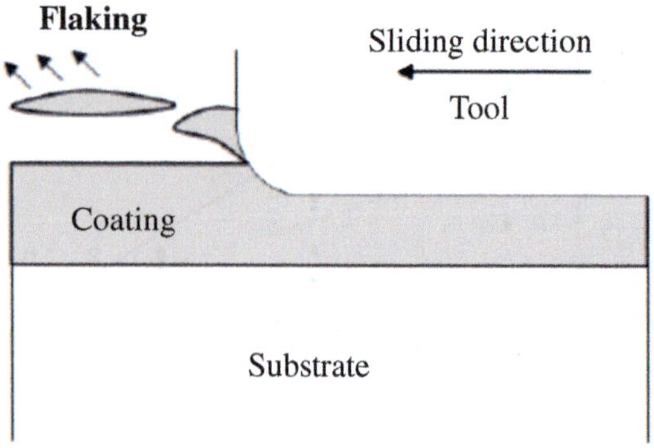

FIG. 12.42 Schematic figure of typical surface flaking behavior on GA, EG, and GI [56].

together with the microstructural details of the coating. The fracture toughnesses of the Γ and δ phases, which usually comprise much of the GA coating thickness, are in the same range as silica glass [57]; therefore, one would not expect GA-coated steel to be formable. However, under compression, several of these Fe-Zn phases, especially the gamma (Γ-Fe_3Zn_{10}), delta palisade (δ_{1p}-$Fe_{13}Zn_{126}$), and zeta (ζ-$FeZn_{13}$) phases, have sufficient ductility to allow forming of GA-coated sheet, notably automobile body panels; this was demonstrated by micropillar compression testing of these phases [57]. However, the Γ phase fractured in a brittle manner during microbeam bending [58]. The K_{1c} fracture toughness of δ_{1p} and Γ is in the range of 0.6–0.7 MPa/$\sqrt{m}$, whereas for δ_{1k}-$FeZn_7$ it is in the range 0.2–0.4 MPa/$\sqrt{m}$. All of these phases exhibit Hall-Petch behavior, i.e., increasing fracture toughness with decreasing grain size [59].

12.2.2.1 Flaking

Cheng [60] related the dependence of GA flaking behavior (separation at the GA coating/steel substrate interface illustrated in Fig. 12.41B) to galvanizing bath soluble Al level and Γ-Fe_3Zn_{10} layer thickness for IF steels, Fig. 12.43. The thickness of Γ-Fe_3Zn_{10} layer at the interface generally is regarded as an indicator of the degree of galvannealing. No ζ-$FeZn_{13}$ was present on the surface of these samples. Liquid Zn penetration into the IF steel grain boundary and further embrittlement were concluded to be the dominant factor in the flaking performance of GA IF steels. Incomplete coverage of the Al rich inhibition layer because of insufficient Al in the medium-Al and low-Al bath is the reason for liquid Zn penetration. The result increases the growth of the Γ-Fe_3Zn_{10} layer that is associated with poor flaking resistance. The flaking failure mode was observed to occur primarily in GA IF system, particularly in overgalvannealed conditions, and rarely occurred in GA AHSS or other highly alloyed steel systems.

FIG. 12.43 Effect of galvanizing bath-soluble Al content and degree of galvannealing (thickness of gamma layer) on GA flaking severity [60].

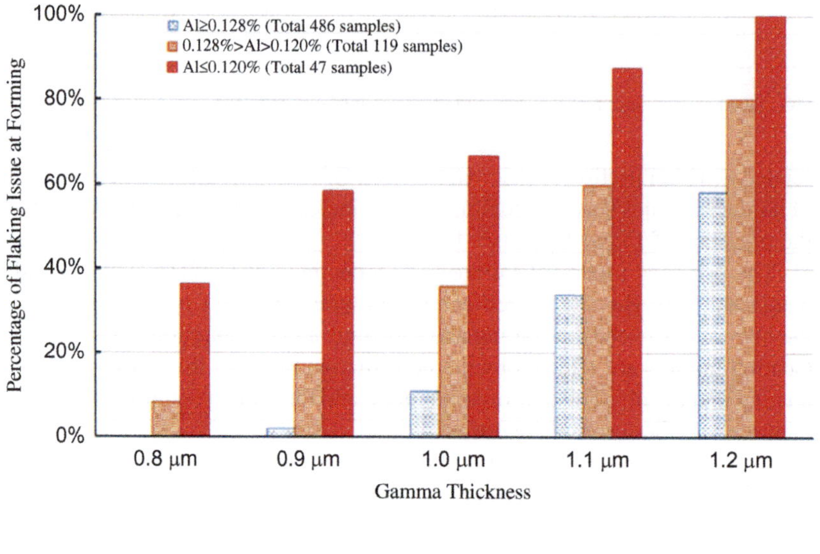

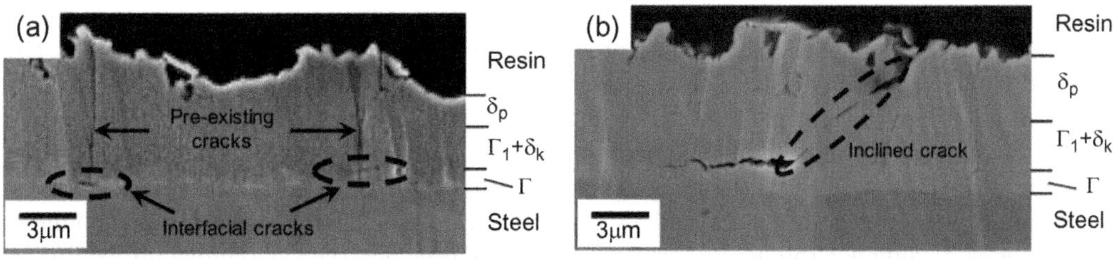

FIG. 12.44 Crack propagation path in the coating layer of the commercial-grade sample with poor powdering performance under compressive stress. a) Crack initiated from preexisting crack along either Fe/Γ or Γ/$Γ_1$ interface after 5% strain, and b) inclined and delamination crack propagating through $δ_{1k}$ phase found after 10% strain [59].

12.2.2.2 Powdering of GA

Powdering (cracking of portions of the GA coating with the release of fractured portions as shown in Fig. 12.41A.) is related to the presence of Γ-Fe_3Zn_{10} and $Γ_1$-$Fe_{11}Zn_{40}$ phases; the δ-$FeZn_7$ layer within the GA coating is often found at the cracked interface with the other phases but is often the powder that is spalled from the GA coating. He [61] determined that the gamma phase has the highest elastic modulus and also the highest yield stress of any of the Fe-Zn phases and the tested DP980 substrate. Consistent with fracture toughness results described before, the Γ-Fe_3Zn_{10} phase is observed to crack first when GA-coated steels are observed in an in-situ three-point bending test with the bending direction perpendicular to the sheet normal direction. Inoue's observation [59] that Γ-Fe_3Zn_{10} has the weakest grain boundaries explains powdering behavior observed in industrial samples and depicted in Fig. 12.44. Cracks first nucleate from preexisting vertical cracks in the GA coating that are created from thermal stresses in the manufacturing process. These cracks propagate along the $Γ_1$/Γ and Γ/steel substrate phase boundaries. The cracks along the $Γ_1$/Γ boundaries, under sufficient stress, deflect into the low toughness $δ_{1k}$-$FeZn_7$ phase, ultimately cracking pieces of this phase to the point of separation, which is manifest as powdering. In samples with better powdering performance, the weak Γ grain boundary was observed to deflect cracks into the Γ/steel boundary and thus to limit powdering.

The critical resolved shear stress (CRSS) for slip in these FeZn phases was determined for each slip system. The $δ_{1p}$-$Fe_{13}Zn_{126}$ phase was determined to have the highest CRSS. Synchrotron observations indicated that all phases were based on an icosahedron crystal structure, each with a body-centered Fe atom. As Fe content increased, the connection geometry of the Fe-centered icosahedra changed from exclusively vertex-sharing in ζ-$FeZn_{13}$ to a mix of vertex and face-sharing in $δ_{1p}$-$Fe_{13}Zn_{126}$ to exclusively face-sharing in Γ-Fe_3Zn_{10}. The occupancy mode at the face-sharing sites changes from exclusive occupancy by Zn atoms to mixed occupancy by Zn and Fe atoms as Fe content increased, and the fraction of dangling Zn atoms decreased with increasing Fe content [57]. In practical terms, this means that the good formability of GA steels is achieved when the coating layer consists mostly of the $δ_1$ ($δ_{1k}$/$δ_{1p}$) phase [57] with optimum formability seen when there is a thin layer of Γ1 phase at the δ1-substrate interface, meaning that the overall coating Fe content should be about 10% [62]. This optimum interfacial structure was observed with TRIP steel compositions of Fe-2% Mn-(1–2)%Si when complete inhibition later coverage was achieved before galvannealing. Then, the first phase formed at the Fe-Zn interface was δ with a concurrent formation of $Γ_1$ at the grain boundaries of the α-grains. The $Γ_1$ grain exhibits a cube-on-cube orientation relationship with the α-grain [63]. Use of an internal oxidation heat treatment, described in Chapter 2, helps provide a coatable steel surface; however, too clean of a surface can result in high reactivity during formation of the GA coating, and therefore formation of Γ phase at the steel-coating interface [64].

The observation of ductile compression behavior of several of the Fe-Zn phases is stated as enabling acceptable stamping behavior of GA-coated steel, even though in bending or tension these phases can fail in brittle fashion [65]. (why this is so is a gap in our knowledge).

12.2.2.3 Surface flaking and galling

The third formability-related failure mode, shown in Fig. 12.42, is flaking of the outer surface of the coating, which is related to friction properties of the forming die and the zinc coating. It is distinct from the "flaking" of GA described in Figs. 12.41 and 12.43, which is failure of the coating near the steel-coating interface. Friction properties are influenced by hardness and roughness of the die and coating together with details of lubrication. Galling, the buildup of the zinc coating on the die surface is related to flaking, but is usually only noticed after at least 20 parts have been formed in the same die, as die surface temperature increases with frictional heating. Galling incidences have become more frequent with the increasing use of zinc-coated high-strength steels in automotive applications. Flaking is more prevalent in GI and ZnAlMg coatings; galling has been observed in all zinc coatings. Galling is related to the surface temperature of the die, which can heat to typically 70–80°C during rapid stroke forming operations, even with good lubricant practice [66]. Higher temperatures are observed with higher strength steels that require increase forming loads. As the die temperature increases with successive strokes, it raises the temperature of the coated steel workpiece, which is also undergoing adiabatic heating during forming. Forming operations typically involve compression of the coating surface, which Kim [67] showed similarly flattens the coating surface of both GI and GA coatings, even though the typical hardnesses of a mixed phase ζ-δ GA surface and the η-Zn GI surface are typically HV = 280 and 55, respectively. Less lubricant can be retained on the smoother surfaces as forming progresses, although roll-applied dry lubricants have been observed to improve formability [68]. The conditions reached with the smoother, hotter, less-lubricated workpiece at the end of the forming stroke are most favorable for the incidence of galling. The harder GA coating was less susceptible to galling than the softer GI. As indicated by Marder [69], GA coatings with a surface layer of the softer zeta phase are occasionally found; the zeta-rich surface is known to increase surface friction while reducing powdering [70] and would therefore be expected to be more susceptible to galling than a GA coating with a delta-rich surface. A harder die surface was also observed to decrease the incidence of galling during both drawing and the more severe ironing operation where typical maximum stresses are 110 and 410 MPa, respectively. The hardest die coating, TiCN (HV = 2500) was observed to give better performance with GI and GA, than the softer CrN and XNP (HV = 1300–1330) coatings [71]. The influence of GA surface roughness and benefit of a preapplied dry lubricant is shown in Fig. 12.45 and Tables 12.7 and 12.8. Using a hardened (HV = 728) Cr-Mo-V steel flat plate tool lubricated only with anti-rust oil, a GA coating having a delta-rich surface gave coefficients of friction in the range 1.6–2.0 while the prelubed GA gave friction values of 0.15–0.16; they both increased with increasing roughness. Regardless of initial roughness, the normal GA has a lower fraction of plateau area (area flattened by the flat plate test) at a relatively low pressure of 7.8 MPa and a low sliding speed of 0.2 m/min, Table 12.7. The reduction of roughness was much less for the prelubricated GA, Table 12.8, and for both types of samples, this reduction of roughness was greater for the samples with higher starting roughness. This is consistent with the higher friction coefficients measured for the rougher samples in

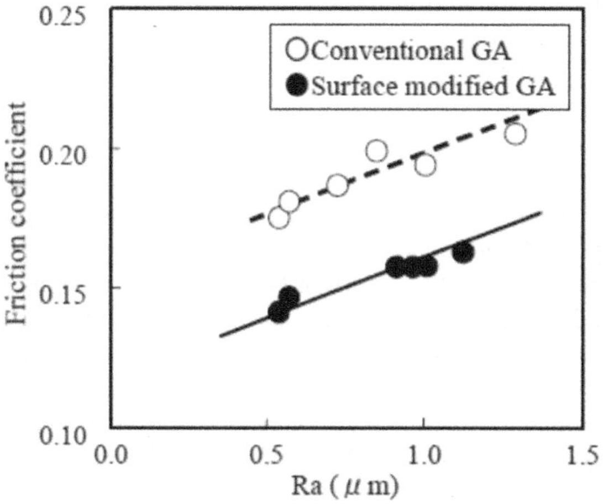

FIG. 12.45 Effect of dry lubrication film (surface modified GA) and GA coating surface roughness of coefficient of friction measured in flat die test (normal stress 7.8 MPa, sliding speed 0.2 m/min, anti-rust oil lubricant [68].

TABLE 12.7 (left) Plateau area ratio (portion of GA coating surface that is flattened) after flat die test of Fig. 12.45: GA is the untreated GA coating; SG is pretreated with a dry lubricant [68].

Test specimens	Plateau area ratio (%)
GA2	41.6
SG2	22.2
GA5	43.9
SG5	21.3

TABLE 12.8 (right) Effect of dry lubricant (SG samples) and initial roughness on final GA coating roughness after flat die test of Fig. 12.45 [68].

Test specimens	Ra before being slid (μm)	Ra after being slid (μm)	Difference between before/after being slid
GA2	0.71	0.52	0.19
SG2	0.69	0.56	0.13
GA5	1.39	0.92	0.47
SG5	1.50	1.07	0.33

Fig. 12.45, because more deformation of coating asperities is required. Surface friction can also be reduced by a light application of a phosphate conversion coating (100–400 mg/m^2), increasing the delta phase fraction in GA increases phosphate pickup and therefore lowering the coefficient of friction [72]. When subjected to cup deformation testing, improved formability to substrate failure was seen with prephosphate panels, although powdering increased [73]. The amount of powdering is therefore related to the total (i.e., Von Mises) state of stress; increasing compressive stress can compensate for the reduction of frictional (and therefore shear) stress and resulting strain states.

A working group of European steelmakers evaluated ZnAlMg coatings over the range for 1–3.7% Al and 1–3%Mg. Because of their proximity to each other on the ZnAlMg phase diagram, the nature of solidification for all these coatings was similar and produced similar microstructures [74]. Coating hardness increases with alloy content as shown in Fig. 12.46 [75]. The higher hardness of the Zn-1.7Al-1.7Mg coating than the GI coating contributes to a more stable coefficient of friction successive passes in the strip drawing test results shown in Fig. 12.47; its lower roughness contributes to its lower coefficient of friction than the GA coating [43,76]. Similar work shows that a zinc electrogalvanized (EG) coating exhibited

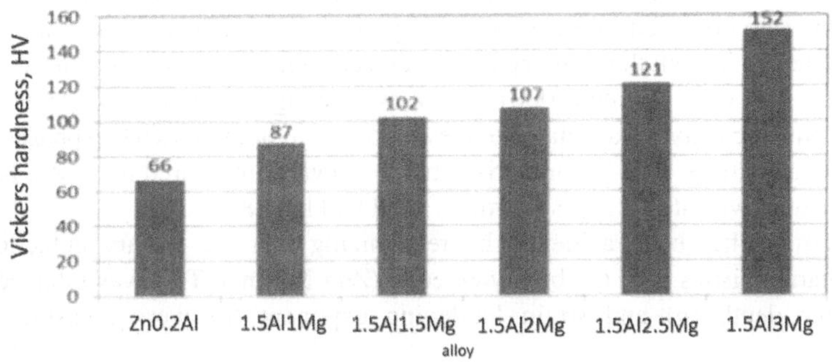

FIG. 12.46 Vickers hardnesses of several Zn-Al-Mg hot dip coatings [75]. Compare with values in Table 12.6.

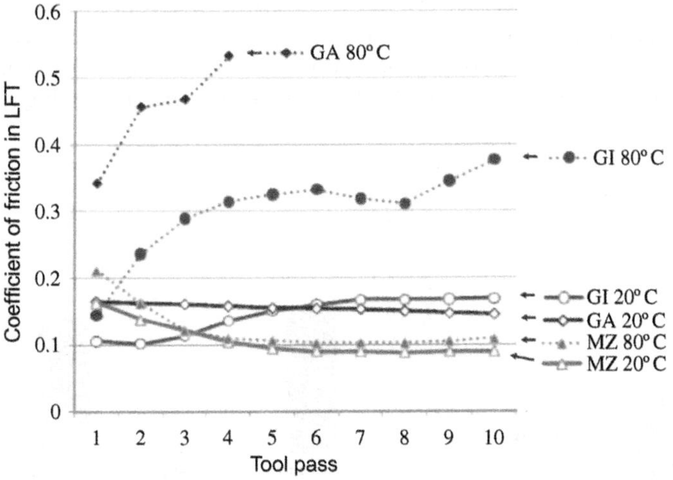

FIG. 12.47 Friction coefficients measured in multiple strip drawing test using a formable steel for GI, GA, and Zn-1.5%Al-1.5%Mg (MZ) coatings [43].

about half the galling severity of a GI coating, with less coating release and fewer, smaller flakes. The coating debris produced by the ZM in this work was one-third of GI, in the form of mainly fine powder. Regarding the drawing tool surfaces, no coating transfer to the tool was observed for the Zn-Al-Mg coating; this was observable for the GI coating and to a lesser extent for EG. Furthermore, no abrasive tool wear could be detected for the ZnAlMg coating [77].

As noted above, the higher-strength steels used in automobiles result in higher stamping forces and therefore higher die surface temperatures. These higher temperatures reduce lubricant effectiveness and soften the zinc coatings, increasing friction coefficients that can lead to forming splits, die pickoff, and galling. An Auto/Steel Partnership study [78] showed that the die and strip temperatures were higher with GA than GI, with more coating adhesive pickup being found with the softer GI while more abrasion was seen with the harder GA, consistent with Fig. 12.47. Overall, the base steel type and strength, the die radius, and sheet thickness were more significant factors in determining local die temperatures than the coating [78]. Higher die temperatures were observed to decrease GA powdering. Powdering with a 100°C die temperature was found to be a quarter of the amount seen with room temperature forming and became almost negligible at 200°C. This is because the GA coating layer was found to soften as temperature increased [79].

12.2.3 Coating cracking and formability

ZnAlMg coatings have been observed to crack more easily during deformation, this has been related to their higher hardness and multiphase structure compared with GI [80]. A Zn-6Al-3Mg coating was observed to be absent of thermal stress cracks associated with cooling of the coated steel during and after solidification, with cracks only appearing in subsequent bending tests. In contrast, a Zn-2.5Al-3Mg coating developed cracks in the as-coated state, which continued growing during bending test. There was little difference in thermal expansion properties of these two coatings in comparison with the steel substrate, therefore the cracking resistance of the higher alloyed coating, in which Al-rich primary phase was formed rather than the Zn-rich primary phase of the lower-alloyed coating, was concluded to be higher. The *c*-axis of the primary zinc crystals of the lower alloyed coating was determined to be the minimum Young's modulus direction, therefore deformation in the coating was concentrated in grains whose c-axis was parallel to the rolling direction. Within the eutectic of both coatings, an orientation relationship between the Zn and $MgZn_2$ eutectic phases was found, with a common rotation about the $\langle 10\bar{1}0\rangle$ axis. No cracks were observed in the $MgZn_2$ component of the eutectic phase; most cracks were in the Zn component. The cracking frequency in both the primary Zn and eutectic Zn phase was concentrated along the basal plane of these crystals when these crystals were oriented $\langle 0001\rangle \perp$ ND, even though the major texture was (0001)||ND. Anisotropy of Young's modulus in the eutectic phases was related to the *c*/*a* axis ratio of these two hcp phases; the $MgZn_2$ phase *c*/*a* of 1.64 is much nearer to the ideal value of 1.63 in comparison to the *c*/*a* value measured for η-Zn. In contrast, the Al fcc primary phase is relatively isotropic with its 12 slip systems of {111} {110} [81].

Improvement to the ductility of ZnAlMg coatings has been achieved by recognizing that the ternary eutectic ($Zn+MgZn_2+Al$) has much better deformation characteristics than the binary eutectic ($Zn+MgZn_2$). This was inferred from nanoindentation results, where the indenter depth (*p*) and strain hardening exponent (*n*) were measured, Table 12.9 [82].

The lower *p* and *n* values measured for the binary eutectic, together with its higher hardness, indicate that this phase has deformation capacity than the primary zinc and ternary eutectic phases. This was confirmed by in-situ uniaxial tensile testing that showed that the first cracks to form were in the binary eutectic. Increasing the Al content while lowering the Mg content was found to decrease the portion of binary eutectic; Fig. 12.48 shows the portions of phases present for a Zn-2.9%Al-1.9%Mg coating compared with a Zn-1.72%Al-1.72%Mg coating that was very similar to the composition characterized in Table 12.9. Uniaxial tensile testing of coated steel samples showed the binary eutectic to begin cracking at

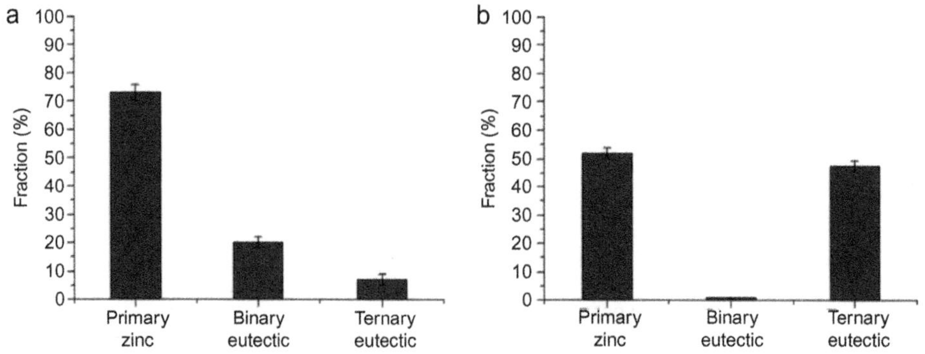

FIG. 12.48 Portions of phases present in a Zn-1.72%Al-1.72%Mg coating (left) and a Zn-2.9%Al-1.9% Mg coating right). Coating thickness range 13–15 μm [83].

TABLE 12.9 Mechanical properties of phases in Zn-1.8%Al-1.8%Mg coating determined by nanoindentation [82].

Phase	Young's modulus (GPa)	Hardness (GPa)	Max. penetration depth, *p* (nm)	Strain hardening exponent, *n*
Primary zinc	108	1.5	328	0.34
Binary eutectic	92	2.7	242	0.08
Ternary eutectic	98	2.4	304	0.24

a true strain of 0.05; the sample containing a minimal amount of binary eutectic exhibited insignificant cracking at true global strains as high as 0.45, and most of the cracks seen were associated with the small portion of binary eutectic. The much finer ternary eutectic was found to better accommodate strain and moreover contributed to strengthening because of dislocation buildup as straining proceeded [83].

Because the Zn-55%Al-1.5%Si coated steel is mainly used in architectural applications, most formability testing has focused on tension bend performance. This alloy coating is susceptible to age hardening at ambient temperatures that increases hardness and lowers ductility [84]. One route to improving the tension bend performance is to heat treat the coating; this reduced its hardness to near the those of GI and Zn-5%Al coatings, as shown in Fig. 12.49. A heat treatment capable of providing an improvement in ductility is to hold the coating at 205°C for 3h followed by cooling in air [86]. When combined with a higher-flexibility primer than conventionally used, noticeably fewer cracks were observed in 3T bend testing, Fig. 12.50. Another way to improve the bending performance of the Zn-55%Al-1.5%Al coating is by grain refinement [87]. An addition of 0.5% TiB_2 grain refiner to a Zn-55%Al-1.5%Si bath reduced the spangle size (mean intercept diameter) from 1400 to 200 μm. This changed the deformation mechanism of the coating from the conventional cracking within the spangle facets to interdendritic cracking. At low strains in the grain-refined coating, the interdendritic

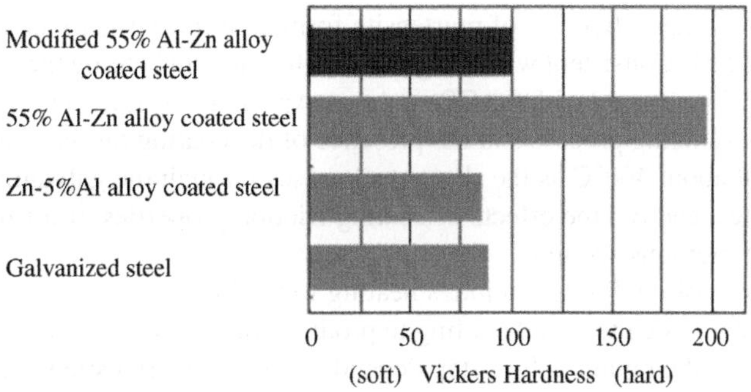

FIG. 12.49 Effect of postcoating heat treatment on hardness of Zn-55%Al-1.5%Si, compared with as-produced Zn-5%Al and GI-coated steels [85].

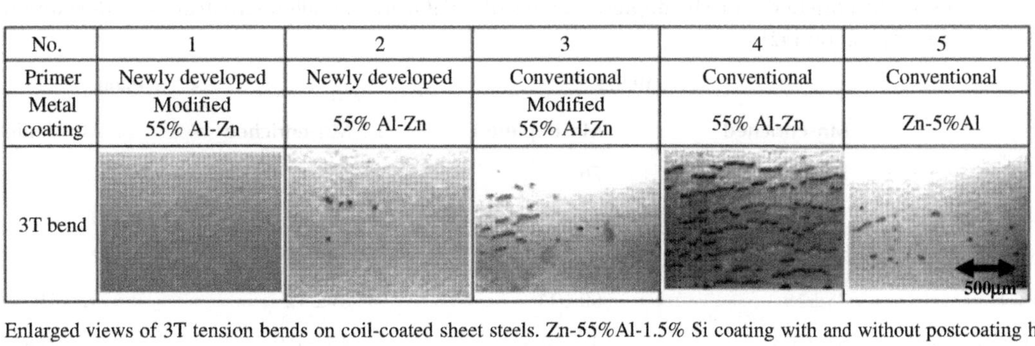

FIG. 12.50 Enlarged views of 3T tension bends on coil-coated sheet steels. Zn-55%Al-1.5% Si coating with and without postcoating heat treatment shown, in comparison with Zn-5%Al coated steel, all 0.35 mm steel thickness. Traditional chromate pretreatment applied, followed by 4 μm prereacted epoxy/polyester primer and 13 μm polyester topcoat [85].

cracks appear at spangle facets, but as strain increases, these cracks also appear within the spangle facets. Therefore, at larger strains, the grain-refined coating exhibits a higher number of smaller cracks than the conventional coating with its larger spangle size. The smaller cracks are more likely to be successfully bridged by a paint system, so that they are not visible and barrier corrosion protection is maintained.

A tension test was developed that permitted in-situ observation of crack initiation and growth in this product in interdendritic Si particles and Zn-rich regions, in areas of porosity and solidification shrinkage voids. These cracks could progress to the surface with sufficient bending severity. Cracks also initiated in the intermetallic interface layer at low bending strains but usually did not progress through the thickness of the coating. Consistent with the ZnAlMg observations above, the Al-rich primary phase deformed rather than exhibiting cracks. A lower Si coating composition (1.3% vs 1.7%) was found to exhibit less cracking [88].

12.2.4 CGL processing effects on formability

Advances in AHSS processing permit a wide range of AHSS microstructures to be produced that influence formability behavior. Although the range of heat treatments possible on the hot-dip galvanizing line cannot match those possible on continuous annealing lines (CAL), galvanized dual-phase steels and multiphase DP-type steels with strengths between 500 and 1000 MPa are produced with varieties of deformation behaviors [89]. An example is a family of types of 780 or 980DP with either high global formabilities (high true uniform strain values and high terminal strain hardening ("n") values) that exceed in stretch formability, or grades with microstructures that provide high true fracture strain (local formability) that provides high flanging, edge stretching, and forgiveness to blanked edge imperfections and high hole expansion ratio ("λ") values [90,91]. The differences between microstructures that give good local versus global formability are thought to be related to the hardness differences between the constituent phases, with higher hardness differences associated with inferior local formability. But rather than hardness differences, it appears that compositional heterogeneity, especially the Mn distribution, is a stronger influence on local formability than a correlation of the hardness differences of microstructural constituents [92]. Table 12.10 shows the local hardness of Mn-enriched bands vs. Mn-depleted bands in DP800 and CP800 microstructures developed from a single 0.12%C-2.00%Mn-0.25%Si-0.55%Cr-0.02%Ti composition using process routes that each included a hot-dip galvanizing step. The hardness of the Mn-enriched bands is clearly higher than that of the Mn-depleted bands. Furthermore, hardness differences between Mn bands are more pronounced in DP800, which can be associated with the higher fraction of martensite formed in the Mn-enriched zones. Stress-strain curves for these grades are shown in Fig. 12.51. Consistent with the greater differences in heterogeneity, the DP grade exhibits a lower hole expansion ratio, 22% vs 42% compared with the CP grade, as expected, this is associated with a higher total elongation.

Influences of the hot-dip galvanizing process and the presence of the coating on sheet steel formability include (A) the effects of the isothermal hold at about 460°C as the sheet steel passes through the galvanizing bath, (B) the constraints of CGL line speeds on cooling rates, and (C) the effects of coating friction properties of formability. Representative coating friction values are given in the previous sections.

Conventional CGL processing of a DP steel includes heating to the intercritical annealing temperature to produce the desired balance of ferrite and new austenite, rapid cooling to produce martensite, and isothermal overaging (equalization). The ideal terminal temperature of the rapid cooling step depends upon the martensite start and finish temperatures of the

TABLE 12.10 Phase fraction, semiquantitative carbon intensity from compositional image maps and hardness of constituent phases in Mn-enriched and Mn-depleted zones of DP800 and CP800 microstructures developed from the same C-Mn-Si-Cr composition [92].

Material	DP800		CP800	
Area	Mn-enriched	Mn-depleted	Mn-enriched	Mn-depleted
Ferrite (vol%)	52	76	37	62
Bainite (vol%)	16	22	48	30
Martensite (vol%)	30	2	15	7
C intensity (–)	75±15	60±11	61±12	57±12
Hardness (GPa)	5.0±1.3	3.7±0.6	4.1±0.9	3.7±0.6

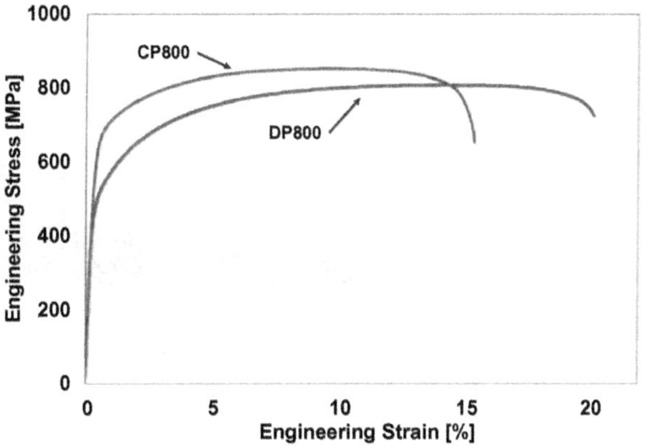

FIG. 12.51 Engineering stress-strain curves for DP800 and CP800 grades made from the same C-Mn-Si-Cr composition [92].

specific grade. If these values are below the zinc pot temperature of 460°C, the microstructure and consequent mechanical properties will be affected. For example, during CGL processing of a 0.15C-1.9Mn-0.41Cr-0.26Si-0.1Mo-0.05Nb grade, which had a martensite start temperature of 365°C, it was found that interruption of cooling by holding at 460°C transformed the small amount of available metastable residual austenite to bainite. This transformation had a significant impact on the mechanical properties of the galvanized steel, Fig. 12.52 In this figure, CR_1 is the cooling rate from the intercritical annealing temperature to the galvanizing temperature, and CR_2 is the cooling rate from the galvanizing temperature to room temperature. Increasing the isothermal hold time associated with hot-dip galvanizing decreased both ultimate tensile and yield strengths, while increasing elongation. In this case, to ensure the desired mechanical properties (UTS > 1100 MPa, YS between 550 and 750 MPa and elongation to fracture greater than 10%), the time of interruption of cooling for hot-dip galvanizing must be between 13 and 17 s; other relevant variables are a heating rate to the annealing temperature of 35°C/s, intercritical annealing at 800°C for 60 s to give a 12% ferrite-88% austenite microstructure; CR_1 = 13°C/s, hot-dip galvanizing at 460°C, and CR_2 = 25°C/s [93].

Thicker GA coatings (50 g/m^2 nominal coating weight per side) require hotter peak metal temperatures (PMT) to fully galvanneal than thinner GA coating types (e.g., 38 g/m^2 nominal). Hotter PMT typically degrades retained austenite (RA) in third-generation steels, with the loss of carbon content in the retained austenite to carbides. Fig. 12.53 shows the effect of GA coating weight and therefore galvanneal processing PMT on tensile and hole expansion properties for an industrial third-generation steel [94]. These hot-dip galvanized steel grades compete with similar 3G grades processed on continuous annealing and electrogalvanizing lines, which can optimize the microstructure to provide hole expansion ratios around 45%, higher than the hole expansion ratio (λ values) shown on the right side vertical axis of Fig. 12.53.

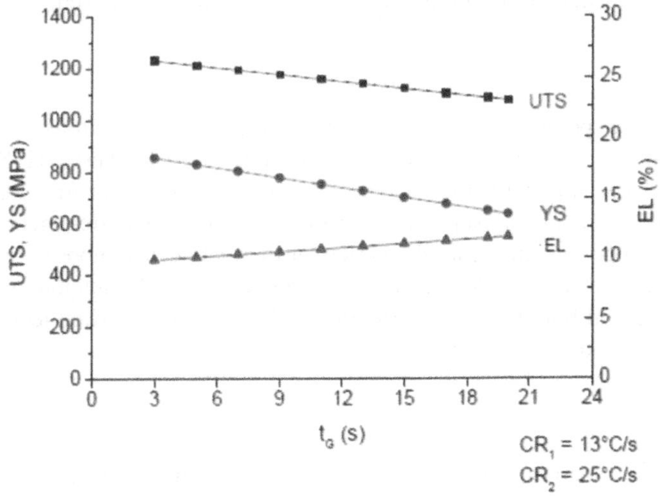

FIG. 12.52 Effect of isothermal hold time at the galvanizing pot temperature on tensile properties of a dual phase steel [93].

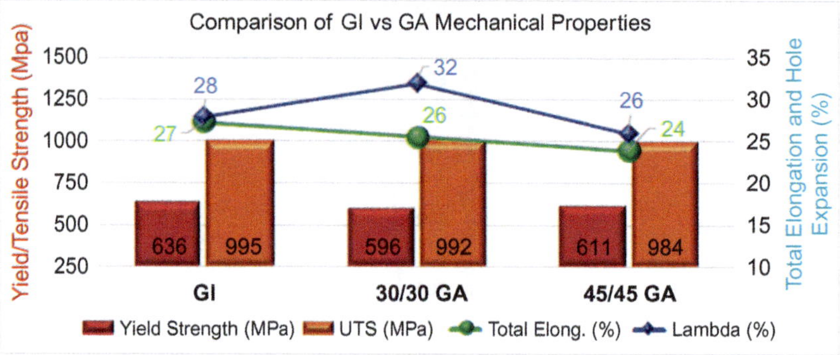

FIG. 12.53 Effect of GA processing on tensile properties and hole expansion ratio for a third-generation 980 MPa UTS steel with a GI coating (left), after galvannealing to produce a GA coating with 30 g/m^2 per side (center), and after galvannealing to produce a GA coating with 45 g/m^2 per side (right) [94].

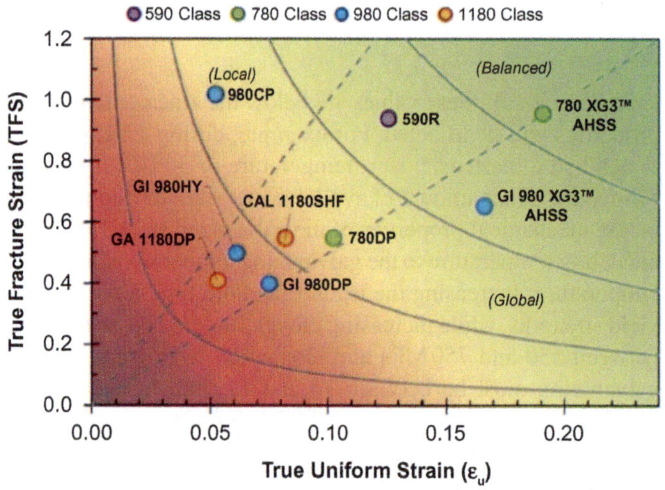

FIG. 12.54 Formability concept map ("Hance Diagram") showing true fracture versus true uniform strains for several hot-dip coated and uncoated 980 MPa AHSS, which originated from application-specific material designs [94].

The local (true fracture strain) and global (true uniform strain) elongations for several galvanized and uncoated AHSS are shown in the Hance diagram in Fig. 12.54 [94]. As noted above, high local strains are associated with high hole expansion ratios. Consistent with the above, AHSS grades with greater differences in strengths between the hard and soft phases would exhibit greater global formability, while those processed to reduce these differences, such as the DP grades shown in this figure, would have a balance between global and local formability. Several quench and partition third-generation grades with high global formability are shown, together with a complex phase grade that has more gradual differences between the properties of microstructural constituents that permit higher local elongations to be achieved.

12.3 Joining zinc-coated steels

Fusion joining processes for zinc-coated steel include resistance spot welding (RSW or spot welding) and laser welding, the latter has come into prominence with the widespread use of tailor welded blanks in vehicle manufacturing. Because AHSS grades have higher electrical resistivities than conventional steels, they require less current to weld; their higher strength means that a higher electrode force is required [47]. The basic design philosophy of AHSS means that thermal treatments are relied on much more than alloying to achieve desired properties, therefore the effects of thermal cycles imposed by welding and brazing process on weld properties are more important to consider than for conventional grades. Fusion joining of zinc-coated sheet steels can also require adjustments to avoid liquid metal embrittlement. The use of nonfusion techniques has become more important, these include brazing and adhesive bonding.

12.3.1 Resistance spot welding

The addition of a zinc coating to sheet steels that will be spot welded together introduces a volume of material much lower in melting point than the steel, in fact the boiling point of zinc, 907°C, is lower than the lowest melting point of steels, 1370°C, that would be considered in practical applications. The beginning of the spot welding cycle therefore usually begins with heat and force applied by the welding electrodes so that the coating can be squeezed out of the current path as it melts. This force is higher than that used for uncoated steels, and this higher force lowers resistivity so that a higher welding current may be required than when welding the same grades in uncoated condition. The remaining zinc can react with the welding electrodes, even if they are, as customary, water-cooled. This results in the formation a softer brass zone in comparison with the initial electrode tip alloy that is normally a Cu-Cr alloy or an Al_2O_3 dispersion-strengthened copper alloy. Oxides can also form on the electrode tip. These tend to be rich in Al that originates from the galvanized coating, raising local surface resistance and therefore temperature when current is applied. TiC-coated copper electrode tips are also used that inhibit the brassing reaction with the electrode. Their lower wettability also lowers the formation of oxides on the electrode tip [95]. Because the GA coating contains zinc in alloyed form with Fe, its melting and boiling points are much higher, and the reactivity of zinc with the electrode surface is reduced; the lower Al content of the GA coating also contributes to longer electrode tip life.

Fig. 12.55 shows the effect of welding current, welding time, and hold time after welding on nugget diameter for a GA-coated HSLA grade. An acceptable range of diameters is generally seen between 9 and 11 kA. Below this current range there is insufficient power to form an acceptable nugget diameter, above it, expulsion of molten material from the workpiece commences [96]. Two weld times are shown, 16 and 18 cycles, these are measured in terms of the frequency of the AC voltage welder that was used, in this case 60 cycles/s (Hz), so that 18 cycles are 300 ms. Because these times are so close, there they have little effect on the shown nugget sizes. The heat input to the nugget is determined from

$$Q = I^2Rt \tag{12.3}$$

where Q is the heat generated in joules, I is the total input current in amperes, R is the total average resistance across the materials and interfaces in ohms, and t is the total welding time, seconds. In Fig. 12.55 there is a significant difference in hold time after the weld nugget is formed, this influences the cooling rate of the workpiece that can greatly influence the microstructure and mechanical properties of spot-welded AHSS. In the case of the referenced HSLA steel, a 5s hold time gave a microstructure containing ferrite and bainite, while the 30s hold time gave a bainitic microstructure. Rather than a single-pulse application of current, multiple pulses can be used, with intervals of about five cycles between pulses; this has been shown to help remove the zinc coating from the faying surfaces that are joined together during spot welding. If the

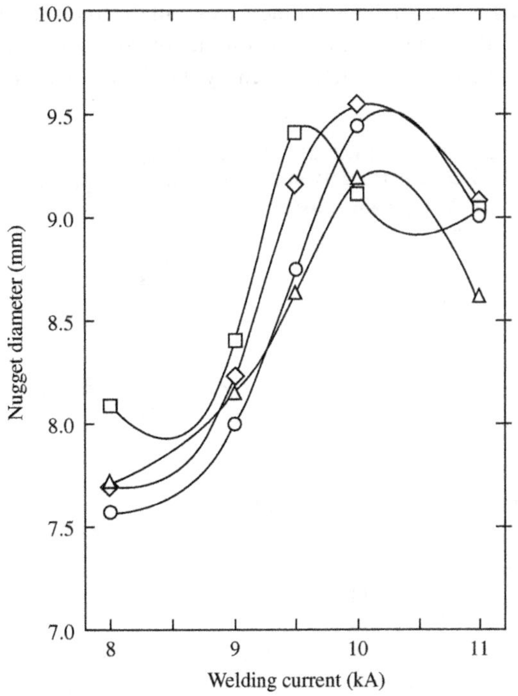

FIG. 12.55 Effect of welding current on nugget diameter for a GA-coated HSLA steel. Diamond: weld time of 16 cycles, hold time of five cycles. Square: weld time of 18 cycles, hold time of five cycles. Circle: weld time of 16 cycles, hold time of 30 cycles. Triangle: weld time of 18 cycles, hold time of 30 cycles [96].

TABLE 12.11 Spot welding parameters for a fully martensitic cold-rolled zinc electrogalvanized grade.

Steel grade (MPa)	Steel thickness (mm)	Coating weight (g/m^2/side)	Welding current type	Electrode tip diameter (mm)	Electrode force (kN)	Weld time	Hold time	Permissible current range (kA)
900	1.2	100	MFDC	6	3.5	340 ms	200 ms	2
900	1.5	75	MFDC	6	3.5	320 ms	200 ms	2.5
1200	2	75	AC	8	5	24 cycles	10 cycles	2.5

Derived from P. Hurtig, T. Nilsson, J. Larsson, Development of cold rolled martensitic electrogalvanized steels, in: Proc. Intl. Conf. on New Developments in Advanced High-Strength Sheet Steels, June 15–18, 2008, Orlando, Association for Iron & Steel Technology, Warrendale, PA, pp. 137–146.

coating is not removed, weld current will need to be higher because of the lower contact resistance of the zinc layer; this is less of an issue for the higher-resistivity GA coating.

Another way of providing energy to the workpieces to be joined, rather than AC, is termed mid-frequency direct current (MFDC). In this case, the AC input is converted to a higher frequency (400–4000 Hz) input to a transformer that provides direct current (DC) with short pulses, typically 8 ms versus the 16 ms of a complete 60 Hz sine wave. With the AC supply, voltage goes to zero it alternates between the positive and negative values of this sine wave. The intervals of low to zero voltage in the sinusoidal voltage pattern permitting the weld to cool; additional cooling can occur because of phase shifts. The MFDC configuration heats continuously, permitting more energy to be provided in less time [97]. Spot welding parameters for a fully martensitic zinc electrogalvanized sheet steel composition of 0.05%–0.11%C, 1.6%–2.0%Mn, 0.2%Si (carbon equivalent CEV = 0.38) are shown in Table 12.11. The minimum current value for each of the permissible current ranges shown in this table was determined as the value required to form the minimum acceptable nugget diameter from the empirical formula

$$d_{\min} = 4\sqrt{t} \tag{12.4}$$

where d is the minimum nugget size in mm, and t is the thickness of the steel substrate, also in mm. The load-bearing capability of the welded join increases as nugget diameter increases, but this requires increased heat input to form the larger nugget that can increase the effect of the welding heat cycle on the microstructure and mechanical properties of the base metals. The diameter of the formed nugget is also determined by the diameter of the electrode tip. Fig. 12.56 shows the

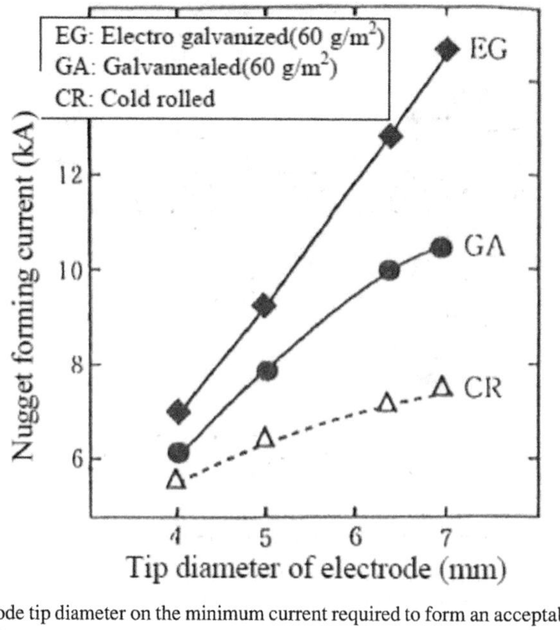

FIG. 12.56 Effect of spot welding electrode tip diameter on the minimum current required to form an acceptable size nugget for cold-rolled, galvanneal, and zinc electrogalvanized sheet steels [98].

relationship between electrode tip diameter and the current required to obtain the minimum acceptable nugget diameter for GA, zinc electrogalvanized and uncoated cold-rolled sheet steels [98]. Consistent with the above, the highest resistivity base metal is the uncoated condition. The addition of the GA coating decreases resistivity less than the EG coating. These lower resistivity conditions require the use of higher currents to obtain the same heat input per Eq. (12.3) to obtain the minimum acceptable nugget diameter.

Many AHSS require the use of relatively high dewpoint pretreatments (e.g., −5°C at 830°C) to sequester more reactive alloying elements such as Si into internal oxides that would otherwise interfere with hot-dip coatability; these internal oxides reside a few microns below the surface where they can influence the electrical properties of the steel and therefore spot welding behavior. To investigate the effects of this galvanizing treatment and the effect of the zinc coating, samples were prepared in as-cold rolled, internally oxidized, and zinc coated conditions using a 1.6mm DP 1000 substrate containing 2.55%Mn and 2%Si, illustrated in Fig. 12.57 [99]. These samples were galvanized on the McMaster University hot-dip process simulator in an atmosphere of N_2-5%H_2 with a dewpoint of −5°C. The annealed-bare samples were heated to 500°C at 15°C/s, followed by heating to the intercritical annealing temperature of 828°C at 5°C/s, and held at this temperature for 120s. Samples were then cooled to room temperature by N_2 jet cooling. The annealed-coated samples received an extra step, dipping in a Fe-saturated 0.2wt% dissolved Al galvanizing bath at 460°C for 4s before final cooling. Panels were then spot welded together using the AWS D 8.9 [100] standard schedule using 7mm Cu-Cr domed radius electrodes and a weld current of 9kA, which formed an acceptable nugget size without incidences of metal expulsion during welding. The effect of steel condition on weld nugget size is shown in Fig. 12.58. The origin of the differences in nugget size is the resistivity of the steel as the welding progresses. The as-rolled samples are covered in oxides that increase resistivity, therefore more heat is initially generated in these samples at the same level of welding current. The coated steel has the lowest resistivity and therefore forms the smallest nugget at constant current. The internal oxides formed in the as-annealed condition prevents external oxides from being formed, resulting in intermediate resistivity. Because of the lower

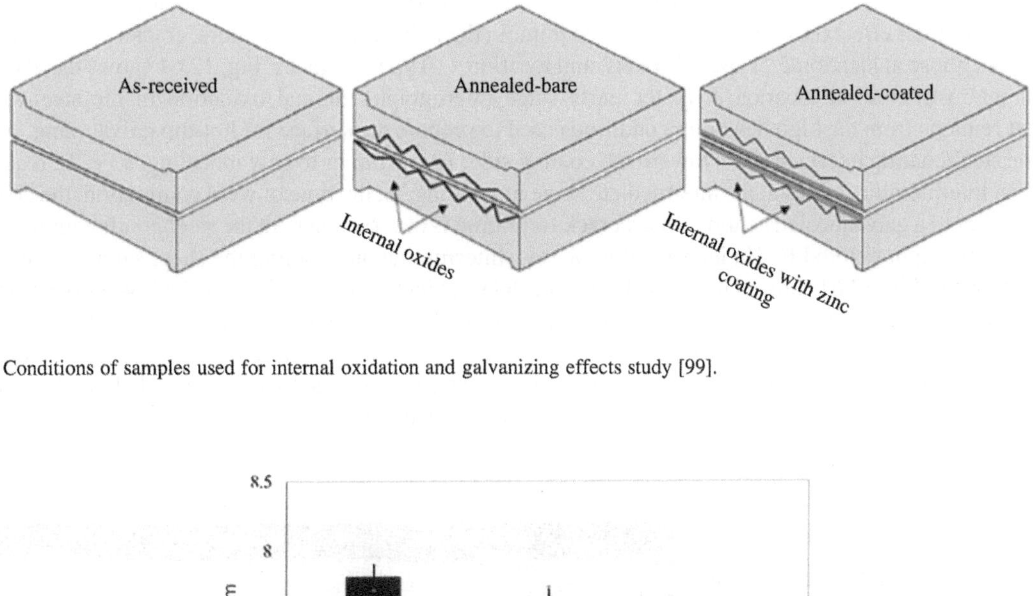

FIG. 12.57 Conditions of samples used for internal oxidation and galvanizing effects study [99].

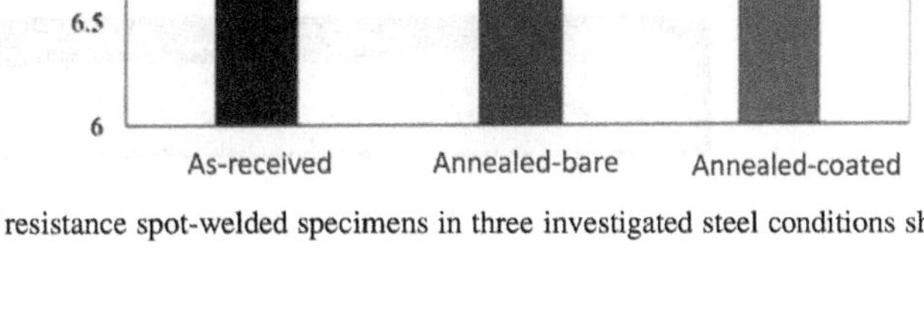

FIG. 12.58 Average nugget size of resistance spot-welded specimens in three investigated steel conditions shown in Fig. 12.57 [99].

resistivity of the coated steel sheet with internally oxidized subsurface layer, operators compensate by increasing weld current to obtain an acceptable nugget size; however, this can be much closer to the expulsion limit. Increased incidences of expulsion have been reported for steels welded in this condition.

12.3.1.1 *Liquid metal embrittlement*

Under certain metallurgical conditions, zinc-coated advanced high-strength steels have been found to be sensitive to liquid metal embrittlement (LME) arising from joining processes in the presence of liquid zinc when a sufficient stress is imposed. Because the melting point of zinc, 419°C, is well below the temperatures required to weld steel, liquid zinc can participate in many reactions occurring during welding of zinc-coated steels including steel phase transformations and alloy reactions that influence weld quality. The heat-affected zone surrounding the weld nugget is one of the most susceptible regions: because this zone is cooler, its strength is also higher, increasing the local state of stress. The cooler temperature also prolongs the presence of liquid zinc rather than zinc vapor; after the weld is completed, it also is not in contact with the water-cooled RSW electrodes, so additional time for liquid Zn exposure is possible [101]. LME is related to liquid zinc alloy penetration in the steel subsurface area under the presence of critical amount of tensile stress in steel microstructures sensitive to this phenomenon; the most sensitive microstructures generally tend to contain appreciable quantities of austenite. Cohesion between grains of these steels decreases and causes embrittlement.

Fig. 12.59 shows the progression of electrode indentation during the spot welding process after early stage (83 ms of weld time), intermediate stage (250 ms), and after complete formation of the weld nugget (383 ms) for a three-sheet stackup shown in Fig. 12.60 in which the top sheet was a 1.5 mm 3G (Si, Mn) 1000 MPa grade, with a 9 μm coating of Zn-0.2%Al (GI) coating on each side, the bottom two sheets were GI-coated extra deep drawing quality steel (EDDS) [102]. The early stage includes the application of electrode force that squeezes the zinc coating out from under the 5 mm electrode. Remnants of the coating can be seen around the perimeter of the electrode indentation. The zinc remains around the perimeter as the weld progresses to completion. Conditions sufficient to cause LME cracking can exist as welding progresses in each of the five locations indicated in Fig. 12.60. Location 2 cracks, termed shoulder cracking, are at the location of maximum tensile stress and are not in contact with the water-cooled electrode. This location, together with location 4, is of the greatest concern because of their effect on strength of the welded joint. In the industrial nomenclature, cracks at location 1 are called "Type A" cracks, those at location 2 "Type B" cracks, and location 4 "Type C" cracks. Fig. 12.61 shows the evolution of the coated steel microstructure at location 2. In the early-stage micrograph, internal oxidation in the steel subsurface is observed that remains from the high dewpoint conditions used to prepare the surface for hot-dip galvanizing. As weld time increases, the steel/coating interface shifts toward the coating side. Also, similar to galvannealing, a Fe-Zn reaction begins, creating Fe-Zn intermetallic phases at the intermediate stage of welding. At the time of weld completion, the coating microstructure is similar to a galvannealed coating. Microcrack formation is observed in both the steel/coating interface and in the coating through the light-colored Fe-Zn intermetallic phases. Interrupting and cooling the sheet samples after each of the three stages shown in Fig. 12.61 showed that it each case the 3G steel had been heated above the lower critical temperature, A_{c1}, of this composition, and the cooling rate that was experienced after while the sheet stackup was still clamped in place by the force of the electrodes was sufficient to form martensite. A small amount of martensite was formed after the early stage (83 ms), while more than 80% martensite is observed in the later stages. Cracks are observed along the prior austenite grain boundaries. Fracture surfaces such as those in Fig. 12.62 typically show the presence of Zn [103].

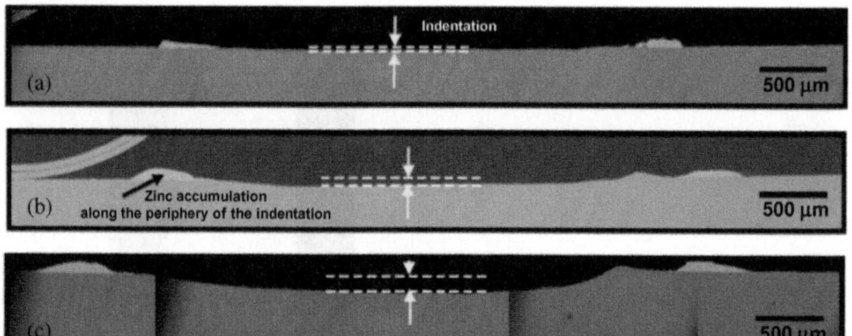

FIG. 12.59 Cross-sectional SEM micrographs of top surface of spot-welded sheet welded in (A) early (83 ms), (B) intermediate 250 ms), and (C) complete (383 ms) stages [102].

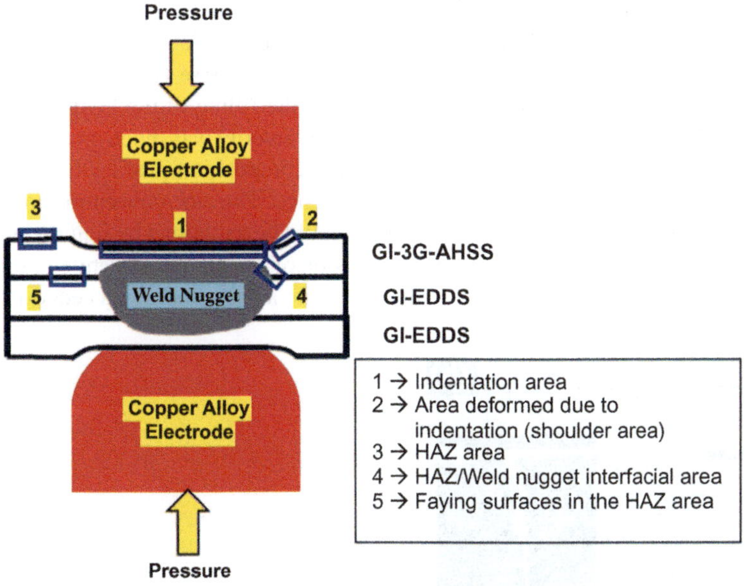

FIG. 12.60 Schematic of three-sheet stackup for LME study using a GI-coated 1000MPa 3G steel grade (top sheet). Probable locations for occurrence of LME cracking on surface and in heat-affected zone (HAZ) are shown as locations 1–5 [102].

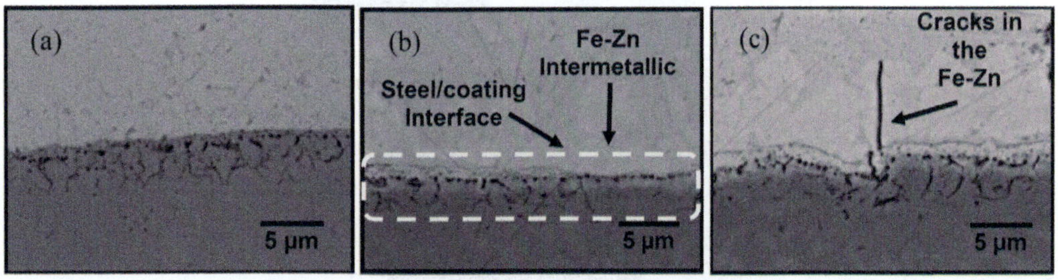

FIG. 12.61 Cross-sectional SEM micrographs of steel/coating interfacial area near the upper surface of location 2 in Fig. 12.60 after welding in (A) early (83 ms), (B) intermediate (250 ms), and (C) complete (383 ms) stages [102].

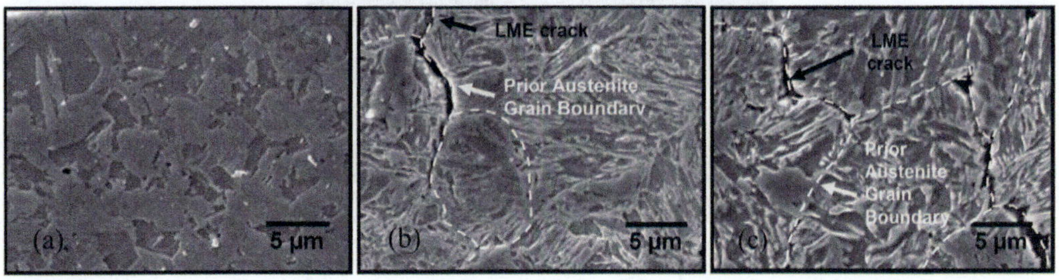

FIG. 12.62 Cross-sectional SEM micrographs of steel subsurfaces corresponding to Fig. 12.61. Note increasing martensite fraction with time, and cracks along prior austenite grain boundaries [102].

The severity of LME has been linked to the relative orientation of the grain boundaries along which LME is observed, and therefore the energies associated with the misorientation of these grain boundaries. Zinc penetration of grain boundaries is possible when

$$\gamma_{GB} > 2\gamma_{SL} \tag{12.5}$$

where γ_{GB} is the grain boundary energy of the steel surface, and γ_{SL} is the solid steel-liquid zinc interface energy. Direct observation of this wettability was observed by Arndt [104]. The factor 2 is present because zinc must wet both sides of the grain boundary. Depending upon misorientation, γ_{GB} is within the range 0.4–1.4 J/m^2 [105]. The mechanism of zinc penetration in austenite-containing microstructures was shown using full penetration laser welds in which an imposed stress was varied [106] joining a GI-coated 10% Mn steel with 50% retained austenite to an uncoated fully austenitic 14%Mn TWIP steel. The procedure is shown in Fig. 12.76 and accompanying text later in this chapter. These two sheets were welded together so that zinc was supplied to the TWIP steel from the coated 10%Mn steel; the filling of the LME crack on the TWIP side of the welded joint is shown in Fig. 12.63A, this also shows that the cracking severity was much greater on the TWIP side. LME severity correlated well with stress level. Fig. 12.63D shows that the composition profiles for zinc within the LME crack together with the composition of the Γ-(Fe,Mn)$_3$Zn$_{10}$ phase for reference. The zinc concentration is observed to decrease near the tip of the crack. The diffusion length was much longer than could be explained by liquid zinc diffusion alone, and therefore, Zn penetration was associated with the stress-assisted diffusion of Zn into the austenite grain boundaries and further liquid-Zn formation by the peritectic reaction at 782°C when the zinc accumulation exceeded its

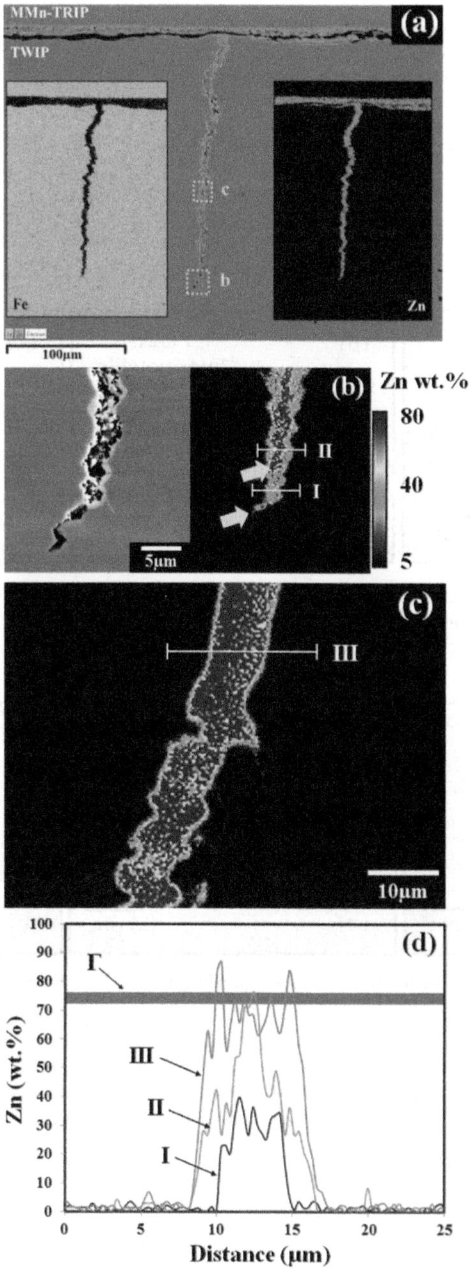

FIG. 12.63 (A) EDS elemental map of an LME crack on the TWIP side of a TRIP-TWIP welded joint showing Zn penetration, EPMA analysis at (B) tip, (C) middle part of the LME crack, and (D) Zn concentration profiles along indicated traverses of the LME crack in (B) and (C), Γ-phase Zn-content indicated in its composition stability range [106].

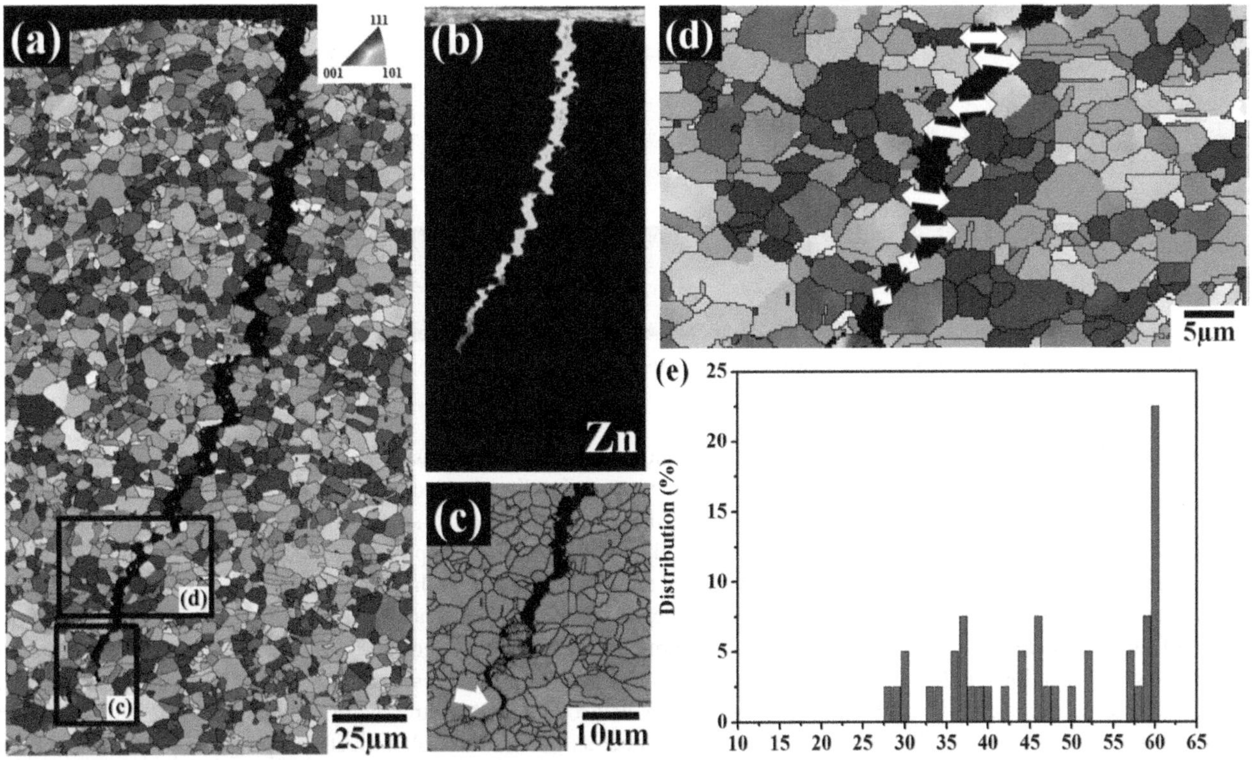

FIG. 12.64 (A) Overview of EBSD analysis of a representative LME crack, (B) Zn penetration along the LME crack, (C) image quality map at the tip of the LME crack showing the propagation along the austenite grain boundaries, (D) comparing the matching grains on either side of the LME crack, and (E) distribution of misorientation angles of matched grains along the LME crack [106].

solid solubility in austenite. This weakened the grain boundary, opening it up to admit liquid zinc [106]. The distribution of misorientation angles, Fig. 12.64, in the LME crack confirmed that Zn penetration only occurs at high-angle grain boundaries; however, Σ3 twinning (60°) boundaries also play a role in Zn penetration in the TWIP substrate [106].

The occurrence of LME is usually associated with locations in the steel that are increased to temperatures above the lower critical temperature A_{c1}, usually around 700°C, but not above 900°C where liquid zinc is no longer present because of its vaporization. LME occurrences in portions of the weldment that do not exceed the lower critical temperature however have been reported, sometimes as low as 500°C. These occurrences are associated with the presence of much higher stresses than normally experienced in the spot welding process. Fig. 12.65 shows a cross section of a two sheets of TRIP 1100 grade containing 2.2%Mn and 1.5%Si with a minimum GI coating weight of 50 g/m^2 per side welded to each other. The electrode was misoriented at angle of 4 degrees. As welding progressed, the electrode moved relative to the lower-right side shoulder of the weld, resulting in the crack observed in the subcritical heat-affected zone (SCHAZ). The upper critical heat-affected zone exceeded 900°C during the welding process and was not affected by the presence of liquid zinc. The intercritical heat-affected zone, where LME is customarily seen under conditions conducive to appearance, is crack-free [108].

FIG. 12.65 Cross sections of a spot weld made with 4 degrees of electrode misalignment showing weld nugget and heat-affected zinc shape with LME cracks highlighted in *yellow* [108].

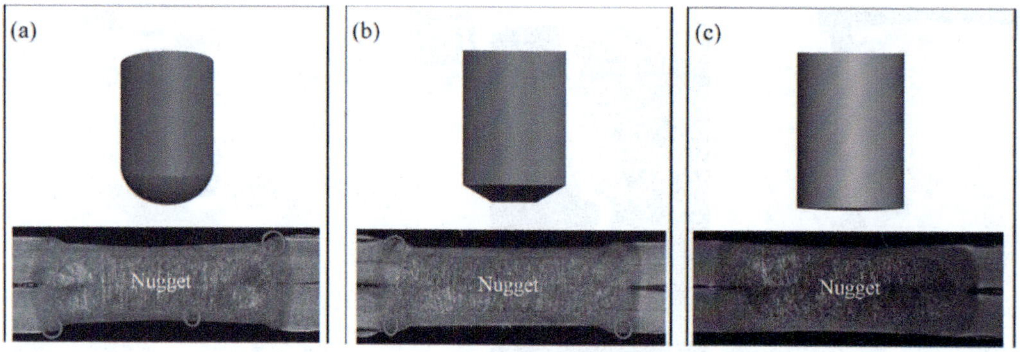

FIG. 12.66 Electrode geometry overall shape and weld cross section for (A) B-nose electrode type, (B) truncated cone, (C) radius tip electrode [109].

The effect of electrode geometry on the occurrence and severity of LME was investigated. Three types of electrodes, shown in Fig. 12.66, were investigated: B-nose, truncated cone, and radius tip electrodes. The face diameter of the first two was 7 mm while a gentle radius of 101.6 mm was used on the latter over its cross section of 16 mm. Spot welding was conducted on the same GI-coated TRIP 1100 grade described above. The micrographs of cross sections shown in Fig. 12.66 show that much less LME cracking was observed with the radius electrode, and that the truncated cone showed more LME cracks than the B-nose. Concentrations of cracks are indicated in the circled areas in each micrograph. The better performance of the radius electrode was related to the lower temperatures that this electrode was able to impart to the shoulder area of the weld (location 2 in Fig. 12.60), and the lower compressive stresses, these variables were all more consistent over the entire welding process compared with either of the two electrodes. Higher stresses in the shoulder area with the other two electrodes were related to the sudden cooling of the sheet as the water-cooled electrode presses into as indentation becomes deeper when the fusion zone is liquid. This cooling results in contraction of the periphery, including the shoulder area and buildup of stress. The points of maximum stress with the truncated cone electrode were found around the periphery while with the B-nose they were found within the indented area. These were the result of the different patterns of contact positions as welding proceeded [109].

The type of coating also influences LME susceptibility. GA-coated steels are observed to be the least susceptible because the zinc is alloyed with iron. GI-coated steels are observed to be more susceptible because the Al constituent forms oxides that can attach to the welding electrode surface, increase local resistivity and therefore temperature. Also, the presence of the inhibition layer in GI means that Zn has a slower alloying rate with Fe until this inhibition layer is broken down, making more Zn available for LME reactions. EG-coated steels are reported as less susceptible to LME because of the absence of this inhibition layer [110,111].

The effect of LME on weld strength was investigated for GI-coated DP980, TRIP690, and TRIP1100 substrates with 0%, 10%, and 16% retained austenite respectively [112]. The loss of tensile lap shear strengths, comparing uncoated vs. zinc-coated samples, was 7.1%, 2.6%, and 43.5%, respectively. The LME cracks in the steels with lower strength losses were typically less than 100 μm, whereas cracks exceeding 500 μm were associated with larger strength losses.

Alterations in welding cycles have been proven to reduce LME severity on the same TRIP 1100 steel described above. In this example [113], a three-sheet stackup was used, with the TRIP 1100 steel on top, joined to two IF steels underneath using a B-nose electrode tip shape. Three different weld current pulse patterns were used, comprising two welding pulses with a two cycle off time between them

a. Short pulse followed by a long pulse: 10-2-14 cycles
b. Equal length pulses: 12-2-12 cycles
c. Long pulse followed by a short pulse: 14-2-10 cycles

The first two gave weld nugget diameter of 7.3 mm while the diameter of the third was 7.5 mm, indicating the total heat input was about the same in these three cases. Cracking patterns in the weld cross sections are shown in Fig. 12.67. The cracking patterns made with these three different pulse patterns are different. The 10-2-14 cycle pulse pattern created very long LME cracks in the center of the weld, Fig. 12.67A). Welds made using the 12-2-12 cycle pulse pattern have cracks that are smaller and less numerous than those seen in the 10-2-14 cycle pulse pattern, Fig. 12.67B). When a longer first pulse was used, with the 14-2-10 cycle welding schedule, a larger number of cracks were again observed; however, they tended to be smaller than those generated by the other pulse patterns, Fig. 12.67C). Because the nuggets are about the same diameter, the difference in LME severity is caused by a difference in thermal history in the weld and not differences in total heat input.

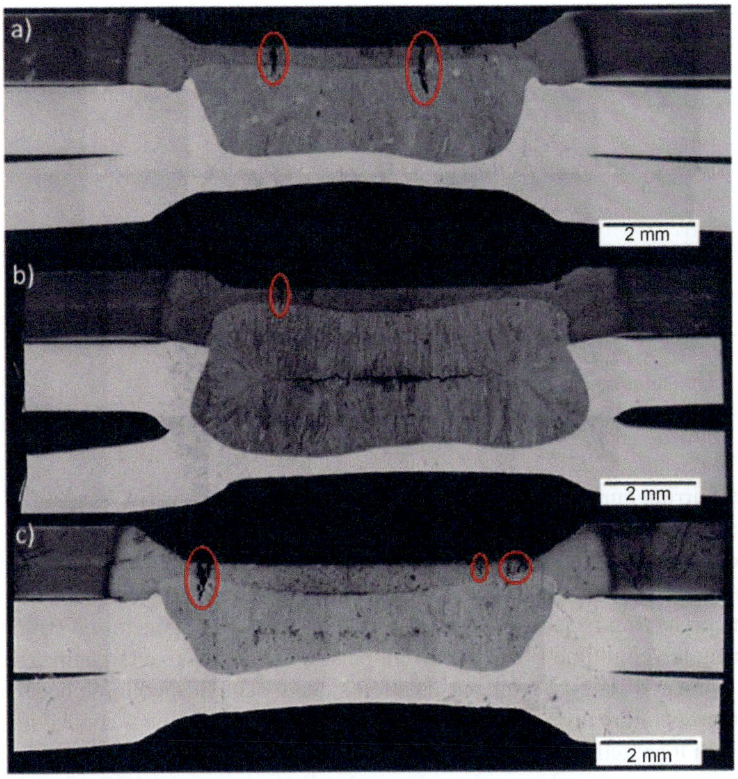

FIG. 12.67 Spot welds in a three-sheet stackup with TRIP 1100 on top together with two sheets of IF steel using on-off-on pulse current pulse patterns of (A) 10-2-14 cycles, (B) 12-2-12 cycles, and (C) 14-2-10 cycles, LME cracks circled [113].

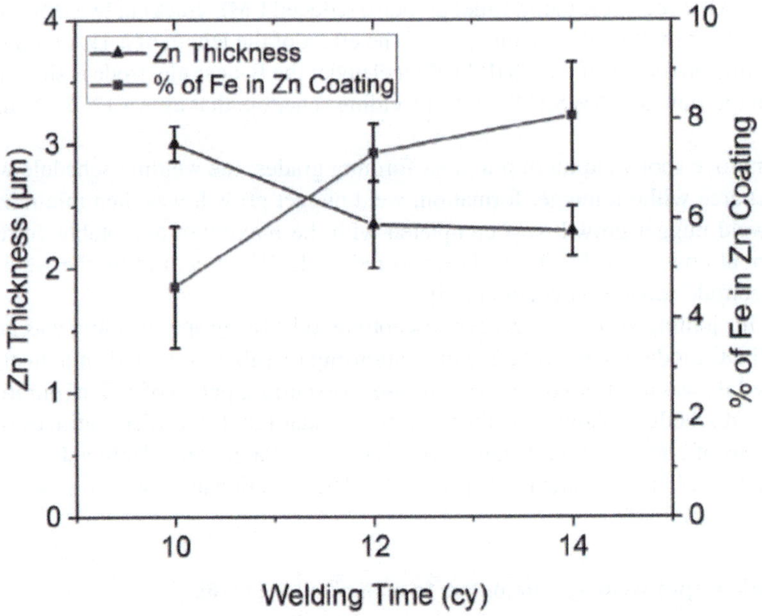

FIG. 12.68 Average thickness and composition of zinc coating remaining after first pulse for the three pulse patterns shown in Fig. 12.67, pulse duration measured in cycles (60 Hz) [113].

EDS analysis on these welds showed that as the length of the first pulse increased, the zinc coating thickness decreased and the amount of iron in the coating increased, Fig. 12.68. The diffusion of iron into the zinc coating stabilized the remaining zinc, increasing the melting temperature of the coating, consequently reducing the risk of formation of LME cracks. Consistent with this, the welds made using the 10-2-14 cycle welding schedule had the most severe LME, Fig. 12.67A). In this case, there is more zinc available for LME and the zinc is less stable due to the lower iron content

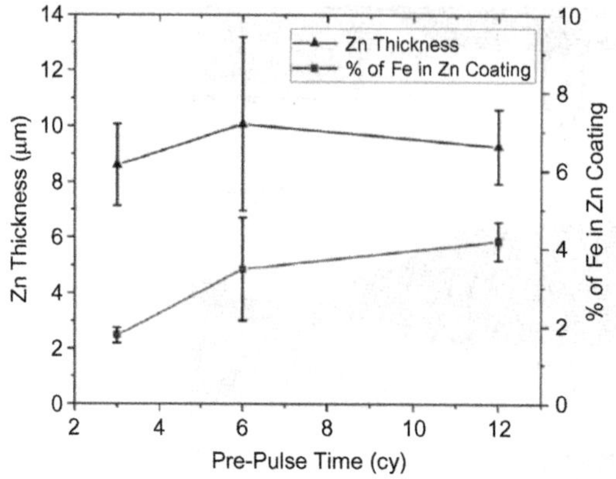

FIG. 12.69 Average thickness and composition of zinc coating remain after 4 kA prepulses of different time durations, measured in cycles (60 Hz) [113].

in the remaining coating [113]. The thermal cycles occurring during the pulse patterns also affect LME severity because of the effects of local thermal patterns on stress, especially at the weld shoulder, location 2 in Fig. 12.60. For the 14-2-10 cycle pulse pattern, at the end of the first pulse, the electrode pushes into the steel, creating an indentation and generating a new thermal contact between the weld shoulder and the electrode, this is known as mechanical collapse. The new thermal contact allows the water-cooled electrodes to begin conducting heat away from the shoulder area. Heat generation (from to the welding current) also stops for two cycles during the cool time between current pulses. This leads to a large drop in temperature beginning at 14 cycles. For the 12-2-12 cycle and 10-2-14 cycle pulse patterns, the mechanical collapse occurs during the second current pulse, causing two smaller temperature drops, the first during the cool time, and the second at the point of mechanical collapse. Therefore, the changes in local temperature at the shoulder are not as severe for these pulse pattens. The coincidence of the start of the cooling time between pulses and mechanical collapse using the 14-2-10 cycle pulse pattern results in higher shoulder stresses and consequently higher LME severity. Therefore, the use of pulse patterns to minimize LME has the objective of stabilizing the zinc through alloying without creating excessive tensile stress.

In the same work, the use of a low current (4 kA), three-cycle (short time) prepulse reduced LME cracking by stabilizing the zinc by diffusing it into the steel substrate without melting the coating [113]. The effect of the length of a 4 kA prepulse on the thickness and iron content of the zinc coating on the top of the TRIP 1100 steel under the B-nose electrode is shown in Fig. 12.69. The rest of the weld was made with the standard AWS D 8.9 [100] welding schedule that uses a 12-2-12 pulse pattern.

Multiple pulse welding was also used to improve spot welding of hot press forming grades; the welding schedule was designed to first stabilize the electrode contact area without nugget formation, weld nugget growth was then minimized while maximizing contact area, after which weld nugget growth was completed with the maximum acceptable current (below expulsion) [114]. The use of weld current downslope has been shown to reduce LME by permitting the weld to gain strength at a cooler temperature when electrode separation occurs [115].

An industrially practical welding schedule for joining sheet steel grades susceptible to LME by spot welding was proposed by Pearson [116]. This work used the MFDC mode described before in relationship to Table 12.11 and an industrial C-type welding gun. The AWS D 8.9 [100] schedule was used as a reference; this uses two current pulses of 167 ms duration separated by a 34 ms cooling time. The optimized schedule, shown together with the standard in Table 12.12, utilizes five pulses of 67 ms duration, each separated by 34 ms of cooling time. A galvanized 3G 1000 MPa grade, 1.4 mm automotive grade galvanized steel sheet was used; its Yurioka carbon equivalent [117] was 0.66. The current was chosen to produce a

TABLE 12.12 Optimized and AWS D 8.9 standard spot welding parameters to minimize LME [116].

Welding schedule	Electrode geometry	Electrode force (kN)	Squeeze time (ms)	Pulses	Pulse length (ms)	Hold time (ms)	Cooling time between pulses (ms)	Current (kA)
Standard	6R50	4.9	167	2	167	167	34	7.7
Optimized	8R50	4.9	167	5	67	100	34	9.3

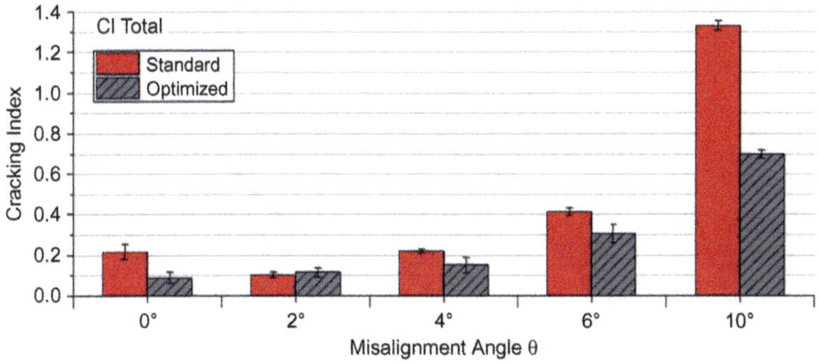

FIG. 12.70 Effect of welding electrode misalignment angle shown in Fig. 12.71 on cracking index (CI) for all locations in the weld, using standard and optimized pulsed welding schedules of Table 12.12 [116].

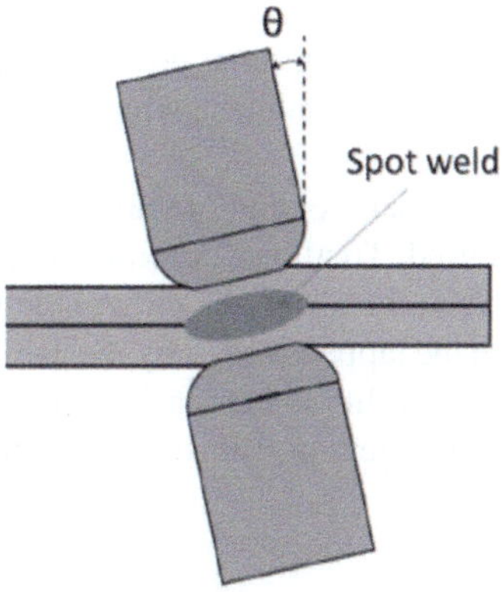

FIG. 12.71 Misalignment angle Θ of welding electrodes used as primary variable for experimental results shown in Fig. 12.68 [116].

6 mm nugget using B-nose electrodes with a tip radius of 50 mm. A range of variations normally encountered in industrial production were then imposed to verify that the schedule is industrially viable including electrode misalignment, tensile loading, and gapped joints. Fig. 12.70 shows one example of this work, the effect a range of joint misalignments on LME severity using the standard and optimized welding schedules. The results are expressed in terms of the cracking index CI [118].

$$CI = \frac{nL}{\tau} \tag{12.6}$$

where n is the mean number of cracks per sample, L is the lognormal median crack depth, and τ is the sheet thickness. The cracking index can be thought of as the ratio of the total length of cracks to the sheet thickness and has shown to correlate to the decrease in joint strength caused by LME cracks for a RSW loaded in tensile shear [118]. Overall, under ideal conditions that exclude misalignment and other factors, use of the optimized schedule resulted In a 51% reduction in LME occurrence.

In summary, LME can be encountered when three simultaneous conditions are present: stress, zinc, and a sensitive steel. Means of minimizing LME include the minimization of local stresses by optimizing electrode shape, minimizing force by ensuring accurate alignment of electrodes and workpieces, and use of weld current cycles that help stabilize the zinc without imposing stress levels required for LME. Understanding the role of austenite grain boundaries, especially high angle boundaries, as pathways for introduction of liquid metal into the microstructure and consequent weakening of these grain boundaries is key to understanding and minimizing LME.

12.3.2 Laser welding

12.3.2.1 Lap welding

In fiber laser welding, the laser light is generated in an active fiber that serves as the optical cavity and guided to the work piece by means of a flexible delivery fiber, which acts as a light guide. Fiber laser welding is widely used to join together ungalvanized coil ends in the strip entry section of continuous galvanizing lines and for preparation of tailor welded blanks.

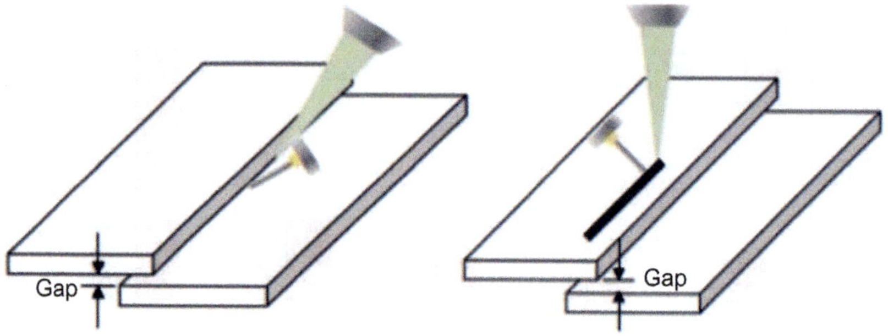

FIG. 12.72 Lap-welded joint configurations for zinc-coated steel (left) fillet (right) full penetration, showing use of gap to permit exit of vaporized zinc coating [121].

In both of these applications, butt welded joints are used. Lap welded joints are also used in many applications. Fiber laser welding is growing in importance because, in comparison with the CO_2 laser, it is more robust, compact, and more stable, especially when welding sheet thicker than 3mm. It has also been found to provide more consistent results when welding higher-strength AHSS grades. Its higher efficiency and minimal power consumption in standby mode can result in energy savings of 90% compared with CO_2 lasers. It improves stability, which also permits less shielding gas to be used in many situations [119].

Both butt and lap configurations are used for laser welding of galvanized steel. In lap welding, both lap fillet weld and conventional full penetration lap welds can be used. To allow for removal of the melting and vaporizing zinc coating from the weld joint, a small gap (0.1–0.2mm) between the sheets on either side of the joint can be used [120]. This has been provided by dimpling the portion of the sheet in the lapped area or by the use of spacers. In this way, pores and other defects can be avoided because the zinc vapor has the possibility to escape from the weld. Fillet welding can utilize a suitable filler metal, while full penetration welding is usually autogenous. Illustrations of the two types of lap-welded joints are shown in Fig. 12.72.

12.3.2.2 Butt welding

A typical laser butt welding configuration is shown in Fig. 12.73. The edges of the sheets to be joined are carefully prepared to form close-fitting interfaces that are joined together by autogenous welding.

An example of nature of butt welding of GA and GI-coated steels is found in the work of Cui [123] who examined these on the DP steels whose compositions are shown in Table 12.13. These have similar carbon equivalents, 0.236% and 0.245%, respectively. Sheets (100 × 100mm) of each type were joined to each other to form 100 × 200mm specimens for study. The 4kW fiber laser with a laser spot diameter of 0.6mm and a focal length of 200mm was run at a 20.5m/min travel speed. No shielding gas was used, but air was blown during welding to remove vapors. The initial

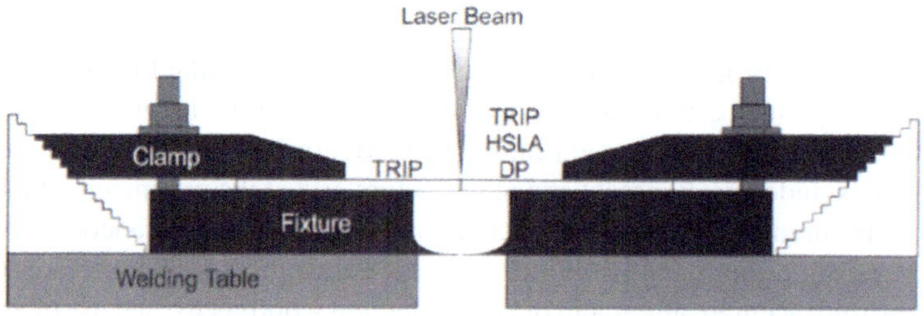

FIG. 12.73 Fixturing and sheet positioning for laser butt welding [122].

TABLE 12.13 Compositions of GA and GI-coated DP steels for coating study.

Steel composition (wt%)	C	Mn	Si	Al	Cr	Mo	Nb	Ti	C_{eq}
GI	0.15	1.45	0.33	0.05	0.02	0.002	0.001	0.002	0.245
GA	0.1	2.19	0.33	0.04	0.21	0.238	0.016	0.022	0.236

From Q.L. Cui, D. Parkes, D. Westerbaan, S.S. Nayak, Y. Zhou, D. Liu, F. Goodwin, S. Bhole, D.L. Chen, Effect of coating on fiber laser welded joints of DP980 steels, Mater. Des. 90 (2016) 516–523.

microstructures were the expected matrix of ferrite and islands of martensite, but some of the martensite in the GA-coated steel had spheroidized because of the additional heating in the GA furnace. As in the case of spot-welded joints with AHSS shown in Fig. 12.60, there were distinct HAZ variations in the steel, including an upper-critical HAZ, an intercritical HAZ, and a subcritical HAZ. In the upper-critical HAZ, the temperature was above the A_{c3} point for each composition, leading to a microstructure after completion of welding that consisted of mainly martensite with a very small amount of ferrite. In the intercritical HAZ, the temperature was between the A_{c1} and A_{c3} points, resulting in the coexistence of ferrite and austenite, the latter of which subsequently transformed into martensite during postweld cooling. In the subcritical HAZ, the temperature was below the A_{c1} line. This caused the martensite that was normally present in the base to temper and decompose into partially tempered martensite. A "soft zone" was thus created where the steel was the weakest. There are significant differences in hardnesses of these zones, Fig. 12.74, the GI hardness profile is higher because of the higher carbon equivalent of this steel.

Coating cross sections were taken at various distances from the fusion zone (weld nugget) whose dimensions are indicated in Fig. 12.74. Figs. 12.75A and B are from a distance 0.74 and 0.6 mm, respectively, from the edge of the fusion zone. These appear as typical as-produced coatings. The 12 μm GI coating in Fig. 12.75A exhibits the customary Zn-Al overlay, separated from the steel by its inhibition layer. The 7 μm GA coating in Fig. 12.75B exhibits a thick δ-$FeZn_7$ layer over an

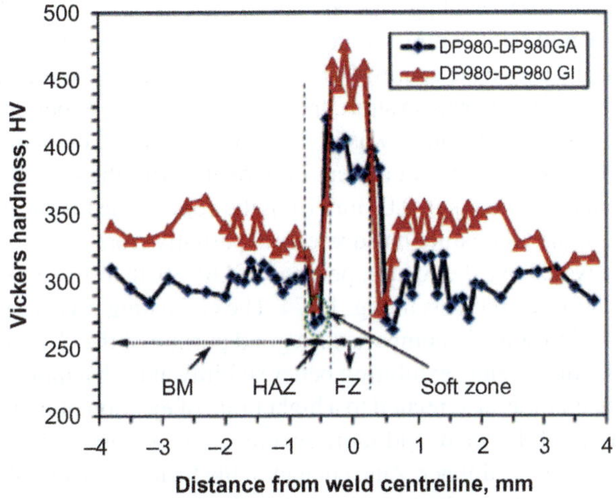

FIG. 12.74 Microhardness curves of GA and GI-coated DP980-DP980 welded joints [123].

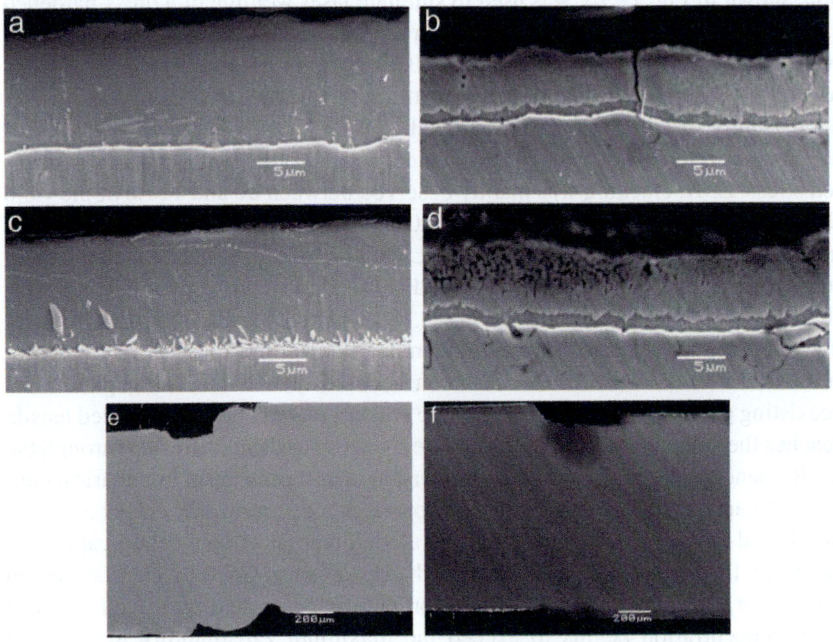

FIG. 12.75 (A) Normal GI coating outside of HAZ, 0.74 mm from the edge of FZ; (B) normal GA coating outside of HAZ, 0.60 mm from the edge of FZ; (C) GI coating inside HAZ, 0.29 mm from the edge of FZ; (D) GA coating inside HAZ, 0.29 mm from the edge of fusion zone; (E) weld of GI-coated DP980; and (F) weld of GA-coated DP980 [123].

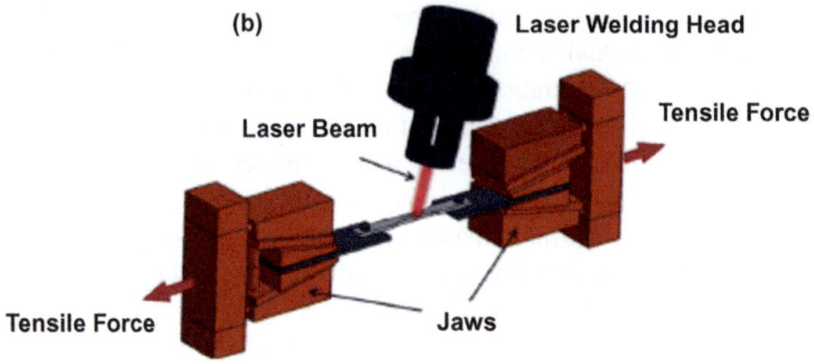

FIG. 12.76 Loaded bead-on-plate laser welding configuration used to determine LME susceptibility [107].

unidentified layer expected to be richer in Fe, the Fe content of the δ layer was 11.2% Figs. 12.75C and D were taken within the HAZ, 0.29 mm away from the edge of the fusion zone. Fig. 12.75C was taken in the HAZ, 0.29 mm from the edge of fusion zone. At this location, the heat from welding was enough to melt the GI coating shown in Fig. 12.75C. Fine Fe-Zn intermetallic precipitates can be seen in faint contrast along the boundaries of the zinc grains in this coating. Fe-Zn intermetallic particles are also found at and above the coating/substrate interface because the inhibition layer has begun to break down. Because the heat cycle for this reaction was so short, the peak temperature experienced by the coating is estimated to be around 550°C. This was also because the location was in the soft zone shown in the hardness profile of Fig. 12.74. The GA coating at this location, Fig. 12.75D, exhibits a gradient in porosity, with the side closer to the fusion zone on the left being more porous. Figs. 12.75E and F) were taken 0.04 mm from the edge of the fusion zone and show the entire thickness of the coated steel. The thickness of the GI coating has decreased, partially from evaporation of zinc as the boing point of zinc is exceeded. This temperature was above the A_{c3} temperature of DP980, therefore this location was in the upper-critical HAZ, corresponding to the hardness measurements in Fig. 12.74. The GI coating was observed to be completely removed at the distance of 0.01 mm from the fusion zone. In comparison, a good portion of the GA coating survived direct contact with molten steel right to the edge of the fusion zinc, exhibiting better stability at high temperatures. The higher loss of material in the GI-coated specimen in the fusion was also related to a higher rate of ejection of metal from the weld than with the GA coating, resulting in larger concavity of the weld, and more severe welding defects. The lower loss of metal from the GA coating is very likely because this coating is thinner. Zinc content in the fusion zone of both GI and GA specimens could not be detected, indicating that essentially all zinc had evaporated in the fusion zone.

Usually, there is little force applied to laser-welded joints during the welding process, in contrast with spot welding where significant force is applied. Therefore, incidences of LME, which require force to occur, are rarely seen with laser welds. A loaded bead-on-plate configuration, shown in Fig. 12.76, was used to simulate laser welding of a press-hardened steel grade under load. A 1.2 mm 0.23%C-1.19%Mn-0.25%Si-0.2%Cr-0.003%B grade was used for these tests, with a 9 μm GA coating consisting of 7 μm δ phase overlaying 1–2 μm mixed Γ/Γ_1 phases [107]. Steel to be used in press-hardened applications is hot-dip galvanized in a ferrite-pearlite condition. The heat treatments and deformations of press hardening occur after the part is cut to shape, and increasingly, used in tailor welded blank configurations prepared by laser butt welding. During laser welding, the load was varied up to 120% of the 390 MPa yield stress of this steel. Fig. 12.77 shows the dependence of LME cracking on applied stress. LME cracking was mainly observed between 80 and 110% of the yield stress. LME-free joints could be achieved if the applied tensile stress is below 80% of the yield stress.

These cracks are associated with the upper critical HAZ. The mechanism of LME cracking in this case is depicted in Fig. 12.78. As seen, the microcrack initiates from the coating-steel interface and subsequently propagates into the steel substrate. In this model, first, the δ and Γ/Γ_1 phases initially in the coating Fig. 12.78 (I) partially melt in the upper critical HAZ temperature range as they pass through their respective peritectic reaction temperatures of 672°C and 782°C, respectively. These phases respectively contain about 75 and 90 wt% of Zn that then becomes available, reaching the steel interface mainly by flowing through the preexisting cracks in the GA coating or the cracks created due to the applied tensile stresses, Fig. 12.78 (II). As the liquid-Zn reaches the substrate, it starts to penetrate by stress-assisted diffusion through the austenite grain boundaries, Fig. 12.78 (III). By penetration of Zn along the higher-angle austenite grain boundaries, consistent with Fig. 12.64, and under the effect of the applied tensile stress, decohesion takes place and LME cracks open. As the cracks open, it is possible for additional liquid-Zn to flow into the crack by a combination of convection capillarity phenomena, further opening the crack. Therefore, the local propagation path of the crack is governed by the distribution of high-angle austenite grain boundaries; however, the overall crack propagation path is controlled by the direction of the applied tensile stress. As the local zinc concentration begins to exceed the maximum Zn-solubility in austenite (7 wt%), α-Fe(Zn) (alpha ferrite) begins to form locally at the crack opening area. The diffusion coefficient of Zn in

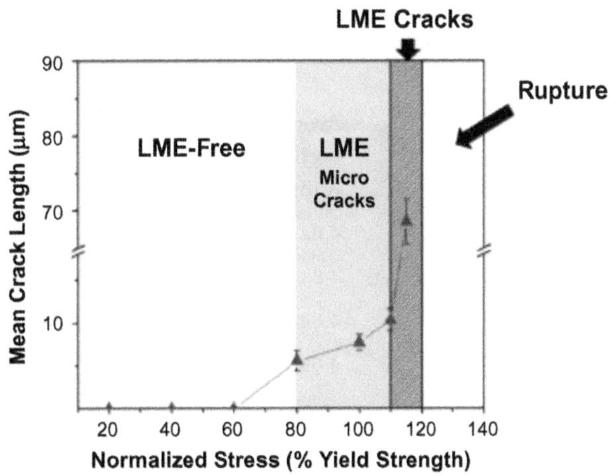

FIG. 12.77 Effect of applied stress on LME crack lengths [107].

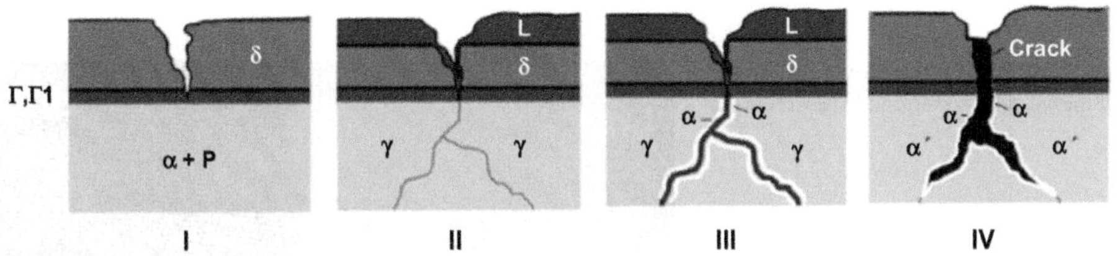

FIG. 12.78 The schematic of LME mechanism during laser welding of GA-coated press-hardening steel (α: Ferrite, P: Pearlite, γ: Austenite, α′: Martensite, δ: Fe-Zn delta-phase, L: Liquid, Γ/Γ_1: Gamma/Gamma$_1$) [107].

austenite and ferrite at 850°C is 1.11×10^{-17} and 3.53×10^{-15} m^2/s, respectively [124]. Therefore, the formation of ferrite along the LME-crack path promotes Zn penetration due to the faster Zn transport in this phase. As the end of the welding heat cycle is reached, the remaining austenite transforms into a lath martensitic structure; however, α-Fe(Zn) fine grains adjacent to the LME-crack remain stable due to high Zn content, Fig. 12.78(IV).

12.3.3 Brazing

Brazing of zinc-coated steel joints can be carried out using gas metal arc or laser energy sources. Gas metal arc is the most widely used technique. This process delivers significantly lower heat input to the substrate that both reduces the amount of zinc coating removed from the area of the joint, it also reduces the HAZ softening phenomenon described before for welding processes that affects steel mechanical properties and also reduces welding defects such as porosity and blowholes. The strength of an arc-brazed joint depends on the spreading of the molten filler material over the joint to create a bond between the parts as the molten braze solidifies. GI and GA coatings have been found to give different spreading behaviors for the Cu-Si brazing alloys typically used to join galvanized steel sheet. For example, 1.4mm thick GI and GA-coated DP600 steel sheets were joined using a Cu-3%Si-1%Mn filler wire in a bead-on-plate configuration, with and without prior plasma cleaning [125]. Currents ranged between 98 and 106 A, voltages between 15.9 and 16.2V, and heat inputs between 79.7 and 85.4J/mm. The result of braze bead measurements is shown in Table 12.14. The as-received GA (ARGA) sample showed the worst wetting with the largest bead height, smallest bead width, and largest contact angle; in contrast, the plasma cleaned (PCGI) sample had the best wetting with the smallest bead height, largest bead width, and smallest contact angle. The PMZ for the different samples confirms that the PC samples were subjected to a higher HI since these samples have larger partially melted zones as compared with the AR samples. Evidence of the better brazing wettability of GI is seen in the "tail" on the bottom on the brazed joint, Fig. 12.79. Also evident is the interface layer that is rich in Cu-Si intermetallics. Thicker Cu-Si layers are associated with higher joint strength and greater heat input during brazing, measurements of interface layers are shown in Table 12.15. In this example, the surface preparation of the Zn-coatings, i.e., plasma-cleaned vs. as-received, had a significant effect on the average thickness of the intermetallic layer, with the plasma cleaned samples showing a 40% reduction in intermetallic layer thickness as compared with the as-received samples.

TABLE 12.14 Measured bead geometries for as-received GI and GA specimens (ARGI, ARGA) and plasma cleaned GI and GA specimens (PCGI, PCGA).

Sample	Average bead height, *H* (mm)	Average bead width, *W* (mm)	Average cross-sectional area, *A* (mm^2)	Average contact angle (*θ* degrees)	PMZ area, PMZ_A (mm)
ARGI	1.15	5.02	4.82	57.86	0.55
PLG1	1.11	5.65	4.83	43.88	0.78
ARGA	1.29	4.54	4.84	60.57	0.58
PLGA	1.21	4.95	4.89	48.20	0.70

PMZ refers to the partially melted zone [125].

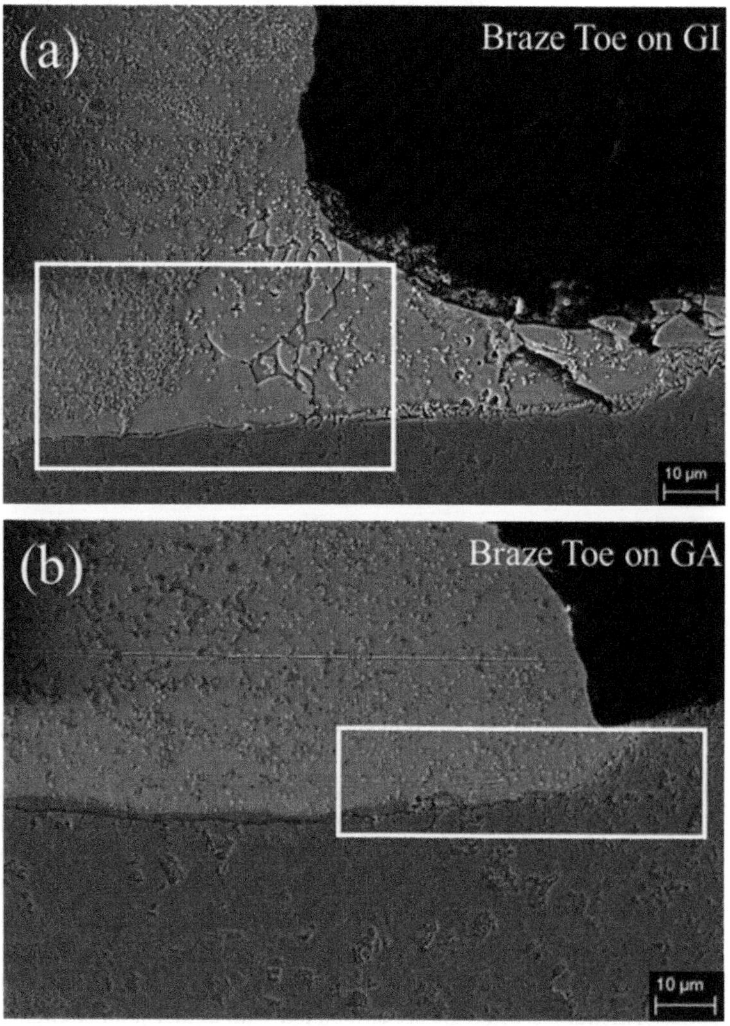

FIG. 12.79 Toe region of (A) GI and (B) GA-coated DP600 brazed joints. The toe of the GI joint was found to be rich in Zn [125].

TABLE 12.15 Intermetallic layer thickness measurements for as-received GI and GA specimens (ARGI, ARGA) and plasma cleaned GI and GA specimens (PCGI, PCGA). PMZ refers to the partially melted zone [125].

	IMC layer	
Sample	Avg. thickness (μm)	95% confidence level
ARGI	11.81	2.28
PCGI	7.18	1.59
ARGA	15.52	4.73
PCGA	9.25	2.33

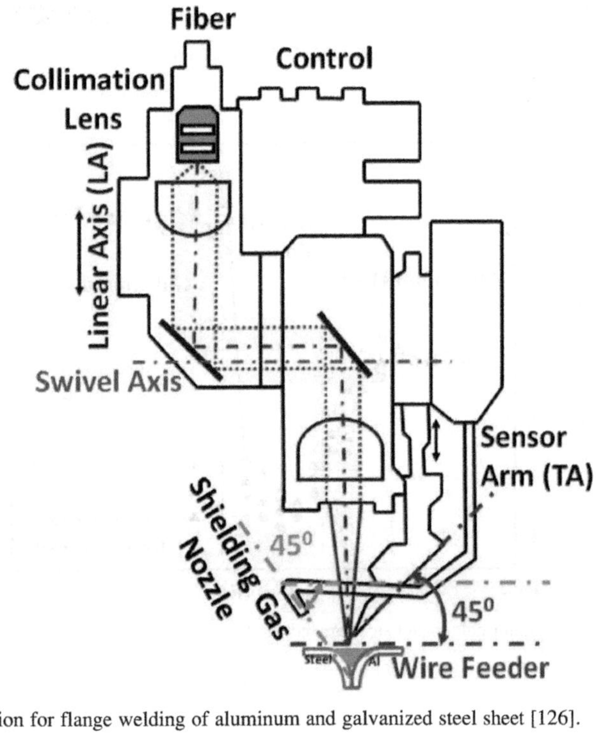

FIG. 12.80 Laser brazing configuration for flange welding of aluminum and galvanized steel sheet [126].

Laser brazing is also used to join zinc-coated steels and to join multi-metal configurations. For example, a flange joint between aluminum alloy 6082 (an Al-Mg-Si composition) and a GI-coated low carbon steel was made using aluminum alloy 4047 (Al-12.6%Si), which is the Al-Si eutectic composition [126]. The process configuration is shown in Fig. 12.80. Laser powers between 2.5 and 4.5kW provided for melting and fusing of the joint, the travel speed was 2m/min, and the feed rate of the wire was 2.5m/min. The laser spot was focused by covering approximately 90% of the filler and 10% of the base material so that it would activate the surface and improve wetting. Argon was fed as shielding gas through a 5mm diameter nozzle in reverse configuration with a gauge pressure of 1 bar and a nozzle spacing distance of 3mm. The filler metal was found to smoothly fuse with the Al base metal while on the galvanized steel side an interface of Fe-Al-Si intermetallics formed. A representative image is shown in Fig. 12.81. Microhardness profiles across the brazed joint for several laser powers are shown in Fig. 12.82. A significant HAZ is observed on the Al side because of coarse columnar dendrites that form there as solidification occurs; in contrast, there is a minimal HAZ on the galvanized steel side. Consistent with this, tensile and fatigue test samples always failed on the aluminum side of the joint.

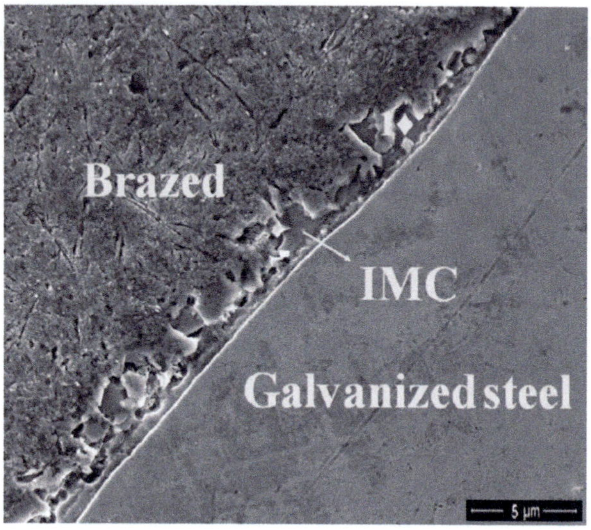

FIG. 12.81 Microstructure of galvanized steel side of laser brazed joint, 4.5 kW laser power [126].

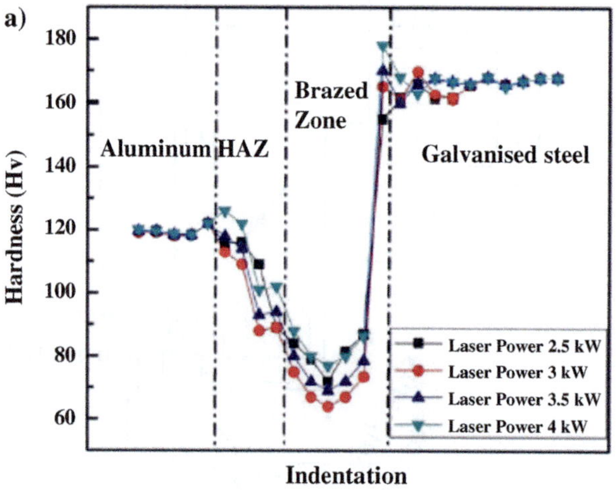

FIG. 12.82 Microhardness traces across aluminum-galvanized steel brazed joints made with a range of laser powers [126].

References

[1] M. Zapponi, E. Di Libero, H. Lazzarino, S. Bruno, J. Pedraza, F. Actis, Prepainted with enhanced performance in high-corrosion environments, in: Proc. 10th International Conference on Zinc and Zinc Alloy Coated Steel Sheet (Galvatech 2015), May 31–June 4, Association for Iron & Steel Technology, Warrendale, PA, 2015, pp. 224–230. Toronto.

[2] ASTM G85-19, Standard Practice for Modified Salt Spray (Fog) Testing, ASTM International, West Conshohocken, PA, 2019.

[3] ASTM B117-19, Standard Practice for Operating Salt Spray (Fog) Apparatus, ASTM International, West Conshohocken, PA, 2019.

[4] ASTM D2247-20, Standard Practice for Testing Water Resistance of Coatings in 100% Relative Humidity, ASTM International, West Conshohocken, PA, 2020.

[5] ASTM G87-02-2018, Standard Practice for Conducting Moist SO2 Tests, ASTM International, West Conshohocken, PA, 2018.

[6] K.M. de Souza, ASTM prohesion test predicts service performance of prepainted sheet steel, in: Proc. 6th International Conference on Zinc and Zinc Alloy Coated Steel Sheet (Galvatech 2004), April 4–7, Association for Iron & Steel Technology, Warrendale, PA, 2004, pp. 339–345. Chicago, IL.

[7] K. Pélissier, D. Thierry, Multiscale and multi-technical approach to characterize the hot-dip galvanized steel surface and its consequence(s) on paint adhesion and tendency to blistering, Coatings 11 (6) (2021) 704, https://doi.org/10.3390/coatings11060704.

[8] GalvInfo Note 2.12, Pretreatments for Metallic-Coated Sheet, Rev. 1.3, GalvInfo Center, International Zinc Association, Durham, NC, 2017. https://www.galvinfo.com/wp-content/uploads/sites/8/2017/05/GalvInfoNote_2_12.pdf.

[9] J. Favilla, D. Fudge, Non-Cr passivation treatment options, in: Presented at 6th International Conference on Zinc and Zinc Alloy Coated Steel Sheet (Galvatech 2004), April 4–7, Association for Iron & Steel Technology, Warrendale, PA, 2004. Chicago, IL.

[10] A. Matsuzaki, M. Yamashita, N. Hara, Effect of pretreatment film composition on adhesion of organic film on zinc coated steel sheet, Tetsu-to-Hagané 73 (6) (2009) 475–483 (in Japanese).

[11] J. De Strycker, F. Hörzenberger, C. Xhoffer, S. Siau, Three years of industrial production with Cr(VI)-free passivation treatments, in: Proc. 7th International Conference on Zinc and Zinc Alloy Coated Steel Sheet, (Galvatech 2007), November 18–22, Iron and Steel Institute of Japan, Tokyo, 2007, pp. 762–767. Osaka.

[12] E. Kudo, Y. Kinoshita, J. Uchida, The development of new inorganic chromium-free chemical for galvanized steel sheets, in: Proc. 10th International Conference on Zinc and Zinc Alloy Coated Steel Sheet (Galvatech 2015), May 31–June 4, Association for Iron & Steel Technology, Warrendale, PA, 2015, pp. 384–390. Toronto.

[13] A. Matsuzaki, T. Matsuda, R. Kaneko, K. Tsuchimoto, Development of next generation chromate-free coatings by noble inorganic polymer, in: Proc. 11th International Conference on Zinc and Zinc Alloy Coated Steel Sheet (Galvatech 2017), November 12–16, Iron and Steel Institute of Japan, Tokyo, 2017, pp. 349–355. Tokyo.

[14] T.C. Kim, I.-R. Sohn, Y.H. You, M.S. Kim, J.S. Kim, Investigation of hot-dip ZnMgAl coatings for automotive applications, in: Proc. 12th International Conference on Zinc and Zinc Alloy Coated Steel Sheet (Galvatech 2021), June 22–24, Austrian Society for Metallurgy and Materials, 2021, pp. 100–107. Virtual Conference.

[15] K. Ishizuka, K. Hayashi, H. Shindo, K. Nishimura, The effect of Mg addition to the phosphate film on the properties of prephosphated electrogalvanized steel sheets, in: Proc. 7th International Conference on Zinc and Zinc Alloy Coated Steel Sheet, (Galvatech 2007), November 18–22, Iron and Steel Institute of Japan, Tokyo, 2007, pp. 219–223. Osaka.

[16] T. Takata, et al., Corrosion resistance of chromate-free pre-painted hot-dip 55%Al-2%Mg-Zn alloy coated steel sheet, CAMP-ISIJ 28 (2015) 904.

[17] H. Kajiyama, S. Fujita, K. Fujii, M. Sakai, Development of new accelerated corrosion test for zinc-coated steel sheets used in electrical appliances, in: Proc. 6th International Conference on Zinc and Zinc Alloy Coated Steel Sheet (Galvatech 2004), April 4–7, Association for Iron & Steel Technology, Warrendale, PA, 2004, pp. 149–159. Chicago, IL.

[18] Y. Suzuki, A. Morishita, S. Nishinomiya, M. Fuda, H. Nakano, Effect of phosphoric acid added to the chemical conversion coating layer on corrosion protection in scribed areas of zinc coated steels, in: Proc. 12th International Conference on Zinc and Zinc Alloy Coated Steel Sheet (Galvatech 2021), June 22–24, Austrian Society for Metallurgy and Materials, 2021, pp. 191–198. Virtual Conference.

[19] Y. Kawamura, H. Furukawa, M. Matsumoto, K. Ueda, Influence of the anticorrosive pigment on the cut edge corrosion of chromate-free pre-painted steel sheets in rain water, in: Proc. 11th International Conference on Zinc and Zinc Alloy Coated Steel Sheet (Galvatech 2017), November 12–16, Iron and Steel Institute of Japan, Tokyo, 2017, pp. 622–626. Tokyo.

[20] S. Yamamoto, F. Kumon, T. Taomoto, H. Entani, Corrosion resistance of pre-painted Zn-6%Al-3%Mg alloy coated steel sheet, in: Proc. 7th International Conference on Zinc and Zinc Alloy Coated Steel Sheet, (Galvatech 2007), November 18–22, Iron and Steel Institute of Japan, Tokyo, 2007, pp. 659–664. Osaka.

[21] J. Schulz, O. Bendick, B. Werner, R. Wittkowski, Zn-Mg coatings—innovation for coil coated steel sheets, in: Proc. 8th International Conference on Zinc and Zinc Alloy Coated Steel Sheet (Galvatech 2011), June 21–25, Associazione Italiana di Metallurgia, Milan, 2011. Genoa, Italy.

[22] J.S. Kim, T.C. Kim, S.K. Lee, Recent progress of zinc and zinc alloy coated steel sheets in Korea, in: Proc. 12th International Conference on Zinc and Zinc Alloy Coated Steel Sheet (Galvatech 2021), June 22–24, Austrian Society for Metallurgy and Materials, 2021, pp. 9–15. Virtual Conference.

[23] R.P. Edavan, R. Kopinski, Corrosion resistance of painted zinc alloy coated steels, Corros. Sci. 51 (2009) 2429–2442.

[24] K. Ueda, A. Takahashi, Y. Kubo, Investigation Of corrosion resistance of pre-painted Zn-11%Al-3%Mg-0.2%Si alloy coated steel sheet through outdoor exposure test in Okinawa, in: Proc. 8th International Conference on Zinc and Zinc Alloy Coated Steel Sheet (Galvatech 2011), June 21–25, Associazione Italiana di Metallurgia, Milan, 2011. Genoa, Italy.

[25] D. Penney, J. Sullivan, D. Worsley, Investigation into the effect cutting technique has on the cut edge corrosion resistance of organically coated steel, in: Proc. 11th International Conference on Zinc and Zinc Alloy Coated Steel Sheet (Galvatech 2017), November 12–16, Iron and Steel Institute of Japan, Tokyo, 2017, pp. 679–686. Tokyo.

[26] T. Prosek, A. Nazarov, D. Thierry, Effect of steel and zinc coating thickness on edge creep of coil coated materials, in: Proc. 10th International Conference on Zinc and Zinc Alloy Coated Steel Sheet (Galvatech 2015), May 31–June 4, Association for Iron & Steel Technology, Warrendale, PA, 2015, pp. 255–262. Toronto.

[27] N. LeBozec, D. Thierry, T.T.X. Hang, P.T. San, Assessment of coil coated steel materials after long-term exposure in different natural weathering sites worldwide, in: Proc. 10th International Conference on Zinc and Zinc Alloy Coated Steel Sheet (Galvatech 2015), May 31–June 4, Association for Iron & Steel Technology, Warrendale, PA, 2015, pp. 271–278. Toronto.

[28] Q. Liu, J. Williams, D. Nolan, B. Shedden, R. Smith, A. Neufeld, Development of Mg-containing 55%Al-Zn coated steel for building applications, in: Proc. 10th International Conference on Zinc and Zinc Alloy Coated Steel Sheet (Galvatech 2015), May 31–June 4, Association for Iron & Steel Technology, Warrendale, PA, 2015, pp. 208–215. Toronto.

[29] AS 2331.3.13-2006, Methods of Test for Metallic and Related Coatings - Corrosion and Related Property Tests—Wet (salt fog)/Dry/Humidity, Cycle E, Standards Australia, Sydney, 2006.

[30] P.R. Seré, C. Deyá, C.I. Elsner, A.R. Di Sarli, Corrosion of painted galvanneal steel, Proc. Mater. Sci. Forum 8 (2015) 1–10.

[31] T. Steck, J. Gerdenitsch, J. Faderl, Impact of surface topography of electro galvanized steel sheet on paint appearance, in: Proc. 12th International Conference on Zinc and Zinc Alloy Coated Steel Sheet (Galvatech 2021), June 22–24, Austrian Society for Metallurgy and Materials, 2021, pp. 318–322. Virtual Conference.

[32] ASTM E284-17, Standard Terminology of Appearance, ASTM International, West Conshohocken, PA, 2017.

[33] R. Sato, S. Yamamoto, T. Kanazawa, M. Omoda, S. Ando, D. Mizuno, A.R. Setiawan, Atmospheric corrosion behaviours of automotive steel sheets exposed in East and South-East Asia regions, in: Proc. 12th International Conference on Zinc and Zinc Alloy Coated Steel Sheet (Galvatech 2021), June 22–24, Austrian Society for Metallurgy and Materials, 2021, pp. 356–364. Virtual Conference.
[34] D. Santos, H. Raminhos, M. Costa, T. Diamantino, F. Goodwin, Performance of conductive pre-primers applied on galvanized steel sheets for automotive bodies, in: SAE Congress and Exposition, April 2–6, SAE, Warrendale, PA, 2007. Detroit, MI. SAE Paper No. 2007-01-1750.
[35] ASTM D3170-2014, Standard Test Method for Chipping Resistance of Coatings, ASTM International, West Conshohocken, PA, 2017.
[36] ISO 11997-1, Paints and Varnishes—Determination of Resistance to Cyclic Corrosion Conditions—Part 1: Wet (Salt Fog)/Dry/Humid, International Organization for Standardization, Geneva, 2017.
[37] D. Santos, H. Raminhos, M. Costa, T. Diamantino, F. Goodwin, Performance of finish-coated galvanized steel sheets for automotive bodies, in: SAE International Congress, April 13–17, SAE, Warrendale, PA, 2008. Detroit, MI. SAE Paper No. 2008-01-1465.
[38] DIN 50014, Climates and Their Technical Application; Standard Atmospheres, DIN e. V., Dussseldorf, Germany, 1985.
[39] ISO 12944, Paints and Varnishes—Corrosion Protection of Steel Structures by Protective Paint Systems - Part 6: Laboratory Performance Test Methods, International Organization for Standardization, Geneva, 2018.
[40] ISO 4628-2, Paints and Varnishes—Evaluation of Degradation of Coatings- Designation of Quantity and Size of Defects, and of Intensity of Uniform Changes in Appearance—Part 2: Assessment of Degree of Blistering, International Organization for Standardization, Geneva, 2016.
[41] H. Shige, M. Iwai, Development of new prepainted galvannealed steel sheet for automotive industry, in: Proc. 6th International Conference on Zinc and Zinc Alloy Coated Steel Sheet (Galvatech 2004), April 4–7, Association for Iron & Steel Technology, Warrendale, PA, 2004, pp. 263–270. Chicago, IL.
[42] A. Schnell, F.M. Androsch, K.-H. Stellnberger, Weldable corrosion protection primers for the automotive industry, in: Proc. 6th International Conference on Zinc and Zinc Alloy Coated Steel Sheet (Galvatech 2004), April 4–7, Association for Iron & Steel Technology, Warrendale, PA, 2004, pp. 279–289. Chicago, IL.
[43] M. Vlot, N. van Landschoot, R. Bleeker, ZnAlMg coatings versus galvannealed, in: Proc. 8th Intl. Conf. on Zinc and Zinc Alloy Coated Steel Sheet (Galvatech 2011), June 21–25, AIM, Genoa, Italy, 2011.
[44] N. LeBozec, N. Blandin, D. Thierry, Accelerated corrosion tests in the automotive industry: a comparison of the performance towards cosmetic corrosion, Mater. Corros. 59 (11) (2008) 879–894.
[45] R. Autengruber, G. Luckeneder, S. Kolnberger, J. Faderl, A.W. Hassel, Corrosion behavior of press-hardened hot-dipped galvanized steel sheet, in: Proc. 8th International Conference on Zinc and Zinc Alloy Coated Steel Sheet (Galvatech 2011), June 21–25, Associazione Italiana di Metallurgia, Milan, 2011. Genoa, Italy.
[46] K. Tsuchimoto, T. Matsuda, A. Matsuzaki, N. Yoshimi, Development of multi-functional chromate-free coated steel sheets for electrical appliances, in: Proc. 10th International Conference on Zinc and Zinc Alloy Coated Steel Sheet (Galvatech 2015), May 31–June 4, Association for Iron & Steel Technology, Warrendale, PA, 2015, pp. 216–223. Toronto.
[47] R.P. Krupitzer, D.M. Urban, D.A. Witmer, Designing and manufacturing vehicles with advanced high-strength steels, in: Proc. 6th International Conference on Zinc and Zinc Alloy Coated Steel Sheet (Galvatech 2004), April 4–7, Association for Iron & Steel Technology, Warrendale, PA, 2004, pp. 31–50. Chicago, IL.
[48] S. Hironaka, K. Hosomi, H. Tanaka, T. Matsumoto, Development of Si bearing high strength galvannealed steel sheets with superior formability, in: Proc. 7th International Conference on Zinc and Zinc Alloy Coated Steel Sheet, (Galvatech 2007), November 18–22, Iron and Steel Institute of Japan, Tokyo, 2007, pp. 357–362. Osaka.
[49] T. Iwama, F. Kitano, Y. Nagataki, H. Matsuda, Y. Tanaka, Y. Hosoya, Development of galvannealed dual-phase ultra-high strength steel sheets with superior formability and spot-weldability, in: Proc. 6th International Conference on Zinc and Zinc Alloy Coated Steel Sheet (Galvatech 2004), April 4–7, Association for Iron & Steel Technology, Warrendale, PA, 2004, pp. 539–545. Chicago, IL.
[50] B. Mintz, A review on the hot-dip galvanizability and properties of TRIP and dual-phase strip steels with 590 MPa tensile strength, in: Proc. 5th International Conference on Zinc and Zinc Alloy Coated Steel Sheet (Galvatech 2001), June 26–28, Verlag Stahleisen, Duesseldorf, 2001, pp. 551–559. Brussels.
[51] O. Yakubovsky, N. Fonstein, C. Cheng, D. Bhattacharya, Effect of alloying elements on mechanical properties and coatability of multi phase steels with ferrite-bainite and ferrite-bainite plus retained austenite microstructure, in: Proc. 6th International Conference on Zinc and Zinc Alloy Coated Steel Sheet (Galvatech 2004), April 4–7, Association for Iron & Steel Technology, Warrendale, PA, 2004, pp. 547–558. Chicago, IL.
[52] F. Huber, W. Bleck, S. Scheik, M. Schleser, U. Reisgen, Investigation of galvanneal coating adherence in adhesive bonded joints using lap shear test, in: Proc. 8th International Conference on Zinc and Zinc Alloy Coated Steel Sheet (Galvatech 2011), June 21–25, Associazione Italiana di Metallurgia, Milan, 2011. Genoa, Italy.
[53] L. Bordignon, F. Goodwin, Galvanizing of TRIP-Si steels, in: Proc. 12th International Conference on Zinc and Zinc Alloy Coated Steel Sheet (Galvatech 2021), June 22–24, Austrian Society for Metallurgy and Materials, 2021, pp. 394–403. Virtual Conference.
[54] K. Han, I. Ohnuma, K. Okuda, R. Kainuma, Hardness of intermetallic compounds in the Zn-Fe binary system, in: Proc. 11th International Conference on Zinc and Zinc Alloy Coated Steel Sheet (Galvatech 2017), November 12–16, Iron and Steel Institute of Japan, Tokyo, 2017, pp. 83–85. Tokyo.
[55] L.G. Garza, C.J. Van Tyne, The effect of ζ-phase on the frictional behavior of galvannealed interstitial free sheet steels, in: Proc. 7th International Conference on Zinc and Zinc Alloy Coated Steel Sheet, (Galvatech 2007), November 18–22, Iron and Steel Institute of Japan, Tokyo, 2007, pp. 248–253. Osaka.
[56] Y. Yamasaki, Past and current trends in research on press forming of zinc-based coated sheet steels for automotive use, in: Proc. 11th International Conference on Zinc and Zinc Alloy Coated Steel Sheet (Galvatech 2017), November 12–16, Iron and Steel Institute of Japan, Tokyo, 2017, pp. 424–427. Tokyo.

[57] H. Inui, N.L. Okamoto, Crystal structure and mechanical properties of intermetallic compounds in the Fe-Zn system in the coat layer of Galvanized steel, in: Proc. 11th International Conference on Zinc and Zinc Alloy Coated Steel Sheet (Galvatech 2017), November 12–16, Iron and Steel Institute of Japan, Tokyo, 2017, pp. 6–10. Tokyo.

[58] N.L. Okamoto, Y. Hashizume, S. Michishita, H. Inui, Plastic deformation and fracture behavior of single crystals of the intermetallic compounds in the Fe-Zn system, in: Proc. 11th International Conference on Zinc and Zinc Alloy Coated Steel Sheet (Galvatech 2017), November 12–16, Iron and Steel Institute of Japan, Tokyo, 2017, pp. 200–203. Tokyo.

[59] J. Inoue, Microstructure and toughness of Fe-Zn intermetallic coatings of galvannealed steel sheet, in: Proc. 11th International Conference on Zinc and Zinc Alloy Coated Steel Sheet (Galvatech 2017), November 12–16, Iron and Steel Institute of Japan, Tokyo, 2017, pp. 109–114. Tokyo.

[60] C. Cheng, V. Krishnardula, H. Hahn, The effect of Al content in the coating on the flaking resistance of GA IF steels, in: Proc. 10th International Conference on Zinc and Zinc Alloy Coated Steel Sheet (Galvatech 2015), May 31–June 4, Association for Iron & Steel Technology, Warrendale, PA, 2015, pp. 96–103. Toronto.

[61] J. He, J. Lian, A. Aretz, N. Vajragupta, U. Hangen, F. Goodwin, S. Münstermann, Fracture properties of zinc coating layers in a galvannealed steel and an electrolytically galvanized steel, Mater. Sci. Eng. A 732 (2018) 320–325, https://doi.org/10.1016/j.msea.2018.05.084.

[62] C.S. Lin, M. Meshii, Effect of steel chemistry on the microstructure and mechanical properties of the commercial galvanneal coatings, in: Proc. 3rd International Conference on Zinc and Zinc Alloy Coated Steel Sheet (Galvatech 1995), September 17–21, Association for Iron & Steel Technology, Warrendale, PA, 1995, pp. 477–484. Chicago, IL.

[63] K.-K. Wang, C.-W. Hsu, L. Chang, D. Gan, L.-J. Chiang, Formation of Fe-Zn intermetallic phases in galvannealed Mn-Si TRIP steels, in: Proc. 10th International Conference on Zinc and Zinc Alloy Coated Steel Sheet (Galvatech 2015), May 31–June 4, Association for Iron & Steel Technology, Warrendale, PA, 2015, pp. 81–87. Toronto.

[64] J.P. Guimarães, A.H.A. Barbosa, B.M. Gonzalez, R.R. Vieira, Effect of the atmosphere dew point of continuous annealing furnaces on the quality of GA coating on dual phase steel, in: Proc. 10th International Conference on Zinc and Zinc Alloy Coated Steel Sheet (Galvatech 2015), May 31–June 4, Association for Iron & Steel Technology, Warrendale, PA, 2015, pp. 88–95. Toronto.

[65] N.L. Okamoto, D. Kashioka, M. Inomoto, H. Inui, H. Takebayashi, S. Yamaguchi, Compression deformability of Γ and ζ Fe–Zn intermetallics to mitigate detachment of brittle intermetallic coating of galvannealed steels, Scr. Mater. 69 (2013) 307–310, https://doi.org/10.1016/j.scriptamat.2013.05.003.

[66] H. Kim, S. Han, Q. Yan, T. Altan, Evaluation of tool materials, coatings and lubricants in forming galvanized advanced high strength steels (AHSS), CIRP Ann. Manuf. Technol. 57 (2008) 299–304, https://doi.org/10.1016/j.cirp.2008.03.029.

[67] H. Kim, J. Sung, F.E. Goodwin, T. Altan, Investigation of galling in forming galvanized advanced high strength steels (AHSSs) using the twist compression test (TCT), J. Mater. Process. Technol. 205 (2008) 459–468, https://doi.org/10.1016/j.jmatprotec.2007.11.281.

[68] K. Hoshino, Y. Yamasaki, W. Tanimoto, M. Nagoshi, S. Taira, N. Yoshimi, Frictional behavior of galvannealed steel sheet depending on tool material, in: Proc. 10th International Conference on Zinc and Zinc Alloy Coated Steel Sheet (Galvatech 2015), May 31–June 4, Association for Iron & Steel Technology, Warrendale, PA, 2015, pp. 104–111. Toronto.

[69] A.R. Marder, The metallurgy of zinc-coated steel, Prog. Mater. Sci. 45 (2000) 191–271.

[70] L.G. Garza, C. Van Tyne, Friction and formability of galvannealed interstitial free sheet steel, J. Mater. Process. Technol. 187 (2007) 164–168,-https://doi.org/10.1016/j.jmatprotec.2006.11.062.

[71] H. Kim, S. Han, F.E. Goodwin, K. Kim, T. Altan, Evaluation of die coatings and stamping lubricants in forming galvanized advanced/ultra high strength steels (A/UHSS) using the strip drawing test (SDT), in: Proc. International Deep Drawing Research Group IDDRG 2009 International Conference, 1–3 June, Institute for Metal Forming Technology, Stuttgart, 2009. Golden, CO, USA.

[72] D.-S. Han, J.-H. Ahn, M.-H. Hong, H.-W. Lee, Effect of phosphate amount and δ1/ζ phase ratio on formability of GA sheets for automotive outer panels, in: Proc. 7th International Conference on Zinc and Zinc Alloy Coated Steel Sheet, (Galvatech 2007), November 18–22, Iron and Steel Institute of Japan, Tokyo, 2007, pp. 266–271. Osaka.

[73] D.J. Paik, G.M. Michal, The effects of phosphate treatment on coating delamination of galvannealed steel subjected to biaxial tensile and plane strain deformation, in: S. Seal (Ed.), Proc. Symposium on Surface Engineering in Materials Science II, San Diego, March, Met. Soc. AIME, Pittsburgh, PA, 2003, pp. 123–131.

[74] G. Angeli, R. Brisberger, M. Bülter, L. Diez, C. Filthaut, W. Fischer, T. Koll, T. Maiwald, C. Pesci, A. Richter, J. Schulz, U. Stellmacher, N. van Landschoot, Zinc-magnesium-aluminium for automotive industry, in: Proc. 9th International Conference on Zinc and Zinc Alloy Coated Steel Sheet (Galvatech 2013), September 23–27, Chinese Society for Metals, Metallurgical Industry Press, Beijing, 2013, pp. 632–636. Beijing.

[75] X. Hao, C. Yue, S. Jiang, Q. Zhang, Y. Bai, L. Li, Effect of Mg content on the microstructure and properties of a Zn-Al-Mg alloy, in: A. Patty, P. Zhou (Eds.), Advances in Materials Science, Energy Technology and Environmental Engineering, Taylor & Francis Group, London, 2017, pp. 405–408, ISBN: 978-1-138-19668-1.

[76] B. Berkhout, C. Dane, Advanced zinc coatings for improved performance and press shop efficiency, in: Proc. 5th Intl. Conf. on Steel in Cars and Trucks 2017, June 19–21, Steel Institute VDEh, Dusseldorf, 2017. Amsterdam.

[77] T. Koll, F. Beier, T. Boeddeker, U. Eggers, C.F.M. Pohl, S. Schulz, C. Schlegel, Application properties of ZM coatings for car production, in: Proc. 9th International Conference on Zinc and Zinc Alloy Coated Steel Sheet (Galvatech 2013), September 23–27, Chinese Society for Metals, Metallurgical Industry Press, Beijing, 2013, pp. 625–631. Beijing.

[78] A.D. Pearson, G. Dalton, A/SP tribology team die wear study, in: Proc. Great Designs in Steel 2007, March 7, American Iron and Steel Institute, Southfield, MI, 2007. Livonia, MI.

[79] T. Nakamura, H. Takebayashi, M. Nakata, T. Toki, Effect of warm forming on GA steel sheet powdering, in: Proc. 7th International Conference on Zinc and Zinc Alloy Coated Steel Sheet, (Galvatech 2007), November 18–22, Iron and Steel Institute of Japan, Tokyo, 2007, pp. 260–265. Osaka.

[80] M. Vlot, M. Zuijderwijk, M. Toose, L. Elliott, R. Bleeker, T. Maalman, Hot dip ZnAlMg coatings: microstructure and forming properties, in: Proc. 7th International Conference on Zinc and Zinc Alloy Coated Steel Sheet, (Galvatech 2007), November 18–22, Iron and Steel Institute of Japan, Tokyo, 2007, pp. 574–579. Osaka.

[81] Y.B. Park, I.G. Kim, S.G. Kim, W.T. Kim, T.C. Kim, M.S. Oh, J.S. Kim, Orientation dependence of cracking in hot-dip Zn-Al-Mg alloy coatings on a sheet steel, Metall. Mater. Trans. A 48A (2017) 1013–1020.

[82] M. Ahmadi, B. Salgın, B.J. Kooi, Y. Pei, Mechanical properties and cracking behavior of hot-dip galvanized ZnAlMg coatings, in: Proc. 12th International Conference on Zinc and Zinc Alloy Coated Steel Sheet (Galvatech 2021), June 22–24, Virtual Conference, Austrian Society for Metallurgy and Materials, 2021, pp. 877–884.

[83] M. Ahmadi, B. Salgın, M. Ahmadi, B.J. Kooi, Y. Pei, Unraveling dislocation mediated plasticity and strengthening in crack-resistant ZnAlMg coatings, Int. J. Plast. 144 (2021), 103041.

[84] T.E. Torok, P.W. Shin, A.R. Borzillo, Method of Improving the Ductility of the Coating of al Aluminum-Zinc Alloy Coated Ferrous Product, US Patent No. 4,287,008, 1981.

[85] K. Yoshida, T. Ooi, C. Kato, New prepainted 55% Al-Zn alloy coated steel sheet with high formability and high corrosion resistance, in: Proc. 6th International Conference on Zinc and Zinc Alloy Coated Steel Sheet (Galvatech 2004), April 4–7, Association for Iron & Steel Technology, Warrendale, PA, 2004, pp. 319–325. Chicago, IL.

[86] D.J. Willis, M.S. Salon, Method of Enhancing the Ductility of Aluminum-Zinc Alloy Coating on Steel Strip, US Patent No. 5,049,202, 1991.

[87] H.N. Hahn, T.C. Simpson, Development of SLEEK AZ coated sheet product, in: Proc. 6th International Conference on Zinc and Zinc Alloy Coated Steel Sheet (Galvatech 2004) Chicago, April, 2004, pp. 327–337. AIST Proc. 6th International Conference on Zinc and Zinc Alloy Coated Steel Sheet (Galvatech 2004), April 4–7, 2004, Chicago, IL, Association for Iron & Steel Technology, Warrendale, PA, pp. 327–337.

[88] J.X. Li, D.R. Johnston, M.C. Palastro, Crack mechanisms of 55%Al-Zn coatings, in: Presented at 6th International Conference on Zinc and Zinc Alloy Coated Steel Sheet (Galvatech 2004), April 4–7, Association for Iron & Steel Technology, Warrendale, PA, 2004. Chicago, IL.

[89] F.E. Goodwin, E.A. Silva, An overview of North American zinc-based sheet steel coatings production: status and opportunities, in: Proc. 7th International Conference on Zinc and Zinc Alloy Coated Steel Sheet (Galvatech 2007), November 18–22, Iron and Steel Institute of Japan, Osaka, 2007, pp. 6–13. Proc. 7th International Conference on Zinc and Zinc Alloy Coated Steel Sheet, (Galvatech 2007), November 18–22, 2007, Osaka, Iron and Steel Institute of Japan, Tokyo, pp. 6–13.

[90] B. Hance, Advanced high-strength steel (AHSS) performance level definitions and targets, SAE Int. J. Mater. Manuf. 11 (4) (2018) 505–516, https://doi.org/10.4271/2018-01-0629.

[91] N. Fujita, T. Nonaka, T. Tomokiyo, H. Taniguchi, K. Goto, K. Yamazaki, Development of ultra-high strength steel sheets with tensile strength of 980MPa, in: SAE 2007 World Congress, April 16–19, SAE, Warrendale, PA, 2007. Detroit, MI. Paper Number 2007-01-0341.

[92] Y. Chang, C. Haase, D. Szeliga, L. Madej, U. Hangen, M. Pietrzyk, W. Bleck, Compositional heterogeneity in multiphase steels: characterization and influence on local properties, Mater. Sci. Eng. A 827 (2021), 142078.

[93] P. Costa, G. Altamirano, A. Salinas, S. Gonzalez-Gonzalez, F. Goodwin, Optimization of the continuous galvanizing heat treatment process in ultra-high strength dual phase steels using a multivariate model, Metals 9 (2019) 703.

[94] M.M. McCosby, E.A. Silva, D.P. Hoydick, U. S. Steel 3rd generation steel: customer-friendly chemistry, properties, and manufacturability enabled by unique processing, in: Proc. 12th International Conference on Zinc and Zinc Alloy Coated Steel Sheet (Galvatech 2021), June 22–24, Austrian Society for Metallurgy and Materials, 2021, pp. 134–144. Virtual Conference.

[95] R. Michelena, Resistance spot welding of galvanized/coated materials, Met. Form. (2021) 28–30.

[96] M. Delgado, M.P. Guerrero-Mata, V.H. Baltazar, R. Garza, Study of the resistance spot welding on a galvannealed HSLA steel for automotive applications, in: Proc. 8th International Conference on Zinc and Zinc Alloy Coated Steel Sheet (Galvatech 2011), June 21–25, Associazione Italiana di Metallurgia, Milan, 2011. Genoa, Italy.

[97] Resistance Welding Manufacturing Alliance (RWMA), Resistance Welding Manual, revised, fourth ed., American Welding Society, Miami, FL, 2003, ISBN: 978-0-9624382-0-2.

[98] N. Fujita, Recent developments and outlook on zinc and zinc alloy coated steel sheets in Japan, in: Proc. 12th International Conference on Zinc and Zinc Alloy Coated Steel Sheet (Galvatech 2021), June 22–24, Austrian Society for Metallurgy and Materials, 2021, pp. 16–23. Virtual Conference.

[99] A. Ghatei Kalashami, S. Zhang, M. Shoajee, A.R.H. Midawi, F. Goodwin, N.Y. Zhou, Failure behavior of resistance spot welded advanced high strength steel: the role of surface condition and initial microstructure, J. Mater. Process. Technol. 299 (2022) 117370.

[100] AWS D8.9M:2012, Test Methods for Evaluating the Resistance Spot Welding Behavior of Automotive Sheet Steel Materials, American Welding Society, Miami, FL, 2012.

[101] R. Ashiri, M.A. Haque, C.-W. Ji, M. Shamanian, H.R. Salimijazi, Y.-D. Park, Supercritical area and critical nugget diameter for liquid metal embrittlement of Zn-coated twining induced plasticity steels, Scr. Mater. 109 (2015) 6–10, https://doi.org/10.1016/j.scriptamat.2015.07.006.

[102] A. Chakraborty, H. Ghassemi-Armaki, Evaluation of initiation and propagation of LME cracks on the galvanized 3G-AHSS using interrupted resistance spot-welding method, in: Proc. 12th International Conference on Zinc and Zinc Alloy Coated Steel Sheet (Galvatech 2021), June 22–24, Austrian Society for Metallurgy and Materials, 2021, pp. 829–837. Virtual Conference.

[103] Y.G. Kim, I.J. Kim, J.S. Kim, Y.I. Chung, D.Y. Choi, Evaluation of surface crack in resistance spot welds of Zn-coated steel, Mater. Trans. 55 (1) (2014) 171–175.

[104] M. Arndt, P. Kürnsteiner, T. Truglas, J. Duchoslav, K. Hingerl, D. Stifter, C. Commenda, J. Haslmayr, S. Kolnberger, J. Faderl, H. Groiss, Nanoscale investigation of microcracks and grain boundary wetting in press hardened galvanized 20MnB8 steel, arXiv (2020). arXiv:2007.01136.

[105] J. Wang, G.K.H. Madsen, R. Drautz, Grain boundaries in bcc-Fe: a density-functional theory and tight-binding study, Model. Simul. Mater. Sci. Eng. 26 (2018), 025008.

[106] M.H. Razmpoosh, E. Biro, D.L. Chen, F. Goodwin, Y. Zhou, Liquid metal embrittlement in laser lap joining of TWIP and medium manganese TRIP steel: the role of stress and grain boundaries, Mater Charact 145 (2018) 627–633, https://doi.org/10.1016/j.matchar.2018.09.018.

[107] M.H. Razmpoosh, A. Macwan, E. Biro, D.L. Chen, Y. Peng, F. Goodwin, Y. Zhou, Liquid metal embrittlement in laser beam welding of Zn-coated 22MnB5 steel, Mater. Des. 155 (2018) 375–383.

[108] C. DiGiovanni, A. Ghatei Kalashami, F. Goodwin, E. Biro, N.Y. Zhou, Occurrence of sub-critical heat affected zone liquid metal embrittlement in joining of advanced high strength steel, J. Mater. Process. Technol. 288 (2021), 116917.

[109] C. DiGiovanni, L. He, U. Pistek, F. Goodwin, E. Biro, N.Y. Zhou, Role of spot weld electrode geometry on liquid metal embrittlement crack development, J. Manuf. Process. 49 (2020) 1–9.

[110] D. Kim, S.-J. Kim, J.-H. Kang, D.-Y. Choi, Effect of Zn-coating process on liquid metal embrittlement of TRIP steel, in: Proc. 12th International Conference on Zinc and Zinc Alloy Coated Steel Sheet (Galvatech 2021), June 22–24, Austrian Society for Metallurgy and Materials, 2021, pp. 859–866. Virtual Conference.

[111] D. Bhattacharya, Liquid metal embrittlement during resistance spot welding of Zn-coated high-strength steels, Mater. Sci. Technol. 34 (15) (2018) 1809–1829, https://doi.org/10.1080/02670836.2018.1461595.

[112] C. DiGiovanni, E. Biro, N.Y. Zhou, Impact of liquid metal embrittlement cracks on resistance spot weld static strength, Sci. Technol. Weld. Join. 24 (3) (2019) 218–224, https://doi.org/10.1080/13621718.2018.1518363.

[113] E. Wintjes, C. DiGiovanni, L. He, S. Bag, F. Goodwin, A. Macwan, E. Biro, Y. Zhou, Effect of multiple pulse resistance spot welding schedules on liquid metal embrittlement severity, J. Manuf. Sci. Eng. 141 (10) (2019) 101001. 9 pp. https://doi.org/10.1115/1.4044099.

[114] J.Y. Cheon, V. Vijayan, S. Murgun, Y.D. Park, J.H. Kim, J.Y. Yu, C. Ji, Optimization of pulsed current in resistance spot welding of Zn-coated hot-stamped boron steels, J. Mech. Sci. Technol. 33 (4) (2019) 1615–1621.

[115] C. DiGiovanni, S. Bag, C. Mehling, K.W. Choi, A. Macwan, E. Biro, N.Y. Zhou, Reduction in liquid metal embrittlement cracking using weld current ramping, Weld. World 63 (6) (2019) 1583–1591, https://doi.org/10.1007/s40194-019-00790-5.

[116] K. Pearson, B. Barber, M. Shojaee, A.R.H. Midawi, O. Sherepenko, H. Ghassemi-Armaki, E. Biro, Industrially viable methods to reduce liquid metal embrittlement severity in resistance spot-welded third-generation advanced high-strength steel, in: Sheet Metal Welding Conference XIX November 2–4, American Welding Society, Livonia, MI, Miami, FL, 2021.

[117] N. Yurioka, H. Suzuki, S. Ohshita, S. Saito, Determination of necessary preheating temperature in steel welding, Weld J. 62 (6) (1983) 147-s–153-s.

[118] E. Wintjes, C. DiGiovanni, L. He, E. Biro, N.Y. Zhou, Quantifying the link between crack distribution and resistance spot weld strength reduction in liquid metal embrittlement susceptible steels, Weld. World 63 (2019) 807–814, https://doi.org/10.1007/s40194-019-00712-5.

[119] J. Szonn, C. Dornscheidt, J. Artel, C. Sasse, Welding of high-strength steel grades with X-Pro® laser welder in continuous strip processing lines, in: AISTech 2020-Proceedings of the Iron & Steel Technology Conference, August 31–September 2, Association for Iron & Steel Technology, Warrendale, PA, 2020, https://doi.org/10.33313/380/124. Virtual Conference.

[120] P. Hurtig, T. Nilsson, J. Larsson, Development of cold rolled martensitic electrogalvanized steels, in: Proc. Intl. Conf. on New Developments in Advanced High-Strength Sheet Steels, June 15–18, Association for Iron & Steel Technology, Warrendale, PA, 2008, pp. 137–146. Orlando.

[121] A. Sharma, U.K. Mohanty, M. Tanaka, T. Suga, Mechanism of gap bridgeability in lap-fillet laser-arc hybrid welding, Lasers Manuf. Mater. Process. 8 (2021) 355–371, https://doi.org/10.1007/s40516-021-00150-6.

[122] N. Lun, D.C. Saha, A. Macwan, H. Pan, L. Wang, F. Goodwin, N. Zhou, Microstructure and mechanical properties of fiber laser welded medium manganese TRIP steel, Mater. Des. 131 (2017) 450–459.

[123] Q.L. Cui, D. Parkes, D. Westerbaan, S.S. Nayak, Y. Zhou, D. Liu, F. Goodwin, S. Bhole, D.L. Chen, Effect of coating on fiber laser welded joints of DP980 steels, Mater. Des. 90 (2016) 516–523.

[124] H. Kang, L. Cho, C. Lee, B.C. De Cooman, Zn penetration in liquid metal embrittled TWIP steel, Metall. Mater. Trans. A 47 (2016) 2885–2905.

[125] M. Shehryar Khan, Y.-H. Cho, M. Alfano, F. Goodwin, E. Biro, Y. Zhou, A systematic study on the effect of coating type and surface preparation on the wettability of Si-bronze brazing filler metal on GI and GA-coated DP600, Surf. Coat. Technol. 425 (2021), 127735.

[126] D. Narsimhachary, K. Dutta, S.M. Shariff, G. Padmanabham, A. Basu, Mechanical and microstructural characterization of laser weld-brazed AA6082-galvanized steel joint, J. Mater. Process. Technol. 263 (2019) 21–32.

Chapter 13

Zn-coated high-strength steel

13.1 High-temperature Zn-coated steel processing

13.1.1 Advanced high-strength steel (AHSS)

According to Mallick [1], approximately 55% of a typical US car's weight is due to steel. The advantages of steel over other materials being considered for use in automotive body structures and panels are as follows:

1. Low cost and high modulus (207 GPa), which is much higher than all other competitive structural metals.
2. Wide range of strength and ductility, which can be achieved by alloying, work hardening, and heat treatment.
3. Excellent formability, which makes them highly suitable for high production rate forming operations.
4. Excellent weldability by resistance spot welding.
5. Improved corrosion resistance can be achieved by new types of zinc alloy coatings.

It is not the objective of this section to review in detail the development of high-strength steels for automotive applications. There are many excellent reviews on the subject, e.g., [2–6]. What follows is a brief overview of the types of steel being used for automotive use. Fig. 13.1 illustrates a typical body-in-white use of high-strength steels, and Fig. 13.2 plots the elongation/strength relationship for high-strength steel (AHSS) applications. Table 13.1 lists the steel types used for automotive applications. Missing from this table is third-generation advanced high-strength steel that is also discussed below.

13.1.1.1 Basic principles of heat-treatable AHSS

AHSS steels began as a result of some significant early work on Dual-Phase steels in the late 1970s and early 1980s. Fig. 13.3, a classic figure taken from the original work of Rashid and Rao [9], shows the effects of intercritical annealing followed by quenching on the mechanical properties of a conventional HSLA steel. They annealed the steels in the two phase $\alpha + \gamma$ region in the equilibrium phase diagram, between the Ac_1 and the Ac_3 temperature for the specific steel composition. Fig. 13.4 shows a schematic representation of an intercritical annealing heat treatment. As will be shown later, the selected intercritical annealing temperature (IAT) will determine the amount of austenite in the steel at temperature, and thus the amount of second phase on quenching. The results in Fig. 13.3 show stress-strain data for a plain carbon steel, an HSLA steel (SAE 980X), and the same SAE 980X steel after intercritical annealing and quenching to produce a dual-phase steel (referred to as GM980X). In contrast to the HSLA steel, the dual-phase (DP) steel exhibits continuous yielding and a significant increase in elongation with essentially the same ultimate tensile strength. The surprising result of continuous yielding with an increase in tensile strength while maintaining or improving ductility generated extensive research on DP steels [11–13]. It was found that the fraction second phase of intercritically annealed DP steel increased the UTS, Fig. 13.5, and reduced the uniform elongation, Fig. 13.6. One important finding, as illustrated in the data in Fig. 13.7, was the contribution of retained austenite on the deformation behavior of DP steels. Specifically, it was observed [15] that DP steels contain retained austenite and that the ductility of DP steels increased with greater retained austenite content. Not surprisingly, it has been reported that up to 5% retained austenite was found in low-carbon martensite (lath martensite) [16]. The data in Fig. 13.3 illustrate the basis on which new developments in AHSS for automotive applications, including transformation-induced plasticity (TRIP) steels, occurred.

It has been well established that controlled mixed microstructure, e.g., ferrite/pearlite, ferrite/martensite, bainite/martensite, and ferrite/bainite/martensite, offers unique property possibilities [17]. Typically, volume fraction second phase increases the tensile strength and reduces the uniform elongation, see e.g., Figs. 13.5 and 13.6; however, it has been shown that increased retained austenite can improve ductility [15]. The well-known Hall-Petch relationship for steels shows that strength will increase with a finer grain size although ductility will decrease.

The Metallurgy of Zinc Coated Steels. https://doi.org/10.1016/B978-0-323-99984-7.00016-6

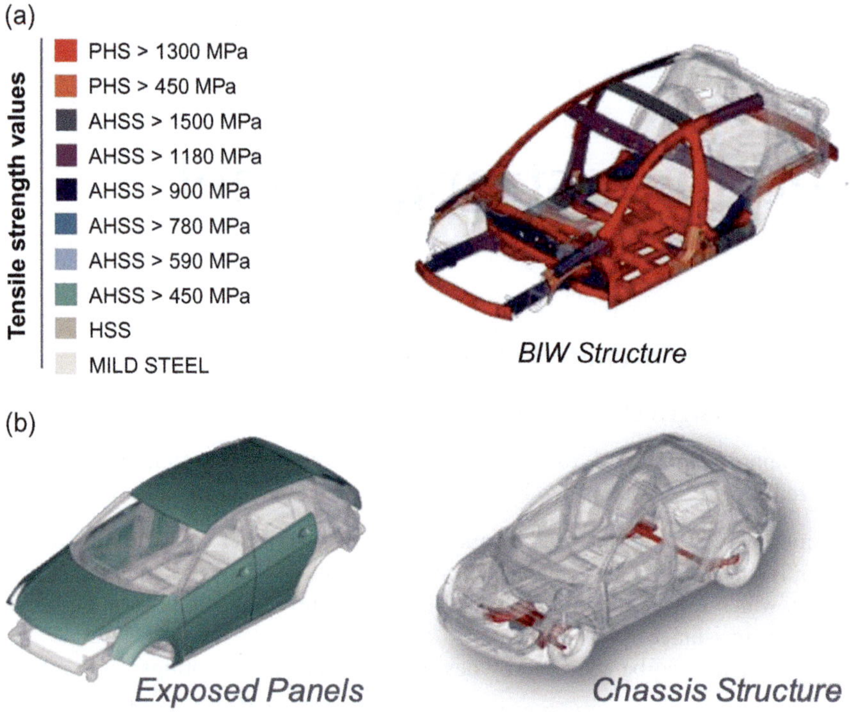

FIG. 13.1 Various high-strength steels used in Arcelor Mittal Body-in-White (BIW) structure, exposed panels, and chassis structure [7].

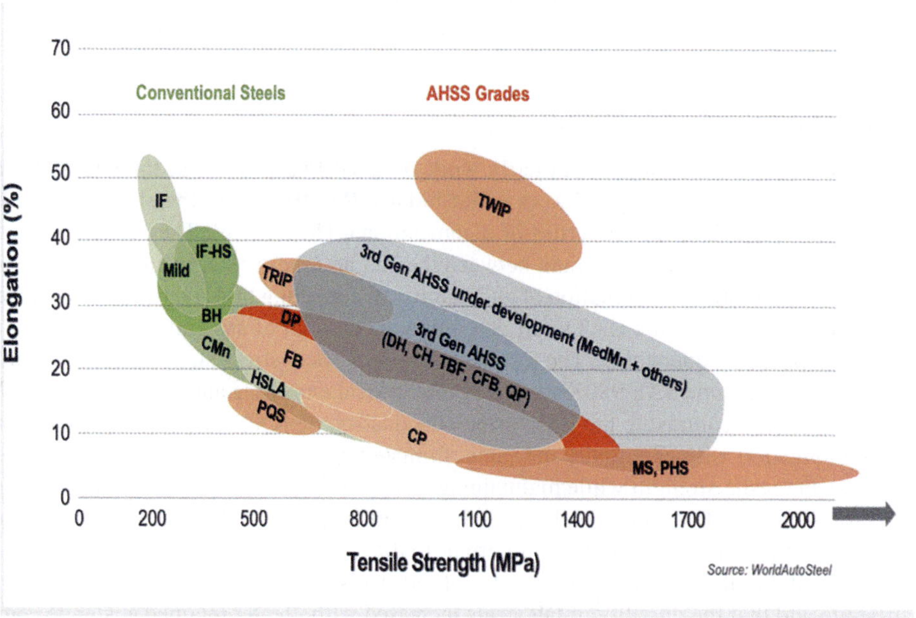

FIG. 13.2 Conventional, current, and prospective automotive steels and their corresponding elongations and strengths [8].

$$\sigma_y = \sigma_i + k_y d^{-1/2}$$

It has also been shown that martensite packet size controls the strength of martensite for a given composition [16]. Fig. 13.8 shows the microstructural morphology of low-carbon martensite. A sodium bisulfite etched light optical micrograph is seen in Fig. 13.8 (top) where the various structural features, blocks and packets, are shown within a prior austenite grain. Laths are difficult to see at this magnification, but it is apparent that within each packet, there is a substructure that has been identified as laths in TEM studies [16]. Fig. 13.8 (bottom) is a schematic of the low-carbon martensite microstructure that

TABLE 13.1 Types of automotive steels.

Low-carbon and conventional high-strength steel
Mild low-carbon steel (LC) Solid solution strengthened (SSS) Bake hardenable (BH) High-strength low alloy (HSLA) Ferrite/bainite (FB) Interstitial Free (IF)
First-generation advanced high-strength steels
Dual phase (DP) Transformation-induced plasticity (TRIP) Complex phase (CP) Fully martensitic (MS)
Second-generation advanced high-strength steels
Twinning-induced plasticity (TWIP) Lightweight steels with induced plasticity (L-IP)
Third-generation advanced high-strength steels
Enhanced DP Modified TRIP High Mn—TRIP Lower Mn—TWIP/TRIP TRIP-assisted bainitic ferrite (TBF) Quench and partition (QP)

(Adapted and updated C.D. Horvath, Advanced steels for lightweight automotive structures, in: Chapter 2, Materials, Design and Manufacturing for Lightweight Vehicles, Woodhead Publishing, 2021.)

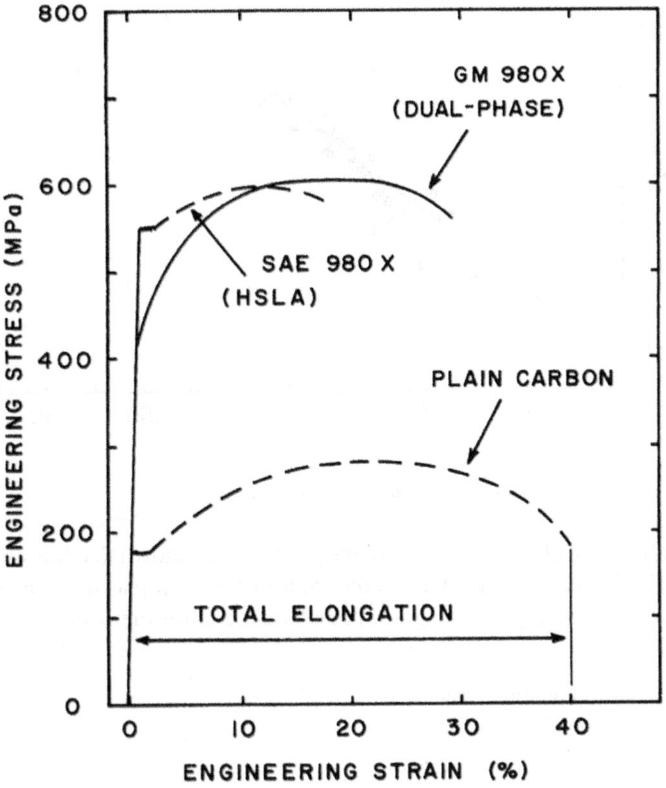

FIG. 13.3 Comparison of the stress-strain curve for a dual-phase steel with a plain carbon steel and an HSLA steel [9].

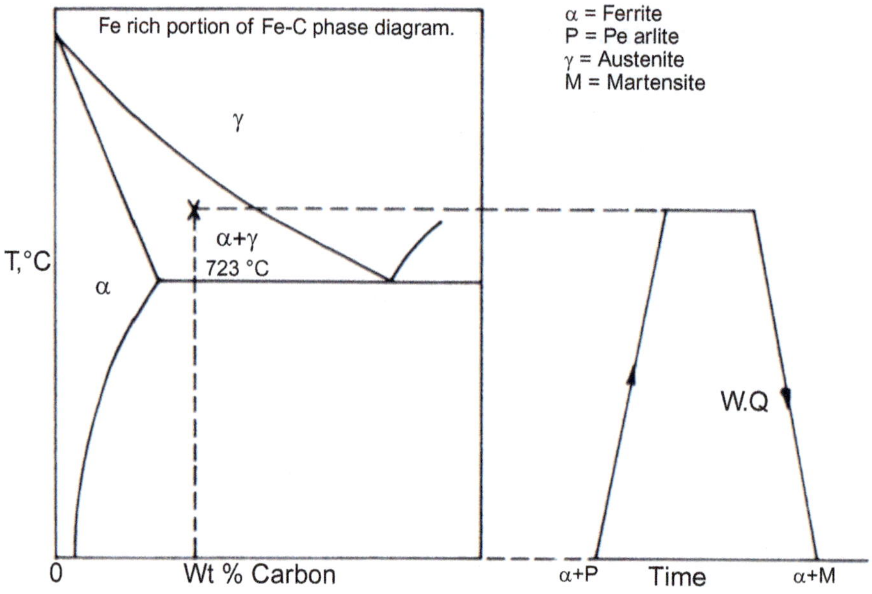

FIG. 13.4 Schematic representation of an intercritical annealing heat treatment cycle to produce dual-phase steel [10].

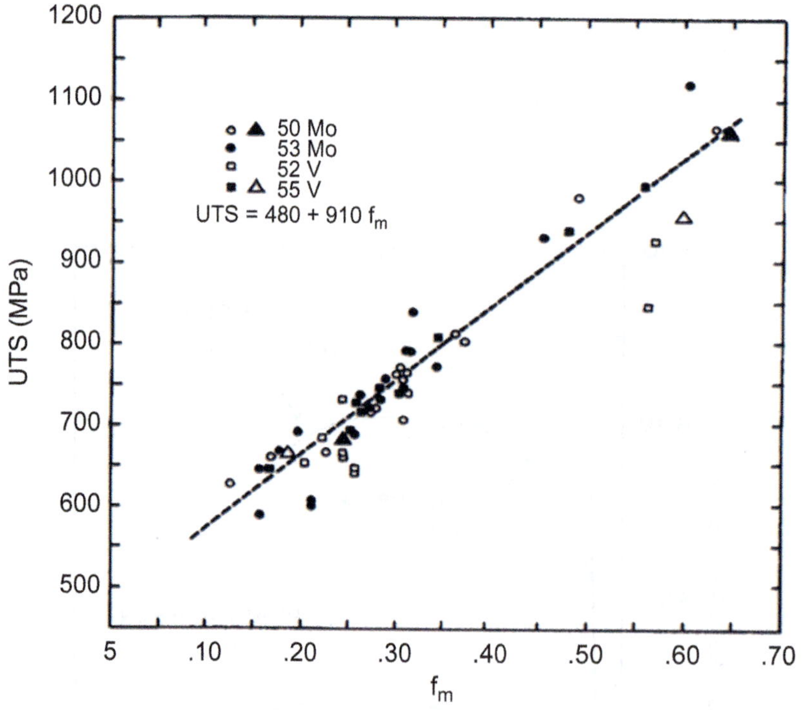

FIG. 13.5 The effect of volume fraction martensite on the ultimate tensile strength [14].

shows laths within packets or blocks within the prior austenite grain. It was shown that the structural feature that controlled strength was the packet size [16]. In effect, when the martensite transforms, it partitions the original prior austenite grain, and these smaller entities, packets with high-angle grain boundaries, control the strength of the martensite.

To sum up:

- increased strength in martensitic containing steels is controlled by:
  1. increased volume fraction second phase
  2. decreased ferrite grain size
  3. decreased packet size

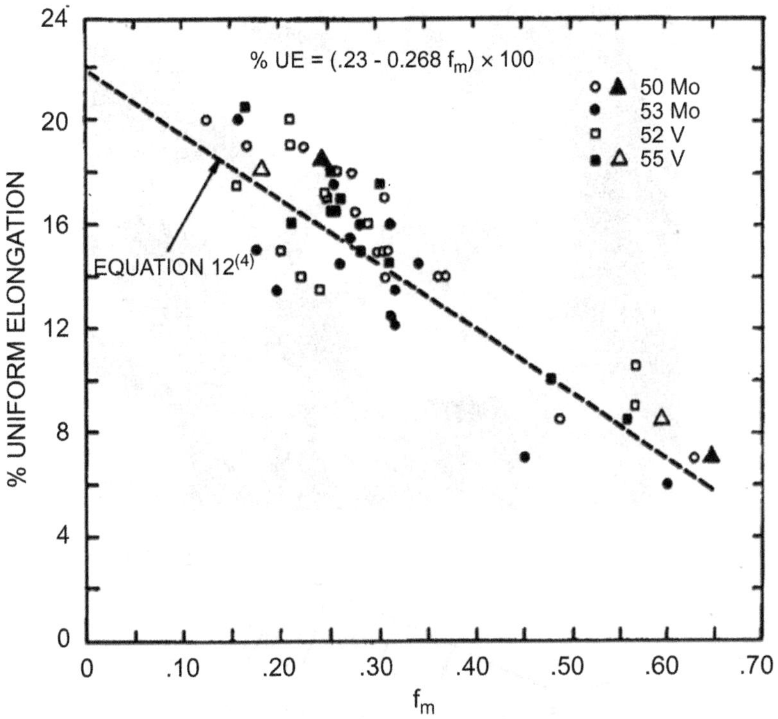

FIG. 13.6 The effect of volume fraction martensite on uniform elongation [14].

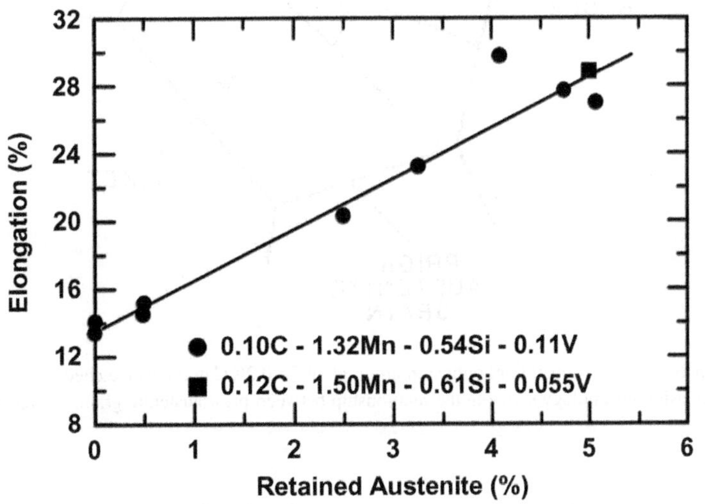

FIG. 13.7 The effect of retained austenite on the ductility of two low-carbon dual-phase steels [15].

- increased uniform elongation is controlled by:
  1. decreased volume fraction second phase
  2. increased ferrite grain size
  3. increased packet size
  4. most significantly, increased retained austenite

Therefore, in order to obtain high strength with improved ductility, it is important to refine the structure and increase the amount of retained austenite. This can be achieved by the use of an intercritical anneal, see Fig. 13.4. A typical dual-phase structure is seen in Fig. 13.9 [18]. Fig. 13.10 shows a pseudo binary Fe(Me)-C phase diagram with Me representing a second alloying element in Fe. It is seen that as the temperature is increased in the two phases $\alpha + \gamma$ region, by the lever rule, the amount of γ increases, but the composition of γ, i.e., % C content, decreases. In order to determine the ability to form both martensite and retained austenite, it is necessary to determine the composition of the γ phase at temperature and its

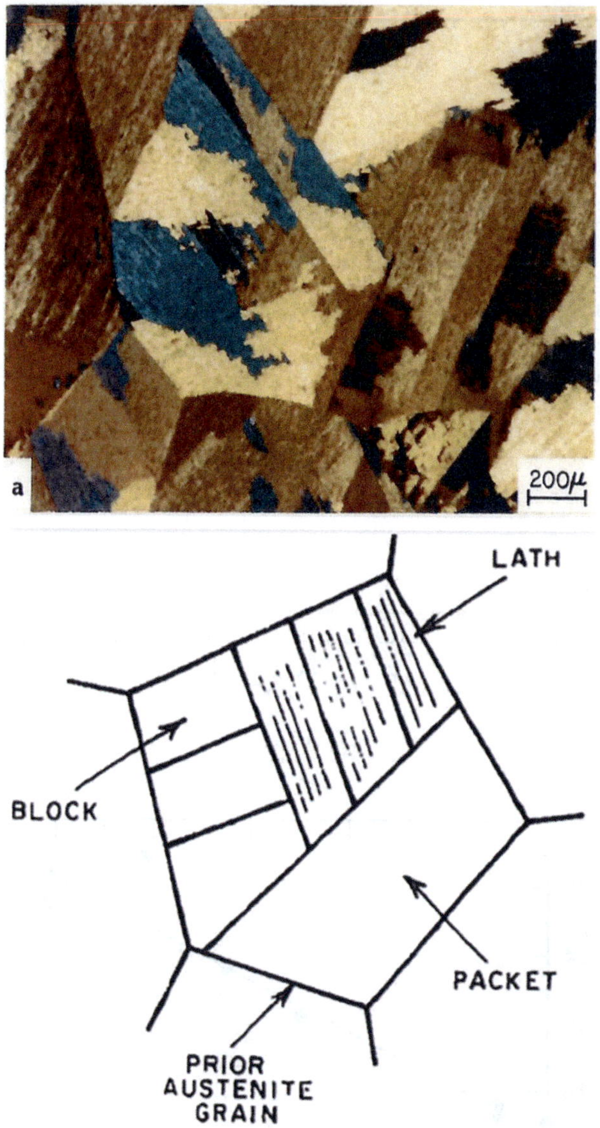

FIG. 13.8 Lath martensite morphology. Top: Actual light optical micrograph of Fe-0.2%C martensite etched with sodium bisulfite [16], and Bottom: a schematic of the low-carbon martensite morphology showing the relationship between prior austenite grain, packet, block, and lath [17].

associated Cooling Transformation (CT) diagram or Continuous Cooling Transformation (CCT) diagram, see Fig. 13.11. Then the cooling rate of the component can be overlayed to determine the microstructure.

Fig. 13.11 schematically shows the interrelationship between heat treating cycle and the CT diagram for a hypothetical Fe-C steel and the austenite composition obtained prior to cooling. It must be emphasized that the CT diagram represents the composition of the austenite at the intercritical annealing temperature, not the overall composition of the alloy. Superimposed are the heat-treating cycles for several AHSS steels such as: Quench and Tempered (QT), Dual Phase (DP), Quench and Partition (QP), and Transformation Induced Plasticity (TRIP). Also included in the diagram is how different alloy additions will affect both the Ac_1 and Ac_3 boundaries of the two phase $\alpha + \gamma$ region, as well as how alloy additions effect the C curve stability of ferrite and bainite, along with their effect on the martensite start temperature (M_s). For example, additions of Al in the steel will raise both the Ac_1 and Ac_3 temperatures while moving the ferrite C curve to the left, making it more difficult to quench the austenite in the steel to martensite or bainite. On the other hand, additions of C, Mn, Cr, Nb move both the ferrite and bainite C curves to the right, making it easier to form martensite; in addition, these same element additions lower the M_s temperature so that it will be easier to retain austenite. Fig. 13.12 is a sketch

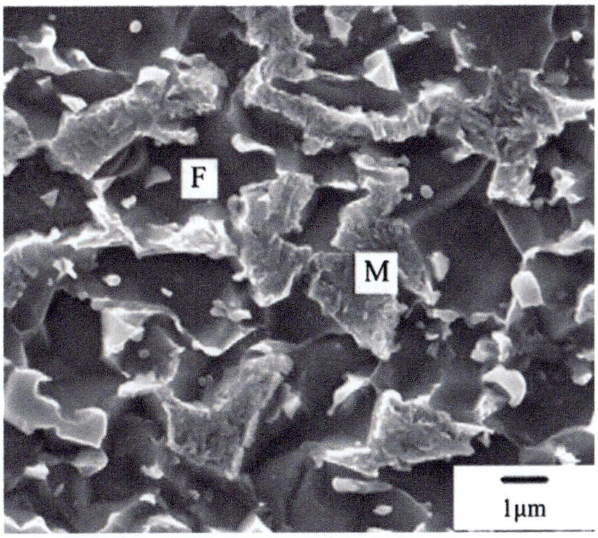

FIG. 13.9 Scanning electron image of a dual-phase steel microstructure showing ferrite (F) and martensite (M) [18].

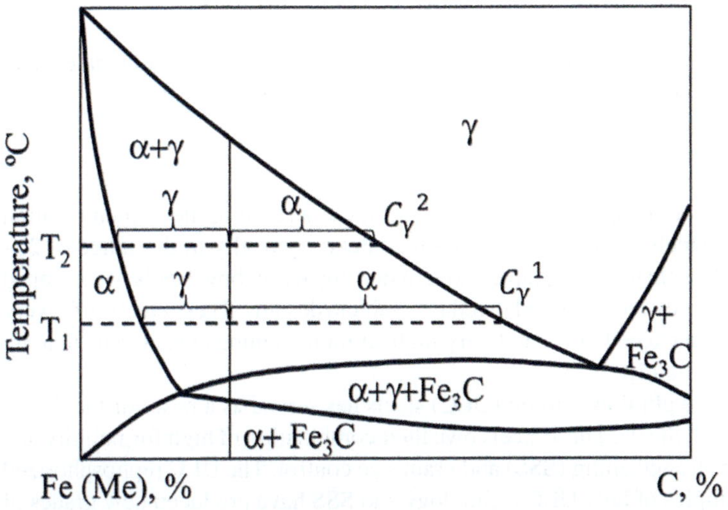

FIG. 13.10 Pseudo binary Fe(Me)-C phase diagram with Me representing a second alloying element in Fe [6].

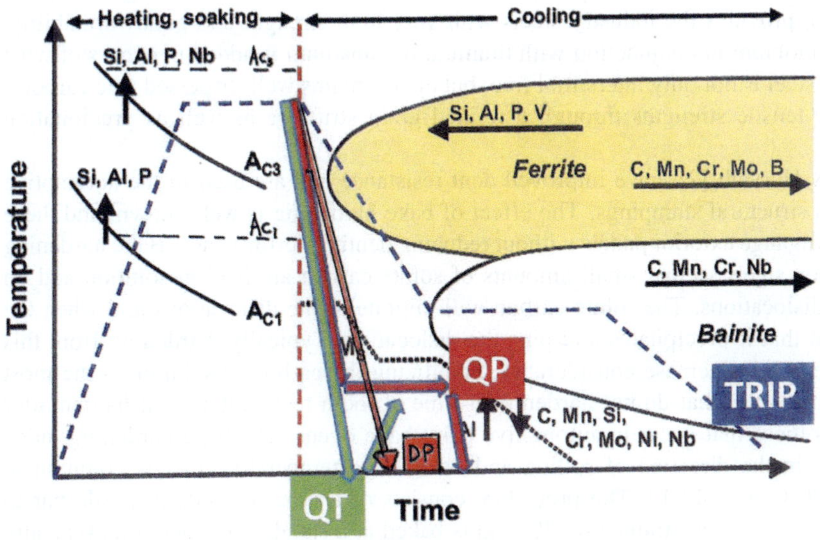

FIG. 13.11 Relationship between heat treating cycle and the CT diagram for various AHSS heat treatments to produce QT, DP, TRIP, QP. *(Adapted and modified from W. Bleck, K. Phiu-On, Effects of microalloying in multi phase steels for car body manufacture, Microstruct. Text. Steels Springer, London, 2009, pp. 145–163.)*

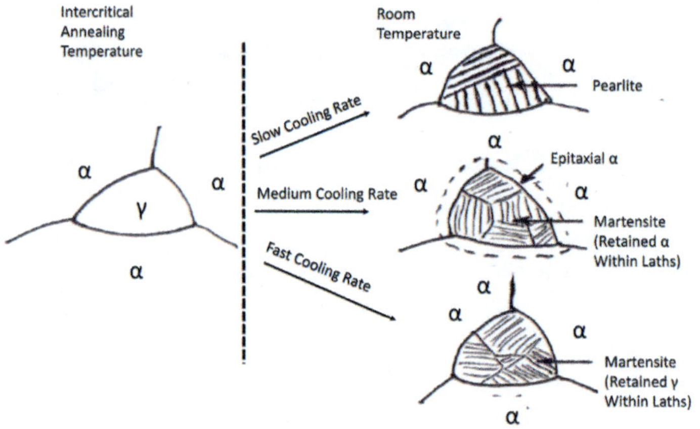

FIG. 13.12 Sketch diagram showing microstructure produced by cooling austenite in intercritically annealed steel at various carbon contents.

diagram showing the microstructures produced by cooling austenite in intercritically annealed steel at various rates. Not obvious is the retained austenite that exists between martensite laths but detected by X-ray and other means.

13.1.1.2 Interstitial free (IF) steels

Homogeneous alloys of very formable steels are now commonly available to the industry as a result of the implementation continuous casting technologies. Interstitial free (IF) steels with very low carbon contents, typically in the range of 20–50 ppm (0.002%–0.005%), and manganese levels of less than 0.25%, have had a major impact on how steels are stamped and have enabled the use of the complex features commonly incorporated in today's vehicle designs. These steels offer very high elongations that can approach and sometimes even exceed 50%, and very high strain hardening coefficients that are advantageous when stamping very difficult-to-form parts [2].

The use of vacuum degassing technology to produce ultralow-carbon (ULC) steels has served as a base for the development of new classes of highly formable High-Strength Steels. These steels owe their combination of high formability and high strength largely through the use of solid solution strengthening (SSS) and grain size control. The ULC rephosphorized steels and some bake-hardenable steels are good examples of how ULC technology and SSS have produced new grades of steel that meet many of the needs of modern automotive designs. In some cases, multiple strengthening mechanisms are used in conjunction with the ULC technology. The addition of microalloying elements such as niobium and/or titanium and vanadium, in conjunction with the SSS elements such as phosphorus, manganese, and silicon, along with grain sizes of 8 or finer (based on ASTM E112), provides the industry steels with very high strength and good formability. With respect to microalloying, niobium or niobium in conjunction with titanium or vanadium is added in excess of what is needed to stabilize carbon. The resulting steel is not only interstitial free, but also contains well-dispersed fine carbonitride precipitates that contribute to higher tensile strengths through a refined grain structure as well as precipitation hardening.

Bake-hardenable steels, as described by Horvath [2], have improved dent resistance and are used in the automotive industry for exposed body panels and even structural stampings. The effect of bake hardening is well known, and these steels have been used for many years to downgauge exterior panels without reducing denting performance. Bake hardening involves thermally treating these low-carbon steels so that small amounts of solute carbon are held in solution and on heating, causes carbon to diffuse and pin dislocations. The solute carbon will migrate to the dislocations and when the carbon concentration increases to the point that it precipitates and pins the dislocation. Typically, hardening from this bake-hardening effect causes the yield strength to increase considerably. Stabilizing the carbon in solution is the most important step in producing bake-hardenable steels that do not harden over time at room temperature, but harden after exposure to elevated temperatures, such as those that occur in automotive paint-bake ovens. The bake hardening index is measured after the steel is strained 2% in the direction of rolling and baked at a temperature that is common to OEM paint-bake ovens, usually around 170°C, Fig. 13.13. The procedure consists of testing a standard tensile bar to determine its yield strength. Another tensile bar is then strained by 2% and is baked at a standard temperature, typically 170°C for 20 min, at which time the tensile bar is pulled again until it breaks. The bake-hardening index is determined by the

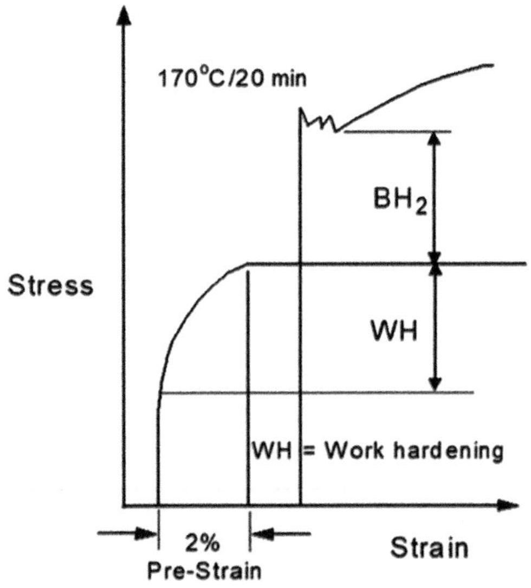

FIG. 13.13 Bake-hardening measurement [2].

TABLE 13.2 Mechanical properties and chemistries of selected solid solution strengthened and bake-hardenable steels [2].

	Mechanical properties					Chemistry				
	YS (MPa)	TS (MPa)	El (%)[a]	*n*-value	*r*-value	C (%)	Mn (%)	Si (%)	P (%)	Nb (%)
180MPa SSS	186	340	37	0.22	1.6	0.005 max	0.4	n/a	0.05	–
210MPa SSS	223	335	35	0.2	1.5	0.005 max	0.4	n/a	0.07	–
240MPa SSS	252	363	34	0.19	1.4	0.005 max	0.4	n/a	0.09	–
180MPa BH	193	342	37	0.21	1.5	0.005 max	0.5	n/a	0.04	–
210MPa BH	218	337	35	0.18	1.3	0.005 max	0.5	n/a	0.05	–
240MPa BH	245	344	34	0.18	1.3	0.005 max	0.6	n/a	0.06	–
440 IF HSS	290	446	34	n/a	1.9	0.005	0.62	0.01	0.04	0.068

difference in the yield strength of the original tensile bar and the bar that was strained and then baked. It is important to note that the hardening effect from baking and from ambient age hardening will occur with less than 2% strain; tests show that strains of 0.5% are sufficient to activate the hardening mechanism. Some of these solid solution strengthened and bake-hardenable steels are shown in Table 13.2 [2].

13.1.1.3 High-strength low-alloy steels (HSLA)

High-strength low-alloy steels (HSLA) are steels that have higher strengths relative to plain low-carbon steels with ferrite/pearlite microstructures. Fig. 13.14 illustrates a comparison of stress-strain curves for plain carbon mild steel, HSLA steel, and DP steel, and shows that the yield strength of HSLA steel is double that of plain carbon mild steel. Fine ferrite grain size is the major strengthening component of HSLA steels, Fig. 13.15. The alloying in these steels is considered microalloyed and contains a carbon content between 0.05 and 0.25%, and includes other alloying elements up to 2.0% manganese and small quantities of copper, nickel, niobium, nitrogen, vanadium, chromium, molybdenum, titanium, calcium, rare-earth elements, or zirconium. Copper, titanium, vanadium, and niobium are added for strengthening purposes and are intended to change the microstructure of carbon steels, usually ferrite-pearlite, to a very fine dispersion of alloy carbides in an almost pure ferrite matrix. To increase corrosion resistance, copper, silicon, nickel, chromium, and phosphorus are added. Zirconium, calcium, and rare-earth elements are added for sulfide-inclusion shape control, which increases formability, and are needed because most HSLA steels have directionally sensitive properties.

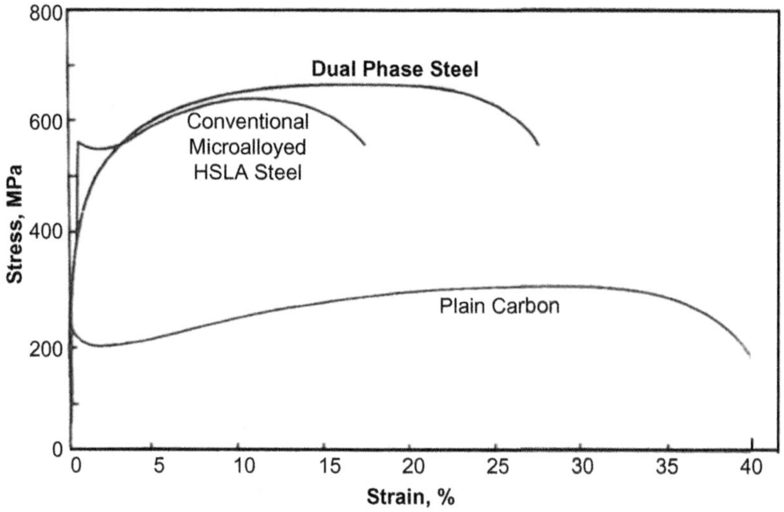

FIG. 13.14 Stress-strain curves for plain carbon, HSLA, and dual-phase steel [19].

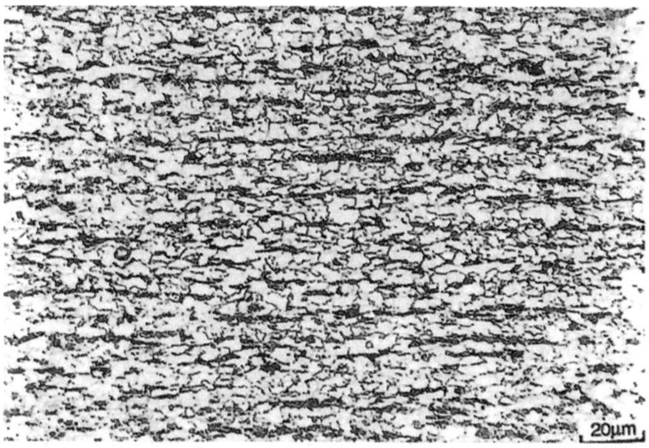

FIG. 13.15 Typical ultrafine-grained ferrite produced from deformed unrecrystallized austenite [19].

Several special terms are used to describe various types of HSLA steels [19]:

- **Weathering steels**: Steels that have better corrosion resistance.
- **Control-rolled steels**: Hot-rolled steels, which have a highly deformed austenite structure that will transform to a very fine equiaxed ferrite structure upon cooling.
- **Pearlite-reduced steels**: Strengthened by precipitation hardening, these very low-carbon content steels have little or no pearlite, but rather a very fine grain ferrite matrix.
- **Acicular ferrite steels**: These steels have very low carbon (typically 0.03–0.06%) with enough hardenability (by Mn, Mo, Nb, and B additions) to transform on cooling to a very fine high-strength acicular ferrite structure (also called low-carbon bainite), rather than polygonal ferrite. They also exhibit continuous yielding behavior and good toughness.
- **Microalloyed steels**: These steels contain very small additions of niobium, vanadium, and/or titanium to obtain a refined grain size and/or precipitation hardening and exhibit discontinuous yielding behavior.

13.1.1.4 First-generation advanced high-strength steels (1st GEN)

These steels contain a mixture of ferrite and martensite obtained by heat treatment in the two-phase austenite plus ferrite region of the equilibrium phase diagram called intercritical annealing heat treatment [20,21]. Included in the first-generation advanced high-strength steels are dual phase (DP), TRIP, complex phase (CP), fully martensitic steels (MS), and boron heat-treatable steels. Dual-phase steels (typically 0.1% C and 1.0% Mn) have a better combination of strength and ductility than do manganese strengthened or HSLA steels. TRIP steels have even better ductilities at higher

TABLE 13.3 Mechanical properties of various intercritical annealed steels [2].

	Yield strength (MPa)	Tensile strength (MPa)	Total elongation (%)[a]	Uniform elongation (%)[a]	n-value
Mild steel	181	296	42.4	21.4	0.208
HSLA 340	375	480	24	13	0.15
Dual-Phase 600	355	623	26	19	0.18
Dual-Phase 800	440	825	17	18	0.13
TRIP800	503	831	27.6	21.5	0.236

[a]*Elongation is in 50mm with an ISO #1 tensile bar.*

strength. DP steels grades, DP 600, DP 800, and DP 1000, can have tensile strengths in the range of 600–1000 MPa with associated ductility from 21 to 8% total elongation, respectively. While Dual-Phase steels are cooled directly to room temperature from the intercritical temperature, TRIP steels are isothermally transformed at subcritical temperatures after intercritical heating, Fig. 13.11. The austenite retained after the subcritical isothermal transformation is capable of strain-induced transformation to martensite, or **TR***ansformation*-**I***nduced* **P***lasticity*. The increased carbon and manganese in TRIP steel (typically 0.18%C and 1.5% Mn) are necessary to stabilize the austenite at room temperature. A comparison of mechanical properties of various steels is seen in Table 13.3. Note the increased total elongation for the TRIP 800 over the DP 800, even though the tensile strength is equivalent.

Stretch flangeable or high hole expansion, measured as the hole expansion ratio HER in various steels, is an important property for the automotive application of AHSS. S.K. Paul [22] recently reviewed HER for these steels. Dual-Phase steel HER depends upon the volume fraction martensite, the carbon content in the martensite, and the difference in hardness between the ferrite and martensite. Ferrite/bainite steels were reported to have a higher resistance to edge cracking compared with dual-phase ferrite martensite steels [2]. Further, it was reported that complex-phase (CP) steels and TWIP steels show better HER than DP steels [22].

Complex-Phase steels (CP) are similar to TRIP steels in composition except they are adjusted to contain less retained austenite. In place of the large volume fraction of retained austenite in TRIP steels, complex-phase (CP) steels contain fine precipitates along with a fine microstructure of ferrite and a high-volume fraction of hard phases. Precipitation hardening in these steels occurs because of the addition of niobium, titanium, and/or vanadium in addition to the normal alloying elements that DP and TRIP steels utilize. As a result, these steels have higher tensile strength and lower formability with lower alloy contents, which improves weldability at lower cost.

Martensitic steels (MS or QT in Fig. 13.11), which consist of a matrix rich in martensite with smaller amounts of bainite and/or ferrite, are produced by quenching at very high rates from the austenite ferrite region. In the automotive industry, martensitic steels are produced by several different methods; one method is to develop the microstructure by heat treatment after the part is stamped, and the other is to make the steel with the desired strength during steelmaking. In the case of heat treatment after stamping, the microstructure is developed by heat treatment of a high hardenability boron-bearing steel in water-cooled dies. In coil form, cold-rolled products are produced on water-quenched continuous annealing lines with high cooling rates while hot-rolled products are produced by water quenching on the run-out table in the hot mill. In the water-quenching process, steels of varying carbon content are used to produce martensitic steels with different strength levels. The use of lower carbon content steels of around 0.09% carbon and 0.50% manganese will result in martensitic steels with tensile strengths around 900–1000 MPa. If higher strengths are desired, the carbon content of the steel can be increased. When carbon contents are increased to about 0.23% carbon, the martensitic structure is strengthened such that the tensile strength is increased to 1300–1400 MPa. All of this strength comes at a cost, however, in that the ductility of the steel is reduced such that total elongations are now 5% or less. Even so, there is enough ductility for this steel to be roll formed or bent into the necessary shape. Martensitic steels are very useful in automotive applications for door beams, bumpers, very lightweight and high-strength lower side members (rocker panels), and cross car bars and beams that are designed to prevent intrusion into the passenger compartment [2].

13.1.1.5 Second-generation advanced high-strength steels (2nd GEN)

Second-generation Advanced High-Strength Steels (2nd AHSS) include Twinning-induced plasticity (TWIP) and Light-weight steels with induced plasticity(L-IP). In contrast to the ferrite-based structure, which characterizes the first-generation AHSS, the second-generation steels consist of a face-centered cubic austenite structure at room temperature that is stable because of a high manganese content (ranging from 15% to 30%) and a carbon content of 0.6%. As TWIP steels are strained [23], they deform by dislocation glide and secondary deformation mechanisms such as α_{bcc} to ε_{hcp} martensite formation and/or deformation-nucleated twins [24] that typically form in the austenitic structure because of a low stacking fault energy (SFE), Fig. 13.16. The twins act as dislocation barriers and reduce the effective mean free path of dislocations, which increases the flow stress. These twins are quite thin, and there is a continuous nucleation of new, increasingly smaller deformation twins. The tensile strength ranges of TWIP steels are similar to those of DP and TRIP steels (600–1000 MPa), but the tensile ductility range is significantly higher at 40%–80%, Fig. 13.17. Ductility and strength are related to the manganese content in that smaller manganese additions typically show higher strength and lower ductility. The high total elongation levels present a challenge for automotive usage because significantly higher strains need to be imparted to the component than those that normally occur in traditional stamping processes to get similar final component strength levels in TRIP and DP steels [25].

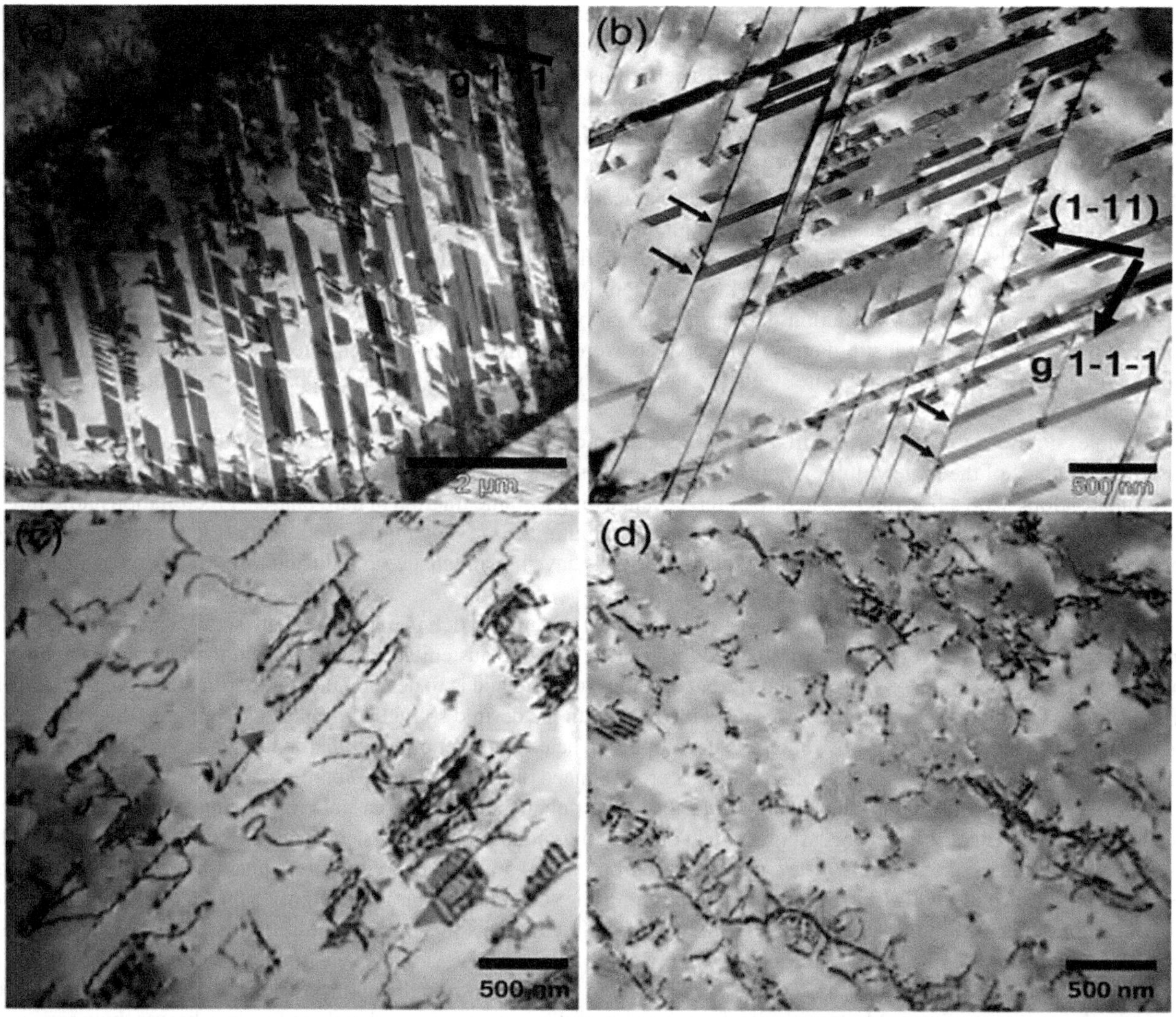

FIG. 13.16 BF TEM micrographs after 0.03 plastic true strain. (A) The microstructure of a grain in the 22%Mn alloy exhibiting large stacking faults on two slip systems. (B) Image of the microstructure in the 22%Mn alloy obtained with a beam direction near [110] using a 1-1-1 diffracting (g)-vector. (C) Grain exhibiting a defect structure with both planar and wavy characteristics in the 25%Mn alloy. (D) Image of the microstructure in the 28%Mn alloy showing wavy dislocations with localized areas of dislocation tangles [23].

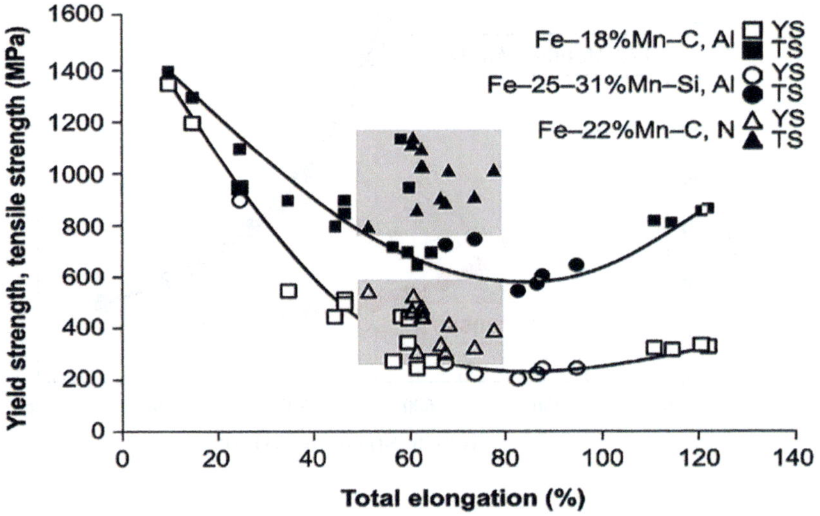

FIG. 13.17 Strength versus ductility ranges for common TWIP steels [25].

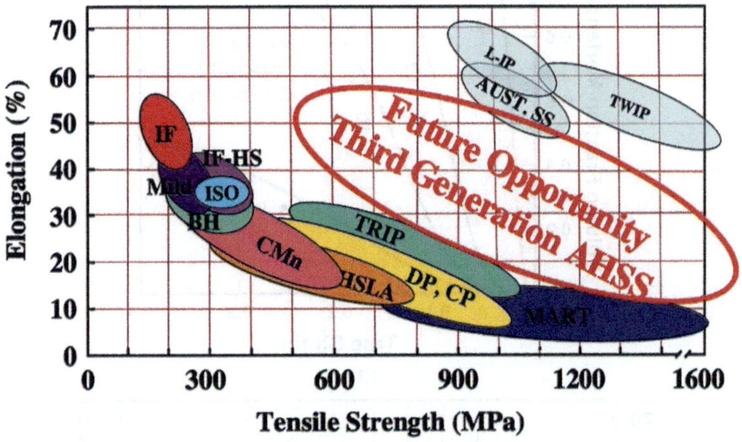

FIG. 13.18 Automotive steel ductility ladder [26].

13.1.1.6 Third-generation advanced high-strength steels (3rd GEN)

The new third-generation steels fill the gap in the ductility ladder between the first-generation and second-generation AHSS as seen in the "Automotive Steel Strength/Ductility Ladder," Fig. 13.18. Besides their improved ductility, the third-generation steels have improved energy absorption and performance as compared with the higher strength, but lower ductility than dual-phase (DP) and complex-phase (CP) steels of similar tensile strengths. These steels generally fall into the following classifications: enhanced DP, modified TRIP steels, Quench and Partition (Q&P) steels, lower manganese TWIP/TRIP, ultrafine bainite, Flash processing, and TRIP-assisted Bainite-Ferrite steels (TBF). These steel types utilize varied approaches to alloying and processing to achieve their special combination of strength and ductility that separate them from conventional carbon steels and first-generation AHSS [2,26–28].

For third-generation AHSS, the product of ultimate tensile strength and total elongation (UTS x TEL) is required to be larger than 30GPa%. A model was used to identify specific combinations of microstructural constituents that lead to properties within the opportunity band shown in Fig. 13.18. Design considerations for third-generation AHSS have employed a systematic design methodology using simplified composite model [29]. Both martensite/ferrite and martensite/austenite microstructures were evaluated, and the results of the calculations are shown in Fig. 13.19. The predicted tensile properties for the hypothetical ferrite/martensite microstructures overlap with the properties exhibited by the first-generation AHSS. The property band corresponding to the austenite/martensite mixtures is situated between the bands of the first and second generation AHSS, i.e., within the desired "third generation" regime. Transformation of metastable austenite with strain was included [28] as a second step of the model. Four hypothetical austenite stability conditions were considered as illustrated in Fig. 13.20a. The resulting strength/elongation combination predictions are given in Fig. 13.20b where the different austenite stability conditions are indicated by letters A through D and the individual data points on each curve correspond

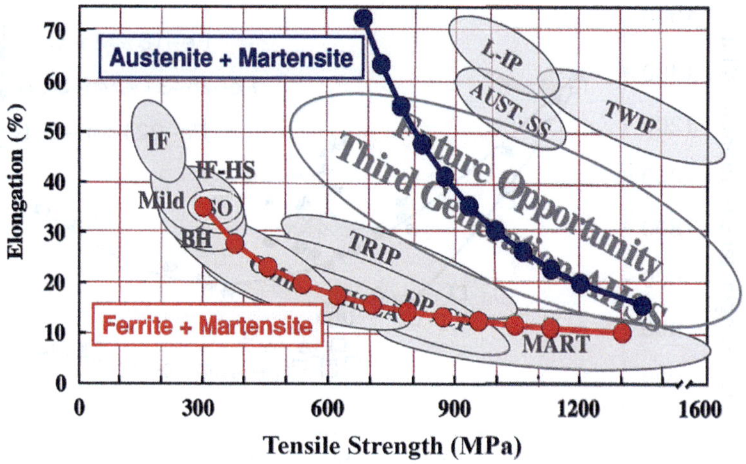

FIG. 13.19 Superposition of predicted strength/ductility combinations of hypothetical ferrite/martensite and austenite/martensite mixtures [26].

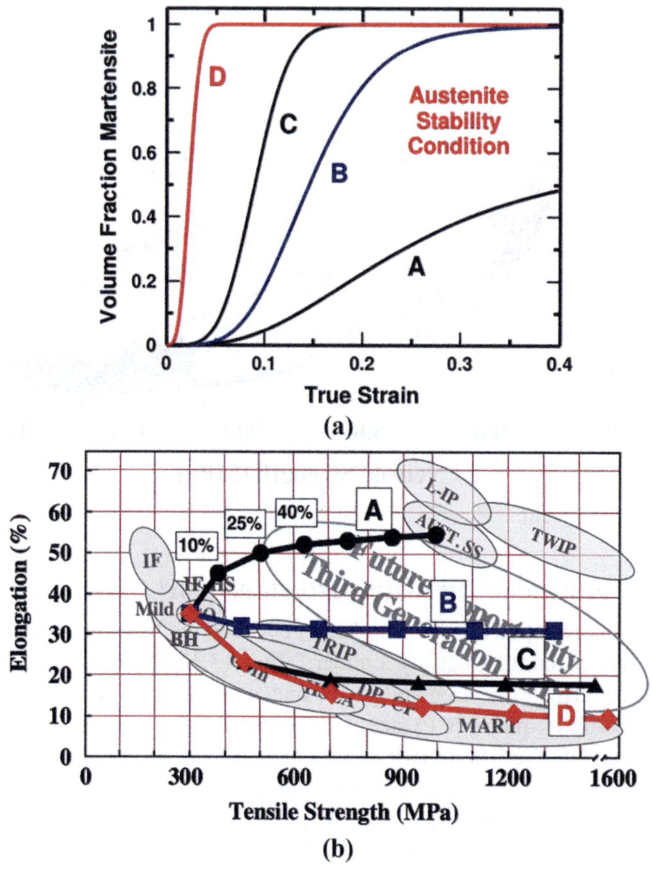

FIG. 13.20 Effect of austenite stability on predicted mechanical property combinations for (a) four different austenite stabilities identified as A through D and (b) predicted mechanical property combinations corresponding to ferrite plus austenite with the different austenite stabilities shown in (a). In (b), each data point on each curve corresponds to an initial austenite fraction, ranging from 0% to 85% [26].

to different initial austenite contents with the remainder of the microstructure being ferrite. Austenite stability clearly has a pronounced effect on the predicted properties. The best combinations of strength and ductility are predicted for high volume fractions of relatively stable austenite (curve B in Fig. 13.20b). As a result of this work, an understanding of the contribution of the individual constituents is realized. However, in order to obtain the next generation of AHSS steels, complex microstructures are needed consisting of significant fractions of high strength phases, which may be martensite, bainite or ultrafine-grained ferrite, in combination with highly ductile austenite with controlled stability against transformation of austenite to martensite with strain. Tables 13.1 and 13.4 list the types of third-generation automotive steels and processing

TABLE 13.4 Processing routes and chemical composition employed in different studies [27].

Process	Composition	Sample geometry
Enhanced DP	0.17C-1.63Mn-0.28Si	3.5 × 5 mm, GL: 10 mm
Modified TRIP	0.39C-1.16Si-1.20Mn	JIS-5 25 mm wide, GL: 50 mm
	0.38C-1.46Si-1.20Mn	JIS-5 25 mm wide, GL: 50 mm
	0.39C-1.99Si-1.20Mn	JIS-5 25 mm wide, GL: 50 mm
	0.23C-1.65Mn-0.45Si-1.14Al-0.076P-0.085Ti	A80: 20 × 1 mm, GL: 80 mm
	0.24C-1.67Mn-0.45Si-1.38Al-0.080P-0.081Nb	A80: 20 × 1 mm, GL: 80 mm
	0.22C-1.55Mn-1.55Si-0.035Nb	Small specimen test technique based on punching a 3 mm disk out of a 300–350 μm thick specimen [20]
	0.186C-1.53Mn-1.48Si-24ppmTi-1081ppmV-81ppmN	no properties reported
	0.225C-1.58Mn-1.60Si-25ppmTi-1550ppmV-90ppmN	no properties reported
	0.20C-1.51Mn-1.51Si	JIS-13B specimens: 12.5 × 1.2 mm, GL: 50 mm
	0.20C-1.71Mn-1.48Si-0.02Nb	JIS-5 specimens, 25 × 1.2 mm, GL: 50 mm
	0.185C-2.3Mn-1.5Si-0.25Mo	2.5 × 6.25 mm, GL: 150 mm
	0.285C-2.3Mn-1.5Si-0.25Mo	2.5 × 6.25 mm, GL: 150 mm
Bainite	0.79C-1.94Mn-1.59Si-1.33Cr-0.30Mo-0.11V	no properties reported
	0.98C-1.89Mn-1.46Si-1.26Cr-0.26Mo-0.09V	no properties reported
	0.83C-1.98Mn-1.57Si-1.02Cr-0.24Mo-1.54Co	no properties reported
	0.78C-1.95Mn-1.49Si-0.97Cr-0.24Mo-1.60Co-0.99Al	no properties reported
	0.80C-2.01Mn-1.59Si-1Cr-0.24Mo-1.51Co	5 mm diameter, GL: 25 mm
	0.79C-1.98Mn-1.56Si-1.01Cr-0.24Mo-1.51Co-1.01Al	5 mm diameter, GL: 25 mm
	0.43C-3.00Mn-2.02Si	Hounsfield no. 11: 3 mm diameter, GL: 50 mm
	0.19C-2.90Mn-2.13Si	Hounsfield no. 11: 3 mm diameter, GL: 50 mm
	1.1C-3.00Mn-2.02Si	Hounsfield no. 11: 3 mm diameter, GL: 50 mm
	0.22C-3.00Mn-2.03Si	Hounsfield no. 11: 3 mm diameter, GL: 50 mm
	0.39C-2.05Si-4.08Ni	Hounsfield no. 11: 3 mm diameter, GL: 50 mm
	0.39C-2.05Si-4.09Ni	Hounsfield: 3.5 mm diameter, GL: 20 mm
	0.40C-2.01Si-4.15Ni	Hounsfield: 3.5 mm diameter, GL: 20 mm
	0.22C-3.00Mn-2.03Si	Hounsfield: 3.5 mm diameter, GL: 20 mm
	0.22C-3.00Mn-1.99Si	Hounsfield: 3.5 mm diameter, GL: 20 mm
	0.32C-1.97Mn-1.45Si-1.26Cr-0.26Mo-0.10V	BS EN 10002–1:1990

Continued

TABLE 13.4 Processing routes and chemical composition employed in different studies—cont'd

Process	Composition	Sample geometry
	0.31C-1.51Si-1.44Cr-0.25Mo-3.52Ni-0.10V	BS EN 10002–1:1990
	0.30C-1.51Si-1.42Cr-0.25Mo-3.53Ni	BS EN 10002–1:1990
	0.31C-1.51Si-1.44Cr-3.52Ni-0.25Mo-0.10V	3mm diameter, GL: 19mm
	0.30C-1.51Si-1.42Cr-3.53Ni-0.25Mo	3mm diameter, GL: 19mm
	0.29C-2.25Mn-1.50Si-0.26Mo	3mm diameter, GL: 19mm
	0.29C-1.97Mn-1.46Si-0.46Cr-0.25Mo	3mm diameter, GL: 19mm
	0.29C-1.56Mn-1.49Si-1.47Cr-0.25Mo	3mm diameter, GL: 19mm
	0.27C-1.53Mn-1.71Si-1.47Ni-0.17Cr-0.24Mo	3mm diameter, GL 19mm
	0.22C-2.23Mn-1.50Si-0.25Mo	3mm diameter, GL: 19mm
	0.21C-1.56Mn-1.46Si-1.49Cr-0.24Mo	3mm diameter, GL: 19mm
Quenching and partitioning	0.20C-1.71Mn-1.48Si-0.02Nb	JIS-5 specimens, 25 × 1.2mm, GL: 50mm
	0.19C-1.59Mn-1.63Si	subsize ASTM: 6.34 × 1.2mm, GL: 25.4mm
	0.20C-1.63Mn-1.63Si	A50: 12.5 × 1mm, GL: 50mm
	0.24C-1.61Mn-1.45Si-0.30Al	A50: 12.5 × 1mm, GL: 50mm
	0.25C-1.70Mn-0.55Si-0.69Al	A50: 12.5 × 1mm, GL: 50mm
	0.24C-1.60Mn-0.12Si-1.41Al-0.17Mo	A50: 12.5 × 1mm, GL: 50mm
	0.21C-1.96Mn-1.49Si-0.25Mo	A50: 12.5 × 1mm, GL: 50mm
	0.17C-1.65Mn-0.38Si-1.11Al-0.08P	A50: 12.5 × 1mm, GL: 50mm
	0.2C-1.5Si-1.8Mn	A50: 12.5 × 1mm, GL: 50mm
	0.2C-1.6Si-3Mn	ASTM E8: 12.7 × 1mm, GL: 57.15mm
	0.29C-1.59Si-2.95Mn	ASTM E8: 12.7 × 1mm, GL: 57.15mm
	0.28C-1.64Si-4.95Mn	ASTM E8: 12.7 × 1mm, GL: 57.15mm
	0.41C-1.27Si-1.30Mn-1.01Ni-0.56Cr	16 × 1mm, GL: 80mm
Flash processing	0.16C-0.16 Si-0.79Mn-0.13Cr-0.08Ni-0.13Cu	Not reported
	0.15C-0.21 Si-1.46Mn	Not reported
	0.04C-0.16Mn-0.02Cr-0.09Ni	Not reported
	0.21C-0.22 Si-0.63Mn-0.33Cr-0.01Ni-0.04Cu	12.7 × 3.1mm, GL: 50mm
	0.22C-0.26 Si-1.05Mn-0.22Cr	12.7 × 3.1mm, GL: 50mm
	0.21C-0.27Si-0.73Mn-0.48Cr-0.48Ni-0.16Mo-0.18Cu	6.3 × 1.6mm width, GL: 25mm

TABLE 13.4 Processing routes and chemical composition employed in different studies—cont'd

Process	Composition	Sample geometry
Lower Mn TWIP/ TRIP	0.02C-15.8Mn-3.0Si-2.9Al	not reported
	0.04C-20.1Mn-2.8Si-2.9Al	not reported
	0.03C-25.6Mn-3.0Si-2.8Al	not reported
	1.13C-11.4Mn-0.2Si-0.17Ni-0.08Mo	6.25 mm diameter, GL: 22 mm
High Mn TRIP	0.10C-5.2Mn-0.12Si	subsize ASTM: 6.34 × 1.2 mm, GL: 25.4 mm
	0.10C-5.8Mn-0.13Si	subsize ASTM: 6.34 × 1.2 mm, GL: 25.4 mm
	0.10C-7.1Mn-0.13Si	subsize ASTM: 6.34 × 1.2 mm, GL: 25.4 mm

routes and composition, respectively. What follows is a brief description of each of the third-generation steels that are more detailed in [2,26–28].

Enhanced DP

A strength increase over DP 980 dual-phase steels can be obtained by microstructural refinement resulting from special hot deformation practices [28,30–37] referred to as Deformation-Induced Ferrite Transformation (DIFT) [27], schematically seen in Fig. 13.21. By deforming the material between the Ar_3 and Ae_3 temperatures, a strain-induced transformation from austenite to ferrite was found to produce ultrafine ferrite with a grain size on the order of 1 μm. A tensile strength of 893 MPa combined with 21% elongation was obtained, but properties keep the steel within the first-generation band.

Modified TRIP

Early TRIP steel research was performed on grades with higher carbon contents than currently used in commercially available grades that result in properties required for third-generation AHSS objectives. Grain refinement of TRIP steels by microalloying has also been investigated [38–40], and tensile strength levels up to 1 GPa with ductility levels of about 20% total elongation have been obtained. Other TRIP steel development with modified heat treatments includes isothermal holding following full austenitization. Significant amounts of austenite are retained in these microstructures as a result of the bainite transformation.

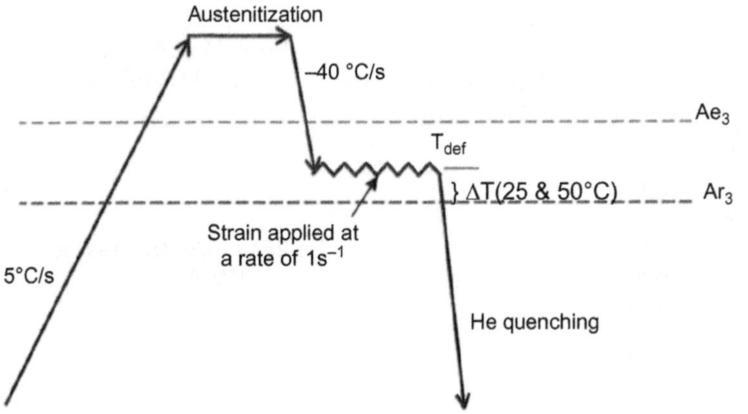

FIG. 13.21 Thermomechanical processing path for deformation-induced ferrite transformation (DIFT) [27].

High Mn—TRIP

Fine-grained or ultra-fine-grained duplex ferrite-austenite microstructures based on "medium" manganese (5–7 wt%), low-carbon (0.1 wt%) compositions are seen in Table 13.4. Manganese enrichment of austenite during intercritical annealing was applied to a cold-rolled 0.1-C 7.1-Mn (wt%) steel to produce a range of microstructures with varying austenite fraction and stability. Conventional hot rolling, cold rolling, and batch annealing processes were used. As a result of long holding times at the peak intercritical annealing temperature allows manganese partitioning to occur. Ultra-fine ferrite and manganese austenite are obtained at room temperature with austenite amounts up to 17% in the 0.10C-5.2Mn steel and austenite levels up to 38% as the Mn level increases to 7.1% Mn.

Lower Mn—TWIP/TRIP

Table 13.4 shows the variety of compositions that have been proposed for the second-generation austenitic steel grades, including (in wt%): 15,20,25,30 Mn. However, some research has focused on reducing the Mn content in these grades, see Table 13.4. The effect of lowering the Mn content from 25 to 15 wt% on the stress-strain behavior shows that if the true strain exceeds 0.15, there is a pronounced increase in strain hardening, Fig. 13.22. This was believed to be associated with the α martensite transformation.

Trip-assisted bainitic ferrite (TBF)

Trip-assisted bainitic ferrite (TBF) steels can be produced by heat treatment using a multistep thermal processing route, automatizing in the fully austenite γ phase and fast cooling to a temperature between the bainite-start (B_S) and the martensite-start (M_S) temperatures of the steels in order to isothermal transform (or austempering) the product [41–44]. The microstructure for TBF steels is stable retained austenite in a bainitic ferrite matrix with retained austenite, Fig. 13.23. Typical chemical compositions of TBF steels contain C, Si, and Mn as major alloying elements. Alloy modifications include variations of the Al, Nb, and Cr content [46]. During isothermal transformation, the transformation of austenite to bainitic ferrite without any carbides and the carbon enrichment to austenite occur simultaneously. On final cooling to room temperature, most of the carbon- enriched austenite is retained with a small amount of α'-martensite. The Si content suppresses the cementite formation normally formed in the bainitic transformation, and the added Si enhances the C content and stabilizes the retained austenite. High Si contents of 1.5 wt% are used in these types of steels. Although Si has major importance to prevent carbide precipitation during annealing of the cold-rolled material, it causes problems during processing in continuous annealing lines. Therefore, other alloying elements having similar effect of suppressing carbide formation have to be considered. The use of Cr- and Mo-bearing TBF steels in the manufacture of such hot-forged parts as drivetrains, engine common rail systems, and others is expected to contribute to the manufacture of lightweight automobiles with lower fuel consumption. The TBF steel parts produced by hot forging have the advantage of improved impact toughness due to their refined microstructure and increased retained austenite fraction, compared with cold-rolled TBF sheet components. Therefore, the TBF steels have the potential to replace conventional low alloy structural steels such as Cr-Mo and Cr-Mo-Ni quenching and tempered steels with low Si content [47].

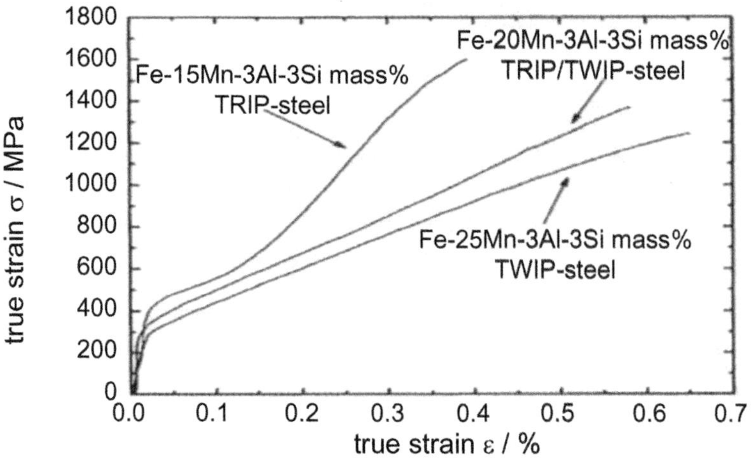

FIG. 13.22 True stress-true strain curves showing the effect of Mn content on strain hardening [24].

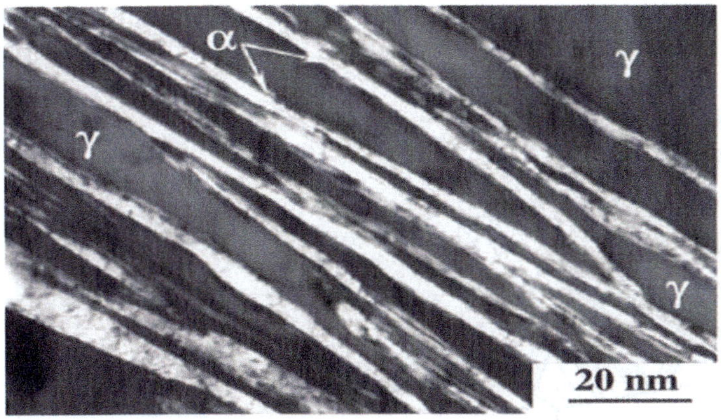

FIG. 13.23 Transmission electron micrograph of very fine bainitic microstructure developed in 0.98C-1.89Mn-1.46Si-1.26Cr-0.26Mo-0.09V alloy after heat treatment at 200°C for 15 days [45].

Quench and partition (Q&P)

A schematic of the Q&P heat treatment is shown in Fig. 13.24. The procedure, which was described by Speer et al. [30], consists of the austenitization of the steel followed by quenching to, and briefly holding at, a predetermined quenching temperature (QT) between the martensite start (M_s) and finish (M_f) temperatures. Martensite is formed during the initial quenching but because QT is above M_f, a certain amount of residual austenite remains in the partially martensitic microstructure. The aim of the next step, the so-called partitioning treatment (PT), is to stabilize the residual austenite by carbon enrichment via carbon depletion of the supersaturated martensite and diffusive transport into the adjacent austenite, which has a much higher carbon solubility. This microstructural process has been identified as dominant mechanism for austenite stabilization by Clarke et al. [31]. As the austenite is enriched in carbon, the M_s-temperature decreases, once the M_s-temperature reaches room temperature, the austenite present at QT is also preserved as residual austenite even upon subsequent cooling to room temperature (RT). Jirkova ́ et al. [32] could prove that the carbon-enriched austenite and the carbon-depleted martensite are primarily responsible for the high ductility combined with the excellent strength of Q&P steels. Steels produced in this manner can have significant amounts of retained austenite, which greatly increases the ductility and elongation of the steel [2,33].

The concept of partitioning austenite to a finer effective grain size to improve steel properties had been suggested as early as the 1970s and documented in [17]. Ohtani et al. [34] proposed that the austenite grain can be partitioned by the bainite transformation prior to the formation of martensite, Fig. 13.25. In this case, the partitioning of austenite by bainite improves the toughness of these structures by causing a reduction in the mean free path for crack propagation, called the "unit crack length." Crack propagation is altered at each bainite lath and thus improves the strength of the steel. It has been found that for optimum impact and tensile properties, the bainite constituent should be around 10–25 vol.% [35,36]. Ohmari et al. [37] reported that the initially formed bainite laths partition the austenite grains and reduce the "effective grain size."

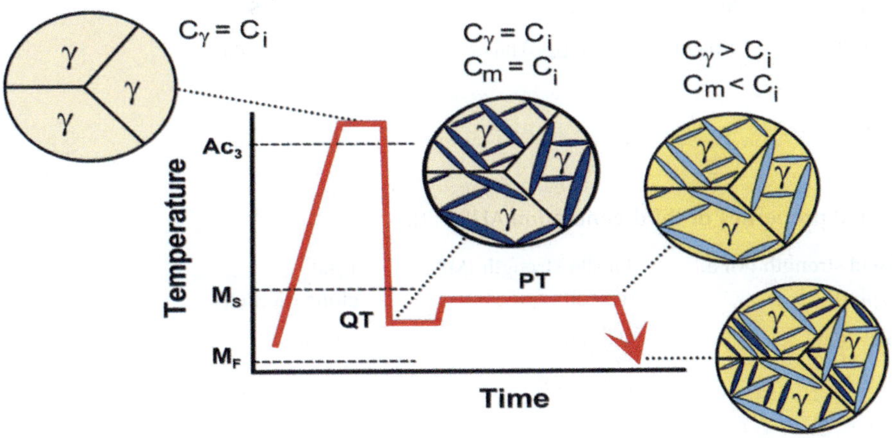

FIG. 13.24 Heat treatment suitable for processing Q&P steels, according to [26]. It consists of the austenitization of the steel followed by quenching to, and briefly holding at, a predetermined quenching temperature (QT) between the M_s- and M_f-temperatures. The last step is the partitioning treatment (PT) for the stabilization of the residual austenite by carbon enrichment.

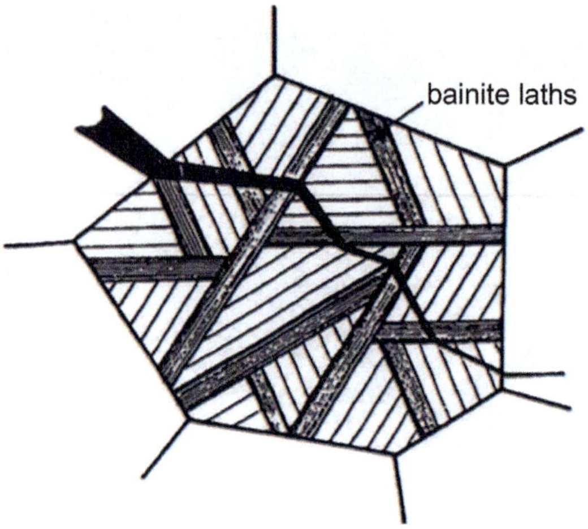

FIG. 13.25 Schematic representation of a bainite/martensite microstructure where the bainite has partitioned the prior austenite with the remaining austenite transformed to martensite [17].

These results may offer another processing path, similar to austempering, for the Q&P AHSS product that may fit more easily into the CGL line parameters. However, it may be necessary to modify the alloy to cause the CT diagram to accommodate the formation of a small amount of bainite prior to quenching to martensite while in the Zn pot. The bainite could then partially transform, and on removal, the remaining austenite would be quenched to form martensite.

Strain induced by deformation during a subsequent forming operation or during a crash event will transform retained austenite to martensite by the TRIP affect. As strains increase, the austenite to martensite transition continues to strengthen the steel and distribute loads, further increasing the strength and load carrying capacity of the steel. It is the high amount of retained austenite combined with the TRIP effect that gives these steels their combination of strength and ductility. Typical mechanical properties of selected third-generation steels and their chemistries are shown in Tables 13.5 and 13.6, respectively.

The chemistry of these steels has been developed for the stabilization of austenite with carbon and manganese and to utilize silicon to suppress carbide formation, which would reduce the availability of carbon for austenite stabilization. The chemistry in combination with specific thermomechanical processing enables the formation of steels with approximately 10%–30% retained austenite. Concentrations that are thermally stable at the operating temperatures of today's automobiles. Applications for third-generation steels are still evolving and generally fall into two categories; those parts that are designed to absorb impact energy and those that are designed to resist deformation in crash events [2].

TABLE 13.5 Chemical composition of typical third-generation AHSS [2].

C	Mn	Si	Al	P	S
0.15–0.30	1.5–3.0	1.0–2.0	0.02–0.06	<0.015	<0.01

TABLE 13.6 Selected properties of third-generation AHSS [2].

Grade	Yield strength (MPa, typical)	Tensile strength (MPa, typical)	Total elongation (%)
Q&P 980	600–850	980–1100	19 min
Q&P 1180	850–1150	1180–1300	12 min

13.2 Zn coating formation

13.2.1 AHSS processing and product challenges

There are several challenges in the continuous galvanizing process for AHSS steels. These involve (1) the relationship between hot strip processing and subsequent processing in the continuous annealing line (CAL) and hot dip galvanizing (HDG), (2) the formation of surface oxides during hot strip processing and continuous annealing, (3) spot weldability, and (4) liquid metal embrittlement.

13.2.1.1 Effect of hot strip processing

Prior to Zn coating AHSS steels, it is necessary to optimize the chemical composition, rolling condition, and cooling rate on the run-out table of the hot strip mill to obtain the proper mechanical properties to prepare the hot strip steel for cold rolling. Not much research is found in the literature pertaining to the effect of the hot-rolled structure on the final cold-rolled and zinc-coated product. For example [48], Q&P Mn hot-rolled sheets with a finish rolling temperature of 900°C were tempered in a batch-annealing-like furnace for 16h at 550°C to produce an equilibrium microstructure of ferrite and second phase (pearlite, bainite, or martensite). Subsequently, the 4mm hot-rolled sheet was cold rolled to a final thickness of 1mm in preparation for the CGL and hot dip Zn coating step. In another example [49], a Trip 800 steel was hot coiled at 600°C to produce a starting microstructure of ferrite and pearlite, Fig. 13.26, prior to cold rolling to 1mm in the full hard condition prior to HDG. McDermid et al. [50] evaluated the effect of hot-rolled microstructure of several prototype C-6Mn-xSi-yAl-zSn 3G-AHSS steels. A difference in microstructure was seen between one group of steels that had a martensitic microstructure with small volume fractions of ferrite in a lath martensitic matrix and a steel with as cold-rolled martensitic laths with some evidence of carbide precipitation resulting from a tempering heat treatment. They suggested that:

1. the starting microstructure has a significant effect on the as-annealed microstructure. The martensite and martensite/ferrite starting microstructures result in a more lath-like martensite and retained austenite morphology as opposed to a more globular morphology for the tempered martensite starting structure,
2. the microstructural constituents are extremely finely divided. The majority of the retained austenite is inter-lath for the martensite and martensite/ferrite starting microstructure in contrast to a more globular structure for the tempered martensite starting microstructure,
3. the tempered martensite structures resulted in undesirably rapid austenite to martensite transformation kinetics during tempering and were not found to meet the required 3G-AHSS property window using CGL-compatible thermal processing.

It is extremely important to be able to process these high-strength hot-rolled grades on a cold mill prior to coating in the CGL. All of this recent research shows the importance of correlating the alloy composition, hot rolling microstructure with CGL processing, to produce the desired final mechanical properties for zinc coated AHSS steels, and much more research is necessary. Most continuous annealing lines use rapid gas jet cooling, gas-water spray cooling, and cold or hot water quenching [51]. Recently, hydrogen quench technology has been introduced on the CGL for AHSS processing [52].

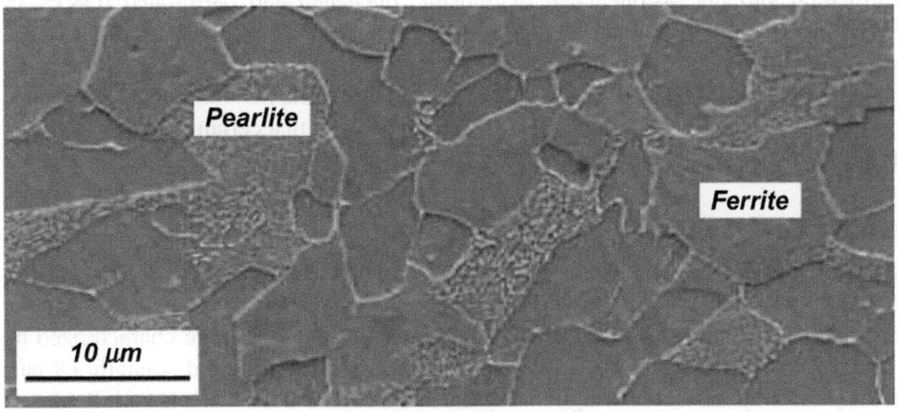

FIG. 13.26 Hot-rolled microstructure of the 0.15% C, 2% Mn, 0.3% Si, 0.8% Al, and 500ppm P TRIP steel [49].

13.2.1.2 Oxide formation

The presence of alloying elements such as Mn and Si is responsible for improving the mechanical properties of advanced high-strength steels (AHSS).

First, for example, the heat treatments conventionally used to produce the desired TRIP steel multiphase microstructure are not necessarily compatible with the hot-dip galvanizing thermal cycle, since the isothermal bainite transformation (IBT) time for hot-dip galvanizing lines is short and the range of IBT temperature is limited due to the requirements for the Zn bath [53–56]. Moreover, selective oxidation of alloying elements such as Si during the annealing prior to galvanizing can result in poor reactive wetting and unacceptable bare spots in the Zn coating.

These elements could form oxide particles as MnO at the steel surface during hot rolling and undergoing continuous annealing prior to the Zn pot step. The forming of oxides on the steel surface could affect the quality of the coating and wettability between the coating and steel [57]. Surface oxides and wettability are discussed in more detail in Chapter 5, Interface Reactions. The dissolved Al in the Zn bath promotes the aluminothermic reduction of MnO, improving the surface quality of the coating, as shown in the equation [58].

$$3\text{MnO (solid)} + 2\text{Al (solution)} = \text{Al}_2O_3 \text{ (solid)} + 3\text{Mn (solution)}$$

13.2.1.3 Spot weldability

Resistance spot weldability is the main method of joining used in the automotive industry with each vehicle containing several thousand welds [59] Understanding the effect of these coatings on the resistance spot welding behavior of these steel grades is important. The two most commonly used grades of AHSS in automotive body-in-white (BIW) welded applications are dual-phase and transformation-induced plasticity (TRIP) steels [60]. Although other AHSS grades such as complex-phase and twinning-induced plasticity steels are commercially available, they have limited applicability in the automotive industry because of their high cost and limited global availability. Tumuluru [61] has reviewed the spot weldability of advanced high-strength steels. He concluded that as of 2015, the available AHSS grades of DP and TRIP steels were easily weldable.

13.2.1.4 Liquid metal embrittlement

An important challenge encountered with Zn-coated AHSS is their susceptibility to Zn-assisted liquid metal embrittlement (LME) [62–64]. LME can be activated when Zn-coated steels are heated to elevated temperatures at which the Zn melts on the surface of the (solid) steel, and there is a simultaneous presence of a tensile stress [62–64]. "Intergranular" cracks are formed with liquid Zn penetration along the grain boundary interfaces of a steel, often resulting in premature fracture of Zn-coated AHSS. The critical conditions for triggering LME can be met during resistance spot welding [65].

Beal [64] showed that Zn-coated steels experience Zn LME-induced ductility loss over a specific range of temperature; the width of the LME "ductility trough" is an important indicator of LME susceptibility of a steel alloy. Fig. 13.27 illustrates the LME "ductility trough" [66]. Research has indicated that a generally higher LME-susceptibility (wider ductility trough) for AHSS compared with conventional mild or high-strength low-alloy (HSLA) steels, the severity of Zn-assisted LME might be inferred to be "stress"-controlled, i.e., LME susceptibility is greater at elevated strength levels [67–69].

On the other hand, the increased LME sensitivity of AHSS compared with lower strength steels may also be associated with the microstructure or alloy composition [70]. A dual-phase (DP) microstructure exhibited lower LME susceptibility compared with other AHSS microstructural variants (as-quenched martensitic, quenched and partitioned (Q&P), or transformation-induced plasticity (TRIP)-assisted bainitic ferrite (TBF)), indicating that the microstructure of an AHSS can influence its LME susceptibility [53]. In addition, previous studies by Hong et al. [71], Tumurulu et al. [72], and Kalashami et al. [73] compared the LME susceptibility of AHSS having different Si contents, while maintaining a similar base microstructure. These authors reported aggravated LME behavior at high Si levels, revealing that certain alloying elements can play a role in the LME cracking mechanism, which was confirmed by Bhattacharya et al. [66].

13.2.1.5 Hydrogen embrittlement sensitivity

Hydrogen embrittlement can be a severe type of failure mode for high-strength steels and heat-treated steels. It is characterized by a loss of ductility and strength from the absorption of hydrogen by the metal, which eventually diffuses to the grain boundaries and weakens the material [2]. Materials that suffer from hydrogen embrittlement are characterized by existence of intergranular cracking and can exhibit fractures far below the nonaffected materials yield strength. Of particular concern to the automotive industry in advanced and ultrahigh-strength steels and heat-treated steels with Rockwell

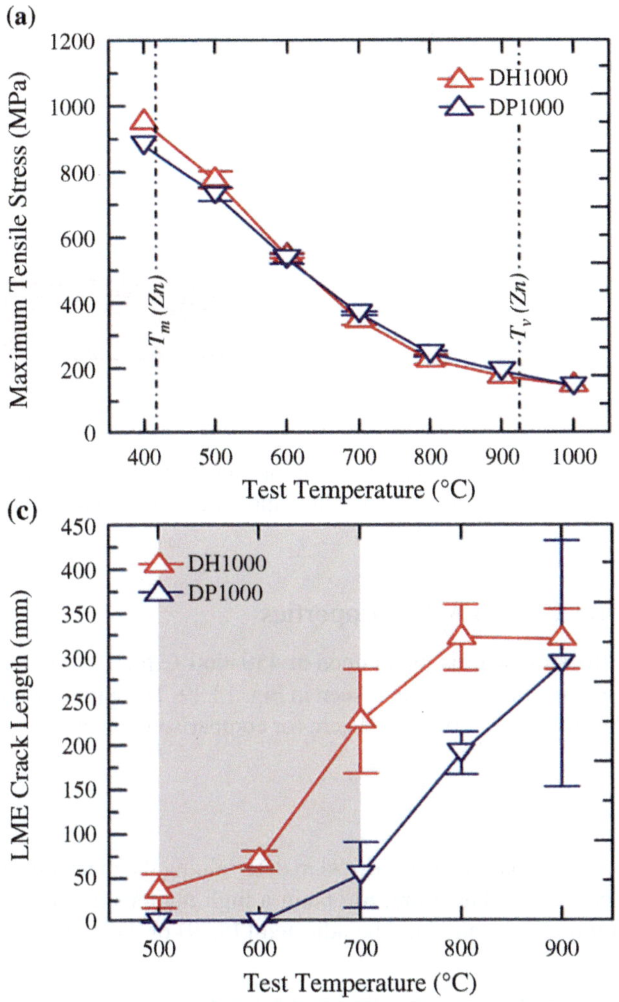

FIG. 13.27 Summary of hot tension test results for the DH1000 and DP1000 steels at different temperatures. (A) Maximum tensile stress, (B) LME-associated ductility loss, (C) average LME crack length, and (D) number of LME cracks as a function of the deformation temperature at a constant displacement rate (150 mm/s) [66].

hardness values above Rc 38, which translates to a tensile strength of about 1000 MPa. Hydrogen embrittlement is simply described by the absorption and migration of hydrogen or hydrogen ions by the metal, which then can migrate to the grain boundaries and weaken them, resulting in intergranular cracking. For cracking to occur, three factors must be present:

(1) The material must have a tensile strength of around 1000 MPa or higher;
(2) There must be a continuous stress on the part of sufficient amplitude to cause cracking; and
(3) There must be hydrogen available of sufficiently high concentration to cause cracking [2].

These conditions are appropriately shown in the Venn diagram in Fig. 13.28. Hydrogen can be introduced in several ways: through steel manufacturing, assembly process, coating processes such as EG coating, and through environmental exposure. A common way for hydrogen to occur in steel manufacturing is during acid cleaning of the steel during the pickling operation. For sheet steels, the most common steel manufacturing process that can introduce hydrogen is during the plating operation of an electrogalvanizing line [2].

In the hot-dip galvanizing process, hydrogen is mainly absorbed during the high-temperature operations in hydrogen containing environments before hot dipping. After hot dipping, the metallic zinc coatings impede hydrogen diffusion out of the steel. Thus, excess hydrogen can remain in the steel substrate that will lead to a possible embrittlement. Mandy et al. [75] recently showed that galvannealing promotes hydrogen degassing as compared with GI and ZnAlMg (ZM1212) coatings. GI coatings show a slight degassing at 60°C after 10 days. In a study of DP and IF steels, Georges et al. [76] found that differences in martensite fraction were important. Dislocations can serve as reversible traps and can vary dependent

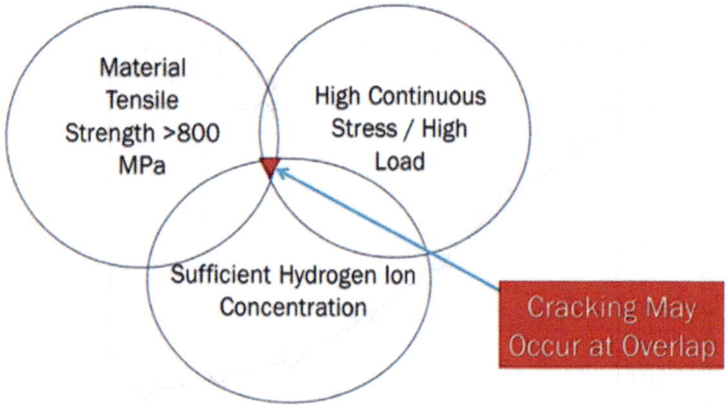

FIG. 13.28 Venn diagram showing conditions favorable for hydrogen cracking [74].

upon the dislocation distribution in the ferrite of a particular alloy. It was found that an overaging step between 100°C and 200°C for around 30 s had a large influence on the final diffusible hydrogen content, depending upon the alloy composition. Integrating the overaging step into the CGL will be challenging.

13.2.2 Effect of Zn coating temperature cycle (GI and GA) on AHSS properties

In the CGL hot-dip galvanizing line, after the annealing step, the strip steel is maintained at 450–460°C for 10–40 s for galvanization as it is hot-dipped into the Zn bath. A typical galvanizing thermal cycle is seen in Fig. 13.29. The galvanizing thermal cycle is described in more detail in Chapter 2, but the thermal cycle is presented here for comparison to the various AHSS coating thermal cycles.

13.2.2.1 Dual-phase (DP) steels

In hot-dip galvanized DP steel, as shown above, the volume fraction of martensite is critical in determining the mechanical properties, including YS, UTS, and elongation, among other properties. Therefore, to obtain a high martensite volume fraction and comprehensive mechanical properties, the steel composition can either be adjusted [78,79] or the annealing

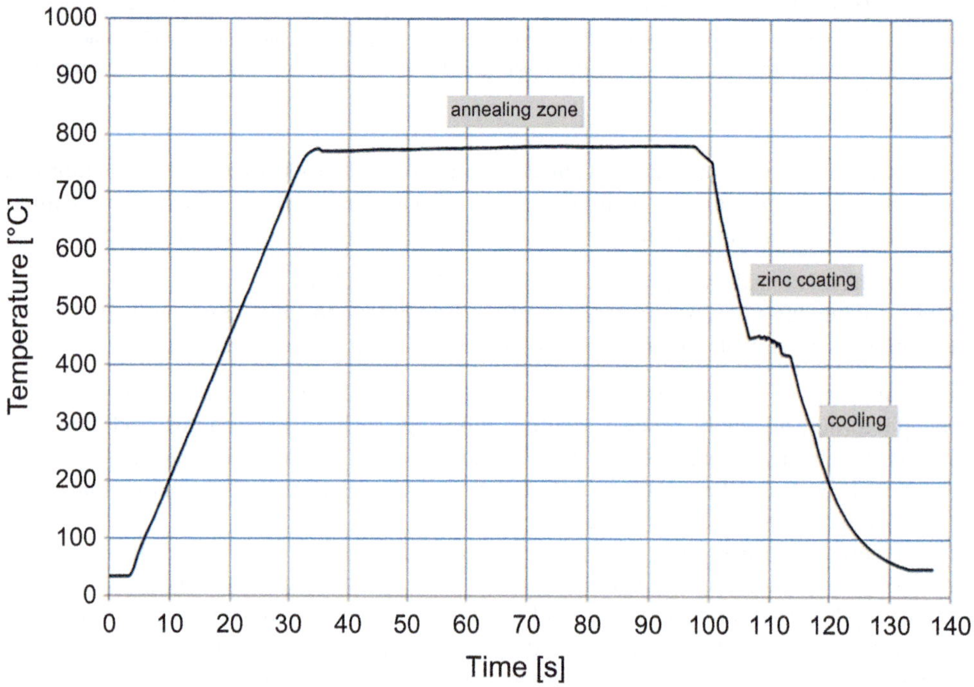

FIG. 13.29 The galvanizing thermal cycle [77].

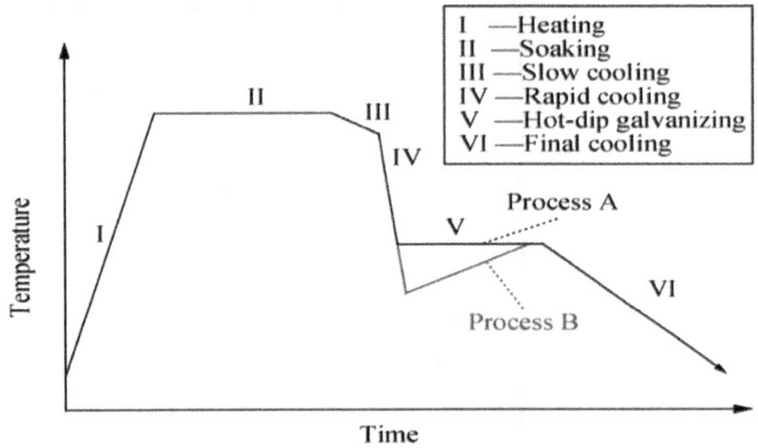

FIG. 13.30 Schematic of the hot-dip galvanizing annealing process [80].

process of DP steel can be altered. Kuang et al. [80] studied two processes on the subsequent microstructure and properties of a DP Zn-coated steel. Fig. 13.30 shows a schematic of the hot-dip galvanizing annealing process. Although similar to the standard CGL process, there were a few differences. The steels were heated to 800°C for 90 s, then either rapidly cooled to 450°C and held for 30 s (process A) or rapidly cooled to 350°C and then reheated to 450°C (process B) to simulate the hot-dip galvanizing process. In both processes, the samples were finally cooled to room temperature in air. Table 13.7 compares the YS and yield ratio of the DP600 two processes. Both process A and process B showed excellent properties in comparison to the standard DP properties for DP600, although the microstructures were slightly different. For process A, the microstructure was composed of ferrite, martensite, and a small amount of bainite. By contrast, the microstructure for process B was composed of ferrite, martensite, bainite, and cementite because of the transformation of supercooled austenite into bainite and the decomposition of martensite into ferrite and carbide [80].

Pan et al. [81] found that increasing the annealing temperature from 790°C to 850°C in a simulated CGL process promoted the austenite transformation and increases the volume fraction of martensite at sufficient cooling rates. It was also found that the total elongation was not affected by annealing temperature. Recently, Moraes et al. [82] evaluated both the soaking temperature (ST) and the quench temperature (QT) in a simulated intercritical annealing cycle based on their industrial CGL line at Usiminas, Fig. 13.31. They found that there was an increase in yield strength and tensile strength with increasing ST and tensile values were greater than 980 MPa when the ST was greater than or equal to 750°C. However, elongation decreased with increasing ST due to the increase of second-phase bainite, martensite, and MA constituent. On the other hand, quenching start temperature (QT) showed minimum influence on the microstructure and mechanical properties.

Lin et al. [83] studied GI and GA behaviors in two kinds of DP steels with 2 wt% Mn, 0.5 wt% Si, and 0.5 wt% Cr. They found that when the precoating annealing process was conducted at an extremely low dew point, external selective oxidation prevailed, resulting in various oxides on the steel surface. The oxides did not fully cover the steel surface under the reducing atmosphere and the molten Zn(Al) reacted locally with the steel surface to form Fe-Al inhibition alloys. The Al in the molten Zn(Al) bath effectively removed the oxides by aluminothermic reduction, which is an important factor contributing to the formation of the Fe-Al inhibition layer. The Fe-Al inhibition alloy was then gradually transformed to Al-bearing Fe-Zn phases. The results showed that the dual-phase steels were easily galvanized because no bare spots were observed, and the Fe-Zn phases were readily formed at the interface. During GA process, the DP steels showed slower GA kinetics compared with the IF steels. The solubility of Si in the ζ phase was extremely low. With continued GA reaction,

TABLE 13.7 Mechanical properties of hot-dip galvanized DP steels [80].

Process	YS / MPa	UTS / MPa	A_{80} / %	n	Yield ratio	UTS × A_{80}
A	362	638	24.3	0.17	0.57	15,500
B	411	629	23.5	0.13	0.65	14,780
Standard	340–420	≥600	≥20.0	≥0.14	–	–

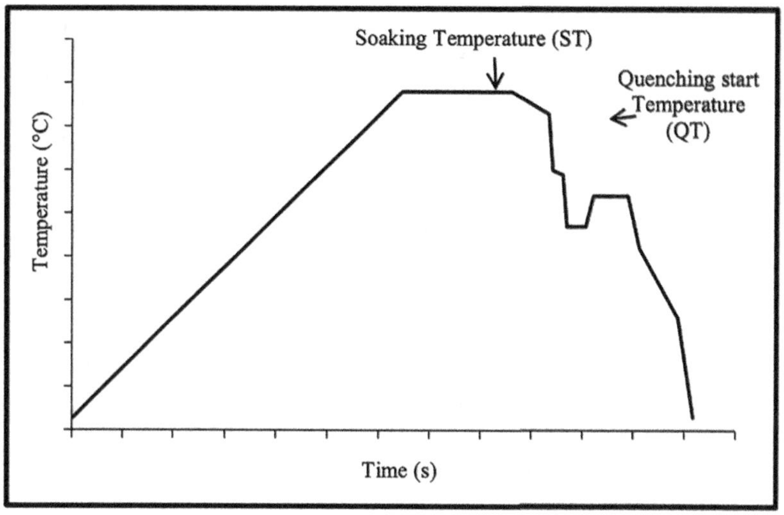

FIG. 13.31 Schematic representation of the cycles of intercritical annealing and galvannealing [82].

the ζ phase transformed into the δ phase, which contained approximately 1.0 at.%. Si. The Si also diffused into the Zn layer during the GA reaction. Hence, the ζ phase did not homogeneously nucleate at the steel substrate/Zn coating interface but was found at the area away from the interface. Therefore, the Fe-Zn phases on the CMnSiCr dual-phase steels were relatively nonuniform compared with those on interstitial-free steel.

Recently, Bleich and Cheng [84] reported that under-galvanneal, or under-alloy, of AHSS is one of the major quality concerns for the continuous galvanizing line. Based on several case studies, they reviewed various mechanisms that have caused under-alloy in MnSiCr Dual-Phase steel. Metallography results showed that nonuniform zinc coating and silicon segregation at the strip edge are two major contributing factors for edge under-alloy, Fig. 13.32. Both internal oxidation and preoxidation practices can produce abnormally heavy Zn-Fe buildup on the strip in the zinc pot during a line stop. It is believed that when the tensile strength for dual phase is over 1180 MPa, the austenite volume fraction plays a critical role in the galvannealing process. The high resistivity and low magnetic permeability of the steel could potentially max out power voltage of a longitudinal galvannealing induction furnace, causing unalloyed coated steel; thus, coating weight uniformity control across the width is critical for galvannealing quality.

13.2.2.2 Transformation-induced plasticity (TRIP) steels

In general, Si-bearing advanced high-strength steels (AHSS) possess excellent mechanical properties but poor galvanizability. The galvanizability of a transformation-induced plasticity (TRIP) steel, 1180TRIP, containing 2.2% Mn and 1.7%

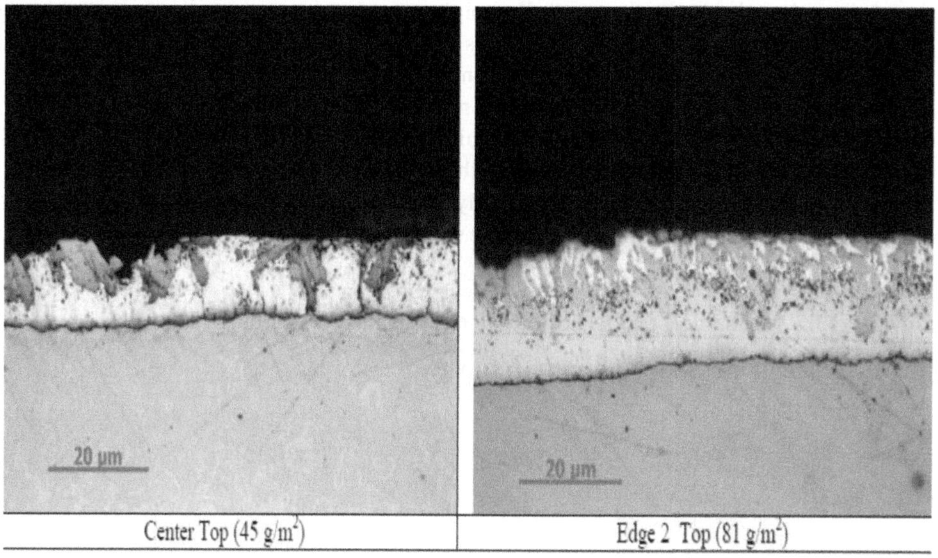

FIG. 13.32 Coating cross sections of DP590 (A) center and (B) edge locations [84].

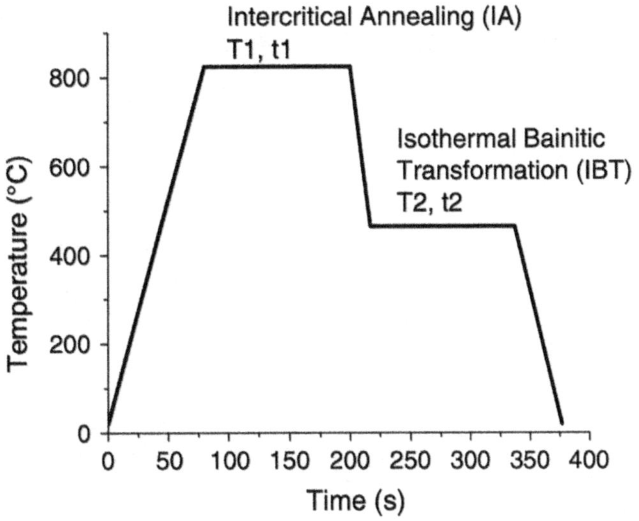

FIG. 13.33 Schematic of the experimental heat treatments [86].

Si was studied by Kim et al. [85]. They found that various oxides formed on the surface of these steels, depending on the steel composition and on the dew point control. Coating quality was good at 0°C dew point but deteriorated as the dew point decreased to 35°C and 65°C. The results showed that to ensure good coating quality on 1180TRIP steel, the formation of SiO_2 must be prevented. This can be achieved by either slightly increasing the Mn:Si ratio in the bulk concentration (for instance, slightly increasing Al content and accordingly reducing Si content) or by further increasing the oxygen partial pressure at the initial stage of annealing.

In another study by Bellhouse and McDermid [86], heat treatments were performed using an isothermal bainitic transformation (IBT) temperature compatible with continuous hot-dip galvanizing on two high Al-low Si transformation induced plasticity (TRIP)-assisted steels. Both steels had 0.2 wt% C and 1.5 wt% Mn; one had 1.5 wt% Al and the other had 1 wt% Al and 0.5 wt% Si. Fig. 13.33 is a schematic of the heat treatment cycle used. Two different intercritical annealing temperatures (ITA) were used resulting in intercritical microstructures of 50% ferrite (α)-50% austenite (γ) and 65% α-35% γ. The effect of time on the isothermal bainite transformation (IBT) temperature of 465°C, was evaluated. Increasing IBT time resulted in an increase in the uniform elongation and an increase in retained austenite. The best combinations of strength and ductility and their corresponding heat treatments to produce 50% retained austenite are found in Table 13.8.

A similar study using an IBT of 460°C with the same thermal cycle and alloy composition confirmed that elongation and retained austenite increased with increasing IBT time [87]. Later, Ding et al. [88] also illustrate the potential for replacing Si by additions of Al and P without lowering the strength and ductility in similar TRIP steels. They reported that the IBT was critical in determining the mechanical properties of the steel since it controls the fraction of bainite. They found that with increasing bainite fraction, the fraction of retained austenite increases while the fraction of martensite decreases. The desired microstructure containing ferrite, bainite, retained austenite, and martensite was obtained after the various heat treatments, Fig. 13.34. Three heat treatment parameters can be controlled during hot-dip galvanizing, i.e., intercritical annealing (IA) temperature, isothermal bainitic transformation (IBT) temperature, and IBT time. It was found that the IA temperature has a negligible influence on the proeutectoid ferrite formation since the cooling rate of 20°C/s was slow enough to allow for further proeutectoid transformation upon cooling, resulting in similar quantities of proeutectoid ferrite

TABLE 13.8 Summary of optimal mechanical properties [86].

Alloy	IA (°C)	IBT (s)	UTS (MPa)	Uniform elongation	UTS × uniform elongation (MPa)
1.5% Al	862 (50% γ)	90	895	0.26	233
1.5% Al	862 (50% γ)	120	880	0.27	238
1.0% Al	825 (50% γ)	120	1009	0.22	222

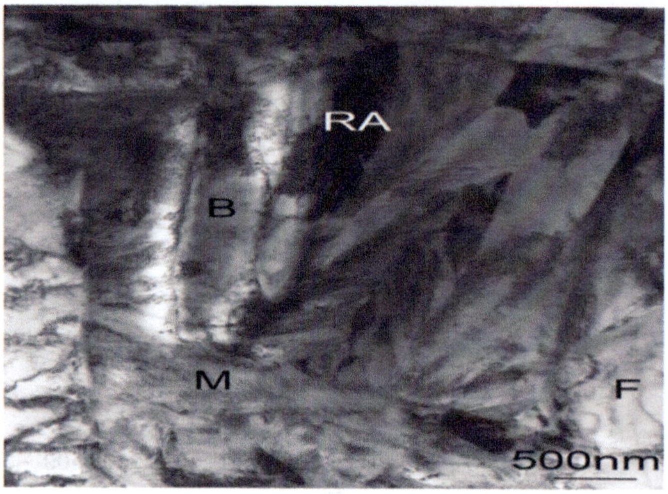

FIG. 13.34 TEM micrograph of the C-Mn-Al-P steel [88].

for IA temperatures between 820°C and 880°C. The IBT treatment on the other hand was found to have a significant effect on microstructure and mechanical properties. The most significant parameter was the IBT time where prolonged times gave larger bainite fractions and also changed the characteristics of the retained austenite, especially the stability of the retained austenite. The large effect of IBT time on the mechanical properties was mainly due to the differences in the fraction of retained austenite [88].

A different thermal cycle was evaluated by Lee et al. [49] that focused on the design of a robust GI/GA-compatible 800MPa TRIP steel with the following composition: 0.15% C, 2.0% Mn, 0.3% Si, 0.8% Al, and 500ppm P, Fig. 13.35. This TRIP steel with a reduced amount of Si could be galvanized and galvannealed in normal HDG line conditions. No amorphous SiO_2 layer was formed at the surface of this TRIP steel after an annealing simulation in a standard low dew point galvanizing line gas atmosphere. A thermal cycle with an IBT temperature of 465°C resulted in a higher retained austenite content than a cycle with an IBT temperature of 400°C. The volume fraction of retained austenite decreased and the amount of C in the retained austenite was reduced by galvannealing. Processing at a temperature of 560°C for a duration of 30s resulted in a GA coating with the δ phase making up to 80% of the coating thickness and the Fe content of the GA coating was approximately 10%, Fig. 13.36.

13.2.2.3 Twinning-induced plasticity

The high Mn content of TWIP steels presents great difficulty in processing Zn coated product. Besides the high cost of Mn, difficulties arise because the austenitic structure results in low yield strengths, complications due to surface oxide formation, and liquid metal embrittlement of Zn-coated TWIP steel [89]. A summary of the steel compositions and the

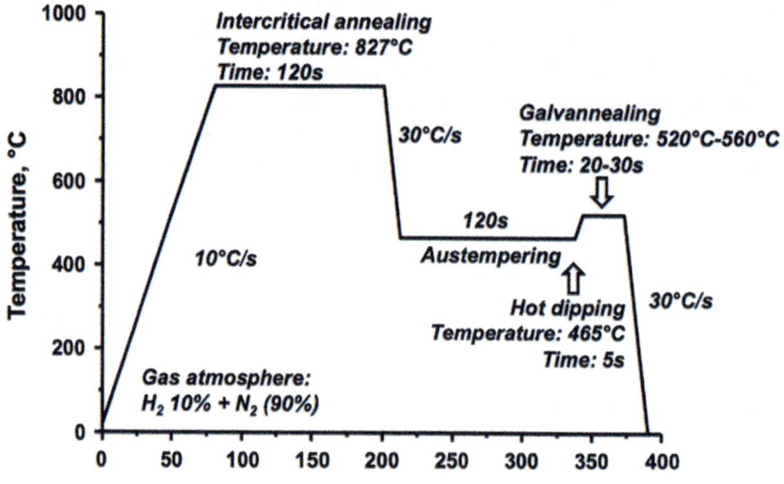

FIG. 13.35 Example of a thermal cycle used in the course of the present project to simulate continuous galvanizing and galvannealing. Note that the austempering was carried out at 465°C [49].

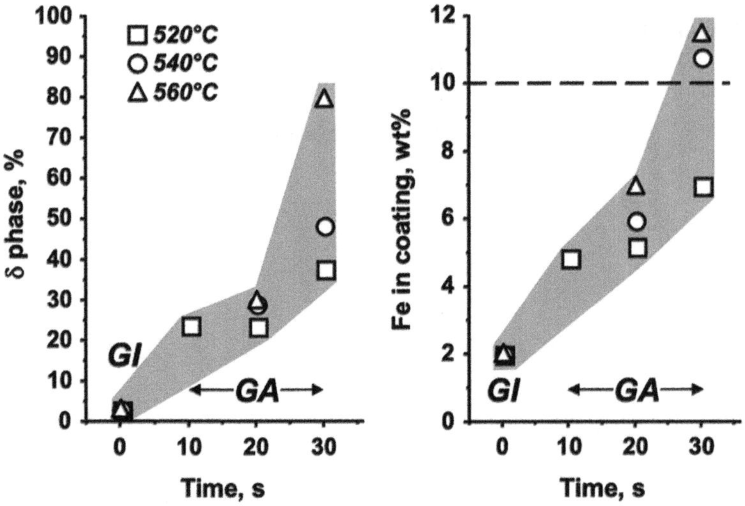

FIG. 13.36 Increase of the δ-phase volume percentage and Fe concentration of the coating during galvannealing [49].

TABLE 13.9 A summary of the steel composition and key annealing parameters for the TWIP steel selective oxidation studies in literature [91–98].

No.	Composition (mass%)	Soaking temperature (°C)	Soaking time (s)	H_2 content (vol%P)	Dew point (°C)	Note
1	18Mn-0.1Si-0.6C	800	162	10	−17, −3	
2	18Mn-0.1Si-0.6C	800	162	10	−17	
3	18Mn-0.2Si-1Al-0.5C	800	162	10	−17	
4	17Mn-0.1Si-1.2Al-0.5C	800	43	15	−40	
5	17Mn-0.1Si-1.2Al-0.5C	800	43	15	−20	
6	17Mn-0.1Si-1.2Al-0.5C	800	43	15	−20	Ni precoating
7	23Mn-0.3Si-0.6C	800	60	5	−30	Preoxidation
8	17Mn-0.06Si-1.95Al-0.66C	500–650	0–180	0–20	−20	

key annealing parameters in TWIP steel selective oxidation studies is compiled by Jin et al. [90] in Table 13.9. Due to these issues and their poor welding characteristics, use of these steels has been very limited and their future as an automotive material is uncertain [2], especially Zn-coated TWIP steels.

13.2.2.4 Third-generation AHSS

Initial research by McDermid et al. [99] studied two prototype 3G AHSS compositions that were thermally processed with the objective of attaining high volume fractions of retained austenite and properties within the target 30,000 MPa% $\leq$ UTS $\times$ %elong $\leq$40,000 MPa% property envelope such that the thermal treatments were CGL-compatible. The prototype steels were based on the medium-Mn C-6Mn-Si-Al system, and both steels achieved properties consistent with the 3G AHSS property envelope. The generic thermal profile is pictured in Fig. 13.37. Both steels achieved properties within the 3G AHSS property envelope where high volume fractions of relatively stable retained austenite were present. The best

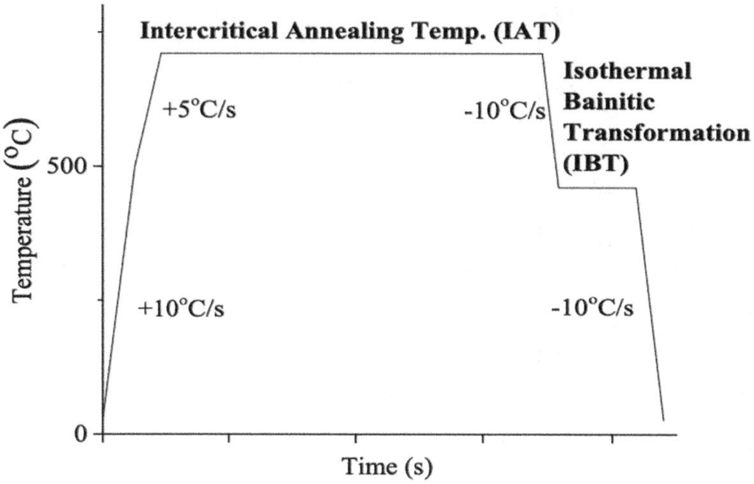

FIG. 13.37 Generic experimental thermal profile [99].

heat treatment conditions tested, which are compatible with the continuous galvanizing process, were found to be at 675°C IAT with a 120 s holding time for a martensitic starting microstructure [100].

A parallel study [82] sought to evaluate thermal processing routes to obtain enhanced mechanical properties and the desired complex microstructure comprising significant volume fractions of bainitic ferrite and transformable retained austenite. The development of surface chemistry during intercritical annealing and its subsequent effect on reactive wetting during galvanizing were also characterized. An ultimate tensile strength of 1415 MPa and 10% elongation were achieved for the experimental 3G AHSS steel (12C, 6.4 Mn, 0.5 Si, 1.68 Al), using intercritical (60% γ) annealing for 120 s followed by cooling at 20°C/s to the 460°C galvanizing temperature, holding for 60 s, and finally cooling to room temperature at 10°C/s. Surfaces that could be successfully reactively wet by the Zn(Al,Fe) bath were produced using a −30°C dew point and 95%N_2–5%H_2 process atmosphere. Fig. 13.38 shows a schematic of the experimental heat treatment cycle, which consisted of heating to the intercritical annealing temperature (IAT) of 670°C, holding at the IAT for either 60 or 120 s followed by cooling at 20°C/s to the 460°C galvanizing temperature, holding at 460°C for either 60 s or 120 s, and finally cooling to room temperature at 10°C/s. Using the intercritical heat treatment, which achieved the maximum volume fraction of retained austenite (i.e., 670°C + 120 s followed by 460°C + 60 s resulting in 24 vol% retained austenite), the target ultimate tensile strength of 1200 MPa (minimum) was achieved; however, the target % total elongation (10%) did not meet the target value of 20% (minimum).

In a subsequent study, McDermid et al. [50] continued the study of the prototype C-6Mn-xSi-yAl-zSn alloys to determine if third-generation advanced high-strength steel (3G-AHSS) target properties (i.e., $24{,}000 \leq \text{UTS} \times \text{TE} \leq 40{,}000$ MPa%) could be achieved using Zn coating CGL-compatible heat treatments. A schematic thermal profile for their intercritical heat treatment is shown in Fig. 13.39. It was found that the mechanical properties consistent with 3G-AHSS targets were obtained using the prototype C-6Mn-xSi-yAl-zSn system and CGL-compatible heat treatments when using either a martensitic or martensite-ferrite starting microstructure. The 3G-AHSS properties observed

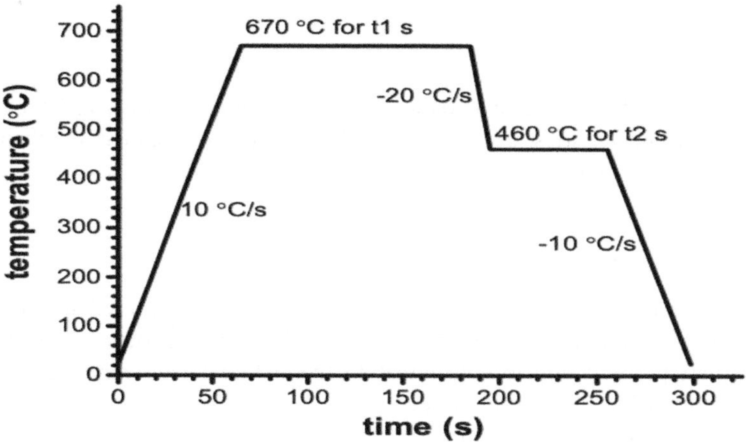

FIG. 13.38 Schematic of heat-treatment schedule [101].

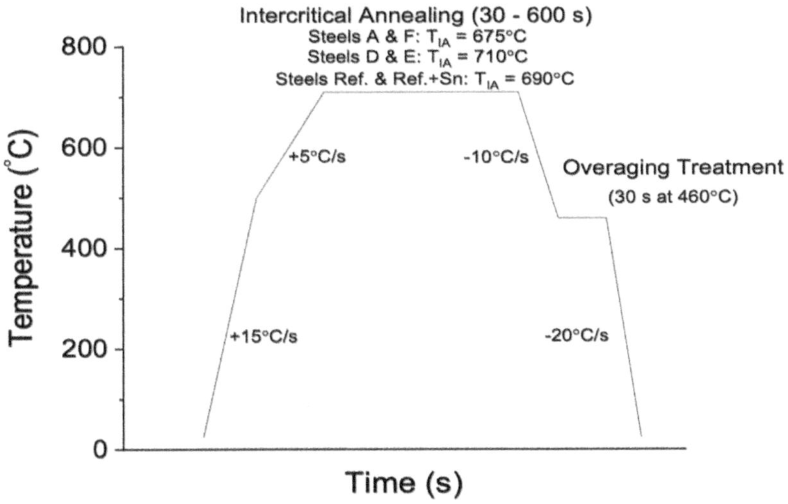

FIG. 13.39 Schematic of intercritical annealing heat treatments [50].

were obtained by having high volume fractions of chemically stable retained austenite, which gradually transformed to martensite during deformation. There was also evidence that additional plasticity enhancement was obtained by twinning-induced plasticity in the retained austenite.

The "quenching and partitioning" (Q&P) process [30,102] is a unique technological route for the production of high-strength steels with significant amounts of retained austenite that provides a better combination of strength and ductility. In this work, intercritically annealed specimens followed by a Q&P treatment have been applied to low-carbon steel with chemical composition typical for conventional TRIP-assisted steels. The hot-galvanized Q&P treatment involves dipping the steel in a molten zinc bath at a temperature around 460°C with a duration of less than 30s. Sun and Yu [103] developed two schemes for a hot-dipping galvanization process, Fig. 13.40. Their results indicate that both hot-dip galvanizing schemes present a limited reduction in tensile strength and a slight enhancement of ductility. The scheme of galvanizing and partitioning after the quenching progress showed a better combination of strength and ductility.

The Q&P microstructure was mainly composed of ferrite, lath martensite, martensite-austenite islands, retained austenite, and a small amount of bainite formed during partitioning. The fraction of bainite formed during partitioning was proportional to the quenching temperature. Quenching and partitioning (Q&P) treatment was an effective way for retaining a significant fraction of austenite, with a maximum austenite fraction of approximately 10%. In comparison with the conventional Q&T process, the steel treated by Q&P process exhibited an improved combination of strength and ductility due to the presence of retained austenite. Both hot-dip galvanizing schemes presented a limited reduction in tensile strength and a slight enhancement of ductility. However, the scheme of galvanizing and partitioning after the quenching progress shows a better combination of strength and ductility [103].

In more recent research, a novel hot-galvanized quenching and partitioning (Q&P) steel (0.225C-0.85Si-2.02Mn-0.91Al, in wt%) with excellent strength-ductility balance was fabricated by intercritical annealing followed by a high-temperature short-time overaging treatment [104]. The Q&P process was compared with the conventional quenching and austempering (QAT) process, Fig. 13.41. The combined effects of structural subdivision by primary martensite and carbon partitioning simultaneously from martensite and bainite lead to the retention of larger amounts of retained austenite (RA) constituents with different morphologies, sizes, and stabilities at room temperature. On the contrary, the QAT samples have coarser martensite-austenite (M-A) islands and smaller fraction of RA due to relatively insufficient partition and bainitic transformation kinetics Fig. 13.42. As compared with the QAT treatment, the better mechanical properties with a yield strength (YS) of 736 MPa, tensile strength up to 1020 MPa, total elongation (TEL) of around 24%, and product of strength and elongation (PSE) above 24 GPa% can be achieved after the Q&P process, and particularly, more ductile tempered martensite and refined M-A particles result in a simultaneous increase of YS and TEL. The instantaneous work-hardening exponent vs. true strain curve of Q&P samples displays a small peak followed by a long plateau region before break, which can be attributed to a less active and sustained TRIP effect [104].

In further work, Wang et al. [105] determined the effects of quenching temperature on phase transformation kinetics, microstructure evolutions, and mechanical properties of a hot-galvanized Q&P steel (0.225C-0.85Si-2.02Mn-0.91Al, in wt %). The thermal cycle used is seen in Fig. 13.43. Their results indicated that a larger fraction of primary martensite at lower quenching temperature could strongly promote subsequent bainite transformation kinetics, which is attributed to more martensite-austenite interfaces and defect density. They were able to develop a model to predict RA fractions by taking

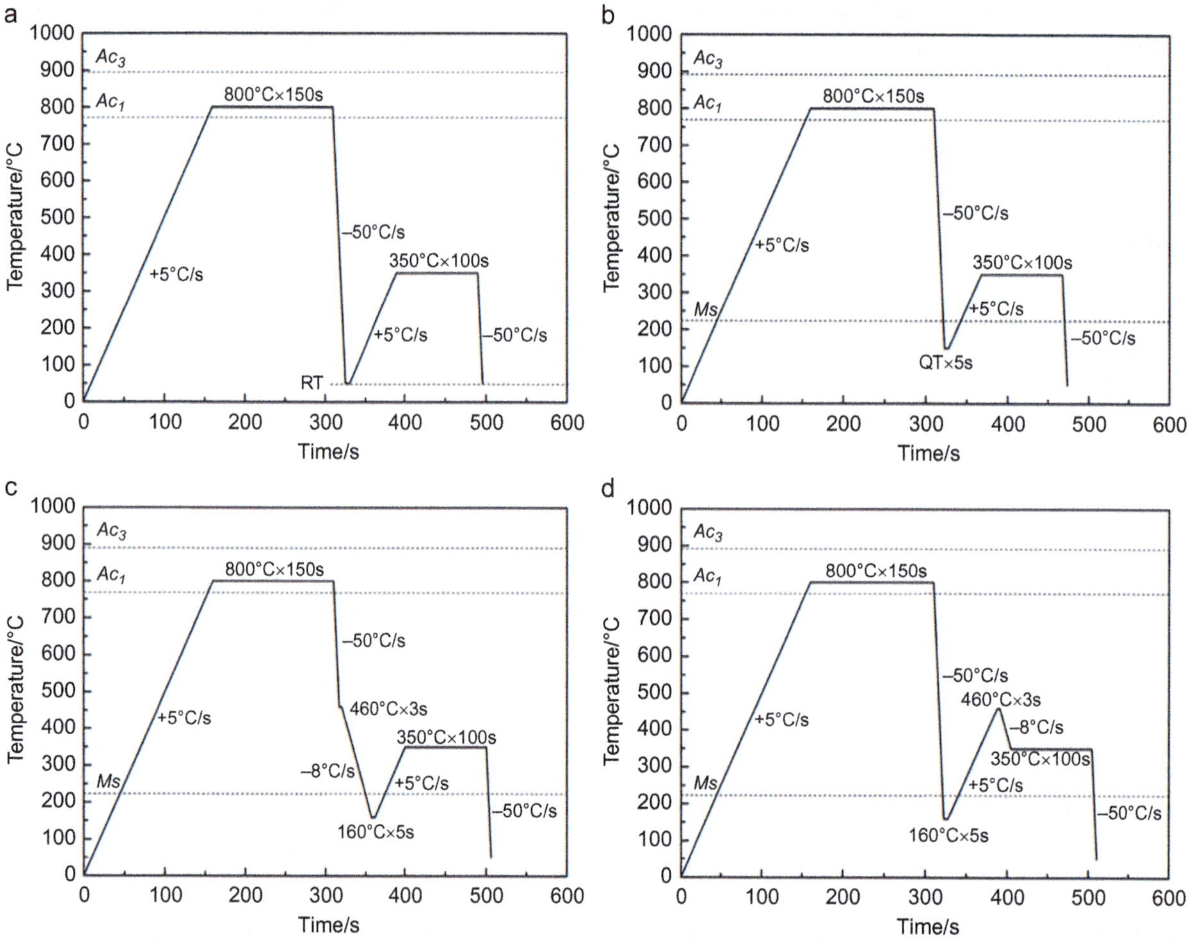

FIG. 13.40 Schematic illustration of heat treatments applied to the steels: (A) Q&T process; (B) Q&P process; (C) galvanizing before quenching and partitioning progress; (D) galvanizing and partitioning after the quenching progress [103].

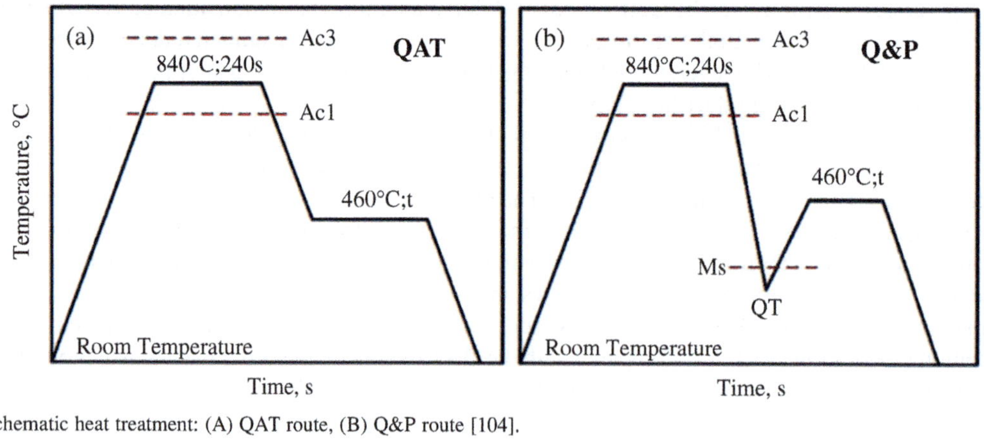

FIG. 13.41 Schematic heat treatment: (A) QAT route, (B) Q&P route [104].

into account intercritical ferrite and the short time bainite transformation. They found that as the quenching temperature decreases, small-sized blocky RA along martensite boundaries and filmy RA between martensite laths increase, while coarse lath/blocky RA inside bainite structures or at bainite boundaries decreases gradually. In addition, the YS decreases from 763 to 431 MPa with the increase of quenching temperature, while the UTS varies in a narrow range between 967 and 1036 MPa. A stable TEL of 22.6%–25.7% was obtained at a wide range of quenching temperature (150–275°C), which was attributed to the joint effects of the TRIP effect and the multiphase structure.

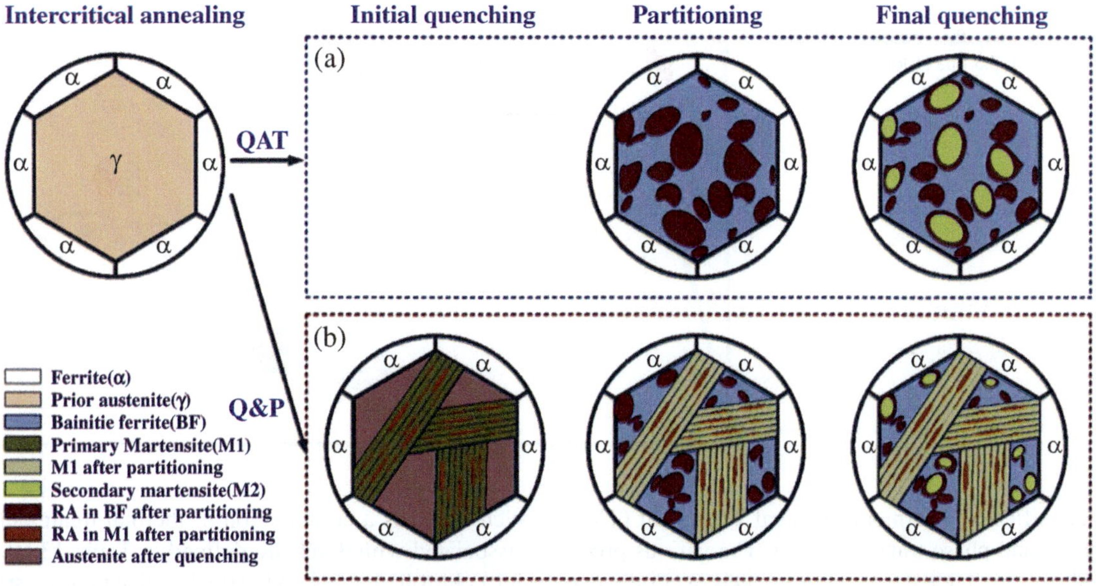

FIG. 13.42 Schematic diagrams representing microstructural evolutions [104].

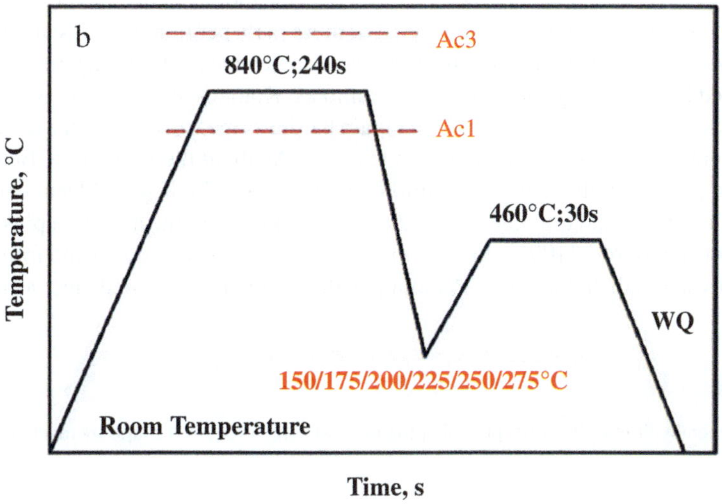

FIG. 13.43 The thermal cycle heat treatment process [105].

Most recently [106], it was reported that Arcelor Mittal has taken the new SIDGAL3.5 process section into service. They were able to achieve the requirements for thermal treatment homogeneity and obtain coating properties. A schematic of the possible Q&P annealing cycle used in the SIDGAL 3.5 line is shown in Fig. 13.44.

13.3 Hot press forming of Zn-coated steel

13.3.1 Hot press forming

In 1920, Camp and Francis (Making, Shaping and Treating of Steel, Carnegie Steel Company) classified hot working as working at a temperature above the upper critical range, with the temperature varying inversely with C content from 700°C to 900°C. They identified three types of hot working processes: hammering, pressing, and rolling. Later, sheet forming methods were developed, which comprise deformation processes in which a metal blank is shaped by tools and dies, primarily under the action of tensile stresses. The design and control of such processes depend upon many factors, including the characteristics of the workpiece material and tooling temperature. In more recent times, these processes have been

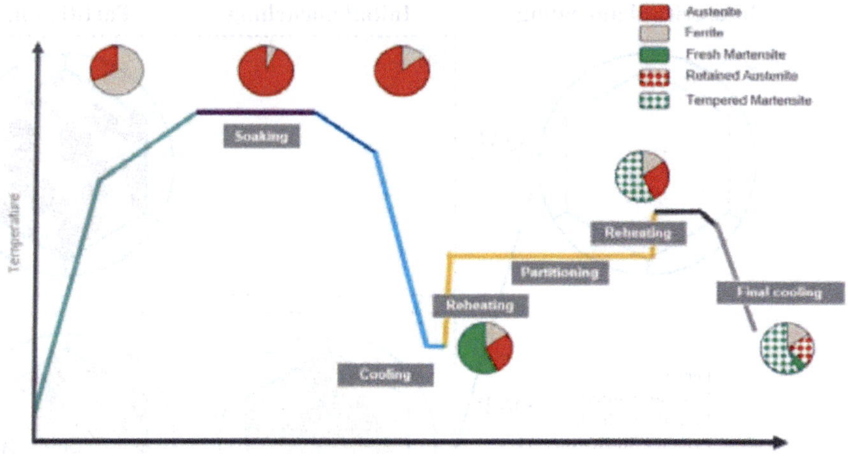

FIG. 13.44 The typical annealing cycle for Q&P processing on the Arcelor Mittal ANDRITZ SIDGAL 3.5 line [106].

broadly classified as Thermomechanical Treatment (TMT), which involves a combination of thermal treatment, plastic deformation, and phase transformation. The various processes associated with TMT are defined in Table 13.10.

The advantage of hot-stamped boron steels over fully martensitic cold-rolled or hot-rolled steels is that they are stamped at high temperatures where ductility is much higher than at room temperature. This allows for the formation of much more complicated stampings than could be formed from conventionally stamped, cold-rolled martensitic steel. The downside is that the cycle time required to stamp and cool the steel in the die is long and the tooling is quite expensive. Compared with stamping conventional steels, where it is not uncommon to produce 12 or more parts per minute, hot stamping generally is restricted to about one per minute, although this production rate can be increased somewhat by designing dies that stamp more than one part.

TMT processes involved both pressing and rolling. As examples, Komatsu et al. [107] invented a process and apparatus for press-forming and quenching steel. They were able to obtain hardness measurements in excess of R_c 60, indicative of the formation of martensite from the high-temperature austenite phase. At about the same time, Bethlehem Steel was utilizing TMT to produce hot-rolled product with significantly improved properties. Melloy and Dennison [108] produced an unexpected combination of high strength and toughness by rolling the workpiece in the austenite plus ferrite region, followed by cooling. Similarly, Bramfitt and Marder [109] were able to hot roll austenite as it cooled and formed bainite, Fig. 13.45. Hot stamping was developed by a Swedish company, Plannja [110], in which a hardened steel article is formed by heating a

TABLE 13.10 Definitions.

The following definitions generally fit into the concept of shaping or producing a required shape by mechanical force often times using elevated temperature.
***Hot* forming.**
Deforming metal plastically at such a temperature and strain rate that recrystallization takes place simultaneously with the deformation, thus avoiding any strain hardening. Also known as ***hot working***.
***Hot* press forging.**
Plastically deforming metals between dies in presses at temperatures high enough to avoid strain hardening.
***Hot* working.**
Deforming metal plastically at such a temperature and strain rate that recrystallization takes place simultaneously with the deformation, thus avoiding any strain hardening.
***Hot*-cold working.**
(1) A high-temperature thermomechanical treatment consisting of deforming a metal ***above its transformation temperature*** and cooling it quickly enough to preserve some or all of the deformed structure. (2) A general term synonymous with *warm working*.
Warm working.
Plastically deforming metal at a temperature above ambient (room) temperature but below the temperature at which the material undergoes recrystallization.

TABLE 13.10 Definitions—cont'd

Forging.
Plastically deforming metal, usually ***hot***, into desired shapes with compressive force, with or without dies.
Forge-Shaping
Includes the steps of heating the sheet-shaped material to a predetermined temperature, inserting the heated material into dies, pressing and continuously quenching the material while being kept pressed.
Press forming
Any sheet metal-forming operation performed with tooling by means of a mechanical or hydraulic press.
Shaping
Producing flat surfaces using single-point tools. The work is held in a vise or fixture, or is clamped directly to the table. The ram supporting the tool is reciprocated in a linear motion past the work.
Stamping
A general term covering almost all sheet metal press operations. It includes blanking, shearing, ***hot*** or cold forming, drawing, bending, and coining.
Hot stamping
A thermo-mechanical forming process with intended phase transformations

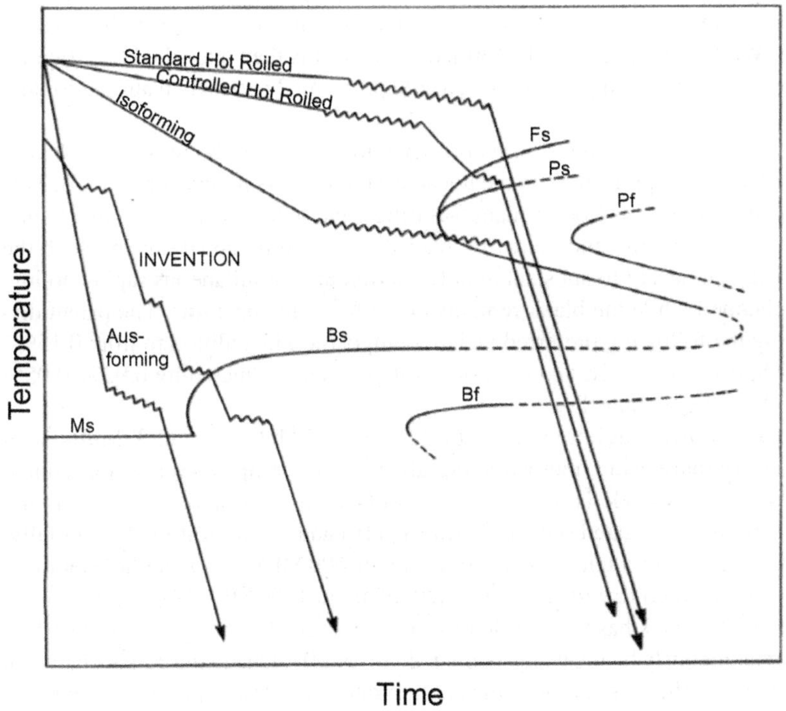

FIG. 13.45 Schematic of various hot rolling processes superimposed on a CCT diagram. Invention refers to bainitic rolling to produce a complete bainitic structure [109].

hardenable steel blank to a hardening temperature and then placed in a forming apparatus in which the blank is deformed and simultaneously rapidly cooled to obtain a martensitic and/or bainitic structure. Okunishi et al. [111] invented a method for forge-shaping sheet. A martensitic stainless steel is used to hot-forge, hot-press shaped, and press-quenched. On quenching they were able to form mixed structures of ferrite and martensite. The development of TMT continued and as an example, Jonsson of SSAB developed a press hardening technique (austenitize the blank, hot press, and quench) to produce bumper beams [112] and a B-pillar for a vehicle body [113].

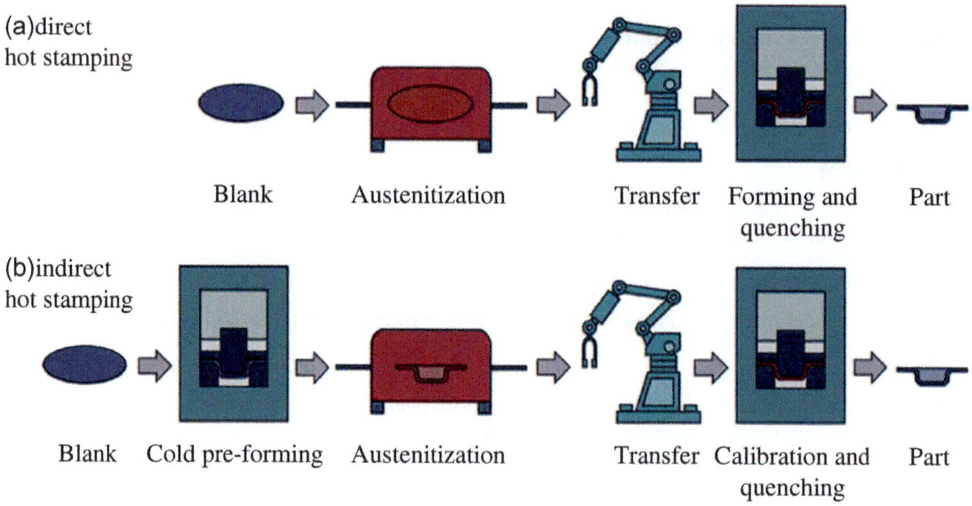

FIG. 13.46 Schematic of the direct and indirect hot stamping process chain [114].

Hot stamping, also known as press hardening, is the practice of thermally treating steel either after it has been shaped or at the same time as shaping. With hot stamping, the steel parts ultimately may achieve a high mechanical resistance after the thermal treatment transforms the steel into martensite and/or bainite (depending on the desired tensile strength). Because the steel is shaped in the press before the conversion to martensite, they retain some ductility during the shaping process, which allows the formation of complex parts. Fig. 13.46 shows two exemplary hot-stamping operations. In the direct hot stamping operation, the steel sheet is first heated to a temperature above the austenitization temperature. Then, the hot steel sheet is placed into a die and formed into a part. While still in the die, the formed steel sheet is then quenched (i.e., rapidly cooled) to achieve a martensitic structure. In the indirect hot stamping method, a cold preformed part is then austenized and quenched.

Hot stamping was pioneered by the Swedish company Norrbottens Järnverk, AB in the 1970s. Norbottens Järnverk later became Swedish Steel. Norbottens Järnverk filed a patent application on the hot stamping process in Sweden in 1973. This patent was then filed in Great Britain [110] and discloses a method of manufacturing a hardened steel article, in which a blank of hardenable steel is heated to a hardening temperature and thereafter placed in a forming apparatus in which the blank is formed to the desired final shape by being subjected to substantial deformation and simultaneous rapid cooling, such that a martensitic or bainitic structure is obtained while the blank remains in the forming apparatus. The patent discloses an exemplary hot-stamping steel as having the following preferred weight composition in addition to iron: 0.15%–0.25% C, 0.7%–1.5% Mn, <0.03% P, <0.04% S, 0.1%–0.3% Cr, and/or various other elements, including 0.02%–0.05% Ti, 0.03%–0.07% Al, and 0.03%–0.07% B, with no significant Si.

Investigations on ultrahigh-strength steels have shown that boron alloys of 22MnB5, 27MnCrB5, and 37MnB4 steel grades are the only steel grades that produce a fully martensitic microstructure after hot stamping when a water- cooled tool is used, Table 13.11 [115]. Although 22MnB5 steel grade is the most commonly used steel grade in hot stamping processes, other modifications of this grade have been investigated such as 20MnB8 [116] and 22MnMoB8 [117]. Initially, the material exhibits a ferritic-pearlitic microstructure with a tensile strength of about 600 MPa. After the hot stamping process, the component finally has a martensitic microstructure with a total strength of about 1500 MPa. In order to achieve such a microstructure and hardness transformation, the blank has to be austenitized in a furnace for at least 5 min at 950°C. Afterward, the blank is formed and quenched simultaneously by the water-cooled die for 5–10 s. Due to the contact between the hot blank and the cold tool, the blank is quenched in the closed tool. If the cooling rate exceeds a minimum cooling rate of approximately 27°K/s, at a temperature of around 400°C, a diffusionless martensitic transformation will be induced, which finally is responsible for the resulting high strength of the part [118]. A typical temperature vs. process time for hot press forming is presented in Fig. 13.47.

13.3.2 Zn coatings

Zinc-based coating systems have been developed in association with hot forming technologies for oxidation protection during hot forming and corrosion resistance during service. An important processing step in direct hot forming is to heat

TABLE 13.11 Chemical components and mechanical properties of boron steels [115].

Steel	Al	B	C	Cr	Mn	N	Ni	Si	Ti
20MnB5	0.04	0.001	0.16	0.23	1.05	–	0.01	0.40	0.034
22MnB5	0.03	0.002	0.23	0.16	1.18	0.005	0.12	0.22	0.040
8MnCrB3	0.05	0.002	0.07	0.37	0.75	0.006	0.01	0.21	0.048
27MnCrB5	0.03	0.002	0.25	0.34	1.24	0.004	0.01	0.21	0.042
37MnB4	0.03	0.001	0.33	0.19	0.81	0.006	0.02	0.31	0.046

			Yield stress in MPa		Tensile strength in MPa	
Steel	Martensite start temperature in °C	Critical cooling rate in K/s	As delivered	Hot stamped	As delivered	Hot stamped
20MnB5	450	30	505	967	637	1354
22MnB5	410	27	457	1010	608	1478
8MnCrB3	–[a]	–[a]	447	751	520	882
27MnCrB5	400	20	478	1097	638	1611
37MnB4	350	14	580	1378	810	2040

[a]*Unknown value.*

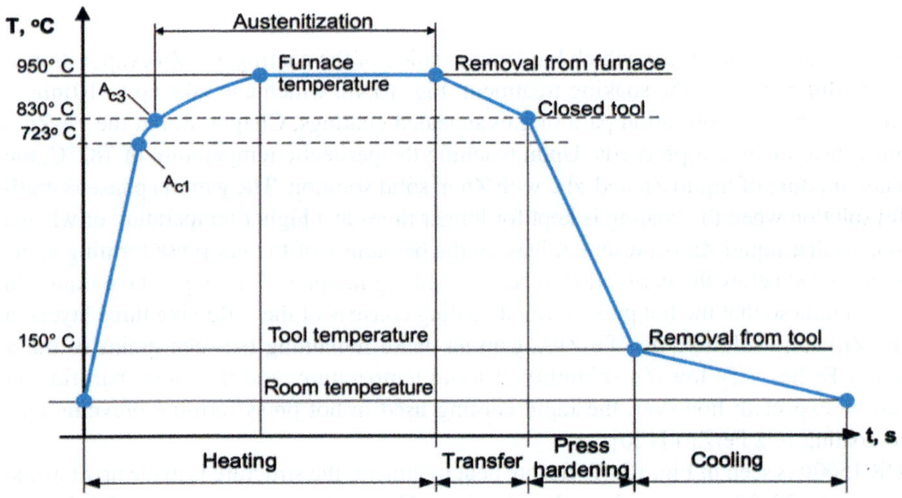

FIG. 13.47 Temperature vs. process time for hot press forming [119].

the steel blanks above the austenitizing temperature for about 3–10 min, which results in surface oxidation and occasionally decarburization of the surface substrate. The surface oxides are detrimental to heat transfer, tool and steel friction behavior, and increase wear of the dies. In earlier technologies, the scale was removed by chromium shot blasting. The present processing technology utilizes coated steel sheet, in particular Zn coatings, instead of bare steel. The coatings prevent surface oxidation and decarburization and enhance corrosion resistance [120]. Several comprehensive reviews of the hot forming process can be found, e.g., [114,120].

Both GI and GA coatings are used as starting materials for hot stamping. The lower Al content of the GA coating results in a faster Fe-Zn reaction during the hot stamping process because of the incomplete coverage of the Fe-Al inhibition layer at the steel-coating interface, as described in Chapter 5. For both GI and GA precursor coatings, much of the Al in the

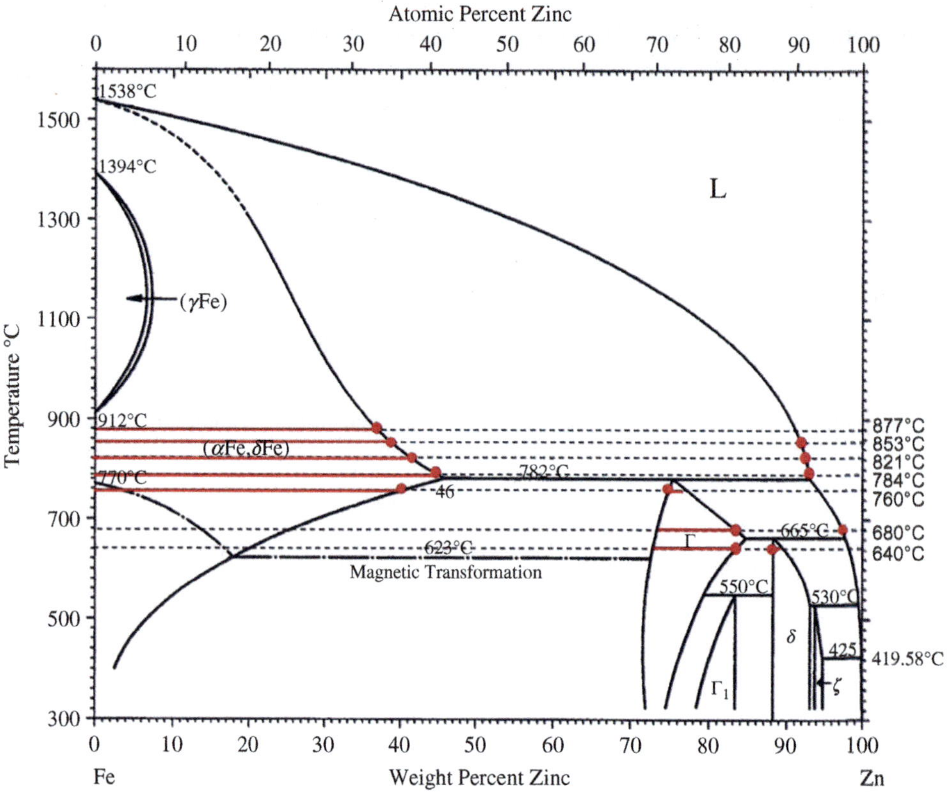

FIG. 13.48 Fe-Zn binary-phase diagram with the phase assumed to exist in the state at each temperature [121].

coating overlay is used to form an Al_2O_3 oxide surface layer, which hinders oxidation of the coating and Zn evaporation at the soaking temperature required for 22MnB5. During the soaking treatment, Fig. 13.48, which can take up to 10min at 880–930°C [121], the coating first follows the transformation path of galvannealed coatings, Chapter 8, but the Fe_3Zn_{10} gamma phase then continues to grow as heat treatment proceeds. Upon reaching the peritectic temperature of 782°C, the gamma phase transforms to a two-phase mixture of liquid Zn and αFe with Zn in solid solution. The gamma phase is gradually consumed by the α Fe(Zn) solid solution when the coating is kept for longer times at a higher temperature or when a thinner coating is used [122]. This means that liquid Zn is present at least at the beginning of the hot press forming operation. After soaking, the coated sheet is cooled below the peritectic temperature during hot press forming and the liquid, in equilibrium, should transform back to gamma so that the hot press-formed coating consists of the following three layers: a surface oxide layer, a matrix of α Fe(Zn) solid solution, and Fe_3Zn_{10} gamma phase remaining between grains of the α Fe(Zn) solid solution. In equilibrium, α Fe has very low Zn solubility at room temperature, and therefore, partitioning of Zn into other Fe-Zn phases would be expected; however, the rapid cooling used in hot press forming prevents this, so that most of the hot press formed coating is α Fe(Zn) [120].

The microstructure of the USIBOR 1500P is seen in Fig. 13.49. Before heat treatment, the structure is made up of a zinc coating with a 6 μm intermetallic layer and a 20–22 μm typical overlay structure. The steel substrate consists of deformed ferrite and pearlite. After heat treatment, the coating has developed a layered coating of zinc rich Zn(Fe) over a layer of α Fe(Zn). Fig. 13.50 shows the effect of time on the coating structure. The total coating layer increases from 34 to 47 μm, the interdiffusion layer increases from 6 to 25 μm, and porosity increases as the time at temperature increases from 8 min to 28 min.

Fig. 13.51 shows the schematic illustration of the structural change of the GA coating during hot stamping heating [124].

- At room temperature to 640°C, the double layers, in which the upper layer was the δ_1 phase and the lower one was the Γ phase, were formed. As the liquid Zn phase did not exist during heating, the Γ phase grew through solid-state diffusion.
- At 680°C, the part that was only the δ_1 in the upper layer at 640°C separated to the pillar-shaped Γ and the δ_1 phase.
- At 760°C, the upper layer changed to only the Γ phase and the Fe-Zn solid solution formed at the interface between the Γ phase and the substrate.

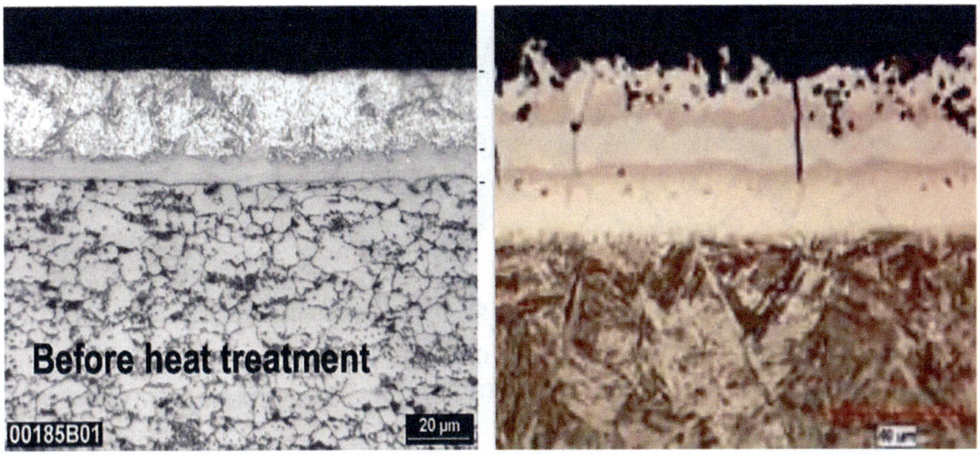

FIG. 13.49 The change in microstructure after heat treatment [123].

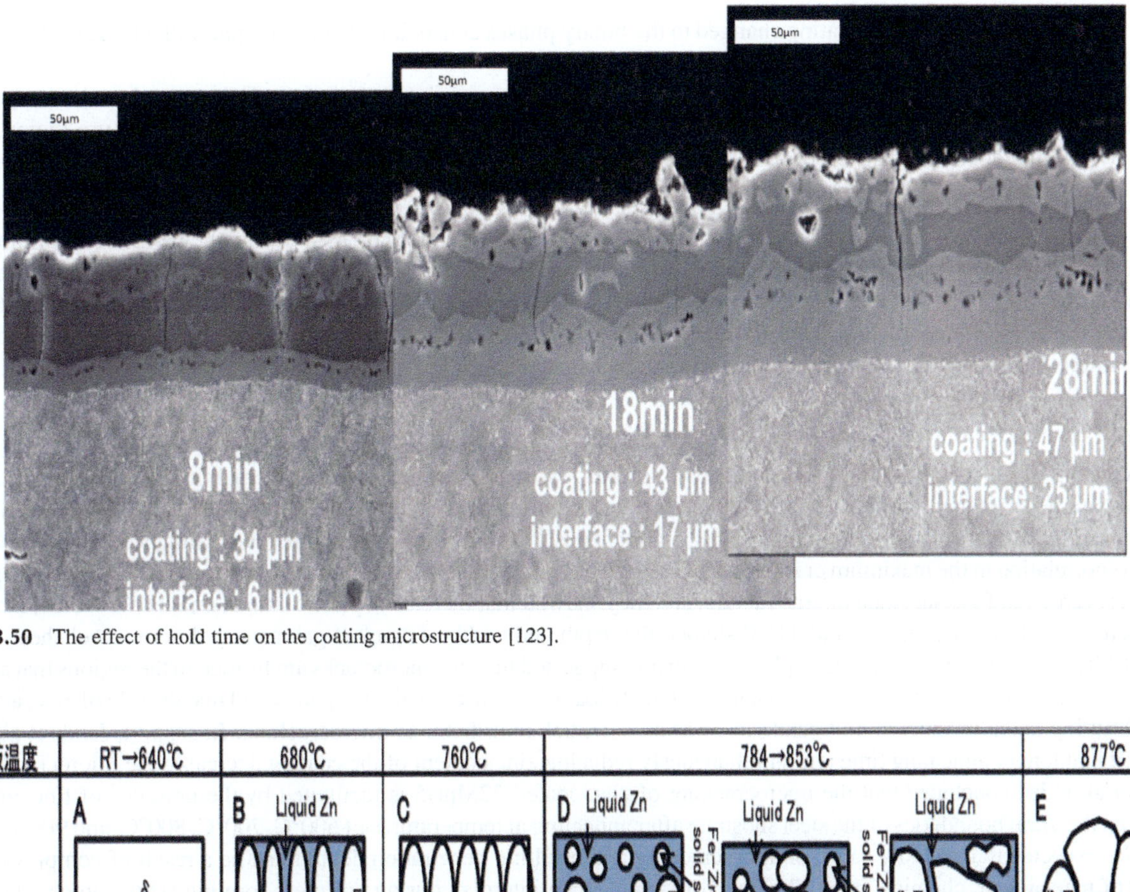

FIG. 13.50 The effect of hold time on the coating microstructure [123].

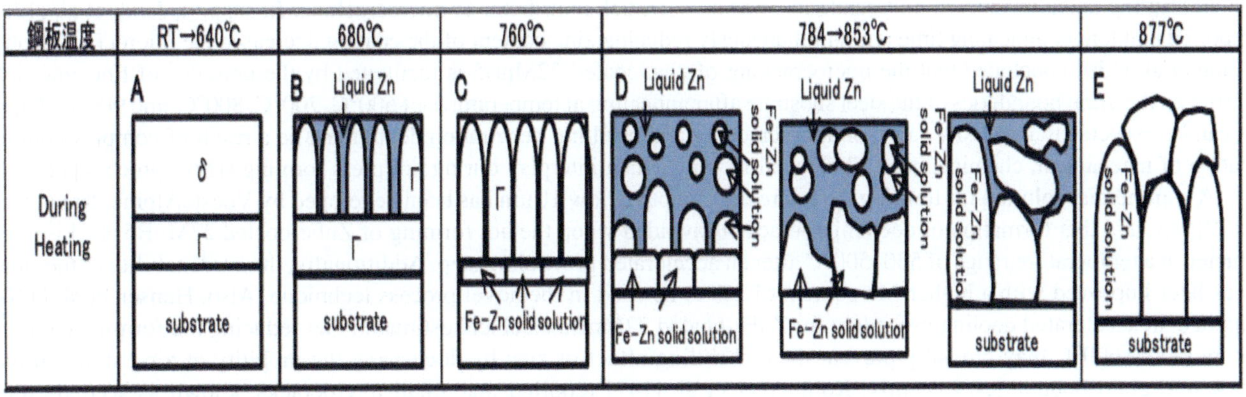

FIG. 13.51 Schematic illustration of the structural change of the GA coating rapidly cooled after heating [124].

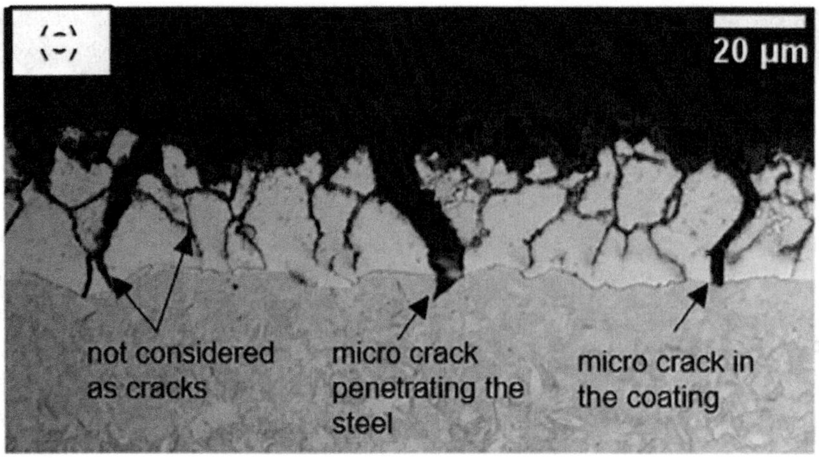

FIG. 13.52 Typical microcracks [117].

- At 784°C, the Γ phase in the coating changed to the binary phases consisting of both the spheroidal Fe-Zn solid solution and the Γ phase.
- At 821–853°C, the volume ratio of the Fe-Zn solid solution increased and that of the Γ decreased as the temperature increased. Both the Fe-Zn solid solution and the liquid Zn phase were able to exist in this temperature region.
- At 877°C, the coating consisted of only the Fe-Zn solid solution. It means that the liquid Zn phase did not exist.

Although typical hot forming occurs around 900°C (see Fig. 13.36, above), it can be seen that at this austenitization temperature, a Zn(Fe) liquid is in equilibrium with a high Zn-content ferrite, α-Fe(Zn), as shown in the Fe-Zn equilibrium phase diagram in Fig. 13.37. As a result, microcracking of the α-Fe(Zn) has been reported during deformation [125].There has been a number of studies that reported that microcracks appeared in the steel substrate during hot stamping [113,126–128] or resistance spot welding [129–131] due to the Zn coating, and most studies suggest that liquid Zn alloys were responsible for the microcracking [127,132–134]. Typical microcracks are seen in Fig. 13.52 [117].

13.3.2.1 Microcracking

Microcracking has been attributed to several causes. In 2017, a major research project on crack initiation mechanisms and process parameter interactions in Zn coated Boron steel was reported by the European Commission [135]. Drillet et al. [136] emphasized the strong role of high friction forces simultaneously with ferrite formation below coating/steel interface and its possible correlation to the maximum crack depth. Kim et al. [126] concluded that microcracks with equal penetration depth are formed regardless of the substrate microstructure, and they showed that increasing plastic strain (smaller die radius) increases the microcrack depth. Takahashi et al. [133] studied the depth and width of cracked galvannealed boron steel sheet. Köyer et al. [137] pointed out the strong role of plastic strain and suggested that the microcracks are formed in the regions that are first subjected to compressive stress and subsequently to tensile loading. Hensen et al. [138] proposed that the role of zinc, acting as an embrittling agent on steel grain boundaries, was an essential contributor to microcracking. In support, Janik et al. [128] reported that longer annealing time and simultaneously reducing zinc content of the coating decreased the microcrack depth. Kang et al. [139] concluded that the microcracking of zinc-coated 22MnB5 is facilitated by the presence of fine zinc-rich ferrite on the grain boundaries of the steel substrate after annealing at temperatures of 600°C, 700°C, 800°C, and 900°C. Thus, it can be expected that the root cause of microcracking and resulting penetration depth must be a result of complex combination of mechanical, chemical, and tribological factors, which interplay during hot press forming (HPF) process [117].

A commercial solution for limiting the microcrack depth below 10 µm has been developed by VoesteAlpine. Kurz et al. [122] reported that formation of deep microcracks is avoided when the hot forming of ZnFe-coated 22MnB5 steel is conducted at a temperature range of 550–500°C after an accelerated precooling step. Additionally, the hardenability of the steel was later improved with a higher Mn content of 2% to be used in the novel process technique. Also, Hensen et al. [140] applied an accelerated cooling procedure for ZnFe-coated 22MnB5 steel and estimated that reducing the forming temperature below 600°C may crucially prevent the embrittling effect of zinc by decreasing the mobility of zinc atoms to the austenite grain boundaries. Recently, Kolnberger et al. [141] reported that small microcracks, known as second-order microcracks ("with depths up to several 10 µm"), may even occur at forming temperatures well below 650°C and are due to vapor metal embrittlement (VME).

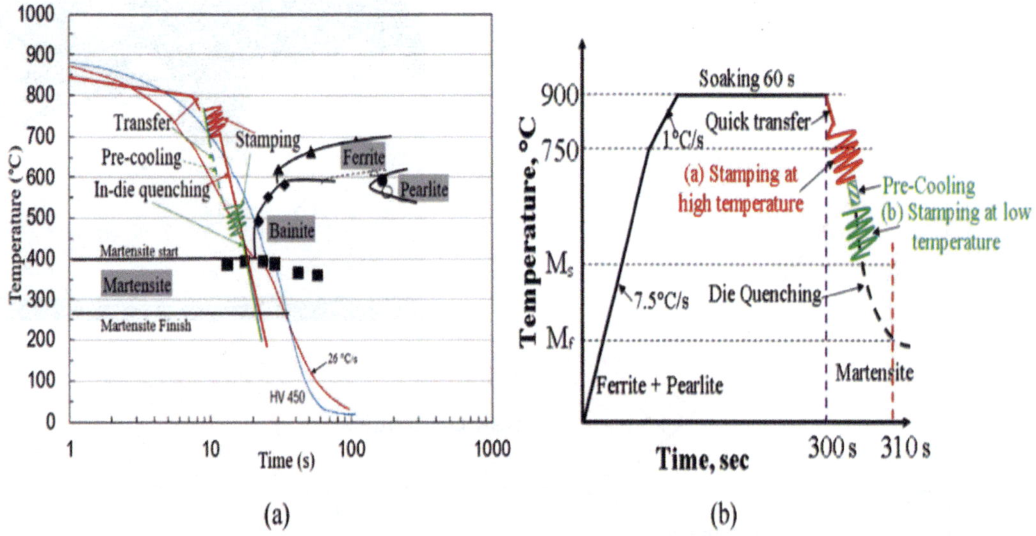

FIG. 13.53 The conventional hot stamping process (*red*—without precooling) and with precooling (*green*—a new process), illustrated in the (A) CCT diagram and (B) temperature profile [142].

A typical continuous cooling transformation (CCT) diagram for 22MnB5 is seen in Fig. 13.53. As the CCT diagram shows, for the as-delivered material, a minimum cooling rate is 27°C/s to ensure complete martensitic phase transformation. The martensite transformation begins at 400°C (M_s, the martensite start temperature) and ends at about 260°C (M_f, martensite finish temperature). Fig. 13.53 illustrates the typical thermal cycle for stamping at high temperature. Superimposed in both figures is a precooling step, which allows for stamping at a lower temperature just above the martensitic transformation start temperature, M_s. As a result, the material strain hardening is improved, which results in better drawability [143]. Critical to the direct hot forming process is the transfer time between the furnace and the precooling unit (transfer 1) and the transfer time between precooling and the start of hot forming (transfer 2), which maintains the part in the austenite region so that deformation is much easier, Fig. 13.54. In the development of USIBOR 1500P,Arcelor Mittal reported that to avoid microcracks it was important to stay below 780°C during stamping [123]. Fig. 13.55 illustrates the "risk of cracking" region in the Fe-Zn equilibrium phase diagram that must be avoided during stamping to prevent cracking. In a recent study, Thomsen et al. [145] developed a prototype press-hardened steel (PHS) with increased Mn content relative to 22MnB5 to enable stamping below the 782°C peritectic temperature, eliminating the liquid Zn phase essential for Liquid Metal Embrittlement, see Figs. 13.48 and 13.55.

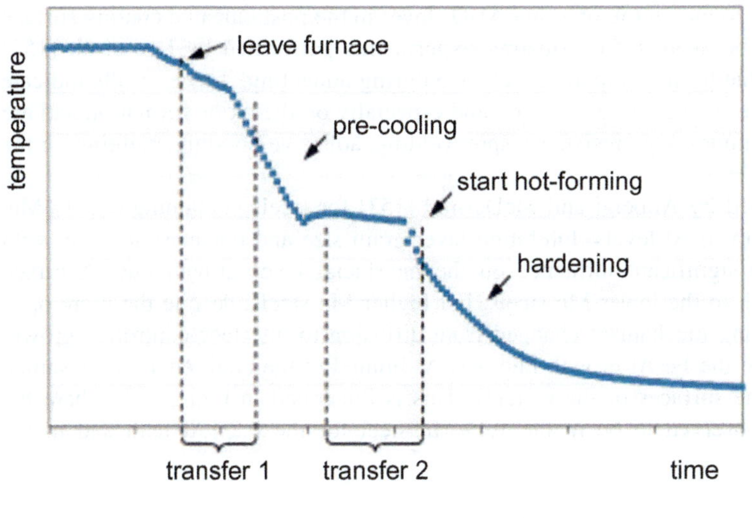

FIG. 13.54 Processing steps of the direct hot forming process with precooling shown by a typical time-temperature graph [144].

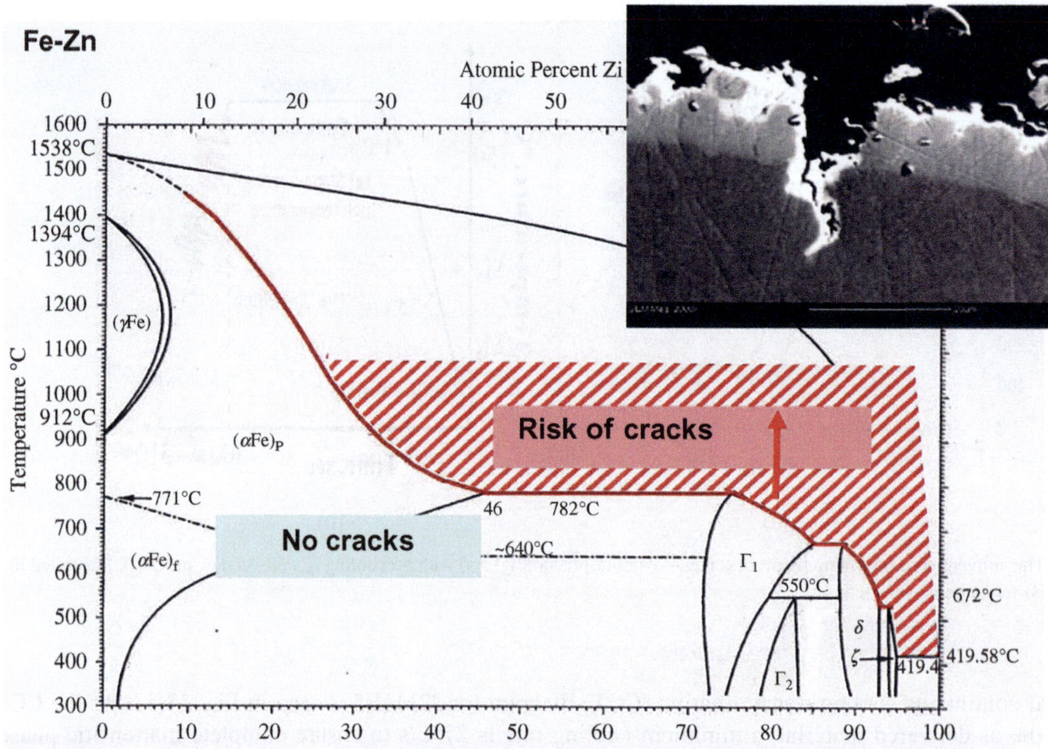

FIG. 13.55 The "risk of cracking" region in the Fe-Zn equilibrium phase diagram [123].

13.3.3 Oxides

AHSS can be covered with oxides of Si, Mn, and other alloying elements that cannot be reduced, leading to the use of higher dewpoint gases or oxidation-reduction treatments to limit the surface concentration of such oxides. Sagl [146] showed that if the (Si,Mn) oxide morphology on the steel surface is granular in nature rather than filmlike, a $Fe_2Al_{(5-x)}Zn_x$ inhibition layer can be successfully formed, permitting galvanization. Two mechanisms for inhibition layer formation were observed: First, in the spaces between surface granules, the inhibition layer was observed to grow. Then, as the inhibition layer grew, the oxides surrounding the inhibition layer either cracked and broke off or, especially near the inhibition layer-steel interface, were incorporated inside or interspersed with the inhibition layer. A schematic of this process is shown in Fig. 13.56 [127]. Arndt et al. [147] confirmed the huge influence of preoxidized alloying elements (mainly Mn) on the wetting behavior and proposed a model for the wetting process. An overview of the coating evolution during annealing was reported by Kang et al. [139] and Järvinen et al. [148,149]. The diffusion process for the Fe-Zn phases depends on the Zn concentration in GA-coated 22MnB5 described by Wang et al. [150]. A mixture of mainly ZnO and Mn_3O_4 was found on top of a thin Al_2O_3 layer in the postannealed coating surface of GI-coated press-hardened steel by Autengrube et al. [151]. Heating experiments performed by Lee et al. [152] described the sequence of the oxide formation due to the temperature change during annealing. These results indicate a strong influence of the alloying elements on the final coating structure and especially on the oxide formation and are consequential for postannealing processing techniques, e.g., resistance spot welding, adhesive bonding, painting, or the application of further (organic) coatings.

The kinetics of this mechanism were described by Alibeigi and McDermid [153] for steels containing 0.2–3%Mn, using galvanizing baths with 0.2 and 0.3% dissolved Al levels. Inhibition layer grain size and openness together with the presence of unreduced surface MnO had a significant influence on the interfacial layer growth rate. A mixed diffusion-model described inhibition layer growth in the lower Mn steels. For higher Mn steels, despite the more open interfacial layer microstructure, the rate-controlling mechanism changed from diffusion to interface-controlled growth predicting the interfacial layer Al uptake because the Fe-Al growth rate was Al limited to the additional Al consumed in reducing the thicker MnO layer present on the surfaces of these steels. This is illustrated in Fig. 13.57, where the highest value of Al in the inhibition layer is observed to be in the 1.4%Mn steel for the 0.2%Al bath and in the 2.5%Mn steel for the 0.3%Al bath.

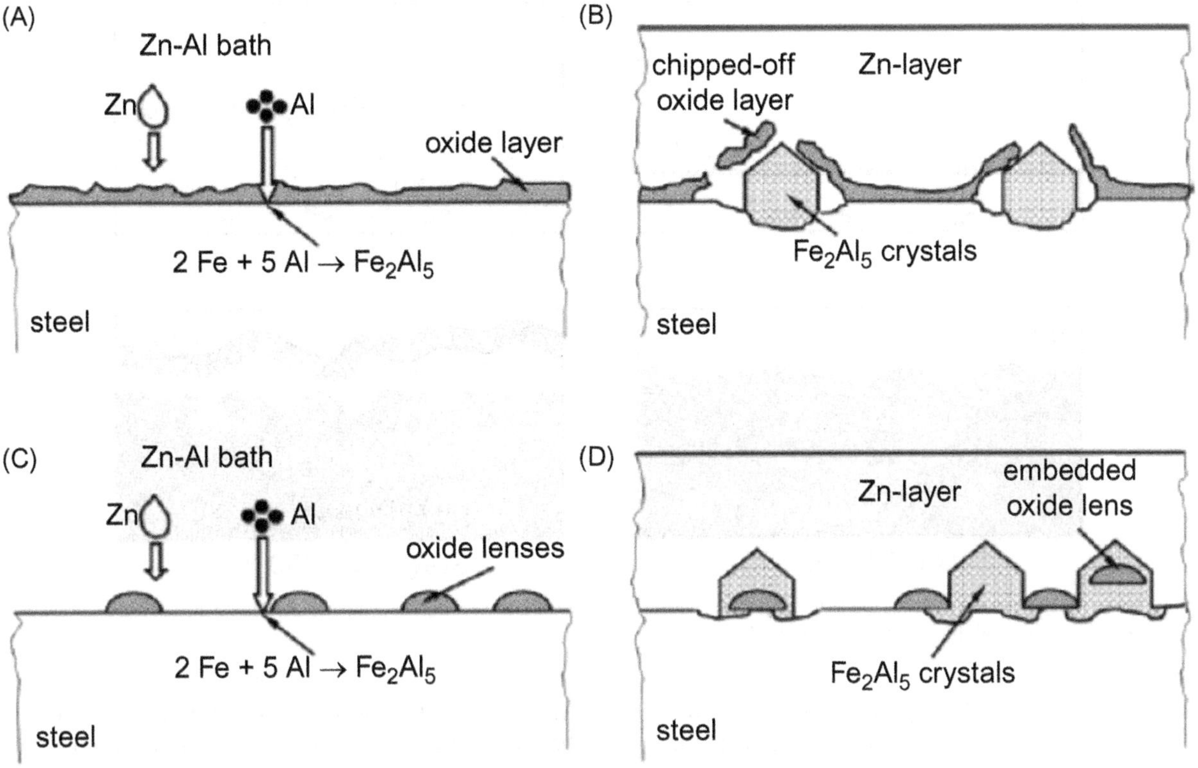

FIG. 13.56 Development on inhibition layer on AHSS with granular surface oxide morphology [146].

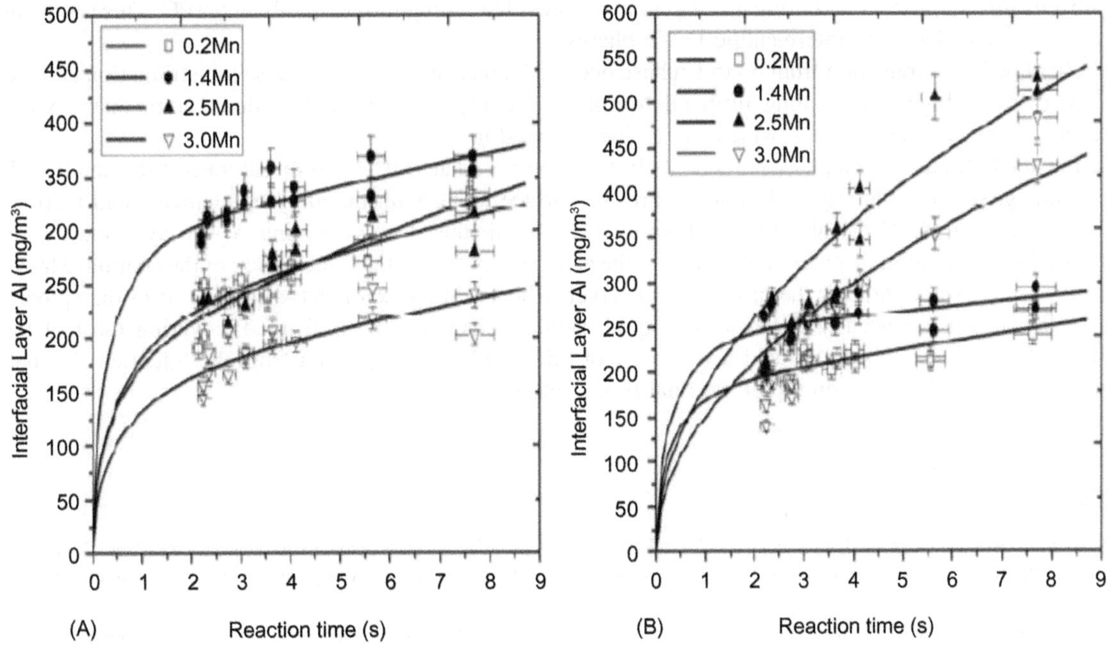

FIG. 13.57 Experimental data of interfacial layer Al uptake as a function of reaction time and steel Mn content for (A) 0.20 wt% and (B) 0.30 wt% dissolved Al baths shown with modeling results [153].

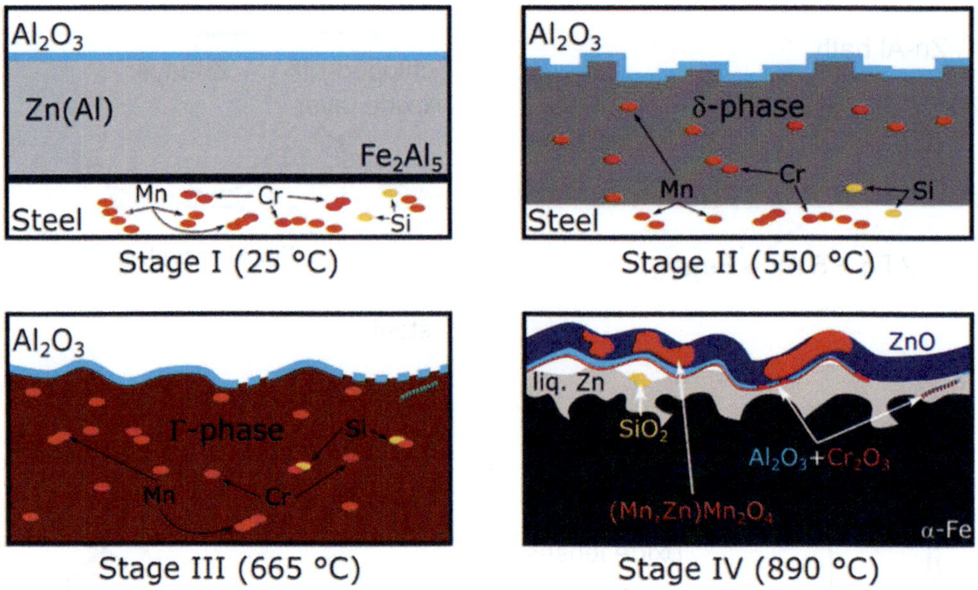

FIG. 13.58 Schematic of four important oxide-forming stages during press hardening annealing showing only the dominant Fe-Zn phases and the main oxide-formation of Al, Zn and the main alloy elements [154].

Gaderbauer et al. [154] proposed a schematic model to describe the oxide formation and is seen Fig. 13.58. The model for GI-coated specimens is adaptable for GA-coated samples, where the initial structure is similar to a fully δ-phase transitioned Stage II instead of Stage I. The stages of the model are:

- Stage I: at 25°C, the layered structure consists of the steel substrate, containing alloying elements such as Si, Cr, or Mn and the coating, consisting of Al and Zn. These substrate elements form oxides near the substrate/coating interface as a result of selective oxidation from the hot-dip galvanizing annealing process. The steel matrix is separated from the Zn coating by a thin Fe_2Al_5 inhibition layer, which acts as a diffusion barrier. Due to the low Al additions in the liquid Zn bath and the high affinity of Al to O, a several nanometers thick Al_2O_3 layer covers the coating.
- Stage II: at 550°C, the diffusion of Fe and alloying elements into the Zn coating, either through galvannealing (in the hot-dip galvanizing process) or during heating up in the press hardening furnace to about 550°C. This results in a transformation of Fe and Zn into intermetallic Fe-Zn phases.
- Stage III: at 665°C, a transition from δ - to Γ-phase occurs. A consequence of the phase transformation and the heating rate is a volume change of the coating, further deforming the Al_2O_3 layer. Due to the induced strain on the oxide layer, it breaks apart into smaller particles, which can be incorporated in the Fe-Zn phase.
- Stage IV: at 890°C, the Fe-Zn phase has already been decomposed at 780°C into Zn saturated α-Fe and liquid Zn. The α-Fe grains grow from the steel substrate toward the surface along with the alloying elements and their oxides. In GI-coated steel strips, Al dissolved in the liquid Zn can repair the already existing Al_2O_3 layer and the remaining Al can lead to a formation of a new Al_2O_3 layer in the early stages during the liquefaction of the coating. Mn containing liquid Zn can protrude through the broken Al_2O_3 layer, where Mn and Zn react with O to form the spinel (Mn,Zn) Mn_2O_4. When Mn has been depleted in the liquid, another layer of ZnO is formed, covering most of the spinel. The high thermodynamic stability and low density of SiO_2 lead to an expulsion from the liquefied Zn coating, and therefore, SiO_2 can be found on top of the final Fe-Zn intermetallic phase.

References

[1] P.K. Malick, Overview, Chapter 1, Materials, Design and Manufacturing for Lightweight Vehicles, Woodhead Publishing, 2021.

[2] C.D. Horvath, Advanced steels for lightweight automotive structures, Chapter 2, in: Materials, Design and Manufacturing for Lightweight Vehicles, Woodhead Publishing, 2021.

[3] D.K. Matlock, J.G. Speer, E. De Moor, Recent AHSS developments for automotive applications: processing, microstructures, and properties, in: Proceedings of the Addressing Key Technology Gaps in Implementing Advanced High-Strength Steels for Automotive Lightweighting, USCAR Offices, Southfield, MI, USA, 2012, pp. 9–10.

[4] X.J. Jin, T.Y. Hsu, Y.H. Rong, X.D. Wang, H.P. Liu, Y. Wang, R.M. Wu, X.W. Lu, L. Wang, H. Dong, J. Shi, J.F. Wang, X.C. Xiong, High Strength Steels Treated by Quenching and Partitioning Process, NSF China, University of Cambridge, 2012. June 28.

[5] W. Bleck, K. Phiu-On, Effects of microalloying in multi phase steels for car body manufacture, in: Microstructure and Texture in Steels, Springer, London, 2009, pp. 145–163.

[6] N. Fonstein, Dual-phase steels, Chapter 7, in: Automotive Steels, Elsevier, 2017, pp. 169–216.

[7] A. Belhadj, Latest 3rd GEN AHSS and 2nd GEN Press Hardened Steels: Solutions to New Automotive Challenges, St Quentin En Yvelines, October, Arcelor Mittal, 2019.

[8] Courtesy of WorldAutoSteel.

[9] M.S. Rashid, B.V.N. Rao, Tempering characteristics of a vanadium containing dual-phase steel, in: R.A. Kot, B.L. Bramfitt (Eds.), Fundamentals of Dual-Phase Steels, TMS-AIME, Warrendale, PA, 1981, pp. 246–264.

[10] I.A. El-Sesy, Z.M. El-Baradie, Influence carbon and/or iron carbide on the structure and properties of dual-phase steels, Mater. Lett. 57 (2002) 580–585.

[11] A.T. Davenport (Ed.), Formable HSLA and Dual-Phase Steels, TMS AIME, Warrendale, PA, 1979.

[12] R.A. Kot, J.W. Morris (Eds.), Structure and Properties of Dual-Phase Steels, TMS AIME, Warrendale, PA, 1979.

[13] R.A. Kot, B.L. Bramfitt (Eds.), Fundamentals of Dual-Phase Steels, TMS AIME, Warrendale, PA, 1981.

[14] A.R. Marder, Deformation characteristics of dual-phase steels, Metall. Mater. Trans. A 13A (1982) 85–92.

[15] A.R. Marder, Factors affecting the ductility of "dual-phase" alloys, in: A.T. Davenport (Ed.), Formable HSLA and Dual-Phase Steels, TMS AIME, Warrendale, PA, 1979, pp. 87–98.

[16] A.R. Marder, The Morphology and Strength of Iron-Carbon Martensite, PhD dissertation, Lehigh University, 1968.

[17] A.R. Marder, Structure- -property relationships in ferrous transformation products, in: A.R. Marder, J.I. Goldstein (Eds.), Phase Transformations in Ferrous Alloys, TMS-AIME, 1983, pp. 11–41.

[18] Y.L. Kang, Q.H. Han, X.M. Zhao, M.H. Cai, Influence of nanoparticle reinforcements on the strengthening mechanisms of an ultrafine-grained dual phase steel containing titanium, Mater. Des. 44 (2013) 331–339.

[19] A.K. Sinha, Ferrous Physical Metallurgy, Butterworths Publishers, 1989.

[20] S. Hayarai, T. Furukawa, Proceedings Microalloying Conference, Washington, D.C., 1975, pp. 78–87.

[21] K. Araki, K. Nakaoka, M. Abe, N. Ohashi, 9th Biennial Conf. of IDDRG Research Group Proceedings, 1976, pp. 39–53.

[22] S.K. Paul, A critical review of hole expansion, Dent. Mater. 9 (2020), 100566.

[23] D.T. Pierce, J.A. Jimenez, J. Bentley, D. Raabe, J.E. Wittig, The influence of stacking fault energy on the microstructural and strain-hardening evolution of Fe-Mn-Al-Si steels during tensile deformation, Acta Mater. 100 (2015) 178–190.

[24] G. Frommeyer, U. Brux, P. Neumann, Supra-ductile and high-strength manganese TRIP/TWIP steels for high energy absorption purposes, ISIJ Int. 43 (3) (2003) 438–446.

[25] T.B. Hilditch, T. De Souza, P.D. Hodgson, Properties and automotive applications of advanced high-strength steels (AHSS), in: Welding and Joining of Advanced High Strength Steels (AHSS), Woodhead Publishing, 2015, pp. 9–28.

[26] D.K. Matlock, J.G. Speer, E. De Moor, P.J. Gibbs, Recent developments in advanced high strength sheet steels for automotive applications: an overview, JESTECH 15 (2012) 1–12.

[27] E. De Moor, P.J. Gibbs, J.G. Speer, D.K. Matlock, Strategies for third-generation advanced high-strength steel development, AIST Transactions 7 (2010) 133–144.

[28] B.C. De Cooman, K. Chin, J. Kim, High Mn TWIP steels for automotive applications, in: M. Chiaberge (Ed.), New trends and Developments in Automotive System Engineering, InTech, 2011, ISBN: 978-953-307-517-4.

[29] D.K. Matlock, J.G. Speer, in: A. Halder, S. Suwas, D. Bhattacharjee (Eds.), Third Generation of AHSS: Microstructure Design Concepts, Microstructure and Texture in Steels and Other Materials, Springer, London, 2009, pp. 185–205.

[30] J. Speer, D.K. Matlock, B.C. De Cooman, J.G. Schroth, Carbon partitioning into austenite after martensite transformation, Acta Mater. 51 (2003) 2611–2622.

[31] A.J. Clarke, J.G. Speer, M.K. Miller, R.E. Hackenberg, D.V. Edmonds, D.K. Matlock, F.C. Rizzo, K.D. Clarke, E. De Moor, Carbon partitioning to austenite from martensite or bainite during the quench and partition (Q&P) process: a critical assessment, Acta Mater. 56 (2008) 16–22.

[32] H. Jirkova, B. Masek, M.F.X. Wagner, D. Langmajerová, L. Kučerová, R. Treml, D. Kiener, Influence of metastable retained austenite on macro and micromechanical properties of steel processed by the Q&P process, J. Alloys Compd. 615 (2014) S163–S168.

[33] J.G. Speer, A.F.C. Rizzo, D.K. Matlock, D.V. Edmonds, The quenching and partitioning process: background and recent progress, Mater. Res. 8 (2005), 417423.

[34] H. Ohtani, F. Terasaki, T. Kunitake, Trans. Iron Steel Inst. Jpn 12 (1972) 118.

[35] T. Kunitake, F. Terasaki, Y. Ohmori, H. Ohtani, The bainite and the bainite-Martensite duplex structure in low-carbon low-alloy high strength steels, Iron Steel 45 (1972) 647–653.

[36] Y. Tomita, K. Okabayashi, Improvement in lower temperature mechanical properties of 0.40 Pct C-Ni-Cr-Mo ultrahigh strength steel with the second phase lower Bainite, Mettal. Trans 14A (1983) 485–492.

[37] Y. Ohmori, H. Ohtani, T. Kunitake, Tempering of the bainite and the bainite/martensite duplex structure in a low-carbon low-alloy steel, Metal Sci. 8 (1974) 357.

[38] D. Krizan, B.C. De Cooman, J. Antonissen, Retained Austenite stability in the cold rolled CMnAlSiP Micro-alloyed TRIP steels, in: Proc. of the Int. Conf. on Advanced High Strength Sheet Steels for Automotive Applications, June 6–9, AIST, Warrendale, PA, 2004, pp. 205–216. Winter Park, CO.

[39] A. Zarei Hanzaki, P.D. Hodgson, S. Yue, Hot deformation characteristics of Si-Mn TRIP steels with and without Nb microalloy additions, ISIJ Intl. 35 (1995) 324–331.
[40] F. Perrard, C. Scott, Vanadium precipitation during intercritical annealing in cold rolled TRIP steels, ISIJ Intl. 47 (2007) 1168–1177.
[41] K. Sugimoto, M. Tsunezawa, T. Hojo, Ductility of 0.1–0.6C–1.5Si–1.5Mn ultra high-strength low-alloy TRIP-aided sheet steels with bainitic ferrite matrix, ISIJ Int. 44 (2004) 1608–1614.
[42] K. Sugimoto, T. Iida, J. Sakaguchi, Retained austenite characteristics and tensile properties in TRIP type bainitic sheet steels, ISIJ Int. 40 (2000) 902–908.
[43] K. Sugimoto, J. Sakaguchi, T. Iida, Stretch-flangeability of TRIP type bainitic sheet steels, ISIJ Int. 40 (2000) 920–926.
[44] K. Hausmann, D. Krizan, K. Spiradek-Hahn, The influence of Nb on the transformation behavior and mechanical properties of TRIP-assisted bainitic-ferrite sheet steels, Mater. Sci. Eng. A 588 (2013) 142–150.
[45] A. Bachmaier, K. Hausmann, D. Krizan, A. Pichler, Development of TBF steels with 980 MPa tensile strength for automotive applications, in: Proceedings of International Conference on New Developments in Advanced High Strength Steels, Colorado, 2013, pp. 23–27.
[46] K.-I. Sugimoto, T. Hojo, J. Kobayashi, Critical assessment 29: TRIP aided bainitic ferrite steels, Mater. Sci. Technol. 33 (2017) 2005–2009.
[47] F.G. Cabellero, H.K.D.H. Bhadesia, Very strong bainite, Curr. Opin. Solid State Mater. Sci. 8 (2004) 251–257.
[48] Y.S. Jin, Development of advanced high strength steels for automotive applications, La Metallurgia Italiana (2011) 43–48.
[49] S.J. Lee, Y. Lee, F.E. Goodwin, B.C. De Cooman, Galvanizing and galvannealing of an 800MPa TRIP steel grade, in: Proc. of 8th Inter. Conf. on Zinc and Zinc Coated Steel Sheet, Genova, 2011, pp. 87–94.
[50] J.R. McDermid, K.M. Bhadhon, D.M. Pallisco, V. Patel, M. Pourmajidian, F.E. Goodwin, Galvanizing of third generation advanced high strength steels, in: 110th Meeting of the Galvanizer's Association, 2018.
[51] T. Ohara, H. Iida, M. Iwaki, Development of a new coiling technology for continuous annealing, Trans. ISI Japan 25 (1985) 1156–1162.
[52] F. Barrado, T. Zhou, C. Cathcart, P. Badgley, S. Zhang, D. Overby, Development of advanced high strength steels using hydrogen quench continuous annealing technology, Mater. Eng. Res. 2 (2020) 106–112.
[53] P. Drillet, Z. Zermout, D. Bouleau, J. Mataigne, S. Claessens, Galvatech'04, 2004, pp. 1123–1134.
[54] J. Mahieu, B.C. De Cooman, J. Maki, S. Claessens, Iron Steelmaker, 2002, pp. 29–34.
[55] J. Mahieu, S. Claessens, B.C. De Cooman, Galvanizability of high-strength steels for automotive applications, Metall. Mater. Trans. A 32A (2001) 2905–2908.
[56] N. Ebrill, Y. Durandet, L. Strezov, Galvatech'01, 2001, pp. 329–336.
[57] H.L. Eleuterio, Influence of Al content in the Zn bath on galvanized coating formation of high strength steels, in: Galvatech '21, 2021, pp. 635–644.
[58] S. Alibeigi, R. Kavitha, R.J. Meguerian, J.R. McDermid, Reactive wetting of high Mn steels during continuous hot-dip galvanizing, Acta Mater. 59 (2011) 3537–3549.
[59] M. Tumuluru, An overview of the resistance spot welding of coated high strength dual phase steel, Weld J. 85 (2006) 31–37s.
[60] Ducker Worldwide Report, 2007. www.duckerworldwide.com.
[61] M. Tumuluru, Resistance spot welding techniques for advanced high-strength steels (AHSS), in: Welding and Joining of Advanced High Strength Steels (AHSS), Woodhead Publishing, 2015, pp. 55–70.
[62] G. Jung, I.S. Woo, D.W. Suh, S.J. Kim, Liquid Zn assisted embrittlement of advanced high strength steels with different microstructures, Met. Mater. Int. 22 (2016) 187–195.
[63] J. Barthelmie, A. Schram, V. Wesling, Liquid metal embrittlement in resistance spot welding and hot tensile tests of surface-refined TWIP steels, IOP Conf. Ser., Mater. Sci. Eng. 118 (2016), 012002.
[64] C. Beal, Mechanical Behavior of a New Automotive High Manganese TWIP Steel in the Presence of Liquid Zinc, Ph.D. thesis, INSA, Diss. Lyon, 2011.
[65] D. Bhattacharya, Liquid metal embrittlement during resistance spot welding of Zn- coated high-strength steels, Mater. Sci. Technol. 34 (2018) 1809–1829.
[66] D. Bhattacharya, L. Cho, D. Marshal, M. Walker, E. van der Aa, A. Pichler, H. Ghassemi-Armaki, K.O. Findley, J.G. Speer, Liquid metal embrittlement susceptibility of two Zn-coated advanced high strength steels of similar strengths, Mater. Sci. Eng. A 823 (2021), 141569.
[67] Z. Ling, T. Chen, L. Kong, M. Wang, H. Pan, M. Lei, Liquid metal embrittlement cracking during resistance spot welding of galvanized Q&P980 steel, Metall. Mater. Trans. 50 (2019) 5128–5142.
[68] R. Sierlinger, M. Gruber, A cracking good story of liquid metal embrittlement during spot welding of advanced high strength steels, in: 5th International Conference On Steels in Cars and Trucks, Amsterdam-Schiphol, The Netherlands, 2017.
[69] D. Bhattacharya, L. Cho, E. Van der Aa, H. Ghassemi-Armaki, A. Pichler, K.O. Findley, J.G. Speer, Quantitative assessment of the characteristics of liquid metal embrittlement during resistance spot welding of Zn-coated high strength steels, in: Sheet Metal Welding Conference XVII, Detroit, Michigan, USA, 2018.
[70] D. Bhattacharya, L. Cho, E. van der Aa, A. Pichler, N. Pottore, H. Ghassemi-Armaki, K.O. Findley, J.G. Speer, Influence of the starting microstructure of an advanced high strength steel on the characteristics of Zn-assisted liquid metal embrittlement, Mater. Sci. Eng. A 804 (2021) 140391.
[71] S.H. Hong, J.H. Kang, D. Kim, S.J. Kim, Si effect on Zn-assisted liquid metal embrittlement in Zn-coated TWIP steels: importance of Fe-Zn alloying reaction, Surf. Coat. Technol. 393 (2020), 125809.
[72] M. Tumurulu, Effect of silicon and retained austenite on the liquid metal embrittlement cracking behavior of Gen3 and high strength automotive steels, Weld J. 98 (2019) 351–362.
[73] A.G. Kalashami, C. DiGiovanni, M.H. Razmpoosh, F. Goodwin, N.Y. Zhou, The role of internal oxides on the liquid metal embrittlement cracking during resistance spot welding of dual phase steel, Metall. Mater. Trans. A 51 (2020) 2180–2191.

[74] C.D. Horvath, E.S. Batt, Delayed cracking of advanced high strength steel solutions, in: Proc. Great Designs in Steel 2017, May 17, Livonia, MI, American Iron and Steel Institute, Southfield, MI, 2017.
[75] M. Mandy, B. Nabi, M. LARNICOL, X. Vanden Eynde, C. Georges, F. Goodwin, Influence of Zn-based coating alloys on hydrogen diffusion and corroision resistance in DP steel, in: GALVATECH '21, 2021, pp. 986–993.
[76] C. Georges, X. Vanden Eynde, Hydrogen solubility effects in galvanized advanced high strength steels, SAE Int. J. Mater. Manuf. 9 (2016). 2016-01-0360.
[77] C.S. Li, Z.X. Li, Y.M. Cen, B. Ma, G. Huo, Microstructure and mechanical properties of dual phase strip steel in the overaging process of continuous annealing, Mater. Sci. Eng. A 627 (2015) 281.
[78] A.A. Sayed, S. Kheirandish, Effect of the tempering temperature on the microstructure and mechanical properties of dual phase steels, Mater. Sci. Eng. A 532 (2012) 21.
[79] J.P. Mogeritsch, A. Ludig, B. Böttger, G. Angeli, C.K. Riener, R. Ebner, Phase-field modelling of ternary eutectic solidification in hot dip galvanization, in: ICOUPLED VII: Proceedings of the VII International Conference on Computational Methods for Coupled Problems in Science and Engineering, 2017, pp. 1171–1182.
[80] C.F. Kuang, Z.W. Zheng, M.L. Wang, Q. Xu, S.G. Zhang, Effect of hot-dip galvanizing processes on the microstructure and mechanical properties of 600-MPa hot-dip galvanized dual-phase steel, Int. J. Miner. Metall. Mater. 24 (2017) 1379–1383.
[81] E. Pan, H. Di, G. Jiang, C. Bao, Effect of heat treatment on microstructures and mechanical properties of hot-dip galvanized DP steels, Acta. Metall. Sin. (English Lett.) (2014) 469–475.
[82] E.A. Moraes, R.O. Rocha, F.S. Costa, D.B. Santos, Continuous annealing temperature influence on the microstructure and mechanical properties of galvanized dual phase steel (DP980), in: Galvatech '21, 2021, pp. 1216–1224.
[83] K.-C. Lin, P.-W. Chu, C.-S. Lin, H.-B. Chen, Galvanizing and galvannealing behavior of CMnSiCr dual-phase steels, Metall. Mater. Trans. A 44 (2013) 2690–2698.
[84] F. Bleich, C. Cheng, Challenges in galvannealing AHSS at an industrial hot dip galvanizing line, in: Galvatech '21, 2021, pp. 122–133.
[85] M.S. Kim, J.H. Kwak, J.S. Kim, Y.H. Liu, N. Gao, N.Y. Tang, Galvanizability of advanced high-strength steels 1180TRIP and 1180CP, Metall. Mater. Trans. A 40A (2009) 1903–1910.
[86] E.M. Bellhouse, J.R. McDermid, Effect of continuous galvanizing heat treatments on the microstructure and mechanical properties of high Al-low Si transformation induced plasticity steels, Metall. Mater. Trans. A 41A (2010) 1460–1473.
[87] W. Ding, D. Tang, H. Jiang, W. Huang, Influence of isothermal bainite transformation time on microstructure and mechanical properties of hot-dip galvanized TRIP steel, J. Mater. Eng. Perform. 20 (2011) 997–1002.
[88] W. Ding, P. Hedström, Y. Li, Heat treatment, microstructure and mechanical properties of a C–Mn–Al–P hot dip galvanizing TRIP steel, Mater. Sci. Eng. A 674 (2016) 151–157.
[89] D. Kim, J.-H. Kang, S.-J. Kim, Heating rate effect on liquid Zn-assisted embrittlement of high Mn austenitic steel, Surf. Coat. Technol. 347 (2018) 157–163.
[90] X. Jin, Y. Zhong, L. Wang, H. Wang, Effect of annealing temperature on the surface and subsurface microstructure of Al-added TWIP steel, Surf. Coat. Technol. 386 (2020) 125479.
[91] Y.F. Gong, B.C. De Cooman, Kirkendall void formation during selective oxidation, Metall. Mater. Trans. A 41A (2010) 2180–2183.
[92] Y.F. Gong, B.C. De Cooman, Selective oxidation and sub-surface phase transformation of TWIP steel during continuous annealing, Steel Res. Int. 82 (2011) 1310–1318.
[93] L. Cho, B.C. De Cooman, Selective oxidation of TWIP steel during continuous annealing, Steel Res. Int. 83 (2012) 391–397.
[94] Y. Kim, J. Lee, K.S. Shin, S. Jeon, K.G. Chin, Effect of dew point on the formation of surface oxides of twinning-induced plasticity steel, Mater Charact 89 (2014) 138–145.
[95] Y. Kim, K.S. Shin, S.H. Jeon, K.G. Chin, J. Lee, The influence of the dew point on the wettability of twinning-induced-plasticity steels by liquid Zn–0.23-wt% Al, Corros. Sci. 85 (2014) 364–371.
[96] Y. Kim, J. Lee, K.S. Shin, S.H. Jeon, K.G. Chin, Effect of nickel precoating on wettability of twinning-induced plasticity steels by liquid Zn-0.23 wt. % Al, Metall. Mater. Trans. A 47A (2016) 4960–4969.
[97] M. Blumenau, M. Norden, F. Friedel, K. Peters, Use of pre-oxidation to improve reactive wetting of high manganese alloyed steel during hot-dip galvanizing, Surf. Coat. Technol. 206 (2011) 559–567.
[98] W. Chen, A Study on Interface Condition in High Mn TWIP Steel After Initial Stage of Oxidation Process, Seoul National University, 2019.
[99] J.R. McDermid, K.M. Bhadhon, E.A. McNally, D.M. Pallisco, F.E. Goodwin, N.W. Resort, Composition and Processing of Third Generation Steels using CGL-compatible Thermal Cycles, in: Proc. of the 108th Galvanizers Association Meeting, Farmington, PA, 2016.
[100] K.M.H. Bhadhon, J.R. McDermid, F.E. Goodwin, Effect of galvanizing heat treatment on the microstructure and mechanical properties of a 6Mn-1. 5Si third generation advanced high strength steel, in: Galvatech '15, 2015, pp. 936–943.
[101] K. Ranganathan, J.R. McDermid, F.E. Goodwin, Continuous hot-dip galvanizing of a third generation (3G) advanced steel, in: Galvatech '15, 2015, pp. 182–189.
[102] J.G. Speer, D.V. Edmonds, F.C. Rizzo, D.K. Matlock, Solid State Mater. Sci. 8 (2004) 219–237.
[103] J. Sun, H. Yu, Microstructure development and mechanical properties of quenching and partitioning (Q&P) steel and an incorporation of hot-dipping galvanization during Q&P process, Mater. Sci. Eng. A 586 (2013) 100–107.
[104] Y. Wang, Y. Xu, R. Liu, F. Peng, X. Gu, T. Zhang, X. Hou, W. Sun, Microstructure evolution and mechanical behavior of a novel hot-galvanized Q&P steel subjected to high-temperature short-time overaging treatment, Mater. Sci. Eng. A 789 (2020), 139665.
[105] Y. Wang, Y. Xu, T. Zhang, J. Li, X. Hou, W. Sun, Effects of quenching temperature on bainite transformation, retained austenite and mechanical properties of hot-galvanized Q&P steel, Mater. Sci. Eng. A 822 (2021), 141643.

[106] K. Kahoul, M. Hamman, B. Caekaert, A. De Boeck, C. Bollaert, J. Gijs, Arcelor Mittal Belgium produces 3rd generation AHSS incl. Q&P on the unique annealing furnace from Andritz Metals, in: Galvatech '21, 2021, pp. 533–542.
[107] N. Komatsu, T. Suzuki, T. Ito, Y. Hara, K. Asakura, Process and Apparatus for Press-Forming and Quenching a Steel Stock, US Pat. No. 3668917, June 13, 1972.
[108] G.F. Melloy, J.D. Dennison, Rolled Steel Having High strength and Low Impact Transition Temperature and Method of Producing Same, US Pat. No. 3753796, Aug. 21, 1973.
[109] B.L. Bramfitt, A.R. Marder, As-Worked Bainitic Ferrous Alloy and Method, US Pat. No. 3806378, April 23, 1974.
[110] A.B. Norbottensjaernverk, Manufacturing a Hardened Steel Article, GB Pat. No. 1490535A, Nov. 2, 1974.
[111] H. Okunishi, H. Nakaji, H. Suwa, H. Sato, Method and Apparatus for Forge-Shaping Sheet Members, US Pat. No. 4356717, Nov. 2, 1982.
[112] M. Jonsson, A Method of Producing a Hardened Sheet Steel Product, WO99/07492, Feb. 18, 1999.
[113] M. Jonsson, Pillar for a Vehicle Body, US Pat. No. 6267438 B1, Jul. 31, 2001.
[114] H. Karbasian, A.E. Tekkaya, A review on hot stamping, J. Mater. Process. Technol. 210 (2010) 2103–2118.
[115] M. Naderi, Hot Stamping of Ultra High Strength Steels, Doctoral Thesis, RWTH, Aachen, 2007.
[116] T. Kurz, P. Larour, J. Lackner, T. Steck, G. Jesner, Press-hardening of zinc coated steel-characterization of a new material for a new process, IOP Conf. Ser. Mater. Sci. Eng. 159 (2016), 012025.
[117] H. Järvinen, J. Rämö, A. Sabr, S. Järn, E. Lepikko, M. Järvenpää, P. Peura, Effect of steel composition and processing parameters on the penetration depth of microcracks in ZnFe-coated boron steels, Steel Res. Int. 92 (2021) 2100112.
[118] M. Merklein, J. Lechler, T. Stöhr, Characterization of tribological and thermal properties of metallic coating for hot stamping boron-manganese steels, in: Proceedings of the 7th International Conference Coatings in Manufacturing 2008. Engineering, Chalkidiki, Greece, 2008, pp. 219–222.
[119] M. Tisza, Hot forming of boron alloyed manganese steels, Mater. Sci. Forum 885 (2016) 25–30.
[120] D.W. Fan, B.C. De Cooman, State-of-the-knowledge on coating systems for hot stamped parts, Steel Res. Int. 83 (2012) 412–433.
[121] J.N. Hall, J.P. Singh, P. Belanger, J. Cantalin, G. Zywicki, J. Coryell, M. Liu, An evaluation of Zn-coated press hardened steels, in: MS&T, AIST, Pittsburgh, PA, 2012.
[122] T. Kurz, G. Luckeneder, H. Schwinghammer, T. Manzenreiter, A. Sommer, Zinc coated press-hardening steels: challenges and solutions, in: Proc. 5th Int. Conf. Hot Sheet Metal Forming of High Performance Steel, Toronto, 2015, pp. 345–353.
[123] ArcelorMittal, Hot stamping with USIBOR1500P® AP&T Hot Stamping Seminar, Detroit, September 15, 2010.
[124] A. Sengoku, H. Takebayashi, K. Matsumura, Microstructural and phase evolution of galvannealed coating during hot stamping heating, in: 5th International Conference on Proceedings on Hot Sheet Metal Forming of High-Performance Steel, 363, Association for Iron & Steel Technology, Toronto, 2015.
[125] J.R. McDermid, Y. Kim, C. Dever, J. Kish, F.E. Goodwin, Effect of Zn content on the corrosion properties of hot dip zinc galvanized coatings on press hardened steels, in: Proc EUROCORR, 2017, pp. 3–7.
[126] S. Kim, I. Son, J. Kim, S.-J. Kim, Cracking issues of Zn coated press hardening steel in direct hot press forming, in: M. Oldenburg, K. Steinhoff (Eds.), Proceedings of 4th International Conference on "Hot Sheet Metal Forming of High-Performance Steel. CHS2", Verlag Wissenschaftliche Scripten, 2013, pp. 537–544.
[127] C. Lee, W. Choi, L. Cho, Y. Cho, B. De Cooman, Liquid-metal-induced embrittlement related microcrack propagation on Zn-coated press hardening steel, ISIJ Int. 55 (2015) 264–271.
[128] V. Janik, Y. Lan, P. Beentjes, D. Norman, G. Hensen, S. Sridhar, Zn diffusion and α- Fe(Zn) layer growth during annealing of Zn-coated B steel, Metall. Mater. Trans. A 47A (2016) 400–411.
[129] Y.G. Kim, I.J. Kim, J.S. Kim, Y.I. Chung, D.Y. Choi, Evaluation of surface crack in resistance spot welds of Zn-coated steel, Mater. Trans. 55 (2014) 171–175.
[130] R. Ashiri, M.A. Haque, C.-W. Ji, M. Shamanian, H.R. Salimijazi, Y.-D. Park, Supercritical area and critical nugget diameter for liquid metal embrittlement of Zn- coated twining induced plasticity steels, Scr. Mater. 109 (2015) 6–10.
[131] R. Ashiri, M. Shamanian, H.R. Salimijazi, M.A. Haque, J.-H. Bae, C.-W. Ji, K.-G. Chin, Y.-D. Park, Liquid metal embrittlement-free welds of Zn-coated twinning induced plasticity steels, Scr. Mater. 114 (2016) 41–47.
[132] C.W. Lee, D.W. Fan, I.R. Sohn, S.-J. Lee, B.C. De Cooman, Liquid-metal-induced embrittlement of Zn-coated hot stamping steel, Metall. Mater. Trans. A 43A (2012) 5122–5127.
[133] M. Takahashi, M. Nakata, K. Imai, N. Kojima, N. Otsuka, Liquid metal embrittle- ment of hot stamped galvannealed boron steel sheet - effect of heating time on crack formation, ISIJ Int. 57 (2017) 1094–1101.
[134] H. Kang, L. Cho, C. Lee, B.C. De Cooman, Zn penetration in liquid metal embrittled TWIP steel, Metall. Mater. Trans. A 47 (2016) 2885–2905.
[135] D. Casellas, I. Pujante, A. Ademaj, U. Weidig, K. Strinhoff, H. Schwinghammer, T. Kurz, D. Norman, G.C. Hensen, Exploring of crack initiation mechanisms and process parameter interactions in press hardening boron steel components actively protected by Zn-based coatings, Final Report, European Commission, 2017. EUR 28448.
[136] P. Drillet, R. Grigorieva, G. Leuillier, T. Vietoris, Study of cracks propagation inside the steel on press hardened steel zinc based coatings, La Metallurgia Italiana 1 (2012) 3–8.
[137] M. Köyer, J. Banik, S. Graff, J.F. Lenze, G. Parma, S. Sikora, in: M. Oldenburg, K. Steinhoff, K.B. Prakash (Eds.), Proc. of the 4th Int. Conf. on Hot Sheet Metal Forming of High-Performance Steel, Verlag Wissenschaftliche Scripten, Auerbach, Germany, Luleå, 2013, pp. 363–370.
[138] G. Hensen, W. Melfo, S. Chen, R. Bleeker, W. Verloop, in: M. Oldenburg, K. Steinhoff, K.B. Prakash (Eds.), Proc. of the 4th Int. Conf. on Hot Sheet Metal Forming of High-Performance Steel, Verlag Wissenschaftliche Scripten, Auerbach, Germany, Luleå, 2013, pp. 463–470.
[139] J.H. Kang, D. Kim, D.H. Kim, S.J. Kim, Fe-Zn reaction and its influence on microcracks during hot tensile deformation of galvanized 22MnB5 steel, Surf. Coat. Technol. 357 (2019) 1069–1075.

[140] G. Hensen, P. Beentjes, M. Abspoel, J. Loiseaux, in: M. Oldenburg, K. Steinhoff, K.B. Prakash (Eds.), Proc. of the 5th Int. Conf. on Hot Sheet Metal Forming of High-Performance Steel, Verlag Wissenschaftliche Scripten, Auerbach, Germany, Toronto, 2015, pp. 85–92.

[141] S. Kolnberger, J. Faderl, T. Kurz, J. Haslmayr, Avoiding Zinc Induced Cracking in Hot Forming. BHM Berg-und Hüttenmännische, Monatshefte (2021) 1–7.

[142] M. Ganapathy, N. Li, J. Lin, M. Abspoel, D. Bhattacharjee, Experimental investigation of a new low-temperature hot stamping process for boron steels, Int. J. Adv. Manuf. Technol. 10 (5) (2019) 669–682.

[143] S. Yang, L. Wang, Study on selective cooling of ultra-high strength steel car body in hot forming, Int. J. Adv. Manuf. Technol. 97 (2018) 1583–1590.

[144] T. Kurz, G. Luckeneder, T. Manzenreiter, H. Schwinghammer, A. Sommer, Zinc coated press-hardening steel-challenges and solutions, SAE Technical Paper, 2015-01-0565, April 14, 2015.

[145] C. Thomsen, J. McDermid, Development of a new grade of Zn-coated direct press-hardenable steel, in: GALVATECH '21, 2021, pp. 1106–1115.

[146] R. Sagl, A. Jarosik, D. Stifter, G. Angeli, The role of surface oxides on annealed high-strength steels in hot-dip galvanizing, Corros. Sci. 70 (2013) 268–275.

[147] M. Arndt, J. Duchoslav, R. Steinberger, G. Hesser, C. Commenda, L. Samek, E. Arenholz, D. Stifter, Nanoscale analysis of the influence of pre-oxidation on oxide formation and wetting behavior of hot-dip galvanized high strength steel, Corros. Sci. 93 (2015) 148–158.

[148] H. Järvinen, M. Isakov, T. Nyyssönen, M. Järvenpää, P. Peura, The effect of initial microstructure on the final properties of press hardened 22MnB5 steels, Mater. Sci. Eng. A 676 (2016) 109–120.

[149] H. Järvinen, M. Honkanen, M. Patnamsetty, S. Järn, E. Heinonen, H. Jiang, P. Peura, Press hardening of zinc-coated boron steels: role of steel composition in the development of phase structures within coating and interface regions, Surf. Coat. Technol. 352 (2018) 378–391.

[150] K. Wang, B. Zhu, Z. Wang, Y. Liu, L. Wang, Y. Zhang, S. Li, Successive phase and morphology evolution of galvannealed coating in hot stamping and diffusion modeling of α-Fe(Zn)/steel system considering the effect of Zn concentration, Surf. Coat. Technol. 380 (2019), 125036.

[151] R. Autengruber, G. Luckeneder, S. Kolnberger, J. Faderl, A.W. Hassel, Surface and coating analysis of press-hardened hot-dip galvanized steel sheet, Steel Res. Int. 83 (2012) 1005–1011.

[152] C.W. Lee, W.S. Choi, Y.R. Cho, B.C. De Cooman, Surface oxide formation during rapid heating of Zn-coated press hardening steel, ISIJ Int. 54 (2014) 2364–2368.

[153] S. Alibeigi, J.R. McDermid, Modeling of interfacial layer growth kinetics on Mn-containing steels during continuous hot-dip galvanizing, in: Galvatech '15, 2015, pp. 334–341.

[154] W. Gaderbauer, M. Arndt, T. Truglas, T. Steck, N. Klingner, D. Stifter, J. Faderl, H. Groiss, Effects of alloying elements on surface oxides of hot–dip galvanized press hardened steel, Surf. Coat. Technol. 404 (2020), 126466.

Chapter 14

Defect identification and remediation in zinc coated steel sheet

14.1 Introduction

Although in many cases a defect in a sample of zinc coated steel sheet at first glance appears to be related to a coating-related cause, careful examination in many cases will reveal that many other line operating conditions can be involved, including conditions existing before and after the sheet steel is processed on the continuous galvanizing line. The occurrence of a defect can only be cured by identifying and controlling all factors relevant to its occurrence. Many coating defects that, when examined by the unaided eye, appear to be "dross-related" have little or nothing to do with dross [1]. The resolution of the human eye for detecting visual defects is about 100 μm. Recent advances in automated surface inspection have improved this resolution and also the identification of defects which is now content-based rather than contrast-based, further improving our ability to correctly and identify coating-related defects [2].

The identification of the cause of a defect can sometimes be very difficult. It is very helpful to document the exact production conditions during the time of occurrence, also the location on the strip: which side, end of coil or central part, close to edge or not. Almost always, it is helpful to "take a look" with surface and cross-section microscopy and by obtaining at least qualitative readings on metal elements and compounds, such as intermetallics and oxides, believed to be playing a role. Many of the examples shown in this chapter show this approach. Also, details of how the customer is using the coated sheet and exposure environments before and after handling by the customer may reveal information to help identify the nature of the defect.

Zinc coated sheet steel coating defects can be divided into six general categories [3,4]:

1. Defects from steel substrate that originate in upstream processing
2. Defects associated with chemical cleaning
3. Defects originating from the annealing furnace
4. Defects related to the zinc bath
5. Defects originating from the gas jet wiping
6. Defects related to temper rolling

In addition, strip shape and recoiling conditions can affect the quality of coil at the delivery end of the line. Handling during transport and storage also affect the quality of the coil ultimately processed by the customer.

14.2 Defects related to strip processing

As pointed out by Chen and Patil, the galvanized coating does not hide any steel defect, no matter how minute [4]. An example of this is an incident of residual gas entrapped in the steel after casting. Modern steels are degassed by the addition of "killing" additions such as aluminum and silicon, But on rare occasions, entrapped gas can pass unnoticed through upstream thermomechanical processing, only to be noticed after galvanizing. Fig. 14.1A shows the appearance of the galvanneal coating covering this defect. After removal of the coating and light polishing, the defect is manifest as a raised blister, Fig. 14.1B. Further polishing reveals gas-related voids and a central crack, Fig. 14.1C. Later, another casting-related defect, slivers, with an electrogalvanizing-related effect is discussed.

Fig. 14.2A shows a defect related to rolled-in scale that was traced to the hot rolling mill. This is exhibited as a faint, short line defect oriented vertically. A more detailed examination of this kind of defect will be discussed in the galvanneal-specific section later. While immersed in the galvanizing bath, the zinc can flow into any openings under the scale, reacting

The Metallurgy of Zinc Coated Steels. https://doi.org/10.1016/B978-0-323-99984-7.00001-4

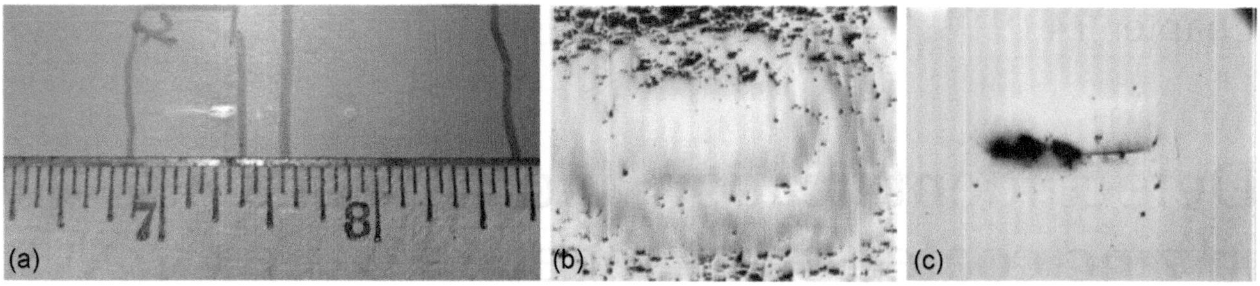

FIG. 14.1 (A) Galvanneal-coated steel sheet showing an entrapped gas defect. Dimensions in inches. (B) Entrapped gas defect sample after removal of galvanneal coating and light polishing, occurrence of raised blister. Bottom axis of photo is 2700 μm long. (C) Entrapped gas defect sample after removal of galvanneal coating and further polishing, entrapped gas features evident. Bottom axis of photo is 72 μm long. *(Reproduced with permission from (A–C) F. Chen, R. Patil, An in-depth analysis of various subtle coating defects of the 2000's, in: Proceedings of 6th International Conference on Zinc and Zinc Alloy Coated Steel Sheet (Galvatech 2004), Chicago, IL, April 4-7, 2004, Association for Iron & Steel Technology, Warrendale, PA, pp. 1055–1066.)*

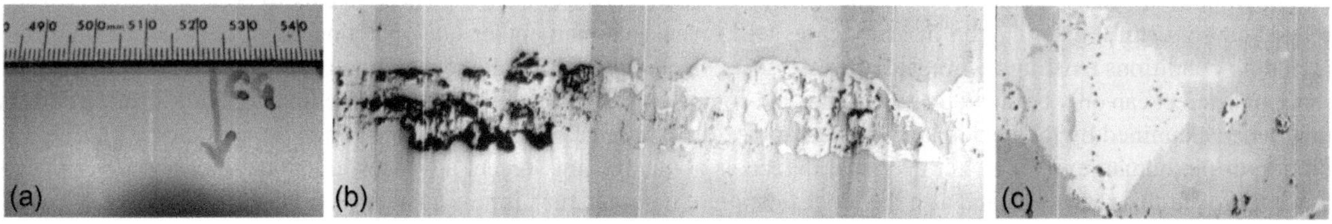

FIG. 14.2 (A) Hot mill scale defect, shown as *vertical light-colored line* in upper center under galvanneal coating. (B) Hot mill scale defect, oriented horizontally after removal of galvanneal coating and polishing. Bottom axis of photo is 9350 μm long. (C) Hot mill scale defect after further polishing, showing remaining scale and zinc that reacted with substrate under the scale. Bottom axis of photo is 300 μm long. *(Reproduced with permission from (A–C) F. Chen, R. Patil, An in-depth analysis of various subtle coating defects of the 2000's, in: Proceedings of 6th International Conference on Zinc and Zinc Alloy Coated Steel Sheet (Galvatech 2004), Chicago, IL, April 4-7, 2004, Association for Iron & Steel Technology, Warrendale, PA, pp. 1055–1066.)*

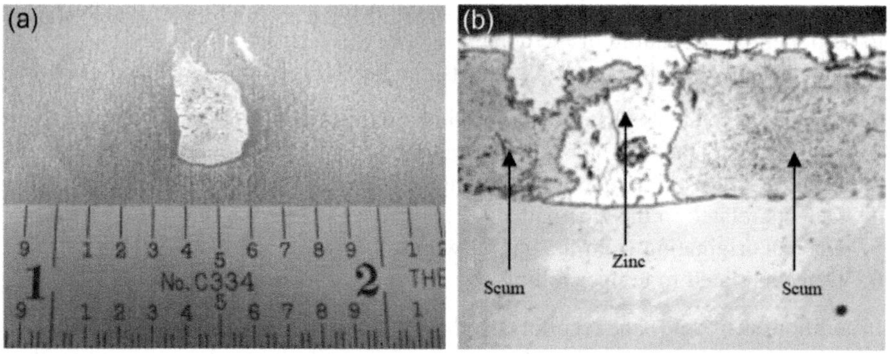

FIG. 14.3 (A) Rolled-in residue (scum) originating in a cold rolling tandem mill, under a galvanneal coating. Dimensions in inches. (B) Cross-section of galvanneal coating surrounding rolled-in residue (scum) originating in a cold rolling tandem mill, under a galvanneal coating. Bottom axis of photo is 270 μm long. *(Reproduced with permission from (A, B) F. Chen, R. Patil, An in-depth analysis of various subtle coating defects of the 2000's, in: Proceedings of 6th International Conference on Zinc and Zinc Alloy Coated Steel Sheet (Galvatech 2004), Chicago, IL, April 4-7, 2004, Association for Iron & Steel Technology, Warrendale, PA, pp. 1055–1066.)*

with the surrounding steel that was reduced prior to zinc immersion. Removal of the coating and polishing reveals the scale, Fig. 14.2B, and zinc that had reacted with the substrate under the scale (Fig 14.2C).

Residue (scum) pickup from the cold rolling mill can also contribute to defect formation. If roll surfaces and other coil handling equipment that touches the steel surface is not kept clean, these residues can adhere. Incomplete removal of rolling lubricants and residues can also contribute to the type of defect shown in Fig. 14.3.

14.3 Defects relating to chemical cleaning

As indicated in Chapter 2, less than 20 mg/m^2/side of contaminants on the strip surface is desirable to control iron fines carryover to the zinc bath to minimize dross generation from this source and also to minimize contaminant-related zinc wettability issues. The contaminants are mainly oil-related substances and iron fine residues from mechanical working

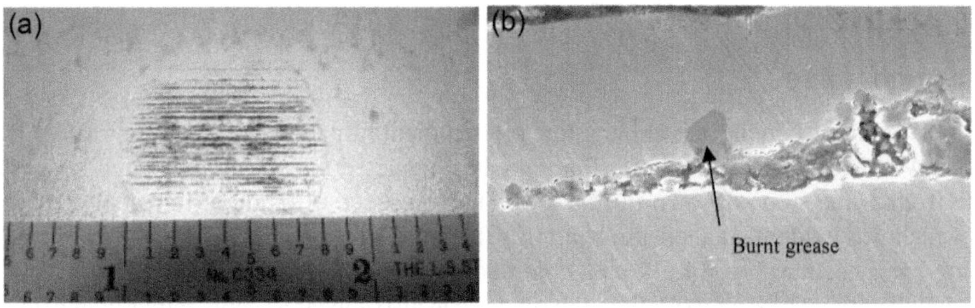

FIG. 14.4 (A) GI-coated area over the remnants of a grease spot present on an incoming full hard cold rolled strip. Dimensions in inches. (B) Cross-section of a GI-coated area over the remnants of a grease spot present on an incoming full hard cold rolled strip. Bottom axis of photo is 110 μm long. *(Reproduced with permission from (A, B) F. Chen, R. Patil, An in-depth analysis of various subtle coating defects of the 2000's, in: Proceedings of 6th International Conference on Zinc and Zinc Alloy Coated Steel Sheet (Galvatech 2004), Chicago, IL, April 4-7, 2004, Association for Iron & Steel Technology, Warrendale, PA, pp. 1055–1066.)*

operations. On coating lines than mainly produce construction product, a wet cleaning section is not used, instead a direct fired furnace, described in Chapter 2, is used to burn off oils and other carbonaceous contamination, however iron fines then still remain. These iron fines can play a role in defect formation later in processing, for example by forming dross in the galvanizing bath, as described below. The surface emissivity of the strip is also affected by its degree of cleanliness, therefore affecting the strip temperature up to the point where it is immersed in the galvanizing bath, this also can increase the amount of dross formed in the pot, as detailed in Chapter 4.

Fig. 14.4 shows a GI-coated grease spot from a line using a wet cleaning section, it could not be removed by the alkaline treatment in this section. Passing through the pretreatment furnace before dipping in zinc, the grease was carbonized, or burned, by removing its volatile components, the remainder was then partially covered by the zinc coating.

Insufficient strip cleanliness is also related to alloy outbursts in the GI coating, Fig. 14.5 [5]. These outbursts affect coating adherence and appearance. Excessive local levels of iron fines, by producing dross, decrease local effective aluminum levels in the bath, making the FeZn equilibrium more stable than the FeAlZn, so that only partial steel surface coverage by the inhibition layer is possible. There also appears to be unconnected intermetallics in the Fig. 14.5 coating that do not have the characteristic monoclinic shape of the FeZn zeta phase that is expected to form the outburst. These defects could originate from iron fines suspended in the bath.

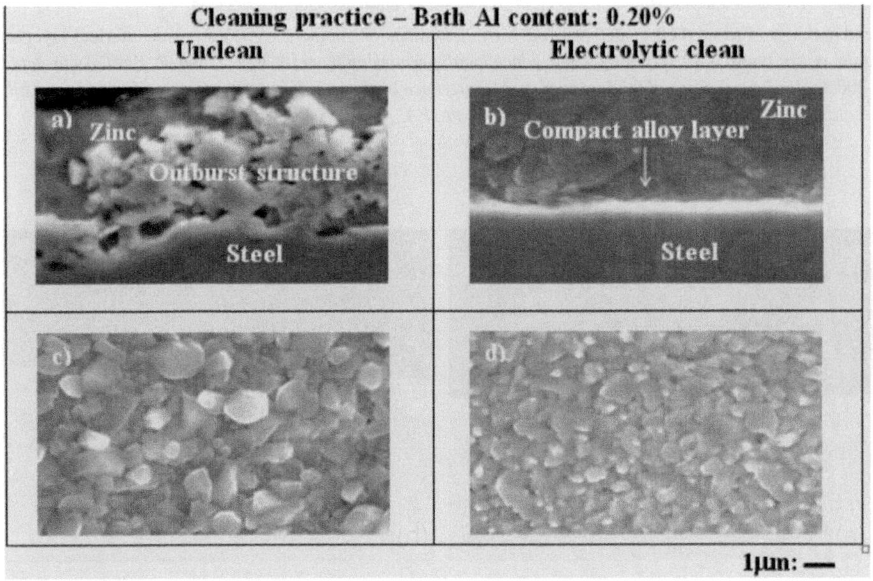

FIG. 14.5 Effect of strip cleanliness of Fe-Zn on GI coating interface stability: FeZn outbursts on unclean steel substrate (A) cross-section (C) surface view, uniform FeAlZn inhibition layer formed on clean substrate (B) cross-section (D) surface view. *(Reproduced with permission from W. Zhong, T. Le, I. O'Reilly, B. Nelson, Experience of Arcelor Mittal Dofasco in automotive exposed GI and GA production, in: Proceedings of 10th International Conference on Zinc and Zinc Alloy Coated Steel Sheet (Galvatech 2015), Toronto, May 31-June 4, 2015, Association for Iron & Steel Technology, Warrendale, PA, pp. 129–136.)*

14.4 Defects related to annealing furnace

14.4.1 Furnace atmosphere

Uncontrolled oxidation of the strip in the galvanizing line pretreatment furnace can arise from furnace leaks, line stops, and actual temperatures lower that target temperatures that may be connected to emissivity-related temperature measurement errors. An example from a line stop is shown in Fig. 14.6. The presence of oxide at the steel-coating interface reduced coating adherence which was evident on a tension bend in a flange of a formed part, Fig. 14.6A. The oxide layer is clearly visible in the coating cross-section, Fig. 14.6B, as a continuous black line.

If there is only a small amount of strip oxidation, the oxides can also appear as spots rather than being continuous. These result in partial coverage of the inhibition layer, with the coating developing only over the surface covered by the inhibition layer. The resultant bare spots can be bridged and result in an interior pore if it is small enough so that a reduction in surface energy results from bridging, or if the surface oxide can be undercut by the zinc bath-strip surface reaction, Fig. 14.7A. Or, the bare spot can remain open, Fig. 14.7B. In the former case a high spot results that can result in "comet tails," Fig. 14.8. However, these surface oxides are only one cause of "comet tails." It is necessary to undertake a more thorough examination to determine the exact cause.

Oxides on the surface of the strip can also be traced to insufficient scale removal and pickling before the hot strip mill, and before the cold rolling mill. In this case, the surface oxides are usually associated with rolled-in scale, discussed in the prior section and later in the galvanneal defect section. An example of strip surface oxides related to hot coiling practice is also described later in the galvanneal defect section.

Fig. 14.9 shows a defect related to hearth roll bearing grease than leaked in the pretreatment furnace, creating a local carburizing atmosphere as it decomposed. This increased the carbon content, the carbon interstitials in the steel then inhibited dislocation movement during forming, resulting in yield point elongation behavior and the appearance of coil breaks shown in this figure.

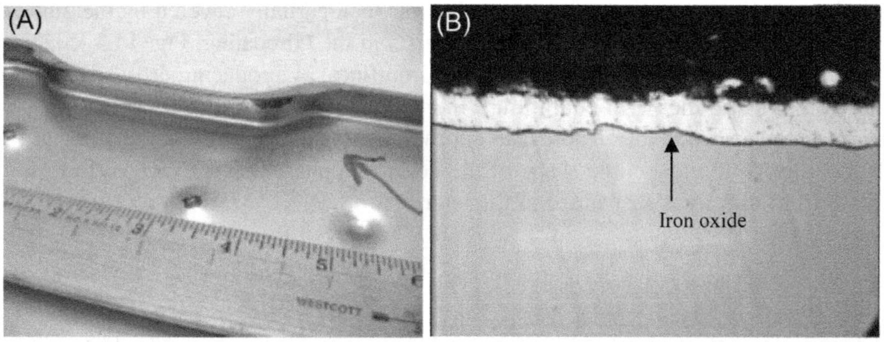

FIG. 14.6 (A) Loss of adherence on tension bend in flange of formed GI-coated part *(dark area)*. Dimensions in inches. (B) Iron oxide interface layer on GI-coated strip that encountered a line stop. Bottom axis of photo is 125 μm long. *(Reproduced with permission from (A, B) F. Chen, R. Patil, An in-depth analysis of various subtle coating defects of the 2000's, in: Proceedings of 6th International Conference on Zinc and Zinc Alloy Coated Steel Sheet (Galvatech 2004), Chicago, IL, April 4-7, 2004, Association for Iron & Steel Technology, Warrendale, PA, pp. 1055–1066.)*

FIG. 14.7 (A) Pore in GI coating resulting in high spot (B) open bare spot caused by strip surface oxide. *(Reproduced with permission from (A, B) Galvanized Autobody Partnership Program ZCO-34-1, Origins of Dross in High Quality Free Zinc Automotive Coatings—Preliminary Examination of Coating Samples, Supplement to Progress Report 1, International Zinc Assn., Durham, NC, July 4, 2000.)*

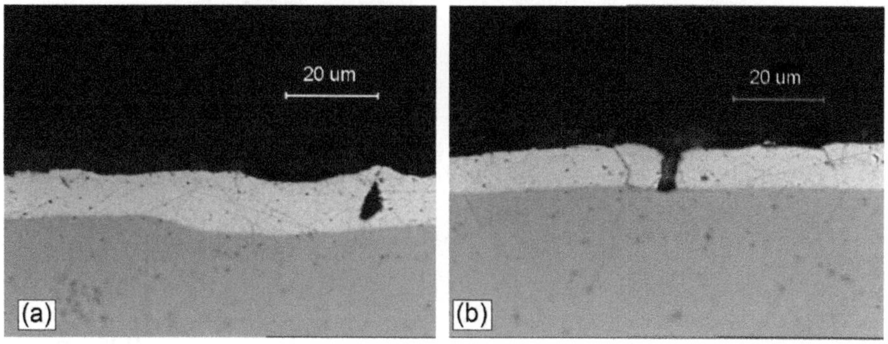

FIG. 14.8 Appearance of "comet tails" on strip surface resulting from oxides on surface of the strip before immersion in the zinc bath. *(Reproduced with permission from M. Dubois, private communication, March 2022, with permission.)*

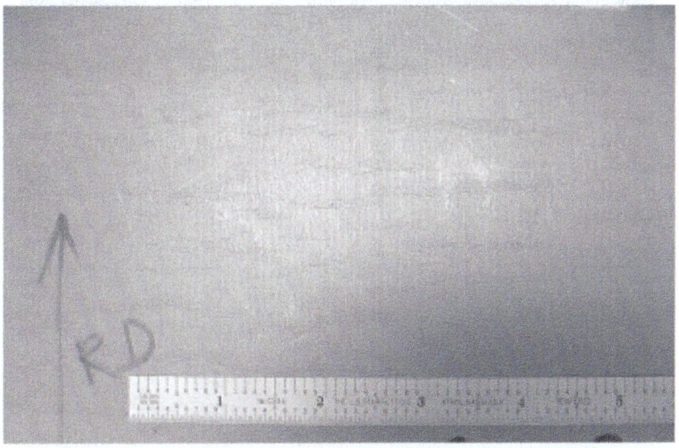

FIG. 14.9 Coil breaks (fine vertically oriented lines of nonuniform elongation, in the direction of strip travel) in GA-coated strip carburized by decomposition of grease in the pretreatment furnace. Dimensions in inches. *(Reproduced with permission from F. Chen, R. Patil, An in-depth analysis of various subtle coating defects of the 2000's, in: Proceedings of 6th International Conference on Zinc and Zinc Alloy Coated Steel Sheet (Galvatech 2004), Chicago, IL, April 4-7, 2004, Association for Iron & Steel Technology, Warrendale, PA, pp. 1055–1066.)*

14.4.2 Other furnace conditions

Friction scratches can occur in the annealing furnace when the surface speed of a process roll does not match strip speed or when a roll is not aligned with the strip pass line. Poor or misaligned roll bearings can result in side-to-side roll movement that can also cause scratches. In the case of a GI coating, the coating surface usually has the appearance of small balls of coating along the line of the scratch, Fig. 14.10A. In the case of GA, bright spots are observed, Fig. 14.10B, these form because of the thicker coating present in the depression of the scratch, so that the galvanneal reaction does not run to completion, leaving unalloyed zinc on the surface above the scratch. In this case, the raised surface in the vicinity of the scratch, which was on the nonprime side, produced a print-through defect on the prime side after stamping. Fig 14.10C shows the scratch in the steel substrate after removal of the GA coating shown in Fig. 14.10B. Although not shown, such scratches usually show some evidence of cold work, manifest as much smaller or elongated grains, in the vicinity of the scratch. Roll coatings have shown benefit in reducing the frequency and severity of these scratches.

There are many other causes of damage to the surface of the steel surface, including nonuniform or nonoptimized rolling conditions, the presence of scale and handling. Fig. 14.11A shows a surface irregularity on the surface that is about the same dimensions of the shown GI coating, about 25 μm. Here, the intermetallic layer has grown through the overlay portion of the coating at its thinnest point. The layer in the vicinity of this growth through to the surface is a FeZn intermetallic rather than the AlFeZn intermetallic normally associated with a GI coating. This occurred because enough of the local Al has been

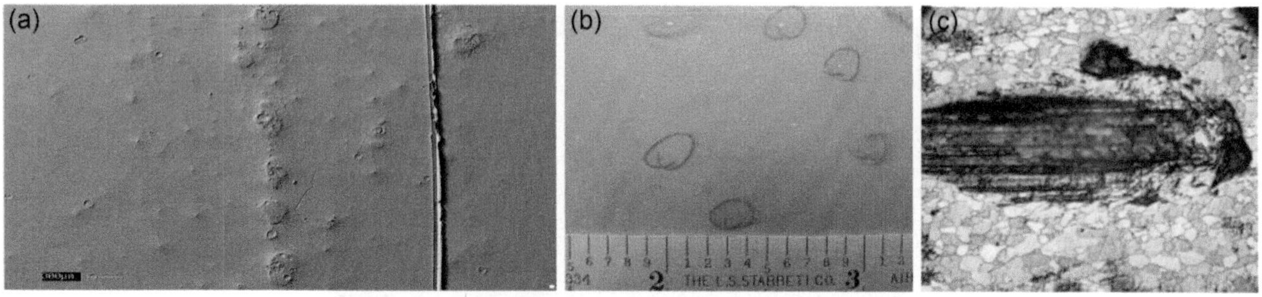

FIG. 14.10 (A) Balled material in a GI coating covering a scratch originating in the galvanizing pretreatment furnace. The gouge on the right side is an artifact. (B) Appearance of GA coating covering a scratch originating in the galvanizing pretreatment furnace. Dimensions in inches. (C) Scratch associated with Fig. B after removal of the GA coating. Bottom axis of photo is 640 μm long. *(Reproduced with permission from (A) M. Rogers, Hot dip galvanize products coating defects & their avoidance, in: Presented at Hot-Dip Galvanized Sheet Production Course, Merrillville, IN, June 9-11, 2010, GalvInfo Center of International Zinc Association, Durham, NC. Reproduced with permission from (B, C) F. Chen, R. Patil, An in-depth analysis of various subtle coating defects of the 2000's, in: Proceedings of 6th International Conference on Zinc and Zinc Alloy Coated Steel Sheet (Galvatech 2004), Chicago, IL, April 4-7, 2004, Association for Iron & Steel Technology, Warrendale, PA, pp. 1055–1066.)*

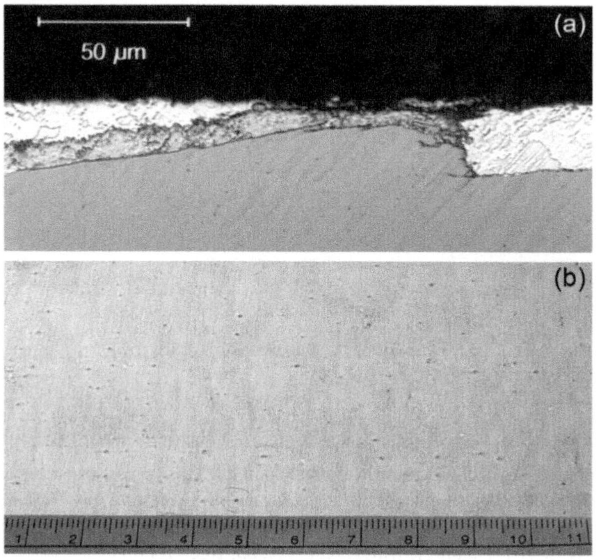

FIG. 14.11 (A) Cross-section of black or gray spot in GI coating related to mechanical damage of steel substrate, caused by through-thickness growth of FeZn phases. (B) Appearance of GI coated sheet with defect shown in Fig. 14.10A. *(Reproduced with permission from (A, B) N.-Y. Tang, F.E. Goodwin, A study of defects in galvanized coatings, in: Proceedings of 5th International Conference on Zinc and Zinc Alloy Coated Steel Sheet (Galvatech 2001), Brussels, June 26-28, 2001, Verlag Stahleisen, Duesseldorf, pp. 49–55.)*

consumed so that the FeZn phase is stable and appears as black or gray spots on the surface, whose appearance is shown in Fig. 14.11B.

Heat buckles result in an irregular wave profile on the strip, with the peaks and valleys of the waves oriented in the direction of strip travel. An example is shown in Fig. 14.12. They are caused by localized yielding of the strip surface, under conditions where the yield strength of the steel at the local temperature of the strip is less than the applied strip tension. To avoid heat buckles, strip tension and tracking must be set so that the stress on the cross-section of the strip never exceeds the yield stress of the steel at temperatures it will experience in the furnace zone. It is important to maintain the correct strip shape, usually a "full center" profile, with the center thicker than the edges, to minimize the chance of the strip being "pulled out" under tension. Furnace roll shape can also change with time and can be checked as a cause of this defect. The "heat index," defined as the width of the strip divided by the thickness, is a useful way of determining the capabilities of a galvanizing line to produce strip free of heat buckles. Scheduling of coils with similar heat indices can also help minimize heat buckle occurrence.

Proper strip cooling, especially in the jet cooling zone, will help minimize temperature differences both across the strip width and in the direction of strip travel so that nonuniform thermal contraction that can influence strip shape will be

FIG. 14.12 Appearance of heat buckles on strip after exiting the wiping zone. *(Reproduced with permission from M. Dubois, private communication, March 2022, with permission.)*

minimized. Good strip cleaning is related to this, because of the effect of strip surface emissivity on heat transfer. Nonuniform heat transfer across the strip width during heating and soaking can also cause local thermal expansion/contraction effects that affect shape. Phase transformations in the steel also affect shape because ferrite is less dense than austenite. Nonuniform temperatures across the strip width can also influence shape as the strip contracts during the formation of austenite and expands as ferrite and hard phases are formed upon cooling.

Hearth roll residue pickup is more frequently found when the strip is supported on uncoated steel rolls as it travels through the galvanizing pretreatment furnace than when these rolls are coated. A deposit of oxides or other contamination on the roll surface repeats this defect on the strip at distances related to the circumference of the roll. It is useful to maintain a library of all roll circumferences used in the galvanizing line, so that when periodic defects occur their spacing on the strip surface can used to identify rolls whose circumference may be connect them to the cause of the defect. Fig. 14.13 shows a hearth roll defect resulting from imprinting of a hearth roll onto the steel strip in the annealing section. The plastic

FIG. 14.13 Hearth roll pickup. Contaminant or roll material released from a hearth roll attached to a steel sheet before dipping in the zinc bath. Bottom axis of photo is 380 μm long. *(Reproduced with permission from M. Rogers, Hot dip galvanize products coating defects & their avoidance, in: Presented at Hot-Dip Galvanized Sheet Production Course, Merrillville, IN, June 9-11, 2010, GalvInfo Center of International Zinc Association, Durham, NC.)*

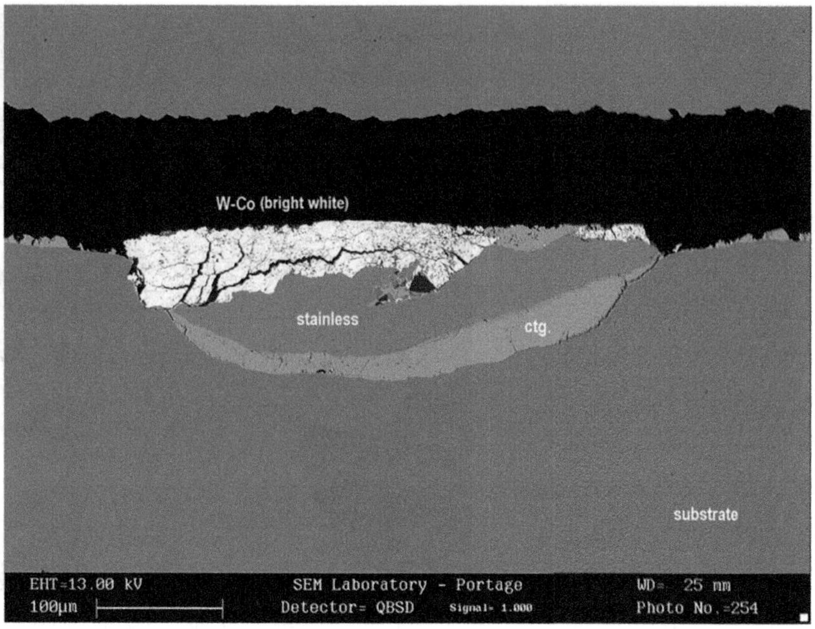

FIG. 14.14 Hearth roll pickup. Contaminant or roll material released from a WC-Co coated stainless steel hearth roll attached to a steel sheet and then hot-dip galvanized. Remnants of zinc coating below the pickup debris can be seen, after removal of the coating covering the strip. *(Reproduced with permission from M. Rogers, Hot dip galvanize products coating defects & their avoidance, in: Presented at Hot-Dip Galvanized Sheet Production Course, Merrillville, IN, June 9-11, 2010, GalvInfo Center of International Zinc Association, Durham, NC.)*

deformation arising from this imprinting results in a smaller, more oriented grain structure in the steel substrate under the imprinted foreign material if the denting occurred after the annealing section of the pretreatment furnace. If it occurred before, these small grains would have grown during annealing. The use of more durable coatings such as tungsten carbide-cobalt cermets thermal sprayed onto the roll surfaces can reduce the frequency of hearth roll defects. However, eventually flaking from these coated rolls can occur, as shown in Fig. 14.14. Practices associated with minimizing hearth roll defects include proper strip cleaning, operation of furnaces to avoid scale and smut buildup, i.e., operating burners with a CO/CO_2 ratio that avoids soot generation, matching roll speed to strip speed, utilizing coatings on rolls that are repeatedly connected to hearth roll defects and increasing tension to avoid slippage between roll surfaces and the strip.

14.5 Defects related to the snout, zinc bath condition, and bath hardware

The occurrence of outbursts of zeta phase at the steel-coating interface was discussed above in connection with strip cleanliness. Another cause of outbursts is the condensation of zinc vapor on the steel strip as it passes through the entry snout to the galvanizing bath. Because this condensed vapor is free of aluminum, a AlFeZn inhibition layer cannot be formed on strip surfaces where the vapor has condensed. These outbursts manifest as raised spots on the coating. Fig. 14.15 shows this defect, including the appearance of exposed zeta phase at the surface.

In many other situations, the dewpoint is the snout is not high enough to maintain the zinc in vapor form; it combines with the partial pressure of oxygen in the snout to form zinc oxide. This is the objective of wet snout practice described in Chapters 2 and 4. Fig. 14.16A and B shows a dark line defect on galvanized and galvanneal coatings, respectively, that was traced to zinc oxide deposition on the strip as it passed through the snout. Fig. 14.16C shows a cross-section of Fig. 14.16B. defect, showing how the oxide prevented the reaction of the zinc with the steel surface.

As discussed in Chapter 4, the sink roll in the zinc pot is grooved to permit passage of liquid zinc between the strip and roll surfaces without roll slippage that would damage the strip surface. However, marks from strip roll grooves can also occur. The marks run in the direction of strip travel and their pattern can be related to the spacing of the roll grooves. For the galvanneal coating, these marks can interfere with the development of uniform interface layers and therefore the growth of the galvanneal coating, Fig. 14.17, but can still be seen with higher aluminum content galvanizing baths, even these exceeding 0.25% soluble aluminum [6]. These marks tend to be more severe with more reactive steels, such as certain ultralow-carbon grades [4]. These marks can be related to the sink roll surface conditions, such as poorly chamfered groove edges, or wear on the sink roll. Overly high strip tension for the strength and thickness of the steel being processed can also contribute to the appearance of the grooves, a high strip tension can give a slightly corrugated strip profile that is accentuated in appearance after temper rolling. The marks do not seem to be dross-related [7]. As discussed in Chapter 4, much more dross can build up on the roll surface that is outside the range of strip contact, but conditions can be favorable for dross

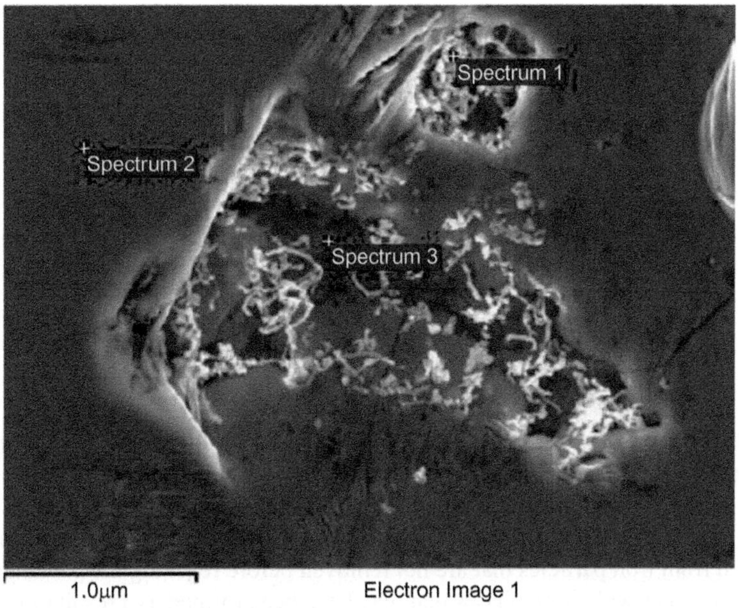

FIG. 14.15 Deposition of metallic zinc and zinc oxide from the snout onto the strip surface resulted in the formation of outbursts and nonwetted areas. *(Reproduced with permission from W. Zhong, T. Le, I. O'Reilly, B. Nelson, Experience of Arcelor Mittal Dofasco in automotive exposed GI and GA production, in: Proceedings of 10th International Conference on Zinc and Zinc Alloy Coated Steel Sheet (Galvatech 2015), Toronto, May 31-June 4, 2015, Association for Iron & Steel Technology, Warrendale, PA, pp. 129–136.)*

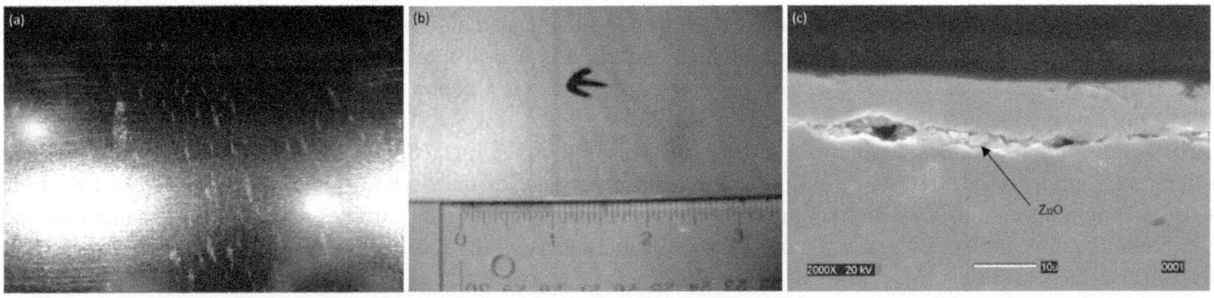

FIG. 14.16 (A) Surface appearance of galvanized coating over zinc oxide deposited on strip in the snout. (B) Surface appearance of galvanneal coating over zinc oxide deposited on strip in the snout. Dimensions in inches. (C) Zinc oxide deposited on strip surface in snout, at steel/galvanneal coating interface. *(Reproduced with permission from (A) M. Dubois, private communication, March 2022, with permission. Reproduced with permission from (B, C) F. Chen, R. Patil, An in-depth analysis of various subtle coating defects of the 2000's, in: Proceedings of 6th International Conference on Zinc and Zinc Alloy Coated Steel Sheet (Galvatech 2004), Chicago, IL, April 4-7, 2004, Association for Iron & Steel Technology, Warrendale, PA, pp. 1055–1066.)*

FIG. 14.17 Sink roll groove marks on galvanneal coating, strip traveling in horizontal orientation. *(Reproduced with permission from P. Sippola, private communication, February 2022, with permission.)*

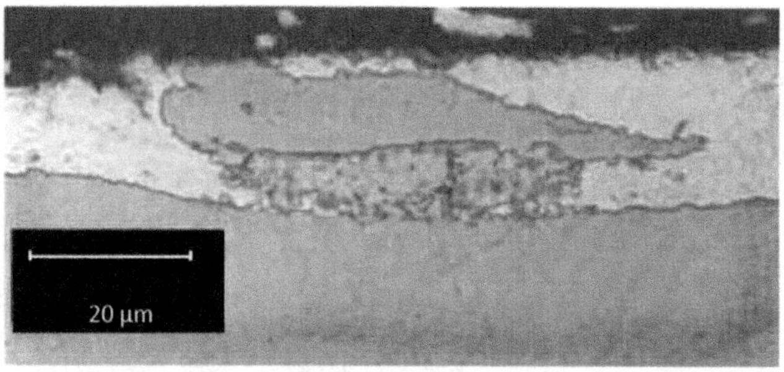

FIG. 14.18 Iron particle entrapped in a GI coating. *(Reproduced with permission from N.-Y. Tang, F.E. Goodwin, A study of defects in galvanized coatings, in: Proceedings of 5th International Conference on Zinc and Zinc Alloy Coated Steel Sheet (Galvatech 2001), Brussels, June 26-28, 2001, Verlag Stahleisen, Duesseldorf, pp. 49–55.)*

buildups in the grooves in the portion of the roll contacted by the strip. In this case, dross buildup has been reduced by lowering the pot temperature, maintaining a correct strip entry temperature relative to the pot temperature, using roll coatings such as the WC-Co cermet and maintaining correct strip tension for the strip product being coated.

The amount of dross present in the zinc bath is influenced by many factors as discussed in Chapter 4. As described above, dross can be formed from iron particles that are not removed before reaching the galvanizing bath. An iron particle entrapped in a GI coating, together with FeZn phases that may have been formed as a result of reaction of the bath with the iron from the coating below and the iron particle above it, is shown in Fig. 14.18. Because of its elongated shape and size, it is likely that this iron particle originated as a microflap during cold rolling. Iron fines transmitted to the bath as a result in incomplete cleaning are usually 3–5 μm in size.

The visual appearance of a dross defect on a galvanneal coating is shown in Fig. 14.19A. This is similar in appearance to the "scratch" defect described before. This kind of defect is frequently seen when transitioning the zinc bath chemistry from GI to GA operation. The AlFeZn dross found in a GI bath, which has a density considerably less than liquid zinc as discussed in Chapter 4, is converted to ZnFe dross that is only slightly denser than the zinc bath. This results in this "bottom dross" requiring much more time to settle to the bottom, as described in Chapter 4. Dross suspended in the bath can adhere to the strip, causing the defect shown in Fig. 14.19A.

After polishing, the surface, FeZn dross particles are apparent on the surface, Fig. 14.19B. If coatings containing these particles are stamped, the entrapped dross will result in the appearance of an auto body panel shown in Fig. 14.19C. Even if the dross-containing coating is on the "nonprime" unexposed side, the hard dross particles can print through upon stamping, giving a similar appearance. There are can be other causes of defects giving this appearance, more causes are shown in the galvanneal defects section.

When the strip is withdrawn from the zinc bath, it may entrain top skimmings, which are a mixture of oxides, top dross and bath metal as described in Chapter 4. Fig. 14.20 shows a coating defect containing top skimmings.

Sometimes the top dross dragged out with top skimmings appears without the oxide portion of the skimmings. Fig. 14.21 shows a GI coating, with its fully formed inhibition layer, in which top AlFeZn top dross particles are suspended. Here, the oxide portion of the top skimmings was present on the coating surface, giving it a dull appearance. Only top dross particles

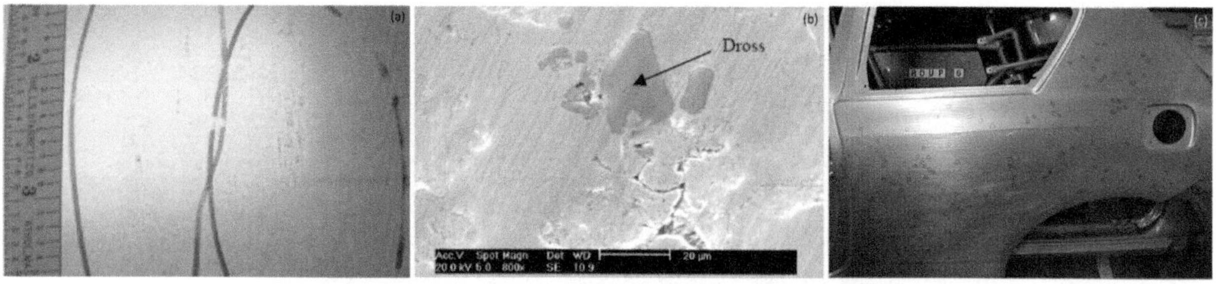

FIG. 14.19 (A) Appearance of dross defect on galvanneal coating. Dimensions in inches. (B) View of dross particles after the surface shown in Fig. A was polished. (C) Appearance of dross defects *(circled)* on stamped galvanized body panel. *(Reproduced with permission from (A, B) F. Chen, R. Patil, An in-depth analysis of various subtle coating defects of the 2000's, in: Proceedings of 6th International Conference on Zinc and Zinc Alloy Coated Steel Sheet (Galvatech 2004), Chicago, IL, April 4-7, 2004, Association for Iron & Steel Technology, Warrendale, PA, pp. 1055–1066. Reproduced with permission from (C) M. Dubois, private communication, March 2022, with permission.)*

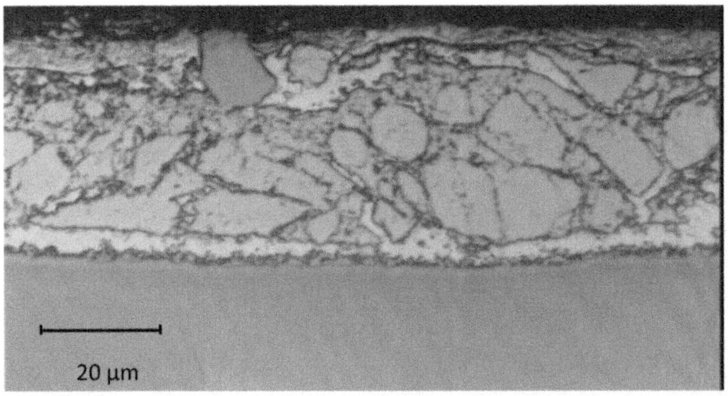

FIG. 14.20 Top skimmings entrained in GI coating, showing mixtures of oxides and top dross particles. Also note incompletely inhibited interface layer. *(Reproduced with permission from Galvanized Autobody Partnership Project ZCO-34, Origins of Dross in High Quality Free Zinc Automotive Coatings, Final Report, November 7, 2001 International Zinc Assn., Durham, NC)*

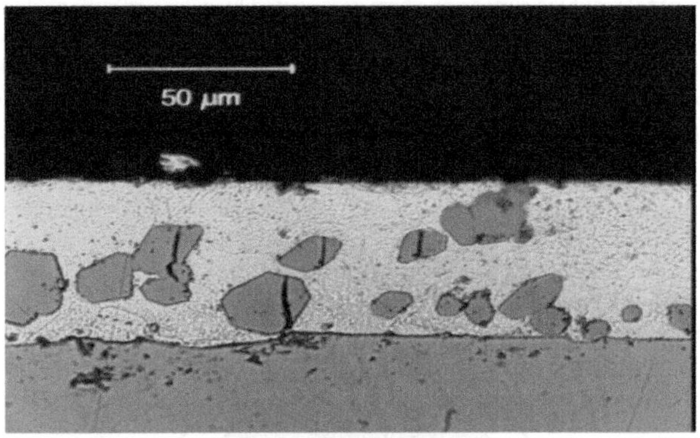

FIG. 14.21 Top dross (AlFeZn) particles entrapped in GI coating. *(Reproduced with permission from N.-Y. Tang, F.E. Goodwin, A study of defects in galvanized coatings, in: Proceedings of 5th International Conference on Zinc and Zinc Alloy Coated Steel Sheet (Galvatech 2001), Brussels, June 26-28, 2001, Verlag Stahleisen, Duesseldorf, pp. 49–55.)*

that had been dragged out of the bath with the top skimmings were present in the interior of this very thick (50 μm) coating that presumably was produced with a low wiping pressure.

Bare spots are caused by incomplete wetting during dipping. For well-cleaned strip, the most frequent reason for bare spot occurrence is the presence of oxides on the strip surface. One source of these oxides is from entrained surface skimmings. If a film of oxide, normally rich in aluminum oxide, contacts the strip surface during its period of entry, immersion and withdrawal from the bath, it can form a nonwettable area that can result in a bare spot. Views of this bare spot on a Ti-IF steel are shown in Fig. 14.22A. After stripping of the GI coating in fuming nitric acid, the discontinuous Fe_2Al_5 inhibition layer is seen in Fig. 14.22B, and the aluminum oxide layer interrupting this inhibition layer in Fig. 14.22C.

The mechanism of skimmings-caused bare spot formation is slightly different for a Mn-Si TRIP steel because of its lower reactivity than the Ti-IF grade. Also, portions of the TRIP surface are covered with granular (Mn,Si) oxides, shown schematically in Fig. 14.23. Whereas the Ti-IF grade reactivity can be sufficient to consume enough aluminum to drive the local equilibrium into the Fe-Zn range and produce zeta outbursts, the TRIP grade remains in the aluminum composition range where Fe_2Al_5 is stable. A schematic of the bare spot formation mechanisms in these steels is shown in Fig. 14.23 [8].

Bare spots are often related to oxidation in the furnace, most likely resulting from an oxygen-rich location that remains after purging, from a furnace leak, from improper combustion in the direct fired furnace (DFF) because of an improperly adjusted air/fuel ratio or damage to the DFF burner nozzles. Jet coolers are also critical part of temperature control in the pretreatment furnace. For AHSS with their higher contents of Mn, Si, Al and other elements that preferentially oxidize during processing, several conditions of dewpoints and temperatures may be required during the pretreatment process, together with rapid cooling, making the preparation of a galvanizable surface more challenging, as described in Chapter 2. An example of a bare spot caused by the presence of a nonwettable surface oxide is shown in Fig. 14.24. The sensitivity of coating quality to the presence of oxides increases with increasing bath aluminum content [1].

Too low of an aluminum content in the GI bath can produce an "orange peel" appearance, Fig. 14.25. This defect has been related to iron oxide coverage on steels that were subjected to an excessively high coiling temperature after hot rolling. Cracks in the oxide layer formed during cold rolling resulted in the initiation of growth of zinc coating grains at crack

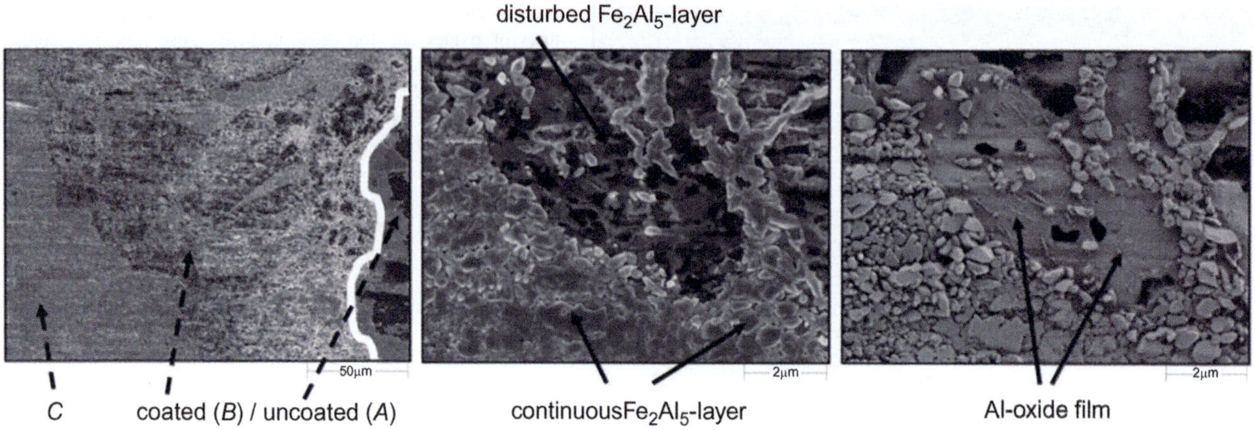

FIG. 14.22 Scanning electron microscope views of the surface of a bare spot caused by entrained surface skimmings (A). overall view of the bare spot (B). view after stripping of GI overlay coating of discontinuous inhibition layer (C). view after stripping of GI overlay coating of aluminum oxide layer interrupting the inhibition layer shown in (B). *(Reproduced with permission from M. Norden, F. Macherey, F. Friedel, M. Peters, Influence of thin zinc-melt oxide films on galvanizing of If and advanced high strength steels, in: Proceedings of 8th International Conference on Zinc and Zinc Alloy Coated Steel Sheet (Galvatech 2011), Genoa, June 21-25, 2011, Italy, Associazione Italiana di Metallurgia, Milan.)*

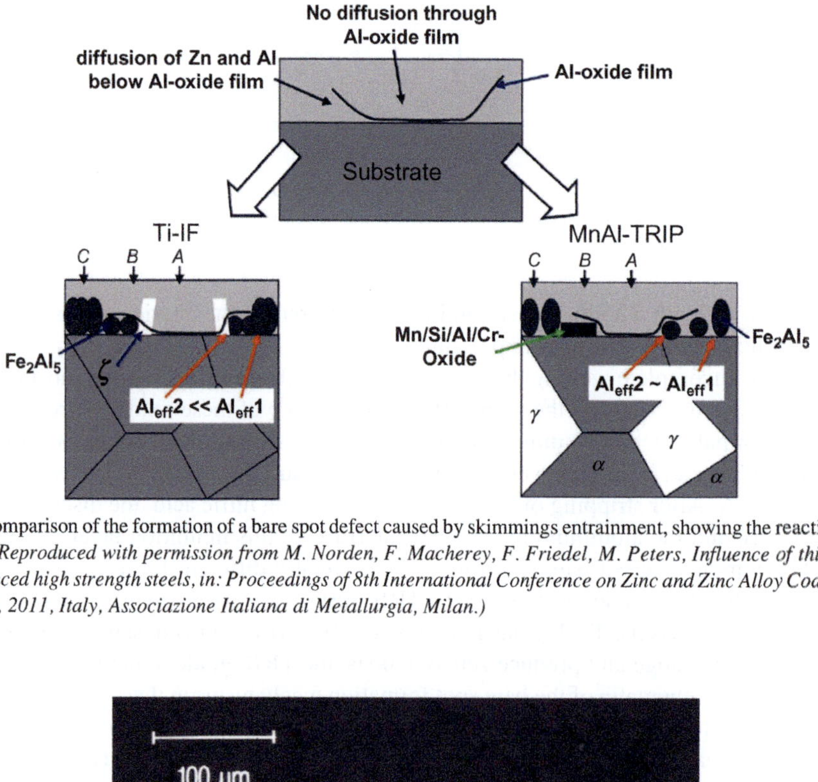

FIG. 14.23 Schematic comparison of the formation of a bare spot defect caused by skimmings entrainment, showing the reaction paths observed in Ti-IF and Mn-Al TRIP grader. *(Reproduced with permission from M. Norden, F. Macherey, F. Friedel, M. Peters, Influence of thin zinc-melt oxide films on galvanizing of If and advanced high strength steels, in: Proceedings of 8th International Conference on Zinc and Zinc Alloy Coated Steel Sheet (Galvatech 2011), Genoa, June 21-25, 2011, Italy, Associazione Italiana di Metallurgia, Milan.)*

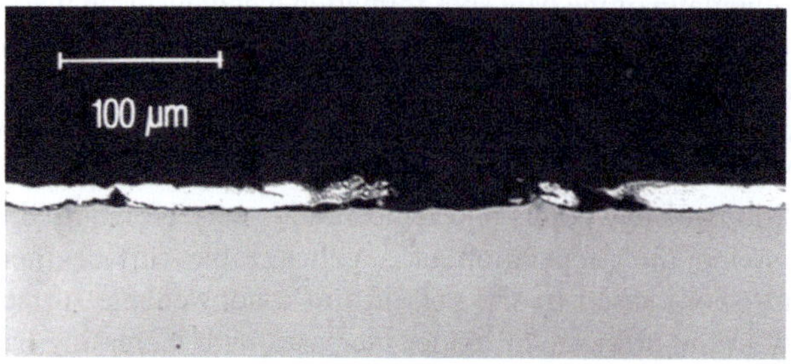

FIG. 14.24 Bare spot resulting from nonwettable surface oxide. *(Reproduced with permission from N.-Y. Tang, F.E. Goodwin, A study of defects in galvanized coatings, in: Proceedings of 5th International Conference on Zinc and Zinc Alloy Coated Steel Sheet (Galvatech 2001), Brussels, June 26-28, 2001, Verlag Stahleisen, Duesseldorf, pp. 49–55.)*

FIG. 14.25 Orange peel appearance of GI coating. *(Reproduced with permission from E. Ma, M. Peng, Z. Yang, L. Cui, H. Zhan Y. Liu, Orange peel surface defect on high strength galvanizing steel sheet, in: Proceedings of 12th International Conference on Zinc and Zinc Alloy Coated Steel Sheet (Galvatech 2021), June 22-24, 2021, Virtual Conference, Austrian Society for Metallurgy and Materials, pp. 1354–1360.)*

locations, these were then sites of outbursts as coating formation progressed [9]. The edges of the fully formed zinc coating grains then contract during solidification, causing grain boundary depressions. This problem can be minimized by increasing bath aluminum content and therefore improve inhibition layer development. In Fig. 14.25, each of the grain diameters is about 500 μm, whereas the coating roughness is about 2 μm, so that the orange peel appearance is on the scale of customary waviness measurements.

14.6 Wiping related defects

The condition of the wiping knife nozzle slot greatly influences coating quality. The use of automatic nozzle lip cleaners described in Chapter 7 can help minimize defects relates to this. Problems with nozzle condition are most frequently manifest as "flow lines" in the coating, Fig. 14.26. The surface of the lines are usually raised because of the lower wiping force applied across the line because of the local wiping nozzle constriction. The frequency of this defect can be prevented by maintaining good strip shape and position stability as it passes between the air knives that may require the used use of strip

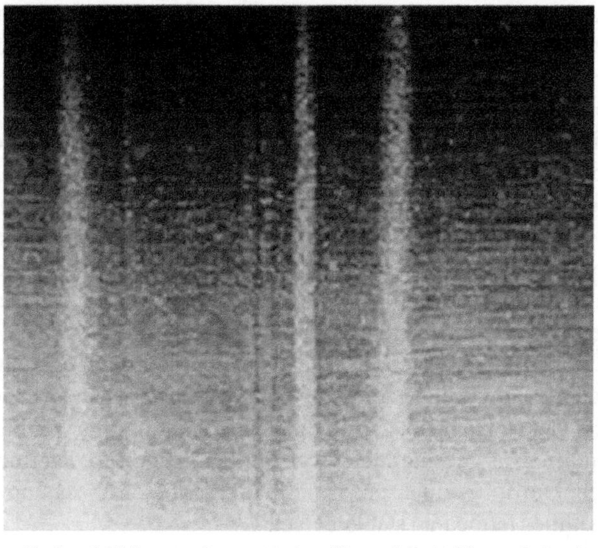

FIG. 14.26 "Flow lines" in coating caused by local wiping nozzle constriction. *(Reproduced with permission from M. Rogers, Hot dip galvanize products coating defects & their avoidance, in: Presented at Hot-Dip Galvanized Sheet Production Course, Merrillville, IN, June 9-11, 2010, GalvInfo Center of International Zinc Association, Durham, NC.)*

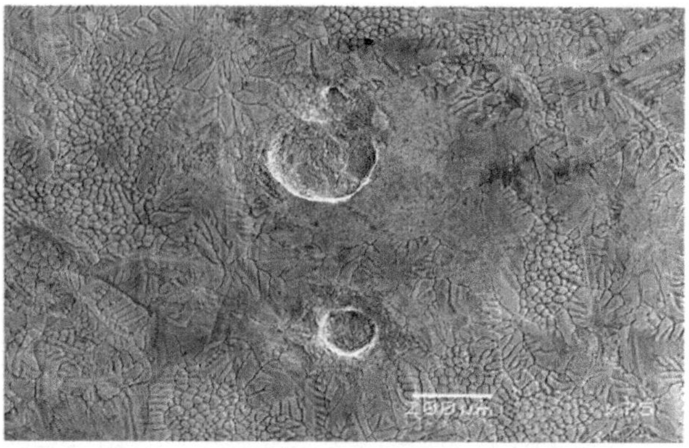

FIG. 14.27 "Dent" caused by impingement of solid particle on wiped coating surface while the coating was partially solidified. *(Reproduced with permission from N.-Y. Tang, F.E. Goodwin, A study of defects in galvanized coatings, in: Proceedings of 5th International Conference on Zinc and Zinc Alloy Coated Steel Sheet (Galvatech 2001), Brussels, June 26-28, 2001, Verlag Stahleisen, Duesseldorf, pp. 49–55.)*

stabilization equipment described in Chapter 7, keeping the wiping knives in good condition, and keeping the strip-to-knife distance and wiping pressure in the correct range for the targeted coating thickness.

"Dents" can result from the impingement of solid zinc particles on the liquid coating surface as a result of turbulent gas flow in the wiping zone and can be related to splashing when wiping conditions are at or beyond the splashing limit described in Chapter 7. In many cases, the dents result from redeposition of the splashed material. Fig. 14.27 shows an example, the coating in this figure is thinner in the dent area, in other cases there can be areas of local thickness. The shape of the dent may be circular or elongated, depending upon the progress of coating solidification at the time of particle impingement. An elongated dent would indicate that liquid zinc flowed around the particle as the coating cooled, a circular dent would indicate impingement of the particle after some coating solidification had occurred. In one case, it was found that an iron fine contained in the coating was blown out during wiping and then re-deposited on the coating surface as it traveled out of the wiping zone [1].

"Dents" can also be caused by conditions encountered later in the continuous galvanizing process. These include adherence of flakes from nonadherent coating, from small zinc particles released from the coating surface, from damaged or cut process rolls, side-to-side movement of strip relative to rolls, and "pick-off" of the zinc coating that then adheres to the process roll. If the strip temperature exceeds 300°C when it reaches the top roll, the risk of coating transfer to the roll surface is increased. For exposed quality coatings, the top roll should be kept in the range of 200–220°C. "Dents" from these causes can be either raised of they were a result of particle impingement of the reverse side of the strip, or depressed if impingement was on the examined side. Defects related to passline rolls will normally show a repeating distance/pattern related to roll circumference. As noted in reference to hearth roll defects before, roll speeds need to match strip speed, proper strip tension should be maintained, and cut or damaged rolls replaced. Further discussion of this kind of defect is also included in the temper mill section later in this chapter.

Sag lines and coating waviness are also encountered as the coating exits the wiping zone and cools above it. An example of sag line appearance is shown in Fig. 14.28. The light and dark areas are unoxidized and oxidized areas, respectively. The formation of oxides is related to the reaction of these areas with air from the wiping gas. It is more prevalent at low line speeds with coatings not containing Pb or Sb and can be remedied by closer-strip-to-knife distances together with reduced pressure. It can also be found on very smooth steel substrates, where zinc flow is easier, enabling breakage of the oxide skin that results in reflectivity differences. High strip and bath temperatures can also contribute to its severity. Nitrogen wiping can help eliminate this condition. This wiping condition closer to the strip will also reduce a contributing factor to waviness, discussed in Chapter 7. A "wavelet defect" related to excessive strip-to-knife distance, higher wiping pressures and lower line speeds is shown in Fig. 14.29. This defect is more severe at high aluminum contents.

Coating sag lines are also related to flow around defects in or under the coating. An example for a thick coating is shown in Fig. 14.30. The surface defects can be bare spots or oxidized zinc that inhibits wetting and uniform flow of the wiped coating. Which of these two conditions is causing the defect can be difficult to determine. This appearance can also be associated with top skimmings pickup at low line speeds, in that case the cross-section of the defect will contain dross, similar to Fig. 14.20 and is sometimes termed a "staple defect."

Related to this, Fig. 14.31 shows a "caterpillar" defect resulting from flow of zinc around a nonwetted area after passing through the wiping zone. An overall view of the coating and this defect are shown in Fig. 14.31A, a scanning electron microscope image is shown in Fig. 14.32B.

FIG. 14.28 Sag lines appearance related to nonuniform oxidation of coating surface. *(Reproduced with permission from M. Dubois, private communication, March 2022, with permission.)*

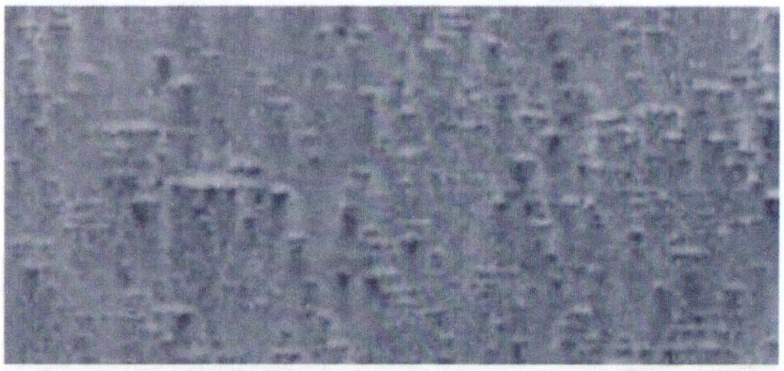

FIG. 14.29 "Wavelet defects" associated with lower line speeds, low GI effective Al contents, high strip-to-knife distance and high plenum pressure. *(Reproduced with permission from F.E. Goodwin, M. Dubois, Review on wiping, a key process limiting CGL productivity, in: AISTech 2012-Proceedings of the Iron & Steel Technology Conference, Atlanta, GA, May 7-12, 2012, Association for Iron & Steel Technology, Warrendale, PA)*

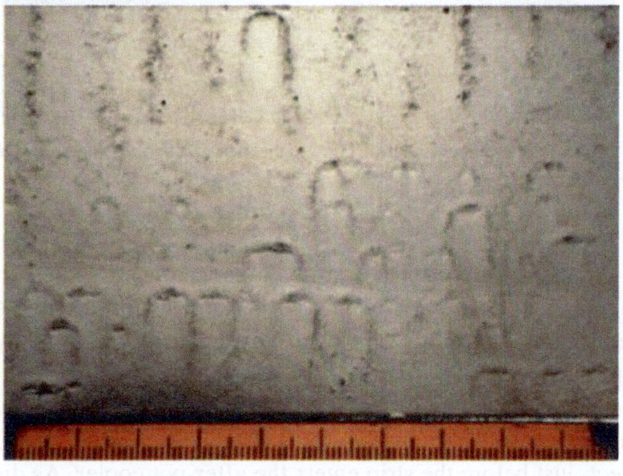

FIG. 14.30 Sag lines associated with steel substrate oxides, zinc particle impingement on a 20 μm thick GI coating. Dimensions in mm. *(Reproduced with permission from M. Dubois, Production of heavy coatings at low line speed, in: Proceedings of 12th International Conference on Zinc and Zinc Alloy Coated Steel Sheet (Galvatech 2021), June 22-24, 2021, Virtual Conference, Austrian Society for Metallurgy and Materials, pp. 733–745.)*

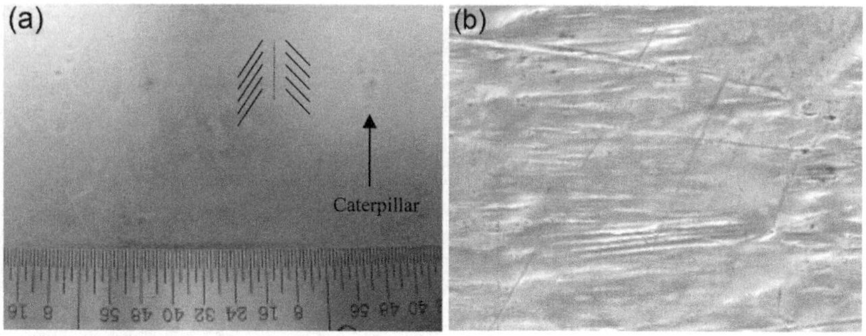

FIG. 14.31 "Caterpillar defect" resulting from nonwetted surface (A) Overall view of coating, dimension in inches. (B) Scanning electron microscope image of surface. Bottom axis of photo is 100 μm long. *(Reproduced with permission from (A, B) F. Chen, R. Patil, An in-depth analysis of various subtle coating defects of the 2000's, in: Proceedings of 6th International Conference on Zinc and Zinc Alloy Coated Steel Sheet (Galvatech 2004), Chicago, IL, April 4-7, 2004, Association for Iron & Steel Technology, Warrendale, PA, pp. 1055–1066.)*

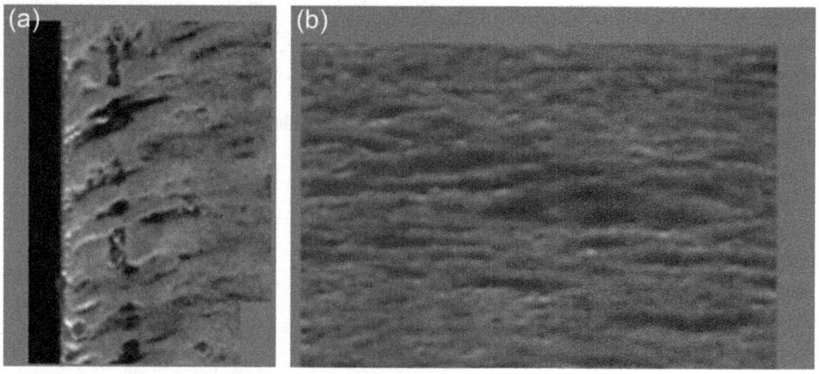

FIG. 14.32 Coating sags related to coating runback (A) at edge (B) general view of strip surface. *(Reproduced with permission from M. Rogers, Hot dip galvanize products coating defects & their avoidance, in: Presented at Hot-Dip Galvanized Sheet Production Course, Merrillville, IN, June 9-11, 2010, GalvInfo Center of International Zinc Association, Durham, NC.)*

FIG. 14.33 Waviness defect related to wiping jet flapping, just before entry to the after-pot cooler. *(Reproduced with permission from M. Dubois, private communication, March 2022, with permission.)*

Coating sags are also related to coating runback. This washboard appearance, shown in Fig. 14.32 is related to control of solidification, meaning that pot and strip temperatures need to be kept within a range so that solidification time is not excessive. It can also be affected by too high of a strip to knife distance and accompanying higher gas wiping pressure, also too high of a bath aluminum content.

The "check mark" waviness defect was discussed in Chapter 7, where the time to relax surface waves was longer than the local solidification time. Another cause of waviness is jet flapping, also discussed in Chapter 7. Fig. 14.33 shows the appearance of this waviness defect, just before the strip enters the after-pot cooler. As discussed in Chapter 7, flapping is related to impingement of opposing jets. Tilting or slightly offsetting jet positions has been found to reduce the intensity of jet flapping, mainly affecting undulations near the strip edges.

The "angel wings" defect is related to the withdrawal of thicker strip running at lower line speeds from the zinc bath and typically forms in a diagonal pattern from the strip edge, Fig. 14.34. Withdrawal of metal from the bath produces a liquid layer covered by an oxide film, explained in Chapter 7, Section 7.8. As this film approaches the after-pot cooler, it is broken up, so that bright patterns of less oxidized metal are formed. Fig. 14.34A also shows an area of poor coating adherence, the dark area that is covered with a coating less than 10 μm thick. A more severe case near a strip edge is shown in Fig. 14.34B.

The "angel hair" defect is shown in Fig. 14.35, this defect is the appearance of thin white oxide lines running transverse to the direction of strip travel. It is found when producing thin coatings at high line speeds and is associated with high aluminum content. Also it can occur when Pb and/or Sb are being used to create a spangled appearance.

The appearance of dendrites on the surface of the coating, Fig. 14.36, indicates that solidification is occurring in the forced cooling zone of the after-pot cooler rather than before it. Slower cooling rates produce planar rather than dendritic solidification. These kinds of defects are usually associated with coatings containing Pb or Sb.

Lead and antimony are added to the zinc coating to produce a spangle appearance, as described in Chapter 6. Variations in spangle size within a single coating, Fig. 14.37, are usually related to variations in the steel substrate that influence coating growth such as the presence of surface oxides, or improper feeding of ingots containing lead or antimony that results in composition variations within the bath. The density of liquid antimony is very near that of liquid zinc, 6.53 vs 6.57 g/cm^3,

FIG. 14.34 (A) "Angel wings" defect related to coating runback with heavier coatings and thicker strip. (B) A location next the strip edge on right-hand side. *(Reproduced with permission from (A, B) M. Dubois, private communication, March 2022, with permission.)*

FIG. 14.35 "Angel hair" defect on a GI coating. *(Reproduced with permission from M. Dubois, private communication, March 2022, with permission.)*

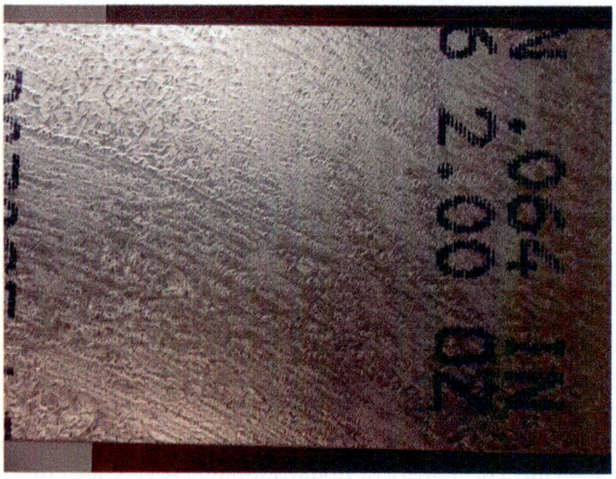

FIG. 14.36 Dendritic appearance of coating exposed to rapid cooling rate in after-pot cooler before solidification was complete. *(Reproduced with permission from M. Dubois, private communication, March 2022, with permission.)*

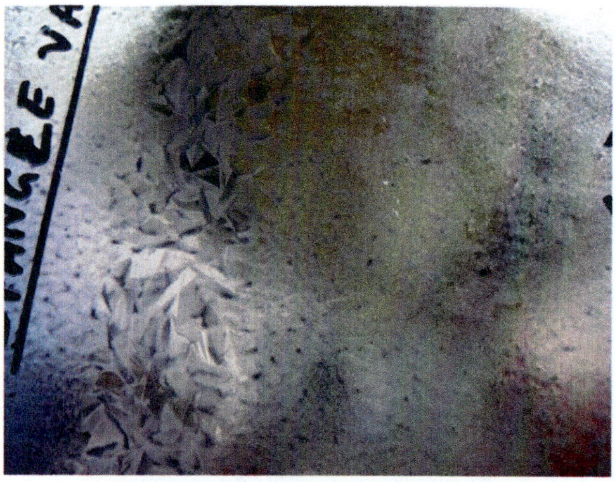

FIG. 14.37 Variation in spangle appearance caused by substrate condition. *(Reproduced with permission from M. Dubois, private communication, March 2022, with permission.)*

so there will be no density-induced segregation of antimony in the galvanizing bath. By contrast, the density of liquid lead is about 10.6 g/cm^3, so that good mixing is required to maintain uniform composition in the bath when lead is added. Because antimony reacts with aluminum to form intermetallic compounds, it is recommended that zinc-antimony alloys be used to make antimony additions. Lead is usually added using zinc-lead alloys; because lead does not react with zinc, aluminum or iron over the range of conditions used in galvanizing, pure lead can be added without affecting bath equilibrium, other than local temperature deviation considerations. Health and safety precautions should always be taken when being exposed to lead and antimony.

A defect related to the presence of spangles is spangle cracking. When lead is added to the coating to create a spangled appearance, its low solubility in zinc causes it to segregate to the spangle boundaries and surfaces as shown in Fig. 14.38. With heat and moisture, bimetallic corrosion cells form, and the coating can detach, starting at where the boundaries meet the steel. Fig. 14.39 also illustrates intergranular cracking that occurs for this reason. On the outside of the 2T bend the coating cracks at the spangle boundaries and then peels off the steel surface when the substrate is bent. With lead-free zinc there is no lead to segregate to grain boundaries, so coating adhesion failure does not occur over time.

Touch rolls are occasionally used to stabilize strip to be galvannealed as it passes between the wiping knives and the galvanneal furnace. Thermal conditions between the zinc bath and the galvanneal furnace produce a "chimney" of rising hot air. This rising air can convey zinc vapor from the surface of the bath upward in the direction of strip travel. Zinc dust can condense and grow needle-shaped dust particles on the touch rolls, Fig. 14.40. The appearance of the strip resulting from this is shown in Fig. 14.41.

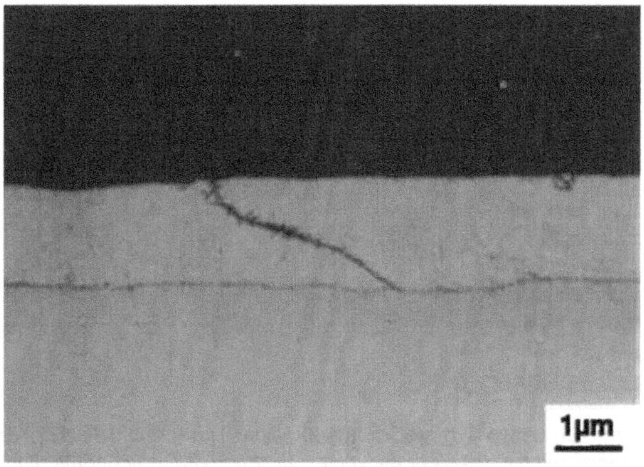

FIG. 14.38 Crack at a spangle boundary resulting from intergranular corrosion after exposure of a Zn-0.3%Al-0.083%Pb coating to 90°C, 75% relative humidity for 1 week. *(Reproduced with permission from D. Willis, A. De Liseo, Intergranular corrosion in continuously galvanized steel, in: Proceedings of 3rd International Conference on Zinc and Zinc Alloy Coated Steel Sheet (Galvatech 1995), Chicago, IL, September 17-21, 1995, Association for Iron & Steel Technology, Warrendale, PA, pp. 145–153.)*

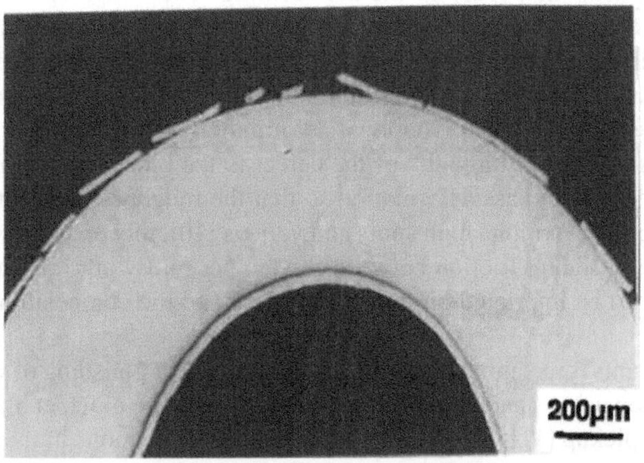

FIG. 14.39 Cross section of tension bend of coating with same composition as in Fig. 14.38, exposed at 90°C, 75% humidity for 2 weeks. Cracks occur at spangle boundaries. *(Reproduced with permission from D. Willis, A. De Liseo, Intergranular corrosion in continuously galvanized steel, in: Proceedings of 3rd International Conference on Zinc and Zinc Alloy Coated Steel Sheet (Galvatech 1995), Chicago, IL, September 17-21, 1995, Association for Iron & Steel Technology, Warrendale, PA, pp. 145–153.)*

FIG. 14.40 Microscopic view of condensed zinc dust on touch rolls. Bottom axis of photo is 120 μm long. *(Reproduced with permission from F. Chen, R. Patil, An in-depth analysis of various subtle coating defects of the 2000's, in: Proceedings of 6th International Conference on Zinc and Zinc Alloy Coated Steel Sheet (Galvatech 2004), Chicago, IL, April 4-7, 2004, Association for Iron & Steel Technology, Warrendale, PA, pp. 1055–1066.)*

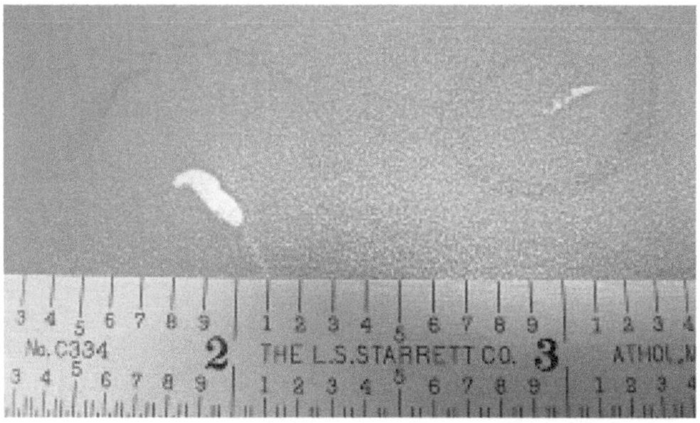

FIG. 14.41 Surface appearance of touch roll defect caused by condensing zinc vapor. Dimensions in inches. *(Reproduced with permission from F. Chen, R. Patil, An in-depth analysis of various subtle coating defects of the 2000's, in: Proceedings of 6th International Conference on Zinc and Zinc Alloy Coated Steel Sheet (Galvatech 2004), Chicago, IL, April 4-7, 2004, Association for Iron & Steel Technology, Warrendale, PA, pp. 1055–1066.)*

14.7 Temper mill related defects

Together with shape control, surface texturing and the imposition of a small amount of plastic deformation in the steel substrate, temper rolling also flattens high spots on the zinc coating. These high spots originate from substrate surface roughness, variations in steel reactivity and coating control variations. After temper rolling, the high spots are flattened and roughened with the texture imparted by the work roll. The "valleys" that have not been roughened are more reflective and therefore appear brighter than the high spots. The resulting thickness profile is shown in Fig. 14.42 for a soft IF steel substrate. Here, the high spot has been flattened. The local deformation on the substrate associated with flattening of the high spot on the top side results in a "print through" of the defect to the other side. Elimination of this kind of defect is possible by increasing the degree of skin pass deformation so that the roughness transfer from the skin pass roll is comparable to that of the incoming coating with its high spots and valleys [10], this practice is more effective with galvanneal than GI coatings. Because there is sliding friction between the temper mill work roll and the zinc coating surface, zinc pickup can occur. This pickup can be imprinted on the strip surface and the composition of the zinc flake in the pickup will be the same as the coating.

Use of wet temper mill lubricants is recommended together with frequent polishing of work roll surfaces. Zinc pickup is more frequently encountered when using stand-alone temper mills [4]. The use of wiper blades to maintain roll surfaces can also help control the severity of pickup [11]. To enable the pinning of dislocations in the steel that will help control yield point elongation, the strip entry temperature to the temper mill should never exceed 38°C.

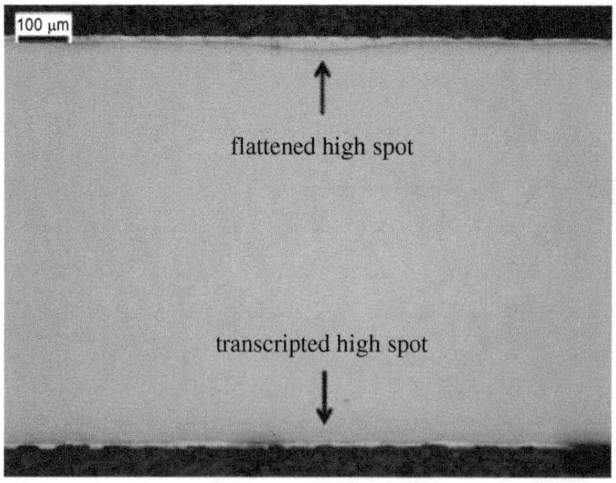

FIG. 14.42 Cross-section of a temper rolled IF steel with a GI coating. Flattening of the high spot on the top side results in an area of high reflectivity. Moreover, with this soft substrate the high spot is printed through to the opposite side. *(Reproduced with permission from T. Koll, M. Bretschneider, T. Klinkberg, F. Luther, B. Maas, On optimizing the zinc coating surface by an improved temper rolling process, in: Proceedings of 10th International Conference on Zinc and Zinc Alloy Coated Steel Sheet (Galvatech 2015), Toronto, May 31-June 4, 2015, Association for Iron & Steel Technology, Warrendale, PA, pp. 611–618.)*

FIG. 14.43 "Fish bone" or skin pass folds in a galvanized strip where the work roll crown was incompatible with roll force and strip tension. *(Reproduced with permission from M. Dubois, private communication, March 2022, with permission.)*

Improper temper mill work roll profiles or mill settings can lead to irregular deformation in the strip, Fig. 14.43. Here, the crown on the work roll was insufficient to provide uniform elongation of the strip. This was not assisted by the strip tension or entry and exit intermesh roll settings in this case.

A very high strip entry tension to the temper mill can result in yield point elongation and the formation of Lüders lines (coil breaks) in the skin passed sheet. This is controlled by adjusting the tension imparted by the entry bridle.

14.8 Galvanneal-specific defects

"Low spot" and "high spot" defects are encountered in galvanneal coatings, where conditions impair or accelerate growth of the coating. A "high spot defect" related to dross entrapment is shown in Fig. 14.44. The temper rolled surface, showing flattening of the high spot, is shown in Fig. 14.44A, and the corresponding cross-section in Fig. 14.44B. Stamping of this sheet resulted in print-through of the defect, Fig. 14.45. It was found that the bath composition during production of the coating with this defect was operation in the two-phase region just above the AlFeZn phase diagram "knee point," 0.135% Al at 460°C, where both FeZn delta and $Fe_2Al_{(5-x)}Zn_x$ phases are present. The incidence of high spots was reduced by lowering the aluminum composition into the field where only delta is stable [12].

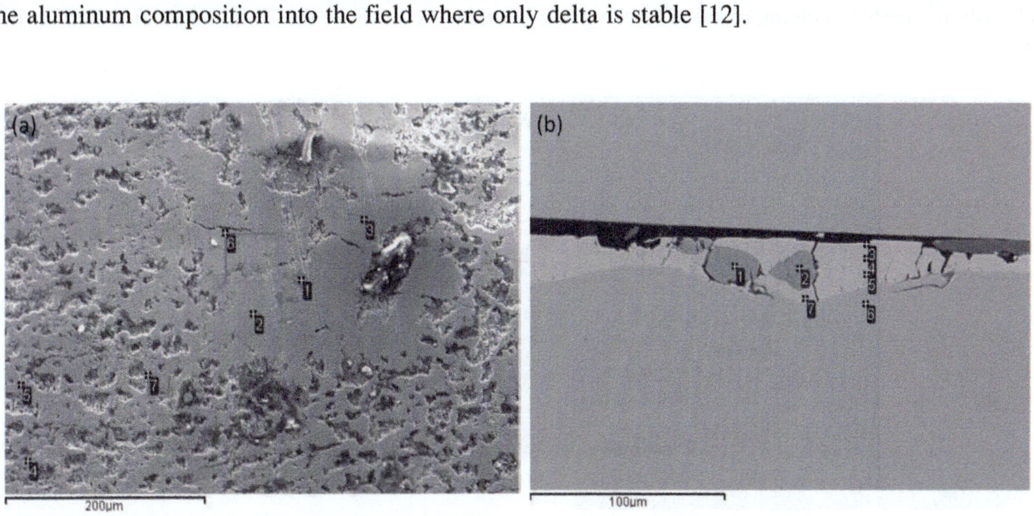

FIG. 14.44 High spot defect in galvanneal coating (left) temper rolled surface (right) cross section showing dross entrapment. *(Reproduced with permission from R. Pais, S. Agnihotri, S. Roy, M. Kadarbhai, P. Narang, Investigation of "high spots" defects in galvannealed automotive outer panels, in: Proceedings of 10th International Conference on Zinc and Zinc Alloy Coated Steel Sheet (Galvatech 2015), Toronto, May 31-June 4, 2015, Association for Iron & Steel Technology, Warrendale, PA, pp. 137–144.)*

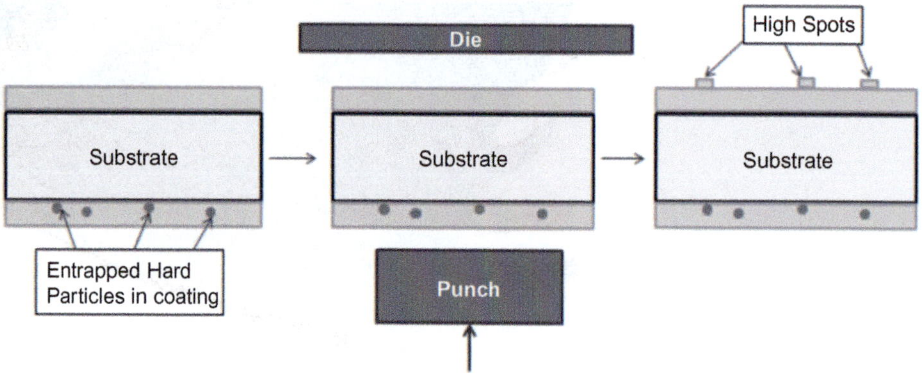

FIG. 14.45 Mechanism of high spot print-through during stamping. *(Reproduced with permission from R. Pais, S. Agnihotri, S. Roy, M. Kadarbhai, P. Narang, Investigation of "high spots" defects in galvannealed automotive outer panels, in: Proceedings of 10th International Conference on Zinc and Zinc Alloy Coated Steel Sheet (Galvatech 2015), Toronto, May 31-June 4, 2015, Association for Iron & Steel Technology, Warrendale, PA, pp. 137–144.)*

Another "high spot" defect has the appearance of a "white spot." In this case, MnO formed on the surface of a 2.5% Mn dual phase steel transferred to the surface of a hearth roll in the galvanizing line pretreatment furnace. The deposit grew to a diameter of about 100 μm before it was transferred to the strip surface. The oxide layer was successfully coated in the zinc bath but prevented the formation of the inhibition layer. Therefore, the FeZn compounds comprising the galvanneal coating grew quickly in this area, even before leaving the zinc bath. The surface of these compounds blocked a uniform wiping action as it passed through the gas knives. Not all of remaining zinc was consumed as the strip passed through the galvannealing furnace. The local coating iron content was about 7.7% versus 9.9% and the coating was twice as thick in this area. The appearance of the strip surface and a magnified view are shown in Fig. 14.46 [13].

A "low spot" defect in a galvanneal coating was reported to change in orientation, from being transverse to the direction of strip travel near the strip center to an inclination of 45 degrees to this direction closer to the strip edge, Fig. 14.47 (left). A coating cross-section showed a dent in which the galvanneal coating followed the contours of the dent, Fig. 14.47 (right). Its average depth was about 25 μm. The defect was found on both sides of the strip. Because of the elongated grains in the strip surrounding the defect, and because the coating is fully developed over the entire surface, it is judged that the dents were made after the recrystallization furnace treatment for this IF steel. The differences in surface friction along the width, caused by strip or roll shape and movement details, relate to the differences in defect orientation from center to edge. Bath hardware roll surface interactions may also play a role [14].

Streaky marks on galvanneal coatings are related to uneven coating growth. Early work with a "wood grain" defect on galvanneal showed that uneven coating growth was its cause and recommended more thorough pickling or a longer dipping time in the zinc pot as a cure [15]. Since then several other reasons for this have been identified. Both "linear streaks" and "tiger streaks," shown in Fig. 14.48, were related to uneven growth from grains and grain boundaries in Ti-stabilized IF steels. This family of steels has clean, reactive ferrite grain boundaries, because the Ti in the steel is tied up in TiO_2, TiN,

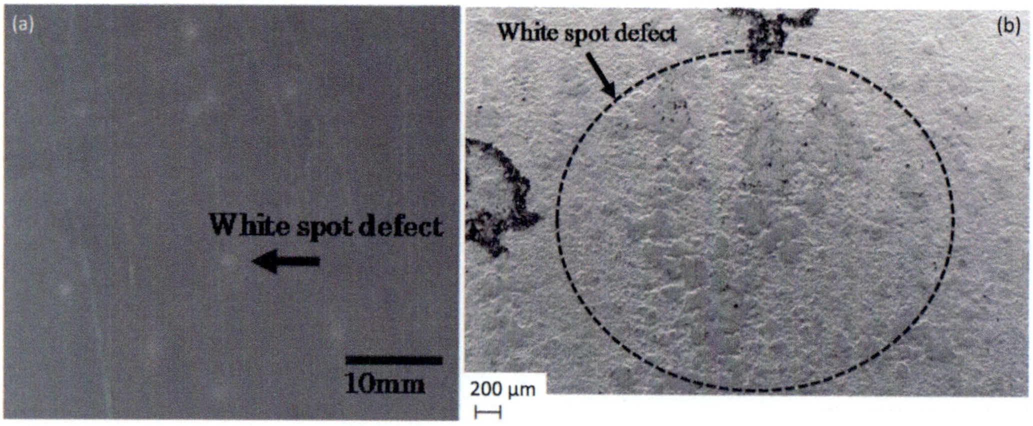

FIG. 14.46 "White spot defect" in galvanneal coating caused by hearth roll pickup. (right) overall view (left) scanning electron micrograph of coating surface. *(Reproduced with permission from (left, right) T. Konishi, H. Irie, White spot defect on high-tensile strength galvannealed steel sheet, in: Proceedings of 11th International Conference on Zinc and Zinc Alloy Coated Steel Sheet (Galvatech 2017), Tokyo, November 12-16, 2017, Iron and Steel Institute of Japan, Tokyo, pp. 840–845.)*

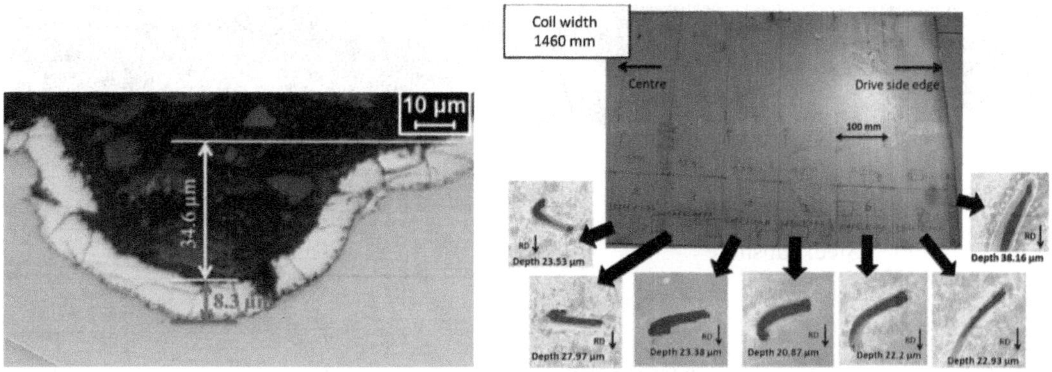

FIG. 14.47 "Low spot defect" in galvanneal coating. (left) relative orientation to direction of strip travel (right) cross section of defect. *(Reproduced with permission from (left, right) A. Mondal, S. Agnihotry, A. Chakraborty, A.K. Halder, P. Mohanta, S. Nayak, A. Kumar, R. Pias, M. Dutta, Study of a surface defect on galvannealed sheet, in: Proceedings of 11th International Conference on Zinc and Zinc Alloy Coated Steel Sheet (Galvatech 2017), Tokyo, November 12-16, 2017, Iron and Steel Institute of Japan, Tokyo, pp. 853–860.)*

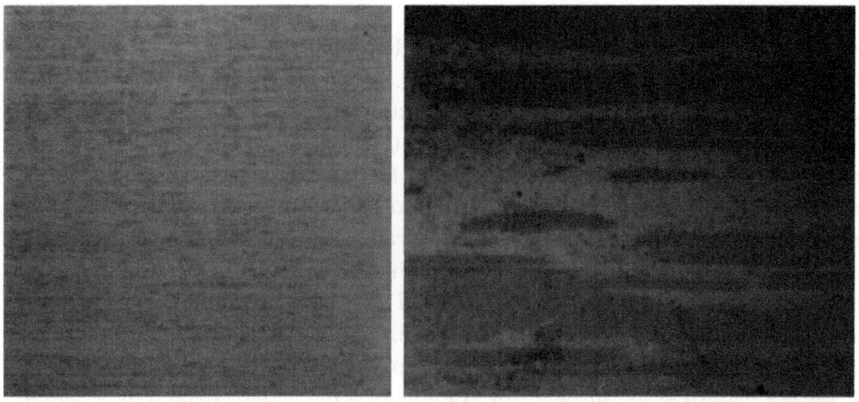

FIG. 14.48 Streak features in galvanneal coatings (left) linear streaks on a 50 g/m^2 coating on Ti-IF steel (right) tiger streak. Bottom axis of each photo is 55 mm long. *(Reproduced with permission from (left, right) W. Zhong, R. Dziuba, Formation of GA streaky defects simulated by using lab hot dip simulator, in: Proceedings of 10th International Conference on Zinc and Zinc Alloy Coated Steel Sheet (Galvatech 2015), Toronto, May 31-June 4, 2015, Association for Iron & Steel Technology, Warrendale, PA, pp. 642–649.)*

$Ti_4C_2S_2$ and/or TiC precipitates and inclusions. As a result of the differences in reactivity between grains and grain boundaries intersecting the steel strip surface, the iron oxide surfaces produced in these grades during upstream processing are therefore nonuniform. This also produces nonuniform reactions during galvannealing, resulting in uneven development of the galvanneal coating, including the development of outbursts and the resultant streaky marks. In IF steels stabilized with both Ti and Nb and also low carbon aluminum-killed grades, Nb and C segregate at the ferrite grain boundaries and the steel surface, slowing the Fe-Zn reaction and formation of FeZn outbursts. This results in more a more uniform GA coating and greatly reduced the incidence of streaky marks [16]. It has been found that decreasing bath aluminum content or decreasing snout entry temperature can reduce the thickness of the $Fe_2Al_{(5-x)}Zn_x$ inhibition layer, promoting more uniform GA reactions, so that less outbursts and craters are formed, reducing streaky mark intensity and frequency. The use of high galvanneal furnace temperatures can result nonuniform galvanneal coating reactions, producing cratered and porous coatings. The development of craters in galvanneal is described in Chapter 8. High coating weights, more than 50 g/m^2 per side can better mask these surface defects. Streak marks are also found to be less intense and frequent when the strip is prepared for galvanizing at a very low dew point, confirming that streaky marks are related to steel surface chemistry [5].

Streaky marks in galvanneal were also identified in an 1180 dual phase grade with a composition of 0.18%C-1.7%Si-2.7%Mn. However, they only occurred on the edges of the strip. This was related to difference in cooling rates for the coiled strip after hot rolling. This led to internal oxidation of Mn and Si near the strip center, while within 20 mm of the strip edge external oxidation with these elements was found. This resulted in reduced galvanneal reactivity near the edges and consequent dark streaks in the coating. Cross-sections of coatings from these two locations are shown in Fig. 14.49, showing the nonuniform development on the galvanneal coating near the strip edge [17]. The reverse effect was seen with a DP980 grade. Only external oxides were observed on the strip surface because of the likely lower Si content of this grade, and they

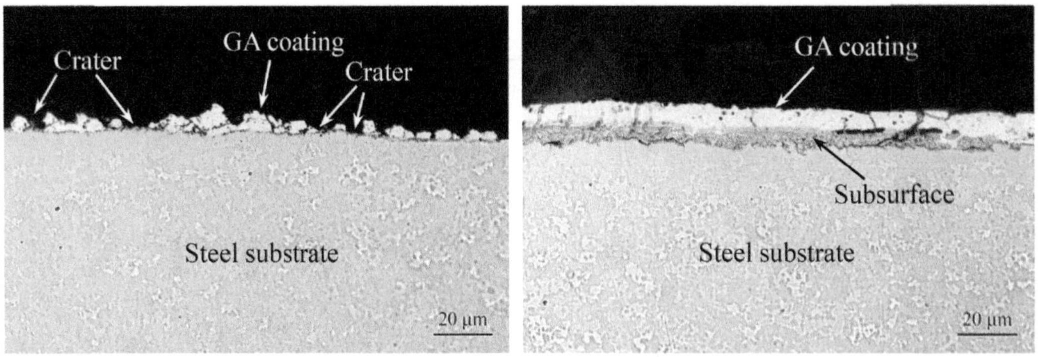

FIG. 14.49 Development of galvanneal coating on a 0.18%C-1.7%Si-2.7%Mn steel substrate (left) near edge, showing craters related to presence of external Mn and Si oxides (right) near center, uniform coating development over internally oxidized substrate. *(Reproduced with permission from W. Bi, X. Jin, X. Wei, Formation of a dark streaky edge defect on galvannealed ultra-high strength steel, Surf. Coat. Technol. 433 (2022) 128162, https://doi.org/10.1016/j.surfcoat.2022.128162.)*

were thicker at the coil center than at the edge because of the slower cooling rate of the coil center. Therefore, the streaky marks appeared at the strip center rather than at the edge [18].

Underalloying of the galvanneal coating also results in appearance differences. For a DP590 grade, this was related to differences in coating weights from center to edge caused by strip crossbow [18]. Fig. 14.50 shows that on the top side of the strip shown, the center showed a coating weight of 45 g/m^2, while the edge had a coating weight of 81 g/m^2. Moreover, like the example described immediately above, the edges of the strip exhibited more external oxidation, resulting in uncoated areas. It is believed that the poorly-adhering heavier coating weight at the narrow face of the edge adhered to the baffle rollers in-line with the gas wiping knife contributing to the heavy coating weight observed at the strip edges.

Microflaps originate from rolled-in-scale during hot rolling. If they are not removed before cold rolling, the scale can be rolled into the surface. When a microflap and subsurface scale, partially reduced in the galvanizing pretreatment furnace, is dipped into the zinc pot, liquid zinc penetrates the gap between the steel surface and rolled-in scale. This liquid zinc solidifies in the gap, thus expanding the gap. It also lifts up the micro flap, causing a localized higher coating thickness, typically greater than 10 μm versus 5–6 μm for a normally specified coating. The local area of thicker coating does not completely galvanneal, resulting in an underalloyed streaky pattern at the surface. An example of a microflap covered by a partially galvannealed coating is shown in Fig. 14.51. The scale in this case was mainly a silicon oxidation product [18]. Similar microflap-caused defects can be found in galvanized coatings. For both, they are a critical defect for exposed quality automotive sheet.

The higher Si content in TRIP grades presents challenges for meeting the simultaneous objective of preparing a strip with the desired microstructure together with a surface that can be coated with zinc. Here, surface oxides affecting coating quality can originate in the galvanizing pretreatment furnace rather than during prior rolling stages. The partially alloyed galvanneal coating on a TRIP 590 grade shown in Fig. 14.52 coincides with "white band" areas on the strip surface. In this area, liquid zinc did not wet the surface because of the presence of surface oxides, resulting in a thin coating and nonuniform galvanneal coating development, such as the FeZn outbursts shown. An elevated Si composition was found in these white band areas [18].

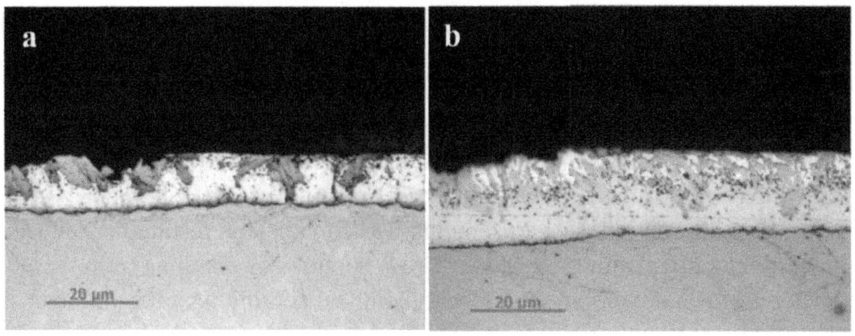

FIG. 14.50 Variations in coating weight and galvanneal coating development related to crossbow in a DP590 grade (A) Strip center, 45 g/m^2 (B) strip edge 81 g/m^2. *(Reproduced with permission from F. Bleich, C. Cheng, Challenges in galvannealing AHSS at an industrial hot dip galvanizing line, in: Proceedings of 12th International Conference on Zinc and Zinc Alloy Coated Steel Sheet (Galvatech 2021), June 22-24, 2021, Virtual Conference, Austrian Society for Metallurgy and Materials, Vienna, pp. 122–133.)*

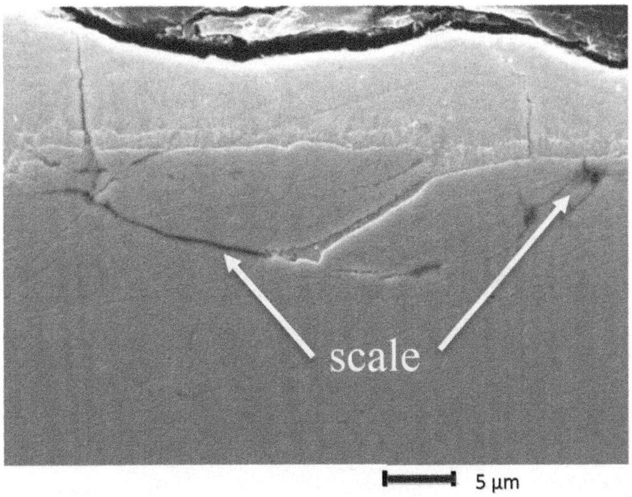

FIG. 14.51 Microflap covering rolled-in scale in DP590 substrate and partially galvannealed coating on its surface. *(Reproduced with permission from F. Bleich, C. Cheng, Challenges in galvannealing AHSS at an industrial hot dip galvanizing line, in: Proceedings of 12th International Conference on Zinc and Zinc Alloy Coated Steel Sheet (Galvatech 2021), June 22-24, 2021, Virtual Conference, Austrian Society for Metallurgy and Materials, Vienna, pp. 122–133.)*

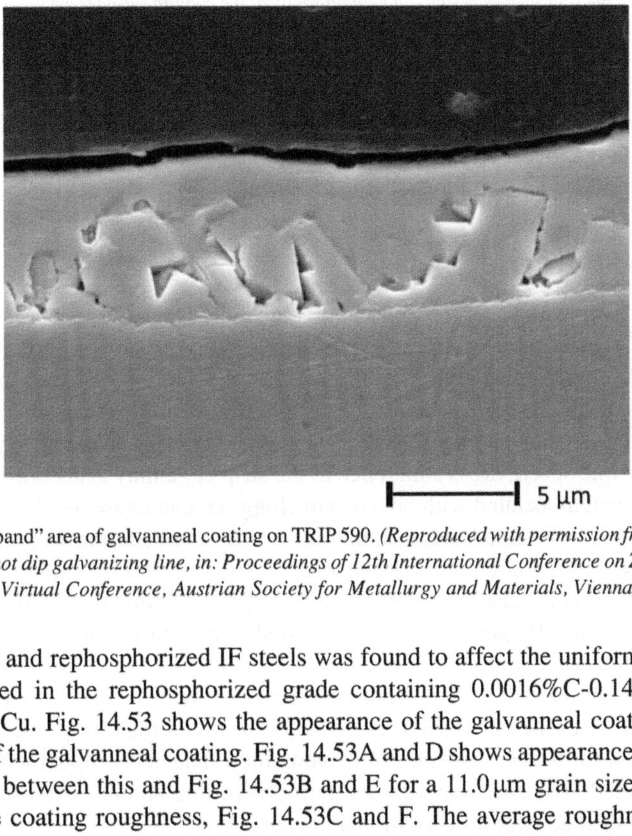

FIG. 14.52 Cross section of "white band" area of galvanneal coating on TRIP 590. *(Reproduced with permission from F. Bleich, C. Cheng, Challenges in galvannealing AHSS at an industrial hot dip galvanizing line, in: Proceedings of 12th International Conference on Zinc and Zinc Alloy Coated Steel Sheet (Galvatech 2021), June 22-24, 2021, Virtual Conference, Austrian Society for Metallurgy and Materials, Vienna, pp. 122–133.)*

The grain size of Ti-Nb IF and rephosphorized IF steels was found to affect the uniformity of galvanneal appearance. The effect is more pronounced in the rephosphorized grade containing 0.0016%C-0.143%Mn-0.0487%P, 0.053%Si-0.033%Al-0.003%Nb-0.053%Cu. Fig. 14.53 shows the appearance of the galvanneal coating and counterpart steel substrate surfaces after removal of the galvanneal coating. Fig. 14.53A and D shows appearances for a 9.5 μm grain size. There was little difference observed between this and Fig. 14.53B and E for a 11.0 μm grain size. Increasing steel grain size to 16.5 μm significantly increase coating roughness, Fig. 14.53C and F. The average roughnesses of the substrates before coating were 2.30, 2.82 and 3.02 μm, respectively. The coating iron contents of the locations 3 and 4 in Fig. 14.53C were 12.37% and 23.93%, respectively. The reason for this difference is the slower galvannealing reaction on top of the ferrite grain surfaces, compared to the grain boundaries. During the galvannealing reaction at the grain surfaces, unreacted liquid zinc moves to grain boundaries to react with iron atoms available from the high diffusion path there. This results in a depletion of zinc that is more severe at larger ferrite grain surfaces, resulting in the difference in uniformity of appearance shown in Fig. 14.53 [19].

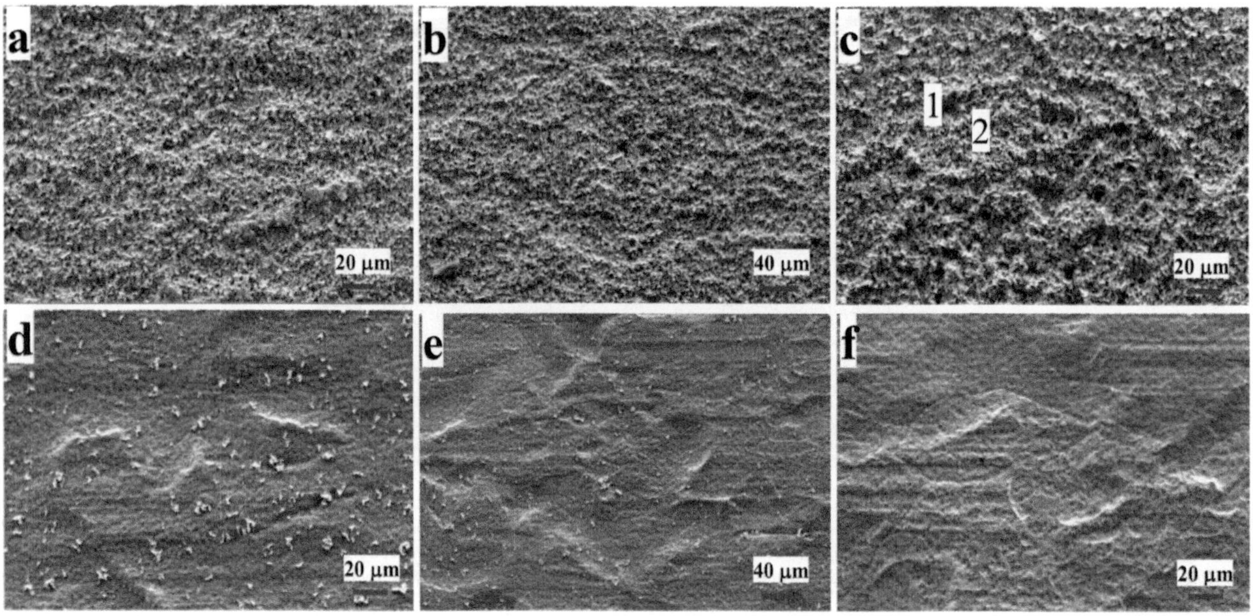

FIG. 14.53 Scanning electron micrographs of galvanneal coating surfaces at different substrate grain sizes: (A–C) Before and (D–F) after removing galvanneal coating layers at 9.5, 11.0, and 16.5 μm, respectively. *(Reproduced with permission from D.-J. Paik, C.-W. Kim, H.-S. Hwang, J.-W. Lee, D.-G. Kang, S.-Y. Choun, M.-H. Hong, Effects of steel substrates on nonuniformity of the coating surfaces during galvannealing reaction, in: Proceedings of 11th International Conference on Zinc and Zinc Alloy Coated Steel Sheet (Galvatech 2017), Tokyo, November 12-16, 2017, Iron and Steel Institute of Japan, Tokyo, pp. 846–852.)*

Finally, "bright edges" on galvanneal are a result of overcoating of strip edges in the gas wiping stage. If it this cannot be controlled by improving coating thickness control practices, edge heaters can be used to fully alloy the coating. This typically occurs over the 5–10 mm of width nearest each edge [20].

14.9 Delivery end and other defects

Defects encountered during recoiling at the delivery end of the line include spooling, where the edges of the coil produce a larger diameter than the center. This is usually related to edge overcoating. As discussed in Chapter 7, crossbow and edge effects associated with gas knife operation are primary causes. If spooling is found on one coil and not on the following coil with otherwise identical characteristics, the shape profile originating in the rolling mill steps may be a cause. If side trimming was performed on hot rolled coil and not later, edge burrs may contribute to built-up edge occurrences. On heavier steel gauges processed at lower line speed, dross adherence to the strip edge may also contribute to incidences of spooling.

The "flow lines" wiping defect, associated with nozzle gap clogging, can cause local areas of higher coating thickness. When a finished strip with this defect is coiled, ridge on the finish coil can occur that will result in a poor flatness when the strip is uncoiled. The appearance of a coil containing this defect is shown in Fig. 14.54.

Staggered winding, telescoping coil shapes and coil collapse, Fig. 14.55, are all related to proper regulation and control of strip tension as the coil is wound. Proper alignment and good maintenance of equipment is essential—the winding capstan must be parallel with the alignment of the previous support roll. If the surface is oiled, the correct oil coverage rate on the strip surface must be maintained. Coil winding practice is more critical when the strip is coated with a thin organic coating because of the low coefficient of surface friction of these coatings. To prevent coil collapse, usually a higher tension is applied for the first few turns, then lower tension maintained. The belt wrapper tension must also be controlled so that excessive force is not applied. The strip temperature during recoiling should be no more than 30°C higher than the ambient temperature, or shape changes during cooling may occur.

"Fretting corrosion" can occur after transportation of coils during which there is excessive movement, typically when transporting over rough roads. Its appearance is shown in Fig. 14.56, where aggregated areas of stripes or blotches are observed. A mirror image of this appearance is seen on surfaces that contacted the sample during transport although they may vary in intensity. In coiled material, the appearance of this defect is more pronounced in the outer 15–25 coil wraps, after which it becomes less severe. There is always a high oxygen content at the fretting locations. It is expected that oxidation is either present on the coil as it is wrapped or builds up as the coil is transported and relative motion between the

FIG. 14.54 Appearance of a coil with local ridges caused by local coating overthickness originating from flow line defect. *(Reproduced with permission from M. Dubois, private communication, March 2022, with permission.)*

FIG. 14.55 Collapsed coils. *(Reproduced with permission from M. Rogers, Hot dip galvanize products coating defects & their avoidance, in: Presented at Hot-Dip Galvanized Sheet Production Course, Merrillville, IN, June 9-11, 2010, GalvInfo Center of International Zinc Association, Durham, NC.)*

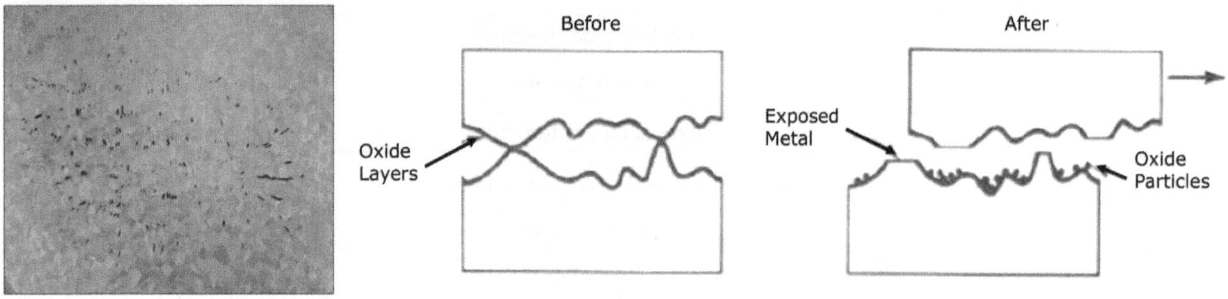

FIG. 14.56 Fretting corrosion (left) example and (right) proposed mechanism. The bottom axis of the photo is 20cm long. *(Reproduced with permission from (left) M. Dubois, private communication, March 2022, with permission. Reproduced with permission from (right) S. Chakrabarti, R.S. Pais, B. Van Veldhuizen, Fretting corrosion: a chronic surface problem in transported galvanised coils, in: Proceedings of 8th International Conference on Zinc and Zinc Alloy Coated Steel Sheet (Galvatech 2011), Genoa, Italy, June 21-25, 2011, Associazione Italiana di Metallurgia, Milan.)*

adjacent steel surfaces occurs. In either case, the abrasion of the oxidized surface exposes fresh metal, further movement results in the detachment of metal particles from the coating surface that then oxidize, giving the nonuniform appearance [21].

"White rust" or "storage stain" was described in Chapter 11 and is related to restricted oxygen access to the coating surface under conditions when moisture is present. Mild and severe cases of this condition are shown in Fig. 14.57. Its appearance begins as the formation of zinc hydroxide, a white corrosion product, Fig. 14.57 (left). After prolonged exposure

FIG. 14.57 Appearance of "white rust" or "storage stain" in (left) early stage and (right) late stage of attack by moisture in oxygen-restricted environment. *(Reproduced with permission from Steel Technology Services Inc., private communication, March 2022, with permission.)*

to moist conditions, it can result in loss of coating protection for the steel and the formation of red rust, Fig. 14.57 (right). The problem is controlled by use of correct levels of passivation treatments and storage in locations where conditions for condensing moisture formation will not be favorable. Coil wrapping and packing materials impregnated with moisture absorbing chemicals such as Vapor Corrosion Inhibitors (VCI) have been found to be helpful. There are many passivating treatments available to control white rust, further details are given in Chapters 11 and 12.

14.10 Defect modeling and control

Analytical models have been developed to search for and solve process-related defects in organized and powerful ways. For example, an edge waviness issue found on galvanized coils was analyzed using a Bayesian network at ArcelorMittal's Vega do Sul plant [22]. Normally a "full center" strip profile is desired, so that edges are thinner than center, but if the edges are too thin then edge waviness results. A flow diagram for a typical Bayesian process is shown in Fig. 14.58. Applying this to the edge waviness problem it was found to be connected to the operation of the cold rolling tandem mill that had both roll bending and continuously variable crown (CVC), provided by the continuous S-shaped profiles of the work rolls. This gave the capability of varying crown between 0 and 0.45 μm. By applying this analysis technique to the process line involved in producing galvanized sheet steel, it was possible develop presets for both roll bending and CVC for the rolling mill stands involved in this issue. This improved production tonnage and reduced the material loss rate from 0.54% to 0.19%, Fig. 14.59.

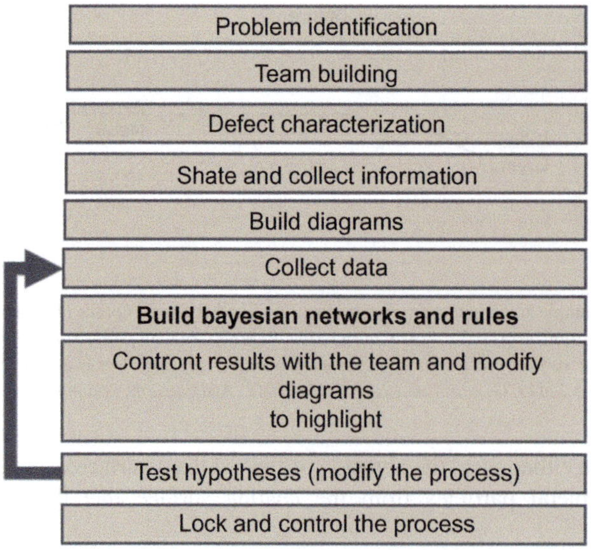

FIG. 14.58 Flow diagram for defects analysis. *(Reproduced with permission from F.B.P. de Souza, L. Colzani, R.S. de Medeiros, D.G. de Oliveira, R. Patil, Defect modeling: an advanced approach to solve quality problems and support decision making, in: Proceedings of 8th International Conference on Zinc and Zinc Alloy Coated Steel Sheet (Galvatech 2011), Genoa, Italy, June 21-25, 2011, Associazione Italiana di Metallurgia, Milan.)*

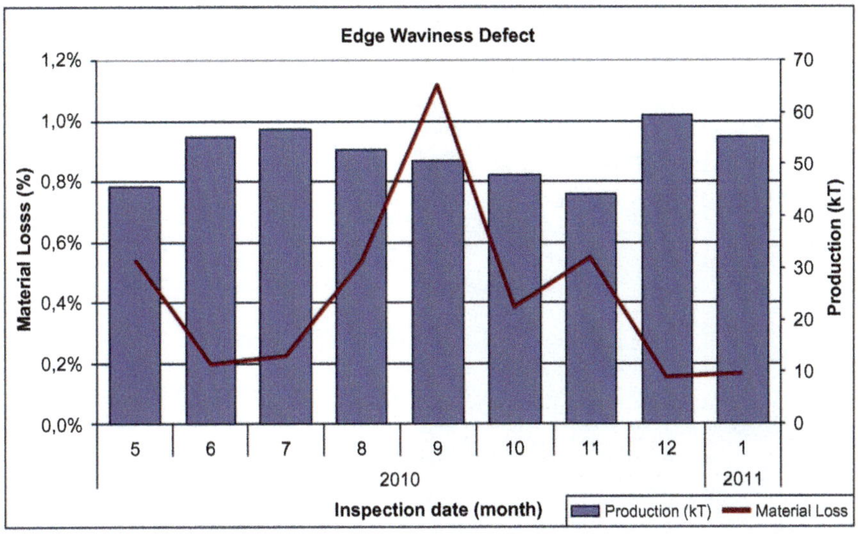

FIG. 14.59 Improvement in production and materials loss measures resulting from Bayesian defects analysis. *(Reproduced with permission from F.B.P. de Souza, L. Colzani, R.S. de Medeiros, D.G. de Oliveira, R. Patil, Defect modeling: an advanced approach to solve quality problems and support decision making, in: Proceedings of 8th International Conference on Zinc and Zinc Alloy Coated Steel Sheet (Galvatech 2011), Genoa, Italy, June 21-25, 2011, Associazione Italiana di Metallurgia, Milan.)*

This approach was further generalized in development of a root cause analysis for quality deviations in hot-dip galvanized coils with a Zn-Al coating. In this case, the mechanical properties of the coating were occasionally out of specification; variations in coating microstructure appeared to be related to this. Using data provided by sensors from the coating section through after-pot cooling, a principal component analysis was developed [23]. The domains of the principal components are shown in Fig. 14.60 (left). Here, Principal Component 1 includes outcomes mainly related to microstructural details, while Principal Component 3 includes outcomes related to process details and therefore provides clues for process improvement. The result of the analysis was an adjustment of the after-pot cooling profile, shown in Fig. 14.60 (right). The locations of the two after-pot coolers (APC1 and APC2) in this curve are shown.

Integrated process and quality management software for the hot strip mill, cold rolling mill and electrogalvanizing line was used to determine the cause of an electrogalvanized coating "spear head" defect [24]. The machine vision image of this defect captured by the automatic inspection system is shown in Fig. 14.61. This defect was normally found on strip tails and edges, on both strip sides. There was a greater frequency of defects in coils scheduled toward the end of the rolling mill

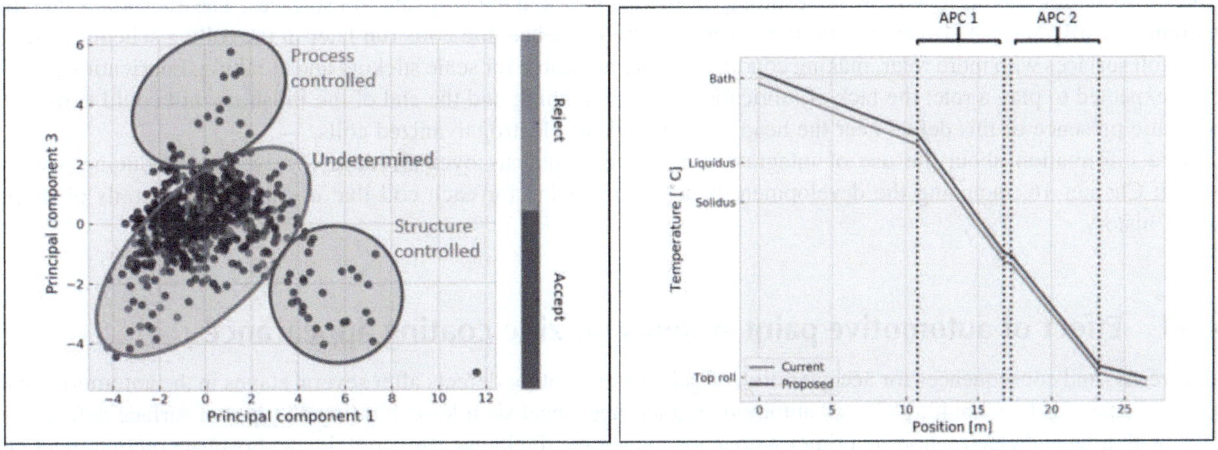

FIG. 14.60 Use of Principal Components Analysis to determine suitable temperature profiles for targeted strip mechanical properties (left) maps of Principal Components 1 and 2 showing interactions relating to steel microstructure and galvanizing line process variables (right) temperature profile from galvanizing bath exit to top roll, effect of after-pot cooler (APC) settings. *(Reproduced with permission from W. Dal'Maz Silva, H. Saint-Raymond, C. Dulcy, An integrated methodology for the root cause analysis of mechanical and metallurgical defects of hot-dip galvanized coatings, in: Proceedings of 12th International Conference on Zinc and Zinc Alloy Coated Steel Sheet (Galvatech 2021), June 22-24, 2021, Virtual Conference, Austrian Society for Metallurgy and Materials, pp. 955–963.)*

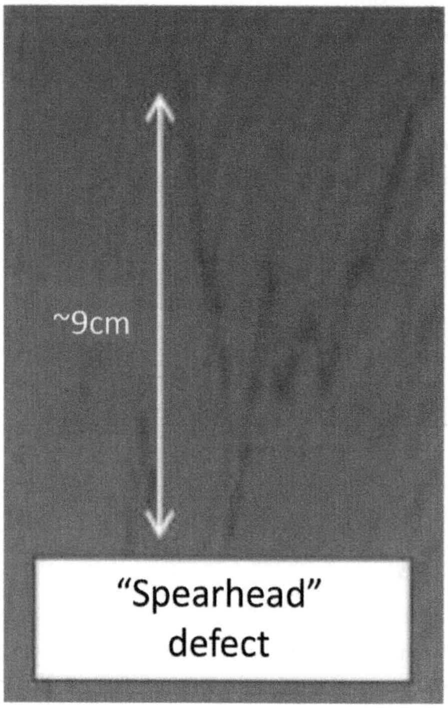

FIG. 14.61 "Spear head" defect image from automatic inspection system. *(Reproduced with permission from M. Monnoyer, R. Biland, E. Menigault, D. Seguret, Fives' Eyeron™, All-in-one quality management software, in: Proceedings of 12th International Conference on Zinc and Zinc Alloy Coated Steel Sheet (Galvatech 2021), June 22-24, 2021, Virtual Conference, Austrian Society for Metallurgy and Materials, pp. 797–805.)*

campaign, most of the affected coils were from slabs whose reheating time exceeded 4 h. Also, all the coils where this defect was exhibited showed a sudden increase of finishing mill temperature at tail end and an increase of roughing mill temperature. Using analysis software, the relationship between the positions of "spear head" defects and the sudden increase of finishing mill temperature was confirmed. It was hypothesized that these defects are caused by the formation of microflaps on the surface of the hot rolled coils. Long slab reheating times and high roughing and finishing mill temperatures are favorable for the development of a thick and brittle scale layer on strip surface. When the strip reaches the finishing stands of the hot strip mill, this scale will break, resulting in sticking between the strip and roll surfaces. The microflaps then fold over onto the strip surface. Together with the reheating time influence, the coils run later in the rolling schedule are also seeing roll surfaces with more wear, making conditions more favorable for scale sticking and breaking. Lubrication practice is also expected to play a role: the lack of lubrication at the beginning and the end of the finishing mill could explain the systematic presence of this defect near the head and tail ends of electrogalvanized coils.

More information about the use of integration of process analysis over all steps involved in producing a coil is given in Chapter 16, including the development of a "digital twin" for each coil that documents all details of its processing history.

14.11 Effect of automotive paint systems on zinc coating appearance

The severity, and consequences for acceptability, of galvanized coating defects after several stages in the automotive paint cycle was analyzed by Choi [25]. Coated automotive grade steel sheet samples exhibiting 13 types of surface defects were coated with a typical electrophoretic primer with a 30 μm average thickness after applying a phosphate pretreatment. The roughness of both unpainted and primer painted surfaces was evaluated using noncontact optical profilometry and visual techniques. Two different topcoat systems were then used. The first was as base coat-clear coat system with average layer thicknesses of 23 μm and 48 μm respectively, the second was a high-build antichip powder system with an initial powder

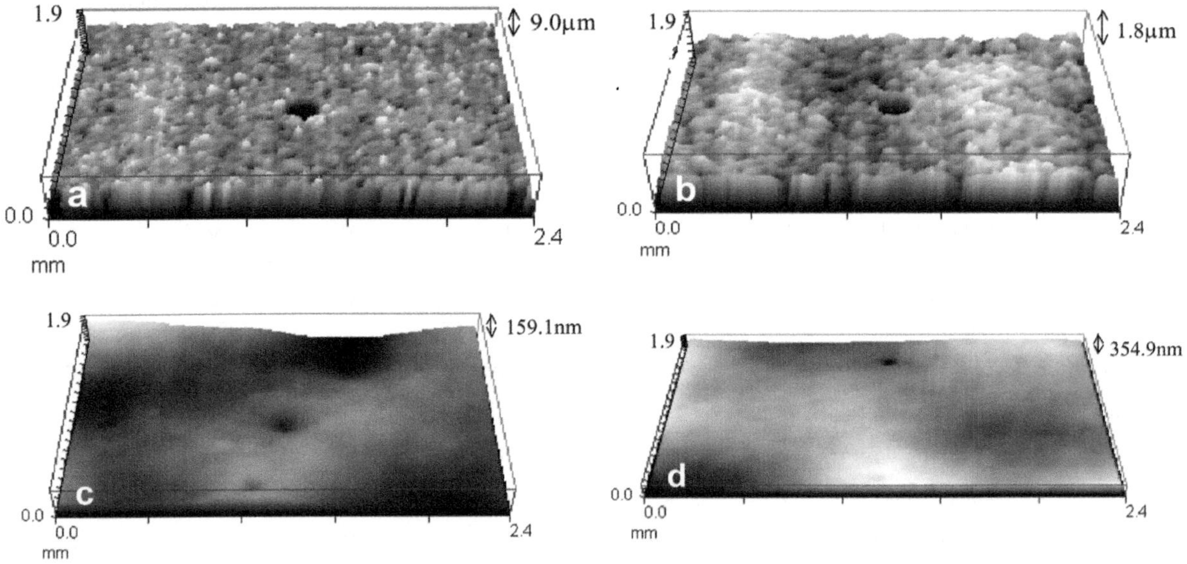

FIG. 14.62 Optical profilometry images of "small white spot" imperfection (13 in Table 14.1) during various stages of painting (A) unpainted (B) after electrophoretic primer (C) after final top coating, using base coat-clear coat (D) after final top coating using powder-liquid coating. *(Reproduced with permission from Y.-M. Choi, Topography assessment of surface imperfections evolution after painting in automotive sheet steel, in: AISTech 2005-Proceedings of the Iron & Steel Technology Conference, Charlotte, May 9-12, 2005, Association for Iron & Steel Technology, Warrendale, PA.)*

coating layer average thickness of 76 μm and a top liquid layer with an average thickness of 43 μm. For some types of defects, the number of samples was insufficient for a full analysis, so only the powder topcoat system was evaluated. The appearance of a "white spot" galvanneal defect after the various painting steps, measured with optical profilometry, is shown in Fig. 14.62. Note that the vertical and horizontal scales are quite different. The dimensions of the galvanneal defect were measured as 0.29 mm wide and 5.7 μm deep. Painting with the electrophoretic primer reduces the depth to 1.8 μm, but the width is has not significantly changed. Application of the base coat-clear coat system further reduces the defect depth; use of the alternative powder topcoat system renders the defect almost invisible. A summary of results for each of the 13 studied defects is shown in Table 14.1. It was found that the studied paint systems were far more effective at attenuating the heights or depths of coating defects while their widths were little changed. In some cases, even though the total thickness of the paint system exceeded 100 μm, defects between 5 and 10 μm in height depth were still visible after curing. Such defects were not visible after initial paint application, before curing. The difference between these two conditions is that these paints shrink typically 5% during curing. If the paint cannot move laterally to accommodate shrinkage, the change in volume must come from the vertical direction, so that in the vicinity of a defect the paint cannot mask it as effectively as if there was no paint shrinkage.

Cheng [26] considered a larger population of sample defects than Choi, dividing them into raised spot, indented spot and line defects. Only a full paint system using a base coat-clear coat topcoat was included, and only contact profilometry was used to assess roughness. Figs. 14.63 and 14.64 show typical characteristics and dimensions of raised spot and line defects. Cheng confirmed Choi's findings that defect widths are more difficult to mask by painting and that smaller defect widths are easier to detect than defect heights of the same dimensions. For example, Fig. 14.65 shows the relative visibility of defects on panels with high spot defects after painting. Both the width and depth of the defect influence the slope of the paint surface covering the imperfection. Higher slopes reflect more light and hence are more visible. Also, the larger the defect, the more reflective and visible the painted surface will be. Initially, the applied paint will flow and level the surface, but shrinkage after curing will result in a more faithful replication of the morphology of the defect. Paint can bridge and therefore hide small imperfections. Therefore, the border between visible and invisible changes slope in Fig. 14.65 for small dimensions. Fig. 14.66 shows the relative visibility of defects on panels with line defects after painting. The borderline between visible and invisible is much lower than that of spot defects. This is related both the details of measurement and the fact that the visibility detection limit for a line is much lower than that of a spot.

TABLE 14.1 Summary of optical profilometry and visual inspection assessments of defect visibility on panels after different stages of painting.

			Visual inspection			Profilometry		
				After topcoat			After topcoat	
ID	Coating	Imperfection type	After E-coat	Liquid primer	Powder primer	After E-coat	Liquid primer	Powder primer
1	GI	Light sink roll dross mark	V	BV	I	V	V	I
2	GI	Light sink roll dross line	V	BV	I	V	V	I
3	GI	Light drag dross line	V	BV	I	V	V	I
4	GI	Splash mark	V	–	BV	V	–	BV
5	GI	Light dross	V	–	I	V	–	I
6	GI	Mid/heavy dross	V	–	BV	V	–	BV
7	GI	Rough Surface	V	–	I	V	–	I
8	GI	Rub/gouge dross	V	V	I	V	V	I
9	GI	Chatter mark	V	V	V	V	V	V
10	GA	White spot	V	BV	I	V	V	BV
11	GA	Dent	I	I	I	I	I	I
12	GA	Cold rolled spot	I	I	I	I	I	I
13	GA	Small white spot	V	BV	I	V	V	I

BV, barely visible; *I*, invisible; *V*, visible.
Reproduced with permission from Y.-M. Choi, Topography assessment of surface imperfections evolution after painting in automotive sheet steel, in: AISTech 2005-Proceedings of the Iron & Steel Technology Conference, Charlotte, May 9-12, 2005, Association for Iron & Steel Technology, Warrendale, PA.

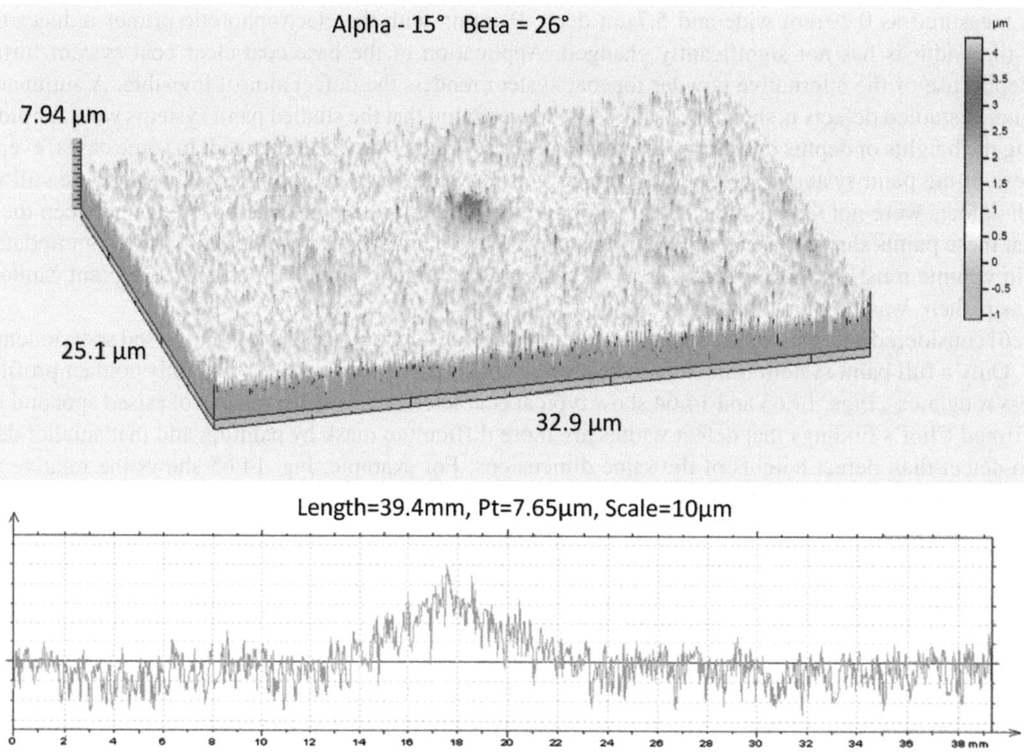

FIG. 14.63 (top) Contact profilometry surface profile for a high spot defect (bottom) profile across the peak of a typical high spot defect. *(Reproduced with permission from (top, bottom) X. Cheng, S. Snopek, D. Pineau, Effect of the dimension of a spot and line type surface imperfection on its visibility after automotive painting, in: Proceedings of 10th International Conference on Zinc and Zinc Alloy Coated Steel Sheet (Galvatech 2015), Toronto, May 31-June 4, 2015, Association for Iron & Steel Technology, Warrendale, PA, pp. 658–667.)*

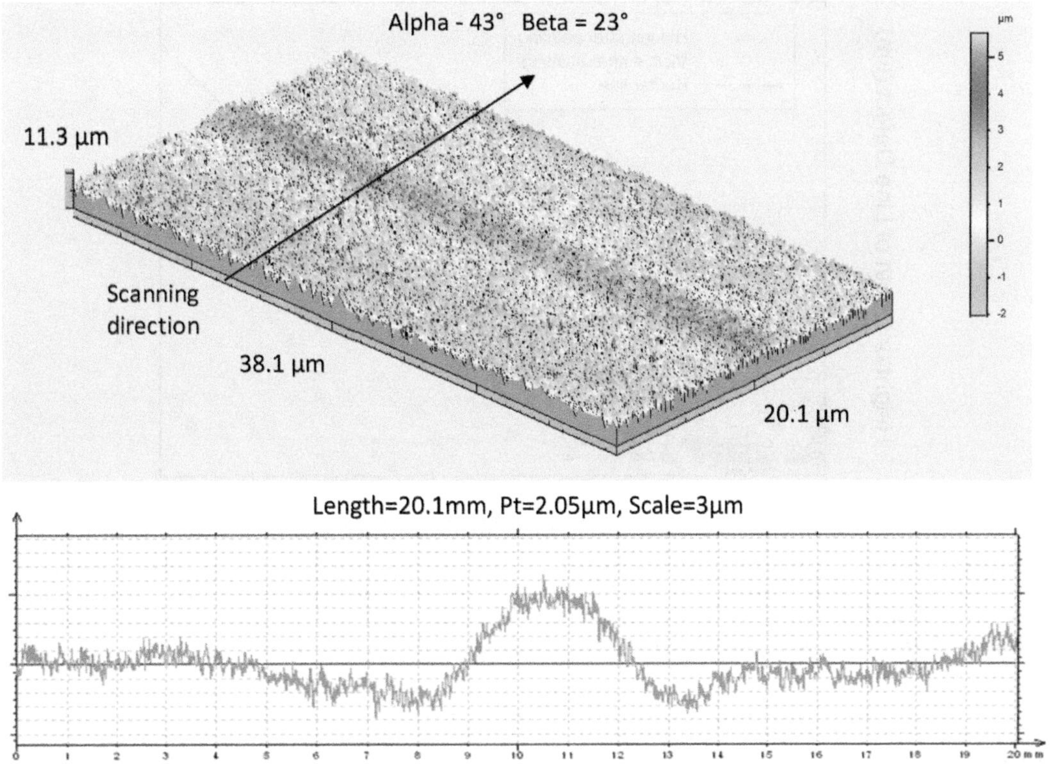

FIG. 14.64 (top) Contact profilometry surface profile for a line defect (bottom) profile across the peak of a typical line defect. *(Reproduced with permission from X. Cheng, S. Snopek, D. Pineau, Effect of the dimension of a spot and line type surface imperfection on its visibility after automotive painting, in: Proceedings of 10th International Conference on Zinc and Zinc Alloy Coated Steel Sheet (Galvatech 2015), Toronto, May 31-June 4, 2015, Association for Iron & Steel Technology, Warrendale, PA, pp. 658–667.)*

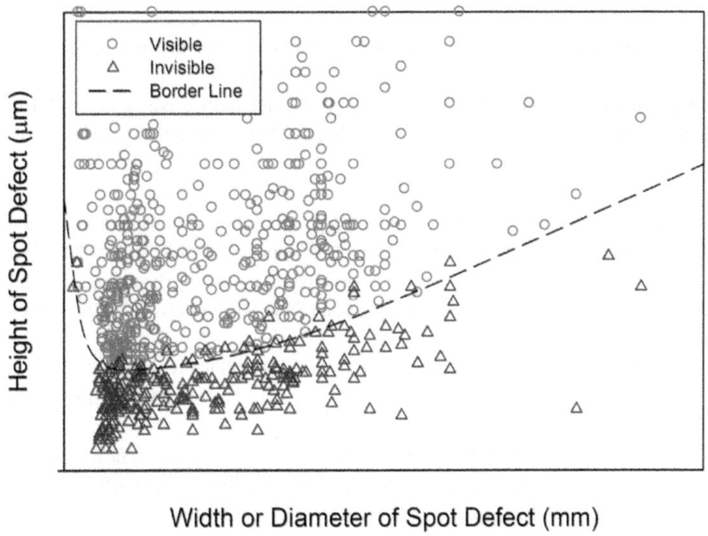

FIG. 14.65 Visibility of high spot defects after receiving full automotive paint system, relative comparison of defect width and depth. *(Reproduced with permission from X. Cheng, S. Snopek, D. Pineau, Effect of the dimension of a spot and line type surface imperfection on its visibility after automotive painting, in: Proceedings of 10th International Conference on Zinc and Zinc Alloy Coated Steel Sheet (Galvatech 2015), Toronto, May 31-June 4, 2015, Association for Iron & Steel Technology, Warrendale, PA, pp. 658–667.)*

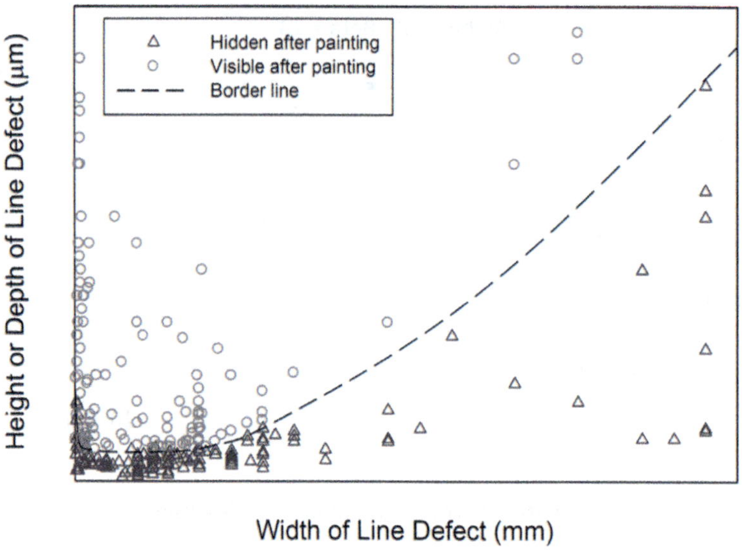

FIG. 14.66 Visibility of line defects after receiving full automotive paint system, relative comparison of defect width and depth. Note difference in scales. *(Reproduced with permission from X. Cheng, S. Snopek, D. Pineau, Effect of the dimension of a spot and line type surface imperfection on its visibility after automotive painting, in: Proceedings of 10th International Conference on Zinc and Zinc Alloy Coated Steel Sheet (Galvatech 2015), Toronto, May 31-June 4, 2015, Association for Iron & Steel Technology, Warrendale, PA, pp. 658–667.)*

References

[1] N.-Y. Tang, F.E. Goodwin, A study of defects in galvanized coatings, in: Proceedings of 5th International Conference on Zinc and Zinc Alloy Coated Steel Sheet (Galvatech 2001), Brussels, June 26-28, Verlag Stahleisen, Duesseldorf, 2001, pp. 49–55.

[2] M. Raus, N. Kotthoff, D. March, New developments in the surface inspection of galvanized strip, in: Proceedings of 5th International Conference on Zinc and Zinc Alloy Coated Steel Sheet (Galvatech 2001), Brussels, June 26-28, Verlag Stahleisen, Duesseldorf, 2001, pp. 474–481.

[3] J. Mahieu, M. De Meyer, B.C. De Cooman, Galvanizability of high strength steels for automotive application, in: Galvanized Steel Sheet Forum-Automotive, London, May 15-16, Institute of Metals, London, 2000, pp. 185–196.

[4] F. Chen, R. Patil, An in-depth analysis of various subtle coating defects of the 2000's, in: Proceedings of 6th International Conference on Zinc and Zinc Alloy Coated Steel Sheet (Galvatech 2004), Chicago, IL, April 4-7, Association for Iron & Steel Technology, Warrendale, PA, 2004, pp. 1055–1066.

[5] W. Zhong, T. Le, I. O'Reilly, B. Nelson, Experience of Arcelor Mittal Dofasco in automotive exposed GI and GA production, in: Proceedings of 10th International Conference on Zinc and Zinc Alloy Coated Steel Sheet (Galvatech 2015), Toronto, May 31-June 4, Association for Iron & Steel Technology, Warrendale, PA, 2015, pp. 129–136.

[6] M. Dubois, private communication, March 2022, with permission.

[7] M. Dubois, Analysis of Zn flow inside the sink roll grooves, in: Proceedings of 11th International Conference on Zinc and Zinc Alloy Coated Steel Sheet (Galvatech 2017), Tokyo, November 12-16, Iron and Steel Institute of Japan, Tokyo, 2017, pp. 226–233.

[8] M. Norden, F. Macherey, F. Friedel, M. Peters, Influence of thin zinc-melt oxide films on galvanizing of If and advanced high strength steels, in: Proceedings of 8th International Conference on Zinc and Zinc Alloy Coated Steel Sheet (Galvatech 2011), Genoa, June 21-25, Italy, Associazione Italiana di Metallurgia, Milan, 2011.

[9] E. Ma, M. Peng, Z. Yang, L. Cui, H. Zhan, Y. Liu, Orange peel surface defect on high strength galvanizing steel sheet, in: Proceedings of 12th International Conference on Zinc and Zinc Alloy Coated Steel Sheet (Galvatech 2021), June 22-24, Austrian Society for Metallurgy and Materials, 2021, pp. 1354–1360. Virtual Conference.

[10] T. Koll, M. Bretschneider, T. Klinkberg, F. Luther, B. Maas, On optimizing the zinc coating surface by an improved temper rolling process, in: Proceedings of 10th International Conference on Zinc and Zinc Alloy Coated Steel Sheet (Galvatech 2015), Toronto, May 31-June 4, Association for Iron & Steel Technology, Warrendale, PA, 2015, pp. 611–618.

[11] M. Rogers, Hot dip galvanize products coating defects & their avoidance, in: Presented at Hot-Dip Galvanized Sheet Production Course, Merrillville, IN, June 9-11, GalvInfo Center of International Zinc Association, Durham, NC, 2010.

[12] R. Pais, S. Agnihotri, S. Roy, M. Kadarbhai, P. Narang, Investigation of "high spots" defects in galvannealed automotive outer panels, in: Proceedings of 10th International Conference on Zinc and Zinc Alloy Coated Steel Sheet (Galvatech 2015), Toronto, May 31-June 4, Association for Iron & Steel Technology, Warrendale, PA, 2015, pp. 137–144.

[13] T. Konishi, H. Irie, White spot defect on high-tensile strength galvannealed steel sheet, in: Proceedings of 11th International Conference on Zinc and Zinc Alloy Coated Steel Sheet (Galvatech 2017), Tokyo, November 12-16, Iron and Steel Institute of Japan, Tokyo, 2017, pp. 840–845.

[14] A. Mondal, S. Agnihotry, A. Chakraborty, A.K. Halder, P. Mohanta, S. Nayak, A. Kumar, R. Pias, M. Dutta, Study of a surface defect on galvannealed sheet, in: Proceedings of 11th International Conference on Zinc and Zinc Alloy Coated Steel Sheet (Galvatech 2017), Tokyo, November 12-16, Iron and Steel Institute of Japan, Tokyo, 2017, pp. 853–860.

[15] K.-C. Chou, V. Jagannathan, Surface quality of galvanneal coated sheet—an analysis of wood-grain surface appearance, in: Proceedings Galvanizers Association 86th Meeting, Dearborn, MI, October 23-26, Galvanizers Association, Marine City, MI, 1994.

[16] W. Zhong, R. Dziuba, Formation of GA streaky defects simulated by using lab hot dip simulator, in: Proceedings of 10th International Conference on Zinc and Zinc Alloy Coated Steel Sheet (Galvatech 2015), Toronto, May 31-June 4, Association for Iron & Steel Technology, Warrendale, PA, 2015, pp. 642–649.

[17] W. Bi, X. Jin, X. Wei, Formation of a dark streaky edge defect on galvannealed ultra-high strength steel, Surf. Coat. Technol. 433 (2022) 128162, https://doi.org/10.1016/j.surfcoat.2022.128162.

[18] F. Bleich, C. Cheng, Challenges in galvannealing AHSS at an industrial hot dip galvanizing line, in: Proceedings of 12th International Conference on Zinc and Zinc Alloy Coated Steel Sheet (Galvatech 2021), June 22-24, Austrian Society for Metallurgy and Materials, Vienna, 2021, pp. 122–133. Virtual Conference.

[19] D.-J. Paik, C.-W. Kim, H.-S. Hwang, J.-W. Lee, D.-G. Kang, S.-Y. Choun, M.-H. Hong, Effects of steel substrates on nonuniformity of the coating surfaces during galvannealing reaction, in: Proceedings of 11th International Conference on Zinc and Zinc Alloy Coated Steel Sheet (Galvatech 2017), Tokyo, November 12-16, Iron and Steel Institute of Japan, Tokyo, 2017, pp. 846–852.

[20] J. Vellente, Induction strip edge heating for galvanneal product, in: Proceedings Galvanizers Association 85th Meeting, Baltimore, MD, October 10-13, Galvanizers Association, Marine City, MI, 1993.

[21] S. Chakrabarti, R.S. Pais, B. Van Veldhuizen, Fretting corrosion: a chronic surface problem in transported galvanised coils, in: Proceedings of 8th International Conference on Zinc and Zinc Alloy Coated Steel Sheet (Galvatech 2011), Genoa, Italy, June 21-25, Associazione Italiana di Metallurgia, Milan, 2011.

[22] F.B.P. de Souza, L. Colzani, R.S. de Medeiros, D.G. de Oliveira, R. Patil, Defect modeling: an advanced approach to solve quality problems and support decision making, in: Proceedings of 8th International Conference on Zinc and Zinc Alloy Coated Steel Sheet (Galvatech 2011), Genoa, Italy, June 21-25, Associazione Italiana di Metallurgia, Milan, 2011.

[23] W. Dal'Maz Silva, H. Saint-Raymond, C. Dulcy, An integrated methodology for the root cause analysis of mechanical and metallurgical defects of hot-dip galvanized coatings, in: Proceedings of 12th International Conference on Zinc and Zinc Alloy Coated Steel Sheet (Galvatech 2021), June 22-24, Austrian Society for Metallurgy and Materials, 2021, pp. 955–963. Virtual Conference.

[24] M. Monnoyer, R. Biland, E. Menigault, D. Seguret, Fives' Eyeron™, All-in-one quality management software, in: Proceedings of 12th International Conference on Zinc and Zinc Alloy Coated Steel Sheet (Galvatech 2021), June 22-24, Austrian Society for Metallurgy and Materials, 2021, pp. 797–805. Virtual Conference.

[25] Y.-M. Choi, Topography assessment of surface imperfections evolution after painting in automotive sheet steel, in: AISTech 2005-Proceedings of the Iron & Steel Technology Conference, Charlotte, May 9-12, Association for Iron & Steel Technology, Warrendale, PA, 2005.

[26] X. Cheng, S. Snopek, D. Pineau, Effect of the dimension of a spot and line type surface imperfection on its visibility after automotive painting, in: Proceedings of 10th International Conference on Zinc and Zinc Alloy Coated Steel Sheet (Galvatech 2015), Toronto, May 31-June 4, Association for Iron & Steel Technology, Warrendale, PA, 2015, pp. 658–667.

Chapter 15

General galvanizing

15.1 Introduction

General galvanizing, also termed after-fabrication galvanizing or batch galvanizing, is very different from the continuous galvanizing process considered in most of this book. Whereas continuous galvanizing treats steel products that are almost always subjected to joining and forming operations after processing, general galvanizing treats steel products that are in essentially final form. General galvanizing is therefore used to treat workpieces that include welds and formed portions and can include sections of different thickness and composition in the articles to be immersed in liquid zinc or zinc alloys. Articles to be general galvanized also enter the zinc bath at much lower temperatures because furnace pretreatments to prepare the steel surface for hot-dip coating are not used; therefore, the residence time in the zinc bath for the galvanizing reaction can be generally several minutes in duration rather than the 4–6 s of immersion time typical for continuous galvanizing. Since 2000, the general galvanizing sector has seen significant improvements in process capability and control, efficiency, and environmental performance.

As described in Marder [1], the classic general galvanizing process consists of the steps shown in Fig. 15.1.

15.2 General galvanizing pretreatment steps

As shown in Fig. 15.1, the classic cleaning step involved degreasing in alkaline solutions such as NaOH, operating at 50–80°C, to which surfactants are added to reduce surface tension and emulsifiers to disperse the droplets of oil that are lifted from the surface. Incoming steelwork usually has surface carbon contents in the range of 500–1000 g/m^2, this must be reduced to <130 g/m^2 for hot-rolled steel and < 50 g/m^2 for cold-rolled steel. Contamination of the downstream pickling baths with insufficiently cleaned surfaces can make pickling rates up to 100 times slower [3]. The emulsifiers often included phosphates that because of their eutrophication properties have been strictly regulated in household detergents [4]. Recently, phosphate-free alkali cleaners have been developed; however, the lifetime of the solution is shorter, and there is an added expense for disposal of the alkaline waste, although for many types of works the bath temperature can be lowered to 30°C, reducing both evaporation and energy consumption [5].

Alkaline degreasing has been replaced in many plants by acid degreasing. About 20% of North American galvanizers have now replaced alkaline degreasing with acid degreasing [6]. The benefits are the ability to operate the degreasing bath at 20–35°C and to eliminate the need for rinsing before pickling. If Fe sludge is well managed, there is no need to dispose of the bath, the bath life is on the order of years, versus months for an alkaline bath. Acid degreasing is not as effective as alkaline for heavily soiled articles and takes longer, typically 15–60 min vs. 5–15 min for alkaline degreasing. The first generation treatment was based on phosphoric acid; new treatments free of phosphates have now been developed. The advantages of a phosphate-free acid degreasing bath are a reduction in buildup on tank heaters and no phosphate carryover to the subsequent pickling tank; the latter is more critical with elimination of the intervening rinse step [5]. Carryover rates depend on the roughness of the process work and range from 0.04–0.08 L/m^2 of steel surface for smooth steel surfaces to 0.12–0.2 L/m^2 for rough and textured components [3].

Following cleaning, the work must be rinsed if an alkaline bath is used. The pickling step then follows. Most picking is performed with HCl, although H_2SO_4 is also used. In North America, HCl is used exclusively by 55% of galvanizers, H_2SO_4 by 38%, and both by 7% [6]. HCl pickling is normally performed at room temperature and dissolves each surface oxide layer on the steel from the outside surface according to the reaction

$$Fe_xO_y + HCl \rightarrow FeCl_2 + H_2O$$

The Metallurgy of Zinc Coated Steels. https://doi.org/10.1016/B978-0-323-99984-7.00020-8

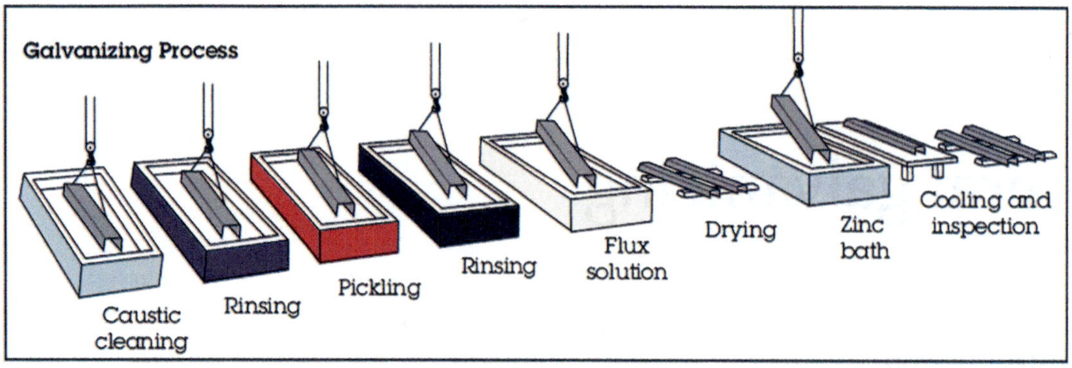

FIG. 15.1 Customary steps of the general galvanizing process, involving immersion of steel in a sequence of treatments. Steel is first degreased in a caustic solution, then rinsed. Oxide scale is then removed by pickling, followed by rinsing. The flux solution in the "dry process" shown protects the surface from oxidation during dipping in Zn and further cleans the steel surface. The flux is dried and heated before the steel is immersed in the Zn bath. In the "wet process," the flux floats on top of a portion of the Zn bath, the steel, after pickling and rinsing, is immersed through this flux layer and then removed from a flux-free portion of the bath when the galvanizing reaction is complete [2].

H_2SO_4 pickling is performed at 60–80°C, it first penetrates the oxide layer and then dissolves the interface between the surface oxide and the underlying steel. This dissolution generates hydrogen gas, lifting off the surface oxide according to the reaction

$$H_2SO_4 + Fe_2O_3 \rightarrow FeSO_4 + H_2O$$

An excellent description of HCl pickling is given in Kuklik [7].

Both pickling processes have their comparative advantages and disadvantages. For H_2SO_4 it is easy to regenerate the pickling solution by cooling and precipitation of oversaturated $FeSO_4$, whereas the HCl picking solution is difficult to regenerate and is usually sent to a third party for treatment; there is some value in the contained Fe and Zn of the spent HCl pickling solution. The elevated temperatures required for H_2SO_4 pickling result in higher equipment and energy costs. HCl pickling is less aggressive to the steel surface, resulting in a smoother surface and therefore less carryover of liquids to downstream processing tanks; HCl also has a slower pickling rate. Increasing the temperature of the HCl bath by 10°C can double the picking rate, but Cl fume emission is noticeably higher, resulting in an increased occupational hazard. HCl also produces fewer residues, therefore less thorough rinsing is required after pickling. The decision to use one pickling process versus the other is usually based on a combination of economic and technical factors together with experience from past practices.

The rate of pickling depends upon the oxides present on the steel surface. These are usually a mixture of FeO (wüstite), which pickles rapidly because of its porous structure; Fe_3O_4 (magnetite), which if on the surface will give the surface a bluish-black appearance; and Fe_2O_3 (hematite), which gives the surface a reddish appearance. Higher proportions of magnetite or hematite in the surface slow the pickling rate; much of the pickling action with HCl consists of dissolution of FeO, the other oxides then lose their adherence and fall away. A high proportion of FeO is found on steel articles that are heat-treated above 577°C, below which FeO is not stable. Steel heat-treated below 577°C has oxide layers consisting mainly of Fe_3O_4 and Fe_2O_3 that are more difficult to remove.

General galvanizing of high-strength steels with more than 2% of combined alloying elements can be more difficult to pickle because of the surface-selective oxidation of alloying elements such as Mn, Si, Cr, and Al. A successful pickling route involving an intermediate immersion in a halide salt bath has been used to treat such steels [8]. In this process, a 20s dip in an HCl pickling bath containing 190/200 g/L total chloride at 85°C was first used. This concentration and temperature were higher than those of normal galvanizing practice, which typically uses a concentration of 50–150 g/L and a bath temperature between room temperature and 30°C, which enables a 20s pickling time. The 85°C bath temperature would be associated with much higher levels of Cl fume emissions, but enabled a 20s dipping time, compatible with continuous strip processing. Use of conventional pickling solution strengths and temperatures would likely permit comparable results to be obtained at longer picking times.

When processing steel or iron that has been sand cast, any remaining sand particles on the workpiece surface must be removed because they will not be wetted during the hot-dip galvanizing step. It is possible to remove such sand by pickling with hydrofluoric acid; however, this treatment is very hazardous and must be practiced with extreme caution, including strict control of the workplace environment. Rather than using this pickling method, most galvanizers use abrasive cleaning to prepare such surfaces. Abrasive cleaning is also used instead of conventional pickling to prepare high-strength fasteners to avoid hydrogen pickup, as discussed later.

After pickling, the steelwork is again rinsed. A main purpose of this rinsing step is to minimize Fe salts carryover from the pickling step into the flux tank and to help control the pH of the flux between 3.5 and 5.5—too low of a pH will result in more Fe dissolution in the flux that can then be carried over to the galvanizing bath and can increase the risk of hydrogen pickup as discussed later in this chapter. The purpose of the flux step is to remove any remaining oxide film and provide a temporary protective covering to the steel surface and improve wettability as it comes into contact with the zinc bath. Flux compositions are based on zinc ammonium chloride together with a wide variety of additives. Three flux compositions are used:

- Mono salt: ZnCl-NH_4Cl: This is used for the "wet process"
- Double salt: ZnCl-2NH_4Cl: This is the most common "dry process" flux used in Europe; control of the process is more forgiving than when the triple salt is used
- Triple salt: ZnCl-3NH_4Cl: This is the most common "dry process" flux used in North America

The higher NH_4Cl compositions emit more fumes. The flux tank concentration is 400–600 g/L for double salt and 300–400 g/L for triple salt. Higher concentrations are used to treat heavier steel sections. Drying is also faster at higher concentrations. The flux bath temperature is between 40°C and 70°C, with increased evaporation experienced above 50°C. Fe content is maintained below 10 g/L and preferably below 5 g/L; 1 g of Fe carried over to the galvanizing bath creates 40 g of bottom dross. Kuklik provides further details on flux practice and regeneration of spent fluxes that mainly involves the removal of Fe [7]. After fluxing, the steelwork is dried. Drying must be at a high enough temperature to remove all moisture (so that a furnace air temperature of 100°C is desirable) but not so high that the NH_4Cl thermally dissociates (so a steel surface temperature of 80°C cannot be exceeded). A drying time of 20 min is typical. Drying times are more sensitive to the volume of air flowing through the drying oven than temperature [3]. Thorough drying is especially important for galvanizing baths containing Al; moisture in the flux will result in Al loss because of the oxidation of Al [9].

15.3 Hot-dip coating with zinc and zinc alloys

15.3.1 Galvanizing alloys: Overview

General galvanizing alloys can be divided into two categories: those that influence the reaction of liquid Zn with the steel surface and those that influence the physical properties of liquid Zn. This is shown schematically in Fig. 15.2.

A detailed view of the effects and uses of these alloy additions is given in Sections 15.3.4–15.3.7. Traditionally, galvanizing baths contained about 1% Pb because it reduces the surface tension of pure Zn by more than 40%, while not changing viscosity. The surface tension of pure Zn at 440°C was found to be 780 dynes/cm, decreasing to 500 dynes/cm at 1% Pb [11]. This greatly aids draining of zinc as the steel article is removed from the zinc bath. Excess Pb in a supersaturated bath sinks to the bottom. Because all Zn-Fe intermetallics are denser than liquid Zn, the intermetallics float on top of this bottom layer, easing the removal of bottom dross. Environmental and occupational health regulations together with social demand have restricted and/or complicated the usage of lead in general galvanizing processes and products, so its use has declined in many countries. For examples, the 2019 AGA process survey reported that 10% of responding plants used Pb additions, whereas 66% used Bi additions, which, as Fig. 15.1 indicates, also influences Zn physical properties [6]. This is a large reduction from 2005, when 70% of these plants used Pb-containing baths [12].

15.3.2 Immersion of steel in the galvanizing bath

As indicated in the previous section, steel work exiting the flux drying step does not exceed 80°C to avoid flux dissociation. Although this steel is then slowly immersed into the zinc bath, its temperature is still significantly below the melting point of zinc galvanizing alloys when immersed. Depending upon steel section thickness, it takes a varying amount of time for the steel to be heated to the point where the galvanizing reaction commences. For thin sections such as sheet and wire, this may be a matter of several seconds, whereas for large heavy sections such as transmission poles and structural sections, the galvanizing reaction may be delayed as long as 30 min after immersion.

The immersion of this steelwork ideally at about 80°C into a galvanizing bath held initially at in the temperature range 445–450°C results in cooling of the bath, especially in the volume of bath metal immediately surrounding the steelwork. In this volume, especially for thicker steel sections, a solidified shell of zinc around the initially immersed steel is formed

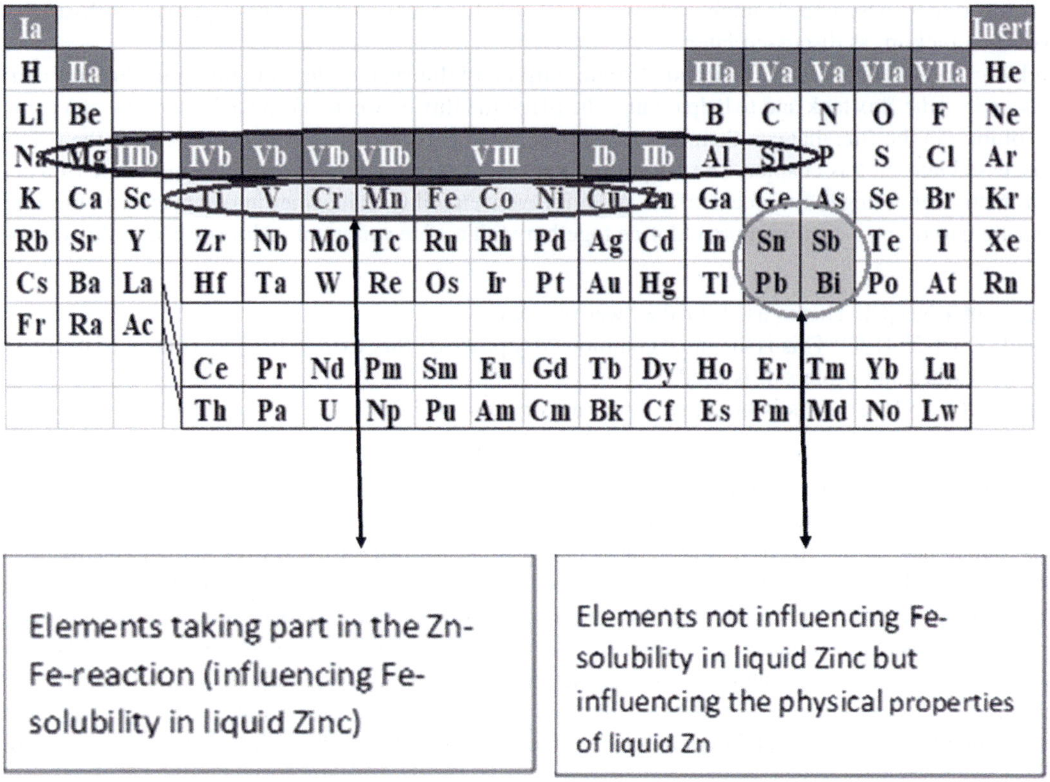

FIG. 15.2 Classification of zinc bath alloying additions according to two characteristics. Ten elements have been shown to influence the Zn-Fe reaction during hot-dip galvanizing. Four others can be added to influence the physical properties of liquid zinc such as surface tension [10].

[13,14]. Even for a thin (3 mm) steel section, it takes 20s for the section to reach the 419.5°C melting point of zinc when dipping room temperature samples, prepared by fresh surface grinding of the surface, in a 450°C bath, as shown in Fig. 15.3 [13].

Because in all cases the zinc is not pure, but alloyed at least with Fe, which after sustained production saturates the Zn bath at a level of about 0.03% at 450°C, this solidified layer does not have a single melting point, but solidifies over a temperature range, depending on the bath alloy composition. Solidification with all commonly used galvanizing alloys begins near the melting point of pure Zn, 419°C, the usual alloy solidification behavior through a semisolid or mushy zone then proceeds, including microsegregation of the lower-melting-point constituents within this shell. Removal of the 3 mm thick sample described above after 4 s of immersion revealed a 2 μm thick interface layer to which the surrounding liquid was attached as the sample was withdrawn from the bath; the constituents and composition of this thin interface layer were not conclusively identified (Fig. 15.4).

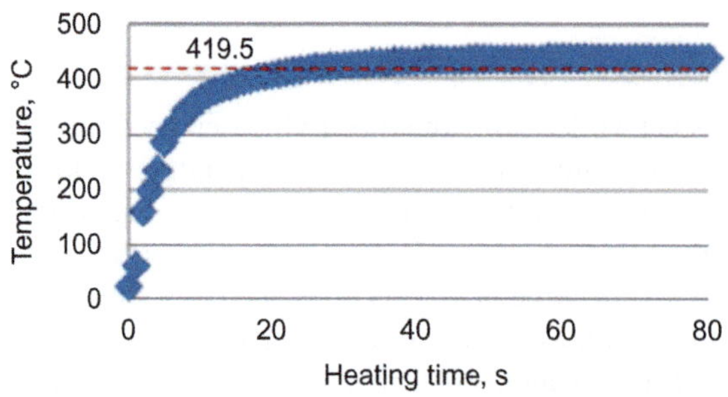

FIG. 15.3 Time dependence of the rise in temperature of a 3 mm thick steel sample dipped in a 450°C galvanizing bath. The melting point of pure zinc is indicated [13].

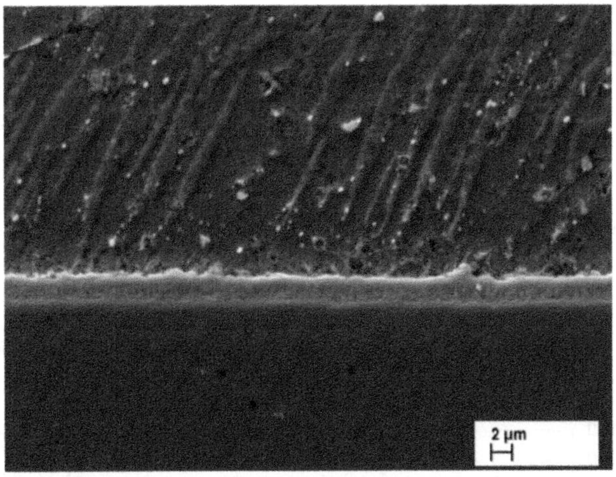

FIG. 15.4 Microstructure of the coating layer of the 3mm sample of the previous figure after 4s of dipping time indicating a 2μm thick interface layer over which a liquid zinc layer was formed with withdrawal from the galvanizing bath [13].

The liquid phase containing these constituents can be in contact with the steel for a longer period of time and therefore react with the steel during this period before the intended formation of the galvanized coating commences [14]. The flux, applied in the previous step, decomposes in the galvanizing bath; traditional fluxes were designed to decompose at the same time the lead in the solidified steel became molten, around 327°C. With the advent of lead-free compositions, fluxes with higher decomposition temperatures were introduced. One of these is described in US Patent 5,437,738, "Fluxes for lead-free galvanizing" and is capable of providing protection up to 500°C, well above the usual galvanizing temperature [15]. The reaction of the solute-rich, low-melting-point liquid at the interior of the solidified shell, when in contact with sensitized areas such as precracked or improperly prepared weldments, has been associated with liquid-metal-assisted cracking [16]. To reduce the risk of such cracking to a minimal level, bath composition limits have been set; for example, the British Constructional Steelwork Association limits the sum of lead and tin contents in the bath to less than 1.3wt% and additionally Bi $\leq$0.1wt% [17]. German limits have been set as a maximum of Sn<0.1wt% and Pb +10Bi <1.5wt% [18]. An example where such practices were followed showed that for a production of more than 1,000,000 metric tons of steelwork, of which 560,000 was classified as structural steel, such cracking occurrences are quite rare. The total overall cracked tonnage was found to be 0.033%, with 0.026% of the structural steel tonnage having cracks and 0.043% of the nonstructural category containing cracks. The large majority of cracks were associated with weld zones. The underlying causes of cracking fell into well-known categories of suboptimum design, involving either weld details or design for galvanizing [19].

The galvanizing reaction begins by wettability of the prepared steel surface, followed by reaction and dissolution of iron into liquid zinc. The rate of this Zn-Fe reaction depends upon temperature. Fig. 15.5 shows the rate of attack of zinc on iron containing varying Si levels, the vertical axis provides the constant "a" in the equation

$$m^2 = at \tag{15.1}$$

where m is Fe lost in g/m^2, and t is immersion time in minutes. Fig. 15.5. shows that the kinetics of Zn attack is parabolic for almost pure iron (0.002%Si) between 450°C and 490°C and above 520°C; between these ranges attack is linear, not slowing with time [20]. This range of linear attack broadens, depending on steel Si content, the reasons for this will be discussed later in this chapter. The intermetallics formed in the parabolic reaction domains are adherent and slow further diffusion, this also means they contribute little Fe to the zinc bath. In the linear range, the Fe-Zn reaction products, especially zeta phase [21], are discontinuous and nonadherent and can result in significant dross in the bath.

The bath temperature also influences the rate of attack on the galvanizing kettle surface and therefore influences kettle life. Many galvanizing kettles are made from Armco iron, which has a minimum Fe content of 99.85% and a maximum C content of 0.01% and minimal Si [22]. To maximize kettle life and optimize energy efficiency, it is recommended that the bath temperature be maintained in the range 445–460°C, below the 475°C temperature indicated in Fig. 15.5A and B where more rapid attack begins for a 0.0002%Si composition [23].

An iron balance of the galvanizing reaction was carried out for steels containing 0.02wt%Si and 0.6wt%Si [24]. As explained later, these steels have very different characteristics of reaction in the zinc bath. The iron balance was conducted for zinc bath temperatures of 450°C, 550°C, and 570°C. The latter two temperatures are of interest for the process

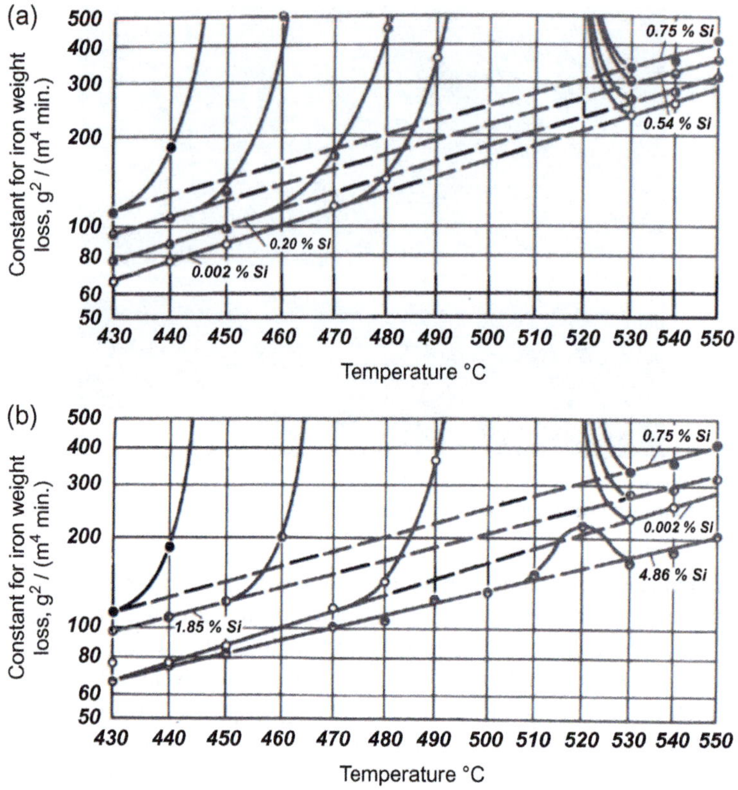

FIG. 15.5 Temperature dependence of weight loss of iron samples with Si levels between 0.002% and 4.86% dipped in liquid zinc [20]. (A) Effect of temperature on parabolic rate constant "a" in Eq. (15.1) for 0.002% < Si < 0.75%. (B) Effect of temperature on parabolic rate constant "*a*" in Eq. (15.1) for 0.002% < Si < 4.86%.

of high-temperature galvanizing, in the range where the FeZn zeta phase is not stable, as described in Chapter 4, and can be used to galvanize high Si content steels as described in [7]. To perform the mass balance, the average Fe content of each coating was determined along with the total coating mass. The product of these two measurements is the total Fe present in the coating. The total iron loss from the steel sample was then measured by subtracting the mass of the stripped sample from the sample mass before hot dipping. The "excess iron" (EI), that is, the amount of iron expected to report to dross, is then

$$\mathrm{EI} = \mathrm{TIL} - (\%\mathrm{Fe})(\mathrm{CW}) \tag{15.2}$$

where TIL = total iron loss from the steel coupon, CW = coating weight.

Table 15.1 shows the results of these calculations, based on coating of 5 mm plate samples. These results for the low Si steel fall in the broad range of dross production rates reported by general galvanizers, which the 2019 American Galvanizers Association process survey reported as ranging from 0.4 to 27 kg of dross produced per ton of galvanized product [6] with normal steady-state operations for conventional jobbing work producing 6–7 kg/ton [25]. The high Si steel can be seen to have dross rates much higher than the low Si steel, regardless of bath temperature. The high Si steel dross rate can be improved by use of alloy coatings, described later in this chapter.

Dross is also produced by precipitation in the bath when it decreases in temperature as steel is dipped into it. It is typical for a galvanizing bath to cycle between a setpoint of 440°C to a minimum temperature of 426°C and then reach as maximum temperature of 448°C during the reheating cycle [26]. These temperature cycles can take 1 h to complete, depending upon production rate (mass of steel being galvanized per hour) and the thickness of the steel being galvanized, which will require more mass heating as thickness increases. Fig. 15.6 shows a typical pattern of temperature cycles in a galvanizing bath over a 12-day period of 8-h production shifts, for periods of high, medium, and low demands for mass heating.

Because the typical galvanizing bath is saturated with iron during periods of constant temperature when steelwork is not being galvanized or makeup zinc being added to the bath, any decrease in temperature will result in dross precipitation. For baths operating in a temperature range up to 490°C, the stable Fe-Zn phase that will precipitate during a decrease in bath temperature is zeta phase, as given by the Fe-Zn phase diagram in Chapter 3. Studies on continuous galvanizing baths [27]

TABLE 15.1 Calculated mass balance for two steel silicon contents at temperatures representative of conventional and high-temperature galvanizing [24].

	10 min immersion			
	Excess Fe	Fe in dross[a]	Calculated dross	
	g/m^2	%	g/m^2	kg/ton
0.02 wt% Si steel				
450°C	9	3	301	15
550°C	14	5	280	14
570°C	0	5	0	0
0.60 wt% Si steel				
450°C	52	3	1732	87
550°C	71	5	1431	72
570°C	62	5	1237	62

[a]*Assumes dross is 50% Fe-Zn iron crystals (zeta for 450°C and delta for 550°C, 570°C) and 50% Zn metal.*

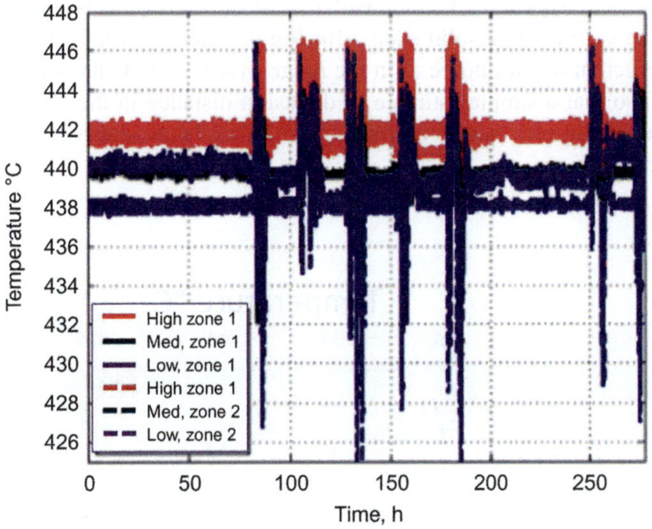

FIG. 15.6 Galvanizing bath temperature cycles for seven working shifts within a 12-day period [26].

have determined that when bath temperature is dropped below the local Fe saturation temperature at Al contents less than 0.1%, zeta phase immediately nucleates and can grow very quickly to a size of 35 μm. If the bath continues to decrease in temperature, particle sizes can reach 50 μm after 1 h. However, it would be unusual for a general galvanizing bath to remain significantly below its setpoint temperature for more than 30 min, by which time heating of the bath to raise the temperature would be well underway. As the bath is heated, the zeta dross particles have been found to dissolve at the same rate as they precipitated. Therefore, the contribution of temperature excursions to dross generation is only transient and does not contribute to the overall iron mass balance that adds dross to the galvanizing bath.

15.3.3 Galvanized coating formation

The stages of galvanized coating development have received much study for many years and proved difficult to explain. The heats of formation and also free energies at 460°C of the Fe-Zn phases, shown Table 15.2, do not differ significantly

TABLE 15.2 Enthalpies of formation and free enthalpies at 460°C for Fe-Zn phases [28].

Phase	Crystal structure	Formula	wt% Fe at 450°C	ΔH^* kJ/mol	$\Delta G^*_{450°C}$
η (Zn)	Hexagonal	Zn	<0.03	–	–
ζ	Monoclinic	$FeZn_{13}$	5–6	−11.7	−2.8
δ_1	Hexagonal	$FeZn_7$	7–12	−11.5	−3.5
Γ_1	Face centered cubic	Fe_5Zn_{21}	17–19	−11.7	−4.1
Γ	Body centered cubic	Fe_3Zn_{10}	23–28	−10.9	−4.2

Data from J.-Y. Dauphin, P. Perrot, U.G. Tchissambot, Thermodynamic description of the iron-zinc system, Mèm. Sci. Rev. Mèt. 84 (6) (1987) 329–336.

[28], so that small additions to either the steel or zinc alloys used in the hot-dip galvanizing process or small variations in the galvanizing process details can lead to differing results.

The rate constants for the zeta-delta, delta-zeta, gamma-one-delta, delta-gamma-one, and delta or gamma-one to Fe_2Al_5 phases are shown in the Arrhenius plot of Fig. 15.7 [29]. These range over four orders of magnitude for units of k in units of s^{-1}. The rate constant for the delta-zeta reaction ranges from 3 to 6×10^{-5}/s for temperatures exceeding 450°C, whereas the reverse zeta-delta reaction rate constant is in the range of $6–10 \times 10^{-7}$/s at temperatures from 478 up to 490°C.

During the initial dipping period, when a steel article with low Si and P contents is surrounded by solid zinc, it has been proposed that the initial coating is formed by solid-state diffusion of zinc into iron, forming thin layers of zeta and delta phase [13]. The diffusion coefficient of Zn into Fe at in the range of 400–500°C has been reported to be on the order of magnitude of 10^{-17} m^2/s [30] so that a simple estimate of diffusion distance in the first 4s gives an intermetallic layer thickness of 0.07 μm, which is negligible; therefore, the 2 μm interface layer observed in Fig. 15.4 must mainly consist of the solidified metallic zinc shell.

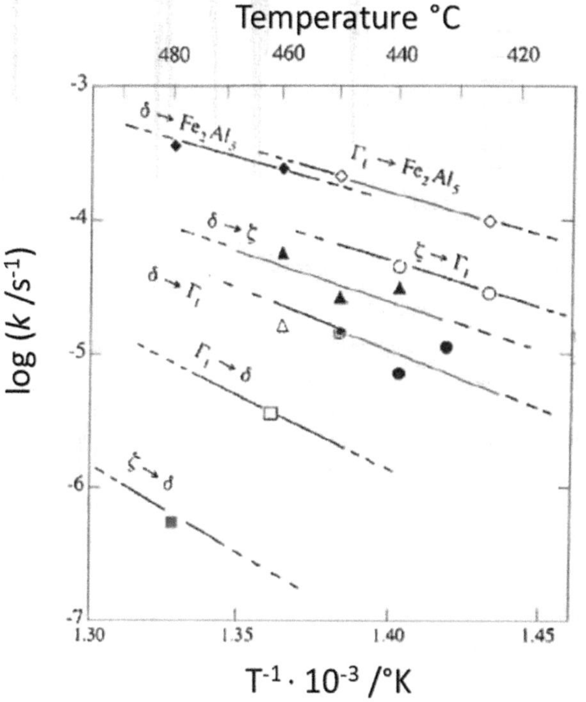

FIG. 15.7 Arrhenius plot for transformation reactions in several FeZn phases and Fe_2Al_5 [29].

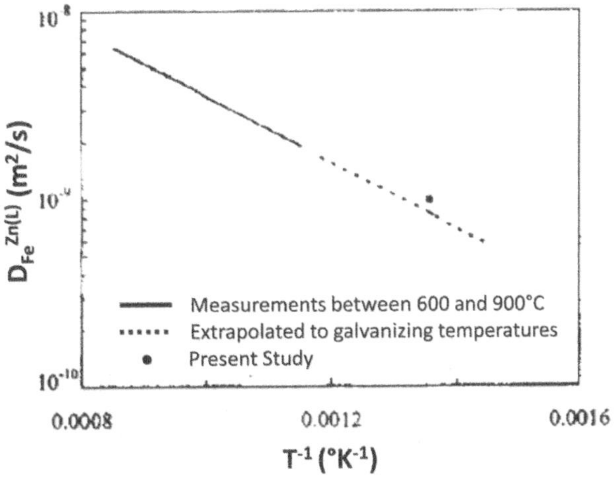

FIG. 15.8 Temperature dependence of the diffusion coefficient of Fe in liquid Zn. The black dot was experimentally determined in [31] and is consistent with past studies [31].

Far more rapid than solid-state diffusion is the diffusion of solid Fe into liquid Zn [31]. The diffusion coefficient for solid Fe into liquid Zn has been experimentally determined to be $9.8 \pm 0.1 \times 10^{-10}$ m²/s. This diffusion coefficient can be used to describe the flux of Fe into the Zn bath and is consistent with measurements made at higher temperatures, between 600°C and 900°C. The relationship between diffusion coefficient and system temperature is shown in Fig. 15.8.

The dissolution of Fe into liquid Zn marks the beginning of the galvanizing reaction. Kuklik provides a detailed description of the effect of steel chemistry effects of coating microstructure, especially Si and P, termed the Sandelin effect [7]. For steels below the Sandelin reactivity range, i.e., <0.03%Si, the galvanizing reaction of the dissolution of Fe into liquid Zn creating a mixture of very fine crystals of zeta (ξ) and delta (δ) phases in compact layers under the unreacted zinc, termed eta (η) phase. A typical microstructure for this type of steel is shown in Fig. 15.9.

The steel reactivity depends upon the local Fe concentration in the liquid, which is influenced by steel roughness. Roughness peaks expose a larger area of Fe to the Zn bath, and therefore, local Fe supply from dissolution is higher at these dissolving peaks. It has been reported that the transient Fe solubility near the liquid Zn/steel surface interface could reach 5 at.%, more than 150 times higher than the equilibrium saturation of 0.03 at.% [32]. Such high Fe solution conditions therefore permit delta phase to be grown.

Shortly after forming, zeta "outbursts" appear, in which the long axes of the monoclinic zeta crystals are generally normal to the plane of the steel surface [28]. These outbursts appear to be coincident with steel grain boundaries that intersect the surface of the ferritic steel substrate. Further dissolution enriches the delta layer on the surface, transforming the base of the zeta outburst crystals at the coating-substate interface to delta phase, so that this phase proceeds to grow as a uniform layer, limited by Fe diffusion through it, so that parabolic growth behavior is observed. The columnar zeta crystals

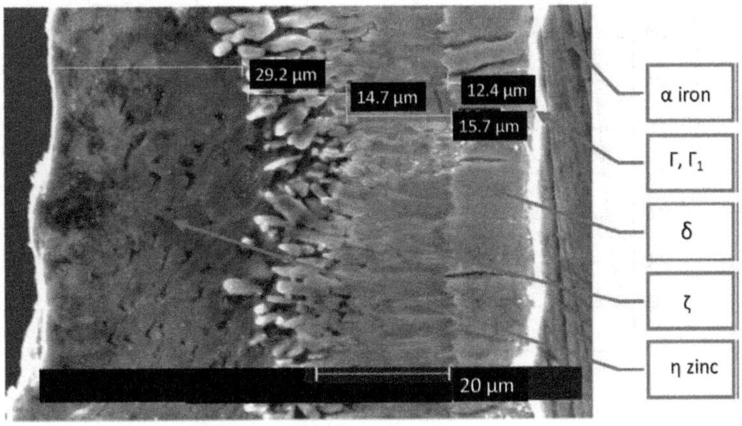

FIG. 15.9 Galvanized coating on a nonreactive low-Si steel, dipped in a 460°C bath for 10 min [7].

continue to be present in the liquid-solid region covering the growing delta phase in the field of thermodynamically stable compositions given by the Fe-Zn phase diagram (Chapter 3), but the limitation of Fe flux through the delta phase limits further zeta phase growth. The zeta columns then fully cover the underlying delta phase at the interface between these phases so that there is no contact between the liquid Zn and the delta phase.

Zeta growth continues into the liquid zinc, regulated by Fe supply provided in the intercrystalline liquid channels that extend down to the solid front of the previously grown zeta phase. As noted before, the kinetics for the transformation of zeta to delta phase is 100 times lower than the zeta-delta transformation reaction. Meanwhile, at the delta phase-steel interface, solid-state diffusion of Zn from the delta phase into Fe creates first a solid solution of Zn in ferrite, termed αFe(Zn). This phase and associated Kirkendall vacancies, made obvious by the porous appearance of the steel, are shown in Fig. 4.5 in reference [7]. As this αFe(Zn) enriches in Zn, it reaches the boundary of the two-phase region, αFe(Zn) + Γ, when a thin gamma layer is formed at the steel-coating interface. This can grow until it forms a continuous layer at the interface. Further Zn enrichment of Γ leads to growth of Γ_1 phase. This phase, with twice the lattice parameter of Γ and a face-centered cubic structure rather than the body-centered cubic structures of ferrite and Γ, was only discovered in 1974 [33]. Γ_1 can also be formed from delta phase during slow cooling of the galvanized article when it is withdrawn from the bath.

Higher levels of Si and P in the steel result in thicker coatings caused by increased steel reactivity in the zinc bath. The effect of steel Si content on coating thickness is given by the well-known Sandelin curve, reproduced as Fig. 15.10 from Fig. 5.6 in [7]. Phosphorus has an effect on steel reactivity in the galvanizing bath, so that a customary indication of the lower bound of the Sandelin reactivity range is $Si + 2P < 0.09$ wt% with $Si < 0.04$ wt% [34].

Representative microstructures for coatings at different compositions over this reactivity range are shown in Figs. 15.9 and 15.12.

Increasing steel Si results in a destabilization of coating growth, as explained by Tang [32], who described two regimes of reactive steel behavior: Si levels between 0.035 wt% and 0.2 wt% are termed Sandelin peak behavior, Si levels greater than 0.2 wt% are termed hyper-Sandelin steels. Within and just beyond the Sandelin peak, $0.15 < Si < 0.25\%$ is termed the Sebisty range; the Sebisty effect [35] that reduces coating reactivity in this range will be explained later. Fig. 15.11, the Zn-rich corner of the Zn-Fe-Si phase diagram, helps explain the behavior of reactive steels.

Referring to the ternary section in Fig. 15.11, as diffusion of the steel into the liquid Zn begins, the interface composition moves from the lower left-hand corner (zero content of both Fe and Si) up the left-hand border as Fe content in the liquid Zn increases. Zeta begins to be formed as the 0.03 at.% point is crossed, moving into the two-phase liquid-zeta mixture. Because of the narrowness of the Zn composition in the zeta single-phase stability range at 460°C (94–94.8 wt%), delta phase can also be formed, as with the growth of nonreactive coatings in regions of locally high Fe concentration in the liquid Zn solution. Si in solid solution with Fe in the steel also diffuses into the liquid Zn solution, enriching it with Si

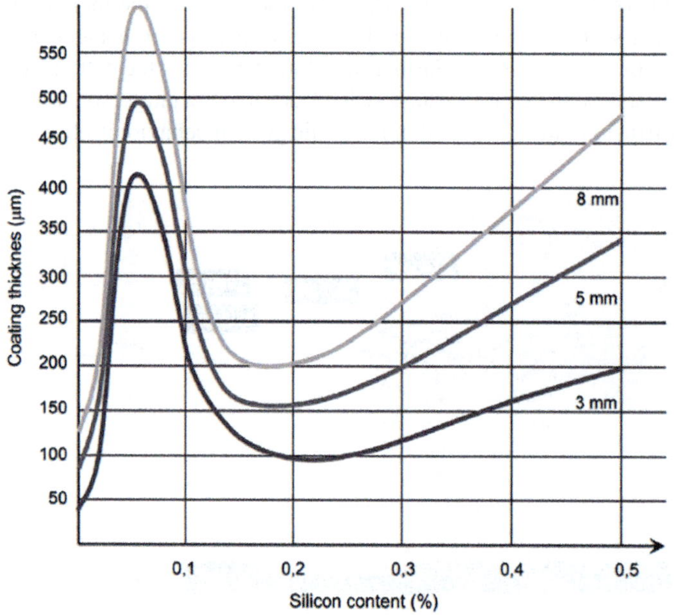

FIG. 15.10 Example of Sandelin galvanizing reactivity curve, showing coating weights for three steel thicknesses [7].

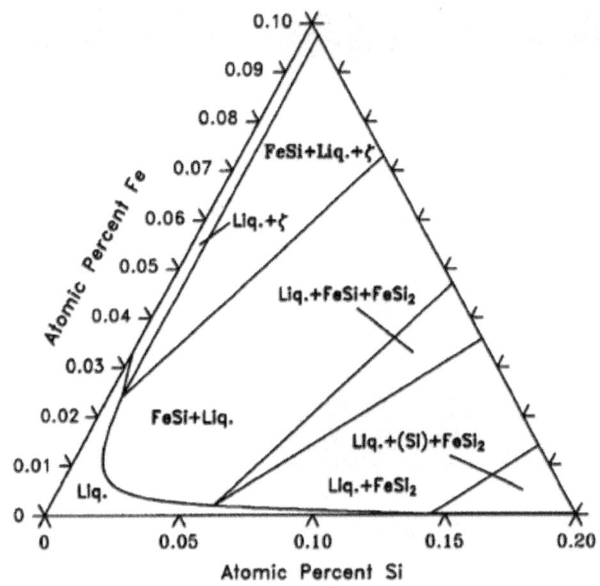

FIG. 15.11 The Zn-rich corner of the 450°C isothermal section of the Zn-Fe-Si phase diagram. Note the coexistence of the liquid and δ phases is thermodynamically prohibited [36].

in a proportion equal to its composition in the steel at this dilute concentration. Until the local liquid composition exceeds the solubility limit of Si in zeta phase according to the FeZnSi ternary diagram, normal galvanizing behavior will be observed. However, zeta phase practically cannot contain any Si in solution; Si is rejected into surrounding liquid and fine FeSi particles begin to grow, as evidenced in [32]. The rate of growth (precipitation) of the zeta phase determines the shape of the zeta crystals. Rapid growth of zeta phase proceeds together with rejection of Si from the precipitating crystals, results in interface stability similar to dendritic solidification rather than growth of the expected regular monoclinic zeta crystals. The zeta layer is then no longer continuous over the underlying delta layer, permitting liquid Zn access to the surface of the delta phase. According to Fig. 15.11, it is not thermodynamically possible for the liquid Zn-Fe-Si solution to exist together with delta phase, and so the delta phase is destabilized. The re-dissolution of the delta phase that begins at the liquid-delta interface, where Si has already been rejected from the zeta phase and is also present in the delta phase, permits liquid to again access the steel-coating interface. Because the rate of coating growth is then not limited by Fe diffusion through the delta phase, the coating growth rate is no longer parabolic with time, but linear, resulting in greater coating thickness because of rapid growth of the zeta phase. Because the gamma-one and gamma phases are also not stable when in contact with the Si-containing liquid Zn, they too destabilize.

The result for a steel with $0.035\% < Si < 0.12\%$, in the Sandelin peak, is a coating with thick zeta outburst structures, and thinner but still compact delta, gamma-one and gamma phases that can be discontinuous [32]. A representative microstructure is shown in Fig. 15.12A. At Si contents higher than 0.2%, the growing zeta phase over the initially stable delta phase again rejects Si into the liquid Zn, opening liquid channels to the underlying delta phase, which destabilize it. This results in formation of what Tang terms "diffused delta phase," which is a mixture of liquid Zn, newly formed zeta crystallites, tiny FeSi particles that can serve as nuclei for further solid phase growth and surviving delta particles. Within the surviving delta layer, two composition regions are observed, one with about 8 at.%Fe and the other, nearer to the steel surface, with about 12 at.%Fe. The lower Fe content portion is associated with the delta palisade structure that can be regarded as an allotriomorph of the higher Fe compact delta phase. Even lower Fe content, about 7 at.% metastable delta phase compounds can be observed in this diffused phase. The gamma-one and gamma phases are also destabilized in this high Si content environment [32].

The portion of the Sandelin curve with lower coating thickness beyond the Sandelin peak, $0.12 < Si < 0.2$ wt% termed Sebisty behavior, was experimentally confirmed by Che [37]. The microstructure of a galvanized coating on a Sebisty steel is shown in Fig. 15.12B. Returning again to Fig. 15.11, the liquid domain narrows as Si content increases, ceasing to be a single-phase domain above 0.14%Si. This corresponds to the valley in the Sandelin curve, termed Sebisty behavior. In this portion of the Sandelin curve, liquid Zn, with Si and Fe contained in liquid solution, is only present in a multiphase mixture involving $FeSi_2$ and Si. Therefore, the Si in this mixture is largely partitioned to these compounds at the start of the galvanizing reaction, leaving a limited amount of liquid Zn with lower Si content capable of producing Fe-Zn phases in early

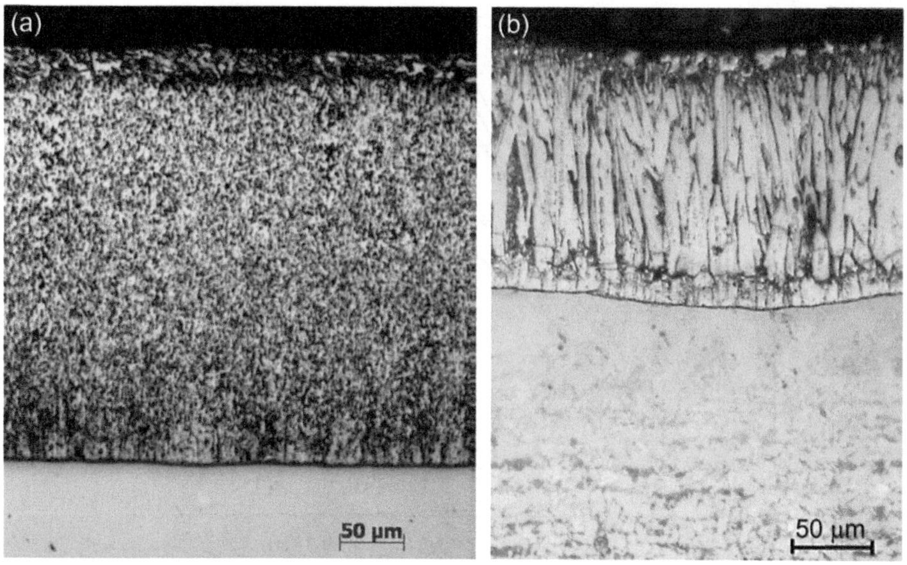

FIG. 15.12 (A) Galvanized coating on a highly reactive steel with a composition near the Sandelin peak of about 0.08%Si shown in Fig. 15.10, dipped in a 460°C bath for about 10min [7]. (B) Galvanized coating on a steel with a composition in the Sebisty range of about 0.21%Si shown in Fig. 15.10 dipped in a 445°C bath for about 10min [7].

stages of coating growth. The solid portions of this mixture can form compounds completely covering the steel surface, permitting diffusion-limited kinetics to again control growth of the galvanized coating. Because of the intolerance of Γ and $Γ_1$ phases for Si, as detailed by [32], these phases can only be formed in initial stages of coating growth, becoming unstable as the system moves toward the state of equilibrium described by the phase diagram. As described by Sebisty [35], increasing the temperature of the Zn-Fe-Si section of Fig. 15.11 to higher temperatures increases Fe solubility.

For steels in the Sebisty range, increasing the bath temperature increased Fe solubility in the bath from 0.045 wt% at 450°C to 0.07% at 480°C. This was found sufficient to shift the Zn-Fe-Si equilibrium to keep the gamma phase stable, so that parabolic kinetics governed coating growth for the investigated 0.14%Si grade investigated by Che [37]. The zeta phase still formed the bulk of the coating thickness but grows at a parabolic rate that decreases with increasing temperature. Further growth of the zeta phase is limited by two factors. First, Fe diffusive transport through the growing coating and its subsequent consumption in the constituent phases limit supply of Fe content at the front of the zeta crystals, these can even reduce in size by redissolution, further moderating coating weight in the Si composition range. Second, the increasing Si content in the liquid results in additional precipitation of FeSi particles. In galvanizing baths some convection of liquid zinc resulting from thermal differences between the kettle surface and the zinc bath is usual [26], convection at the growing front of the galvanized coating has been observed to influence coating growth rate and phase makeup [31]. For this reason, it is common practice to slowly raise and lower the steel being galvanized in the zinc bath during dipping to improve coating uniformity.

For the purposes of understanding steel reactivity, the nominal Si content of the steel may not always correspond to the composition of Si present near the surface being treated for galvanization, because of the surface-selective oxidation behavior of Si during hot working and heat treatment processes. One study found that near-surface Si contents were reliably three times higher than bulk compositions [38]. Sebisty [35] found that it is possible to induce normal reactivity in steels over the reactive composition range by vacuum annealing. This was attributed to internal oxidation of near-surface Si that prevented it from participating in the galvanizing reaction.

A large-scale study by International Lead Zinc Research Organization, involving 172 steels compositions including more than 6000 test specimens, developed the following categorization system for galvanized steel reactivity:

The coating layer structure becomes independent of Si content when the galvanizing bath temperature is held at 490–530°C. In this temperature range, the coating consists of a homogeneous microstructure containing fine crystals of δ and ζ phases [39]. Table 15.1 indicates that lower amounts of dross are associated with these higher temperatures.

The final step of coating formation is the withdrawal of the steel from the bath. Forces working to reduce coating thickness are the gravitational force and the surface tension on of the zinc bath; those working to increase coating thickness are the viscosity of the liquid coating that contributes to a viscous shear force, inertial forces, and the surface tension between the coating and substrate. Most of these, in turn, are dependent upon bath temperature. The greatest influence on the thickness of the zinc overlay coating is the speed of withdrawal from the bath. The overall coating thickness includes

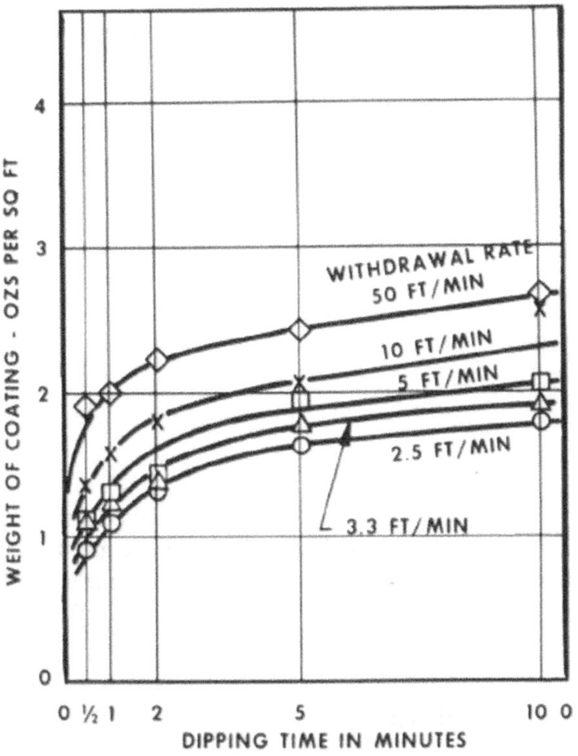

FIG. 15.13 Effect of withdrawal speed on coating weight for a range of typical dipping times for a steel of normal reactivity, bath temperature of 435°C [40].

the thickness of the intermetallic layers, which depend upon dipping time. Fig. 15.13 shows the resulting total coating weight for a range of dipping times and withdrawal speeds for a bath containing 1.2% Pb at a temperature of 435°C when galvanizing a steel of normal reactivity (Classes 1 and 2 in Table 15.3).

TABLE 15.3 Reactivity classification for general galvanized steels.

Steel reactivity classification—Hot-dip galvanizing in conventional zinc baths					
Class	**Description**	**Si wt%**	**P wt%**	**Steel reactivity**	**Coating appearance**
1	Normal/hypo-sandelin	0–0.035	0–0.025	Generally normal but occasionally low	Few defects
2	Semi-sandelin	0–0.04	0.025–0.035	Normal	Localized defects from zeta oubursts
3	Nonclassical sandelin	0–0.04	>0.035	High, especially with high P	Pronounced surface defects, i.e., "tree bark", high flaking tendency
4a	Sandelin low-P	0.04–0.135	<0.01	Moderate, increasing with Si	May appear normal with few defects
4b	Sandelin high-P	0.04–0.135	>0.01	High	Generally, few defects
5a	Hyper-sandelin low P	0.135–.35	<0.03	High, but generally thinner coatings than Class 5b	May appear normal with few defects
5b	Hyper-sandelin high P	0.135–0.35	>0.03	High	Tendency to flake, especially with high P
6	Extreme hyper-sandelin	>0.35	>0	High, increasing with Si	Tendency to flake, increasing with P content

Derived from R.W. Richards, H. Clarke, F.E. Goodwin, Analysis of galvanized coatings, in: Proc. 17th International Galvanizing Conf. (Intergalva 1994), June 6–8, 1994, Paris, European General Galvanizers Assn., Solihull, UK.

TABLE 15.4 Minimum coating weight and equivalent thickness requirements of ISO 1461 [41].

	Local coating (minimum)		Mean coating (minimum)	
Articles and its thickness	g/m²	μm	g/m²	μm
Steel >6 mm	505	70	610	85
Steel >3 mm to ≤6 mm	395	55	505	70
Steel ≥1.5 mm to ≤3 mm	325	45	395	55
Steel <1.5 mm	250	35	325	45
Castings ≥6 mm	505	70	575	80
Castings <6 mm	430	60	505	70

ISO 1461 [40] includes a table of the permissible minimum coating weights, in terms of both local and mean minima, for different thicknesses of wrought and cast steel articles. The challenge to the galvanizer is to produce a coating weight as close as possible to the minimum, to avoid unnecessary consumption of Zn, while balancing materials and process variables including steel reactivity, bath temperature, dipping time, and withdrawal speed (Table 15.4).

15.3.4 Alloy coatings to control steel reactivity

The variations in steel reactivity in an unalloyed zinc galvanizing bath described above can result in rough coatings with thicknesses much higher than the 85 μm mean minimum required by the ISO 1461 standard. This can result in excessive and unnecessary consumption of zinc. As indicated in Fig. 15.2, many alloy additions can be made to the zinc bath to influence steel reactivity. Ti and V have been investigated and found to be of limited practical use, in the case of Ti excessive dross precipitation occurs and with V the toxicity of V salts limits use [42]. A combination of Al, Sn, and Mg was developed in the 1970s as the Polygalva alloy, but difficulties with bath management and related flux reactivity with the Al and Mg additions resulted in only brief industrial use. Cr, Mn, and Cu additions were associated with unacceptable levels of dross generation; Cu is also undesirable when the steel is eventually recycled because it cannot be removed in normal steel mill melt shop practices [43].

As described by Tang [32], the addition of Ni to the zinc bath at a level of 0.055 wt% introduces the T phase that is reported to possess the same crystal structure at the Zn-Fe Γ_1 phase [44] and has the equilibrium formula $Fe_6Ni_5Zn_{89}$ [42]. The T phase begins to form at a Ni content of 0.06 wt% at 450°C, a much lower content than Fe in the counterpart Zn-Fe Γ or Γ_1 phases. In these Zn-Fe phases, there is significant Ni solubility, 11 and 4 wt%, respectively. As in conventional galvanizing, the zeta phase forms first, but no outbursts occur. The T phase is the next to form, giving coating thickness and appearance similar to conventional galvanizing. Ni additions help control steel reactivity up to a Si level of 0.25%, over which the reactive steel growth patterns resume. The effect of bath Ni content on coating thickness is shown in Fig. 15.14. for a range of steel Si contents over the Sandelin peak ($0.05 < Si < 0.21$ wt%), in the Sebisty range ($Si = 0.21$ wt%) and for a hyper-Sandelin composition ($Si = 0.29$ wt%). In practice, the amount of Ni in the bath is maintained below 0.06 wt% to obtain the benefits of lower coating thickness shown in Fig. 15.14 but avoiding excessive amounts of T phase [25].

The quality of the Ni-containing bath is further improved by the addition of Al. Guttman and Niessen [46] found that an addition of 0.05wt%Al to the Zn bath modified the expected Zn-Al intermetallic layer to Zn-Al-Si for steels in the Sandelin activity range and delayed the formation of this layer, helping to moderate steel reactivity. As mentioned before, Al additions to the bath react with the pretreatment flux, so their addition must be limited, the high limit for Al in both Ni-containing and Ni-free galvanizing baths using conventional fluxes is 0.01 wt% [42]. As with other bath compositions, this small Al addition improves the brightness of the otherwise satin-appearing Zn-Ni coating and also helps moderate oxidation from the bath surface and therefore reduces top ash formation, by covering it with a thin, protective layer of Al oxide. The 2019 AGA process survey reported that 97% of respondents used Al additions, and 55% used Ni additions [6]. Al is always added to the bath as a prealloyed Zn-Al ingot, usually using Zn-5%Al, which has the lowest melting point and most uniform microstructure of any Zn-Al alloy. Overall, the amount of top ash produced during general galvanizing production is about the same as the amount of dross produced, about 6–7 kg per ton of steel galvanized [25].

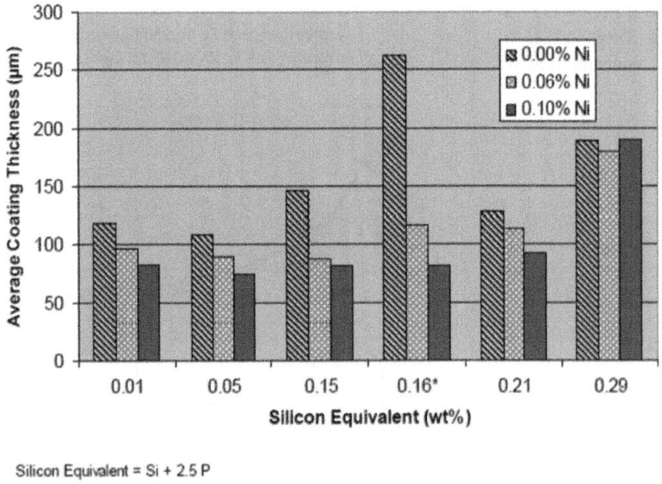

FIG. 15.14 Effect of bath Ni level on coating thickness of six steels [45].

The dross produced by eventual saturation of the ZnNiAl bath as a result of extended production has been reported to be the T phase indicated above, with a typical content of 4.6wt%Fe-2.3wt%Ni. Because of its low content in production baths (0.002%–0.005%), Al does not play a noticeable role in Zn-Ni dross. Because the density of this dross phase is not much different than that of the liquid Zn bath, it can be suspended for many hours before floating to the top of the bath, giving opportunities for dross pickup in the galvanized coatings [47]. Dross associated with bath alloy additions is minimized use of Zn-Ni ingots with a Ni composition less than the Zn-Ni eutectic composition of 0.25%Ni. Ingot microstructures in this composition range feature Zn primary crystals surrounded by Zn-Ni eutectic that has a slightly lower melting point, 418.5°C, than pure Zn [48]; therefore this range of compositions will melt, rather than dissolve, in the galvanizing bath. During typical galvanizing operations, the Ni content of the coating will be 0.1%–0.2%, depending upon steel reactivity, the dross will contain 0.3%–0.8% and the top ashes 0.04% [49]. Therefore, an ingot addition of 0.15% should compensate for the Ni consumed during galvanizing. As indicated before, the galvanizing bath is expected to be always saturated in Fe. As shown in Fig. 15.15, the addition of Ni lowers Fe solubility in Zn over the range of temperatures of interest. It is especially

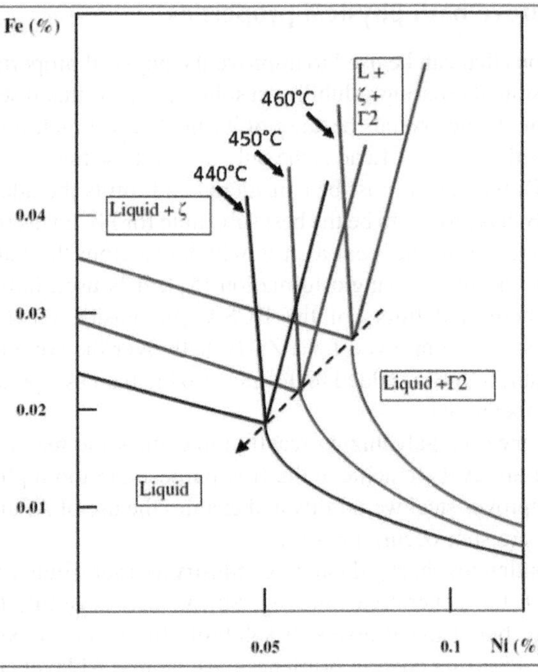

FIG. 15.15 Zn-rich corner of the Zn-Ni-Fe phase diagram, indicating solubility lines for 440°C, 450°C, and 460°C [49].

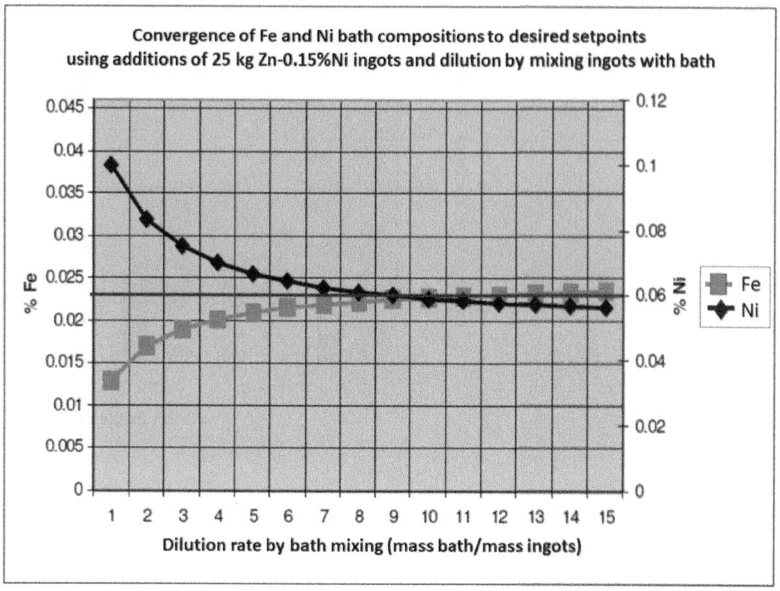

FIG. 15.16 Evolution of local Ni and Fe levels in a 450°C galvanizing bath into which a 25kg Zn-0.15% Ni ingot has been added. The aim concentration is 0.05% Ni [49].

important to avoid the formation of T phase, as indicated before this is the dross phase in the ZnNiAl bath saturated with Fe. Operating at 460°C rather than 440°C increases the solubility of Ni from 0.05% to 0.07% with a corresponding increase in Fe solubility. Therefore, the addition of a Zn-0.15% Ni ingot would be expected to produce T phase dross. In practice, this does not happen if dilution by mixing into the surrounding bath occurs as the Zn-Ni ingot gradually melts. This dilution behavior is shown in Fig. 15.16 for a bath held at 450°C. The local volume surrounding the ingot is diluted in Zn, lowering the local Fe content; however, the local Ni content is raised. As both ingot melting and bath mixing continue, local Fe content increases while local Ni content decreases. Good practice permits the Fe and Ni levels to remain below the solubility limit, even when a Zn-0.15% Ni ingot is used [49].

15.3.5 Alloy additions to improve bath physical properties

Fig. 15.2 shows the four alloy additions that can be used to improve the physical properties of the zinc bath. As mentioned before, lead reduces bath surface tension. Its sparing solubility in solid Zn means that it segregates to grain boundaries upon solidification, it also limits nucleation of zinc crystals because of its much lower melting point and therefore contributes to formation of large spangles on the coating surface. Lead is the only one of these three additions that is nonreactive with the steel. The reactivity of Sn, Sb, and Bi with steel must be considered and limits the additions of these alloys before steel properties can be affected. Of these, Bi has proven to be the best substitute for Pb. It was reported that an addition of 0.1wt% Bi has the effect on improving draining from the steel as it is withdrawn from the bath as 1wt%Pb [50]. Bismuth also reduces surface energy and, like Pb, promotes spangle formation [51]. It is used industrially as an alloy with 0.1wt% Bi, 0.4wt%Sn. This is within the permitted limits of the BCSA but outside the limits of the German regulations [17,18]. Kettle life was found to be slightly improved for a Zn-Bi bath over the composition range $0.15 < Bi < 4.5$ wt%, compared with a traditional Zn-Pb bath [52]. Bi is added to the galvanizing bath as a prealloyed ingot or Bi-shots to reduce melt losses and simply galvanizing operations.

Tin can be used both to the influence the galvanizing reaction to control the reactivity of Si-containing steels and to improve surface appearance, but the Sn levels to achieve the first purpose are too high to meet BSCA and German regulations [17,18]. At low levels, Sn improves steel wettability and enables the use of Al in Bi-Sn-Al galvanizing alloys with conventional fluxes by eliminating bare spot occurrences [45].

Antimony has been used in the continuous sheet galvanizing industry as a substitute for Pb when spangled appearance is desired and is described in Chapter 4. For general galvanizing, Sb can also substitute for Pb, and, like continuous sheet galvanizing, is used together with Al, but at lower levels. If solubility limits are exceeded, around 0.2wt%Sb at 450°C, precipitation of Sb_3Zn_4 can occur [53] causing coating brittleness. At the low Al levels used in general galvanizing baths, the Sb-Al precipitation that can occur in continuous galvanizing baths is well beyond this Sb composition.

15.3.6 Alloys to improve corrosion resistance

Because the Zn-Al alloy system does not form intermetallics but rather a series of solid solutions of each element in the other below the liquidus at any temperature, Zn alloys containing Al have been developed for many applications. This includes a wide range of Zn-Al coating alloys, mainly for continuous galvanizing of steel sheet, which offer improved corrosion resistance in comparison with coatings composed of nearly pure zinc. For general galvanizing, the development of Zn-Al alloy coatings has been limited by the constraints associated with the zinc-ammonium chloride flux; at low Al concentrations, Al forms an oxide that can interfere with flux action, and at higher concentrations can form appreciable quantities of aluminum chloride fume. Al-tolerant fluxes have been developed that have been proven to work with a wide range of Al compositions but are limited in industrial usage to Al compositions less than around 5% [54]. The improvement of corrosion resistance with Al additions has been ascribed to several effects [55]. First, Fe solubility in Zn decreases as Al increases. Zn-Fe precipitates coming from the customary Fe-saturated galvanizing bath are cathodic to the more Zn-rich coating constituents. The Fe solubility is reduced below 1 ppm for Al compositions around 5%. Second, in the multiphase coating structure produced in Zn-Al coatings, the eutectic portion, a finely layered structure of Al and Zn lamellae, plays an important role. The highest corrosion resistance was observed in coatings where the primary Zn-rich phase was enveloped by the Al-rich lamella in the eutectic phase. Corrosion resistance for coatings with different Al contents is shown in Fig. 15.17; corrosion testing was conducted in a cyclic corrosion test that included salt fog exposures. Higher Al coatings also benefited from the development of an aluminum oxide film in addition to other corrosion products considered in more detail in Chapter 11.

Upon dipping steel in a Zn-Al galvanizing bath, Fe-Al intermetallics are the first to grow from the steel/coating interface because of Fe diffusion into the liquid. Fe_2Al_5 is the most energetically favorable intermetallic compound to form. Although it is only expected to be found in the range $0.12 < Al < 1.7$ wt% in Fe-saturated baths at 460°C, it can be found in coatings produced in baths between 0.05 and 100% Al [28]. As dipping time proceeds, the thin interface layer of Fe_2Al_5 is destabilized by liquid Al diffusion and as noted before Fe solubility in the Zn melt is very low. This results in rapid growth of $FeAl_3$. Therefore, precise control of the galvanizing process is required. Galvanizing with higher Al content alloys is usually practiced in industry with series of steel articles of uniform section thickness and mass such as automobile underbody components. The varying residence times customary for mixed steels that make up jobbing work do not lend themselves to production of high Al content coatings when it is required that the coating microstructure can be consistently produced with uniform, thin FeAl intermetallic layers and metallic overlay coatings [56]. Related to this process is the semicontinuous production of galvanized reinforcing bar for concrete. Here, an Al-compatible flux is used, and rebar transported through a trough containing a Zn bath with an Al content between 0.14% and 0.3% at 460°C. The coating produced is similar to that produced on galvanized sheet with a thin Fe_2Al_5 inhibition layer and a zinc overlay coating [57].

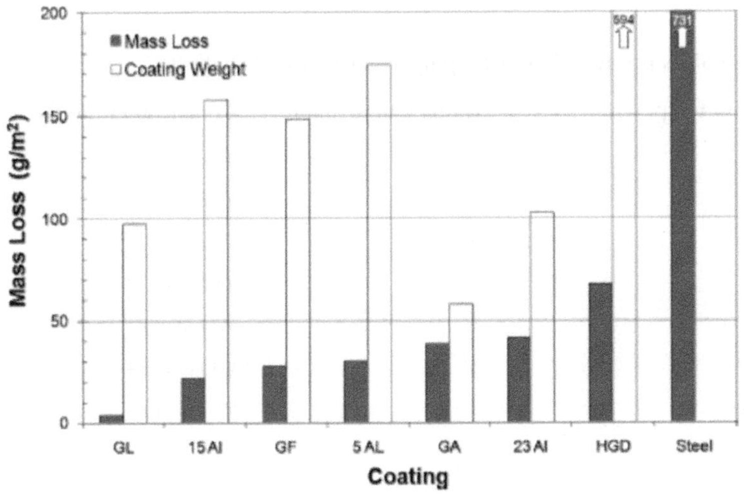

FIG. 15.17 Mass losses of various Zn-Al coatings in g/m^2 after 4 weeks of cyclic testing with a total fog spray time of 221 h. GL = Zn-55%Al, GF = Zn-5%Al, HDG = conventional hot-dip galvanized coating [55].

The corrosion resistance of ZnAlMg coatings is described in Chapter 11. While these alloy coatings for steel sheet have reached full industrial status, they are only in early stages of development for the general galvanized process. Approaches have included dipping steel articles in an Al-Mg fluidized bed after first dipping in liquid Zn, and a double dip process where steel articles are first dipped in liquid Zn and then into a Zn-5%Al-1%Mg bath without intermediate application of a flux [58,59]. A European patent has been issued for a dry process flux compatible with ZnAlMg alloys, but this is not known to be used in industrial practice [60].

In practice, general galvanized articles and their coatings are customarily produced to be in conformance with the ISO 1461 standard specification, which requires that the hot-dip galvanizing bath shall primarily contain molten zinc. This standard also states that other elements added to the bath, as described in standards ISO 752, EN 1179, or EN 13283, excluding Sn and Fe, shall not exceed 1.5% by mass. As described in Chapter 7, the viscosity and density of alloys with appreciable Al additions are lower than those of pure zinc, making it difficult if not impossible to meet the 85 μm mean minimum coating thickness required by ISO 1461 with these alloys. Therefore, the alloys described in this section have so far only been used in niche applications.

15.3.7 Alloying effects on surface color

Alloying Zn with 0.2% Ti has been found to change the color of the coating from the usual silver-gray of zinc to a variety of others [61]. The usual pretreatments were used, but bath temperatures ranged between 560°C and 640°C. When a typical nonreactive steel (0.014wt%Si, 0.005wt%P) of either 2 or 4 mm thickness was dipped in the bath for 3 min, colored surfaces were obtained that depended upon bath temperature as shown in Table 15.5.

All of the produced coatings have the same phase makeup and other bulk coating features. The color difference is thought to be caused by the development of surface oxides of different thicknesses that give an optical interference effect.

15.4 Hydrogen effects in general galvanized articles

Most steels that are batch galvanized are lower-strength (<550 MPa tensile strength) hot-rolled grades with microstructures consisting of ferrite and pearlite. Because these microstructures have minimal harder phases such as martensite or bainite, and few contained dislocations or high-angle grain boundaries, they therefore do not have a tendency to trap hydrogen when it is introduced during the pickling and fluxing steps. Hydrogen can also be introduced during the decomposition of the ammonium chloride component of the flux. The hydrogen diffusion coefficient in ferrite at 460°C is around 10^{-7} m^2/s [62]. This is 1000 times faster than the diffusion of solid Fe into liquid Zn described before, which is a key part of the galvanizing reaction. Therefore, hydrogen diffuses into steel quickly during pickling and fluxing and can exit quickly during flux drying and dipping in zinc unless it is caught in traps. Its exit during the galvanizing process can be manifest in a foamy coating appearance and can also be associated with separation of the gamma phase from its interface with the steel [7]. This phenomenon is especially associated with steel compositions within the Sandelin peak [63].

TABLE 15.5 Effect of bath temperature on surface colors produced after dipping in Zn-0.2wt%Ti bath [61].

Color hues	Sample dimensions (mm)	Bath temperature (°C)	Dipping time (min)	Cooling after dipping
Golden yellow	50 × 20 × 2	560	3	WQ after 3 s
Purple	50 × 20 × 2	580	3	WQ after 3 s
Blue	50 × 20 × 2	600	3	WQ after 3 s
Light Blue	50 × 20 × 2	610	3	WQ after 3 s
Light Yellow	50 × 20 × 2	620	3	WQ after 4 s
Pink	50 × 20 × 2	630	3	WQ after 5 s
Green	50 × 20 × 4	640	3	WQ after 6 s

WQ, water quench.

Exposure of higher-strength steels to pickling and fluxing treatments can result in trapping of hydrogen at the sites of dislocations and dislocation-rich constituents such as martensite and bainite lath boundaries. The thermal activation required to release this hydrogen requires temperatures higher than those associated with the galvanizing process. If martensite or bainite is present in the steel, there is also a risk of tempering if the process temperature exceeds the tempering temperature. For several grades of quench and tempered steels such as ASTM A193 Grade B7 and F1554 Grade 105, the minimum tempering temperature is 590°C, so that there is only a risk of lowering strengths if the high-temperature galvanizing process, with a bath temperature of 550°C, is used. Many other quenched and tempered steels such as ASTM A325, A349, and A354 Grade BC have minimum tempering temperatures around 425°C. If these are hot-dip galvanized, some effect on strength reduction must be anticipated. Many higher-strength grades can be hot-dip galvanized without risk of hydrogen embrittlement. Avoiding the pickling and fluxing process steps prevents exposure to hydrogen, the steel surface in this case is cleaned by abrasive blasting. This has been successfully applied in European plants to bolts covered by ASTM A490 with tensile strengths between 1034 and 1192 MPa [64], but hot-dip galvanizing is still not permitted according to this standard.

15.5 Management of general galvanizing baths

Good composition control of the galvanizing bath depends upon good bath analysis and ingot addition practices. The patterns of temperature and composition within the bath vary greatly as steel is processed through it because of the need to heat the steel in the bath before the galvanizing reaction can commence. The bath heating system reacts to restore temperature, resulting in an upward flow of the bath at the heated wall surface and downward in the cooled volume of zinc surrounding the steel being galvanized. The distribution of bath alloying elements is influenced by this thermal convection and also their reactions with the steel being galvanized and, when supersaturated, the formation of dross. It is therefore important to take bath samples at times when the bath is stable. Fig. 15.18 shows the distribution of four common elements in a 2.6 m deep general galvanizing bath: Pb, Bi, Al, and Ni, after a "rest period" when no work was processed, and the bath was allowed to stabilize. This is for a typical Ni-containing bath used to process a mix of reactive and nonreactive steels. The stratification of Pb and Bi is expected because of the higher densities of these elements compared with Zn, but there are also regions within the bath cross section where there is depletion of these elements. Ni is closer in density to Zn and shows intermediate stratification, while Al concentrations are higher in the top half of the bath.

The composition of the galvanizing bath during actual operations is also important to understand, to help correlate the condition of produced coatings with galvanizing process conditions. Fig. 15.19 shows a systematic way of measurement: bath samples are taken (1) just before zinc and alloy additions are made to the bath, (2) just after the completion of melting of all additions, (3) after three cycles of hot-dip galvanizing, and (4) just before the next round of zinc and alloy additions are made to the bath. This usually occurs over an 8-h working shift. This high sampling frequency is normally only done to help understand the behavior of a specific bath or to aid in troubleshooting; a more typical sampling frequency for a bath operating under normal conditions is once each month [65].

Results of chemical analysis made be inductively coupled plasma spectroscopy (ICP) are shown in Table 15.6 for three locations (A, B, C) in the bath. There are slight variations in moving from the first through to the fifth measurement for each of the elements but overall, the bath is quite stable.

The variation of Al and Ni composition from another series of measurements is shown in Fig. 15.20; the reader will note that the scale of Ni measurements is much finer that that shown in Table 15.5, meaning that the composition is quite stable. The composition readings agree with expectations, i.e., an increase in composition after all metal additions have been melted, followed by depletion as work is processed through the bath (Fig. 15.20).

Moving out to a longer timescale, the variations in bath composition over a period of 15 months is shown in Fig. 15.21. Some drifts in composition, followed by the results of corrective actions, are seen.

Chemical analysis of general galvanizing baths is conducted either with ICP or Optical Emission Spectroscopy (OES). OES is generally less accurate but can provide satisfactory results if reliable standards near to the unknow composition are used. A major source of analysis error is the presence of dross in the sample subjected to chemical analysis. This can give readings for Fe, Ni, and Al that are far above expectations. Fig. 15.22 shows coincident high readings for Fe and Ni that were taken when a dross particle was included in the chemical analysis. Good sampling procedures and repeated analysis to obtain confidence in true readings will improve accuracy and better inform any needed changes to the customary zinc and alloy addition schedule.

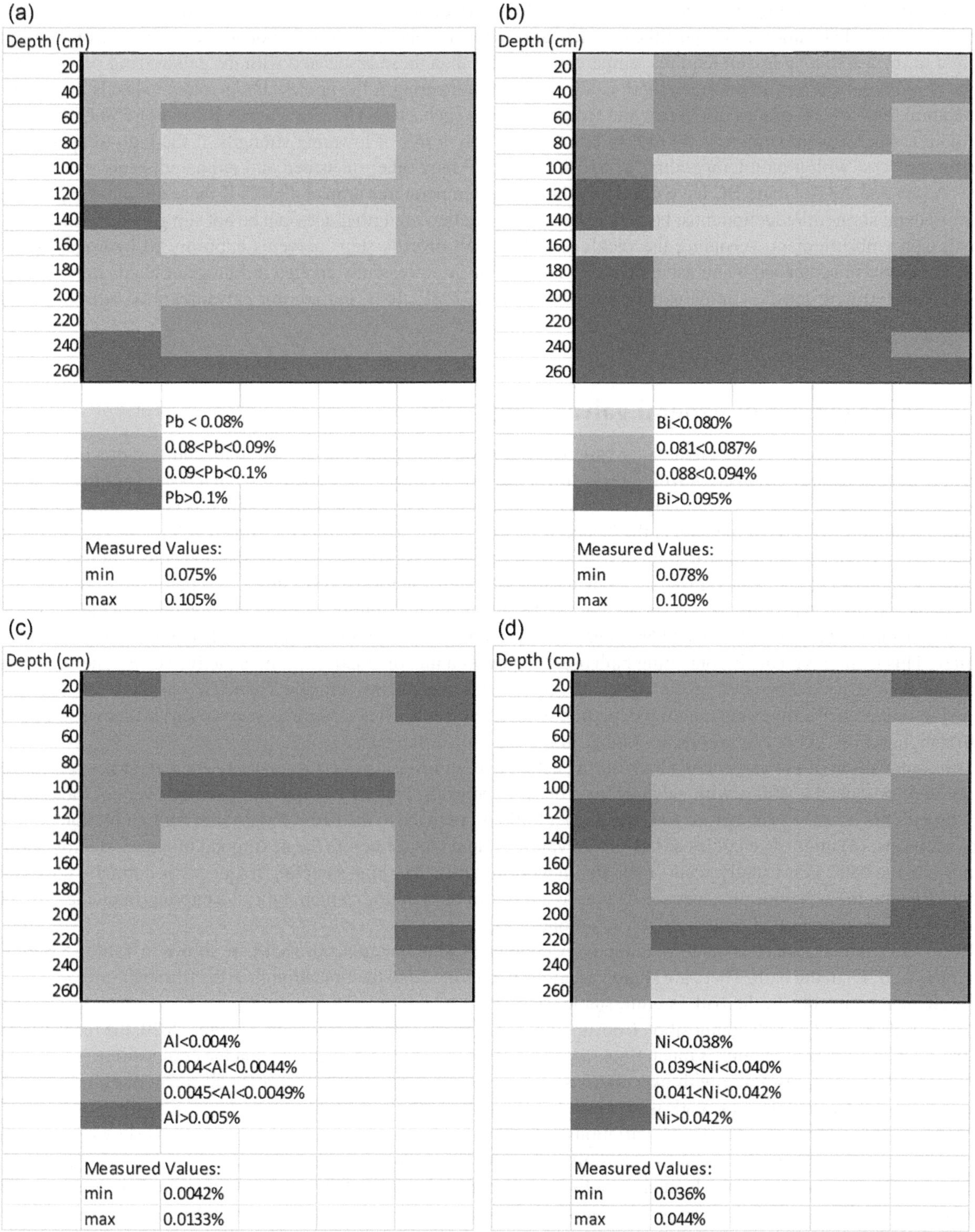

FIG. 15.18 Distribution of alloy elements in a 2.6 m deep galvanizing bath containing Ni, Al, Bi, and Pb after a "rest period" when no work was processed [65]. (A, B) Pb and Bi distributions and (C, D) Al and Ni distributions.

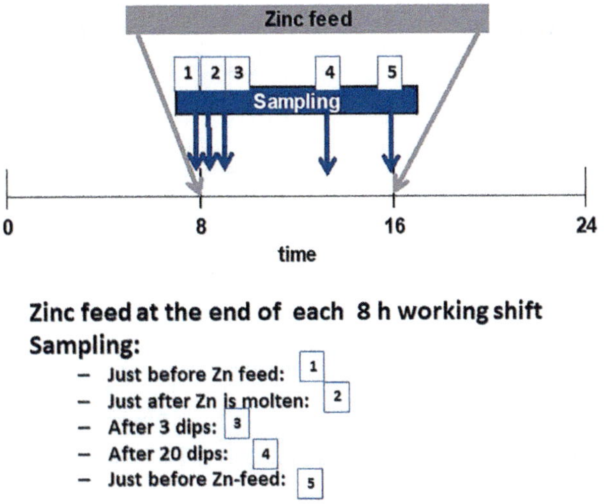

FIG. 15.19 Bath sampling schedule over an 8-h working shift of hot-dip galvanizing, with zinc and alloy additions before and after the shift [65].

TABLE 15.6 Results of ICP analysis for three bath locations over an 8-h working shift, for the time points 1 through 5 described in Fig. 15.19 [65].

Alloy element	Al	Fe	Ni	Cu	Cd	Sn	Pb	Bi
Sample	Composition (100 × weight percent)							
A1	0.45	2.4	4.1	1.3	<0.05	0.15	0.71	8.8
A2	0.56	2.5	4.1	1.3	<0.05	0.15	0.72	8.8
A3	0.47	2.5	4.1	1.3	<0.05	0.15	0.72	8.9
A4	0.44	2.5	4.1	1.3	<0.05	0.15	0.71	8.8
A5	0.44	2.5	4.1	1.3	<0.05	0.15	0.72	8.9
B1	0.43	2.3	4.1	1.3	<0.05	0.15	0.7	8.8
B2	0.46	2.6	4.2	1.4	<0.05	0.15	0.72	8.9
B3	0.45	2.7	4.1	1.3	<0.05	0.15	0.71	8.7
B4	0.43	2.4	4.1	1.3	<0.05	0.14	0.72	8.7
B5	0.43	2.5	4.2	1.3	<0.05	0.15	0.73	8.9
C1	0.44	2.2	4.2	1.3	<0.05	0.14	0.71	8.6
C2	0.44	2.4	4.2	1.3	<0.05	0.15	0.73	8.8
C3	0.47	2.4	4.2	1.3	<0.05	0.15	0.72	8.9
C4	0.44	2.3	4.1	1.3	<0.05	0.14	0.72	8.8
C5	0.43	2.3	4.1	1.3	<0.05	0.15	0.72	8.8

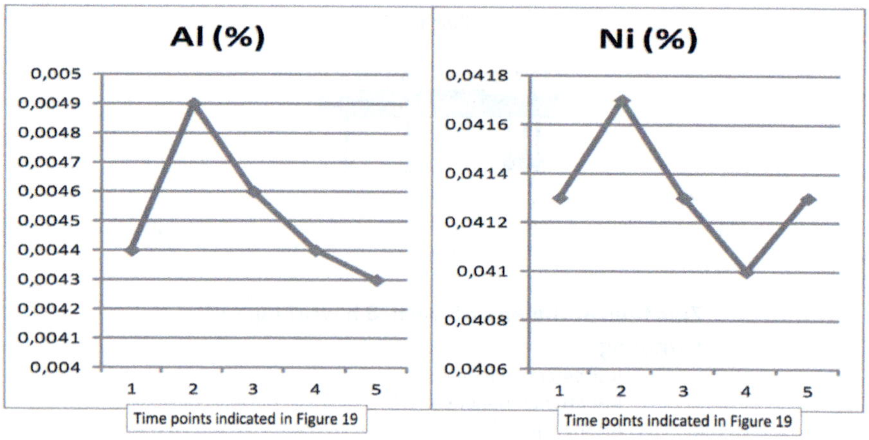

FIG. 15.20 Variations in Al and Ni compositions in a general galvanizing bath over an 8-h working shift, for the time points 1 through 5 described in Fig. 15.19 [65].

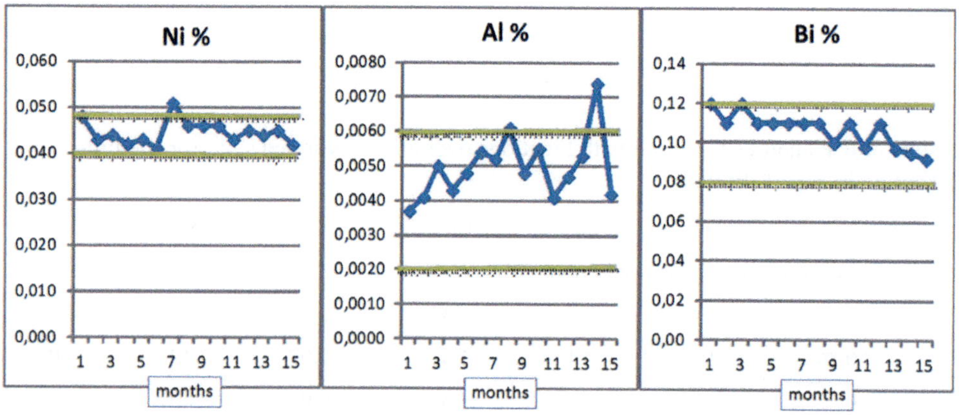

FIG. 15.21 Trends of bath composition over a 15-month period [65].

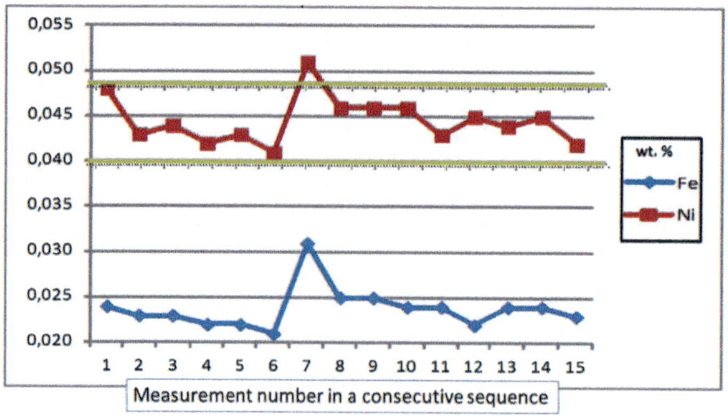

FIG. 15.22 High reading 7 indicates that a dross particle containing Ni and Fe was included in the bath chemical analysis [65].

References

[1] A.R. Marder, The metallurgy of zinc-coated steel, Prog. Mater. Sci. 45 (2000) 191–271.

[2] American Galvanizers Assn., Zinc Coatings – A Comparative Analysis of Process and Performance Characteristics, American Galvanizers Assn., Centennial, CO, 2011.

[3] F. Schmelz, Chemical pre-treatment of steel in the hot dip galvanizing process with focus of flux and drying conditions, in: Proc. 25th International Galvanizing Conf. (Intergalva 2018), June 17–22, European General Galvanizers Assn., Solihull, UK, 2018, pp. 137–148. Berlin.

[4] ISSA State Phosphate Survey, National Sanitary Supply Assn., Rosemont, IL, 2015 https://www.issa.com/wp-content/uploads/phosphate_survey2015.pdf.

[5] J. Kader, New environmentally friendly treatment for hot dip galvanizing, in: Proc. 24th International Galvanizing Conf. (Intergalva 2015), June 7–12, European General Galvanizers Assn, Solihull, UK, 2015. Liverpool.

[6] American Galvanizers Association, 2019 Galvanizing Process Survey, American Galvanizers Assn., Centennial, CO, 2019.

[7] V. Kuklík, J. Kudláček, Hot-Dip Galvanizing of Steel Structures, Butterworth-Heinemann, Oxford, UK, 2016.

[8] D. Sequeira, D. Balcer, L.R. Pignotti, J.C. Malloy, D. Price, O. Lanzi, Improved descaling method for advanced high-strength steels (AHSS) using fused conditioning salt, Iron Steel Technol. 15 (2) (2018) 32–38.

[9] H. Bablik, Galvanizing (Hot-Dip), E.F. Spon, London, 1950.

[10] R. Pankert, Influence of alloying elements on galvanizing abnormalities with a special view of surface defects, in: Proc. 25th International Galvanizing Conf. (Intergalva 2018), June 17–22, European General Galvanizers Assn., Solihull, UK, 2018, pp. 181–190. Berlin.

[11] H.R. Thresh, D.W.G. White, J.O. Edwards, J.W. Meier, Properties of molten zinc and zinc alloys—final report, in: International lead Zinc Research Organization Project ZM-54, October 15, 1964. Available from: International Zinc Assn., Durham, NC.

[12] B. Dugan, private communication, April 2021.

[13] S. Sepper, P. Peetsalu, P. Kulu, M. Saarņa, V. Mikli, The role of silicon in the hot dip galvanizing process, Proc. Est. Acad. Sci. 65 (2) (2016) 159–165. https://kirj.ee/public/proceedings_pdf/2016/issue_2/proc-2016-2-159-165.pdf.

[14] D. Balloy, A.-L. Cristol, C. Niclaeys, P. Quaegebeur, L. Néel, Experimentation and modelling of thermo-mechanical phenomena that appear during hot dip galvanizing, in: Proc. 22nd International Galvanizing Conference (Intergalva 2009), June 7–11, European General Galvanizers Assn., Solihull, UK, 2009. Madrid, Spain.

[15] Y. Gerenrot, D. Leychkis, T.L. Ranck, J.L. Griffin, G. Stefanick, Fluxes for Lead-Free Galvanizing, US Patent 5,437,748, issued August 1, 1995.

[16] T. Kinstler, Current knowledge of the cracking of steels during galvanizing—a synthesis of the available technical literature and collective experience, in: Proc. 21st International Galvanizing Conference (Intergalva 2006), June 11–16, European General Galvanizers Assn., Solihull, UK, 2006. Naples, Italy.

[17] Galvanizing Structural Steelwork – An Approach to Management of Liquid Metal Assisted Cracking, British Constructional Steelwork Association and UK Galvanizers Association, London, 2005.

[18] M. Huckshold, Improving design guidance to avoid cracking of galvanized structural steelwork, in: Proc. of the 22nd International Galvanizing Conference (Intergalva 2009), June 8–12, European General Galvanizers Assn., Solihull, UK, 2009. Madrid.

[19] C. Leighfield, M.N. James, The conclusion from one million tonnes of experience in galvanizing steel – LMAC is not a primary instigator of cracking, Eng. Fail. Anal. 106 (2019) 104151.

[20] D. Horstmann, Reactions between liquid zinc and silicon-free and silicon-containing steels, in: Proc. Seminar on Galvanizing of Silicon Containing Steels, May 21–22, Centre de Recherches Metallurgiques and International Lead Zinc Research Org., Inc., 1975, pp. 86–95. Liège, Belgium.

[21] Z.W. Chen, N.F. Kennon, J.B. See, M.A. Barter, Technigalva and other developments in batch hot-dip galvanizing, J. Met. 44 (1992) 22–26.

[22] AK Steel International Product Bulletin, Armco® Pure Iron High Quality Iron, December 2018. www.aksteel.eu.

[23] Galvanizing Kettles—Recommendations for the User, English translation of German brochure, Institut für Stahlbau, Leipzig, 1991.

[24] R.C. Amistadi, R.G. Konecny, C.S. Young, Experiments in galvanizing silicon-killed steels, in: Proc Seminar on Galvanizing of Silicon Containing Steels, May 21–22, Centre de Recherches Metallurgiques and International Lead Zinc Research Org., Inc., Liège, Belgium, 1975, pp. 150–167.

[25] R. Pankert, private communication, September 2021.

[26] M. Ubiali, What advanced computerized simulation is showing us about high velocity furnaces, in: Presented at 24th International Galvanizing Conf. (Intergalva 2015), June 7–12, European General Galvanizers Assn., Solihull, UK, 2015. Liverpool.

[27] F. Ajersch, P. Bouchard, M. Sabsabi, F.E. Goodwin, Al and Fe solubility limits and characterization of dross particles during galvanneal and galvanized operations, in: Proceedings Galvanizers Association 99th Meeting, October 21–14, Galvanizers Association, Marine City, MI, 2007. Montreal, Quebec.

[28] M. Guttmann, M., Diffusive phase transformations in hot dip galvanizing, Mater. Sci. Forum 155–156 (1994) 527–548.

[29] S. Yamaguchi, Thermochemical stability and precipitation behavior of dross phases in CGL bath, in: Proc. 4th International Conference on Zinc and Zinc Alloy Coated Steel Sheet (Galvatech 1998), September 20–23, Iron and Steel Institute of Japan, Tokyo, 1998, pp. 84–89. Makuhari, Chiba, Japan.

[30] J.S. Dohie, J.R. Cahoon, W.F. Caley, The grain-boundary diffusion of Zn in α-Fe, J. Ph. Equilibria Diffus. 28 (4) (2007) 322–327, https://doi.org/10.1007/s11669-007-9093-y.

[31] M.-L. Giorgi, P. Durighello, R. Nicolle, J.-B. Guillot, Dissolution kinetics of Iron in liquid zinc, J. Mater. Sci. 39 (2004) 5803–5808.

[32] N.-Y. Tang, Control of silicon reactivity in general galvanizing, J. Ph. Equilibria Diffus. 29 (4) (2008) 337–344.

[33] G.F. Bastin, F.J.J. van Loo, G.D. Rieck, A new compound in the Iron-zinc system, Z. Metallkd. 65 (1974) 657–660.

[34] J. Pelerin, J. Hoffmann, V. Leroy, The influence of silicon and phosphorus on the commercial galvanization of mild steel, Metall 35 (1981) 870–873.

[35] J.J. Sebisty, G.E. Ruddle, Pretreatment effects on galvanizing silicon-containing steels, in: Proc Seminar on Galvanizing of Silicon Containing Steels, May 21–22, Centre de Recherches Metallurgiques and International Lead Zinc Research Org., Inc., Liège, Belgium, 1975, pp. 219–245.
[36] X.P. Su, N.-Y. Tang, J.M. Toguri, 450°C isothermal section of the Zn-Fe-Si ternary phase diagram, Can. Metall. Q. 40 (3) (2001) 377–384.
[37] C.-S. Che, J.-T. Lu, G. Kong, Interpretation on sebisty effect of hot-dip galvanized steels, Trans. Nonferrous Met. Soc. China 5 (6) (2005) 1275–1279.
[38] O.S. Bondereva, A.A. Melnikov, Effect of the silicon content in steel on the hot-dip zinc coating microstructure formation, Mater. Sci. Eng. 156 (2016), 012015, https://doi.org/10.1088/1757-899X/156/1/012015.
[39] W.-D. Schulz, M. Thiele, General Hot-Dip Galvanizing Materials-Technologies-Layer Formation-Properties-Defects, Leuze Verlag, Bad Salgau, Germany, 2012.
[40] The Galvanizing Manual, St. Joe Minerals Corp., Monaca, PA, 1974.
[41] ISO 1461, Hot Dip Galvanized Coatings on Fabricated Iron and Steel Articles – Specifications and Test Methods, third ed., International Organization for Standardization, Geneva, Switzerland, 2009.
[42] H. Kania, J. Mendala, J. Kozuba, M. Saternus, Development of bath chemical composition for batch hot-dip galvanizing—a review, Materials 13 (18) (2020) 4168, https://doi.org/10.3390/ma13184168.
[43] K.E. Daehn, A.C. Serrenho, J.M. Allwood, How will copper contamination constrain future steel recycling, Environ. Sci. Technol. 51 (11) (2017) 6599–6606, https://doi.org/10.1021/acs.est.7b00997.
[44] V. Raghavan, Fe-N-Zn (iron-nickel-zinc), J. Ph. Equilibria Diffus. 28 (4) (2007) 394, https://doi.org/10.1007/s11669-007-9107-9.
[45] J. Zervoudis, G. Anderson, A review of bath alloy additives and their impact on the quality of the galvanized coating, in: Proc. 6th Asia Pacific General Galvanizing Conference, May 29-June 2, Asia Pacific General Galvanizing Association, Melbourne, Australia, 2005. Cairns, Australia.
[46] H. Guttman, P. Niessen, Galvanizing of silicon steels in aluminum containing baths, in: Proc Seminar on Galvanizing of Silicon Containing Steels, May 21–22, Centre de Recherches Metallurgiques and International Lead Zinc Research Org., Inc., Liège, Belgium, 1975, pp. 198–216.
[47] M. Taylor, S. Murphy, A decade of technigalva, in: Proc. 18th International Galvanizing Conf. (Intergalva 1997), June 8–11, European General Galvanizers Assn., Solihull, UK, 1997. Birmingham, UK.
[48] P. Nash, Y.Y. Pan, The Ni-Zn (nickel-zinc) system, J. Ph. Equil.: Bull. Alloy Phase Diagr. 8 (5) (1987) 422–430.
[49] R. Pankert, Use of tailor-made alloys in general galvanizing, in: Proc. 22nd International Galvanizing Conference (Intergalva 2009), June 7–11, European General Galvanizers Assn., Solihull, UK, 2009. Madrid, Spain.
[50] M. Gagne, Hot-dip galvanizing with zinc-bismuth alloys, Metall 53 (1999) 269–271.
[51] M. Gagne, The effect of bath alloy additions on the grain size (spangle) of galvanized steel coatings, in: American Galvanizers Association Tech Forum 2002, September 25–27, American Galvanizers Assn., Centennial, CO, 2002. Toronto.
[52] J. Treibert, T. Klein, S. Six, Lead-free and Bismuth-containing zinc melts and kettles in hot-dip galvanizing, in: Proc. 25th International Galvanizing Conf. (Intergalva 2018), June 17–22, European General Galvanizers Assn., Solihull, UK, 2018, pp. 201–212. Berlin.
[53] S. Peng, S.-K. Xie, Y.-M. Yang, G.Y. Zhao, J.-T. Lu, L.-C. Zhang, Aluminum and antimony segregation on a batch hot-dip galvanized Zn-0.05Al-0.2Sb coating, J. Alloys Compd. 694 (2017) 1004–1010.
[54] D. Leychkis, J. Zervoudis, Flux and Process for Hot Dip Galvanization, U.S. Patent 7,811,389, issued October 12, 2010.
[55] N.-Y. Tang, Y. Liu, Corrosion performance of aluminum-containing zinc coatings, ISIJ Int. 50 (3) (2010) 455–462.
[56] M. Memmi, M. Cecchini, C. Colibri, F.E. Goodwin, New batch Zn-5% hot dip coating process for fabricated steel parts, in: Proc. 21st International Galvanizing Conf. (Intergalva 2006), June 11–16, European General Galvanizers Assn, Solihull, UK, 2006. Naples, Italy.
[57] M. Stroia, D. Liu, A continuous galvanizing process for long steel products, AISTech 2020, Proceedings of the Iron & Steel Technology Conference pp. 1284–1299. DOI 10.33313/380/13.
[58] Y. Kiyokawa, New zinc-aluminum-magnesium galvanizing technology, in: Proc. 25th International Galvanizing Conf. (Intergalva 2018), June 17–22, European General Galvanizers Assn., Solihull, UK, 2018, pp. 213–217. Berlin.
[59] T. Morooka, Application of hot dip Zn-Al-Mg alloy coatings to steel structures, in: Proc. 24th International Galvanizing Conf. (Intergalva 2015), June 7–12, European General Galvanizers Assn., Solihull, UK, 2015. Liverpool.
[60] D. Warichet, J. Balduyck, C. Masquelier, Continuous single-dip process for galvanization of steel long products into Zn-Al-Mg alloys, European Patent EP 2725116, issued March 14, 2018.
[61] S. Natali, V. Volpe, L. Zortea, C. Burattini, V. Di Cocco, F. Iacoviello, Mechanical and structural characterization of Zn-Ti colored coatings, Proc. Eng. 109 (2015) 105–112.
[62] H.K.D.H. Bhadeshia, Prevention of hydrogen embrittlement in steels, ISIJ Int. 56 (1) (2016) 24–36.
[63] W.-D. Schulz, P. Schubert, M. Thiele, An alternative approach to explaining the effect of silicon on the galvanizing reaction, in: Proc. 20th International Galvanizing Conf. (Intergalva 2003), June 1–3, European General Galvanizers Assn, Solihull, UK, 2003. Amsterdam.
[64] T. Langill, Hydrogen Embrittlement Study of Hot-Dip Galvanized High-Strength Bolts, Fasten. Technol. Int. (April 2013) 40–43.
[65] R. Pankert, M. Jungar, Bath homogeneity in general galvanizing baths, in: Proc. 23rd International Galvanizing Conf. (Intergalva 2012), June 10–15, European General Galvanizers Assn., Solihull, UK, 2012. Paris.

Further reading

J.-Y. Dauphin, P. Perrot, U.G. Tchissambot, Thermodynamic description of the iron-zinc system, Mèm. Sci. Rev. Mèt. 84 (6) (1987) 329–336.
R.W. Richards, H. Clarke, F.E. Goodwin, Analysis of galvanized coatings, in: Proc. 17th International Galvanizing Conf. (Intergalva 1994), June 6–8, European General Galvanizers Assn., Solihull, UK, 1994. Paris.

Chapter 16

New opportunities

16.1 Introduction

New opportunities for zinc-coated steel will depend upon the continued development of the capabilities of existing coating process, a more complete integration of these processes with the overall steel sheet manufacturing process, the implementation of entirely new processes, and the improvement of products to meet increasingly challenging user requirements. Environmental performance challenges will increase in importance, both in improving energy efficiency and waste reduction and a significant reduction in carbon emissions.

16.2 Environmental performance

In 2011, the Council of the European Commission published a roadmap for attaining a competitive low-carbon economy by 2050. According to this, European industry, of which the steel industry is only one part, would have to reduce its CO_2 emissions below 1990 levels 80–95% by the year 2050. This is an enormous challenge. There are two strategies for reaching massive CO_2 mitigation in the steel industry: "Smart Carbon Usage" (SCU) for processes that continue to utilize carbon in fuels and steel processing reactions and "Carbon Direct Avoidance" (CDA), which uses electricity and hydrogen instead of carbon. All processes and scenarios required to reach the 95% CO_2 mitigation target by 2050 compared with 1990 involve an enormous increase in consumption of "green" electric energy, "green" hydrogen and biomass are necessary. A significant increase in use of these sources of energy will also be required for CO_2 mitigation measures that are expected to be achieved by applying so-called "end-of-pipe" technologies such as CCS (CO_2 Capture and Storage) and CCU (Carbon Capture and Usage) [1]. Other jurisdictions have also published roadmaps and reporting requirements with the intention of reducing the carbon footprint of the steel manufacturing process, for example, the USA [2].

It has been estimated that 50–60 kg CO_2 is produced from the customary steps of furnace pretreatment, hot-dip galvanizing, and allied processes in the continuous hot-dip galvanizing line [3]. In continuous galvanizing, CO_2 is directly produced as a result of combustion of gases to heat furnaces and indirectly in the use of electricity produced by combustion of carbonaceous fuels. Opportunities to improve the environmental performance of hot-dip galvanizing exist in improving the efficiency of combustion processes within the gas-fired burners used to heat furnaces. For example, when high percent silica content ceramic inserts made in a twisted channel geometry using additive manufacturing were installed in radiant heating tubes, improvements of up to 15% in productivity were gained versus tuned burners without inserts at the same level of fuel consumption and emissions [4]. Replacement of gas-fired radiant tubes with electrically heated tubes can reduce the carbon burden of the process if energy is generated from "green" sources. Also, induction heating can replace many heating requirements in the continuous galvanizing processes and can be especially efficient when the sheet steel is in the ferromagnetic state. For example, a 3.3 MW transverse flux inductor was used at an Arcelor Mittal facility in the annealing furnace for advanced high-strength steels with ultimate tensile strengths in the range of 980–1180 MPa [5].

There are continuing opportunities for zinc-coated steel sheet to contribute to improving the performance of automobiles. Since the 1980s, zinc-coated steels have played a role in greatly extending the service life of vehicles by enabling corrosion-resistant bodies and structures to be produced rapidly and at low cost. During succeeding years, the development of zinc-coated AHSS enabled steel-based vehicles to achieve significant mass reduction without compromises in safety, durability, and cost. The advent of electric vehicle mass production has opened new opportunities for zinc-coated sheet steel. The addition of traction batteries has resulted in significant increases in vehicle mass, for example, the Ford F-150 Lightning, and all-electric pickup truck, increased its mass by 656–818 kg compared with the same model powered by an internal combustion engine [6]. A similar trend is seen in the 2022 Ford Mustang Mach-E that realized a 36% mass gain. Equally important, the allowable crush space used to accommodate crash-related deformations is 10% less in the Mustang [7]. These developments will require steels that can absorb the much higher crash energies associated with

The Metallurgy of Zinc Coated Steels. https://doi.org/10.1016/B978-0-323-99984-7.00004-X

increased vehicle mass in a significantly smaller space. This will require steels with both higher yield strengths and energy absorption capabilities, and for long-term protection, these steels must be protected from corrosion. Opportunities to develop zinc-coated steel sheet grades that meet these increasingly demanding requirements will need to be met to ensure that steel remains the material of choice for such vehicles. Alternatives such as composite materials as of now cannot meet existing requirements for manufacturing economics, including speed of vehicle assembly and materials costs, which are expected from zinc-coated sheet steel [8]. Examples of emerging steel grades that will provide challenges for coating with zinc will be discussed later in this chapter.

16.3 Process integration and quality assurance—"Industry 4.0"

Fig. 16.1 shows the levels of controls used in manufacturing processes [9]. In recent years, controls at each level have increasingly relied on monitoring and corrective actions involving computers. Increasingly, each measurement and control detail results in the generation of data concerning the status of that detail can be input to models at the same or higher levels to improve productivity and reduce the production of defective or out-of-specification material.

In the hot-dip galvanizing line, Level 2 systems for coating thickness control were described in Section 7.10 of Chapter 7. An example of a Level 2 system for control of the galvanizing pretreatment furnace is based on the inputs shown in Fig. 16.2. Here, heat exchange calculations permit strip temperature to be measured at various points as it passes through the furnace zone. It is sufficient to model a single value for temperature in the direction of strip travel. The accuracy of the model is confirmed by running a wide variety of strip temperatures, widths, and speeds. The difference between a Level 1 controller such as a PID (proportional–integral–derivative) feedback controller and Level 2 is that the Level 2 controller has the capability of adjusting model variables to adjust for inaccuracies caused by perturbations and the fact that the system is only partially described by the input measurements, there are many other variables influencing the results that are not measured. This permits predictive control to be realized. An example is given in Fig. 16.3, where changes in line speed are shown in the top graph, the total power delivered to the furnace is shown in the lower graph. The Level 2 system takes furnace inertia into consideration when adjusting furnace power, the power adjustments are not as sudden as in the case of a PID controller [10].

An example of a Level 3 system is the control of line variables in response to a coil change, shown in Fig. 16.4. Here, the strip width and thickness, heat cycle requirements, emissivity, and line speed might all change. As in the above example, predictive control is used to minimize the disturbance of temperature during the transition. The effect of heat exchange between the furnace rolls and the strip is also noted.

This same model also includes a coil scheduling capacity, taking it to a Level 4 capability. Level 4 tasks include the scheduling of coils for production, grouping compatible grades, widths, and thicknesses [11].

Of course, the processing of the steel used in a galvanized sheet occurs not only on the galvanizing line, but in the prior steps including the refining of the liquid steel, casting, soaking, and treatments in the hot and cold mills. Streams of processing data are becoming available from each step of the processing sequence, enabling the "genealogy" of each coil to be tracked and recorded. This has enabled variations in different process steps to be related to out-of-specification and defect occurrences. For example, an electrogalvanized sheet steel coating "spear head" defect was related to details of slab

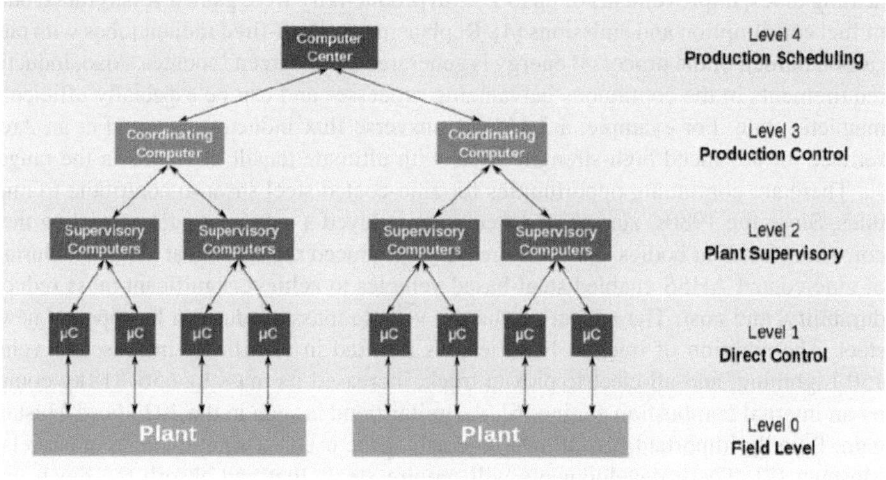

FIG. 16.1 Levels of control used in manufacturing processes [9]. Level 0 refers to measurements made in the process, Level 1 to the microcontrol of machinery or apparatus that affects the Level 0 measurements. Level 2 involves algorithms, including predictive control programs, which influence the actions of Level 1 controls. Level 3 involves higher levels of production control, including conditions associated with incoming material. Level 4 includes production scheduling that determines the nature of incoming material used in Level 3 controls [9].

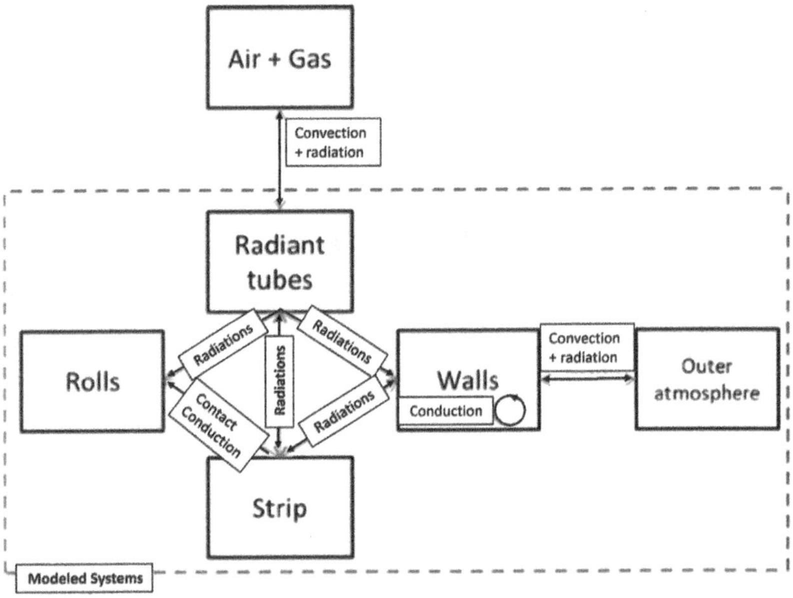

FIG. 16.2 Modeled components of a Level 2 furnace control system [10]. *(C. Perottet, How model predictive control enhances strip temperature regulation, in: Proceedings Galvanizers Association 111th Meeting, October 6–9, 2019, Charleston, SC, Galvanizers Association, Marine City, MI.)*

reheating time, coil position in the hot strip mill schedule and temperatures during roughing and finishing mill processing [12], the system display and rules definitions for each of these steps are shown in Fig. 16.5.

Increasingly, additional streams of data from quality control processes are used to guide process control. Online mechanical properties measurements have been possible to make for several years because of the relationship between the mechanical properties and the magnetic properties of ferromagnetic steel sheet [13]. The results of these measurements are now sent to Level 3 process controls. An example of concurrent line speed, furnace temperature, and residual magnetism (remanence) measurements is shown in Fig. 16.6. Process line data has also been used in analytic models to predict mechanical properties. A correlation of tensile strength and elongation data derived from such a model and laboratory data is shown in Fig. 16.7. These kinds of techniques have been able to qualify coils more efficiently for release to customers by eliminating the need for intensive gathering of samples and making laboratory tests [14].

Image processing is now widely used to detect and classify coating defects. Typically, an 8-h working shift galvanizes about 25 coils, where the average dimensions of the coils being processed would be roughly 0.8 mm thickness, 1200 mm width, 3000 m length, which sums up to about 90,000 m^2 or 12 soccer fields of inspected surface for these 565 metric tons of coated steel. Cameras such as those shown in Fig. 16.8 improve consistency of inspection and provide for a digital flow of processing data that can be integrated with other information flowing from process equipment and other quality measures. The smallest feature needs to cover at least four pixels of the detected surface to provide reliable data [15]. As an example of integration with other plant data, Cleveland-Cliffs' Cleveland, OH continuous galvanizing line utilizes cameras at the entry end of the line that provide images that support quality assessments at the delivery end [16].

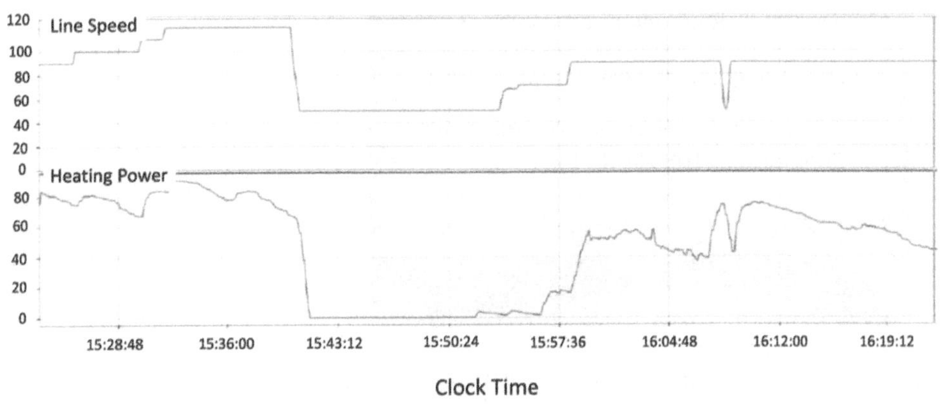

FIG. 16.3 Response of a Level 2 control system controlling furnace power (lower graph) to a change in line speed (upper graph) [10]. *(C. Perottet, How model predictive control enhances strip temperature regulation, in: Proceedings Galvanizers Association 111th Meeting, October 6–9, 2019, Charleston, SC, Galvanizers Association, Marine City, MI.)*

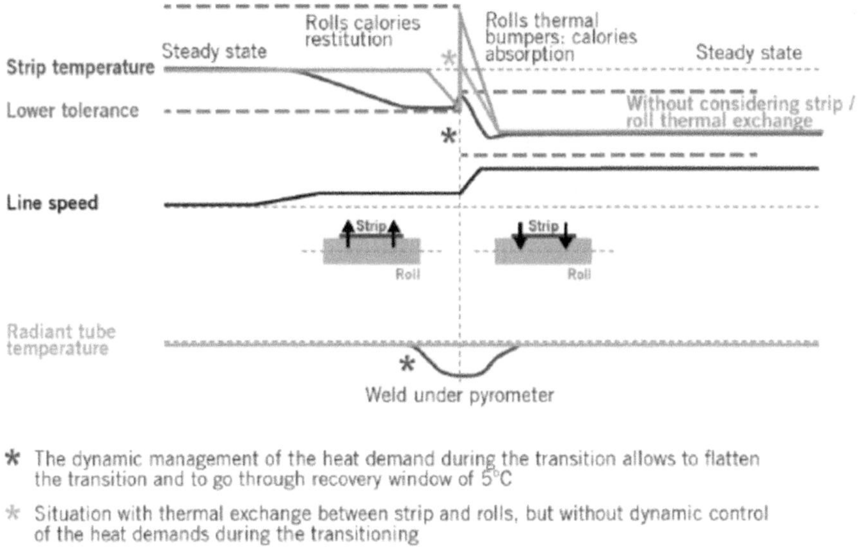

FIG. 16.4 Process variable changes during a thick to thin sheet thickness transition [11]. *(M. Volpato, D. Garassino, A. Alessandrini, Efficient and effective operation of a hot dip galvanizing line, in: Proc. 12th International Conference on Zinc and Zinc Alloy Coated Steel Sheet (Galvatech 2021), June 22–24, 2021, Virtual Conference, Austrian Society for Metallurgy and Materials, pp. 806–817.)*

One of the challenges for dealing with line data is the large amount of computational processing required. For an artificial neural network dealing with images acquired during coating optical inspection, the effect of strip speed on computing requirements is shown in Fig. 16.9. This is for an automated surface inspection system based on 8 × 8 pixel sizes treated with a full convolution to give a two-dimensional map of the defect. The central processing unit (CPU), graphics processing

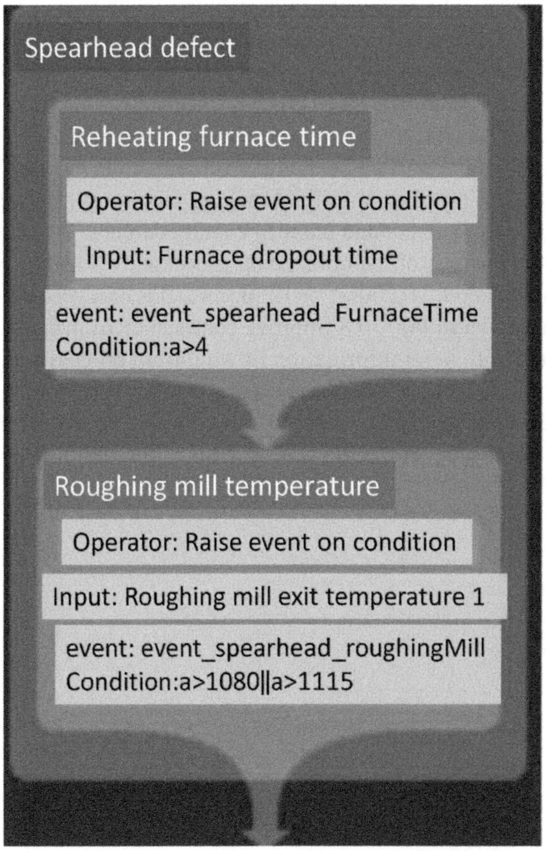

FIG. 16.5 Data and rules generated by a process control system including a continuous galvanizing line and upstream processes. Actions required in the slab reheating furnace and hot strip roughing mill are noted [12]. *(M. Monnoyer, R. Biland, E. Menigault, D. Seguret, Fives' Eyeron™, all-in-one quality management software, in: Proc. 12th International Conference on Zinc and Zinc Alloy Coated Steel Sheet (Galvatech 2021), June 22–24, 2021, Virtual Conference, Austrian Society for Metallurgy and Materials, pp. 797–805.)*

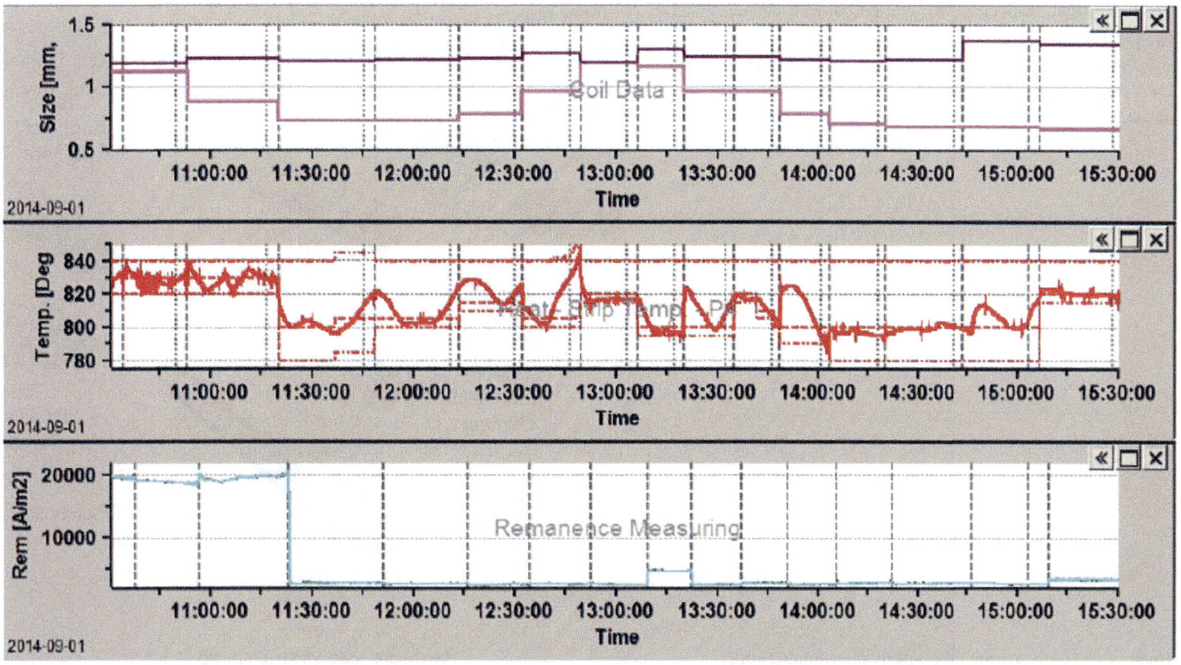

FIG. 16.6 Measurements from continuous galvanizing line process data display during strip thickness changes (top) and resultant changes in temperature (middle) and residual magnetism or remanence (bottom) [13]. *(A. Lhoest, M. Bärwald, U. Sommers, M. Biglari, E. Montagna, W. Beugeling, Online quality monitoring of IF and high strength steels on continuous galvanizing lines controlled by furnace mathematical model, in: Proc. 10th International Conference on Zinc and Zinc Alloy Coated Steel Sheet (Galvatech 2015), May 31–June 4, 2015, Toronto, Association for Iron & Steel Technology, Warrendale, PA, pp. 650–657.)*

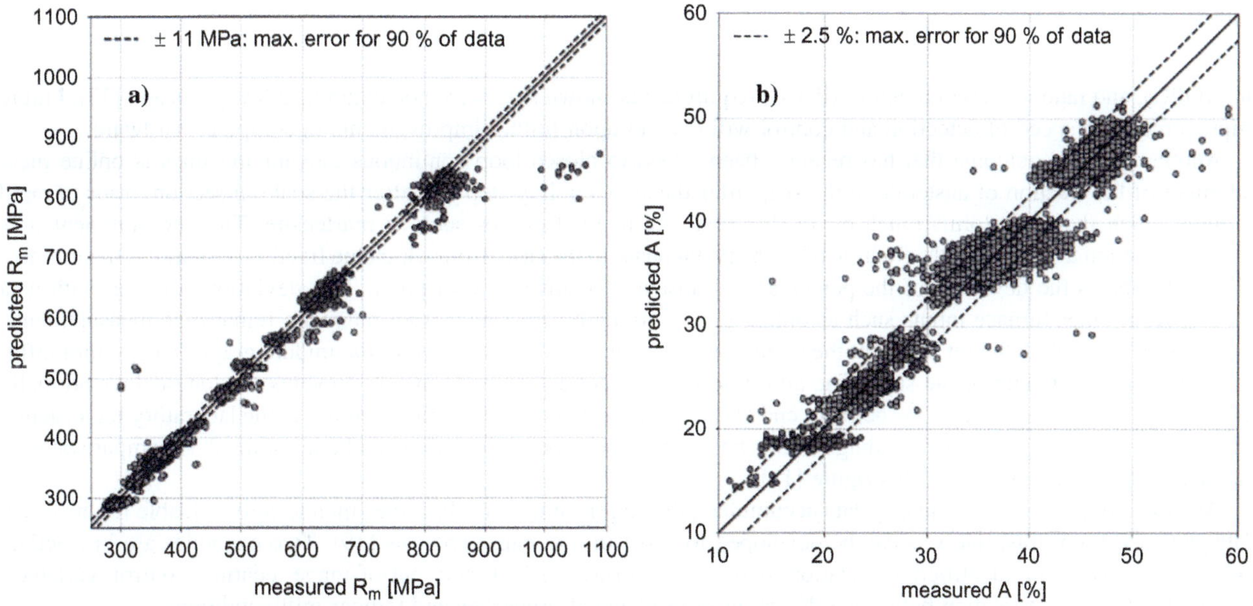

FIG. 16.7 Correlation of predicted (A) ultimate tensile strength and (B) total elongation of galvanized steel sheet from analytical models based on process line data and measured results from counterpart laboratory tensile specimens [14]. *(S. Bragin, J. Klaus, T. Pfatschbacher, J.-F. Plaul, L. Bonhomme, B. Grenier, B. Jalard, Through-process optimization (TPO)—a modern approach to rapid product development, in: Proc. 12th International Conference on Zinc and Zinc Alloy Coated Steel Sheet (Galvatech 2021), June 22–24, 2021, Virtual Conference, Austrian Society for Metallurgy and Materials, pp. 818–828.)*

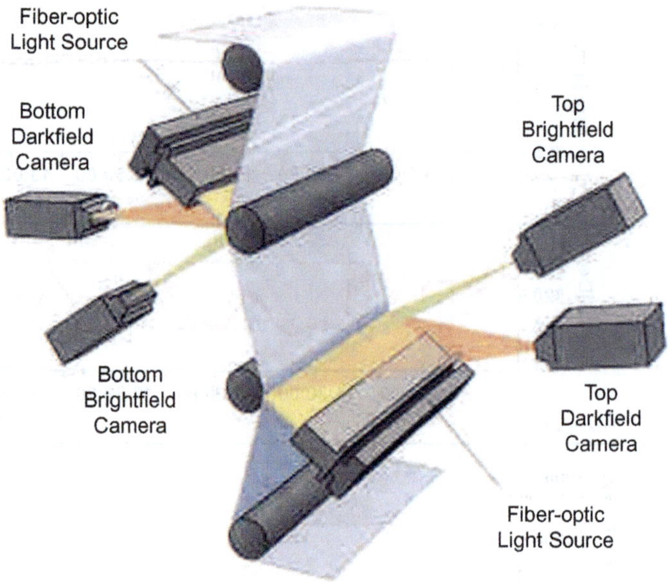

FIG. 16.8 Optical inspection by charge-coupled device (CCD) cameras [15]. *(T. Koll, New Testing Procedures and Quality Control, Proc. 8th International Conference on Zinc and Zinc Alloy Coated Steel Sheet (Galvatech 2011), June 21-25, 2011, Genoa, Italy, Associazione Italiana di Metallurgia, Milan.)*

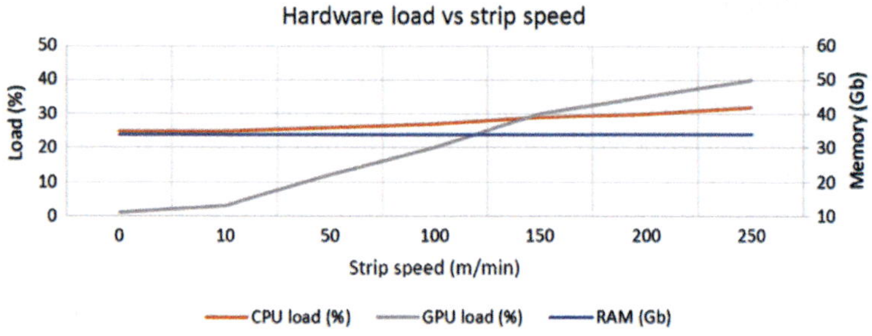

FIG. 16.9 Effect of coating line speed on computational requirements of Central Processing Unit (CPU), Graphics Processing Unit (GPU), and Random Access Memory (RAM) for image processing of surface defects [17]. *(Y. Fontaine, L. Dorel, B. Jalard, A.Perrichon, G. Fricout, O. Bichler, SIAS® 4.0 the next generation of automatic surface inspection system using convolutional neuron network for detection and classification of surface defects, in: Proc. 12th International Conference on Zinc and Zinc Alloy Coated Steel Sheet (Galvatech 2021), June 22–24, 2021, Virtual Conference, Austrian Society for Metallurgy and Materials, pp. 940–954.)*

unit (GPU), and random access memory (RAM) requirements shown here were not available a few years ago [17]. Future improvements in speed of detection and control will depend upon further improvements in computer hardware.

Another process outcome that has recently been added to closed-loop continuous galvanizing lines is online measurement of the fraction of austenite in the strip after the slow cooling step just after the soaking section, prior to rapid cooling which then transforms much or all this austenite to hard phases such as martensite. This measurement uses X-ray diffraction to determine the fraction of each crystal phase in the strip, from which the fraction austenite is determined. Fig. 16.10 shows the deviation of the percentage of austenite relative to a set point. This deviation, together with data coming from other furnace inputs such as online tensile strength assessment from magnetic remanence measurements, is then used to regulate the set point for the completion of slow cooling, as shown in the middle graph. This in turn influences the tensile strength of the steel. The ability to obtain fraction austenite in-line prevents the long delays normally involved in taking samples from the delivery end of the galvanizing line and sending samples to the laboratory for austenite measurements. In this way, downgrading of coils because of this process deviation can be minimized. A comparison with manual control is also shown in this figure [18].

As production and product quality data accumulate, the larger sample size becomes increasingly valuable for historical data analysis. From this, rule sets can be developed to assist in root cause analysis [19]. Two examples are instructive: Fig. 16.11 shows an identified correlation between a surface defect and out-of-range coating control variables. Fig. 16.12 shows a correlation between a deviation in mechanical properties and temper mill conditions.

Integrating historical data, together with expert knowledge that helps establish rules related to product specifications and process details, goes beyond the physical models traditionally used to control modules in the continuous galvanizing line such as coating control. Increasingly, these computational tasks are assumed by neural network models that are the basis of artificial intelligence. The neural network program is fed information from many data sources throughout the production process and links this to specific products and their locations in the stages of production. Measures of quality,

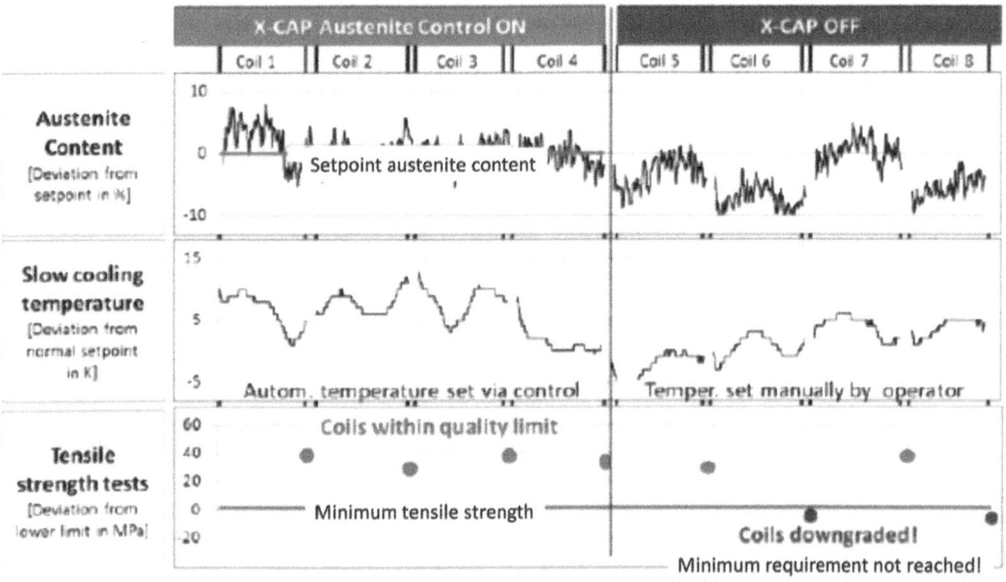

FIG. 16.10 Use of online austenite fraction measurements during the slow cooling section just after the annealing step (top) to control the slow cooling step (middle) and minimize downgrading because of below-specification tensile strength. Automatic control (left), manual control (right) [18]. *(H. Behrens, T. Daube, L. Kümmel, C. Sasse, Hot-dip galvanizing lines—challenges and solutions for the industrial production of coated 3rd gen automotive steels, in: Proc. 12th International Conference on Zinc and Zinc Alloy Coated Steel Sheet (Galvatech 2021), June 22–24, 2021, Virtual Conference, Austrian Society for Metallurgy and Materials, pp. 787–796.)*

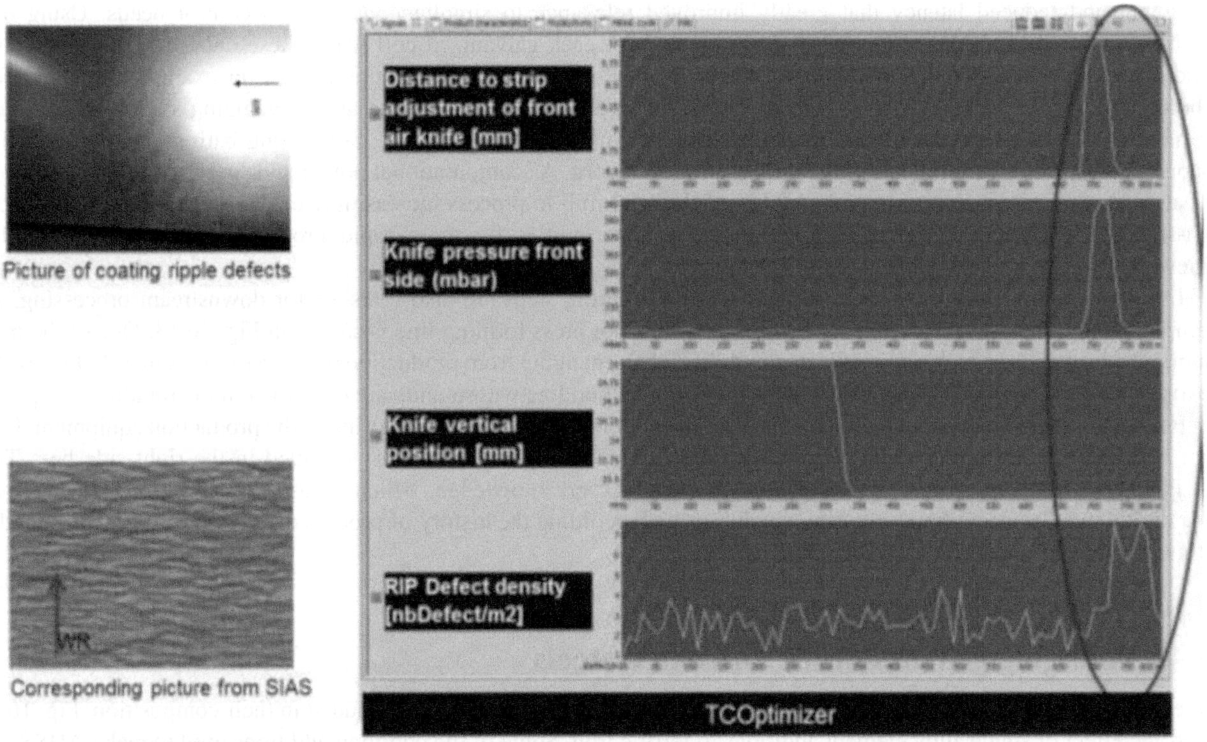

FIG. 16.11 Use of data from advanced process control systems to identify a coating defect. Out-of-range coating control system variables were identified as the root cause [20]. *(B. Grenier, L. Bonhomme, Y. Fontaine, S. Georges, B. Jalard, Y. Kai, S. Maillard, Quality intelligence package dedicated to state-of-the-art processing lines for a comprehensive approach, in: Proc. 11th International Conference on Zinc and Zinc Alloy Coated Steel Sheet (Galvatech 2017), November 12–16, 2017, Tokyo, Iron and Steel Institute of Japan, Tokyo, pp. 322–329.)*

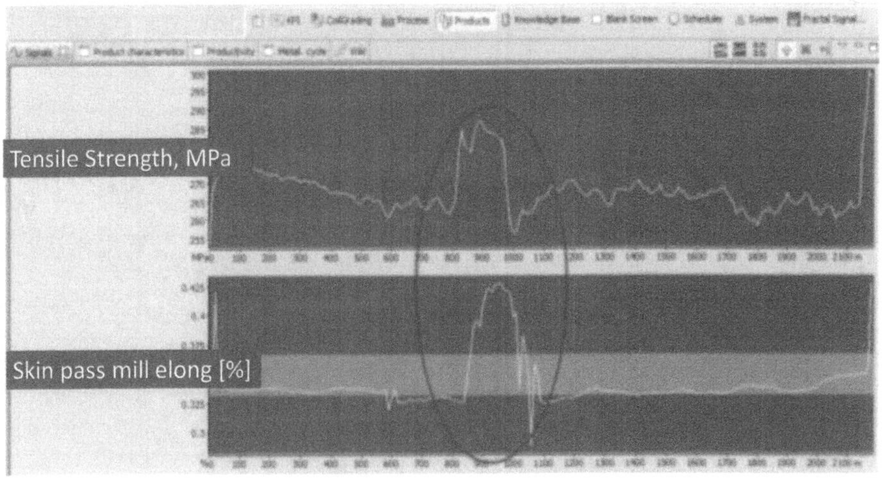

FIG. 16.12 Use of data from advanced process control systems to identify nonconforming tensile strength. Out-of-range skin pass mill elongation was identified as the root cause [20]. *(B. Grenier, L. Bonhomme, Y. Fontaine, S. Georges, B. Jalard, Y. Kai, S. Maillard, Quality intelligence package dedicated to state-of-the-art processing lines for a comprehensive approach, in: Proc. 11th International Conference on Zinc and Zinc Alloy Coated Steel Sheet (Galvatech 2017), November 12–16, 2017, Tokyo, Iron and Steel Institute of Japan, Tokyo, pp. 322–329.)*

including coating, strip shape, and mechanical properties also are fed to the neural network so that it can be trained. With this information, the machine learning system can be built, enabling the system to ultimately determine optimal process settings and recognize process deviations that affect quality. The computational power must be high enough to permit the system to run in real time. An example of such as system is operating at the ArcelorMittal Eisenhüttenstadt plant, where sliver defects originating in continuous casting of exposed automotive quality ultralow-carbon (ULC) interstitial-free (IF) steel were often not detected until they reached the continuous galvanizing line inspection stage. Deep convolutional neural network (CNN) classifiers were used rather than traditional rule-based image analysis that greatly increased analysis throughput and reduced latency that greatly improved relevance to simultaneous process control needs. Using data matching and quality results, a "digital twin" is created for each galvanized coil [21]. These data integrate with Level 2 and 3 controls and enable the matching of defects as they pass through production stages, for example, a scale defect in hot rolling can be picked up on the same coil as it passes through cold rolling and galvanizing stages. Online measurement of galvanizing bath chemistries, using data streams from the LIBS and galvanizing bath sensors described in Chapter 4, is expected to form part of the overall process record. As computational power and speed increase and the historical and technical databases grow, the ability of the steel mill to process increasingly challenging grades of advanced high-strength steels will be enhanced, opening new opportunities for more rapid process development and earlier achievement of commercial status for new products.

High-level control and interaction with prior manufacturing steps are also foreseen for downstream processing. An example of an "Industry 4.0" design for control tasks in a hot press forming line is shown in Fig. 16.13. One of the most important advantages of such a system is its flexibility when changing from production of one part to another. In Fig. 16.13, the top box is the integrated system that controls the production line whose units are described in the middle box, quality monitoring functions are also contained in this box. The bottom box involves monitoring of the production equipment. Data from both the top and bottom boxes are sent to the central management system described in the right side box. This box includes the expert system, based on historical and expert knowledge, which adjusts the process when required. The "digital twin" concept for monitoring conditions and recording the history of process details for each part is included in this control and documentation scheme [22].

16.4 Emerging generations of Zn-coated steels

Lower-density steels are possible to make if lower-density alloying elements are included in their composition. Fig. 16.14 shows the effect of single alloy element additions to ferritic iron. Some of the common additions used to make AHSS, Cr, Mn and Si contribute to a reduction in density, but Al, with its specific gravity of 2.7, is observed to have the biggest influence. Steels containing significant quantities of lighter elements such as Al are termed "low-density steels" (LDS) [23]. There are four families of LDS, their microstructures and typical compositions are shown in Fig. 16.15. Al is limited in these compositions to less than 12%. Densities of the ferritic and austenitic phases are related to alloy content in these steels by the formulas [23]

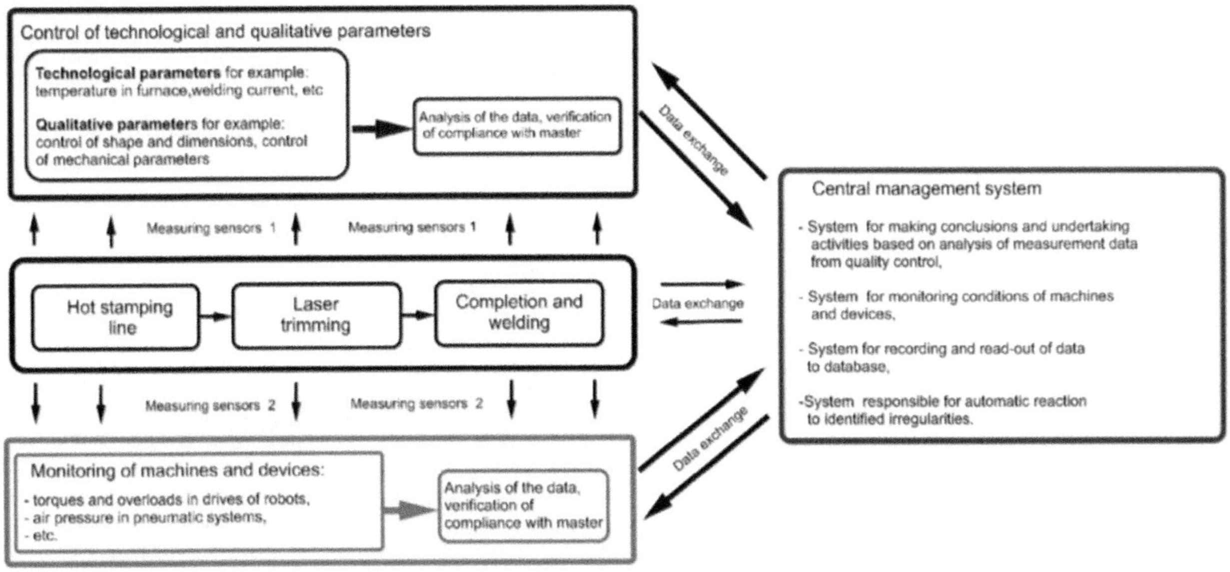

FIG. 16.13 An "Industry 4.0" concept for integrated control and management of the production of hot press-formed steel sheet pieces [22]. *(I. Wrobel, M. Sidzina, Design study for automatic production line of a sub-assemblies of new generation car body structures compliant with the 'Industry 4.0' concept, Sensors 21 (7) (2021) 2434.)*

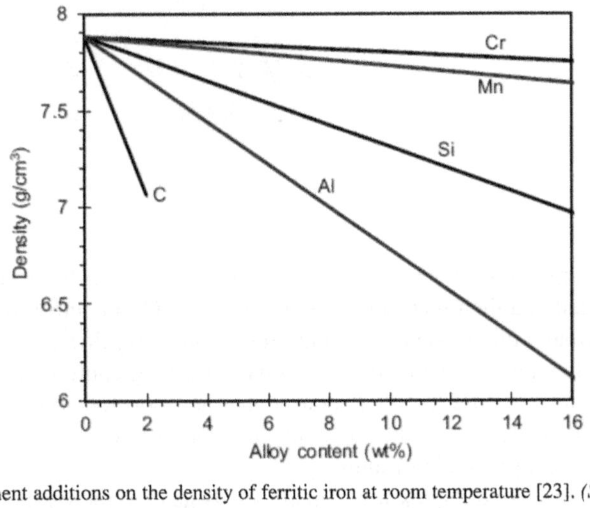

FIG. 16.14 Effect of single alloy element additions on the density of ferritic iron at room temperature [23]. *(S. Chen, R. Rana, Low-density steels, in: R. Rana (Ed.), High-Performance Ferrous Alloys, Springer, Cham.)*

$$\rho_{\text{ferritic (g/cm}^3)} = 7.874 - 0.098\left(x_{\text{Al (wt\%)}}\right) \tag{16.1}$$

$$\rho_{\text{austenitic (g/cm}^3)} = 8.15 - 0.101\left(x_{\text{Al (wt\%)}}\right) - 0.41\left(x_{\text{C (wt\%)}}\right) - 0.085\left(x_{\text{Mn (wt\%)}}\right) \tag{16.2}$$

where x is the composition in weight percent of Al, C, or Mn. As in the case of TRIP and 3G steels, austenite in the austenitic grades is not stable unless appreciable quantities of C and Mn are added and maintained in austenite by selected processing routes. At the levels shown in Fig. 16.15, Al is a ferrite stabilizer unless its effect is countered by C and Mn. Al also plays a role in formation of carbides, generally $(Fe,Mn)_3AlC$ with its perovskite structure that can contribute to strengthening. During hot working, these carbides are above their solvus temperature and do not interfere with microstructural development. At lower temperatures, these and other carbides need to be prevented from forming coarse precipitates at grain boundaries that would reduce mechanical properties. Possible rolling schedules for ferritic LDS are much more limited than

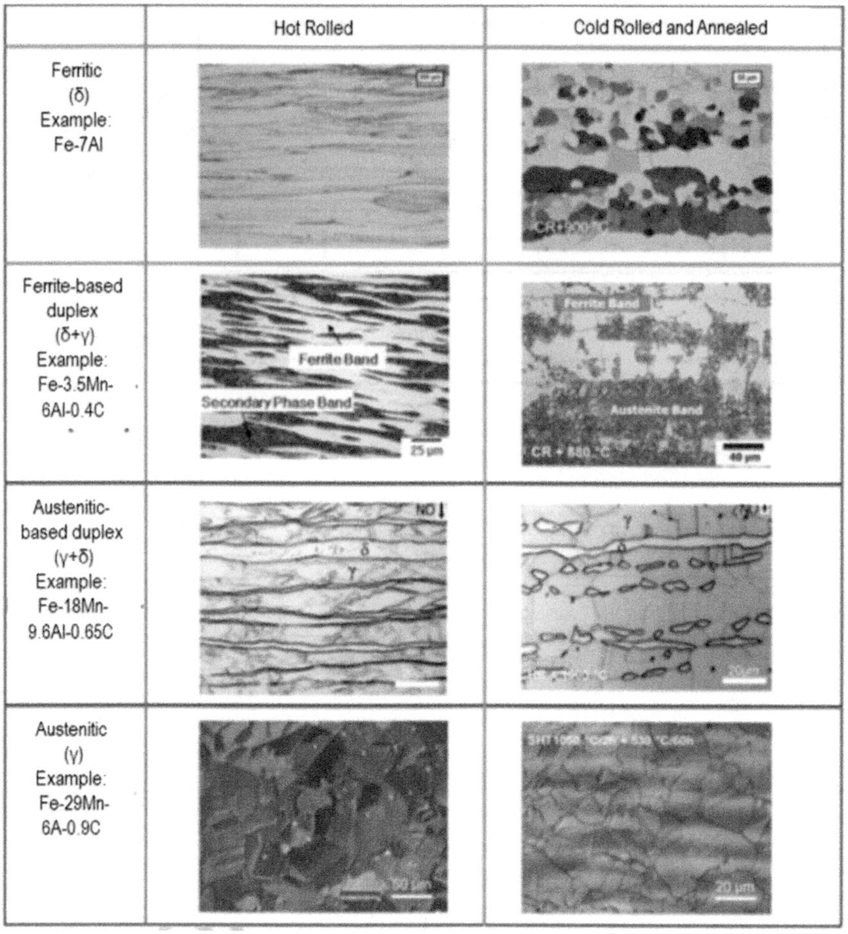

FIG. 16.15 Optical micrographs and typical compositions for the four types of low-density steels [23]. *(S. Chen, R. Rana, Low-density steels, in: R. Rana (Ed.), High-Performance Ferrous Alloys, Springer, Cham.)*

austenitic grades because of the large grain size of delta ferrite that persists after hot rolling the ferritic grades. After hot rolling, these grades are cold-rolled and then recrystallized at 700–1000°C to control texture and also to precipitate carbides, the treatments are compatible with conventional continuous annealing lines. Some of the pretreatment routes for austenitic LDS are also compatible with conventional continuous processing equipment, these include routes that rapidly cool the strip from the austenitic temperature to either room temperature or 500–700°C after which it is slow cooled or isothermally held. Age-hardenable austenitic grades generally include holding at 450–650°C for 5–20h and appear to be less compatible with typical AHSS processing routes.

As indicated in Fig. 16.15, ferrite-based duplex low-density steels have a microstructure of austenite and delta ferrite with a delta ferrite volumetric portion greater than 50%. Austenite-based duplex low-density steels have a microstructure of austenite and delta ferrite at hot working temperatures with the austenitic phase forming the continuous matrix. As expected, the stability of austenite in ferrite-based duplex LDS is relatively low in comparison to the austenite-based compositions; therefore, in the ferrite-based duplex grades, the austenite will transform to other phases at lower temperatures. In austenite-based duplex LDS, the austenite phase is largely stable at room temperature owing to the high content of austenite-stabilizing elements. Stabilization is assisted in the austenite-based duplex grades by the formation of small, dispersed $(Fe,Mn)_3AlC$ particles.

In investigating the galvanizability of steels containing significant amounts of Al, Maki [24] noted that the equilibrium oxygen potential of Al was lower than Si or Mn; therefore, it should be easier to develop internal oxides. Maki's work, however, was limited to TRIP steels with less than 2%Al. Jin [25] examined the galvanizability a steel with a composition of 0.38%C-4%Al-1.26%Mn-0.29%Si using annealing furnace atmosphere dewpoints of −50°C and −10°C, soaking temperatures of 750–870°C, and soaking times between 30 and 150s. With the −10°C dewpoint, an internally oxidized layer consisting of Mn Al was observed. Increasing the dew point from −50°C to −10°C resulted in formation of a very different surface morphology. Whereas the −50°C dewpoint sample was covered with surface oxides, Fig. 16.16, the morphology

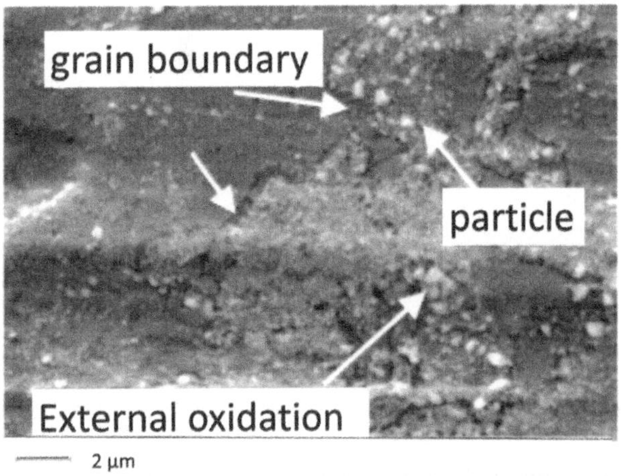

FIG. 16.16 Surface appearance of C-0.38%C-4%Al-1.26%Mn-0.29%Si steel after soaking at 815°C for 150 s in a −50°C dewpoint atmosphere [25]. *(X. Jin, Y. Xie, H. Wang, Selective oxidation of high aluminium low density steel, in: Proc. 11th International Conference on Zinc and Zinc Alloy Coated Steel Sheet (Galvatech 2017), November 12–16, 2017, Tokyo, Iron and Steel Institute of Japan, Tokyo, pp. 506–513.)*

FIG. 16.17 Surface appearance of C-0.38%C-4%Al-1.26%Mn-0.29%Si steel after soaking at 815°C for 150 s in a −10°C dewpoint atmosphere [25]. *(X. Jin, Y. Xie, H. Wang, Selective oxidation of high aluminium low density steel, in: Proc. 11th International Conference on Zinc and Zinc Alloy Coated Steel Sheet (Galvatech 2017), November 12–16, 2017, Tokyo, Iron and Steel Institute of Japan, Tokyo, pp. 506–513.)*

after soaking with the −10°C dewpoint indicates that most of the surface is covered with a layer of polygonal particles with individual sizes less than 1 μm, Fig. 16.17. Some of the particles appear to be partially connected to each other while others are separated. The particles consist mainly of reduced iron, only a small amount of oxygen, and only a trace of aluminum was detected. Higher levels of oxygen and aluminum were detected in the gaps between the particles, indicating the presence of external and/or internal Al-rich oxides. These experiments point the way to opportunities for producing other Al-rich LDS grades, using appropriate combinations of compositions and galvanizing process variables.

One can speculate about the nature of fourth-generation steels. It seems clear that greater ductility capabilities will be required even as strength increases. Possible processing routes will need to provide a balance between the quantity, size, and relative hardness of the hard and soft phases that make up steels in this new generation. TWIP (twinning-induced plasticity) grades with tensile elongation products of 40,000 MPa-% would be expected to rely upon refined grain sizes and precipitation hardening, together with the use of TRIP (transformation-induced plasticity) and TWIP effects to reach these property goals.

16.5 Extending capabilities of existing coating processes

As steels increase in complexity and their properties reach higher levels, the capabilities of processes used to prepare them for hot-dip galvanizing will also need to increase. Starting with the treatment of slabs or strands from continuous casters,

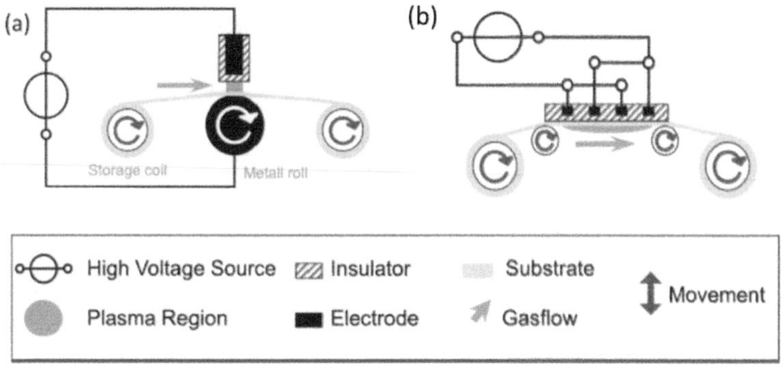

FIG. 16.18 Schematics of possible methods for strip cleaning utilizing the dielectric barrier discharge (DBD) plasma process. (A) A treatment currently used for fabric and (B) a coplanar arrangement [27]. *(R. Brandenburg, Dielectric barrier discharges: progress on plasma sources and on the understanding of regimes and single filaments, Plasma Sources Sci. Technol. 26 (2017) 053001.)*

higher alloy contents are associated with oxide scale coatings that are more difficult to remove. "Red scale" layers are associated with steels containing more than 0.5%Si [26], as higher levels of alloy additions are made that have surface-selective oxidation behavior, removal of these layers, normally accomplished with high-energy water jets, will be become more challenging. Blast slurry scale cleaning and plasma cleaning have been investigated as alternatives [16]. Whereas many plasma processes are only possible to operate under vacuum, the dielectric barrier discharge (DBD) process can operate under normal atmospheric conditions. These have widespread application in the surface treatment of plastic foils but appear to have opportunities for cleaning of steels [27]. Two configurations for treatment of sheet in coil processing lines are shown in Fig. 16.18. The left side configuration produces an electric field between an upper electrode and the support roll with the dielectric and strip surface between them. The right side produces an electric field between electrodes contained within the upper surface, the electric field extends beyond this surface through the strip surface to the dielectric placed on the bottom side of the strip surface.

Increased coating control capabilities will be required to support opportunities for new alloy coatings than can be expected to be applied at lower coating weights. Strip shape control will become more challenging as steel mechanical properties continue to increase. Crossbow modeling and active crossbow correction will become part of the active process control tools available to manage strip shape prior to gas wiping [28]. More accurate process control will also increase use of "hot" coating weight gauges, located as close as possible to the coating control equipment. An emerging technology to help stabilize strip prior to the coating control operation involves hydrodynamic stabilization. This is particularly effective at higher line speeds, in excess of 2 m/s, and utilizes a "cushion" of liquid metal that is entrained upward from the zinc bath to help stabilize strip motion [29]. Fig. 16.19 shows how this stabilization is arranged above the surface of the galvanizing bath. Rigid hydrodynamic pads (6) pivot around an axis (7) that exerts a force to contain the mass of liquid metal entrained with the strip. After passing through the wiping knives above this apparatus (not shown), a runback flow of wiped metal passes over the outer surface of the pads. The hydrodynamic pads can contribute to the wiping operation if their top ends extend into the wiping zone described in Chapter 7.

Strip shape stabilization is a key enabler for increasing line speeds in the unending effort to improve plant productivity. By altering the quantity of liquid metal delivered to the wiping zone, the gas flow in the vicinity of the wiping knives, and the runback flow, the hydrodynamic device shown in Fig. 16.19 may increase the splashing limit described in Chapter 7. Another factor that will be important in enabling faster line speeds is improvement in furnace heating and cooling capabilities. Induction heating, discussed earlier in this chapter, has the potential of both reducing carbon emissions per ton of coated product and providing the increased heating rates that will be required at faster line speeds. It was noted in Chapter 2 that the thermal conductivity of H_2-N_2 gas increases with increasing H_2 content. By injecting pure H_2 into the cooling section after the soaking zone, cooling rates in excess of 100°C/s, corresponding to heat transfer rates of 100–200 $W/m^{2}°C$, can be achieved for strip thicknesses of 1 mm or less [30]. After the galvanized strip exits the coating section, it travels through the after-pot cooler that will also require increased cooling capability without increasing strip vibration, the coupling of these factors was described in Chapter 7. Use of a galvanneal furnace increases the need for cooling so that when the strip reaches the top roll above the galvanizing bath, its temperature is less than 360°C. Otherwise, the warm coating will be torn from the strip and picked up on the top roll surface.

The demand for defect-free coatings at higher line speeds will increase the need for bath management and dross control. As described in Chapter 4, dross removal from zinc baths relies upon gravity, with top dross floatation and removal by skimming the bath surface, and bottom dross removed by pumping or mechanical scoops after it settles to the bottom of the bath. The European Union "CentriClean" project is examining techniques for pumping and then centrifuging

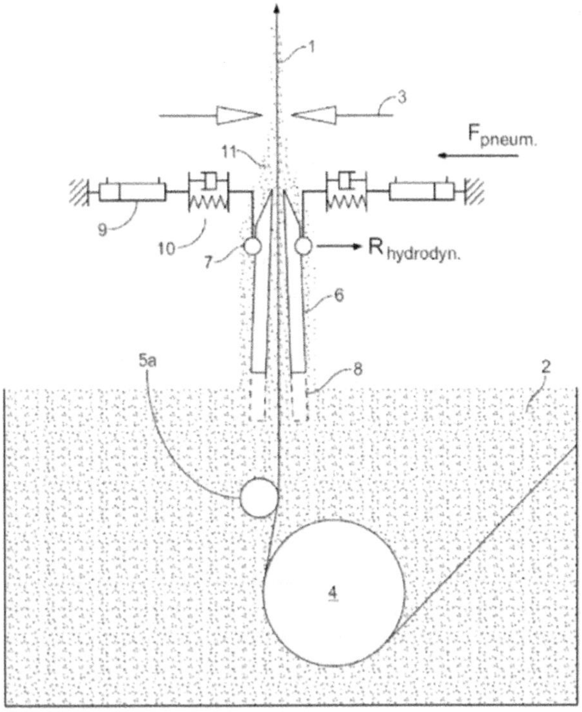

FIG. 16.19 Method of hydrodynamically stabilizing a steel strip upon exit from the galvanizing bath [29]. *(M. Larnicol, Y. Hardy, O. Bregand, P. Gerkens, Device for Hydrodynamic Stabilization of a Continuously Travelling Metal Strip, U.S. Patent 10,550,459, February 4, 2020.)*

dross-liquid mixtures from zinc baths [31]. If successful, this technique will enable continuous removal of dross from the bath and return to the bath of the separated liquid, reducing the amount of liquid metal lost to dross. Another method of dross management is to maintain a cold surface in the zinc bath, below the solubility temperature of the bath composition, so that dross is grown on the cold surface. This is used in the ZQ-Dynamic Galvanizing (trademark of Psitec Oy) process as part of its management of bath temperatures that accommodate high strip entry temperatures, Fig. 16.20 [32]. Recovery of dross from the cold surface must be able to keep up with the rate of dross production; this is shown as reaching 20 kg/h in Table 4.4

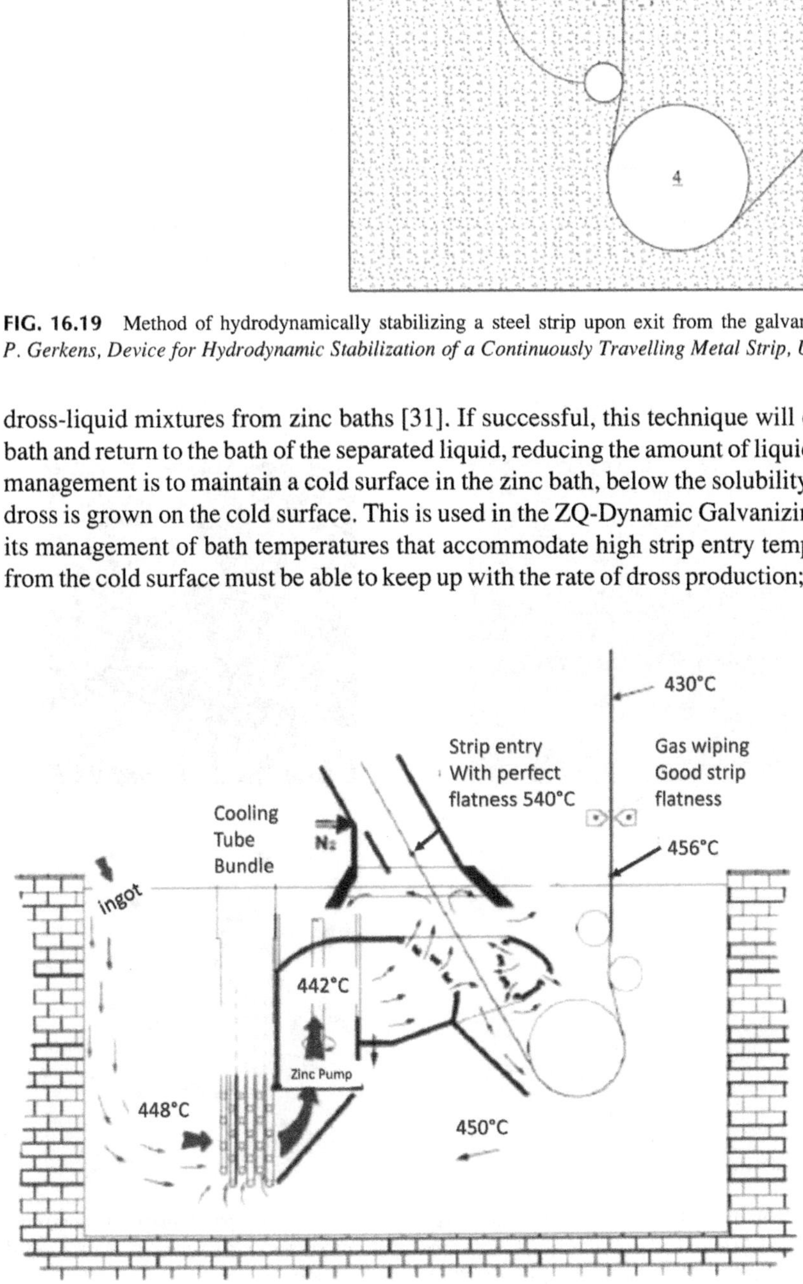

FIG. 16.20 Use of air-cooled tube bundle in the ZQ-Dynamic Galvanizing process to maintain bath temperature and iron content [32]. *(P. Sippola, Unique simultaneous rapid cooling and zinc coating technology—no bainite before zinc pot, in: Proc. 12th International Conference on Zinc and Zinc Alloy Coated Steel Sheet (Galvatech 2021), June 22–24, 2021, Virtual Conference, Austrian Society for Metallurgy and Materials, pp. 426–435.)*

of Chapter 4. This is much less than the typical top skimmings production rate cited as 72–102 kg/h for GI and GA operations, respectively. Top skimmings not only include the dross resulting from bath supersaturation with iron, it also includes the oxidized metal from reactions with air such the runback flow from wiping and exposure of the top surface of the bath to air. Reducing the exposure of the zinc bath surface to air has already been achieved by wiping with nitrogen and/or shrouding of the bath and emerging strip as described in Chapter 7. Further improvements in dross management, together with improved strategies for bath temperature and composition control that minimize dross, will support new opportunities for expanded production of defect-free coatings.

Temper rolling mill enhancements have improved the consistence of surface texture of galvanized sheet. Despite the many advancements made with "Industry 4.0" initiatives, much of temper mill control is surprisingly manual, further instrumentation and control developments are needed to permit creation of a "digital twin" of this processing unit. Improvements in the reliability and consistence of electrical discharge textured (EDT) temper roll surfaces have been made by improving both the power delivery system and the electrode positioning system of the EDT machine [33]. This enables sharper, more consistent craters to be eroded on the skin pass roll, Fig. 16.21. Normally there is a relationship between the average roughness and the peak count in the texture of the strip surface, i.e., larger craters result in higher average roughness (R_a) and fewer peaks per unit length (peak count R_{Pc}). Because of the improved control in crater shape on the temper mill roll, the textures exhibit a significant increase in the highest R_{Pc} values measured in the lower R_a range, peaking at a 60% increase at 1.0R_a. This is illustrated in Fig. 16.22, where the circular points show readings from improved temper roll textures versus conventionally processed rolls where typical values are in the band between the two lower curves. By further optimizing the setup of the EDT machine and texturing parameters, R_{Pc} is significantly increased in the mid-range R_a, where rolls for automotive strip are typically produced at 2.5–3.0R_a. The curve connecting the triangular points is representative of optimized conditions. Increasing peak count increases the exposure surface area of the temper rolled galvanized sheet, increasing reactivity with passivates or paint pretreatments, and contributes to improved paint adhesion.

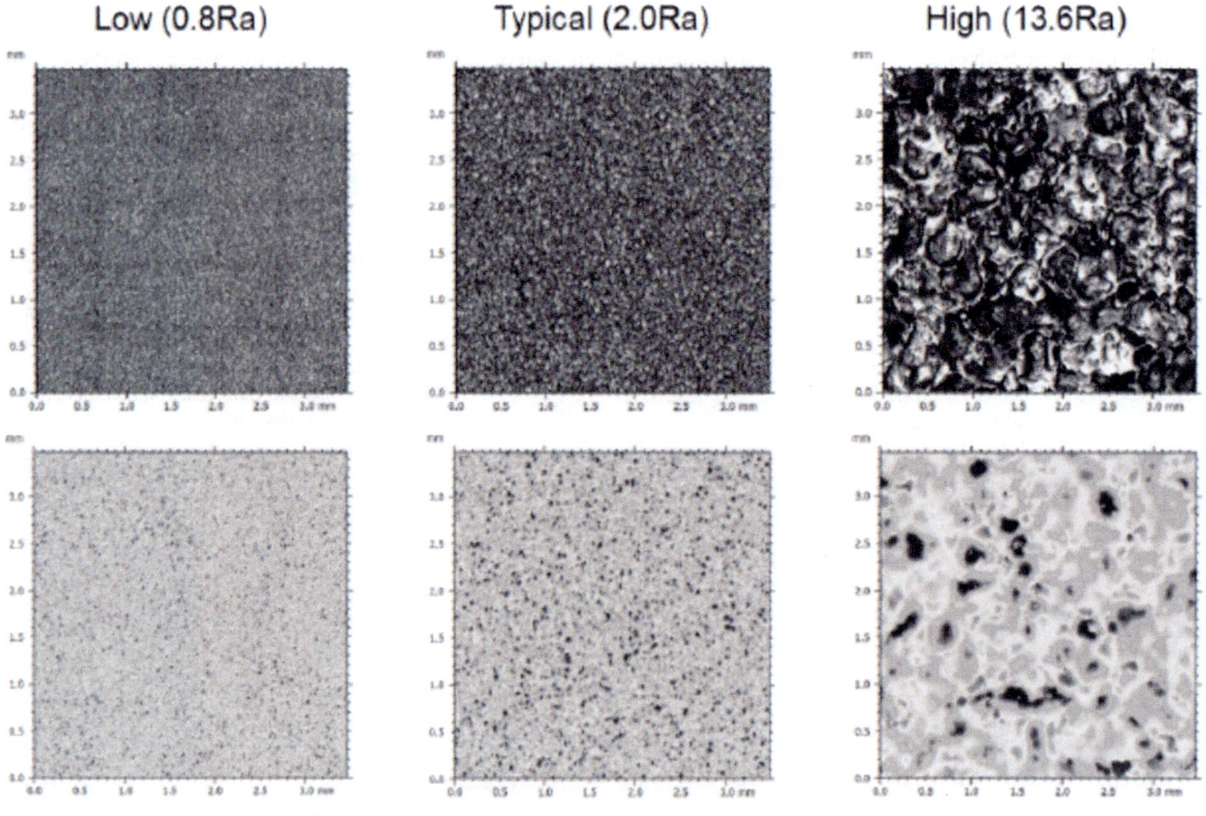

FIG. 16.21 Temper roll roughness profiles produced by improved EDT roll texturing for three typical roughnesses (top) optical images (bottom) corresponding roughness maps [33]. *(C. Childs, Next-generation EDT technology, in: Proc. 12th International Conference on Zinc and Zinc Alloy Coated Steel Sheet (Galvatech 2021), June 22–24, 2021, Virtual Conference, Austrian Society for Metallurgy and Materials, pp. 71–78.)*

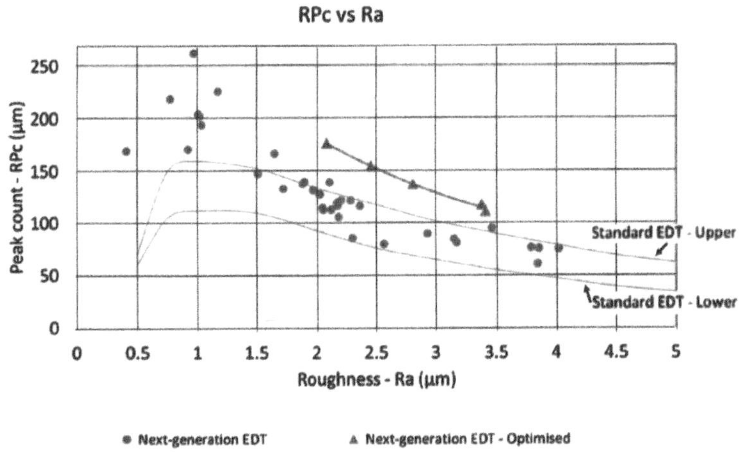

FIG. 16.22 Relationship between average roughness R_a and peak count R_{Pc} for conventional and improved temper mill EDT rolls [33]. *(C. Childs, Next-generation EDT technology, in: Proc. 12th International Conference on Zinc and Zinc Alloy Coated Steel Sheet (Galvatech 2021), June 22–24, 2021, Virtual Conference, Austrian Society for Metallurgy and Materials, pp. 71–78.)*

16.6 New coating processes

New coating processes can be considered that overcome the shortcomings of existing processes. For electrogalvanizing, these include the high consumption of electricity associated with production of thicker coatings, usually 40 g/m^2 per side is considered as an economic limit. Also, the line speeds typically used with hot-dip continuous galvanizing are not possible with electrogalvanizing because the coating deposition rate is limited by current density, meaning that excessively long cells would be required to increase production rates [34]. With regard to continuous hot-dip galvanizing, line speed is limed by many factors, the first of which is the capacity of the pretreatment furnace, this limits productivity when processing thick (>3 mm) strip. Productivity is also limited by the life of submerged bath hardware, including corrosion of the hardware and buildup of dross on roll surfaces that impact coating quality. Moving to higher line speeds also runs the risk of roll slippage because of inertial, hydrodynamic, and frictional forces, this can result in scratching of the strip surface. Line speed is also limited by the splashing limit discussed in Chapter 7 on coating control and after-pot cooling, as indicated above.

The development of physical vapor coating (PVD) processes is described in Chapter 10. Many new opportunities for zinc-coated sheet steel are expected to come from further development of this processes. Advances in chemical vapor deposition (CVD) may make high deposition rates economically possible. This is done by using a magnetron cathode as a source of electrons; these electrons are then forced to exit the source chamber through a reduced dimension magnetic nozzle. As electrons are accelerated through this nozzle, they are joined by the precursor gas, which is ionized to form a plasma. The internal magnetron can be extended to several meters in length. To obtain high deposition rates, two of these nozzles are set up opposing each other, with one being the anode and the other the cathode for each half-cycle of an AC sine wave. But as the electrodes move between cathode and anode, a mirror magnetic field in the opposing nozzle impedes electron flow and sets up an electric field outside this nozzle. Gas atoms encountering the electrons in this electric field and are accelerated away from it. This results in a stream of ions exiting one nozzle or the other during each AC sine wave half-cycle. Ion energy is 30–60 eV, giving typical deposition rates of 100 mm m/min under pressures between 0.001 and 0.025 mbar and temperatures between 50°C and 300°C. Until now, this process has been used to coat flexible conductive foil. [35].

Fluidized bed coating techniques have been investigated for both pure zinc and zinc alloy coatings. An example of a continuous zinc coating line utilizing a fluidized bed is shown in Fig. 16.23 [34]. In this figure, the coating section is between the lower sealing chamber (225) and the upper sealing chamber (215), both of which isolate the coating chamber (220) by sealing rolls. Zinc powder is injected by nozzles (223) and is fluidized by the N_2-H_2 gas injection nozzles below them (226). Excess powder is recovered by suction tubes (227) that feed a cyclone device (not shown). The coating chamber must be kept below 250 °C to prevent agglomeration of the zinc particles. Electrodes (228) electrostatically charge the zinc powder so that it is attracted to the grounded strip surface (201). Voltages around 90 V permitted 200 g/m^2 coatings to be produced. The strip is heated to a temperature between 420°C and 730°C during fluidized bed coating and can be reheated after leaving the coating chamber, consolidating the powder. In this example, the average powder diameter was 45 μm and the Al content varied between 0.1% and 0.7%.

Zinc coatings with higher alloy contents have until now only been achieved using batch processes. A vibratory fluidized bed containing either corundum (Al_2O_3) powder with an addition of 1.5% Zn or corundum mixed with 1.5%Zn-0.5%Al-0.03%Ti-0.03% copper oxide (CuO) powders was used with an ammonia atmosphere to coat high-strength steel bolts [36]. In both cases, the coating was applied at 450–550°C for 5–15 min. The lower temperature limit of the case of the Zn-Al-Ti-

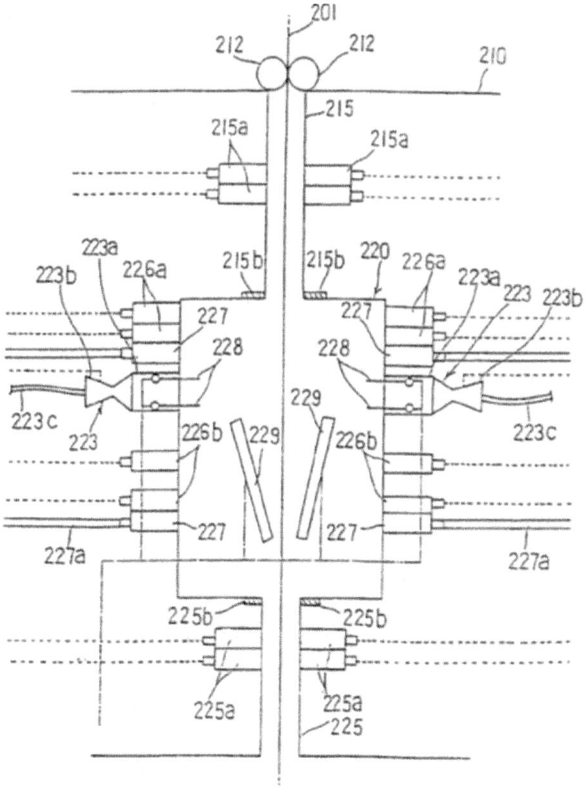

FIG. 16.23 Continuous fluidized bed apparatus for applying zinc to steel strip [34]. *(S.H. Kim, J.H. Chung, N.H. Cho, Apparatus for Coating Zinc on Steel Sheet, and Method Therefor, U.S Patent 6,042,892 March 28, 2000.)*

CuO coating was 500°C because of the formation of Fe-Al coatings that inhibited growth of the coating. In the ZnFe coating produced with pure Zn powder, Γ (Fe_3Zn_{10}), $Γ_1$ (Fe_5Zn_{21}), $δ_1$ ($FeZn_7$), and pure zinc (Zn) phases were detected, but no ζ ($FeZn_{13}$) or η (pure Zn). The coating structure of the Zn-Al-Ti-Cu coating was similar, except for a surface layer of mixed $Al_{0.71}Zn_{0.29}$ and $Zn_{0.68}Ti_{0.32}$ compounds that were associated with improved wear and corrosion resistance.

A batch coating process for coating of ZnAlMg coatings onto steel has been developed. This process, utilizing the traditional general galvanizing pretreatment processes and initial hot-dip coating in zinc before inserting in the fluidized bed, is described further in Chapter 15.

Zn-Al and Zn-Cr coatings have been applied to steel using the pack cementation technique [37]. Powder mixtures containing 50 wt% zinc, 10 wt% Al or Cr, and 2% wt% of the activator NH_4Cl were prepared, Al_2O_3 made up the balance of the powder mixture. Mild steel plates were then buried in the powder. The thermal cycle consisted of a 1 h degas at 50°C and 4 h of cementation at 400°C. Cross sections of the coatings produced are shown in Fig. 16.24. In Zn-Al, Fig. 16.24A, most of the Al segregated to surface producing a 25 μm Al-rich layer that included $Fe_2(Al_{1-x}Zn_x)_5$, where $x < 0.15$. Underneath this, a 42 μm layer consisting of Fe-Zn $δ_1$ ($FeZn_7$) and Γ (Fe_3Zn_{10}) phases made up the balance of the coating. In the Zn-10%Cr coating, Fig. 16.24B, a 5 μm thick layer contained $(Fe_{1-x}Cr_x)Zn_{13}$. Oxidation tests, exposing samples to air at 400°C 24 h, shows both coatings for were quite oxidation resistant, especially the Cr containing coating. The mixed oxides on the surface were mixed coatings included $ZnCr_{0.8}Fe_{1.2}O_4$ and $ZnCrFeO_4$.

Just as tailor-welded blanks enable optimization of auto body materials utilization by placing the right steel grade and thickness as the right place, one can speculate that three-dimensional coatings can be developed to put a coating with the right thickness and alloy composition at the right place. Use of wire or powder feeds together with local melting by lasers, plasma, or electron beams would enable composition and thickness variations over the width and length of the strip, taking advantage of three-dimensional printing process advances already demonstrated for zinc components [38].

16.7 New performance capabilities for zinc-coated steel

As described in Chapter 12, corrosion inhibitors are added to passivate and paint pretreatments to improve the corrosion resistance of coil-coated product. The addition of inhibitors contained within microcapsules that can be co-deposited with

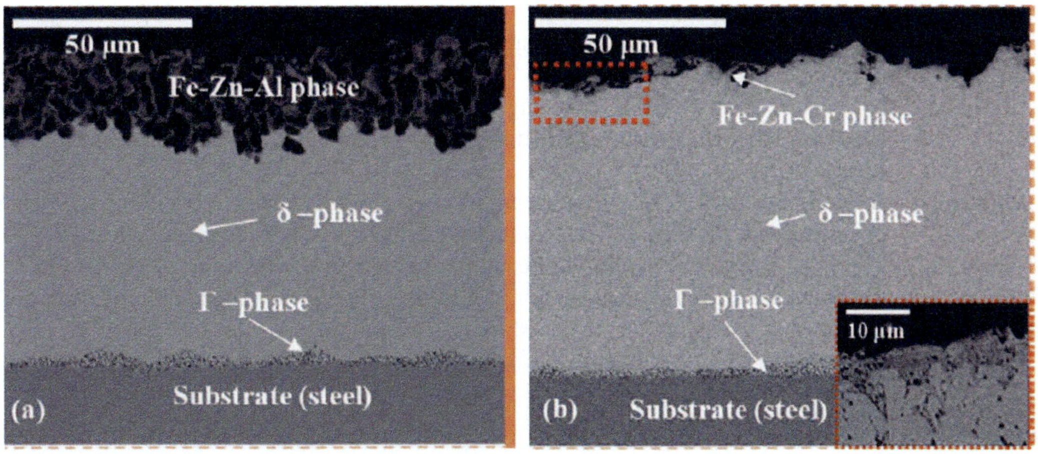

FIG. 16.24 Cross sections of coatings produced by pack cementation (A) Zn-10%Al and (B) Zn-10%Cr [37]. *(D. Chaliampalias, M. Papazoglou, S. Tsipas, E. Pavlidou, S. Skolianos, G. Stergioudis, G. Vourlias, The effect of Al and Cr additions on pack cementation zinc coatings, Appl. Surf. Sci. 256 (11) (2010) 3618–3623.)*

zinc electrogalvanized coatings has shown that the corrosion resistance of these coatings can be improved. The microcapsules are opened, and corrosion-inhibiting species released when either a pH threshold is reached or when the coating is mechanically damaged, such as when it is scratched [39]. There are many challenges in producing a composite electrogalvanized coating on an industrial scale, including the management of particle charge and dispersion. Despite this, Zn-graphite, Zn-SiC, and other additions have been investigated [40]. There are also significant challenges in managing the release of the inhibitor based on pH triggering. The microcapsule must be stable in the acidic electrogalvanizing bath and release inhibitors only under conditions when it is needed. However, corrosion of steel, zinc, and corrosion under paint occur at different pH levels, so that the local corrosion potential would be a more reliable trigger. Two kinds of microcapsules have been used to produce composite zinc electrogalvanized coatings: one with a shell of poly(styrene-*stat*-methacrylic acid) (PSMAA), the other is meso- or microporous silica in which inhibitors can be embedded. A sketch of the PSMAA particle is shown in Fig. 16.25. When the local pH exceeds 4, the surface loses protons, resulting in negative charging of the surface (Fig. 16.25, left), resulting in capsule rupture. The contents are then released (Fig. 16.25, right). Tailoring of the surface of these particles makes them suitable for use in co-deposition during zinc electrogalvanizing. Silica particles have the advantage of compatibility with hydrophilic inhibitors rather than the hydrophobic ones required for synthesis with PSMAA. Most corrosion inhibitors used with galvanized sheet are hydrophilic. These can be introduced into the pores that have typical sizes in the range of 2–50 nm. For example, a pore size of 1.29 nm gives a pore volume of $0.5\,cm^3/g$. There are challenges in determining when the inhibitor release should occur, it is expected that some zinc corrosion will initiate before the pH level is high enough to release the inhibitor, and the optimum quantity of inhibitor to release. With painted product, coordination with inhibitors already utilized in pretreatments and paint layers will also be required to optimize performance. However, the development of this additional means of corrosion protection would extend the capabilities of electrogalvanized coatings whose shortcomings in production of heavier coatings were noted at the beginning of this section.

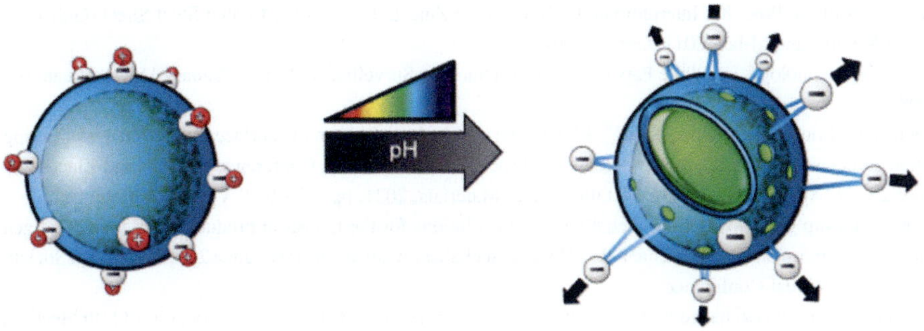

FIG. 16.25 (left) Poly(styrene-*stat*-methacrylic acid) (PSMAA) microcapsule shell losing protons in corrosion environment with pH > 4, resulting in shell rupture (right) release of inhibitor chemicals from microcapsule. Overall diameter is in the submicron range [39]. *(M. Rohwerder, T.H. Tran, Novel zinc-nanocontainer composite coatings for intelligent corrosion protection, in: Proc. 11th International Conference on Zinc and Zinc Alloy Coated Steel Sheet (Galvatech 2017), November 12–16, 2017, Tokyo, Iron and Steel Institute of Japan, Tokyo, pp. 169–176.)*

Many other capabilities opening new opportunities for galvanized steel sheet are being developed using new functional coil coating treatments. Reduction of heat absorption of roof surfaces is being accomplished by application of chemical and organic treatments that increase thermal emissivity, termed "cool roofs." High emissivity steel sheet roofing is already contributing to the control of urban "heat islands," lowering energy consumption by 20%–50% [41] and contributing to quality of life. Improved hydrophobic and surface texturing properties have led to "self-cleaning" architectural panels [42].

Further development of new opportunities will also depend upon advances in characterization techniques to improve insights into corrosion mechanisms and other zinc coating metallurgy effects on user properties such as formability, weldability, and paintability [43]. For example, although the Scanning Kelvin Probe was initially developed to assess the behavior of coatings and phenomena related to paint delamination, its use has been expanded to improve our understanding of the behavior of alloying elements in metallic coatings and the role of hydrogen [44].

References

[1] H.B. Lüngen, Ways to reduce CO_2 emissions in iron and steelmaking in Europe, in: Proc. 12th International Conference on Zinc and Zinc Alloy Coated Steel Sheet (Galvatech 2021), June 22–24, 2021, Virtual Conference, Austrian Society for Metallurgy and Materials, 2021, pp. 603–614.

[2] US Environmental Protection Agency, Greenhouse Gas Reporting Rule, 40 CFR Part 98, Federal Register, vol. 81 (237), December 9, 2016.

[3] M. Dubois, private communication, October 19, 2021.

[4] T. Briselden, B. Nakanishi, Additive manufacturing for improved galvanizing sustainability, in: Proceedings Galvanizers Association 111th Meeting, October 6–9, Galvanizers Association, Marine City, MI, 2019. Charleston, SC.

[5] J. Lovens, P. Weber, M. Anderhuber, Inductive strip heating, in: Proc. 11th International Conference on Zinc and Zinc Alloy Coated Steel Sheet (Galvatech 2017), November 12–16, Iron and Steel Institute of Japan, Tokyo, 2017, pp. 291–298. Tokyo.

[6] B. Visnic, Ford details all-electric Ford Lightning, Automot. Eng. 8 (5) (2021) 13. https://www.sae.org/news/2021/05/ford-f-150-lightning-detailed.

[7] M. Mikolaiczik, 2021 Ford Mustang Mach-E, in: Presented at Great Designs in Steel 2021, Virtual Conference, May 19, American Iron and Steel Institute, Washington, DC, 2021. https://www.steel.org/wp-content/uploads/2021/06/GDIS-2021-_-Track-1_01_Mikolaiczik_Mustang-Mach-E-.pdf.

[8] B.K. Zuidema, On the role of body-in-white weight reduction in the attainment of the 2012–2025 US EPA/NHTSA fuel economy mandate, in: Presented at Great Designs in Steel 2013, Livonia, MI, May 1, American Iron and Steel Institute, Washington, DC, 2013.

[9] Supervisory Control. https://en.wikipedia.org/wiki/Supervisory_control. (Accessed 6 February 2022).

[10] C. Perottet, How model predictive control enhances strip temperature regulation, in: Proceedings Galvanizers Association 111th Meeting, October 6–9, Galvanizers Association, Marine City, MI, 2019. Charleston, SC.

[11] M. Volpato, D. Garassino, A. Alessandrini, Efficient and effective operation of a hot dip galvanizing line, in: Proc. 12th International Conference on Zinc and Zinc Alloy Coated Steel Sheet (Galvatech 2021), June 22–24, Austrian Society for Metallurgy and Materials, 2021, pp. 806–817. Virtual Conference.

[12] M. Monnoyer, R. Biland, E. Menigault, D. Seguret, Fives' Eyeron™, all-in-one quality management software, in: Proc. 12th International Conference on Zinc and Zinc Alloy Coated Steel Sheet (Galvatech 2021), June 22–24, Austrian Society for Metallurgy and Materials, 2021, pp. 797–805. Virtual Conference.

[13] A. Lhoest, M. Bärwald, U. Sommers, M. Biglari, E. Montagna, W. Beugeling, Online quality monitoring of IF and high strength steels on continuous galvanizing lines controlled by furnace mathematical model, in: Proc. 10th International Conference on Zinc and Zinc Alloy Coated Steel Sheet (Galvatech 2015), May 31–June 4, Association for Iron & Steel Technology, Warrendale, PA, 2015, pp. 650–657. Toronto.

[14] S. Bragin, J. Klaus, T. Pfatschbacher, J.-F. Plaul, L. Bonhomme, B. Grenier, B. Jalard, Through-process optimization (TPO)—a modern approach to rapid product development, in: Proc. 12th International Conference on Zinc and Zinc Alloy Coated Steel Sheet (Galvatech 2021), June 22-24, Austrian Society for Metallurgy and Materials, 2021, pp. 818–828. Virtual Conference.

[15] T. Koll, New testing procedures and quality control, in: Proc. 8th International Conference on Zinc and Zinc Alloy Coated Steel Sheet (Galvatech 2011), June 21–25, Associazione Italiana di Metallurgia, Milan, 2011. Genoa, Italy.

[16] G.W. Dallin, Emerging hot-dip sheet galvanizing technology, in: White Paper for AISI Technology Surveillance Project, January 15, American Iron and Steel Institute, Washington, DC, 2008.

[17] Y. Fontaine, L. Dorel, B. Jalard, A. Perrichon, G. Fricout, O. Bichler, SIAS® 4.0 the next generation of automatic surface inspection system using convolutional neuron network for detection and classification of surface defects, in: Proc. 12th International Conference on Zinc and Zinc Alloy Coated Steel Sheet (Galvatech 2021), June 22–24, Austrian Society for Metallurgy and Materials, 2021, pp. 940–954. Virtual Conference.

[18] H. Behrens, T. Daube, L. Kümmel, C. Sasse, Hot-dip galvanizing lines—challenges and solutions for the industrial production of coated 3rd gen automotive steels, in: Proc. 12th International Conference on Zinc and Zinc Alloy Coated Steel Sheet (Galvatech 2021), June 22–24, Austrian Society for Metallurgy and Materials, 2021, pp. 787–796. Virtual Conference.

[19] M.F. Peintinger, Cross-site data integration, unification and exchange for historical analysis, in: Proceedings Galvanizers Association 111th Meeting, October 6–9, Galvanizers Association, Marine City, MI, 2019. Charleston, SC.

[20] B. Grenier, L. Bonhomme, Y. Fontaine, S. Georges, B. Jalard, Y. Kai, S. Maillard, Quality intelligence package dedicated to state-of-the-art processing lines for a comprehensive approach, in: Proc. 11th International Conference on Zinc and Zinc Alloy Coated Steel Sheet (Galvatech 2017), November 12–16, Iron and Steel Institute of Japan, Tokyo, 2017, pp. 322–329. Tokyo.
[21] F.-F. Henrich, O. Jannasch, H. Shafy, J. Dalldrop, S. Arikan, M. Rückl, Systematic application of AI to quality optimization from steelmaking to galvanizing, Iron Steel Technol. (2021) 44–50.
[22] I. Wrobel, M. Sidzina, Design study for automatic production line of a sub-assemblies of new generation car body structures compliant with the 'Industry 4.0' concept, Sensors 21 (7) (2021) 2434, https://doi.org/10.3390/s21072434.
[23] S. Chen, R. Rana, Low-density steels, in: R. Rana (Ed.), High-Performance Ferrous Alloys, Springer, Cham, 2021, https://doi.org/10.1007/978-3-030-53825-5_6.
[24] J. Maki, J. Mahieu, B.C. De Cooman, Galvanizability of silicon free CMnAl TRIP steels, Mater. Sci. Technol. 19 (1) (2003) 125–131.
[25] X. Jin, Y. Xie, H. Wang, Selective oxidation of high aluminium low density steel, in: Proc. 11th International Conference on Zinc and Zinc Alloy Coated Steel Sheet (Galvatech 2017), November 12–16, Iron and Steel Institute of Japan, Tokyo, 2017, pp. 506–513. Tokyo.
[26] H. Okada, T. Fukugawa, H. Ishihara, A. Okamoto, M. Azuma, Y. Matsuda, Prevention of red scale formation during hot rolling of steels, ISIJ Int. 35 (7) (1995) 886–891.
[27] R. Brandenburg, Dielectric barrier discharges: progress on plasma sources and on the understanding of regimes and single filaments, Plasma Sources Sci. Technol. 26 (2017) 053001, https://doi.org/10.1088/1361-6595/aa6426.
[28] O. Pearcey, H. Brown, I. Marincic, J. Tham, The evolution of coating control systems as the need for new coated products expands, Berg. Huetenmaenn Monastch. 166 (11) (2021) 525–534.
[29] M. Larnicol, Y. Hardy, O. Bregand, P. Gerkens, Device for Hydrodynamic Stabilization of a Continuously Travelling Metal Strip, U.S. Patent 10,550,459, February 4, 2020.
[30] N. Lagios, M. Renard, J.-P. Crutzen, J.-M. Raick, New lines for processing advanced high strength steels, in: Proceedings Galvanizers Association 106th Meeting, October 5–8, Galvanizers Association, Marine City, MI, 2014. Jackson MS.
[31] European Union RFCS-2018 Research Project N° 847341, CentriClean—Metallic coating bath cleaning by centrifugation, Technical Report, May 2021.
[32] P. Sippola, Unique simultaneous rapid cooling and zinc coating technology—no bainite before zinc pot, in: Proc. 12th International Conference on Zinc and Zinc Alloy Coated Steel Sheet (Galvatech 2021), June 22–24, Austrian Society for Metallurgy and Materials, 2021, pp. 426–435. Virtual Conference.
[33] C. Childs, Next-generation EDT technology, in: Proc. 12th International Conference on Zinc and Zinc Alloy Coated Steel Sheet (Galvatech 2021), June 22–24, Austrian Society for Metallurgy and Materials, 2021, pp. 71–78. Virtual Conference.
[34] S.H. Kim, J.H. Chung, N.H. Cho, Apparatus for Coating Zinc on Steel Sheet, and Method Therefor, U.S Patent 6,042,892 March 28, 2000.
[35] M.A. George, H. Chandra, P. Mórse, J. Madocks, Latest innovations in large area web coating technology via plasma enhanced chemical vapor deposition source technology, J. Vac. Sci. Technol. A 27 (4) (2009) 970, https://doi.org/10.1116/1.3077287.
[36] Y.A. Balandin, A.S. Kolpakov, E.V. Zharkov, Diffusion multi-component zinc-plating of steel 40kh in vibrofluidized bed, Met. Sci. Heat Treat. 51 (1–2) (2009) 45–48.
[37] D. Chaliampalias, M. Papazoglou, S. Tsipas, E. Pavlidou, S. Skolianos, G. Stergioudis, G. Vourlias, The effect of Al and Cr additions on pack cementation zinc coatings, Appl. Surf. Sci. 256 (11) (2010) 3618–3623, https://doi.org/10.1016/j.apsusc.2009.12.165.
[38] C. Hawk, E. Sullivan, C. Yu, S. Liu, Developments in metal additive manufacturing, Weld. J. (2020) 24–30.
[39] M. Rohwerder, T.H. Tran, Novel zinc-nanocontainer composite coatings for intelligent corrosion protection, in: Proc. 11th International Conference on Zinc and Zinc Alloy Coated Steel Sheet (Galvatech 2017), November 12–16, Iron and Steel Institute of Japan, Tokyo, 2017, pp. 169–176. Tokyo.
[40] S. Claessens, S.K. Ghosh, L. Stappers, D. Soccol, J. Fransaer, Design of functional properties in electrogalvanized coatings, in: Presented at Galvanized Steel Sheet Forum—Functional Properties, May 30–31, International Zinc Assn., Durham, NC, 2006. Dusseldorf.
[41] C. Murguia, D. Valles, Y.-H. Park, S. Kuravi, Effect of high aged albedo cool roofs on commercial buildings energy savings in U.S.A. climates, Int. J. Renew. Energy Res. 9 (1) (2019) 65–72, https://doi.org/10.20508/ijrer.v9i1.8600.g7565.
[42] L.T. Jandel, Systematic research & development for new coil coating systems for construction, in: Presented at Galvanized Sheet Steel Forum—Construction, April 24–25, The Institute of Metals, London, 2003. London.
[43] Z. Yu, Z.J. Hu, H. Meng, A review of recent developments in coating systems for hot-dip galvanized steel, Front. Mater. 7 (2020) 74, https://doi.org/10.3389/fmats.2020.00074.
[44] M. Rohwerder, Scanning kelvin probe techniques for the investigation of corrosion, hydrogen uptake and permeation properties of zinc alloy coatings on steel, in: Proc. 12th International Conference on Zinc and Zinc Alloy Coated Steel Sheet (Galvatech 2021), June 22–24, Austrian Society for Metallurgy and Materials, 2021, pp. 244–253. Virtual Conference.

Index

Note: Page numbers followed by *f* indicate figures and *t* indicate tables.

CPI Antony Rowe
Eastbourne, UK
February 25, 2023